Chemistry of Protein and Nucleic Acid Cross-Linking and Conjugation

Second Edition

Chemistry of Protein and Nucleic Acid Cross-Linking and Conjugation

Second Edition

Shan S. Wong
David M. Jameson

CRC Press
Taylor & Francis Group
Boca Raton London New York

CRC Press is an imprint of the
Taylor & Francis Group, an **informa** business

First published in paperback 2024

First published 2012 by CRC Press
2385 NW Executive Center Drive, Suite 320, Boca Raton FL 33431

and by CRC Press
4 Park Square, Milton Park, Abingdon, Oxon, OX14 4RN

CRC Press is an imprint of Taylor & Francis Group, LLC

© 2012, 2024 Taylor & Francis Group, LLC

Library of Congress Cataloging-in-Publication Data

Wong, Shan S.
  Chemistry of protein and nucleic acid cross-linking and conjugation / Shan S. Wong and David M. Jameson. -- 2nd ed.
        p. ; cm.
    Rev. ed. of: Chemistry of protein conjugation and cross-linking. c1991.
    Includes bibliographical references and index.
    ISBN 978-0-8493-7491-3 (hardcover : alk. paper)
    1. Proteins--Crosslinking. I. Jameson, David M. II. Wong, Shan S. Chemistry of protein conjugation and cross-linking. III. Title.
    [DNLM: 1. Proteins--chemistry. 2. Cross-Linking Reagents. 3. Immunoconjugates--chemistry. 4. Immunoconjugates--therapeutic use. 5. Nucleic Acids--chemistry. QU 55]

QP551.W596 2011
572'.633--dc22   2011010494

ISBN: 978-0-8493-7491-3 (hbk)
ISBN: 978-1-03-291817-4 (pbk)
ISBN: 978-0-429-12723-6 (ebk)

DOI: 10.1201/b11175

Visit the Taylor & Francis Web site at
http://www.taylorandfrancis.com

and the CRC Press Web site at
http://www.crcpress.com

Contents

Preface

It has been 20 years since the publication of *Chemistry of Protein Conjugation and Cross-Linking*. During this period, great advances have been made in the area of cross-linking of biological molecules. New cross-linking reagents, including multifunctional cross-linkers, have been developed and synthesized. The completion of the human genome project has opened a new area for studying nucleic acid and protein interactions using nucleic acid cross-linking reagents. Advances have also been made in the area of biosensors and microarray biochips for the detection and analysis of genes, proteins, and carbohydrates. In addition, physical techniques, especially novel mass spectrometry approaches with unprecedented sensitivity and resolution, have facilitated the analysis of cross-linked products. All these advances warrant a new edition of the old text.

This book offers an overview of the chemical principles underlying the processes of cross-linking and conjugation. Attempts have been made to list all, or at least most, cross-linking reagents published in the literature up to now, covering monofunctional, homobifunctional, heterobifunctional, and multifunctional as well as zero-length cross-linkers. A general methodology for experimental applications of these cross-linkers is provided. This book also includes reviews on the use of these reagents in studying protein tertiary structures, geometric arrangements of subunits within complex proteins and nucleic acids, near neighbor analysis, protein-to-protein or ligand–receptor interactions, and conformational changes of biomolecules. In addition, applications in the area of immonoconjugation for immunoassays, immunotoxin for targeted therapy, microarray technology for analysis of various biomolecules, and solid state chemistry for immobilizations are presented. Therefore, this book is intended to be a valuable reference for multidisciplinary approaches.

It has taken a long time to prepare this book, and the authors thank the publishers for their enormous patience. Shan Wong, as always, is indebted to Lee-Jun Wong for her patience and understanding during the entire book project (yet again!) as well as for her patience and understanding in day-to-day life. David Jameson wishes to thank Marcin Bury and Nicholas James, in his laboratory, for proofreading much of the book, and Dudley Williams and Don Laudicina from Allergan, Inc. for helpful discussions on mass spectrometry. In addition, he wishes to thank Sandra Kopels for her unwavering support in all aspects of his life!

Authors

Shan S. Wong, PhD, has recently retired from the National Institutes of Health, where he served as a scientific review administrator and a program officer. In his latter capacity, he oversaw scientific programs in the area of alternative and complementary medicine. Previously, he served as director of clinical chemistry at Hermann Hospital and Lyndon B. Johnson General Hospital in Houston, Texas, and as a faculty at the University of Texas Health Science Center at Houston, Texas. Before joining the University of Texas, Dr. Wong was a full professor of chemistry at the University of Massachusetts at Lowell, Lowell, Massachusetts. In addition to teaching at the University of Massachusetts at Lowell, he also taught chemistry courses at Denison University, Granville, Ohio, and Ohio State University, Columbus, Ohio.

Dr. Wong graduated in 1970 from the Oregon State University, Corvallis, Oregon, with a BS in chemistry and received his PhD in 1974 from the Department of Chemistry at Ohio State University. After doing postdoctoral work at Temple University, Philadelphia, and Ohio State University, he joined the University of Massachusetts at Lowell.

Dr. Wong has published extensively in various scientific journals in the area of enzymology and clinical chemistry. He has received numerous honors and awards and was active in various professional societies.

David M. Jameson, PhD, joined the Department of Cell and Molecular Biology at the John A. Burns School of Medicine at the University of Hawaii in 1989, where he is presently a full professor. Before moving to Hawaii he was on the faculty of the Pharmacology Department at the University of Texas Southwestern Medical School in Dallas.

Dr. Jameson received his BS in chemistry from Ohio State University in 1971 and his PhD in biochemistry from the University of Illinois at Urbana-Champaign in 1978. His thesis advisor was Gregorio Weber, who laid the foundations of modern fluorescence spectroscopy. He carried out postdoctoral research at the Université Paris-Sud at Orsay before returning to the University of Illinois for a postdoctoral period in Gregorio Weber's laboratory. In 1983, he joined the Pharmacology Department at the University of Texas Southwestern Medical Center at Dallas as an assistant professor. In 1989, he moved to the University of Hawaii.

Dr. Jameson's primary research focus has always been the development and application of fluorescence approaches for the study of biomolecular interactions, in particular protein–protein and protein–ligand interactions. He has published extensively in this area (~130 publications to date) and has received funding from the National Science Foundation, the American Heart Association, and the National Institutes of Health. He has also received the Established Investigator Award from the American Heart Association and the 2004 Gregorio Weber Award for Excellence in Fluorescence Theory and Application. He lectures at numerous fluorescence workshops around the world and is co-organizer of the International Weber Symposium on Innovative Fluorescence Methodologies in Biochemistry and Medicine held every three years in Hawaii.

1 Overview of Protein Conjugation

Completion of the human genome project has opened up tremendous opportunities for the study of complex biological processes at the molecular level.[1] We now know that only about 1%–2% of the genome encodes for proteins.[2] These gene products perform all cellular functions from metabolism to developmental control to apoptosis and cell death. In order to comprehend how the cell works and thus the whole organism, it is important to know the detailed functions of these proteins. From the start, we need to elucidate their three-dimensional (3D) structures and their relationships and interactions with other proteins. Some proteins, such as myoglobin, exist freely in the cytosol as monomers. Others associate into protein complexes, the simplest of which are dimers, either with another identical protein subunit (homodimer), for example, malate dehydrogenase, or a different protein subunit (heterodimer), for example, creatine kinase.[3] Still others may associate into higher architectural organizations such as tetramers, pentamers, hexamers, and larger multicomponent aggregates. Examples of these organizations are shown in Figure 1.1.[4–8] As the number of components increases, so do the complexities of the protein interactions. It then becomes more difficult to elucidate the sites of protein contacts and the 3D dispositions of the individual subunits.

Some proteins associate to regulate or alter their activities. For example, bovine galactosyltransferase normally transfers galactose from UDP-galactose to *N*-acetylglucosamine, either free or as the terminal sugar of glycoproteins.[9] However, when it binds with bovine α-lactalbumin, glucose becomes the preferred galactose acceptor leading to the formation of lactose.[10] The protein–protein interactions become an important aspect of the regulatory process.

Association of proteins as a regulatory process is seen practically in all signaling pathways. An obvious example is that of the hedgehog (Hh) signaling pathway, which is depicted in Figure 1.2 for *Drosophila*.[11–13] In *Drosophila*, the Hh signaling molecules associate with the Patched (Ptc) receptor, a 12-pass membrane protein. This interaction activates the Smoothened (Smo) G-protein leading to the release of active CI155 from a microtubule, Cos2, Fu, SuFu, and CI protein complex.[14] The active CI155 ultimately controls the transcription of specific target genes. In the absence of Hh, Ptc interacts with and inhibits Smo, a seven-pass membrane protein, and Fu, Cos2, and SuFu bind to CI, preventing its activation and retaining it in the cytoplasm. CI in the complex is cleaved to yield CI75 upon phosphorylation by Adenylate Cyclase (AC)-induced protein kinase A (PKA), which involves Slimb and GSK3H. This culmination of protein binding events leads to inhibition of transcription. It is obvious that Cos2, Fu, and SuFu play multiple and complex roles in CI control.[15] In order to understand the details of the signal transduction pathway, it is necessary to reveal exactly how the individual proteins in the assembly of complex protein networks interact with each other. In this example, it would be of interest to understand how the association of Hh with Ptc alters its protein structure such that Smo is activated. It would also be of interest to know the structural organization of the microtubule, Cos2, Fu, SuFu, and CI complex. Even the activation of PKA through CA is an interesting regulatory process.

Cyclic AMP-dependent PKA consists of two regulatory and two catalytic subunits.[16] In its tetrameric holoenzyme form, the catalytic subunits are inactive. However, binding of cAMP to the regulatory subunits results in the dissociation of the ternary complex into a regulatory dimer and two active catalytic monomers as represented in Figure 1.3, demonstrating another level of regulation through protein–protein interactions. Using the lysine-specific bifunctional cross-linking reagent

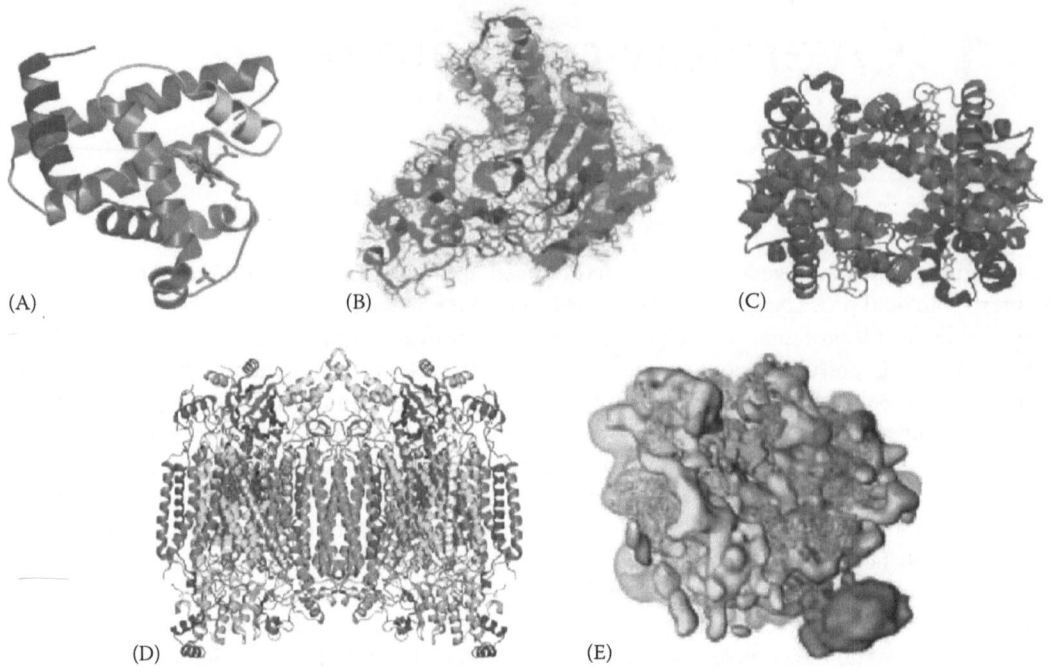

FIGURE 1.1 **(See color insert.)** Examples of different molecular structures of proteins. (A) Myoglobin molecule. (After Phillips, S. E. V. *J. Mol. Biol.*, 142, 531, 1980.) (B) Dimeric creatine kinase. (After Shen, Y. Q. et al., *Acta Crystallogr. D Biol. Crystallogr.*, 57, 1196, 2010.) (C) Tetrameric hemoglobin. (After Paoli, M. et al., *J. Mol. Biol.*, 256, 775, 1996.) (D) Bovine cytochrome C oxidase with 2 copies of 13 different components. (From Shinzawa-Itoh, K. et al., *EMBO J.*, 26, 1713, 2007. With permission.) (E) Yeast 80S ribosome of multicomponent proteins and RNA. (Reprinted from *Cell*, 107, Beckmann, R. et al., Architecture of the protein-conducting channel associated with the translating 80S ribosome, 361, Copyright 2001, with permission from Elsevier.)

dimethyl suberimidate, Charlton et al.[17] have demonstrated the dynamics of dissociation of the tetramer in the presence of cAMP and MgATP. Other information on the 3D architecture should be available using the same technique.

There are numerous other ways in which proteins interact with each other in complex biological processes. In addition, proteins also interact with nucleic acids. As we have seen above in the Hh signaling pathway, CI regulates the cell cycle by binding to nuclear DNA to modulate gene expression. Also, in the structure of ribosomes, protein–RNA interactions are of paramount importance.[18] Such protein–nucleic acid interactions are significant in diverse genetic networks and protein pathways. Determining the interactions of protein–protein and protein–nucleic acid systems is crucial to understanding how biological systems function and how they contribute to cellular and organismal phenotypes.

There are many methods to study protein structures and their interactions. X-ray crystallography has been successfully used to elucidate tens of thousands of protein structures, from monomers to multicomponents complexes. However, proteins in the biological environment are dynamic, and x-ray structures, being restricted by crystal packing, are inherently static, although some measures of the elasticity of these crystal structures are available.[19] This powerful technique has even been able to elucidate fairly high-resolution structures of ribosomes.[18] Because it is based on crystallography, the technique is limited in studying protein interactions that occur transiently as in signal transduction pathways. In recent years, nuclear magnetic resonance (NMR) has become a powerful method for elucidating protein structures in solution, but is limited to relatively small proteins, for example, proteins less than about 30 kDa.[20] The field of computationally based protein structure

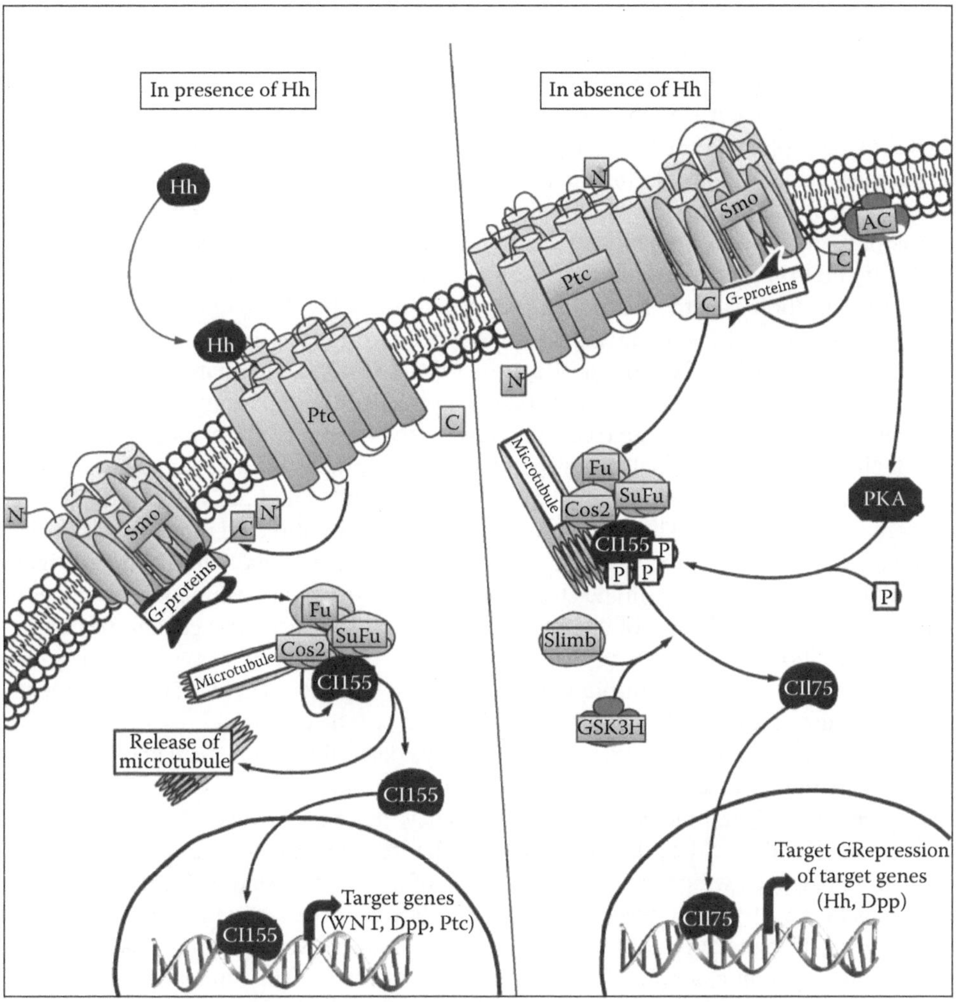

FIGURE 1.2 Hedgehog signaling pathway in *Drosophila*.

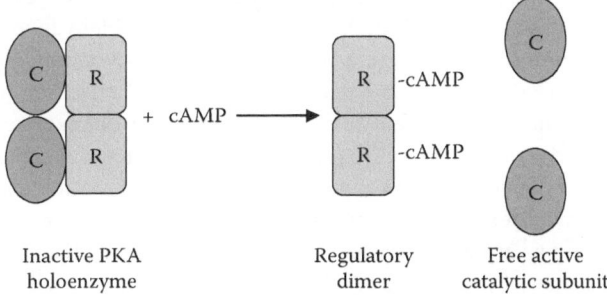

FIGURE 1.3 Activation of PKA by cAMP. Cyclic-AMP binds to the regulatory subunits causing the release of active free catalytic subunits.

prediction has made significant advances in recent years but is not likely to replace experimental structure determinations in the near future.[21] Other proteomic approaches include mass spectrometry,[22] microarrays,[23] the two-hybrid system,[24,25] coprecipitation, and computational statistical analysis.[26] Modern molecular biology techniques of gene knockout,[27] knock-down approaches,[28] and use of small molecular inhibitors[29] have facilitated the determination of protein functions. Of these methods, however, chemical cross-linking provides detailed characterization of protein–protein interactions and enables non-covalent interactions to be captured *in vivo*.[30]

Chemical cross-linking of biological components is a special application of chemical modification of proteins.[31] The history of chemical modifications dates back to the 1920s when enzymes were shown to be proteins.[32] At this early stage, the technique was used mainly to identify the particular amino acids responsible for their catalytic activity. Due to the lack of instrumentation and analytical methods, progress in the development of new procedures and reagents was initially slow. It was only during and immediately after World War II that significant advances started to take place. The first reviews on the subject were published in 1947.[33,34] Since then, the application of chemical modification has grown exponentially hand-in-hand with the development of biochemical analysis techniques. For example, an effective procedure for the determination of amino acid sequences was introduced by Edman and Begg in 1956.[35] In 1960, automated amino acid analyzers became available.[36] Various new ion exchange and gel filtration chromatography media were developed and different forms of gel electrophoresis also became commonplace during the same time period. Numerous review articles and specialized monographs began to appear during the last 30 years.[36–44] New directions such as photoaffinity labeling[40,45,46] and active-site-directed affinity reagents were developed.[40,47–49] The very powerful method of site-directed mutagenesis is also providing results beyond those attainable by chemical modification, especially for the determination of the function of individual residues in the structure and reactivity relationship of a molecule.[50]

Cross-linking as a special form of chemical modification has particular capabilities of its own, unparalleled by *in vitro* mutagenesis.[41,51,52] The process involves joining of two molecular components by a covalent bond achieved through the use of cross-linking reagents. The components may be proteins, peptides, drugs, nucleic acids, or solid particles. The chemical cross-linkers are multi-functional reagents containing reactive functional groups derived from classical chemical modification agents. Bifunctional cross-linkers are the most common. These reagents are capable of reacting with the side chains of the amino acids of proteins. They may be classified into homobifunctional, heterobifunctional, and zero-length cross-linkers. The zero-length cross-linkers are essentially group-activating reagents, which cause the formation of a covalent bond between the components without incorporation of any extrinsic atoms. Thus, dicyclohexyl carbodiimide, which has been used extensively to bring about the formation of amide bonds between carboxyl and amino groups in peptide synthesis, is an example of a zero-length cross-linker. The homobifunctional reagents consist of two identical functional groups and the heterobifunctional reagents contain two different types of reactive functional moieties. They therefore form bridges between the reactive amino acid side chains in proteins. Homobifunctional reagents, such as dialkyl halides and *bis*-imidoesters, were among the early cross-linkers developed,[53–56] although formaldehyde and other reagents had been used in the tanning industry many years prior without known chemical reactions. Since the first application of a bifunctional reagent by Zahn in the 1950s,[53–55] research in this area has flourished, particularly during the last two decades. The introduction of photoactivatable aryl azides marked the beginning of heterobifunctional reagents.[46,57–59] Further advancement in the application of these reagents has led to the design and synthesis of cleavable bifunctional compounds.[60,61] Over 300 cross-linkers have now been synthesized, and more are forthcoming. The diversity of these molecules is as complicated as organic chemistry itself, limited only by the creative imaginations of the researchers. The presence of the *Journal of Bioconjugate Chemistry* will attest to the complexity of these reagents and their usefulness in various applications.

The application of chemical cross-linking is multidisciplinary, ranging from basic protein biochemistry to applied biotechnology and engineering, and from immunology to medicine. These reagents have been used to stabilize tertiary structures of proteins,[62,63] to study protein–protein interactions of subunits in oligomeric proteins,[64] and in complex structures such as ribosomes,[65] to determine distances between reactive groups within or between protein subunits,[56,66] to attach ligands to solid supports,[67] and to identify membrane receptors.[68,69] Applications in the pharmaceutical area have led to the coupling between target-specific proteins and metal-chelating agents for diagnostics in *in vivo* imaging of human patients,[70,71] as well as toxins and enzymes for therapeutics.[72–76] With the advent of enzyme-linked immunoassay and genetic probe technology, chemical cross-linking provides a means for preparation of enzyme-immunoglobulin conjugates and DNA probes.[77–79] These applications will be reviewed and summarized in the second half of this book.

Although the terms cross-linking and conjugation are often used interchangeably, there is a fine distinction in connotation between them. Cross-linking usually refers to the joining of two molecular species that have some sort of affinity between them, that is, they either exist as an aggregate or can associate under certain conditions. Thus, the chemical bonding between a ligand and its receptor is usually referred to as cross-linking. Similarly, cross-linking is used for the covalent bonding between subunits of enzymes. Conjugation, on the other hand, denotes the coupling of two unrelated species. For example, the linking between an enzyme and an immunoglobulin is conjugation. The product is referred to as a conjugate, and in this case, an immunoconjugate.

No matter whether it is conjugation or cross-linking, two types of products usually result from a cross-linking reaction. One is derived from intramolecular cross-linking, the other as a consequence of intermolecular joining of two or more species. The possible chemical reactions of a cross-linking reagent with a protein dimer are illustrated in Figure 1.4. As an example, suppose a protein exists as a monomer in dilute solutions and associates or interacts with another protein molecule at higher concentrations. When the monomeric form reacts with the chemical reagent, intramolecular cross-linking will take place since protein molecules are, on average, far apart. At high concentrations, the molecules will be in closer proximity or will associate to form dimeric or oligomeric aggregates. Under these conditions, the reagent will provide intermolecular cross-linking. Thus, at very low concentrations, intramolecular bonding prevails, whereas intermolecular coupling is important at high concentrations. Cross-linkers have been used to determine distances between two reactive groups in a protein that is close in space, particularly those at the active sites of enzymes.[80–83] Intermolecular cross-linking may conjugate molecules of the same kind or of different kinds to form homopolymers or heteropolymers, respectively, thus providing a means for the preparation of high-molecular-weight complexes. Intermolecular coupling of different kinds of proteins also provides a tool for the study of antibody–antigen interactions, multienzyme complexes, membrane

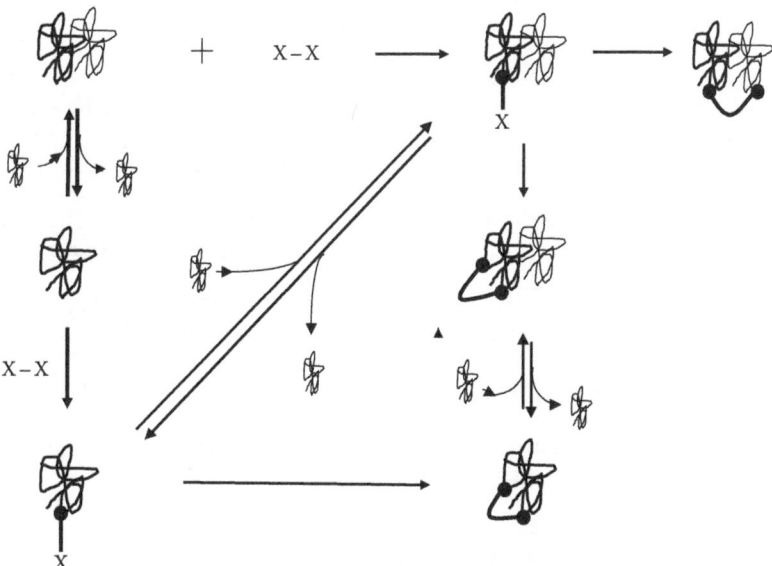

FIGURE 1.4 Cross-linking reactions of a hypothetical protein dimer. Reaction of the cross-linker, X–X, with the monomeric form usually yields intramolecular cross-linking such as in the case of dilute protein solutions. At high protein concentrations or at conditions where the protein molecules associate, intermolecular cross-linking will result.

protein structures, hormone and receptor recognitions, and other protein–protein interactions at the quaternary structural level.

To help understand the basic principles of reactions of these cross-linking reagents, this book begins with a review of the chemical reactivity of amino acid side chains and their reactions with specific chemical reagents. With the background of chemical modification, bifunctional reagents are introduced. All the existing cross-linking reagents are surveyed and classified into homo-, hetero-, and zero-length cross-linkers. Various specific applications of these reagents are mentioned. Examples of conjugation are provided as well as the conditions for the reaction. The reader should find this book useful not only as a reference for the basic information about cross-linking reagents but also as a handbook for experimental application of these reagents.

REFERENCES

1. Collins, F. S., Green, E. D., Guttmacher, A. E., and Guyer, M. S., A vision for the future of genomics research: A blueprint for the genomic era, *Nature*, 422, 835, 2003.
2. The International Human Genome Sequencing Consortium, Initial sequencing and analysis of the human genome, *Nature*, 409, 860, 2001.
3. Bais, R. and Edwards, J. B., Creatine kinase, *CRC Crit. Rev. Clin. Lab. Sci.*, 16, 291, 1982.
4. Phillips, S. E. V., Structure and refinement of oxymyoglobin at 1.6 A resolution. *J. Mol. Biol.*, 142, 531, 1980.
5. Shen, Y. Q., Tang, L., Zhou, H. M., and Lin, Z. J., Structure of human muscle creatine kinase. *Acta Crystallogr. D Biol. Crystallogr.*, 57, 1196, 2001.
6. Paoli, M., Liddington, R., Tame, J., Wilkinson, A., and Dodson, G., Crystal Structure of T State Haemoglobin with Oxygen Bound at All Four Haems. *J. Mol. Biol.*, 256, 775, 1996.
7. Shinzawa-Itoh, K., Aoyama, H., Muramoto, K., Terada, H., Kurauchi, T., Tadehara, Y., Yamasaki, A. et al., Structures and physiological roles of all the integral lipids of bovine heart cytochrome *c* oxidase, *EMBO J.*, 26, 1713, 2007.
8. Beckmann, R., Spahn, C. M. T., Eswar, N., Helmers, J., Penczek, P. A., Sali, A., Frank, J., and Blobel, G., Architecture of the protein-conducting channel associated with the translating 80S ribosome, *Cell*, 107, 361, 2001.
9. Ebner, K. E. and Magee, S. C., *Subunit Enzymes: Biochemistry and Functions*, Ebner, K. E. (ed.), Marcel Dekker, New York, 1975, p. 137.
10. Morrison, J. F. and Ebner, K. E., Studies on galactosyltransferase: Kinetic effects of α-lactalbumin with N-acetylglucosamine and glucose as galactosyl group acceptors, *J. Biol. Chem.*, 246, 3992, 1971.
11. Riobo, N. A. and Manning, D. R., Pathways of signal transduction employed by vertebrate hedgehogs, *Biochem J.*, 403, 369, 2007.
12. Østerlund, T. and Kogerman, P., Hedgehog signalling: How to get from Smo to Ci and Gli, *Trends Cell. Biol.*, 16, 176, 2006.
13. Jacob, L. and Lum, L., Hedgehog signaling pathway in Drosophila, *Sci STKE.*, 2007, cm7, 2007.
14. Hochman, E., Castiel, A., Jacob-Hirsch, J., Amariglio, N., and Izraeli, S., Molecular pathways regulating pro-migratory effects of hedgehog signaling, *J. Biol. Chem.*, 281, 33860, 2006.
15. Frank-Kamenetsky, M., Zhang, X. M., Bottega, S., Guicherit, O., Wichterle, H., Dudek, H., Bumcrot, D. et al., Small-molecule modulators of hedgehog signaling: Identification and characterization of smoothened agonists and antagonists, *J. Biol.*, 1, 10, 2002.
16. Krebs, E. G., Role of the cyclic AMP-dependent protein kinase in signal transduction, *JAMA*, 262, 1815, 1989.
17. Charlton, J. P., Huang, C.-H., and Huang, L. C., Chemical cross-linking of cyclic AMP-dependent protein kinase and its dissimilar subunits, *Biochem. J.*, 209, 581, 1983.
18. Bashan, A. and Yonath, A., Correlating ribosome function with high-resolution structures, *Trends Microbiol.*, 16, 326, 2008.
19. Hinsen, K., Structural flexibility in proteins: Impact of the crystal environment, *Bioinformatics*, 24, 521, 2008.
20. Billeter, M., Wagner, G., and Wüthrich, K., Solution NMR structure determination of proteins revisited, *J. Biomol. NMR*, 42, 155, 2008.
21 Zhang, Y., Protein structure prediction: When is it useful? *Curr. Opin. Struct. Biol.*, 19, 145, 2009.
22. Cravatt, B. F., Simon, G. M., and Yates III, J. R., The biological impact of mass-spectrometry-based proteomics, *Nature*, 450, 991, 2007.

23. Hall, D. A., Ptacek, J., and Snyder, M., Protein microarray technology, *Mech. Ageing Dev.*, 128, 161, 2007.
24. Fields, S. and Song, O.-K., A novel genetic system to detect protein–protein interactions, *Nature*, 340, 245, 1989.
25. MacDonald, P. N. (ed.), *Two-Hybrid Systems, Methods in Molecular Biology*, Humana Press, Totowa, NJ, 2001.
26. Sardiu, M. E., Cai, Y., Jin, J., Swanson, S. K., Conaway, R. C., Conaway, J. W., Florens, L., and Washburn, M. P., Probabilistic assembly of human protein interaction networks from label-free quantitative proteomics, *Proc. Natl. Acad. Sci. U. S. A.*, 105, 1454, 2008.
27. Van der Weyden, L., Adams, D. J., and Bradley, A., Tools for targeted manipulation of the mouse genome, *Physiol. Genomics*, 11, 133, 2002.
28. Hannon, G. J., RNA interference, *Nature,* 418, 244–251 (2002).
29. Stockwell, B. R., Chemical genetics: Ligand-based discovery of gene function, *Nature Rev. Genet.*, 1, 116, 2000.
30. Trakselis, M. A., Alley, S. C., and Ishmael, F. T., Identification and mapping of protein–protein interactions by a combination of cross-linking, cleavage, and proteomics, *Bioconj. Chem.*, 16, 741, 2005.
31. Means, G. E. and Feeney, R. E., Chemical modification of proteins: History and applications, *Bioconj. Chem.*, 1, 2, 1990.
32. Summer, J. B. and Graham, V. A., The nature of insoluble urease, *Proc. Soc. Exp. Biol. Med.*, 22, 504, 1925.
33. Olcott, H. S. and Fraenakel-Conrat, H., Specific group reagents for proteins, *Chem. Rev.*, 41, 151, 1947.
34. Herriott, R. M., Reactions of native proteins with chemical reagents, *Adv. Protein Chem.*, 3, 161, 1947.
35. Edman, P. and Begg, G., A protein sequenator, *Eur. J. Biochem.*, 1, 80, 1967.
36. Moore, S. and Stein, W. H., Chromatographic determination of amino acids by use of automatic recording equipment, *Methods Enzymol.*, 6, 819, 1963.
37. Means, G. E. and Feeney, R. E., *Chemical Modification of Proteins*, Holden-Day, San Francisco, CA, 1971.
38. Hirs, C. H. W. and Timasheff, S. N. (eds.), Enzyme structure (Part B), in *Methods in Enzymology*, Vol. 25, Academic Press, 1972.
39. Glazer, A. N., DeLange, R. J., and Sigman, D. S., *Chemical Modification of Proteins: Selected Methods and Analytical Procedures*, North-Holland, Amsterdam, the Netherlands, 1975.
40. Jakoby, W. B. and Wilcheck, M. (eds.), Affinity Labeling, in *Methods in Enzymology*, Vol. 46, Academic Press, New York, 1977.
41. Eyzaguirre, J. (ed.), *Chemical Modification of Enzymes: Active Site Studies*, Ellis Horwood, Chichester, U.K., 1987.
42. Feeney, R. E., Chemical modification of proteins: Comments and perspectives. *Int. J. Peptide Protein Res.*, 29, 145, 1987.
43. Lundblad, R. L., *Techniques in Protein Modification*, CRC Press, Boca Raton, FL, 1995.
44. Lundblad, R. L., *Chemical Reagents for Protein Modification*, 3rd edn., CRC Press, Boca Raton, FL, 2004.
45. Dhowdhry, V. and Westheimer, F. H., Photoaffinity labeling of biological systems, *Annu. Rev. Biochem.*, 48, 293, 1979.
46. Baley, H., *Photogenerated Reagents in Biochemistry and Molecular Biology*, Elsevier, New York, 1983.
47. Silverman, R. B., Mechanism-based enzyme inactivators for medical uses, in *Protein Tailoring for Food and Medical Uses*, Feeney, R. E. and Whitaker, J. R. (eds.), Marcel Dekker, New York, 1986, p. 215.
48. Baker, B. R., *Design of Active-Site-Directed Irreversible Enzyme Inhibitor: The Organic Chemistry of the Enzymic Active Site*, John Wiley & Sons, New York, 1967.
49. Walsh, C. T., Suicide substrates, mechanism-based enzyme inactivation: Recent developments, *Annu. Rev. Biochem.*, 53, 493, 1984.
50. Wetzel, R., Medical applications of protein engineering, in *Protein Tailoring for Food and Medical Uses*, Feeney, R. E. and Whitaker, J. R. (eds.), Marcel Dekker, New York, 1986, p. 181.
51. Friedman, M. (ed.), *Protein Crosslinking, Biochemical and Molecular Aspects*, Plenum Press, New York, 1977.
52. Scouten, W. H. (ed.), *Solid Phase Biochemistry, Analytical and Synthetic Aspects*, Wiley & Sons, New York, 1983.
53. Zahn, H., Bridge reactions in amino acids and fibrous proteins, *Angew. Chem.*, 67, 561, 1955.
54. Zahn, H., Cross-linking reactions with amino acid and fibrous proteins, *Makromol. Chem.*, 18, 201, 1955.
55. Zahn, H. and Meienbofer, J., Reactions of 1,5-difluoro-2,4-dinitrobenzene with insulin. 1. Synthesis of medial compounds, *Makromol. Chem.*, 26, 126, 1958.
56. Hartmann, F. C. and Wold, F., Cross-linking of bovine pancreatic RNase-A with dimethyl-adipimidate, *Biochemistry*, 6, 2439, 1967.

57. Fleet, G. W. J., Porter, R. R., and Knowles, J. R., Affinity labeling of antibodies with aryl nitrene as reactive group, *Nature*, 224, 511, 1969.

58. Fleet, G. W. J., Knowles, J. R., and Porter, R. R., The antibody binding site: Labeling of a specific antibody against the photo-precursor of an aryl nitrene, *Biochem. J.*, 128, 499, 1972.

59. Knowles, J. R., Photogenerated reagents for biological receptor-site labels, *Acc. Chem. Res.*, 5, 155, 1972.

60. Wang, K. and Richards, F. M., Reaction of dimethyl-3,3'-dithio-bis-propionimidate with intact human erythrocytes, *J. Biol. Chem.*, 250, 6622, 1975.

61. Coggins, J. R., Hooper, E. A., and Perham, R. N., Use of DMS and novel periodate-cleavable *bis*-imidioesters to study the quaternary structure of the pyruvate dehydrogenase multienzyme complex of *E. coli*, *Biochemistry*, 15, 2527, 1976.

62. Alber, T., Mutational effects on protein stability, *Annu. Rev. Biochem.*, 58, 765, 1989.

63. Wold, F., The reaction of bovine serum albumin with the bifunctional reagent *p,p'*-difluoro-*m,m'*-dinitro-dipbenylsulfone, *J. Biol. Chem.*, 236, 106, 1961.

64. Davies, G. E. and Stark, G. R., Use of dimethyl suberimidate, a cross-linking reagent, in studying the subunit structure of oligomeric proteins, *Proc. Natl. Acad. Sci. U. S. A.*, 66, 651, 1970.

65. Sun, T. T., Traut, R. R., and Kahan, L., Protein–protein proximity in the association of ribosomal subunits of *Escherichia coli*. Crosslinking of 30-S protein S16 to 50-S proteins by glutaraldehyde or formaldehyde, *J. Mol. Biol.*, 87, 509, 1974.

66. Fasold, H., Decomposition of azoglobin, separation and identification of single peptide bridges, *Biochem. Z.*, 342, 295, 1965.

67. Faib, R. D., Covalent linkage. I. Enzyme immobilization by covalent linkage on insolubilized support, in *Biomedical Applications of Immobilized Enzymes and Proteins*, Vol. 1, Chang, T. M. S. (ed.), Plenum Publishing, New York, 1977, p. 7.

68. Ji, T. H., Bifunctional reagents, *Methods Enzymol.*, 91, 580, 1983.

69. Ji, T. H., Crosslinking of lectins and receptors in membranes with heterobifunctional cross-linking reagents, in *Membranes and Neoplasia: New Approaches and Strategies*, Marchisi, V. T. (ed.), Alan R. Lias, New York, 1976, p. 171.

70. Meares, C. J., Attaching metal ions to antibodies, in *Protein Tailoring for Food and Medical Uses*, Feeney, R. E. and Whitaker, J. R. (eds.), Marcel Dekker, New York, 1986, p. 339.

71. Meares, C. F., McCall, M. J., Deshpande, S. V., DeNardo, S. J., and Goodwin, D. A., Chelate radiochemistry: Cleavable linkers lead to altered levels of radioactivity in the liver, *Int. J. Cancer*, 2, 99, 1988.

72. Neville, D. M., Jr., Immunotoxins: Current use and future prospects in bone marrow transplantation and cancer treatment, *CRC Crit. Rev. Therapeu. Drug Carrier Syst.*, 2, 329, 1986.

73. Poznansky, M. J. and Juliano, R. L., Biological approaches to the controlled delivery of drugs: A critical review, *Pharm. Rev.*, 36, 277, 1984.

74. Maeda, H., Matsumura, Y., Oda, T., and Sasamoto, K., Cancer selective macromolecular therapeutics: Tailoring of an antitumor protein drug, in *Protein Tailoring for Food and Medical Uses*, Feeney, R. E. and Whitaker, J. R. (eds.), Marcel Dekker, New York, 1986, p. 353.

75. Marsh, J. W. and Neville, D. M., Jr., lmmunotoxins: Chemical variables affecting cell killing efficiencies, in *Proteins Tailoring for Food and Medical Uses*, Feeney, R. E. and Whitaker, J. R. (eds.), Marcel Dekker, New York, 1986, p. 291.

76. Frankel, A. E. (ed.), *Immunotoxins*, Kluwer Academic Publishers, Boston, MA, 1988.

77. Ngo, T. T. and Leahoff, H. M. (eds.), *Enzyme-Mediated Immunoassay*, Plenum Press, New York, 1985.

78. Ngo, T. T. (ed.), *Nonisotopic Immunoassay*, Plenum Press, New York, 1988.

79. Keller, G. H. and Manak, M. M., *DNA Probes*, Stockton Press, New York, 1989.

80. Moore, J., Jr. and Feoselau, A., Reaction of glyceraldehyde-3-phosphate dehydrogenase with dibromoacetone, *Biochemistry*, 11, 3753, 1972.

81. Husain, S. S., Ferguson, J. B., and Fruton, J. S., Bifunctional inhibitors of pepsin, *Proc. Natl. Acad. Sci. U. S. A.*, 68, 2765, 1971.

82. Reichert, A., Heintz, D., Echner, H., Voelter, W., and Faulstich, H., Identification of contact sites in the actin-thymosin beta 4 complex by distance-dependent thiol cross-linking, *J. Biol. Chem.*, 271, 1301, 1996.

83. Vollmer, S. H. and Colman, R. F., Cysteinyl peptides labeled by dibromobutanedione in reaction with rabbit muscle pyruvate kinase, *Protein Sci.*, 1, 678, 1992.

2 Review of Protein and Nucleic Acid Chemistry

2.1 INTRODUCTION

Before we can discuss the chemical cross-linking and conjugation of proteins, we must understand some basic protein chemistry, as the cross-linking reagents depend on the reactivities of the constituents of proteins. In most cases, the biological activities of the individual proteins in the conjugated products have to be preserved. This condition dictates that those amino acids involved in the biological functions must be conserved and only those residues not involved in the biological activities be modified. In addition, the three-dimensional (3D) structure of a protein should remain as invariant as possible during the process of chemical modification. Disturbances of protein structures and properties may occur with reagents that change the charge, size, and other characteristics of the modified amino acid residues. For example, rat liver glycine methyltransferase is completely inactivated on introduction of a large and anionic 2-nitro-5-thiobenzoate, while a smaller and neutral cyano group has no effect.[1] Similar results have been observed for the modification of cysteine residues of many proteins.[2] Thus, only those amino acid residues that are not situated at the active centers or settings critical to the integrity of the tertiary structures of proteins may be targets for chemical cross-linking. Such amino acids are ideally located on the surface of the molecule. It follows, therefore, that the identity of the reactive functional groups on the exterior of a protein is often the most important factor controlling the protein's reactivity toward cross-linking reagents. By knowing which functional groups are located at the protein–solvent interface, one may modify the protein without sacrificing its biological activity. However, this strategy is not always as straightforward as one would like. Proteins vary in their 3D structures as well as their surface compositions. A particular amino acid may occur both buried in a protein's interior and exposed on the protein's surface. This duality may or may not be true in another protein. In addition, the chemical properties of an amino acid side chain may be influenced by the nearby residues with which it interacts. In fact, such differences in reactivity may be used to evaluate the microenvironment of the residue.[3,4] On the other hand, some studies looking at the modulation of biological activities on interaction with other proteins may specifically involve the inhibition of its biological activity. One end of a bifunctional cross-linker would react with the active site and the other end would attach to the nearby modulating protein. The reactivity of the active site residues and residues on neighboring proteins need to be well defined for the modifying agents to work as planned. In order to understand the principles that govern the reactivity of a protein toward chemical reagents, it is necessary to consider the general properties of the amino acid side chains. With the advent of genomic projects, protein–nucleic acid interactions have become increasingly crucial to our understanding of gene regulation and expression. Chemical modifications and cross-linking of protein–nucleic acids have provided valuable information in many biological systems. It is therefore important to review the chemical reactivities of nucleic acids toward chemical reagents as well.

2.2 PROTEIN COMPOSITION

2.2.1 AMINO ACIDS

All proteins are composed of amino acids. Some proteins may contain, in addition to amino acids, other groups such as carbohydrates, lipids, other organic moieties, and metal ions. There are 20 common amino acids with side chains of different sizes, shapes, charges, and chemical reactivity (although selenocysteine is sometimes considered to be the 21st amino acid). These amino acids join through an amide or a peptide bond between the amino and carboxyl groups. The peptide bond is stabilized by resonance, as shown in Figure 2.1, and is not reactive toward chemical reagents except to undergo hydrolysis. The number of amino acids that join together can be as few as two in a dipeptide, as many as over 4000 as in apolipoprotein B,[5] or much more in the case of titan, a nearly 4 MDa single polypeptide chain.[6] As a newly formed protein emerges from the ribosome and as the number of its amino acids increases, the protein chain begins to fold into a specific 3D structure. The degree of hydrophobicity and hydrophilicity of the amino acid side chain composition is one of the major determinants of the 3D structure of proteins (Table 2.1).[7,8] Glycine, alanine, valine, leucine, isoleucine, methionine, and proline have nonpolar aliphatic side chains while phenylalanine and tryptophan have nonpolar aromatic side groups. These hydrophobic amino acids are generally found in the interior of proteins forming the hydrophobic core. The forces involved in such structures are van der Waals forces and so-called hydrophobic forces (considered by some to be a misnomer since these are not "forces" per se but rather the consequence of entropic considerations), although the former is weaker than the latter.[9] Other amino acids, such as arginine, aspartic acid, glutamic acid, cysteine, histidine, lysine, and tyrosine, have ionizable side chains. Together with asparagine, glutamine, serine, and threonine, which contain nonionic polar groups, they are usually located on the protein surface where they can interact strongly with the aqueous environment. The side chains of these amino acids also interact with each other through electrostatic forces and hydrogen bonding. While it can be assumed in general that hydrophobic side chains are buried within the protein and that hydrophilic amino acids are exposed, nonpolar groups may be found on the surface and polar groups may be buried.[9] This scenario is particularly true for amino acids such as methionine, tryptophan, and tyrosine that have both hydrophilic and hydrophobic moieties. Thus, the reactivity of a given protein, in terms of its ability to be chemically modified, will be determined largely by its amino acid composition and the location of the individual amino acids in the 3D structure of the protein. The diversity of amino acid composition and its conformation imparts many of the different chemical reactivities of a protein. Invariably, however, since lysine residues are usually the most abundant amino acid found in proteins, nucleophilic amino groups will be found on the surface of a protein. The question is, then: How many?

2.2.2 PROSTHETIC GROUPS

In addition to the various amino acids, some proteins also contain tightly bound prosthetic groups. These include metal ions, porphyrin groups, coenzymes such as biotin, and other nonpeptidyl moieties. Although many of these structures may contribute to the chemical reactivity of the protein, the most important prosthetic group in the consideration of the protein conjugation is the carbohydrate.

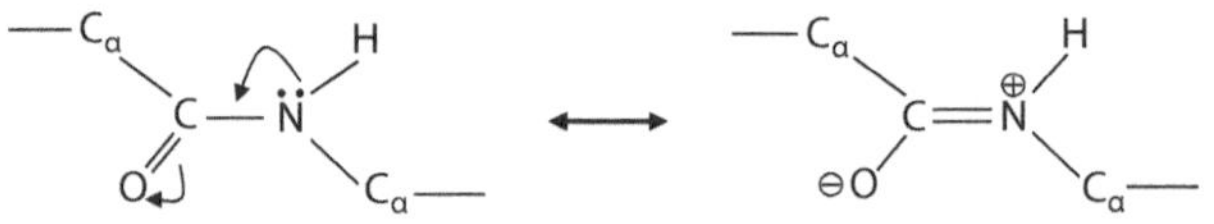

FIGURE 2.1 Resonating hybrids of the peptide bond. Resonance of the peptide bond makes the amide group inactive to chemical reagents.

TABLE 2.1
Side Chain Chemical Structure and Hydrophobicity
of 20 Common Amino Acids

Amino Acid	Side Chain Structure	Hydrophobicity
Phenylalanine	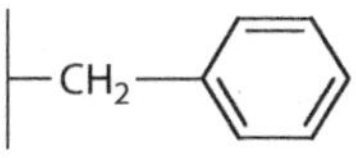 —CH₂—⬡ (benzene ring)	1.000
Leucine	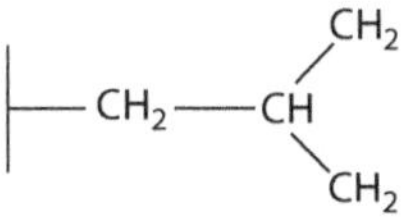 —CH₂—CH(CH₂)(CH₂)	0.943
Isoleucine	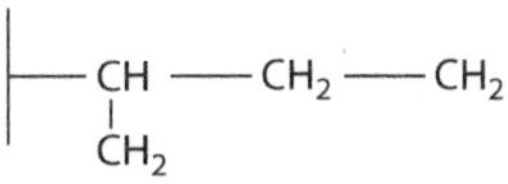 —CH(CH₂)—CH₂—CH₂	0.943
Tyrosine	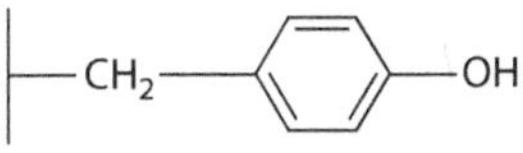 —CH₂—⬡—OH	0.880
Tryptophan	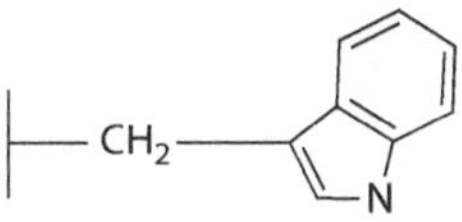 —CH₂—indole (N)	0.878
Valine	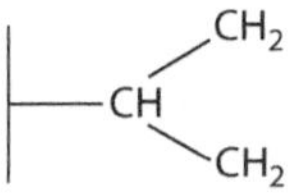 —CH(CH₂)(CH₂)	0.825
Methionine	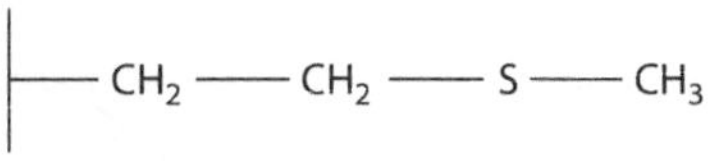 —CH₂—CH₂—S—CH₃	0.738
Proline	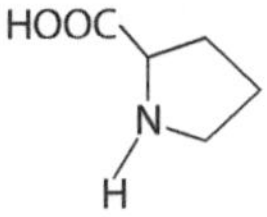 HOOC—pyrrolidine (N–H)	0.711
Cysteine	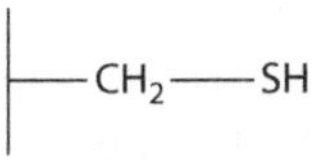 —CH₂—SH	0.680
Alanine	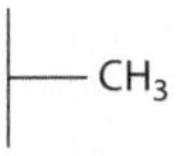 —CH₃	0.616
Glycine	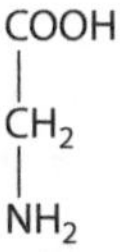 COOH—CH₂—NH₂	0.501
Theonine	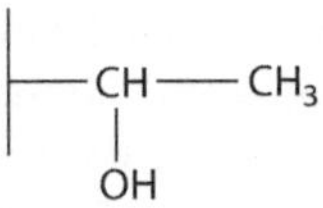 —CH—CH₃ (OH)	0.450
Serine	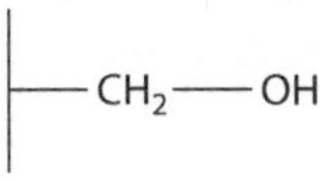 —CH₂—OH	0.359

(continued)

TABLE 2.1 (Continued)
Side Chain Chemical Structure and Hydrophobicity
of 20 Common Amino Acids

Amino Acid	Side Chain Structure	Hydrophobicity
Lysine	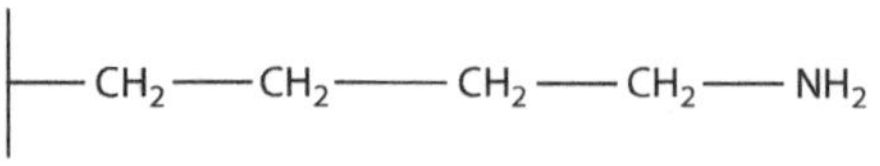 $-CH_2-CH_2-CH_2-CH_2-NH_2$	0.283
Glutamine	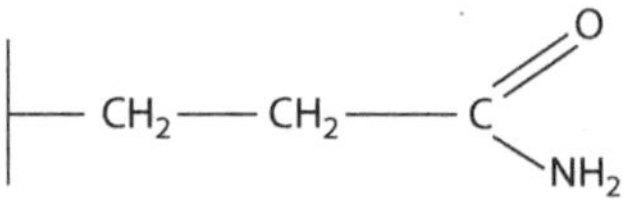 $-CH_2-CH_2-C(=O)NH_2$	0.251
Asparagine	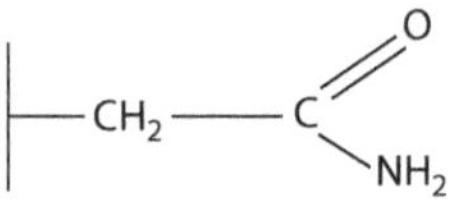 $-CH_2-C(=O)NH_2$	0.236
Histidine	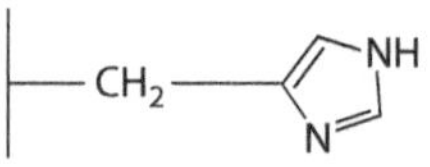 $-CH_2-$ imidazole (NH, N)	0.165
Glutamate	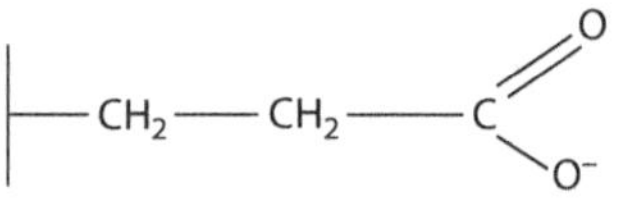 $-CH_2-CH_2-C(=O)O^-$	0.043
Aspartate	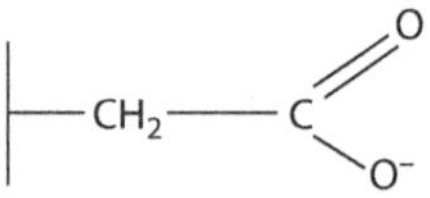 $-CH_2-C(=O)O^-$	0.028
Arginine	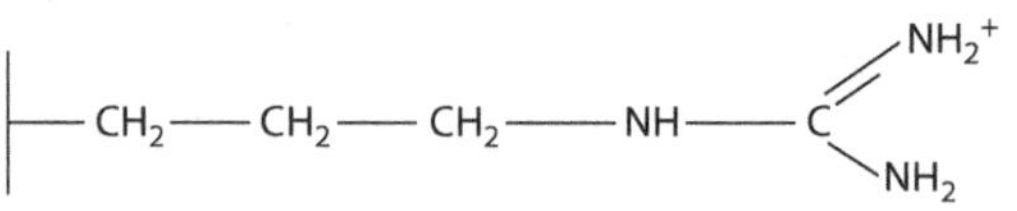 $-CH_2-CH_2-CH_2-NH-C(=NH_2^+)NH_2$	0.000

The hydrophobicity is presented in a scale ranging from 1.000 for the most hydrophobic to 0.000 for the least hydrophobic, according to Black and Mould.[7]

Glycoproteins may contain up to 50% or more of carbohydrates by weight as in proteoglycans.[10] Generally, the carbohydrates in glycoproteins are short, frequently branched sugar chains of 15 residues or less. They are covalently attached to proteins through *O*-glycosidic linkages to the hydroxyl groups of serine, threonine, or hydroxylysine or through *N*-glycosidic linkages to the amide nitrogen of asparagine. The asparagine-linked oligosaccharides are better understood and seem to be more common in glycoproteins.

All asparagine-linked oligosaccharides have in common a mannose-*N*-acetylglucosamine (Man-GlcNac) core of three mannose (Man) and two *N*-acetylglucosamine (GlcNAc) residues through which the oligosaccharide is linked to asparagine.[11] The anomeric carbon of GlcNAc forms the β-*N*-glycosidic linkage with the amide nitrogen of asparagines as shown in Figure 2.2. Additional mannose residues may be attached to this common core to form the high mannose type oligosaccharide. In the complex type, sialic acid (Sia), galactose (Gal), GlcNAc, and L-fucose (Fuc) residues are built on the core. A hybrid type is formed when these sugars and mannose residues are added to the Man-GlcNAc core (Figure 2.3).

Although the precise function of the carbohydrate moiety of most glycoproteins is unknown, oligosaccharide units provide a useful site for chemical modification and cross-linking of proteins. The principle and method of these reactions will be considered later in this chapter.

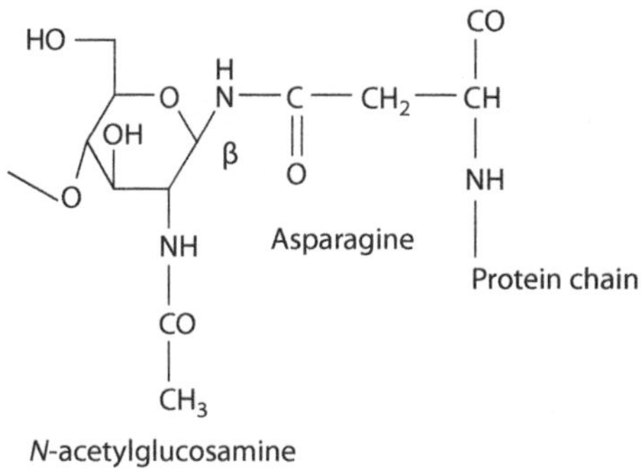

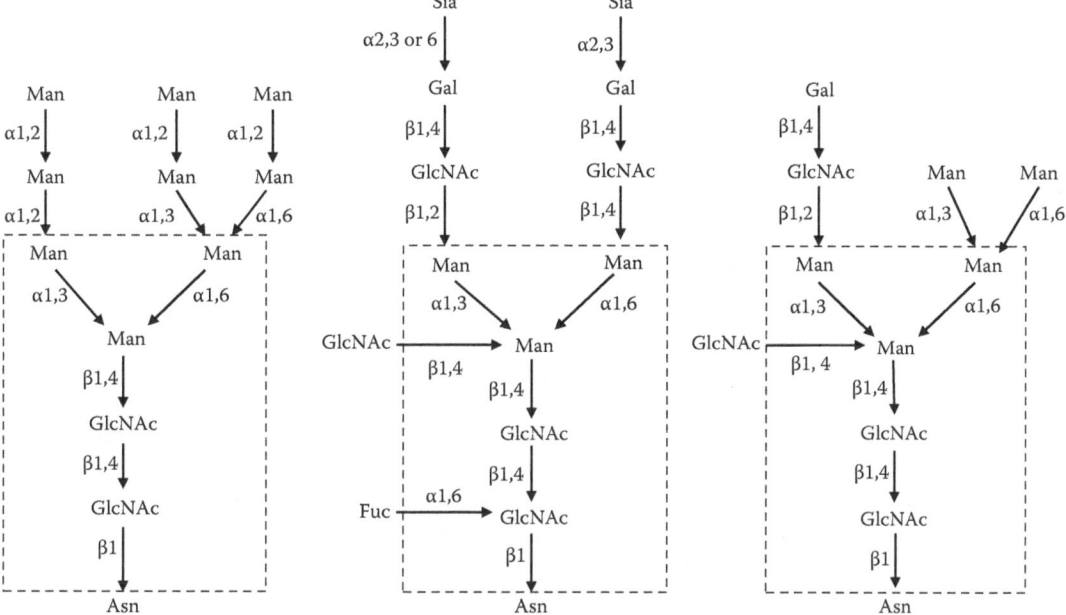

FIGURE 2.2 The β1-*N*-acetylglucosamine-asparagine linkage.

FIGURE 2.3 Types of asparagine-linked oligosaccharides in glycoproteins. The common core of three mannose (Man) and two *N*-acetylglucosamine (GlcNAc) residues linked to Asn are boxed. When additional sugars such as Man, sialic acid (Sia), galactose (Gal), GlcNAc, and Fuc (fucose) are attached to the core, many different patterns are formed, referred to as complex, high mannose, or hybrid types. The sugar linkages are also indicated in the figure. (Adapted from Kornfeld, R. and Kornfeld, S., *Annu. Rev. Biochem.*, 54, 631, 1985.)

2.3 PROTEIN FUNCTIONAL GROUPS

2.3.1 REACTIVE AMINO ACID SIDE CHAINS

In the final analysis, the chemical reactivities of proteins depend on the side chains of their amino acid compositions as well as the free amino and carboxyl groups of the N- and C-terminal residues, respectively. The terminal residues, however, contribute little of significance to chemical modification since they are of limited number compared to the overall large number of amino acids in the protein. Furthermore, only a few of the amino acid side chains are really reactive.[12] Of the 20 amino acids, the alkyl side chains of the hydrophobic residues are for all intents and purposes chemically inert. The aliphatic hydroxyl groups of serine and threonine can be considered as water derivatives and therefore have a low reactivity. Only eight of the hydrophilic side chains are chemically active.[2]

TABLE 2.2
Chemical Reactions of Active Amino Acid Side Chains

Amino Acid	Active Side Chain	Alkylation or Arylation	Acylation	Oxidation	Other Reactions[a]
Cysteine	$-CH_2SH$	+	+	+	a,d,f,h
Lysine	$-NH_2$	+	+	–	c,e,g
Methionine	$-S-CH_3$	+	–	+	i
Histidine	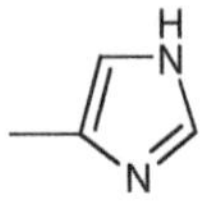 (imidazole structure)	+	+	+	a,c
Tyrosine	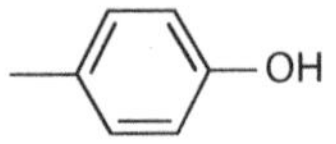 (phenol structure)—OH	+	+	+	a,b,c,d
Tryptophan	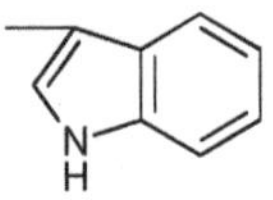 (indole structure)	+	–	+	h
Arginine	$-NH-C(=NH)-NH_2$	–	–	–	g
Aspartic and glutamic acid	$-COOH$	–	+	–	d,e

[a] Other reactions include (a) iodination, (b) nitration, (c) diazotization, (d) esterification, (e) amidation, and reaction with (f) mercurials, (g) dicarbonyls, (h) sulfenyl halides, and (i) cyanogen bromide.

These are the guanidinyl group of arginine, the β- and γ-carboxyl groups of aspartic and glutamic acids, respectively, the sulfhydryl group of cysteine, the imidazolyl group of histidine, the ε-amino group of lysine, the thioether moiety of methionine, the indolyl group of tryptophan, and the phenolic hydroxyl group of tyrosine (Table 2.1). Table 2.2 summarizes the various chemical reactions that can occur with these active side chains. The most important reactions are alkylation and acylation.[13] In alkylation, an alkyl group is transferred to the nucleophilic atom, whereas in acylation, an acyl group is bonded. Since methionine and tryptophan are generally buried in the interior of proteins and are thereby protected from reagents dissolved in the solvent, they show only some selected reactivity in intact proteins. The other ionizable groups are normally exposed on the surface of proteins. They are therefore the target of protein cross-linking and conjugation.

2.3.1.1   Relationship between Nucleophilicity and Reactivity

Most of the protein modification reactions are nucleophilic reactions, involving a direct displacement of a leaving group by the attacking nucleophile, which in this case is the amino acid side chain (Figure 2.4). The rate of such a bimolecular nucleophilic substitution reaction, the S_N2 mechanism, depends on at least two factors: the ability of the leaving group to leave and the nucleophilicity of the attacking group. The more facilely the leaving group comes off, the faster will be the reaction.[13,14]

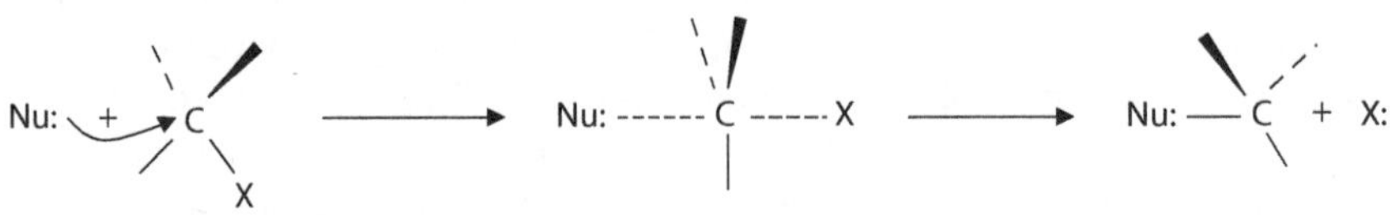

FIGURE 2.4   The nucleophilic substitution reaction, the S_N2 reaction mechanism. The nucleophile (Nu:) attacks an electron-deficient center displacing the good leaving group, (X:).

Similarly, the greater the nucleophilicity of the attacking group, the more expeditiously will the product be formed.[15,16] In terms of protein modification, the relative chemical reactivity is basically a function of the nucleophilicity of the amino acid side chains.

A nucleophile is any species, which has an unshared pair of electrons, whether it is neutral or negatively charged, that is, any Lewis base. The relative availability of these electrons for attacking the positive centers of the substrate determines the nucleophile's relative reactivity.[17] Although nucleophilicity is influenced by various factors such as solvation, size, and bond strength,[15] there are three basic rules of thumb that govern the nucleophilicity of a chemical species[15,16]:

1. A negatively charged nucleophile is always more powerful than its conjugate acid. Thus, ArO^- is more powerful than $ArOH$, and HO^- more powerful than HOH.
2. Going across the same row of the periodic table, nucleophilicity is roughly proportional to the basicity.[18–20] An approximate order of nucleophilicity is $NH_2^- > RO^- > OH^- > ArO^- > RNH_2 > NH_3 > H_2O$.
3. Going down the same column of the periodic table, nucleophilicity increases. Thus, sulfur is a more powerful nucleophile than its oxygen analogs.

Although these rules do not always hold, Edwards and Pearson[16] have formulated an overall nucleophilicity order: $RS^- > ArS^- > I^- > CN^- > HO^- > N_3^- > Br^- > ArO^- > Cl^- >$ pyridine $> AcO^- > H_2O$. With this formulation, it may be deduced that the sulfhydryl group of cysteine is the most potent nucleophile in the protein, particularly in its thiolate form. Nitrogen, as in the amino group, is considerably less potent followed by oxygen and carbon. It should be pointed out that the aliphatic hydroxyl groups of serine and threonine, having about the same nucleophilicity as water, are generally unreactive in aqueous solutions.[21] This lack of reactivity must also be considered in the context of the high concentration of water molecules (55 M) against which the aliphatic alcohols will have to compete. A very reactive reagent will favorably undergo hydrolysis before it reacts with the hydroxyl group containing amino acids. However, the aliphatic hydroxyl group could be activated to its hydroxylate ion as discussed below. In this case, it would be more reactive than water. Similarly, in nonaqueous solvents, the hydroxyl group may react effectively. Finally, we note that more recent approaches to nucleophilicity scales have been proposed and an excellent review of the topic was presented by Mayr and Patz,[22] but a complete consideration of this field is beyond the scope of this book.

2.3.1.2 Effects of pH

Because protonation decreases the nucleophilicity of a species, the pH of the medium affects the rate of many nucleophilic reactions. The relationship between protonation and pH depends on the pK_a of the nucleophile. Table 2.3 lists the pK_as of the reactive groups in free amino acids and in model peptides. Because the pK_a is a function of temperature, ionic strength, and microenvironment of the ionizable group,[23,24] this table reflects only the approximate values of these groups in proteins. Using these values, the ratio of protonated to deprotonated species at a certain pH can be calculated with the Henderson–Hasselbalch equation:[25]

$$pH = pK_a + \log\left\{\frac{[A^-]}{[AH]}\right\}$$

It can be shown from such calculations that the following general rules hold:

1. At one pH unit below its pK_a, the species is 91% protonated.
2. At two pH units below its pK_a, the species is 99% protonated.
3. At one pH unit above its pK_a, the species is 91% deprotonated.
4. At two pH units above its pK_a, the species is 99% deprotonated.
5. When the pH is the same as pK_a, 50% protonation occurs.

TABLE 2.3

pK$_a$ of Reactive Groups in Proteins

Functional Group	Amino Acid Residue	pK$_a$ in Free Amino Acid	pK$_a$ in Model Peptides
α—COOH	C-terminal	1.8–2.6	3.1–3.7
β—COOH	Aspartic acid	3.9	4.4–4.6
γ—COOH	Glutamic acid	4.3	4.4–4.6
α—NH$_3^+$	N-terminal	8.8–10.8	7.6–8.0
ε—NH$_3^+$	Lysine	10.8	10.0–10.2
—SH	Cysteine	8.3	8.5–8.8
—OH	Serine and threonine	>13	—
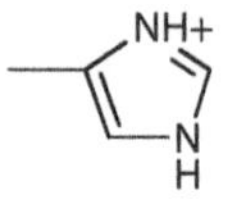 (imidazole, NH+)	Histidine	6.0	6.5–7.0
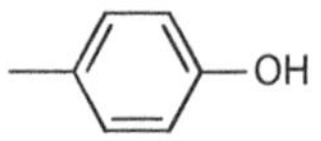 (phenol, —OH)	Tyrosine	10.9	9.6–10.0
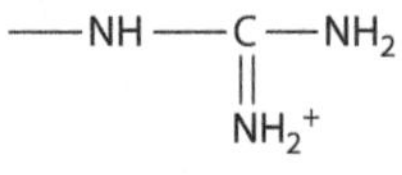 —NH—C(=NH$_2^+$)—NH$_2$	Arginine	12.5	>12

From earlier discussions, it can be shown that uncharged (deprotonated) species have greater nucleophilicity. Thus, at a fixed pH, the most reactive group is usually the one with the lowest pK$_a$, since it is more likely to be deprotonated. Because of their differences in pK$_a$ values, the degree of protonation of different amino acid side-chain groups at a certain pH provides a basis for differential modification. For example, at neutrality, the amino group of lysine is protonated (pK$_a$ about 10) rendering them unreactive. On the other hand, carboxyl and imidazolyl groups are deprotonated and thus would be more reactive. At pH 5, the imidazolyl group will be over 90% protonated (pK$_a$ = 6), leaving only the carboxyl group in the ionic form. For a selective reaction with the carboxyl group, such as with diazoacetate, the condition of an acidic pH should be selected. At higher pHs, other nucleophiles, particularly the sulfhydryl group, will react. As a consequence, it should be obvious that changing the pH also provides a means to control the course of a chemical reaction. An example of such pH control is provided by the iodination reaction of tyrosine. In this reaction, the phenolate anion has been shown to be the reactive species.[26,27] The unionized residue reacts very slowly or not at all. Thus, the rate of iodination of tyrosine increases with increasing pH as the tyrosine anion concentration increases. Consequently, the state of relative reactivity of an individual amino acid side chain to modifying reagents might be considered as the overriding control of the route and extent of modification.

Since the thioether group of methionine is usually not protonated, pH has little effect on its reactivity. Also, the high pK$_a$ of the hydroxyl groups of serine and threonine residues provides an explanation as to why these residues are normally inert at neutral pH, unless they are activated by neighboring groups as in serine proteases discussed below.

2.3.1.3 Effects of Microenvironment

The pK$_a$ values of the reactive amino acid side chains, shown in their free states in Table 2.3, may be influenced by the local microenvironments of the amino acid side chains. As a consequence, their pK$_a$ values in certain peptides are different (Table 2.3). When an ionizable group is in a hydrophobic environment, it will tend not to ionize as the resulting charged group in such an environment will not be stable. Thus, its pK$_a$ will become apparently higher. On the other hand, if such an ionizable group is in a polar or charged environment, it will tend to ionize, resulting in an apparently lower pK$_a$.

TABLE 2.4
Alteration of pK$_a$ as a Result of the Local Environment

Local Microenvironment	Effect on pK$_a$
Hydrophobic environment	Increase
Polar environment	Decrease
Adjacent to like charge	Increase
Adjacent to opposite charge	Decrease
Environment capable of forming salt bridges and hydrogen bonds	Decrease

For example, the pK$_a$ of acetic acid is affected by the presence of ethanol. In water alone, its pK$_a$ is 4.7. In absolute ethanol, it is 10.3. In 80% aqueous ethanol, it is 6.9.[28] It is obvious that the hydrophobicity of the environment increases the pK$_a$ value. Table 2.4 reflects some of the effects of microenvironment on pK$_a$. In proteins, the amino acid side chains have different microenvironments depending on their locations in the protein sequence and hence their spatial locations in the protein. Thus, there is great variation in their reactivity. Few, if any, of the amino acid side chains are totally free from interacting with their neighboring groups. These interactions are particularly true for those residues situated in the interior of the protein where they interact through hydrogen bonding, electrostatic attraction, and van der Waals forces. Those groups on the surface may also be involved in such interactions with their neighbors or with solvent molecules. Consequently, surface polarities around a functional group may affect its chemical and physical properties. Of particular importance is the effect on the pK$_a$ of the dissociable side chains. Differences in local microenvironment will render different effects on pK$_a$ values for identical groups at various sequence positions.[3,4] As a consequence, the pK$_a$ of a group in one protein may not be the same in another protein or, in fact, in the same protein but at a different sequence location. This aspect is revealed by the ranges of pK$_a$ values of the ionizable groups in peptide model compounds as shown in Table 2.3. Thus, the pK$_a$ values of the free amino acids are not definitive but only indicative of their reactivity in a protein. In general, however, most of the pK$_a$s of a protein fall within those expected values shown in Table 2.3, but there are dramatic exceptions. Table 2.5 depicts some examples.[29,30]

One of the most remarkable cases of deviation from the expected pK$_a$s is the activation of hydroxyl side chain in serine proteases.[31] As a free entity, serine is relatively inert because of its high pK$_a$. However, at the active site of serine proteases, chymotrypsin, for example, the hydroxyl proton of serine is hydrogen bonded to the nitrogen of an imidazolyl moiety of a histidine residue, which is polarized by a buried aspartic acid in a charge-relay system as shown in Figure 2.5.[32] The abstraction of the proton activates the hydroxyl group to an excellent nucleophilic alkoxide. The inactive hydroxyl group can now participate as a nucleophile in chemical catalysis.

In addition to the effect of the microenvironment on pK$_a$ values, the local environment can hinder the accessibility to a reactive group. If a chemical modification agent is bulky and the reactive group is located in protein pocket, the reaction may not occur since the groups may not be able to approach close enough for the reaction to occur. Such is the case, of course, for those functional groups buried inside protein molecules. Steric hindrance by neighboring groups will obviously reduce accessibility and therefore the reactivity of a particular group. Such steric effects should be taken into consideration when designing cross-linking reagents.

2.3.2 CHEMICALLY INTRODUCED REACTIVE GROUPS

In addition to the side chains of the 20 amino acids, special reactive groups can be introduced into proteins. These extrinsic moieties are obtained by chemical modification of the existing functional groups.

TABLE 2.5

pK$_a$s of Ionizable Groups in Some Proteins

Group	pK$_a$ (Expected)	Protein	pK$_a$ (Found)
—COOH	3.1–4.6	Ribonuclease	4.7
		β-Lactoglobulin	4.8 (49 of 51 groups)
			7.3 (2 of 51 groups)
		Insulin	4.7
		Serum albumin	4.0
ε—NH$_3^+$	10.0–10.2	Serum albumin	9.8
		Lysozyme	10.4
		Ribonuclease	10.2
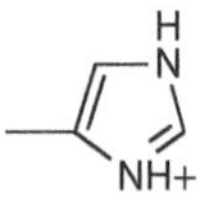 (imidazole)	6.5–7.0	Ribonuclease	6.5
		β-Lactoglobulin	7.4
		Insulin	6.4
		Myoglobin	6.6 (6 of 12 groups)
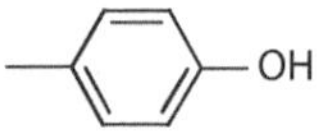 (phenol) —OH	9.6–10.0	Ribonuclease	9.95 (3 of 6 groups)
		Insulin	>12 (3 of 6 groups)
		Serum albumin	9.6
		Chymotrypsinogen	10.4
			9.7 (1 of 4 groups)
			10.4 (1 of 4 groups)

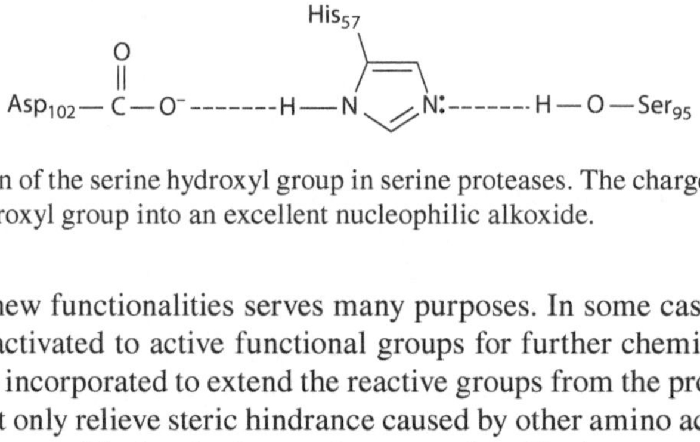

FIGURE 2.5 Activation of the serine hydroxyl group in serine proteases. The charge relay system of chymotrypsin converts the hydroxyl group into an excellent nucleophilic alkoxide.

The addition of such new functionalities serves many purposes. In some cases, inactive units such as carbohydrates are activated to active functional groups for further chemical reactions. In other cases, spacer arms are incorporated to extend the reactive groups from the protein into the medium. This extension will not only relieve steric hindrance caused by other amino acid side chains but will also decrease the influence of the local microenvironment. In still other cases, functional groups are converted into one another to either change their specificity or increase their reactivity. The importance of these manipulations is clearly demonstrated by many examples found in the literature.

2.3.2.1   Reduction of Disulfide Bonds

Disulfide bonds formed from two cysteine residues cross-link two portions of a protein where these amino acids are located. This oxidized form of sulfur is relatively unreactive but can be easily activated by reduction to free sulfhydryl groups. Any thiol-containing compound such as dithiothreitol, dithioerythreitol, 2-mercaptoetbanol, or 2-mercaptoethylamine can serve as a reducing agent.[33,34] The reaction is specific for disulfide bonds and involves disulfide interchange as illustrated in Figure 2.6. Complete conversion of disulfide to free thiol can be achieved with excess reducing agents. With dithiothreitol, a low level is enough to drive the reaction to completion because of the thermodynamically favored formation of a six-membered ring product.[35] It should be mentioned that these mild reagents will generally reduce only the exposed disulfide bonds but not those buried inside the protein. In this case, the integrity of the protein's 3D structure will be preserved. Other stronger reducing

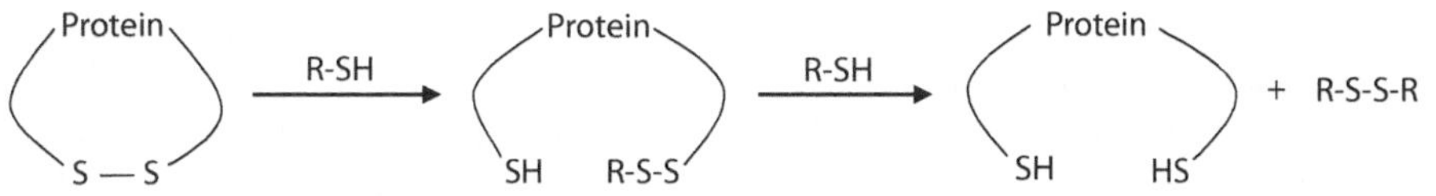

FIGURE 2.6 Reduction of disulfide bonds. The reaction involves protein disulfide interchange reaction with a free thiol.

agents such as sodium borohydride and lithium aluminum hydride will also reduce disulfide bonds, but these reagents are usually used for complete reduction of proteins after denaturation.

The reduction of disulfide bonds has been used to prepare immunoglobulin fragments for various conjugation reactions.[36,37] These reactions will be discussed later in this book.

2.3.2.2 Interconversion of Functional Groups

Introduction of new functional groups through the modification of existing amino acid side chains provides additional diversity in the application of cross-linking reagents. Under certain circumstances, for example, in the preparation of immunotoxins (Chapter 13), it is desirable to convert one functional group into another. This conversion will either increase its nucleophilicity or change its specificity toward a reagent. For immunotoxins, converting an amino group into a sulfhydryl group enables the preparation of cleavable conjugates, which are desired for *in vivo* toxicity. The art of introduction of new functional groups is multifarious, limited only by the imagination of the researcher. The following sections will illustrate the voluminous methods used in the transformation of various functionalities.

2.3.2.2.1 Conversion of Amines to Carboxylic Acids

Amino groups are probably the most abundant hydrophilic group on the surface of a protein. The constituents are the ε-amino group of lysines and the α-amino group of N-terminal amino acids. These exposed functional groups on the surface of proteins are susceptible for conversion to other functionalities. Reaction with dicarboxylic acid anhydrides will convert these amines to carboxylic acids. Succinic and maleic anhydrides are the two most commonly used dicarboxylic acid anhydrides. Although the mechanisms of reaction are similar, the products are different (Figure 2.7). The product of maleylation is stable at neutral pH but rapidly hydrolyzes at acidic pHs.[38] Thus, succinic anhydride is the reagent of choice for introduction of carboxyl groups for the purpose of chemical cross-linking, though it also reacts with tyrosyl, histidyl, cysteinyl, seryl, and threonyl side chains. The tyrosyl and histidyl derivatives formed are reversible, and are either hydrolyzed spontaneously at alkaline pH or rapidly decomposed by hydroxylamine. Ester and thioester derivatives are also susceptible to hydroxylamine cleavage at pH 10. Thus, specific succinylation of amino groups is possible.[39]

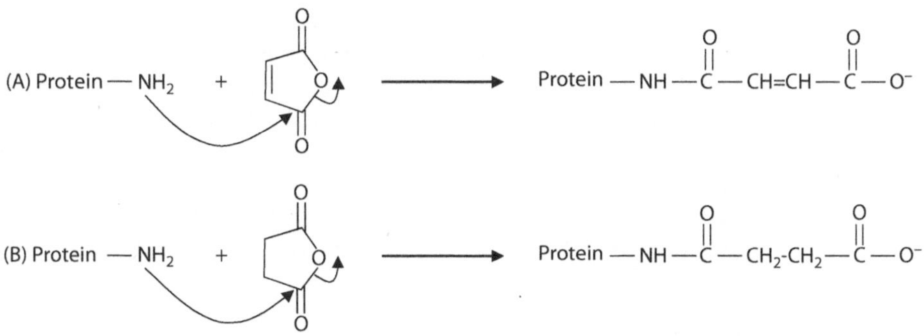

FIGURE 2.7 Conversion of an amine to a carboxylic acid. (A) Reaction with maleic anhydride; (B) reaction with succinic anhydride.

Succinylation and maleylation of a protein result in a net change of a cationic amino group to an anionic acid. Such electrostatic alternation frequently brings about conformational changes and provides a means to dissociate protein complexes.[40,41] Succinylation has been used to hinder tryptic attacks at lysine residues, thus giving rise to peptides with only arginine residue at the carboxy terminus. Its use in chemical cross-linking of proteins is limited. However, such conversion has provided a means of diversifying the surface chemistry of solid supports for the immobilization of proteins (Chapter 14).

2.3.2.2.2 Conversion of Amino to Sulfhydryl Groups

There are several approaches by which a free thiol can be linked to the amino group. Amino-specific reagents containing a free sulfhydryl group have been developed. However, since the thiol is a potent nucleophile, most of the effective reagents generally contain protected sulfhydryl groups, which can be activated to generate free thiols after the amino groups have been modified. These reactions are shown in Figure 2.8 and will be discussed below.[42]

2.3.2.2.2.1 Thiolation with N-Acetylhomocysteine Thiolactone   N-acetylhomocysteine thiolactone was introduced by Benesch and Benesch as a thiolating agent.[43,44] The thiol group is masked in the thiolactone ring. Nucleophilic attack on the carbonyl group of thiolactone opens the ring, liberating a free thiol. Direct reaction of the compound with an amino group of a protein proceeds very slowly except at a rather alkaline pH of 10–11. The reaction can be catalyzed by adding Ag^+ so that it can be carried out near neutral pH. The resulting thiolate moiety contains an acetamido group, which may interfere with the reactivity of the thiol (Figure 2.8A).

2.3.2.2.2.2 Thiolation with S-Acetylmercaptosuccinic Anhydride   Proteins can be thiolated with S-acetylmercaptosuccinic anhydride in a reaction involving a two-step process wherein the amino group first reacts with the thiol-blocked reagent.[45] The reaction occurs adequately at pH 7. At the end of the reaction, the free sulfhydryl group is generated by treatment with 0.01–0.05 M hydroxylamine at pH 7–8 (Figure 2.8B). Similar to the succinylation reaction discussed above, the resulting thiolated protein contains a side chain bearing a negative charge, which may affect the reactivity of the free thiol.

2.3.2.2.2.3 Thiolation with Thiol-Containing Imidoesters   Several imidoesters have been used to thiolate proteins. Perham and Thomas[46] prepared methyl 3-mercaptopropionimidate hydrochloride (Figure 2.8C) and Traut et al.[47,48] synthesized methyl 4-mercaptobutyrimidate (Figure 2.8D). The latter compound gradually cyclizes on storage with elimination of methanol to form 2-iminothiolane (Traut's reagent) (Figure 2.8E).[49,50] The cyclic imidothioester has been used for thiolation of proteins (Figure 2.8E),[49,51] although it is less reactive than the corresponding open-chain methyl 4-thiobutyrimidate. It is, however, more stable and can be stored in the cold for many months. Other imidoesters slowly decompose and can only be stored for a limited amount of time, except pyridylthio-protected methyl 3-mercaptopropionimidate, which can be used to thiolate proteins and other disulfide compounds (Figure 2.8F). These imidoesters react readily with amino groups at pH 7–10 and are excellent thiolating agents since they are amine-specific and do not change the net charge of the reacted protein. Although the primary target of thiolation is the amino group, there is a side reaction with hydroxyl group, which may become prominent with glycoproteins such as antibodies. The number of hydroxyls on the polysaccharide may increase the rate of reaction. With the Traut's reagent, the reaction with amines is 100 times faster than with hydroxyl groups.[52]

2.3.2.2.2.4 Thiolation with Thiol-Containing Succinimidyl Derivatives   In addition to methyl 3-(4-pyridyldithio)propionimidate hydrochloride mentioned above, amine-specific disulfide compounds such as N-succinimidyl 3-(2-pyridyldithio)propionate (SPDP)[53] (Figure 2.9A) and dithiobis(succinimidyl

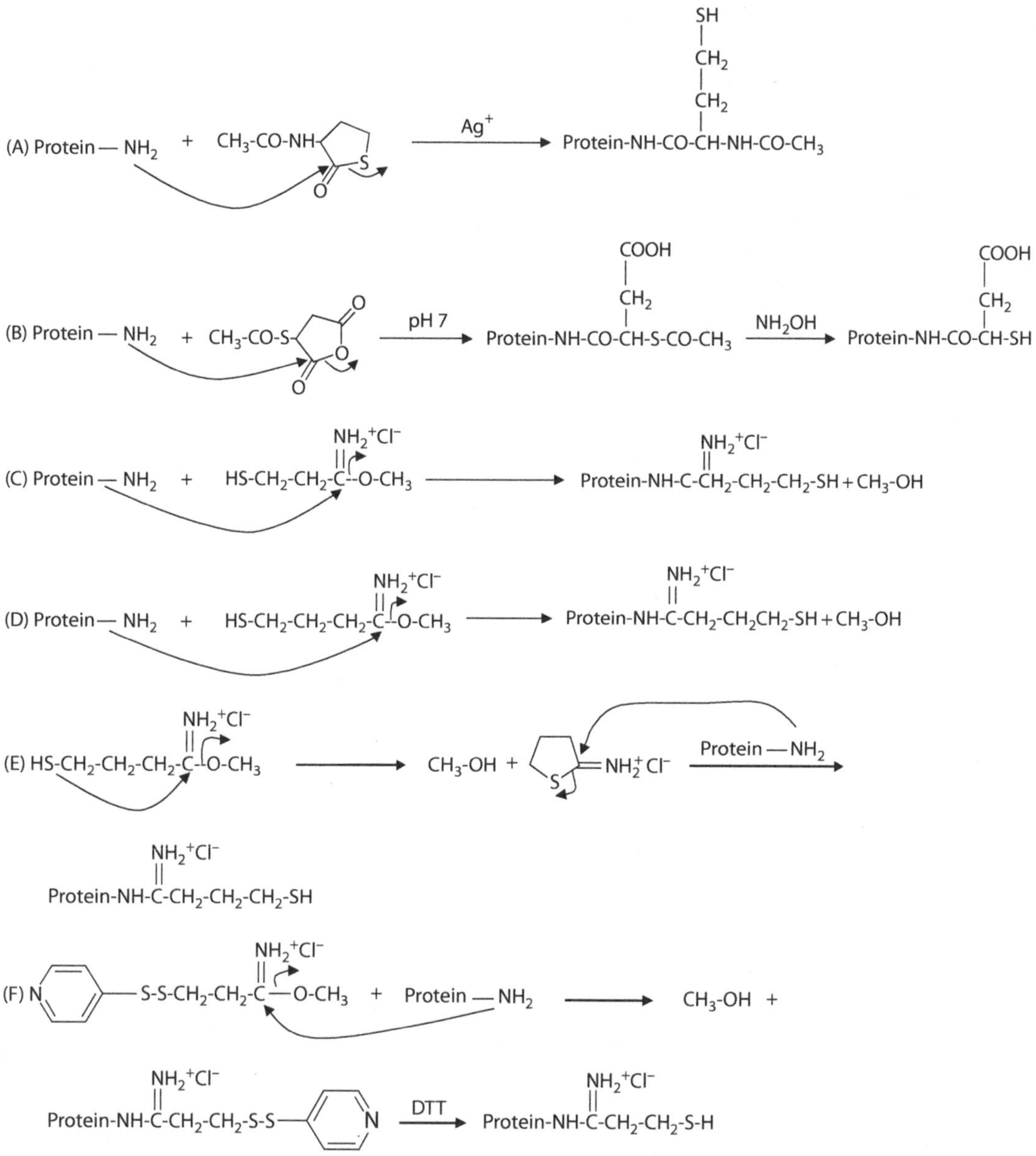

FIGURE 2.8 Methods of protein thiolation. (A) Thiolation with *N*-acetylhomocysteine thiolactone; (B) reaction with *S*-acetylmercaptosuccinic anhydride; (C) thiolation with methyl 3-mercaptopropionimidate; (D) reaction with methyl 4-mercaptobutyrimidate; (E) cyclization of methyl 4-mercaptobutyrimide to form iminothiolane (Traut's reagent) and its thiolation reaction; (F) thiolation with methyl 3-(4-pyridyldithio) propionimidate hydrochloride.

propionate) (DSP)[54] (Figure 2.9B) have been used in the same reaction scheme as thiolating agents, providing additional alternatives for the introduction of spacers. Reduction of the introduced disulfide bond with dithiothreitol (DTT) produces the free sulfhydryl group. Other succinimidyl esters useful for thiolation of proteins are *N*-succinimidyl *S*-acetylthioacetate (Figure 2.10A)[55] and *N*-succinimidyl *S*-acetylthiopropionate (Figure 2.10B).[56] After reaction with amino groups, these protected thioesters are treated with hydroxylamine to liberate the free thiol group as shown in Figure 2.10. The *N*-hydroxysuccinimide esters are stable, crystalline compounds that react cleanly with amines.

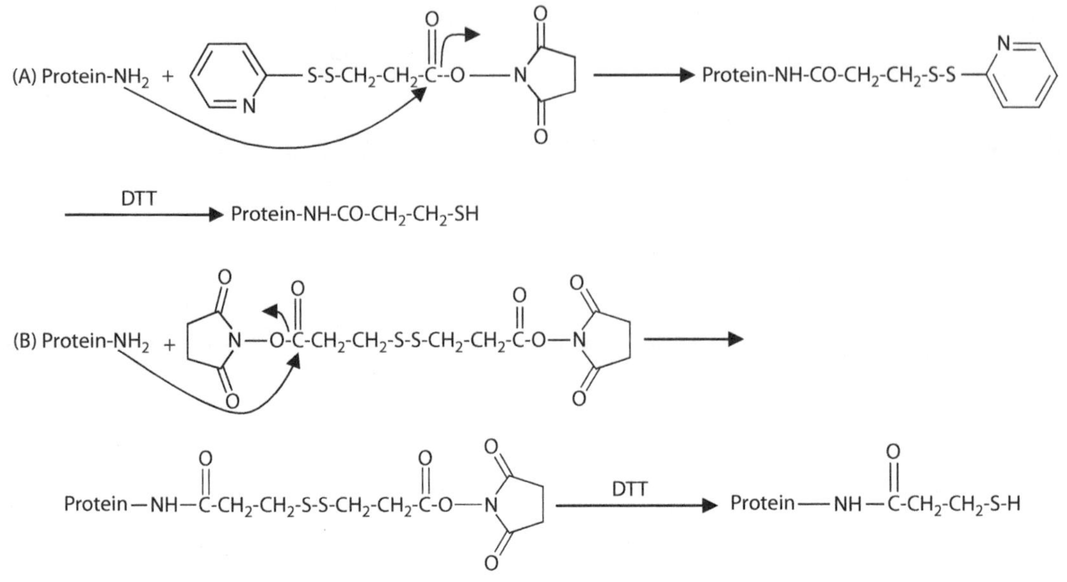

FIGURE 2.9 Thiolation using succinimidyl ester-containing disulfide compounds. (A) Reaction with succinimidyl 3-(2-pyridyldithio)propionate (SPDP); (B) reaction with dithiobis(succinimidylpropionate) (DSP). Dithiothreitol (DTT) reduces the disulfide bond.

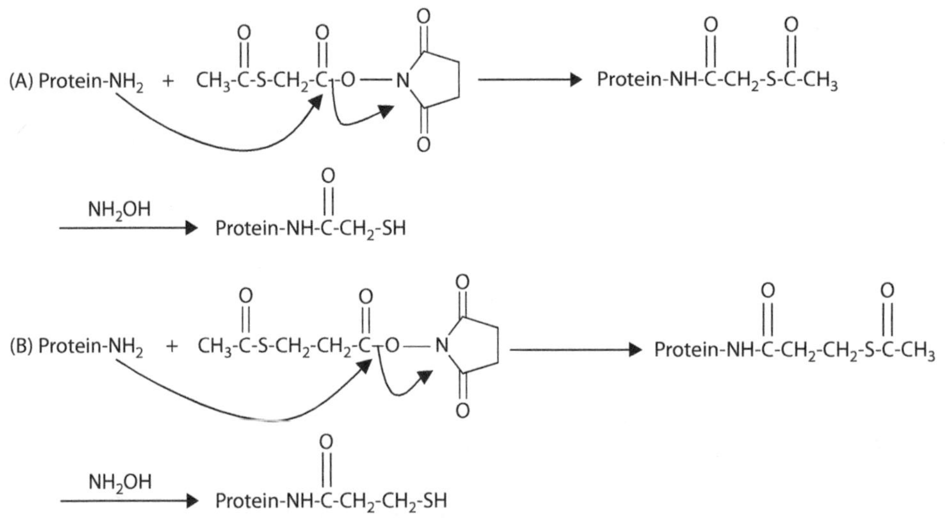

FIGURE 2.10 Thiolation with thiol-protected succinimidyl esters. (A) Succinimidyl acetylthioacetate; (B) succinimidyl acetylthiopropionate.

2.3.2.2.2.5 Other Reactions   Water-soluble carbodiimides have been used by Jou and Bankert to thiolate erythrocytes with dithiodiglycolic acid.[57] An amide bond is formed between the carboxyl group of the reagent and an amino group of the protein. The coupled component is then reduced with excess dithiothreitol to generate the free thiol group (Figure 2.11).

Other reagents that have been used to introduce thiol groups are 3-(3-acetylthiopropionyl)thiazolidine-2-thione and 3-(3-*p*-methoxybenzylthiopropionyl)thiazolidine-2-thione.[56,58] Amino groups readily attack the carbonyl carbon displacing the good leaving group, thiazolidine-2-thione as shown in Figure 2.12. Aminolysis can be easily monitored by the disappearance of the yellow color. The S-protecting groups can be quantitatively and quickly removed: acetyl group by incubating

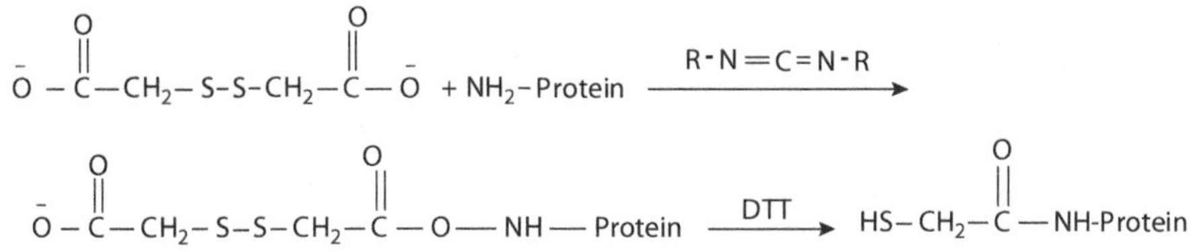

FIGURE 2.11 Thiolation by coupling dithioglycolic acid to proteins with water-soluble carbodiimides.

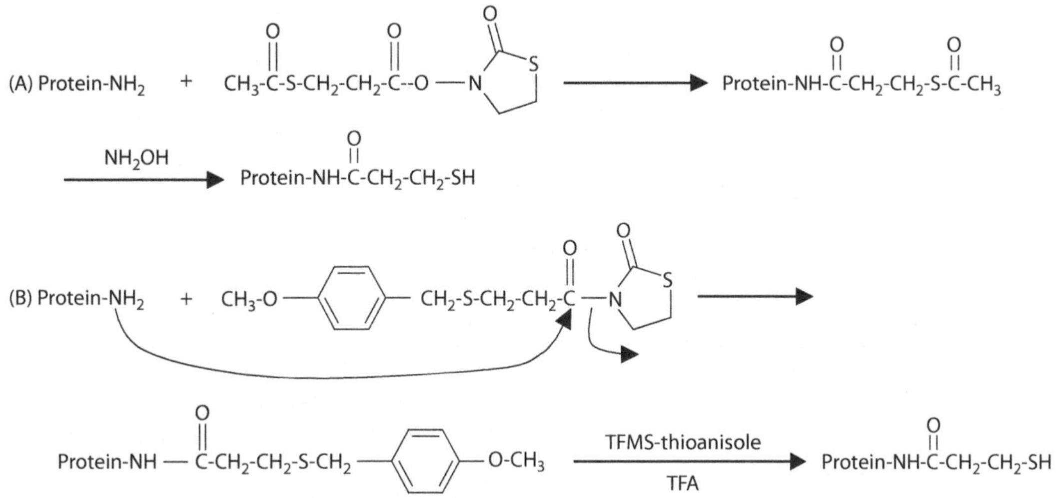

FIGURE 2.12 Thiolation of amino groups with (A) 3-(3-acetylthiopropionyl)thiazolidine-2-thione and (B) 3-(3-*p*-methoxybenzylthiopropionyl)thiazolidine-2-thione.

with hydroxylamine[55] and the methoxybenzyl group by 1 M trifluoromethanesulfonic acid (TFMS)-thioanisole in trifluoroacetic acid (TFA).[59]

The introduction of a sulfhydryl group has many applications. Not only has this approach been used extensively in the preparation of immunotoxins (see Chapter 13), it has also been used to prepare immunoconjugate for immunoassays (see Chapter 12). For example, insulin, after thiolation, has been conjugated to galactosidase with a thiol-specific reagent.[60,61]

2.3.2.2.3 Conversion of Thiols to Carboxylic Acids

Although succinic anhydride also reacts with sulfhydryl groups to yield carboxylic acid derivatives, the thioester bond is susceptible to hydrolysis. Carboxyalkylation is generally preferred for the modification of sulfhydryl groups (Figure 2.13). α-Haloacetates are commonly used as alkylating agents with the highest reactivity for sulfhydryl groups followed, in order of reactivity, by imidazolyl, thioether, and amino groups.[62]

S-carboxymethylation has been used for amino acid quantification of cysteine and cystine residues after reduction. The reaction occurs at neutral pH but the reactivity increases with alkalinity since the sulfur anion is the reactive species. The reactivity of the haloacids is a function of the halogen in the order of I > Br > Cl ≫ F. Iodoacetate reacts about twice as fast as bromoacetate and 20- to 100-fold faster than chloroacetate. Fluoroacetate is quite unreactive. Thus, by careful choice of pH and haloacetate, the sulfhydryl group can be specifically modified.

FIGURE 2.13 S-carboxymethylation of thiol group with iodoacetate.

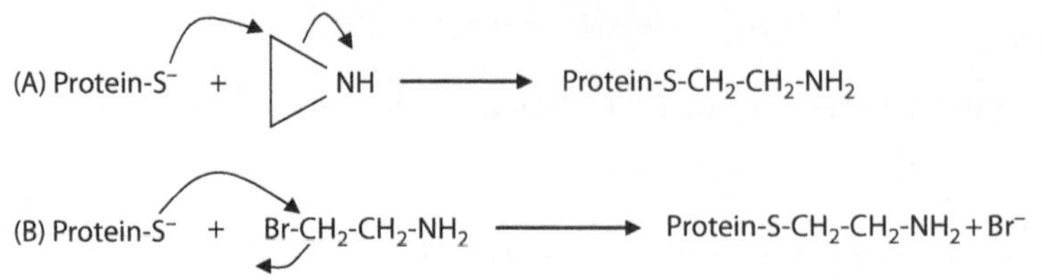

(A) Protein-S⁻ + [NH] ⟶ Protein-S-CH₂-CH₂-NH₂

(B) Protein-S⁻ + Br-CH₂-CH₂-NH₂ ⟶ Protein-S-CH₂-CH₂-NH₂ + Br⁻

FIGURE 2.14 Conversion of thiols to amines. Reaction of the sulfhydryl group with (A) ethylenimine and (B) 2-bromoethylamine.

2.3.2.2.4 Conversion of Thiols to Amines

Conversion of sulfhydryl group to an amine has been achieved with ethylenimine (Figure 2.14A)[63] and 2-bromoethylamine (Figure 2.14B).[64] The reaction is quite specific for the thiol group at slightly alkaline conditions.

Aminoethylation of cysteine residues has been used to introduce new targets for trypsin cleavage into polypeptide chains, although the rate of hydrolysis is much slower.[65] It may also be used for the modification of surface chemistry of solid supports.

2.3.2.2.5 Conversion of Carboxylic Acids to Amines

Water-soluble carbodiimides, as will be discussed in detail later in this book, have been used to promote amide formation. Using carbodiimides such as 1-ethyl-3-(3-dimethylaminopropyl)-carbodiimide (EDC) and a diamine such as ethylenediamine, the carboxyl group of proteins can be modified to a cationic amine (Figure 2.15).[66] The reaction is relatively mild and is best carried out at pHs between 4.5 and 5.0.

Amination of the carboxyl group has the advantage of converting it into a more potent nucleophile and at the same time provides an extended arm from the protein to avoid steric hindrance. This method has also been used to change the surface chemistry of solid particles for protein immobilization (see Chapter 14).

2.3.2.2.6 Conversion of Hydroxyl Group to Sulfhydryl Group

The relatively unreactive hydroxyl group of serine and threonine can be converted to the highly reactive thiol by the reactions shown in Figure 2.16. Activation of the hydroxyl group is first

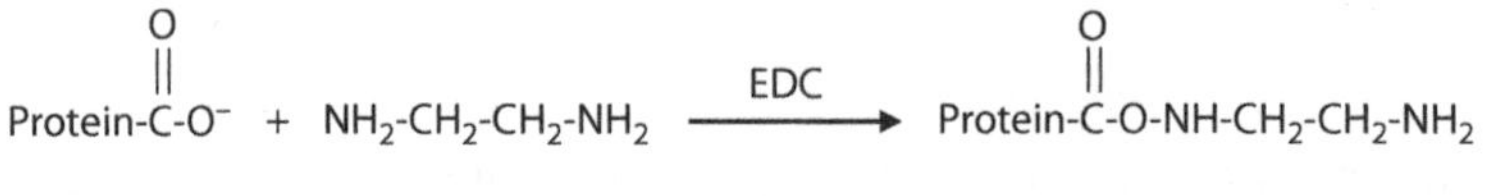

$$\text{Protein-}\overset{\overset{\displaystyle O}{\|}}{C}\text{-O}^- + NH_2\text{-}CH_2\text{-}CH_2\text{-}NH_2 \xrightarrow{EDC} \text{Protein-}\overset{\overset{\displaystyle O}{\|}}{C}\text{-O-NH-}CH_2\text{-}CH_2\text{-}NH_2$$

FIGURE 2.15 Amination of carboxylic acids. The reaction involves coupling of diamines to carboxyl groups of proteins with a water-soluble carbodiimide such as 1-ethyl-3-(3-dimethylaminopropyl)-carbodiimide, EDC.

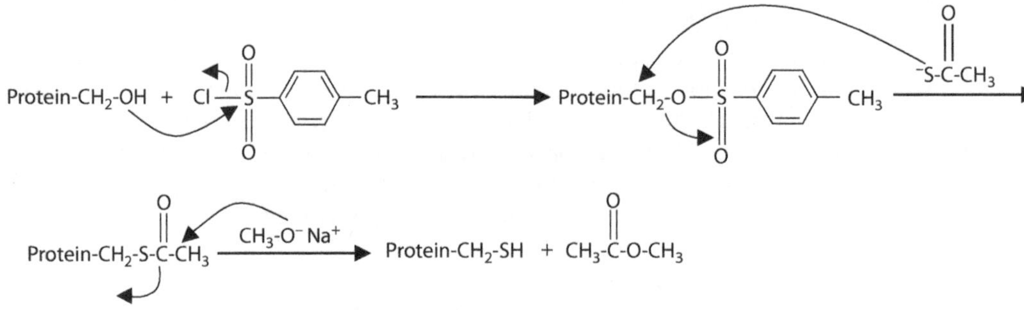

FIGURE 2.16 Conversion of hydroxyl group to sulfhydryl group. Hydroxyl groups are first activated by tosyl chloride and then reacted with thioacetate. Subsequent hydrolysis yields free thiol.

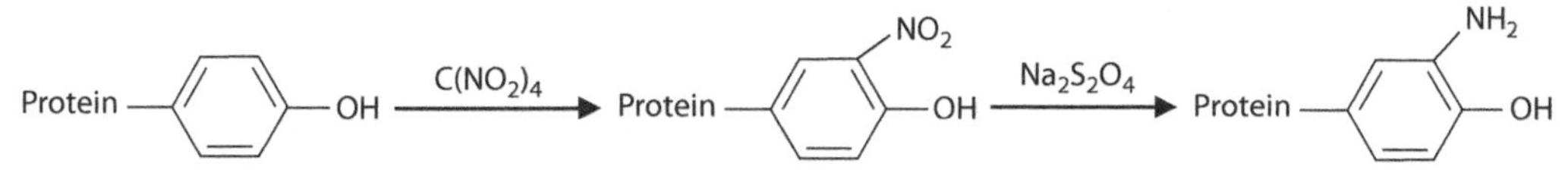

FIGURE 2.17 Conversion of tyrosine to aminotyrosine.

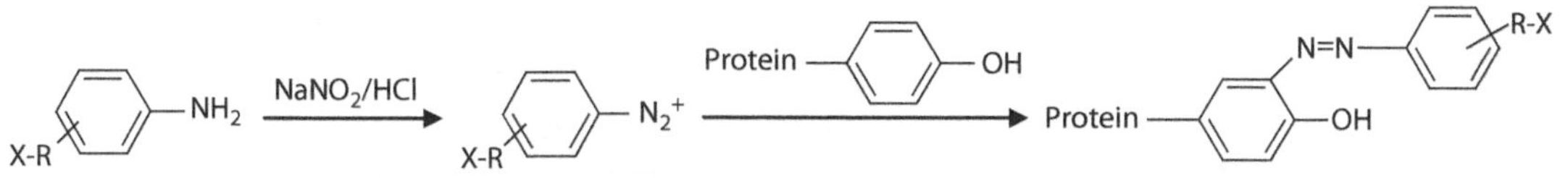

FIGURE 2.18 Introduction of a functional group (X) through diazotization of tyrosine.

achieved by tosyl chloride (toluenesulfonyl chloride). This tosylation step may be carried out in an aqueous phosphate buffer at pH 8–9 containing dioxane or pyridine. Subsequent transesterification is obtained in 0.5 M thioacetate solution at pH 5.5. Hydrolysis of the thioester to generate a free thiol is accomplished with 0.5 N sodium methanoate. Using this procedure, Ebert et al.[67] were able to form disulfide cross-links in various proteins. This procedure should also be useful for altering the surface chemistry of many carbohydrate-based solid supports.

2.3.2.2.7 Conversion of Tyrosine to Aminotyrosine and Other Derivatives

Modification of tyrosine in proteins with tetranitromethane results in the nitration at the 3 position of the benzene ring.[12] The phenolic moiety of 3-nitrotyrosine has a pK_a of about 7, which is much lower than that of tyrosine itself, resulting in an increase in the ionized form at neutral pH.[68] Nitrotyrosine can be further reduced to aminotyrosine by sodium hydrosulfite (Figure 2.17).[69] The amino group of 3-aminotyrosine has a pK_a near 4.8 and is susceptible to a number of nucleophilic reactions at low pHs where other protein groups are inactive.

As will be discussed in a later chapter, diazotization of proteins provides another useful means of modification of tyrosine residues.[12] Various functional groups may be introduced through the reaction with diazonium salts. The most common precursor of diazonium salts are the derivatives of phenylamine. On reaction with nitrous acid, phenylamines will be converted to aryl diazonium salts, which react with tyrosine (Figure 2.18). Such process enables the introduction of many functional groups into tyrosine residues of proteins. For example, various carbohydrate moieties, such as p-aminophenyl galactoside or lactoside have been introduced.[70,71] Nucleotides such as 3′- and 5′-thymidine-*p*-aminophenylphosphate have also been coupled.[72]

2.3.2.3 Introduction of Carbohydrate Prosthetic Groups

Simple proteins lacking nonproteous groups can be covalently linked to prosthetic moieties. Carbohydrate, in particular, can be introduced through covalent bonds to proteins. Plotz and Rifai[73] have modified the soluble carbohydrate polymer Ficoll with chloroacetate, ethylenediamine, glutaric anhydride, and 2,4-dinitrophenol as shown in Figure 2.19A. The modified Ficoll can be covalently attached to antibodies through displacement of the dinitrophenol by an amino group of the protein.

Although carbohydrates can be covalently attached to protein amino groups via reductive amination,[74] they have also been coupled to proteins by diazonium, phenylisothiocyanate, and mixed anhydride reactions.[70,75] Aminophenyl glycosides have been synthesized and diazotized with nitrous acid (Figure 2.19B1) or converted to isothiocyanate with thiophosgen (Figure 2.19B2). These reactive carbohydrate intermediates will react with proteins on mixing. Alternatively, sugars may be activated with isobutylchloroformate. The mixed anhydride formed will react with proteins to form oligosaccharide-protein conjugates (Figure 2.19B3). More recently, Lee et al.[76] have used a masked heterobifunctional reagent to link glycopeptides to proteins. The amino-containing carbohydrate first reacts with an acylazide to introduce an aldehyde. The second step involves reductive alkylation

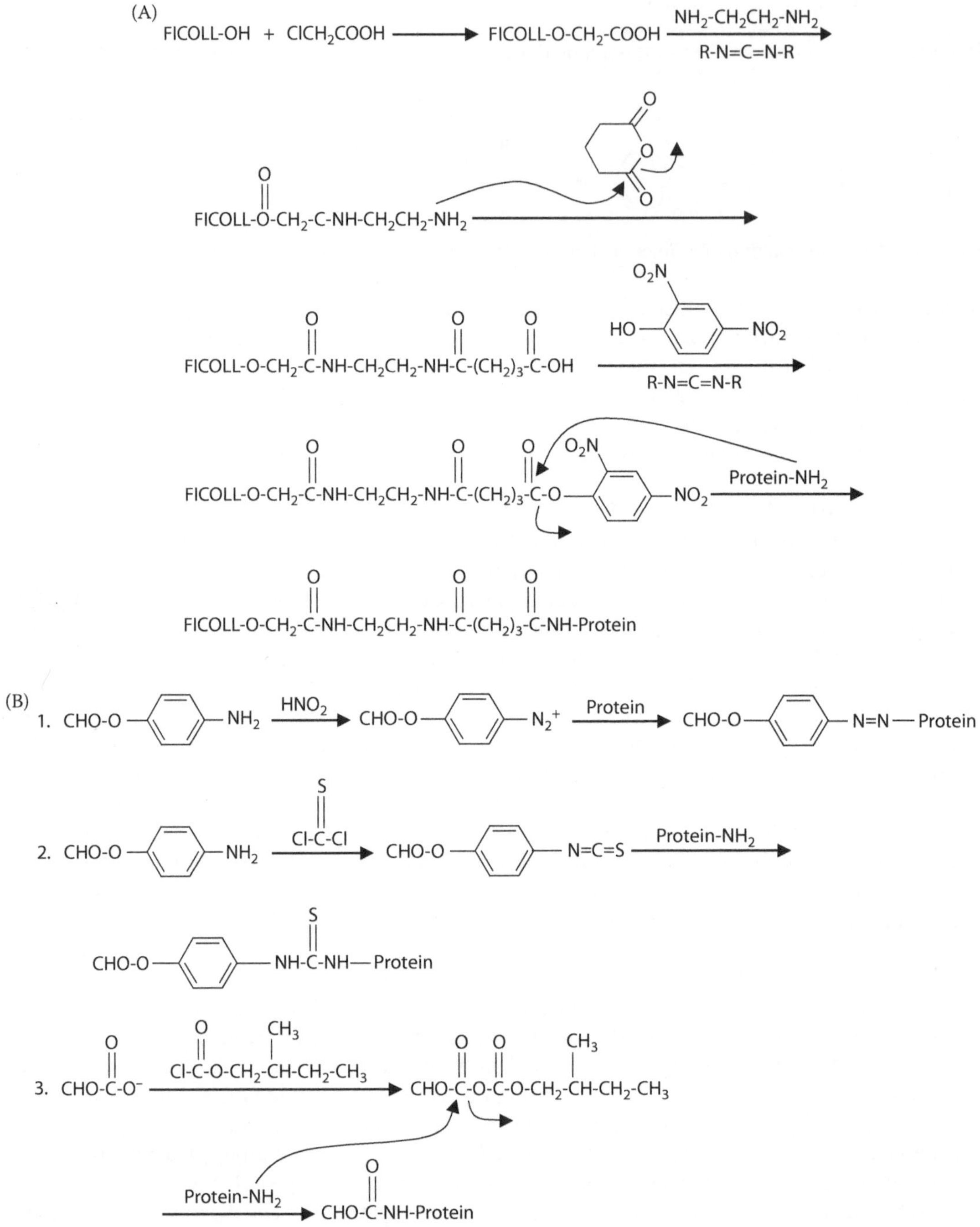

FIGURE 2.19 Introduction of carbohydrate prosthetic group into proteins. (A) Modification of Ficoll for covalent attachment to proteins; (B) coupling of carbohydrates, CHO, to proteins through diazonium, thiocyanate, and mixed anhydride.

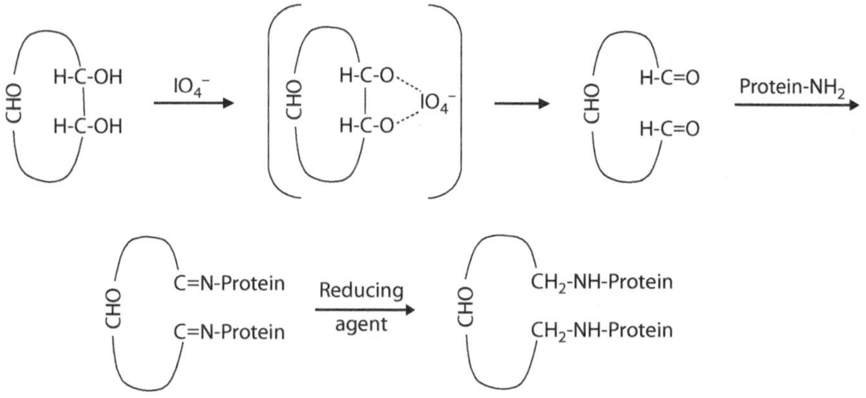

FIGURE 2.20 Reductive alkylation of amino groups with oxidized carbohydrates (CHO).

of the aldehyde with protein. Further modification of the attached carbohydrate is possible to provide additional functional groups for protein cross-linking.

Another method of attaching carbohydrates to proteins is by reductive alkylation. As will be discussed below as well as in a later chapter, aldehydes derived from oxidized carbohydrates form Schiff bases with amino groups of proteins, which can be reduced to form stable covalent bonds.

2.3.2.4 Activation of Carbohydrates by Periodate

In glycoproteins, the carbohydrate moiety contains Sia, Gal, and relatively high content of Man as alluded to earlier. These sugar constituents contain vicinal hydroxyl groups, which are susceptible to oxidation with periodic acid, sodium or potassium periodate. Periodate oxidation cleaves C–C bonds bearing adjacent –OH groups, converting them to dialdehydes.[77] *Cis* vicinal glycols react more rapidly than *trans* isomers because of the formation of a cyclic intermediate during oxidation (Figure 2.20). Diaxial diols rigidly held at 180° are not oxidized because the cyclic intermediate is sterically impossible to form.

After periodate treatment of glycoproteins, the dialdehyde formed can react with a variety of reagents, notably with amine to form imines or Schiff bases (Figure 2.20). The Schiff bases may be further stabilized on reduction with sodium borohydride, sodium cyanoborohydride, or pyridine borane. The choice of the latter is preferred since it reduces only Schiff bases and is nontoxic, whereas borohydrides also reduce aldehydes and disulfides.[78,79] In addition, Gildersleeve et al.[80] have improved the yield of reductive amination by changing the condition of the reaction.

2.4 NUCLEIC ACID CHEMISTRY

2.4.1 PHOTOCHEMICAL REACTIVITIES OF NUCLEIC ACIDS

Nucleic acids are composed of heterocyclic nitrogenous bases, a pentose sugar, and a phosphoryl group. There are five bases, two purines, adenine (A) and guanine (G) and three pyrimidines, thymine (T), uracil (U), and cytosine (C). The sugar component can be ribose or deoxyribose. The former forms the ribonucleic acid (RNA) and the latter deoxyribonucleic acid (DNA). In RNA, the major bases are A, U, G, and C. In DNA, the major bases are A, T, G, and C. The phosphoryl group links to the 5′ position of ribose or deoxyribose to the 3′ position of another ribose or deoxyribose forming polymers of RNA or DNA. These polymers are capable of hydrogen bonding through base pairs of purine and pyrimidine (AT, AU, and GC) transforming the nucleic acids into tertiary structures containing double helices.[81,82] The hydrogen bonds between other pairs of the purines and pyrimidines such as GU are also possible in RNA. The bases have highly conjugated double bonds and the molecules exist in many different resonance structures. Because of the resonance double bonds, nucleic acids show absorption maxima in the near-ultraviolet (UV) region. In the helical structure, the bases are stacked on the top of one another. Such base-stacking interactions decrease the efficiency of absorption of UV light, a phenomenon referred to as the hypochromic effect.[83] The photochemistry of nucleic acids has been studied for many years in the context of UV-induced photodamage to DNA. One of the best known photoreactions involves the cross-linking of adjacent thymine residues in DNA to form a dimer.[84] In addition to the photochemistry induced by normal, low-intensity light sources, high-intensity UV lasers can be used to produce interesting photochemical reactions. For example, excitation of a nucleic acid base by a UV laser can lead to excitation from the singlet S_o ground state to the first excited S_1 manifold.[85] From here, if the light density is sufficiently strong, as in the case of UV laser illumination, the excited S_1 state can absorb another photon to be promoted to the higher singlet level, S_N, above the ionization limit (I.L. in Figure 2.21). The high-energy species can react with other molecules to form photoproducts. However, the excited singlet can release some energy by intersystem crossing to the triplet T_1 manifold, which has a longer lifetime than the singlet S_1. A second photon can promote the triplet T_1 to a higher lying T_N level above the ionization limit. The ions produced can also initiate reaction with other molecules leading to cross-linking products.[86] When DNA is photoactivated, thymine dimers are produced as shown in Figure 2.22.[87] Mechanistically, the UV radiation activates thymine into a diradical,

FIGURE 2.21 Photochemistry of nucleic acid bases.

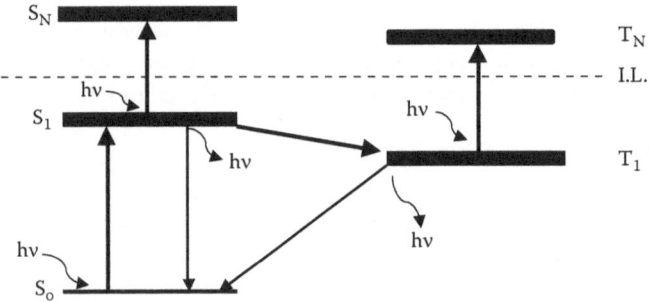

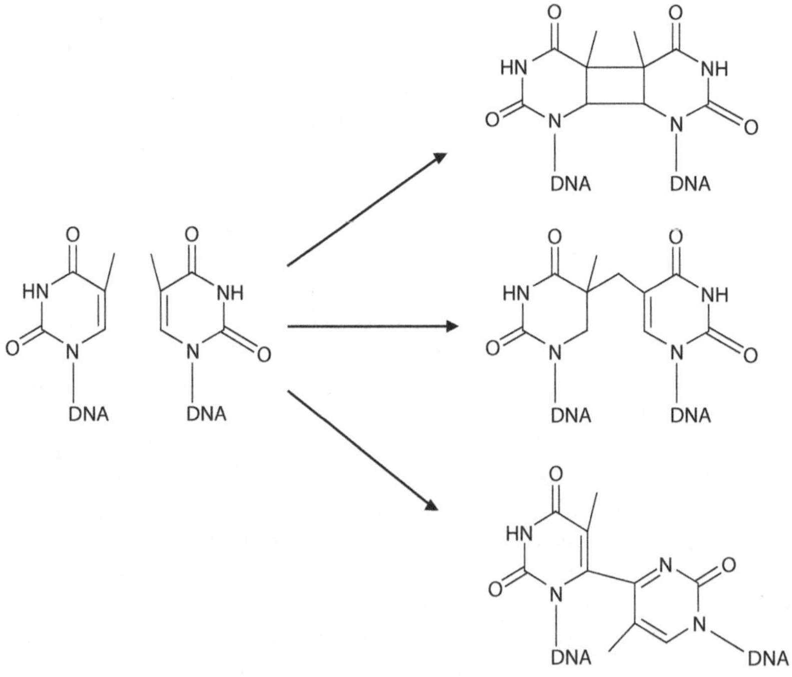

FIGURE 2.22 Photochemical dimerization of thymine in DNA.

which transfers an electron to a neighboring thymine by abstracting a proton. The two radicals then combine to form a covalent bond. Similar photoactivated reactions can cross-link the nucleic acids to proteins as will be discussed later in this book.

2.4.2 CHEMICAL REACTIVITIES OF NUCLEIC ACIDS

Although slow, the nucleic bases undergo spontaneous deamination reactions, where C is converted to U, A to hypoxanthine, and G to xanthine (Figure 2.23). These spontaneous reactions are of great importance in mutagenesis[88] but of little or no application in cross-linking nucleic acids to proteins. Deamination, however, can be achieved by chemical reagents such as nitrous acid, sodium nitrite, nitrosamines, and sodium bisulfite. Since nucleic acids form hydrogen bonding between purines and pyrimidines (Figure 2.24), the hydrogen bond controls some of the chemical reactivities of the bases. For example, sodium bisulfite is unreactive in converting C to U in double-helical polynucleotides.[89]

Another spontaneous reaction of nucleic acids is depurination. In this process, the *N*-glycosyl bond between the base and the pentose is hydrolyzed, which occurs faster for purines than for pyrimidines.

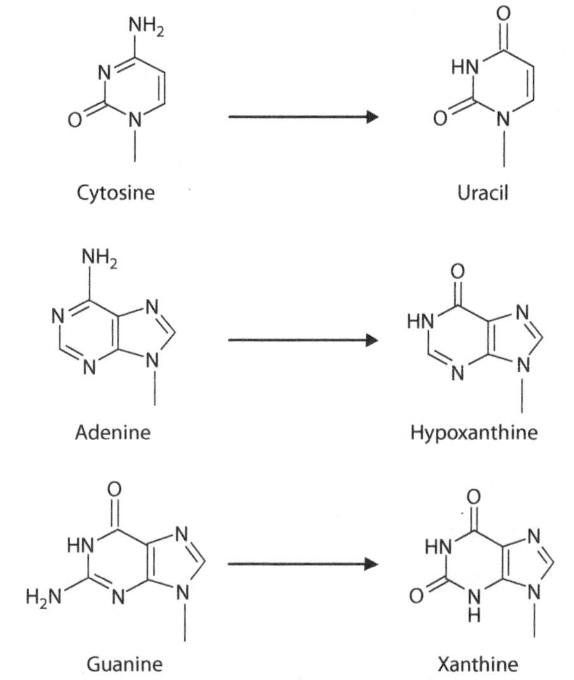

FIGURE 2.23 Deamination of nucleic acids. Cytosine is deaminated to uracil, adenine to hypoxanthine, and guanine to xanthine.

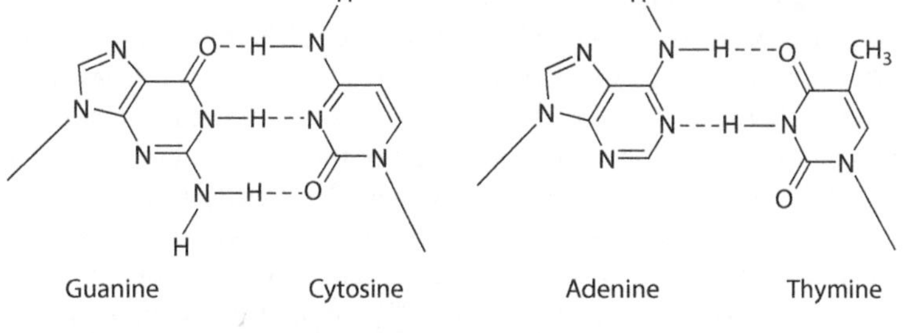

FIGURE 2.24 Hydrogen bonding between base pairs in DNA. Left, a GC base pair with three hydrogen bonds; right, an AT base pair with two hydrogen bonds. Hydrogen bonds are shown as dashed lines.

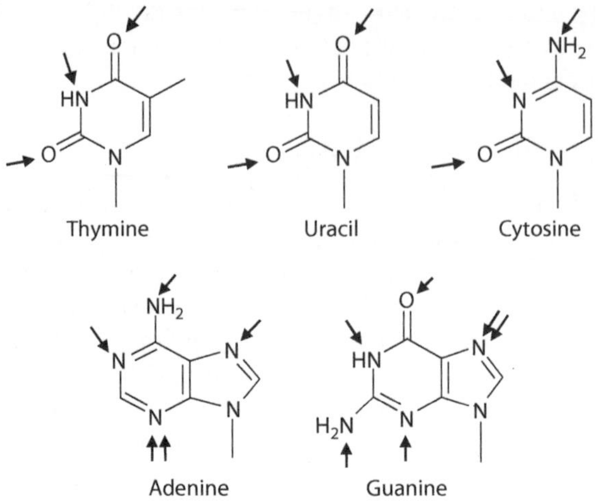

FIGURE 2.25 Nucleophilic sites of nucleic acid bases. The most reactive positions are indicated by double arrows.

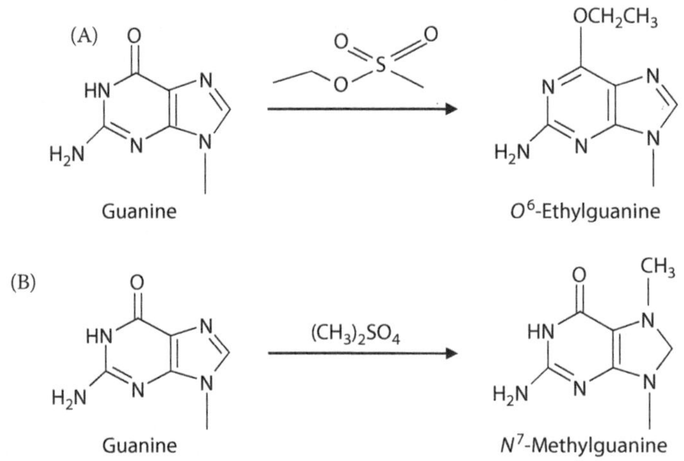

FIGURE 2.26 Alkylation of nucleic acid bases. (A) Reaction of guanine with ethyl methanesulfonate; (B) reaction with dimethysulfate.

This process is relatively slow at ambient temperatures and neutral pH; however, the rate of the reaction is increased in acidic pH. For example, at pH 3, DNA quickly loses its purines to form apurinic acid.[90] It has been estimated that about 1000 nucleotides are depurinated in a typical mammalian cell each day; however, DNA repair mechanisms have evolved to help mitigate this problem. Interestingly, the toxic protein ricin catalyzes adenosine depurination of the 28S eukaryotic ribosomal RNA at position 4256, which results in total loss of the ribosome's ability to synthesize proteins, thus leading to cell death.

Perhaps the most important chemical reaction of nucleic acids for cross-linking is alkylation. As shown in Figure 2.25, there are many nucleophilic sites on the nucleic acid bases.[91] The site of alkylation depends on the alkylating agents, structure of the nucleic acid, as well as the pH.[92] Weak carcinogens such as ethyl methane sulfonate and N-methyl-N-nitrosourea produce significant levels of O-atom alkylation,[93] whereas alkylating agents such as dimethylnitrosamine and dimethylsulfate react predominately with the nitrogen sites of nucleic acids, for example, with guanine residue to yield N^7-methylguanine as shown in Figure 2.26.[94] The numerous nucleophilic sites on the bases therefore provide excellent opportunities for cross-linking nucleic acids and proteins with bifunctional alkylating agents.

REFERENCES

1. Fujioka, M., Takata, Y., Kooishi, K., and Ogawa, H., Function and reactivity of sulfhydryl groups of rat liver glycine methyltransferase, *Biochemistry*, 26, 5696, 1987.
2. Lundblad, R. L., *Chemical Reagents for Protein Modification*, 3rd edn., CRC Press, Boca Raton, FL, 2004, p. 139.
3. Duggleby, K. G. and Kaplan, H., A competitive labeling method for the determination of the chemical properties of solitary functional groups in proteins, *Biochemistry*, 14, 5168, 1975.
4. Schewale, J. G. and Brew, K., Effects of Fe^{3+} binding on the microenvironments of individual amino groups in human serum transferrin as determined by different kinetic labeling, *J. Biol. Chem.*, 257, 9406, 1982.
5. Cladaras, C., Hadzopoulou-Cladaras, M., Nolte, R. T., Atkinson, D., and Zannis, V. I., The complete sequence and structural analysis of human apolipoprotein B-100: Relationship between apoB-100 and apoB-48 forms, *EMBO J.*, 5, 3495, 1986.
6. Tskhovrebova, L. and Trinick, J., Titan: Properties and family relationships, *Nat. Rev. Mol. Cell Biol.*, 4, 679, 2003.
7. Black, S. D. and Mould, D. R., Development of hydrophobicity parameters to analyze proteins which bear post- or cotranslational modifications, *Anal. Biochem.*, 193, 72, 1991.
8. Cornette, J., Cease, K. B., Margalit, H., Spouge, J. L., Berzofsky, J. A., and DeLisi, C., Hydrophobicity scales and computational techniques for detecting amphipathic structures in proteins, *J. Mol. Biol.*, 195, 659, 1987.
9. Chothia, C., Principles that determine the structure of proteins, *Ann. Rev. Biochem.*, 53, 537, 1984.
10. Hassell, J. R., Kimina, J. H., and Cantly, L., Proteoglycan core protein families, *Ann. Rev. Biochem.*, 55, 539, 1986.
11. Kornfeld, R. and Kornfeld, S., Assembly of asparagine-linked oligosaccharides, *Annu. Rev. Biochem.*, 54, 631, 1985.
12. Means, G. E. and Feeney, R. E., *Chemical Modification of Proteins*, Holden-Day, San Francisco, CA, 1971.
13. Loudon, J. D. and Shulman, N., Mobility of groups in chloronitrodiphenyl sulfones, *J. Chem. Soc.*, 1941, 722, 1941.
14. Suhr, H., Effect of the leaving group on the velocity of nucleophilic aromatic substitutions, *Berichte*, 97, 3268, 1964.
15. Bunnett, J. F., Nucleophilic reactivity, *Annu. Rev. Phys. Chem.*, 14, 271, 1963.
16. Edwards, J. O. and Pearson, R. G., The factors determining nucleophilic reactivities, *J. Chem. Soc.*, 84, 26, 1962.
17. Jencks, W. P., *Catalysis*, in *Chemistry and Enzymology*, Dover Publications, New York, 1987.
18. Bruice, T. C. and Lapinski, R., Imidazole catalysis. IV. The reaction of general bases with p-nitrophenyl acetate in aqueous solution, *J. Am. Chem. Soc.*, 80, 2265, 1958.
19. Freedman, R. B. and Radda, G. K., The reaction of 2,4,6-trinitrobenzeuesulfonic acid with amino acids, peptides, and proteins, *Biochem. J.*, 108, 383, 1968.
20. Hudson, R. F., Nucleophilic reactivity, in *Chemical Reactivity and Reaction Paths*, Klopman, G. (ed.), John Wiley & Sons, New York, 1974, Chapter 5.
21. Stark, G. R., Reactions of cyanate with functional groups of proteins. IV. Inertness of aliphatic hydroxyl groups. Formation of carbamyl- and acetylhydantoins, *Biochemistry*, 4, 2363, 1965.
22. Mayr, H. and Patz, M., Scales of nucleophilicity and electrophilicity: A system for ordering polar organic and organometallic reactions, *Angew. Chem. Int. Ed. Engl.*, 33, 938, 1994.
23. Mooz, E. D., Data on the naturally occurring amino acids, in *Handbook of Biochemistry and Molecular Biology*, Vol. I, Fasman, G. D. (ed.), 3rd edn., CRC Press, Cleveland, OH, 1975, p. 111.
24. Cantor, C. R. and Schimmel, P. R., *Physical Chemistry. Part 1. The Conformation of Biological Macromolecules*, W. H. Freeman, San Francisco, CA, 1980.
25. de Levie, R., The Henderson–Hasselbalch equation: Its history and limitations, *J. Chem. Educ.*, 80, 146, 2003.
26. Berliner, E., Kinetics of the iodination of phenol, *J. Am. Chem. Soc.*, 73, 4307, 1951.
27. Mayberry, W. E. and Bertoli, D. A., Kinetics of the iodination. II. General base catalysis in the iodination of N-acetyl-L-tyrosine and N-acetyl-3-iodo-L-tyrosine, *J. Org. Chem.*, 30, 2029, 1965.
28. Lundblad, R. L. and Noyes, C. M., *Chemical Reagents for Protein Modification*, Vol. I, CRC Press, Boca Raton, FL, 1984, p. 16.
29. Tanford, C. and Hauenstein, J. D., Hydrogen-ion equilibriums of ribonuclease, *J. Am. Chem. Soc.*, 78, 5287, 1956.

30. Haschemeyer, R. H. and Haschemeyer, A. E. V., *Proteins: A Guide to Study by Physical and Chemical Methods*, Wiley-Interscience, New York, 1973.
31. Kraut, J., Serine proteases: Structure and mechanism of catalysis, *Annu. Rev. Biochem.*, 46, 331, 1977.
32. Blow, D. M., Birktoft, J. J., and Hartley, B. S., Role of a buried acid group in the mechanism of action of chymotrypsin, *Nature (London)*, 221, 337, 1969.
33. White, F. H., Jr., Regeneration of enzymic activity by air oxidation of reduced ribonuclease with observations of thiolation during reduction with thioglycolate, *J. Biol. Chem.*, 235, 383, 1960.
34. Zahler, W. L. and Cleland, W. W., A specific and sensitive assay for disulfides, *J. Biol. Chem.*, 243, 716, 1968.
35. Konigsberg, W., Reduction of disulfide bonds in proteins with dithiothreitol, *Methods Enzymol.*, 25B, 185, 1972.
36. Ishikawa, E., Hashida, S., Kohno, T., and Tanaka, K., Methods for enzyme-labeling of antigens, antibodies and their fragments, in *Nonisotopic Immunoassays*, Ngo, T. T. (ed.), Plenum Press, New York, 1988, p. 45.
37. Imagawa, M., Hashida, S., Ishikawa, E., and Freytag, J. W., Preparation of a monomeric 2,4-dinitrophenyl Fab'-β-galactosidase conjugate for immunoenzymometric assays, *J. Biochem.*, 96, 1727, 1984.
38. Butler, P. J. G., Harris, J. I., Hartley, B. S., and Leberman, R., Use of maleic anhydride for the reversible blocking of amino groups in polypeptide chains, *Biochem. J.*, 112, 679, 1969.
39. Gounaris, A. D. and Perlman, G. E., Succinylation of pepsinogen, *J. Biol. Chem.*, 242, 2739, 1967.
40. Klotz, I. M., Succinylation, *Methods Enzymol.*, 11, 576, 1967.
41. Klapper, M. H. and Klotz, I. M., Hybridization of chemically modified proteins, *Methods Enzymol.*, 25, 536, 1972.
42. White, F. B., Jr., Thiolation, *Methods Enzymol.*, 25, 541, 1972.
43. Benesch, R. and Benesch, R. E., Thiolation of proteins, *Proc. Natl. Acad. Sci. U. S. A.*, 44, 848, 1958.
44. Benesch, R. and Benesch, R. E., Formation of peptide bonds by aminolysis of homocysteine thiolactones, *J. Am. Chem. Soc.*, 78, 1597, 1956.
45. Klotz, I. M. and Heiney, R. E., Introduction of sulfhydryl groups into proteins using acetyl-succinic anhydride, *Arch. Biochem. Biophys.*, 96, 605, 1962.
46. Perham, R. N. and Thomas, J. O., Reaction of tobacco mosaic virus with a thiol-containing imido ester and a possible application to x-ray diffraction analysis, *J. Mol. Biol.*, 62, 415, 1971.
47. Traut. R. R., Bollen, A., Sun, T. T., Hershey, J. W. B., Sundberg, J., and Pierce, L. R., Methyl 4-mercaptobutyrimidate as cleavable crosslink reagent and its application to the *Escherichia coli* 30s ribosome, *Biochemistry*, 12, 3266, 1973.
48. Kenny, J. W., Sommer, A., and Traut, R. R., Cross-linking studies on the 50S ribosomal subunit of *Escherichia coli* with methyl 4-mcrcaptobutyrimidate, *J. Biol. Chem.*, 250, 9434, 1975.
49. King, T. P., Li, Y., and Kochoumian, L., Preparation of protein conjugates via intermolecular disulfide bond formation, *Biochemistry*, 17, 1499, 1978.
50. Jue, R., Lambert, J. M., Pierce, L. R., and Traut, R. R., Addition of sulfhydryl group of *Escherichia coli* ribosomes by protein modification with 2-iminothiolate (methyl 4-mercaptobutyrimidate), *Biochemistry*, 17, 5399, 1978.
51. Schramm, H. J. and Dulffer, T., The use of 2-iminothiolane as a protein crosslinking reagent, *Hoppe-Seyler's Z. Physiol. Chem.*, 358, 137, 1977.
52. Bush, D. A. and Winkler, M. A., Isoelectric focusing of cross-linked monoclonal antibody heterodimers, homodimers and derivatized monoclonal antibodies, *J. Chrom.*, 489, 303, 1989.
53. Carlsson, J., Drevin, H., and Axen, R., Protein thiolation and reversible protein–protein conjugation. *N*-Succinimidyl 3-(2-pyridylthio)propionate, a new heterobifunctional reagent, *Biochem. J.*, 173, 723, 1978.
54. Loment, A. J. and Fairbanks, G., Chemical probes of extended biological structure: Synthesis and properties of the cleavable protein cross-linking reagent [^{35}S]-dithiobis(succinimidyl propionate), *J. Mol. Biol.*, 104, 243, 1976.
55. Duucan, R. J. S., Weston, P. D., and Wrigglesworth, R., A new reagent which may be used to introduce sulfhydryl groups into proteins, and its use in the preparation of conjugates for immunoassay, *Anal. Biochem.*, 132, 68, 1983.
56. Fuji, N., Akaji, K., Hayashi, Y., and Yajima, H., Studies on peptides. CXXV. 3-(3-p-methoxybenzylthiopropionyl)-thiazolidine-2-thione and its analog as reagents for the introduction of the mercapto group into peptides and proteins, *Chem. Pharm. Bull.*, 33, 362, 1985.
57. Jou, Y. H. and Bankert, R. B., Coupling of protein antigens to erythrocytes through disulfide bond formation: Preparation of stable and sensitive target cells for immune hemolysis, *Proc. Natl. Acad. Sci. U. S. A.*, 78, 2493, 1981.

58. Fuji, N., Hayashi, Y., Katakura, S., Akaji, K., Yajma, H., Inouye, A., and Segawa, T., Studies on peptides. CXXVIII. Application of new heterobifunctional crosslinking reagents for the preparation of neurokinin (A and B)-BSA (bovine serum albumin) conjugates, *Int. J. Pept. Protein Res.*, 26, 121, 1985.

59. Yajima, H. and Fuji, N., Studies on peptides. 103. Chemical synthesis of a crystalline protein with full enzymatic activity of ribonuclease A, *J. Am. Chem. Soc.*, 103, 5867, 1981.

60. Kato, K., Hamaguchi, Y., Fukui, H., and Ishikawa, I., Enzyme-linked immunoassay. I. Novel method for synthesis of the insulin-β-galactosidase conjugate and its applicability for insulin assay, *J. Biochem.*, 78, 235, 1975.

61. Ishikawa, E., Imagawa, M., Hashida, S., Yoshitake, S., Hamaguchi, Y., and Ueno, T., Enzyme labeling of antibodies and their fragments for enzyme immunoassay and immunohistochemical staining *J. Immunoassay*, 4, 209, 1983.

62. Gurd, F. R. N., Carboxymethylation, *Methods Enzymol.*, 11, 532, 1967.

63. Raftery, M. A. and Cole, R. D., Tryptic cleavage at cysteinyl peptide bonds, *Biochem. Biophys. Res. Commun.*, 10, 967, 1963.

64. Lindley, H., A new synthetic substrate for trypsin and its application to the determination of the amino acid sequence of proteins, *Nature*, 178, 647, 1956.

65. Wang, S. S. and Carpenter, F. H., Kinetic studies at high pH of the trypsin-catalyzed hydrolysis of N-alpha-benzoyl derivatives of L-arginamide, L-lysinamide and S-2-aminoethyl-L-cyteinamide and related compounds, *J. Biol. Chem.*, 243, 3702, 1968.

66. Kurzer, F. and Douraghi-Zadeh, K., Advances in the chemistry of carbodiimides, *Chem. Rev.*, 67, 107, 1967.

67. Ebert, C., Ebert, G., and Knipp, H., On the introduction of disulfide crosslinks into fibrous proteins and bovine serum albumin, in *Protein Cross-Linking: Nutritional and Medical Consequences*, Friedman, M. (ed.), Plenum Press, New York, 1977, p. 235.

68. Vincent, J. P., Lazdunski, M., and Delaage, M., Use of tetranitromethane as a nitration reagent. Reaction of phenol side chains in bovine and porcine trypsinogens and trypsins, *Eur. J. Biochem.*, 12, 250, 1970.

69. Sokolovsky, M., Riordan, J. F., and Vallee, B. L., Conversion of 3-nitrotyrosine to 3-aminotyrosine in peptides and proteins, *Biochem. Biophys. Res. Commun.*, 27, 20, 1967.

70. McBroom, C. R., Samanen, C. H., and Goldstein, I. J., Carbohydrate antigens: Coupling of carbohydrates to proteins by diazonium and phenylisothiocyanate reactions, *Methods Enzymol.*, 22, 212, 1976.

71. Gopalakrishnan, P. V., Zimmerman, U. J., and Karush, F., Labeling of antilactose antibody, *Methods Enzymol.*, 46, 516, 1977.

72. Cuatrecasas, P. and Wilchek, M., Staphylococcal nuclease, *Methods Enzymol.*, 46, 516, 1977.

73. Plotz, P. H. and Rifai, A., Stable, soluble, model immune complexes made with a versatile multivalent affinity-labeling antigen, *Biochemistry*, 21, 301, 1982.

74. Gray, G. R., The direct coupling of oligosaccharides to proteins and derivatized gels, *Arch. Biochem. Biophys.*, 163, 426, 1974.

75. Ashwell, G., Carbohydrate antigens: Coupling of carbohydrates to proteins by a mixed anhydride reaction, *Methods Enzymol.*, 22, 219, 1976.

76. Lee, R. T., Wong, T.-C., Lee, R., Yue, L., and Lee, Y. C., Efficient coupling of glycopeptides to proteins with a heterobifunctional reagent, *Biochemistry*, 28, 1856, 1989.

77. Bobbitt, J. M., Periodate oxidation of carbohydrates, *Adv. Carbohyd. Chem.*, 11, 1, 1956.

78. Dottavio-Martin, D. and Ravel, J. M., Radiolabeling of proteins by reductive alkylation with [^{14}C]-formaldehyde and sodium cyanoborohydride, *Anal. Biochem.*, 87, 562, 1978.

79. Cabacuugan, J. C., Ahmed, A. I., and Feeney, R. E., Amine boranes as alternative reducing agents for reductive alkylation of proteins, *Anal. Biochem.*, 124, 272, 1982.

80. Gildersleeve, J. C., Oyelaran, O., Simpson, J. T., and Allred, B., Improved procedure for direct coupling of carbohydrates to proteins via reductive amination, *Bioconjug. Chem.*, 19, 1485, 2008.

81. Watson, J. and Crick, F., Molecular structure of nucleic acids: A structure for deoxyribose nucleic acid, *Nature*, 171, 737, 1953.

82. Saenger, W., *Principles of Nucleic Acid Structure*, Springer-Verlag, New York, 1983.

83. Volkov, S. N., Some aspects of the DNA hypochromic effect theory, *Int. J. Quantum Chem.*, 16, 119, 1979.

84. Lamola, A., Molecular mechanisms in nucleic acid photochemistry. I. Sensitized photochemical splitting of thymine dimer, *J. Am. Chem. Soc.*, 88, 6113, 1966.

85. Pashev, I. G., Dimitrov, S. I., and Angelov, D., Crosslinking proteins to nucleic acids by ultraviolet laser irradiation, *Trends Biochem. Sci.*, 16, 323, 1991.

86. Nikogosyan, D. N., Two-quantum UV photochemistry of nucleic acids: Comparison with conventional low-intensity UV photochemistry and radiation chemistry, *Int. J. Radiat. Biol.*, 57, 233, 1990.

87. Begley, T. P., Thymine dimer photochemistry: A mechanistic perspective, *Compr. Nat. Prod. Chem.*, 5, 371, 1999.
88. Parry, T. E., On the mutagenic action of adenine, *Leuk. Res.*, 31, 1621, 2007.
89. Shaprio, R., Braverman, B., Louis, J. B., and Servis, R. E., Nucleic acid reactivity and conformation: II. Reaction of cytosine and uracil with sodium bisulfite, *J. Biol. Chem.*, 248, 11, 4060, 1973.
90. Lindahl, T. and Nyberg, B., Rate of depurination of native deoxyribonucleic acid, *Biochemistry*, 11, 3610, 1972.
91. Singer, B., Sites in nucleic acids reacting with alkylating agents of differing carcinogenicity of mutagenicity, *J. Toxicol. Environ. Health*, 2, 1279, 1977.
92. Di Capua, E. and Muller, B., The accessibility of DNA to dimethylsulfate in complexes with recA protein, *EMBO J.*, 6, 2493, 1987.
93. Fox, M. and Brennand, J., Evidence for the involvement of lesions other than O6-alkylguanine in mammalian cell mutagenesis, *Carcinogenesis*, 1, 795, 1980.
94. Hoffmann, G. R., Genetic effects of dimethyl sulfate, diethyl sulfate, and related compounds, *Mutation Res.*, 75, 63, 1980.

3 Reagents Targeted to Specific Functional Groups

3.1 INTRODUCTION

Since cross-linking reagents are chemical modifying agents, their efficiencies in linking two moieties depend on their specificity toward the particular functional groups in the components to be linked. For proteins, these are the reactive groups of amino acid side chains; for nucleic acids, the heterocyclic base nitrogen, the amino groups, and the carbonyl oxygen. In order to link two species together, the choice of reagent is critical to the success of the cross-linking and conjugation. In cases wherein a specific group is known to react, a logical choice of cross-linker may be made. Such a choice requires an understanding of the specificity of the reagents. In this connection, this chapter will review the basic factors of protein modification and agents specific for functional groups.

From the considerations presented in the Chapter 2, it is not surprising to find that only a few reagents are absolutely specific toward a particular functional group. The lack of specificity has limited the usefulness of many reagents. However, the same consideration has also provided the basis of selectivity of some reagents. Since cross-reactivity of a reagent toward various groups is due to the nucleophilicities of these groups, as well as the effect of their microenvironment, some reagents may react selectively faster with a particular group. For example, at neutral pH, N-ethylmaleimide reacts much faster with sulfhydryl groups than with amino groups. For other reagents, the stability of the product may play an important role. Cyanic acid, for instance, reacts with amino, sulfhydryl, imidazolyl, tyrosyl, and carboxyl groups of proteins; but only the reaction with amino groups results in the formation of a stable product. Other adducts are readily reversible. Thus, under certain conditions, the relative reactivity of these reagents can be employed to selectively modify a particular group. Table 3.1 gives a summary of different classes of compounds and their reactivity toward the various functional groups found in proteins. These reactions will be discussed to illustrate the principle of chemical applications of cross-linking reagents. There are many excellent monographs and review articles that deal specifically with protein modifications.[1–9] Readers who are interested in this area are encouraged to consult these texts for more detailed coverage.

3.2 SULFHYDRYL REAGENTS

The sulfhydryl moiety, with the thiolate ion as the active species, is the most reactive functional group in a protein. With a pK_a of about 8.6, the reactivity of the thiol is expected to increase with increasing pH, toward and above its pK_a. There are many reagents that react faster with the thiol than any other groups.[10,11] The major classes of compounds that are particularly pertinent to the discussion of cross-linking reagents are presented here.

3.2.1 α-HALOACETYL COMPOUNDS

As we have discussed in Chapter 2, α-haloacetates can be used to react with free sulfhydryl groups to convert them to carboxylic acids. In the S-carboxylation reaction, the thiolate ion acts as a nucleophile. In addition to the commonly used iodoacetate, there are other α-acetyl compounds.

TABLE 3.1

Examples of Group-Specific Reagents

Reagent	Group Specificity
α-Haloacetyl compounds (e.g., ICH$_2$COOH)	-SH, -S-CH$_3$, -NH$_2$, 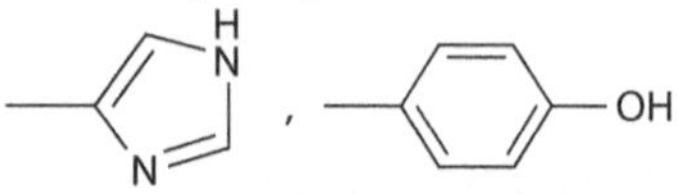 (imidazole), (phenol -OH)
N-Maleimides (e.g., N-ethylmaleimide)	-SH, -NH$_2$
Mercurials 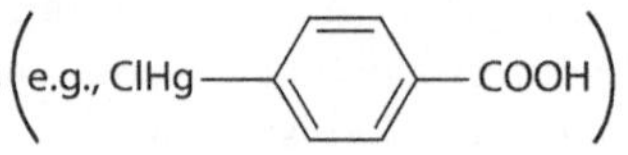 (e.g., ClHg—C$_6$H$_4$—COOH)	-SH
Disulfides (e.g., O$_2$N-, HOOC—C$_6$H$_3$—S—S—C$_6$H$_3$—NO$_2$, COOH) 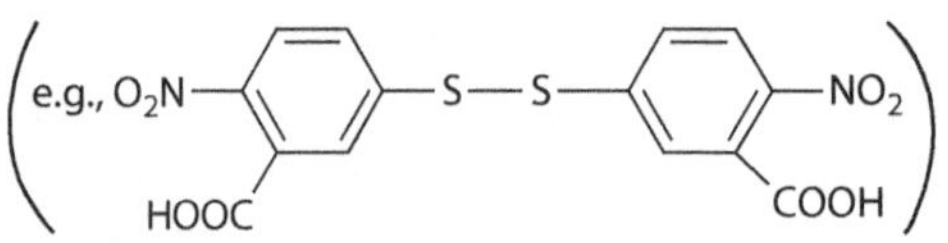	-SH
Aryl halides 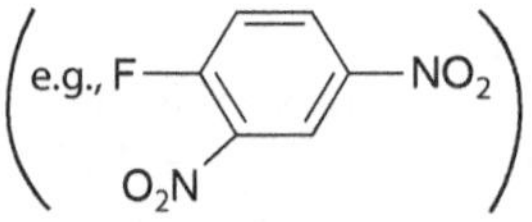 (e.g., F—C$_6$H$_3$(O$_2$N)—NO$_2$)	-SH, -NH$_2$, 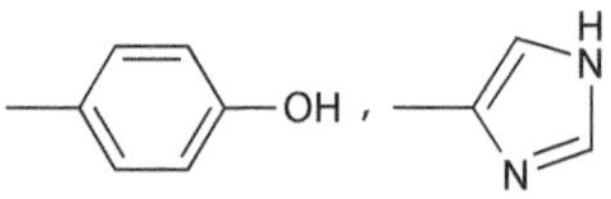 (phenol -OH), (imidazole)
Acid anhydrides (e.g., succinic anhydride) 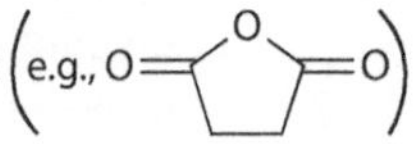	-NH$_2$, (phenol -OH) 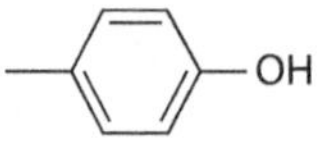
Isocyanates (e.g., H-N=C=O)	-NH$_2$
Isothiocyanates 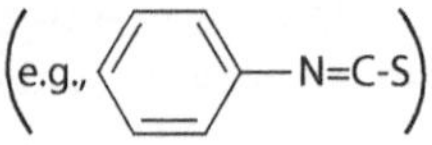 (e.g., C$_6$H$_5$—N=C-S)	-NH$_2$
Sulfonyl halides 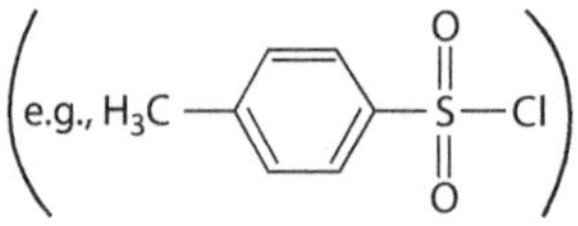 (e.g., H$_3$C—C$_6$H$_4$—SO$_2$—Cl)	-NH$_2$
Imidoesters 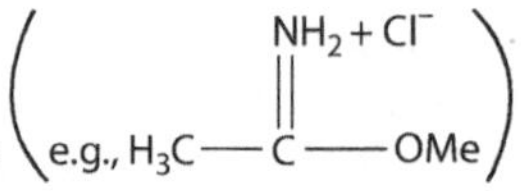 (e.g., H$_3$C—C(=NH$_2^+$ Cl$^-$)—OMe)	-NH$_2$
Diazoacetates 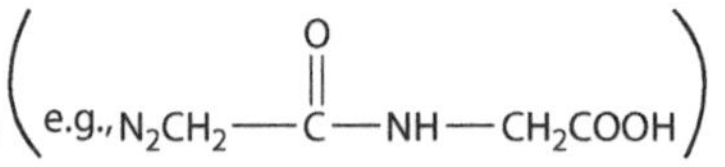 (e.g., N$_2$CH$_2$—C(=O)—NH—CH$_2$COOH)	-SH, -COOH
Diazonium salts 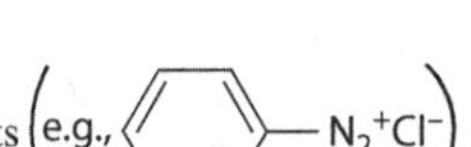 (e.g., C$_6$H$_5$—N$_2^+$Cl$^-$)	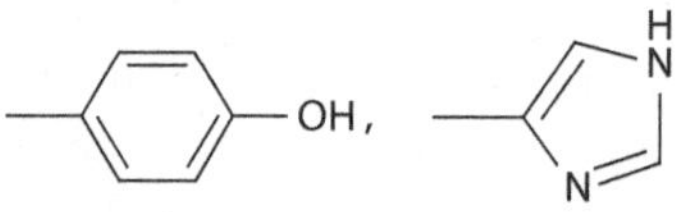 (phenol -OH), (imidazole)
Dicarbonyl compounds (e.g., 1,2-cyclohexanedione) 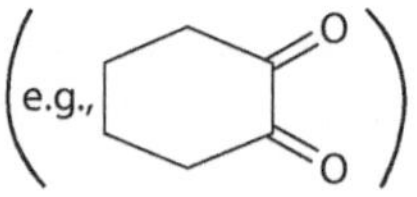	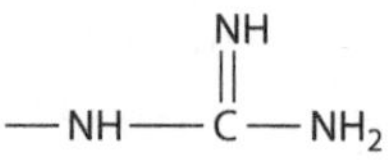 —NH—C(=NH)—NH$_2$

For example, the amide derivatives, α-haloacetamides, follow the same reactivity trend. In all cases, the reaction involves nucleophilic attack of the thiolate ion resulting in a displacement of the halide as shown in Figure 3.1. The reactive haloacetyl moiety, X–CH$_2$CO–, has been incorporated into compounds for various purposes. For instance, bromotrifluoroacetate (compound [3] in Figure 3.1) has been used for the incorporation of ^{19}F,[12] and N-chloroacetyliodotyramine (Figure 3.1 [4]) has

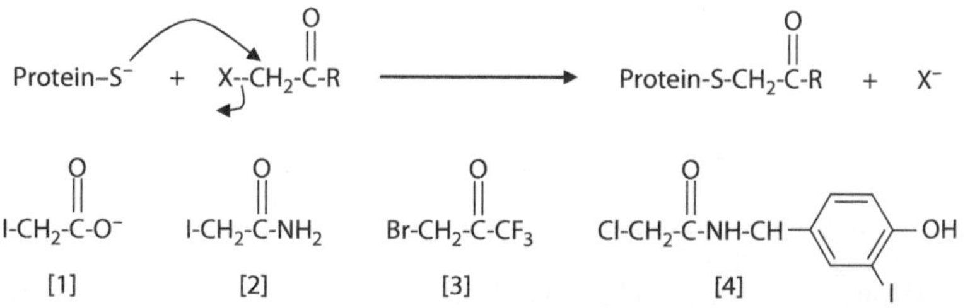

FIGURE 3.1 Reaction of α-haloacetyl compounds with the thiolate ion. The equation shows nucleophilic substitution of the halide. Some examples of α-haloacetyl compounds are: [1] iodoacetate, [2] iodoacetamide, [3] bromotrifluoroacetate, and [4] *N*-chloroacetyliodotyramine.

been employed for the introduction of radioactive iodine into proteins.[13] Such α-haloacetyl groups have also been incorporated into cross-linking reagents as will be discussed later. Although the α-haloacetyl compounds are not completely sulfhydryl specific, and other nucleophiles such as amines will also react, they are most selective for thiols.

3.2.2 *N*-Maleimide Derivatives

Maleimides such as *N*-ethylmaleimide (NEM) (Figure 3.2) are considered fairly specific for the sulfhydryl group, especially at pHs below 7 where other nucleophiles are protonated.[14–16] In acidic and near neutral solutions, the reaction rate with simple thiols is about 1000-fold faster than with the corresponding simple amines. Although the rate increases with pH, the reaction with the amino group also becomes significant at high pHs.[7] The other major competing reaction is the hydrolysis of maleimides to maleamic acids (Figure 3.3). However, at pH 7, the apparent rate of hydrolysis is only 3.2×10^{-4} min^{-1} in 0.1 M sodium phosphate buffer at 20°C, which is too slow to interfere with the reaction with sulfhydryl groups.[17] The rate of decomposition becomes significant only at pHs above neutrality.[15] At pH 7, it is estimated that the half-life of the reaction between millimolar concentrations of mercaptan and a maleimide is on the order of 1 s. Thiols undergo the Michael reaction with maleimides to yield exclusively an adduct as shown in Figure 3.2. The resulting succinimidyl thioether bond is very stable and cannot be cleaved under physiological conditions. However, their imido groups may undergo spontaneous hydrolysis and produces isomeric succinamic acids as water can attack either of the two carbonyl carbons of the imido group.[18] As a result, the isomeric complexity may complicate analysis of the reacted protein. Nevertheless, the quantitative reaction of maleimides with thiol groups provides the basis for a spectrophotometric assay

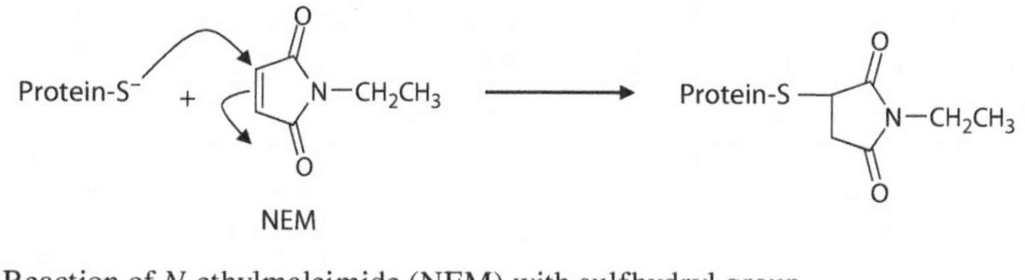

FIGURE 3.2 Reaction of *N*-ethylmaleimide (NEM) with sulfhydryl group.

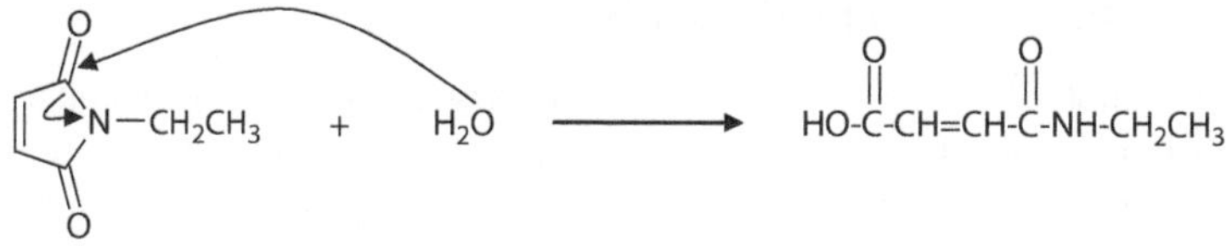

FIGURE 3.3 Hydrolysis reaction of *N*-ethylmaleimide.

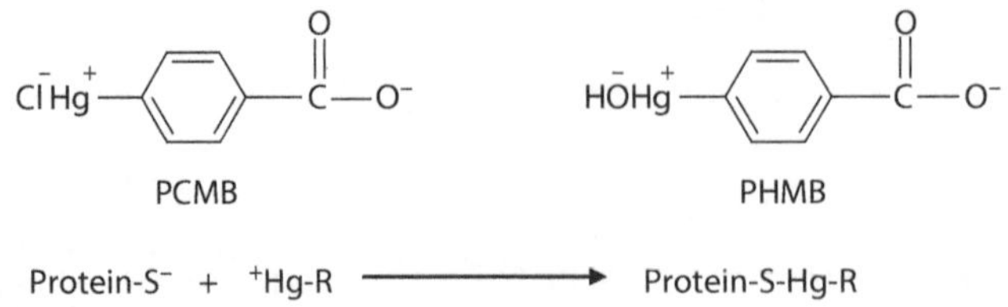

FIGURE 3.4 Mercurial reagents, *p*-chloromercuribenzoate (PCMB) and *p*-hydroxymercuribenzoate (PHMB), and their reactions with a thiol.

of the latter. At pH 6, the loss in absorption of NEM at 300 nm can be used to quantify sulfhydryl groups in the presence of excess NEM.[19] However, the use of Ellman's reagent affords a greater sensitivity (see below).

Many N-substituted maleimides have been synthesized for sulfhydryl group modification. The maleimide group has also been incorporated into cross-linking reagents to selectively react with the thiol group.

3.2.3 MERCURIAL COMPOUNDS

Mercurials are the most specific of all sulfhydryl reagents. Organomercurials such as *p*-chloromercuribenzoate (PCMB) (converted to *p*-hydroxymercuribenzoate (PHMB) in aqueous solution) (Figure 3.4) have been widely used.[20] The optimum rate of reaction is usually around pH 5, although the reaction rate for different mercurials and different sulfhydryl groups within one or different proteins can vary greatly. While the total thiolate ion increases with pH, a faster reaction rate does not result due to the competition of OH⁻ for the reagent.

The affinity of mercurials for sulfhydryl groups is very strong. The dissociation constants of thiol–mercurial complexes are on the order of 10^{-20}, which should allow specific modification of these groups in proteins. However, removal of mercurials from this complex is possible by competitive displacement with high concentrations of low-molecular-weight mercaptans. Because of the specificity of the reaction at low pH toward the thiol group, the mercurial ion has been incorporated into many reagents such as 2-(2′-pyridylmercapto)mercuri-4-nitrophenol.[21,22]

3.2.4 DISULFIDE REAGENTS

Disulfide interchange occurs when sulfhydryl groups react with disulfides as denoted in Section 2.3.2.1 (Figure 2.6). The most commonly used disulfide in the modification of protein sulfhydryl groups is Ellman's reagent, S,S′-dithiobis-(2-nitrobenzoic acid) (DTNB) (Figure 3.5). Other disulfides, such as 4,4′-dithiodipyridine (DTP), methyl-3-nitro-2-pyridyl disulfide (MNPD), and methyl-2-pyridyl disulfide (MPD) (Figure 3.5), have also been synthesized.[23] The protein disulfides formed are readily reversible on incubation with another free mercaptan such as 2-mercaptoethanol (ME) or dithiothreitol (DTT) (Figure 3.5). As pointed out in the Section 2.3.2.1, protein disulfides are easily reduced to free sulfhydryl groups. After reduction, the reduced thiols may react with other sulfhydryl reagents as mentioned above.

3.3 AMINO GROUP–SPECIFIC REAGENTS

The amino group is another strong nucleophile in the protein. Moreover, because of its abundance and omnipresence in proteins, it is the most important target for chemical modification, particularly in cases where cysteine residues are absent.[24] Since the protonated species is not reactive, the reaction rate increases with increasing pH as the free amine is formed. Due to the relatively high pK_a of the ammonium ion, most of the reagents that react with the amino group also react with other

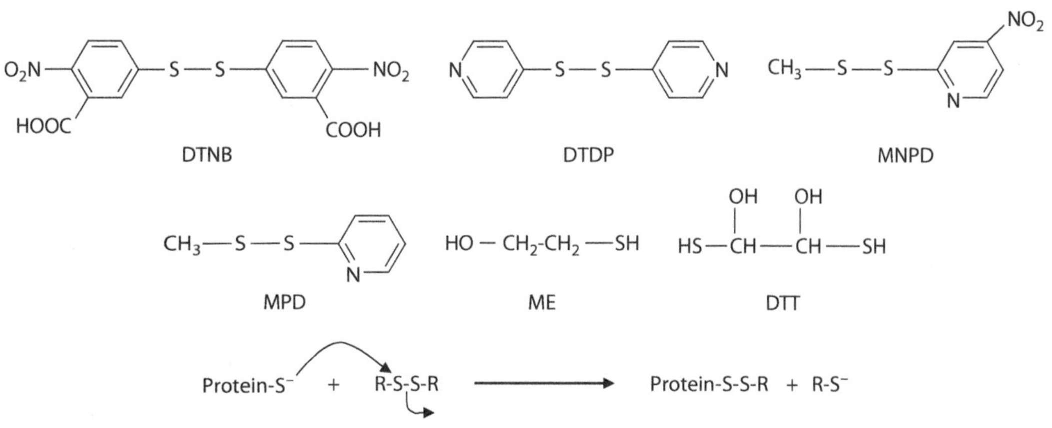

FIGURE 3.5 Disulfide reagents and their reaction with sulfhydryl groups. DTNB: Ellman's reagent, S,S'-dithiobis(2-nitrobenzoic acid); DTDP: 4,4'-dithiodipyridine; MNPD: methyl-3-nitro-2-pyridyl disulfide; MPD: methyl-2-pyridyl disulfide; ME: 2-mercaptoethanol; DTT: dithiothreitol.

functionalities. However, many stable acylated products are formed only with the amino groups, providing a basis of selectivity. The most common reactions of amines are alkylation and acylation reactions.[25] Some of the important reagents are described in the following text.

3.3.1 ALKYLATING AGENTS

3.3.1.1 α-Haloacetyl Compounds

Compared with sulfhydryl groups, amino groups of proteins react with haloacetates and haloacetamides much more slowly and only at high pHs, where they are unprotonated. Even then, the reaction rate is only one-hundredth of that of the thiol. The reaction is generally observed as a by-product of extensive S-carboxymethylation and dialkylation may occur during the process (Figure 3.6). However, in proteins where there is no free sulfhydryl group or where the thiol is buried and unreactive, the reaction with the amino group becomes important. Under these circumstances, alkyl halides containing the reactive structure of α-haloacetyl moiety, $-COCH_2-X$, react with the amino and other groups of proteins. These compounds have been used as affinity labels.[26] The haloacetyl group can be incorporated into active-site-directed irreversible enzyme inhibitors to study the mechanism of enzyme action. For example, a complex UDP-galactose affinity analog (Figure 3.7) was synthesized to investigate UDP-galactose 4-epimerase.[27] Such haloacetyl groups have also been incorporated into many cross-linking reagents.

3.3.1.2 N-Maleimide Derivatives

Like haloacetates, the reaction of maleimides with amino groups becomes significant only at alkaline pHs. Because of their greater reactivity with sulfhydryl groups, this reaction is useful only for proteins containing no cysteine residues.[28] Amino groups can conceivably react with maleimides in

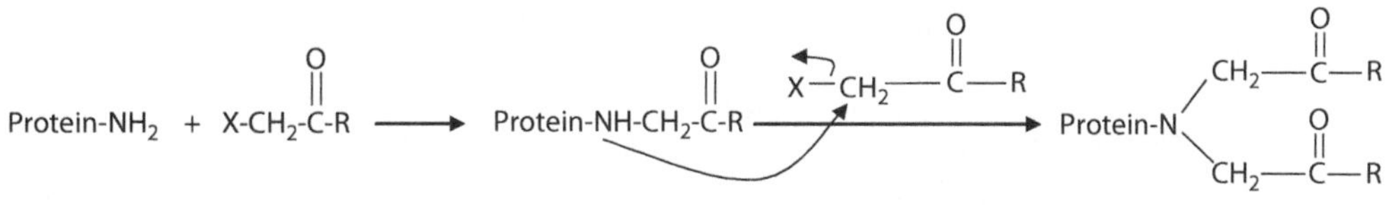

FIGURE 3.6 Dialkylation of amino groups. Reaction with α-haloacetyl compounds is shown with possible dialkylation.

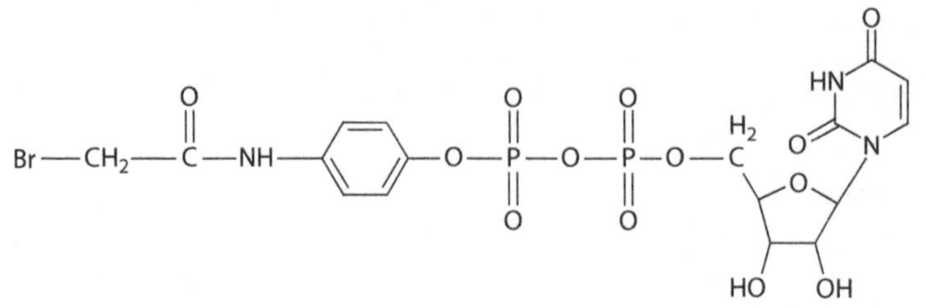

FIGURE 3.7   UDP-galactose affinity analogue: bromoacetylaminophenyl uridylyl pyrophosphate.

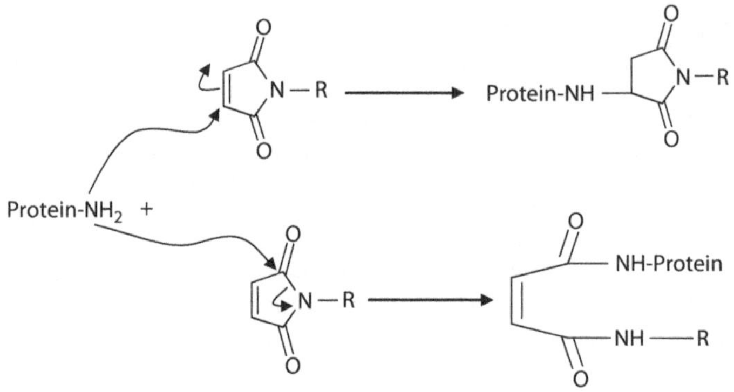

FIGURE 3.8   Reaction of *N*-maleimides with protein amino groups.

one of the two ways (Figure 3.8). The amine nitrogen can attack the double bond of the maleimide ring (Michael reaction analogous to the addition of SH)[29] or undergo acylation by addition to the carbonyl carbon group followed by ring opening. Regardless of the reaction pathway, the reaction with amines is significant only above neutral pH.

3.3.1.3   Aryl Halides

Aryl halides are usually regarded as amino-group reagents. Dinitrofluorobenzene (DNFB) (Figure 3.9), for example, has been used to identify the N-terminal amino acid residues of poly-peptides and proteins, and trinitrobenzenesulfonate (TNBS) (Figure 3.9) has been used to quan-tify amino groups in proteins because of their bright yellow-colored products. The reaction involves bimolecular nucleophilic substitution at the halogen-bonded carbon and is accelerated at alkaline pH since only the unprotonated amino groups react. The reactivity of the aryl halides depends on their leaving group. Fluoro-compounds are the most reactive followed by chloro- or

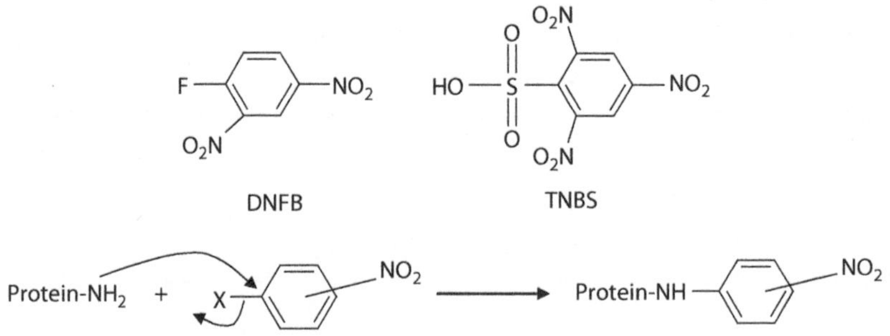

FIGURE 3.9   Arylation of amino groups. Reaction with aryl halide, either 2.4-dinitrofluorobenzene (DNFB) or trinitrobenzenesulfonate (TNBS) is shown.

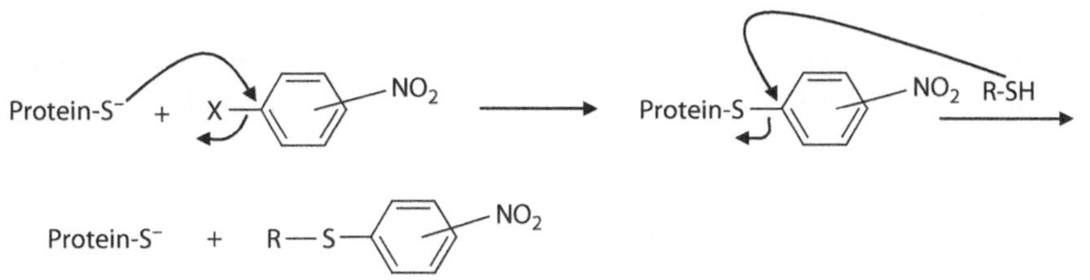

FIGURE 3.10 Reversible thioarylation product. The reaction with thiol is reversed by free thiol reagents.

bromo-derivatives in the order: F > Cl ~ Br > SO$_3$H.[30] Trinitroaryl compounds are more reactive than the corresponding dinitro compounds.

While aryl halides also react readily with other nucleophiles such as thiolate, phenolate, and imidazole, the products of these latter reactions are unstable at alkaline pH. In the case of sulfhydryl groups, the reaction is reversible in the presence of excess β-mercaptoethanol (Figure 3.10).[31]

Several reactive nitrohaloaromatic compounds have been used as cross-linking reagents. These will be further discussed later.

3.3.1.4 Aldehydes and Ketones

Carbonyl compounds such as aliphatic ketones and aldehydes react readily and reversibly with amino groups of proteins to form Schiff bases.[32] As usual, the reaction is pH-dependent with the best result obtainable near pH 9. The adducts so formed can be stabilized with sodium borohydride or sodium cyanoborohydride in a process known as reductive alkylation (Figure 3.11).[33] Amine boranes such as dimethylamine, trimethylamine, t-butylamine, morpholine, and pyridine boranes are also effective.[34] Reduction is best carried out at 0°C in 0.2 M borate buffer, pH 9.0. The concentration of sodium borohydride used for stabilization of the Schiff base is usually low to prevent the reduction of disulfide bonds, in which case, sodium cyanoborohydride may be a better choice. Pyridine borane has been reported to be an even superior reducing agent for reductive alkylation and is less toxic.[34] Under the mild, slightly alkaline conditions required for the extensive alkylation of amino groups, other protein side chains do not give stable derivatives, constituting a high selectivity for the amino group.

Practically any carbonyl compounds can be used to alkylate amino groups of proteins. Reductive methylation of bovine pancreatic ribonuclease, ovomucoid, chymotrypsin, and other proteins has been carried out with formaldehyde (Figure 3.11).[32] Pyridoxal phosphate (PLP) has also been used to label lysine residues at or near PLP binding sites of proteins and enzymes.[35,36] While some

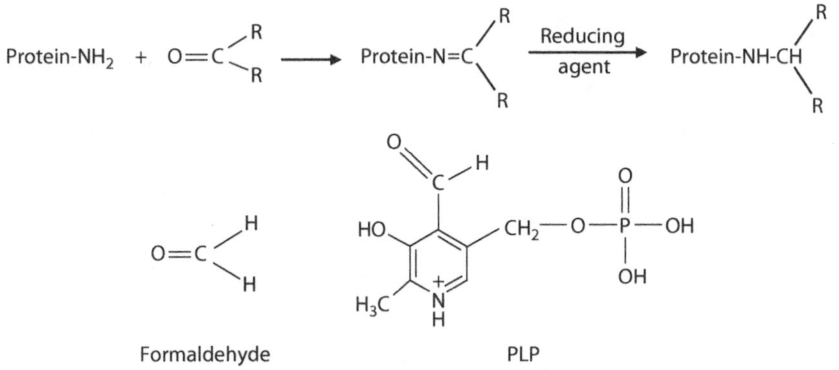

FIGURE 3.11 Reaction of amino groups with aldehydes and ketones such as formaldehyde and pyridoxal phosphate (PLP). Schiff base formed is reversible but can be stabilized with reducing agents.

simple dialdehydes, notably glutaraldehyde, have been used for cross-linking of proteins, the dialdehydes produced upon periodate cleavage of carbohydrate as discussed in Section 2.3.2.4, provide an important and versatile means for protein conjugation. Examples of such reactions will be discussed later.

3.3.2 ACYLATING AGENTS

Acylating agents are compounds that contain an activated acyl group where the nucleophile attacks at the carbonyl carbon displacing a leaving group (Figure 3.12). The rate of acylation of a nucleophile depends on its pK_a, which, as discussed in Chapter 2, determines the nucleophilicity. Other factors such as steric and proximity effects of other residues on a protein (microenvironment) also influence its reactivity. Since water is also a nucleophile, hydrolysis of the acylating agent may be an important side reaction, consuming considerable portions of the reagent. Thus, even if a large excess of the reagent is used, complete acylation of a protein may be difficult to achieve.

All nucleophiles in proteins are susceptible to acylation. The tyrosine phenolic group is generally less reactive than amino groups, partly because of its higher pK_a and partly because it is

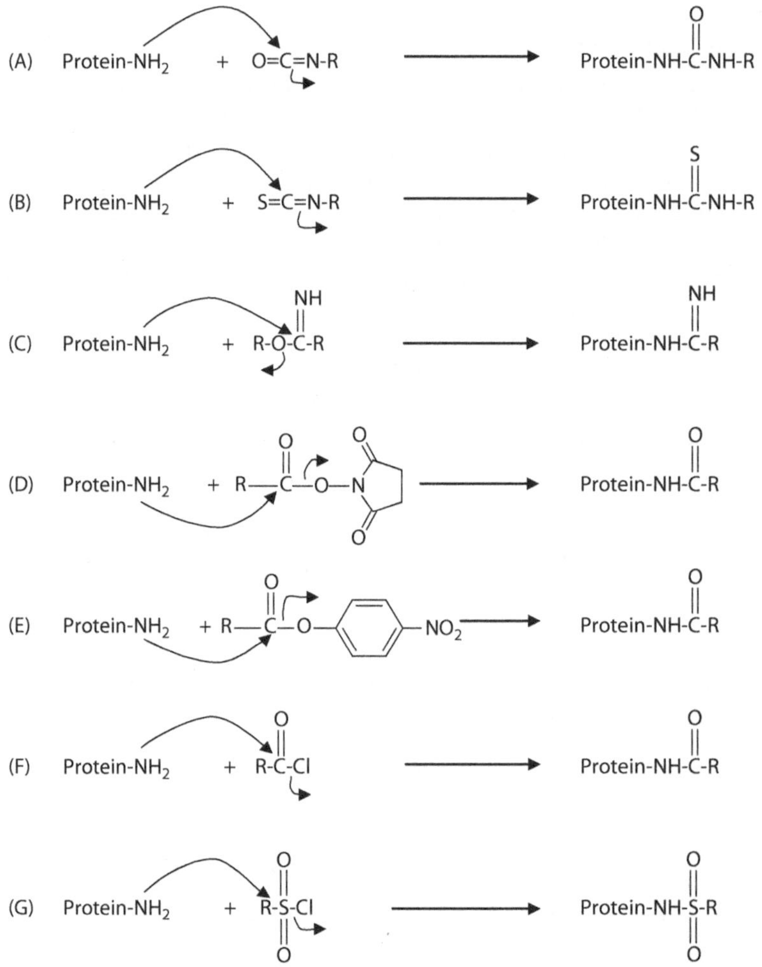

FIGURE 3.12 Acylation reactions of amino groups. Reaction with: (A) isocyanate; (B) isothiocyanate; (C) imidoesters; (D) N-hydroxysuccinimidyl ester; (E) p-nitrophenyl ester; (F) acyl chloride; and (G) sulfonyl chloride.

generally shielded in proteins. Another important difference is that acylation of tyrosine residues is easily reversible. Hydroxylamine at neutral pH or simply mild alkaline conditions are sufficient to deacylate acylated tyrosines.[37] Similar ease of reversibility of acylated products of other nucleophiles, such as sulfhydryl, imidazolyl, and carboxyl, makes most acylating agents amino group selective. Thus, succinic anhydride has been used to convert amino groups to carboxylic acids (Section 2.3.2.2.1).

In addition to acid anhydrides, there are many other acylating agents. Of particular importance are the isocyanates, isothiocyanates, imidoesters, acid halides, N-hydroxysuccinimidyl, and other activated esters. Isocyanates and isothiocyanates react with amino, sulfhydryl, imidazolyl, tyrosyl, and carboxyl groups of proteins. However, only amino groups yield stable products (Figure 3.12A and B). Isocyanates are generally more reactive than isothiocyanates. Such differences in reactivity are reflected by the reaction rates of these groups in 2-isocyanato-4-isothiocyanatotolulene, which contains both isocyanate and isothiocyanate functionalities.[38]

Imidoesters are the most specific reagents for amino groups among the acylating compounds. At mild alkaline pHs (between 7 and 10), imidoesters react only with amines to form imidoamides, the so-called amidines (Figure 3.12C). The products, like the original imidoester, carry a positive charge at physiological pH. Amidination, therefore, retains the net charges of the protein, minimizing the effect of charge on protein conformation.[39] For this reason, imidoesters are favorably incorporated into cross-linking reagents. Amidines are stable in acidic and neutral solutions, but slowly hydrolyze at high pHs. To preserve the cross-linked conjugates, it is necessary to lower the pH of the solution.

Similar to imidoesters, there are many other activated esters that react with amino groups to form stable amide bonds. Common examples include N-hydroxysuccinimidyl (NHS) esters and p-nitrophenyl esters (Figure 3.12D and E). These compounds contain stable leaving groups, N-hydroxysuccinimide and p-nitrophenol, respectively, and are reactive toward nucleophilic substitution. Such groups have been used frequently in the construction of cross-linking reagents. NHS cross-linkers are more reactive and stable than their imidate counterparts at physiological pH.[40]

Like acid anhydrides, acyl halides are very reactive compounds and react with all the nucleophiles of a protein. Reaction with amino groups results in the formation of stable amide bonds (Figure 3.12F). The adducts with other nucleophiles, for example, O-acyltyrosine, are unstable and hydrolyze under moderately alkaline conditions. The major competitive reaction is hydrolysis, since water molecules are present in high concentrations. Sulfonyl halides react with proteins in much the same way as acyl halides (Figure 3.12G). They are, however, more stable and therefore react more slowly. The reaction products are also more stable. Even S-sulfonylcysteine, O-sulfonyltyrosine, and imidazole-sulfonylhistidine are stable in neutral solutions, unlike the corresponding acyl compounds. Thus, the reaction of sulfonyl chlorides and fluorides with proteins is not group specific. Many sulfonyl fluorides have been found to react with serine hydroxyl groups at the active sites of proteases.[41]

3.4 REAGENTS DIRECTED TOWARD CARBOXYL GROUPS

3.4.1 DIAZOACETATE ESTERS AND DIAZOACETAMIDES

Like diazomethane, which has been used to esterify carboxyl groups,[42] diazoacetyl esters and diazoacetamides react with high specificity with carboxyl groups of proteins under mild acid conditions.[43] The reaction is thought to involve protonation of the diazoacetyl group followed by nucleophilic displacement of nitrogen gas (Figure 3.13) and occurs optimally at a pH near 5.[44] At this pH, alkylation of sulfhydryl groups is an important side reaction. At higher pHs, other nucleophiles also react. Diazoacetates and diazoacetamides are highly reactive compounds and react rapidly with water and many simple inorganic anions, which reduce their efficiency in reaction with proteins. At lower pHs, hydrolysis of the reagent is even more significant.

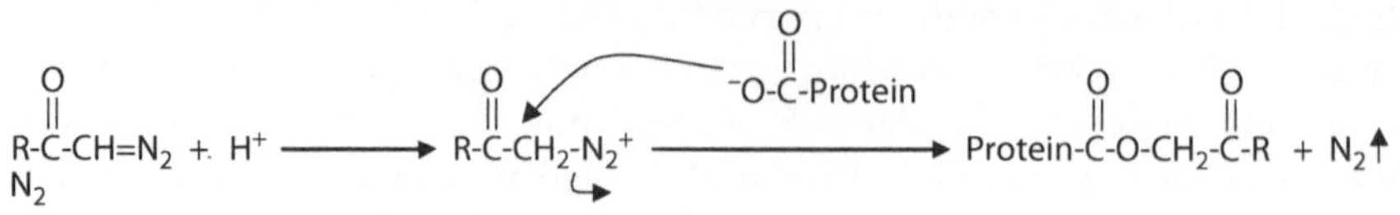

FIGURE 3.13 Reaction of diazoacetyl esters and amides with carboxyl groups. A protonated intermediate is indicated.

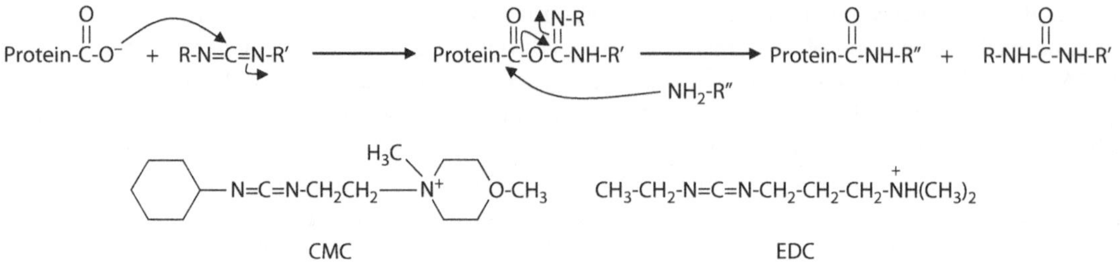

FIGURE 3.14 Carbodiimide-mediate modification of carboxyl groups. The reaction mechanism is shown. Also shown are the structures of 1-cyclohexyl-3-(2-morpholinyl-4-ethyl)carbodiimide (CMC); 1-ethyl-3-(3-dimethylaminopropyl)carbodiimide (EDC).

3.4.2 CARBODIIMIDES

The most important chemical modification reactions of carboxyl groups utilize the carbodiimide-mediated process. In the presence of an amine, carbodiimides promote the formation of an amide bond in two steps. In the initial reaction, the carboxyl group adds to the carbodiimide to form an O-acylisourea intermediate. Subsequent reaction of the intermediate with an amine yields the corresponding amide (Figure 3.14).[45] The reactive intermediate also undergoes hydrolysis slowly and may react with other nucleophiles to form different carboxylated derivatives. With proteins, the optimum pH of the reaction is about 5.[45] Carbodiimides react not only with carboxylic acids but also with alcohols, amines, water, and other nucleophiles. O-arylisourea is formed with the phenolic group of tyrosine and S-cysteinylisourea with cysteine. These reactions may cause inactivation of the modified protein and decrease the coupling efficiency.

Practically any carbodiimides can be used to facilitate the amide bond formation. There are several water-soluble derivatives that are most useful for this application.[46] The most distinct are 1-cyclohexyl-3-(2-morpholinyl-4-ethyl)carbodiimide (CMC) and 1-ethyl-3-(3-dimethylaminopropyl) carbodiimide (EDC) (Figure 3.14). The use of these compounds to convert a carboxyl group to a cationic amine has been discussed in Section 2.3.2.2.5. In the absence of an added amine, cross-linking between proteins may also occur.[47] These reactions will be discussed later.

3.5 TYROSINE SELECTIVE REAGENTS

3.5.1 ACYLATING AGENTS

The phenolate ion of tyrosine reacts similarly to amino groups toward acylating agents. However, the tyrosyl group is generally perceived as having a lower reactivity, not because the phenolate ion has lower nucleophilicity, but because tyrosine residues are usually buried in proteins inaccessible for reactions due to their hydrophobicity. Such steric hindrance usually gives the amino groups the leading edge of selectivity. In addition, many O-acyl-tyrosine products are unstable, even at neutral pH, as has been noted earlier. The most important acylating reagent for the phenolic moiety of tyrosine is N-acetylimidazole (Figure 3.15). Although it reacts with both amino and tyrosyl

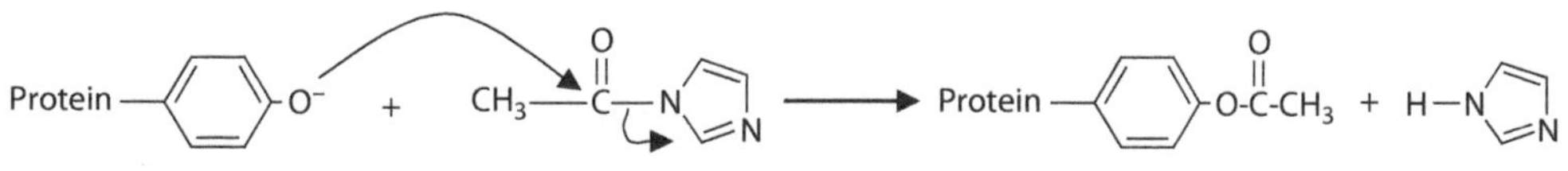

FIGURE 3.15 Reaction of phenolic moiety of tyrosine with *N*-acetylimidazole.

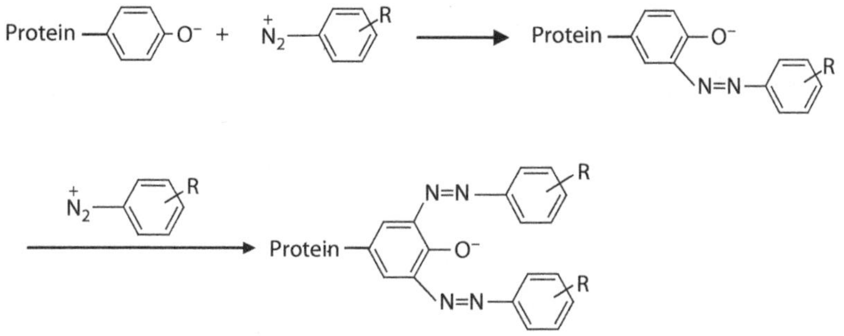

FIGURE 3.16 Reaction of phenolic moiety of tyrosine with diazonium salts.

groups of proteins, it is more selective for tyrosine. Such selectivity enables it to be used for determining the distribution of free and buried tyrosine residues in a protein.[48] As with other acylating agents, the *O*-acetylphenol is reversible at high pHs or in the presence of hydroxylamine.[49]

3.5.2 ELECTROPHILIC REAGENTS

Tyrosine, histidine, and other aromatic residues of proteins are rich in electrons. These residues undergo electrophilic substitution reactions at the aromatic ring. The most important and useful electrophiles for reaction with tyrosine and histidine in proteins are diazonium compounds.[50] These reagents can be easily obtained by treating aromatic amines with nitrous acid at a low temperature. Since diazonium ions are generally rather unstable, they are usually prepared fresh before use. The reaction with tyrosine involves an electrophilic attack of the diazonium ion at the ortho position of the phenol ring displacing a proton without disrupting its aromaticity (Figure 3.16). Bisazotization of tyrosine is possible in the presence of excess reagent. A similar mechanism takes place at the imidazole moiety of histidine. The rate of the reaction increases with increasing alkalinity, with an optimal pH near 9. Other protein components such as lysine, tryptophan, cysteine, and arginine residues react very slowly, such that diazonium reagents can be regarded as tyrosine selective.

3.6 ARGININE-SPECIFIC REAGENTS

A predominant reaction of the guanidinyl moiety of arginine residue is with 1,2-dicarbonyl reagents.[51,52] Commonly used α-dicarbonyls include glyoxal, phenylglyoxal, 2,3-butanedione, and 1,2-cyclohexanedione (Figure 3.17). Under mild alkaline conditions, these compounds condense with the guanidinyl group in an initial reaction very similar to the Schiff base formation, which undergoes further rearrangement to different products with different compounds (Figure 3.17). With phenylglyoxal, a trimeric adduct may be formed.[53] Borate ion may stabilize the adduct as in the case for 1,2-cyclohexanedione.[51] The principal side reaction of modification of arginine with α-dicarbonyls is the reaction with lysine residues, which gives unknown products. However, under slightly alkaline conditions, the reaction with arginine predominates.

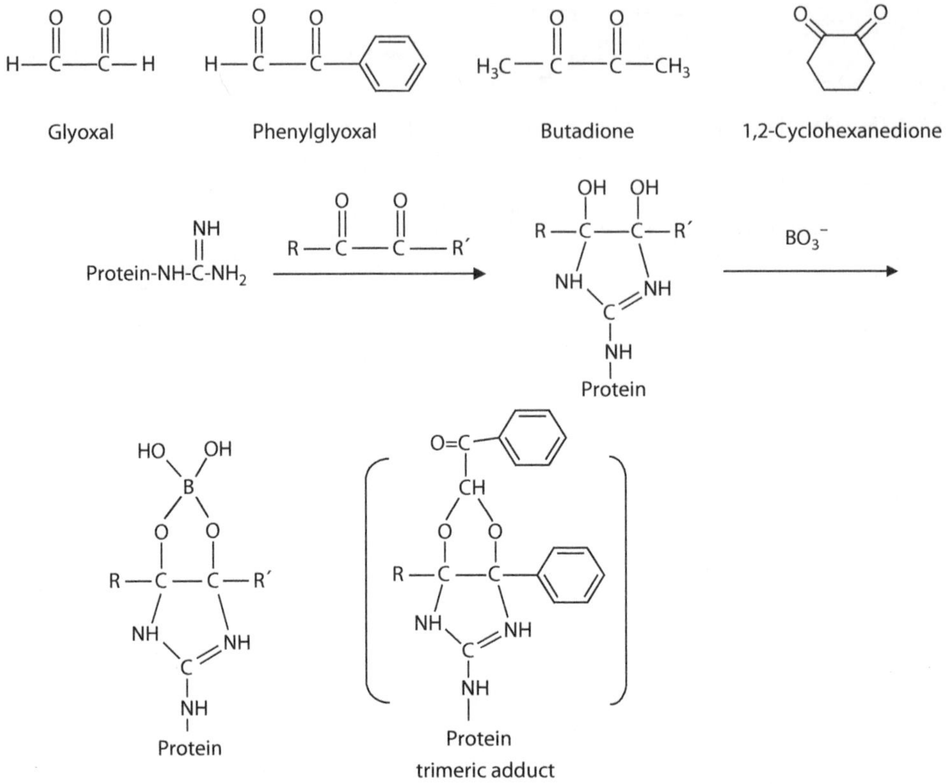

FIGURE 3.17 Modification of arginine residues with dicarbonyl compounds. Some common reagents are: glyoxal; phenylglyoxal; 2,3-butanedione; 1,2-cyclohexanedione. The general reaction is shown. Phenylglyoxal may form a trimeric adduct.

3.7 HISTIDINE-SELECTIVE REAGENTS

While a number of alkylating and acylating agents react with the imidazolyl moiety of histidine, as mentioned earlier, the rates of these reactions are generally inferior to those characteristic of other nucleophiles. Even with α-haloacetate, N-carboxymethylation is generally slow in comparison with sulfhydryl groups. However, when such reactive α-halocarbonyl groups are incorporated into affinity labels, specific reactions may be achieved. For example, p-toluenesulfonylphenylalaninechloromethyl ketone (TPCK) (Figure 3.18) specifically alkylates an

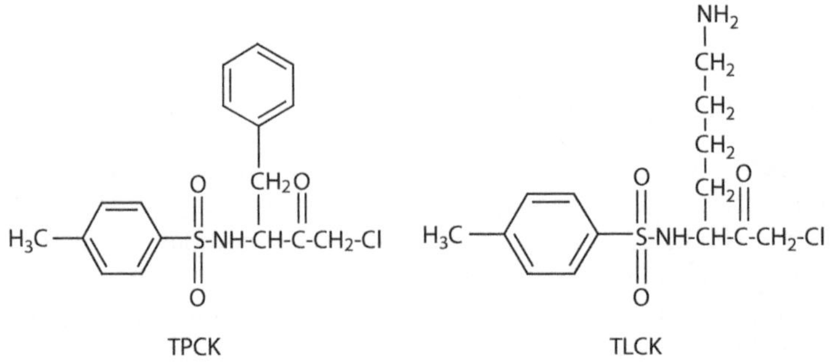

FIGURE 3.18 Active site-directed histidine specific reagents: p-toluenesulfonylphenylalaninechloromethyl-ketone (TPCK) and p-toluenesulfonyllysinechloromethylketone (TLCK).

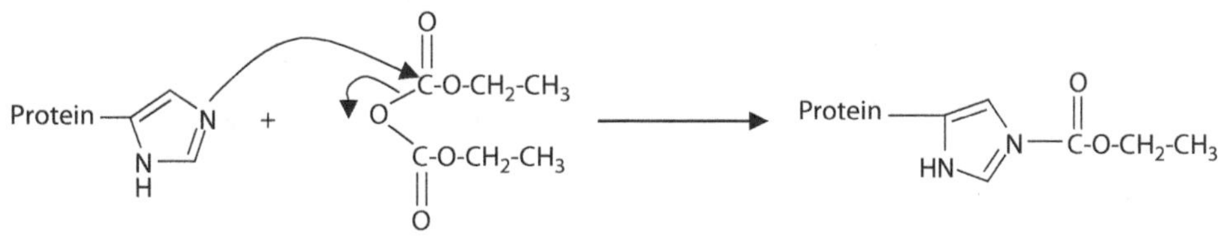

FIGURE 3.19   Acylation reaction of histidine residue with diethylpyrocarbonate.

active-center histidine in α-chymotrypsin.[54] Similarly, p-toluenesulfonyllysinechloromethyl ketone (TLCK) (Figure 3.18) reacts specifically with a histidine at the active site of trypsin.[55] Such active-site-directed affinity labels may be useful in the design of active-site-directed cross-linking reagents. Besides α-haloacetyl groups, other alkylating agents are not reactive toward histidine.

With acylating reagents, histidine forms acylated products that are generally unstable and may undergo spontaneous hydrolysis. The most important acylating agent commonly used for the modi-fication of histidines is diethylpyrocarbonate (ethoxyformic anhydride) (Figure 3.19).[56] At pH 4, this reagent shows good selectivity for histidine, resulting in N-ethoxyformulation of one of the imid-azole nitrogens. Amino groups are the major competitor of the reaction. The acylated imidazole is hydrolyzed at alkaline pH, resulting in the recovery of histidine. Deacylation can be achieved at neutral pH very rapidly with hydroxylamine.

3.8   METHIONINE-ALKYLATING REAGENTS

The major chemical modification reactions of methionine are oxidation and alkylation, although cyanogen bromide has been used extensively to cleave peptide bonds formed with methionine resi-dues.[57] Oxidation of methionine to methionine sulfoxide can be achieved with hydrogen perox-ide,[58] but this reaction has no implication to protein cross-linking and therefore will not be further discussed.

Alkylation of methionine, for example, with α-haloacids, is independent of pH and is most selec-tive at pH 3 or less where other groups are protonated.[59,60] However, because methionine is often sit-uated in the hydrophobic interior of proteins, some disruption of the tertiary structure is often necessary for rapid reaction. For the same reasons, cross-linking reagents are generally not designed with methionine in mind. However, one should bear in mind that in recent years, with the progress in molecular biology, more and more recombinant proteins are constructed and many or even most of these may have methionine as the N-terminal amino acid.

3.9   TRYPTOPHAN-SPECIFIC REAGENTS

Due to its hydrophobicity, tryptophan residues are often buried in the interior of proteins. When exposed, they can be modified with N-bromosuccinimide (NBS)[61] and 2-hydroxy-5-nitrobenzyl bromide (HNBB) (Figure 3.20).[62] Whereas NBS selectively cleaves tryptophanyl peptide bonds in peptides and proteins, HNBB forms an adduct with tryptophan residues. However, practical use of these reagents in protein cross-linking has not been found.

A distinct reagent, p-nitrophenylsulfenyl chloride, has been used for the modification of the indolyl moiety, giving rise to the 2-thioether derivative (Figure 3.21).[63] The reaction is selective for tryptophan and cysteine residues and has been incorporated into a bifunctional cross-linking reagent.[64]

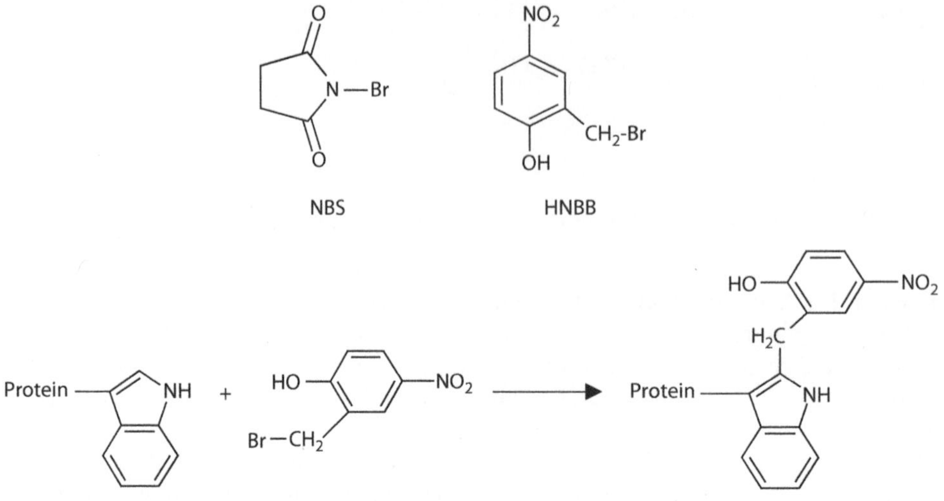

FIGURE 3.20 Tryptophan modifying reagents: *N*-bromosuccinimide (NBS) and 2-hydroxy-5-nitrobenzyl bromide (HNBB). The reaction with HNBB is shown.

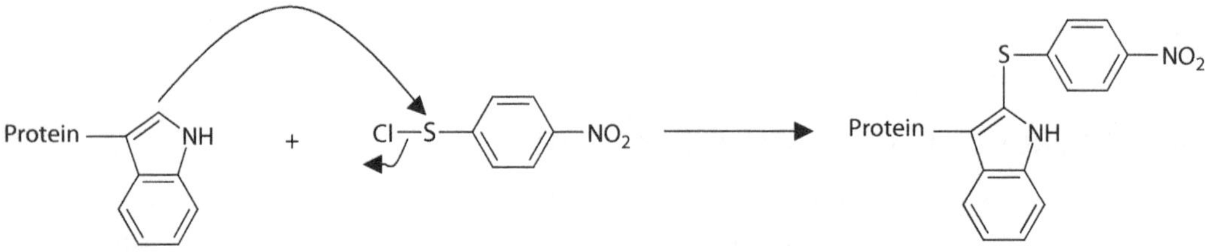

FIGURE 3.21 Reaction of a tryptophan residue with *p*-nitrophenylsulfenyl chloride.

3.10 SERINE-MODIFYING REAGENTS

As discussed in the Section 2.3.1.3, alkyl alcohols are generally inert and normally not subject to chemical modification in aqueous solutions unless highly reactive reagents are used. In organic solvents, hydroxyl groups of solid matrices can react with various reagents for the conjugation of proteins. These reactions will be discussed later. In proteins, hydroxyl groups of serines and threonines undergo modification only under certain circumstances where the hydroxyl group is activated by neighboring groups such as that at the active site of α-chymotrypsin and other serine proteases (see Section 2.3.1.3). Many reactive reagents such as diisopropylfluorophosphate, phenylmethylsulfonylfluoride, and other arylsulfonyl fluorides have been found to react with the active-site serine.[41] But because of the strong competitive reaction of hydrolysis, these groups are generally not targets for chemical modification.

REFERENCES

1. Means, G. E. and Feeney, R. E., *Chemical Modification of Proteins*, Holden-Day, San Francisco, CA, 1971.
2. Lundblad, R. L., *Chemical Reagents for Protein Modification*, 3rd edn., CRC Press, Boca Raton, FL, 2004.
3. Lundblad, R. L., *Techniques in Protein Modification*, CRC Press, Boca Raton, FL, 1994.
4. Francis, M. B., Antos, J. M., Bains, S., Crochet, S. P., Gilmore, J. M., Hooker, J. M., Joshi, N. S. et al., New Chemical tools for protein modification, *ChemInform*, 37(4), DOI: 10.1002/chin.200604251, 2006.

5. Eyzaguirre, J. (ed.), *Chemical Modification of Enzymes: Active Site Studies*, John Wiley & Sons, New York, 1987.

6. Glazer, A. N., Deiange, R. J., and Sigman, D. S., *Chemical Modification of Proteins*, North-Holland/ Elsevier, Amsterdam, the Netherlands, 1975.

7. Glazer, A. N., The chemical modification of proteins by group-specific and site specific reagents, in *The Proteins*, 3rd edn., Neurath, H. and Hill, R. L. (eds.), Academic Press, New York, 1976, Chapter 2.

8. Hirs, C. H. W. and Timasheff, S. N. (eds.), Enzyme structure, in *Methods in Enzymology*, Vol. 91, Academic Press, New York, 1983.

9. Hermanson, G. T., *Bioconjugate Techniques*, Academic Press, San Diego, CA, 1996.

10. Crankshaw, M. W. and Grant, G. A., Modification of cysteine, in *Current Protocols in Protein Science*, Wiley Interscience, Hoboken, NJ, 2001, Chapter 15.

11. Kim, Y., Ho, S. O., Gassman, N. R., Korlann, Y., Landorf, E. V., Collart, F. R., and Weiss, S., Efficient site-specific labeling of proteins via cysteines, *Bioconjug. Chem.*, 19, 786, 2008.

12. Huestis, W. H. and Raftery, M. A., A study of cooperative interactions in hemoglobin using fluorine nuclear magnetic resonance, *Biochemistry*, 11, 1648, 1972.

13. Holowka, D., N-chloroacetyl-[^{125}I]-iodotyramine: An alkylating agent with high specific activity, *Anal. Biochem.*, 117, 390, 1981.

14. Gorin, G., Martin, P. A., and Doughty, G., Kinetics of the reaction of N-ethylmaleimide with cysteine and some congeners, *Arch. Biochem. Biophys.*, 115, 593, 1966.

15. Gregory, J. D., The stability of N-ethylmaleimide and its reaction with sulfhydryl groups, *J. Am. Chem. Soc.*, 77, 3922, 1955.

16. Smyth, D. G., Blumenfeld, O. O., and Konigsberg, W., Reaction of N-ethylmaleimide with peptides and amino acids, *Biochem. J.*, 91, 589, 1964.

17. Heitz, J. R., Anderson, C. D., and Anderson, B. M., Inactivation of yeast alcohol dehydrogenase by N-alkylmaleimides, *Arch. Biochem. Biophys.*, 127, 627, 1968.

18. Kalia, J. and Rainesa, R. T., Catalysis of imido group hydrolysis in a maleimide conjugate, *Bioorg. Med. Chem. Lett.*, 17, 6286, 2007.

19. Benesch, R. and Benesch, R. E., Determination of -SH groups in proteins, *Methods Biochem. Anal.*, 10, 43, 1962.

20. Hirano, J., Miyamoto, K., and Ohta, H., Purification and characterization of aldehyde dehydrogenase with a broad substrate specificity originated from 2-phenylethanol-assimilating *Brevibacterium* sp. KU1309, *Appl. Microbiol. Biotechnol.*, 76, 357, 2007.

21. Stefanini, S., Chiancone, E., McMurray, C. H., and Antonini, E., Dissociation human hemoglobin by different organomercurials, *Arch. Biochem. Biophys.*, 151, 28, 1972.

22. Scott-Ennis, R. J. and Noltmann, E. A., Differential response of cysteine residues in pig muscle phosphoglucose isomerase to seven sulfhydryl-modifying reagents, *Arch. Biochem. Biophys.*, 239, 1, 1985.

23. Kimura, T., Matsueda, R., Nakagawa, Y., and Kaiser, E. T., New reagent for the introduction of the thiomethyl group at sulfhydryl residue of proteins with concomitant spectrophotometric titration of the sulfhydryl: methyl-3-nitro-2-pytidyl disulfide and methyl 2-pyridyl disulfide, *Analyt. Biochem.*, 122, 274, 1982.

24. Geoghegan, K. F., Modification of amino groups, in *Current Protocols in Protein Science*, Wiley Interscience, Hoboken, NJ, 2001, Chapter 15.

25. Lundblad, R. L., Modification of amino groups, in *Chemical Reagents for Protein Modification*, 3rd edn., Lundblad, R. L. (ed.), CRC Press, Boca Raton, FL, 2004, Chapter 2.

26. Baker, B. R., *Design of Active-Site-Directed Irreversible Enzyme Inhibitors: The Organic Chemistry of the Enzymic Active Site*, John Wiley & Sons, New York, 1967.

27. Wong, Y.-H. H. and Frey, P. A., Uridine diphosphate galactose 4-epimcrase. Alkylation of enzyme-bound diphosphopyridine nucleotide by p-(bromoacetamido)phenyl uridyl pyrophosphate, an active-site-directed irreversible inhibitor, *Biochemistry*, 24, 5337, 1979.

28. Brewer, C. F. and Riehm, J. P., Evidence for possible nonspecific reactions between N-ethylmaleimide and proteins, *Anal. Biochem.*, 18, 248, 1967.

29. Smyth, D. G., Nagamatsu, A., and Fruton, J. S., Reactions of N-ethylmaleimide, *J. Am. Chem. Soc.*, 82, 4600, 1960.

30. Elsen, H. N., Belman, S., and Carsten, M. E., The reaction of 2,4-dinitrobenzenesulfonic acid with free amino groups of proteins, *J. Am. Chem. Soc.*, 75, 4583, 1953.

31. Shaltiel, S., Thiolysis of some dinitrophenyl derivatives of amino acids, *Biochem. Biophys. Res. Commun.*, 29, 178, 1967.

32. Means, G. E. and Feeney, R. E., Reductive alkylation of amino groups in proteins, *Biochemistry*, 7, 2192, 1968.

33. Dottavio-Martin, D. and Ravel, J. M., Radiolabeling of proteins by reductive alkylation with [^{14}C] form-aldehyde and sodium cyanoborohydride, *Analyt. Biochem.*, 87, 562, 1978.

34. Cabacungan, J. C., Ahmed, A. I., and Feeney, R. E., Amine boranes as alternative reducing agents for reductive alkylation of proteins, *Anal. Biochem.*, 124, 272, 1982.

35. Schirch, L. G. and Mason, M., A study of the properties of a homogeneous enzyme preparation and of the nature of its interaction with substrates and pyridoxal 5-phospbate, *J. Biol. Chem.*, 238, 1032, 1963.

36. Klein, S. M. and Sagers, R. D., Effect of borohydride reduction on the pyridoxal phosphate-containing glycine decarboxylase from *Peptococcus glycinophilus*, *J. Biol. Chem.*, 242, 301, 1967.

37. Nakamura, K., Nomura, E., and Taniguchi, H., Chemical properties of acyl groups on phenolic hydroxyl group of tyrosine, *Pept. Sci.*, 2001, 43, 2002.

38. Engvall, E. and Perlmann, P., Enzyme-linked immunosorbent assay, II. Quantitative assay of protein antigen, immunoglobulin G, by means of enzyme-labeled antigen and antibody-coated tubes, *Immunochemistry*, 8, 871, 1971.

39. Wofsy, L. and Singer, S. J., Effects of the amidination reaction on antibody activity and on the physical properties of some proteins, *Biochemistry*, 2, 104, 1963.

40. Staros, J. V., *N*-Hydroxysulfosuccinimide active esters: *Bis*(*N*-hydroxysulfosuccinimide) esters of two carboxylic acids are hydrophilic, membrane-impermeant, protein cross-linkers, *Biochemistry*, 21, 3950, 1982.

41. Wong, S. S., Quiggle, K., Triplett, C., and Berliner, L. J., Spin-labeled sulfonyl fluorides as active site probes of protease structure: II. Spin label synthesis and enzyme inhibition, *J. Biol. Chem.*, 249, 1678, 1974.

42. Herriott, R. M., Reactions of native proteins with chemical reagents, *Adv. Protein Chem.*, 3, 169, 1947.

43. Doscher, M. S. and Wilcox, P. E., Chemical derivatives of chymotrypsinogen. IV. A comparison of the reactions of α-chymotrypsinogen and of simple carboxylic acid with diazoacetamide, *J. Biol. Chem.*, 236, 1328, 1961.

44. Riehm, J. P. and Scheraga, H. A., Structural studies of ribonuclease. XVII. A reactive carboxyl group in ribonuclease, *Biochemistry*, 4, 772, 1965.

45. Kurzer, F. and Douraghi-Zadeh, K., Advances in the chemistry of carbodiimides, *Chem. Rev.*, 67, 107, 1967.

46. Gilles, M. A., Hudson, A. Q., and Borders, C. L., Jr., Stability of water-soluble carbodiimides in aqueous solution, *Anal. Biochem.*, 184, 244, 1990.

47. Goodfriend, T. L., Levine, L., and Fasman. G. D., Antibodies to bradykinin and angiotensin: A use of carbodiimides in immunology, *Science*, 144, 1344, 1964.

48. Riordan, J. F., Wacker, W. E. C., and Vallee, B. L., N-Acetylimidazole: A reagent for determination of "free" tyrosyl residues of proteins, *Biochemistry*, 4, 1758, 1965.

49. Smyth, D. G., Acetylation of amino and tyrosine hydroxyl groups: Preparation of inhibitors of oxytocin with no intrinsic activity on the isolated uterus, *J. Biol. Chem.*, 242, 1592, 1967.

50. Riordan, J. F. and Vallee, B. L., Diazonium salts as specific reagents and probes of protein conformation, *Methods Enzymol.*, 25, 521, 1972.

51. Pathy, L. and Smith, E. L., Reversible modification of arginine residues: Application to sequence studies by restriction of tryptic hydrolysis to lysine residues, *J. Biol. Chem.*, 250, 557, 1975.

52. Yankeelov, J. A., Jr., Modification of arginine by diketones, *Methods Enzymol.*, 25, 566, 1972.

53. Takahashi, K., The reaction of phenylglyoxal with arginine residues in proteins, *J. Biol. Chem.*, 243, 6167, 1968.

54. Hayashi, R., Bai, Y., and Hata, T., Evidence for an essential histidine in carboxypeptidase Y. Reaction with the chloromethyl ketone derivative of benzyloxycarbonyl-L-phenylalanine, *J. Biol. Chem.*, 250, 5221, 1975.

55. Shaw, E. and Springhorn, S., Identification of the histidine residue at the active center of trypsin labelled by TLCK, *Biochem. Biophys. Res. Commun.*, 27, 391, 1967.

56. Miles, E. W., Modification of histidyl residues in proteins by diethylpyrocarbonate, *Methods Enzymol.*, 47, 431, 1977.

57. Crimmins, D. L., Mische, S. M., and Denslow, N. D., Chemical cleavage of proteins in solution, in *Current Protocols in Protein Science*, Wiley Interscience, Hoboken, NJ, 2005, Chapter 11.

58. Caldwell, P., Luk, D. C., Weissbach, H., and Brot, N., Oxidation of the methionine residues of *Escherichia coli* ribosomal protein L12 decreases the protein's biological activity, *Proc. Natl. Acad. Sci. U. S. A.*, 75, 5349, 1978.

59. Marks, R. H. L. and Miller, R. D., Chemical modification of methionine residues in azurin, *Biochem. Biophys. Res. Commun.*, 88, 661, 1979.

60. Gundlach, H. G., Moore, S., and Stein, W. H., The reaction of iodoacetate with methionine, *J. Biol. Chem.*, 234, 1761, 1959.

61. Spande, T. F. and Witkop, B., Determination of the tryptophan content of protein with *N*-bromosuccinimide, *Methods Enzymol.*, 11, 498, 1967.

62. Loudon, G. M. and Koshland, D. E., Jr., The chemistry of a reporter group: 2-Hydroxy-5-nitrobenzylbromide, *J. Biol. Chem.*, 245, 2247, 1970.

63. Fontana, A. and Scoffone, E., Sulfenyl halides as modifying reagents for polypeptides and proteins, *Methods Enzymol.*, 25, 482, 1972.

64. Demoliou, C. D. and Epand, R. M., Synthesis and characterization of a heterobifunctional photoaffinity reagent for modification of tryptophan residues and its application to the preparation of a photoreactive glycogen derivative, *Biochemistry*, 19, 4539, 1980.

4 How to Design and Choose Cross-Linking Reagents

4.1 INTRODUCTION

The joining of two or more components, either biological molecules or otherwise, can be achieved by enzyme catalysis, caused by activating agents, or facilitated by cross-linking reagents containing two or more reactive groups. The last strategy seems to be the most common and logical approach. With two group-specific reagents linked through a spacer, bifunctional cross-linkers will react with specific functional groups within a molecule or between two different molecules resulting in a bond between these two components. In addition, multifunctional cross-linkers can also be designed. The reactive groups in a cross-linker can be identical or different, providing a diversity of reagents that can bring about covalent bonding between a wide range of chemical species, either intra- or intermolecularly.

The choice and design of a cross-linking reagent depend on its specific application. In each instance, the molecular species to be cross-linked may be different. These moieties may be located in disparate environments and the cross-linked products may require a certain configuration. Some of the conditions and requirements to be considered when choosing a cross-linker include the following:

1. *Reaction specificity toward a particular group, for example, amino, sulfhydryl, carboxyl, guanidinyl, imidazolyl, and other amino acid side chains.* If there is a special functional group on the protein to which another molecule will be linked, the cross-linking reagent must be specific to that group.
2. *Polarity of the reagent.* A protein in hydrophobic environment may require a nonpolar reagent in order to access the target residue. For example, membrane permeability of a reagent may be necessary for labeling transmembrane or intramembrane proteins.
3. *Cleavability of the reagent.* It may be desirable in some cases to separate cross-linked proteins for identification of the linked components. In this case, the use of cleavable reagents will enable the cross-linking to be reversed.
4. *Size and geometry of the reagent.* The length of bridge between the reactive groups of cross-linkers may be used to measure the distance between the two cross-linked species for intramolecular topology studies. The cross-linkers can also serve as spacers between two conjugated proteins.
5. *Photosensitivity of the reagent.* A photoactivatable agent is essential for photoaffinity labeling studies. This criterion is particularly important when one of the species is unknown, for example, protein receptors of a hormone. Photoaffinity labels have been used to detect and identify membrane acceptors.
6. *Presence of tracer (or reporter) group.* Sometimes it may be necessary to follow a cross-linking reaction, detect molecular conformation after conjugation, or isolate and analyze the labeled product. In these cases, a tracer or reporter group attached to the cross-linking reagent, for example, a fluorescent probe, a spin label, or a radioisotope, will facilitate the process.

Again, cross-linking is essentially a chemical modification process. The reactive groups are located at the ends of the reagent with a connecting backbone where various desirable functionalities can

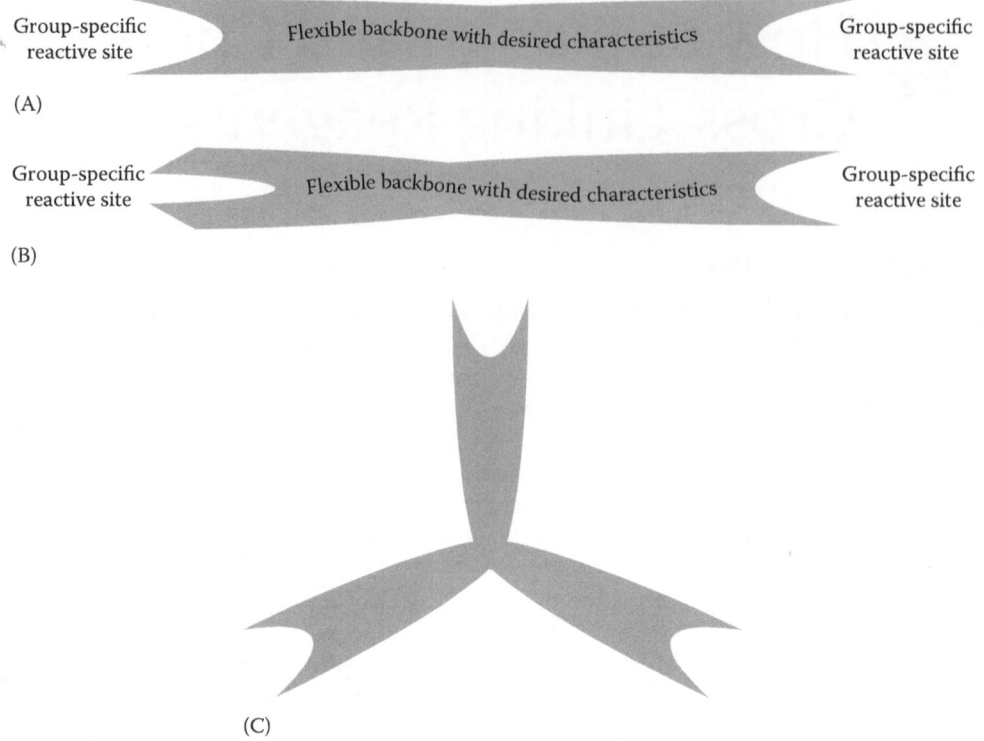

FIGURE 4.1 General features of cross-linking reagents. The group-specific reagents are connected by a bridge of different characteristics. (A) A homobifunctional reagent with two identical reactive groups. (B) A heterobifunctional reagent with two different group-specific moieties. (C) A homotrifunctional reagent containing three identical reactive groups.

be introduced (Figure 4.1). A simple cross-linker can be made by connecting any of the group directed reagents discussed in Chapter 3 with a carbon chain. As represented in Figure 4.1, if there are two identical reactive centers, these reagents are referred to as homobifunctional reagents. If the two reactive centers are different, they are heterobifunctional reagents. A trifunctional reagent is one that contains three reactive groups that can be the same or different, that is, homofunctional or heterofunctional.

An example of the design of a cross-linking reagent is provided by Chong and Hodges,[1] who have synthesized a heterobifunctional reagent, *N*-(4-azidobenzoylglycyl)-*S*-(2-thiopyridyl)-cysteine, for the study of biological interactions (Figure 4.2). The rationale is as follows. The two ends of the compound contain reactive functionalities that will react with proteins. One end contains a pyridyl

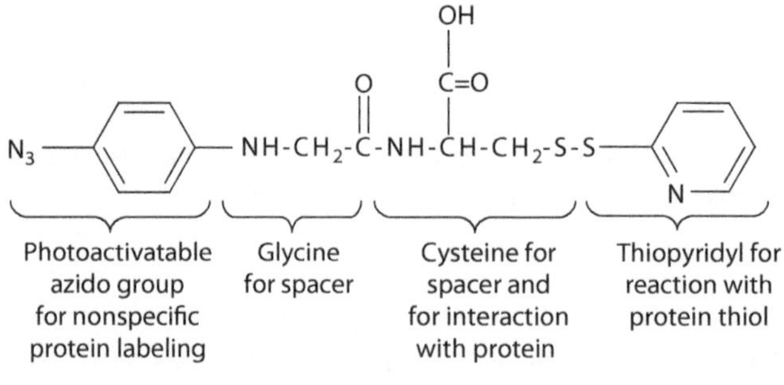

FIGURE 4.2 Rationale for the design of *N*-(4-azidobenzoylglycyl)-*S*-(2-thiopyridyl)-cysteine.

disulfide moiety, which will react with a protein sulfhydryl group through disulfide interchange since the displaced product, pyridine-2-thione, is a good leaving group. It also has the advantage that the progress of the reaction may be followed spectrophotometrically at 343 nm. The other end contains a photoactivatable aryl azido group that will nonspecifically insert into C–H or N–H bonds in the protein on photolysis. Joining the two reactive ends are glycine and cysteine residues, which provide a spacer between the interactive groups as well as providing some affinity for binding the cross-linker to the protein.

The specificity of the cross-linking reagent for a protein depends on the relative reactivity of the amino acid side chains toward the reactive ends of the cross-linking compounds. Many of these groups are derived from the reagents discussed in Chapter 3. Since most of the chemical modifications are nucleophilic reactions with the amino acid side chains acting as the nucleophile, the modifying agent must contain certain characteristics for the nucleophilic attack. The design of cross-linking reagents involves the incorporation of these various special functional groups.

4.2 USE OF NUCLEOPHILIC REACTIONS

4.2.1 THE BASIC REACTION

In nucleophilic substitution reactions, the amino acid side chains with a lone pair of electrons attack the electrophilic centers of the substrate. The rate of the reaction depends on the nucleophilicity of the attacking group and the reactivity of the substrate.[2] In Chapter 2, the relative nucleophilicity of the amino acid side chains is discussed. The reactivity of the substrate, on the other hand, depends on its electrophilicity as well as its leaving group. This concept will be elaborated upon below.

4.2.1.1 Electrophilicity of the Substrate

Electrophilic centers are atoms deficient in electrons. This electron deficiency is generally induced by more electronegative atoms or positive charges bonded directly to or next to the atom, which in most cases is carbon. The electrons of the carbon atom will be drawn to the more electronegative atom, inducing a slight positive charge. The examples in Figure 4.3 show that oxygen is probably the most commonly used electronegative atom. The arrows in the examples indicate the attraction of the electrons to the more electronegative atoms inducing a slight positive charge on the neighboring atom, which becomes an electrophile. These electrophilic centers are targets of nucleophilic attack. The greater the positive aspect, the greater will be the electrophilicity.

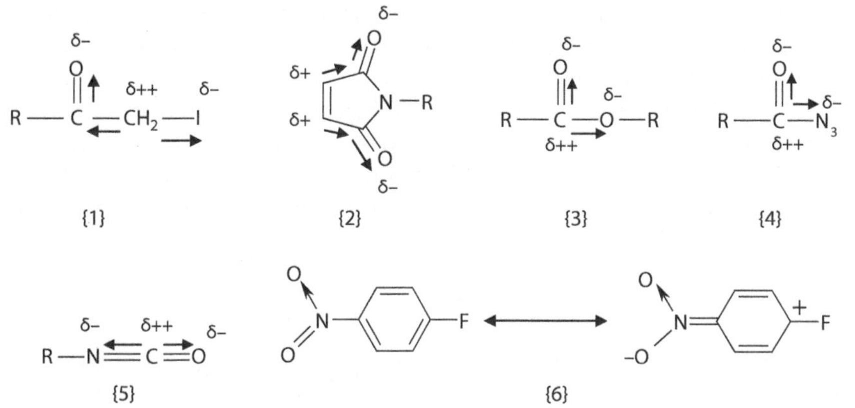

FIGURE 4.3 Examples of induction of electrophilic centers. Electrons are drawn to the more electronegative atoms as shown by the arrows inducing slight positivity at the carbon atom. {1} Iodoacetyl group; {2} maleimido group; {3} ester bond; {4} azidocarbonyl group; {5} isocyanate group; {6} p-nitroflurobenzene.

Depending on whether the positive center is located on an alkyl or an acyl carbon, these electrophiles will either alkylate or acylate the nucleophile, respectively. These reagents can therefore be broadly classified into alkylating and acylating agents. Thus, in Figure 4.3, {1} and {2} are alkylating agents, while {3}, {4}, and {5} are acylating agents. Compound {6} is an arylating agent, specifically an alkylating reagent with an aromatic ring.

4.2.1.2   Leaving Group Reactivity

In the nucleophilic substitution reactions, the nucleophile attacks the electrophilic center and finally displaces a leaving group. The capability of a leaving group to leave affects the reactivity of the substrate toward the nucleophile.[3,4] The more stable the leaving group is as a free entity, the easier it will come off. Thus, the best leaving group is probably N_2 gas from aliphatic diazonium ions, R-N_2^+. Nitrogen gas can be quantitatively evolved from such diazonium compounds. Usually the capability of a leaving group to leave is inversely proportional to its basicity. The best leaving groups are usually the weakest bases. In this context, iodide is the best leaving group among the halide ions and fluoride the poorest. Also, OH$^-$ and RO$^-$ do not ordinarily leave since they are strong bases. They can, however, be converted to better leaving groups after protonation, that is, OH$_2$ and ROH, as shown in Figure 4.4. This fact may explain why alcohols and esters are stable at neutral pHs and usually undergo acid-catalyzed reactions. In general, the approximate order of leaving ability for groups attached to saturated carbons is (leaving group indicated in bold): **R-N$_2^+$** > R-**OTs** (tosyl), R-**OSO$_2$OR** > R-**I** > R-**Br** > R-**OH$_2^+$** > R-**Cl** > R-**O(H$^+$)R′** > R-**ONO$_2$** > R-**NR$_3'^+$** > R-**OCOR′**. For groups bound to the carbonyl carbons the order is (leaving group indicated in bold): R-CO-**Cl** (acyl halides) > R-CO-**OCOR′**. (anhydrides) > R-CO-**OAr** (aryl esters) > R-CO-**OH** (carboxylic acids) > R-CO-**OR′** (alkyl esters) > R-CO-**NH$_2$** (amides). It is important to note that the –OH and –NH$_2$ groups of carboxylic acids and amides, respectively, are very poor leaving groups unless they are first protonated. For these groups the entities after leaving are water and ammonia, respectively. This list provides a general guide to the reactivity of the reagents. The better the leaving group present, the greater will be the reactivity of the reagent.

Some of the common leaving groups that have been incorporated in bifunctional reagents are *N*-hydroxysuccinamide, methanol, iodine, and thiopyridine. These will be further illustrated in the next two sections.

4.2.2   Alkylation

The attack of a nucleophile at an alkyl or aryl carbon results in alkylation of the nucleophile. As discussed in Chapter 2, the best nucleophilic amino acid side chain is the thiolate anion.[5] Thus, alkylation of the thiol is the most prevalent of all bifunctional reagents. Alkylation of the amino, thioether, imidazolyl, phenolate, and indolyl moieties will also occur under certain conditions as discussed in Chapter 3. For some reagents, such as imidoesters, the amino group is highly preferred. In order to promote nucleophilic substitution reactions, good leaving groups must be incorporated in the reagent, which should contain electron withdrawing atoms to induce a positive charge at the desired center of attack. Some of such alkylating groups commonly found in cross-linking reagents are shown in Table 4.1.

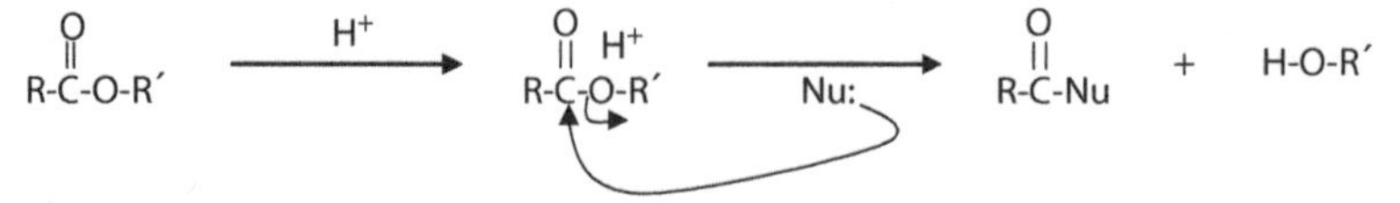

FIGURE 4.4   Example of protonation to convert a poor leaving group into a good leaving group. An ester is first protonated so that the alcohol can leave.

TABLE 4.1

Common Alkylating Groups That Have Been Incorporated into Cross-Linking Reagents

Alkylating Group	Example of Cross-Linking Reagent

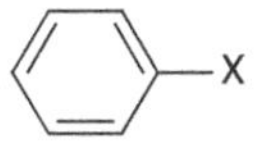

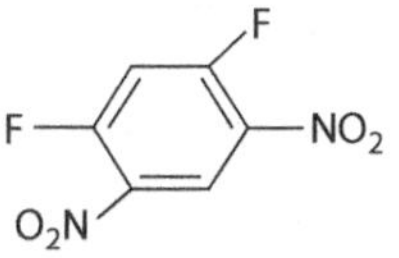

The haloacetate and maleimide cross-linkers, such as dibromoacetone and bismaleimidohexane, react fastest with the thiol and are considered thiol specific.[6,7] Under conditions of higher alkaline pH, excess reagent, and prolonged reaction time, other nucleophiles (i.e., amino, thioether, imidazolyl, phenolate) will also react. In the case of haloacetyl group, the halogen is the leaving group and the leaving ability follows the order: $I > Br > Cl > F$, which is the order of reactivity for the haloacetyl compounds. The homobifunctional cross-linker, dibromoacetone, reacts with two thiol nucleophiles, linking them together (Figure 4.5A). In the case of the bismaleimide derivative, the sulfur nucleophile attacks the double bond, a Michael addition reaction, again resulting in a linkage between the two species (Figure 4.5B). In this reaction, there is no formal leaving group. The electrons in the double bond are shifted to stabilize the product.

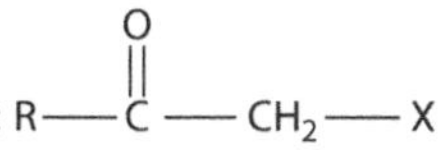

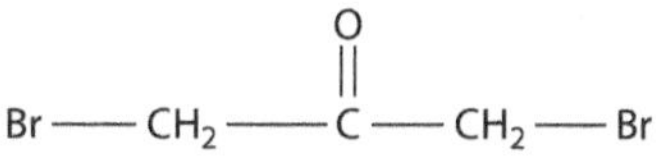

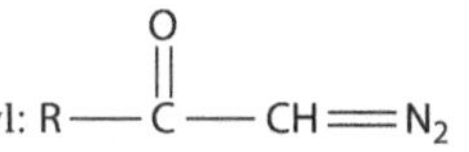

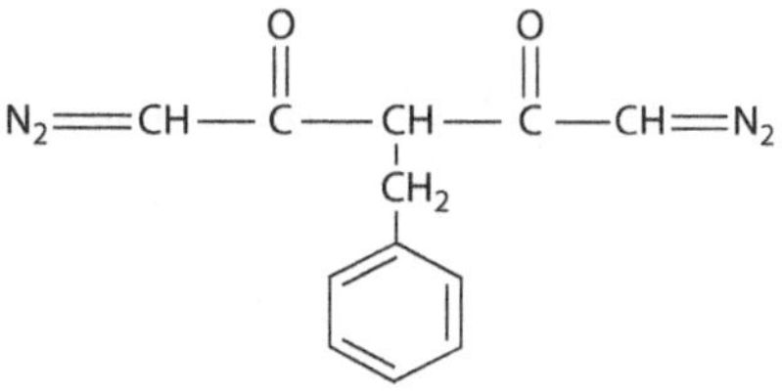

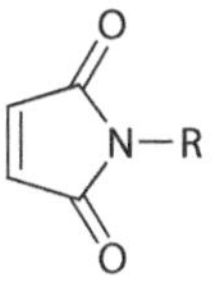

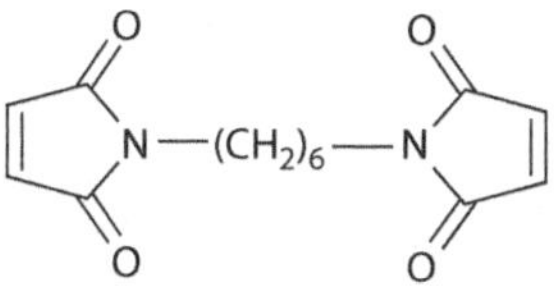

Although usually considered an amino group reagent, aryl halides also readily react with the thiol, phenolate, and imidazolyl groups in proteins. A simple example of the cross-linking reagent derived from such a functionality is 1,5-difluoro-2,4-dinitrobenzene.[8] At neutral pH, the reagent reacts preferentially with sulfhydryl groups. At higher pHs, it also modifies amino groups and other nucleophiles. The strong electronegative nitro groups create a positive center at the carbon where the fluorine atom is bonded (see Figure 4.3). Nucleophilic attack at the positive carbon eliminates the fluorine, which itself is also very electronegative (Figure 4.5C). The adduct formed with the thiol is reversible in the presence of an excess free thiol, for example, 2-mercaptoethanol, which displaces the bonded nucleophile.[9]

The synthesis of a bifunctional reagent containing the diazoacetyl group has been reported, namely, 1,1-*bis*(diazoacetyl)-2-phenylethane,[10] which has been shown to react with carboxyl groups of pepsin.[10,11] During the reaction, the diazoacetyl group is protonated and N_2 is eliminated. Thus, a low pH of about 5 is preferred for the reaction (Figure 4.5D). Other nucleophiles presumably can also react.

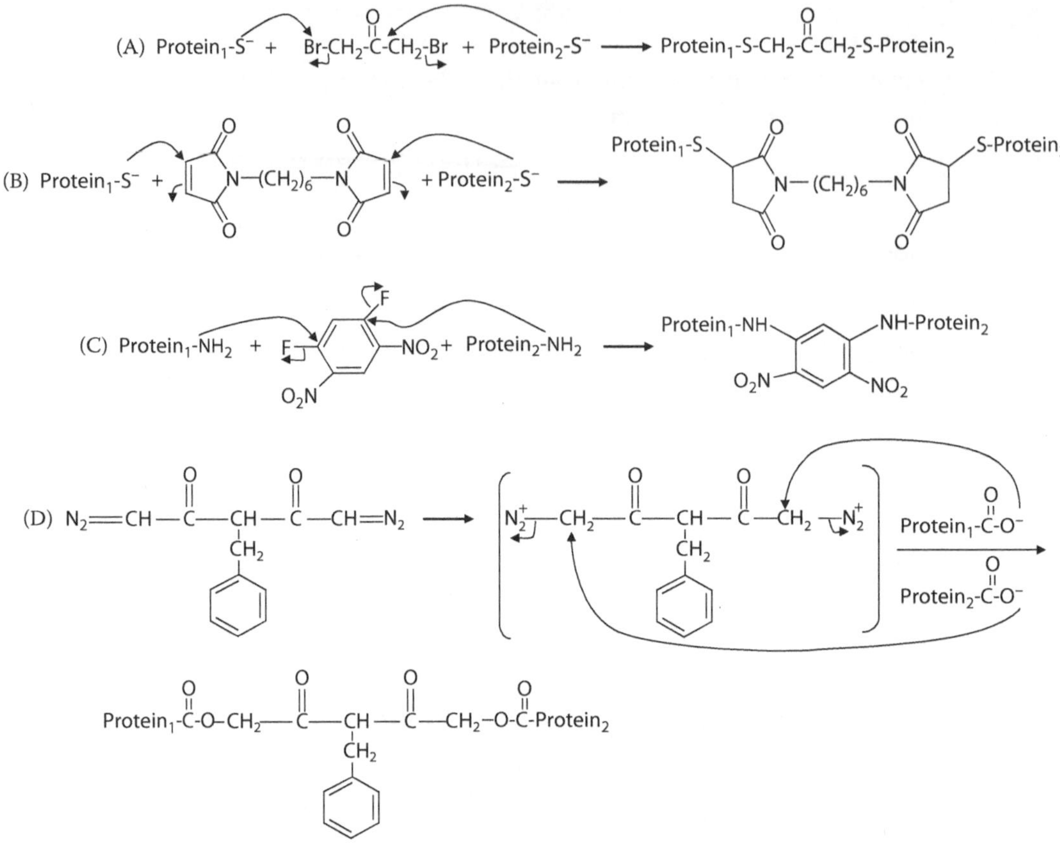

FIGURE 4.5 Alkylation reactions of the cross-linking reagents. Attack of the protein nucleophiles at the electrophilic center cross-links the proteins. Cross-linking with (A) Dibromoacetone. (B) Bismaleimidohexane. (C) 1,5-difluro-2,4-dinitrobenzene. (D) 1,1-*bis*(diazoacetyl)-2-phenylethane.

4.2.3 ACYLATION

In acylation, as in alkylation, the activated acyl compounds contain good leaving groups and are attacked by all amino acid nucleophiles. However, acylation primarily yields amino and phenolate derivatives. As discussed in Chapter 3, the adducts of sulfhydryl and imidazolyl groups are unstable, being rapidly hydrolyzed in aqueous solutions. Since tyrosine residues are often buried in proteins and are not accessible for modification, the acylation reagents are often regarded as amine-specific.

Acylating agents can be easily derived from either carboxylic or sulfonic acids. In fact, any of the aliphatic or aromatic dicarboxylic acids or disulfonic acids can be activated to give rise to bifunctional acylating reagents capable of reacting under mild conditions to form covalent bonds with proteins. The activating groups are generally good leaving groups with electronegative atoms to further withdraw electrons from the carbonyl carbon such as *p*-nitrophenol and halides. Some commonly used leaving groups in bifunctional acylating agents are shown in Table 4.2. The products of these leaving entities are also shown in this table. Of the carboxylic acid esters, the most commonly employed leaving groups are *N*-hydroxysuccinimide and alcohol, although *p*-nitrophenol has also been used. *N*-hydroxysuccinimidyl esters are popular because they can be easily synthesized, are highly activated, and react with amino groups under mild conditions (pH 7–9). The synthesis of these compounds can be achieved, as for other esters such as *p*-nitrophenyl ester, by coupling *N*-hydroxysuccinimide to dicarboxylic acids using dicyclohexylcarbodiimide (Figure 4.6).[12] Another advantage of the hydroxysuccinimidyl functionality

TABLE 4.2
Activated Acyl Groups in Bifunctional Reagents

Activated Acyl Functional Group	Leaving Group

Succinimidyl: R—C(=O)—O—N(succinimide)

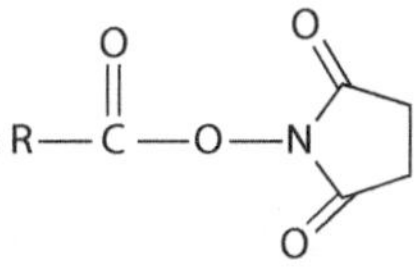

N-Hydroxysuccinimide: HO—N(succinimide)

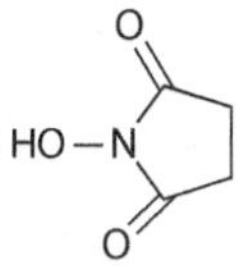

Sulfosuccinimidyl: R—C(=O)—O—N(sulfosuccinimide)—SO$_3$Na

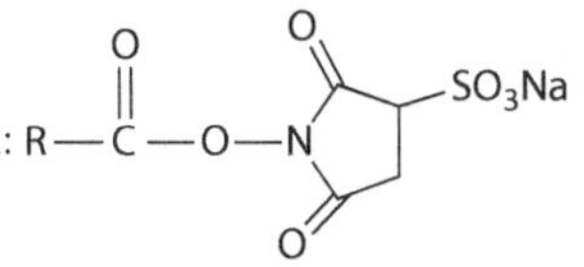

N-Hydroxysulfosuccinimide: HO—N(sulfosuccinimide)—SO$_3$Na

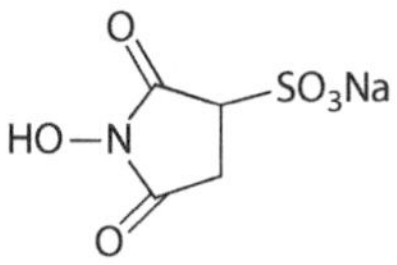

Imidoester: R—C(=NH)—O——CH$_3$

Methanol: H—O—CH$_3$

Carbonyl chloride: R——C(=O)——Cl

Chloride ion: Cl$^-$

Azidocarbonyl: R——C(=O)——N$_3$

Azide ion: N$_3^-$

Sulfonyl chloride: R—S(=O)(=O)—Cl

Chloride ion: Cl$^-$

p-Nitrophenyl: R—C(=O)—O—C$_6$H$_4$—NO$_2$

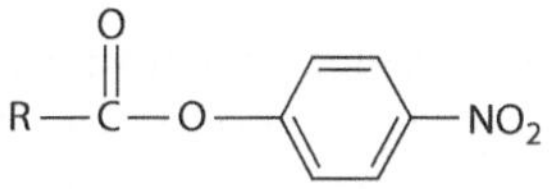

p-Nitrophenol: H—O—C$_6$H$_4$—NO$_2$

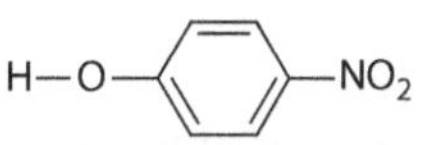

Isothiocyanate: R-N=C=S

None—electron shift with protonation

Isocyanate: R-N=C=O

None—electron shift with protonation

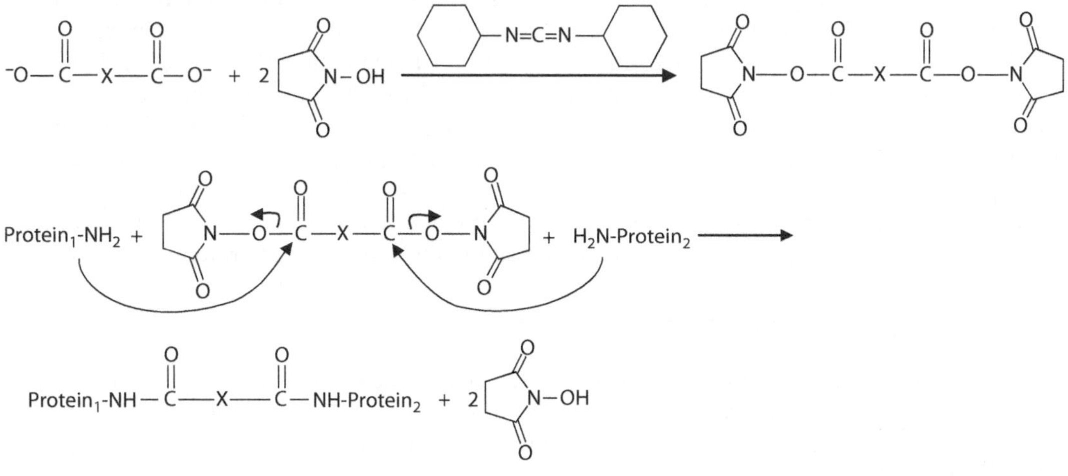

FIGURE 4.6 Synthesis of *bis-N*-hydroxysuccinimidyl eaters and their cross-linking reaction with protein amino groups.

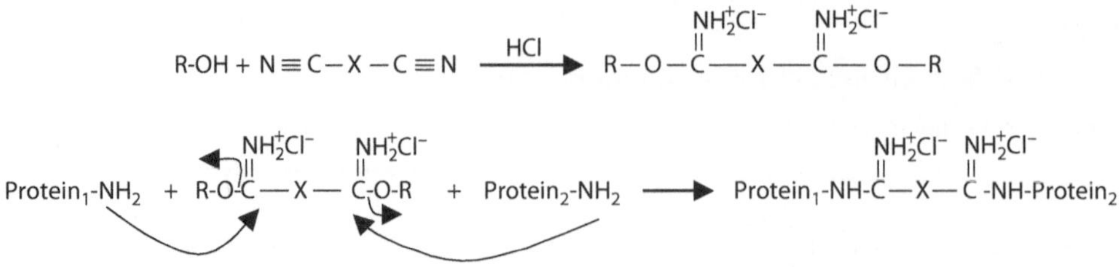

FIGURE 4.7 Synthesis of *bis*-imidoesters and their cross-linking reaction with protein amino groups.

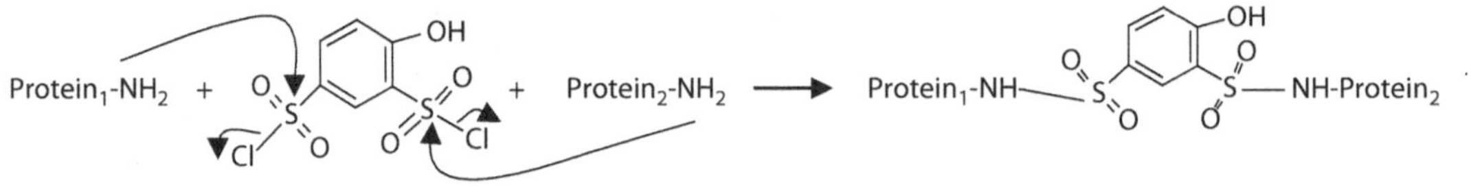

FIGURE 4.8 Cross-linking with a disulfonyl chloride: phenol-2,4-di(sulfonyl chloride).

is the water solubility of the reaction product.[12] The *N*-hydroxysuccinimidyl product has been further sulfonylated to render it even more water soluble.

In the case of imidoesters, these compounds are very soluble in water and react under mild conditions (pH 7–10) with a high degree of specificity toward amino groups. As discussed in Chapter 2, the side reactions with thiol, phenolate, carboxyl, imidazolyl, and guanidinyl groups are negligible. *Bis*-imidoesters can be readily synthesized by a variety of procedures. The most common is the Pinner method where a primary alcohol is allowed to react with a nitrile in the presence of dry HCl (Figure 4.7).[13–15] These compounds are usually isolated as hydrochloride salts. After reaction of the imidoesters with amino groups, the resulting amidines are quite stable and resistant to acid hydrolysis. Total amino acid hydrolysis can be performed on the modified proteins for quantification.

Acid halides are very reactive compounds with the halogen ions as good leaving groups. Although these compounds can be synthesized easily in the absence of water, acyl halides are rarely used in the design of cross-linkers because of their instability in aqueous solutions. Similarly, acylazides are reactive by virtue of the good leaving azido group, and *bis*-acylazides have been used to cross-link proteins.[16] As discussed in Chapter 3, sulfonyl halides are more stable and have been incorporated into cross-linking reagents, such as phenol-2,4-di(sulfonyl chloride) (Figure 4.8). These compounds react with all nucleophiles of the protein and are generally not group specific.

The isocyanates and isothiocyanates do not contain a leaving entity. During the nucleophilic attack at the electrophilic carbon, the electrons are shifted to the nitrogen to pick up a proton. With nitrogen as the nucleophile, isocyanates form stable urea derivatives and isothiocyanates form thiourea derivatives. As pointed out in Chapter 3, these compounds are considered amino group specific. Aryl di-isocyanates and di-isothiocyanates cross-linking reagents can be easily prepared from the corresponding diamines on reaction with phosgene or thiophosgene (Figure 4.9).[17] Condensation with amino groups of proteins will result in cross-linking the proteins.

4.3   USE OF ELECTROPHILIC REACTIONS

In electrophilic reaction, the protein amino acid side chain attacks an electron-rich center of the substrate. This reaction is relatively rare since the amino acid side chains are mostly nucleophilic in nature, unless they are protonated as in the phenolic group of tyrosine and imidazolyl group of histidine. As the aromatic benzene and imidazole rings are relatively rich in electrons, the most important electrophilic reaction in protein modification is the reaction of diazonium salts

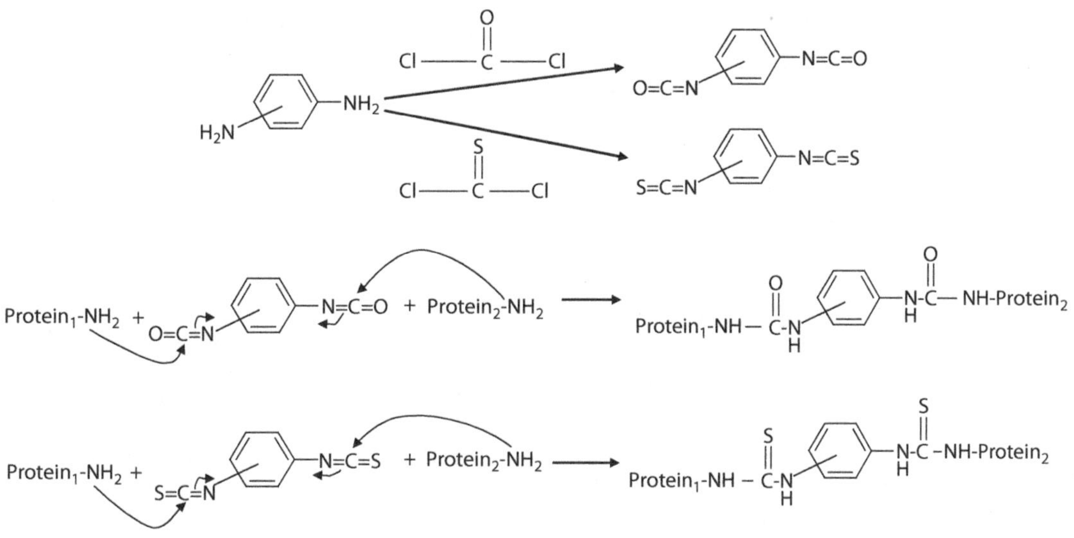

FIGURE 4.9 Synthesis of di-isocyanates and di-isothiocyanates and their cross-linking reactions with protein amino groups.

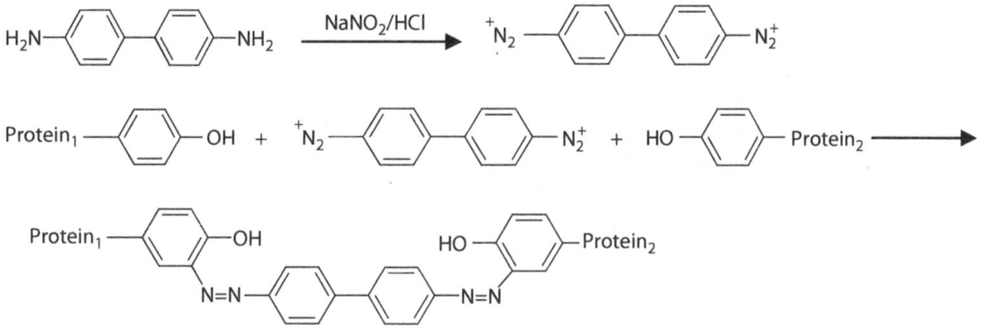

FIGURE 4.10 Diazotization of diaminobenzidine and the cross-linking reaction of bisdiazobenzidine.

with tyrosines and histidines. Such diazonium salts, particularly aryl diazonium compounds, may be incorporated into cross-linking reagents. An example is bisdiazobenzidine, which has been used to cross-link papain[18,19] and other proteins.[20] The aryl diazonium compounds are easily prepared from the corresponding aryl amines and nitrous acid, which is generated *in situ* from sodium nitrite and hydrochloric acid. Diazotization of diaminobenzidine will yield bisdiazobenzidine as a cross-linker (Figure 4.10). Thus, cross-linking reagents may be designed with a phenylamine at the end of the bridge.

4.4 INCORPORATING GROUP-DIRECTED REAGENTS

There are relatively few reactive moieties that are really specific for certain amino acid side chains. The degree of specificity depends on the relative nucleophilicity of the amino acid side chains. Any of the group-directed compounds discussed in Chapter 3 may be incorporated into cross-linking reagents. And a few have been successfully implemented. These are mercurials and disulfides for thiols, carbonyls for reductive alkylation with amines, and vicinal dicarbonyls for arginine. The incorporation of these reactive groups into cross-linking reagents is illustrated below.

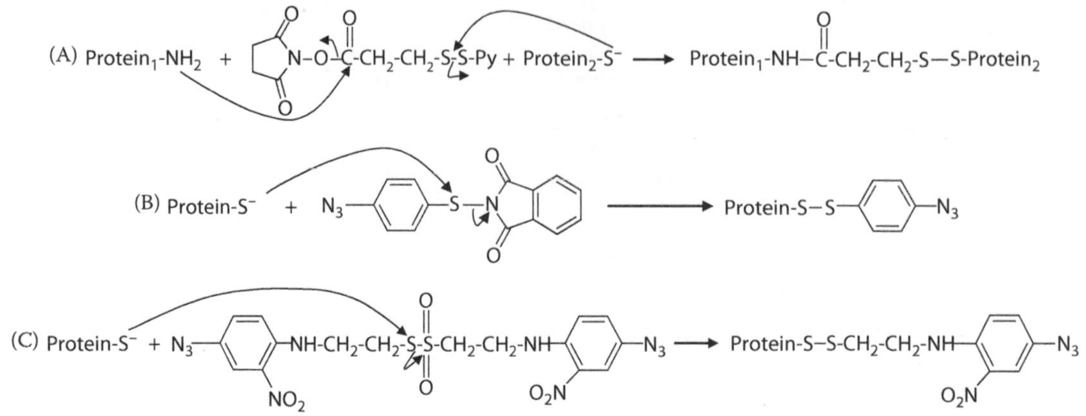

FIGURE 4.11 Disulfide cross-linkers. (A) Cross-linking reaction with *N*-succinimidyl-3-(2-pyridyldithio) propionate; (B) reaction with *N*-(4-azidophenylthio)phthalimide; (C) reaction with *N,N'*-(4-azido-2-nitrophenyl)cystamine dioxide.

4.4.1 Disulfide Reagents

The specific thiol–disulfide interchange reaction is illustrated by the cross-linkers derived from pyridyl disulfide. The thiopyridine is an excellent leaving group, being stabilized by several resonance forms. An example of such a cross-linking reagent is *N*-succinimidyl-3(2-pyridyldithio)propionate (Figure 4.11A).[21] The reagent reacts with an amino group of a protein at the *N*-succinimidyl ester bond and a thiol of another protein at the disulfide bond. The reaction with thiol involves a direct displacement of the thiopyridine, generating a new disulfide bond (Figure 4.11A).

Thiophthalimides and compounds containing $-S-SO_2-$ are also thiol-specific reagents resulting in the formation of disulfide bonds.[22] The phthalimide and sulfinic acid provide excellent leaving groups and, at the same time, activate the thiol in the compound. Attack of a protein sulfhydryl group at the sulfur atom eliminates the leaving group and a disulfide bond is formed as illustrated in Figure 4.11B and C for the cross-linking reagents *N*-(4-azidophenylthio)phthalimide and *N,N'*-(4-azido-2-nitrophenyl)cystamine dioxide, respectively. The disulfide bonds formed can be cleaved by reducing agents providing a reversible conjugate, which will be discussed below.

4.4.2 Mercurial Reagents

While mercuric ion itself can reversibly cross-link two sulfhydryl groups,[23,24] insertion of the ion into protein disulfide bonds makes this reagent undesirable for general conjugation.[25,26] However, mercurial derivatives have been synthesized where the mercurial ion is monovalent. One example is 3,6-bis(mercurimethyl)dioxane (Figure 4.12). It has been used to prepare mercaptalbumin dimers.[27]

4.4.3 Reductive Alkylation

As has been discussed in Chapter 3, reductive alkylation is a special reaction for the amino group. The process involves the formation of a Schiff base followed by reduction with a reducing agent. It should be noted that either the amino group or the carbonyl group may be contributed by the

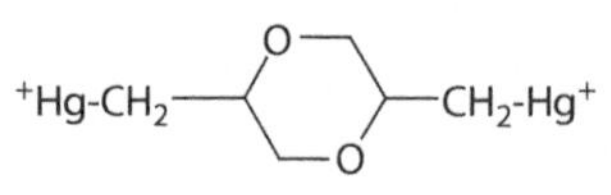

FIGURE 4.12 Mercury-derived thiol-specific reagent: 3,6-*bis*(mercurimethyl)dioxane.

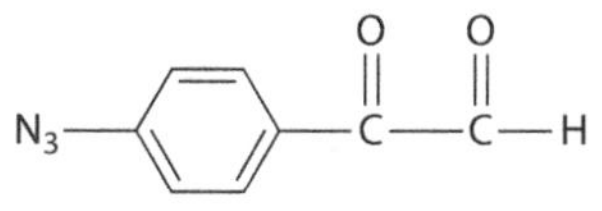

FIGURE 4.13 An arginine-specific dicarbonyl compound: azidophenyl glyoxal.

protein. Aldehydes may be derived from the carbohydrate moieties of glycoproteins by periodate treatment. In this case, glycoproteins may be cross-linked with diamines. On the other hand, proteins may be cross-linked through the amino groups by means of dialdehydes.[28] Examples of such dialdehydes can be found in Chapter 5.

4.4.4 VICINAL DICARBONYL REAGENTS

Vicinal dicarbonyl reagents are specific for the modification of guanidinyl group of arginine. Incorporation of such groups into bifunctional reagents will cross-link arginine moieties. Such a bifunctional derivative of phenylglyoxal, azidophenyl glyoxal (Figure 4.13), has been synthesized to study enzymes that contain an arginine at the active site.[29] This compound has also been used for cross-linking RNA to proteins.[30]

4.5 INCORPORATING PHOTOACTIVATABLE NONSPECIFIC GROUPS

Nonspecific reactions involve reagents that react indiscriminately with chemical entities in their very close proximity, such as amino acid side chains. One class of such compounds that have evolved for biochemical studies are photoaffinity labels. The chemistry of these compounds has been reviewed by Knowles and Bayley, and others.[31–40] Readers who are not familiar with this technique are encouraged to consult these publications for detailed treatment. Basically, photoaffinity labels are compounds that, upon irradiation, generate reactive intermediates such as carbenes and nitrenes. The most widely used photoreactive groups are azides, diazo moiety, benzophenone, diazonium salts, and diazirines. Absorption of light of certain wavelength activates these photophores to an excited state, resulting in nitrenes for azides, carbenes for diazo derivatives and diazirines, biradicals for benzophenones, and aryl cation for diazonium group.[34–40] These highly reactive species react immediately with their nearest molecule. Common nitrene precursors are aryl azides, which have absorbance in the long ultraviolet region. With an additional nitro group on the phenyl ring, the absorbance of dinitroarylazides is further shifted away from the protein absorption band. Figure 4.14 shows the photoactivation of an aryl azide and outlines the principle reactions for nitrenes. Carbenes undergo similar reactions. The major reaction of interest is the insertion into C–H and N–H bonds, resulting in a covalent bond with the photophore. Carbenes are generally generated from diazoacetyl compounds and diazirines as shown in Figure 4.15.[41,42] Trifluoromethyl derivatives of diazoacetyl compounds are more stable and give a better yield of carbenes. The diazirine unit is small, nonbulky, lipophilic, and has an absorbance that extends significantly into the 300 nm range, away from protein and nucleic acid absorption bands. Most common diazirines are derivatives of the trifluoroethyldiazirinephenyl group.[42]

In addition to the reactive nitrenes and carbenes, other reactive intermediates are found on photoactivation of benzophenones and diazonium salts (Figure 4.16). Benzophenones and acetophenones, for example, are activated to carbonyl diradical triplet states when irradiated with 350 nm wavelength light. The excited state can either decay back to the starting material or abstract a proton to form a carbon radical, which reacts irreversibly with the nearby molecules.[43] Benzophones react preferentially with unreactive C–H bonds (Figure 4.16A).[37] Diazonium salts have been used to photolabel biological systems that bind cationic groups such as nicotinic receptor.[44] On photoactivation at wavelengths greater 350 nm, aryldiazonium salts generate a highly active aryl carbocations,

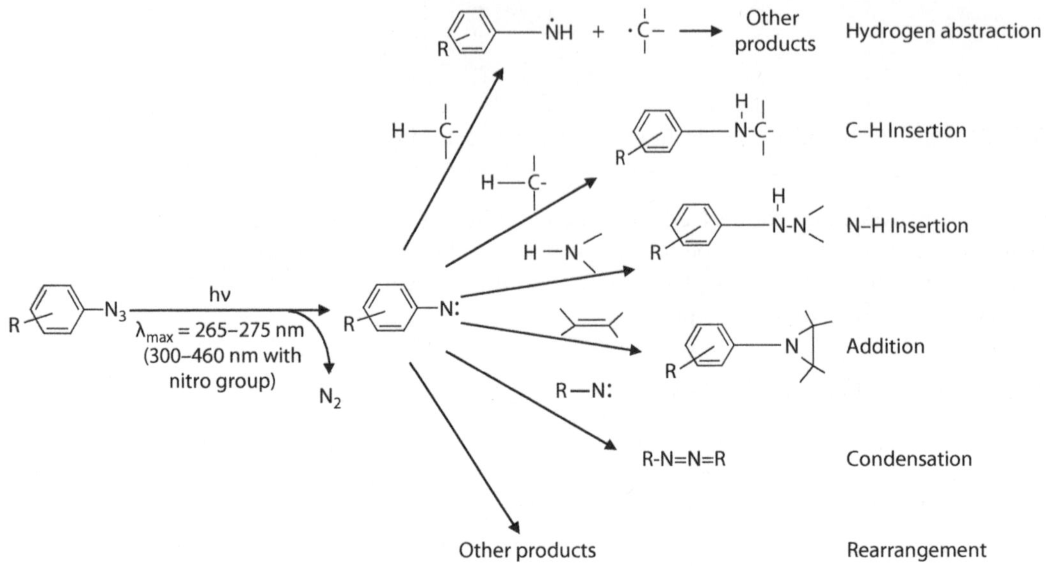

FIGURE 4.14   Photoactivation of arylazides and some important reactions of nitrenes.

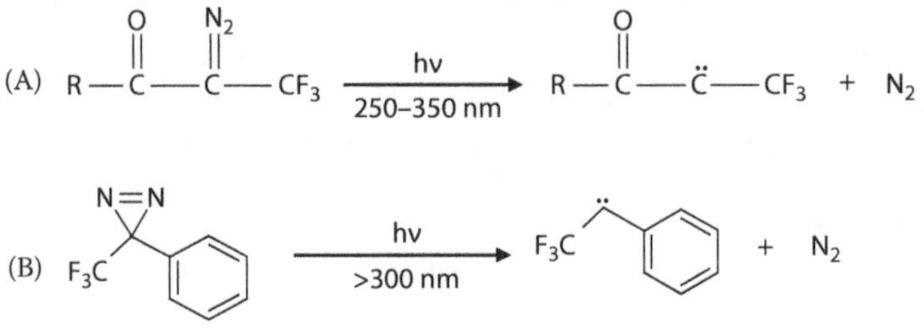

FIGURE 4.15   Photoactivation of (A) a diazoacetyl compound and (B) 3-phenyl-3-(trifluoromethyl)diazirine to generate carbenes.

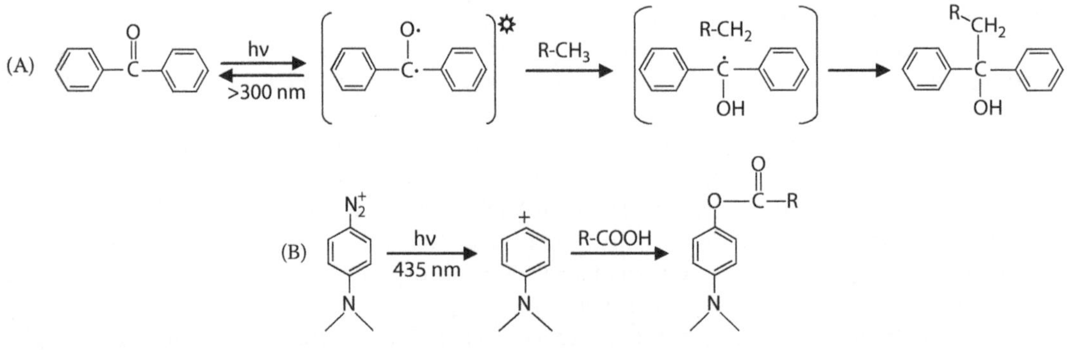

FIGURE 4.16   Photochemical reactions of (A) benzophenone and (B) aryldiazonium salt.

which react with the surrounding molecule (Figure 4.16B). Together with nitrene- and carbene-generating photophores, these components may be incorporated into bifunctional reagents for nonspecific cross-linking of proteins. Many reagents have been synthesized for this purpose.[45] Azidophenyl derivatives are, however, by far the most commonly used. Examples of these series of compounds can be found in Chapter 6.

4.6 CHANGING THE WATER SOLUBILITY OF CROSS-LINKERS

By virtue of their constituents, bifunctional cross-linking reagents vary in their water solubilities. The degree of water solubility may be enhanced by incorporating hydrophilic moieties into the molecules. For example, the hydrophobicity of *N*-hydroxysuccinimides can be increased by adding a sulfonate group to the succinimide ring.[46] The resulting molecule with increased solubility in water has, however, decreased membrane permeability. Similarly, introducing a sulfonate group into the bridge increases the polarity of the compound and decreases the membrane solubility.[47,48] The hydrophobicity and hydrophilicity of the reagents therefore determine the depth of probe penetration into the membrane. 4-Azidoiodobenzene, for example, was able to penetrate into membranes and cross-link phospholipids and proteins.[49] As shown in Table 4.3, other means of changing the hydrophobicity and hydrophilicity of the compounds are possible.[50–53] These methods include introduction of ether-oxygen groups, hydroxyl group, ester and amide bonds, and other water-soluble functionalities into the bridging spacer.

Not only does an increase in hydrophilicity of the reagent increase its solubility in water, the change of hydrophobicity also affects the mode and extent of intermolecular cross-linking. This fact was vividly demonstrated by Fasold et al.[53] who compared the efficiency of cross-linking of hemoglobulin using tartryl-*bis*-glycinazide and *bis*-(ureido)prolylazobenzene. These two compounds have similar spans of 30 Å but the former is distinctly more hydrophilic than the latter. The results showed that the tartryl derivative gave a higher percentage of interchain cross-linking. It can be reasoned that the hydrophilic reagents tend to stick out from the surface of the modified protein to react with a second molecule, whereas the hydrophobic reagent, in contrast, may tend

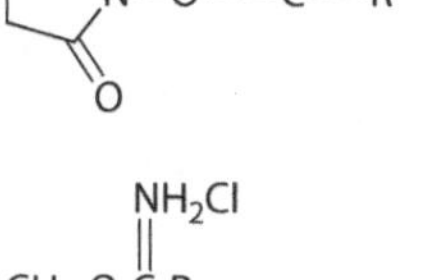

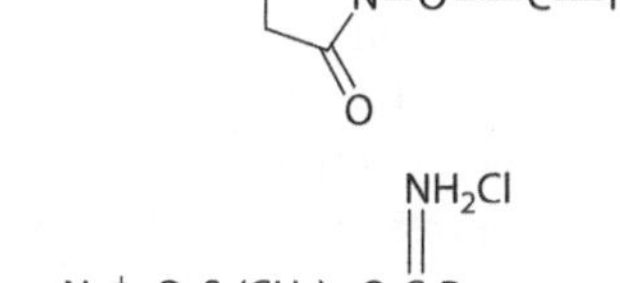

TABLE 4.3
Modifying the Water Solubility Characteristics of Cross-Linkers

Increased Hydrophobicity	Increased Hydrophilicity
$CH_3\text{-O-C}(=NH_2Cl)\text{-R}$	$Na^+\ {}^-O_3S\text{-}(CH_2)_2\text{-O-C}(=NH_2Cl)\text{-R}$
$R\text{-O-}(CH_2)_5\text{-O-R}$	$R\text{-O-}(CH_2)_2\text{-S}(=O)_2\text{-}(CH_2)_2\text{-O-R}$
$R\text{-}(CH_2)_n\text{-R}$	$R\text{-[O-}(CH_2)_n\text{-O-}(CH_2)_n]_x\text{-O-R}$
$R\text{-}(CH_2)_n\text{-R}$	$R\text{-}CH_2\text{-}(CH(OH))_n\text{-}CH_2\text{-R}$
$R\text{-}(CH_2)_n\text{-R}$	$R\text{-}[C(=O)\text{-O-}(CH_2)_n\text{-C}(=O)\text{-O}]_x\text{-R}$

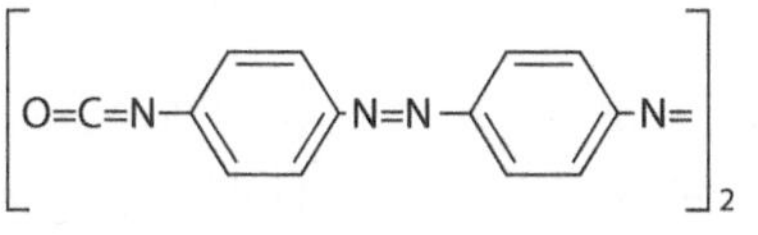

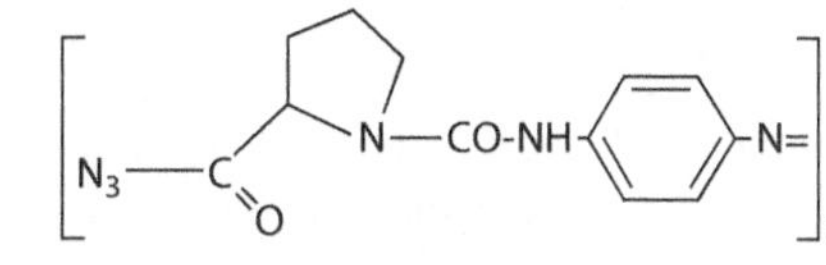

to stick to the hydrophobic portions of the protein favoring internal cross-linking. In addition to decreased efficiency of intermolecular cross-linking, hydrophobic reagents also have a higher tendency to denature proteins.[53]

4.7   INCORPORATING SPECIAL CHARACTERISTICS IN THE BRIDGE SPACER

4.7.1   INCORPORATION OF CLEAVABLE BONDS

The bridge spacer arm that joins the bifunctional reactive groups provides opportunities for incorporating various functionalities into the cross-linker. The inclusion of cleavable bonds enables the cross-linked moieties to be separated for analysis. The advantages include confirmation of the cross-linking reactions, establishment of the amino acids modified, monitoring of the effect of cross-linking, and identification of the cross-linked species. A number of cleavable bonds are available for this purpose. These include disulfide bonds,[54–56] mercurial group,[57] vicinal glycol,[58] azo,[59] sulfone,[60–62] selenoethylene,[63,64] ester,[65,66] and thioester[67] linkages as shown in Table 4.4 and briefly discussed below.

4.7.1.1   Disulfide Bond

As mentioned above, disulfide bonds can be cleaved by any reducing agent, such as free mercaptans,[68] sodium borohydride,[69] sodium phosphorothioate,[70] and sulfite.[71,72] Cleavage is generally achieved under mild conditions with 2-mercaptoethanol, dithiothreitol, or dithioerythritol at concentrations between 10 and 100 mM. The reaction is a simple two-step thiol–disulfide interchange as discussed in Chapter 2.

In the cross-linking reagents, the disulfide can be intrinsically built into the molecule as shown in Table 4.4.A or generated during the modification reactions. The latter has been demonstrated in Figure 4.11. These reagents must be used under nonreducing conditions and in the complete absence of free thiols to prevent disulfide–thiol interchange.

4.7.1.2   Mercurial Group

The reaction of mercurial compounds with sulfhydryl groups is reversible in the presence of free thiols.[57] Thus, mercurial cross-linked proteins may be separated in the presence of 2-mercaptoethanol or dithiothreitol (Table 4.4.B). The reaction is similar to thiol–disulfide interchange. Because of the very small dissociation constant of thiol–mercurial complexes, relatively high concentrations of free thiol may be required for complete dissociation of the cross-linked proteins.

4.7.1.3   Vicinal Glycol Bond

As in the carbohydrate moiety of glycoproteins (discussed in Chapter 2), vicinal glycol bonds can be cleaved with sodium periodate converting the diols into dialdehydes.[73,74] Although the rate of cleavage is lower than disulfide bonds, the oxidation reaction takes place under mild conditions, at neutral pH and ambient temperatures, for instance, 15 mM sodium periodate at pH 7.5 and 25°C for 4–5 h. Proteins coupled by reagents containing such a vicinal glycol linkage are therefore reversible. Disuccinimidyl tartarate shown in Table 4.4.C is such a reagent. The cleavage reaction of cross-linked proteins produces aldehydes. It should be remembered that the carbohydrate moieties of cross-linked glycoproteins are also susceptible to oxidation. In addition, the reactivity of the aldehydes formed during the oxidation reaction may cause Schiff base formation with amino groups on the proteins, rendering potential complications.

4.7.1.4   Azo Linkage

Azo linkages can be cleaved with 0.1 M sodium dithionite at pH 8.0 for 25 min.[59] The reduction produces two free amines. An example of such a cleavable reagent is N-[4-(p-azidophenylazo)benzoyl]-3-aminohexyl-N′-oxysuccinimide ester shown in Table 4.4.D.

TABLE 4.4
Cleavable Groups in Bifunctional Reagents

Cleavable Bond in Cross-Linked Product	Example of Cross-Linker	Cleavage Conditions	Cleaved Products	References
A. P~S–S~P'	CH_3–O–C–$(CH_2)_2$–S–S–$(CH_2)_2$–C–O–CH_3 (with $NH_2^+Cl^-$ groups)	Reducing agent, e.g., mercaptoethanol	P–SH + HS–P'	36–38,48–52
B. P~Hg–x~Hg~P'	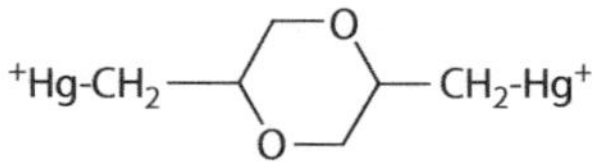 ^+Hg–CH_2 ... CH_2–Hg^+	Free thiol, e.g., mercaptoethanol	P–SH + HS–P'	39
C. P~CH–CH~P' (OH OH)	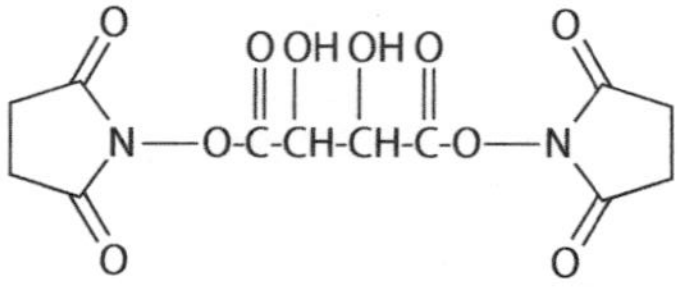 (maleimide–O–C–CH–CH–C–O–maleimide)	Periodate	P–CHO + OHC–P'	40,53,54
D. P–N=N–P'	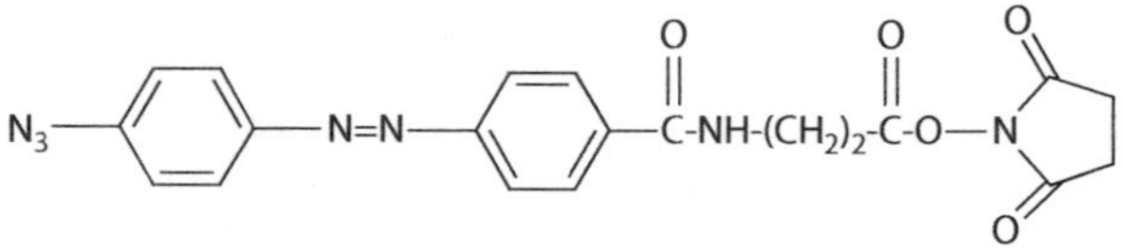 (structure with C–NH–$(CH_2)_2$–C–O–succinimide, azo $N=N$, N_3)	Dithionite	P–NH_2 + H_2N–P'	41
E. P~S–P' (O=S=O)	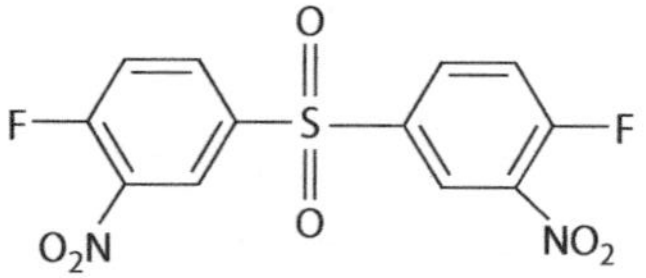 (sulfone structure with F, NO_2, O_2N)	Base	P–SO_3 + HO–P'	42–44
F. P–$(CH_2)_2$–Se–$(CH_2)_2$–~P'	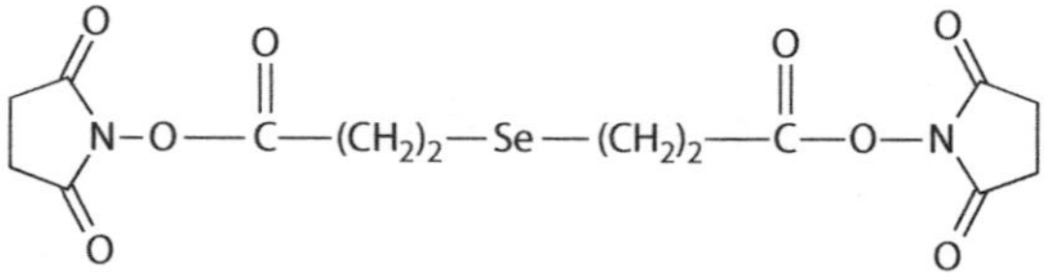 (succinimide–O–N–C–$(CH_2)_2$–Se–$(CH_2)_2$–C–O–succinimide)	Periodate or iodobeads	Unknown	63,64

(continued)

TABLE 4.4 (Continued)
Cleavable Groups in Bifunctional Reagents

Cleavable Bond in Cross-Linked Product	Example of Cross-Linker	Cleavage Conditions	Cleaved Products	References
G. 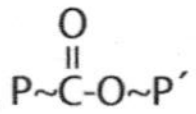 $\overset{O}{\overset{\|}{P\text{-}C\text{-}O\text{-}P'}}$	Maleimide–N–CH$_2$–O–C(=O)–(CH$_2$)$_2$–N–maleimide	Base or NH$_2$OH	P-COOH + HO-P' or P-CO-N-R + OH-P'	45 46
H. 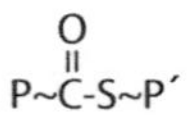 $\overset{O}{\overset{\|}{P\text{-}C\text{-}S\text{-}P'}}$	N$_3$–(NO$_2$-phenyl)–C(=O)–S–CH$_2$–CH$_2$–C(=O)–OH	R-NH$_2$	P-CO-NH-R + HS-P'	47
I. 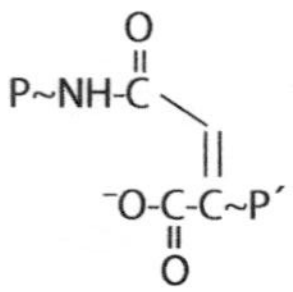 $P{\sim}NH\text{-}C\overset{O}{\overset{\|}{}}$ ... $^{-}O\text{-}C\text{-}C{\sim}P'$ ($C=O$)		H$^+$	P~NH$_2$ + X-P'	57
J. 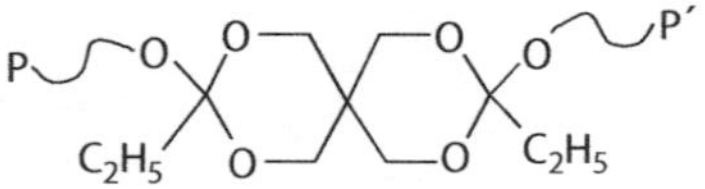 (spiro ketal cross-linker with C$_2$H$_5$ groups, maleimide termini)	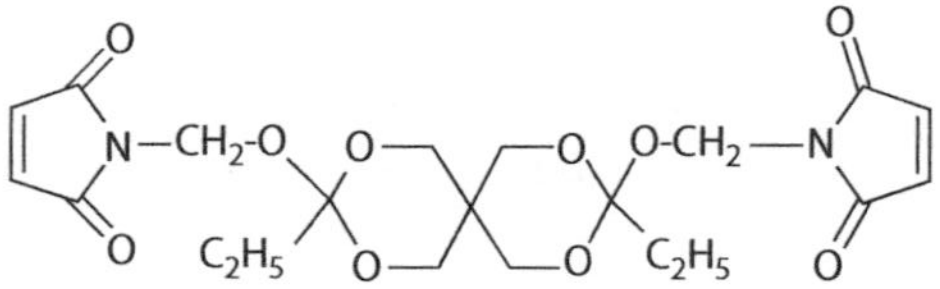 (spiro bis-dioxane, N–CH$_2$–O– groups, C$_2$H$_5$ substituents, maleimide)	H$^+$	P-OH + HO-P'	58–61
K. 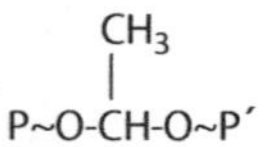 $P{\sim}O\text{-}\overset{CH_3}{\overset{\|}{CH}}\text{-}O{\sim}P'$	Maleimide–N–CH$_2$–O–CH(–CH$_3$)–O–CH$_2$–N–maleimide	H$^+$	P-OH + HO-P'	58–61

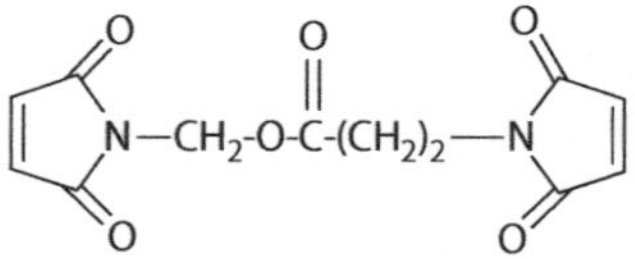

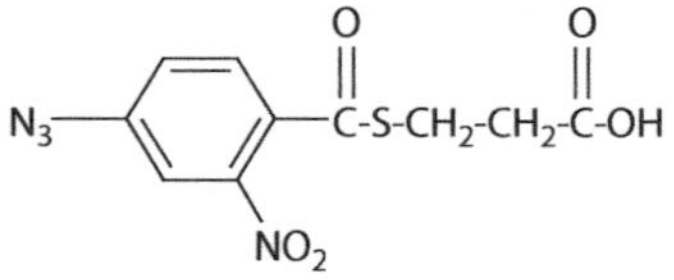

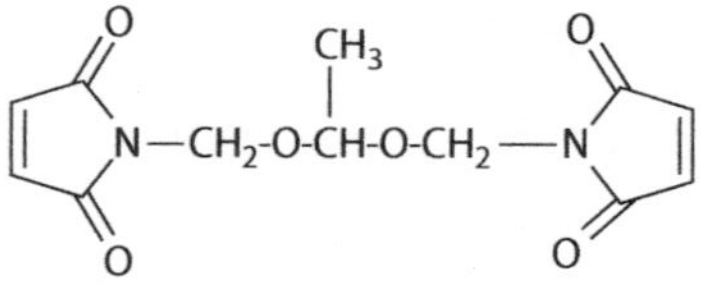

4.7.1.5 Sulfone Linkage

Sulfone linkages provide another means for cleavage. The bond is easily hydrolyzed with base, at pH 11–12 for 2 h at 37°C, to the sulfonic acid.[60-62] 4,4'-Difluoro-3,3'-dinitrophenyl-sulfone shown in Table 4.4.E is an example of a cleavable cross-linking reagent containing such a group.

4.7.1.6 Selenoethylene Group

The selenoethylene group in cross-linking reagents can be easily cleaved by mild oxidation. For example, it can be cleaved with 2 mM sodium periodate within minutes at 20°C or with N-chlorobenzenesulfonamide immobilized on polystyrene beads (Iodo-Beads).[63] The mechanism of cleavage is thought to be β-elimination of an oxidized seleno derivative, but the final products are unknown. An example of selenoethylene group containing reagent is shown in Table 4.4.F.[64]

4.7.1.7 Ester Bond

Theoretically esters can be hydrolyzed under acidic or basic conditions. Generally this reaction is done with 0.1 N NaOH at ambient temperature.[65] The products of the hydrolysis are a carboxylic acid and an alcohol. The ester bond may also be cleaved by incubating with 1 M hydroxylamine at pH 7.5–8.5 and at 25°C to 37°C for 3–6 h.[66] This reaction produces an amide and an alcohol. However, the use of hydroxylamine is incompatible with proteins containing Asn–Gly bonds, which are known to be attacked by this chemical.[75] A bifunctional dimaleimide with such a cleavable ester bond, maleimidomethyl-3-maleimido propionate, synthesized by Sato and Nakao[65] is shown in Table 4.4.G.

4.7.1.8 Thioester Bond

As in ester bonds, thioesters can be hydrolyzed in acid or base. They are preferably cleaved with an amine, which displaces the sulfide and forms an amide bond. A cross-linking reagent that forms a cleavable thioester on reaction with sulfhydryl groups of proteins is 3-(4-azido-2-nitrobenzoylthio) propionic acid, shown in Table 4.4.H.[67]

4.7.1.9 Maleylamide Linkage

As indicated in Chapter 2, maleylation of amino groups with maleic anhydride is reversible. With 2-methylmaleic anhydride and 2,3-dimethylmaleic anhydride, the acyl derivatives are even more easily hydrolyzed under acidic conditions.[76] Blättler et al.[77] have incorporated the 2-methylmaleic anhydride in the synthesis of an acid labile heterobifunctional reagent (Table 4.4.I). Upon reaction with an amino group, the maleylamide formed is stable above neutral pH, but hydrolyzes readily at mildly acidic pHs (pH 4–5). The reaction is shown in Figure 4.17.

4.7.1.10 Acetals, Ketals, and Ortho Esters

Acetals, ketals, and ortho esters have long been established as acid labile and base stable.[78,79] Of these, ortho esters are the most susceptible to acid-catalyzed hydrolysis, followed by ketals and acetals. Ketals undergo faster hydrolysis than acetals because of enhanced stabilization of the intermediate carbonium ion by an additional alkyl group.[80] Among the ketals, acyclic ketals of acetone are the most acid labile. Srinivasachar and Neville[81] have incorporated these functionalities in the synthesis of acid-cleavable protein cross-linking reagents, bismaleimide ortho ester and acetal and ketal cross-linkers, as shown in Table 4.4.J and K. These compounds are stable toward base and can be stored and manipulated at pH 8–9. They can be hydrolyzed at pH 5.5 with half-lives less than an hour for ortho esters and ketals.[81]

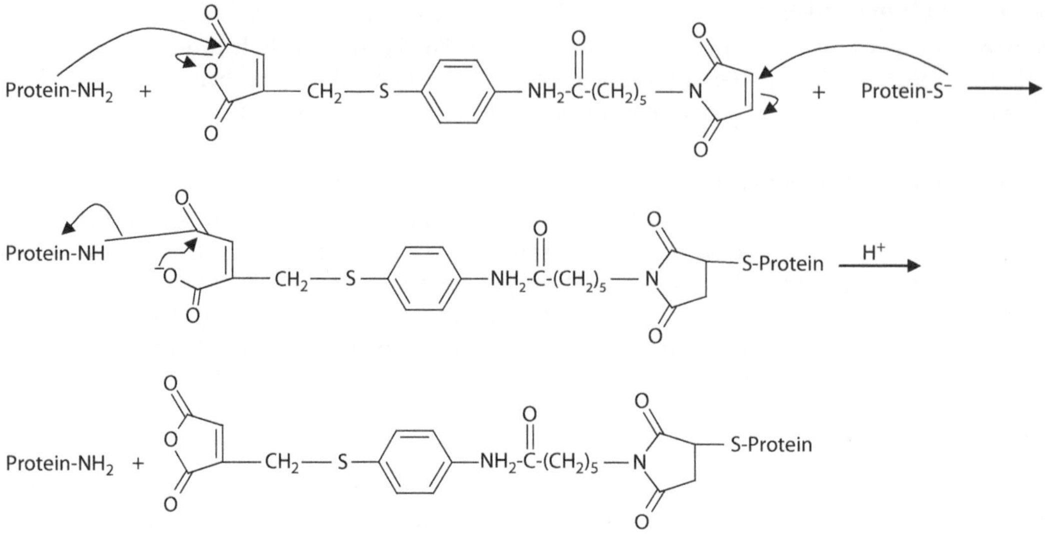

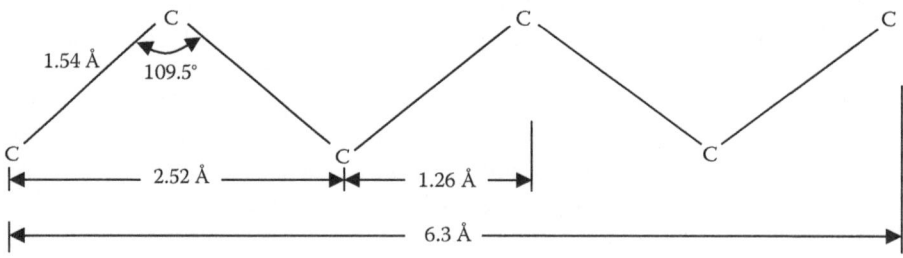

FIGURE 4.17 Cross-linking reaction of a maleimide derivative of 2-methylmaleic anhydride and hydrolysis of the cross-linked product.

4.7.2 INCORPORATING MOLECULAR DISTANCE RULERS

In addition to cleavable functionalities as described above, the spacer arm can be used to measure the distance between two targets where the cross-linker reacts as a molecular ruler.[82] The two ends of a cross-linker are where the reagent attaches to the target object. So, the longer the bridge, the more distant will be the attached points. Carbon chains are the best means for constructing different lengths of the bridge. The carbon atom is unique in that its hybridized orbitals are tetrahedral in shape with a bond angle of 109.5° and a C–C bond length of 1.54 Å.[83,84] As shown in Figure 4.18, the distance between alternate carbons in a carbon chain is about 2.52 Å, which translates into 1.26 Å for the "projected" C–C bond. From these values, it can be calculated that the length of a bridge of six carbon chains will have a distance of 6.3 Å. Thus, cross-linkers with different carbon chains in the bridge will give an estimate of the distance between two linked groups. As an example, a series of *bis*-imidoester with different number of carbon atoms is shown in Table 4.5.[85] In this example, the maximum distance includes the bonds between the reactive atom of the cross-linker and the attached atom of the protein. Since the C–C single bond can rotate freely, the molecule can assume various conformations. Many of these conformational states will have spans much less than the estimated maximum distance. It is therefore possible for a cross-linker to join two groups within the calculated distance but not beyond.

It should be noted that these values are calculated from methylene carbons. Any substitution of the carbon atom that affects the bond angle and bond length will change the projected distance. Introduction of oxygen or sulfur as ether or thioether in the carbon chain will shorten or lengthen the bond distance, respectively, since C–O bond is shorter (1.43 Å) and the C–S is longer (1.81 Å). The projected distance

FIGURE 4.18 Various molecular measurements of carbon–carbon chains.

TABLE 4.5

Imidoester Cross-Linkers of Different Chain Lengths

Reagent	Maximum Distance between Cross-Linked Group (Å)
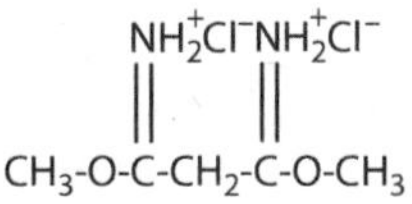 $NH_2^+Cl^- \quad NH_2^+Cl^-$ $CH_3\text{-}O\text{-}C\text{-}CH_2\text{-}C\text{-}O\text{-}CH_3$	5
$-(CH_2)_2-$	6
$-(CH_2)_4-$	9
$-(CH_2)_5-$	10
$-(CH_2)_6-$	11
$-(CH_2)_8-$	14

for the C–O bond and the C–S bond is, respectively, 1.17 and 1.44 Å. Introduction of disulfide bonds will further extend the chain length since the S–S bond length is 1.89 Å, giving a projected distance of about 1.50 Å as shown in Figure 4.19. For aromatic compounds, the distance between two groups bonded to the ortho positions is about 4.8 Å, and that between the *para*-positions is about 5.3 Å (Figure 4.20).[86] Thus, using the various bond lengths, the distance span of any cross-linked species can be estimated. For example, *bis*[2-(succinimidyloxycarbonyloxy)ethyl]sulfone (BSOCOES) can be estimated to have a maximum span of 10.4 Å between the reactive carbons (Figure 4.21).[62] The distance between two cross-linked groups will be about 13 Å, since a bond is formed between the reactive carbon of the reagent and a reactive group in a protein. Thus, the maximum distance between two coupled groups is that of the span of the reagent plus two bond lengths.[82]

It can be inferred from the above discussion that the bridge spacer arm can be designed to include various atoms of different chain lengths for cross-linking groups in the target molecule at different distances. By doing so, the molecular distance of the cross-linked atoms can be measured. An experimentalist can design different molecular rulers to construct a three-dimensional picture of biological aggregates. Such an approach has been taken by various researchers.[87–90] Since there are many

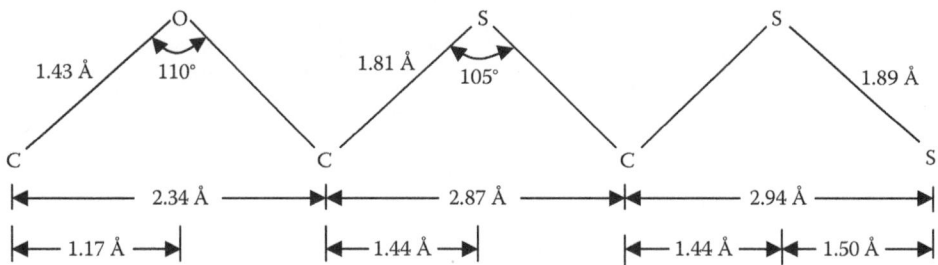

FIGURE 4.19 Various molecular measurements of carbon–oxygen, carbon–sulfur, and sulfur–sulfur bonds.

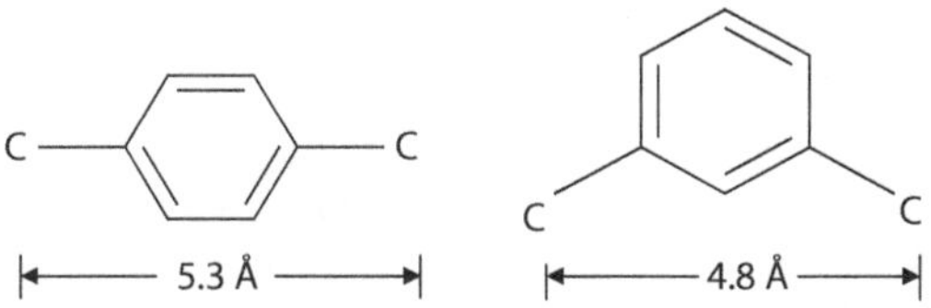

FIGURE 4.20 Bond lengths of aromatic compounds.

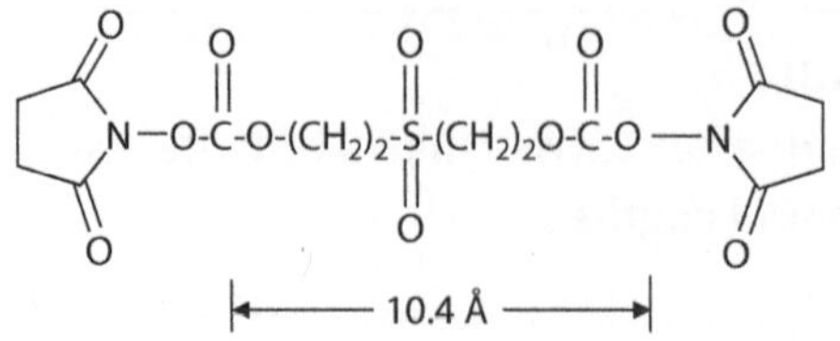

FIGURE 4.21   Maximum bond distance of a cross-linking agent, *bis*[2-(succinimidyloxycarbonyloxy)ethyl] sulfone (BSOCOES).

cross-linkers available commercially, one can chose suitable reagents of various lengths to study protein–protein and subunit–subunit interactions in solutions or crystals and in membranes.[82,88,91–93]

4.7.3   Incorporating Reporter Groups

There are many different reporter groups that are useful for the study of protein subunits and protein environment interactions,[93–95] for monitoring the extent of chemical modification,[34–36,59,96,97] and for isolation and analysis of cross-linked components. The common reporter groups that have been employed are described below. Researchers with different needs may incorporate new reporter groups into the cross-linking reagents.

4.7.3.1   UV–VIS Absorption Chromophores

UV–VIS absorbing chromophores provide a convenient means to follow a chemical reaction, which can be easily monitored using a simple spectrophotometer. We have seen earlier that the protein sulfhydryl group can be activated through disulfide exchange reaction using Ellman's reagent and other thiopyridine derivatives (Chapter 3). These reactions can be used to follow the extent of the reaction since the resulting thiolate ions have distinct absorbances in the UV-visible region. The molar absorptivities of some of these products are listed in Table 4.6.[98–107] These chromophores can be built into cross-linking reagents to generate chromogenic compounds. For example, in

TABLE 4.6
Molar Absorptivity of Some Good Leaving Groups

Leaving Group	Molar Absorptivity (M⁻¹ cm⁻¹)	References
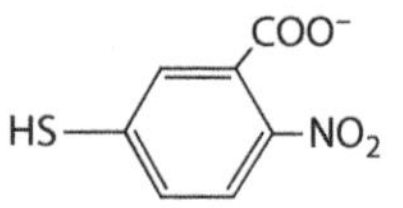 2-Nitro-5-thiobenzoate	14,150 (at 412 nm, pH 7.4) 13,553 (at 412 nm, pH 8.0) 13,600 (at 412 nm, pH 7.5)	98,99 100 101,102
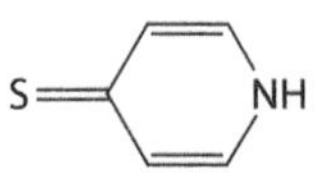 Pyridine-4-thione	19,800 (at 324 nm, pH 7.5)	102,103
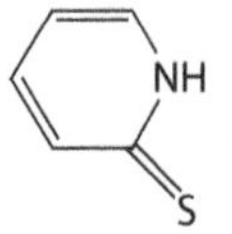 Pyridine-2-thione	7,060 (at 343 nm, pH 3–8) 8,080 (at 343 nm, pH 3–8)	103–105 106
 4-Nitrophenol	17,700 (at 402 nm, pH 8.5)	107

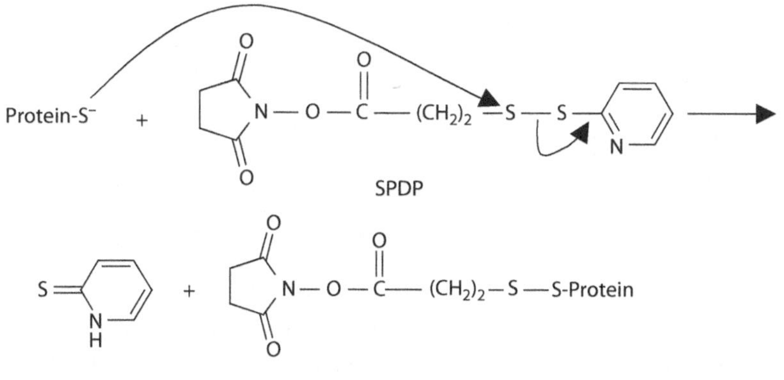

FIGURE 4.22 A chromogenic cross-linking reagent, *N*-succinimidyl-3-(2-pyridyldithio)propionate and its reaction to generate the chromophore, pyridine 2-thione.

N-succinimidyl-3-(2-pyridyldithio)propionate, chromogenic 2-thiopyridine is incorporated into the disulfide bond and the chromophore, pyridine-2-thione, is released on reaction with a sulf-hydryl group as shown in Figure 4.22. The reaction can be followed at wavelengths near 343 nm. Unfortunately, as can be seen from Table 4.6, the molar absorptivities of the chromophores are relatively low. Therefore, the method does not have the requisite sensitivity enough for monitoring reactions at low concentrations. Although the chromophores can be incorporated in the bridge arm of the cross-linker, such as in 5-isothiolcyanato-1-naphthalene azide,[108] the proteins itself also contain chromophores such as tryptophan and tyrosine and the absorbance of the protein (in the 220–310 nm region) may interfere with the measurement of the cross-linking chromophores. Thus, these chromogenic probes are generally not useful for following the labeled proteins.

4.7.3.2 Infrared-Absorbing Chromophores

Using infrared (IR) absorption of specific cross-linkers to detect cross-linked proteins is an innovative approach. Gardner et al.[109] have developed an IR chromogenic cross-linker, which contains a phosphate moiety as shown in Figure 4.23. The phosphate group has a high absorption efficiency at the wavelength of a continuous wave CO_2 laser, 10.6 μm. Thus, labeled phosphopeptides will undergo dissociation upon IR irradiation far more readily than unmodified peptides. This differential reaction and the use of mass spectrometry enable identification of labeled peptides. The idea of this approach presents a new way for creating IR-sensitive cross-linkers.

4.7.3.3 Fluorescent Probes

Fluorescent probes are much more sensitive than UV–VIS absorption chromophores and can be incorporated into the cross-linker bridge arm. In fluorophores, the light energy excites the molecular electrons to higher energy states. Regardless of whether the molecule is excited into the first electronic excited state or higher energy levels, within a very short time the molecule reverts to the lowest

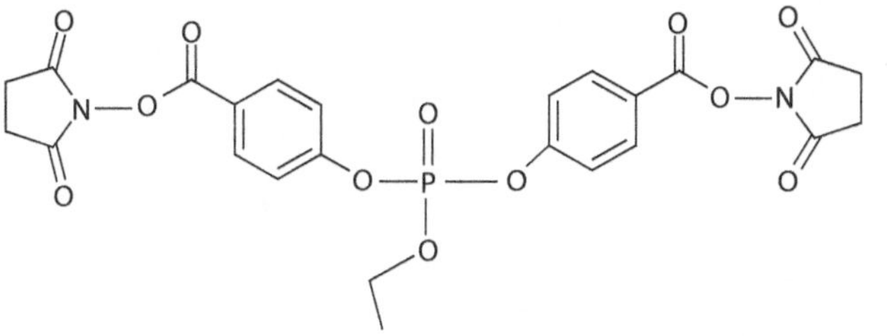

FIGURE 4.23 IR chromogenic cross-linker, dibenzoyloxysuccinimidyl ethyl phosphate.

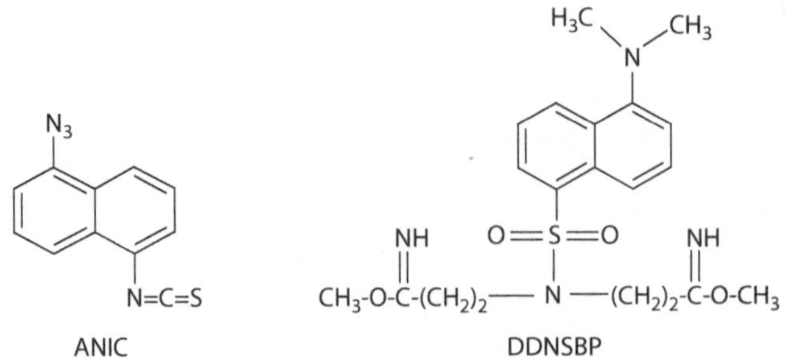

FIGURE 4.24 Examples of cross-linking reagents containing fluorescent probes: 1-azido-5-naphthaleneiso-thiocyanate (ANIC) and dimethyl-3,3′-[*N*-(5-(*N,N*-dimethylamino)naphthyl)sulphonyl]-*bis*propion-imidate" (DDNSBP).

(first) electronic energy level and eventually, after a time on the order of nanoseconds to tens of nano-seconds, returns to the ground state with concomitant emission of light (fluorescence). In addition to simple intensity measurements to quantify the extent of incorporation of the fluorescent moiety, more refined fluorescence measurements, such as polarization (anisotropy) and lifetimes can be used to gain information about the mobility of the fluorophore and aspects of its microenvironment.[110] If two fluorophores can be incorporated into a protein, it may also be possible to gain information on their spatial dispositions using Förster resonance energy transfer (FRET), although precise distance deter-minations are extremely difficult to achieve due to uncertainties regarding the so-called orientation factor, kappa square (κ^2).[110] The molecular structure that fluoresces usually contains aromatic compo-nents and is therefore sometimes bulky and hydrophobic in nature.[94] Thus, the purposes of the cross-linking agent need to be carefully considered before embarking on the design and synthesis of such a cross-linker. To incorporate a fluorescent probe, the fluorophore may be introduced as the bulk of the bridge spacer such as in 1-azido-5-naphthaleneisothiocyanate (ANIC)[108] (Figure 4.24). Fluorophores can also be attached to the bridge spacer arm as in dimethyl-3,3′-[*N*-(5-(*N,N*-dimethylamino)naphthyl) sulfonyl]-*bis*propionimidate (DDNSBP) shown in Figure 4.24. Literally thousands of fluorophores are now available commercially, which permits free range to the imagination of the chemist to syn-thesize novel fluorescent cross-linkers to probe diverse molecular properties.

4.7.3.4   Spin Labels

Spin labels are paramagnetic reporter groups with a nitroxide moiety containing an unpaired electron localized on the nitrogen and oxygen atoms.[111] A general nitroxide structure, together with those in pyrrolidine and piperidine, is shown in Figure 4.25. These nitroxide spin labels can be detected using an electron paramagnetic resonance spectrometer. In solution, they exhibit a spectrum that is differ-ent from those bound to proteins as shown in Figure 4.26. As reporter groups, these compounds can be used to study the conformations of proteins[112] and membrane structure.[94,95] They may also be used to determine the interspin distances of two or more spin labels attached to specific sites in proteins.[113] For designing cross-linking reagents, these nitroxide entities may be incorporated into the bridge backbone as shown in Figure 4.27 for *bis*(sulfo-*N*-succinimidyl)-doxyl-2-spiro-5′-axelate (BSSDA).

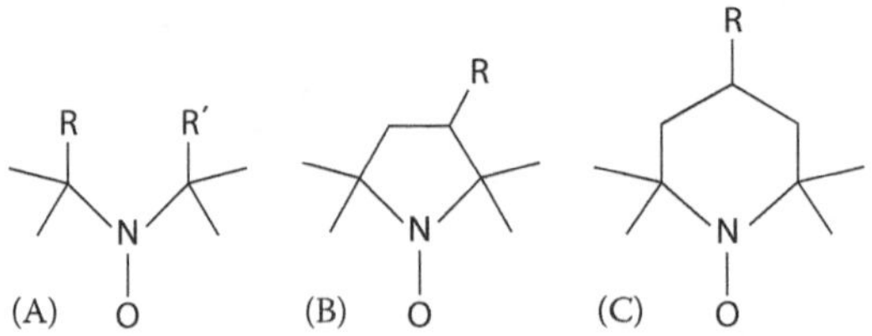

FIGURE 4.25 A general nitroxide spin label structure (A) and those in pyrrolidine (B) and piperidine (C) form.

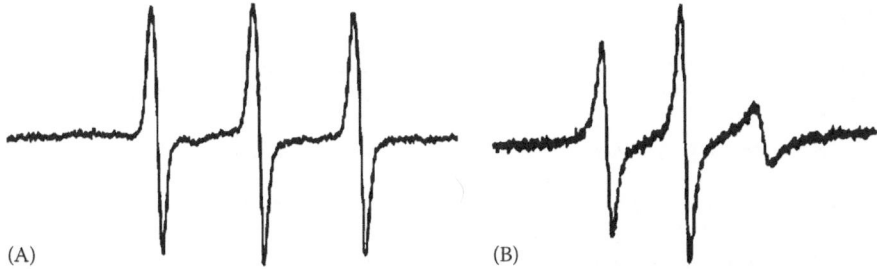

FIGURE 4.26 Electron spin resonance spectrum for a nitroxide spin label: (A) free in solution; (B) bound to a protein.

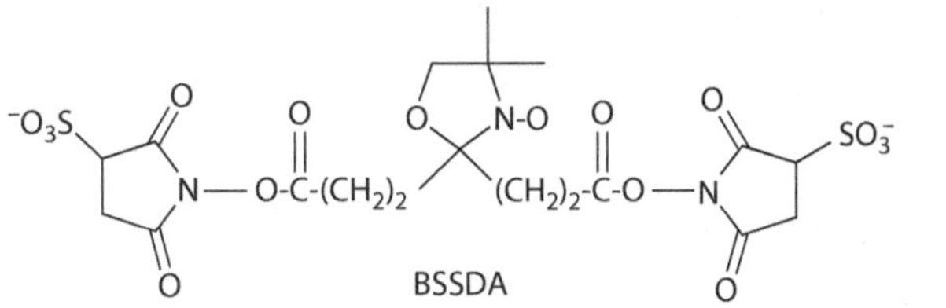

FIGURE 4.27 A spin labeled cross-linking reagent, *bis*(sulfo-*N*-succinimidyl)-doxyl-2-spiro-5′-axelate (BSSDA).

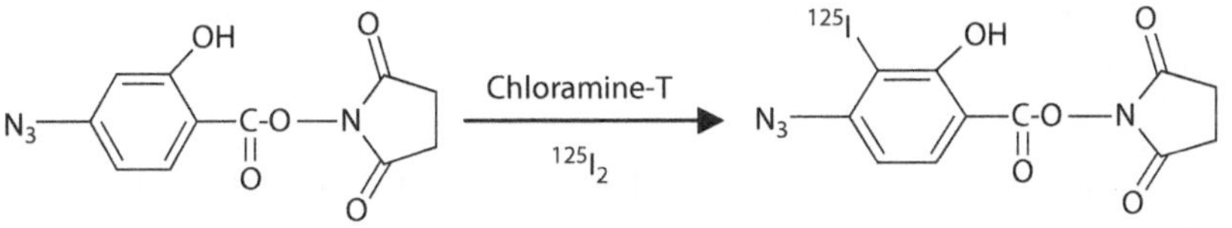

FIGURE 4.28 Introduction of radioactive iodine isotope, ^{125}I, into a bifunctional cross-linking reagent containing a hydroxyphenyl group, *N*-(4-azidocarbonyl-3-hydroxyphenyl)maleimide.

4.7.3.5 Radioactive and Nonradioactive Isotopes

Radioactive isotopes are the most sensitive means for following cross-linked components. In particular, when the cross-linked species are separated using cleavable cross-linking reagent, the individual components can be identified with the help of the radioactive tracer. Although radioactive cross-linking reagents can be synthesized de novo from compounds labeled with ^{3}H, ^{14}C, ^{35}S, or other radioactive isotopes,[114] the ability to radiolabel an already-made cross-linker in a simple one-step process is of great appeal. Iodination has generally been used for such a purpose particularly for proteins.[115,116] There are several reagents available for the introduction of ^{125}I. These include chloramine-T (*N*-chlorotoluenesulfonamide)[117,118] and its immobilized analogue, IODO-BEADS,[119] iodine monochloride,[120] iodogen (1,3,4,6-tetrachloro-3α,6α-diphenylglycouril),[121] and lactoperoxidase.[122] The phenolic side chain of tyrosine provides a convenient major site of substitution. The incorporation of such hydroxyphenyl groups into cross-linking reagent has also been used for this purpose. An example is *N*-(4-azidocarbonyl-3-hydroxyphenyl)maleimide. As shown in Figure 4.28, radioactive iodine can be introduced readily with chloramine-T.[123]

With the advent of mass spectrometry for the analysis of cross-linked products, nonradioactive isotopes have been incorporated into bifunctional cross-linkers. Heavy isotopes like ^{18}O and ^{2}H have been used.[124] Collins et al.[125] used ^{18}O labeled *bis*(sulfosuccinimidyl)suberate to cross-link FGF-2. The cross-linked protein was then analyzed using a method called mass spectrometry in three dimensions to solve the protein structures.[125] For deuterium, Petrotchenko et al.[126] synthesized D$_{12}$–12-ethylene glycol *bis*(sulfosuccinimidylsuccinate) to study protein–protein interactions in various proteins.

REFERENCES

1. Chong, P. C. S. and Hodges, R. S., A new heterobifunctional cross-linking reagent for the study of biological interactions between proteins. I. Design, synthesis and characterization, *J. Biol. Chem.*, 256, 5064, 1981.
2. March, J., *Advanced Organic Chemistry: Reactions. Mechanism and Structure*, John Wiley & Sons, New York, 1985.
3. Loudon, J. D. and Schulman, N., Mobility of groups in chloronitrodiphenyl sulfones, *J. Chem. Soc.*, 722, 1941.
4. Suhr, B., Effects of the leaving group on the velocity of nucleophilic aromatic substitutions, *Chemische Berichte*, 97, 3268, 1964.
5. Edwards, J. O. and Pearson, R. G., The factors determining nucleophilic reactivities, *J. Am. Chem. Soc.*, 84, 26, 1962.
6. Gurd, F. R. N., Carboxymethylation, *Methods Enzymol.*, 11, 532, 1967.
7. Gorin, G., Martin, P. A., and Doughty, G., Kinetics of the reaction of *N*-ethylmaleimide with cysteine and some congeners, *Arch. Biochem. Biophys.*, 115, 539, 1966.
8. Zahn, H. and Meinhoffer, J., Reactions of 1,5-difluoro-2,4-dinitrobenzene with insulin, *Makromol. Chem.*, 26, 153, 1958.
9. Shaltiel, S., Thiolysis of some dinitrophenyl derivatives of amino acids, *Biochem. Biophys. Res. Commun.*, 29, 178, 1967.
10. Husain, S. S., Ferguson, J. B., and Fruton, J. S., Bifunctional inhibitors of pepsin, *Proc. Natl. Acad. Sci. U. S. A.*, 68, 2765, 1971.
11. Lundblad, R. L. and Stein, W. H., On the reaction of diazoacetyl compounds with pepsin, *J. Biol. Chem.*, 244, 154, 1969.
12. Anderson, G. W., Zimmerman, J. E., and Callahan, E. M., The use of esters of N-hydroxysuccinimide in peptide synthesis, *J. Am. Chem. Soc.*, 86, 1839, 1964.
13. McElvain, S. M. and Schroeder, J. P., Ortho esters and related compounds from malono- and succino-nitriles, *J. Am. Chem. Soc.*, 71, 40, 1949.
14. Peters, K. and Richards, F. M., Chemical cross-linking: Reagents and problems in studies of membrane structure, *Annu. Rev. Biochem.*, 46, 523, 1977.
15. Hunter, M. J. and Ludwig, M. L., Amidination, *Methods Enzymol.*, 25, 585, 1972.
16. Lutter, L. C., Ortanderl, F., and Fasold, H., Use of a new series of cleavable protein-crosslinkers on the *Escherichia coli* ribosome, *FEBS Lett.*, 48, 288, 1974.
17. Rifai, A. and Wong, S. S., Preparation of phosphorylcholine-conjugated antigens, *J. Immunol. Methods*, 94, 25, 1986.
18. Silman, I. H., Albu-Weissenberg, M., and Katchalski, E., Some water-insoluble papain derivatives, *Biopolymers*, 4, 441, 1966.
19. Silman, I. H. and Katchalski, E., Water-insoluble derivatives of enzymes, antigens, and antibodies, *Annu. Rev. Biochem.*, 35, 873, 1966.
20. Gordon, J., Rose, B., and Schon, A. H., Detection of "non-precipitating" antibodies in sera of individuals allergic to ragweed pollen by an *in vitro* method, *J. Exp. Med.*, 108, 37, 1958.
21. Carlsson, J., Drevin, H., and Axén, R., Protein thiolation and reversible protein–protein conjugation. *N*-Succinimidyl 3-(2-pyridyldithio)propionate, a new heterobifunctional reagent, *Biochem. J.*, 173, 723, 1978.
22. Ji, T. H., The application of chemical crosslinking for studies on cell membranes and the identification of surface reporters, *Biochem. Biophys. Acta*, 559, 39, 1979.
23. Hughes, W. L., Albumin fraction isolated from human plasma as a crystalline mercuric salt, *J. Am. Chem. Soc.*, 69, 1836, 1947.
24. Jovin, T. M., Englund, P. T., and Kornberg, A., Enzymatic synthesis of deoxyribonucleic acid. XXIIV. Chemical modification of deoxyribonucleic acid polymerase, *J. Biol. Chem.*, 244, 3009, 1969.
25. Arnon, R. and Shapira, E., Crystalline papain derivative containing an intramolecular mercury bridge, *J. Biol. Chem.*, 244, 1033, 1969.
26. Sperling, R., Burstein, Y., and Steinberg, I. Z., Selective reduction and mercuration of cystine IV-V in bovine pancreatic ribonuclease, *Biochemistry*, 8, 3810, 1969.
27. Kay, C. M. and Edsall, J. T., Dimerization of mercaptalbumin in the presence of mercurials. III. Bovine mercaptalbumin in water and in concentrated urea solutions, *Arch. Biochem. Biophys.*, 65, 354, 1956.
28. Means, G. E. and Fereney, R. E., Reductive alkylation of amino groups in proteins, *Biochemistry*, 7, 2192, 1968.

29. Ngo, T. T., Yam, C. F., Lenhoff, H. M., and Ivy, J., *p*-Azidophenylglyoxal: A heterobifunctional photoactivable cross-linking reagent selective for arginyl residues, *J. Biol. Chem.*, 256, 11313, 1981.

30. Politz, S. M., Noller, H. F., and McWhiter, P. D., Ribonucleic acid-protein cross-linking in *Escherichia coli* ribosomes: (4-Azidophenyl)glyoxal, a novel heterobifunctional reagent, *Biochemistry*, 20, 372, 1981.

31. Knowles, J. R., Photogenerated reagents for biological receptor-site labeling, *Acc. Chem. Res.*, 5, 155, 1972.

32. Bayley, H. and Knowles, J. R., Photoaffinity labeling, *Methods Enzymol.*, 46, 69, 1977.

33. Bayley, H., Photogenerated reagents in biochemistry and molecular biology, in *Laboratory Techniques in Biochemistry and Molecular Biology*, Vol. 12, Work, T. S. and Burdon, R. H. (eds.), Elsevier, Amsterdam, the Netherlands, 1984, p. 208.

34. Vodovozova, E. L., Photoaffinity labeling and its application in structural biology, *Biochem. (Moscow)*, 72, 1, 2007.

35. Hatanaka, Y. and Sadakane, Y., Photoaffinity labeling in drug discovery and developments: Chemical gateway for entering proteomic frontier. *Curr. Top. Med. Chem.*, 2, 271, 2002.

36. Dorman, G. and Prestwich, G. D., Using photolabile ligands in drug discovery and development, *Trends Biotechnol.*, 18, 64, 2000.

37. Dorman, G. and Prestwich, G. D., Benzophenone photophores in biochemistry, *Biochemistry*, 33, 5661, 1994.

38. Fleming, S. A., Chemical reagents in photoaffinity labeling, *Tetrahedron*, 51, 12479, 1995.

39. Kotzyba-Hibert, F., Kapfer, I., and Goeldner, M., Recent trends in photoaffinity labeling, *Angew. Chem. Int. Ed.*, 34, 1296, 1995.

40. Brunner, J., New photolabeling and crosslinking methods, *Ann. Rev. Biochem.*, 62, 483, 1993.

41. Moss, R. A., Diazirines: Carbene precursors par excellence, *Acc. Chem. Res.*, 39, 267, 2006.

42. Hashimoto, M. and Hatanaka, Y., Recent progress in diazirine-based photoaffinity labeling, *Eur. J. Org. Chem.*, 15, 2513, 2008.

43. Walling, C. and Gibian, M. J., Hydrogen abstraction reactions by the triplet states of ketones, *J. Am. Chem. Soc.*, 87, 3361, 1965.

44. Chatrenet, B., Kotzyba-Hibert, F., Mulle, C., Changeux, J. P., Goeldner, M. P., and Hirth, C., Photoactivatable agonist of the nicotinic acetylcholine receptor: Potential probe to characterize the structural transitions of acetylcholine binding site in different states of the receptor, *Mol. Pharmacol.*, 41, 1100, 1992.

45. Schafer, H.-J., Divalent azido-ATP analog for photoaffinity cross-linking of F_1 subunits, *Methods Enzymol.*, 126, 649, 1986.

46. Staros, J. V., N-Hydroxysulfosuccinimide active esters: *bis*(N-hydroxysulfosuccinimide) esters of two dicarboxylic acids are hydrophilic, membrane-impermeant, protein cross-linkers, *Biochemistry*, 21, 3950, 1982.

47. Staros, J. V., Morgan, D. G., and Appling, D. R., A membrane-impermeant, cleavable cross-linker: Dimers of human erythrocyte band 3 subunits cross-linked at the extra-cytoplasmic face, *J. Biol. Chem.*, 256, 5890, 1981.

48. Staros, J. V., Membrane impermeant cross-linking reagents: Probes of the structure and dynamics of membrane proteins, *Acc. Chem. Res.*, 12, 435, 1988.

49. Harris, R. and Findlay, J. B., Investigation of the organization of the major proteins in bovine myelin membranes. Use of chemical probes and bifunctional cross-linking reagents, *Biochim. Biophys. Acta*, 732, 75, 1983.

50. Tesser, G. I., De Hoog-Declerck, R. A., and Westerhuis, L. W., Some new bifunctional reagents for the conjugation of a protein with an amino compound in water. *Hoppe Seylers Z. Physiol. Chem.*, 356, 1625, 1975.

51. Staros, J. V., Lee, W. T., and Conrad, D. H., Membrane impermeant cross-linking reagent application to studies of the cell surface receptor for IgE, *Methods Enzymol.*, 150, 503, 1988.

52. Schramm, H. J. and Dülffer, T., Synthesis and application of cleavable and hydrophilic crosslinking reagents, in *Protein Crosslinking: Biochemical and Molecular Aspects*, Friedman, M. (ed.), Plenum Press, New York, 1976, p. 197.

53. Fasold, H., Bäumert, H., and Fink, G., Comparison of hydrophobic and strongly hydrophilic cleavable cross-linking reagents in intermolecular bond formation in aggregates of proteins or protein–RNA, in *Protein Cross-Linking: Biochemical and Molecular Aspects*, Friedman, M. (ed.), Plenum Press, New York, 1976, p. 207.

54. Traut, R. R., Bollen, A., Sun, T. T., Hershey, J. W. B., Sundberg, J., and Pierce, L. R., Methyl-4-mercaptobutyrimdate as a cleavable cross-linking reagent and its application to the *Escherichia coli* 30S ribosome, *Biochemistry*, 12, 3266, 1973.

55. Sun, T. T., Bollen, A., Kahan, L., and Traut, R. R., Topography of ribosomal proteins of the *E. coli* 30S subunit as studied with the reversible cross-linking reagent methyl-4-mercaptobutyrimidate, *Biochemistry*, 13, 2334, 1974.

56. Bragg, P. D. and Hoa, C., Subunit composition, function, and spatial arrangement in the Ca^{2+} and Mg^{2+} activated adenosine triphosphatase of *Escherichia coli* and *Salmonella typhimurium*, *Arch. Biochem. Biophys.*, 167, 311, 1975.

57. Webb, J. L., *Enzyme and Metabolic Inhibitors*, Vol. 2, Academic Press, New York, 1966, 729.

58. Smith, R. J., Capaldi, R. A., Muchmore, D., and Dahlquist, F., Cross-linking of ubiquinone-cytochrome-c-reductase with periodate-cleavable bifunctional reagent, *Biochemistry*, 18, 3719, 1978.

59. Jaffe, C. L., Lis, B., and Sharon, N., New cleavable photoreactive heterobifunctional cross-linking reagents for studying membrane organization, *Biochemistry*, 19, 4423, 1980.

60. Wold, F., Reaction of bovine serum albumin with the bifunctional reagent p,p'-difluoro-m,m'-dinitrodiphenylsulfone, *J. Biol. Chem.*, 236, 106, 1961.

61. Wold, F., Bifunctional reagents, *Methods Enzymol.*, 25, 623, 1972.

62. Zahling, D. A., Watson, A., and Bach, F. H., Mapping of lymphocyte surface polypeptide antigens by chemical cross-linking with BSOCOES, *J. Immunol.*, 124, 913, 1980.

63. Buchardt, O., Eisner, H. I., Nielsen, P. E., Petersen, L., and Suenson, E., Protein crosslinking reagents containing a selenoethylene linker are cleaved by mild oxidation, *Anal. Biochem.*, 158, 87, 1986.

64. Birkelund, S., Lundemose, A. G., and Christiansen, G., Chemical cross-linking of *Chlamydia trachomatis*, *Infect. Immun.*, 56, 654, 1988.

65. Sato, S. and Nakao, M., Cross-linking of intact erythrocyte membrane with a newly synthesized cleavable bifunctional reagent, *J. Biochem.*, 90, 1177, 1981.

66. Abdella, P. M., Smith, P. K., and Royer, G. P., A new cleavable reagent for cross-linking and reusable immobilization of proteins, *Biochem. Biophys. Res. Commun.*, 87, 734, 1979.

67. Friebel, K., Buth, B., Jany, K. D., and Trummer, W. E., Semireversible cross-linking synthesis and application of a novel heterobifunctional reagent, *Z. Physiol. Chem.*, 362, 421, 1981.

68. Cleland, W. W., Dithiothreitol: A new protective reagent for SH groups, *Biochemistry*, 3, 480, 1964.

69. Stockmayer, W. B. D., Rice, D. W., and Stephenson, C. C., Thermodynamic properties of sodium borohydride and aqueous borohydride ion, *J. Am. Chem. Soc.*, 77, 1980, 1955.

70. Neumann, H., Goldberger, R. F., and Sela, M., Interaction of phosphorothioate with disulfide bonds of ribonuclease and lysozyme, *J. Biol. Chem.*, 239, 1536, 1964.

71. Cecil, R. and McPhee, J. R., Kinetic study of the reactions on some disulfides with sodium sulfite, *Biochem. J.*, 60, 496, 1955.

72. McPhee, J. R., Further studies on the reactions of disulfides with sodium sulfite, *Biochem. J.*, 64, 22, 1956.

73. Bobbitt, J. M., Periodate oxidation of carbohydrates, *Adv. Carbohydr. Chem.*, 11, 1, 1956.

74. Coggins, J. R., Hooper, E. A., and Perham, R. N., Use of dimethyl suberimidate and novel periodate cleavable *bis*(imidoesters) to study the quaternary structure of the pyruvate dehydrogenase multienzyme complex of *Escherichia coli*, *Biochemistry*, 15, 2527, 1976.

75. Bornstetn, P. and Balian, G., Cleavage of Asn-Gly bonds with hydroxylamine, *Methods Enzymol.*, 47, 132, 1977.

76. Dixon, H. B. F. and Perham, R. N., Reversible blocking of amine groups with citraconic anhydride, *Biochem. J.*, 109, 312, 1968.

77. Blättler, W. A., Kuenzi, B. S., Lambert, J. M., and Senter, P. D., New heterobifunctional protein cross-linking reagent that forms an acid-labile link, *Biochemistry*, 24, 1517, 1985.

78. Cordes, E. H., Mechanism and catalysis for the hydrolysis of acetals, ketals, and orthoesters, *Prog. Phys. Org. Chem.*, 4, 1, 1967.

79. Cordes, E. H. and Bull, H. G., Mechanism and catalysis for hydrolysis of acetals, ketals and orthoesters, *Chem. Rev.*, 74, 581, 1974.

80. Cordes, E. H., Ortho Esters in *The Chemistry of Carboxylic Acids and Esters*, Patai, S. (ed.), Interscience Publishers, New York, 1969, p. 623–667.

81. Srinivasachar, K. and Neville, D. M., Jr., New protein cross-linking reagents that are cleaved by mild acid, *Biochemistry*, 28, 2501, 1989.

82. Green, N. S., Reisler, E., and Houk, K. N., Quantitative evaluation of the lengths of homobifunctional protein cross-linking reagents used as molecular rulers, *Protein Sci.*, 10, 1293, 2001.

83. Pauling, L., *The Nature of the Chemical Bond*, Cornell University Press, Cornell, NY, 1960.

84. Lide, D. R. (ed.), *CRC Handbook of Chemistry and Physics*, 88th edn., CRC Press, Boca Raton, FL, 2008.

85. Ji, T. H., Cross-linking of glycolipids in erythrocyte ghost membrane, *J. Biol. Chem.*, 249, 7841, 1974.

86. Nitao, L. K. and Reisler, E., Probing the conformational states of the SH1–SH2 helix in myosin: A cross-linking approach, *Biochemistry*, 37, 16704, 1998.

87. Nagy, J. K., Lau, F. W., Bowie, J. U., and Saunders, C. R., Mapping the oligomeric interface of diacylglycerol kinase by engineered thiol crosslinking: Homologous sites in the transmembrane domain, *Biochemistry*, 39, 4154, 2000.

88. Kwaw, I., Sun, J., and Kaback, H. R., Thiol cross-linking of cytoplasmic loops in the lactose permease of *Escherichia coli*, *Biochemistry*, 39, 3134, 2000.

89. Zecherle, G. N., Oleinikov, A., and Traut, R. R., The proximity of the C-terminal domain of *Escherichia coli* ribosomal protein L7/L12 to L10 determined by cysteine site-directed mutagenesis and protein–protein crosslinking, *J. Biol. Chem.*, 267, 5889, 1992.

90. Sun, J. and Kaback, H. R., Proximity of periplasmic loops in the lactose permease of *Escherichia coli* determined by site-directed crosslinking, *Biochemistry*, 36, 11959, 1997.

91. Ishiguro, A., Kimura, M., Yasui, K., Iwata, A., Ueda, S., and Ishihama, A., Two large subunits of the fission yeast RNA polymerase II provide platforms for the assembly of small subunits, *J. Mol. Biol.*, 279, 703, 1998.

92. Rappsilber, J., Siniossoglou, S., Hurt, E. C., and Mann, M., A generic strategy to analyze the spatial organization of multi-protein complexes by cross-linking and mass spectrometry, *Anal. Chem.*, 72, 267, 2000.

93. Gonzalez-Ros, J. M., Farach, M. C., and Martinez-Carrion, M., Ligand-induced effects at regions of acetylcholine receptor accessible to membrane lipids, *Biochemistry*, 22, 3807, 1983.

94. Gaffney, B. J., Willingham, G. L., and Schepp, R. S., Synthesis and membrane interactions of spin label bifunctional reagents, *Biochemistry*, 22, 881, 1983.

95. Willingham, G. L. and Gaffney, B. J., Reactions of spin-label cross-linking reagents with RBC proteins, *Biochemistry*, 22, 892, 1983.

96. Ueno, T., Hikita, S., Muno, D., Sato, E., Kanaoka, Y., and Sekine, T., New fluorogenic, photoactivable heterobifunctional crosslinking thiol reagents, *Anal. Biochem.*, 140, 63, 1984.

97. Maassen, J. A., Cross-linking of ribosomal proteins by 4-(6-formyl-3-azidophenoxy)butyrimidate, a heterobifunctional, cleavable cross-linker, *Biochemistry*, 18, 1288, 1979.

98. Eyer, P., Worek, F., Kiderlen, D., Sinko, G., Stuglin, A., Simeon-Rudolf, V., and Reiner, E., Molar absorption coefficients for the reduced Ellman reagent: Reassessment. *Anal. Biochem.*, 312, 224, 2003.

99. Riddles, P. W., Blakeley, R. L., and Zerner, B., Reassessment of Ellman's reagent, *Methods Enzymol.*, 91, 49, 1983.

100. Hosseinimehr, S. J., Ebrahimi, P., Hassani, N., Mirzabeigi, P., and Amini, M., Spectrophotometric determination of captopril with DTNB reagent in pharmaceutical formulation, *Boll. Chim. Farm.*, 143, 249, 2004.

101. Habeeb, A. F. S. A., Reaction of protein sulfhydryl group with Ellman's reagent, *Methods Enzymol.*, 25, 457, 1972.

102. Talgoy, M. M., Bell, A. W., and Duckworth, H. W., The reactions of *Escherichia coli* citrate synthase with the sulfhydryl reagents 5,5'-dithiobis-(2-nitrobenzoic acid) and 4,4'-dithiodipyridine, *Can. J. Biochem.*, 57, 822, 1979.

103. Grassetti, D. R. and Murray, J. F., Determination of sulfhydryl groups with 2,2'- or 4,4'-dithiodipyridine, *Arch. Biochem. Biophys.*, 119, 41, 1967.

104. Brocklehurst, K. and Little, G., Reactions of papain and of low-molecular-weight thiols with some aromatic disulphides. 2,2'-Dipyridyl disulphide as a convenient active-site titrant for papain even in the presence of other thiols, *Biochem. J.*, 133, 67, 1973.

105. Kimura, T., Matsueda, R., Nakagawa, Y., and Kaiser, E. T., New reagents for the introduction of the thiomethyl group at sulfhydryl residues of proteins with concomitant spectrophotometric titration of the sulfhydryl: Methyl 3-nitro-2-pyridyl disulfide and methyl 2-pyridyl disulfide, *Anal. Biochem.*, 122, 274, 1982.

106. Stuchbury, T., Shipton, M., Norris, R., Malthouse, J. P. G., and Brocklehurst, K., Reporter groups delivery system with both absolute and selective specificity for thiol groups and an improved fluorescent probe absolute and selective specificity for thiol groups and an improved fluorescent probe containing the 7-nitrobenz-2-oxa-1,3-diazole moiety, *Biochem. J.*, 151, 417, 1975.

107. Shikita, M., Fahey, J. W., Golden, T., R., Holtzclaw, W. D., and Talalay, P., An unusual case of 'uncompetitive activation' by ascorbic acid: Purification and kinetic properties of a myrosinase from *Raphanus sativus* seedlings, *Biochem. J.*, 341, 725, 1999.

108. Sigrist, B., Allegrini, P. R., Kempf, C., Schnippering, C., and Zahler, P., 5-isothiocyanato-1-phthalene azide and p-azidophenylisothiocyanate. Synthesis and application in hydrophobic heterobifunctional photoactive cross-linking of membrane proteins, *Eur. J. Biochem.*, 125, 197, 1982.

109. Gardner, M. W., Vasicek, L. A., Shabbir, S., Anslyn, E. V., and Brodbelt, J. S., Chromogenic cross-linker for the characterization of protein structure by infrared multiphoton dissociation mass spectrometry, *Anal. Chem.*, 80, 4807, 2008.

110. Valuer, B., *Molecular Fluorescence: Principles and Applications*, Wiley-VCH, New York, 2002.

111. Stone, T. J., Buckman, T., Nordio, P. L., and McConnell, H. M., Spin-labeled biomolecules, *Proc. Natl. Acad. Sci. U. S. A.*, 54, 1010, 1965.

112. Berliner, L. J. and Wong, S. S., Evidence against two "pH locked" conformations of phosphorylated trypsin, *J. Biol. Chem.*, 248, 1118, 1973.

113. Steinhoff, H. J., Radzwill, N., Thevis, W., Lenz, V., Brandenburg, D., Antson, A., Dodson, G., and Wollmer, A., Determination of interspin distances between spin labels attached to insulin: Comparison of electron paramagnetic resonance data with the X-ray structure, *Biophys. J.*, 73, 3287, 1997.

114. Schwartz, M. S., Das, O. P., and Hynes, R. O., A new radioactive cross-linking reagent for studying the interactions of proteins, *J. Biol. Chem.*, 257, 2343, 1982.

115. Hughes, W. L., The chemistry of iodination, *Ann. N.Y. Acad. Sci.*, 70, 3, 1957.

116. Seevers, R. B. and Counsell, R. E., Radioiodination techniques for small organic molecules, *Chem. Rev.*, 82, 575, 1982.

117. Greenwood, F. C., Hunter, W. M., and Glover, J. S., The preparation of ^{131}I labeled human growth hormone of high specific radioactivity, *Biochem. J.*, 89, 114, 1963.

118. Wilbur, D. S., Radiohalogenation of proteins: An overview of radionuclides, labeling methods, and reagents for conjugate labeling, *Bioconjug. Chem.*, 3, 433, 1992.

119. Markwell, M. A. K., A new solid-state reagent to iodinate proteins: Conditions for the efficient labeling of antiserum, *Anal. Biochem.*, 125, 427, 1982.

120. McFarlane, A. S., Efficient trace-labeling of proteins with iodine, *Nature (Lond.)*, 182, 53, 1958.

121. Fraker, P. J. and Speck, J. C., Jr., Protein and cell membrane iodinations with a sparingly soluble chloroamide, 1,3,4.6-tetrachloro-3a,6a-diphenylglycouril, *Biochem. Biophys. Res. Commun.*, 80, 849, 1978.

122. Marchalonis, J. J., Enzymic method for the trace iodination of immunoglobulins and other proteins, *Biochem. J.*, 113, 299, 1969.

123. Ji, T. H. and Ji, I., Macromolecular photoaffinity labeling with radioactive photoactivable heterobifunctional reagents, *Anal. Biochem.*, 121, 286, 1982.

124. Seebacher, J., Mallick, P., Zhang, N., Eddes, J. S., Aebersold, R., and Gelb, M. H., Protein cross-linking analysis using mass spectrometry, isotope-coded cross-linkers, and integrated computational data processing, *J. Proteome Res.*, 5, 2270, 2006.

125. Collins, C. J., Schilling, B., Young, M., Dollinger, G., and Guy, R. K., Isotopically labeled crosslinking reagents: Resolution of mass degeneracy in the identification of crosslinked peptides, *Bioorg. Med. Chem. Lett.*, 13, 4023, 2003.

126. Petrotchenko, E. V., Olkhovik, V. K., and Borchers, C. H., Isotopically coded cleavable cross-linker for studying protein–protein interaction and protein complexes, *Mol. Cell. Proteomics*, 4, 1167, 2005.

5 Homobifunctional Cross-Linking Reagents

5.1 INTRODUCTION

Homobifunctional reagents are compounds that contain two reactive functional groups that have equal reactivity toward the same amino acid side chain or DNA base. In all cases, these two reactive groups are identical. There may, however, be exceptions where the reaction mechanism indicates that the two identical reactive groups react differently. For example, in the case of nitrosourea derivatives, it is thought that one of the two reactive groups spontaneously forms a carbonium intermediate which elicits the cross-linking reaction. In such cases, these reagents are referred to as heterobifunctional if two different groups are cross-linked. A reagent, although classified as a homobifunctional cross-linker, may cross-link two different groups when the second identical group is not available. For example, a sulfhydryl-targeted reagent may cross-link a thiol and an amino group when a second sulfhydryl group is not available for reaction. This situation is particularly so for protein–DNA cross-linking and for DNA strand cross-linking where two different bases are cross-linked due to the complementary nature of the helical structure. Similarly, nonspecific photoactive compounds containing two identical photosensitive groups are regarded as homobifunctional compounds. In addition, a compound with obscure functional groups is classified as homobifunctional if it cross-links two identical groups in proteins.

The homobifunctional reagents have been shown to induce cross-linking intramolecularly between two locales within a macromolecule and intermolecularly between two molecules. Intramolecular cross-linking has been used to increase the stability of proteins against thermal and mechanical denaturation to determine the distances between two reactive groups, to study protein surface topology, and to detect different conformational states.[1-5] Intermolecular cross-linking has been employed to characterize the nature and extent of protein–protein interactions, to couple two or more different proteins, to determine interacting protein neighbors, and to prepare various bioconjugates for medicinal and industrial applications.[1,2,6-9] These reagents are also useful for identification of receptors within membranes.[10-13]

A variety of reactive groups have been incorporated into these compounds. These include group-selective functionalities such as those presented in the last three chapters for thiols, amines, carboxylates, phenols, and other amino acid side chains. Nonspecific photosensitive functional groups have also been synthesized. In addition, various cleavable bonds, as discussed in the last chapter, have been incorporated into these chemicals. In this chapter, the homobifunctional reagents will be presented according to their selectivity toward a particular amino acid side chain. An additional section is devoted to DNA cross-linking reagents. It should be clear from the previous chapters that there are no absolutely group-specific reagents. For alkylating and acylating agents, the nucleophiles encompass the sulfhydryl, amino, imidazolyl, phenolate, carboxylate, and hydroxyl groups. The conditions and the reactants, as discussed in Chapter 3, determine the selectivity of the reaction. Thus, the potential exists that two different groups may be cross-linked. The categories listed in this chapter are for general references only and should not be taken as absolute classification of these reagents.

It may be noted that the nomenclature of these bifunctional reagents is inconsistent with that recommended by the International Union of Pure and Applied Chemistry. Since these compounds are generally complex and complicated, systemic names are usually long and difficult. The common names used in the literature are adopted here and used throughout this book. Those readers interested in the systemic names are referred to the Chemical Abstracts.

5.2   AMINO GROUP–DIRECTED CROSS-LINKERS

There are some specific functionalities that preferentially react with amino groups of proteins as discussed in Chapter 3. Those functional groups that have been incorporated into homobifunctional cross-linking reagents are shown in Table 5.1. A comprehensive listing of these amino group–directed cross-linking reagents that have been used and reported in the literature can be found in Appendix A. Many of these reagents are commercially available. Table 5.1 only provides an example of a cross-linking reagent for each of the various classes of the cross-linkers, which will be discussed below. The imidates and *N*-succinimidyl derivatives are generally considered the most selective for amino groups followed by aryl halides. Many of the alkylating and acylating agents that seemingly react preferentially with amino groups at alkaline pHs because of their abundance also react with other nucleophiles.

5.2.1   BISIMIDOESTERS (BISIMIDATES)

Imidoesters (see Table 5.1.A for general structures) readily react with amino groups eliminating an alcohol to form amidines (Figure 4.7). Because of ease of synthesis,[9,10,14] a large number of bisimidoester reagents with different spacer arms are available as shown in Appendix A. A series of compounds with different chain lengths of the spacer arm from 5 to 14 Å have been synthesized (II through X in Appendix A).[15-23] In addition to dimethyl imidates, diethyl imidates, such as diethyl malonimidate (DEM, I in Appendix A), has also been prepared.[24] To increase the hydrophilicity of the bridge, ether-oxygen-groups have been incorporated into the spacer arm (see XI through XV in Appendix A), which also increases the chain span to 17 Å.[25] In general, these bis-imidoesters are water soluble and react with amino groups with a high degree of specificity under mild conditions (pH 7–10) to form amidine derivatives.[14] A general reaction mechanism is shown in Figure 4.7. The reaction is favored at alkaline pHs, which offset the major competitive hydrolysis at neutral pH.[24,26] In order to maximize the reaction, incremental additions of the reagent are recommended.[11] The side reactions with other groups are negligible.[10] Since this class of compounds react with amino groups, Tris and imidazole buffers should be avoided.

A potential advantage of imidoesters over other amino group–directed cross-linkers is the retention of positive charge on the original amino group. The products are normally stable to mild acid hydrolysis[27] but are reported to be susceptible to hydrolysis by hydrazine[28] or ammonia.[26]

Several cleavable bis-imidoesters (compounds XVI through XXIII in Appendix A) containing the disulfide bond have been synthesized.[25,29-32] These compounds can be cleaved in the presence of thiol containing reagents such as mercaptoethanol. Another series of imidoesters (XXIV through XXVI in Appendix A) containing vicinal diols cleavable by periodate have also been used.[33-35] Other characteristic reporter groups have also been incorporated in the spacer arm. In addition to the cleavable homobifunctional cross-linkers, there are chromogenic reagents and fluorescent compounds with labels at the spacer arm, such as compounds XXVII through XXIX shown in Appendix A.[31,32] The chromogenic compounds, dimethyl-3,3′-(*N*-2,4-dinitrophenyl)bispropionimidate (DDPB) and dimethyl-3,3′-(*N*-2,4-dintro-5-carboxyphenyl)bispropionimidate (DDCB), containing dinitrophenyl group in the spacer were used in studies of tertiary structures of proteins.[36] The reagent, dimethyl-3,3′-[*N*-(5-(*N*,*N*′-dimethylamino)naphthyl)sulfonyl]bispropionimidate (DDNBP), is fluorescent and was demonstrated to cross-link lactate dehydrogenase.[37]

TABLE 5.1
Examples of Amino Group–Directed Homobifunctional Cross-Linking Reagents

A. Bisimidoesters (bisimidates)

General structure:

$$R-O-\overset{\overset{\overset{+}{NH_2Cl^-}}{\|}}{C}-X-\overset{\overset{\overset{+}{NH_2Cl^-}}{\|}}{C}-O-R$$

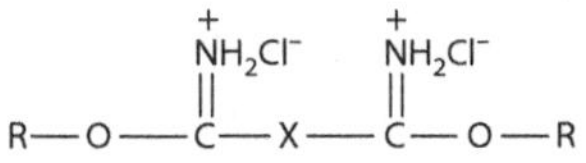

Example: Dimethyl malonimidate·2HCl

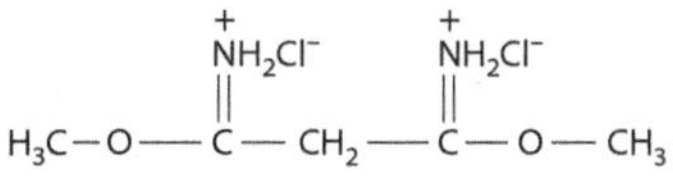

$$H_3C-O-\overset{\overset{\overset{+}{NH_2Cl^-}}{\|}}{C}-CH_2-\overset{\overset{\overset{+}{NH_2Cl^-}}{\|}}{C}-O-CH_3$$

B. Bis-*N*-succinimidyl derivatives

General structure:

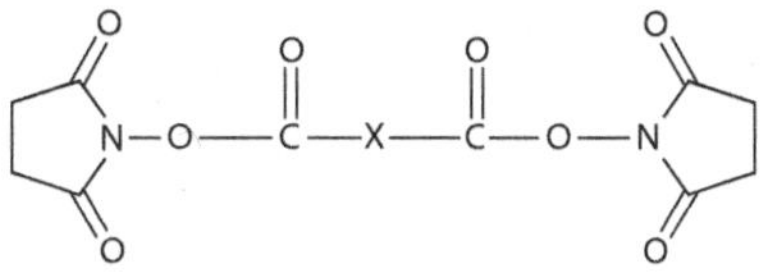

Example: Disuccinimidylsuberate (DSS) (or *N*-hydroxysuccinimidylsuberate (NHS-SA)

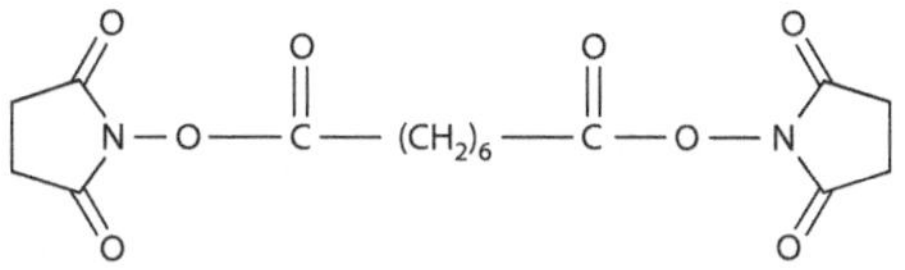

Example: Bis(sulfosuccinimidyl)suberate (BSSS)

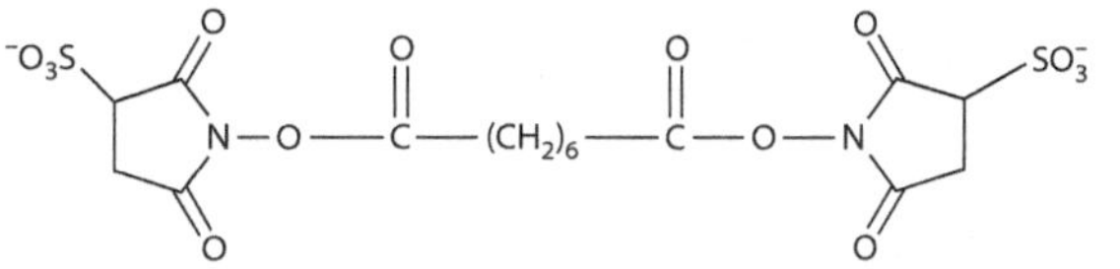

C. Bifunctional aryl halides

General structure:

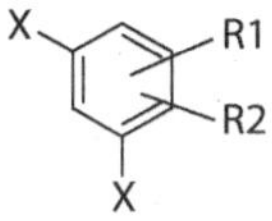

(X = F, Cl, Br; R1, R2 = other substituents)

Example: 1,5-Difluoro-2,4-dinitrobenzene (DFDNB, DFNB)

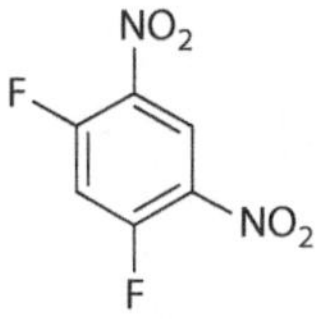

Example: Bis(3-nitro-4-fluorophenyl)sulfone

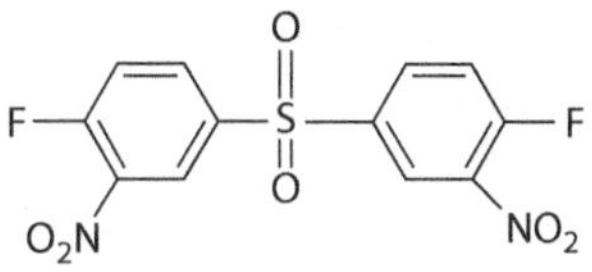

D. Diisocyanates and Diisothiocyanates

General structure: O=C=N-X-N=C=O and S=C=N-X-N=C=S

Example: 1,6-Hexamethylene diisocyanate   O=C=N-(CH$_2$)$_6$-N=C=O

(*continued*)

TABLE 5.1 (Continued)
Examples of Amino Group–Directed Homobifunctional Cross-Linking Reagents

E. Bifunctional sulfonyl halides

General structure:

$$X-\underset{O}{\overset{O}{S}}-R-\underset{O}{\overset{O}{S}}-X \quad (X = halogen; R = other\ groups)$$

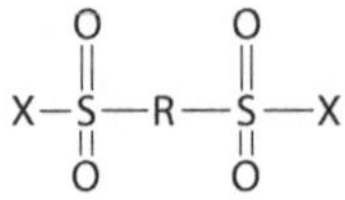

Example: Phenol-2,4-disulfonyl chloride

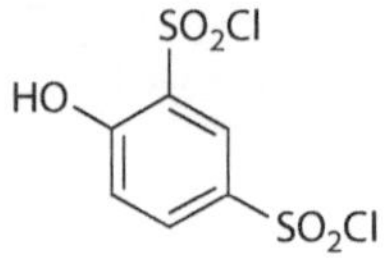

(structure: phenol ring with SO_2Cl at positions 2 and 4, HO on ring)

F. Bisnitrophenol esters

General structure:

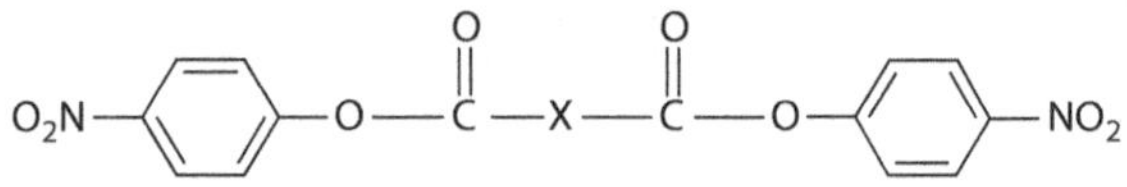

$$O_2N-\!\!\!\bigcirc\!\!\!-O-\overset{O}{\overset{\|}{C}}-X-\overset{O}{\overset{\|}{C}}-O-\!\!\!\bigcirc\!\!\!-NO_2$$

Example: Bis(p-nitrophenyl)adipate

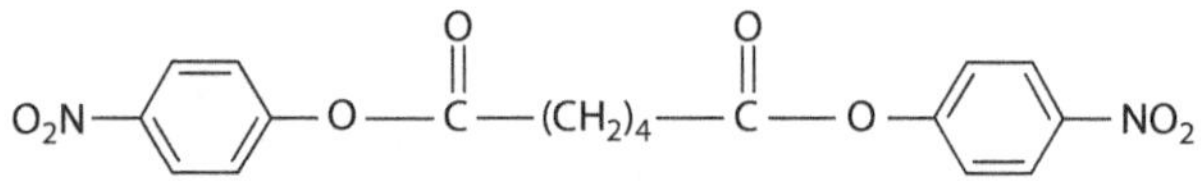

$$O_2N-\!\!\!\bigcirc\!\!\!-O-\overset{O}{\overset{\|}{C}}-(CH_2)_4-\overset{O}{\overset{\|}{C}}-O-\!\!\!\bigcirc\!\!\!-NO_2$$

G. Bifunctional acylazides

General structure:

$$N_3-\overset{O}{\overset{\|}{C}}-X-\overset{O}{\overset{\|}{C}}-N_3$$

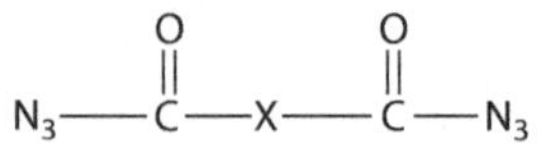

Example: Tartryl diazide

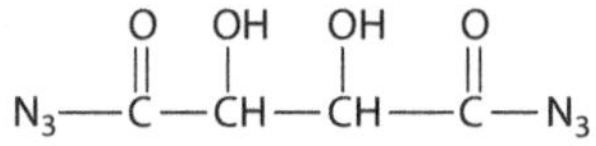

$$N_3-\overset{O}{\overset{\|}{C}}-\overset{OH}{\overset{|}{CH}}-\overset{OH}{\overset{|}{CH}}-\overset{O}{\overset{\|}{C}}-N_3$$

H. Dialdehydes

General structure:

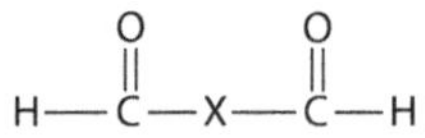

$$H-\overset{O}{\overset{\|}{C}}-X-\overset{O}{\overset{\|}{C}}-H$$

Example: Malondialdehyde

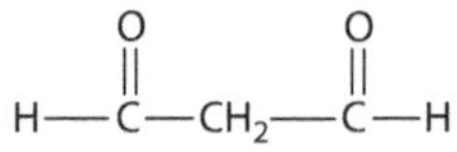

$$H-\overset{O}{\overset{\|}{C}}-CH_2-\overset{O}{\overset{\|}{C}}-H$$

I. Diketones

General structure:

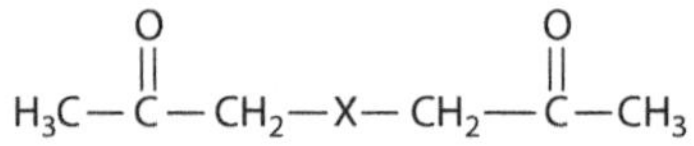

$$H_3C-\overset{O}{\overset{\|}{C}}-CH_2-X-CH_2-\overset{O}{\overset{\|}{C}}-CH_3$$

Example: 2,5-Haxanedione

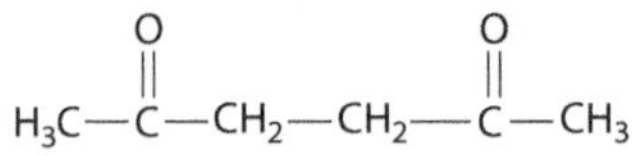

$$H_3C-\overset{O}{\overset{\|}{C}}-CH_2-CH_2-\overset{O}{\overset{\|}{C}}-CH_3$$

It is impossible to cite all of the proteins that have been cross-linked using bisimidoesters. The homologous series with different chain lengths has been used for studying α-crystalline.[23] Cross-linkers, such as dimethyl adipimidate (DMA) and dimethyl suberimidate (DMS), were used to construct low-resolution three-dimensional structures of cytochrome *c* and lysozyme.[38] DMS, dimethyl pimelimidate (DMP), and dimethyl 3,3′-dithiobisproprionimidate (DTBP) were used to study macromolecular self-association of ADP-ribosyltransferase.[39] Other reagents have been used to study proteins in erythrocytes,[11,40,41] synaptosomal membranes,[42] Mengo virus capsids,[43] Rous sarcoma virus,[44] ion-channels,[45] transcription termination factor rho,[46] and ribosomes.[47] Many enzymes such as glycogen phosphorylase b,[48] transhydrogenase,[49] phosphorylase-kinase,[50] lactose synthase,[51] phosphofructokinase,[52] HIV-1 reverse transcriptase,[53] ribonuclease,[54] aldose,[13] glyceraldehyde-3-phosphate dehydrogenase,[13] tryptophan synthetase B protein,[13] L-arabinose isomerase,[13] (Na,K) ATPase,[55] monooxygenases,[56,57] and aspartate transcarbamylase[13] have been cross-linked as well as a series of other proteins, for example, nucleosomal histones,[58] tubulin,[59] IgE,[60] α-tropomyosin,[61] low-density lipoprotein,[62] concanavalin A,[63] hemoglobin,[64,65] collagen,[66,67] acetylcholine receptor,[68] and simian virus induced antigen.[69] These reagents have also been used to introduce cross-links between immunoglobulins and other proteins,[17,70,71] bovine leukemia virus RNA and proteins,[72] copolymers and proteins,[73] lipids and proteins in retrovirus,[74] and DNA and DNA-binding proteins.[75]

5.2.2   Bis-Succinimidyl Derivatives (N-Hydroxysuccinimidyl Esters, NHS Esters)

Bis-N-hydroxysuccinimide esters (NHS) are another group of homobifunctional cross-linking reagents that are amino group selective. A general structure of these compounds is shown in Table 5.1.B. An extensive list of these reagents can be found in Appendix A (compounds XXX through LXXXIII).[76–98] These compounds are easily synthesized by condensing N-hydroxysuccinimide with the corresponding dicarboxylic acids in the presence of dicyclohexylcarbodiimide (Figure 4.6).[99] Theoretically, any homologous series such as those in imidoesters may be synthesized from the available dicarboxylic acids. Thus, a variety of NHS esters have been prepared. These include spacer arms with different numbers of methylene groups (XXXIII through XLV in Appendix A), polyethyleneglycol (compounds L through LIII), and phenyl rings (LXVII through LXXI). Reporter groups have also been incorporated into the spacer arm, such as fluorophores (BiPS, compounds LXXV and LXXVI in Appendix A),[81] spin labels (compounds LXXVII through LXXXII),[82–86] and deuterated isotopes (XXXVI, XL, XLIII, and XLVI).[79,80] Dibenzoyloxysuccinimidylethylphosphate (IRCX, LXXXIII in Appendix A) is a new chromogenic cross-linker for infrared studies.[87] Various bonds have been used for making cleavable NHS esters. As in bis-imidoesters, NHS esters containing disulfide bonds (LVII through LXI in Appendix A) are cleaved by free thiol compounds, those containing vicinal hydroxyl groups (LXII through LXIV) are cleavable by periodate, and those with sulfonyl group by base (LV and LVI). The bimane derivatives, BiPS and BiPS-D8 (LXXV and LXXVI), are photocleavable.[81] Of interest are the two compounds, disuccinimidyl selenodiacetate (DSSDA; LXV in Appendix A) and disuccinimidyl selenodipropionate (DSSDP; LXVI), which contain seleno groups cleavable by mild oxidation with hydrogen peroxide.[97,98] To increase the solubility of NHS esters in aqueous solutions, sulfonate groups have been introduced into the succinimidyl rings as in bis(sulfosuccinimidyl)suberate (BS^3) shown in Table 5.1.B. Additional sulfonated reagents can be found in Appendix A. Incorporation of hydrogen bonding atoms in the spacer arm, such as ester oxygens and hydroxyl groups, also increases their hydrophilicity. Bis-succinimidyl cross-linkers that are sparingly soluble in water may be initially dissolved in an organic solvent such as acetone, dimethylformamide, or dimethyl sulfoxide before addition to the protein solution.[89] A 10-fold molar excess of reagent is generally sufficient to modify all accessible protein amino groups.

The NHS cross-linkers react preferentially with amino groups eliminating N-hydroxysuccinimide as the leaving group. The product of the reaction with an amino group is an amide and therefore resists mild acid hydrolysis. The reaction mechanism is shown in Figure 4.6. The NHS cross-linkers do not react with the thioether of methionine, the indole amine of tryptophan, or the guanidine group of arginine, and there is little reactivity with the hydroxyl groups of serine and tyrosine.[100,101] However, a recent article indicated that there may be undesired side reactions with serine and tryrosine at slightly acidic conditions.[102] The NHS group released has a molar extinction coefficient, at 260 nm, of $8.2 \times 10^3 M^{-1} cm^{-1}$ in pH 9.0 buffers and may be used to monitor the extent of the reaction.[103] The reaction is usually complete within 10–20 min at pH 6–9.[77,89,100–104] Since unprotonated amino groups are the reactive species, an increase in pH will increase the rate of the reaction. However, rapid hydrolysis at alkali pHs effectively competes against the reaction. For example, the half-life of hydrolysis at pH 7.5 and 8.6 is 4–5 h and less than 10 min, respectively.[93] The NHS esters are more stable below pH 6. Although the imidazolyl group of histidine competes with amines, the product is unstable and is readily hydrolyzed. In this manner imidazolyl groups accelerate the hydrolysis.[100] NHS esters, however, are stable for months under anhydrous conditions.

There are widespread applications of these NHS esters. Bis-succinimidyl esters of various chain lengths have been used in studies of the spatial arrangements of muscle proteins and stabilization of α-helices.[78,105] Spin-labeled reagents were used for the study of membranes.[82–86] Cross-linkers with deuterium and ^{15}N isotopes were used in studies of human erythrocytes[86] and for mass spectrometry analysis of cross-linked calmodulin and an adenylyl cyclase peptide.[80] BiPS-D8, a fluorescent photocleavable reagent, was used to investigate the protein structure of RNase S protein complex with MALDI-mass spectrometry.[81] Other reagents have been used to study prothrombin and thrombin,[106] thrombin receptor,[107] platelet proteins,[108] membrane proteins,[109] transhydrogenase,[49] P450 and P450 reductase,[90] ubiquinone-cytochrome *c* reductase,[96] vasoactive intestinal peptide receptor,[110,111] microtubule,[112] DNA topoisomerase II,[113] lymphotoxin receprors,[114] galanin receptor,[115] gonadotropin-releasing hormone,[116] F_1-adenosine triphosphatase,[94] virus induced antigen,[69] HIV Type 1 gp41 glycoprotein,[117] viral proteins,[43,118] prion protein,[119] von Willebrand factor,[120] chorionic gonadotropin receptor,[121] epidermal growth factor-induced receptor,[122,123] endoplasmic reticulum translocase,[124] asialoorosomucoid receptor,[125] and to study conformational changes in serine/threonine kinase Akt activation.[126] These reagents have also been used to enhance fluorescence in histochemical preparations of cytoskeletal proteins.[88]

5.2.3   BIFUNCTIONAL ARYL HALIDES

A general structure of bifunctional aryl halide is shown in Table 5.1.C. The aryl halogens react preferentially with amino and tyrosine phenolic groups, but they also react with the thiol and imidazolyl groups as mentioned in Chapter 3.[127–132] The electron-withdrawing carboxyl and nitro groups on the benzene ring activate the halogen for nucleophilic substitution (see Figure 4.3). However, rapid reaction requires relatively high pH conditions. Fluoro-compounds are the most reactive, followed by chloro-, and then bromo-derivatives. The reaction can be followed by the release of hydrogen halide using a pH stat, or by the characteristic visible and UV spectra of the product. Unfortunately, most of the aryl halides are insoluble in water. They can, nevertheless, be added to an aqueous protein solution after first being dissolved in acetone, dioxane, alcohol, or other water-miscible organic solvents. The lysine and tyrosine adducts are resistant to acid hydrolysis, but alkaline hydrolysis will liberate the reacted amino acids.[133,134] Among the aryl halides, the dibromosalicyl derivatives (LXXXIX through XCV in Appendix A) and bis(3-nitro-4-fluorophenyl)sulfone (XCVI) are cleavable by base or by Ni-catalytic reduction.[27,130–136]

The reactivities of the two halogens on the phenyl ring are different. The first halogen generally reacts faster than the second.[27] For example, the replacement of one of the fluorine atoms of 1,5-difluoro-2,4-dinitrobenzene (DFDNB, Table 5.1.C) decreased the reactivity of the remaining fluorine atom.

A number of proteins have been cross-linked with the bifunctional aryl halides. The series of bis(3,5-dibromosalicyl) derivatives with difference lengths have been used to study hemoglobin.[137–140] Other proteins that have been studied include collagen[135] bovine serum albumin,[132,136] aspartate aminotransferase,[141] (Na,K)-ATPase,[142] thymidylate synthetase,[143] silk,[133] wool,[133] insulin,[144] ribonuclease,[145,146] staphylococcal nuclease,[147] erythrocyte membranes,[148] and retrovirus proteins.[118] These compounds have also been used to form covalently linked conjugates of several pairs of proteins, for example, IgG and ferritin,[149] IgG and horse radish peroxidase,[150] and the coupling of antigens to erythrocytes.[151]

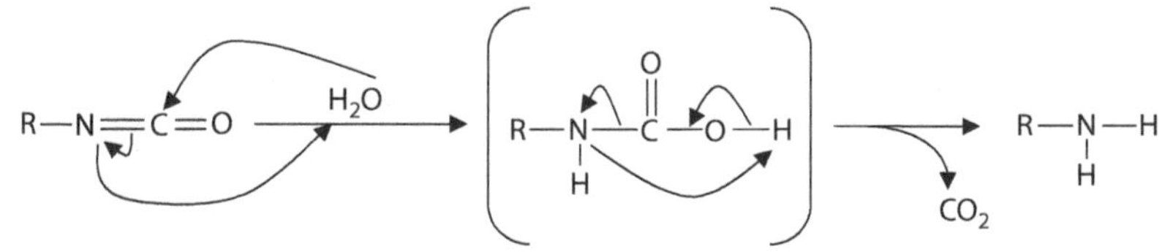

FIGURE 5.1 Mechanism of hydrolysis of isocyanates to free amines.

5.2.4 DIISOCYANATES AND DIISOTHIOCYANATES

Isocyanates and isothiocyanates have similar structures (Table 5.1.D). Most of the diisocyanate and diisothiocyanate cross-linkers are aromatic derivatives (see XCVIII through CVI and CVIII through CXVIII in Appendix A).[152–163] These reagents generally react with amino groups to form stable urea and thiourea derivatives, respectively (see Figure 4.9). Their reactions with sulfhydryl, imidazolyl, and phenolic groups give relatively unstable bonds that undergo spontaneous hydrolysis.[19,27] The isocyanates are not stable in aqueous solutions. Hydrolysis to form free amines is shown in Figure 5.1. The half-life of aliphatic isocyanates at pH 7.6 is less than 2 min.[27] Aromatic isocyanates are somewhat more reactive than the aliphatic ones. There is also a differential reactivity between the two functional groups in mixed isocyanate asymmetric compounds such as toluene-2-isocyanate-4-isothiocyanate (CI in Appendix A). In this case, the isocyanate group is more reactive than the isothiocyanate. Steric hindrance also affects the reactivity. For example, in 3-methoxydiphenylmethane-4,4′-diisocyanate (CVI in Appendix A), the 4′-isocyanate moiety is more reactive than the methoxy group hindered 4-isocyanate group.[27]

Most of the isocyanates and isothiocyanates compounds are insoluble in aqueous solutions; only those with carboxyl and sulfonyl derivatives are water soluble. In addition to the hydrophilic moieties, azo bonds can be incorporated in the spacer arm making the cross-linking reversible (CIX through CXIV in Appendix A). Cleavage of the cross-linked product can be achieved in the presence of dithionite. Of interest is the stilbene derivative, 4,4′-diisothiocyanato-2,2′disulfonic acid stilbene (CXVII in Appendix A), which is fluorescent and which has many applications.[163–165] However, its reduced analog, dihydrostibene (CXVIII in Appendix A), is nonfluorescent.

These bifunctional diisocyanate and diisothiocyanate reagents, having a wide variety of reactivities, sizes, and solubilities, have been used for different purposes. They have been used to study myoglobin,[166] chymotrypsin,[152,167] ribonuclease,[152] elastase,[167] and to prepare cross-linked collagen[168] and gelatin.[169] They have been used to conjugate different protein molecules, for example, IgG and bovine serum albumin,[154] and IgG and ferritin.[154,157]

5.2.5 BIFUNCTIONAL SULFONYL HALIDES

As shown in Table 5.1.E, sulfonyl halides are acylating agents in which the halogen is the leaving group and the attacking nucleophile is sulfonated as discussed in Chapter 4. Most bifunctional sulfonyl chloride cross-linkers are aryl derivatives. These compounds (CXIX through CXXI in Appendix A) react with amino groups to form sulfonamide derivatives (see Figure 4.8).[170,171] Although these reagents themselves are not cleavable, the sulfonamide linkage formed can be reversed with HBr in glacial acetic acid without breaking the peptide bonds.[27] Unfortunately, these reagents are quite insoluble in water and must be dissolved in organic solvents before addition to aqueous solutions where hydrolysis is rather rapid. However, they have been used to cross-link lysozyme[170,171] and insulin.[172] There are other bisulfonyl chlorides commercially available, such as biphenyl-4,4′-disulfonyl chloride, diphenyl ether 4,4′-disulfonyl chloride (4-(4-chlorosulfonylphenoxy)benzenesulfonyl chloride), and 4-amino-6-chloro-benzene-1,3-disulfonyl dichloride, but they have not yet been discussed in the literature.

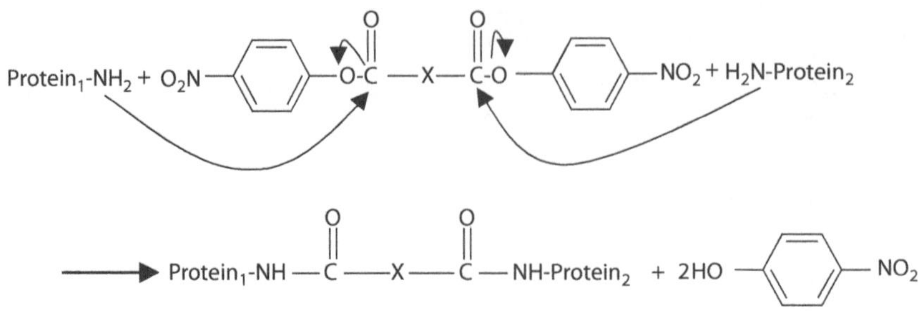

FIGURE 5.2 Cross-linking reaction of bis-nitrophenyl esters.

5.2.6 BIS-NITROPHENYL ESTERS

Nitrophenol is a good leaving group that makes nitrophenyl esters good acylating agents. These esters (see general structure in Table 5.1.F), such as compounds CXXII through CXXV in Appendix A, react with amino groups most rapidly, although their specificity is not very high.[173–177] The reaction involves nucleophilic attack at the ester carbonyl carbons displacing nitrophenol as shown in Figure 5.2. The liberated nitrophenol has a distinct yellow color above pH 7 where the phenolic group is deprotonated. Therefore, absorbance at 410 nm may be used to quantify the reaction. In addition to amino groups, other nucleophiles also react to form acylated derivatives. Hydrolysis is the most important competitive reaction in aqueous solutions.

These compounds can easily be synthesized by coupling nitrophenol to the corresponding dicarboxylic acids and, as with the succinimidyl esters, a large number of diphenyl esters may be prepared.[177] Of particular interest is carbonyl bis(L-methionine p-nitrophenyl ester) (CXXV in Appendix A) synthesized by Busse and Carpenter.[175] This reagent is a derivative of methionine and therefore cleavable by CNBr. It bas been used to cross-link insulin.[176] Other bis-nitrophenyl esters have been used to cross-link silk fibroin, wool, and collagen.[178]

5.2.7 BIFUNCTIONAL ACYLAZIDES

Bifunctional acylazides, like other acylating reagents, react readily with amino groups to produce amide bonds. These compounds (see Table 5.1.G for general structure) were originally designed for photoaffinity labeling but reacted readily in the dark as acylating agents because the azido moiety is a good leaving group. The tartryl diazides with different spacer arm lengths (CXXVI through CXXXI in Appendix A) were developed by Lutter et al.[35] They have different membrane permeabilities and contain vicinal hydroxyl groups that can be cleaved by mild treatment with periodate.[179] These compounds have been used to cross-link various ribosomal proteins,[35] influenza virion,[180] aspartate transcarbamylase,[181] and erythrocyte proteins.[179] The acyl azide derivative, p-bis(ureido) oligoprolylazobenzene (PAPA, compound CXXXII in Appendix A) was synthesized and purified as the hydrazide. Prior to protein cross-linking, the hydrazide was converted to the acylazide by sodium nitrite in 1 N HCl.[182] This compound was tested on hemoglobin for its cross-linking ability. PAPA has also been shown to cross-link ferritin subunits.

5.2.8 DICARBONYL COMPOUNDS

The dicarbonyl compounds as exemplified in Table 5.1.H include dialdehydes, diketones, and compounds containing both aldehyde and ketone moieties. A comprehensive list of these reagents can be viewed in Appendix A (compounds CXXXIII through CLX). The simplest dialdehyde is glyoxal with adjacent carbonyl groups. With various carbon chain spacers inserted between the carbonyls, a homologous series of dialdehydes is formed as represented by compounds CXXXIV through CXXXVII in Appendix A. Methyl glyoxal (CXLVII in Appendix A) is a derivative of glyoxal containing both an aldehyde and a ketone group. These α-dicarbonyl compounds, together with phenylglyoxal, selectively

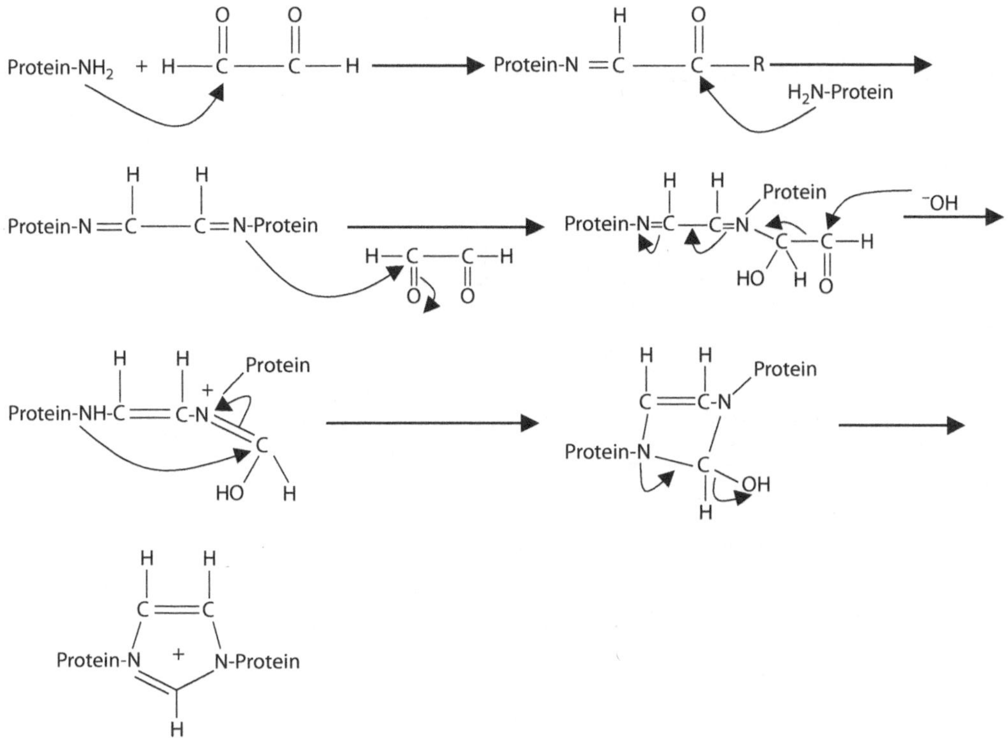

FIGURE 5.3 Speculated protein cross-linking reaction mechanism of glyoxal, an α-carbonyl aldehyde.

react with arginine guanidino groups as discussed in Chapter 3.[183] They may also react with amino groups of lysines to form various cross-linked products. However, their rate of cross-linking reaction varies dramatically such that some, for example, phenylglyoxal, do not yield any cross-linked products.[183] When they do react, these α-dicarbonyl compounds, together with other dialdehydes and diketones, either aliphatic or cyclic, react with amino groups by forming Schiff bases following the initial Maillard reaction as discussed in Chapter 3.[184–186] Since the rates of the reactions vary dramatically and different cross-linked end products are obtained with different reagents, the molecular mechanism of the reaction remains elusive.[184] For α-dicarbonyl compounds, it has been speculated that a second dicarbonyl molecule is added to the initially formed diimine as shown in Figure 5.3 for glyoxal. The end result is a complex cross-linked imidazolium compound.[184,185] On the other hand, in the reaction with spacer-separated dialdehydes, such as 1,6-hexanedial, the molecular mechanism involves cyclization of the initial Schiff base formed with the first amino group. A second amino group then attacks the heterocyclic ring to form a pyridium product as indicated in Figure 5.4.[184] The reaction mechanism is similar for diketones as well as for cycloalkane dicarbonyls.[184] For example, 2,5-hexanedione and its derivatives, after forming a Schiff base with the amino group of lysine, cyclizes to form a pyrrole. A second pyrrole then condenses to effect the cross-linking of the proteins.[187–189]

All the dicarbonyl compounds listed in Appendix A (CXXXIII through CLX) have been shown to cross-link proteins.[184–226] The most extensively used reagent is probably glutaraldehyde.[191–214] Conceivably the reaction would proceed through a Schiff base. However, the product formed is irreversible. Glutaraldehyde has been found to form polymers in solution. At acidic pH, the polymers are cyclic hemiacetals. At neutral or slightly alkaline pHs at which cross-linking is carried out, the α,β-unsaturated aldehyde polymers are formed, which increases in length as the pH is raised.[193,211] Presumably the unsaturated polymer cross-links the amino groups of proteins as illustrated in Figure 5.5.[192,194,212] Interaction of the Schiff base with adjacent double bonds provides the stability toward hydrolysis. With excess amino groups, nucleophilic addition of the ethylenic double

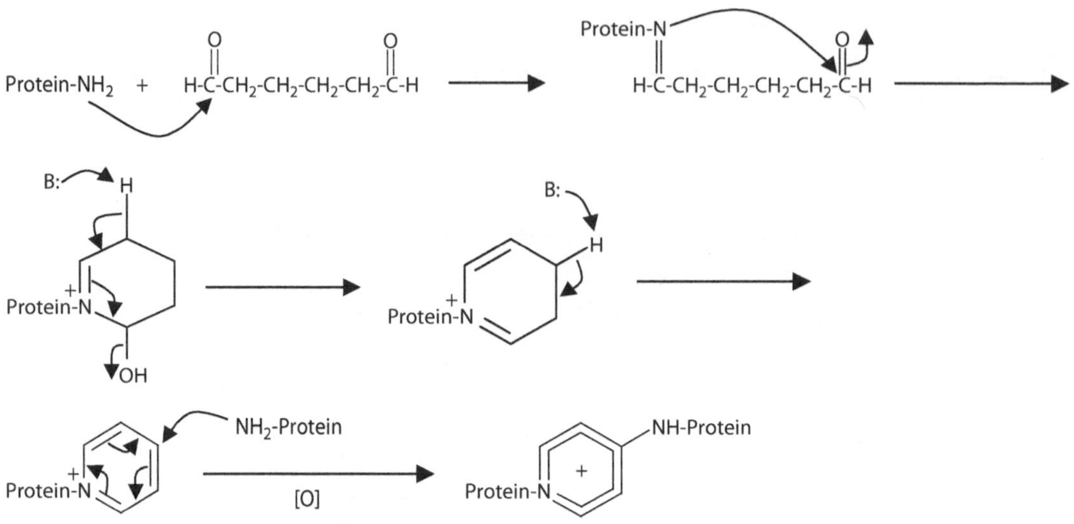

FIGURE 5.4 Speculated protein cross-linking reaction mechanism of a dialdehyde, 1,6-hexanedial.

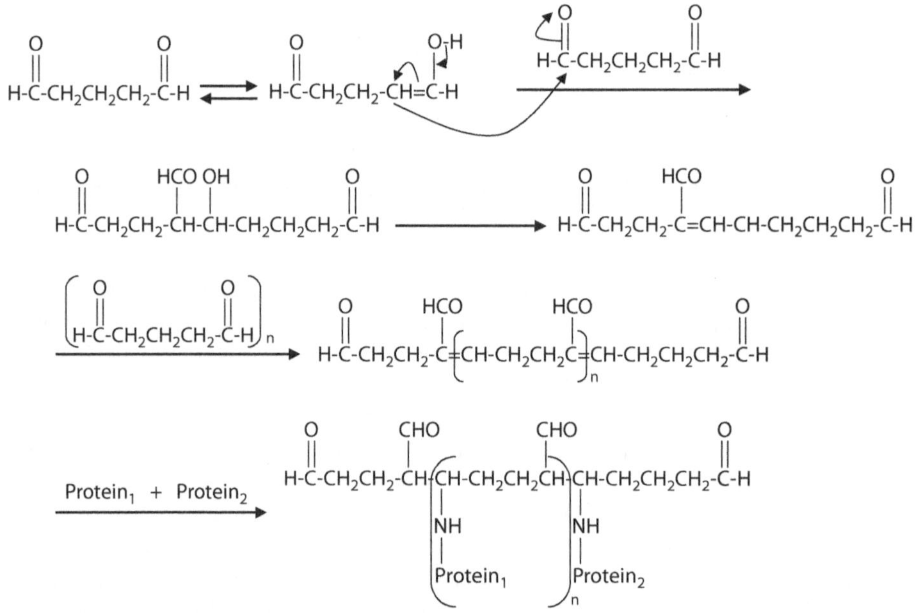

FIGURE 5.5 Polymerization of glutaraldehyde and its cross-linking reaction with proteins.

bond is possible.[10] Because of the different polymeric forms, the distance between two cross-linked groups cannot be estimated. Recently, several cross-linked products have been detected by mass spectrometry,[212,213] which led Wine et al.[212] to suggest that different mechanisms of cross-linking reactions may take place under different alkaline or acidic conditions.

Glutaraldehyde has been used extensively to cross-link and stabilize proteins and enzymes in solution and in crystals and has also been used for the immobilization of proteins.[192–214] Catalase,[199] hemoglobin,[200–202] phosphofructokinase,[203] lactate dehydrogenase,[204] insulin,[205] photosystem polypeptide,[208] gelatin,[213] and glutamate dehydrogenase[214] have been studied with glutaraldehyde. Crystalline carboxypeptidase and β-lactoglobulin have been cross-linked with glutaraldehyde.[195–197] Glutaraldehyde has also been used to form enzyme-immunoglobulin conjugates for immunoassays[206,207] and to immobilize proteins onto solid supports.[209,210]

Other glutaraldehyde derivatives have also been found to cross-link proteins. Two glutaraldehyde derivatives, 3-methylglutaraldehyde and 3-methoxy-2,4-dimethylglutaraldehyde (CXXXIX and CXL in Appendix A), have been used in tanning and to cross-link albumin and casein.[214–217] o-Phthalaldehyde, a fluorogenic aromatic dialdehyde (CXLI in Appendix A), has been demonstrated to cross-link components of the cytochrome b6f complex.[218] Pyridoxal phosphate, which has been used as a monofunctional reagent to modify lysine residues at the phosphate-binding sites, has been dimerized into various forms (see CXLII through CXLVI in Appendix A). These compounds contain the pyrophosphate bond, which may be hydrolyzed in acid or base. They have been used to cross-link glycogen phosphorylase b and hemoglobin.[219,220]

In addition to the specific examples given above, the dicarbonyl compounds, including the diketones, both aliphatic and cyclic, have been shown to cross-link proteins and DNAs.[221–226] Malondialdehyde, for example, has been found to cross-link histones to DNA[190] and its cross-linked proteins such as lysozyme, cytochrome c, and ribonuclease have been studied for their immunologic epitopes.[223] Other bifunctional aldehydes have also been found to interact with DNAs.[221] 2,5-Hexanedione, as mentioned above for studies on ribonuclease, is found to induce neuropathy by cross-linking neurofilaments.[185,186,224] 2,3-Butanedione has also been used to study protein cross-linking by the Maillard reaction.[184] Dehydroascorbic acid (CLX in Appendix A), an oxidized form of ascorbic acid, is actually a triketone and has been found to be involved in the Maillard-type chemistry that leads to protein cross-linking.[226]

Dialdehydes can be generated from carbohydrates. Schoevaart and Kieboom[227] used D-galactose oxidase to oxidize D-galactose to give a dialdehyde that occurs as two bicyclic structures in aqueous solution as established by NMR spectroscopy. The dialdehyde is very reactive and has the potential to cross-link gelatin. Similarly, D-galactose oxidase is able to convert lactose, lactylamine, and lactobionic acid into their corresponding 6'-aldehyde compounds, which can cross-link proteins to yield stable products after reductive amination.[228]

5.2.9   OTHER AMINO GROUP–REACTING CROSS-LINKING REAGENTS

There are several other reagents that do not readily fit into any of the categories of amino directed cross-linkers listed above but which have been implicated to react with amino groups. Among these, benzoquinones such as p-benzoquinone (also known as p-quinone; compound CLXI in Appendix A) have been used in protein–protein and enzyme antibody coupling.[229–231] The quinone can be derived from hydroquinone in the presence of a simple Cu(II) salt.[231] The mechanism of the reaction follows a two-step pathway as shown in Figure 5.6. First, the nucleophile is added to the benzoquinone to form an adduct that undergoes oxidation to generate a substituted benzoquinone. A second nucleophile is added in the subsequent step to yield a cross-linked product. In addition to

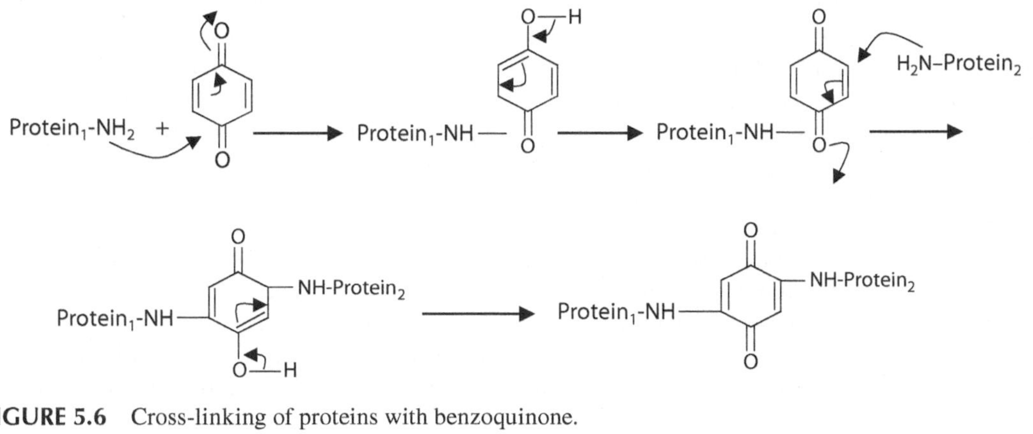

FIGURE 5.6   Cross-linking of proteins with benzoquinone.

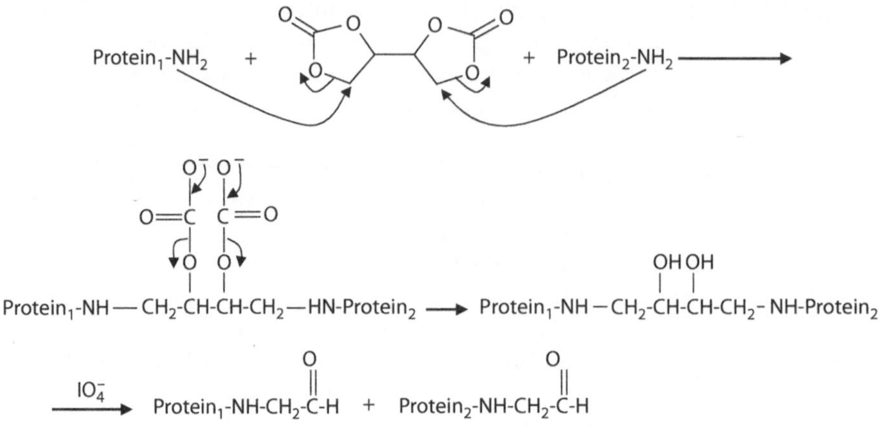

FIGURE 5.7 Reaction mechanism of erythritolbiscarbonate and the cleavage of the cross-linked product.

lysine, hydroxylysine and histidine in gelatin have been found to react with benzoquinone to affect cross-linking.[231] Many other quinone derivatives have been studied including benzoquinone alkylating agents such as benzoquinone mustards. These compounds react differently with thiol groups and will be considered below under sulfhydryl group–directed alkylating agents. Most of them are also involved in DNA cross-linking, which will be discussed in detail later in this chapter.

Erythritolbiscarbonate, erythritol-1,2:3,4-dicarbonate (EBC, CLXII in Appendix A), is commercially available and has been used as a bifunctional reagent that reacts with amino groups at pH ranges 7.5–9.5. The reaction starts with the addition of the amino groups to yield bis-carbonate as shown in Figure 5.7. After decarboxylation to form vicinal diols, the cross-linked product may be cleaved by periodate.[95] The use of this compound is limited except in the preparation of patented cross-links.[232,233]

Mucobromic and mucochloric acids (CLXIII and CLXIV in Appendix A) have been shown to cross-link gelatin.[234] These compounds contain a carboxyl and an aldehyde group that can exist in a cyclic lactone. In this form, amino groups attack the carbonyl carbon to form an amide bond. Schiff base formation of the aldehyde with another amino group completes the cross-linking reaction as shown in Figure 5.8. Michael addition of a nucleophile to the double bond is another possible mechanism for the cross-linking reaction.

Genipin (CLXXI in Appendix A) has recently been shown to cross-link macromolecules.[235–245] It is a hydrolytic product of geniposide, which is found in the fruit of *Gardenia jasminoides* Ellis. Genipin was shown to be effective at improving the stability of collagen-based biomaterials with less cytotoxicity and reduced *in vivo* inflammatory response.[245] The genipin cross-linked material has been used as a bioadhesive, in wound dressing materials, as a bone substitute, and other applications.[236–245] Studies have been conducted to investigate genipin cross-linked gelatin as a conduit for peripheral nerve regeneration,[245] genipin cross-linked chitosan for wound dressing and drug delivery,[238,239] and casein for protein drug delivery.[236] The cross-linking mechanism of genipin is not well understood. It is postulated that primary amines are involved. Butler et al.[246] proposed two separate reactions involving different sites on the genipin molecule for the formation of cross-links between primary amino groups. The fastest reaction occurs with a nucleophilic attack of the macromolecular amino group on the genipin C3 carbon atom, leading to the formation of a heterocyclic compound of genipin linked to the macromolecule, such as gelatin. The slower second reaction is the nucleophilic substitution of the ester group on genipin to release methanol to form a secondary amide link with the macromolecule. Acid catalysis is necessary for one or both of the reactions to proceed. This reaction causes an opening of the dihydropyran ring. An attack on the resulting aldehyde group by the second amino group results in cross-linking between the macromolecules. The reaction is complicated by the oxygen radical-induced polymerization of genipin that occurs once the heterocyclic compound has formed.

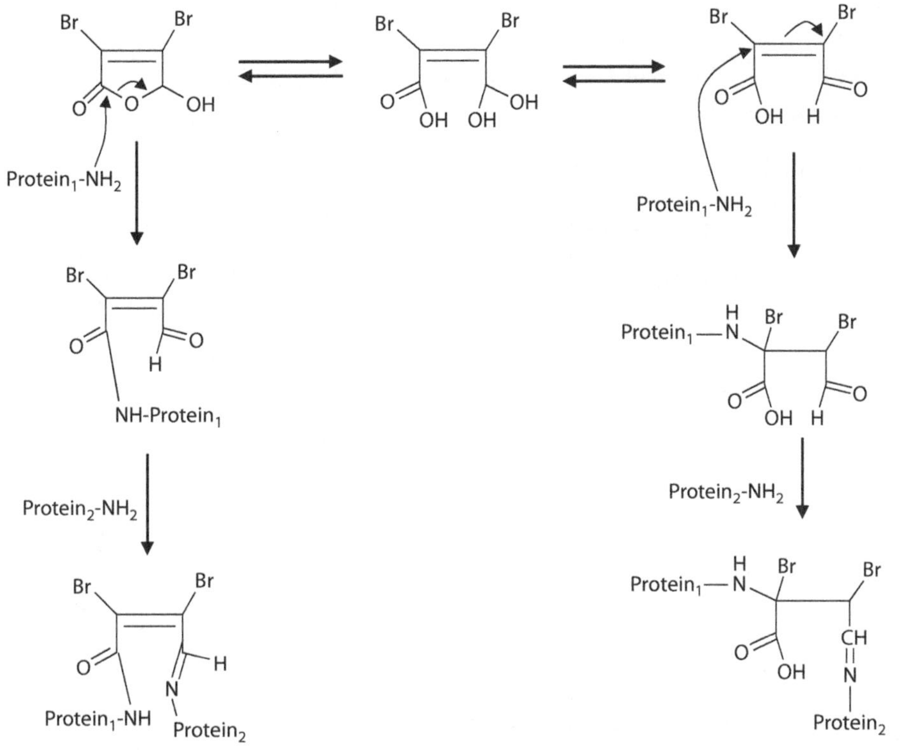

FIGURE 5.8 Cross-linking reaction mechanism of mucobromic acid.

Numerous other natural compounds have been found to cross-link proteins, but their reaction mechanisms are unknown, although amino groups may be involved. These compounds are usually discovered from pathologic conditions and have not been used for chemical studies as cross-linking reagents. For example, various compounds of the intermediate metabolites such as homogentisic acid (CLXVII in Appendix A) have been shown to cross-link collagen preparations.[247–249] However, they have not yet been used as cross-linking agents for chemical studies.

Many compounds discussed below under sulfhydryl group–directed reagents also react with amino groups. For example, a series of polyepoxides like glycerol polyglycidyl ethers have been used in preparing cross-linked collagen and other applications.[250,251] In most cases, adjustment of reaction conditions may direct the reaction from sulfhydryl to amino groups.

5.3 SULFHYDRYL GROUP–DIRECTED CROSS-LINKERS

The sulfhydryl cross-linking reagents consist of derivatives of mercury, maleimide, disulfide, halomethylketone, and other alkylating agents as listed in Table 5.2. A comprehensive list of these compounds that have been reported in the literature can be found in Appendix B. Of these reagents, only the mercurial derivatives, disulfide- and disulfide-forming compounds are truly thiol specific. *N*-maleimido derivatives, although considered as sulfhydryl group directed, may react with other nucleophiles. This qualification is also true for other alkylating agents. Since the thiol is the most potent nucleophile at neutral pH, these reagents are considered sulfhydryl selective at these pHs. It should be borne in mind that they are not thiol specific and that amino groups may be the target of modification, particularly at higher pHs. Thus, adjusting the coupling condition may change the reaction specificity of these reagents. In the absence of a free sulfhydryl group, other amino acid nucleophiles, such as the amino group, become important and would be subject to cross-linking.

TABLE 5.2
**Examples of Sulfhydryl Group–Directed Homobifunctional
Cross-Linking Reagents**

A. Mercurial reagents

 General structure: $^+Hg\text{-}CH_2\text{-}X\text{-}CH_2Hg^+$

 Example: 1,4-Bis(bromomercurial)butane $Br^{-+}Hg\text{-}(CH_2)_4\text{-}Hg^+Br^-$

B. Disulfide–forming agents

 General structure: R-S-S-X-S-S-R and HS-X-SH

 Example: 5,5′-Pentamethylene bis(methanethiosulfonate)

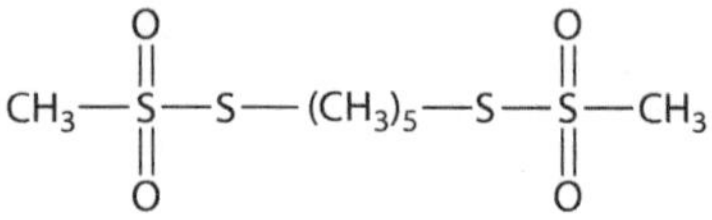

C. Bismaleimides

 General structure:

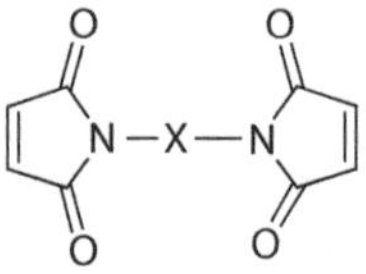

 Example: *N,N′*-Methylenebismaleimide

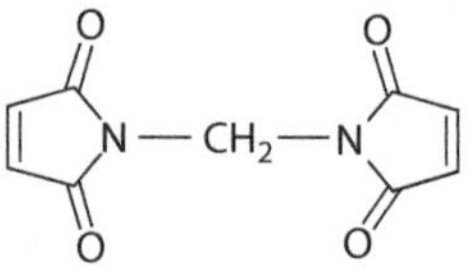

D. Bis-haloacetyl derivatives

 General structure:

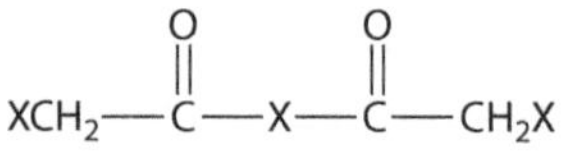

 Example: 1,3-Dibromoacetone

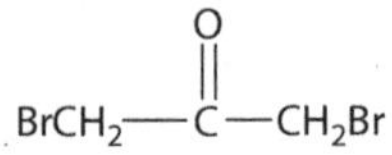

E. Di-alkyl halides

 General structure: $XCH_2\text{—}R\text{——}CH_2X$

 Example: Di(2-chloroethyl)sulfone

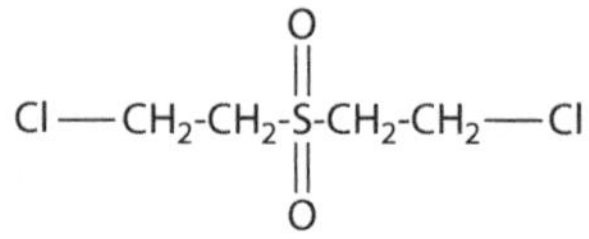

F. Dichloro-*s*-triazines

 General structure:

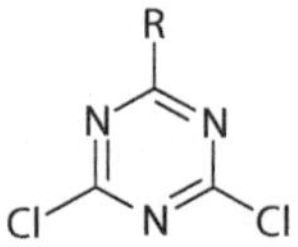

 Example: 2,4-Dichloro-6-methoxy-*s*-triazine

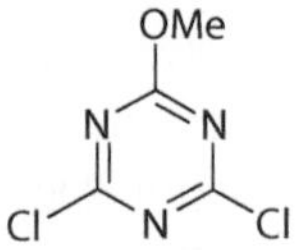

TABLE 5.2 (Continued)
Examples of Sulfhydryl Group–Directed Homobifunctional
Cross-Linking Reagents

G. Aziridines

General structure:

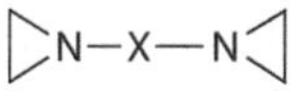

Example: *N,N′*-Ethyleneiminoyl-1,6-diaminohexane

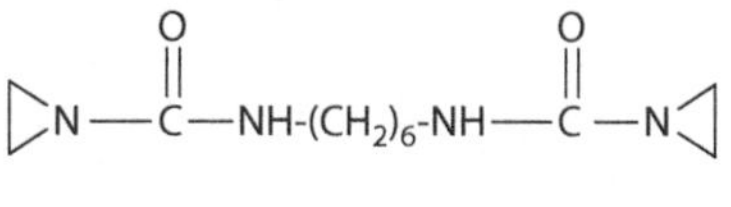

H. Bis-epoxides

General structure:

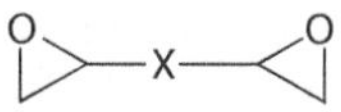

Example: 1,2:3,4-Diepoxybutane

I. Other: Divinyl Sulfone

Structure:

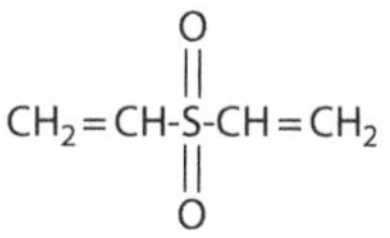

$$CH_2=CH\text{-}S\text{-}CH=CH_2$$

5.3.1 MERCURIAL REAGENTS

Mercuric ion (Hg^{2+}) reacts reversibly with sulfhydryl groups.[252–256] It is essentially a monofunctional reagent. However, because of its divalent nature, it functions as a bifunctional reagent, inserting between two sulfhydryl groups. The first thiol group reacts very quickly followed by a slower reaction with a second sulfhydryl group. Cross-linking occurs when the ratio of Hg^{2+} ion to –SH groups is 0.5 or less. Intramolecular cross-linking has been prepared with reduced papain,[253] albumin,[252] and pancreatic ribonuclease.[254] Intermolecular cross-links have been demonstrated with mercaptalbumin[256] and *Escherichia coli* DNA polymerase.[255]

Di-mercurial compounds have been prepared for cross-linking sulfhydryl groups. 3,6-Bis-(mercurimethyl)dioxane with different counterions of acetate, chloride, or nitrate (see compounds II through IV in Appendix B) has been prepared from allyl alcohol and the corresponding mercury(II) salt.[257] These compounds react similar to but faster than the free mercuric ion. They also introduce a linkage distance of 15 Å versus about 5 Å for the mercuric ion. These bifunctional mercurials have been used to prepare mercaptoalbumin dimers[257,258] and for cross-linking bovine serum albumin, ovalbumin, and ribonuclease with antibodies against these proteins.[259]

Another mercurial derivative, 1,4-bis(bromomercuri)butane (Table 5.2.A), was synthesized by Vas and Caanády by reacting $HgBr_2$ and 1,4-dibromobutane using the appropriate Grignard reagent.[260] The mercurial compound was found to inhibit 3-phosphoglycerate kinase by cross-linking two fast-reacting thiol groups.

5.3.2 DISULFIDE-FORMING REAGENTS

The cross-linking reaction of this group of reagents (see Table 5.2.B) is based on disulfide–thiol interchange. Such an interchange can take place between free sulfhydryl groups on the protein and disulfide bonds of a cross-linking reagent. Conversely, the interchange can also occur between protein disulfides and free sulfhydryls of cross-linkers. Examples of compounds that fall into the first category are compounds VI through XXX listed in Appendix B.[82,83,261–274] There are additional reagents commercially available but they have not yet been used as chemical cross-linkers, for example, 3,6,9-trioxaundecane-1,11-diyl dimethanethiosulfonate and 3,6,9,12-tetraoxatetradecane-1,14-diyl dimethanethiosulfonate. Radioactive analogs labeled with ^{35}S and ^{14}C have also been synthesized.[83,267] The products of the reaction, disulfide derivatives, are easily cleaved in the presence of excess free mercaptans. For the alkylene bis(5,5′-dithio-2-nitrobenzoic acid) series (compounds VI through XIV in Appendix B), the 2-nitro-5-thiobenzoate group (NTB) is replaced by a protein sulfhydryl group. The release of NTB can be used to monitor the reaction at 412 nm where NTB has a molar extinction coefficient of 14,150 M^{-1} cm^{-1}.[262] With 3–12 methylene groups in the spacer arm, these compounds have molecular lengths ranging from 8.6 to 20.3 Å.[262] They have been used as molecular rulers to measure the distance between SH1 and SH2 in rabbit skeletal myosin subfragment 1.[261,262] The spin-labeled compound (compound XIV in Appendix B) has been used to cross-link hemoglobin and to incorporate spin labels into membranes similar to amino group selective spin labels LXXVII through LXXXII shown in Appendix A.[82,83]

Instead of a nitrobenzoic acid, pyridylthio group has been incorporated in the cross-linkers as a good leaving group (XXVIII through XXX in Appendix B). During the disulfide interchange reaction, two pyridine-2-thione molecules are released, which can be used to follow the course of the reaction by monitoring the absorbance at 343 nm. Pyridine-2-thione has a molar extinction coefficient of 8.08 × 10^3 M^{-1} cm^{-1} at 343 nm.[263] 1,4-Di(3′,2′-pyridyldithio)propionamidobutane (DPDPB, compound XXX in Appendix B) has been used to investigate the relationship of ribosomal protein L7/L12 and L10,[263] in the preparation of Filgrastim–Transferrin conjugates,[264] and for immobilizing protein complexes to gold substrates.[265]

In polymethylene bis(methanethiosulfonate) reagents (compounds XVII through XXV in Appendix B), methanethiosulfonate serves as the active moiety of the molecule. The reaction with thiols involves disulfide formation and elimination of methanesulfinate, which is subsequently oxidized to methane sulfonic acid (Figure 5.9). Although cleavable by thiols, the cross-link is stable to CNBr and tryptic digestion. This stability allows identification of the modified residue with routine procedures. These cross-linkers were developed by Bloxham et al. for studying pairs of thiol groups in lactate dehydrogenase, pyruvate kinase, phosphofructokinase, and glyceraldehyde-3-phosphate dehydrogenase.[266,267] They can be synthesized by reacting dibromoalkanes with potassium methane thiosulfonate in refluxing methanol. Radioactive analogs have also been synthesized.[267] The series of bismethanethiosulfonates with varying methylene groups from 1 to 12 (compounds XVII through XXV in Appendix B) as well as reagents containing various ether groups in the spacer (compounds XXVI and XXVII) have been used as molecular rulers to study the oligomeric structure of folate carrier,[268] the transmembrane structure of ATP synthase,[269] conformational dynamics of G- and F-actin,[270] the transmembrane structure of lactose permease,[271,272] the structure of acetoacetyl coenzyme A thiolase,[273] and the cofilin binding site on G-actin.[274]

An example of the second type of disulfide–thiol exchange cross-linker is crabescein (compound XXXI in Appendix B), which is a fluorescent derivative of fluorescein containing two free sulfhydryls. The compound has been shown to add across the disulfide bonds of reduced antibodies.[275] It was used to measure the rotational correlation time of IgG. In addition to crabescein,

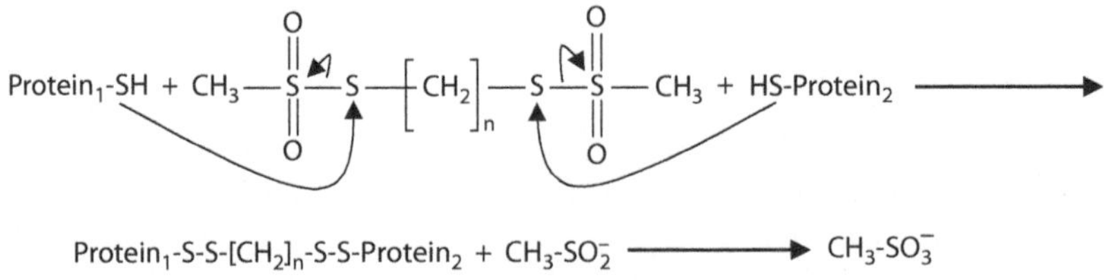

FIGURE 5.9 Cross-linking reaction of polymethylenebis(methanethiosulfonate).

presumably any compound with two free sulfhydryl groups are potential cross-linkers as long as they can undergo a disulfide–thio exchange with disulfides in target molecules.

With the introduction of extrinsic sulfhydryl groups into proteins, as discussed in Section 2.3.2, the scope of disulfide-forming cross-linking reagents can be further broadened. For example, intrinsic protein sulfhydryls and extrinsic thiols may be cross-linked.[10]

5.3.3 BISMALEIMIDES

The most commonly used sulfhydryl reagents are probably the N-substituted bismaleimide derivatives (see general structures in Table 5.2.C).[28,276–310] A comprehensive list of the bismaleimides that have been described in the literature is provided in Appendix B (compounds XXXII through LXI). These homobifunctional cross-linkers are generally synthesized by condensation of the corresponding amine and maleic anhydride with either acetic anhydride or dicyclohexyl carbodiimide.[300,301] They react readily at pH 7–8 with thiol groups to form sulfides through Michael addition. They also react at a much slower rate with amino and imidazolyl groups. At pH 7, for example, the reaction with simple thiols is about 1000-fold faster than with the corresponding amines.[302] The characteristic absorbance change in the 300 nm region associated with the reaction provides a convenient method for monitoring its progression.[304] Since these reagents are generally insoluble in water, they may be added to the aqueous protein solution as solids or after dissolving in water-miscible organic solvents. These compounds are stable at low pHs but are susceptible to hydrolysis at high pHs. Maleimido groups attached to benzene rings are also labile at neutral pH.[304,305]

A few dimaleimide compounds are cleavable. Azophenyldimaleide containing an azo bond (compound LV in Appendix B) can be cleaved by dithionite. Maleimidomethyl-3-maleimido propionate (MMP) (compound XLI) with an ester linkage can be hydrolyzed by base. Dithiobismaleimidoethane (DTME) (compound XLVIII) with a disulfide linkage can be cleaved with free thiols, and 1,4-dimaleimidyl-2,3-dihydroxybutane (BMDB) (compound XLVI) with vicinal hydroxyl groups can be cleaved with periodate. Those with acetal, ketal, and ortho ester bonds (compounds XLII through XLV) are susceptible to hydrolysis at acidic pH values.[284]

There are also compounds containing reporter groups that are fluorogenic. The stilbene compounds, 4,4′-dimaleimidostilbene and 4,4′-dimaleimidylstilbene-2,2′-disulfonic acid (DMSDS) (compounds LIII and LIV in Appendix B), have a low fluorescence until both maleimides have reacted.[282,306] Maximum absorbance is about 316 nm. After reaction with thiols, the emission maximum shifts to around 385–430 nm. With the combined use of FRET and cross-linking with DMSDS, Chantler et al.[307] were able to estimate the distance between the two myosin heads. Similar to the dimaleimidylstilbenes, naphthalene-1,5-dimaleimide (NDM, compound LVI in Appendix B) becomes highly fluorescent upon addition of thiol and has been used to study myosin subfragment 1.[308,309] Keillor et al.[296,298] have developed a series of maleimide fluorogens that contain two maleimide groups attached directly to the fluorescent cores (compounds LVII through LXI in Appendix B). Like NDM, these fluorogens do not fluoresce until the maleimide groups have undergone their typical thiol addition reaction. In this way, the compounds serve as potential fluorescent protein sulfhydryl group labeling agents. The coumarin derivatives have been tested successfully with recombinant proteins.[296]

Compound LV in Appendix B, 4,4′-azobenzene dimaleimide (APDM), is of special interest. It is a photochromic cross-linker and undergoes cis–trans photo-isomerization upon UV/VIS light irradiation.[297] During this transition, there is a change in molecular length between 5 and 17 Å. It has been used to study skeletal muscle myosin.

The dimaleimide cross-linkers have numerous applications. They have been used extensively to study the conformational changes of myosin subfragment 1,[280,282,287,289,292,294,297,307–309] the structure of E. coli ribosome,[279] organization of oligomeric tryptophan synthetase,[277,278] phosducin-induced transducin alteration,[288] Na+/K+ exchanger of rod photoreceptor,[291,299] 3-hydroxy-3-methylglutaryl-CoA lyase,[286] bovine heart mitochondria proteins,[293] erythrocyte membrane,[283,290] chromosome banding,[306] and membrane components.[281,284,291] They have also been used to prepare bispecific antibody dimer against carcinoembryonic antigen and ErbB-2,[295] bovine serum albumin dimer,[276] and coupling of immunoglobulins to enzymes for immunoassays.[285,304] The series of dimaleimides with different number of methylene spacer groups (compounds XXXII through XXXVII in Appendix B) and those with phenyl groups (compounds XLIX through LV) have been used as molecular rulers.[262] The S–S distances with such cross-linkers range from 9.39 Å (for oPDM; L) to 12.31 Å (for NDM; LVI). They are frequently used owing to their high reactivity and specificity and the perceived lack of length overlap among them.[265,292,296]

5.3.4   BIS-HALOACETYL DERIVATIVES

The haloacetyl-containing cross-linkers (compounds LXII through LXXXIII in Appendix B) react primarily with sulfhydryl, imidazolyl, and amino groups.[311–321] For example, 1,3-dibromoacetone (Table 5.2.D, compound LXII in Appendix B) cross-links histidine and cysteine residues at the active sites of ficin and stem-bromelain.[311] At neutral pH, the reagents can be considered as thiol specific, since the sulfhydryl group is the most nucleophilic group at that pH. At higher pHs, however, reactions with amino groups become prominent. The reaction with proteins can be readily followed in a pH-stat by the production of halogen acids. The azo-containing compounds (LXXV through LXXVIII in Appendix B) also have maximum absorbance at 370 nm, which can be used to monitor the reaction.[27] Furthermore, they are cleavable. Quantitative cleavage can be achieved by reduction of the azo group with dithionite.[318] Similarly, the disulfide-containing compound, DIDBE (LXXIX in Appendix B) is cleavable through thio-disulfide exchange with mercaptans.[316] There are several fluorescent diiodoacetyl cross-linkers (compounds LXXX through LXXXIII in Appendix B). The fluorescein derivative (compound LXXX), which fluoresces with an emission at about 552 nm when excited at 496 nm, was once commercially available but does not appear to be available presently.[317] It has not been used in chemical studies. Instead, the fluorescent rhodamine derivative, BSR (compound LXXXII in Appendix B), is available and has been used to determine the orientation of kinesin bound to microtubules in the presence of a nonhydrolyzable ATP analog[319] and to label regulatory light chains of chicken gizzard myosin for fluorescence polarization experiments.[320] Another fluorescent diiodoacetyl reagent is 4,4-difluoro-3,5-di(iodoacetamidomethyl)-4-bora-3a,4a-diaza-s-indacene, BODIPY-FL (compound LXXXIII in Appendix B) which has been used to investigate two nearby cysteine residues of calmodulin by fluorescence anisotropy measurements.[321]

Various proteins have been cross-linked by these bis-haloacetyl compounds for structural studies. These include H-meromyosin,[27,318] chymotrypsin,[313] aldolase,[312,316] IgG,[315] and tubulin.[316]

5.3.5   DI-ALKYL HALIDES

In alkyl halides (see general structure in Table 5.2.E) the halogen leaving group is activated by groups other than the carbonyl in acetyl halides. For benzyl halides (compounds LXXXIV and LXXXV in Appendix B), the halogen is activated by the benzene ring through resonance and reacts in a manner similar to haloacetyl compounds.[322] They have been used to intermolecularly cross-link lysozyme.[322]

The halogen atoms beta to the sulfur and nitrogen form the S- and N-mustards, respectively, and are readily replaced by nucleophiles. Alkylation of thiol groups in proteins is favored at neutral pH, although reactivity toward other nucleophiles increases with increasing pH.[323–327] Compounds containing the sulfone group as in di(2-chloroethyl)sulfone (Table 5.2.E) are cleavable and may be hydrolyzed under basic conditions. These nitrogen and sulfur mustards have been used to cross-link collagen,[328] bacteriophage,[329] wool fibers,[95] hemoglobin,[323–325] serum albumin,[323–325,329] keratine, insulin, gelatin, pepsin, egg albumin, hexokinase, protein components of tobacco mosaic virus, chymotrypsinogen,[323,324] chymotrypsin,[313] serum globulin, fibrinogen,[323,324,330] and ovalbumin.[325] The sulfur mustard, di(2-chloroethyl)sulfide, has been implicated in DNA cross-linking, which may account for its toxicity.[331] Nitrogen mustards attached to benzoquinone are known as benzoquinone alkylating agents or benzoquinone mustards (see Table 5.5.VIII.F.6), which have been shown to induce protein–DNA cross-links in K562 nuclei and cells.[332] These compounds and many other nitrogen mustards are implicated in DNA cross-linking. They will be discussed in detail below under Section 5.11: "Nucleic Acid Cross-linking Reagents."

Dibromobimane (BD, compound CI in Appendix B) is a fluorescent cross-linker. However, like 4,4′-dimaleimidostilbene and 4,4′-dimaleimidylstilbene-2,2′-disulfonic acid (DMSDS) (compounds LIII and LIV in Appendix B), it has minimal fluorescence until both alkyl bromides have reacted generating an absorption maximum at 385 nm and a fluorescent peak at 477 nm.[333] It has been used to cross-link thiols in myosin,[334] hemoglobin,[335] E. coli lactose permease,[336] ATPase,[337] and P-glycoprotein.[338] It has also been shown to intramolecularly cross-link thiols in a complex of nebulin and calmodulin.[339]

5.3.6 Chloro-s-Triazines

Chloro-s-triazines are heterocyclic compounds (see general structure in Table 5.2.F) in which the chlorine atoms are very reactive and can easily be replaced by nucleophiles,[94] including hydroxyl groups of carbohydrates.[95,340–342] Because of their reactivity, they are rapidly hydrolyzed in aqueous solutions resulting in poor cross-linking. In the absence of sulfhydryl and amino groups, the imidazolyl group of histidine, and phenol group of tyrosine, all react.[343] The rates of reaction of the chlorine atoms on the triazine ring are remarkably different. For example, in cyanuric chloride (compound CII in Appendix B), displacement of the first chlorine proceeds rapidly even in the cold with a half-life of 30 s. The second chlorine reacts comparatively slower with a half-life of 30 min at 40°C. The last chlorine is displaced very slowly requiring hours of incubation at elevated temperatures. Thus, cyanuric chloride, which is actually a trifunctional agent, functions as a bifunctional cross-linker. These triazines (compounds CII through CIX in Appendix B) have been used to cross-link collagen[174,342] and to conjugate antigens to erythrocytes.[344] 5-(4,6-Dichloro-s-triazin-2-yl)aminofluorescein (5-DTAF), a fluorescein derivative of dichloro-s-triazine, has been used to modify glycoproteins and to study the rotational correlation time of IgG after intramolecular cross-linking.[345,346] Because of the poor cross-linking properties of these chloro-s-triazines, they are not commonly used in conjugation studies.

5.3.7 Aziridines (Ethyleneimines)

Aziridine or ethyleneimine (see general structures in Table 5.2.G) contains a three-membered heterocycle with one amine and two methylene groups (compounds CX through CXII in Appendix B). The bond angles in aziridine are 60°, considerably less than the bond angle of 109.5° found in ordinary hydrocarbons. This smaller angle results in strain making the compound very reactive. Like chloro-s-triazines, any nucleophile will react with aziridines. The nucleophiles attack one of the methylene groups causing ring opening, thus releasing the bond strain.[322,324,347,348] As may be imagined, the major competitive reaction is hydrolysis. These aziridines effectively cross-link proteins such as albumin,[348] wool fibers, and γ-globulin.[347,348] The oxidized and reduced forms of 2,5-diaziridinyl-1,4-benzoquinone (DZQ, compounds CXI and CXII in Appendix B) have been shown to cross-link guanine bases in DNA with the reduced form being the most studied.[349] DNA cross-linking with these and other diaziridinyl benzoquinone compounds will be further discussed below under Section 5.11.

5.3.8 Bis-Epoxides (Bisoxiranes)

Like aziridines, epoxides contain a strained three member heterocyclic ring with an oxygen and two methylene groups (see general structures in Table 5.2.H). They also undergo ring opening reactions with nucleophiles, including hydroxyl groups. These compounds are reactive and unstable.

The bis-epoxide-containing compounds (see CXIII through CXXI in Appendix B) have been used to cross-link keratin,[350] ribosomal proteins,[351] alkyltransferase and glyceraldehyde 3-phosphate dehydrogenase and DNA,[352,353] elongation factor G and RNA,[72,354] and DNA strands,[355–359] as well as to activate matrices for immobilization of ligands, which will be discussed in Chapter 14.[360,361] Of these reagents, diepoxybutane (compound CXIII in Appendix B) will produce a vicinal diol linkage after reaction. The cross-linked product is therefore cleavable by sodium periodate. Additional epoxide compounds derived from azinomycin are used to cross-link DNAs, which will be discussed in detail under Section 5.11.

The literature contains a wide coverage of diglycidyl and polyglycidyl ethers as cross-linkers for industrial and medical applications. These compounds are commercially available to prepare gels, resins, coating materials, fabrics, tissue grafts, and other cross-linked products.[362] The glycerol diglycidyl ether and other polyglycidyl ethers have been used to cross-link collagen preparations,[363] elastin,[364] and other proteins.[250,365] There are numerous other diglycidyl ethers commercially available such as biphenol A diglycidyl ether and bisphenol A propoxylate diglycidyl ether. However, their application as chemical cross-linkers has not been reported.

5.3.9 SULFONE DERIVATIVES

Divinylsulfone, as shown in Table 5.2.I, is an alkylating agent. It reacts with sulfhydryl, amino, and hydroxyl nucleophiles by addition to the double bond, which is activated by the sulfonyl group. Figure 5.10 shows the reaction mechanism for coupling nucleophilic ligands to solid materials such as agarose.[361,366] At alkaline pHs (e.g., pH 11), divinylsulfone reacts with hydroxyl groups of agarose. Reactions with thiols and amino groups occur at lower pHs and at somewhat higher rates. The product with hydroxyl functions is unstable above pH 9.0 and the product with amino groups is unstable above pH 8.0. Thus, for the immobilization reaction, the pH needs to be well controlled.

Divinylsulfone and 1,6-hexane bisvinylsulfone, compounds CXXIV and CXXV in Appendix B, have been used to prepare cross-linked proteins,[367] protein conjugates for immunoassays,[368] oligo ribonuclease,[369] di-ubiquitin,[370] hydrogels,[371] and modified carbohydrates.[372] In addition to these two vinylsulfones, there are commercial vinylsulfones available, such as bis(vinylsulfone) methane, 1,2-ethane bis(vinylsulfone), and bis(vinylsulfone) benzene derivatives. However, their use in chemical studies has not appeared in the literature.

Poly(2-methacryloyloxyethyl phosphorylcholine)-bis-sulfone (PMPC-bis-sulfone), whose structure is shown in CXXVI in Appendix B, was synthesized by Lewis et al.[373] to introduce poly(2-methacryloyloxyethyl phosphorylcholine) into protein to improve immunogenicity and reduce toxicity of protein-based medicines. PMPC-bis-sulfone cross-links two sulfhydryl groups. For example, it reacts with interferon-α2a whose disulfide bonds were first reduced by dithiothreitol to produce two free thio groups. As shown in Figure 5.11, it is speculated that one of the bis-sulfone end groups can undergo *in situ* elimination of a sulfinyl moiety to give alkenyl monosulfone. After the addition of a thio to the double bond, a second alkenyl function is formed

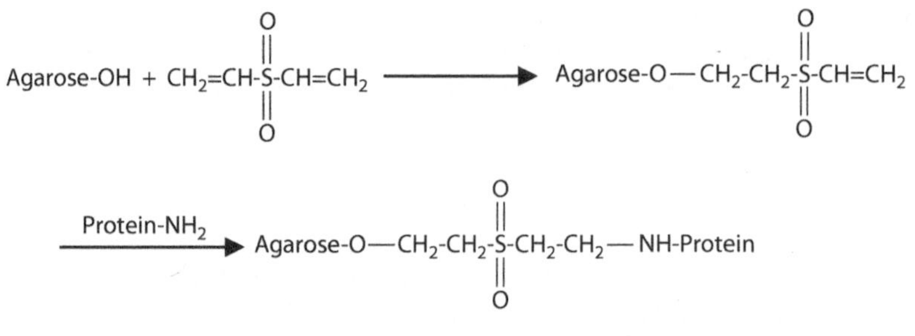

FIGURE 5.10   Reaction mechanism of divinylsulfone in the immobilization of proteins to agarose.

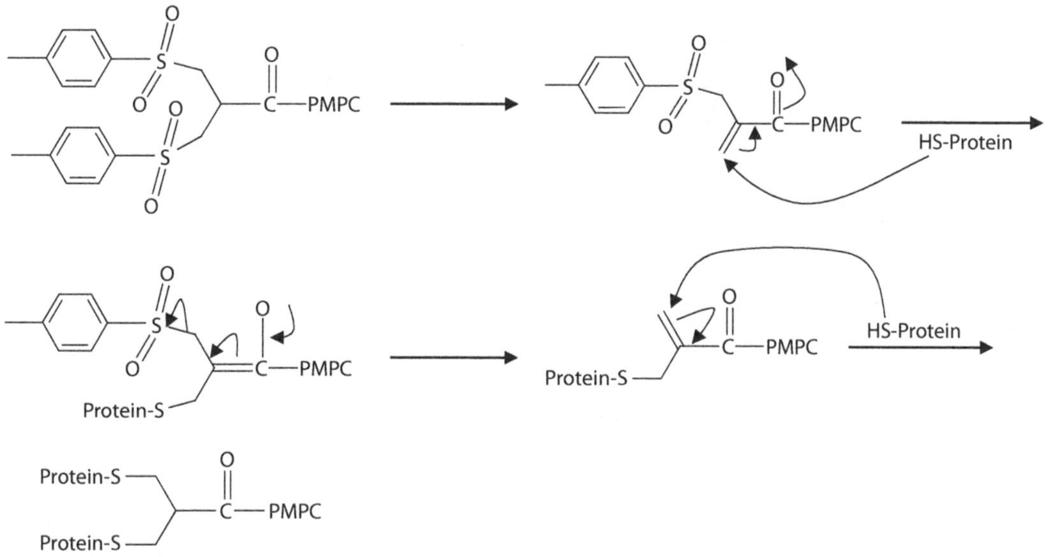

FIGURE 5.11 Speculated reaction mechanism of poly(2-methacryloyloxyethyl phosphorylcholine)-bis-sulfone (PMPC-bis-sulfone).

which a second sulfhydryl attacks completing the cross-linking reaction. Lewis et al.[373] have demonstrated that the compound reacts well with reduced interferon-α2a. It might be useful for conjugating other proteins.

5.4 CARBOXYL GROUP–DIRECTED CROSS-LINKING AGENTS

There are relatively few carboxyl group directed homobifunctional cross-linking reagents. Carbodiimides, which are generally regarded as zero-length cross-linkers (see Chapter 8), can, in the presence of diamines such as ethyl diamine or cleavable cystamine, cross-link protein carboxyl groups as shown in Figure 5.12.[374] In this case, carbodiimides activate the carboxyl groups of proteins to form O-acylisoureas, which react with diamines to from cross-linked products. Theoretically any carbodiimide can be used, but water-soluble compounds such as those shown in Chapter 3 are the most common choices. Other carbodiimides will be discussed in Chapter 8. In addition, there are other carboxyl group activating agents. These will be further discussed in Chapter 8 as monofunctional cross-linkers.

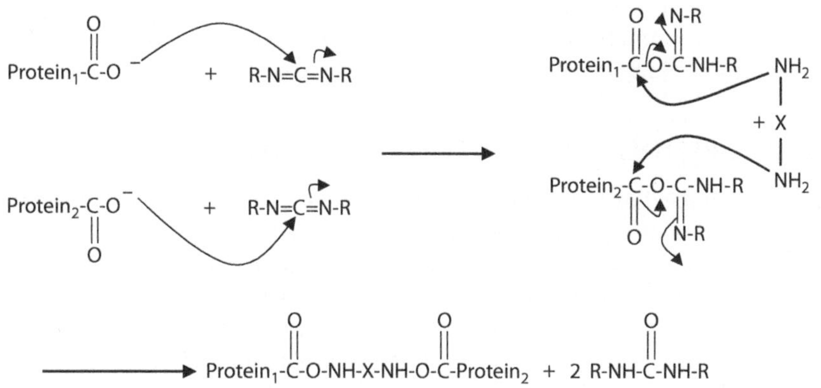

FIGURE 5.12 Cross-linking of proteins with carbodiimides and diamines.

TABLE 5.3
Carboxyl Group–Directed Homobifunctional
Cross-Linking Reagents

1,1-Bis(diazoacetyl)-2-phenylethane:	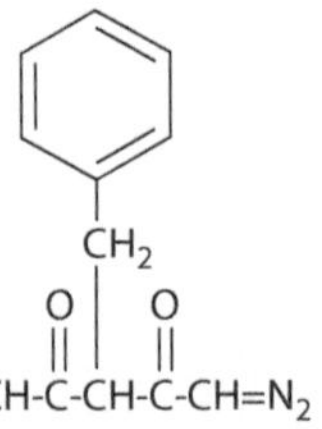 $\begin{array}{c} \text{(phenyl ring)} \\ \text{CH}_2 \\ \text{O} \quad \text{O} \\ \| \quad \| \\ N_2\text{=CH-C-CH-C-CH=}N_2 \end{array}$
Bisdiazohexane:	$N_2\text{=CH-(CH}_2)_4\text{-CH=}N_2$

The carboxylate ion, as a nucleophile, can effectively compete with other nucleophiles at neutral or slightly acidic pH, where nucleophiles such as amino, thiol, and phenolic groups are present in less-reactive protonated forms. Thus, the carboxyl group can react with nitrogen and sulfur mustards and other reagents.[323,324,328,330,331] Similarly, the diazo compounds shown in Table 5.3 have been used to modify carboxyl groups.[375–377] Bisdiazohexane, which is also reactive toward thiols, phenols, and other nucleophiles, has been used to cross-link wool and fibroin,[377] whereas 1,1-bis(diazoacetyl)-2-phenylethane was used to cross-link pepsin intramolecularly.[375,376]

5.5 PHENOLATE AND IMIDAZOLYL GROUP–DIRECTED CROSS-LINKING REAGENTS

The majority of tyrosine- and histidine-directed cross-linkers are diazonium salts. These compounds readily react with the phenolate and imidazolyl groups by electrophilic substitution reactions as discussed in Section 3.5.2.[378,379] Bis-diazonium compounds are easily prepared by treatment of aryl diamines with sodium nitrite in acidic conditions.[380] The diazonium species formed is very reactive and thus intramolecular cross-linking is a major side reaction. The diazonium salts derived from the diamines listed in Appendix C have been used to couple antigens for antibody production and assays,[381–388] to cross-link immunoglubulins,[389–391] to prepare protein aggregates,[392–394] to study protein conformation and antigen properties,[380,395] and to prepare liposome conjugates for drug delivery.[396] Of the diazonium compounds, the ones containing disulfide bonds are cleavable by mercaptan (e.g., compounds IX and X in Appendix C). The resulting reaction product containing the azo bond can also be cleaved by dithionite.[397]

In addition to the diamines listed in Appendix C, other aromatic diamines may be activated to diazonium ions. Poly-diazonium salts derived from polyphenyl amines have also been used. These multifunctional cross-linking reagents will be further discussed in Chapter 7.

5.6 ARGININE RESIDUE–DIRECTED CROSS-LINKERS

Only one kind of arginine directed homobifunctional cross-linker is presently available. As discussed in Chapter 3, vicinal diones react with the guanidino moiety of arginine. Such a reactive functional group has been incorporated into p-phenylenediglyoxal (PDG) and 4,4′-biphenyl-diglyoxal (BDG) as shown in Figure 5.13. These reagents have been shown to react with arginine residues of proteins and guanosine base of RNA.[398–400] The reaction with nucleic acid is reversible in mild alkaline conditions, providing an opportunity to study cross-linked protein in protein–nucleic acid cross-linking reactions. The specificity of PDG was employed to investigate ribosomal protein interactions.[399–401] It has also been used to study protein structures using mass spectrometry,[402] and nucleotide- and actin-induced

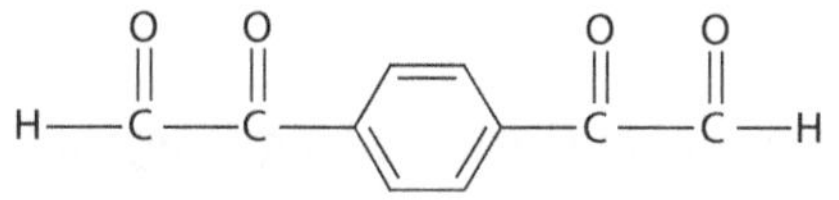

p-Phenylenediglyoxal

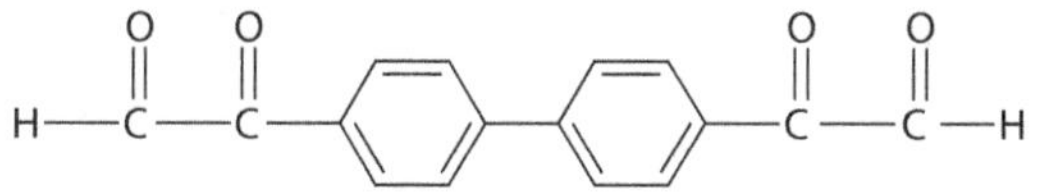

4,4′-Biphenyldiglyoxal

FIGURE 5.13   Chemical structures of homobifunctional arginine reagents.

intramolecular movements of the myosin subfragment-1 molecule.[403] Both PDG and BDG have been used to demonstrate the feasibility of elucidation of biomolecular structures by employing peptide mapping with electrospray ionization Fourier transform ion cyclotron resonance mass spectrometry.[404]

5.7   METHIONINE RESIDUE CROSS-LINKING AGENT

The sulfur atoms of methionine residues in proteins sometimes compete with other nucleophiles in cross-linking reactions as shown in various instances discussed above. No methionine-specific cross-linkers have been published. Although not specifically designed for cross-linking methionine residues, *cis*-diaminedichloroplatinum(II) (Cisplatin, *cis*-DDP) as shown in Figure 5.14, has been found to cross-link α_2-macroglobulin with methionine residues at or near its receptor recognition site.[405–407] The cross-linking can be reversed by diethyldithiocarbamate, which is a potent platinum chelator.[407] *Trans*-diaminedichloroplatinum(II) (*trans*-DDP, Figure 5.14) also reacts with α_2-macroglobulin, but is less efficient at inhibition. Both complexes have been shown to react with albumin and other plasma α-1 proteinase inhibitors, however, cysteines may be involved.[408,409] In addition, they cross-link DNAs as will be discussed more fully below.

5.8   CARBOHYDRATE MOIETY–SPECIFIC REAGENTS

As discussed in Section 3.3.1.4, the monosaccharide unit of glycoproteins can be oxidized by sodium periodate to yield dialdehydes. These dialdehydes will form Schiff bases with amines, which can be stabilized by reducing agents in the process of reductive amination. Thus, amines can be used as cross-linkers after oxidation of the carbohydrate moiety of glycoproteins. Adipic acid dihydrazide (Figure 5.15) has been used to cross-link glycoproteins, such as acid phosphatase and invertase.[410] Other diamines of different chain lengths may also be used in this process. The efficiency of reductive amination can be increased by including 500 mM sodium sulfate in the reaction media.[411]

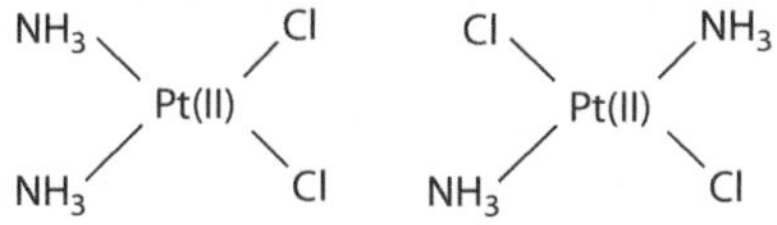

cis-DDP   *Trans*-DDP

FIGURE 5.14   Chemical structure of *cis*-diaminedichloroplatinum (*cis*-DDP) and *trans*-diaminedichloroplatinum (*trans*-DDP).

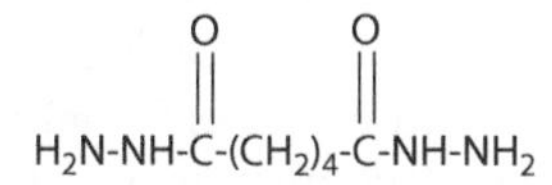

$$H_2N\text{-}NH\text{-}C\text{-}(CH_2)_4\text{-}C\text{-}NH\text{-}NH_2$$

FIGURE 5.15   Chemical structure of adipic acid dihydrazide (AAD).

TABLE 5.4

Photoactivatable Homobifunctional Cross-Linking Reagents

A. 4,4′-Diazidobiphenyl (DABP):

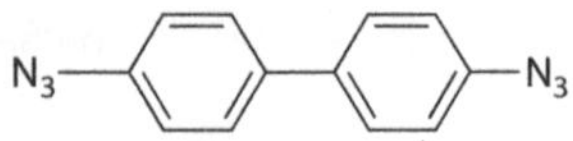

B. 1,5-Diazidonaphthalene (NAD):

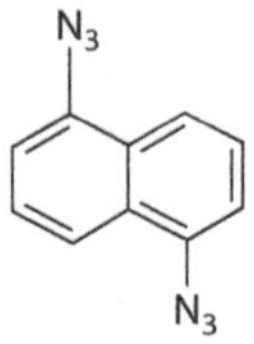

C. *N*,*N*′-Bis(*p*-azido-*o*-nitrophenyl)-1,3-diamino-2-propanol:

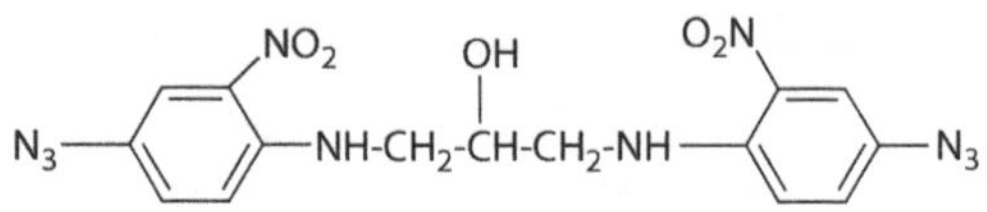

D. 4,4′-Dithiobisphenylazide (DTPA, DTBPA):

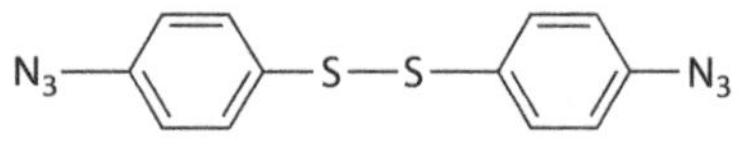

5.9 NONDISCRIMINATORY PHOTOACTIVATABLE CROSS-LINKERS

Among the photoactivatable moieties, diazoalkanes and diazirines have not been incorporated into homobifunctional cross-linkers, although they are found in heterobifunctional reagents, which will be discussed in the next chapter. The more commonly used group is the azido function. Alkyl azides, however, have not been favorably used.[10,412] Acylazides as well as sulfonyl and phosphoryl azides are generally not used as photoactivatable reagents because of their nucleophilic reactivity in the dark.[35] The most common photoactivatable homobifunctional reagents are derived from aryl azides as shown in Table 5.4.[413,414] These aryl azides are photolyzed at wavelengths of 300–400 nm so that the biological component, be it protein or nucleic acid, would not be damaged by photoirradiation. They are used mainly to label erythrocyte membrane proteins.[413,415] 4,4′-Dithiobisphenylazide (DTBPA) containing a disulfide bond was synthesized by Mikkelsen and Wallach[413] as a cleavable reagent in the presence of free thiols. It has been used to identify erythrocyte membrane proteins. It should be noted that the reagent may be used as a heterobifunctional reagent by reaction first with a sulfhydryl group through the disulfide–thiol interchange reaction and followed by photoactivation.

5.10 NONCOVALENT HOMOBIFUNCTIONAL CROSS-LINKING REAGENTS

While chemical cross-linking has the connotation that the reagents react to form covalent bonds, there are species that physically associate with proteins or other entities to form such tight complexes that binding is essentially irreversible. The avidin–biotin complex is a well-known example of such a system. Some immunoglobulins can also bind to antigens with dissociation constants in the order of femtomolar. Lectins are another class of proteins that have specific binding targets. These molecules may be important under certain circumstances and are hence useful for some applications.

The avidin–biotin and the lectin systems will be discussed in the context of multifunctional reagents in Chapter 7. Avidin and streptavidin are proteins with four high affinity binding sites for biotin.[415] They will cross-link four biotin moieties bound to biological molecules. The lectins are complex sugar binding proteins with multivalent binding sites for specific carbohydrates.[416] These proteins will cross-link multiple glycoproteins. In Chapter 7, these systems will be further explored in detail.

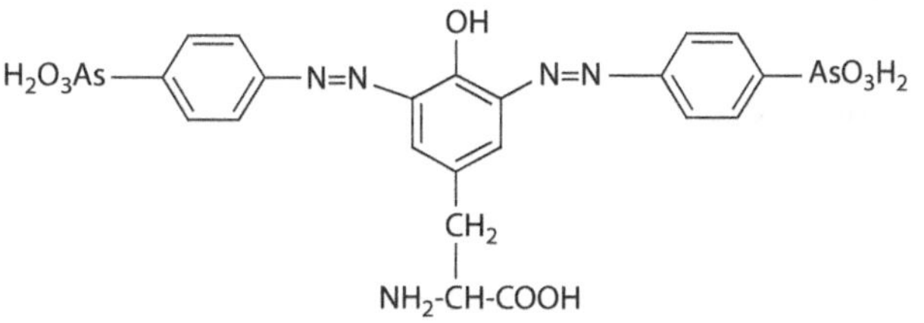

FIGURE 5.16   Structure of a bifunctional antigen: L-tyrosine-bis(p-azobenzenearsenate) (Bis-RAT).

For the immunological systems, some antibodies can be used as homobifunctional cross-linkers while others have been designed to be heterobifunctional.[417] The latter will be discussed in the next chapter. The basic unit (monomer) of immunoglobulins contains two Fab (antigen-binding) fragments and one Fc (crystallizable) fragment. These molecules are therefore bifunctional in that they can bind specifically two antigens, thus cross-linking them. With monoclonal antibodies, the dissociation constant of antibody–antigen complex can be as low as 10^{-15}M, making the complex essentially irreversible. There are five classes of immunoglobulins, designated as IgG, IgA, IgM, IgD, and IgE. All the antibodies exist as monomers except IgM, which is pentameric. In adult animals, the majority of serum immunoglobulins are in the IgG class. IgG is thus the most common monospecific reagent. Treatment of IgG with pepsin yields a F(ab)$_2^1$ fragment, which retains the two antigen binding sites. F(ab)$_2^1$ can be purified and used as a cross-linking reagent.

Homobifunctional antibodies are used in various immunoassays, including agglutination precipitin assays, and more recently in proteomic tissue profiling by antibody-based microarray.[418,419] They are also used to construct immunotoxins.[7] However, their use as cross-linking agents for chemical studies is limited.

Similarly, compounds that contain two antigenic determinants will cross-link antibody molecules. Simple molecules such as L-tyrosine-bis(p-azobenzenearsenate) (Bis-RAT, see Figure 5.16) have been shown to cross-link anti-RAT antibodies antibodies.[420] Similar cross-linking of the immunoglobulins is achieved by bis-azobenzenearsenate separated by various spacers. Other examples of multiepitopic molecules are hapten-conjugated antigens such as dinitrophenyl labeled bovine serum albumin.[421] In addition to proteins, polysaccharides have also been used as carriers. The antigen–antibody complexes have been used in various studies.

5.11   NUCLEIC ACID CROSS-LINKING REAGENTS

From the discussions in earlier sections, it is obvious that there are reagents that not only cross-link proteins but also react with nucleic acids. Reagents like bisimidates, dicarbonyl compounds, mustards, aziridine benzoquinones, diepoxides, and platinum complexes, can also cross-link protein and nucleic acids, and DNAs. However, there are series of compounds that have been reported to mainly cross-link nucleic acids. These DNA cross-linking reagents, listed in Table 5.5, are generally antitumor agents and will be described below.[422]

5.11.1   METAL COMPOUNDS

In addition to *cis*- and *trans*-DDP mentioned in Section 5.7, there are di-, tri-, and poly-nuclear platinum complexes that will react with DNA (Table 5.5.I.A).[423–425] The active form of the reagent is believed to be the aquatic species, formed by the displacement of the chlorides by water. For *cis*-DDP this would be *cis*-[Pt(NH$_3$)$_2$(H$_2$O)$_2$]$^{2+}$.[426] These compounds are thought to exert their antitumor activities through DNA adducts formation with the N7-nitrogen of purines, mainly guanine. The main products of *cis*-DDP and DNA reactions are intrastrand cross-links between guanine residues in 5′-GG-3′, 5′-GNG-3′, and between adenine and guanine in 5′-AG-3′ sequences.[427]

TABLE 5.5
Nucleic Acid Cross-Linking Reagents

I. Metal complexes

A. Platinum complexes

1. Dinuclear platinum complexes

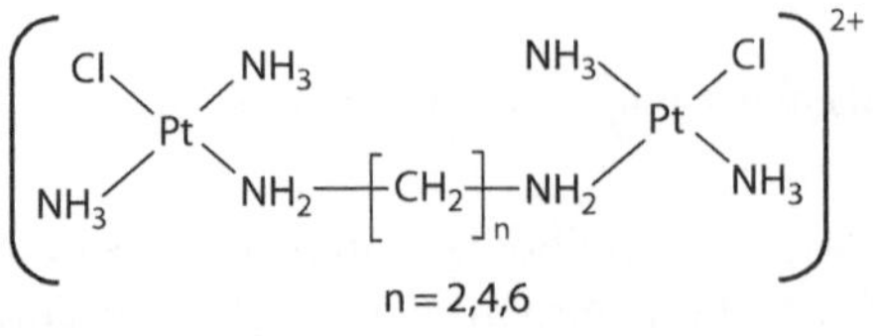

$n = 2,4,6$

2. Trinuclear platinum complex

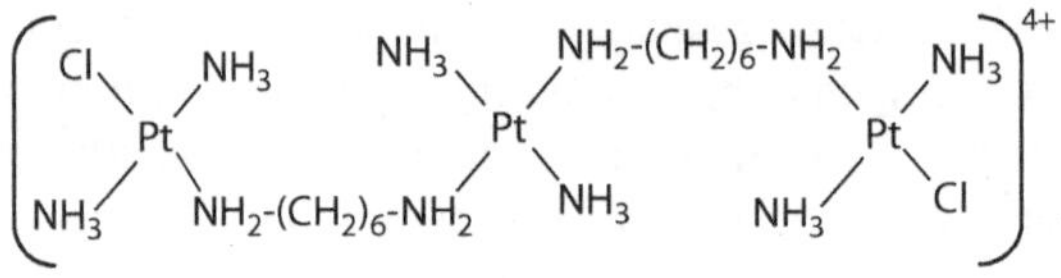

B. Phenyl selenides:

1. 2,6-Bis(phenylselenylmethyl)phenol

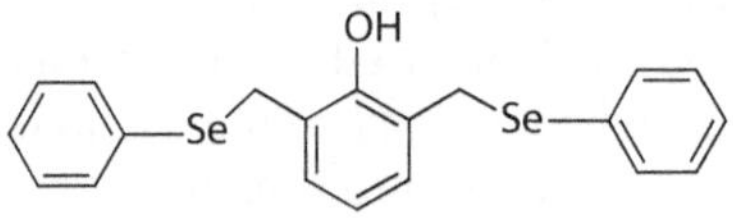

2. 2,5-Bis(phenylselenylmethyl)benzene-1,4-diol

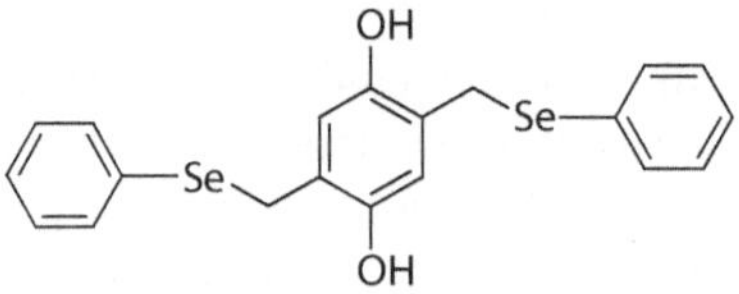

3. 3,3′-Bis(phenylselenylmethyl)-4,4′-dihydroxy-1,1′-biphenyl

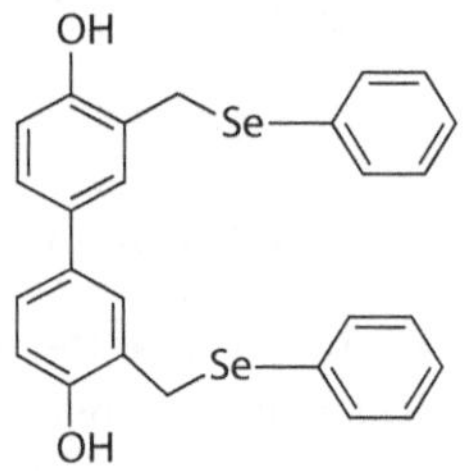

C. Aluminum(III) complex

1. (R,R)-*N,N′*-Bis[5-methyl-3-(4-methylpiperazinyl)-salicylidene]-1,2-diphenylethanediamine aluminum(III) chloride ([SalenAl^{III}]Cl)

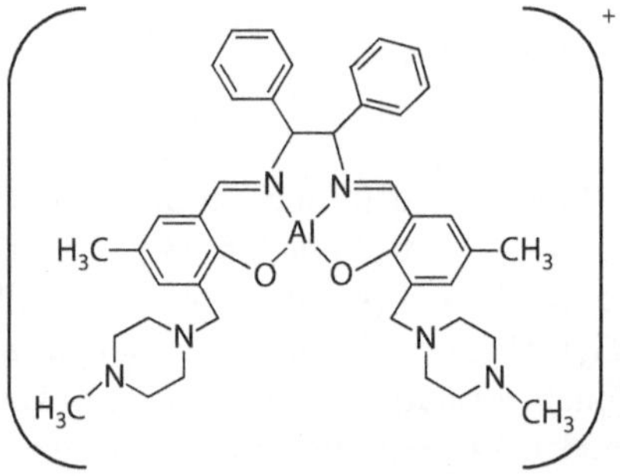

TABLE 5.5 (Continued)
Nucleic Acid Cross-Linking Reagents

II. Azinomycin derivatives

A. Azinomycin bisepoxides with methylene spacer

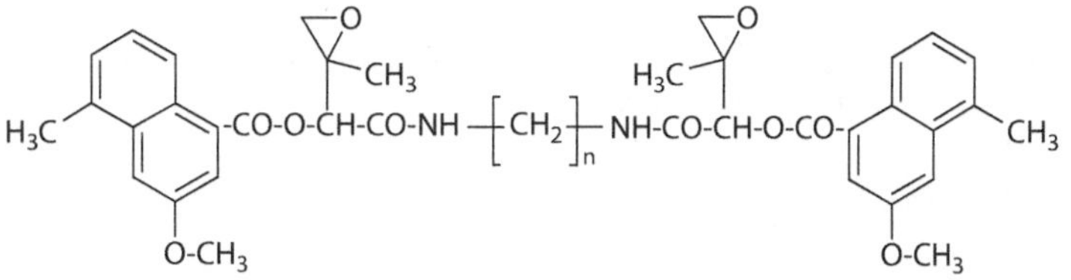

1. n = 2: 1,2-Bis[(2S,3S)-3,4-epoxy-2-(3-methoxy-5-methyl-1-naphthoyloxy)-3-methylbutanamido]ethane
2. n = 3: 1,3-Bis[(2S,3S)-3,4-epoxy-2-(3-methoxy-5-methyl-1-naphthoyloxy)-3-methylbutanamido]propane
3. n = 4: 1,4-Bis[(2S,3S)-3,4-epoxy-2-(3-methoxy-5-methyl-1-naphthoyloxy)-3-methylbutanamido]butane
4. n = 5: 1,5-Bis[(2S,3S)-3,4-epoxy-2-(3-methoxy-5-methyl-1-naphthoyloxy)-3-methylbutanamido]pentane
5. n = 6: 1,6-Bis[(2S,3S)-3,4-epoxy-2-(3-methoxy-5-methyl-1-naphthoyloxy)-3-methylbutanamido]hexane

B. 2,2′-Bis[(2S,3S)-3,4-epoxy-2-(3-methoxy-5-methyl-1-naphthoyloxy)-3-methylbutanamido]diethylmethylamine

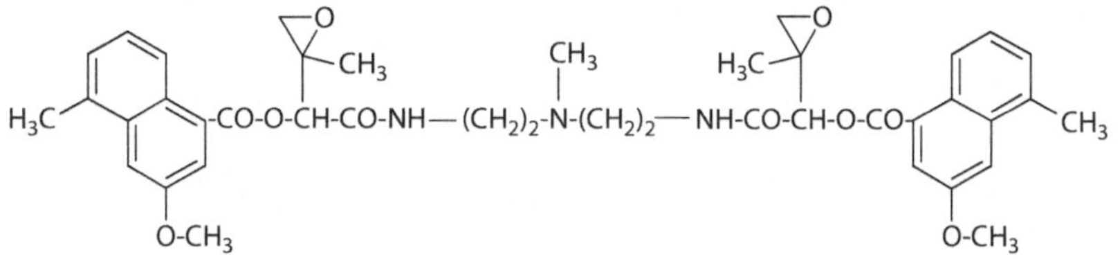

C. Azinomycin bisepoxides with aromatic spacer

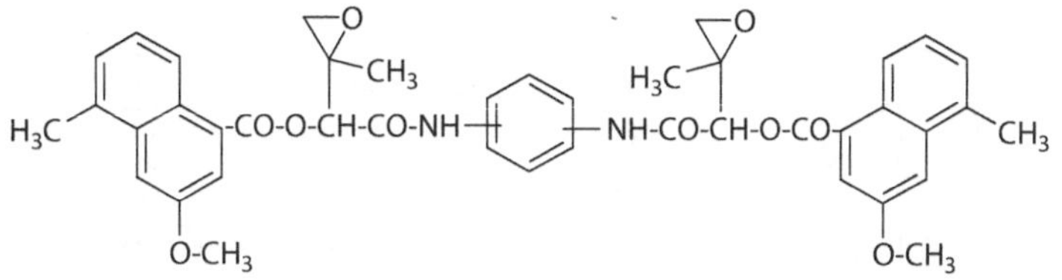

1. *p*-Aryl: 1,4-Bis[(2S,3S)-3,4-epoxy-2-(3-methoxy-5-methyl-1-naphthoyloxy)-3-methylbutanamido]benzene
2. *m*-Aryl: 1,3-Bis[(2S,3S)-3,4-epoxy-2-(3-methoxy-5-methyl-1-naphthoyloxy)-3-methylbutanamido]benzene
3. *o*-Aryl: 1,2-Bis[(2S,3S)-3,4-epoxy-2-(3-methoxy-5-methyl-1-naphthoyloxy)-3-methylbutanamido]benzene

D. 1,4-Bis[(2S,3S)-3,4-epoxy-2-(1-naphthoyloxy)-3-methylbutanamido]benzene

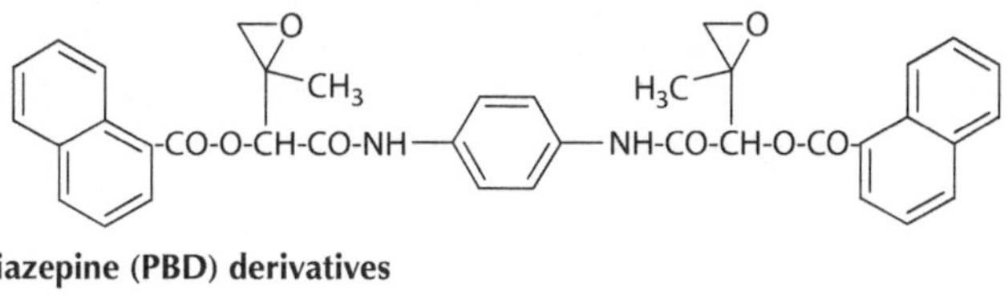

III. Bis-pyrrolobenzodiazepine (PBD) derivatives

A. C7/C7′-Linked PBD dimers

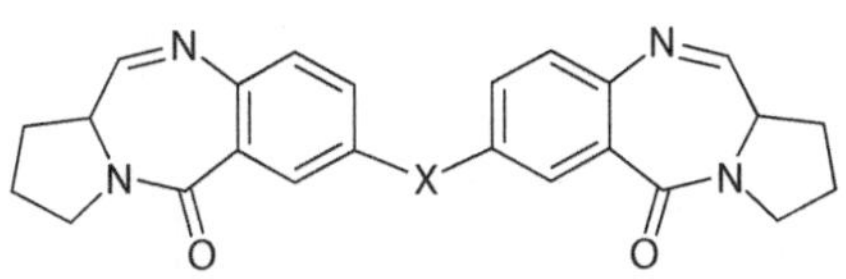

1. X=S-(CH₂)₃-S: 1,3-Bis[7-sulfo-(11a*S*)-1,2,3,11a-tetrahydro-5*H*-pyrrolo[1,2-*c*][1,4]benzodiazepin-5-one]propane

(continued)

TABLE 5.5 (Continued)
Nucleic Acid Cross-Linking Reagents

2. X = O-(CH$_2$)$_2$-N-(CH$_2$)$_2$-O: 2,2′-Bis[7-oxy-(11aS)-1,2,3,11a-tetrahydro-5H-pyrrolo[1,2-c][1,4]benzodiazepin-5-
   | one]ethylmethylamine
   CH$_3$

3. X = NH-(CH$_2$)$_3$-N-(CH$_2$)$_3$-HN: 2,2′-Bis[7-amino-(11aS)-1,2,3,11a-tetrahydro-5H-pyrrolo[1,2-c][1,4]benzo-
   | diazepin-5-one]ethylmethylamine
   CH$_3$

B. C8/C8′-Alkanediyldioxy-linked PBD dimers

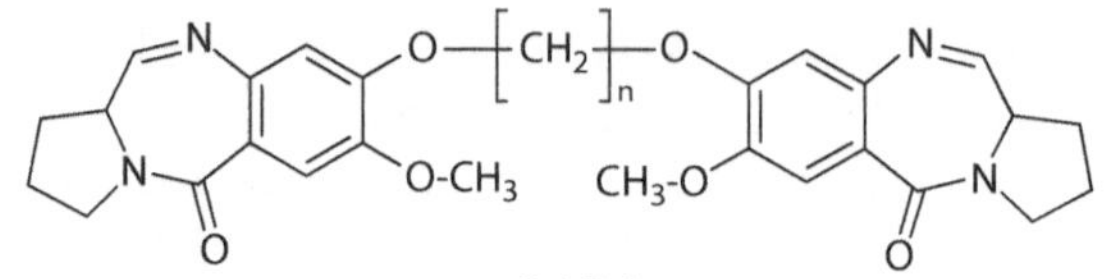

n = 3,4,5,6

1. n = 3: 1,1′-[(Propane-1,3-diyl)dioxy]bis[(11aS)-7-methoxy-1,2,3,11a-tetrahydro-5H-pyrrolo[1,2-c][1,4]
   benzodiazepin-5-one] (DSB-120)
2. n = 4: 1,1′-[(Butane-1,4-diyl)dioxy]bis[(11aS)-7-methoxy-1,2,3,11a-tetrahydro-5H-pyrrolo[1,2-c][1,4]
   benzodiazepin-5-one]
3. n = 5: 1,1′-[(Pentane-1,5-diyl)dioxy]bis[(11aS)-7-methoxy-1,2,3,11a-tetrahydro-5H-pyrrolo[1,2-c][1,4]
   benzodiazepin-5-one]
4. n = 6: 1,1′-[(Hexane-1,6-diyl)dioxy]bis[(11aS)-7-methoxy-1,2,3,11a-tetrahydro-5H-pyrrolo[1,2-c][1,4]
   benzodiazepin-5-one]

C. C8/C8′-Alkanediyldioxy-linked C2-difluorinated PBD dimers

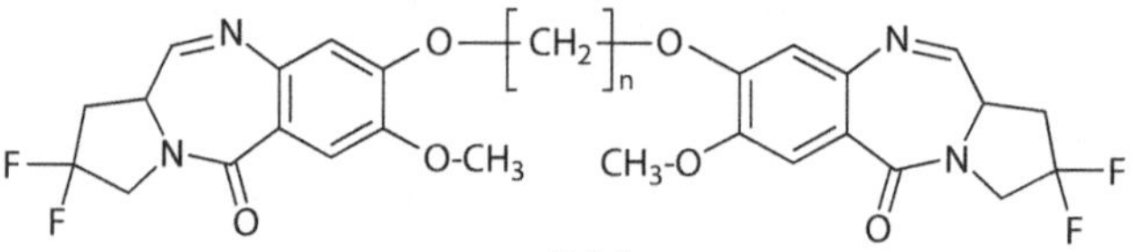

n = 3,4,5

1. n = 3: 1,1′-[Propane-1,3-diyl)dioxy]-bis[(11aS)-7-methoxy-2,2-difluoro-1,2,3,11a-tetrahydro-5H-pyrrolo[2,1-c]
   [1,4]benzodiazepin-5-one]
2. n = 4: 1,1′-[Butane-1,4-diyl)dioxy]-bis[(11aS)-7-methoxy-2,2-difluoro-1,2,3,11a-tetrahydro-5H-pyrrolo[2,1-c]
   [1,4]benzodiazepin-5-one]
3. n = 5: 1,1′-[Pentane-1,5-diyl)dioxy]-bis[(11aS)-7-methoxy-2,2-difluoro-1,2,3,11a-tetrahydro-5H-pyrrolo[2,1-c]
   [1,4]benzodiazepin-5-one]

D. C8/C8′-Alkanediyldioxy-linked C2-$exo/endo$ unsaturated PBD dimers
   1. C2-Exo unsaturated PBD dimer: 1,1′-[[(Propane-1,3-diyl)dioxy]bis[(11aS)-7-methoxy-2-methylidene-1,2,3,11a-
      tetrahydro-5H-pyrrolo[2,1-c][1,4]benzodiazepin-5-one] (SJG-136)

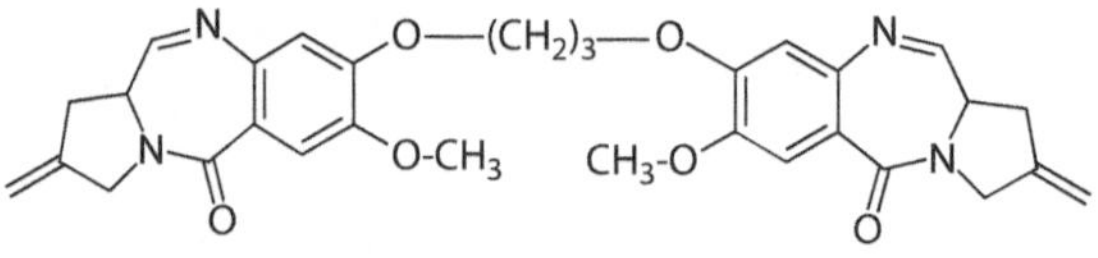

   2. C2-Exo-difluoromethylene PBD dimers

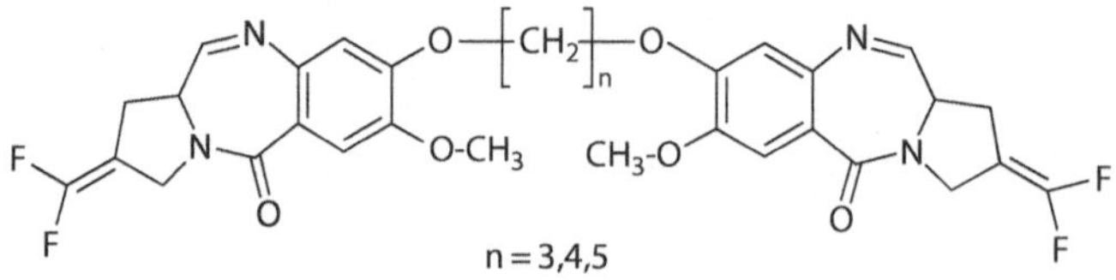

n = 3,4,5

TABLE 5.5 (Continued)

Nucleic Acid Cross-Linking Reagents

    a. n = 3: 1,1'-[[(Propane-1,3-diyl)dioxy]bis[(11aS)-7-methoxy-2-fluoromethylidene-1,2,3,11a-tetrahydro-5H-pyrrolo[2,1-c][1,4]benzodiazepin-5-one]

    b. n = 4: 1,1'-[[(Butane-1,4-diyl)dioxy]bis[(11aS)-7-methoxy-2-fluoromethylidene-1,2,3,11a-tetrahydro-5H-pyrrolo[2,1-c][1,4]benzodiazepin-5-one]

    c. n = 5: 1,1'-[[(Pentane-1,5-diyl)dioxy]bis[(11aS)-7-methoxy-2-fluoromethylidene-1,2,3,11a-tetrahydro-5H-pyrrolo[2,1-c][1,4]benzodiazepin-5-one]

  3. C2-C3/C2'-C3'-*endo* unsaturated PBD dimer

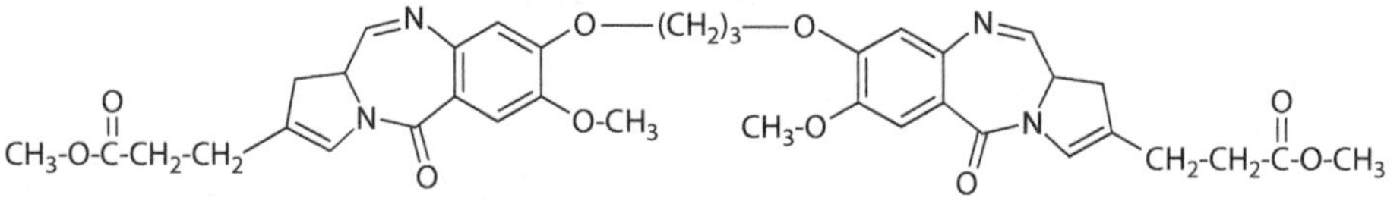

E. C8/C8'-Polyheterocyclic-linked PBD dimers

  1. Tripyrrole-containing PBD dimer (AT-235)

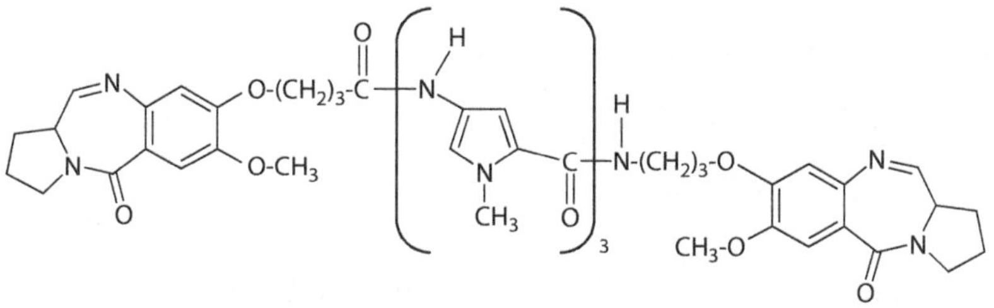

  2. Polypyrrole-polyamide-containing PBD dimers

$$n = 1,2,3$$

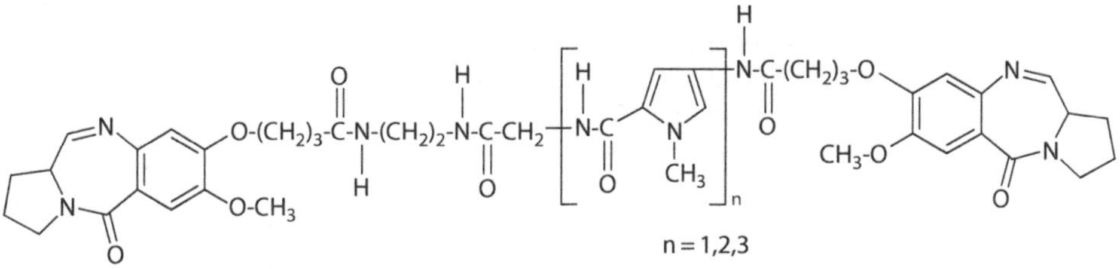

  3. Polyimidazole-polyamide-containing PBD dimers

$$n = 1,2,3$$

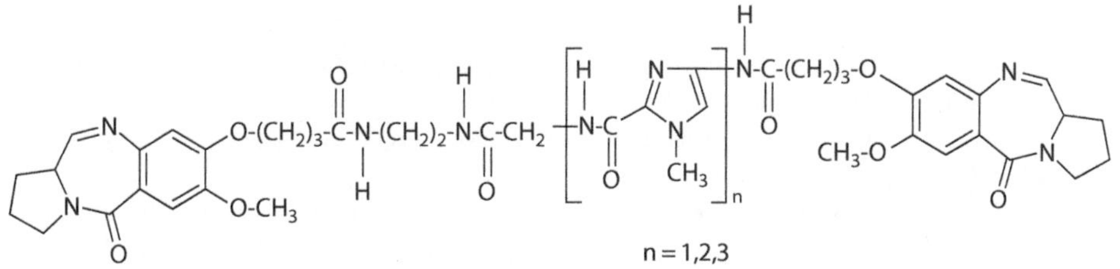

F. C8/C8'-Linked-1,2,3-triazole-containing PBD dimers

$$n = 3,4,5,6,8$$

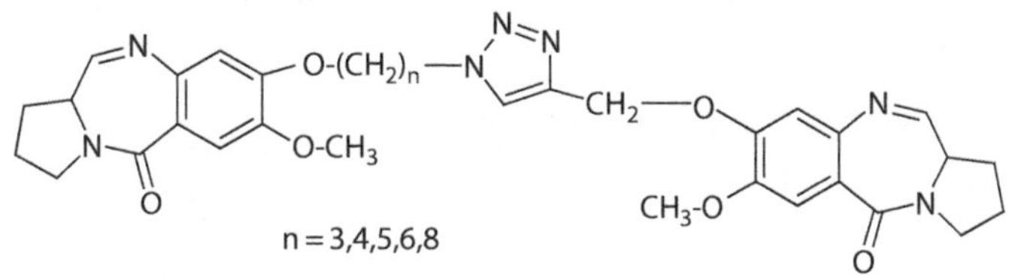

(*continued*)

TABLE 5.5 (Continued)
Nucleic Acid Cross-Linking Reagents

G. C8/C8′-piperazine-linked PBD dimers

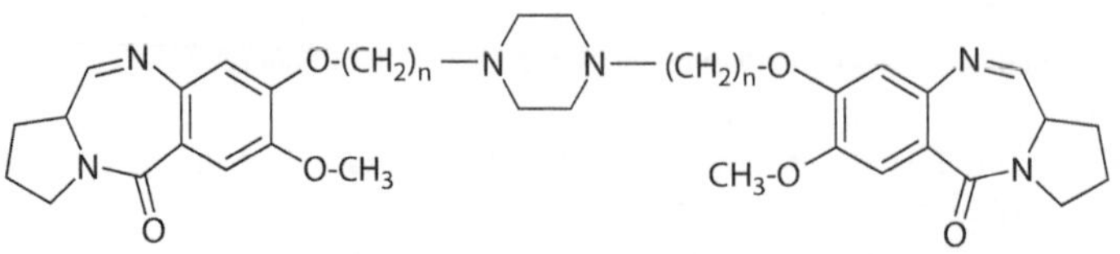

1. n = 2: 1,1′-[[1,4-Di(ethane-1,2-diyl)hexahydropiperazine]dioxy]-bis[(11aS)-7-methoxy-1,2,3,11a-tetrahydro-5H-pyrrolo[2,1-c][1,4]benzodiazepin-5-one]
2. n = 3: 1,1′-[[1,4-Di(propane-1,3-diyl)hexahydropiperazine]dioxy]-bis[(11aS)-7-methoxy-1,2,3,11a-tetrahydro-5H-pyrrolo[2,1-c][1,4]benzodiazepin-5-one]
3. n = 4: 1,1′-[[1,4-Di(butane-1,4-diyl)hexahydropiperazine]dioxy]-bis[(11aS)-7-methoxy-1,2,3,11a-tetrahydro-5H-pyrrolo[2,1-c][1,4]benzodiazepin-5-one]
4. n = 5: 1,1′-[[1,4-Di(pentane-1,5-diyl)hexahydropiperazine]dioxy]bis[(11aS)-7-methoxy-1,2,3,11a-tetrahydro-5H-pyrrolo[2,1-c][1,4]benzodiazepin-5-one]

H. C8/C8′-Piperazine-linked C2-monoflourinated PBD dimers

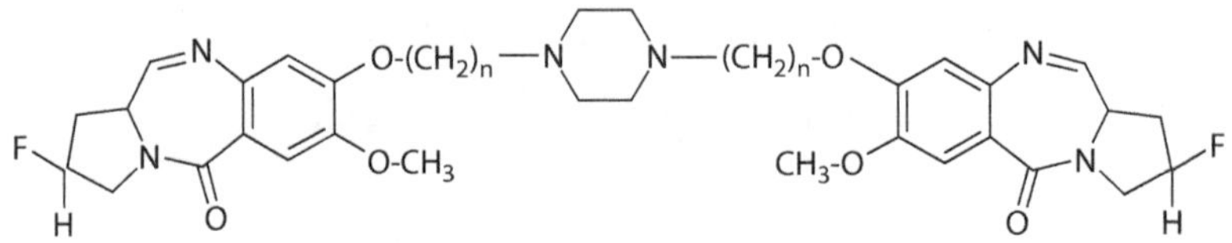

1. n = 3: 1,1′-{[1,4-Di(propane-1,3-diyl)hexahydropiperazine]dioxy}-bis[(11aS)-2-fluoro-7-methoxy-1,2,3,11a-tetrahydro-5H-pyrrolo[2,1-c][1,4]benzodiazepine-5-one]
2. n = 4: 1,1′-{[1,4-Di(butane-1,4-diyl)hexahydropiperazine]dioxy}-bis[(11aS)-2-fluoro-7-methoxy-1,2,3,11a-tetrahydro-5H-pyrrolo[2,1-c][1,4]benzodiazepine-5-one]
3. n = 5: 1,1′-{[1,4-Di(pentane-1,5-diyl)hexahydropiperazine]dioxy}-bis[(11aS)-2-fluoro-7-methoxy-1,2,3,11a-tetrahydro-5H-pyrrolo[2,1-c][1,4]benzodiazepine-5-one]

I. C8/C8′-piperazine-linked C2-difluorinated PBD dimers

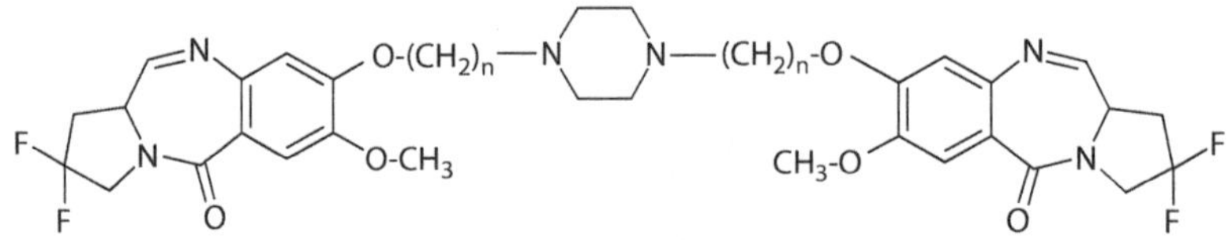

1. n = 3: 1,1′-{[1,4-Di(propane-1,3-diyl)hexahydropiperazine]dioxy}-bis[(11aS)-2,2-di-fluoro-7-methoxy-1,2,3,11a-tetrahydro-5H-pyrrolo[2,1-c][1,4]benzodiazepine-5-one]
2. n = 4: 1,1′-{[1,4-Di(butane-1,4-diyl)hexahydropiperazine]dioxy}-bis[(11aS)-2,2-di-fluoro-7-methoxy-1,2,3,11a-tetrahydro-5H-pyrrolo[2,1-c][1,4]benzodiazepine-5-one]
3. n = 5: 1,1′-{[1,4-Di(pentane-1,5-diyl)hexahydropiperazine]dioxy}-bis[(11aS)-2,2-di-fluoro-7-methoxy-1,2,3,11a-tetrahydro-5H-pyrrolo[2,1-c][1,4]benzodiazepine-5-one]

J. C8/C8′-Piperazine-linked C2-exo-difluoromethylene PBD dimer: 1,1′-{[1,4-Di(propane-1,3-diyl) hexahydropiperazine]dioxy}-bis[(11aS)-2-difluoromethylidine-7-methoxy-1,2,3,11a-tetrahydro-5H-pyrrolo[2,1-c] [1,4]benzodiazepine-5-one]

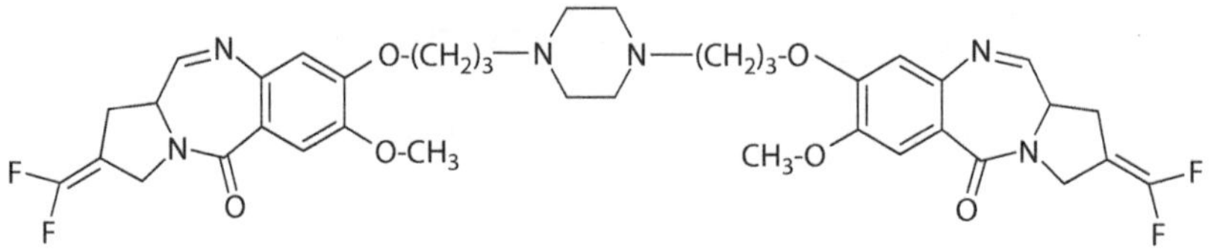

TABLE 5.5 (Continued)
Nucleic Acid Cross-Linking Reagents

K. C8/C8′-anthraquinone-linked PBD dimers

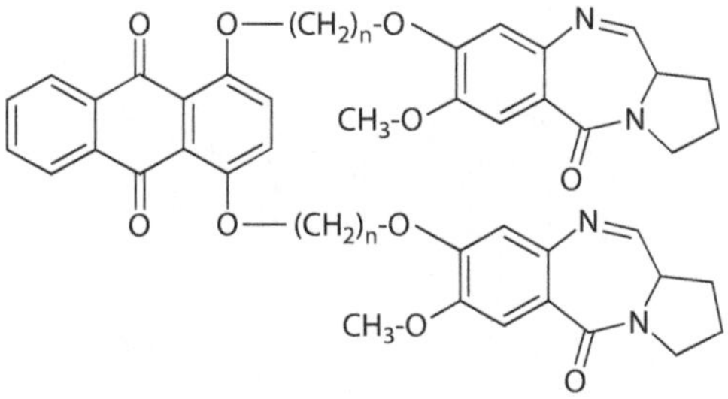

1. n = 3: 1,4-Bis-{3-[7-Methoxy-8-oxy-(11a*S*)-1,2,3,11a tetrahydro-5*H*-pyrrolo [2,1-*c*][1,4]benzodiazepin-5-one] propyloxy}anthracene-9,10-dione
2. n = 4: 1,4-Bis-{4-[7-Methoxy-8-oxy-(11a*S*)-1,2,3,11a tetrahydro-5*H*-pyrrolo [2,1-*c*][1,4]benzodiazepin-5-one] butyloxy}anthracene-9,10-dione
3. n = 5: 1,4-Bis-{5-[7-Methoxy-8-oxy-(11a*S*)-1,2,3,11a tetrahydro-5*H*-pyrrolo [2,1-*c*][1,4]benzodiazepin-5-one] pentyloxy}anthracene-9,10-dione

L. C2/C2′-Linked PBD dimers

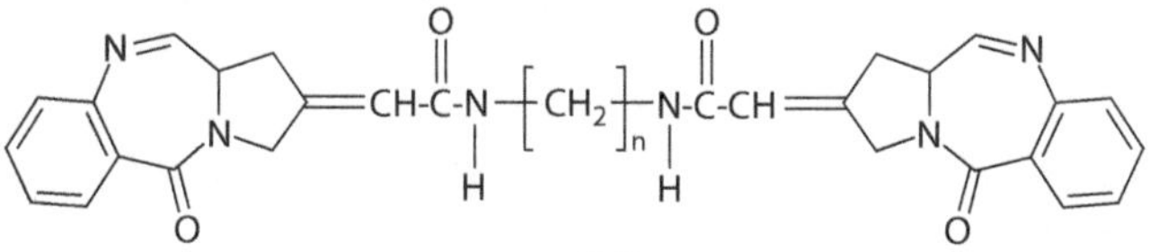

n = 5–7

M. A-C8/C-C2 Amide-linked PBD dimers

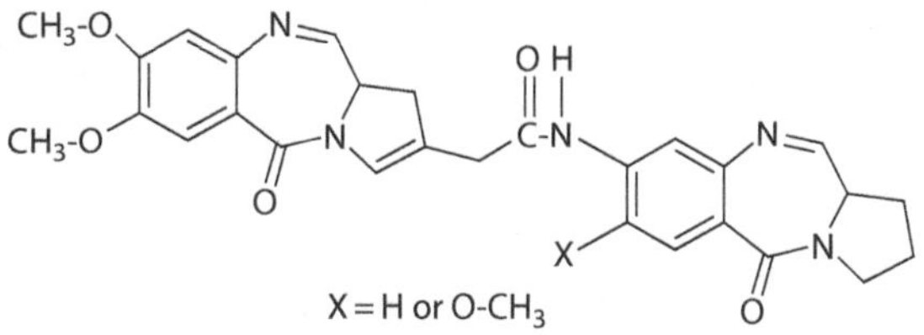

X = H or O-CH₃

N. A-C8/C-C2 Alkoxyamido-linked PBD dimers

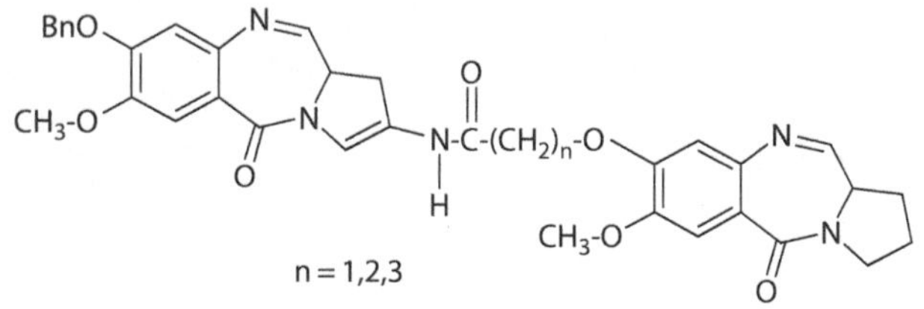

n = 1,2,3

O. C8/C2-Linked-1,2,3-triazole-containing PBD dimers

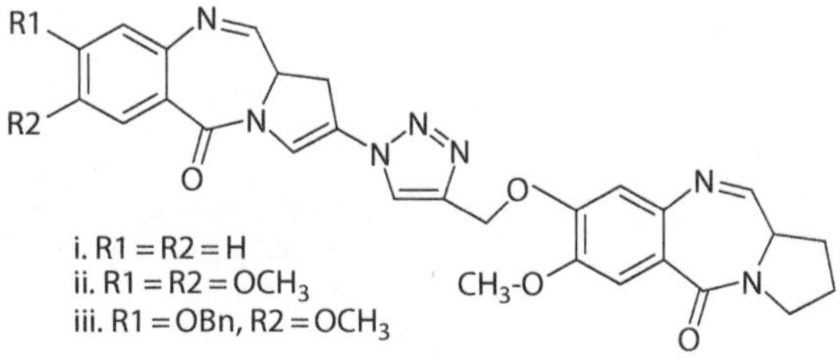

i. R1 = R2 = H
ii. R1 = R2 = OCH₃
iii. R1 = OBn, R2 = OCH₃

(continued)

TABLE 5.5 (Continued)
Nucleic Acid Cross-Linking Reagents

IV. Cyclopropylpyrroloindole (CPI)-based dimers

  A. Bis-Duocarmycin SA enantiomer analogs

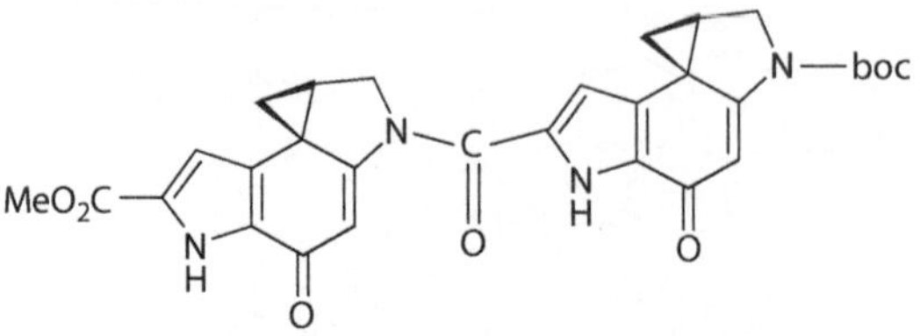

i. (+)(+)-DSA$_2$ ii. (+)(−)-DSA$_2$
iii. (−)(+)-DSA$_2$ iv. (−)(−)-DSA$_2$

  B. Bis(*N*-methylimidazole-*N*-methylpyrrole-CPI)

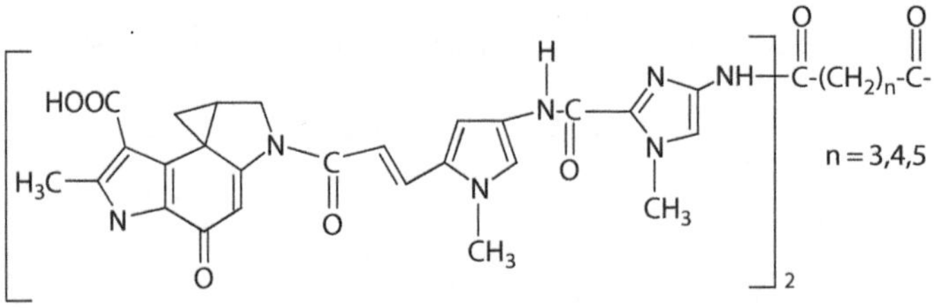

$n = 3,4,5$

  C. *Seco*-CPI-based dimers

   1. Bizelesin, 1,3-bis[2-[(8S)-8-(chloromethyl)-4-hydroxy-1-methyl-7,8-dihydro-3*H*-pyrrolo[3,2-e]indole-6-carbonyl]-1*H*-indol-5-yl]urea

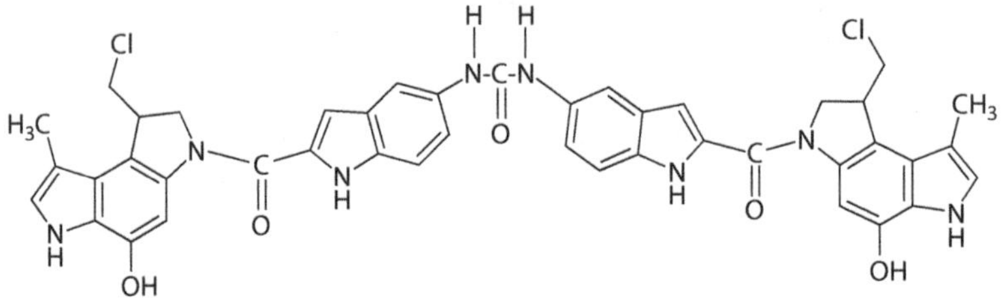

   2. U78779

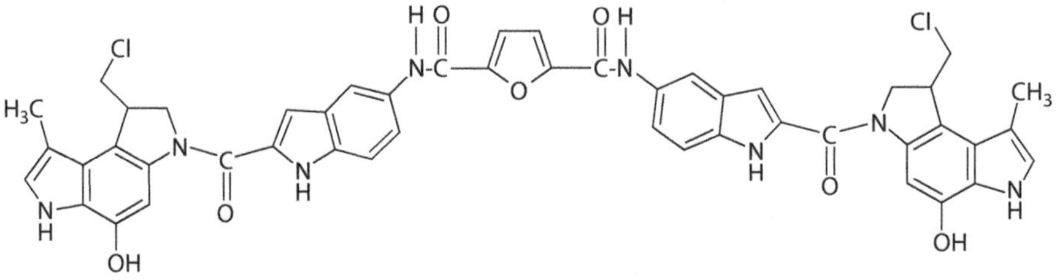

   3. Bis(3-methoxycarbonyl-2-trifluoromethyl-CPI) (MCTFCPI)

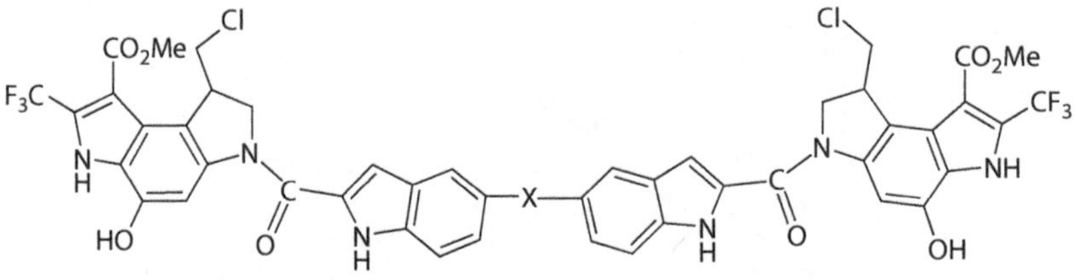

a. X = O   b. X = CH$_2$ c. X = C = C   d. X = NH-CO-NH

TABLE 5.5 (Continued)
Nucleic Acid Cross-Linking Reagents

4. Methoxycarbonyl/methyl-CPI dimer linked by phenylenediacryloyl group

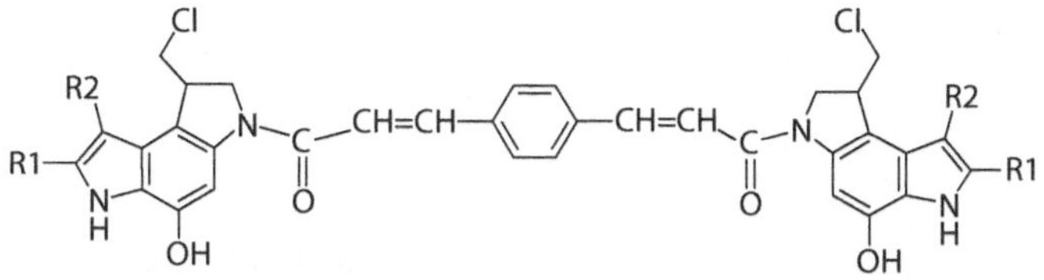

a. R1=CF$_3$, R2=CO$_2$Me   b. R1=Me, R2=CO$_2$Me   c. R1=H, R2=CO$_2$Me
d. R1=CO$_2$Me, R2=H   e. R1=H, R2=Me

5. Phenyl-CPI dimer linked by phenylenediacryloyl group

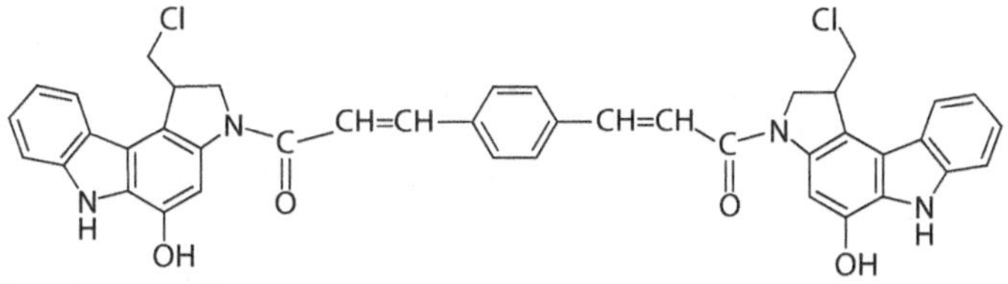

V. *Seco*-CBI (cyclopropanebenz[e]indoline)-based dimers

A. C7-C7 *Seco*-CBI dimers

1. With methylene spacer arm

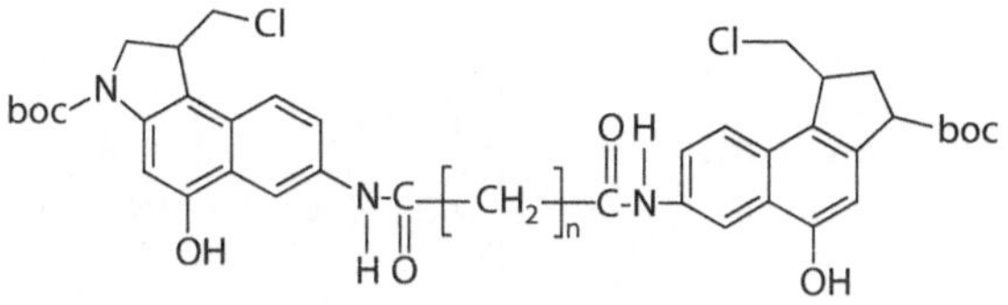

a. n = 3   b. n = 4   c. n = 5   d. n = 6

2. With pyrrole polyamide spacer arm

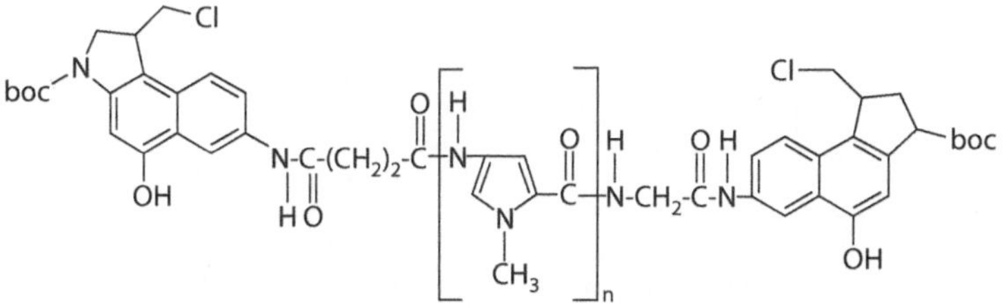

a. n = 1   b. n = 2   c. n = 3

3. With imidazole polyamide spacer arm

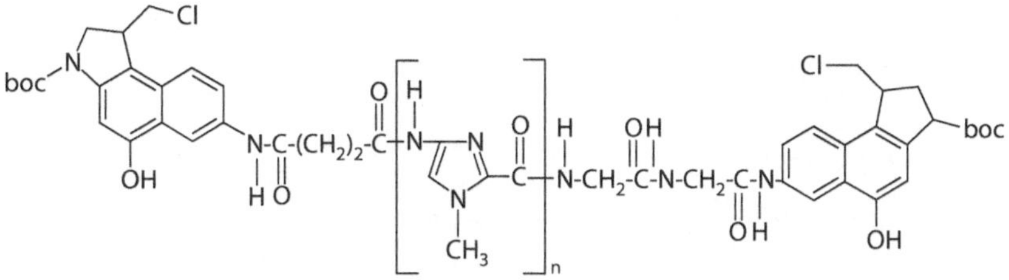

a. n = 1   b. n = 2   c. n = 3

(continued)

TABLE 5.5 (Continued)
Nucleic Acid Cross-Linking Reagents

B. N3-N3 *Seco*-CBI dimers

    1. With methylene spacer arm

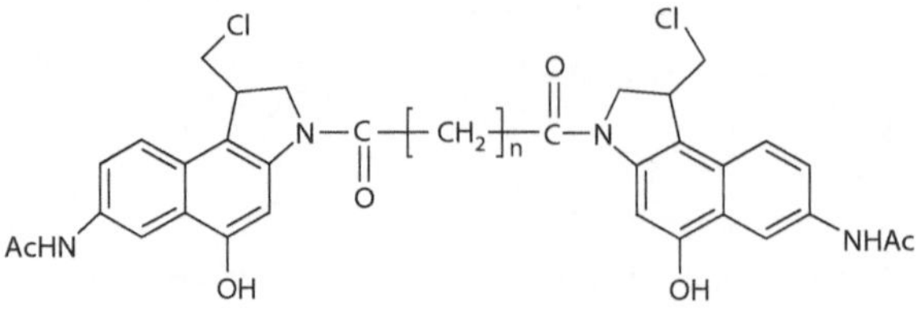

a. n = 3   b. n = 4   c. n = 5   d. n = 6

    2. With pyrrole polyamide spacer arm

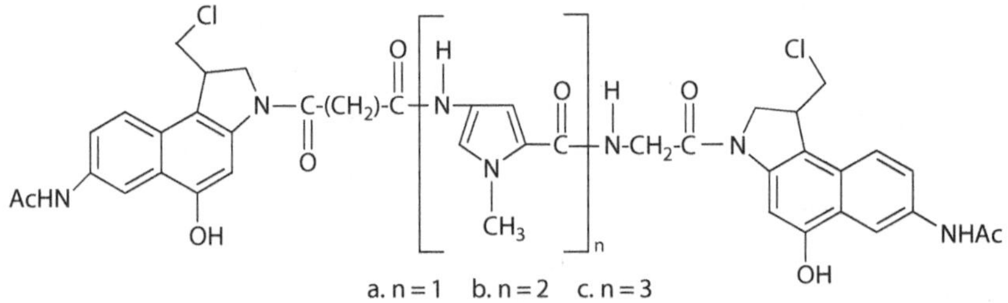

a. n = 1   b. n = 2   c. n = 3

    3. With imidazole polyamide spacer arm

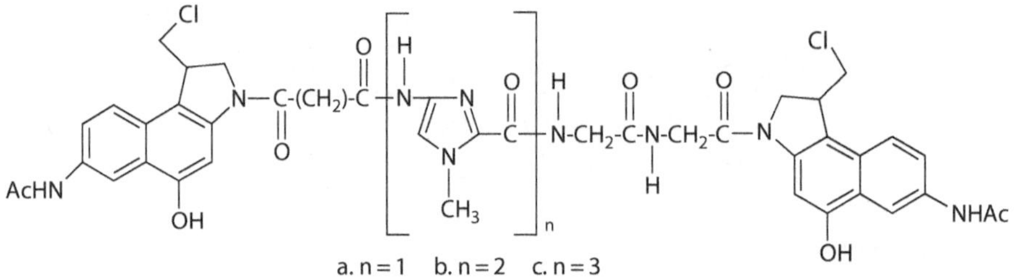

a. n = 1   b. n = 2   c. n = 3

C. C7-N3 *Seco*-CBI dimers

    1. With methylene spacer arm

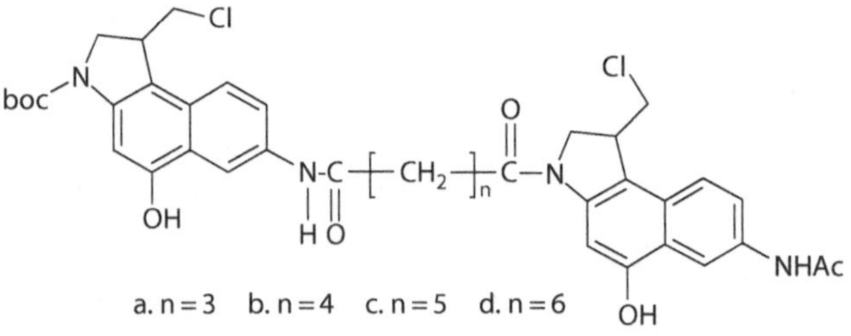

a. n = 3   b. n = 4   c. n = 5   d. n = 6

    2. With pyrrole polyamide spacer arm

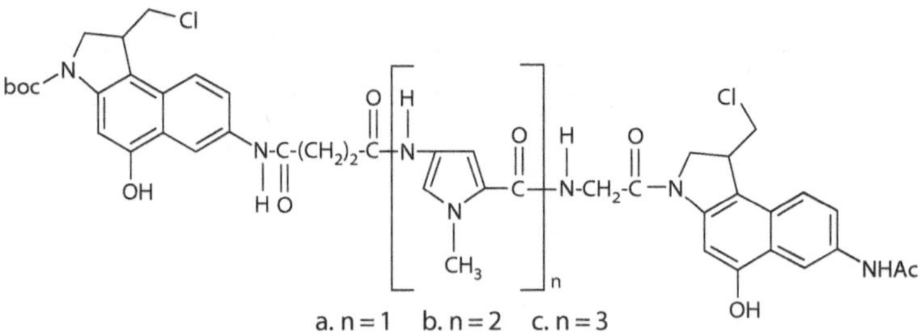

a. n = 1   b. n = 2   c. n = 3

TABLE 5.5 (Continued)
Nucleic Acid Cross-Linking Reagents

3. With imidazole polyamide spacer arm

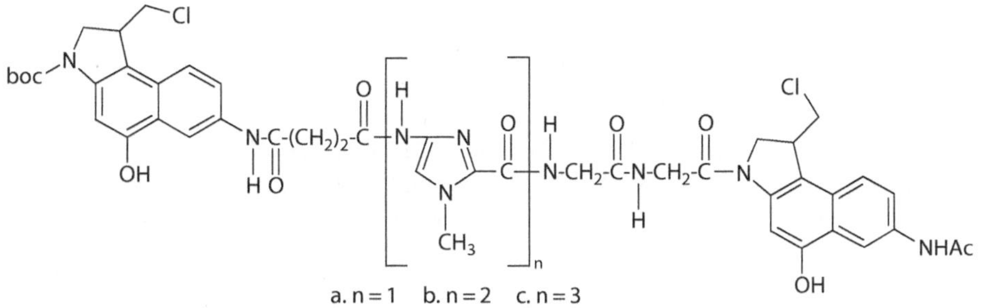

a. n = 1   b. n = 2   c. n = 3

VI. Diaziridinyl benzoquinone derivatives

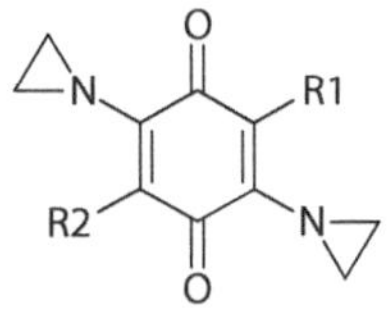

A. R1=R2=H: 2,5-Diaziridinyl-1,4-benzoquinone (DZQ)

B. R1=R2=CH3: 3,6-Dimethyl-2,5-diaziridinyl-1,4-benzoquinone (MeDZQ)

C. R1=CH3, R2=CH2OH: 2,5-Diaziridinyl-3-(hydroxymethyl)-6-methyl-1,4-benzoquinone (RH1)

D. R1=H, R2=phenyl: 2,5-Diaziridinyl-3-phenyl-6-methyl-1,4-benzoquinone (PDZQ)

E. R1=R2=NHCH2CH2OH: 3,6-Bis[(2-hydroxyethyl)amino]-2,5-diaziridinyl-1–4-benzoquinone (BZQ)

F. R1=R2=NHCOCH3: 2,5-Bis(carbomethoxyamino)-3,6-diaziridinyl-1,4-benzoquinone

G. R1=R2=NHCO-ethyl: 2,5-Bis(carboethoxyamino)-3,6-diaziridinyl-1,4-benzoquinone (AZQ)

H. R1=R2=NHCO-n-propyl: 2,5-Bis(carbo-n-propoxyamino)-3,6-diaziridinyl-1,4-benzoquinone

I. R1=R2=NHCO-isopropyl: 2,5-Bis(carboisopropoxyamino)-3,6-diaziridinyl-1,4-benzoquinone

J. R1=R2=NHCO-n-butyl: 2,5-Bis(carbo-n-butoxyamino)-3,6-diaziridinyl-1,4-benzoquinone

K. R1=R2=NHCO-isobutyl: 2,5-Bis(carbomethoxyamino)-3,6-diaziridinyl-1,4-benzoquinone

L. R1=R2=NHCO-sec-butyl: 2,5-Bis(carbo-sec-butoxyamino)-3,6-diaziridinyl-1,4-benzoquinone

M. R1=NHCOCH$_3$, R2=NHCO-n-propyl:
   2-carbomethoxyamino-5-carbo-n-propoxyamino-3,6-diaziridinyl-1,4-benzoquinone

N. R1=NHCOCH$_3$, R2=NHCO-n=butyl:
   2-carbomethoxyamino-5-carbo-n-butoxyamino-3,6-diaziridinyl-1,4-benzoquinone

O. R1=H, R2=

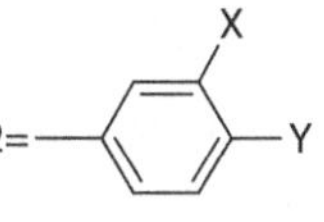

1. X=F, Y=H: 2,5-Diaziridinyl-3-(3′-fluorophenyl)-1,4-benzoquinone

2. X=H, Y=F: 2,5-Diaziridinyl-3-(4′-fluorophenyl)-1,4-benzoquinone

3. X=H, Y=CH2OH: 2,5-Diaziridinyl-3-[4′-(hydroxymethyl)phenyl]-1,4-benzoquinone

4. X=H, Y=OH: 2,5-Diaziridinyl-3-(4′-hydroxyphenyl)-1,4-benzoquinone

5. X=OH, Y=H: 2,5-Diaziridinyl-3-(3′-hydroxyhenyl)-1,4-benzoquinone

6. X=H, Y=CO-O-phenyl: 2,5-Diaziridinyl-3-(4′-phenoxycarbonylphenyl)-1,4-benzoquinone

7. X=H, Y=CH2-CO-O-phenyl: 2,5-Diaziridinyl-3-[4′-(benzoyloxymethyl)-phenyl]-1,4-benzoquinone

8. X=O-Co-phenyl, Y=H: 2,5-Diaziridinyl-3-[3′-(benzoyloxy)phenyl]-1,4-benzoquinone

(*continued*)

TABLE 5.5 (Continued)
Nucleic Acid Cross-Linking Reagents

P. R1=H, R2=

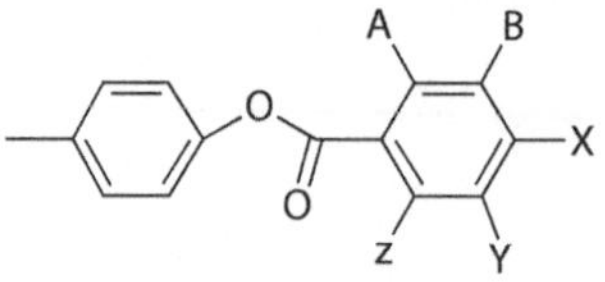

   1. A=B=X=Y=Z=H: 2,5-Diaziridinyl-3-[4'-(benzoyloxy)phenyl]-1,4-benzoquinone
   2. A=Z=H, B=X=Y=OCH3: 2,5-Diaziridinyl-3-[4'-(3″,4″,5″-trimethoxy-benzoyloxy)phenyl]-1,4-benzoquinone
   3. B=X=Y=H, A=Z=OCH3: 2,5-Diaziridinyl-3-[4'-(2″,6″-dimethoxy-benzoyloxy)phenyl]-1,4-benzoquinone

Q. R1=H, R2=

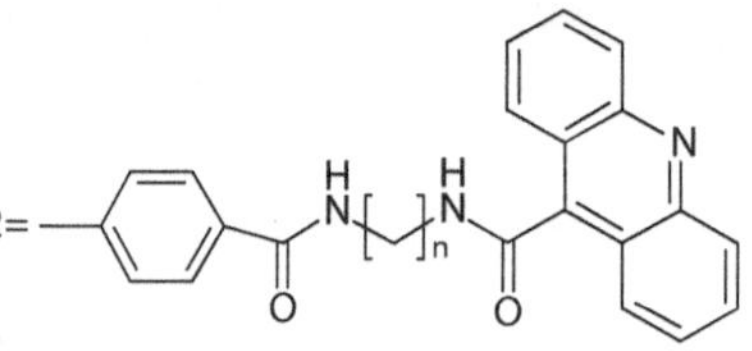

   1. n = 2: 2-{4'-Benzamido-N'-[2″-(acridine-9‴-carboxamido)ethyl]}-3,6-diaziridinyl-1,4-benzoquinone
   2. n = 3: 2-{4'-Benzamido-N'-[3″-(acridine-9‴-carboxamido)propyl]}-3,6-diaziridinyl-1,4-benzoquinone
   3. n = 4: 2-{4'-Benzamido-N'-[4″-(acridine-9‴-carboxamido)butyl]}-3,6-diaziridinyl-1,4-benzoquinone
   4. n = 5: 2-{4'-Benzamido-N'-[5″-(acridine-9‴-carboxamido)pentyl]}-3,6-diaziridinyl-1,4-benzoquinone
   5. n = 6: 2-{4'-Benzamido-N'-[6″-(acridine-9‴-carboxamido)hexyl]}-3,6-diaziridinyl-1,4-benzoquinone

R. R1=CH₃, R2=

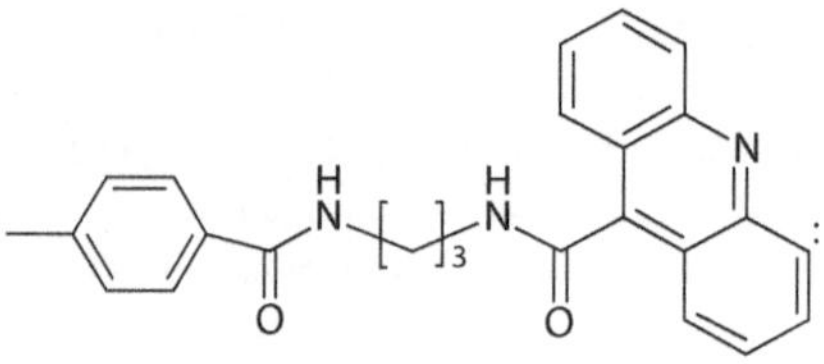

: 3-(4'-(N-(3″-Acridine-9‴-carboxamido)
propyl)-benzamido)-2,5-diaziridinyl-5-methylquinone

VII. Mitomycin C dimers
   A. Bismitomycin C linked by methylene spacer arm

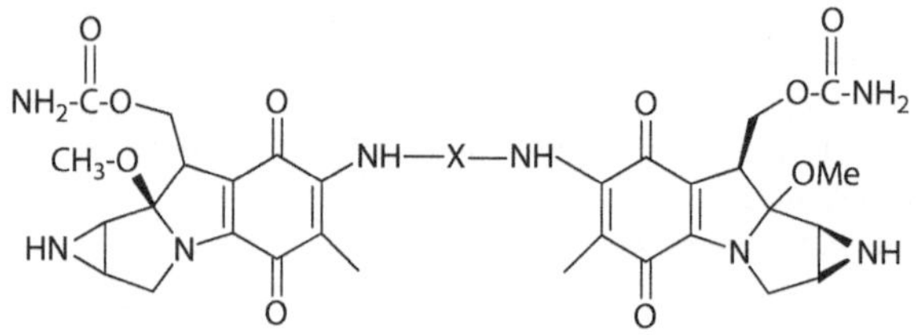

   1. X=(CH₂)₄   2. X=(CH₂)₇   3. X=(CH₂)₁₀   4. X=(CH₂)₂O(CH₂)₂   5. X=(CH₂)₃O(CH₂)₃
   6. X=(CH₂)₂NH(CH₂)₂   7. X=(CH₂)₃NH(CH₂)₃   8. X=(CH₂)₃N(CH₃)(CH₂)₃

   B. Bismitomycin C spacer containing disulfide: 7-N,7'-N'-(1″,2″-Dithianyl-3″,6″-dimethylenyl)bismitomycin C

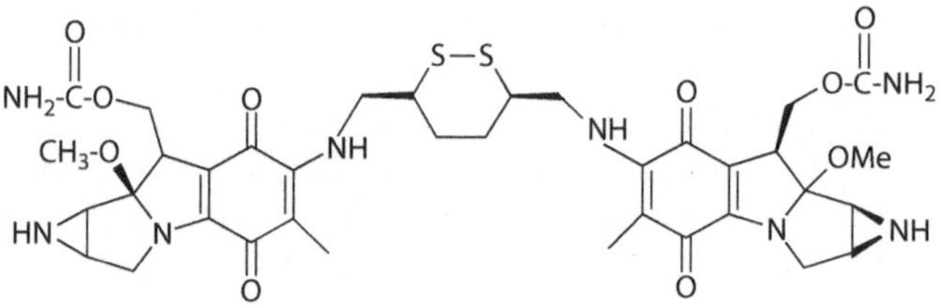

TABLE 5.5 (Continued)
Nucleic Acid Cross-Linking Reagents

C. 7-N,7'-N'-(1″,2″,9″,10″-Tetrathia-cyclohexadecanyl-3″,8″,11″,16″-tetramethylenyl)tetrakismitomycin C

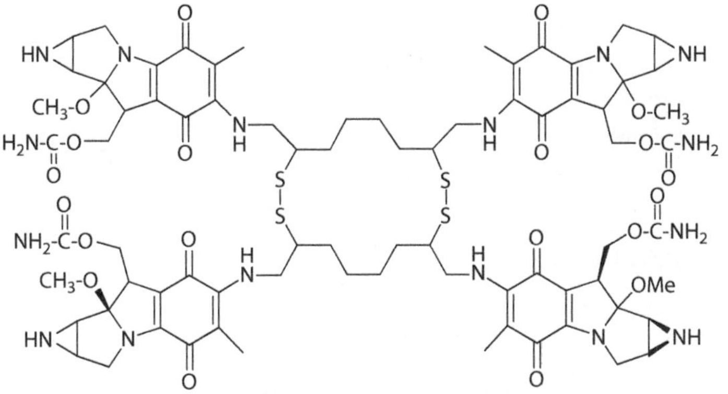

D. Bismitomycin C spacer containing cyclohexane: 7-N,7'-N'-(Cyclohexanyl-trans-1″,4″-dimethylenyl)bismitomycin C

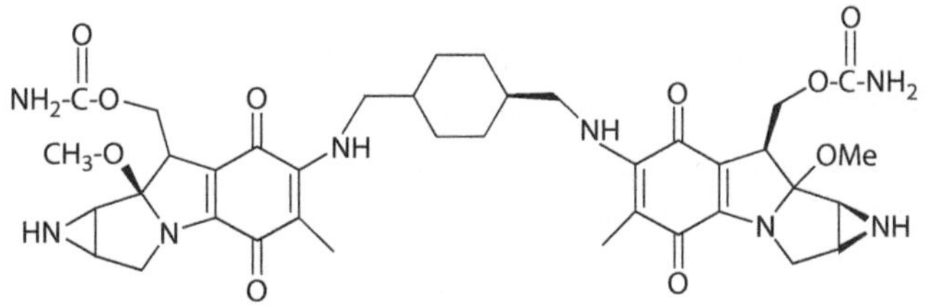

E. Bismitomycin C spacer containing dihydroxy group:

1. 7-N,7'-N'-(2″,5″-Dihydroxy-1″,6″-hexanediyl)bismitomycin C

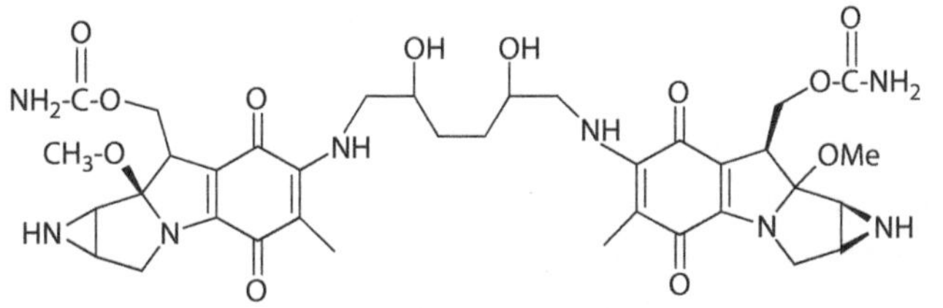

2. 7-N,7'-N'-(2″,7″-Dihydroxy-1″,8″-octanediyl)bismitomycin C

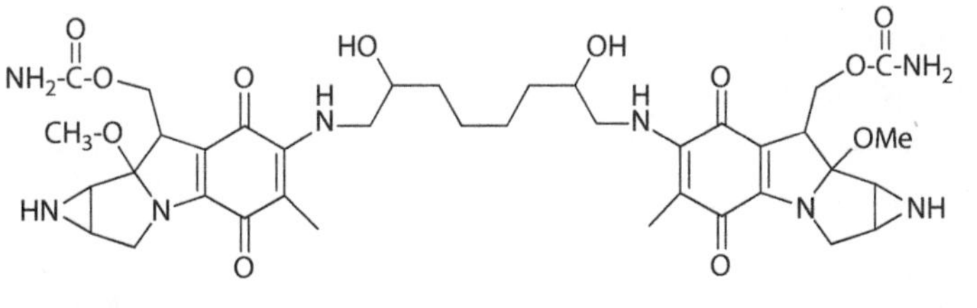

VIII. Chloroethylamine derivatives

A. N,N'-Dichloroethyl-N,N'-dimethylhexane-1,6-diamine

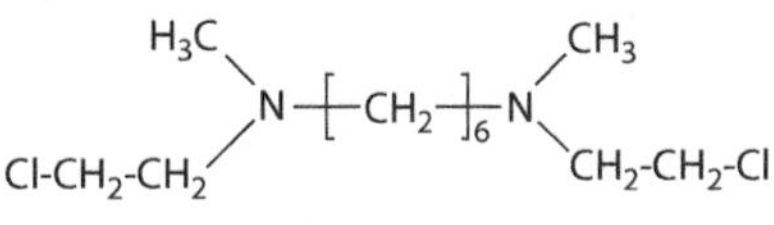

(continued)

TABLE 5.5 (Continued)
Nucleic Acid Cross-Linking Reagents

B. N'-Chloroethylaziridine

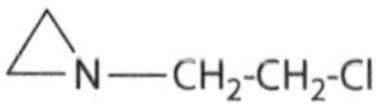

$$\triangle N—CH_2\text{-}CH_2\text{-}Cl$$

C. Dichloroethylpiperazines

    1. N,N'-Dichloroethylpiperazine

$$Cl\text{-}CH_2\text{-}CH_2—N \bigcirc N—CH_2\text{-}CH_2\text{-}Cl$$

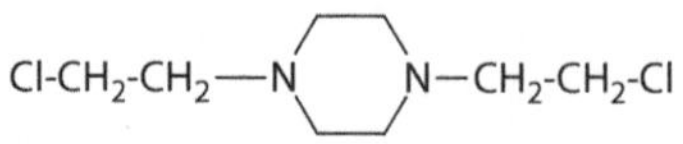

    2. Bis-chloromethylpiperidines

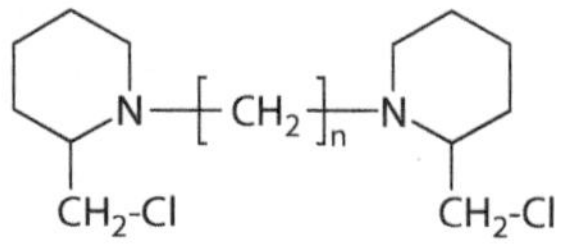

$$N—[CH_2]_n—N \quad CH_2\text{-}Cl \quad CH_2\text{-}Cl$$

        a. n = 2: N,N'-(1,2-Ethanediyl)bis(2-chloromethylpiperidine)

        b. n = 3: N,N'-(1,2-Propanediyl)bis(2-chloromethylpiperidine)

        c. n = 4: N,N'-(1,2-Butanediyl)bis(2-chloromethylpiperidine)

        d. n = 5: N,N'-(1,2-Pentanediyl)bis(2-chloromethylpiperidine)

        e. n = 6: N,N'-(1,2-Hexanediyl)bis(2-chloromethylpiperidine)

D. Bis-chloromethylpyrrolidines

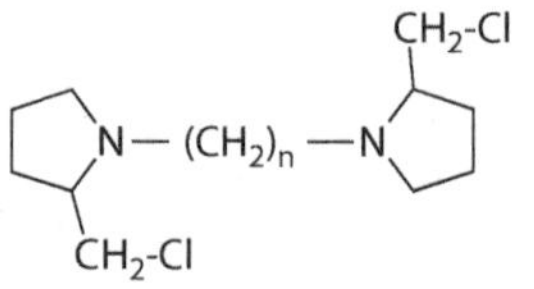

$$N—(CH_2)_n—N \quad CH_2\text{-}Cl \quad CH_2\text{-}Cl$$

    1. n = 2: N,N'-(1,2-Ethanediyl)bis(2S-2-chloromethylpyrrolidine)

    2. n = 3: N,N'-(1,3-Propanediyl)bis(2S-2-chloromethylpyrrolidine)

    3. n = 4: N,N'-(1,4-Butanediyl)bis(2S-2-chloromethylpyrrolidine)

    4. n = 5: N,N'-(1,5-Pentanediyl)bis(2S-2-chloromethylpyrrolidine)

    5. n = 6: N,N'-(1,6-Hexanediyl)bis(2S-2-chloromethylpyrrolidine)

    6. n = 7: N,N'-(1,7-Heptanediyl)bis(2S-2-chloromethylpyrrolidine)

    7. n = 8: N,N'-(1,8-Octanediyl)bis(2S-2-chloromethylpyrrolidine)

E. Polybenzamide half-mustards

    1. *Meta*-anilinocarboxamides

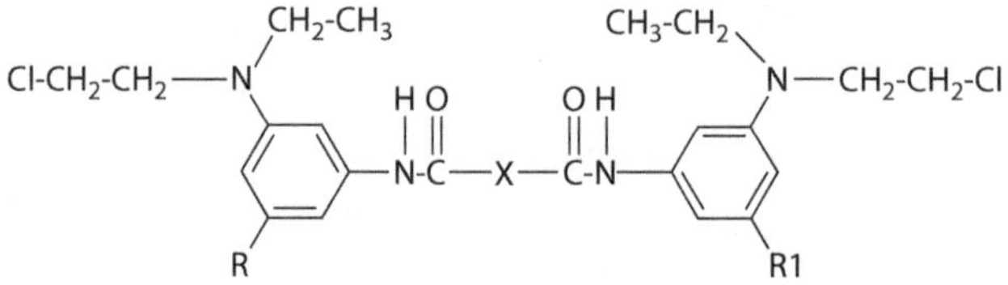

$$Cl\text{-}CH_2\text{-}CH_2—N(CH_2\text{-}CH_3)—[Ar(R)]—N\text{-}C(HO)—X—C(OH)\text{-}N—[Ar(R1)]—N(CH_3\text{-}CH_2)—CH_2\text{-}CH_2\text{-}Cl$$

        a. R=R1=CH_2NMe_2, X=  —⟨ ⟩— : N,N'-Bis[3-[N-(2-chloroethyl)-N-ethylamino]-5-(N,N-dimethylaminomethyl)phenyl]-1,4-benzenedicarboxamide, Alkamin

TABLE 5.5 (Continued)
Nucleic Acid Cross-Linking Reagents

b. R=R1=CH$_2$NMe$_2$, X= : N,N'-Bis[3-[N-(2-chloroethyl)-N-ethylamino]-5-(N,N-dimethylaminomethyl)phenyl]-1,3-benzenedicarboxamide

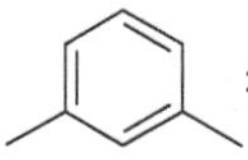

c. R=R1=CH$_2$NMe$_2$, X= : N,N'-Bis[3-[N-(2-chloroethyl)-N-ethylanfino]-5-(N,N-dimethylaminomethyl)phenyl]-1-methyl-3,5-pyrazoledicarboxamide

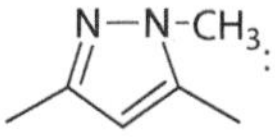

d. R=R1=CH$_2$NMe$_2$, X= : N,N'-Bis[3-[N-(2-chloroethyl)- N,N'-ethylamino]-5-(N,N'-dimethylaminomethyl)phenyl]-1,4-bicyclo[2.2.2] octanedicarboxamide

e. R=CH$_2$NMe$_2$, R1=H, X= : N-[3-[N-(2-Chloroethyl)-N-ethylamino]-5-(N,N-dimethylaminomethyl)phenyl]-N'-[3-(N-(2-chloroethyl)-N-ethylamino)phenyl]-1,4-benzenedicarboxamide

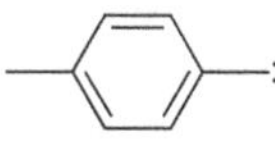

f. R=R1=H, X= : N,N'-Bis[3-[N-(2-chloroethyl)-N-ethylamino]phenyl]-N''-[2-(dimethylamino)ethyl]-1,3,5-benzenetricarboxamide

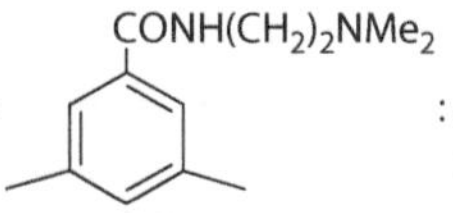

2. *Para*-anilinocarboxamide: N,N'-Bis[4-[N-(2-chloroethyl)-N-ethylamino]phenyl]-N''-[2-(dimethylamino)ethyl]-1,3,5-benzenetricarboxamide

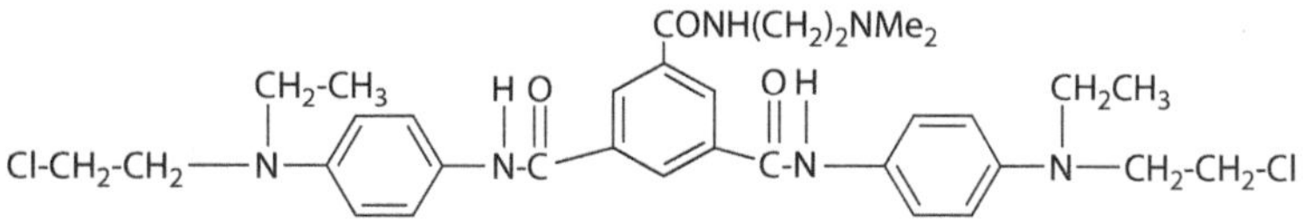

F. Nitrogen mustards

1. Mechlorethamine, chlormethine, chlorethazine, 2-chloro-N-(2-chloroethyl)-N-methylethanamine

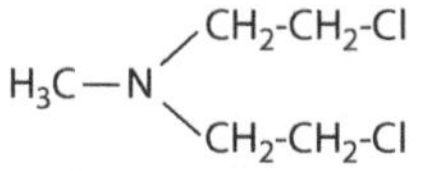

2. Phosphamide mustard, N,N-bis(2-chloroethyl)phosphorodiamidic acid

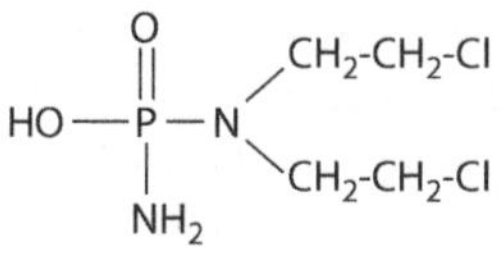

3. Cyclophosphamide, N,N-bis(2-chloroethyl)-1,3,2-oxazaphosphinan-2-amine 2-oxide

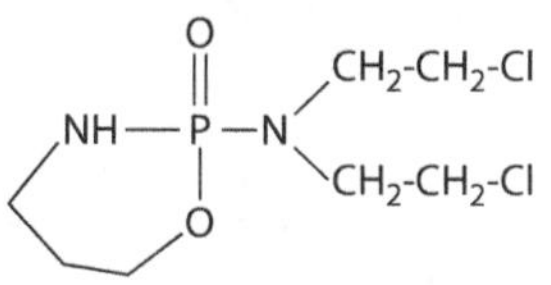

4. Ifosfamide, $N,3$-bis(2-chloroethyl)-1,3,2-oxazaphosphinan-2-amide-2-oxide

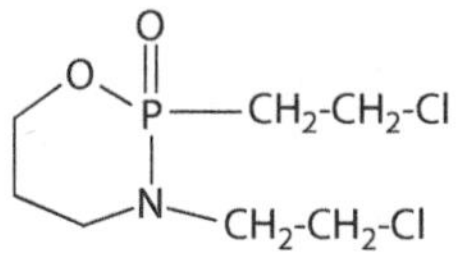

(continued)

TABLE 5.5 (Continued)
Nucleic Acid Cross-Linking Reagents

5. Simple derivatives of aniline mustard

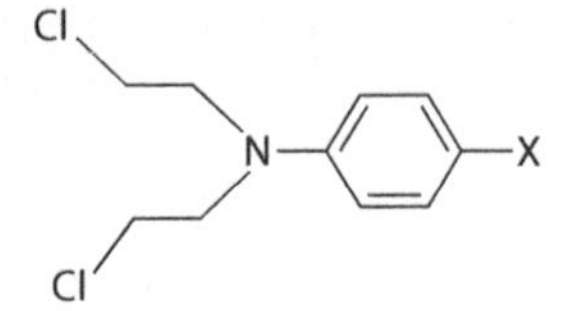

a. X=H: Aniline mustard, *N,N*-bis(2-chloroethyl)aniline
b. X=NH$_2$: 4-Amino-*N,N*-bis(2-chloroethyl)aniline
c. X=CH$_3$: 4-[*N,N*-bis(2-chloroethyl)amino]toluene
d. X=CH$_2$-CH$_2$-CH$_2$-COOH: Chlorambucil, 4-[*N,N*-bis(2-chloroethyl) amino]benzenebutanoic acid
e. X=CH$_2$-CH(NH$_2$)-COOH: Melphalan, 4-[*N,N*-bis(2-chloroethyl) amino]phenylalanine
f. X=O-CH$_3$: 4-[*N,N*-Bis(2-chloroethyl)amino]phenyl methyl ether
g. X=S-CH$_3$: 4-[*N,N*-Bis(2-chloroethyl)amino]phenylmethylsulfide
h. X=SO$_2$-CH$_3$: 4-[*N,N*-Bis(2-chloroethyl)amino]phenylmethylsulfone
i. X=NH-CO-CH$_3$: *N*-4-[*N,N*-Bis(2-chloroethyl)amino]phenyl acetamide
j. X=COOH: Benzoic acid mustard (BAM), 4-[*N,N*-bis(2-chloroethyl) amino]benzoic acid
k. X=CO-CH$_3$: 4-[*N,N*-Bis(2-chloroethyl)amino]phenyl methyl ketone
l. X=CO-N(CH$_3$)$_2$: 4-[*N,N*-Bis(2-chloroethyl)amino]-*N′,N′*-dimethyl benzamide

6. Benzoquinone mustards

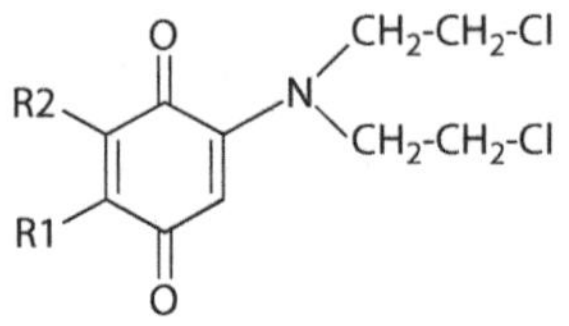

a. R1=R2=H: 2-[Di(chloroethyl)amino]-1,4-benzoquinone (BM)
b. R1=CH$_3$, R2=H: 5-Methyl-2-[di(chloroethyl)amino]-1,4-benzoquinone (MeBM)
c. R1=OCH$_3$, R2=H: 5-Methoxy-2-[di(chloroethyl)amino]-1,4-benzoquinone (BMB)
d. R1=Cl, R2=H: 5-Chloro-2-[di(chloroethyl)amino]-1,4-benzoquinone (CBM)
e. R1=F, R2=H: 5-Chloro-2-[di(chloroethyl)amino]-1,4-benzoquinone (FBM)
f. R1=phenyl, R2=H: 5-Chloro-2-[di(chloroethyl)amino]-1,4-benzoquinone (PBM)
g. R1=H, R2=CH$_3$: 5-Chloro-2-[di(chloroethyl)amino]-1,4-benzoquinone (m-MeBM)
h. R1=H, R2=phenyl: 5-Chloro-2-[di(chloroethyl)amino]-1,4-benzoquinone (m-PBM)
i. R1=H, R2=C(CH$_3$)$_3$: 5-Chloro-2-[di(chloroethyl)amino]-1,4-benzoquinone (mTBM)
j. R1=N(CH$_2$)$_2$Cl, R2=H: 2,5-Bis(di(2′-chloroethyl)amino)-1,4-benzoquinone (DBM)

7. Chlorambucil-spermidine conjugate

a. *N*-{3-[*N*-(3-Aminopropyl)-*N*-(4-aminobutyl)]aminopropyl}-4-[*p*-bis(2-chloroethyl)amino] phenylbutanamide

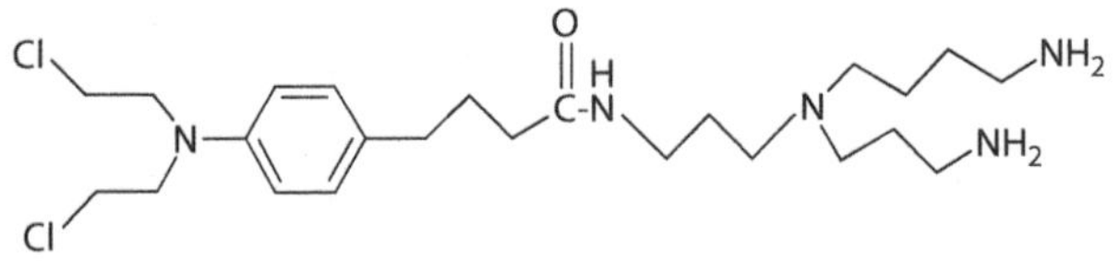

b. *N*-[4-[Bis(2-chloroethyl)amino]phenethyl]-*N*$^{\alpha}$-(3-aminopropyl)-L-ornithinamide

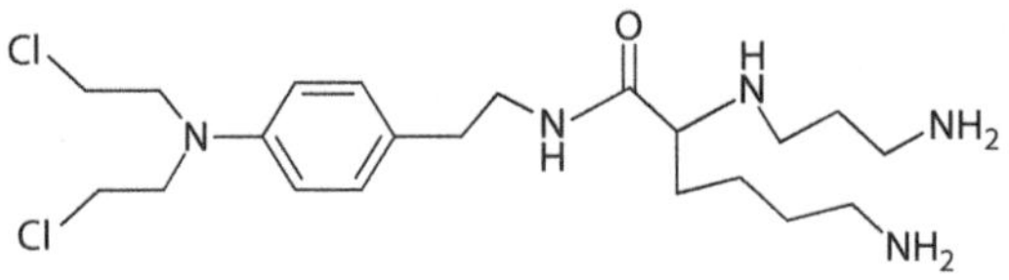

TABLE 5.5 (Continued)
Nucleic Acid Cross-Linking Reagents

8. Aniline mustards containing amidine

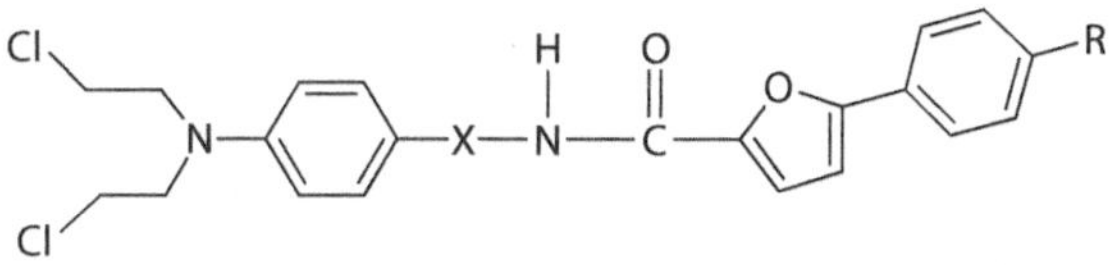

a. X=(CH$_2$)$_3$-CO-NH-CH$_2$-CH$_2$, R=C(=NH)NH$_2$: *N*-(2-(4-(4-Bis(2-chloroethyl)aminophenyl)butyryl) aminoethyl)-5-(4-amidinophenyl)-2-furancarboxamide

b. X=(CH$_2$)$_3$-CO-NH-CH$_2$-CH$_2$, R=C(=NH)NH-CH(CH$_3$)$_2$: *N*-(2-(4-(4-Bis(2-chloroethyl)aminophenyl) butyryl) aminoethyl)-5-[4-[(*N*-isopropyl)amidino)] phenyl-2-furancarboxamide

c. X=(CH$_2$)$_3$-CO-NH-CH$_2$-CH$_2$, R=C(=NH)NH- 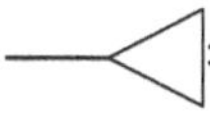 : *N*-(2-(4-(4-Bis(2-chloroethyl)aminophenyl) butyryl)aminoethyl)-5-[4-[(*N*-cyclopropyl) amidino)]phenyl-2-furancarboxamide

d. X=(CH$_2$)$_3$-CO-NH-CH$_2$-CH$_2$, R=C(=NH)NH 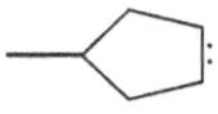 : *N*-(2-(4-(4-Bis(2-chloroethyl)aminophenyl)butyryl) aminoethyl)-5-[4-(*N*-cyclopentylamidino) phenyl]-2-furancarboxamide

e. X=(CH$_2$)$_3$-CO-NH-CH$_2$-CH$_2$, R= 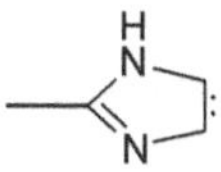 : *N*-(2-(4-(4-Bis(2-chloroethyl)aminophenyl)butyryl)- aminoethyl)-5-[4-(4,5-dihydro-1*H*-imidazol-2-yl) phenyl]-2-furancarboxamide

f. X=(CH$_2$)$_3$-CO-NH-CH$_2$-CH$_2$, R= 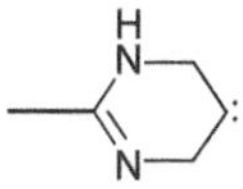 : *N*-(2-(4-(4-Bis(2-chloroethyl)aminophenyl)butyryl) aminoethyl)-5-[4-(3,4,5,6-tetrahydro-1*H*-pyrimidine-2-yl) phenyl]-2-furancarboxamide

g. X=CH$_2$-CH(COOCH$_3$), R=C(=NH)NH$_2$: Methyl-2-(5-(4-amidinophenyl)furan)amido-3-(4-(bis(2-chloroethyl)amino)-phenyl)propanoate

h. X=CH$_2$-CH(COOCH$_3$), R=C(=NH)NH-C(CH$_3$)$_2$: Methyl-3-(4-(bis(2-chloroethyl)amino)phenyl)-2-(2-(4-[(*N*-isopropyl)-amidino)]phenyl)furan-5-carboxamido) propanoate

i. X=CH$_2$-CH(COOCH$_3$), R=C(=NH)NH 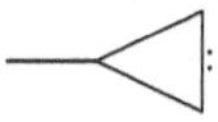 : Methyl-3-(4-(bis(2-chloroethyl)amino)phenyl)-2-((5-((4-(*N*-cyclopropyl)amidino)-phenyl)furan)amido)propanoate

j. X=CH$_2$-CH(COOCH$_3$), R=C(=NH)NH 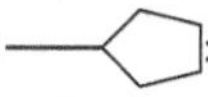 : Methyl-3-(4-(bis(2-chloroethyl)amino)phenyl)-2-((5-((4-(*N*-cyclopentyl)amidino)-phenyl)furan)amido)propanoate

k. X=CH$_2$-CH(COOCH$_3$), R= 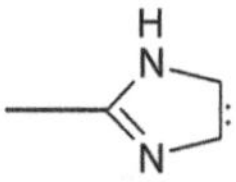 : Methyl-3-(4-(bis(2-chloroethyl)-amino)phenyl)-2-(2-(4-[(4,5-dihydro-1*H*-imidazol-2-yl)]-phenyl)furan-5-carboxamido) propanoate

l. X=CH$_2$-CH(COOCH$_3$), R= 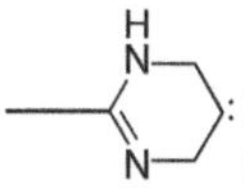 : Methyl-3-(4-(bis(2-chloroethyl)-amino)phenyl)-2-((5-(4-(1,4,5,6-tetrahydro-1*H*-pyrimidine-2-yl)phenyl)furan)amido)propanoate

m. X=(CH$_2$)$_3$, R=C(=NH)NH$_2$: *N*-(3-(4-Bis(2-chloroethyl)amino-phenyl) propyl)-5-(4-amidinophenyl)-2-furancarboxamide

n. X=(CH$_2$)$_3$, R=C(=NH)NH-C(CH$_3$)$_2$: *N*-(3-(4-Bis(2-chloroethyl)-aminophenyl)propyl)-5-(4-(*N*-isopropylamidino)phenyl)-2-furancarboxamide

(continued)

TABLE 5.5 (Continued)
Nucleic Acid Cross-Linking Reagents

o. X=(CH$_2$)$_3$, R=C(=NH)NH ——◁ : *N*-(3-(4-Bis(2-chloroethyl)-aminophenyl)propyl)-5-[4-(4,5-dihydro-1*H*-imidazol-2-yl)phenyl]-2-furancarboxamide

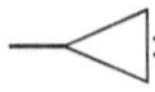

p. X=(CH$_2$)$_3$, R=C(=NH)NH ——⬠ : *N*-(3-(4-Bis(2-chloroethyl)-aminophenyl)propyl)-5-(4-(*N*-cyclopropylamidino)phenyl)-2-furancarboxamide

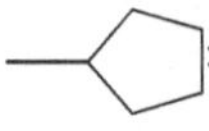

r. X=(CH$_2$)$_3$, R= ——[imidazoline] : *N*-(3-(4-Bis(2-chloroethyl)aminophenyl)-propyl)-5-(4-(*N*-cyclopentylamidino) phenyl)-2-furancarboxamide

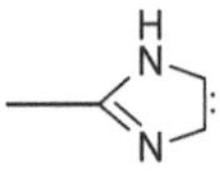

s. X=(CH$_2$)$_3$, R= ——[tetrahydropyrimidine] : *N*-(3-(4-Bis(2-chloroethyl)aminophenyl)-propyl)-5-[4-(3,4,5,6-tetrahydro-1*H*-pyrimidine-2-yl)phenyl]-2-furancarboxamide

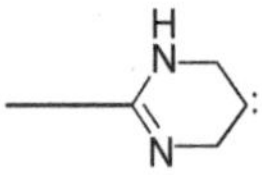

9. Bendamustine, 4-{5-[bis(2-chloro-ethyl)-amino]-1-methyl-1*H*-benzoimidazol-2-yl}-butanoic acid

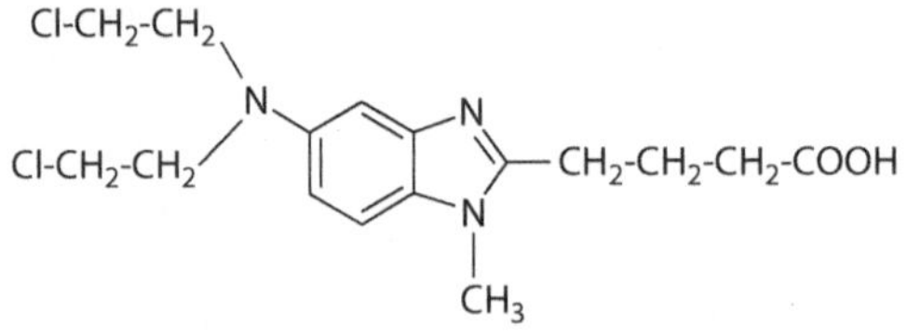

10. Aniline mustards linked to bis-benzimidazole at the para-position

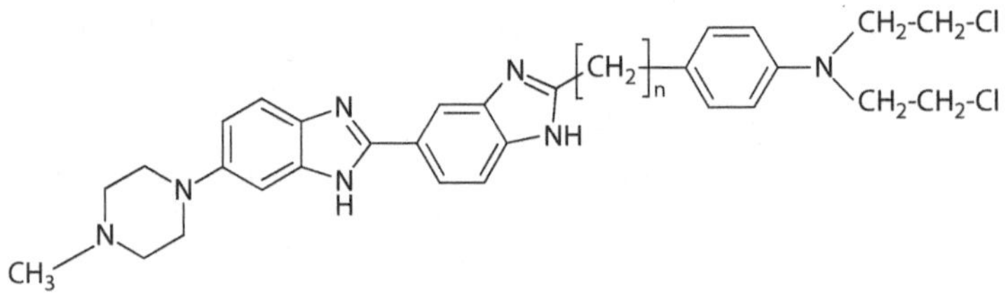

a. n = 0: 2-[2-[4-(*N*,*N*-Bis(2-chloroethyl)amino)phenyl]-5-benzimidazolyl]-5-(1-methyl-4-piperazinyl) benzimidazole

b. n = 1: 2-[2-[[4-(*N*,*N*-Bis(2-chloroethyl)amino)phenyl]methyl]-5-benzimidazolyl]-5-(1-methyl-4-piperazinyl)benzimidazole

c. n = 2: 2-[2-[2-[4-(*N*,*N*-Bis(2-chloroethyl)amino)phenyl]ethyl]-5-benzimidazolyl]-5-(1-methyl-4-piperazinyl)benzimidazole

d. n = 3: 2-[2-[3-[4-(*N*,*N*-Bis(2-chloroethyl)amino)phenyl]propyl]-5-benzimidazolyl]-5-(1-methyl-4-piperazinyl)benzimidazole

e. n = 6: 2-[2-[6-[4-(*N*,*N*-Bis(2-chloroethyl)amino)phenyl]hexyl]-5-benzimidazolyl]-5-(1-methyl-4-piperazinyl)benzimidazole

11. Aniline mustards linked to bis-benzimidazole carrier at the *ortho*- and *meta*-position

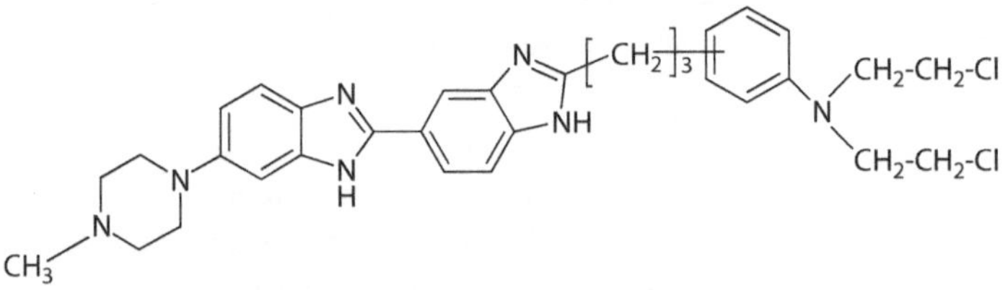

a. *Ortho*-: 2-[2-[3-[2-(*N*,*N*-Bis(2-chloroethyl)amino)phenyl]propyl]-5-benzimidazolyl]-5-(1-methyl-4-piperazinyl)benzimidazole

b. *Meta*-: 2-[2-[3-[3-(*N*,*N*-Bis(2-chloroethyl)amino)phenyl]propyl]-5-benzimidazolyl]-5-(1-methyl-4-piperazinyl)benzimidazole

TABLE 5.5 (Continued)
Nucleic Acid Cross-Linking Reagents

12. Aniline mustard linked to bis-benzimidazole analog benzoxazole carrier

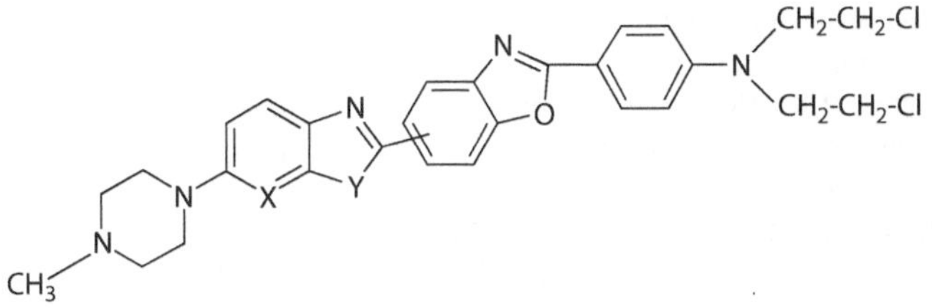

a. X=N, Y=NH, 6-yl: 2-[4-*N,N*-Bis(2-chloroethyl)aminophenyl] benzoxazol-6-yl-6-(4-methylpiperazin-1-yl) imidazolo[4,5-b]pyridine

b. X=N, Y=NH, 5-yl: 2-[4-*N,N*-Bis(2-chloroethyl)aminophenyl] benzoxazol-5-yl-6-(4-methylpiperazin-1-yl) imidazolo[4,5-b]pyridine

c. X=N, Y=O, 6-yl: 2-[4-*N,N*-Bis(2-chloroethyl)aminophenyl] benzoxazol-6-yl-6-(4-methylpiperazin-1-yl) oxazole[4,5-b]pyridine

d. X=CH, Y=NH, 6-yl: 2-[4-*N,N*-Bis(2-chloroethyl)aminophenyl] benzoxazol-6-yl-5-(4-methylpiperazin-1-yl) benzimidazole

13. Aniline mustard analogs of distamycin containing pyrrole

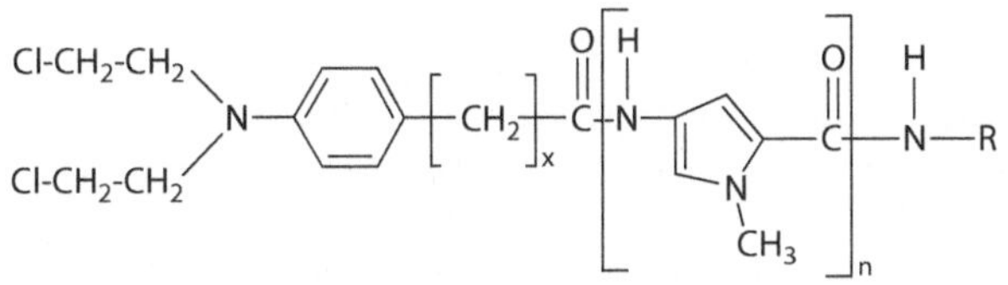

a. x = 0, n = 1, R=(CH$_2$)$_2$-N(CH$_3$)$_2$: 2-[1-Methyl-4-{4-bis(2-chloroethyl)-aminobenzamido} pyrrole-2-carboxamido]-dimethylaminoethane

b. x = 0, n = 2, R=(CH$_2$)$_3$-N(CH$_3$)$_2$: 3-[1-Methyl-4-{1-methyl-4-{4-bis(2-chloroethyl)aminobenzamido} pyrrole-2-carboxamido}pyrrole-2-carboxamido]dimethylaminopropane

c. x = 0, n = 3, R=(CH$_2$)$_3$-N(CH$_3$)$_2$: melphalan, 3-[1-methyl-4-{1-methyl-4-(1-methyl-4-(4-bis(2-chloroethyl) aminobenzamido)-pyrrole-2-carboxamido)pyrrole-2-carboxamido}pyrrole-2-carboxamido] dimethylaminopropane

d. x = 0, n = 3, R=(CH$_2$)$_2$-C(NH)NH$_2$ ·HCl: FCE 24517, tallimustine, 3-[1-methyl-4-[1-methyl-4-[1-methyl-4-[4-[*N,N*-bis(2-chloroethyl)amino]benzenecarboxamido]pyrrole-2-carboxamido]pyrrole-2-carboxamido] pyrrole-2-carboxamido]-propionamidine hydrochloride

e. x = 0, n = 3, R=(CH$_2$)$_2$-NH$_2$: *N*-(2-Aminoethyl)-1-methyl-4-[1-methyl-4-(1-methyl-4-[*p-N,N*-bis-(2-chloroethyl)benzamido]pyrrole-2-carboxamido)pyrrole-2-carboxamido]pyrrole-2-carboxamide

f. x = 0, n = 3, R=(CH$_2$)$_3$-NH$_2$: *N*-(3-Aminopropyl)-1-methyl-4-[1-methyl-4-(1-methyl-4-[*p-N,N*-bis-(2-chloroethyl)benzamido] pyrrole-2-carboxamido)pyrrole-2-carboxamido]pyrrole-2-carboxamide

g. x = 0, n = 3, R=(CH$_2$)$_4$-NH$_2$: *N*-(4-Aminobutyl)-1-methyl-4-[1-methyl-4-(1-methyl-4-[*p-N,N*-bis-(2-chloroethyl)benzamido] pyrrole-2-carboxamido)pyrrole-2-carboxamido]pyrrole-2-carboxamide

h. x = 0, n = 3, R=(CH$_2$)$_3$-NH-(CH$_2$)$_2$-NH$_2$: 3-[*N*-(2-aminoethyl) aminopropyl]-1-methyl-4-[1-methyl-4-[1-methyl-4-[*p-N,N*-bis(2-chloroethyl)aminobenzamido]pyrrole-2-carboxamido] pyrrole-2-carboxamido]-pyrrole-2-carboxamide

i. x = 0, n = 3, R=p-Ar-NH$_2$: *N-p*-Aminophenyl-1-methyl-4-[1-methyl-4-(1-methyl-4-[*p-N,N*-bis-(2-chloroethyl)aminobenzamido] pyrrole-2-carboxamido]pyrrole-2-carboxamido]pyrrole-2-carboxamide

j. x = 3, n = 1, R=(CH$_2$)$_2$-N(CH$_3$)$_2$: 2-[1-Methyl-4-{4-bis(2-chloroethyl)-aminophenylbutanamido} pyrrole-2-carboxamido]-dimethylaminoethane

k. x = 3, n = 2, R=(CH$_2$)$_2$-N(CH$_3$)$_2$: 2-[1-Methyl-4-{1-methyl-4-{4-bis(2-chloroethyl) aminophenylbutanamido}pyrrole-2-carboxamido}pyrrole-2-carboxamido]dimethylaminoethane

(*continued*)

TABLE 5.5 (Continued)
Nucleic Acid Cross-Linking Reagents

l. x = 3, n = 3, R=(CH$_2$)$_3$-N(CH$_3$)$_2$: 3-[1-Methyl-4-{1-methyl-4-(1-methyl-4-{4-bis(2-chloroethyl) aminophenylbutanamido}-pyrrole-2-carboxamido)pyrrole-2-carboxamido}pyrrole-2-carboxamido] dimethylaminopropane

m. x = 3, n = 3, R=(CH$_2$)$_2$-C(NH)NH$_2$·HCl, MEN 10569, 3-[1-methyl-4-{1-methyl-4-(1-methyl-4-{4-bis(2-chloroethyl)aminophenyl-butanamido}pyrrole-2-carboxamido)pyrrole-2-carboxamido} pyrrole-2-carboxamido]propionamidine hydrochloride

n. x = 3, n = 4, R=(CH$_2$)$_2$-C(NH)NH$_2$·HCl, MEN 10710, 3-[1-Methyl-4-{1-methyl-4-(1-methyl-4-(1-methyl-4-{4-bis(2-chloroethyl)-aminophenylbutanamido}pyrrole-2-carboxamido)pyrrole-2-carboxamideo) pyrrole-2-carboxamido}pyrrole-2-carboxamido] propionamidine hydrochloride

14. Nitrogen mustard analogs of distamycin containing pyrrole

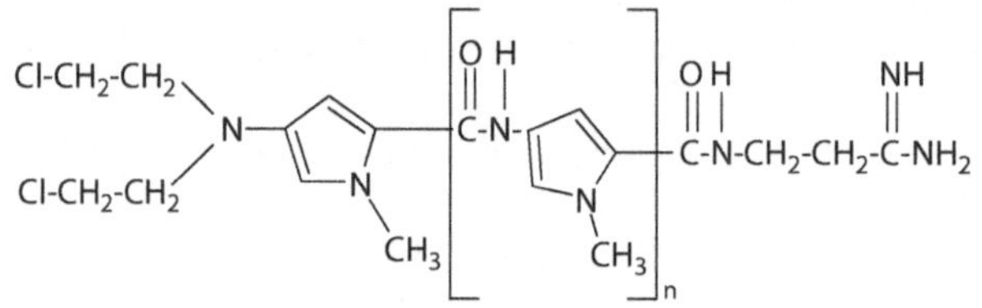

a. n = 1: 3-[1-Methyl-4-[1-methyl-4-[N,N-bis(2-chloroethyl)-amino]pyrrole-2-carboxamido] pyrrole-2-carboxamido]-propionamidine

b. n = 2: 3-[1-Methyl-4-[1-methyl-4-[1-methyl-4-[N,N-bis(2-chloroethyl)amino]pyrrole-2-carboxamido] pyrrole-2-carboxamido]pyrrole-2-carboxamido]propionamidine

c. n = 3: 3-[1-Methyl-4-[1-methyl-4-[1-methyl-4-[1-methyl-4-[N,N-bis(2-chloroethyl)amino]pyrrole-2-carboxamido]pyrrole-2-carboxamido]pyrrole-2-carboxamido]pyrrole-2-carboxamido]propionamidine

15. 2-Trimethylsilyethyl-1-methyl-4-[1-methyl-4-[N,N-bis(2-chloroethyl)amino]pyrrole-2-carboxamino] pyrrole-2-carboxylate

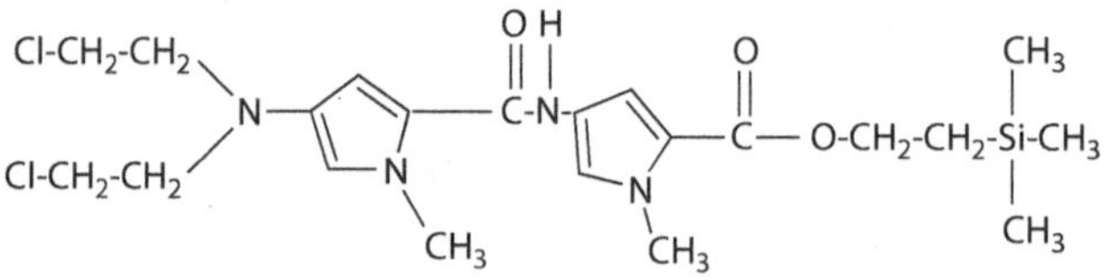

16. 3-[1-Methyl-4-[1-methyl-4-[1-methyl-4-[4-[N,N-bis(2-chloroethyl)amino]thiophene-2-carboxamido] pyrrole-2-carboxamido]pyrrole-2-carboxamido]pyrrole-2-carboxamidolpropionamidine

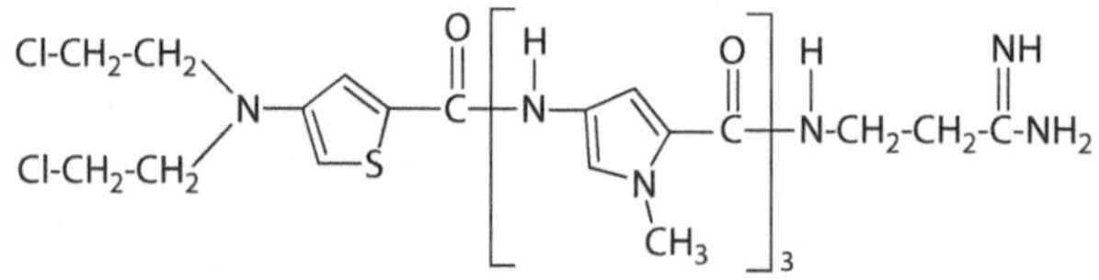

17. Mustard analogs of distamycin containing imidazole

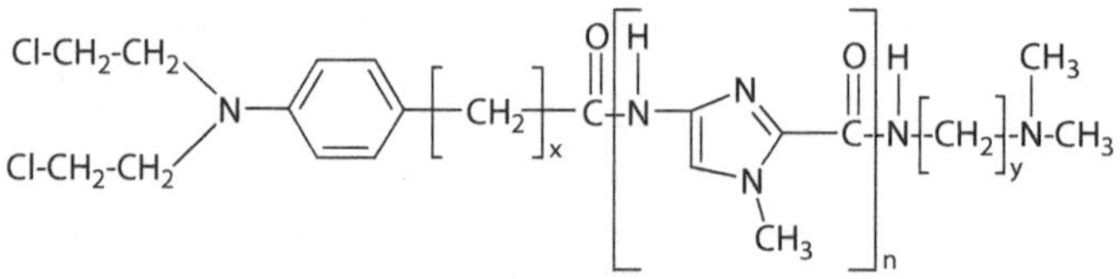

a. x = 0, n = 1, y = 2: 4[4-[Bis(2-chloroethyl)amino]benzamido]-N-[2-(dimethylamino) ethyl]-1-methylimidazole-2-carboxamide

b. x = 0, n = 2, y = 2: 4[[4-4-[Bis(2-chloroethyl)amino]benzamido]-1-methylimidazol-2-yl]carboxamido]-N-[2-(dimethylamino) ethyl]-1-methylimidazole-2-carboxamide

c. x = 0, n = 3, y = 2: 4[[4-[[4-4-[Bis(2-chloroethyl)amino]benzamido]-1-methylimidazol-2-yl] carboxamido]-1-methylimidazol-2-yl]carboromido]-N-[2-(dimethylamino) ethyl]-1-methylimidazole-2-carboxamide

TABLE 5.5 (Continued)
Nucleic Acid Cross-Linking Reagents

    d. x = 3, n = 1, y = 2: *N*-(*N'*,*N'*-Dimethylaminoethyl)-1-methyl-4-[4-(4-(bischloroethyl)aminophenyl) butanamido]imidazole-2-carboxamide

    e. x = 3, n = 2, y = 2: *N*-(*N'*,*N'*-Dimethylaminoethyl)-1-methyl-4-[4-(4-(bischloroethyl)aminophenyl) butanamido-1-methylimidazole-2-carboxamido]imidazole-2-carboxamide

    f. x = 3, n = 3, y = 2: *N*-(*N'*,*N'*-Dimethylaminoethyl)-1-methyl-4-{1-methyl-4-[4-(4-(bischloroethyl) aminophenyl)butanamido]-1-methylimidazole-2-carboxamido]imidazole-2-carboxamido} imidazole-2-carboxamide

18. Benzyl-mustard: *N*-(*N'*,*N'*-Dimethylaminoethyl)-1-methyl-4-{1-methyl-4-[4-(4-(bischloroethyl)aminobenzyl) amino-1-methylimidazole-2-carboxamido]imidazole-2-carboxamido}imidazole-2-carboxamide

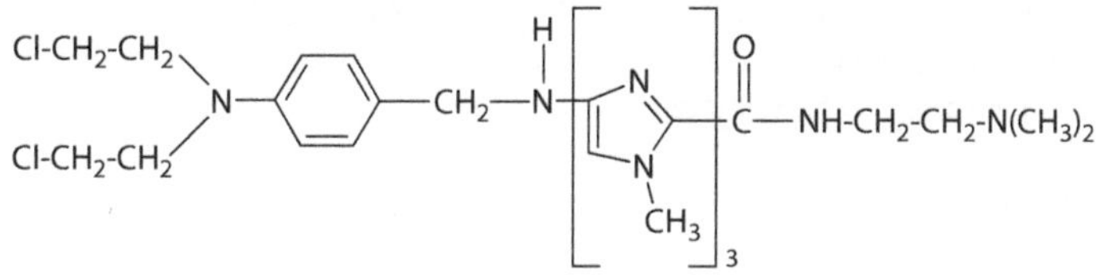

19. Mustard analogs of distamycin containing pyrrole and imidazole

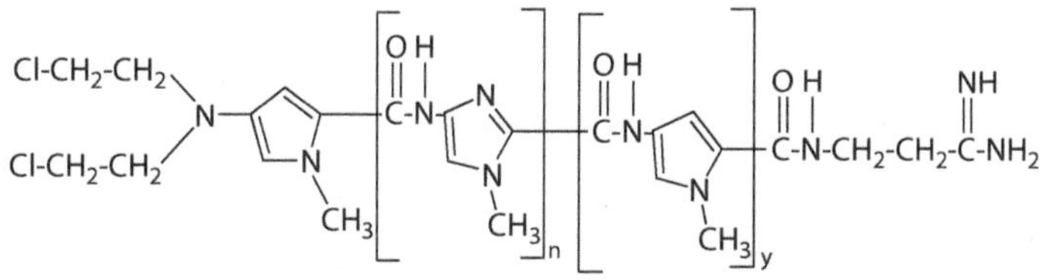

    a. n = 1, y = 0: 3-[1-Methyl-4-[1-methyl-4-[*N*,*N*-bis(2-chloroethyl) amino]pyrrole-2-carboxamido]imidazole-2-carboxamido]propionamidine

    b. n = 1, y = 1: 3-[1-Methyl-4-[1-megthyl-4-[1-methyl-4-[*N*,*N* -bis(bis(2-chloroethyl)amino)pyrrole-2-carboxamido]imidazole-2-carboxamido]pyrrole-2-carboxamido]propionamidine

    c. n = 2, y = 0: 3-[1-Methyl-4-[1-methyl-4-[1-methyl-4-[*N*,*N* -bis(2-chloroethyl)amino]pyrrole-2-carboxamido]imidazole-2-carboxamido]imidazole-2-carboxamido]propionamidine

20. Benzoyl mustard analogs of distamycin containing pyrazole

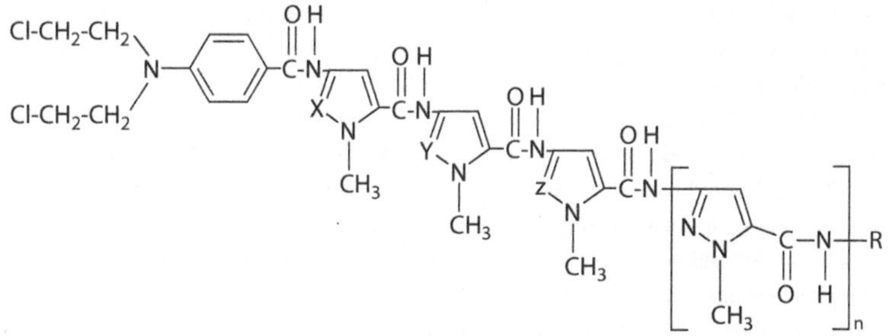

    a. X=CH, Y=CH, Z=N, n = 0, R=CH₂-CH₂-C(NH)NH₂: 3-[1-Methyl-3-[1-methyl-3-[1-methyl-4-[4-bis(2-chloroethyl)aminophenyl amido)]pyrrole-2-carboxamido]pyrrole-2-carboxamido] pyrazole-5-carboxamido] propionamidine

    b. X=CH, Y=N, Z=N, n = 0, R=CH₂-CH₂-C(NH)NH₂: 3-[1-Methyl-3-[1-methyl-3-[1-methyl-4-[4-bis(2-chloroethyl)aminophenyl amido)]pyrrole-2-carboxamido]pyrazole-5-carboxamido] pyrazole-5-carboxamido]propionamidine

    c. X=N, Y=CH, Z=CH, n = 0, R=CH₂-CH₂-C(NH)NH₂: 3-[1-Methyl-4-[1-methyl-4-[1-methyl-3-[4-bis(2-chloroethyl)aminophenyl amido]pyrazole-5-carboxamido]pyrrole-2-carboxamido] pyrrole-2-carboxamido] propionamidine

(*continued*)

TABLE 5.5 (Continued)

Nucleic Acid Cross-Linking Reagents

  e. X=N, Y=N, Z=CH, n = 0, R=CH₂-CH₂-C(NH)NH₂: 3-[1-Methyl-4-[1-methyl-3-[1-methyl-3-[4-bis(2-chloroethyl)aminophenyl amido]pyrazole-5-carboxamido]pyrazole-5-carboxamido] pyrrole-2-carboxamido]propionamidine

  f. X=N, Y=N, Z=N, n = 0, R=CH₂-CH₂-C(NH)NH₂: 3-[1-Methyl-3-[1-methyl-3-[1-methyl-3-[4-bis(2-chloroethyl)aminophenyl amido]pyrazole-5-carboxamido]pyrazole-5-carboxamido] pyrazole-5-carboxamido]propionamidine

  g. X=N, Y=N, Z=N, n = 0, R=CH₂-CH₂-CH₂-C(CH₃)₂: 3-[1-Methyl-3-[1-methyl-3-[1-methyl-3-[4-bis(2-chloroethyl)aminophenyl amido]-pyrazole-5-carboxamido]pyrazole-5-carboxamido] pyrazole-5-carboxamido]dimethylaminopropane

  h. X=CH, Y=CH, Z=N, n = 1, R=CH₂-CH₂-C(NH)NH₂: 3-[1-Methyl-3-[1-methyl-3-[1-methyl-4-[1-methyl-4-[4-bis-(2-chloroethyl) aminobenzenamido]pyrrole-2-carboxamido]pyrrole-2-carboxamido]pyrazole-5-carboxamido]pyrazole-5-carboxamido]propionamidine

  i. X=N, Y=CH, Z=N, n = 1, R=CH₂-CH₂-C(NH)NH₂: 3-[1-Methyl-3-[1-methyl-3-[1-methyl-4-[1-methyl-4-[4-bis-(2-chloroethyl) aminobenzenamido]pyrrole-2-carboxamido]pyrrole-2-carboxamido]pyrazole-5-carboxamido]pyrazole-5-carboxamido]propionamidine

 21. Benzoyl and cinnamoyl nitrogen mustards containing benzoheterocycle and pyrrole

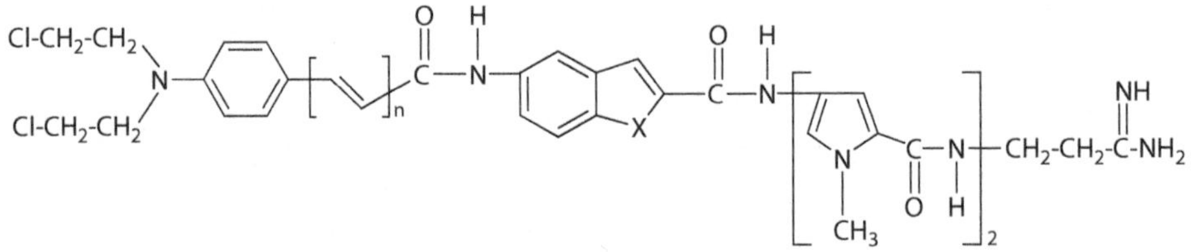

  a. n = 0, x=NH: 3-[1-Methyl-4-[1-methyl-4-[5-[4-*N*,*N*-bis(2-chloroethyl) aminobenzene-1-carboxamido]indole-2-carboxamido] pyrrole-2-carboxamido]pyrrole-2-carboxamido] propionamidine

  b. n = 0, x=N-CH₃: 3-[1-Methyl-4-[1-methyl-4-[1-methyl-5-[4-*N*,*N*-bis(2-chloroethyl)aminobenzene-1-carboxamido]indole-2-carboxamido]pyrrole-2-carboxamido]pyrrole-2-carboxamido] propionamidine

  c. n = 0, x = O: 3-[1-Methyl-4-[1-methyl-4-[5-[4-*N*,*N*-bis(2-chloroethyl)aminobenzene-1-carboxamido]benzofuran-2-carboxamido]pyrrole-2-carboxamido]pyrrole-2-carboxamido]propionamidine

  d. n = 0, x = S: 3-[1-Methyl-4-[1-methyl-4-[5-[4-*N*,*N*-bis(2-chloroethyl)aminobenzene-1-carboxamido]benzothiophene-2-carboxamido]pyrrole-2-carboxamido]pyrrole-2-carboxamido]propionamidine

  e. n = 1, x=NH: 3-[1-Methyl-4-[1-methyl-4-[5-[4-*N*,*N*-bis(2-chloroethyl)aminocinnamoyl]aminoindole-2-carboxamido]pyrrole-2-carboxamido]pyrrole-2-carboxamido]propionamidine

  f. n = 1, x=N-CH₃: 3-[1-Methyl-4-[1-methyl-4-[1-methyl-5-[4-*N*,*N*-bis(2-chloroethyl)aminocinnamoyl]aminoindole-2-carboxamido]pyrrole-2-carboxamido]pyrrole-2-carboxamido]propionamidine

  g. n = 1, x=O: 3-[1-Methyl-4-[1-methyl-4-[5-[4-*N*,*N*-bis(2-chloroethyl)-aminocinnamoyl]amino benzofuran-2-carboxamido]pyrrole-2-carboxamido]pyrrole-2-carboxamido]propionamidine

  h. n = 1, x=S: 3-[1-Methyl-4-[1-methyl-4-[5-[4-*N*,*N*-bis(2-chloroethyl)-aminocinnamoyl]amino benzothiophene-2-carboxamido] pyrrole-2-carboxamido]pyrrole-2-carboxamido] propionamidine

 22. Chlorambucil linked to pyrrole–imidazole polyamide

  a. ImPy-β-ImPy-(R)CHL-γ-ImPy-β-ImPy-β-Dp

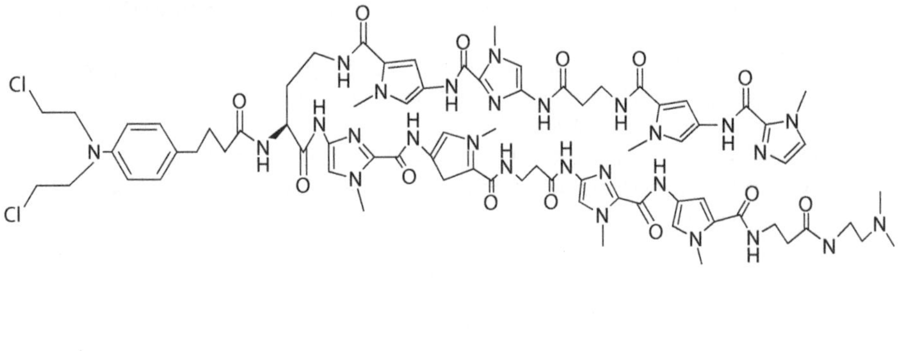

TABLE 5.5 (Continued)

Nucleic Acid Cross-Linking Reagents

b. ImIm-β-Im-(R/S)CHL-α-PyPyPyPy-β-Dp

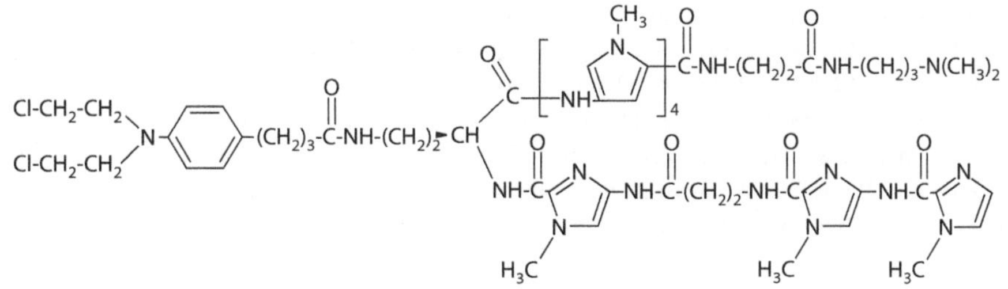

c. ImIm-β-Im-(R/S)CHL-γ-PyPyPyPy-β-Dp

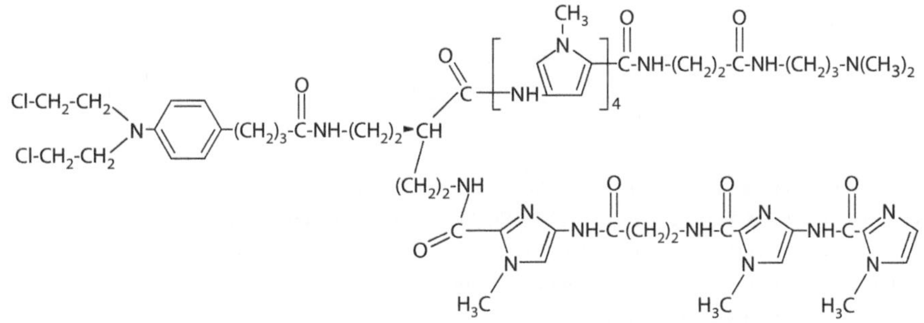

23. Anilinoquinoline-based aniline mustards

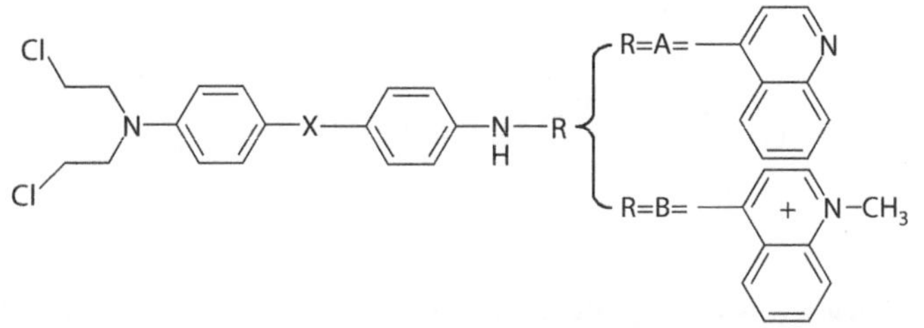

a. R=A, X=O: 4-(4-Aminoquinolyl)phenyl 4′-[bis(2-chloroethyl)amino]phenyl ether

b. R=A, X=CH₂: 4-(4-Aminoquinolyl)phenyl 4′-[bis(2-chloroethyl)amino]phenylmethane

c. R=A, X=S: 4-(4-Aminoquinolyl)phenyl 4′-[bis(2-chloroethyl)amino]phenylsulfide

d. R=A, X=CONH: N-4-(4-Aminoquinolyl)phenyl 4′-[bis(2-chloroethyl)amino]benzamide

e. R=A, X=CO: 4-(4-Aminoquinolyl)phenyl 4′-[bis(2-chloroethyl)amino]phenylketone

f. R=A, X=SO₂: 4-(4-Aminoquinolyl)phenyl 4′-[bis(2-chloroethyl)amino]phenyl sulfone

g. R=B, X=O: 4-(N-Methyl-4-aminoquinolinium)phenyl 4′-[bis(2-chloroethyl)amino]phenyl ether

h. R=B, X=CH₂: 4-(N-Methyl-4-aminoquinolinium)phenyl 4′-[bis(2-chloroethyl)amino]phenylmethane

i. R=B, X=CONH: N-4-(N-Methyl-4-aminoquinolinium)phenyl 4′-[bis(2-chloroethyl)amino]benzamide

j. R=B, X=CO: 4-(N-Methyl-4-aminoquinolinium)phenyl 4′-[bis(2-chloroethyl)amino]phenylketone

24. Anthraquinone-linked alkyl mustards

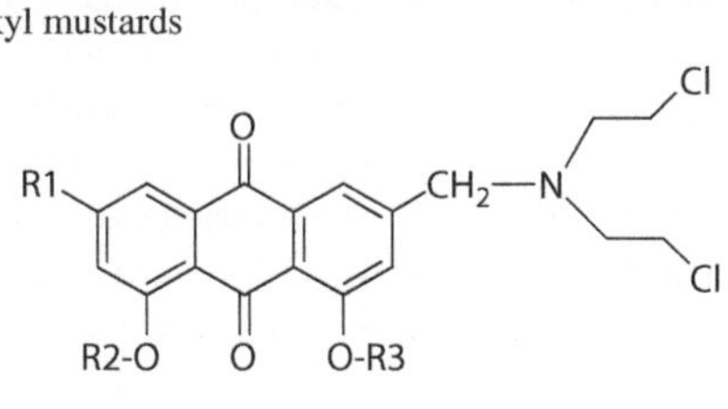

(continued)

TABLE 5.5 (Continued)
Nucleic Acid Cross-Linking Reagents

a. R1=H, R2=R3=Me: 3-[bis(2-chloroethyl)amino]-1,8-di-*O*-methylchrysophanol

b. R1=H, R2=Me, R3=H: 3-[Bis(2-chloroethyl)amino]-8-*O*-methylchrysophanol

c. R1=R2=H, R3=Me: 3-[Bis(2-chloroethyl)amino]-1-*O*-methylchrysophanol

d. R1=R2=R3=H: 3-[Bis(2-chloroethyl)amino]chrysophanol

e. R1=OMe, R2=R3=Me: 6-[Bis(2-chloroethyl)amino]-1,3,8-tri-*O*-methylemodin

f. R1=OMe, R2=Me, R3=H: 6-[Bis(2-chloroethyl)amino]-1,3-di-*O*-methylemodin

g. R1=OMe, R2=H, R3=Me: 6-[Bis(2-chloroethyl)amino]-3,8-di-*O*-methylemodin

h. R1=OMe, R2=R3=H: 6-[Bis(2-chloroethyl)amino]-3-*O*-methylemodin

25. Alkyl mustards linked to acridine

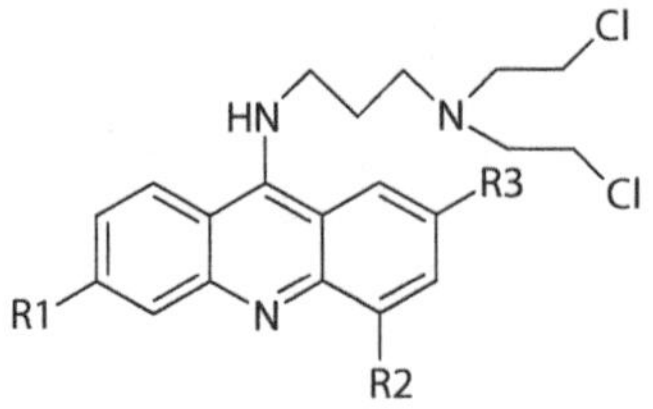

a. R1=R2=R3=H: 9-[3-Bis(2-chloroethyl)aminopropylamino]acridine

b. R1=R2=H, R3=OMe: 2-Methoxy-9-[3-bis(2-chloroethyl)amino propylamino]acridine

c. R1=R3=H, R2=OMe: 4-Methoxy-9-[3-bis(2-chloroethyl)amino propylamino]acridine

d. R1=Cl, R2=H, R3=Ø: 2-Phenyl-6-chloro-9-[3-bis(2-chloroethyl) aminopropylamino]acridine

26. Aniline mustards linked to 9-aminoacridine

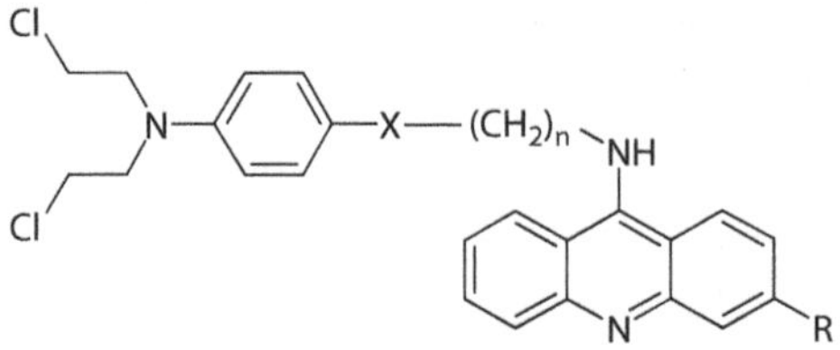

a. R=H, X=CH$_2$, n = 2: 9-[3-[4-[*N*,*N*-Bis(2-chloroethyl)amino]phenyl]-propylamino]acridine

b. R=H, X=CH$_2$, n = 3: 9-[4-[4-[*N*,*N*-Bis(2-chloroethyl)amino]phenyl]-butylamino]acridine

c. R=H, X=CH$_2$, n = 4: 9-[5-[4-[*N*,*N*-Bis(2-chloroethyl)amino]phenyl]-pentylamino]acridine

d. R=H, X=CH$_2$, n = 5: 9-[6-[4-[*N*,*N*-Bis(2-chloroethyl)amino]phenyl]-hexylamino]acridine

e. R=H, X=O, n = 2: 9-[2-[4-[*N*,*N*-Bis(2-chloroethyl)amino]phenoxy]-ethylamino]acridine

f. R=H, X=O, n = 3: 9-[3-[4-[*N*,*N*-Bis(2-chloroethyl)amino]phenoxy]-propylamino]acridine

g. R=H, X=O; n = 4: 9-[4-[4-[*N*,*N*-Bis(2-chloroethyl)amino]phenoxy]-butylamino]acridine

h. R=H, X=O, n = 5: 9-[5-[4-[*N*,*N*-Bis(2-chloroethyl)amino]phenoxy]-pentylamino]acridine

i. R=H, X=S, n = 2: 9-[2-[[4-[*N*,*N*-Bis(2-chloroethyl)amino]phenyl]-thio]ethylamino]acridine

j. R=H, X=S, n = 3: 9-[3-[[4-[*N*,*N*-Bis(2-chloroethyl)amino]phenyl]-thio]propylamino]acridine

k. R=H, X=S, n = 4: 9-[4-[[4-[*N*,*N*-Bis(2-chloroethyl)amino]phenyl]-thio]butylamino]acridine

l. R=H, X=S, n = 5: 9-[5-[[4-[*N*,*N*-Bis(2-chloroethyl)amino]phenyl]-thio]pentylamino]acridine

m. R=H, X=SO$_2$, n = 2: 9-[2-[[4-[*N*,*N*-Bis(2-chloroethyl)amino]phenyl]-sulfonyl]ethylamino]acridine

n. R=H, X=SO$_2$, n = 3: 9-[3-[[4-[*N*,*N*-Bis(2-chloroethyl)amino]phenyl]-sulfonyl]propylamino]acridine

o. R=H, X=SO$_2$, n = 4; 9-[4-[[4-[*N*,*N*-Bis(2-chloroethyl)amino]phenyl]-sulfonyl]butylamino]acridine

p. R=H, X=SO$_2$, n = 5: 9-[5-[[4-[*N*,*N*-Bis(2-chloroethyl)amino]phenyl]-sulfonyl]pentylamino]acridine

q. R=H, X=CO, n = 5: 9-[5-[4-[*N*,*N*-Bis(2-chloroethyl)amino]benzyol]-pentylamino]acridine

r. R=H, X=NHCO, n = 4: *N*,*N*-Bis(2-chloroethyl)-4-[5-[*N*-(9-acridine)-amino]pentanamido]aniline

s. R=H, X=CONH, n = 4: *N*-[4-(9-Acridinylamino)butyl]-4-[*N*,*N*-bis(2-chloroethyl)amino]benzamide

t. R=C(CH$_3$)$_3$, X=CH$_2$, n = 2: 3-*t*-Butyl-9-[3-[4-[*N*,*N*-bis(2-chloroethyl)amino]phenyl]propylamino]acridine

TABLE 5.5 (Continued)
Nucleic Acid Cross-Linking Reagents

27. Aniline mustards linked to 1'-position of 9-anilinoacridine

a. X=CH$_2$, n = 2: N-[3-Methoxy-4-(9-acridinylamino)phenyl]-4-[4-(N,N-bis-(2-chloroethyl)amino)phenyl] butanamide

b. X=O, n = 2: N-[3-Methoxy-4-(9-acridinylamino)phenyl]-3-[4-(N,N-bis-(2-chloroethyl)amino)phenoxy] propanamide

c. X=O, n = 5: N-[3-Methoxy-4-(9-acridinylamino)phenyl]-5-[4-(N,N-bis-(2-chloroethyl)amino)phenoxy] pentanamide

d. X=S, n = 2: N-[3-Methoxy-4-(9-acridinylamino)phenyl]-3-[4-(N,N-bis(2-chloroethyl)amino)thiophenyl] propanamide

e. X=SO$_2$, n = 2: N-[3-Methoxy-4-(9-acridinylamino)phenyl]-3-[4-(N,N-bis(2-chloroethyl)amino) sulfonylphenyl]propanamide

28. Aniline mustards linked to 4-position of 9-anilinoacridine

a. R=MeSO$_2$, X=CH$_2$, n = 2: 4-[4-(N-(3-(4-(N,N-Bis(2-chloroethyl)amino)phenyl)propyl) carboxamido)-9-acridinylamino]-3-methoxymethanesulfonanilide

b. R=MeCO, X=CH$_2$, n = 2: N-[3-Methoxy-4-(4-(N-(3-(4-(N,N-bis-(2-chloroethyl)amino)phenyl)propyl) carboxamido)-9-acridinylamino)phenyl]acetamide

c. R=MeCO, X=O, n = 2: N-[3-Methoxy-4-(4-N-(2-(4-(N,N-bis-(2-chloroethyl)amino)phenoxy)ethyl) carboxamido)-9-acridinylamino)]acetamide

d. R=MeCO, X=O, n = 5: N-[3-Methoxy-4-(4-N-(5-(4-(N,N-bis-(2-chloroethyl)amino)phenoxy)pentyl) carboxamido)-9-acridinylamino)]acetamide

e. R=MeCO, X=S, n = 2: N-[3-Methoxy-4-(4-(N-(2-(4-(N,N-bis(2-chloroethyl)amino)thiophenyl)ethyl) carboxamido)-9-acridinylamino)phenyl]acetamide

f. R=MeCO, X=SO$_2$, n = 2: N-[3-Methoxy-4-(4-(N-(2-(4-(N,N -bis(2-chloroethyl)amino)sulfonylphenyl) ethyl)carboxamido)-9-acridinylamino)phenyl]acetamide

(continued)

TABLE 5.5 (Continued)
Nucleic Acid Cross-Linking Reagents

29. Aniline mustards linked to naphthoate

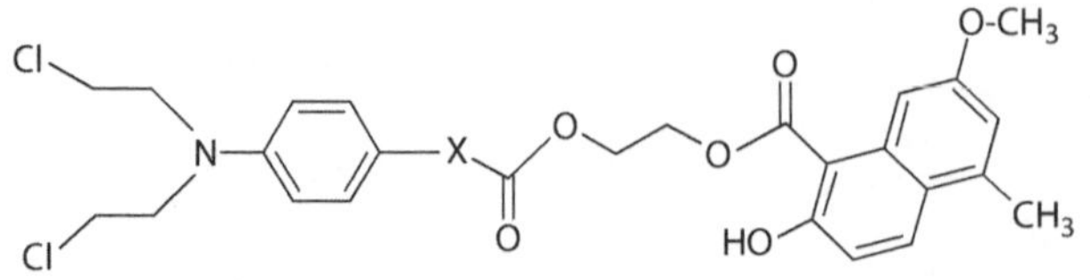

     a. X=(CH$_2$)$_3$: 2-Hydroxy-7-methoxy-5-methyl-naphthalene-1-carboxylic acid 2-(4-{4-[bis-(2-chloro-ethyl)-amino]-phenyl}butyryloxy)ethyl ester

     b. X=CH$_2$-CH(NH$_2$): 2-Hydroxy-7-methoxy-5-methyl-naphthalene-1-carboxylic acid 2-(2-amino-3-{4-[bis-(2-chloroethyl)amino]-phenyl}propionyloxy)ethyl ester

30. Aniline mustards linked to napthalimide

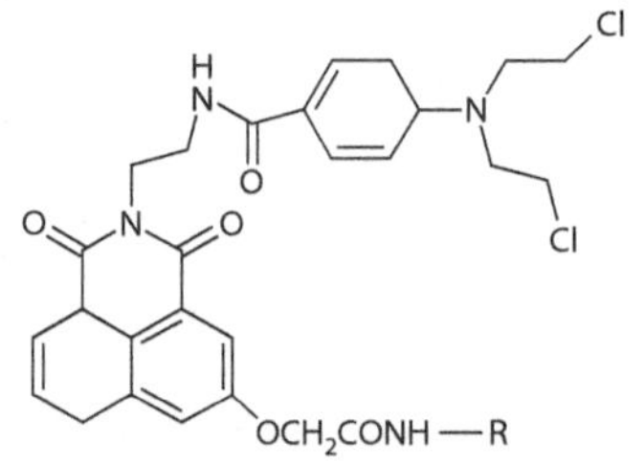

a. R 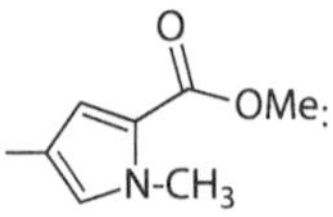 =   *N-p*-[*N,N*-Bis(2-chloroethyl)]aminobenzamidoethyl-3-(1-methyl-2-methoxycarbonyl-4-pyrrolyl)aminocarbonyl-methoxy-1,8-naphthalimide

b. R = 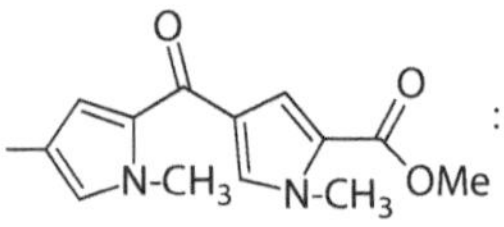   : *N-p*-[*N,N*-Bis(2-chloroethyl)]aminobenzamido-ethyl-3-[1-methyl-2-[(1-methyl-2-methoxy-carbonyl-4-pyrrolyl)aminocarbonyl]-4-pyrrolyl]aminocarbonyl-methoxy-1,8-naphthalimide

c. R = 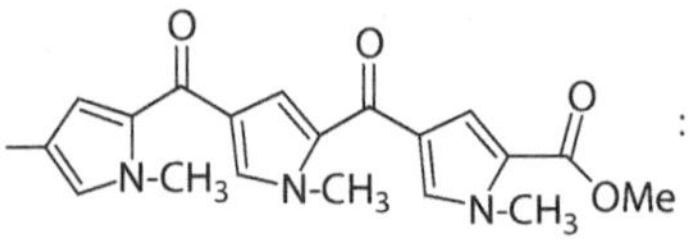   : *N-p*-[*N,N*-Bis(2-chloroethyl)]aminobenzamido-ethyl-3-[1-methyl-2-[[1-methyl-2-[(1-methyl-2-methoxycarbonyl-4-pyrrolyl)aminocarbonyl]-4-pyrrolyl]amino-carbonyl]4-pyrrolyl]aminocarbonylmethoxy-1,8-naphthalimide

d. R = 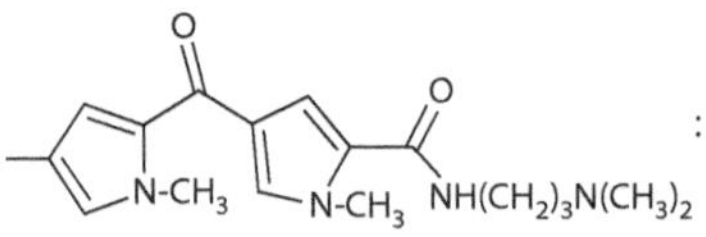   : *N-p*-[*N,N*-Bis(2-chloroethyl)]aminobenz-amido-ethyl-3-[1-methyl-2-[[1-methyl-2-(3-*N,N*-dimethylaminopropylaminocarbonyl)-4-pyrrolyl]amino-carbonyl]-4-pyrrolyl]aminocarbonylmethoxy-1,8-naphthalimide

e. R = 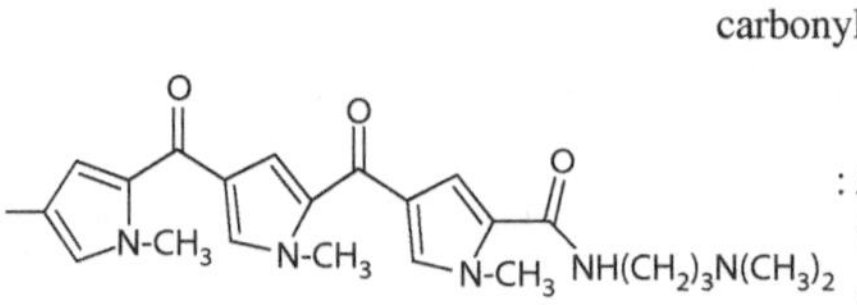   : *N-p*-[*N,N*-Bis(2-chloroethyl)]aminobenzamido-ethyl-3-[1-methyl-2-[[1-methyl-2-(3-*N,N*-dimethylaminopropylaminocarbonyl)-4-pyrrolyl]amino-carbonyl]-4-pyrrolyl]aminocarbonyl]-4-pyrrolyl]aminocarbonyl-methoxy-1,8-naphthalimide

f. R = 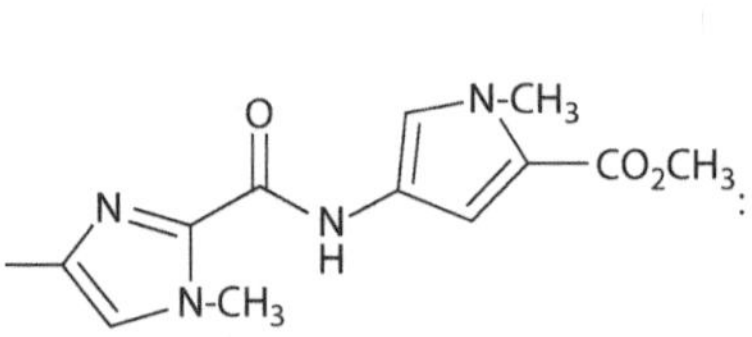   : *N-p*-[*N,N*-Bis(2-chloroethyl)]aminobenzamido-ethyl-3-[1-methyl-2-[(1-methyl-2-methoxy-carbonyl-4-pyrrolyl)aminocarbonyl]-4-imidazolyl]aminocarbonyl-methoxy-1,8-naphthalimide

TABLE 5.5 (Continued)
Nucleic Acid Cross-Linking Reagents

g. R=(CH$_2$)$_3$N(CH$_3$)$_2$: *N*-*p*-[*N*,*N*-Bis(2-chloroethyl)]aminobenzamido-ethyl-3-(3-*N*,*N*-dimethylaminopropyl) aminocarbonyl-methoxy-1,8-naphthalimide

31. Aniline mustards as hypoxia-selective cytotoxins (HSC)

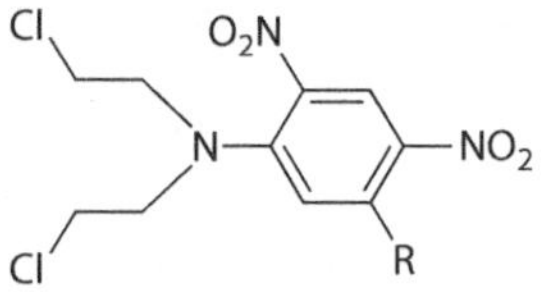

a. R=CONH$_2$: 5-[*N*,*N*-Bis(2-chloroethyl)amino]-2,4-dinitrobenzamide, SN 23862

b. R=CONHMe: *5*-[*N*,*N*-Bis(2-chloroethyl)amino]-*N*-methyl-2,4-dinitrobenzamide

c. R=CONMe$_2$: 5-[*N*,*N*-Bis(2-chloroethyl)amino]-*N*,*N*-dimethyl-2,4-dinitrobenzamide

d. R=COOH: 5-[*N*,*N*-Bis(2-chloroethyl)amino]-2,4-dinitrobenzoic acid

e. R=CO——N⟩O: 5-[*N*,*N*-Bis(2-chloroethyl)amino]-2,4-dinitrobenzoylmorpholide

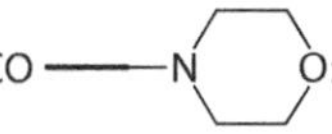

f. R=CONH(CH$_2$)$_2$-——N⟩O: 5-[*N*,*N*-Bis(2-chloroethyl)amino]-*N*-[2-(4-morpholino) ethyl]-2,4-dinitrobenzamide

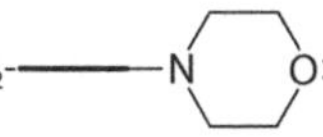

g. R=CONHCH$_2$CH$_2$OH: 5-[*N*,*N*-Bis(2-chloroethyl)amino]-*N*-(2-hydroxyethyl)-2,4-dinitrobenzamide

h. R=CONHCH$_2$CH(OH)CH$_2$OH: 5-[*N*,*N*-Bis(2-chloroethyl) amino]-*N*-(2,3-dihydroxypropyl)-2,4-dinitrobenzamide

I. R=CONH(CH$_2$)$_2$COOH: 5-[*N*,*N*-Bis(2-chloroethyl)amino]-*N*-(2-carboxyethyl)-2,4-dinitrobenzamide

j. R=CN: 5-[*N*,*N*-Bis(2-chloroethyl)amino]-2,4-dinitrobenzonitrile

k. R=CSNH$_2$: 5-[*N*,*N*-Bis(2-chloroethyl)amino]-2,4-dinitrothiobenzamide

l. R=CONH(CH$_2$)2N(O)Me$_2$ ·HCl: 5-[*N*,*N*-Bis(2-chloroethyl)amino]-*N*-([*N*′,*N*′-dimethylamino)ethyl]-2,4-dinitrobenzamide *N*′-oxide

m. R=SO$_2$NH$_2$: 5-[*N*,*N*-Bis(2-chloroethyl)amino]-2,4-dinitrobenzenesulfonamide

n. R=SO$_2$NHSO$_2$Me: 5-[*N*,*N*-Bis(2-chloroethyl)amino]-*N*-(methylsulfonyl)-2,4-dinirobenzenesulfonamide

o. R=CONH(CH$_2$)$_2$NMe$_2$ HCl: 5-[*N*,*N*-Bis(2-chloroethyl)amino]-*N*-([*N*′,*N*′-dimethylamino) ethyl]-2,4-dinitrobenzamide

32. 5-[*N*,*N*-Bis(2-chloroethyl)amino]-1-methyl-2-nitroimidazole

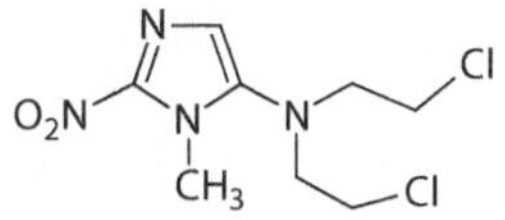

33. PR-104, 2-((2-bromoethyl)-2-{[(2-hydroxyethyl)amino]carbonyl}-4,6-dinitroanilino)ethyl methanesulfonate phosphate ester

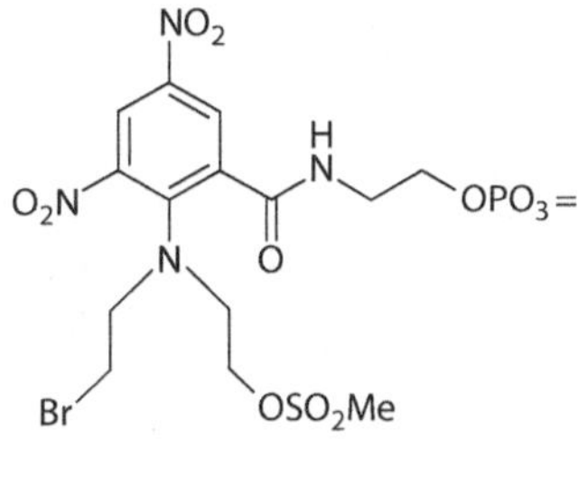

(continued)

TABLE 5.5 (Continued)

Nucleic Acid Cross-Linking Reagents

34. Metal mustards: Co(III) N,N'-bis(2-chloroethyl)ethylenediamine (BCE) alkylpentanedionato complex

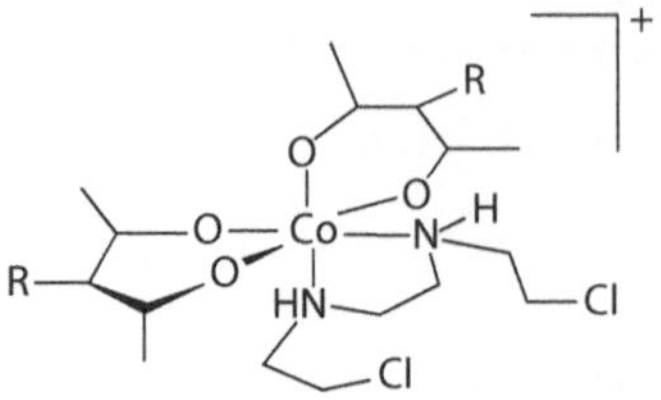

    a. R=H: Bis(2,4-pentanedionato)(N,N'-bis(2-chloroethyl)ethylene-diamine)cobalt(III) perchlorate, Co(acac)BCE.

    b. R=Me: Bis(3-methyl-2,4-pentanedionato)(N,N'-bis(2-chloroethyl)-ethylenediamine)cobalt(III) perchlorate, Co(Meacac)BCE

    c. R=Et: Bis(3-ethyl-2,4-pentanedionato)(N,N'-bis(2chloroethyl)-ethylenediamine)cobalt(III) perchlorate, Co(Etacac)BCE

    d. R=Pr: Bis(3-n-propyl-2,4-pentanedionato)(N,N'-bis(2-chloroethyl)ethylenediamine)cobalt(III) perchlorate, Co(Pracac)BCE

    e. R=Cl: Bis(3-chloro-2,4-pentanedionato)(N,N'-bis(2-chloroethyl)-ethylenediamine)cobalt(III) chloride, Co(Clacac)BCE

35. Metal mustards: Co(III) N,N'-bis(2-chloroethyl)ethylenediamine (DCE) alkylpentanedionato complex

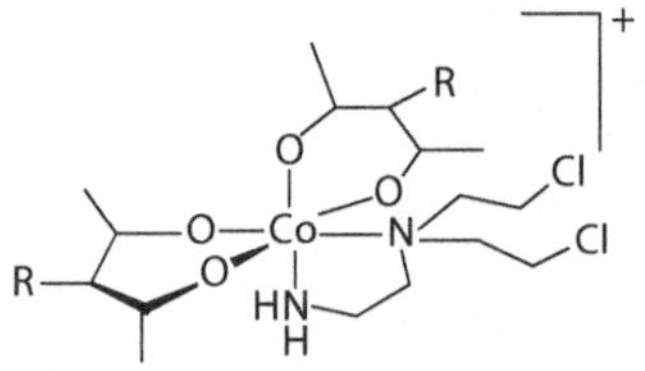

    a. R=H: Bis(2,4-pentanedionato)(N,N-bis(2-chloroethyl)ethylene-diamine)cobalt(III) perchlorate, Co(acac) DCE

    b. R=Me: Bis(3-methyl-2,4-pentanedionato)(N,N-bis(2-chloroethyl)-ethylenediamine)cobalt(III) perchlorate, Co(Meacac)DCE

    c. R=Et: Bis(3-ethyl-2,4-pentaneddionato)(N,N-bis(2-chloroethyl)-ethylenediamine)cobalt(III) perchlorate, Co(Etacac)DCE

    d. R=Pr: Bis(3-n-propyl-2,4-pentanedionato)(N,N-bis(2-chloroethyl)-ethylenediamine)cobalt(III) perchlorate, Co(Pracac)DCE

36. Metal mustard: Bis(tropolonato)-N,N'-bis(2-chloroethyl)ethylene-diamine cobalt(III) perchlorate, Co(trop)BCE

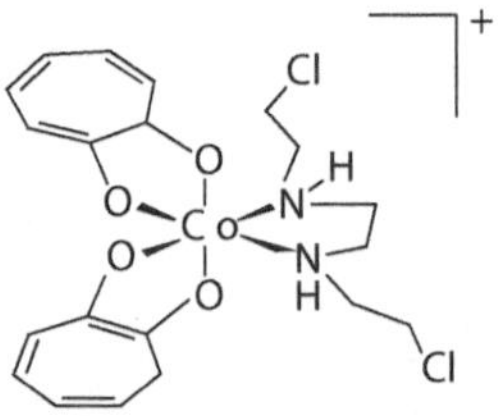

37. Metal mustard: Bis(tropolonato)-N,N-bis(2-chloroethyl)ethylene-diamine cobalt(III) perchlorate, Co(trop)DCE

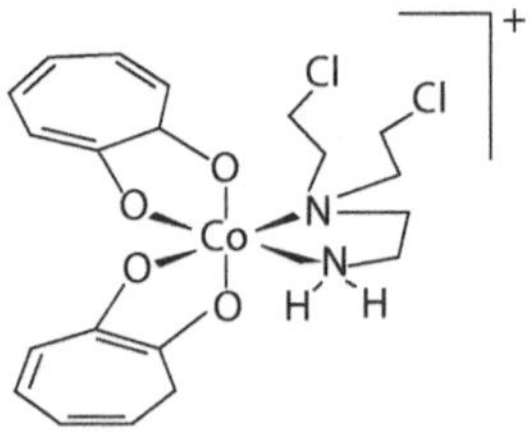

TABLE 5.5 (Continued)
Nucleic Acid Cross-Linking Reagents

38. Metal mustard: Potassium bis(carbonato)-N,N-bis(2-chloroethyl)-ethylenediamine cobaltate(III), K[Co(CO$_3$)$_2$(DCE)]

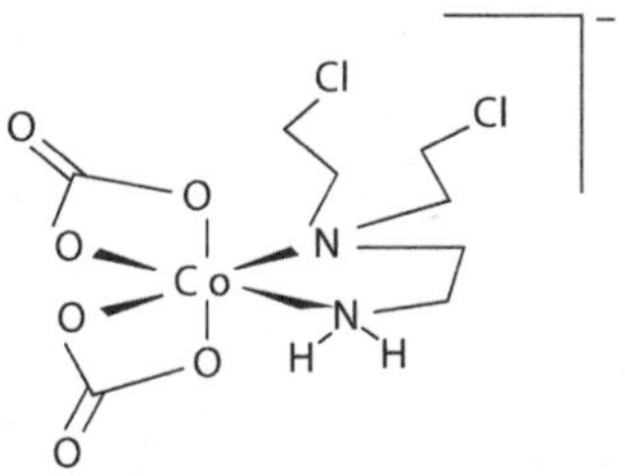

39. Metal mustard: Potassium bis(oxalate)-N,N-bis(2-chloroethyl)-ethylenediamine cobaltate(III), K[Co(ox)$_2$(DCE)]

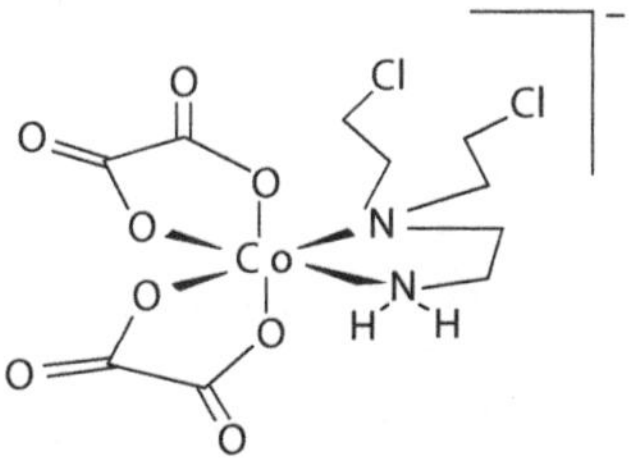

40. Metal mustards: Co(III) N,N'-bis(2-chloroethyl)ethylenediamine dithiocarbamate complex

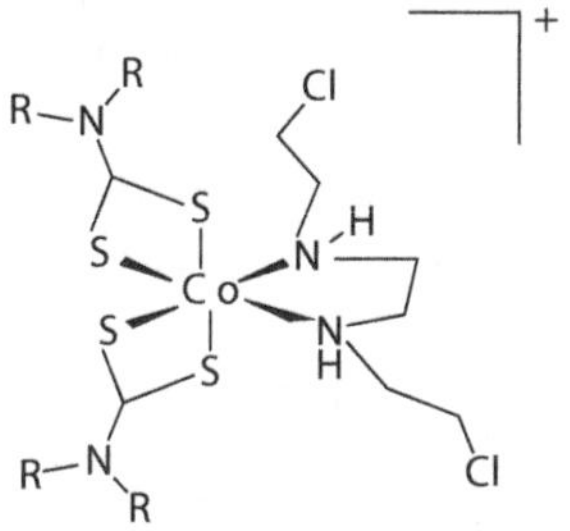

    a. R=Me: Bis(dimethyldithiocarbamato)-N,N'-bis(2-chloroethyl)-ethylenediamine cobalt(III) perchlorate, [Co(Me2dtc)BCE]

    b. R-R=pyrrole: Bis(pyrrolidinedithiocarbamato)-N,N'-bis(2-chloroethyl)ethylenediaminecobalt(III) perchlorate, [Co(pyrrndtc)BCE]

41. Metal mustards: Co(III) N,N-bis(2-chloroethyl)ethylenediamine dithiocarbamate complex

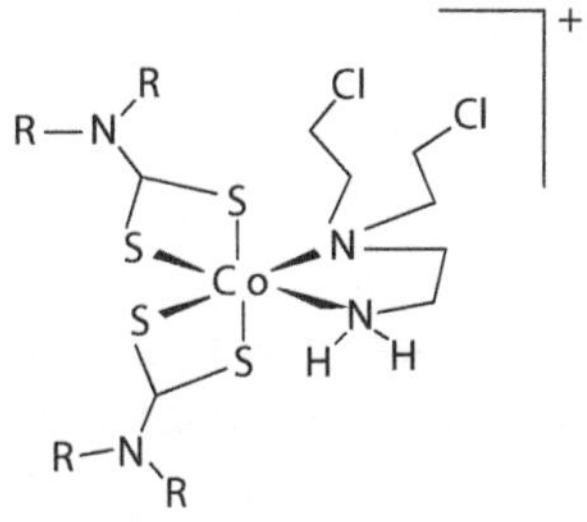

    a. R=Me: Bis(dimethyldithiocarbamato)-N,N-bis(2-chloroethyl)-ethylenediamine cobalt(III), [Co(Me2dtc)DCE]

    b. R=Et: Bis(diethyldithiocarbamato)-N,N-bis(2-chloroethyl)-ethylenediamine cobalt(III), [Co(Et2dtc)DCE]

(*continued*)

TABLE 5.5 (Continued)
Nucleic Acid Cross-Linking Reagents

IX. Bis-carbamate derivatives

A. Bispiperidine nitrophenylcarbamates

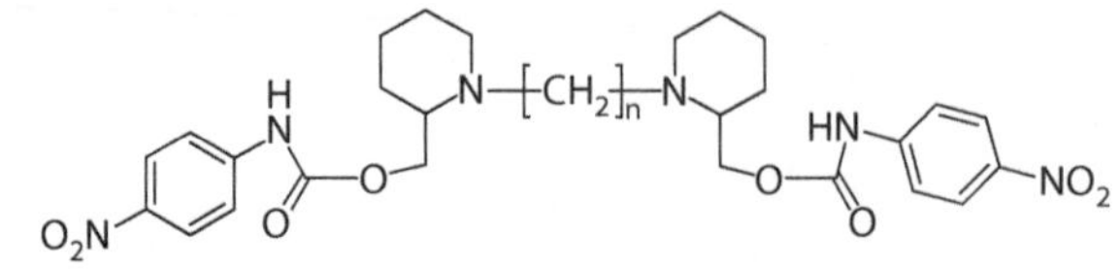

1. n = 2: N,N'-(1,2-Ethanediyl)bis(2-p-nitrophenylaminocarbonyloxymethyl-piperidine)
2. n = 3: N,N'-(1,3-Propanediyl)bis(2-p-nitrophenylaminocarbonyloxymethyl-piperidine)
3. n = 4: N,N'-(1,4-Butanediyl)bis(2-p-nitrophenylaminocarbonyloxymethyl-piperidine)
4. n = 5: N,N'-(1,5-Pentanediyl)bis(2-p-nitrophenylaminocarbonyloxymethyl-piperidine)
5. n = 6: N,N'-(1,6-Hexanediyl)bis(2-p-nitrophenylaminocarbonyloxymethyl-piperidine)

B. Vinylogous carbinolamines

1. Imidazole bis-carbamates

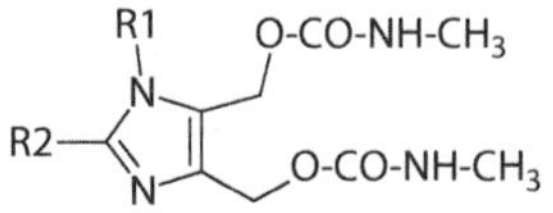

a. R1=CH$_3$, R2=CH$_3$S-: Carmethizol, 1-Methyl-2-(methylthio)-4,5-bis(hydroxymethyl)imidazole bis(N-methylcarbamate)

b. R1=CH$_3$, R2=CH$_3$CH$_2$S-: 4,5-Bis[[(N-methylcarbamoyl)oxy]-methyl]-1-methyl-2-(ethylthio)imidazole

c. R1= CH$_3$, R2=CH$_3$CH$_2$CH$_2$S-: 4,5-Bis[[(N-methylcarbamoyl)oxy]-methyl]-1-methyl-2-(propylthio) imidazole

d. R1= 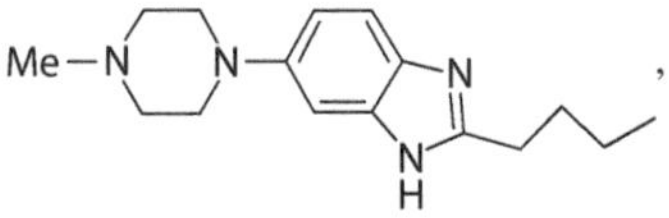 , R2=H: 2-[3-[N-(4,5-Bis(methylcarbamoyl)methyl)-imidazolyl] propyl]-5-(1-methyl-4-piperazinyl)-benzimidazole

e. R1= 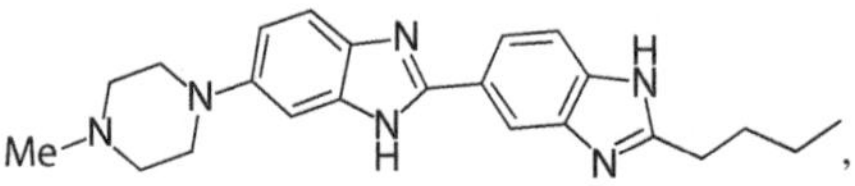 , R2=H: 2-[2-[3-[N-(4,5-Bis(methylcarbamoyl)-methyl) imidazolyl]propyl]-5-benzimidazolyl]-5-(1-methyl-4-piperazinyl)benzimidazole

2. Pyrrol bis-carbamates

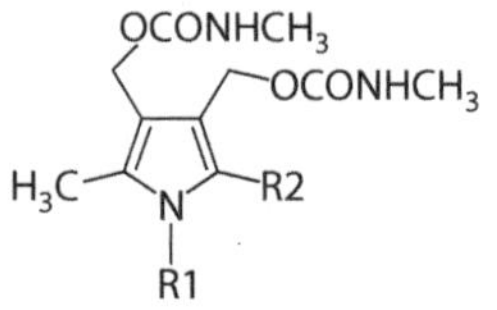

a. R1=phenyl, R2=CH$_3$: N-Phenyl-2,5-dimethyl-3,4-bis(hydroxyl-methyl)pyrrole bis(N-methylcarbamate)

b. R1=p-methoxyphenyl, R2=CH$_3$: N-(p-Methoxyphenyl)-2,5-dimethyl-3,4-bis(hydroxylmethyl)pyrrole bis(N-methylcarbamate)

c. R1=CH$_3$, R2=phenyl: 1,2-Dimethyl-3,4-bis(hydroxymethyl)-5-(phenylpyrrole bis(N-methylcarbamate)

d. R1=CH$_3$, R2=p-methoxyphenyl: 1,2-Dimethyl-3,4-bis(hydroxyl-methyl)-5-(p-methoxyphenyl)pyrrole bis(N-methylcarbamate)

TABLE 5.5 (Continued)
Nucleic Acid Cross-Linking Reagents

3. Pyrrolizine bis-carbamates

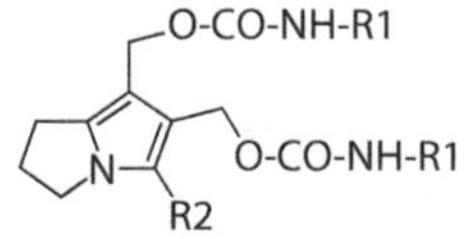

a. R1=CH$_3$, R2=3,4-dichlorophenyl: 2,3-Dihydro-5-(3′,4′-dichlorophenyl)-6,7-bis(hydroxymethyl)-1H-pyrrolizine bis(N-methylcarbamate)

b. R1=CH$_3$, R2=4-fluorophenyl: 2,3-Dihydro-5-(4′-fluorophenyl)-6,7-bis(hydroxymethyl)-1H-pyrrolizine bis(N-methylcarbamate)

c. R1=CH$_3$, R2=4-methoxyphenyl: 2,3-Dihydro-5-(4′-methoxyphenyl)-6,7-bis(hydroxymethyl)-1H-pyrrolizine bis(N-methylcarbamate)

d. R1=C$_2$H$_5$, R2=3,4-dichlorophenyl: 2,3-Dihydro-5-(3′,4′-dichlorophenyl)-6,7-bis(hydroxymethyl)-1H-pyrrolizine bis(N-ethylcarbamate)

e. R1=CH(CH$_3$)$_2$, R2=3,4-dichlorophenyl: 2,3-Dihydro-5-(3′,4′-dichlorophenyl)-6,7-bis(hydroxymethyl)-1H-pyrrolizine bis[N-(2-propyl)carbamate]

f. R1=CH(CH$_3$)$_2$, R2= : 1-Methyl-2-fluoro-4-[2,3-dihydro-6,7-bis[[(N-2-propylcarbamoyl)oxy]methyl]-1H-pyrrolizin-5-yl]pyridinium iodide

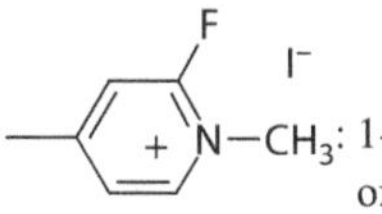

g. R1=CH(CH$_3$)$_2$, R2= : 1-Methyl-2-fluoro-5-[2,3-dihydro-6,7-bis[[(N-2-propylcarbamoyl)oxy]methyl]-1H-pyrrolizin-5-yl]pyridinium iodide

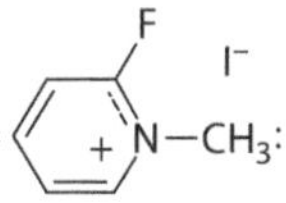

h. R1=CH(CH$_3$)$_2$, R2= : 2,3-Dihydro-5-(1-methyl-1,2-dihydro-2-oxopyridin-4-yl)-6,7-bis(hydroxymethyl)-1H-pyrrolizine bis[N-(2-propyl)carbamate]

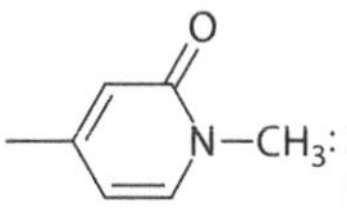

4. Pyrrolo-isoquinoline bis-carbamates

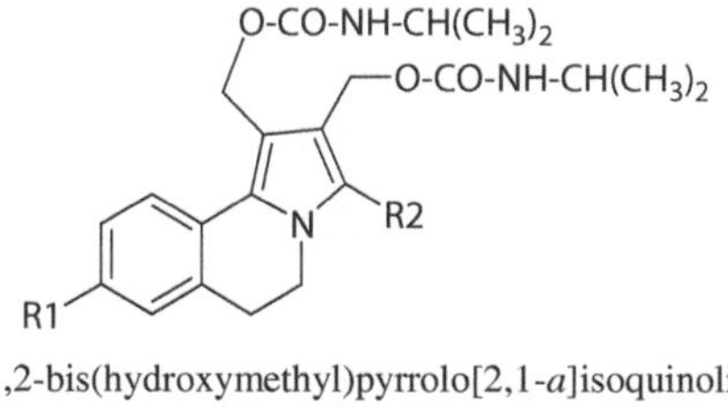

a. R1=R2=H: 5,6-Dihydro-1,2-bis(hydroxymethyl)pyrrolo[2,1-a]isoquinoline bis[N-(2-propyl)carbamate]

b. R1=CH$_3$O, R2=H: 5,6-Dihydro-8-methoxy-1,2-bis(hydroxymethyl)-pyrrolo[2,1-a]isoquinoline bis[N-(2-propyl)carbamate]

c. R1=CH$_3$O, R2=CH$_3$: 1,2-Bis(hydroxylmethyl)-5,6-dihydro-8-methoxy-3-methylpyrrolo[2,1-a]isoquinoline bis[N-(2-propyl)carbamate]

5. Pyrrolo-quinoline bis-carbamates

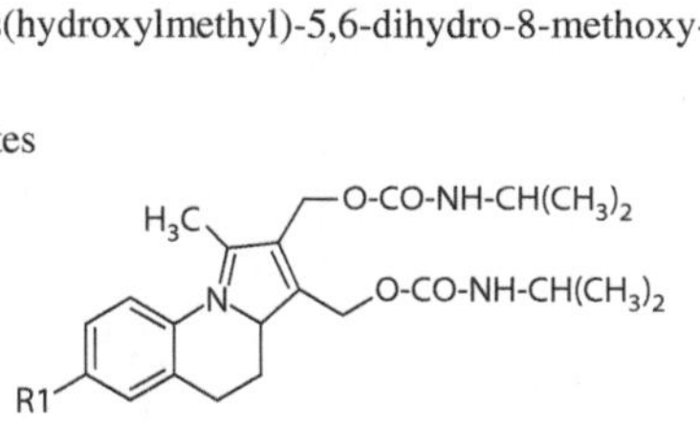

d. R1=H: 1-Methyl-2,3-bis-(hydroxymethyl)-4,5-dihydropyrrolo[1,2-a]quinoline bis[N-(2-propyl)carbamate]

e. R1=CH$_3$O: 7-Methoxy-1-methyl-2,3-bis(hydroxymethyl)-4,5-dihydropyrrolo[1,2-a]quinoline bis[N-(2-propyl)carbamate]

(continued)

TABLE 5.5 (Continued)
Nucleic Acid Cross-Linking Reagents

6. Pyrrolo-benzazepine bis-carbamates

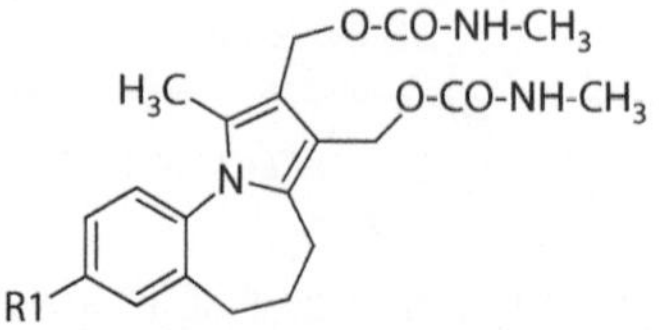

a. R1=H: 1-Methyl-2,3-bis(hydroxymethyl)-5,6-dihydro-4*H*-pyrrolo[1,2-*a*]benzazepine bis(*N*-methylcarbamate)
b. R1=CH₃O: 1-Methyl-8-methoxy-2,3-bis(hydroxymethyl)-5,6-dihydro-4*H*-pyrrolo[1,2-*a*]benzazepine bis(methylcarbamate)
7. Pyrrolo-isobenzazepine bis-carbamate: 3-Methyl-8-methoxy-1,2-bis(hydroxymethyl)-5,6-dihydro-4*H*-pyrrolo[2,1-*a*]isobenzazepine bis(*N*-methylcarbamate)

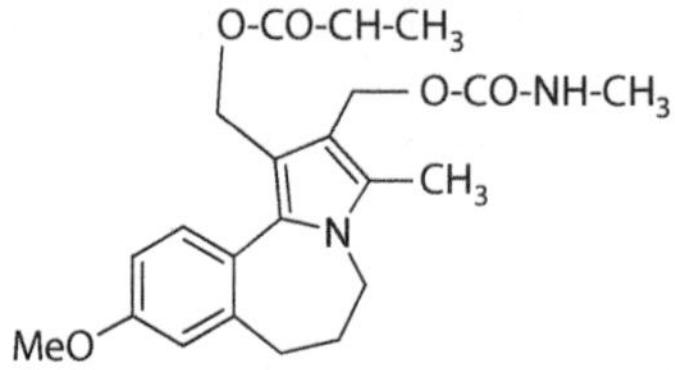

X. Pyrrolizidine alkaloids

A. Retronecine-type: Riddelliine as example

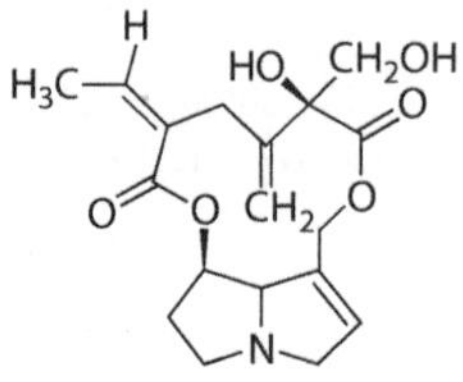

B. Heliotridine-type: Lasiocarpine as example

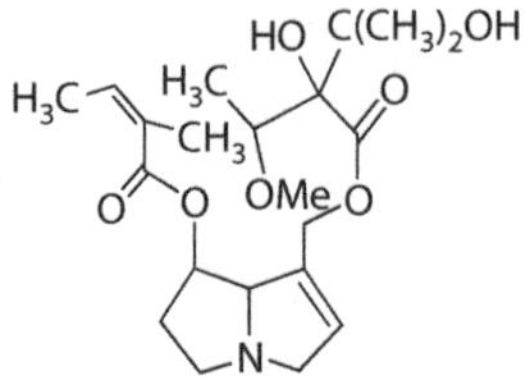

C. Otonecine-type: Clivorine as example

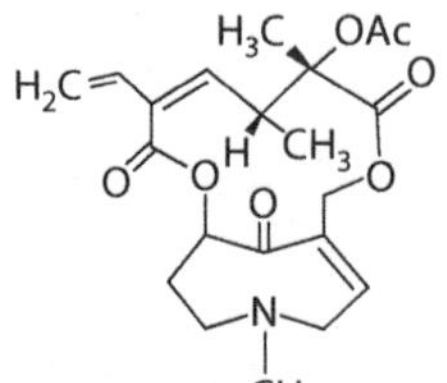

XI. Bis-catechol derivatives

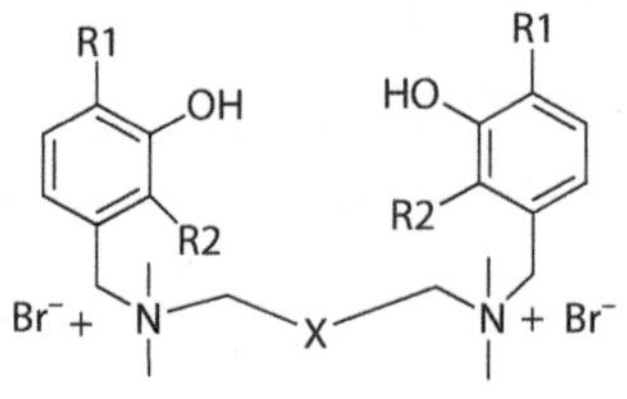

TABLE 5.5 (Continued)
Nucleic Acid Cross-Linking Reagents

    A. R1=H, R2=OH, X=$(CH_2)_2$: N,N'-(3,4-Dihydroxylbenzyl)-N,N,N',N'-tetramethyl-1,4-butanediamine dibromide

    B. R1=H, R2=OH, X=p-phenyl: N,N'-(2,3-Dihydroxylbenzyl)-N,N,N',N'-tetramethyl-p-xylylenediamine dibromide

    C. R1=OH, R2=H, X=$(CH_2)_2$: N,N'-(2,3-Dihydroxylbenzyl)-N,N,N',N'-tetramethyl-1,4-butanediamine dibromide

XII. Quinone methide precursors
    A. Benzo-quinone methides

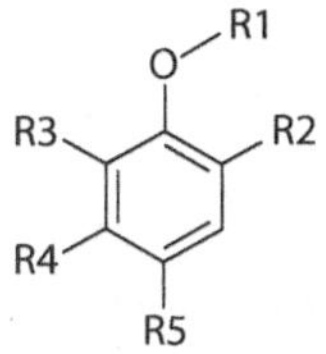

    1. R1=R4=R5=H, R2=R3=$CH_2N(CH_3)_3^+$ I^-: 2-Hydroxy-1,3-di-[(trimethylammonium iodide)methyl]phenyl

    2. R1=R4=H, R2=R3=$CH_2N(CH_3)_3^+$ I^-, R5=CH_3: 2-Hydroxy-5-methyl-1,3-di[(trimethylammonium iodide)methyl] phenyl

    3. R1=R3=H, R2=R4=$CH_2N(CH_3)_3^+$ I^-, R5=OH: 2,5-Dihydroxy-1,3-di-[(trimethylammonium iodide)methyl]phenyl

    4. R1=R4=H, R2=R3=$CH_2N(CH_3)_2$, R5=$COOCH_3$: 3,5-Bis(dimethyl-aminomethyl)-4-hydroxybenzoic acid methyl ester

    5. R1=R4=H, R2=R3=$CH_2N(CH_3)_2$, R5=OCH_3: 2,6-Bis(dimethyl-aminomethyl)-4-methoxyphenol

    6. R1=R4=H, R2=R3=$CH_2N(CH_3)_3^+$ I^-, R5=$COOCH_3$: 3,5-Bis[(trimethyl-ammonium iodide)methyl]-4-hydroxybenzoic acid methyl ester

    7. R1=t-butyldimethylsilyl, R2=R3=CH_2OAc, R4=H, R5=$(CH_2)_n$-CO-NH-(ImPy-β-ImPy)PyIm-β-PyIm-β-Dp: N-polyamide-[4-$tert$-butyldimethylsilyloxy-3,5-bis(acetoxymethyl)phenyl]alkylamide

    8. R1=t-butyldimethylsilyl, R2=R3=CH_2OAc, R4=H, R5=$(CH_2)_2$-CO-NH-$(CH_2)_2$-NH-9-acridine: N-[(9-acridinylamino)ethyl]-3-[4-tert-butyl-dimethylsilyloxy-3,5-bis(acetoxymethyl)phenyl]propanamide

    B. Biphenyl-quinone methides

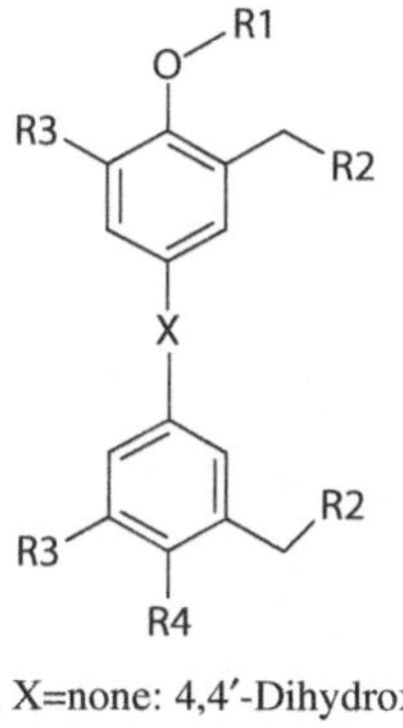

    1. R1=H, R2=$N(CH_3)_3^+$ I^-, R3=H, R4=OH, X=none: 4,4′-Dihydroxy-3,3′-di-[(trimethylammonium iodide)methyl] biphenyl

    2. R1=H, R2=$N(CH_3)_3^+$ I^-, R3=H, R4=OH, X=S: 2,2′-Bis(trimethyl-ammoniumethylene)bis(4-hydroxyphenyl) sulfide iodide

    3. R1=H, R2=$N(CH_3)_3^+$ I^-, R3=H, R4=OH, X= : 2,2′-Bis(trimethyl-ammoniumethylene)-1,3-bis(4-hydroxyphenyloxyl)benzene iodide

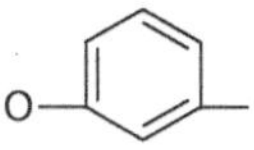

    4. R1=H, R2=$N(CH_3)_3^+$ I^-, R3=$CH_2N(CH_3)_3^+$ I^-, R4=OH, X=none: 2,5,2′,5′-Tetra(trimethylammoniumethylene)-4,4′-dihydroxydiphenyl iodide

    5. R1=H, R2=$N(CH_3)_3^+$ I^-, R3=$CH_2N(CH_3)_3^+$ I^-, R4=OH, X=S: 2,5,2′,5′-Tetra(trimethylammoniumethylene) bis(4-hydroxyphenyl)sulfide iodide

(continued)

TABLE 5.5 (Continued)
Nucleic Acid Cross-Linking Reagents

C. Bipyridine-quinone methides

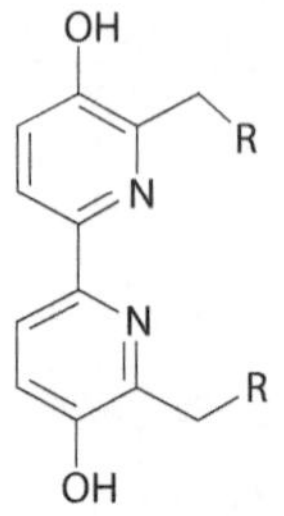

1. R=N(CH₃)₂: 6,6′-Bis-dimethylaminomethyl-[2,2′]bipyridinyl-5,5′-diol

2. R= 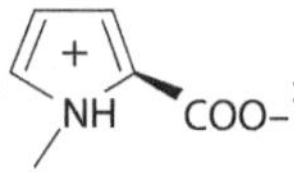 : 6,6′-Bis-L-proline-4-ylmethyl-[2,2′]bipyridinyl-5,5′-diol

D. Bis-binol-quinone methide: 2,2′-Dihydroxy-3,3′-bis[(trimethylammonium iodide)methyl]-1,1′-binaphthyl

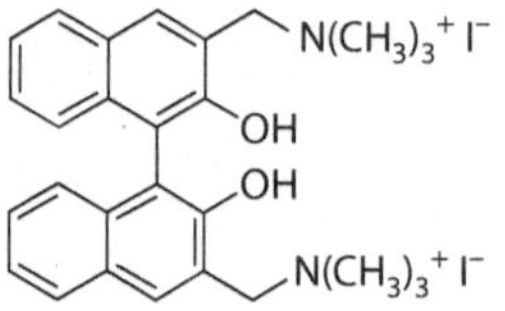

E. (5,5′-(6,7-Dimethoxyquinazoline-2,4-diyl)bis(azanediyl)bis(2-hydroxy-5,1-phenylene))bis(*N,N,N*-trimethylmethanaminium) iodide

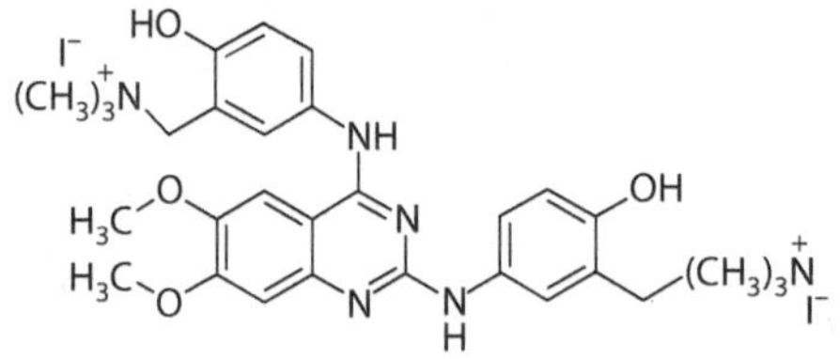

F. Porphyrin-quinone methides

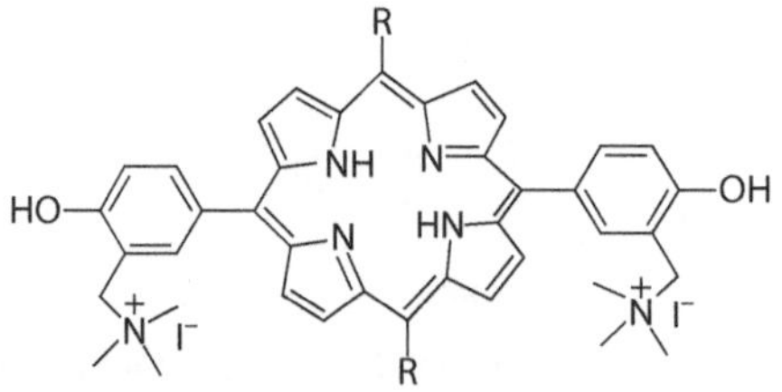

1. R=H: 10,20-{4-Hydroxy-3-(trimethylammonium iodide)methyl}phenyl porphyrin
2. R=CH₂CH₃: 5,15-Diethyl-10,20-{4-hydroxy-3-(trimethylammonium iodide)methyl}phenylporphyrin
3. R=C6H5: 5,15-Diphenyl-10,20-{4-hydroxy-3-(trimethylammonium iodide)methyl}phenylporphyrin

4. R= 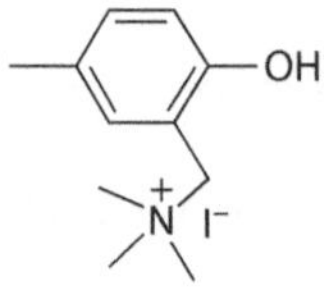 OH: 5,10,15,20-Tetra-{4-hydroxy-3-(trimethylammonium iodide)-methyl}phenylporphyrin

TABLE 5.5 (Continued)
Nucleic Acid Cross-Linking Reagents

XIII. Nitrosoureas

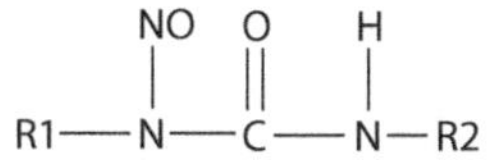

$$R1-\underset{\underset{NO}{|}}{N}-\underset{\underset{O}{\|}}{C}-\underset{\underset{H}{|}}{N}-R2$$

A. R1=R2=-CH₂CH₂Cl: 1,3-Bis(2-chloroethyl)-1-nitrosourea (BCNU, Carmustine)

B. R1=R2= ⬡ OH: 1,3-Bis(*trans*-4-hydroxycyclohexyl)-1-nitrosourea (BHNU)

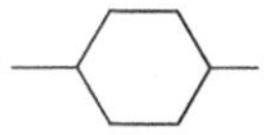

C. R1=-CH₂CH₂Cl, R2= ⬡ : 1-(2-Chloroethyl)-3-cyclohexylnitrosourea (CCNU, Lumustine)

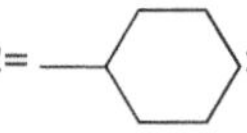

D. R1=-CH₂CH₂Cl, R2=-CH₂CH₂OSO₂N(CH₃)₂: 1-(2-Chloroethyl)-3-[2-(dimethyl-aminosulfonyl)ethyl]-1-nitrosourea (TCNU)

E. R1=-CH₂CH₂Cl, R2= [phosphonate structure]: Fotemustine

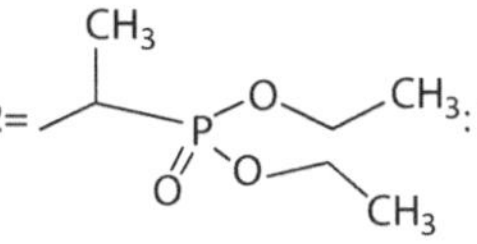

F. R1=-CH₂CH₂Cl, R2= –CH₂CH₂ [pyrrole structure] : [1-Methyl-4-[1-methyl-4-[3-[3′-(2-chloroethyl)-3′-nitrosoureido]-propanamido] pyrrole-2-carboxamido] pyrrole-2-carboxamido]propane, (CENU-lexitropsin)

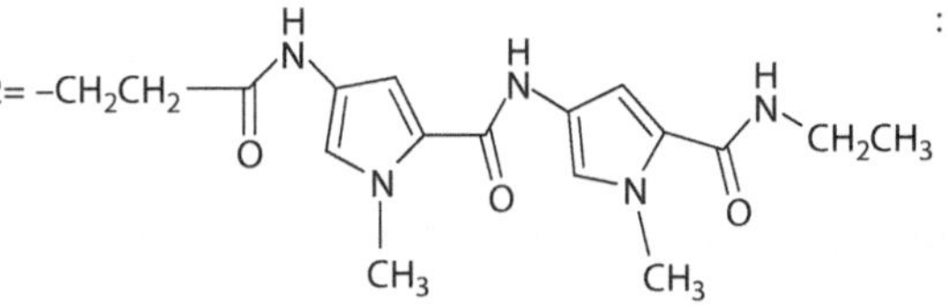

Interstrand *cis*-DDP cross-links are formed between guanines in 5′-GC-sequences to the extent of only 2%–5%.[428] The structures of the cross-linked products are shown in Figure 5.17. There are many factors that affect the interstrand cross-linking.[429,430] However, once formed, these interstrand cross-links result in a significant deformation of the DNA helix, which is the basis of cytotoxicity.

Cisplatin has been attached to various DNA-binding ligands, such as acridine orange, 9-aminoacridine, doxorubicin, etc. Unfortunately, these modifications have relatively little impact on the intrinsic cytotoxicity of the compound. In fact, some analogs become ineffective *in vivo* even against wild-type P388 leukemia.[431]

Other metal complexes that cross-link DNA are phenyl selenides and aluminum complexes (Table 5.5.I.B and I.C). Weng et al.[432] have synthesized a series of phenyl selenides and demonstrated their DNA cross-linking ability. These compounds are activated by periodate oxidation or

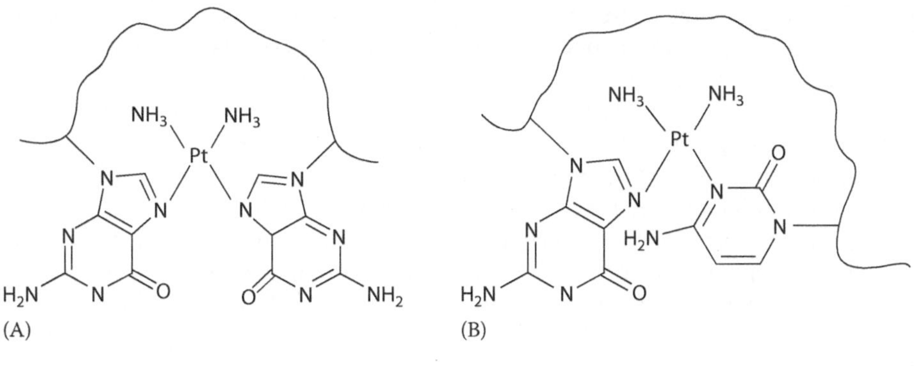

FIGURE 5.17 Structure of *cis*-DDP DNA cross-links: (A) guanine to guanine cross-linked adduct; (B) guanine to cytosine cross-linked adduct.

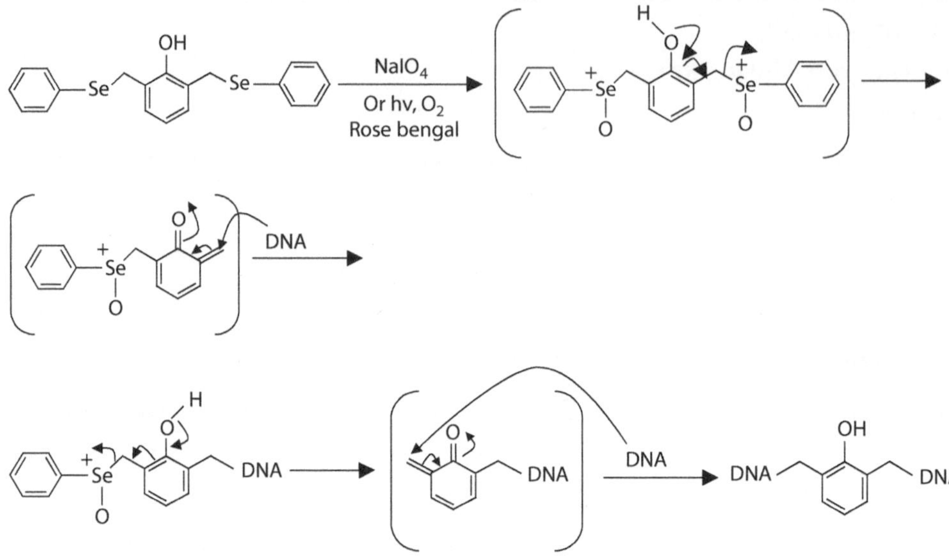

FIGURE 5.18 Proposed reaction mechanism of phenyl selenides cross-linking DNA as exemplified by 2, 6-bis(phenylselenylmethyl)phenol.

light with Rose bengal as the singlet oxygen sensitizer to *o*-quinone methide, which reacts with DNA. Guanine, cytosine, and adenine residues are also thought to be involved. The proposed mechanism of action is shown in Figure 5.18.[432]

An aluminum(III) complex, [SalenAl^{III}]Cl, as shown in Table 5.5.I.C, has been found to cross-link DNA.[433] It is speculated that the phenyl rings in the complex may intercalate between the base pairs of the DNA during the cross-linking process, which actually involves the metal center. The actual mechanism is not known and it is not clear if the reagent shows any base specificity.

There are other metal compounds that may be involved in DNA–DNA and DNA–protein cross-linking. For example, chromine(IV) and chromium(III) have both been shown to form DNA cross-links.[434] However, details of these reactions are lacking. Complexes of iron, cobalt, gold, titanium, ruthenium, and gallium have shown promising cytotoxicity. Their reactions with DNA seem to be different from that of platinum metal. Whether they cause interstrand DNA cross-links remains to be elucidated.[435]

5.11.2 Azinomycin Bis-Epoxides

In Section 5.3.8, we mentioned that many bis-epoxide reagents cause protein, protein–nucleic acid, and DNA cross-linking. The epoxide of azinomycin exhibits similar reactivity (see Chapter 6 for discussion of azinomycin B as a heterobifunctional cross-linker).[436] Using this information, Hartley et al.[437] and others[438] synthesized a series of bis-epoxide analogs containing two azinomycin domains linked by a flexible hydrocarbon spacer arm of varying lengths (Table 5.5.II.A). It was found that a spacer arm with three methylene units was the most efficient in cross-linking DNA. However, the efficiency increased two- to fourfold with the introduction of a methylamino group in the spacer (Table 5.5.II.B).[439] Presumably, cross-linking proceeds with the addition of N7 of guanine to the epoxide centers.[437–439] Recently, rigid aromatic spacer arms have been incorporated into the reagent to link the two azinomycin epoxide centers as shown in Table 5.5.II.C.[440] However, these compounds cross-link DNA with less efficiency than those with flexible spacer arms.

5.11.3 Bis-Pyrrolobenzodiazepines

Pyrrolobenzodiazepines (PBDs) (see Figure 5.19 for core structure) are a family of naturally occurring antitumor antibiotics produced by various *Streptomyces* species and are generally referred to as the anthramycin family. These PBDs differ in the number, type, and position of the substituents in both the aromatic A-ring and the pyrrolo C-ring, and the nature of saturation of the C-ring at the C2–C3 bond or at C2 (*exo* cyclic) bond. All naturally occurring PBDs possess the (S)-configuration at the C11a-position, which provides the molecule with a right-handed twist when viewed from the C-ring toward the A-ring, enabling it to fit into the minor groove of DNA.[441] The antitumor and cytotoxic nature of the PBDs are attributable to their covalent binding to the exocyclic N2 group of guanine via an acid labile amine bond to the electrophilic imine at C-11 position as shown in Figure 5.20 for a PBD dimer.[442] These PBD monomers span three DNA base pairs and have a general preference for 5′-purine-G-purine sequences in the minor groove. Based on these studies, two PBD units have been linked to form dimers to explore their antitumor activity and as interstrand cross-linking agents on opposite DNA strands.[443] A dramatic increase in cytotoxicity and sequence selectivity as cross-linking agents on opposite DNA strands (i.e., interstrand cross-linking) has been achieved. The structures of these compounds are shown in Table 5.5.III. PBD units have been joined with various spacer arms through their different positions such as A-C7/A-C7′, A-C8/A-C8′, C-C2/C-C2′, and A-C8/C-C2′. Farmer et al.[444] first

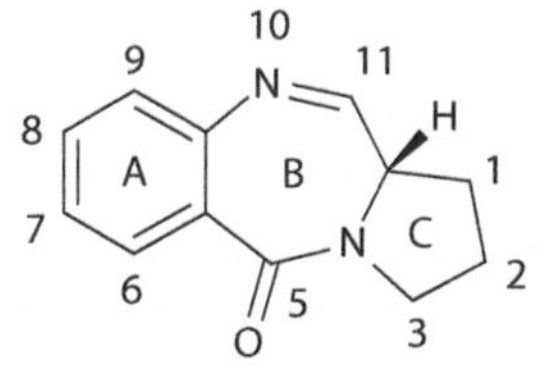

FIGURE 5.19 The core structure of pyrrolobenzodiazepine with the numbering system.

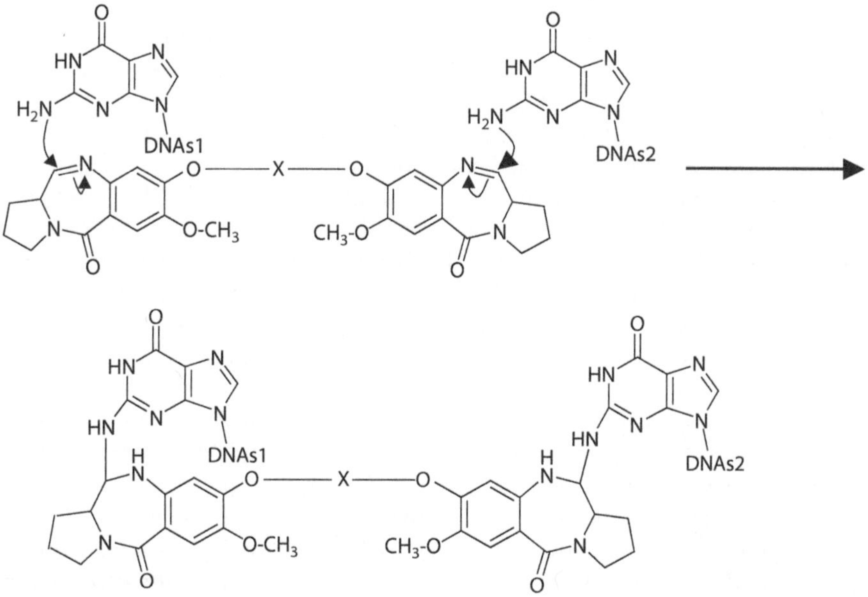

FIGURE 5.20 Speculated mechanism of DNA reaction with a PDB derivative. DNAs1 and DNAs2 are complementary DNA strands showing the guanine base.

synthesized C7/C7′-linked dimers (Table 5.5.III.A.1 through 3). These molecules were found to cross-link plasmid DNA, restriction fragments, and short oligonucleotides. They have specificity for dG-containing duplex DNA.

The C8/C8′-linked PBD dimers with alkane spacer arms were designed and synthesized by Thurston's group as efficient irreversible DNA interstrand cross-linking agents.[445] In this series, the monomeric PBD units were conjugated by diether linkages at their C8 position through alkyl chains of varying lengths, that is, C8-O-(CH_2)n-O-C8′ where n varies from 3 to 6 (Table 5.5.III.B.1 through 4). These reagents were studied for their ability to interact with oligonucleotide duplexes containing potential target-binding sites. The results showed that the PBD dimers form irreversible interstrand cross-links between two guanine bases within the minor groove via their exocyclic N2 atom.[446] DSB-120 (n = 3: Table 5.5.III.B.1) spans 6 bp in the minor groove and actively recognizes and cross-links a 5′-purine(Pu)-GATC-pyrimidine(Py) sequence. It is >300-fold more efficient at cross-linking DNA than the clinically used cross-linking agent melphalan under the same conditions. The more extended PBD dimer (n = 5: Table 5.5.III.B.3) can span an extra base pair and cross-link the 5′-Pu-GA(T/A)TC-Py sequence. While the cross-linking efficiency of 3 and 5 carbon chain–linked dimers is broadly similar, the 4 and 6 carbon chain–linked dimers are approximately 18- and 14-fold less efficient respectively.[442–446] C2-Difluorinated analogs have also been synthesized (Table 5.5.III.C.1 through 3).[447] These fluorinated compounds showed increased DNA-binding ability in comparison with nonfluoro PBD molecules and exhibited significant anticancer activity. Modification of C2 is extended to include *exo/endo* unsaturation of the carbon atom (Table 5.5.III.D.1 through 3). C2-*exo* unsaturated analog with C2/C2′-*exo*-methylene modification (SJG-136: Table 5.5.III.D.1) has been synthesized as a novel cross-linking agent with remarkable DNA-binding affinity and cytoxicity.[448] Like DSB-120, SJG-136 spans six base pairs in the minor groove with a preference for binding to 5′-Pu-GATC-Py sequences and produces DNA interstrand cross-links between two N2 guanine positions on opposite strands and separated by two base pairs.[449] Its potential antitumor activity has led to clinical trials. C2-*exo*-Difluoro-SJG-136 and other analogs have been synthesized (Table 5.5.III.D.2.a through c).[450] These C2/C2′-*exo*-difluoromethylene PBD dimers exhibit remarkable DNA-binding ability and the spacer length in these dimers may modulate the DNA reactivity potential. The C2-C3/C2′-C3′-*endo* unsaturated PDP dimer as shown in Table 5.5.III.D.3 also showed enhanced cytotoxicity and DNA-binding affinity compared to other saturated A-ring-linked PBD dimers. However, its effect is less than that with C2-*exo*-unsaturation.[451]

In addition to the alkanediyldioxy spacers, PBD monomers have also been linked through C8/C8′ by polyheterocyclic polyamide moieties containing pyrroles and imidazoles, by alkane-1,2,3-triazole, alkane-piperazine, and alkane-anthraquinone (Table 5.5.III.E through K). Of the polyheterocyclic spacers, the tripyrrole-containing PBD dimer (AT-235: Table 5.5.III.E.1) was synthesized by Tiberghien et al.[452] It spans 11 DNA base pairs with high affinity binding to a 5′-GCTTATAATGG-3′ sequence in the minor groove and is shown to effectively cross-link interstrand linear plasmid pUC18 DNA. Presumably the PBD moieties bind to the 3′-CGA (opposite 5′-GCT) and 5′-TGG triplet sequences cross-linking G–G with the heterocyclic linker interacting with the central 5′-TATAA base pairs.[452]

PBD dimers containing polypyrrole polyamides or polyimidazole polyamides shown in Table 5.5.III.E.2 and 3, respectively, were synthesized by Kumar and Lown.[453] These compounds were tested against a panel of 60 human cancer cells and were found, in general, potent against many human cancer cell lines. Kamal et al.[454] synthesized C8/C8′ linked PBD dimers containing 1,2,3-triazole (Table 5.5.III.F) attached through various carbon chain lengths. Introduction of the 1,2,3-triazole moiety has a significant effect on DNA-binding ability. As the spacer arm increased from three to four carbons, there was a decrease in the DNA-binding ability. However, upon further increase from five to six carbons in the chain, DNA stabilization was enhanced. The compound with six methylene groups was shown to have enhanced DNA-binding ability compared to the DSB-120 PBD dimer. Similarly, incorporation of a piperazine moiety also significantly enhanced the DNA-binding ability.[455] (For structures, see Table 5.5.III.G.1 through 4.) Presumably, the piperazine ring in the middle of an alkanedioxy spacer arm can enhance hydrophobic interactions and may also achieve a superior isohelical fit within the DNA minor groove. These compounds are significantly more cytotoxic in a number of human cancer cell lines[455] than those dimers with only an alkanedioxy spacer. C2-Mono- and di-fluorinated analogs of these piperazine containing PBD dimers have also been synthesized (Table 5.5.III.H.1 through 3 and I.1 through 3).[447] Kamal et al.[447] found that, in general, these C2-fluorinated compounds have enhanced DNA-binding ability and are biologically more potent than their C2-unsubstituted counterparts. A C2/C2'-exo-difluoromethylene PBD dimer analog of SJG-136 as shown in Table 5.5.III.J has also been studied.[447] It exhibited remarkable DNA-binding affinity. The increased affinity is probably due to a more favorable fit of the molecules in the DNA minor groove.

Alkyloxy-anthroquinone is another spacer arm that links C8 PBD units. Because anthraquinones are anticancer agents, which generally bind to DNA by insertion and stacking between the base pairs, Kamal et al.[456] synthesized a series of such PBD-anthroquinone conjugates as shown in Table 5.5.III.K.1 through 3. These compounds displayed promising anticancer activity in various cell lines. The compound with the three-carbon spacer exhibited the highest DNA-binding ability. As the spacer length increased from three to four carbons, the DNA-binding ability decreased. On further increase from four to five carbons, the DNA-binding ability was enhanced. These results indicate the importance of spacer chain length in such PBD dimers.

While C8/C8'-linked PBD dimers represent the most investigated interstrand DNA cross-linkers, Reddy et al.[457,458] have synthesized PBD dimers joined together at the C2 position of the C ring of PBD through an alkyl amido spacer arm (see Table 5.5.III.L). These compounds exhibited moderate promising cytotoxic potency against different cancer cells. Increases in the chain length of the spacer arm significantly increased the cytotoxic potency.

Head-to-tail linking of C8 to C2 of PBD units to form A-C8/C-C2 PBD dimers has also been investigated. Gregson et al.[459] synthesized the first A-C8/C-C2 amide-linked PBD dimers (Table 5.5.III.M) in 2003. Unfortunately, these compounds demonstrated poor cytotoxicity in a number of human tumor cell lines and poor DNA cross-linking ability compared to A-C8/A-C8' linked PBD dimers. A different spacer arm, alkoxyamido, was used to link A-C8/C-C2 PBD dimers (Table 5.5.III.N).[460] In contrast to amide-linked dimers, these alkoxyamidolinked reagents exhibited significant DNA-binding ability but with moderate anticancer activity. Thus, a correlation between the DNA-binding affinity and cytotoxicity could not be derived from this class of head-to-tail PBD dimers. Another series of A-C8/C-C2-linked PBD dimers containing triazole in the spacer arm has been studied (Table 5.5.III.O).[454] Although these compounds show DNA-binding affinity, it is not that significant, probably due to the structural rigidity at C2 of the PBD unit.

5.11.4   Bis-Cyclopropylpyrroloindole (CPI)-Based Reagents

Many antitumor antibiotics such as CC-1065, adozelesin, duocarmycin A, and (+)duocarmycin SA contain a cyclopropylpyrroloindole (CPI) moiety, 1,2,8,8a-tetrahydro-7-methylcyclopropa[c] pyrrolo[3,2-e]indole-4-one, which alkylates DNA. The antitumor activities of these compounds occur through alkylation of N3 of adenine or guanine in the minor groove of duplex DNA by the cyclopropyl group as shown in Figure 5.21.[461] Numerous analogs have been synthesized to increase DNA regional specificity.[462] For cross-linking DNAs, Boger et al.[463] constructed CPI dimers derived from head-to-tail coupling of the two enantiomers of duocarmycin SA alkylation subunits as shown in Table 5.5.IV.A. These four combinations of enantiomeric dimers showed potent cytotoxicity in assays against the L1210 cell lines and displayed a two- to threefold higher activity than duocarmycin SA. Dimerization the duocarmycin SA subunit presumably alkylates double strand DNA causing interstrand cross-linking, providing two to three times more potency than duocarmycin SA. However, the interstrand cross-linking process has not been investigated.

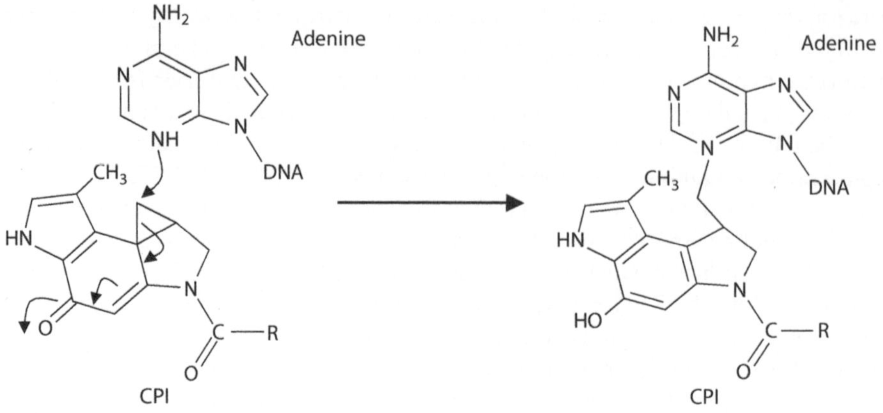

FIGURE 5.21　Proposed mechanism of CPI reaction with N3-adenine of DNA.

It has been shown that CPI with *N*-methylimidazole (Im)-*N*-methylpyrrole (Py) exhibits a remarkable sequence specificity for minor groove of the DNA duplex, particularly in the presence of a triamide, ImImPy partner.[462,464] With this effect in mind, Bando et al.[461,464–467] synthesized CPI dimers containing ImPy linked with a series of spacers. Only those with tri-, tetra-, or penta-methylene spacers (see Table 5.5.IV.B) give cross-linked DNA products in the presence of ImImPy, with tetramethylene as an optimal spacer for the formation of interstrand cross-linking. These ImPy-containing CPI dimers efficiently produce DNA interstrand cross-links at the nine-base-pair sequence, 5′-Py-GGC(T/A)GCC-Pu-3′, reacting with N3 of adenine.[466,467] Their reaction with N3 of guanine in the corresponding oligomers at cross-link sites in the presence of ImImPy occurs with a slightly lower efficiency.[467]

Compounds of the CPI family also include those with a chloromethyl group attached to the 8-position of pyrroloindole moiety instead of the cyclopropyl group. Such *seco*-CPI is converted to CPI as shown in Figure 5.22.[461] Many *seco*-CPI compounds were studied for their antitumor efficiency.[462] Bis-*seco*-CPI dialkylators such as those listed in Table 5.5.IV.C have been designed and studied. Bizelesin (Table 5.5.IV.C.1) displays excellent cytotoxic efficiency in comparison to its CPI mono alkylating parent compound and is currently under clinical trial. Its exceptional potency may be due to its targeting of AT-rich matrix-associated regions, domains of critical importance for replication.[468] It binds mainly to T(A/T)$_4$A sites in cellular DNA, forming inter-strand cross-links through the N3 of two adenines 6 or 7 bp apart, although some monoadducts are also observed.[469] With a sequence of 5′-TTAGTTA-3′, a 7 bp cross-link is overwhelmingly pre-ferred over a possible 6 bp sequence. The unique presence of a G-C base pair in the middle of the sequence causes exocyclic distortion of 2-amino group of guanine, effectively reducing the cross-linked distance.[470] U78779, an analog of bizelesin (Table 5.5.IV.C.2) with a furan spacer rather than urea, also forms interstrand DNA cross-links. Like bizelesin, it is highly preferential for

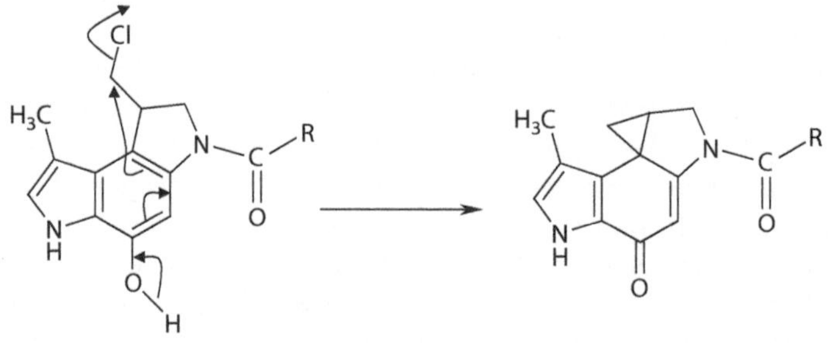

FIGURE 5.22　Conversion of a *seco*-CPI to CPI.

long AT islands and binds at (A/T)$_6$A sequence, although it prefers mixed A/T–G/C motifs.[468,471] There are other bizelesin analogs with various modifications on the CPI moiety and spacers of different flexibility. Fukada et al.[472,473] modified the CPI moiety with methoxycarbonyl, trifluoromethyl, methyl, and phenyl moieties at the 2- and/or 3-position (Table 5.5.IV.C.3 through 5). In MCTFCPI (Table 5.5.IV.C.3) the two groups are connected with rigid spacer arms of varying lengths and types containing 5,5′-bis(2-carbonyl-1H-indole).[472] In others (Table 5.5.IV.C.4 and 5), the modified CPI are linked by 3,3′-(1,4-phenylene)diacrylic acid.[473] The bisalkylators show strong cytotoxicity against HeLaS3 human uterine cervix carcinoma and antitumor activity against colon 26 murine adenocarcinoma. The length of a rigid linker has a significant influence on the cytotoxicity and antitumor activity rather than the type of rigid linker. However, their cross-linking activities are unknown.

5.11.5 BIS-CYCLOPROPANEBENZ[e]INDOLINE (CBI)-BASED REAGENTS

Modification of the cyclopropylpyrroloindole moiety of CPI has led to the synthesis and investigation of derivatives of 1-(chloromethyl)-5-hydroxy-1,2-dihydro-3H-benz[e]indole (seco-CBI) as the basis of alkylation of DNA, which is found to be more potent than CPI itself.[461] Like seco-CPI, the seco-CBI moiety is converted to 1,2,9,9a-tetrahydrocyclopropa[1,2-c]benz[1,2-e]-indol-4-one (CBI) as an alkylating moiety (Figure 5.23). These compounds selectively alkylate the N3 of adenine at the 3′ end of three or more consecutive AT base pairs in DNA in a similar manner as CPI.[464] Lown's group have coupled seco-CBI subunits in three discrete ways with different spacer arms to investigate the structure activity relationship systematically.[474,475] As shown in Table 5.5.V, the seco-CBI units are linked through C7 and N3 to form C7–C7, N3–N3, and C7–N3 dimers that contain two racemic CBI moieties. The spacer arm used is methylene, pyrrole polyamide, or imidazole polyamide. For the bis-seco-CBI dimers liked by a flexible methylene chain of variable lengths (Table 5.5.V.A.1, V.B.1, and V.C.1), all are active against almost all 60 human tumor cell lines derived from leukemia, non-small cell lung cancers, colon cancer, CNS cancer, melanoma, ovarian cancer, renal cancer, prostate cancer, and breast cancer. The antitumor activities of these dimers are strongly related to the position and length of the spacer. In general, the antitumor activity sequence is C7–C7 dimers < C7–N3 dimers < N3–N3 dimers.[474] Among the C7–C7 dimers, the

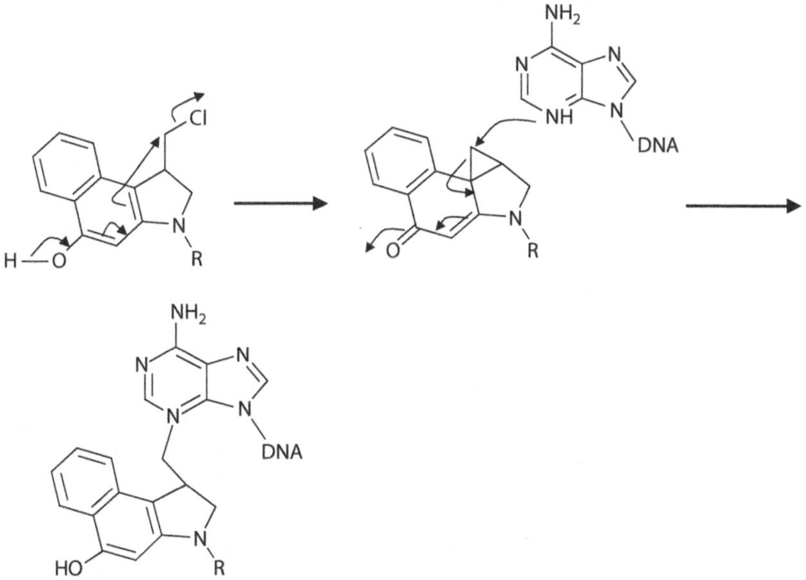

FIGURE 5.23 Conversion of seco-CBI to CBI and its alkylation of N3-adenine of DNA.

potency decreases with the increasing length of the spacer arm. For the C7–N3 dimers, compound V.C.1.d, which possesses the longest spacer (n = 6) in the series, proved to be the most potent with potency decreasing in the order of n = 6 > n = 3 > n = 4 > n = 5. The N3–N3 dimers are the most potent of all methylene linked bis-*seco*-CBI dimers with compound V.B.1.a (n = 1) displaying the highest overall potency. The *seco*-CBI dimers linked with one or more pyrrole and imidazole units (compounds 2 and 3 of Table 5.5.V.A through C) were also evaluated against three human tumor cell lines consisting of MCF7 (Breast), NCI-H460 (Lung), and SF-268 (CNS) cells. All compounds have varying cytotoxic potency activity against these three cancer cell lines. Those with only one pyrrole or imidazole unit give the higher cytotoxicity.[475] However, the cross-linking activities of all bis-*seco*-CBI have not been investigated.

5.11.6   Diaziridinyl Benzoquinones

While aziridine reagents react with specificity toward sulfhydryl groups (Section 5.3.7) and benzoquinone reacts preferentially with amino groups (Section 5.2.9), diaziridinylbenzoquinones are shown to react with DNAs resulting in interstand cross-links. Table 5.5.VI represents some of the diaziridinylbenzoquinone compounds that have been studied, several of which are under clinical trials as antitumor drugs, such as RH1 (Table 5.5.VI.C) and AZQ (Table 5.5.VI.G).[476] Dzielendziak et al.[349] synthesized a series of bis(carboethoxyamino) diaziridinyl benzoquinone analogs (Table 5.5.VI.F through N) and found that their toxicities correlated with their efficiencies of forming cross-links in DNA, which were in turn a function of the ease with which the compounds could be reduced to produce more active forms. Although bioreductive activation may play a role in the DNA cross-linking,[477] a single mechanism cannot fully explain all the observed cytotoxic effects of these quinones. For example, AZQ (Table 5.5.VI.G) and BZQ (Table 5.5.VI.E) both cross-link DNAs, but AZQ can be readily reduced to semiquinone radicals in a biological system under normal physiological conditions, while BZQ is not easily reduced. Butler et al.[478,479] concluded that BZQ functions as a bifunctional alkylating agent by an acid-catalyzed aziridine ring-opening mechanism. On the other hand, Mayalarp et al.[480] proposed that quinone methide may be involved. However, bioreductive activation remains an important aspect of these anticancer drugs.[481] Reduction by ascorbic acid increased the cross-linking, which was particularly striking in the case of DZQ (Table 5.5.VI.A).[481] Reduction also changed DZQ's preferential cross-linking site of N7-guanine in a 5′-GXC (X = any base) to a 5′-GC sequence with particular preference for 5′-TGC-3′ sites.[481,482] Berardini et al.[482] speculated that the altered binding site by the reduced DZQ was due to DNA intercalation of the reduced hydroquinone. While other reduced diaziridinylquinones showed similar inclination particularly those with a hydrogen in position-6 of the quinone,[483] the reduced form of MeDZQ (Table 5.5.VI.B) was found to preferentially cross-link at 5′-GNC sites.[482] Additional minor cross-linking sites such as the 5′-GXXC sequence were possible. Di Francesco et al.[484] synthesized and investigated a series of diaziridinyl benzoquinone phenyl esters (Table 5.5.VI.O and P) and found that certain phenyl esters were significantly more toxic than others in six different human cancer cell lines (H460, H596, HT29, BE, K562, and A2780). These esters were hydrolyzed by esterases to form stable *meta*-phenol and an unstable *para*-phenol. The former was highly cytotoxic. However, all the compounds elicited DNA cross-linking; therefore, the enhanced toxicity could be due to other mechanisms.

Diaziridinylbenzoquinones have been linked to DNA sequence-specific binders. DZQ was linked with a triplex-forming oligonucleotide, which binds in the major groove to specific double-helical DNA sequences and it was shown to have improved sequence-directed cross-linking activity.[485] Di Francesco et al.[486] linked diaziridinlybenzoquinones to acridine as shown in Table 5.5.VI.Q and R. All of the compounds were shown to produce DNA interstrand cross-links at submicromolar. However, they were between three and four times less efficient at cross-linking DNA than RH1 (Table 5.5.VI.C). The reason for the difference is not known. Computer modeling suggested that the acridine moiety could intercalate into DNA while the flexible methylene spacer could allow the aziridines to come

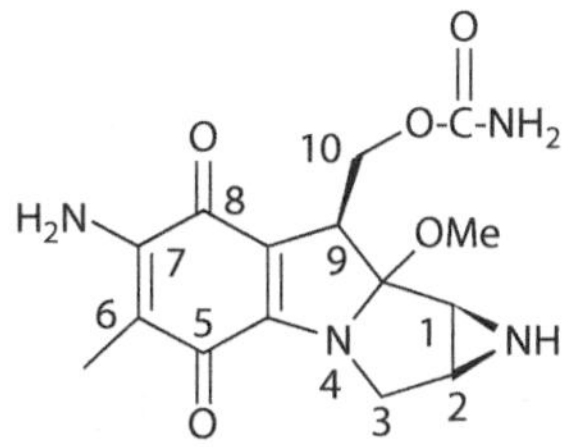

FIGURE 5.24 Structure of mitomycin C and the numbering system.

close to the N7 guanine position for cross-linking. The model, however, is not consistent with the observation that the extent of DNA cross-links produced by these acridine compounds does not vary dramatically between these analogs.[486]

5.11.7 MITOMYCIN C DIMERS

Mitomycin C as shown in Figure 5.24 contains an aziridine ring as well as a benzoquinone moiety. It is a potent antitumor antibiotic, but it does not react with DNA until it undergoes activation under enzymatic or chemical reductive conditions to generate reactive species, which react with DNA to generate both mono- and bis-alkylation DNA adducts with interstrand DNA cross-links.[487] Since mitomycin C itself is a DNA intercolator, its use as a reagent for chemical cross-linking studies of DNA is limited. Dimeric mitomycin C with various spacer arms has been synthesized (see Table 5.5.VII).[488–490] Those with methylene spacers as shown in Table 5.5.VII.A.1 through 8 were thought to react with DNA at C(1) aziridine ring of mitomycin C unit, although reaction at C(10) position was also observed under reductive conditions.[488,489] Thus, these compounds can undergo molecular rearrangement and activation as in the case of diaziridinylbenzoquinone discussed above. There is a specific DNA-binding site for these dimeric compounds, but the reaction seems to have 5′-CpG sequence specificity, reacting with the N2 of guanine. In addition to the methylene group linked mitomycin C dimers, Lee and Kohn[490] have synthesized and studied a dimer with a cyclic disulfide unit spacer (Table 5.5.VII.B). This compound was found to produce DNA interstrand cross-links. The distance between the mitomycin units was calculated to be between 7 and 25 Å depending on their conformation. This distance can cross-link guanines on different DNA strands separated by as many as five base pairs. The cross-linking activity was greatly enhanced by nucleophile phosphine. Apparently phosphine attacks at the disulfide bond to generate thiolate species capable of activating the mitomycin units.[490,491] Of interest is the tetramer in which four mitomycin units are attached to the novel bis-disulfide linker, 3,8,11,16-tetrakis(aminomethyl)-1,2,9,10-tetrathia-cyclohexadecane (Table 5.5.VII.C).[492] On disulfide exchange in the presence of dithiothreitol, the tetramer is reduced to the dimer, providing a DNA interstrand cross-linking capacity similar to the disulfide dimer. In addition, other analogs with cyclohexane and dihydroxy spacer arms (Table 5.5.VII.D and E) were studied and found to react with DNA to produce interstrand cross-links, but with much lower efficiencies.[491,492]

5.11.8 BIS-CHLOROETHYLAMINE DERIVATIVES

The 2-chloroethylamino group is a special functional group that is present in many anticancer drugs, including the nitrogen mustards. This particular group can undergo internal molecular arrangement where the free pair of electrons on nitrogen attacks the β-carbon displacing the halogen ion and forms an active aziridinium cation as shown in Figure 5.25.[493] The aziridinium ion alkylates DNA primarily at the guanine-N7 position to form an N7-alkylated guanine derivative. Compounds containing two 2-chloroethylamino groups exert their biological activity by producing DNA interstrand cross-links. With two chloroethylamine groups, the first guanine

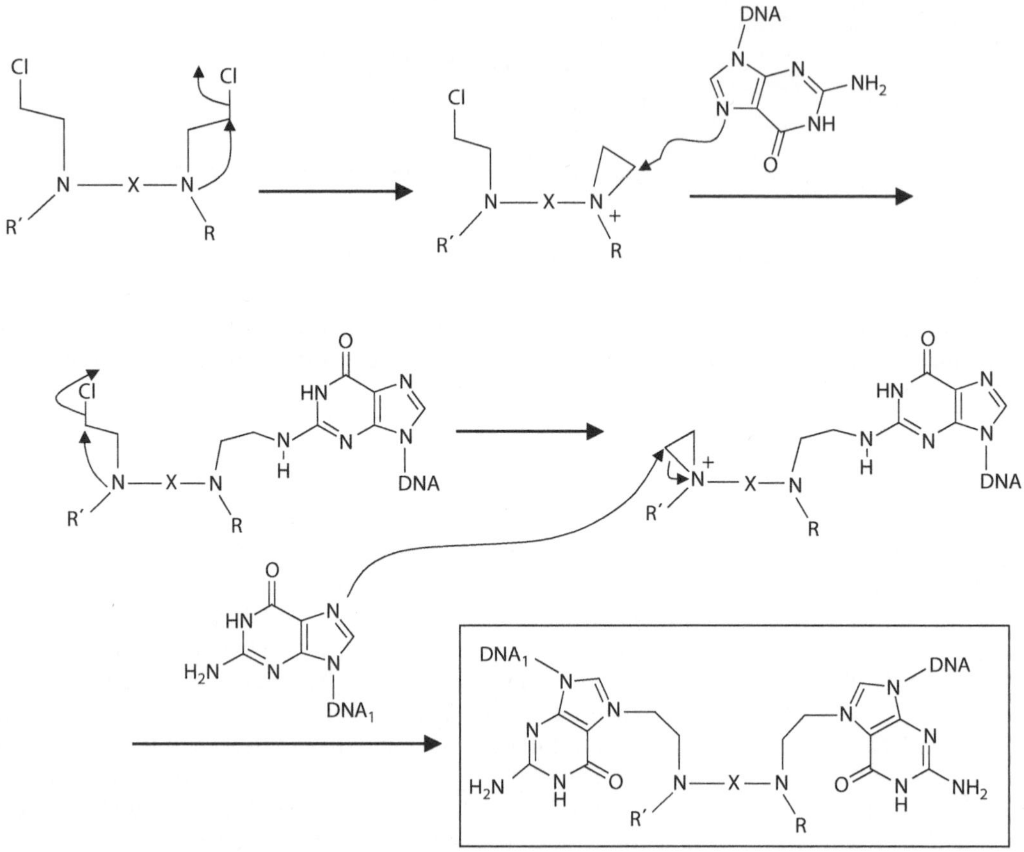

FIGURE 5.25 Formation of aziridinium ion from 2-chloroethylamine and its alkylation of guanine-N7 of DNA, resulting in interstrand cross-linking.

monoadduct produced can form another reactive aziridinium intermediate that can react either with water to form a 2-hydroxyethyl monoadduct or with a second guanine residue to form the interstrand cross-link product (Figure 5.25). Anderson et al.[494] have demonstrated that compounds containing bis-chloroethylamine, such as N,N'-dichloroethyl-N,N'-dimethylhexane-1,6-diamine (Table 5.5.VIII.A), resulted in DNA interstrand cross-links. The corresponding compound with the chlorine atom removed from the nitrogen by one additional carbon, N,N'-dichloropropyl-N,N'-dimethylhexane-1,6-diamine, is not a cross-linker, supporting the notion of an aziridinium cation intermediate in the cross-linking process. Further support for an aziridinium ion intermediate comes from a preformed aziridine, N'-chloroethylaziridine (Table 5.5.VIII.B), which is an efficient cross-linker.[494] The amino group to which chloroethyl is attached can be part of a heterocyclic ring as in piperidine and pyrrolidine. These series of compounds as listed in Table 5.5.VIII.C and D, respectively, have been shown to cross-link interstrand DNA at the guanine-N7 position in the major groove. Of the bispiperidine series, the one with piperidines linked by an ethylene spacer, N,N'-(1,2-ethanediyl)bis(2-chloromethylpiperidine) (Table 5.5.VIII.C.2.a), is the most reactive. On the other hand, in the bispyrrolidine series, the pyrrolidines linked by three and four carbons, that is, N,N'-(1,3-propanediyl)bis(2S-2-chloromethylpyrrolidine) and N,N'-(1,4-butanediyl)bis(2S-2-chloromethylpyrrolidine) (Table 5.5.VIII.D.2 and 3), have little or no DNA cross-linking reactivity.[494]

The binding and alkylation specificity of chloroethylamine-containing compounds depends on the chemical structure to which chloroethylamine is attached. Denny and others[495–499] have synthesized and studied a series of polybenzamide-linked half-mustards where one chloroethyl group is attached to each of the two anilines linked by various spacers (see Table 5.5.VIII.E). The two separate mono-functional mustards appear necessary for maximum cytotoxicity. These polybenzamide mustards are effective interstrand DNA cross-linkers. They preferentially attach to the minor groove of DNA and alkylate adenine-N3 sites in polyA sequences and to a small extent, at 5'-TA and 5'-AT sites. However, the sequence alkylation specificity depends mostly on the H-bonds formed between these compounds and DNA, with factors such as the degree and positioning of cationic charge being less influential. The size and conformation of the spacer in the polybenzamide mustards contribute to this specificity. For example, N,N'-bis[3-[N-(2-chloroethyl)-N-ethylamino]-5-(N,N-dimethylaminomethyl) phenyl]-1,4-benzenedicarboxamide, (alkamin, Table 5.5.VIII.E.1.a), with an annular structure closely matched to the minor groove of duplex DNA, has a strong preference for alkylating adenines in sites containing at least three consecutive adenines and less affinity for 5'-AT and 5'-TA sequences. Gel electrophoresis studies and in vitro cytotoxicity assays against repair-deficient AA8 mutant cell lines showed that this compound has a high degree of DNA interstrand cross-linking ability.[495–499] Using mass spectrometry and model DNA dodecamers d(CGCGAATTCGCG)$_2$, d(CGCAAATTTGCG)$_2$, and d(CGCAAAAAAGCG)·d(CGCTTTTTTGCG), Abdul Majid et al.[500,501] found that alkamin formed a variety of interstrand cross-links between N3-adenines and between adenine and guanine as well as intrastrand cross-links between these bases, demonstrating its ability to cross-link cellular DNA at AT tracts. Compounds VIII.E.1.b, c, and f in Table 5.5 whose annular structure matches that of the minor groove with at least one potential H-bonding carboxamide NH group give the highest specificity.[496–499] They show a high specificity for alkylating various adenines in sequences possessing four or more consecutive adenines, with the consensus sequence being 5'-(A/T)A(G/C)(A/T)N (N being any nucleotide). Their preference to alkylate A preceded by T may be a consequence of their tighter curvature. Compound VIII.E.1.c. also alkylates specific guanines, presumably at the N3 position. The compound with a bicyclo[2.2.2]octane spacer (Table 5.5.VIII.E.1.d) is the least sequence specific. In addition to alkylating adenine N3, it also alkylates guanine-N3. Compounds with a single cationic unit linked to the central aromatic ring (Table 5.5.VIII.E.1.f and E.2) are much more cytotoxic than the corresponding parent compound without the chloroethylamine group. This cytotoxicity is likely due to alkylating events of the half-mustards.[496]

In addition to mono-chloroethylamines, two chloroethyl groups can be attached to a nitrogen atom, that is, N,N-bis(2-chloroethyl)amines. These compounds form a class of nitrogen mustards (see Table 5.5.VIII.F). Nitrogen mustards are the most widely used DNA interstrand cross-linking agents.[423,499] The reaction mechanism involves spontaneous cyclization of each of the chloroethyl chains to form aziridinium ions capable of adding to a nucleophilic site in DNA. The first resulting monoadduct can form a second aziridinium ion which can simply react with solvent or add to another nearby nucleophilic site, resulting in a crosslink either between DNA and protein or between two DNA bases as shown in Figure 5.25.[422,423,493–495] The overall reactivity correlates closely with the basicity of the nitrogen.[499] For aromatic mustards, electron-donating substituents increase reactivity, while for aliphatic mustards the rate-determining step is an SN2 reaction of the aziridinium cation on DNA. The simple aliphatic nitrogen mustards such as mechlorethamine, cyclophosphamide (which metabolizes to phosphamide), and ifosfamide (Table 5.5.VIII.F.1 through 4) have long been explored for their antitumor cytotoxicity and have been made into drugs. Not only do they cross-link interstrand DNA,[502] they also cross-link DNA and protein.[503] They have been shown to react with the N7 of guanines. The preferred targets for interstrand cross-linking was shown to be the 5'-GX'-3' sequences, where X can be any of the four deoxyribonucleotide bases.[504–506] Recently, mechlorethamine was found to cross-link DNA C–C mismatch pairs.[507] It has also been used to study RNA base-specific effects using mass spectrometry.[403] However, the most common nitrogen mustard employed in clinical drugs contain the N,N-bis(2-chloroethyl)aniline functionality. Some of the simple aniline mustards are shown in Table 5.5.VIII.F.5. Studies in Denny's laboratory showed that the rate of alkylation of the simple aniline mustards were much more dependent on the electronic parameter of the substituent.[508,509] The sulfone mustard was least reactive (Table 5.5.VIII.F.5.h), whereas the O-substituted mustard was most reactive (Table 5.5.VIII.F.5.f). The simple aliphatic and aniline mustards have little DNA-binding affinity or specificity. Regiospecificity is largely governed by the electronic and steric properties of the DNA. This specificity results in a high degree of alkylation in the major groove at the N7 position of guanine, which is the most accessible and which has the lowest electrostatic potential.[510] Interstrand cross-link formation occurs almost exclusively with guanine residues in 5'-GNC-3' sequences in DNA, where N is any base.[505,511] The minimal distance between the guanines in this sequence is approximately 6.8 Å, which is more than the 5.1 Å that can span between the five atoms of the mustard cross-link. This constraint produces a distortion of the DNA helix in the region of the cross-link as demonstrated by Rink and Hopkins for mechlorethamine, which produces a static bend of approximately 14° in DNA.[512] N7-alkylated guanines are to some extent unstable and can undergo a further reaction resulting in cleavage of the N-glycosyl bond.[500]

Analogs of aniline mustards can be found in the series of benzoquinone mustards where the nitrogen mustard is attached to a benzoquinone ring. Thus, the molecule can undergo both free radical as well as alkylation reaction with DNA.[513] The DNA cross-linking activities of both benzoquinone mustard (BM) (Table 5.5.VIII.F.6.a) and benzoquinone dimustard (DBM) (Table 5.5.VIII.F.6.j) were enhanced by reduction of the quinone group. It seems that the quinone group may play an important role in modulating the alkylating activity of quinone alkylating agents. To study such an effect, Fourie et al.[514,515] synthesized and investigated BM analogs with electron-donating groups (MeBM, MBM, m-MeBM), electron-withdrawing groups (CBM, FBM), sterically bulky groups (PBM, m-PBM, m-TBM), and positional isomers (MeBM, m-MeBM, PBM, m-PBM), as shown in Table VIII.F.6. After reduction by DT-diaphorase, the BM analogs produced a concentration-dependent increase in DNA crosslink, which was affected by functional groups on the benzoquinone ring. Electron-donating functional group substitutions at the C5 and C6 positions of the quinone seemed to yield greater DNA cross-linking, whereas electron-withdrawing groups and sterically bulky groups at the C6 position had no effect or decreased the ability of the compounds to produce DNA damage. While cytotoxicity of the BM analogs correlated with the maximum levels of DNA crosslinks formed with each BM analog, the half-time of reduction of the BM analogs by DT-diaphorase did not correlate with DNA cross-link formation. Although the reaction mechanism of DNA crosslinking by these BM analogs is not known, it seems that the reaction differs from direct alkylation through aziridium ion but may involve bioreduction alkylation of the quinone ring.

The regiospecificity of DNA alkylation can be modified by attaching the di-chloroethylamine mustard to DNA-affinic carriers.[462,499] Depending on the carrier, the DNA-binding specificity can be greatly enhanced. Holley et al.[516] linked spermidine to chlorambucil (see Table 5.5.VIII.F.7.a) to increase its binding affinity to DNA based on the rationale that under physiological pH, the polycationic spermidine chain would associate with the duplex DNA target. The chlorambucil-spermidine conjugate was found to be approximately 10,000-fold more active than chlorambucil at forming interstrand cross-links with naked DNA and was 35- to 225-fold more toxic than chlorambucil. The increased toxicity of the conjugate compared to chlorambucil was possibly due to enhanced DNA binding and/or facilitated uptake via the polyamine uptake system. In a similar effort, Stark et al.[517] devised a slightly altered conjugate (Table 5.5.VIII.F.7.b), which demonstrated unexpectedly low cytotoxicity against BL6 melanoma cells. Unfortunately, its DNA cross-linking ability was not reported.

In addition to spermidines, other DNA-affinic carriers have been explored, particularly the DNA minor groove binding ligands. Targeting nitrogen mustards to the DNA minor groove region enhances in vitro cytotoxicity and in vivo antitumor activity of most mustards when compared with untargeted mustards of similar structure.[499] Such targeting can also significantly alter the pattern of DNA alkylation. Using 5-[4-(N-alkylamidino)phenyl]furan as a minor groove targeting agent, which is able to bind to double-stranded DNA, preferentially at AT base pairs, along the minor groove by formation of hydrogen bonds, Bielawski et al.[518–522] synthesized three series of analogs of chlorambucil, melphalan, and alkyl aniline mustard (Table 5.5.VIII.F.8). The chlorambucil analogs (Table 5.5.VIII.F.8.a through f) showed DNA minor groove binding characteristics and moderate specificity for AT base pairs, although they also showed weak binding affinity for

GC base pairs.[518,519] These compounds were potent topoisomerase II inhibitors. It was speculated that the combined effect resulting from DNA minor groove alkylation and their potency in topoisomerase II inhibition might be responsible for their increased cytotoxicity. Similarly, the amidine mephalan analogs (Table 5.5.VIII.F.8.g through l) were found to intercalate into the minor groove in AT sequences of DNA and were more active than melphalan.[520,521] These compounds also inhibit topoisomerase II and their cytotoxic properties correlate with their inhibitory effects. Linking 4-(N,N-bis(2-chloroethyl)aminophenyl)propylamine to a 5-(4-N-alkylamidinophenyl)-2-furancarboxylic acid also provided a series of compounds (Table 5.5.VIII.F.8.m through s) with enhanced cytotoxicity. There was a smooth trend of decreasing cytotoxic potency as the size of N-terminal amidine group increased. The increase in the size of the N-alkyl terminal amidine substituents also decreased the DNA-binding affinity, which preferred binding to the minor groove AT regions more than to the GC region. These compounds were also topoisomerase II inhibitors with VIII.F.8.m and VIII.F.8.r being the most potent. Studies with DNA alkylation suggested that DNA cross-link formation was dependent upon the drug orientation prior to alkylation and the deformation of the DNA required to permit 1,3 cross-linking could largely be achieved in the non-covalent intercalated complex.

Bendamustine (see Table 5.5.VIII.F.9) is a drug that has been used since the 1960s.[523] It contains three components, a nitrogen mustard group, a benzimidazole ring, and a butyric acid side chain. The nitrogen mustard group is similar to other alkylators like cyclophosphamide and chlorambucil. As an alkylator, it induces more DNA breaks than does cyclophosphamide. The benzimidazole ring, which replaces the benzene ring present in chlorambucil, is unique and is similar in structure to some purine analogs such as 2-chlorodeoxyadenosine. Hence, it may be involved in DNA binding.

With the incorporation of various specific DNA-affinic carriers into nitrogen mustards, the regiospecificity of DNA can be modified. For example, when the DNA minor groove binding fluorophore pibenzimol (Hoechst 33258) is linked to aniline mustards by variable methylene lengths (Table 5.5.VIII.F.10 and 11), the cytotoxicity is greatly increased across the homologous series due to an altered binding of the mustard to the DNA minor groove, resulting in different patterns of alkylation.[524] As the methylene chain is increased from 0 to 6 (Table 5.5.VIII.F.10.a through e), the equilibrium binding constants decrease.[525] However, the compounds alkylate DNA more rapidly as the chain length increases.[526] There are also significant differences in alkylation patterns with different chain lengths. Compounds with a $(CH_2)_3$ spacer chain (n = 3) (Table 5.5.VIII.F.10.d) show very strong alkylation at adenine sites in poly-AT regions, particularly 5'-GAGAT, 5'-GAGAATA, and 5'-AAAAATA sites, whereas the $(CH_2)_2$ (n = 2) analog (Table 5.5.VIII.F.10.c) alkylates at a number of guanine sites, presumably at the guanine-N7 of the major groove site.[526] Smaill et al.[527] also synthesized bis-benzimidazole analogs bearing o- and m-aniline mustards with $(CH_2)_3$ linker (Table 5.5.VIII.F.11.a and b). These compounds showed broadly similar sequence-specificity of adenine alkylation, with preferred sequences being 5'-TTTAXAXAAXX and 5'-ATTAXAXAAXX (X being any nucleotide). AT-rich sequences are required on both the 5' and 3' sides of the alkylated adenine. The propyl linker between the mustard and benzimidazole has no DNA sequence preference, thus providing the flexibility to orient the mustard for base alkylation after the bis-benzimidazole carrier binding to AT-rich sequences. Different aniline mustards showed similar efficiencies of DNA cross-link formation with little variation in alkylation pattern despite the changes in orientation and positioning of the mustard. Studies on bis-benzimidazoles modified analogs where one of the benzimidazole units is altered by changing the heteroatoms to benzoxazole (Table 5.5.VIII.F.12.a through d) showed that these compounds retain DNA-affinic H-bonding moieties to the minor groove and alkylate predominantly at the 5'-A or 5'-G termini of mixed sequences determined by the sequence-recognizing properties of bis-benzimidazole.[528]

Distamycin, an antiviral compound originally isolated from the cultures of *Streptomyces distallicus*, is endowed with high affinity for AT-rich sequences of B DNA. Since it contains a three-pyrrole skeleton, oligopyrrole is considered a classical DNA minor-groove binder that recognizes consecutive AT base pairs in DNA.[455,500] Mustard analogs of distamycin with varying number of pyrroles and methylene chains have been synthesized and studied (Table 5.5.VIII.F.13).[462,500,529–534] Compounds VIII.F.13.a through i in Table 5.5 are benzoic acid mustards (BAM) whereas compounds VIII.F.13.j through n are aniline mustards containing chlorambucil (CAM). All of these mustards showed highly specific alkylation at N3-adenines of AT tracts and broad-spectrum solid tumor activity. For example, tallimustine (FCE 24517, Table 5.5.VIII.F.13.d) is an AT-specific anticancer drug, alkylating almost exclusively N3 of adenine in the minor groove at the sequence 5′-TTTTGA, and at the present time is in clinical trials.[462,469,500] The number of pyrroleamide units affects the pattern of DNA alkylation. Comparing BAM with one to three pyrroleamide units (compounds VIII.F.13.a through c in Table 5.5), Wyatt et al.[531,532] showed that the BAM monopyrrole compound (Table 5.5.VIII.F.13.a) alkylated several sites, including strong alkylation at the sequences 5′-TTTTAA, 5′-TTAAA, and 5′-TTTTGG. It also alkylated guanine-N7 similar to that of the untargeted mustard. The BAM dipyrrole reagent (Table 5.5.VIII.F.13.b) provided alkylation at fewer sites and strong alkylation at the sequences 5′-TTTTGG, 5′-TTTTAA, and 5′-ATATGA. The BAM tripyrrole conjugate (Table 5.5.VIII.F.13.c) strongly alkylated N3-purine base at the sequence 5′-TTTTGG and 5′-TTTTGA. It also bound noncovalently to AT tracts similarly to distamycin. The BAM dipyrrole and tripyrrole conjugates did not produce detectable guanine-N7 alkylation. Thus, for each increase in the number of pyrroles, there is strong alkylation at fewer sites indicating an increase in sequence specificity in DNA binding. Furthermore, for each increase in the number of pyrrole units there was a corresponding increase in cytotoxicity, with the tripyrrole conjugate being greater than 50-fold more cytotoxic than benzoic acid mustard itself. These observations are similar for CAM containing one to three pyrroles (Table 5.5.VIII.F.13.j through l),[530,531] with the one exception that AG-N3 is strongly alkylated by dipyrrole CAM and G-N3 strongly alkylated by the CAM tripyrrole compound in the sequence 5′-GAAGAT of the minor groove. In general, CAM compounds were able to cross-link plasmid DNA at a 10-fold lower dose than chlorambucil itself.

Brooks et al.[533] studied the effect of various structures of the C-terminus of BAM containing aminoalkyl, diaminoalkyl, and aniline groups (Table 5.5.VIII.F.13.c through i). The results show that all the compounds have a similar sequence-specific alkylation pattern with a preference for the 3′-G residue in the sequence 5′-TTTTGPu-3′ (Pu = G or A). Cytotoxicities of these compounds are related to their relative ability to alkylate the consensus DNA–binding sequence. The effects of the tri-methylene spacer between the aromatic mustard and the oligopyrrole moiety have also been investigated. Ciucci et al.[534] compared mephalan and tallimustine with no spacers (Table 5.5.VIII.F.13.c and d) and MEN 10569 and MEN 10710 with trimethylene spacers (Table 5.5.VIII.F.13.m and n). They found that the flexible trimethylene chain improved the ability to bind DNA and gave a higher alkylation rate and therefore greater interstrand cross-linking. As a consequence, the CAM is more cytotoxic than BAM. Therefore, the presence of a flexible trimethylene chain confers to the CAM distamycin derivatives a particular mode of interaction with DNA.[462]

In addition to BAM and CAM, a dichloroethylamino mustard moiety has been attached directly to pyrrole creating close distamycin derivatives (Table 5.5.VIII.F.14 through 16).[462,499,535,536] All these analogs were able to inhibit tumor cell proliferation *in vitro* and *in vivo*, and showed interstrand cross-linking and strong alkylation of A-N3 residues as compared to G bases, demonstrating specific association with AT-rich regions of DNA. Of all analogs, the silicon-bearing compound (Table 5.5.VIII.F.15) showed the least alkylating reactivity.[536]

In the development of GC sequence-directed alkylating agents to produce more effective anticancer drugs, Lee et al.[537,538] synthesized a series of CAM and BAM distamycin analogs containing varying number of imidazole units as shown in Table 5.5.VIII.F.17 and 18. Compounds VIII.F.17.a through c in Table 5.5 are BAM distamycin derivatives, and compounds VIII.F.17.d through f are CAM distamycin analogs. The imidazole-CAM analogs were less cytotoxic than the corresponding pyrrole-CAM, probably due to lower noncovalent binding affinities.[532] They, together with benzyl-mustard (Table 5.5.VIII.F.18), were found to alkylate guanine-N7 in the major groove and were more effective in producing DNA interstrand cross-links in isolated DNA than were imid-azole-BAM compounds.[537] For the BAM, these imidazole-containing analogs were found to bind to GC-rich DNA sequences in the minor groove, albeit with weaker affinities than distamycin.[532] The monoimidazole–BAM (Table 5.5.VIII.F.17.a) produced guanine-N7 alkylation in a similar pattern to BAM, but at a 100-fold lower dose.[529] The diimidazole- and triimidazole-BAM analogs (Table 5.5.VIII.F.17.b and c, respectively) did not produce detectable guanine-N7 alkylation but only alkylated selected sites in the minor groove in a similar pattern to that seen for the corresponding di- and tripyrrole–BAM compounds. The consensus sequence, 5′-TTTTGPu (Pu = G or A) was strongly alkylated by the triimidazole conjugate in preference to other similar sites including three occurrences of 5′-TTTTAA.[499,529]

Distamycin nitrogen mustard analogs containing mixed pyrroles and imidazoles have also been studied. Xie et al.[536] have synthesized hybrid pyrrole–imidazole-containing analogs of distamycin as shown in Table 5.5.VIII.F.19. These compounds were shown to strongly alkylate adenine-N3 residues in 5′-AATA, 5′-ATAA, 5′-AATG, 5′-ATGG, 5′-AAAT, and 5′-ATTT. Moderate alkylation of guanine residue was also observed at the sequences 5′-GTTA, 5′-GTCA, 5′-GATA, and 5′-GGTT. However, a careful study of interstrand cross-linking was not reported. In additional to these pyrrole–imidazole hybrids, Baraldi et al.[539] have synthesized pyrrole–pyrazole-containing distamycin mustard analogs (Table 5.5.VIII.F.20). These benzoyl mustards were found to have various cytotoxicities. For tallimustine analogs (Table 5.5.VIII.F.20.a through f), the pyrrole position was shown to be critical for the activity. The presence of this heterocycle near the benzoyl nitrogen mustard moiety was necessary for activity both *in vitro* and *in vivo*. Incorporation of an additional pyrrole ring near the alkylating moiety as in compound VIII.F.20.h in Table 5.5 further increased the activity, whereas introduction of a pyrazole ring in the same position (Table VIII.F.20.i) decreased activity. Changing the propionamidino moiety (Table 5.5.VIII.F.20.f) to dimethylaminopropane (Table 5.5.VIII.F.20.g) decreased the activity by twofold. Neither the DNA sequence specificity nor the DNA residues alkylated were reported, however. Further structural-activity studies of compounds containing benzoyl and cinnamoyl nitrogen mustards tethered to different benzoheterocycles, and to oligopyrroles related to netropsin consisting of two pyrrole–amide units and terminating with an amidine moiety (Table 5.5.VIII.F.21), were carried out by Baraldi et al.[540] In Table 5.5, compounds VIII.F.21.a through d are benzoyl nitrogen mustards, whereas VIII.F.21.e through h are vinylogues compounds with a vinylic double bond increasing distance between the alkylating moiety and the DNA–polypyrrolic binding frame. They were 2- to 50-fold less cytotoxic than tallimustine against human K562 leukemia cells. The introduction of a vinylic double bond between the phenyl nitrogen mustard and the benzoheterocycle provided an inconsistent advantage in cytotoxicity. No clear-cut structure–activity relationship can be deduced from the study, and it is difficult to find common physicochemical features among the compounds. All of the mustards showed evidence of noncovalent binding to two AT-rich sequences in the minor groove (AAATAA and TATAT). Sequence-specific alkylation was observed for several of the compounds with a preference for adenine of 5′-TTTTGA sequence similar to tallimustine.

Hairpin polyamides containing pyrrole (Py) and imidazole (Im) amino acids, linked via chiral diaminobutyric acid turn moieties to give a hairpin structure, have an affinity and specificity for DNA comparable to naturally occurring DNA-binding proteins.[461,462,464,499,541] Wurtz and Dervan[542] synthesized an eight-ring hairpin polyamide targeted to the HIV-1 promoter and attached chlorambucil (CHL) to the chiral α-amino group of diaminobutyric acid (DABA) γ-turn (Table 5.5.VIII.F.22). The compound, ImPy-β-ImPy-(R)CHL-γ-ImPy-β-ImPy-β-Dp (where HT = hairpin turn and Dp = 3-(dimethylamino)-propylamine), is shown in Table 5.5.VIII.F.22.a. The modification with mustard did not alter the polyamide-binding affinity and specificity for DNA, and alkylation proceeded in high yield. Adenine residues adjacent on either side of the polyamide-binding site sequence 5′-(A/T)GC(A/T)GC(A/T)-3′ in the HIV promoter region were alkylated. DNA interstrand cross-linking in a cell-free SV40 system was demonstrated and the polyamide mustard was found to be a much more efficient agent than chlorambucil itself.[543] The position of CHL attachment to the polyamide confers important differences in the alkylation activities of these compounds. Tsai et al.[544] synthesized a polyamide of the sequence: ImImβIm-HT-PyPyPyPy-βDp. In one case, 1(R/S)-CHL (Table 5.5.VIII.F.22.b), CHL utilized an α-DABA turn unit, whereas the other, 2(R/S)-Chl (Table 5.5.VIII.F.20.c), employed the standard γ-DABA turn unit. The S-enantiomers of the compounds are similar to their R analogs in their alkylation specificities. However, the alkylation profiles of 1R-Chl and 2R-Chl are dramatically different: 1R-Chl alkylates DNA more specifically than 2R-Chl and appears less reactive. Thus, hairpin polyamides containing the α-DABA turn unit may be an important class of DNA-binding small molecules.

Another minor groove carrier to which an aniline mustard is attached is 4-anilinoquinoline.[499] Gravatt et al.[545] synthesized and evaluated a family of anilinoquinoline-based aniline mustards as shown in Table 5.5.VIII.F.23. Studies on the methyl quaternary anilinoquinolinium analogs showed that the binding of these compounds to DNA was consistent with a minor groove binding mode. The incorporation of the 4-anilinoquinoline moiety has resulted in aniline mustards of enhanced cytotoxic potency. These compounds were much more cytotoxic than the parent diols. Cytotoxicity is consistent with the full mustard killing via a DNA cross-linking mechanism. Comparative cell line studies suggested that the mechanism of cytotoxicity varied with mustard reactivity. The most reactive mustards cross-linked DNA, while cell killing by the less reactive compounds appeared to be by the formation of bulky monoadducts. McClean et al.[546] also showed that these compounds (Table 5.5.VIII.F.23.a, b, and h) nonspecifically and weakly alkylate the N7 of guanine in the major groove, but they strongly alkylate the N3 of adenine and guanine in the minor groove at specific classical AT-rich sites. Other binding sites include AT-rich with GC base pairs. There is evidence that the anilinoquinoline aniline mustards cross-linked interstrand DNA at concentrations as low as 0.05–0.01 µM, which is 60- to 100-fold more effective than melphalan. The quaternary compound (Table 5.5.VIII.F.23.h) is the most effective cross-linking agent in this series. Although the exact cross-linking site is not defined, there is an indication that an AATTA site is involved.

DNA intercalating agents have been sought as DNA-affinic carriers. Koyama et al.[547] prepared a series of anthraquinone-linked alkyl mustards as shown in Table 5.5.VIII.F.24. Compounds VIII.F.24.a through d of Table 5.5 are chrysophanol derivatives and compounds VIII.F.24.e through h are emodin derivatives. With minimal modification of the anthraquinone moiety, that is, compounds VIII.F.24.d and h, the antraquinone mustards showed maximum activity, presumably due to intercalation of the anthraquinone ring into DNA. Introduction of two bulky methoxy groups, that is, compounds VIII.F.24.a and e of Table 5.5, interfered with the intercalation process and reduced the cytotoxicity. The activity of 1- or 8-mono-O-methylanthraquinones, that is, compounds VIII.F.24.b, c, f, and g of Table 5.5, fell between those of the 1,8-dihydroxy and 1,8-dimethoxy analogs. Thus, DNA intercalation plays some role in alkylating the DNA. Another intercalator that has been explored is acridine which is considered to bind to DNA in a preferred conformation in which the long axis of the acridine lies parallel with the base pair long axis. Creech et al.[548] have studied a series of mono and dichloroethyl alkyl mustards linked to various heterocyclic chromophores, particularly acridines, as shown in Table 5.5.VIII.F.25. They showed that these compounds were more potent than the corresponding simple mustards against ascitic tumors *in vivo* and suggested that this potency was due to the high affinity of the chromophores for DNA. To continue this line of investigation, Denny's group synthesized several series of 9-aminooacridine aniline mustard[508,549,550] as shown in Table 5.5.VIII.F.26. The anilines were attached to the DNA-intercalating acridine through -CH$_2$-, -O-, -S-, -SO$_2$-, -CO-, -NHCO-, and -CONH- with varying alkyl chain lengths, thus providing a series of mustards with widely varying electronic properties. Most of the acridine-linked DNA-targeted mustards showed *in vivo* antitumor activity with potency up to 100-fold more than the corresponding untargeted compound.[508] The overall reactivity is controlled primarily by the electronic properties of the para-substituent on the mustard. In the -CH$_2$-, -O-, and -S- series (Table 5.5.VIII.F.26.a through l), there is little change in *in vitro* cytotoxicity with chain length. For the SO$_2$ compounds (Table 5.5.VIII.F.26.m through p), there is a distinct variation in cytotoxicity across the series, with compound VIII.F.26.p being the least reactive.[550] The least reactive SO$_2$ compounds were also the weakest alkylating agents. Studies with reactive compounds (e.g., Table 5.5.VIII.F.26.d and h) showed that the primary cytotoxic effect is due to DNA cross-linking. These compounds have a striking preference for alkylating guanines in 5'-GT sequences, probably in the N7 (major groove) site, and are capable of adenine alkylation at significant levels. The -S- compounds (Table 5.5.VIII.F.26.i through l) showed much lower cross-linking capabilities than the compounds in the -CH$_2$- and -O- series (Table 5.5.VIII.F.26.a through h). Compounds of the -O- and -S- series require thymine and purine bass on the 3'-side of the guanine which is alkylated, whereas the -CH$_2$- series does not show such sequence selectivity.[549] As the chain length of the series is increased there is a switch from guanine alkylation in 5'-GT sequences to adenine alkylation in the complementary 5'-AC sequence. Within each series, cross-linking ability altered with chain length, being maximal with the C4 analog. Overall, these compounds showed marked changes in their patterns of DNA alkylation, switching from alkylation at guanine N7 sites in the major groove to adenine N3 sites in the minor groove and even adenine N1 sites in the intergroove.[501] The compound with tert-butyl-9-aminoacridine chromophores (Table 5.5.VIII.F.26.t) showed an overall decrease in DNA alkylation ability and a marked preference for alkylating guanines in 5'-GT.[508] These tendencies are probably due to steric hindrance of the *t*-butyl group affecting DNA intercalation.

Because of the low level of cross-linking by acridine-linked aniline mustards, which appears to result from the slow rate of the second alkylation event as a result of unfavorable spatial orientation, Gourdie et al.[551] synthesized acridine mustards with two half-mustards separately attached to the chromophore as shown in Figure 5.26. It was hoped that the two mono-mustards on acridine intercalation would be well-placed to cross-link the DNA. Unfortunately the two compounds, *N*,9-bis[[3-[4-[(2-chloroethyl)ethylamino]phenyl]propyl]amino]acridine-4-carboxamide and *N*,9-bis[[2-[[4-[(2-chloroethyl)ethylamino]phenyl]thio]ethyl]amino]acridine-4-carboxamide, gave only monoalkylated DNA.

In addition to acridine itself, anilinoacridines are well-known intercalators.[499] Fan et al.[552] prepared two series of mustard analogs of this carrier, with aniline mustards attached either at the 1'-position of the 9-anilino ring (Table 5.5.VIII.F.27) or off the 4-carboxamide (Table 5.5.VIII.F.28). Compounds of both series cross-linked DNA and showed similar patterns of alkylation-induced cleavage of DNA. The efficiency of the cross-linking process appeared to depend more on the reactivity of the mustard rather than its position on the chromophore. The most efficient cross-linkers are the most reactive compounds (i.e., Table 5.5.VIII.F.27.a through c and VIII.F.28.a through d). All of the mustards alkylated DNA at N7-guanines as well as some adenines similar to the pattern for untargeted mustards. Strong alkylation was observed at all available runs of G residues or guanines in the GC-rich region such as 5'-GCGG, 5'-CGC, 5'-GGC, and 5'-CCGTG. Alkylation of adenines occurred at AT-rich sequences, such as 5'-TTTAAT and 5'-TAA. Within each class, the extent of alkylation varied with the reactivity of the mustard. Drugs with an O- or CH$_2$-linker (Table 5.5.VIII.F.27.a through c and VIII.F.28.a through d) showed fast alkylation. The SO$_2$-containing

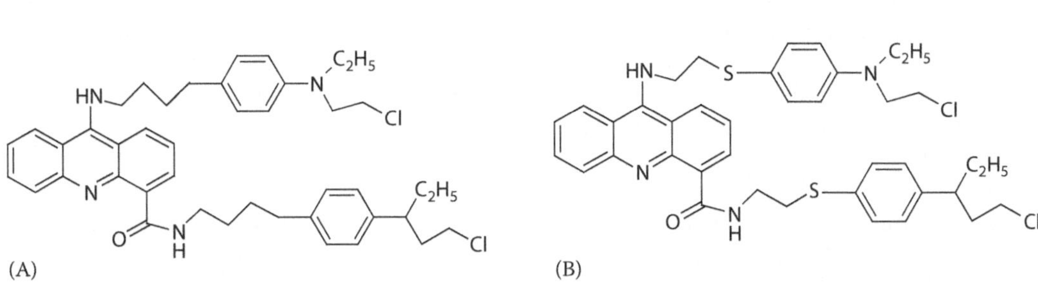

FIGURE 5.26 Acridine with two half-aniline mustards. (A) *N*,9-Bis[[3-[4-[(2-chloroethyl)ethylamino] phenyl]propyl]amino]acridine-4-carboxamide. (B) *N*,9-Bis[[2-[[4-[(2-chloroethyl)ethylamino]phenyl]thio] ethyl]amino]acridine-4-carboxamide.

mustards (Table 5.5.VIII.F.27.e and VIII.F.28.f) showed very slow alkylation. The length of the linker chain also influence alkylation, with the shorter-chain derivatives (Table 5.5.VIII.F.27.b and VIII.F.28.c) appearing to alkylate more rapidly. All of the compounds were considerably more cytotoxic than analogous untargeted mustards. The 4-carboxamide-linked analogs showed slightly higher *in vivo* antileukemic activity than the corresponding 1′-linked analogs. In general, toxicity correlated with DNA cross-linking. The S- and SO_2-linked compounds (Table 5.5.VIII.F.27.d and e, and VIII.F.28.e and f), which showed lower cross-linking capability, were also less cytotoxic. In a different application, Temple et al.[553] used compounds VIII.F.27.b and c to study protein–DNA interaction and found that the two amsacrine-based compounds were extremely sensitive to local variations in protein–DNA structure. Thus, the anilinocritine mustards can be used to probe protein–DNA interactions in the chromatin of intact human cells.

Another DNA intercalator involves naphthalene. Naphthoate as part of neocarzinostatin (NCS), an antitumor antibiotic isolated from *Streptomyces carzinostaticus*, binds to the DNA minor groove. Intercalation of the naphthoate moiety positions the NCS to execute its toxicity. Urbaniak et al.[554] linked chlorambucil and mephalan to naphthoate (Table 5.5.VIII.F.29) and studied the effects. For the melphalan analog (Table 5.5.VIII.F.29.b), the level of DNA damage and *in vitro* cytotoxicity improved, whereas the linkage did not provide any advantage to chlorambucil. The position of DNA damage observed was similar for both compounds, suggesting that DNA alkylation occurred at the N7 position of guanine nucleotides. No significant alkylation of adenine nucleotides was detected. ApoNCS was found to bind significantly to the melphalan derivative, which reduced the extent of hydrolysis of the conjugate. Thus, binding of an apo-protein may increase the stability of a drug.

Napthalimides are a class of intercalating antitumor agents. Gupta et al.[555] linked the naphthalimide moiety with alkylating aniline mustard to DNA minor groove-binding lexitropsin to study its efficacy.[462,555] A series of the naphthalimide lexitropsin aniline mustards were synthesized (Table 5.5.VIII.F.30). Cytotoxicities to KB cells in culture were comparable for compounds VIII.F.30.a, d, and e in Table 5.5). Compounds VIII.F.30.b and c were the least toxic, while the compound lacking the lexitropsin moiety (Table 5.5.VIII.F.30.g) was the most toxic. In contrast to distamycin-bearing nitrogen mustard moieties where DNA alkylation is directed to adenine N3 sites in the minor groove, the naphthalimide nitrogen mustards alkylate DNA at accessible N7-guanine sites within the major groove. Compound VIII.F.30.e in Table 5.5 bearing three pyrrole groups showed greater alkylation potential than compound VIII.F.30.d (Table 5.5). Overall, both of these agents were more effective than other compounds. The compound with an additional imidazole group (Table 5.5.VIII.F.30.f) showed relatively weak alkylation. Greater reactivity was found for guanines flanked on either side by TC, AA, or AT, followed by GT, GC, and GG, which may affect the stability of the DNA–drug formation. The distamycin and netropsin moieties in this series did not divert the compounds to bind to AT-rich sequences. Binding of the intercalative naphthalimide moiety rather than the lexitropsin unit to DNA determines guanine alkylation by the nitrogen mustard moiety. Thus, lexitropsin moiety in conjugation with naphthalimide mustards neither enhanced sequence specificity nor increased cytostatic activity.

Hypoxia activation of aniline mustards provides another means of DNA alkylation.[556] Palmer et al.[557,558] synthesized a series of 5-[*N,N*-bis(2-chloroethyl)amino]-2,4-dinitrobenzamide analogs as hypoxia-selective cytotoxins (HSC) (Table 5.5.VIII.F.31), which appear to be fully active only at extremely low oxygen concentrations. These HSC contained nitro groups which deactivated the aniline mustard. However, bioreduction of either of the electron withdrawing nitro groups to hydroxylamine or amine dramatically activate the mustard.[559] For example, 5-[*N,N*-bis(2-chloroethyl)amino]-2,4-dinitrobenzamide (Table 5.5.VIII.F.31.a, SN 23862) was 60-fold more cytotoxic to hypoxic W4 cells than to the same cells under aerobic conditions. Reactivity and cytotoxicity of these compounds are very dependent on the electronic properties of the substituents on the aromatic ring. In the series listed in Table 5.5.VIII.F.31, analogs with ionizable or dipolar carboxamide side chains showed improved solubility but generally had reduced cytotoxic potency and hypoxic selectivity. However, no general structure–activity relationships for hypoxic selectivity could be discerned. Compound VIII.F.31.o in Table 5.5 acted as the preferred compound overall with respect to solubility, potency, and *in vitro* hypoxic cell selectivity. Radiobiological studies indicated that the compound was equally active against both aerobic and hypoxic cells in KHT tumors. DNA elution studies indicated reductive activation transformed it to a DNA cross-linking agent under hypoxia.

Similar to dinitrobenzamide mustards, 2-nitroimidazoles have significantly higher intrinsic one-electron reduction potentials than nitrobenzene. Lee et al.[560] synthesized 5[*N,N*-bis(2-chloroethyl) amino]-1-methyl-2-nitroimidazole (Table 5.5.VIII.F.32). The 2-nitroimidazole mustard showed comparable hypoxic selectivity in UV4 cells to the dinitrobenzamide mustard SN 23862 (Table 5.5.VIII.F.31.a) and was also significantly hypoxia-selective in repair-competent AA8 cells.

Since HSC are inactive in aerobic cells, they are essentially prodrugs. Another dinitrobenza-mide-related prodrug is PR-104 (Table 5.5.VIII.F.33) studied by Patterson et al.[561] As opposed to 2,4-dinitrobenzamide-5-mustard, PR-104 is a 3,5-dinitrobenzamide-2-mustard, which is more readily reduced in hypoxic cells and its asymmetrical nitrogen mustard contains more reactive leav-ing groups (bromide and mesylate rather than chloride). The water-soluble phosphate "pre-prodrug" is converted efficiently to the more lipophilic alcohol, which is hypoxia-activated to the hydroxyl-amine derivative resulting from reduction of the nitro group para to the mustard moiety. It is further metabolized to various active intermediates including 2-chloroethylamine-containing compounds. The hypoxia-activated metabolites cross-link interstrand DNA with marked activity against human tumor xenografts. Studies on DNA damage suggested that DNA interstrand cross-linking is the major mechanism of cytotoxicity, although the quantitative relationship between cytotoxicity and cross-link formation has yet to be established.

Nitrogen mustards coordinated to Co(III) are potential HSC. Coordination of the nitrogen lone pair to Co(III) should suppress its toxicity since the electron pair is no longer available to act as a nucleophile. On chemical or metabolic one-electron reduction of the inert Co(III) complex, the resulting labile Co(II) species undergoes very facile ligand substitution by water, releasing the cytotoxic free nitrogen mustard. Thus, Co(III) complexes of nitrogen mustards constitute a new class of bioreductive drugs. Ware et al.[562] investigated two series of Co(III) complexes of the bidentate bisalkylating nitrogen mustard ligands *N,N*-bis(2-chloroethyl)ethylenediamine (DCE) and *N,N'*-bis(2-chloroethyl)ethylenediamine (BCE) (Table 5.5.VIII.F.35 and 34, respectively). The complexes also bear two 3-alkylpentane-2,4-dionato (acac) auxiliary ligands. In both series, the cobalt(III) complexes showed moderate hypoxic selectivity. The DCE complexes appeared to have hypoxic selectivity superior to that of the BCE compounds and were one order of magnitude more cytotoxic than the corresponding BCE compounds. For example, the DCE complexes (Table 5.5.IX.F.35) proved much more cytotoxic in aerobic AA8 cultures. However, the unsubstituted cobalt–DCE complex (Table 5.5.VIII.F.35.a) was much less toxic than the parent ligand toward aerobic AA8 cells. On the other hand, the methyl analog (Table 5.5.VIII.F.35.b) was selectively toxic to hypoxic cells. Its cytotoxicity was clearly due to release of the DCE ligand, as shown by its effects on UV4 cells, which are hypersensitive to DNA cross-linking agents, and by the demonstra-tion of DNA cross-links by alkaline elution. Essentially, the patterns of cytotoxicities of the cobalt complexes were broadly similar to those of the respective free ligands.

In addition to 3-alkylpentane-2,4-dionato (acac) as an auxiliary ligand, Ware et al.[563] also synthe-sized tropolonato bidentate mustard complexes (Table 5.5.VIII.F.36 and 37). The tropolonato com-plexes have significantly higher reduction potentials than the corresponding acac complexes, which is likely to facilitate their cellular reduction and release of the mustard ligands. Unfortunately, they did not show any hypoxic selectivity in a clonogenic assay, where the corresponding Meacac DCE complex (Table 5.5.VIII.F.35.b) is significantly hypoxia-selective.

Other auxiliary ligands have also been used, such as carbonate, oxalate, and dithiocarbamate.[564,565] Craig et al.[565] studied cobalt(III) complexes containing *N,N*-bis(2-chloroethyl)-1,2-ethanediamine with carbonato, or oxalato anionic ancillary ligands (Table 5.5.VIII.F.38 and 39, respectively). Both the nitrogen mustard complexes K[Co(CO$_3$)$_2$(DCE)] and K[Co(ox)$_2$(DCE)] were found to be less cytotoxic than free DCE and exhibited hypoxia selectivity. The experiments indicated that the nitrogen mustard ligand was released from the complexes and involved in the cytotoxic event. The carbonato and oxalate ligands are no better than the methyl acac ligand (e.g., Table 5.5.VIII.F.35.b) discussed above. Another ligand that does not fair well is dithiocarbamate. Ware et al.[564] studied cobalt(III) complexes containing two dithiocarbamate ligands and a bidentate nitrogen mustard ligand (Table 5.5.VIII.F.40 and 41). It was speculated that the sulfur donor ligands should lead to both faster electron transfer and a more negative redox potential for the Co(III)/Co(II) couple, relative to complexes containing the oxygen donor acac ligands discussed above. The aerobic cytotoxicities of the [Co(Me2dtc)DCE] and [Co(Et2dtc)DCE] complexes (Table 5.5.VIII.F.41.a and b) against AA8 cells were about 60-fold greater than that of the corresponding methyl acac complex (Table 5.5VIII.F.35.b). However, clonogenic cell killing is not appreciably enhanced under hypoxic conditions for any of the dithiocarbamato complexes. Although DCE was released on Co(III) complex reduction and contributed to clonogenic cell killing, the thiocarbamate ligands may play a larger role in the antiproliferative effects of these complexes. Dialkyldithiocarbamates are known to be potent inhibitors of cell proliferation. This finding, coupled with the instability of the parent complexes, suggests that the dialkyldithiocarbamato ligand system is not suitable for further development.

5.11.9   BIS-CARBAMATE DERIVATIVES

In carbamate-activated compounds, the (carbamoyl)oxylmethyl group serves as a leaving group to generate an active intermediate. Like chlorine in nitrogen mustards, the carbamate must be bonded at a suitable location in relationship to the nitrogen for the group to leave. For example, Henderson et al.[493] have synthesized the corresponding carbamate analogs (Table 5.5.IX.A) of chloromethylpiperazine, replacing chloromethyl group with 2-p-nitrophenylaminocarbonyloxymethyl group. Similar to the nitrogen mustards, the 2-p-nitrophenylaminocarbamate was displaced in the formation of aziridium ion, which is then attacked by DNA nucleophile resulting in interstrand cross-linking as shown in Figure 5.27. Although the bispiperidine carbamate did cross-link naked DNA, they are less efficient than the corresponding mustards. The compounds exhibited a range of *in vitro* cytotoxicity, with compounds IX.A.1, 3, and 4 in Table 5.5 being the most potent.

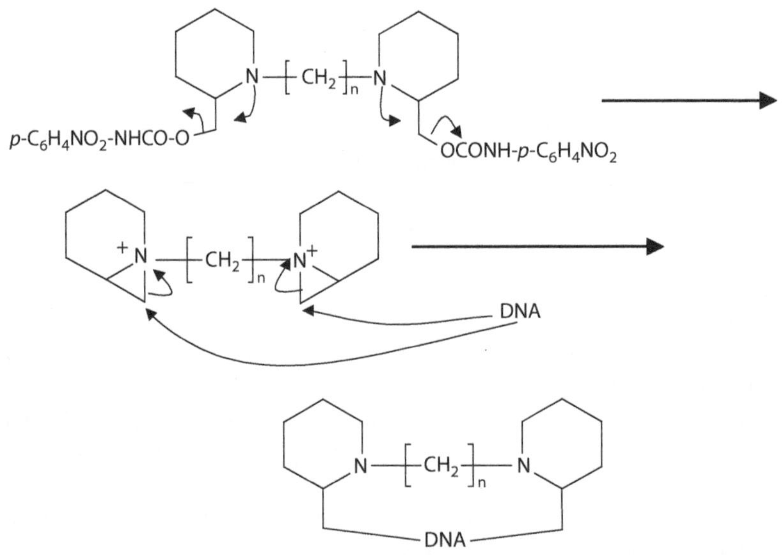

FIGURE 5.27   Formation of aziridinium ion from bispiperidine carbamate and its interstrand cross-linking of DNA.

Another group of bis-carbamate compounds are the vinylogous carbinolamines, which include bis-carbamate-substituted imidazoles, thioimidazoles, pyrroles, pyrrolizines, pyrrolo-quinolines, and pyrrolo-benzazepines (Table 5.5.IX.B). Again, the carbamate moiety serves as leaving groups in an alkyl-oxygen cleavage mechanism yielding methylenic carbons bonded directly to a heteroaromatic nucleus reactive electrophilic center. The mechanism of action is displayed in Figure 5.28 for imidazole bis-carbamate.[566] The reaction is similar for other vinylo-gous carbinolamines. Numerous imidazole bis-carbamates with different substituents on the heterocyclic ring have been synthesized and studied.[566,567] Compounds with antitumor activi-ties are listed in Table 5.5.IX.B.1. As can be seen, only electron-donating substituents on the imidazole ring gave antitumor active compounds. 2-Methylthio-derivatives as in carmethiole (Table 5.5.IX.B.1.a) seem to be most active.[567] 2-(Ethylthio)imidazole has comparable antitumor activity as carmethiole.[566] As the length of the 2-(alkylthio)-imidazole side chain increases, antitumor potency decreases such that the 2-(n-propylthio)-imidazole (Table 5.5.IX.B.1.c) is less active than carmethiole. However, electron-withdrawing groups on the sulfur decreased chemical reactivity such that carmethizole sulfoxide had no antitumor activity.[566] Cell studies showed that the interaction of carmethizole with DNA produces monoadducts, DNA–protein, and DNA–DNA interstrand cross-links at several sites similar to melphalan.[568] Since the imid-azole bis-carbamates are not DNA regiospecific, Smaill et al.[527] linked them to DNA minor groove binding ligands such as benzimidazole as shown in Table 5.5.IX.B.1.d and e. These two compounds showed DNA interstrand cross-linking ability and preferentially alkylated gua-nines at 5′-CG sequences. The pattern of alkylation is very similar to the corresponding untar-geted compounds, with little evidence of additional selectivity imposed by the benzimidazole AT-preferring carrier.

Related to imidazole bis-carbarmates are the pyrrole derivatives. Anderson et al. synthesized a large number of pyrrole bis-carbamate analogs.[569,570] While most of the compounds showed signifi-cant activity in the P388 *in vivo* antileukemic assay, the studies lacked information on their DNA cross-linking properties. Thus, only a few representatives are shown in Table 5.5.IX.B.2. It seems that the phenyl and methoxyphenyl derivatives either at position 1 or 2 of the heterocyclic pyrrole ring provided the most antileukemic activity.

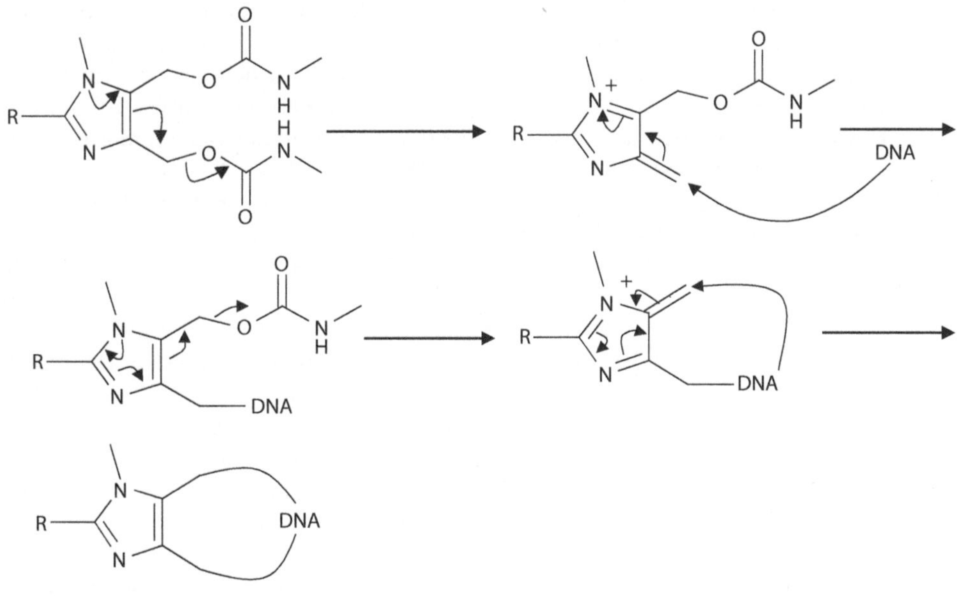

FIGURE 5.28   Proposed mechanism of action for imidazole biscarbamate DNA cross-linking.

A bicyclic analog of pyrrole is pyrrolizine. Anderson and Corey[571] investigated a series of pyrrolizine compounds and found bis(N-methylcarbamate) derivatives showed significant antileukemic activity in the *in vivo* P-388 assay. Some representative compounds are listed in Table 5.5.IX.B.3.a through c. Compound IX.B.3.a is the most potent followed by others. It appears that the *p*-methoxyphenyl substitution at C-5 of pyrrolizine (Table 5.5.IX.B.3.c) is inferior to the 3,4-dichlorophenyl group (Table 5.5.IX.B.3.a). Numerous other pyrrolizine bis-carbamates have also been synthesized.[572–574] The most important ones are listed in Table 5.5.IX.B.3.d through h. Of the pyrrolizinyl-halopyridinium iodides, the α-fluoropyridinium compounds (Table 5.5.IX.B.3.f and g) were active but the α-chloro compounds were not. The pyridone that corresponds to the pyridinium salts (Table 5.5 IX.B.3.h) was also active against P388 lymphocytic leukemia.

In addition to bicyclic compounds, tricyclic bis[[(carbamoyl)oxy]methyl] derivatives have also been explored. These include pyrrolo[2,1-*a*]isoquinolines, pyrrolo[1,2-*a*]quinolines, pyrrolo[2,l-*a*]isobenzazepines, and pyrrolo[1,2-a]benzazepines.[575] Numerous compounds have been synthesized and some examples are listed in Table 5.5.IX.B.4 through 7. In the pyrrolo[2,1-*a*]isoquinoline series (Table 5.5.IX.B.4), the C-3 methyl group had a more pronounced effect on activity and toxicity than the C-7 methoxy group. For example, addition of a C-3 methyl group to compound IX.B.4.b in Table 5.5 (converted to Table 5.5.IX.B.4.c) caused marked decrease in activity. Although addition of C-3 methyl group to IX.B.4.a in Table 5.5 retained its activity, IX.B.4.a is more potent. In the pyrrolo[1,2-*a*]quinoline series (Table 5.5.IX.B.5), the 4,5-dihydro compounds were more potent than the corresponding fully unsaturated analogs. The pyrrolo[1,2-*a*]benzazepine bis(carbamates) (Table 5.5.IX.B.6.a and b) showed approximately equivalent activity to the comparable pyrrolo[1,2-*a*]quinolines (Table 5.5.IX.B.5.a and b). The isobenzazepine (Table 5.5.IX.B.7) is less active and more toxic than the corresponding pyrrolo[2,1-*a*]isoquinoline (Table 5.5.IX.B.4.c). For all these bis-carbamates, DNA binding and cross-linking have not been studied.

5.11.10 PYRROLIZIDINE ALKALOIDS (PAs)

Pyrrolizidine alkaloids (PAs) are naturally occurring phytochemicals found in more than 12 higher plant families.[422,576] More than 660 PAs and N-oxide derivatives have been identified in over 6000 plants. An example of each of the retronecine-type, heliotridine-type, and otonecine-type PAs is shown in Table 5.5.X.A through C (stereochemistry not exactly designated). Retonecine-type PAs (Table 5.5.X.A) are macrocyclic diesters and heleiotridine-type PAs (Table 5.5.X.B) are open diesters. Both types possess a similar necine base. The difference is at the C7 position, with the retronecine-type PAs possessing an R absolute configuration and the heliotridine-type an S stereochemistry. The otonecine-type PAs (Table 5.5.X.C) have a necine base structurally different from the other PAs. However, all PAs require metabolic activation to exert toxicity and tumorigenicity.[576–581] They are all metabolized by cytochromes P-450 (CYP), primarily CYP 3A4, to a pyrrolic ester, the dehydropyrrolizidine alkaloid, which is chemically and biologically reactive. Once formed, the pyrrolic ester metabolites can rapidly bind with DNA, leading to DNA cross-linking, DNA–protein cross-linking, and DNA adduct formation. A potential DNA cross-linking reaction mechanism is shown in Figure 5.29. Because of its high reactivity, the pyrrolic ester metabolites can also react readily with water and other endogenous constituents, such as glutathione, to form the detoxified products. Coulombe et al.[581] used dehydrosenecionine and dehydromonocrotaline (DHMO) derived from oxidation of the corresponding PAs to study DNA–actin cross-linking. They found that the patterns of the proteins cross-linked to DNA were similar to those induced by standard bifunctional alkylating agents such mitomycin C, *cis*-dichlorodiammine platinum(II), and nitrogen mustard. Kim et al.[582] used similar pyrrolic PAs to study DNA cross-linking and found that that structural features, most notably the presence of a macrocyclic diester, confer potent cross-link activity to PAs. In addition to DNA cross-links, all PAs studied so far generated the same set of 6,7-dihydro-7-hydroxy-1-hydroxymethyl-5H-pyrrolizine (DHP)-derived DNA adducts *in vivo* and/or *in vitro*.[577–580]

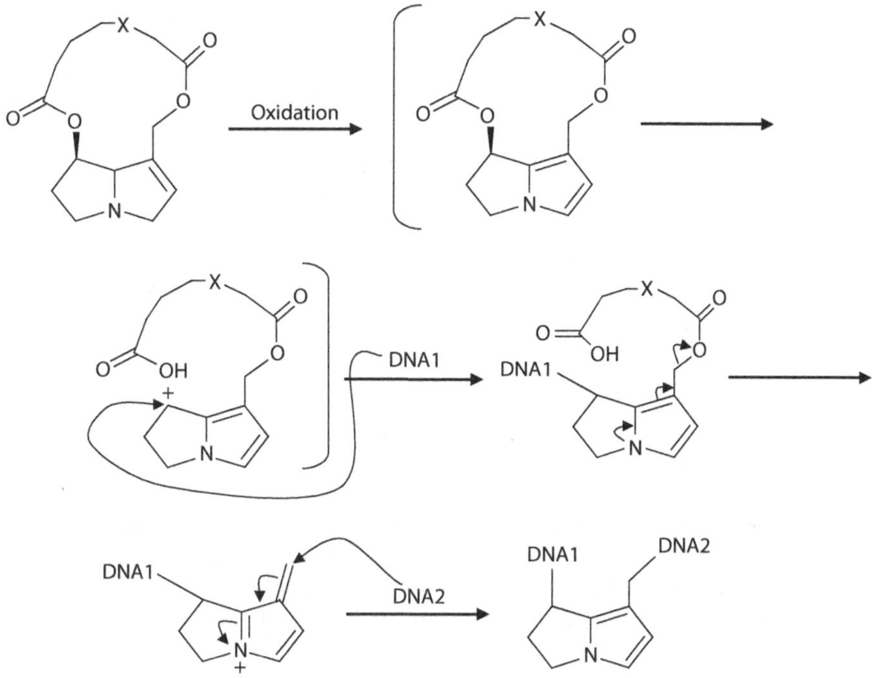

FIGURE 5.29 Speculated DNA cross-linking reaction of PA. Metabolic oxidation activates PA to a carbonium ion intermediate through a pyrrolic ester. Nucleophilic addition of DNA followed by a second addition after pyrrolic activation to a methylenic intermediate completes the cross-linking process.

Thus, DHP–DNA adduct was proposed to be a potential biomarker for PA carcinogenesis. Several DNA bases were implicated to be involved in cross-links by pyrrolic PAs. For example, dehydroretronecine was shown to react with the N2 of deoxyguanosine and N6 of deoxyadenosine. The O2 sites of uridine and deoxythymidine or thymidine have also been identified as targets for linking to DNA. DHMO and dehydroretrorsine have been shown to preferentially cross-link dG-to-dG at a 5′-CG sequence in synthetic duplex DNA. DHMO was shown to cross-link at the N7 position of guanine in a 35 bp fragment of pBR322 with a preference for 5′-GG and 5′-GA sequences. However, Rieben et al.[583] showed that DHMO exhibited no strong base preference when forming crosslinks with DNA. This result is probably expected because the PAs do not carry any specific DNA-affinic moiety.

5.11.11 Bis-Catechol Derivatives

Bis-catechol derivatives are composed of two catechol monomers joined by a spacer arm. Each catechol monomer acts as a DNA cross-linking unit. Song et al.[584] designed and synthesized a new family of bis-catechol quaternary ammonium derivatives as shown in Table 5.5.XI. The positively charged quaternary ammonium was thought to interact with high affinity to DNA and the aliphatic and aromatic chains were used to achieve a wide spread between the monomeric catechols for cross-linking DNA. In order for the catechol moiety to react with DNA, it must be oxidized to a quinone intermediate. Nucleophilic reaction with DNA then forms the cross-link as shown in Figure 5.30. Song et al.[584] used tyrosinase to convert the dihydoxy groups of catechol to the corresponding quinone and found that all the compounds in Table 5.5.XI could cross-link interstrand DNA with compound XI.C, giving the best results. However, when oxidation of the catechol to quinone intermediate was induced by sodium periodate, compound XII.C became ineffective. It was speculated that this might be due

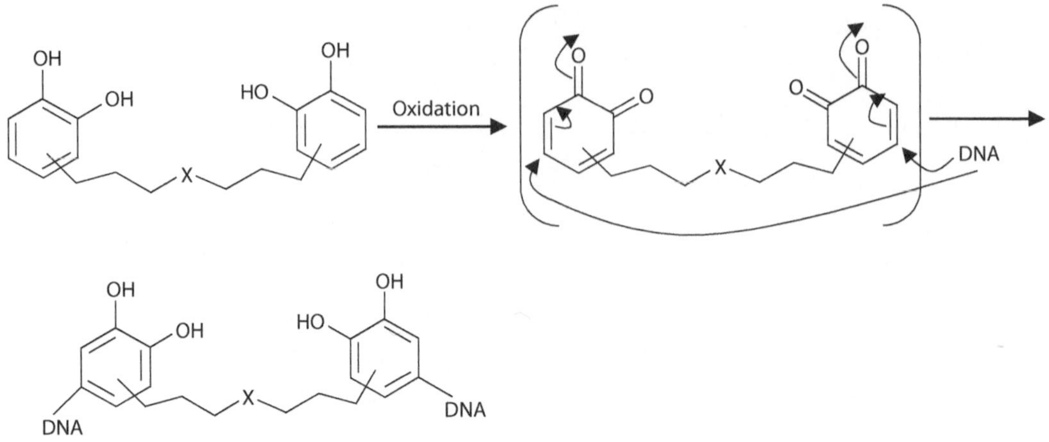

FIGURE 5.30 Proposed mechanism of DNA cross-linking reaction by bis-catechol.

to the instability of the quinone intermediate. McDonald et al.[585] also synthesized many bis-catechols with alkyl spacer arms, some of which were nordihydroguaiaretic acid analogs. The compound with maximum anticancer activity contained a spacer with a four-carbon chain. Those with shorter or longer spacers were less active. Unfortunately, their reactions with DNA were not reported.

5.11.12 QUINONE METHIDES

Related to catechols are the quinone methide precursors. Quinone methides are highly electrophilic and transient intermediates that are implicated in alkylation of DNA by drugs such as mitomycin and tamoxifen, food additives such as butylated hydroxytoluene, and certain natural products.[586] They also react with amines, thiols, water, amino acids, and peptides. There are many methods to generate quinone methides.[586] For DNA interstrand cross-linking, the most common means of activation are photolysis and introduction of fluoride ion, if certain structural requirements are satisfied. The basic structure of the precursors contains a good leaving group, such as a quaternary ammonium, attached through a methylene ortho to a hydroxyl group on a benzene ring. Upon activation by light at wavelengths longer than 350 nm, o-quinone methide is produced. If the hydroxyl group is attached a *tert*-butyldimethylsilyl group, fluoride ion may be used to activate the compound. The reactive quinone methide will react with DNA to produce interstrand cross-linked products as shown in Figure 5.31. Numerous quinone methide precursors have been synthesized. Representative DNA cross-linking agents are listed in Table 5.5.XII. Various modifications on the o-methyltrimethylammonium phenol compounds have been synthesized and studied (Table 5.5 XII.A). All the simple benzo-quinone methide precursors (Table 5.5.XII.A.1 through 6) are activated by light to form the corresponding o-quinone methides, which are found to cross-link DNA. The cross-linking process probably takes two steps with two sequential formation of quinone methides as shown in Figure 5.31.[586–588] Compounds XII.A.7 and 8 in Table 5.5 contain t-butyldimethylsilyl group and conversion to quinone methide can be achieved with fluoride ion.[586,589,590] For biphenyl, bipyridinyl and bis-binol compounds (Table 5.5. XII.B through D), DNA cross-linking can be achieved with one step formation of the corresponding o-quinone methide on both phenyl rings upon irradiation.[586–593] The biphenyl quinone methides are also better DNA cross-linking agents than the monophenol precursors. For example, compound XII.B.1 (Table 5.5) was 100-fold more efficient than compound XII.A.2 at effecting DNA cross-linking.[591] Since the biphenyl groups are flexible and freely rotatable, they can adopt suitable orientations to fit the DNA helical conformation and could be easily localized on DNA, thus resulting in better DNA interstrand cross-linking after photo-illumination.[588] Song et al.[592] further studied the effect of various linkers between the two phenyl groups by investigating two series of quaternary ammonium phenol

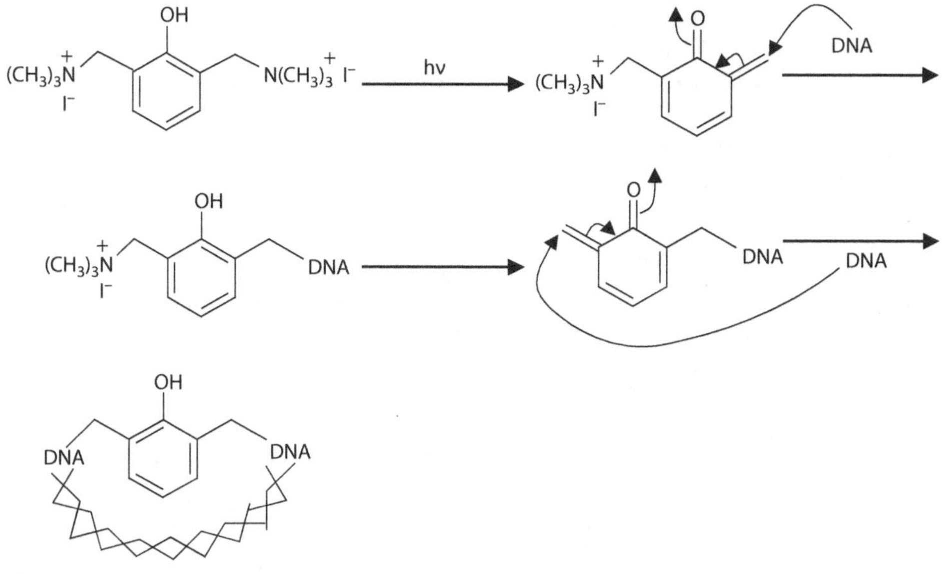

FIGURE 5.31 Interstrand DNA cross-linking reaction of an *o*-quinone methide. The *o*-quinone methide precursor is first activated by light to *o*-quinone methide, which undergoes electrophile reaction with DNA. The reaction is repeated for by the generation of a second *o*-quinone methide.

derivatives (Table 5.5.XII.B). In most cases, the compounds with electron-donating linkers have higher DNA cross-linking abilities than those with electron-withdrawing linkers. The two biphenol compounds that have the highest activities are shown in Table 5.5.XII.B.2 and 3. In addition, the number of charges, which could increase the ability to cross-link DNA, is another main factor to induce DNA cross-linking. Compounds XII.B.4 and 5 in Table 5.5 have four quaternary ammonium groups with four positive charges. They dramatically increased the DNA cross-linking ability compared to the derivatives with two quaternary ammonium salts. Compound XII.B.4 induced 98.2% crosslinking efficiency upon exposure to light for 30 min.

Various structural modifications of biphenyl compounds have been investigated. Richter et al.[587] synthesized a new bino quaternary ammonium derivative containing binaphthyl groups (Table 5.5. XII.D). The binol-quinone methide precursor can be photogenerated at 360 nm. It was shown that the quinone methide produced induced detectable cross-links of both the circular and linear forms of the plasmid. Although the quaternary ammonium positive change of a quinone methide would direct it to bind to DNA, these compounds do not have intrinsic DNA affinity. In order to design agents that can selectively cross-link or alkylate DNA, Zhang et al.[594] studied quinazoline-linked quinone methides (Table 5.5. XII.E), as the quinazoline moiety may be a good intercalator to increase the interactions with DNA. The compound was found to cross-link 28% linear DNA at a concentration of 1 μM and was affected by pH with maximum cross-linking at pH 7.7. Quinone methide precursors have also been linked to other DNA affinic groups. Kumar et al.[589] studied a series of quinone methide precursors linked to a hairpin pyrrole–imidazole polyamide (see Table 5.5. XII.A.7) for selective association with the DNA minor groove. Interstrand cross-linking was found to be more efficient than alkylation but still quite modest and equivalent to that generated by a comparable compound containing the *N*-mustard chlorambucil. The low yield is thought to be due to intramolecular trapping of the quinone methide, which is highly unstable and which has the potential to react with a range of nucleophiles. Varying the length of the linker connecting the polyamide and quinone methide derivative did not greatly affect the yield of DNA cross-linking. When a sily-protected bis(acetoxymethyl)phenol was linked to 9-aminoacridine (see Table 5.5. XII.A.8), two quinone methides were formed upon addition of fluoride.[586,590] DNA was cross-linked with high efficiency and it was found that they formed one cross-link for every four alkylation events.

Comparison with similar analogs of *N*-mustards, the acridine–quinone methide enhanced its cross-linking efficiency by at least 64-fold. Another attempt to direct quinone methides to DNA was made by He et al.[595,596] who synthesized a series of phenol quaternary ammonium porphyrins (see Table 5.5.XII.F). DNA cross-linking was implicated in the process of cytotoxicities of these compounds which might involve both singlet oxygen and *o*-quinone methide intermediate. However, the actual DNA cross-links were not studied.

5.11.13 NITROSOUREA DERIVATIVES

Nitrosourea derivatives (Table 5.5.XIII) were mostly constructed for their antitumor activities, particularly the chloroethylnitrosoureas (CENUs).[597–603] These reagents may contain the chloroethyl functional group as in nitrogen mustards or cyclohexyl group as in 1,3-bis(*trans*-4-hydroxyhexyl)-1-nitrosourea (Table 5.5.XIII.B). They are effective alkylators and cross-linkers of DNA interstrands and DNA and proteins.[559–602,604–611] For example, interstrand DNA cross-links induced by carmustine (BCNU in Table 5.5.XIII.A) were formed rapidly following a 2 h exposure to 50 μM of the compound.[604] More than 14 different types of DNA alkylations were described, leading essentially to the formation of N^1-, N^3-, and N^7-alkyladenines or to N^3-, N^7-, and O^6-alkylguanines.[601] However, O^6-alkylguanine is considered to be the most mutagenic and cytotoxic lesion and leads to DNA interstrand cross-linking. The antineoplastic action of CENUs is correlated to their DNA–DNA cross-linking activity. Although the exact mechanism of cross-linking is not understood, it has been postulated that these compounds decompose to alkylating carbonium ions as well as organic isocyanates under physiological conditions.[599,601,602,606–615] Thus, they may react with any nucleophiles including sulfhydryl, amino, and hydroxyl groups.[613] However, for haloalkyl nitrosourea cross-linking of DNA, for example 1,3-bis(2-chloroethyl)-1-nitrosourea (Table 5.5.XIII.A), it was demonstrated that the path of reaction followed an initial attack of a 2-haloethyl carbonium on the O^6-position of guanine followed by an intramolecular rearrangement and secondary reaction with cytosine resulting DNA interstrand cross-links between the N^1-position of guanine in one strand and the N^3-position of cytosine in the opposite strand of DNA as shown in Figure 5.32. Support for the reaction mechanism comes from the identification of O^6-(2-haloalkyl)guanine,[606] activation of the ATR-Chk1 pathway,[607] prevention of cross-linking by methylguanine-DNA methyltransferase (MGMT),[608,609] inhibition of interstrand cross-linking by ellagic acid (an inhibitor of O^6-guanine alkylation),[602] neutralization of carbamylation reaction by nitric oxide,[613] and the synthesis of a cross-linked duplex containing an ethylene-bridged N^1–2′-deoxyinosine-N^3-thymidine base as well as guanine and cytidine.[614,615] Intrastrand cross-linking may also occur. In this case, the predominant position for DNA alkylation is at N^7-guanine. The N^7-(2-chloroethyl)-guanine adduct may then react with the same position of a neighboring guanine, leading to a N^7-G::N^7-G intrastrand cross-link. The nitroisoureas also cross-link DNA and proteins, such as MGMT.[610,611] In the course of enzymatic transferase reaction, MGMT binds to the 1-O^6-ethanoguanine of CENU-treated DNA forming an MGMT–DNA complex. The 1-O^6-ethanoguanine then reacts with a sulfhydryl group of the protein instead of another base forming a DNA–protein cross-link.[611]

To increase DNA regiospecificity, the CENUs have been tethered to DNA N-terminus minor groove binding dipeptides (lex, information reading peptides) based on *N*-methylpyrrole-carboxamide subunits (e.g., Table 5.5.XIII.F).[612] These CENU-lex's showed significant changes in groove- and sequence-selective DNA alkylation, preferentially induces N^3-(2-chloroethyl)-adenine and N^3-(2-hydroxyethyl)-adenine, at lex binding sites in the minor groove, with a concomitant significant reduction of the alkylation at G residues in the major groove when compared to *N*-(2-chloroethyl)-*N*-cyclohexyl-*N*-nitrosourea.

In addition to the classical nitrosoureas, new analogs have been prepared. Fotemustine (Table 5.5. XIII.E), a third generation nitrosourea, is used both in treatment of melanoma and brain tumors.[603,616] The reaction with DNA seems to be identical to other chloroethylnitrosoureas. However, new fotemustine analogs, the chloroethylnitrososulfamides that contain a sulfonyl group instead of the carbonyl

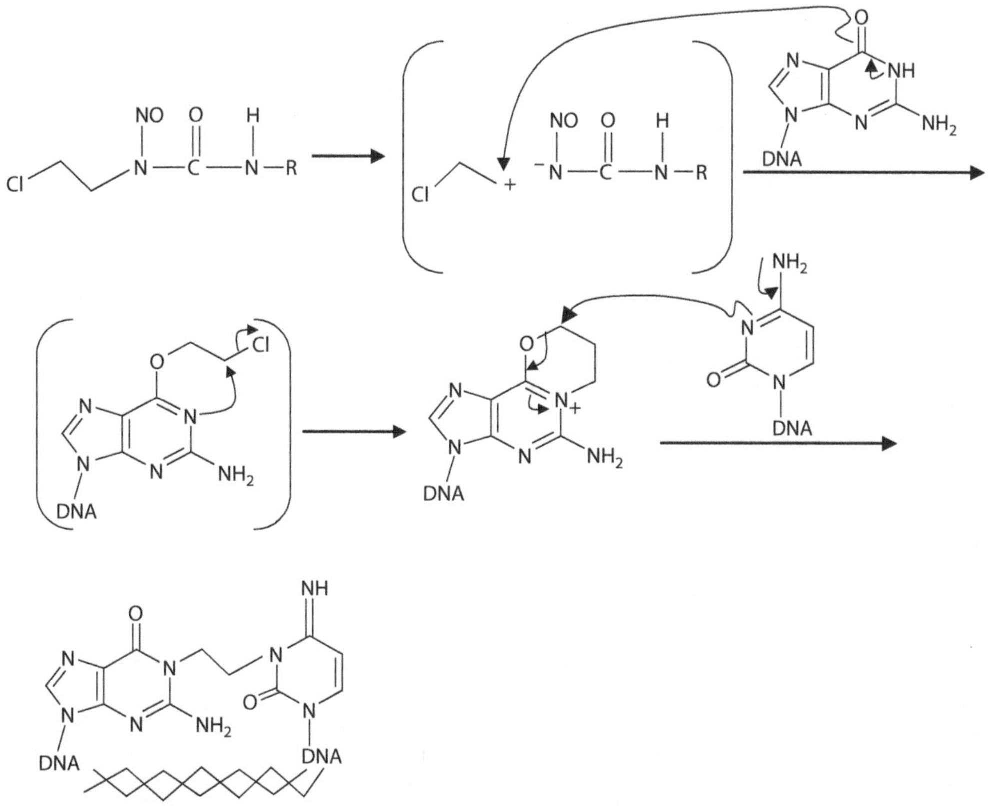

FIGURE 5.32 Speculated DNA cross-linking reaction of a 2-chloroethylnitrosourea. O^6-guanine of DNA is first ethylated which undergoes intramolecular rearrangement to 1-O^6-ethanoguanine. Subsequent reaction with N3 of cytosine in the complementary strand produces an interstrand cross-link.

group, react with DNA differently,[602] although they exhibit antitumor effects on several cancer cell lines, including colon and breast cancers and melanoma. Replacement of the carbonyl group with a sulfonyl group prevents the release of the carbamylating species. Because their DNA cross-linking abilities and efficacies have not been reported, further detailed discussions are not warranted here.

REFERENCES

1. Jin, L. Y., Mass spectrometric analysis of cross-linking sites for the structure of proteins and protein complexes, *Mol. Biosyst.*, 4, 816, 2008.
2. Sinz, A., Chemical cross-linking and mass spectrometry to map three-dimensional protein structures and protein–protein interactions, *Mass Spectrom. Rev.*, 25, 663, 2006.
3. Tyagi, R. and Gupta, M. N., Chemical modification and chemical cross-linking for protein/enzyme stabilization, *Biochemistry (Mosc.)*, 63, 334, 1998.
4. Novak, P. and Kruppa, G. H., Intra-molecular cross-linking of acidic residues for protein structure studies, *Eur. J. Mass Spectrom. (Chichester, Eng.)*, 14, 355, 2008.
5. Shandiz, A. T., Capraro, B. R., and Sosnick, T. R., Intramolecular cross-linking evaluated as a structural probe of the protein folding transition state, *Biochemistry*, 46, 13711, 2007.
6. Patel, R. P. and Price, S., Derivatives of proteins. I. Polymerization of α-chymotrypsin by use of N-ethyl-5-phenylisoxazolium-3′-sulfonate, *Biopolymers*, 5, 583, 1967.
7. Ghetie, V. and Vitetta, E. S., Chemical construction of immunotoxins, *Mol. Biotechnol.*, 18, 251, 2001.
8. Bennett, K. L., Kussmann, M., Björk, P., Godzwon, M., Mikkelsen, M., Sorensen, P., and Roepstorff, P., Chemical cross-linking with thiol-cleavable reagents combined with differential mass spectrometric peptide mapping—A novel approach to assess intermolecular protein contacts, *Protein Sci.*, 9, 1503, 2000.

9. Davies, G. E. and Stark, G. R., Use of dimethyl suberimidate, a cross-linking reagent, in studying the subunit structure of oligomeric proteins, *Proc. Natl. Acad. Sci. U. S. A.*, 66, 651, 1970.

10. Peters, K. and Richards, F. M., Chemical cross-linking: Reagents and problems in studies of membrane structure, *Annu. Rev. Biochem.*, 46, 523, 1977.

11. Ji, T. H., The application of chemical crosslinking for studies on cell membranes and the identification of surface reporters, *Biochim. Biophys. Acta*, 559, 39, 1979.

12. Pilch, P. F. and Czech, M. P., Interaction of cross-linking agents with the insulin effector system of isolated fat cells. Covalent linkage of ^{125}I-insulin to plasma membrane receptor protein of 140,000 daltons, *J. Biol. Chem.*, 254, 3375, 1979.

13. Studdert, C. A. and Parkinson, J. S., In vivo crosslinking methods for analyzing the assembly and architecture of chemoreceptor arrays, *Methods Enzymol.*, 423, 414, 2007.

14. Hunter, M. J. and Ludwig, M. L, Amidination, *Methods Enzymol.*, 25, 585, 1972.

15. Wang, T. W. and Kasseell, B., The preparation of a chemically cross-linked complex of the basic pancreatic trypsin inhibitor with trypsin, *Biochemistry*, 13, 698, 1974.

16. Bartholeyns, J. and Moore, S., Pancreatic ribonuclease: Enzymic and physiological properties of a cross-linked dimer, *Science*, 186, 444, 1974.

17. Dutton, A., Adam, M., and Singer, S. J., Bifunctional imidoesters as cross-linking reagents, *Biochem. Biophys. Res. Commun.*, 23, 730, 1966.

18. Niehaus, W. G., Jr. and Wold, F., Cross-linking of erythrocyte membranes with dimethyl adipimidate, *Biochem. Biophys. Acta*, 196, 170, 1970.

19. Hartman, F. C. and Wold, F., Bifunctional reagents. Cross-linking of pancreatic ribonuclease with a diimido ester, *J. Am. Chem. Soc.*, 88, 3890, 1966.

20. Ji, T. H., Cross-linking of glycolipids in erythrocyte ghost membrane, *J. Biol. Chem.*, 249, 7841, 1974.

21. Hucho, F., Mullner, H., and Sund, H., Investigation of the symmetry of oligomeric enzymes with bifunctional reagents, *Eur. J. Biochem.*, 59, 79, 1975.

22. Tinberg, H. M., Nayudu, P. R. V., and Packer, L., Crosslinking of membranes: The effect of dimethylsuberimidate, a bifunctional alkylating agent, on mitochondrial electron transport and ATPase, *Arch. Biochem. Biophys.*, 172, 734, 1976.

23. Siezen, R. J., Bindels, J. G., and Hoenden, H. J., The quaternary structure of bovine α-crystallin: Chemical crosslinking with bifunctional imidoesters, *Eur. J. Biochem.*, 107, 243, 1980.

24. Browne, D. T. and Kent, B. H., Formation of nonamidine products in the reaction of primary amines with imido esters, *Biochem. Biophys. Res. Commun.*, 67, 126, 1975.

25. Schramm, H. J. and Dülffer, T., Synthesis and application of cleavable and hydrophilic cross-linking reagents, in *Protein Cross-Linking: Biochemical and Molecular Aspects*, Friedman, M. (ed.), Plenum Press, New York, 1976, p. 197.

26. Hunter, M. J. and Ludwig, M. L., The reaction of imidoesters with proteins and related small molecules, *J. Am. Chem. Soc.*, 84, 3491, 1962.

27. Wold, F., Bifunctional reagents, *Methods Enzymol.*, 25, 623, 1972.

28. Means, G. E. and Feeney, R. E., *Chemical Modification of Proteins*, Holden-Day, San Francisco, CA, 1971.

29. Staros, J. V., Morgan, D. G., and Appling, D. R., A membrane-impermeant, cleavable cross-linker. Dimers of human erythrocyte band 3 subunits cross-linked at the extracytoplasmic membrane face, *J. Biol. Chem.*, 256, 5890, 1981.

30. Wang, K. and Richards, F. M., Behavior of cleavable cross-linking reagents based on the disulfide group, *Isr. J. Chem.*, 12, 375, 1974.

31. Ruoho, A., Bartett, P. A., Duttoa, A., and Singer, S. J., Disulfide-bridge bifunctional imidoester as a reversible cross-linking reagent, *Biochem. Biophys. Res. Commun.*, 63, 417, 1975.

32. Aizawa, S., Kurimoto, F., and Yokono, O., Crosslinking studies with different length dithiobisalkylimidates. (I). Solubilized erythrocyte spectrin, *Biochem. Biophys. Res. Commun.*, 75, 870, 1977.

33. Coggins, J. R., Hooper, E. A., and Perham, R. N., Use of dimethyl suberimidate and novel periodate-cleavable bis(imido esters) to study the quaternary structure of the pyruvate dehydrogenase multienzyme complex of *E. coli*, *Biochemistry*, 15, 2527, 1976.

34. Fasold, H., Baumert, H., and Fink, G., Comparison of hydrophobic and strongly hydrophilic cleavable cross-linking reagents in intermolecular bond formation in aggregates of proteins or protein–RNA, in *Protein Cross-Linking: Biochemical and Molecular Aspects*, Friedman, M. (ed.), Plenum Press, New York, 1976, p. 207.

35. Lutter, L. C., Ortanderl, F., and Fasold, H., Use of a new series of cleavable protein-crosslinkers on the *Escherichia coli* ribosome, *FEBS Lett.,* 48, 288, 1974.

36. Schramm, H. J., Synthese von farbigen Nitrilen, Dinitril und bifunktimellen Imidsoureestern, *Hoppe-Seyler's Z. Physiol. Chem.*, 348, 289, 1967.

37. Schramm, H. J., The synthesis of mono- and bifunctional nitriles and imidoesters carrying a fluorescent group, *Hoppe-Seyler's Z. Physiol. Chem.*, 356, 1375, 1975.

38. Dihazi, G. H. and Sinz, A., Mapping low-resolution three-dimensional protein structures using chemical cross-linking and Fourier transform ion-cyclotron resonance mass spectrometry, *Rapid Commun. Mass Spectrom.*, 17, 2005, 2003.

39. Bauer, P. I., Buki, K. G., Hakam, A., and Kun, E., Macromolecular association of ADP-ribosyltransferase and its correlation with enzymic activity, *Biochem. J.*, 270, 17, 1990.

40. Gibson, J. S., Stewart, G. W., and Ellory, J. C., Effect of dimethyl adipimidate on K+ transport and shape change in red blood cells from sickle cell patients, *FEBS Lett.*, 480, 179, 2000.

41. García-Pérez, A. I., Pérez, M. T., Lucas, L., Pinilla, M., Luque, J., Sancho, P., Oxygenation capacity of hypotonized and crosslinked rat erythrocytes, *Life Sci.*, 61, 445, 1997.

42. Smith, A. P. and Loh, H. H., Effect of bisimidate cross-linking reagents on synaptosomal plasma membrane, *Biochemistry*, 17, 1761, 1978.

43. Hordern, J. S., Leonard, J. D., and Scraba, D. G., Structure of the Mengo virion, *Virology*, 97, 131, 1979.

44. Gebhardt, A., Bosch, J. V., Ziemiecki, A., and Griis, R. R., Rous sarcoma virus p19 and gp35 can be chemically crosslinked to high molecular weight complexes: An insight into virus assembly, *J. Mol. Biol.*, 174, 297, 1984.

45. Drews, G. and Rack, M., Modification of sodium and gating currents by amino group specific crosslinking and monofunctional reagents, *Biophys. J.*, 54, 383, 1988.

46. Horiguchi, T., Miwa, Y., and Shigesada, K., The quaternary geometry of transcription termination factor rho: Assignment by chemical cross-linking, *J. Mol. Biol.*, 269, 514, 1997.

47. Uchiumi, T., Terao, K., and Ogata, K., Identification of neighboring protein pairs cross-linked with DTBP in rat liver 40 S ribosomal subunits, *J. Biochem.*, 90, 185, 1981.

48. Hajdu, J., Dombradi, V., Bot, G., and Friedrich, P., Structural changes in glycogen phosphorylase as revealed by cross-linking with bifunctional di-imidates: Phosphorylase-b, *Biochemistry*, 18, 4137, 1979.

49. Anderson, W. M. and Fisher, R. R., The subunit structure of bovine heart mitochondrial transhydrogenase, *Biochim. Biophys. Acta*, 635, 194, 1981.

50. Lambooy, P. K. and Steiner, R. F., The cross-linking of phosphorylase-kinase, *Arch. Biochem. Biophys.*, 213, 551, 1982.

51. Brew, K., Shaper, J. H., Olsen, K. W., Trayer, I. P., and Hill, R. L., Cross-linking of components of lactose synthetase with dimethyl-pimelimidate, *J. Biol. Chem.*, 250, 1434, 1975.

52. Lad, P. M. and Hammes, G. G., Physical and chemical properties of rabbit muscle phosphofructokinase crosslinked with dimethyl suberimidate, *Biochemistry*, 13, 4530, 1974.

53. Debyser, Z. and De Clercq, E., Chemical crosslinking of the subunits of HIV-1 reverse transcriptase, *Protein Sci.*, 5, 278, 1996.

54. Hartman, F. C. and Wold, F., Cross-linking of bovine pancreatic ribonuclease A with dimethyl adipimidate, *Biochemistry*, 6, 2439, 1967.

55. De Pont, J. J. H. H. M., Reversible inactivation of (Na,K)ATPase by use of a cleavable bifunctional reagent, *Biochim. Biophys. Acta*, 567, 247, 1979.

56. Baskin, L. S. and Yang, C. S., Cross-linking studies of monooxygenase enzymes, in *Microsomes, Drug Oxidations, and Chemical Carcinogenesis*, Vol. I, Coon, M. J., Cooney, A. H., Estabrook, R. W., Gelboin, H. V., Gillette, J. R., and O'Brien, P. J. (eds.), Academic Press, New York, 1980, p. 102.

57. Baskin, L. S. and Yang, C. S., Identification of cross-linked cytochrome P-450 in rat liver microsomes by enzyme immunoassay, *Biochem. Biophys. Res. Commun.*, 108, 700, 1982.

58. Suda, M. and Iwai, K., Identification of suberimidate cross-linking sites of four histone sequences in the H1-depleted chromatin. Histone arrangement in nucleosome core, *J. Biochem.*, 86, 1659, 1979.

59. Galella, G. and Smith, D. B., The cross-linking of tubulin with imido-esters, *Can. J. Biochem.*, 60, 71, 1982.

60. Kagey-Sobotka, A., Dembo, M., Goldstein, B., Metzger, H., and Ligbtenstein, L. M., Qualitative characteristics of histamine release from human basophils by covalently cross-linked IgE, *J. Immunol.*, 127, 2285, 1981.

61. Ohara, O., Takanashi, S., and Ooi, T., Cross-linking study on tropomyosin, *J. Biochem.*, 87, 1795, 1980.

62. Ikai, A. and Yanagita, Y., A cross-linking study of apo-low density lipoprotein, *J. Biochem.*, 88, 1359, 1980.

63. Ji, T. H., A novel approach to the identification of surface receptors. The use of photosensitive heterobifunctional cross-linking reagent, *J. Biol. Chem.*, 252, 1566, 1977.

64. Pennathur-Das, R., Heath, R. H., Mentzer, W. C., and Lubin, B. H., Modification of hemoglobin S with dimethyl adipimidate: Contribution of individual reacted subunits to changes in properties, *Biochim. Biophys. Acta*, 704, 389, 1982.

65. Hu, T. and Su, Z., Preparation of well-defined bovine polyhemoglobin based on dimethyl adipimidate and glutaraldehyde cross-linkage, *Biochem. Biophys. Res. Commun.*, 293, 958, 2002.

66. Charulatha, V. and Rajaram, A., Crosslinking density and resorption of dimethyl suberimidate-treated collagen, *J. Biomed. Mater Res.*, 36, 478, 1997.

67. Charulatha, V. and Rajaram, A., Dimethyl 3,3′-dithiobispropionimidate: A novel crosslinking reagent for collagen, *J. Biomed. Mater Res.*, 54, 122, 2001.

68. Watty, A., Methfessel, C., and Hucho, F., Fixation of allosteric states of the nicotinic acetylcholine receptor by chemical cross-linking, *Proc. Natl. Acad. Sci. U. S. A.*, 94, 8202, 1997.

69. Dietrich, J. B., Chemical crosslinking of different forms of the simion virus 40 large T antigen using bifunctional reagents, *FEBS Lett.*, 201, 311, 1986.

70. Sekhar, M. C., Sharma, G. L., Gangal, S., Joshi, A. P., and Sarma, P. U., Dimethyl suberimidate as an effective crosslinker for antibody-enzyme conjugation, *Prep. Biochem.*, 21, 215, 1991.

71. Bons, J. A., Michielsen, E. C., de Boer, D., Bouwman, F. G., Jaeken, J., van Dieijen-Visser, M. P., Rubio-Gozalbo, M. E., and Wodzig, W. K., A specific immunoprecipitation method for isolating isoforms of insulin-like growth factor binding protein-3 from serum, *Clin. Chim. Acta.*, 387, 59, 2008.

72. Uckert, W., Wunderlich, V., Ghysdael, J., Portetelle, D., and Burny, A., Bovine leukemia virus (BLV): A structural model based on chemical cross-linking studies, *Virology*, 133, 386, 1984.

73. Yi, X., Batrakova, E., Banks, W. A., Vinogradov, S., and Kabanov, A. V., Protein conjugation with amphiphilic block copolymers for enhanced cellular delivery, *Bioconjug. Chem.*, 19, 1071, 2008.

74. Uckert, W. and Rudolph, M., Chemical crosslinking of lipids and proteins within mason-pfigen monkey and squirrel monkey type D retrovirus, *Arch. Geschwulstforsch.*, 56, 107, 1986.

75. Dodson, M. S., Dimethyl suberimidate cross-linking of oligo(dT) to DNA-binding proteins, *Bioconjug. Chem.*, 11, 876, 2000.

76. Staros, J. V., N-Hydroxysulfosuccinimide active esters: Bis(N-hydroxysulfosuccinimide) esters two carboxylic acids are hydrophilic, membrane-impermeable, protein cross-linkers, *Biochemistry*, 21, 3950, 1982.

77. Lindsay, D. G., Intramolecular cross-linked insulin, *FEBS Lett.*, 21, 105, 1972.

78. Fujimoto, K., Kajino, M., and Inouye, M., Development of a series of cross-linking agents that effectively stabilize α-helical structures in various short peptides, *Chem. Eur. J.*, 14, 857, 2008.

79. Ihling, C., Schmidt, A., Kalkhof, S., Schulz, D. M., Stingl, C., Mechtler, K., Haack, M., Beck-Sickinger, A.G., Cooper, D. M. F., and Sinz, A., Isotope-labeled cross-linkers and Fourier transform ion cyclotron resonance mass spectrometry for structural analysis of a protein/peptide complex, *J. Am. Soc. Mass Spectrom.*, 17, 1100, 2006.

80. Müller, D. R., Schindler, P., Towbin, H., Wirth, U., Voshol, H., Hoving, S., and Steinmetz, M. O., Isotope-tagged crosslinking reagents. A new tool in mass spectrometric protein interaction analysis, *Anal. Chem.*, 73, 1927, 2001.

81. Petrotchenko, E. V., Xiao, K., Cable, J., Chen, Y., Dokholyan, N. V., and Borchers, C. H., BiPS, a photocleavable, isotopically coded, fluorescent cross-linker for structural proteomics, *Mol. Cell. Proteomics*, 8, 273–286, 2009.

82. Gaffney, B. J., Willingham, G. L., and Schopp, R. S., Synthesis and membrane interactions of spin-label bifunctional reagent, *Biochemistry*, 22, 881, 1983.

83. Willingham, G. L. and Gaffney, B. J., Reactions of spin-label cross-linking reagents with red blood cell proteins, *Biochemistry*, 22, 892, 1983.

84. Anjaneyulu, P. S. R., Beth, A. H., Cobb, C. E., Juliao, S. F., Sweetman, B. J., and Staros, J. V., Bis(sulfo-N-succinimidyl)doxyl-2-spiro-5′-azelate: Synthesis, characterization and reaction with anion-exchange channel in intact human erythrocyte, *Biochemistry*, 28, 6583, 1989.

85. Beth, A. H., Conturo, T. E., Venakataramu, S. D., and Staros, J. V., Dynamic and interactions of the anion channel in intact human erythrocytes: An electron paramagnetic resonance spectroscopic study employing a new membrane-impermeable bifunctional spin label, *Biochemistry*, 25, 3824, 1986.

86. Anjaneyulu, P. S. R., Beth, A. H., Sweetman, B. J., Faulkner, L. A., and Staros, J. V., Bis(sulfo-N-succinimidyl) [^{15}N,^{2}H16]doxyl-2-spiro-4′-pimelate, a stable isotope-substituted, membrane-impermeant bifunctional spin label for studies of the dynamics of membrane proteins: Application to the anion-exchange channel in intact human erythrocytes, *Biochemistry*, 27, 6844, 1988.

87. Gardner, M. W., Vasicek, L. A., Shabbir, S., Anslyn, E. V., and Brodbelt, J. S., Chromogenic cross-linker for the characterization of protein structure by infrared multiphoton dissociation mass spectrometry, *Anal. Chem.*, 80, 4807, 2008.

88. Safiejko-Mroczka, B. and Bell, P. B., Jr., Bifunctional protein cross-linking reagents improve labeling of cytoskeletal proteins for qualitative and quantitative fluorescence microscopy, *J. Histochem. Cytochem.*, 34, 641, 1996.

89. Mattson, G., Conklin, E., Desai, S., Nielander, G., Savage, M. D., and Morgensen, S., A practical approach to crosslinking, *Mol. Biol. Rep.*, 11, 167, 1993.

90. Baskin, L. S. and Yang, C. S., Cross-linking studies of cytochrome P-450 and reduced NADPH-cytochrome-P-450 reductase, *Biochemistry*, 19, 2260, 1980.

91. Abdella, P. M., Smith, P. K., and Royer, G. P., A new cleavable reagent for cross-linking and reversible immobilization of proteins, *Biochem. Biophys. Res. Commun.*, 87, 734, 1979.

92. Zarling, D. A., Watson, A., and Bach, F. H., Mapping of lymphocyte surface polypeptide antigens by chemical cross-linking with BSOCOES, *J. Immunol.*, 124, 913, 1980.

93. Lomant, A. J. and Fairbanks, G., Chemical probes of extended biological structures: Synthesis and properties of the cleavable protein cross-linking reagent [^{35}S]dithiobis(succinimidylpropionate), *J. Mol. Biol.*, 104, 243, 1976.

94. Bragg, P. D. and Hou, C., Chemical cross-linking of α-subunit in the F_1 adenosine triphosphatase of *Escherichia coli*, *Arch. Biochem. Biophys.*, 244, 361, 1986.

95. Han, K.-K., Richard, C., and Delacourte, A., Chemical cross-links of proteins by using bifunctional reagents, *Int. J. Biochem.*, 16, 129, 1984.

96. Smith, R. J., Capaldi, R. A., Muchmore, D., and Dahlquist, F., Cross-linking of ubiquinone-cytochrome-C-reductase with periodate-cleavable bifunctional reagent, *Biochemistry*, 17, 3719, 1978.

97. Buchardt, O., Elsner, H. I., Nielsen, P. E., Perersen, L. C., and Suenson, E., Protein crosslinking reagents containing a selenoethylene linker are cleaved by mild oxidation, *Anal. Biochem.*, 158, 87, 1986.

98. Birkelund, S., Lundemose, A. G., and Christiansen, G., Chemical cross-linking of *Chlamydia trachomatis*, *Infect. Immun.*, 56, 654, 1988.

99. Anderson, G. W., Zimmerman, J. E., and Callahan, F. M., Esters of N-hydroxysuccinimide in peptide synthesis, *J. Am. Chem. Soc.*, 86, 1839, 1964.

100. Cuatrecasas, P. and Parikh, I., Adsorbents for affinity chromatography. Use of N-hydroxysuccinimide esters of agarose, *Biochemistry*, 11, 2291, 1972.

101. Leavell, M. D., Novak, P., Behrens, C. R., Schoeniger, J. S., and Kruppa, G. H., Strategy for selective chemical cross-linking of tyrosine and lysine residues, *J. Am. Soc. Mass. Spectrom.*, 15, 1604, 2004.

102. Kalkhof, S. and Sinz, A., Chances and pitfalls of chemical cross-linking with amine-reactive *N*-hydroxysuccinimide esters, *Anal. Bioanal. Chem.*, 392, 305, 2008.

103. Carlsson, J., Drevin, H., and Axen, R., Protein thiolation and reversible protein–protein conjugation. N-succinimidyl 3-(2-pyridyldithio)propionate, a new heterobifunctional reagent, *Biochem. J.*, 173, 723, 1978.

104. Vanin, E. F. and Ji, T. H., Synthesis and application of cleavable photoactivable heterobifunctional reagents, *Biochemistry*, 20, 6754, 1981.

105. Hill, M., Bechet, J. J., and D'Albis, A., Disuccinimidyl esters as bifunctional cross-linking reagents for proteins, *FEBS Letts.*, 102, 282, 1979.

106. Travers, R. C., Noyes, C. M., Roberts, H. R., and Lundblad, R. L., Influence of metal ions on prothrombin self association, *J. Biol. Chem.*, 257, 10708, 1982.

107. Takamatsu, T., Horne, M. K., III, and Gralnick, H. R., Identification of the thrombin receptor on human platelets by chemical crosslinking, *J. Clin. Invest.*, 77, 362, 1986.

108. Davies, G. E. and Palek, J., Platelet protein organization analysis by treatment with membrane permeable cross-linking reagents, *Blood*, 59, 502, 1982.

109. Wiemken, V., Theiler, R., and Bachofen, R., Lateral organization of proteins in chromatophore membrane of *R. rubrum*, studies by chemical cross-linking, *J. Bioenerg. Biomembr.*, 13, 181, 1981.

110. Laburthe, M., Breant, B., and Rouyen-Fessard, C., Molecular identification of receptors for vasoactive intestinal peptide in rat intestinal epithelium by covalent cross-linking: Evidence for two classes of binding sites with different structural and functional properties, *Eur. J. Biochem.*, 139, 181, 1984.

111. Wood, C. L. and O'Dorisio, M. S., Covalent crosslinking of vasoactive intestinal polypeptide to its receptors on intact human lymphoblasts, *J. Biol. Chem.*, 260, 1243, 1985.

112. Boal, A. K., Tellez, H., Rivera, S. B., Miller, N. E., Bachand, G. D., and Bunker, B. C., The stability and functionality of chemically crosslinked microtubules, *Small*, 2, 793, 2006.

113. Meller, V. H. and Fisher, P. A., Nuclear distribution of Drosophila DNA topoisomerase II is sensitive to both RNase and DNase, *J. Cell Sci.*, 108, 1651, 1995.

114. Stauber, G. and Aggarwal, B. B., Characterization and affinity cross-linking of receptors for human recombinant lymphotoxin (tumor necrosis factor-β) on human histiocytic lymphoma cell line, U-937, *J. Biol. Chem.*, 264, 3573, 1989.

115. Amiranoff, B., Lorinet, A. M., and Laburthe, M., Galanin receptor in the rat pancreatic beta-cell line Rin M-5F: Molecular characterization by chemical cross-linking, *J. Biol. Chem.*, 264, 20714, 1989.

116. Conn, P. M., Rogers, D. C., Stewart, J. M., Niedel, J., and Sheffield, T., Conversion of gonadotropin-releasing hormone antagonist to an agonist, *Nat. (London)*, 296, 653, 1982.

117. McInerney, T. L., Ahmar, W. E., Kemp, B. E., and Poumbourios, P., Mutation-directed chemical cross-linking of human immunodeficiency virus type 1 gp41 oligomers, *J. Virol.*, 72, 1523, 1998.

118. Pinter, A. and Fleissner, E., Structural studies of retroviruses: Characterization of oligomeric complexes of murine and feline leukemia virus envelope and core component formed upon cross-linking, *J. Virol.*, 30, 157, 1979.

119. Onisko, B., Fernández, E. G., Freire, M. L., Schwarz, A., Baier, M., Camiña, F., García, J. R., Villamarín, S. R.-S., and Requena, J. R., Probing PrP^{Sc} structure using chemical cross-linking and mass spectrometry: Evidence of the proximity of Gly90 amino termini in the PrP 27-30 aggregate, *Biochemistry*, 44, 10100, 2005.

120. Andrews, R. K., Gorman, J. J., Booth, W. J., Corino, G. L., Castaldi, P. A., and Berndt, M. C., Cross-linking of a monomeric 39/34-kDa diapase fragment of von Willebrand factor (Ieu-480/val-481-gly-718) to the N-terminal region of the α-chain of membrane glycoprotein Ib on intact platelet with bis(sulfosuccinimidyl)suberate, *Biochemistry*, 28, 8326, 1989.

121. Zhang, Q.-Y. and Menon, K. M. J., Characterization of rat Leydig cell gonadotropin receptor structure by affinity cross-linking, *J. Biol. Chem.*, 263, 1002, 1988.

122. Fanger, B. O., Stephens, J. E., and Staros, J. V., Trapping of epidermal growth factor-induced receptor dimers by chemical crass-linking, *FASEB J.*, 2, A1774, 1988.

123. Fanger, B. O., Stephens, J. E., and Staros, J. V., High-yield trapping of EGF-induced receptor dimers by chemical cross-linking, *FASEB J.*, 3, 71, 1989.

124. Wilkinson, B. M., Esnault, Y., Craven, R. A., Skiba, F., Fieschi, J., Képès, F., and Stirling, C. J., Molecular architecture of the ER translocase probed by chemical crosslinking of Sss1p to complementary fragments of Sec61p, *EMBO J.*, 16, 4549, 1997.

125. Herzig, M. C. and Weigel, P. H., Synthesis and characterization of N-hydroxysuccinimide ester chemical affinity derivatives of asialoorosomucoid that covalently cross-link to galactosyl receptors on isolated rat hepatocytes, *Biochemistry*, 28, 600, 1989.

126. Huang, B. X. and Kim, H.-Y., Interdomain conformational changes in Akt activation revealed by chemical cross-linking and tandem mass spectrometry, *Mol. Cell. Proteomics*, 5, 1045, 2006.

127. Zahn, H., Bridge reactions in amino acids and fiber proteins, *Angew. Chem.*, 67, 56, 1955.

128. Grow, T. E. and Fried, M., Lipoprotein geometry. I. Spatial relations of human HDL (high density lipoproteins) apoproteins studied with a bifunctional reagent, *Biochem. Biophys. Res. Commun.*, 66, 352, 1975.

129. Fraenkel-Conrat, H., The chemistry of proteins and peptides, *Annu. Rev. Biochem.*, 25, 318, 1956.

130. Walder, J. A., Zhang, R. H., Walder, R. Y., Steele, J. M., and Klotz, I. M., Diaspirins that cross-link β chains of hemoglobin: Bis(3,5-dibromosalicyl)succinate and bis(3,5-dibromosalicyl)fumarate, *Biochemistry*, 18, 4265, 1979.

131. Chatterjee, R., Welty, E. V., Walder, R. Y., Pruitt, S. L, Rogers, P. H., Arone, A., and Walder, J. A., Isolation and characterization of new hemoglobin derivative cross-linked between α chains (lysine 99α$_1$ → lysine 99α$_2$), *J. Biol. Chem.*, 261. 9929, 1986.

132. Wold, F., Reaction of bovine serum albumin with the bifunctional reagent p,p′-difluoro-m,m′-dinitrodiphenylsulfone, *J. Biol. Chem.*, 236, 106, 1961.

133. Zahn, H., Zuber, H., Ditscher, W., Wegerle, D., and Meienhofer, J., Reactions of aromatic fluoro compounds with amino acids and proteins. XV. Reaction of p,p′-difluoro-m,m′-dinitrodiphenylsulfone with silk fibroin, *Chem. Ber.,* 89, 407, 1956.

134. Mills, G. L., Identification of dinitrophenylamino acids, *Nat. (London)*, 165, 403, 1950.

135. Zahn, H. and Wegerle, D., Collagen. III. Reactions with bis(3-nitro-4-fluorophenyl)sulfone, *Kolloid-Z.*, 172, 29, 1960.

136. Tawde, S., Ram, J. S., and Iyengar, M. R., Physicochemical and immunochemical studies on the reaction of bovine serum albumin with p,p′-difluoro-m,m′-dinitrophenylsulfone, *Arch. Biochem. Biophys.*, 100, 270, 1963.

137. Vandegriff, K. D., Medina, F., Marini, M. A., and Winslow, R. M., Equilibrium oxygen binding to human hemoglobin cross-linked between the α chains by bis(3,5-dibromosalicyl)fumarate, *J. Biol. Chem.*, 264, 17824, 1989.

138. Huang, H. and Olsen, K. W., Thermal stabilities of hemoglobins crosslinked with different length reagents, *Artif. Cells Blood Substit. Immobil. Biotechnol.*, 22, 719, 1994.

139. Zhang, Q. and Olsen, K. W., The modification of hemoglobin by a long crosslinking reagent: Bis(3,5-dibromosalicyl)sebacate, *Biochem. Biophys. Res. Commun.*, 203, 1463, 1994.

140. Bobofchak, K. M., Tarasov, E., and Olsen, K. W., Effect of cross-linker length on the stability of hemoglobin, *Biochim. Biophys. Acta*, 1784, 1410, 2008.

141. Deyev, S. M., Afanasenko, G. A., and Polyanovsky, O. L., Two steps modification of aspartate aminotransferase with DFNB-cross-linking localization, *Biochim. Biophys. Acta*, 534, 358, 1978.

142. Harris, W. E. and Stahl, W. L., Orginizational of thiol groups of electric-eel electric organ Na, K ion-stimulated ATPase studied with bifunction reagents, *Biochem. J.*, 185, 787, 1980.

143. Munroe, W. A. and Dunlap, R. B., Chemical modification of *Lactobocillus casei* thymidylate synthetase with FDBN and FDNB, *Arch. Biochem. Biophys.*, 214, 742, 1982.

144. Zahn, H. and Meienhofer, J., Reactions of 1,5-difluoro-2,4-dinitrobenzene with insulin. I. Synthesis of model compounds, *Makromol. Chem.*, 26, 153, 1958.

145. Marfey, P. S., Nowak, H., Uziel, M., and Yphantis, D. A., Reaction of bovine pancreatic ribonuclease A with 1,5-difluoro-2,4-dinitrobenzene. I. Preparation of monomeric intramolecularly bridged derivatives, *J. Biol. Chem.*, 240, 3264, 1965.

146. Marfey, P. S. and King, M. V., Chemical modification of ribonuclease A crystals. I. Reaction with 1,5-difluoro-2,4-dinitrobenzene, *Biochim. Biophys. Acta*, 105, 178, 1965.

147. Cuatrecasas, P., Fuchs, S., and Anfinsen, C. B., Cross-linking of aminotyrosyl residues in the active site of Staphylococcal nuclease, *J. Biol. Chem.*, 244, 406, 1969.

148. Berg, H. C., Diamond, J. M., and Marfey, P. S., Erythrocyte membrane: Chemical modification, *Science*, 150, 64, 1965.

149. Ram, J. S., Tawde, S. S., Pierce, G. B., Jr., and Midgley, A. R., Jr., Preparation of antibody-ferritin conjugates for immuno electron microscopy, *J. Cell Biol.*, 17, 673, 1963.

150. Modesto, R. R. and Pesce, A. J., The reaction of 4,4′-difluoro-3,3′-dinitrodiphenyl sulfone with γ-globulin and horse-radish peroxidase, *Biochem. Biophys. Acta*, 229, 384, 1971.

151. Ling, N. R., Coupling of protein antigens to erythrocytes with difluoro-dinitrobenzene, *Immunology*, 4, 49, 1961.

152. Ozawa, H., Bridging reagent for protein. I. The reaction of diisocyantes with lysine and enzyme proteins, *J. Biochem. (Tokyo)*, 62, 419, 1967.

153. Snyder, P. D., Jr., Wold, F., Bernlohr, R. W., Dullum, C., Desnick, R. J., Krivit, W., and Condie, R. M., Enzyme therapy. II. Purified human α-galactosidase A. Stabilization to heat and protease degradation by complexing with antibody and by chemical modification, *Biochem. Biophys. Acta*, 350, 432, 1974.

154. Schick, A. F. and Singer, S. J., On the formation of covalent linkages between two protein molecules, *J. Biol. Chem.*, 236, 2477, 1961.

155. Fasold, H., Synthese und reaktionen eines wasserlöslichen, spaltbaren, starren reagens zur verknüpfung von freien aminopruppen in proteinen, *Biochem. Z.*, 339, 482, 1964.

156. Haimovich, J., Hurwitz, E., Novik, N., and Sela, M., Preparation of protein–bacteriophage conjugates and their use in detection of anti-protein antibodies, *Biochem. Biophys. Acta*, 207, 115, 1970.

157. Borek, F. and Silverstein, A. M., Characterization and purification of ferritin-antibody globulin conjugates, *J. Immunol.*, 87, 555, 1961.

158. Fasold, H. and Lusty, C. J., The application of azo dyes to identify reactive groups and determine distances in proteins, *7th International Congress of Biochemistry, Abstracts*, III, 1, Tokyo, Japan,1967.

159. Manecke, G. and Gunzel, G., Darstellung eines wasserunlöslichen, aktiven papains, *Naturwissenschaften*, 54, 647, 1967.

160. Cabantchik, I. Z., Balshin, M., Breuer, W., and Rothstein, A., Pyridoxal phosphate. An anionic probe for protein amine groups exposed on the outer and inner surfaces of intact human red blood cells, *J. Biol. Chem.*, 250, 5130, 1975.

161. Macara, I. G. and Cantley, L. C., Mechanism of anion exchange across the red cell membrane by band 3: Interactions between stilbene sulfonate and NAP-taurine binding sites, *Biochemistry*, 20, 5695, 1981.

162. Lepke, S. and Passow, H., Inverse effect of dansylation of red blood cell membrane on band 3 protein-mediated transport of sulphate and chloride, *J. Physiol.*, 328, 27, 1982.

163. Wells, J. A. and Yount, R. G., Chemical modification of myosin by active-site trapping of metal-nucleotides with thiol crosslinking reagents, *Methods Enzymol.*, 85, 93, 1982.

164. Rao, A., Martin, P., Reithmeier, R. A. F., and Cantley, L. C., Location of the stilbenesulfonate binding sites of the human erythrocyte anion-exchange system by resonance energy transfer, *Biochemistry*, 18, 4505, 1979.

165. Dissing, S., Jesaitis, A. J., and Fortes, P. A. G., Fluorescence labeling of the human erythrocyte anion transport system. Subunit structure studied with energy transfer, *Biochim. Biophys. Acta*, 553, 66, 1979.
166. Fasold, H., Chemical investigation of the tertiary structure of proteins. I. Cross-linking of myoglobin, checking its molecular weight and native state, *Biochem. Z.*, 342, 288, 1965.
167. Brown, W. E. and Wold, F., Alkyl isocyanates as active site-specific inhibitors of chymotrypsin and elastase, *Science*, 174, 608, 1971.
168. Zeugolis, D. I., Paul, G. R., and Attenburrow, G., Cross-linking of extruded collagen fibers—A biomimetic three-dimensional scaffold for tissue engineering applications, *J. Biomed. Mater. Res. A*, 89, 895, 2009.
169. Bertoldo, M., Bronco, S., Gragnoli, T., and Ciardelli, F., Modification of gelatin by reaction with 1,6-diisocyanatohexane, *Macromol. Biosci.*, 7, 328, 2007.
170. Herzig, D. J., Rees, A. W., and Day, R. A., Bifunctional reagents and protein structure determination. The reaction of phenolic disulfonyl chlorides with lysozyme, *Biopolymers*, 2, 349, 1964.
171. Moore, G. L. and Day, R. A., Protein conformation in solution: Cross-linking of lysozyme, *Science*, 159, 210, 1968.
172. Zahn, H. and Meiehofer, J., Experiments with insulin, *Makromol. Chem.*, 26, 153, 1958.
173. Brandenburg, D., Peptides. 87. Preparation of $N^{\alpha A1}$, $N^{\epsilon B29}$-adipoyl-insulin, an intramolecularly cross-linked derivative of beef insulin, *Hoppe-Seyler's Z. Physiol. Chem.*, 353, 869, 1972.
174. Plotz, P. H., Bivalent affinity labeling haptens in the formation of model immune complexes, *Methods Enzymol.*, 46, 505, 1977.
175. Busse, W. D. and Carpenter, F. H., Carbonyl bis(L-methionine p-nitrophenyl ester). A new reagent for the reversible intramolecular cross-linking of insulin, *J. Am. Chem. Soc.*, 96, 5947, 1974.
176. Busse, W. D. and Carpenter, F. H., Synthesis and properties of CBM-insulin, a pro-insulin analogue which is convertible to insulin by CNBr cleavage, *Biochemistry*, 15, 1649, 1976.
177. Zahn, H. and Schade, F., Nitrophenyl esters, *Chem. Ber.*, 96, 1747, 1963.
178. Zahn, H. and Schade, F., Chemiscbe modifizierung von insulin, seidenfibroin, sehnenkollagen und wollkeratin mit nitrophenylestern, *Angew. Chem.*, 75, 377, 1963.
179. Miyakawa, T., Takemoto, L. J., and Fox, C. F., Membrane permeability of bifunctional, amino site-specific, cross-linking reagents, *J. Supramol. Struct.*, 8, 303, 1978.
180. Dimmock, N. J., Dolbear, H. S., and Guest, A. R., Chemical cross-linking of proteins of the influenza virion. 2. Acid-induced irreversible conformational changes in HA1 and HA2, *Arch. Virol.*, 108, 183, 1989.
181. Chan, W. W. and Enns, C. A., Hybrid aspartate transcarbamoylase containing cross-linked subunits, *Can. J. Biochem.*, 59, 461, 1981.
182. Wetz, K., Fasold, H., and Meyer, C., Synthesis of long, hydrophilic, protein cross-linking reagents, *Anal. Biochem.*, 58, 347, 1974.
183. Riordan, J. F., Arginyl residues and anion binding sites in proteins, *Mol. Cell. Biochem.*, 26, 71, 1979.
184. Meade, S. J., Miller, A. G., and Gerrard, J. A., The role of dicarbonyl compounds in non-enzymatic cross-linking: A structure–activity study, *Bioorg. Med. Chem.*, 11, 853, 2003.
185. Miller, A. G., Meade, S. J., and Gerrard, J. A., New insights into protein crosslinking via the Maillard reaction: Structural requirements, the effect on enzyme function, and predicted efficacy of crosslinking inhibitors as anti-ageing therapeutics, *Bioorg. Med. Chem.*, 11, 843, 2003.
186. Miller, A. G. and Gerrard, J. A., Assessment of protein function following cross-linking by α-dicarbonyls, *Ann. N.Y. Acad. Sci.*, 1043, 195, 2005.
187. Xu, G. and Sayre, L. M., Cross-linking of proteins by 3-(trifluoromethyl)-2,5-hexanedione. Model studies implicate an unexpected amine-dependent defluorinative substitution pathway competing with pyrrole formation, *J. Org. Chem.*, 3, 3007, 2002.
188. Xu, G., Singh, M. P., Gopal, D., and Sayre, L. M., Novel 2,5-hexanedione analogues. Substituent-induced control of the protein cross-linking potential and oxidation susceptibility of the resulting primary amine-derived pyrroles, *Chem. Res. Toxicol.*, 14, 264, 2001.
189. Zhu, M., Spink, D. C., Yan, B., Bank, S., and DeCaprio, A. P., Formation and structure of cross-linking and monomeric pyrrole autoxidation products in 2,5-hexanedione-treated amino acids, peptides, and protein, *Chem. Res. Toxicol.*, 7, 551, 1994.
190. Voitkun, V. and Zhitkovich, A., Analysis of DNA–protein crosslinking activity of malondialdehyde in vitro, *Mutat. Res.*, 97, 424, 1999.
191. Migneault, I., Dartiguenave, C., Bertrand, M. J., and Waldron, K. C., Glutaraldehyde: Behavior in aqueous solution, reaction with proteins, and application to enzyme crosslinking, *Biotechniques*, 37, 790, 2004.
192. Richard, F. M. and Knowles, J. R., Glutaraldehyde as a protein cross-linking reagent, *J. Mol. Biol.*, 37, 231, 1968.

193. Hardy, P. M., Nicholls, A. C., and Rydon, H. N., The nature of the crosslinking of proteins by glutaraldehyde. Part I. Interaction of glutaraldehyde with the amino groups of 6-aminohexanoic acid and of α-N-acetyl-lysine, *J. Chem. Soc. Perkin Trans.*, 1, 958, 1976.

194. Moosan, P., Puzo, G., and Mazarguil, H., Etude du mecanisme d'etablissement des liaisons glutaraldehyde proteins, *Biochimie*, 57, 1281, 1975.

195. Quiocbo, F. A. and Richards, F. M., Intermolecular cross linking of a protein in the crystalline state: Carboxypeptidase A, *Proc. Natl. Acad. Sci. U. S. A.*, 52, 833, 1964.

196. Bishop, W. H. and Richards, F. M., Isoelectric point of a protein in the crosslinked crystalline state: β-Lactoglobulin, *J. Mol. Biol.*, 33, 415, 1968.

197. Reeke, G. N., Hartsuck, J. A., Ludwig, M. L., Quiocho, F. A., Steitz, T. A., and Lipscomb, W. N., The structure of carboxypeptidase A. VI. Some results at 2.0 Å resolution, and the complex with glycyltyrosine at 2.8 Å resolution, *Proc. Natl. Acad. Sci. U. S. A.*, 58, 2220, 1967.

198. Habeeb, A. F. S. A. and Hiramoto, R., Reaction of proteins with glutaraldehyde, *Arch. Biochem. Biophys.*, 126, 16, 1968.

199. Schejter, A. and Bar-Eli, A., Preparation and properties of crosslinked water-insoluble catalase, *Arch. Biochem. Biophys.*, 136, 325, 1970.

200. Tam, J. W. and Cheng, L. Y., Chemical cross-linking of hemoglobin-H, *Biochim. Biophys. Acta*, 580, 75, 1979.

201. Scannon, P. J., Molecular modification of hemoglobin, *Crit. Care Med.*, 10, 261, 1982.

202. Guillochon, D., Esclade, L., Remy, M. H., and Thomas, D., Studies on hemoglobin immobilized by cross-linking with glutaraldehyde, *Biochim. Biophys. Acta*, 670, 332, 1981.

203. Cambou, B., Laurent, M., Hervagault, J. F., and Thomas, D., Modulation of phosphofructokinase behavior by chemical modification during the immobilization process, *Eur. J. Biochem.*, 121, 99, 1981.

204. Hermann, R., Jaenicke, R., and Rudolph, R., Analysis of the reconstitution of oligomeric enzymes by cross-linking with glutaraldehyde, *Biochemistry*, 20, 5195, 1981.

205. Alwan, S., Smith, H. J., Mahbouba, M., Evans, J. C., and Morgan, P. H., Cross-linking of insulin with glutaraldehyde to form macromolecules, *J. Pharm. Pharmacol.*, 33, 323, 1981.

206. Engvall, E., and Perlmann, P., Enzyme-linked immunosorbent assay (ELISA). Quantitative assay of immunoglobulin G, *Immunochemistry*, 8, 871, 1971.

207. Avrameas, S. and Ternynck, T., Peroxidase-labeled antibody and Fab conjugates with enhanced intracellular penetration, *Immunochemistry*, 8, 1175, 1971.

208. Armbrust, T. S., Chitnis, P. R., and Cuikema, J. A., Organization of photosystem I polypeptides examined by chemical cross-linking, *Plant Physiol.*, 111, 1307, 1996.

209. López-Gallego, F., Betancor, L., Mateo, C., Hidalgo, A., Alonso-Morales, N., Dellamora-Ortiz, G., Guisán, J. M., and Fernández-Lafuente, R., Enzyme stabilization by glutaraldehyde crosslinking of adsorbed proteins on aminated supports, *J. Biotechnol.*, 119, 70, 2005.

210. Betancor, L., López-Gallego, F., Hidalgo, A., Alonso-Morales, N., Dellamora-Ortiz, G., Guisán, J. M., and Fernández-Lafuente, R., Preparation of a very stable immobilized biocatalyst of glucose oxidase from *Aspergillus niger*, *J. Biotechnol.*, 121, 284, 2006.

211. Hardy, P. M., Nicholls, A. C., and Rydon, H. N., The nature of glutaraldehyde in aqueous solution, *Chem. Commun.*, 565, 1969.

212. Wine, Y., Cohen-Hadar, N., Freeman, A., and Frolow, F., Elucidation of the mechanism and end products of glutaraldehyde crosslinking reaction by X-ray structure analysis, *Biotechnol. Bioeng.*, 98, 711, 2007.

213. Fuguet, E., van Platerink, C., and Janssen, H.-G., Analytical characterisation of glutardialdehyde crosslinking products in gelatine–gum arabic complex coacervates, *Anal. Chim. Acta*, 604, 45, 2007.

214. Josephs, R., Eisenberg, H., and Reisler, E., Some properties of cross-linked polymers of glutamic dehydrogenase, *Biochemistry*, 12, 4060, 1973.

215. Fein, M. L. and Filachione, E. M., Tanning studies with aldehydes, *J. Am. Leather Chem. Assoc.*, 52, 17, 1957.

216. Seligsberger, L. and Sadlier, C., New developments in tanning with aldehydes, *J. Am. Leather Chem. Assoc.*, 52, 2, 1957.

217. Hopwood, D., Comparison of the crosslinking abilities of glutaraldehyde, formaldehyde, and α-hydroxyadipaldehyde with bovine serum albumin and casein, *Histochemie*, 17, 1.51, 1969.

218. Bhagwat, A. S., Blokesch, A., Irrgang, K. D., Salnikow, J., and Vater, J., Crosslinking of components of the cytochrome b6f complex from spinach thylakoids by o-phthalaldehyde, *Arch. Biochem. Biophys.*, 304, 38, 1993.

219. Benescb, R. E. and Kwong, S., *Bis*-pyridoxal polyphosphates: A new class of specific intramolecular crosslinking agents for hemoglobin, *Biochem. Biophys. Res. Commun.*, 156, 9, 1988.

220. Shimomura, S. and Fukui, T., Characterization of the pyridoxal phosphate site in glycogen phosphorylase b from rabbit muscle, *Biochemistry*, 17, 5359, 1978.
221. Brooks, B. R. and Klamerth, O. L., Interaction of DNA with bifunctional aldehydes, *Eur. J. Biochem.*, 5, 178, 1968.
222. Cater, C. W., The evaluation of aldehydes and other bifunctional compounds as cross-linking agents for collagen, *J. Soc. Leather Trade Chem.*, 47, 259, 1963.
223. Chancerelle, Y., Alban, C., Viret, R., Tosetti, F., and Kergonou, J. F., Immunological relevance of malonic dialdehyde (MDA). IV. Further evidences about the epitope recognized by antibodies obtained from rabbits immunized with MDA-modified lysozyme., *Biochem. Int.*, 24, 157, 1991.
224. Sager, P. R., Cytoskeletal effects of acrylamide and 2,5-hexanedione: Selective aggregation of vimentin filaments, *Toxicol. Appl. Pharmacol.*, 97, 141, 1989.
225. Graham, D. G., St. Clair, M. B., Amarnath, V., and Anthony, D. C., Molecular mechanisms of gamma-diketone neuropathy, *Adv. Exp. Med. Biol.*, 283, 427, 1991.
226. Fayle, S. E., Gerrard, J. A., Simmons, L., Meade, S. J., Reid, E. A., and Johnston, A. C., Crosslinkage of proteins by dehydroascorbic acid and its degradation products, *Food Chem.*, 70, 193, 2000.
227. Schoevaart, R. and Kieboom, T., Galactose dialdehyde as potential protein cross-linker: proof of principle, *Carbohydr. Res.*, 337, 899, 2002.
228. van Wijk, A., Siebum, A., Schoevaart, R., and Kieboom, T., Enzymatically oxidized lactose and derivatives thereof as potential protein cross-linkers, *Carbohydr. Res.*, 341, 2921, 2006.
229. Ternynck, T. and Avrameas, S., Conjugation of p-benzoquinone treated enzymes with antibodies and Fab fragments, *Immunochemistry*, 14, 767, 1977.
230. Avrameas, S., Ternynck, T., and Guesdon, J.-L., Coupling of enzymes to antibodies and antigens, *Scand. J. Immunol.*, (Suppl. 7), 7, 1978.
231. Yamauchi, A., Hatanaka, Y., Muro, T., and Kobayashi, O., Enzyme-free quinone crosslinking reaction for proteins: A macromolecular characterization study using gelatin, *Macromol. Biosci.*, 9, 875, 2009.
232. United States Patent: 6048736, Cyclodextrin polymers for carrying and releasing drugs, 2000.
233. United States Patent: 6835718, Biocleavable micelle compositions for use as drug carriers, 2001.
234. Robinson, I. D., Role of crosslinking of gelatin in aqueous solutions, *J. Appl. Polymer Sci.*, 8, 1903, 1964.
235. Slusarewicz, P., Zhu, K., and Hedman, T., Kinetic characterization and comparison of various protein crosslinking reagents for matrix modification, *J. Mater. Sci. Mater. Med.*, 21, 1175, 2010.
236. Bhrany, A. D., Lien, C. J., Beckstead, B. L., Futran, N. D., Muni, N. H., Giachelli, C. M., and Ratner, B. D., Crosslinking of an oesophagus acellular matrix tissue scaffold, *J. Tissue Eng. Regen. Med.*, 2, 365, 2008.
237. Song, F., Zhang, L. M., Yang, C., and Yan, L., Genipin-crosslinked casein hydrogels for controlled drug delivery, *Int. J. Pharm.*, 373, 41, 2009.
238. Liu, B. S., Yao, C. H., and Fang, S. S., Evaluation of a non-woven fabric coated with a chitosan bi-layer composite for wound dressing, *Macromol. Biosci.*, 8, 432, 2008.
239. Chen, M. C., Liu, C. T., Tsai, H. W., Lai, W. Y., Chang, Y., and Sung, H. W., Mechanical properties, drug eluting characteristics and in vivo performance of a genipin-crosslinked chitosan polymeric stent, *Biomaterials*, 30, 5560, 2009.
240. Chiono, V., Pulieri, E., Vozzi, G., Ciardelli, G., Ahluwalia, A., and Giusti, P., Genipin-crosslinked chitosan/gelatin blends for biomedical applications, *J. Mater. Sci. Mater. Med.*, 19, 889, 2008.
241. Yao, C. H., Liu, B. S., Chang, C. J., Hsu, S. H., and Chen, Y. S., Preparation of networks of gelatin and genipin as degradable biomaterials, *Mater. Chem. Phys.*, 83, 204–208, 2004.
242. Sung, H. W., Huang, D. M., Chang, W. H, Huang, L. L. H., Tsai, C. C., and Liang, I. L., Gelatin-derived bioadhesives for closing skin wounds: An in vivo study, *J. Biomater. Sci. Polymer Edn.*, 10, 751–771, 1999.
243. Chang, W. H., Chang, Y., Lai, P. H., and Sung, H. W., A genipin-crosslinked gelatin membrane as wound-dressing material: In vitro and in vivo Studies, *J. Biomater. Sci. Polymer Edn.*, 14, 481, 2003.
244. Yao, C. H., Lui, B. S., Hsu, S. H., and Chen, Y. S., Cavarial bone response to tricalcium phosphate-genipin crosslinked gelatin composite, *Biomaterials*, 26, 3065, 2005.
245. Chen, Y. S., Chang, J. Y., Cheng, C. Y., Tsai, F. J., Yao, C. H., and Liu, B. S., An in vivo evaluation of a biodegradable genipin-cross-linked gelatin peripheral nerve guide conduit material, *Biomaterials*, 26, 3911, 2005.
246. Butler, M. F., Ng, Y. F., and Pudney, P. D. A., Mechanism and kinetics of the crosslinking reaction between biopolymers containing primary amine groups and genipin, *J. Polym. Sci. A Polym. Chem.*, 41, 3941, 2003.
247. Milch, R. A., Studies of alcaptonuria: Collagenase degradation of homogentisic-tanned hide powder collagen, *Proc. Soc. Exp. Biol. Med.*, 107, 183, 1961.

248. Milch, R. A., Viscometric hardening of gelatin sols in the presence of certain intermediary metabolites, *J. Surg. Res.*, 3, 254, 1963.

249. Aliberti, G., Pulignano, I., Pisani, D., March, M. R., Del Porto, F., and Proietta, M., Bisphosphonate treatment in ochronotic osteoporotic patients, *Clin. Rheumatol.*, 26, 729, 2007.

250. Murayama, Y., Satoh, S., Oka, T., Imanishi, J., and Noishiki, Y., Reduction of the antigenicity and immunogenicity of xenografts by a new cross-linking reagent, *ASAIO Trans.*, 34, 546, 1988.

251. Ichikawa, Y., Noishiki, Y., Soma, T., Ishii, M., Yamamoto, K., Takahashi, K., Mo, M., Kosuge, T., Kondo, J., and Matsumoto, A., A new antithrombogenic RV-PA valved conduit, *ASAIO J.*, 40, M714, 1994.

252. Hughes, W. L., Albumin fraction isolated from human plasma as a crystalline mercuric salt, *J. Am. Chem. Soc.*, 69, 1836, 1947.

253. Arnon, R. and Shapira, E., Crystalline papain derivative containing an intramolecular mercury bridge, *J. Biol. Chem.*, 244, 1033, 1969.

254. Sperling, R., Burstein, Y., and Steinberg, I. Z., Selective reduction and mercuration of cystine IV–V in bovine pancreatic ribonuclease, *Biochemistry*, 8, 3810, 1969.

255. Jovin, T. M., Eaglund, P. T., and Kornberg, A., Enzymatic synthesis of deoxyribonucleic acid, *J. Biol. Chem.*, 244, 3009, 1969.

256. Edelboch, H., Katchalsk, E., Maybury, R. H., Hughes, W. L., Jr., and Edsall, J. T., Dimerization of serum mercaptalbumin in presence of mercurials. I. Kinetic and equilibrium studies with mercuric salts, *J. Am. Chem. Soc.*, 75, 5058, 1953.

257. Edsall, J. T., Maybury, R. H., Simpson, R. B., and Straessle, R., Dimerization of serum mercaptalbumin in the presence of mercurials. II. Studies with a bifunctional organic mercurial, *J. Am. Chem. Soc.*, 76, 3131, 1954.

258. Kay, C. M. and Edsall, J. T., Dimerization of mercaptalbumin in the presence of mercurials. III. Bovine mercaptalbumin in water and in concentrated urea solutions, *Arch. Biochem. Biophys.*, 65, 354, 1956.

259. Singer, S. J., Fothergill, J. E., and Shainoff, J. R., A general method for the isolation of antibodies. *J. Am. Chem. Soc.*, 82, 565, 1960.

260. Vas, M. and Caanády, G., The two fast-reacting thiols of 3-phosphoglycerate kinase are structurally juxtaposed: Chemical modification with bifunctional reagents, *Eur. J. Biochem.*, 163, 365, 1987.

261. Kliche, W., Pfannstiel, J., Tiepold, M., Stoeva, S., and Faulstich, H., Thiol-specific cross-linkers of variable length reveal a similar separation of SH1 and SH2 in myosin subfragment 1 in the presence and absence of MgADP, *Biochemistry*, 38, 10307, 1999.

262. Green, N. S., Reisler, E., and Houk, K. N., Quantitative evaluation of the lengths of homobifunctional protein cross-linking reagents used as molecular rulers, *Protein Sci.*, 10, 1293, 2001.

263. Zecherle, G. N., Oleinikov, A., and Traut, R. R., The proximity of the C-terminal domain of *Escherichia coli* ribosomal protein L7/L12 to L10 determined by cysteine site-directed mutagenesis and protein–protein cross-linking, *J. Biol. Chem.*, 267, 5889, 1992.

264. Widera, A., Kim, K. J., Crandall, E. D., and Shen, W. C., Transcytosis of GCSF-transferrin across rat alveolar epithelial cell monolayers, *Pharm. Res.*, 20, 1231, 2003.

265. Mansuy-Schlick, V., Delage-Mourroux, R., Jouvenot, M., and Boireau, W., Strategy of macromolecular grafting onto a gold substrate dedicated to protein–protein interaction measurements, *Biosens. Bioelectron.*, 21, 1830, 2006.

266. Bloxham, D. P. and Sharma, R. P., The development of S,S′-polymethylenebis(methanethiosulfonates) as reversible cross-linking reagent for thiol groups and their use to form stable catalytically active crosslinked dimers with glyceraldehyde-3-phosphate dehydrogenase, *Biochem. J.*, 181, 355, 1979.

267. Bloxham, D. P. and Cooper, G. K., Formation of a polymethylene bis(disulfide) intersubunit cross-link between cys-281 residues in rabbit muscle glyceraldyde-3-phosphate dehydrogenase using octamethylene bis(methane [35]thiosulfonate), *Biochemistry*, 21, 1807, 1982.

268. Hou, Z. and Matherly, L. H., Oligomeric structure of the human reduced folate carrier: Identification of homo-oligomers and dominant-negative effects on carrier expression and function, *J. Biol. Chem.*, 284, 3285, 2009.

269. Moore, K. J. and Fillingame, R. H., Structural interactions between transmembrane helices 4 and 5 of subunit a and the subunit c ring of *Escherichia coli* ATP synthase, *J. Biol. Chem.*, 283, 31726, 2008.

270. Shvetsov, A., Stamm, J. D., Phillips, M., Warshaviak, D., Altenbach, C., Rubenstein, P. A., Hideg, K., Hubbell, W. L., and Reisler, E., Conformational dynamics of loop 262–274 in G- and F-actin, *Biochemistry*, 45, 6541, 2006.

271. Ermolova, N., Guan, L., and Kaback, H. R., Intermolecular thiol cross-linking via loops in the lactose permease of *Escherichia coli*, *Proc. Natl. Acad. Sci. U. S. A.*, 100, 10187, 2003.

272. Guan, L., Murphy, F. D., and Kaback, H. R., Surface-exposed positions in the transmembrane helices of the lactose permease of *Escherichia coli* determined by intermolecular thiol cross-linking, *Proc. Natl. Acad. Sci. U. S. A.*, 99, 3475, 2002.

273. Salam, W. H. and Bloxham, D. P., Hypolipidemic effect of polymethylenemethane thiosulfonates: Inhibitors of acetoacetyl coenzyme A thiolase, *J. Pharmacol. Exp. Ther.*, 241, 1099, 1987.

274. Grintsevich, E. E., Benchaar, S. A., Warshaviak, D., Boontheung, P., Halgand, F., Whitelegge, J. P., Faull, K. F. et al., Mapping the cofilin binding site on yeast G-actin by chemical cross-linking, *J. Mol. Biol.*, 377, 395, 2008.

275. Packard, B., Edidin, M., and Komoriya, A., Site-directed labeling of a monoclonal antibody: Targeting to a disulfide bond, *Biochemistry*, 25, 3548, 1986.

276. Zahn, H. and Lamper, L., Specificity of bifunctional sulfhydryl reagents and synthesis of a defined dimer of bovine serum albumin, *Hoppe-Seyler's Z. Physiol. Chem.*, 349, 485, 1968.

277. Heilmann, H. D. and Holzner, M., The spatial organization of the active sites of the bifunctional oligomeric enzyme tryptophan synthetase: Cross-linking by a novel method, *Biochem. Biophys. Res. Commun.*, 99, 1146, 1981.

278. Freedberg, W. B. and Hardman, J. K., Structural and functional roles of the cysteine residues in the α subunit of the *Escherichia coli* tryptophan synthetase, *J. Biol. Chem.*, 246, 1439, 1971.

279. Thammana, P. and Cantor, C. R., Studies on ribosome structure and interactions near the m62Am62A sequence, *Nucleic Acids Res.*, 5, 805, 1978.

280. Phan, B. C., Peyser, Y. M., Reisler, E., and Muhlrad, A., Effect of complexes of ADP and phosphate analogs on the conformation of the Cys707–Cys697 region of myosin subfragment 1, *Eur. J. Biochem.*, 243, 636, 1997.

281. Moroney, J. V., Warncke, K., and McCarthy, R. E., The distance between thiol groups in the gamma subunit of coupling factor 1 influences the proton permeability of thylakoid membranes, *J. Bioenerg. Biomembr.*, 14, 347, 1982.

282. Chantler, P. and Bower, S. M., Cross-linking between translationally equivalent sites on the heads of myosin: Relationship to energy transfer results between the same pair of sites, *J. Biol. Chem.*, 263, 938, 1988.

283. Sato, S. and Nakao, M., Cross-linking of intact erythrocyte membrane with a newly synthesized cleavable bifunctional reagent, *J. Biochem.*, 90, 1177, 1981.

284. Srinivasachar, K. and Neville, D. M., Jr., New protein cross-linking reagents that are cleaved by mild acid, *Biochemistry*, 28, 2501, 1989.

285. Weston, P. D., Devries, J. A., and Wrigglesworth, R., Conjugation of enzymes to immunoglobulins using dimaleimides, *Biochim. Biophys. Acta*, 612, 40, 1980.

286. Hruz, P. W. and Miziorko, H. M., Avian 3-hydroxy-3-methylglutaryl-CoA lyase: Sensitivity of enzyme activity to thiol/disulfide exchange and identification of proximal reactive cysteines, *Protein Sci.*, 1, 1144, 1992.

287. Levitsky, D. I., Shnyrov, V. L., Khvorov, N. V., Bukatina, A. E., Vedenkina, N. S., Permyakov, E. A., Nikolaeva, O. P., and Poglazov, B. F., Effects of nucleotide binding on thermal transitions and domain structure of myosin subfragment 1, *Eur. J. Biochem.*, 209, 829, 1992.

288. Lee, R. H., Ting, T. D., Lieberman, B. S., Tobias, D. E., Lolley, R. N., and Ho, Y. K., Regulation of retinal cGMP cascade by phosducin in bovine rod photoreceptor cells. Interaction of phosducin and transducin, *J. Biol. Chem.*, 267, 25104, 1992.

289. Sata, M., Sugiura, S., Yamashita, H., Momomura, S., and Serizawa, T., Dynamic interaction between cardiac myosin isoforms modifies velocity of actomyosin sliding in vitro, *Circ. Res.*, 73, 696, 1993.

290. Khan, M. T. and Saleemuddin, M., Unlike its human counterpart, band 3 anion exchange protein from goat erythrocyte membrane shows a lack of reactivity against various-SH oxidants and protease treatments, *Comp. Biochem. Physiol. B Biochem. Mol. Biol.*, 110, 339, 1995.

291. Schwarzer, A., Kim, T. S., Hagen, V., Molday, R. S., and Bauer, P. J., The Na/Ca-K exchanger of rod photoreceptor exists as dimer in the plasma membrane, *Biochemistry*, 36, 13667, 1997.

292. Nitao, L. K. and Reisler, E., Probing the conformational states of the SH1–SH2 helix in myosin: A cross-linking approach, *Biochemistry*, 37, 16704, 1998.

293. Hashimoto, M., Majima, E., Goto, S., Shinohara, Y., and Terada, H., Fluctuation of the first loop facing the matrix of the mitochondrial ADP/ATP carrier deduced from intermolecular cross-linking of Cys56 residues by bifunctional dimaleimides, *Biochemistry*, 38, 1050, 1999.

294. Himmel, D. M., Gourinath, S., Reshetnikova, L., Shen, Y., Szent-Györgyi, A. G., and Cohen, C., Crystallographic findings on the internally uncoupled and near-rigor states of myosin: Further insights into the mechanics of the motor, *Proc. Natl. Acad. Sci. U. S. A.*, 99, 12645, 2002.

295. Dorvillius, M., Garambois, V., Pourquier, D., Gutowski, M., Rouanet, P., Mani, J. C., Pugnière, M., Hynes, N. E., and Pèlegrin, A., Targeting of human breast cancer by a bispecific antibody directed against two tumour-associated antigens: ErbB-2 and carcinoembryonic antigen, *Tumour Biol.*, 23, 337, 2002.

296. Girouard, S., Houle, M. H., Grandbois, A., Keillor, J. W., and Michnick, S. W., Synthesis and characterization of dimaleimide fluorogens designed for specific labeling of proteins, *J. Am. Chem. Soc.*, 127, 559, 2005.

297. Umeki, N., Yoshizawa, T., Sugimoto, Y., Mitsui, T., Wakabayashi, K., and Maruta, S., Incorporation of an azobenzene derivative into the energy transducing site of skeletal muscle myosin results in photo-induced conformational changes, *J. Biochem.*, 136, 839, 2004.

298. Guy, J., Caron, K., Dufresne, S., Michnick, S. W., Skene, W. G., and Keillor, J. W., Convergent preparation and photophysical characterization of dimaleimide dansyl fluorogens: Elucidation of the maleimide fluorescence quenching mechanism, *J. Am. Chem. Soc.*, 129, 11969, 2007.

299. Chen, H. and Fliegel, L., Expression, purification, and reconstitution of the Na(+)/H(+) exchanger sod2 in *Saccharomyces cerevisiae*, *Mol. Cell. Biochem.*, 319, 79, 2008.

300. Cava, M. P., Deana, A. A., Muth, K., and Mitchell, M. J., N-phenylmaleimide, *Org. Synth.*, 41, 93, 1961.

301. Trommer, W. F. and Hendrick, M., Formation of maleimides by a new mild cyclization procedure, *Synthesis*, 8, 484, 1973.

302. Brewer, C. F. and Riehm, J. P., Evidence for possible nonspecific reactions between N-ethylmaleimide and proteins, *Anal. Biochem.*, 18, 148, 1967.

303. Riordan, J. F. and Vallee, B. L., Reactions with N-ethylmaleimide and p-mercuribenzoate, *Methods Enzymol.*, 25, 449, 1972.

304. Hashida, S., lmagawa, M., Inoue, S., Ruan, K.-H., and Ishikawa, E., More useful maleimide compounds for the conjugation of Fab′ to horseradish peroxidase through thiol groups in the hinge, *J. Appl. Chem.*, 6, 56, 1984.

305. Yoshitake, S., Hamaguchi, Y., and Ishikawa, E., Efficient conjugation of rabbit Fab′ with β-D-galactosidase from *Escherichia coli*, *Scand. J. Immunol.*, 10, 81, 1979.

306. Sumner, A. T., Distribution of protein sulfhydryls and disulfides in fixed mammalian chromosomes, and their relationship to banding, *J. Cell. Sci.*, 70, 177, 1984.

307. Chantler, P. D., Tao, T., and Stafford, W. F., On the relationship between distance information derived from cross-linking and from resonance energy transfer, with specific reference to sites located on myosin heads, *Biophys. J.*, 59, 1242, 1991.

308. Perkins, W. J., Wells, J. A., and Young, R. G., Characterization of the properties of ethenoadenosine nucleotides bound or trapped at the active site myosin subfragment, *Biochemistry*, 23, 3994, 1984.

309. Miller, L., Coopedge, J., and Reisler, E., The reactive SH1 and SH2 cysteines in myosin subfragment 1 are cross-linked at similar rates with reagents of different length, *Biochem. Biophys. Res. Commun.*, 106, 117, 1982.

310. Kwaw, I., Sun, J., and Kaback, H. R., Thiol cross-linking of cytoplasmic loops in the lactose permease of *Escherichia coli*, *Biochemistry*, 39, 3134, 2000.

311. Hasain, S. S. and Lowe, G., Evidence for histidine in the active sites for ficin and stem-bromelain, *Biochem. J.*, 110, 53, 1968.

312. Ozawa, H., Bridging reagent for protein. II. The reaction of N,N′-polymethylenebis(iodoacetamide) with cysteine and rabbit muscle aldolase, *J. Biochem. (Tokyo)*, 62, 531, 1967.

313. Gundlach, H. G., Habilitationsschrift, University of Wurzburg, Würzburg, Germany, 1965, 44 p.

314. Wilchek, M. and Givol, D., Affinity cross-linking of heavy and light chains, *Methods Enzymol.*, 46, 501, 1977.

315. Segal, D. M. and Hurwitz, E., Dimers and trimers of immunoglobulin G covalently cross-linked with a bivalent affinity label, *Biochemistry*, 15, 5253, 1976.

316. Ludueña, R. F., Roach, M. C., Trcka, P. P., and Weintraub, S., N,N-Bis(alpha-iodoacetyl)-2,2′-dithiobis(ethylamine), a reversible crosslinking reagent for protein sulfhydryl groups, *Anal. Biochem.*, 117, 76–80, 1981.

317. Haugland, R. P., *Handbook of Fluorescent Probes and Research Chemicals*, Molecular Probes, Inc., Eugene, OR, 1990, 22 p.

318. Fasold, H., Gröschel-Stewart, U., and Turba, F., Synthese and reaktionen eines wasserlöslichen, spaltbarenreagens zur verknupfung frier SH-gruppen in proteinen, *Biochem. Z.*, 339, 487, 1964.

319. Asenjo, A. B., Weinberg, Y., and Sosa, H., Nucleotide binding and hydrolysis induces a disorder-order transition in the kinesin neck-linker region, *Nat. Struct. Mol. Biol.*, 13, 648, 2006.

320. Corrie, J. E., Craik, J. S., and Munasinghe, V. R., A homobifunctional rhodamine for labeling proteins with defined orientations of a fluorophore, *Bioconjug. Chem.*, 9, 160, 1998.

321. Ehrhardt, A. G., Kang, H. C., Tuft, R. A., Fay, F. S., and Ikebe, M., Labeling of Calmodulin with a new bifunctional BODIPY for improved fluorophore stability in anisotropy measurements, *Biophys. J.*, 74(2 Pt 2), A380, Th-Pos200, 1998.

322. Hiremath, C. B. and Day, R. A., Introduction of covalent cross-linkages into lysozyme by reaction with α,α′-dibromo-p-xylenesulfonic acid, *J. Am. Chem. Soc.*, 86, 5027, 1964.

323. Ross, W. C. J., The chemistry of cytotoxic alkylating agents, *Adv. Cancer Res.*, 1, 397, 1953.

324. Alexander, P., The reactions of carcinogens with macromolecules, *Adv. Cancer Res.*, 2, 1, 1954.

325. Burnop, V. C. E., Francis, G. E., Richards, D. E., and Wormall, A., Nitrogen-15-labeled nitrogen mustard. The combination of bis(2-chloroethyl)methylamine with proteins, *Biochem. J.*, 66, 504, 1957.

326. Boursnell, J. C., Some reactions of mustard gas (β,β′-di-chlorodiethyl sulfide) with proteins, *Biochem. Soc. Symp.*, 2, 8, 1948.

327. Philips, F. S., Recent contributions to the pharmacology of bis(2-haloethyl)amines and sulfides, *J. Pharmacol. Exp. Ther.*, 99, 281, 1950.

328. Goodlad, G. A. J., Cross-linking of collagen by sulfur- and nitrogen-mustards, *Biochim. Biophys. Acta*, 25, 202, 1957.

329. Brookes, P. and Lawley, P. D., Effects of alkylating agents on T_2 and T4 bacteriophages, *Biochem. J.*, 89, 138, 1963.

330. Berenblum, I. and Wormall, A., The immunological properties of proteins treated with β,β′-dichlorodiethyl sulfide (mustard gas) and β,β′-dichlorodiethyl sulfones, *Biochem. J.*, 33, 75, 1939.

331. Balali-Mood, M., Mousavi, S. S., and Balali-Mood, B., Chronic health effects of sulphur mustard exposure with special reference to Iranian veterans, *Emerg. Health Threats J.*, 1, e7, 2008.

332. Hasinoff, B. B., Wu, X., Begleiter, A., Guziec, L. J., Guziec F., Jr., Giorgianni, A., Yang, S., Yu Jiang, Y., and Yalowich, J. C., Structure–activity study of the interaction of bioreductive benzoquinone alkylating agents with DNA topoisomerase II, *Cancer Chemother. Pharmacol.*, 57, 221, 2006.

333. Kim, J. S. and Raines, R. T., Dibromobimane as a fluorescent crosslinking reagent, *Anal. Biochem.*, 225, 174, 1995.

334. Konno, K., Ue, K., Khoroshev, M., Martinez, H., Ray, B., and Morales, M. F., Consequences of placing an intramolecular crosslink in myosin S1, *Proc. Natl. Acad. Sci. U. S. A.*, 97, 1461, 2000.

335. Kosower, N. S., Newton, G. L., Kosower, E. M., and Ranney, H. M., Bimane fluorescent labels. Characterization of the bimane labeling of human hemoglobin, *Biochim. Biophys. Acta*, 622, 201, 1980.

336. Wu, J., Voss, J., Hubbell, W. L., and Kaback, H. R., Site-directed spin labeling and chemical crosslinking demonstrate that helix V is close to helices VII and VIII in the lactose permease of *Escherichia coli*, *Proc. Natl. Acad. Sci. U. S. A.*, 93, 10123, 1996.

337. Bhattacharjee, H. and Rosen, B. P., Spatial proximity of Cys113, Cys172, and Cys422 in the metalloactivation domain of the ArsA ATPase, *J. Biol. Chem.*, 271, 24465, 1996.

338. Loo, T. W. and Clarke, D. M., Identification of residues in the drug-binding domain of human P-glycoprotein. Analysis of transmembrane segment 11 by cysteine-scanning mutagenesis and inhibition by dibromobimane, *J. Biol. Chem.*, 274, 35388, 1999.

339. Sinz, A. and Wang, K., Mapping protein interfaces with a fluorogenic cross-linker and mass spectrometry: Application to nebulin-calmodulin complexes, *Biochemistry*, 40, 7903, 2001.

340. Agarwal, K. L., Grudzinski, S., Kenner, G. W., Rogers, N. H., Sheppard, R. C., and McGuigan, J. E., Immunochemical differentiation between gastrin and related peptide hormones through a novel conjugation of peptides to proteins, *Experientia*, 27, 514, 1971.

341. Nakane, P. K., Simultaneous localization of multiple tissue antigens using the peroxidase-labeled antibody method. A study of pituitary glands of the rat, *J. Histochem. Cytochem.*, 16, 557, 1968.

342. Cater, C. W., The efficiency of dialdehydes and other compounds as cross-linking agents for collagen, *J. Soc. Leather Trades Chem.*, 49, 455, 1965.

343. Gotoh, Y., Tsukada, M., and Minoura, N., Chemical modification of silk fibroin with cyanuric chloride-activated poly(ethylene glycol): Analyses of reaction site by 1H-NMR spectroscopy and conformation of the conjugates, *Bioconjug. Chem.*, 4, 554, 1993.

344. Avrameas, S., Taudou, B., and Chuilon, S., Glutaraldehyde, cyanuric chloride, and tetra-azotized o-dianisidine as coupling reagents in the passive hemogglutination test, *Immunochemistry*, 6, 67, 1969.

345. Wadsworth, P. and Sloboda, R. D., Modification of tubulin with the fluorochrome 5-(4,6-dichlorotriazin-2-yl)aminofluorescein and the interaction of the fluorescent protein with the isolated meiotic apparatus, *Biol. Bull.*, 166, 357, 1984.

346. Mahoney, C. W. and Azzi, A., The synthesis of fluorescent chlorotriazinylaminofluorescein–concanavalin A and its use as a glycoprotein stain on sodium dodecyl sulfate/polyacrylamide gels, *Biochem. J.*, 243, 569, 1987.

347. Onoue, K., Yagi, Y., and Pressmann, D., Immunoadsorbent with high capacity, *Immunochemistry*, 2, 181, 1965.

348. Likhite, V. and Sehon, A. H., Protein–protein conjugation, in *Methods in Immunology and Immunochemistry*, Williams, C. A. and Chase, M. W. (eds.), Vol. 1, Academic Press, New York, 1967, p. 150.

349. Dzielendziak, A., Butler, J., Hoey, B. M., Lea, J. S., and Ward, T. H., Comparison of the structural and cytotoxic activity of novel 2,5-bis(carboethoxyamino)-3,6-diaziridinyl-1,4-benzoquinone analogues, *Cancer Res.*, 50, 2003, 1990.

350. Fearnley, C. and Speakman, J. B., Cross-linkage formation in keratin, *Nat. (London)*, 166, 743, 1950.

351. Schönfeld, H. J. and Foulaki, K., Preoperative isolation of a reversible protein–protein crosslink generated in 50 S subunits of *Escherichia coli* ribosomes and identification of its components, *Anal. Biochem.*, 164, 23, 1987.

352. Loeber, R., Rajesh, M., Fang, Q., Pegg, A. E., and Tretyakova, N., Cross-linking of the human DNA repair protein *O*6-alkylguanine DNA alkyltransferase to DNA in the presence of 1,2,3,4-diepoxybutane, *Chem. Res. Toxicol.*, 19, 645, 2006.

353. Loecken, E. M. and Guengerich, F. P., Reactions of glyceraldehyde 3-phosphate dehydrogenase sulfhydryl groups with bis-electrophiles produce DNA–protein cross-links but not mutations, *Chem. Res. Toxicol.*, 21, 453, 2008.

354. Sköld, S.-E., Chemical crosslinking of elongation factor G to the 23SRNA in 70S ribosomes from *Escherichia coli*, *Nucleic Acid Res.*, 11, 4923, 1983.

355. Park, S., Anderson, C., Loeber, R., Seetharaman, M., Jones, R., and Tretyakova, N., Interstrand and intrastrand DNA–DNA cross-linking by 1,2,3,4-diepoxybutane: Role of stereochemistry, *J. Am. Chem. Soc.*, 127, 14355, 2005.

356. Goggin, M., Swenberg, J. A., Walker, V. E., and Tretyakova, N., Molecular dosimetry of 1,2,3,4-diepoxybutane-induced DNA–DNA cross-links in B6C3F1 mice and F344 rats exposed to 1,3-butadiene by inhalation, *Cancer Res.*, 69, 2479, 2009.

357. Tretyakova, N., Livshits, A., Park, S., Bisht, B., and Goggin, M., Structural elucidation of a novel DNA–DNA cross-link of 1,2,3,4-diepoxybutane, *Chem. Res. Toxicol.*, 20, 284, 2007.

358. Yunes, M. J., Charnecki, S. E., Marden, J. J., and Millard, J. T., 1,2,5,6-Diepoxyhexane and 1,2,7,8-diepoxyoctane cross-link duplex DNA at 5′-GNC sequences. *Chem. Res. Toxicol.*, 9, 994, 1996.

359. Millard, J. T., Katz, J. L., Goda, J., Frederick, E. D., Pierce, S. E., Speed, T. J., and Thamattoor, D. M., DNA interstrand cross-linking by a mycotoxic diepoxide, *Biochimie*, 86, 419, 2004.

360. Sundberg, L. and Porath, J., Preparation of adsorbents for biospecific affinity chromatography. I. Attachment of group-containing ligands to insoluble polymers by means of bifunctional oxiraneas, *J. Chromatogr.*, 90, 87, 1974.

361. Porath, J., General methods and coupling procedures, *Methods Enzymol.*, 34, 13, 1974.

362. Durairaj, R. B., *Resorcinol: Chemistry, Technology and Applications*, Springer-Verlag, Berlin, Germany, 2005, p. 427.

363. Tang, Z. and Yue, Y., Crosslinkage of collagen by polyglycidyl ethers, *ASAIO J.*, 41, 72, 1995.

364. Leach, J. B., Wolinsky, J. B., Stone, P. J., and Wong, J. Y., Crosslinked α-elastin biomaterials: Towards a processable elastin mimetic scaffold, *Acta Biomater.*, 1, 155, 2005.

365. Imamura, E., Sawatani, O., Koyanagi, H., Noishiki, Y., and Teruomiyata, N., Epoxy compounds as a new cross-linking agent for porcine aortic leaflets: Subcutaneous implant studies in rats, *J. Card. Surg.*, 4, 50, 1989.

366. Porath, J. and Axén, R., Immobilization of enzymes to agar, agarose, and sephadex supports, *Methods Enzymol.*, 44, 19, 1976.

367. Sereikait, J., Bassus, D., Bobnis, R., Dienys, G., Bumelien, Z., and Bumelis, V.-A., Divinyl sulfone as a crosslinking reagent for oligomeric proteins, *Russ. J. Bioorg. Chem.*, 29, 227, 2003.

368. Houen, G. and Jensen, O. M., Conjugation to preactivated proteins using divinylsulfone and iodoacetic acid, *J. Immunol. Methods*, 181, 187, 1995.

369. Gotte, G. and Libonati, M., Oligomerization of ribonuclease A: Two novel three-dimensional domain-swapped tetramers, *J. Biol. Chem.*, 279, 36670, 2004.

370. Dickinson, B. C., Varadan, R., and Fushman, D., Effects of cyclization on conformational dynamics and binding properties of Lys48-linked di-ubiquitin, *Protein Sci.*, 16, 369, 2007.

371. Collins, M. N. and Birkinshaw, C., Physical properties of crosslinked hyaluronic acid hydrogels, *J. Mater. Sci. Mater. Med.*, 19, 3335, 2008.

372. Pathak, T. and Bhattacharya, R., A vinyl sulfone-modified carbohydrate mediated new route to aminosugars and branched-chain sugars, *Carbohydr. Res.*, 343, 1980, 2008.

373. Lewis, A., Tang, Y., Brocchini, S., Choi, J.-W., and Godwin, A., Poly(2-methacryloyloxyethyl phosphorylcholine) for protein conjugation, *Bioconjug. Chem.*, 19, 2144, 2008.

374. Carraway, K. L. and Koshland, D. E., Jr., Carbodiimide modification of proteins, *Methods Enzymol.*, 25, 616, 1972.

375. Husain, S. S., Ferguson, J. B., and Fruton, J. S., Bifunctional inhibitor of pepsin, *Proc. Natl. Acad. Sci. U. S. A.*, 68, 2765, 1971.

376. Lundblad, R. L. and Stein, W. H., On the reaction of diazoacetyl compounds with pepsin, *J. Biol. Chem.*, 244, 154, 1969.

377. Zahn, H. and Waschka, O., Reaction of bisdiazohexane with amino acids, mercaptans, wool keratin, and silk fibroin, *Makromol. Chem.*, 18, 201, 1956.

378. Gold, P., Jonson, J., and Freedman, S. O., The effect of the sequence of erythrocyte sensitization on the bis-diazotized benzidine hemagglutination reaction, *J. Allergy*, 37, 311, 1966.

379. Stavitsky, A. B. and Arquilla, E. R., Studies of proteins and antibodies by specific hemagglutination and hemolysis of protein-conjugated erythrocytes, *Int. Arch. Allergy Appl. Immunol.*, 13, 1, 1958.

380. Riordan, J. F. and Vallee, B. L., Diazonium salts as specific reagents and probes of protein conformation, *Methods Enzymol.*, 25, 521, 1972.

381. Howard, A. N. and Wild, F., A two-stage method of cross-linking proteins suitable for in serological techniques, *Br. J. Exp. Pathol.*, 38, 640, 1957.

382. Olovnikov, A. M., Sensitization of erythrocytes by polycondensed protein of immune serum and their use for determining antigen content, *Immunochemistry*, 4, 77, 1967.

383. Arquilla, E. R. and Stavitsky, A. B., The production and identification of antibodies to insulin and their use in assaying insulin, *J. Clin. Invest.*, 35, 458, 1956.

384. Pressman, D., Campbell, D. H., and Pauling, L., The agglutination of intact azo-erythrocytes by antisera homologous to the attached groups, *J. Immunol.*, 44, 101, 1942.

385. Borek, F., A new two-stage method for cross-linking proteins, *Nat. (London)*, 191, 1293, 1961.

386. Kalnins, V. I., Stich, H. F., and Yoha, D. S., Electron microscopic localization of virus-associated antigens in human amnion cells (AV-3) infected with human adenovirus, type 12, *Virology*, 28, 751, 1966.

387. Hermanson, G. T., *Bioconjugate Techniques*, Academic Press, New York, 2008, 773–775.

388. Adrian, T. E., Production of antisera using peptide conjugates, in *Methods in Molecular Biology: Neuropeptide Protocols*, Irvine, G. B. and Williams, C. H. (eds.), Humana Press, Totowa, New Jersey, Vol. 73, 239, 1996.

389. DeCarvalho, S., Lewis, A. J., Rand, H. J., and Uhrick, J. R., Immunochromatographic partition of soluble antigens on columns of insoluble diazo-gamma-globulins, *Nat. (London)*, 204, 265, 1964.

390. Koch, N. and Haustein, D., Association of surface IgM with two membrane proteins on murine B lymphocytes detected by chemical crosslinking, *Mol. Immunol.*, 20, 33, 1983.

391. Spycher, M. O. and Nydegger, U. E., Part of the activating cross-linked immunoglobulin G is internalized by human platelets to sites not accessible for enzymatic digestion, *Blood*, 67, 12, 1986.

392. Bar-Eli, A. and Katchalski, E., Preparation and properties of water-insoluble derivatives of trypsin, *J. Biol. Chem.*, 238, 1690, 1963.

393. Riesel, E. and Katchalski, E., Preparation and properties of water-insoluble derivatives of urease, *J. Biol. Chem.*, 239, 1521, 1964.

394. Ishizaka, K. and Ishizaka, T., Biologic activity of aggregated γ-globulin. II. A study of various methods for aggregation and species differences, *J. Immunol.*, 85, 163, 1960.

395. Anderer, F. A. and Schlumberger, H. D., Antigenic properties of proteins cross-linked by multidiazonium compounds, *Immunochemistry*, 6, 1, 1969.

396. Soni, V., Jain, S. K., and Kohli, D. V., Potential of transferrin and transferrin conjugates of liposomes in drug delivery and targeting, *Am. J. Drug Deliv.*, 3, 155, 2005.

397. Jaffe, C. L., Lis, H., and Sharon, N., New cleavable photoreactive heterobifunctional cross-linking reagents for studying membrane organization, *Biochemistry*, 19, 4423, 1980.

398. Wagner, R. and Garrett, R. A., A new RNA–RNA cross-linking reagent and its application to ribosomal 5S RNA, *Nucleic Acid Res.*, 5, 4065, 1978.

399. Chiam, C. L. and Wagner, R., Composition of the *Escherichia coli* 70S ribosomal interface: A cross-linking study, *Biochemistry*, 22, 1193, 1983.

400. Melançon, P., Boileau, G., and Brakier-Gingras, L., Cross-linking of streptomycin to the 30S subunit of *Escherichia coli* with phenyldiglyoxal, *Biochemistry*, 23, 6697, 1984.

401. Scheibe, U. and Wagner. R., Identification of neighbouring proteins by cross-linking of intact 70S ribosomes from *Escherichia coli*, *Biochim. Biophys. Acta*, 869, 1, 1986.

402. Zhang, Q., Crosland, E., and Fabris, D., Nested Arg-specific bifunctional crosslinkers for MS-based structural analysis of proteins and protein assemblies, *Anal. Chim. Acta*, 627, 117, 2008.

403. Blotnick, E. and Muhlrad, A., Effect of nucleotides and actin on the intramolecular cross-linking of myosin subfragment-1, *Biochemistry*, 33, 6867, 1994.

404. Zhang, Q., Yu, E. T., Kellersberger, K. A., Crosland, E., and Fabris, D., Toward building a database of bifunctional probes for the MS3D investigation of nucleic acids structures, *J. Am. Soc. Mass Spectrom.*, 17, 1570, 2006.

405. Roche, P. A., Jensen, P. E. H., and Pizzo, S. V., Intersubunit cross-linking by cis-dichlorodiammineplatinum (II) stabilizes an α_2-macroglobulin "nascent" state: Evidence that thiol ester bond cleavage correlates with receptor recognition site exposure, *Biochemistry*, 27, 759, 1988.

406. Pizzo, S. V., Roche, P. A., Feldman, S. R., and Gonias, S. L., Further characterization of platinum-reactive component of the α_2-macroglobulin-receptor recognition site, *Biochem. J.*, 238, 217, 1986.

407. Goalas, S. L., Oakley, A. C., Walther, P. J., and Pizzo, S. V., Effect of diethyldithiocarbamate and nine other nucleophiles on the intersubunit protein cross-linking and inactivation of purified human α_2-macroglobulin by cis-diamminedichloroplatinum (II), *Cancer Res.*, 44, 5764, 1984.

408. Pizzo, S. V., Swaim, M. W., Roche, P. A., and Gonias, S. L., Selectivity and stereospecificity of the reactions of dichlorodiammineplatinum(II) with three purified plasma proteins, *J. Inorg. Biochem.*, 33, 67, 1988.

409. Gonias, S. L., Swaim, M. W., Massey, M. F., and Pizzo, S. V., Cis-dichlorodiammineplatinum (II) as a selective modifier of the oxidation-sensitive reactive-center methionine in alpha 1-antitrypsin, *J. Biol. Chem.*, 263, 393, 1988.

410. Kozulic, B., Barbaric, S., Ries, B., and Milder, P., Study of the carbohydrate part of yeast acid phosphatase, *Biochem. Biophys. Res. Commun.*, 122, 1083, 1984.

411. Gildersleeve, J. C., Oyelaran, O., Simpson, J. T., and Allred, B., Improved procedure for direct coupling of carbohydrates to proteins via reductive amination, *Bioconjug. Chem.*, 19, 1485, 2008.

412. Das, M. and Fox, F., Chemical cross-linking in biology, *Annu. Rev. Biophys. Bioeng.*, 8, 165, 1979.

413. Mikkelsen, R. B. and Wallach, D. F. H., Photoactivated cross-linking of protein within the erythrocyte membrane core, *J. Biol. Chem.*, 251, 7413, 1976.

414. Guire, P., Stepwise thermophotochemical cross-linking agents for enzyme stabilization and immobilization, *Fed. Proc.*, 35, 1632, 1976.

415. Elia, G., Biotinylation reagents for the study of cell surface proteins, *Proteomics*, 8, 4012, 2008.

416. Sharon, N., Lectins: Past, present and future, *Biochem. Soc. Trans.*, 36, 1457, 2008.

417. Cao, Y. and Suresh, M. R., Bispecific antibodies as novel bioconjugates, *Bioconjug. Chem.*, 9, 635, 1998.

418. Wood, W. G., Immunoassays & co.: Past, present, future?—A review and outlook from personal experience and involvement over the past 35 years, *Clin. Lab.*, 54, 423, 2008.

419. Wingren, C. and Borrebaeck, C. A., Antibody-based microarrays, *Methods Mol. Biol.*, 509, 57, 2009.

420. Nitecki, D. E., Woods, V., and Goodman, J. W., Crosslinking of antibody molecules by bifunctional antigens, in *Protein Cross-Linking. Biochemical and Molecular Aspects*, Friedman, M. (ed.), Plenum Press, New York, 1976, p. 139.

421. Rifai, A., Experimental models for IgA-associated nephritis, *Kidney Int.*, 31, 1, 1987.

422. Rajski, S. R. and Williams, R. M., DNA cross-linking agents as antitumor drugs, *Chem. Rev.*, 98, 2723, 1998.

423. Noll, D. M., Mason, T. M., and Miller, P. S., Formation and repair of interstrand cross-links in DNA, *Chem. Rev.*, 106, 277, 2006.

424. Hofr, C., Farrell, N., and Brabec, V., Thermodynamic properties of duplex DNA containing a site-specific d(GpG) intrastrand crosslink formed by an antitumor dinuclear platinum complex, *Nucleic Acids Res.*, 29, 2034, 2001.

425. Qu, Y., Tran, M. C., and Farrell, N. P., Structural consequences of a $3' \rightarrow 3'$ DNA interstrand cross-link by a trinuclear platinum complex: Unique formation of two such cross-links in a 10-mer duplex, *J. Biol. Inorg. Chem.*, 14, 969, 2009.

426. Reishus, J. W. and Martin, D. S., cis-Dichlorodiammineplatinum(II). Acid hydrolysis and isotopic exchange of the chloride ligands, *J. Am. Chem. Soc.*, 83, 2457, 1961.

427. Siddik, Z. H., Cisplatin: Mode of cytotoxic action and molecular basis of resistance, *Oncogene*, 22, 7265, 2003.

428. Kartalou, M. and Essigmann, J. M., Recognition of cisplatin adducts by cellular proteins, *Mutat. Res.*, 478, 1, 2001.

429. Ruhayel, R. A., Moniodis, J. J., Yang, X., Kasparkova, J., Brabec, V., Berners-Price, S. J., and Farrell, N. P., Factors affecting DNA–DNA interstrand cross-links in the antiparallel 3'-3' sense. A comparison with the 5'-5' directional isomer, *Chemistry*, 21, 9365, 2009.

430. Zerzankova, L., Suchankova, T., Vrana, O., Farrell, N. P., Brabec, V., and Kasparkova, J., Conformation and recognition of DNA modified by a new antitumor dinuclear Pt(II) complex resistant to decomposition by sulfur nucleophiles, *Biochem. Pharmacol.*, 15, 112, 2010.

431. Palmer, B. D., Lee, H. H., Johnson, P., Baguley, B. C., Wickham, G., Wakelin, L. P., McFadyen, W. D., and Denny, W. A., DNA-directed alkylating agents. 2. Synthesis and biological activity of platinum complexes linked to 9-anilinoacridine, *J. Med. Chem.*, 33, 3008, 1990.

432. Weng, X., Ren, L., Weng, L., Huang, J., Zhu, S., Zhou, X., and Weng, L., Synthesis and biological studies of inducible DNA cross-linking agents, *Angew. Chem. Int. Ed.*, 46, 8020, 2007.

433. Rajendran, A., Magesh, C. J., and Perumal, P. T., DNA–DNA cross-linking mediated by bifunctional [SalenAl^{III}]+ complex, *Biochim. Biophys. Acta*, 1780, 282, 2008.

434. Tsou, T.-C., Lin, R.-J., and Yang, J.-L., Mutational spectrum induced by chromium(III) in shuttle vectors replicated in human cells: Relationship to Cr(III)-DNA interactions, *Chem. Res. Toxicol.*, 10, 962, 1997.

435. Chen, D., Milacic, V., Frezza, M., and Dou, Q. P., Metal complexes, their cellular targets and potential for cancer therapy, *Curr. Pharm. Des.*, 15, 777, 2009.

436. Armstrong, R. W., Salvati, M. E., and Nguyen, M., Novel interstrand cross-links induced by the antitumor antibiotic carzinophilin/azinomycin B, *J. Am. Chem. Soc.*, 114, 3144, 1992.

437. Casely-Hayford, M. A., Pors, K., James, C. H., Patterson, L. H., Hartley, J. A., and Searcey, M., Design and synthesis of a DNA-crosslinking azinomycin analogue, *Org. Biomol. Chem.*, 3, 3585, 2005.

438. Hodgkinson, T. J. and Shipman, M., Chemical synthesis and mode of action of azinomycins, *Tetrahedron*, 57, 4467, 2001.

439. LePla, R. C., Landreau, C. A. S., Shipman, M., Hartley, J. A., and Jones, G. D. D., Azinomycin inspired bisepoxides: Influence of linker structure on in vitro cytotoxicity and DNA interstrand cross-linking, *Bioorg. Med. Chem. Lett.*, 15, 2861, 2005.

440. Finerty, M. J., Bingham, J. P., Hartley, J. A., and Shipman, M., Azinomycin bisepoxides containing rigid aromatic linkers: Synthesis, cytotoxicity and DNA interstrand cross-linking activity, *Tetrahedron Lett.*, 50, 3648, 2009.

441. Hurley, L. H., Reck, T., Thurston, D. E., Langley, D. R., Holden, K. G., Hertzberg, R. P., Hoover, J. R. E. et al., Pyrrolo[1,4]benzodiazepine antitumor antibiotics: Relationship of DNA alkylation and sequence specificity to the biological activity of natural and synthetic compounds, *Chem. Res. Toxicol.*, 1, 258, 1988.

442. Kamal, A., Rao, M. V., Laxman, N., Ramesh, G., and Reddy, G. S. K., Recent developments in the design, synthesis and structure–activity relationship studies of pyrrolo[2,1-c][1,4]benzodiazepines as DNA interactive antitumour antibiotics, *Curr. Med. Chem. Anticancer Agents*, 2, 215, 2002.

443. Cipolla, L., Araújo, A. C., Airoldi, C., and Bini, D., Pyrrolo[2,1-c][1,4]benzodiazepine as a scaffold for the design and synthesis of anti-tumour drugs, *Anticancer Agents Med. Chem.*, 9, 1, 2009.

444. Farmer, J. D., Rudnicki, S. M., and Suggs, J. W., Synthesis and DNA crosslinking ability of a dimeric anthramycin analog, *Tetrahedron Lett.*, 29, 5105, 1988.

445. Thurston, D. E., Bose, D. S., Thompson, A. S., Howard, P. W., Leoni, A., Croker, S. J., Jenkins, T. C., Neidle, S., Hartley, J. A., and Hurley, L. H., Synthesis of sequence-selective C8-linked pyrrolo[2,1-c][1,4]benzodiazepine DNA interstrand cross-linking agents, *J. Org. Chem.*, 61, 8141, 1996.

446. Smellie, M., Bose, D. S., Thompson, A. S., Jenkins, T. C., Hartley, J. A., and Thurston, D. E., Sequence-selective recognition of duplex DNA through covalent interstrand cross-linking: Kinetic and molecular modeling studies with pyrrolobenzodiazepine dimers, *Biochemistry*, 42, 8232, 2003.

447. Kamal, A., Rajender, Reddy, D. R., Reddy, M. K., Balakishan, G., Shaik, T. B., Chourasia, M., and Sastry, G. N., Remarkable enhancement in the DNA-binding ability of C2-fluoro substituted pyrrolo[2,1-c][1,4]benzodiazepines and their anticancer potential, *Bioorg. Med. Chem.*, 17, 1557, 2009.

448. Gregson, S. J., Howard, P. W., Hartley, J. A., Brooks, N. A., Adams, L. J., Jenkins, T. C., Kelland, L. R., and Thurston, D. E., Design, synthesis and evaluation of a novel pyrrolobenzodiazepine DNA-interactive agent with highly efficient crosslinking ability and potent cytotoxicity, *J. Med. Chem.*, 44, 737, 2001.

449. Hartley, J. A., Spanswick, V. J., Brooks, N., Clingen, P. H., McHugh, P. J., Hochhauser, D., Pedley, R. B. et al., SJG-136 (NSC 694501), a novel rationally designed DNA minor groove interstrand cross-linking agent with potent and broad spectrum antitumor activity. Part 1: cellular pharmacology, *In vitro* and initial *In vivo* antitumor activity, *Cancer Res.*, 64, 6693, 2004.

450. Kamal, A., Reddy, P. S., Reddy, D. R., Laxman, E., and Murthy, Y. L., Synthesis of fluorinated analogues of SJG-136 and their DNA-binding potential, *Bioorg. Med. Chem. Lett.*, 14, 5699, 2004.

451. Gregson, S. J., Howard, P. W., Corcoran, K. E., Jenkins, T. C., Kelland, L. R., and Thurston, D. E., Synthesis of the first example of a C2-C3/C2'-C3'-*endo* unsaturated pyrrolo[2,1-c][1,4]benzodiazepine dimer, *Bioorg. Med. Chem. Lett.*, 11, 2859, 2001.

452. Tiberghien, A. C., Evans, D. A., Kiakos, K., Martin, C. R. H., Hartley, J. A., Thurstona, D. E., and Howard, P. W., An asymmetric C8/C80-tripyrrole-linked sequence-selective pyrrolo[2,1-c][1,4]benzo-diazepine (PBD) dimer DNA interstrand cross-linking agent spanning 11 DNA base pairs, *Bioorg. Med. Chem. Lett.*, 18, 2073, 2008.

453. Kumar, R. and Lown, J. W., Design, synthesis and in vitro cytotoxic studies of novel bis-pyrrolo[2,1][1,4] benzodiazepine-pyrrole and imidazole polyamide conjugates, *Eur. J. Med. Chem.*, 40, 641, 2005.

454. Kamal, A., Prabhakar, S., Shankaraiah, N., Reddy, C. R., and Reddy, P. V., Synthesis of C8–C8/C2–C8-linked triazolo pyrrolobenzodiazepine dimers by employing 'click' chemistry and their DNA-binding affinity, *Tetrahedron Lett.*, 49, 3620, 2008.

455. Kamal, A., Reddy, P. S., Reddy, D. R., and Laxman, E., DNA binding potential and cytotoxicity of newly designed pyrrolobenzodiazepine dimers linked through a piperazine side-armed-alkane spacer, *Bioorg. Med. Chem.*, 14, 385, 2006.

456. Kamal, A., Ramu, R., Tekumalla, V., Khanna, G. B. R., Barkume, M. S., Juvekar, A. S., and Zingde, S. M., Synthesis, DNA binding, and cytotoxicity studies of pyrrolo[2,1-c][1,4]benzodiazepine-anthraquinone conjugates, *Bioorg. Med. Chem.*, 15, 6868, 2007.

457. Reddy, B. S. P., Damayanthi, Y., and Lown, J. W., Design and efficient synthesis of novel DNA inter-strand cross-linking agents: C2-linked pyrrolo[2,1-c][1,4]benzodiazepine dimers, *Synlett*, 1112, 1999.

458. Reddy, B. S. P., Damayanthi, Y., Reddy, B. S. N., and Lown, J. W., Design, synthesis and in vitro cyto-toxicity studies of novel pyrrolo[2,1-c][1,4]benzodiazepine (PBD)—Polymade conjugates and 2,2′-PBD dimers, *Anticancer Drug Des.*, 15, 225, 2000.

459. Gregson, S. J., Howard, P. W., and Thurston, D. E., Synthesis of the first examples of A-C8/C-C2 amide-linked pyrrolo[2,1-c][1,4]benzodiazepine dimers, *Bioorg. Med. Chem. Lett.*, 13, 2277, 2003.

460. Kamal, A., Ramulu, P., Srinivas, O., and Ramesh, G., Synthesis and DNA-binding affinity of A-C8/C-C2 alkoxyamido-linked pyrrolo[2,1-c][1,4]benzodiazepine dimers, *Bioorg. Med. Chem. Lett.*, 13, 3955, 2003.

461. Kashiwazaki, G., Bando, T., and Sugiyama, H., Sequence-specific alkylation of DNA by pyrrole–imidazole polyamides through cooperative interaction, *Nucleic Acids Symp. Ser.*, 52, 365, 2008.

462. Reddy, B. S., Sharma, S. K., and Lown, J. W., Recent developments in sequence selective minor groove DNA effectors, *Curr. Med. Chem.*, 8, 475, 2001.

463. Boger, D. L., Searcey, M., Tse, W. C., and Jin, Q., Bifunctional alkylating agents derived from duocar-mycin SA: Potent antitumor activity with altered sequence selectivity, *Bioorg. Med. Chem. Lett.*, 10, 495, 2000.

464. Bando, T. and Sugiyama, H., Synthesis and biological properties of sequence-specific DNA-alkylating pyrrole–imidazole polyamides, *Acc. Chem. Res.*, 39, 935, 2006.

465. Bando, T., Iida, H., Saito, I., and Sugiyama, H., The synthesis of pyrrole(Py)/imidazole(lm) polyamide CPI conjugates which possess DNA interstrand cross-linking activity, *Nucleic Acids Symp. Ser.*, 44, 45, 2000.

466. Bando, T., Narita, A., Saito, I., and Sugiyama, H., Highly efficient sequence-specific DNA interstrand cross-linking by pyrrole/imidazole CPI conjugates. *J. Am. Chem. Soc.*, 125, 3471, 2003.

467. Bando, T., Iida, H., Saito, I., and Sugiyama, H., Sequence-specific DNA interstrand cross-linking by imidazole-pyrrole CPI conjugate, *J. Am. Chem. Soc.*, 123, 5158, 2001.

468. Woynarowski, J. M., Targeting critical regions in genomic DNA with AT-specific anticancer drugs, *Biochim. Biophys. Acta*, 1587, 300, 2002.

469. Woynarowski, J. M., Trevino, A. V., Rodriguez, K. A., Hardies, S. C., and Benham, C. J., AT-rich Islands in genomic DNA as a novel target for AT-specific DNA-reactive antitumor drugs, *J. Biol. Chem.*, 276, 40555, 2001.

470. Thompson, A. S., Fan, J.-Y., Sun, D., Hansen, M., and Hurley, L. H., Determination of the structural role of the internal guanine-cytosine base pair in recognition of a seven-base-pair sequence cross-linked by bizelesin, *Biochemistry*, 34, 11005, 1995.

471. Herzig, M. C. S., Rodriguez, K. A., Trevino, A. V., Dziegielewski, J., Arnett, B., Hurley, L. H., and Woynarowski, J. M., The genome factor in region-specific DNA damage: The DNA-reactive drug U-78779 prefers mixed A/T-G/C sequences at the nucleotide level but is region-specific for long pure AT Islands at the genomic level, *Biochemistry*, 41, 1545, 2002.

472. Fukuda, Y., Furuta, H., Kusama, Y., Ebisu, H., Oomori, Y., and Terashima, S., The novel cyclopropapyrroloindole(CPI) bisalkylators bearing methoxycarbonyl and trifluoromethyl groups, *Bioorg. Med. Chem. Lett.*, 8, 1387, 1998.

473. Fukuda, Y., Seto, S., Furuta, H., Ebisu, H., and Oomori, Y., The novel cyclopropapyrroloindole (CPI) bisal-kylators bearing 3,3′-(1,4-phenylene)diacryloyl group as a linker, *Bioorg. Med. Chem. Lett.*, 8, 2003, 1998.

474. Jia, G. and Lown, J. W., Design, synthesis and cytotoxicity evaluation of 1-chloromethyl-5-hydroxy-1,2-dihydro-3H-benz[e]indole (seco-CBI) dimers, *Bioorg. Med. Chem.*, 8, 1607, 2000.

475. Kumar, R. and Lown, J. W., Synthesis and cytotoxicity evaluation of novel C7–C7, C7–N3 and N3–N3 dimers of 1-chloromethyl-5-hydroxy-1,2-dihydro-3*H*benzo[*e*]indole (*seco*-CBI) with pyrrole and imidazole polyamide conjugates, *Org. Biomol. Chem.*, 1, 2630, 2003.

476. Ward, T. H., Danson, S., McGown, A. T., Ranson, M., Coe, N. A., Jayson, G. C., Cummings, J., Hargreaves, R. H., and Butler, J., Preclinical evaluation of the pharmacodynamic properties of 2,5-diaziridinyl-3-hydroxymethyl-6-methyl-1,4-benzoquinone, *Clin. Cancer Res.*, 11, 2695, 2005.

477. Hargreaves, R. H. J., O'Hare, C. C., Hartley, J. A., Ross, D., and Butler, J., Cross-linking and sequence-specific alkylation of DNA by aziridinylquinones. 3. Effects of alkyl substituents, *J. Med. Chem.*, 42, 2245, 1999.

478. Butler, J., Hoey, B. M., and Ward, T. H., The alkylation of DNA in vitro by 2,5-bis(2-hydroxyethylamino)-3,6-diaziridinyl-1,4-benzoquinone (BZQ), *Biochem. Pharmacol.*, 38, 923, 1989.

479. Hartley, J. A., Berardini, M., Ponti, M., Gibson, N. W., Thompson, A. S., Thurston, D. E., Hoey, B. M., and Butler, J., DNA cross-linking and sequence selectivity of aziridinylbenzoquinones: A unique reaction at 5'-GC-3' sequences with 2,5-diaziridinyl-1,4-benzoquinone upon reduction, *Biochemistry*, 30, 11719, 1991.

480. Mayalarp, S. P., Hargreaves, R. H. J., Butler, J., O'Hare, C. C., and Hartley, J. A., Cross-linking and sequence specific alkylation of DNA by aziridinylquinones. 1. Quinone methides, *J. Med. Chem.*, 39, 531, 1996.

481. Danson, S., Ward, T. H., Butler, J., and Ranson, M., DT-diaphorase: A target for new anticancer drugs, *Cancer Treat. Rev.*, 30, 437, 2004.

482. Berardini, M. D., Souhami, R. L., Lee, C.-S., Gibson, N. W., Butler, J., and Hartley, J. A., Two structurally related diaziridinylbenzoquinones preferentially cross-link DNA at different sites upon reduction with DT-diaphorase, *Biochemistry*, 32, 3306, 1993.

483. Hargreaves, R. H., Mayalarp, S. P., Butler, J., McAdam, S. R., O'Hare, C. C., and Hartley, J. A., Cross-linking and sequence specific alkylation of DNA by aziridinyl quinones. 2. Structure requirements for sequence selectivity, *J. Med. Chem.*, 40, 357, 1997.

484. Di Francesco, A. M., Hargreaves, R. H., Wallace, T. W., Mayalarp, S. P., Hazrati, A., Hartley, J. A., and Butler, J., The abnormal cytotoxicities of 2,5-diaziridinyl-1,4-benzoquinone-3-phenyl esters, *Anticancer Drug Des.*, 15, 347, 2000.

485. Reed, M. W., Wald, A., and Meyer, R. B., Triplex-directed interstrand cross-linking by aziridinylquinone-oligonucleotide complexes, *J. Am. Chem. Soc.*, 120, 9729, 1998.

486. Di Francesco, A. M., Mayalarp, S. P., Kim, S., Butler, J., and Lee, M., Synthesis and biological evaluation of novel diaziridinylquinone-acridine conjugates, *Anticancer Drugs*, 14, 601, 2003.

487. Tomasz, M. and Palom, Y., The mitomycin bioreductive antitumor agents: Cross-linking and alkylation of DNA as the molecular basis of their activity, *Phamacol. Ther.*, 76, 73, 1997.

488. Na, Y., Li, V.-S., Nakanishi, Y., Bastow, K. F., and Kohn, H., Synthesis, DNA cross-linking activity, and cytotoxicity of dimeric mitomycins, *J. Med. Chem.*, 44, 3453, 2001.

489. Paz, M. M., Kumar, G. S., Glover, M., Waring, M. J., and Tomasz, M., Mitomycin dimers: Polyfunctional cross-linkers of DNA, *J. Med. Chem.*, 47, 3308, 2004.

490. Lee, S. H. and Kohn, H., 7-*N*,7'-*N*'-(1″,2″-Dithianyl-3″,6″-dimethylenyl)bismitomycin C: Synthesis and nucleophilic activation of a dimeric mitomycin, *Org. Biomol. Chem.*, 3, 471, 2005.

491. Lee, S. H., Disulfide and multisulfide antitumor agents and their modes of action, *Arch. Pharm. Res.*, 32, 299, 2009.

492. Lee, S. H. and Kohn, H., Nucleophilic activation of a tetra-substituted mitomycin cyclic bis-disulfide, *Chem. Pharm. Bull.*, 57, 149, 2009.

493. Henderson, N. D., Lacy, S. M., O'Hare, C. C., Hartley, J. A., McClean, S., Wakelin, L. P., Kelland, L. R., and Robins, D. J., Synthesis of new bifunctional compounds which selectively alkylate guanines in DNA, *Anticancer Drug Des.*, 13, 749, 1998.

494. Anderson, F., O'Hare, C., Hartley, J., and Robins, D., Synthesis of new homochiral bispyrrolidines as potential DNA cross-linking antitumour agents, *Anticancer Drug Design*, 15, 119, 2000.

495. Prakash, A. S., Valu, K. K., Wakelin, L. P., Woodgate, P. D., and Denny, W. A., Synthesis and anti-tumour activity of the spatially-separated mustard bis-N,N'-[3-(N-(2-chloroethyl)-N-ethyl)amino-5-[N,N-dimethylamino)methyl]-aminophenyl]-1,4-benzenedicarboxamide, which alkylates DNA exclusively at adenines in the minor groove, *Anticancer Drug Des.*, 6, 195, 1991.

496. Atwell, G. J., Yaghi, B. M., Turner, P. R., Boyd, M., O'Connor, C. J., Ferguson, L. R., Baguley, B. C., and Denny, W. A., Synthesis, DNA interactions and biological activity of DNA minor groove targeted polybenzamide-linked nitrogen mustards, *Bioorg. Med. Chem.*, 3, 679, 1995.

497. Turner, P. R., Ferguson, L. R., and Denny, W. A., Polybenzamide mustards: Structure–activity relationships for DNA sequence-specific alkylation, *Anticancer Drug Des.*, 14, 61, 1999.

498. Turner, P. R., Denny, W. A., and Ferguson, L. R., Role of DNA minor groove alkylation and DNA cross-linking in the cytotoxicity of polybenzamide mustards, *Anticancer Drug Des.*, 15, 245, 2000.

499. Denny, W. A., DNA minor groove alkylating agents, *Curr. Med. Chem.*, 8, 533, 2001.

500. Abdul Majid, A. M. S., Smythe, G., Denny, W. A., and Wakelin, L. P. G., Structure of the d(CGCGAATTCGCG)2 complex of the minor groove binding alkylating agent alkamin studied by mass spectrometry, *Mol. Pharmacol.*, 71, 1165, 2007.

501. Abdul Majid, A. M., Smythe, G., Denny, W. A., and Wakelin, L. P., Mass spectrometry studies of the binding of the minor groove-directed alkylating agent alkamin to AT-tract oligonucleotides, *Chem. Res. Toxicol.*, 22, 146, 2009.

502. Kohn, K. W., Spears, C. L., and Doty, P., Interstrand cross-linking of DNA by nitrogen mustard, *J. Mol. Biol.*, 19, 266, 1966.

503. Loeber, R. L., Michaelson-Richie, E. D., Codreanu, S. G., Liebler, D. C., Campbell, C. R., and Tretyakova, N. Y., Proteomic analysis of DNA–protein cross-linking by antitumor nitrogen mustards, *Chem. Res. Toxicol.*, 22, 1151, 2009.

504. Ojwang, J. O., Grueneberg, D. A., and Loechler, E. L., Synthesis of a duplex oligonucleotide containing a nitrogen mustard interstrand DNA–DNA crosslink, *Cancer Res.*, 49, 6529, 1989.

505. Bauer, G. W. and Povirk, L. F., Specificity and kinetics of interstrand and intrastrand bifunctional alkylation by nitrogen mustards at a G-G-C sequence, *Nucleic Acids Res.*, 25, 1211, 1997.

506. Povirk, L. F. and Shuker, D. E. DNA damage and mutagenesis induced by nitrogen mustards, *Mutat. Res.*, 318, 205, 1994.

507. Romero, R. M., Rojsitthisak, P., and Haworth, I. S., DNA interstrand crosslink formation by mechlorethamine at a cytosine ± cytosine mismatch pair: Kinetics and sequence dependence, *Arch. Biochem. Biophys.*, 386, 143, 2001.

508. Gourdie, T. A., Valu, K. K., Gravatt, G. L., Boritzki, T. J., Baguley, B. C., Wakelin, L. P. G., Wilson, W. R., Woodgate, P. D., and Denny, W. A., DNA-directed alkylating agents. 1. Structure–activity relationships for acridine-linked aniline mustards: Consequences of varying the reactivity of the mustard, *J. Med. Chem.*, 33, 1177, 1990.

509. Ferguson, L. R., Palmer, B. D., and Denny, W. A., Relationships between structure, toxicity and genetic effects in *Salmonella typhimurium* and *Saccharomyces cerevisiae* for substituted aniline mustards, *Mutat. Res.*, 224, 95, 1989.

510. Sunters, A., Springer, C. J., Bagshawe, K. D., Souhami, R. L., and Hartley, J. A., The cytotoxicity, DNA crosslinking ability and DNA sequence selectivity of the aniline mustards melphalan, chlorambucil and 4-[bis(2-chloroethyl)amino] benzoic acid, *Biochem. Pharmacol.*, 44, 59, 1992.

511. Struck, R. F., Davis, R, L., Jr., Berardini, M. D., and Loechler, E. L., DNA guanine-guanine crosslinking sequence specificity of isophosphoramide mustard, the alkylating metabolite of the clinical antitumor agent ifosfamide, *Cancer Chemother. Pharmacol.*, 45, 59, 2000.

512. Rink, S. M. and Hopkins, P. B., A mechlorethamine-induced DNA interstrand cross-link bends duplex DNA, *Biochemistry*, 34, 1439, 1995.

513. Begleiter, A., Leith, M., and Pan, S.-S., Mechanisms for the modulation of alkylating activity by the quinone group in quinone alkylating agents, *Mol. Pharmacol.*, 40, 454, 1991.

514. Fourie, J., Guziec, F., Jr., and Guziec, L., Monterrosa, C., Fiterman, D. J., and Begleiter, A., Structure–activity study with bioreductive benzoquinone alkylating agents: Effects on DT-diaphorase-mediated DNA crosslink and strand break formation in relation to mechanisms of cytotoxicity, *Cancer Chemother. Pharmacol.*, 53, 191, 2004.

515. Fourie, J., Oleschuk, C. J., Guziec, F., Jr., Guziec, L., Fiterman, D. J., Monterrosa, C., and Begleiter, A., The effect of functional groups on reduction and activation of quinone bioreductive agents by DT-diaphorase, *Cancer Themother. Pharmacol.*, 49, 101, 2002.

516. Holley, J. L., Mather, A., Wheelhouse, R. T., Cullis, P. M., Hartley, J. A., Bingham, J. P., and Cohen, G. M., Targeting of tumor cells and DNA by a chlorambucil–spermidine conjugate, *Cancer Res.*, 52, 4190, 1992.

517. Stark, P. A., Thrall, B. D., Meadows, G. G., and Abdel-Monem, M. M., Synthesis and evaluation of novel spermidine derivatives as targeted cancer chemotherapeutic agents, *J. Med. Chem.*, 35, 4264, 1992.

518. Bielawska, A., Bielawski, K., Wolczynski, S., and Anchim, T., Structure–activity studies of novel amidine analogues of chlorambucil: Correlation of cytotoxic activity with DNA-binding affinity and topoisomerase II inhibition, *Arch. Pharm. Pharm. Med. Chem.*, 336, 293, 2003.

519. Bielawska, A., Bielawski, K., and Muszynska, A., Synthesis and biological evaluation of new cyclic amidine analogs of chlorambucil, *IL FARMACO*, 59, 111, 2004.

520. Bielawski, K., Bielawska, A., Sosnowska, K., Miltyk, W., Winnicka, K., and Pałka, J., Novel amidine analogue of melphalan as a specific multifunctional inhibitor of growth and metabolism of human breast cancer cells, *Biochem. Pharm.*, 72, 320, 2006.

521. Bielawska, A., Bielawski, K., and Anchim, T., Amidine analogues of melphalan: Synthesis, cytotoxic activity, and DNA binding properties, *Arch. Pharm. Chem. Life Sci.*, 340, 251, 2007.

522. Bielawski, K., Bielawska, A., and Poplawska, B., Synthesis and cytotoxic activity of novel amidine analogues of bis(2-chloroethyl)amine, *Arch. Pharm. Chem. Life Sci.*, 342, 484, 2009.

523. Kalaycio, M., Bendamustine: A new look at an old drug, *Cancer*, 115, 483, 2009.

524. Gravatt, G. L., Baguley, B. C., Wilson, W. R., and Denny, W. A., DNA-directed alkylating agents. 6. Synthesis and antitumor activity of DNA minor groove-targeted aniline mustard analogues of pibenzimol (Hoechst 33258), *J. Med. Chem.*, 37, 4338, 1994.

525. Smaill, J. B., Fan, J.-Y., Papa, P. V., O'Connor, C. J., and Denny, W. A., Mono- and dysfunctional nitrogen mustard analogues of the DNA minor groove binder pibenzimol. Synthesis, cytotoxicity and interaction with DNA. *Anticancer Drug Des.*, 13, 221, 1998.

526. Ferguson, L. R. and Denny, W. A., Microbial mutagenic effects of the DNA minor groove binder pibenzimol (Hoechst 33258) and a series of mustard analogues, *Mutat. Res.*, 329, 19, 1995.

527. Smaill, J. B., Fan, J.-Y., and Denny, W. A. DNA minor groove targeted alkylating agents based on bis-benzimidazole carriers: Synthesis, cytotoxicity and sequence-specificity of DNA alkylation, *Anticancer Drug Des.*, 13, 857, 1998.

528. Gupta, R., Wang, H., Huang, L., and Lown, J. W., Design, synthesis, DNA sequence preferential alkylation and biological evaluation of N-mustard derivatives of Hoechst 33258 analogues, *Anticancer Drug Des.*, 10, 25, 1995.

529. Wyatt, M. D., Lee, M., and Hartley, J. A., The sequence specificity of alkylation for a series of benzoic acid mustard and imidazole-containing distamycin analogues: The importance of local sequence conformation, *Nucleic Acids Res.*, 25, 2359, 1997.

530. Wyatt, M. D., Lee, M., and Hartley, J. A., Alkylation specificity for a series of distamycin analogues that tether chlorambucil, *Anticancer Drug Des.*, 12, 49, 1997.

531. Wyatt, M. D., Garbiras, B. J., Haskell, M. K., Lee, M., Souhami, R. L., and Hartley, J. A., Structure–activity relationship of a series of nitrogen mustard- and pyrrole-containing minor groove-binding agents related to distamycin, *Anticancer Drug Des.*, 9, 511, 1994.

532. Wyatt, M. D., Lee, M., Garbiras, B. J., Souhami, R. L., and Hartley, J. A., Sequence Specificity of alkylation for a series of nitrogen mustard-containing analogues of distamycin of increasing binding site size: Evidence for increased cytotoxicity with enhanced sequence specificity, *Biochemistry*, 34, 13034, 1995.

533. Brooks, N., Hartley, J. A., Simpson, J. E., Jr., Wright, S. R., Woo, S., Centioni, S., Fontaine, M. D., McIntyre, T. E., and Lee, M., Structure–activity relationship of a series of C-terminus modified aminoalkyl, diaminoalkyl- and anilino-containing analogues of the benzoic acid mustard distamycin derivative tallimustine: Synthesis, DNA binding and cytotoxicity studies, *Bioorg. Med. Chem.*, 5, 1497, 1997.

534. Ciucci, A., Manzini, S., Lombardi, P., and Arcamone, F., Backbone and benzoyl mustard carrying moiety modifies DNA interactions of distamycin analogues, *Nucleic Acids Res.*, 24, 311, 1996.

535. Arcamone, F. M., Animati, F., Barbieri, B., Configliacchi, E., D'Alessio, R., Geroni, C., Giuliani, F. C. et al., Synthesis, DNA-binding properties, and antitumor activity of novel distamycin derivatives, *J. Med. Chem.*, 32, 774, 1989.

536. Xie, G., Gupta, R., and Lown, J. W., Design, synthesis, DNA sequence preferential alkylation and biological evaluation of N-mustard derivatives of distamycin and netropsin analogues, *Anticancer Drug Des.*, 10, 389, 1995.

537. Lee, M., Rhodes, A. L., Wyatt, M. D., Forrow, S., and Hartley, J. A., Design, synthesis, and biological evaluation of DNA sequence and minor groove selective alkylating agents, *Anticancer Drug Des.*, 8, 173, 1993.

538. Lee, M., Rhodes, A. L., Wyatt, M. D., D'Incalci, M., Forrow, S., and Hartley, J. A., In vitro cytotoxicity of GC sequence directed alkylating agents related to distamycin, *J. Med. Chem.*, 36, 863, 1993.

539. Baraldi, P. G., Cozzi, P., Geroni, C., Mongelli, N., Romagnolia, R., and Spalluto, G., Novel benzoyl nitrogen mustard derivatives of pyrazole analogues of distamycin A: Synthesis and antileukemic activity, *Bioorg. Med. Chem.*, 7, 251, 1999.

540. Baraldi, P. G., Romagnoli, R., Pavani, M. G., del Carmen Nunez, M., Bingham, J. P., and Hartley, J. A., Benzoyl and cinnamoyl nitrogen mustard derivatives of benzoheterocyclic analogues of the tallimustine: Synthesis and antitumour activity, *Bioorg. Med. Chem.*, 10, 1611, 2002.

541. Sharma, S. K., Jia, G., and Lown, J. W., Novel cyclopropylindole conjugates and dimers: Synthesis and anti-cancer evaluation, *Curr. Med. Chem. Anticancer Agents*, 1, 27, 2001.

542. Wurtz, N. R. and Dervan, P. B., Sequence specific alkylation of DNA by hairpin pyrrole–imidazole polyamide conjugates, *Chem. Biol.*, 7, 153, 2000.

543. Wang, Y.-D., Dziegielewski, J., Wurtz, N. R., Dziegielewska, B., Dervan, P. B., and Beerman, T. A., DNA crosslinking and biological activity of a hairpin polyamide ± chlorambucil conjugate, *Nucleic Acids Res.*, 31, 1208, 2003.

544. Tsai, S. M., Farkas, M. E., Chou, C. J., Gottesfeld, J. M., and Dervan, P. B., Unanticipated differences between α- and γ-diaminobutyric acid-linked hairpin polyamide-alkylator conjugates, *Nucleic Acids Res.*, 35, 307, 2007.

545. Gravatt, G. L., Baguley, B. C., Wilson, W. R., and Denny, W. A., DNA-directed alkylating agents. 4. 4-Anilinoquinoline-based minor groove directed aniline mustards, *J. Med. Chem.*, 34, 1552, 1991.

546. McClean, S., Costelloe, C., Denny, W. A., Searcey, M., and Wakelin, L. P., Sequence selectivity, cross-linking efficiency and cytotoxicity of DNA-targeted 4-anilinoquinoline aniline mustards, *Anticancer Drug Des.*, 14, 187, 1999.

547. Koyama, M., Kelly, T. R., and Watanabe, K. A., Novel type of potential anticancer agents derived from chrysophanol and emodin. Some structure–activity relationship studies, *J. Med. Chem.*, 31, 283, 1988.

548. Creech, H. J., Preston, R. K., Peck, R. M., and O'Connell, A. P., Antitumor and mutagenic properties of a variety of heterocyclic nitrogen and sulfur mustards, *J. Med. Chem.*, 15, 739, 1972.

549. Prakash, A. S., Denny, W. A., Gourdie, T. A., Valu, K. K., Woodgate, P. D., and Wakelin, L. P. G., DNA-directed alkylating ligands as potential antitumor agents: Sequence specificity of alkylation by intercalating aniline mustards, *Biochemistry*, 29, 9799, 1990.

550. Valu, K. K., Gourdie, T. A., Boritzki, T. J., Gravatt, G. L., Baguley, B. C., Wilson, W. R., Wakelin, L. P. G., Woodgate, P. D., and Denny, W. A., DNA-directed alkylating agents. 3. Structure–activity relationships for acridine-linked aniline mustards: Consequences of varying the length of the linker chain, *J. Med. Chem.*, 33, 3015, 1990.

551. Gourdie, T. A., Prakash, A. S., Wakelin, L. P. G., Woodgate, P. D., and Denny, W. A., Synthesis and evaluation of DNA-targeted spatially separated bis(aniline mustards) as potential alkylating agents with enhanced DNA cross-linking capability, *J. Med. Chem.*, 34, 240, 1991.

552. Fan, J.-Y., Valu, K. K., Woodgate, P. W., Baguley, B. C., and Denny, W. A., Aniline mustard analogues of the DNA-intercalating agent amsacrine: DNA interaction and biological activity, *Anticancer Drug Des.*, 12, 181, 1997.

553. Temple, M. D., Cairns, M. J., Denny, W. A., and Murray, V., Protein–DNA interactions in the human beta-globin locus control region hypersensitive site-2 as revealed by four nitrogen mustards, *Nucleic Acid Res.*, 25, 3255, 1997.

554. Urbaniak, M. D., Bingham, J. P., Hartley, J. A., Woolfson, D. N., and Caddick, S., Design and synthesis of a nitrogen mustard derivative stabilized by apo-neocarzinostatin, *J. Med. Chem.*, 47, 4710, 2004.

555. Gupta, R., Lin, J., Xie, G., and Lown, J. W., Novel DNA-directed alkylating agents consisting of naphthalimide, nitrogen mustard and lexitropsin moieties: Synthesis, DNA sequence specificity and biological evaluation, *Anticancer Drug Des.*, 11, 581, 1996.

556. Denny, W. A., Prospects for hypoxia-activated anticancer drugs, *Curr. Med. Chem. Anticancer Agents*, 4, 395, 2004.

557. Palmer, B. D., Wilson, W. R., Atwel, G. J., Schultz, D., Xu, X. Z., and Denny, W. A., Hypoxia-selective antitumor agents. 9. Structure–activity relationships for hypoxia-selective cytotoxicity among analogues of 5-[N,N-Bis(2-chloroethyl)amino]-2,4-dinitrobenzamide, *J. Med. Chem.*, 37, 2175, 1994.

558. Palmer, B. D., Wilson, W. R., Anderson, R. F., Boyd, M., and Denny, W. A., Hypoxia-selective antitumor agents. 14. Synthesis and hypoxic cell cytotoxicity of regioisomers of the hypoxia-selective cytotoxin 5-[N,N-bis(2-chloroethyl)amino]-2,4-dinitrobenzamide, *J. Med. Chem.*, 39, 2518, 1996.

559. Palmer, B. D., van Zijl, P., Denny, W. A., and Wilson, W. R., Reductive chemistry of the novel hypoxia-selective cytotoxin 5-[N,N-bis(2-chloroethyl)amino]-2,4-dinitrobenzamide, *J. Med. Chem.*, 38, 1229, 1995.

560. Lee, H. H., Palmer, B. D., Wilson, W. R., and Denny, W. A., Synthesis and hypoxia-selective cytotoxicity of a 2-nitroimidazole mustard, *Bioorg. Med. Chem. Lett.*, 8, 1741, 1998.

561. Patterson, A. V., Ferry, D. M., Edmunds, S. J., Gu, Y., Singleton, R. S., Patel, K., Pullen, S. M. et al., Mechanism of action and preclinical antitumor activity of the novel hypoxia-activated DNA cross-linking agent PR-104, *Clin. Cancer Res.*, 13, 3922, 2007.

562. Ware, D. C., Palmer, B. D., Wilson, W. R., and Denny, W. A., Hypoxia-selective antitumor agents. 7. Metal complexes of aliphatic mustards as a new class of hypoxia-selective cytotoxins. Synthesis and evaluation of cobalt(III) complexes of bidentate mustards, *J. Med. Chem.*, 36, 1839, 1993.

563. Ware, W. C., Palmer, H. R., Brothers, P. J., Rickard, C. E. F., Wilson, W. R., and Denny, W. A., *Bis*-tropolonato derivatives of cobalt-(III) complexes of bidentate miphatic nitrogen mustards as potential hypoxia-selective cytotoxins, *J. Inorg. Biochem.*, 68, 215, 1997.

564. Ware, D. C., Palmer, H. R., Pruijn, F. B., Anderson, R. F., Brothers, P. J., Denny, W. A., and Wilson, W. R., Bis(dialkyl)dithiocarbamato cobalt(III) complexes of bidentate nitrogen mustards: Synthesis, reduction chemistry and biological evaluation as hypoxia-selective cytotoxins, *Anticancer Drug Des.*, 13, 81, 1998.

565. Craig, P. R., Brothers, P. J., Clark, G. R., Wilson, W. R., Denny, W. A., and Ware, D C., Anionic carbonato and oxalato cobalt(III) nitrogen mustard complexes, *Dalton Trans.*, 4, 611, 2004.

566. Jarosinski, M. A., Reddy, P. S., and Anderson, W. K., Synthesis, chemical reactivity, and antitumor evaluation of congeners of carmethizole hydrochloride, an experimental "acylated vinylogous carbinolamine" tumor inhibitor, *J. Med. Chem.*, 36, 3618, 1993.

567. Anderson, W. K., Bhattacharjee, D., and Houston, D. M., Design, synthesis, antineoplastic activity, and chemical properties of bis(carbamate) derivatives of 4,5-bis(hydroxymethyl)imidazole, *J. Med. Chem.*, 32, 119, 1989.

568. Elliott, W. L., Fry, D. W., Anderson, W. K., Nelson, J. M., Hook, K. E., Hawkins, P. A., and Leopold, W. R., III, In vivo and in vitro evaluation of the alkylating agent carmethizole, *Cancer Res.*, 51, 4581, 1991.

569. Anderson, W. K. and Corey, P. F., Antileukemic activity of derivatives of 1-phenyl-2,5-dimethyl-3,4-bis(hydroxymethyl)pyrrole bis(N-methylcarbarnate), *J. Med. Chem.*, 20, 1691, 1977.

570. Anderson, W. K. and Halat, M. J., Antileukemic activity of derivatives of 1,2-dimethyl-3,4-bis(hydroxymethyl)-5-phenylpyrrole bis(N-methylcarbamate), 22, 977, 1979.

571. Anderson, W. K. and Corey, P. F., Synthesis and antileukemic activity of 5-substituted 2,3-dihydro-6,7-bis(hydroxymethy1)-1 H-pyrrolizine diesters, *J. Med. Chem.*, 20, 813, 1977.

572. Anderson, W. K., New, J. S., and Corey, P. F., Tumor inhibitory agents: Bis-(N-alkylcarbamate) derivatives of 2,3-dihydro-5-(3′,4′-dichlorophenyl)-6,7-bis(hydroxymethyl)-lH-pyrrolizine, *Arzneim-Forsch.*, 30(I), 765, 1980.

573. Anderson, W. K., Dean, D. C., and Endo, T., Synthesis, chemistry, and antineoplastic activity of a-halopyridinium salts: Potential pyridone prodrugs of acylated vinylogous carbinolamine tumor inhibitors, *J. Med. Chem.*, 33, 1667, 1990.

574. Anderson, W. K., Chang, C.-P., and McPherson, H. L., Jr., Synthesis, evaluation of chemical reactivity, and murine antineoplastic activity of 2-hydroxy-5-(3,4-dichlorophenyl)-6,7-bis(hydroxymeth yl)-2,3-dihydro-1H-pyrrolizine bis(2-propylcarbamate) and 2-acyloxy derivatives as potential water-soluble prodrugs, *J. Med. Chem.*, 26, 1333, 1983.

575. Anderson, W. K., Heider, A. R., Raju, N., and Yucht, J. A., Synthesis and antileukemic activity of bis[[(carbamoyl)oxy]methyl]-substituted pyrrolo[2,1-a]isoquinolines, pyrrolo[1,2-a]quinolines, pyrrolo[2,1-a]isobenzazepines, and pyrrolo[1,2-a]benzazepines, *J. Med. Chem.*, 31, 2097, 1988.

576. Fu, P. P., Xia, Q., Lin, G., and Chou, M. W., Pyrrolizidine alkaloids—Genotoxicity, metabolism enzymes, metabolic activation, and mechanisms, *Drug Met. Rev.*, 36, 1, 2004.

577. Yan, J., Xia, Q., Chou, M. W., and Fu, P. P., Metabolic activation of retronecine and retronecine N-oxide—Formation of DHP-derived DNA adducts, *Toxicol. Ind. Health*, 24, 181, 2008.

578. Xia, Q., Yan, J., Chou, M. W., and Fu, P. P., Formation of DHP-derived DNA adducts from metabolic activation of the prototype heliotridine-type pyrrolizidine alkaloid, heliotrine, *Toxicol. Lett.*, 178, 77, 2008.

579. Xia, Q., Chou, M. W., Edgar, J. A., Doerge, D. R., and Fu, P. P., Formation of DHP-derived DNA adducts from metabolic activation of the prototype heliotridine-type pyrrolizidine alkaloid, lasiocarpine, *Cancer Lett.*, 231, 138, 2006.

580. Wang, Y.-P., Yan, J., Fu, P. P., and Chou, M. W., Human liver microsomal reduction of pyrrolizidine alkaloid N-oxides to form the corresponding carcinogenic parent alkaloid, *Toxicol. Lett.*, 155, 411, 2005.

581. Coulombe, R. A., Jr., Drew, G. L., and Stermitz, F. R., Pyrrolizidine alkaloids crosslink DNA with actin, *Toxicol. Appl. Pharmacol.*, 154, 198, 1999.

582. Kim, H.-Y., Stermitz, F. R., Li, J. K.-K., and Coulombe, R. A., Jr., Comparative DNA cross-linking by activated pyrrolizidine alkaloids, *Food Chem. Toxicol.*, 37, 619, 1999.

583. Rieben, W. K., Jr. and Coulombe, R. A., Jr., DNA cross-linking by dehydromonocrotaline lacks apparent base sequence preference, *Toxicol. Sci.*, 82, 497, 2004.

584. Song, Z., Weng, X., Weng, L., Huang, J., Wang, X., Bai, M., Zhou, Y., Yang, G., and Zhou, X., Synthesis and oxidation-induced DNA cross-linking capabilities of bis(catechol) quaternary ammonium derivatives, *Chem. Eur. J.*, 14, 5751, 2008.

585. McDonald, R. W., Bunjobpon, W., Liu, T., Fessler, S., Pardo, O. E., Freer, I. K. A., Glaser, M., Seckl, M. J., and Robins, D. J., Synthesis and anticancer activity of nordihydroguaiaretic acid (NDGA) and analogues, *Anticancer Drug Des.*, 16, 261, 2001.

586. Wang, P., Song, Y., Zhang, L., He, H., and Zhou, X., Quinone methide derivatives: Important intermediates to DNA alkylating and DNA cross-linking actions, *Curr. Med. Chem.*, 12, 2893, 2005.

587. Richter, S. N., Maggi, S., Mels, S. C., Palumbo, M., and Freccero, M., Binol quinone methides as bisalkylating and DNA cross-linking agents, *J. Am. Chem. Soc.*, 126, 13973, 2004.

588. Wang, P., Liu, R., Wu, X., Ma, H., Cao, X., Zhou, P., Zhang, J. et al., A potent, water-soluble and photoinducible dna cross-linking agent, *J. Am. Chem. Soc.*, 125, 1116, 2003.

589. Kumar, D., Veldhuyzen, W. F., Zhou, Q., and Rokita, S. E., Conjugation of a hairpin pyrrole–imidazole polyamide to a quinone methide for control of DNA cross-linking, *Bioconjug. Chem.*, 15, 915, 2004.

590. Veldhuyzen, W. F., Pande, P., and Rokita, S. E., A transient product of DNA alkylation can be stabilized by binding localization, *J. Am. Chem. Soc.*, 125, 14005, 2003.

591. Song, Y., Wang, P., Wu, J., Zhou, X., Zhang, X.-L., Weng, L., Cao, X., and Liang, F., Biological studies of photoinducible phenol quaternary ammonium derivatives, *Bioorg. Med. Chem. Lett.*, 16, 1660, 2006.

592. Song, Y., Tian, T., Wang, P., He, H., Liu, W., Zhou, X., Cao, X., Zhang, X.-L., and Zhou, X., Phenol quaternary ammonium derivatives: Charge and linker effect on their DNA photo-inducible cross-linking abilities, *Org. Biomol. Chem.*, 4, 3358, 2006.

593. Verga, D., Richter, S. N., Palumbo, M., Gandolfi, R., and Freccero, M., Bipyridyl ligands as photoactivatable mono- and bis-alkylating agents capable of DNA cross-linking, *Org. Biomol. Chem.*, 5, 233, 2007.

594. Zhang, L., Ren, L., Bai, M., Weng, L., Huang, J., Wu, L., Deng, M., and Zhou, X., Synthesis and biological activities of quinazoline derivatives with ortho-phenol-quaternary ammonium salt groups, *Bioorg. Med. Chem.*, 15, 6920, 2007.

595. He, H., Tian, T., Wang, P., Wu, L., Xu, J., Zhou, X., Zhang, X., Cao, X., and Wu, X., Porphyrin–DNA cross-linking agent hybrids: Chemical synthesis and biological studies, *Bioorg. Med. Chem. Lett.*, 14, 3013, 2004.

596. He, H., Zhou, Y., Liang, F., Li, D., Wu, J., Yang, L., Zhou, X., Zhang, X., and Cao, X., Combination of porphyrins and DNA-alkylation agents: Synthesis and tumor cell apoptosis induction, *Bioorg. Med. Chem.*, 14, 1068, 2006.

597. Ludlum, D. B., The chloroethylnitrosoureas: Sensitivity and resistance to cancer chemotherapy at the molecular level, *Cancer Invest.*, 15, 588, 1997.

598. Ali-Osman, F., Giblin,J., Berger, M., Murphy, M. J., Jr., and Rosenblum, M. L., Chemical structure of carbamoylating groups and their relationship to bone manow toxicity and antiglioma activity of bifunctionally alkylating and carbamoylating nitrosoureas, *Cancer Res.*, 45, 4185, 1985.

599. Prestayko, A. W., Baker, L. H., Crooke, S. T., Carter, S. L., and Schein, P. S., *Nitrosoureas. Current Status and New Developments*, Academic Press, New York, 1981, Chapter 4.

600. Tew, K. D., Dean, S. W., and Gibson, N. W., The effect of a novel taurine nitrosourea, 1-(2-chloroethyl)-3-[2-(dimethylaminosulfonyl)ethyl]-1-nitrosourea (TCNU) on cytotoxicity, DNA crosslinking and glutathione reductase in lung carcinoma cell lines, *Cancer Chemother. Pharmacol.*, 19, 291, 1987.

601. Gnewuch, C. T. and Sosnovsky, G., A critical appraisal of the evolution of nitrosoureas as anticancer drugs, *Chem. Rev.*, 97, 829, 1997.

602. Hayes, M. T., Bartley, J., Parsons, P. G., Eaglesham, G. K., Prakash, A. S., Mechanism of action of fotemustine, a new chloroethylnitrosourea anticancer agent: Evidence for the formation of two DNA-reactive intermediates contributing to cytotoxicity, *Biochemistry*, 36, 10646, 1997.

603. Merimsky, O., Inbar, M., Gerard, B., and Chaitchik, S., Fotemustine—An advance in the treatment of metastatic malignant melanoma, *Melanoma Res.*, 2, 401, 1992.

604. Ali-Osman, F., Rairkar, A., and Young, P., Formation and repair of 1,3-bis-(2-chloroethyl)-1-nitrosourea and cisplatin induced total genomic DNA interstrand crosslinks in human glioma cells, *Cancer Biochem. Biophys.*, 14, 231, 1995.

605. Fiumicino, S., Martinelli, S., Colussi, C., Aquilina, G., Leonetti, C., Crescenzi, M., and Bignami, M., Sensitivity to DNA cross-linking chemotherapeutic agents in mismatch repair-defective cells *in vitro* and in xenografts, *Int. J. Cancer*, 85, 590, 2000.

606. Tong, W. P., Kirk, M. C., and Ludlum, D. B., Mechanism of action of the nitrosoureas—V. Formation of O^6-(2-fluoroethyl)guanine and its probable role in the crosslinking of deoxyribonucleic acid, *Biochem. Pharmacol.*, 32, 2011, 1983.

607. Cui, B., Johnson, S. P., Bullock, N. H., Ali-Osman, F., Bigner, D. D., and Friedman, H. S., Bifunctional DNA alkylator 1,3-bis(2-chloroethyl)-1-nitrosourea activates the ATR-Chk1 pathway independently of the mismatch repair pathway, *Mol. Pharmacol.*, 75, 1356, 2009.

608. Brent, T. P., Isolation and purification of O^6-alkylguanine-DNA alkyltransferase from human leukemic cells. Prevention of chloroethylnitrosourea-induced cross-links by purified enzyme, *Pharmacol. Ther.*, 31, 121, 1985.

609. Hansen, R. J., Nagasubramanian, R., Delaney, S. M., Samson, L. D., and Dolan, M. E., Role of O^6-methylguanine-DNA methyltransferase (MGMT) in protecting from alkylating agent-induced toxicity and mutations in mice, *Carcinogenesis*, 28, 1111, 2007.

610. Ewig, R. A. G. and Kohn, K. W., DNA–protein cross-linking and DNA interstrand cross-linking by haloethylnitrosoureas in L1210 cells, *Cancer Res.*, 38, 3197, 1977.

611. Gonzaga, P. E., Potter, P. M., Niu, T.-Q., Yu, D., Ludlum, D. B., Rafferty, J. A., Margison, G. P., and Brent, T. P., Identification of the cross-link between human O^6-methylguanine-DNA methyltransferase and chloroethylnitrosourea-treated DNA, *Cancer Res.*, 52, 6052, 1992.

612. Chen, F.-X., Zhang, Y., Church, K. M., Bodell, W. J., and Gold, B., DNA crosslinking, sister chromatid exchange and cytotoxicity of *N*-2-chloroethylnitrosoureas tethered to minor groove binding peptides, *Carcinogenesis*, 14, 935, 1993.

613. Yin, J. H., Yang, D. I., Chou, H., Thompson, E. M., Xu, J., and Hsu, C. Y., Inducible nitric oxide synthase neutralizes carbamoylating potential of 1,3-bis(2-chloroethyl)-1-nitrosourea in c6 glioma cells, *J. Pharmacol. Exp. Ther.*, 297, 308, 2001.

614. Wilds, C. J., Xu, F., and Noronha, A. M., Synthesis and characterization of DNA containing an N^1–2′-deoxyinosine-ethyl-N^3-thymidine interstrand cross-link: A structural mimic of the cross-link formed by 1,3-bis-(2-chloroethyl)-1-nitrosourea, *Chem. Res. Toxicol.*, 21, 686, 2008.

615. Fischhaber, P. L., Gall, A. S., Duncan, J. A., and Hopkins, P. B., Direct demonstration in synthetic oligonucleotides that N,N′-bis(2-chloroethyl)-nitrosourea cross links N^1 of deoxyguanosine to N^3 of deoxycytidine on opposite strands of duplex DNA, *Cancer Res.*, 59, 4363, 1999.

616. Passagne, I., Evrard, A., Winum, J.-Y., Depeille, P., Cuq, P., Montero, J. L., Cupissol, D., and Vian, L., Cytotoxicity, DNA damage, and apoptosis induced by new fotemustine analogs on human melanoma cells in relation to O^6-methylguanine DNA-methyltransferase expression, *J. Pharmacol. Exp. Ther.*, 307, 816, 2003.

6 Heterobifunctional Cross-Linkers

6.1 INTRODUCTION

In contrast to homobifunctional reagents, heterobifunctional reagents contain two dissimilar functional groups of different specificities. These two reactive functionalities may be any combination of the conventional group-selective moieties discussed in earlier chapters. For example, one end of the cross-linker may be selective for an amino group while the other end directed to a sulfhydryl group. As categorized in Table 6.1, a variety of combinations is possible; thus, different amino acids in proteins as well as DNAs can be conjugated together. A comprehensive list of these heterobifunctional cross-linkers can be found in Appendix D. It should be stressed again, however, that with few exceptions, no reagent is absolutely group specific. Cross-reactivity exists between the various classes of heterobifunctional cross-linkers presented in the classification, which is a guide for selectivity rather than specificity. By taking advantage of the differential selectivity of the different functional groups, cross-linking reactions can be controlled selectively and sequentially.

In addition to the conventional group-specific functionalities, one end of the heterobifunctional cross-linker may be a nonspecific photosensitive group as represented in Table 6.2. A comprehensive list of photoactivatable reagents is presented in Appendix E. These photosensitive groups, as elaborated in Chapter 4, react indiscriminately on activation by irradiation. With one end of the cross-linker anchored to an amino acid residue or DNA, the photoreactive moiety can be used to probe its surrounding environment. The photosensitive reagents are most effectively used in macromolecular photoaffinity labeling, in which the reagent is first incorporated into a polypeptide or DNA in the dark. The labeled ligand is then allowed to bind to specific receptors that are cross-linked to the ligand on photolysis.[1,2] The use of this technique for the identification of receptors will be discussed in Chapter 11. This chapter will give a general presentation of the activities of these reagents.

6.2 GROUP-SELECTIVE HETEROBIFUNCTIONAL REAGENTS FOR PROTEIN CROSS-LINKING

6.2.1 AMINO- AND SULFHYDRYL-GROUP–DIRECTED CROSS-LINKERS

This assembly of heterobifunctional cross-linking reagents contains two-headed reactive functionalities directed toward amino and sulfhydryl groups. As may be realized from Appendix D.A, most of the amino group–directed reactive functionalities are acylating agents, while those directed toward sulfhydryl groups are mostly alkylating agents. As shown in Table 6.1, one end these compounds may be an hydroxysuccinimidyl ester moiety, which will undergo nucleophilic substitution reaction with amino groups, liberating N-hydroxysuccinimide.[1,3–15] Other activated esters include p-nitrophenyl esters[13–20] and imidoester.[21–24] Acyl azides,[25,26] acyl chlorides,[27] aryl halides,[21] and epoxides[28] are also reactive toward amino groups. Since they are highly activated, they may also react with hydroxyl groups.[27] Haloketones and alkyl halides, which react with various nucleophiles, are also amino group selective.[26,29,30] These functionalities will, of course, react faster with thiols and are regarded as thiol selective in some reagents (see Table 6.1 and Appendix D).[31,32] Other thiol-directed groups in these two-headed compounds are the maleimido moiety, disulfide bonds, and other alkylating functionalities.[33,34] The majority of the amino- and sulfhydryl-group–directed bifunctional reagents are

TABLE 6.1
Representative Group Selective Heterobifunctional Cross-Linkers

I. Amino- and Sulfhydryl-Group–Directed Bifunctional Reagents

A. Succinimide and thiopyridine. General structure

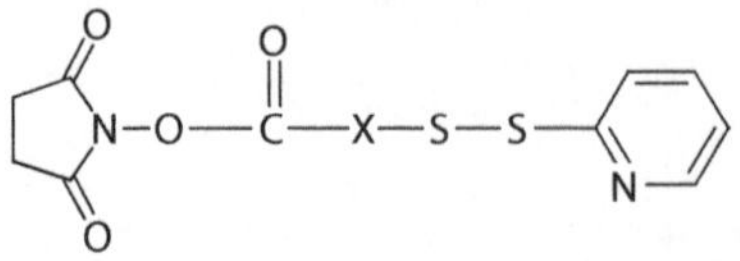

Example: *N*-Succinimidyl-3-(2-pyridyldithio)propionate (SPDP)

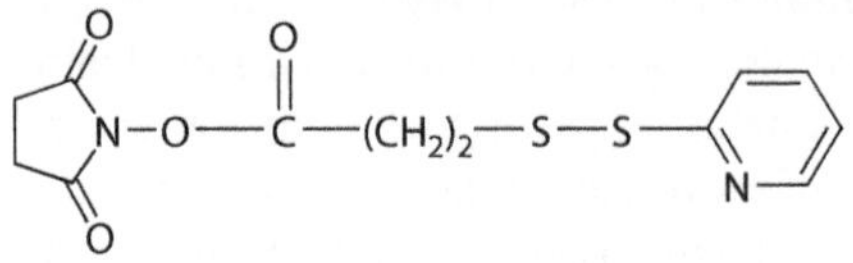

B. Succinimide and maleimide. General structure

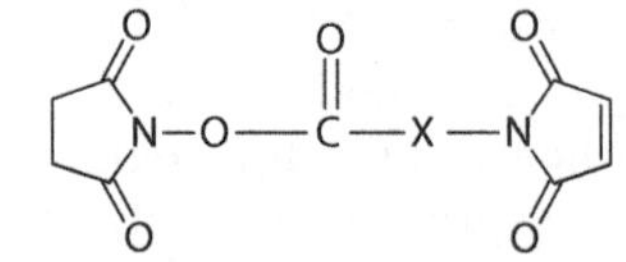

Example: *N*-Succinimidylmaleimidoacetate (AMAS)

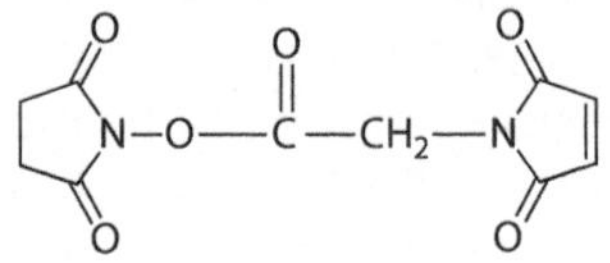

C. Succinimide and alkylhalide. General structure

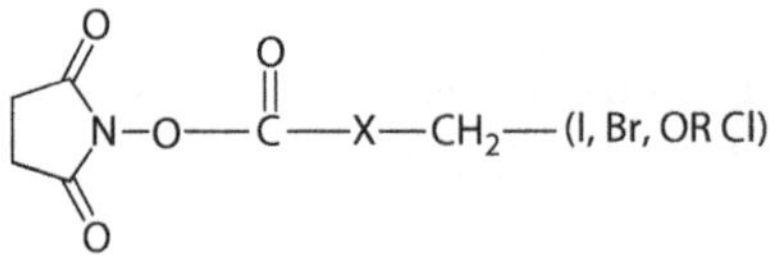

Example: *N*-Succinimidylbromoacetate

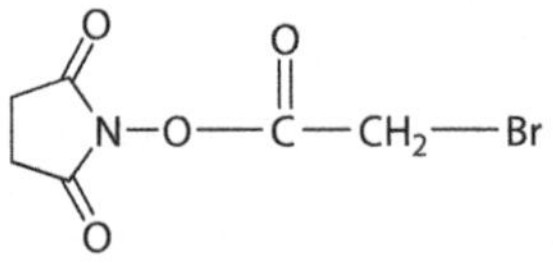

D. Nitrophenol and maleimide. General structure

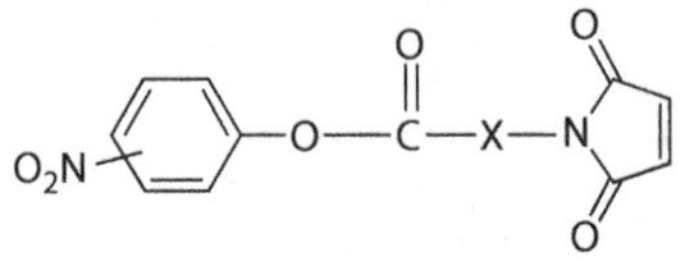

Example: *p*-Nitrophenyl-6-maleimidocaproate

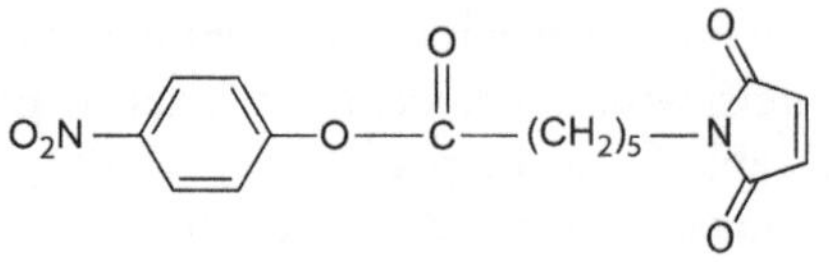

TABLE 6.1 (Continued)
Representative Group Selective Heterobifunctional Cross-Linkers

E. Nitrophenol and alkylhalide. General structure

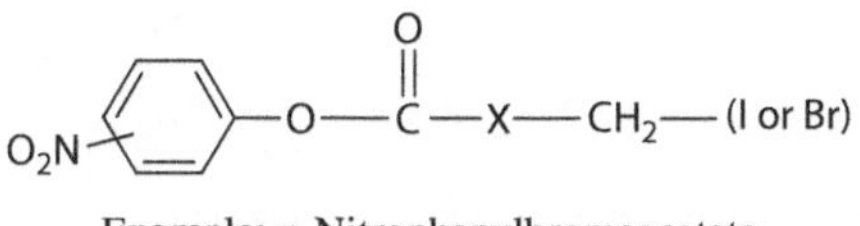

Example: *p*-Nitrophenylbromoacetate

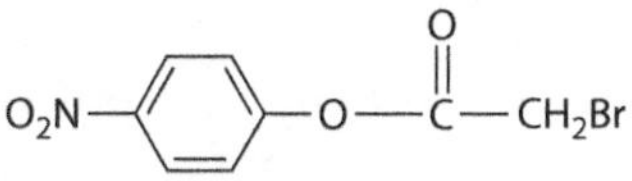

F. Acetimidate and alkylhalide. General structure

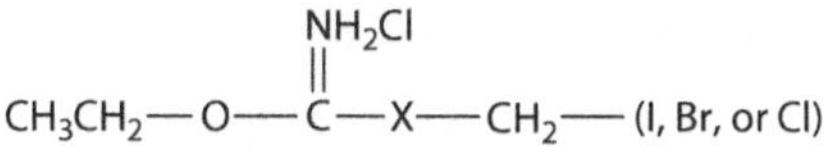

Example: Ethyl iodoacetimidate HCl

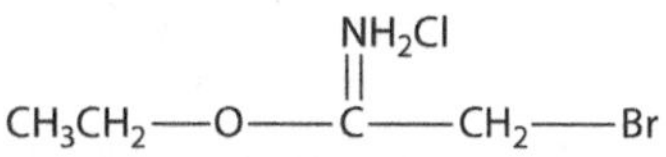

G. Benzoyl derivative and maleimide. General structure

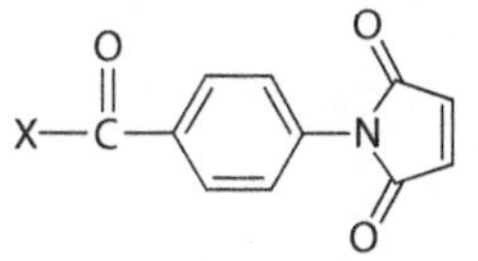

Example: 2-Hydroxy-4-(*N*-maleimido)benzoylazide (HMB)

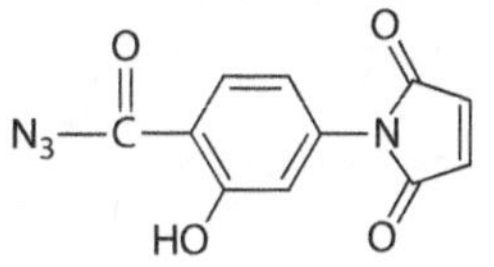

II. Carboxyl- and Either Sulfhydryl- or Amino-Group–Directed Bifunctional Reagents

A. Diazoacetyl and thiopyridine. General structure

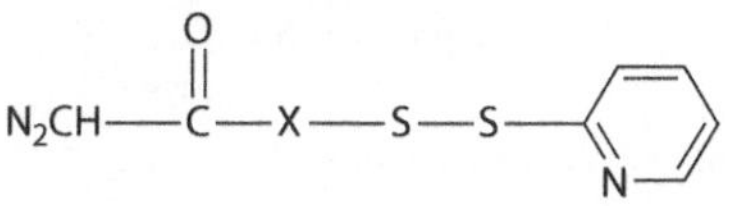

Example: Pyridyl-2-2′-dithiobenzyldiazoacetate (PDD)

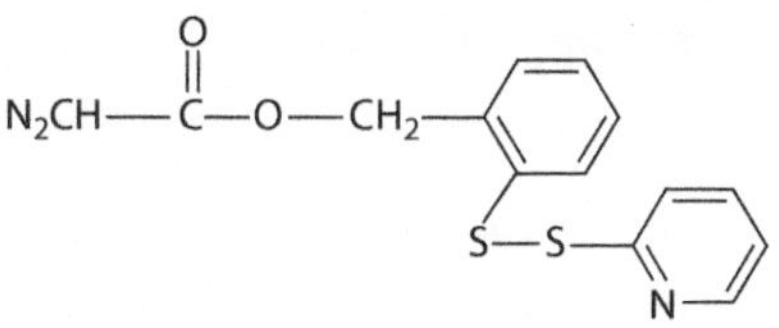

(*continued*)

TABLE 6.1 (Continued)
Representative Group Selective Heterobifunctional Cross-Linkers

B. Diazoacetyl and nitrophenol. General structure

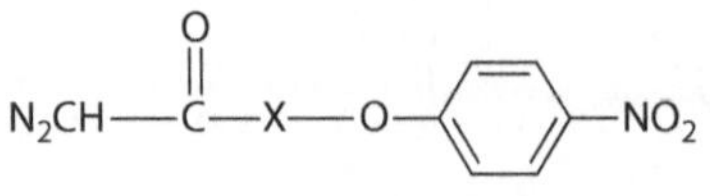

$$N_2CH-\overset{\overset{\displaystyle O}{\|}}{C}-X-O-\langle\text{benzene ring}\rangle-NO_2$$

Example: *p*-Nitrophenyl diazoacetate

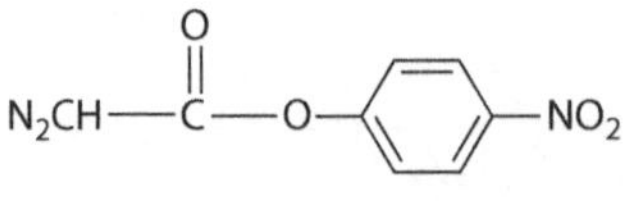

$$N_2CH-\overset{\overset{\displaystyle O}{\|}}{C}-O-\langle\text{benzene ring}\rangle-NO_2$$

III. Carbonyl- and Amino-Group–Directed Bifunctional Reagents

A. Hydrazide and succinimide. General structure

$$H_2N-\overset{\overset{\displaystyle H}{|}}{N}-X-\overset{\overset{\displaystyle O}{\|}}{C}-O-N\langle\text{succinimide}\rangle$$

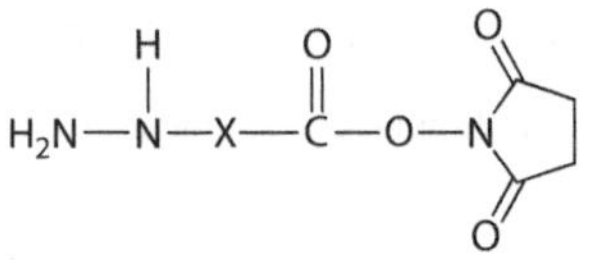

Example: *N*-Succinimidyl-4-hydrazidoterephthalate

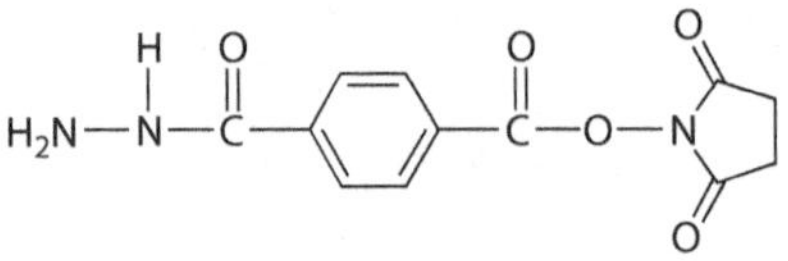

$$H_2N-\overset{\overset{\displaystyle H}{|}}{N}-\overset{\overset{\displaystyle O}{\|}}{C}-\langle\text{benzene ring}\rangle-\overset{\overset{\displaystyle O}{\|}}{C}-O-N\langle\text{succinimide}\rangle$$

IV. Carbonyl- and Sulfhydryl-Group–Directed Bifunctional Reagents

A. Aminooxy and thiopyridine. General structure

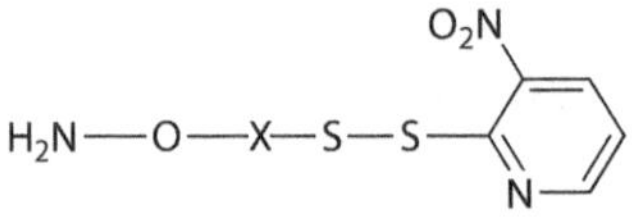

$$H_2N-O-X-S-S-\langle\text{3-nitro-2-pyridyl}\rangle$$

Example: 1-Aminooxy-4-[(3-nitro-2-pyridyl)dithio]butane

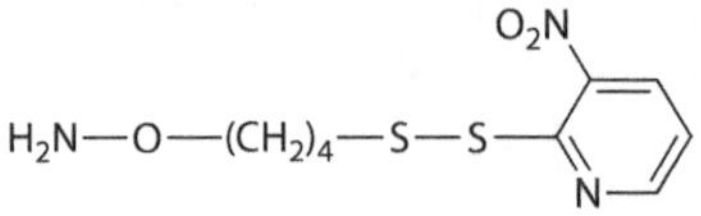

$$H_2N-O-(CH_2)_4-S-S-\langle\text{3-nitro-2-pyridyl}\rangle$$

B. Hydrazide and maleimide: General structure

$$NH_2\text{-}NH-\overset{\overset{\displaystyle O}{\|}}{C}-X-N\langle\text{maleimide}\rangle$$

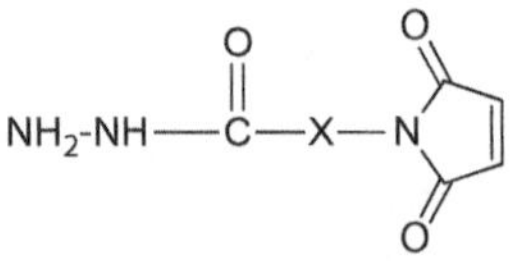

Example: *N*-(β-maleimidopropionic acid) hydrazide

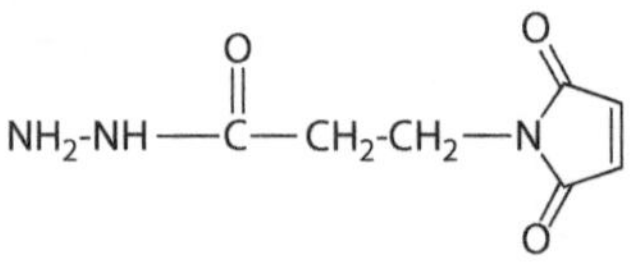

$$NH_2\text{-}NH-\overset{\overset{\displaystyle O}{\|}}{C}-CH_2\text{-}CH_2-N\langle\text{maleimide}\rangle$$

TABLE 6.2

Representative Anchoring Agents of Photoactivatable Heterobifunctional Cross-Linkers

I. Amino Group Anchoring Agents

A. Succinimides. General structure

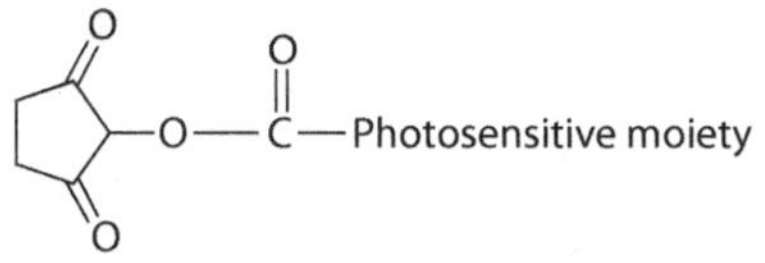

Example: *N*-Succinimidyl-4-azidobenzolyglycinate

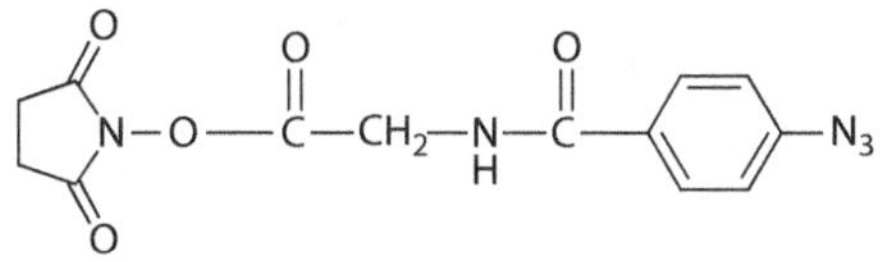

B. Imidates. General structure

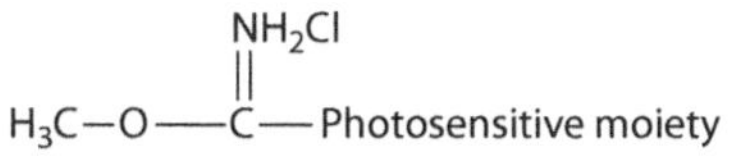

Example: Methyl-[3-(4-azidophenyl)dithio]propionimidate HCl (MADP)

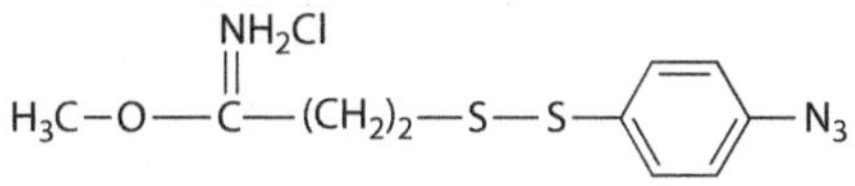

C. Phenylhalides. General structure

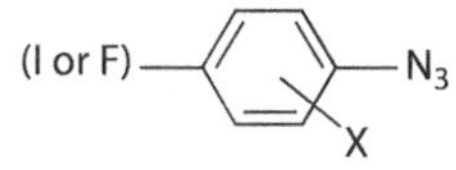

Example: 4-Fluoro-3-nitrophenylazide (FNA)

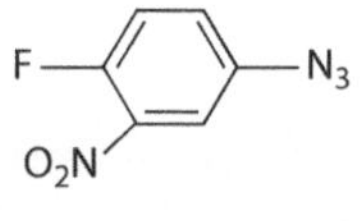

D. Phenylisocyanates. General structure

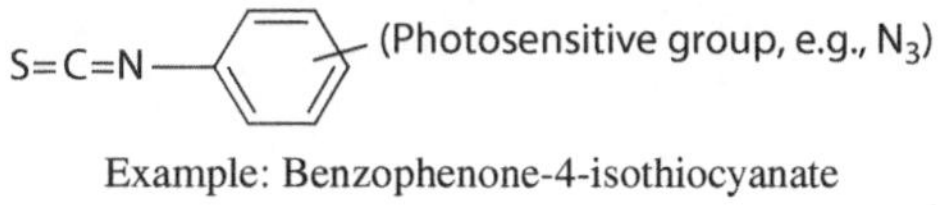

Example: Benzophenone-4-isothiocyanate

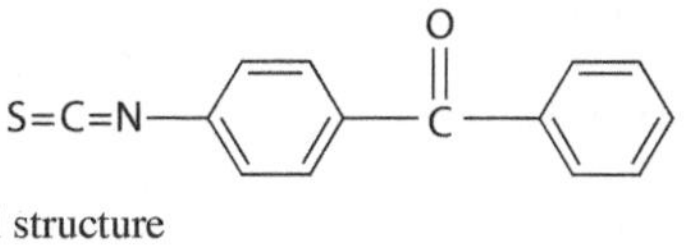

E. Nitrophenols. General structure

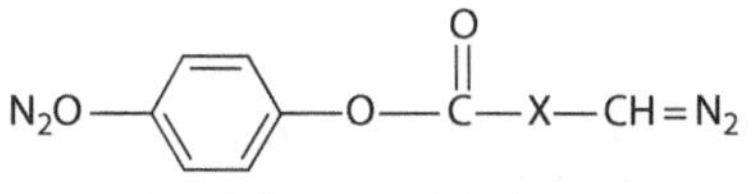

Example: *p*-Nitrophenyl-3-diazopyruvate

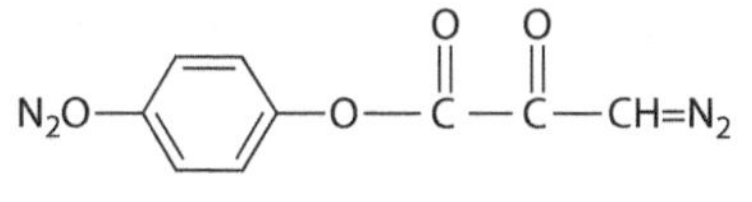

(continued)

TABLE 6.2 (Continued)
Representative Anchoring Agents of Photoactivatable
Heterobifunctional Cross-Linkers

II. Sulfhydryl Group—Anchored Agents

A. Disulfides. General structure

Photosensitive moiety—S—S— Photosensitive moiety

Example: 4,4'-Dithiobisphenylazide

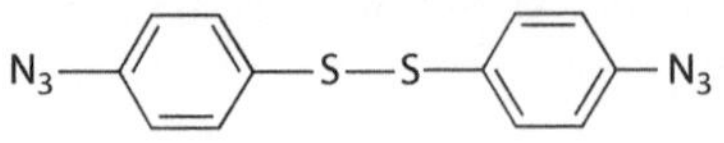

N_3—⟨⟩—S—S—⟨⟩—N_3

B. Sulfenyls. General structure

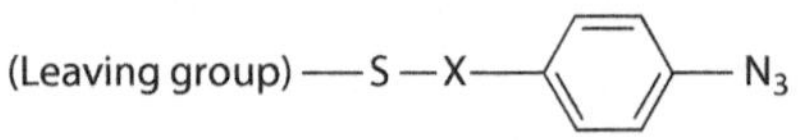

(Leaving group) —S—X—⟨⟩—N_3

Example: 2-Nitro-4-azidophenylsulfenyl chloride

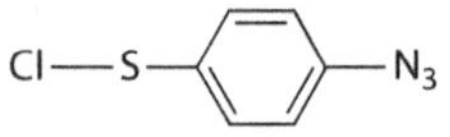

Cl—S—⟨⟩—N_3

C. Seleno derivatives. General structure

R—Se—C(=O)—Photosensitive moiety

Example: 3-(4-Azido-2-nitrobenzoylseleno)propionic acid

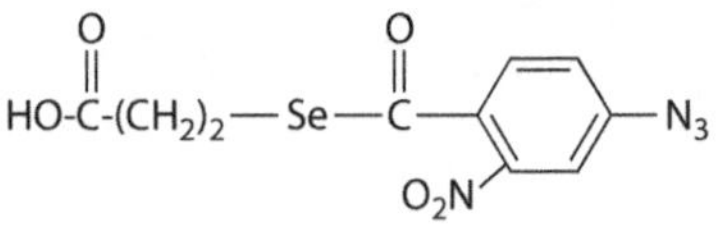

HO-C(=O)-$(CH_2)_2$—Se—C(=O)—⟨⟩—N_3
O_2N

D. Maleimides. General structure

N—X——(Photosensitive group, e.g., N_3)

Example: 4-Azidophenylmaleimide (APM)

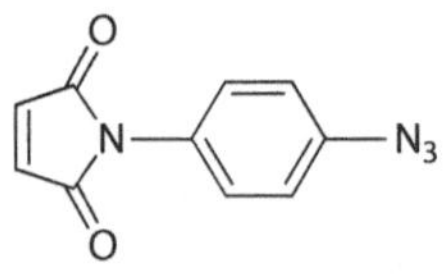

N—⟨⟩—N_3

E. Akylhalides. General structure

(Br or I)—CH_2—X—⟨⟩—N_3

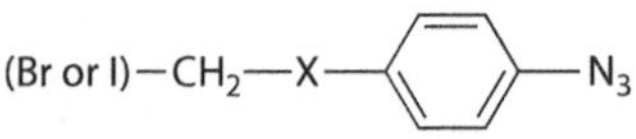

Example: 4-(Bromoaminoethyl)-3-nitrophenylazide

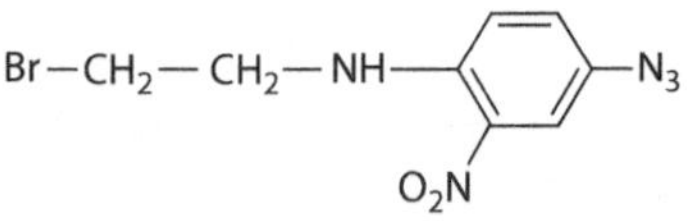

Br—CH_2—CH_2—NH—⟨⟩—N_3
O_2N

TABLE 6.2 (Continued)
Representative Anchoring Agents of Photoactivatable Heterobifunctional Cross-Linkers

III. Guanidinal Group Anchoring Agent

A. Dicarbonyls. General structure

$$R-\overset{\overset{O}{\|}}{C}-\overset{\overset{O}{\|}}{C}-\text{Photosensitive moiety}$$

Example: 4-Azidophenylglyoxal

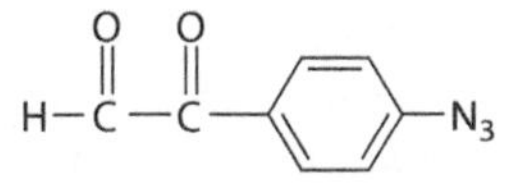

$$H-\overset{\overset{O}{\|}}{C}-\overset{\overset{O}{\|}}{C}-\overset{\text{}}{\bigcirc}-N_3$$

IV. Carboxyl or Carboxamide Groups Anchoring Agent

A. Alkylamines. General structure

$$H_2N-CH_2-\text{Photosensitive moiety}$$

Example: N-[β-(β'-Aminoethyldithioethyl)]-4-azido-2-nitroaniline

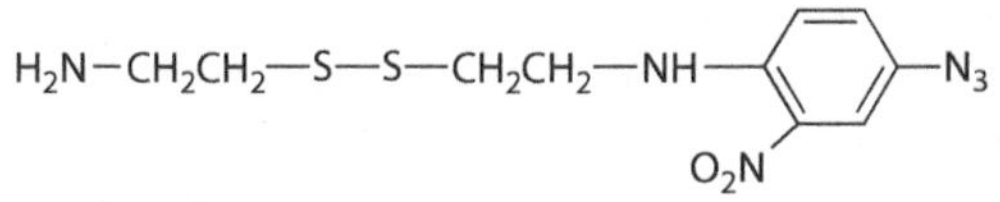

$$H_2N-CH_2CH_2-S-S-CH_2CH_2-NH-\overset{\text{}}{\underset{O_2N}{\bigcirc}}-N_3$$

composed of the hydrosuccinimidyl (amino group selective) and maleimide (sulfhydryl group selective) moieties (XVI through LII in Appendix D), followed by compounds with hydrosuccinimidyl and haloacetyl (sulfhydryl group selective) groups (LIII through LXIII). The rest of the compounds have various combinations of nitrophenol, maleimide, haloacetyl, imidate, haloacyl, azidoacyl, epoxide, disulfhydryl, and other groups. Since the maleimido group attached to aromatic rings is labile at neutral pH, the most stable compound is probably N-succinimidyl-4-(N-maleimidomethyl) cyclohexane-1-carboxylate (SMCC) and its sulfonated analog, sulfo-SMCC (XXXIII and XXXIV in Appendix D).[35-45] Compounds with nitrovinyl groups such as 2,4-dinitrophenyl-p-(β-nitrovinyl) benzoate (LXX in Appendix D) react with the thiol through Michael addition to the double bond more readily than N-maleimido derivatives under acidic conditions.[20] In this particular cross-linker, the dinitrophenyl ester at the other end of the compound also reacts with amines at a much faster rate than N-hydroxysuccinimide esters under basic conditions. The cross-linking reaction is shown in Figure 6.1. Cross-linking of sulfhydryl and amino groups is also achievable with epichlorohydrin (ECH, compound XCIX in Appendix D).[32] Although its main application is in the immobilization of proteins and carbohydrates, it has been shown to cross-link DNA,[46] which will be further discussed below.

These amino- and sulfhydryl-group–directed heterobifunctional cross-linkers are not absolutely specific; many of them can react with different amino acid side chains. Ethyl haloacetimidates

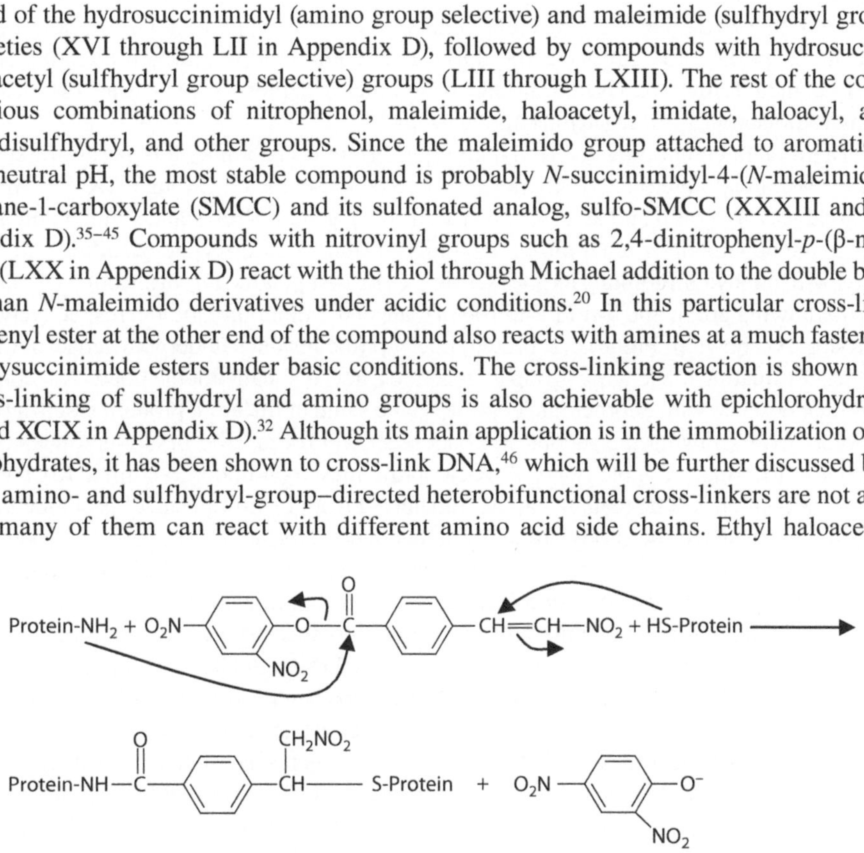

FIGURE 6.1 Cross-linking reaction of 2,4-dinitrophenyl-p-(β-nitrovinyl)benzoate.

(LXXXIX through XCI in Appendix D), for example, have a broad reactivity. Although its imidate moiety reacts quite specifically with lysine residues through the amidination reaction, the haloaetamido group can react with any nucleophile including histidine as shown by Diopoh and Olomucki in cross-linking studies with RNAse.[23,24] Since the thiol is a strong nucleophile, many of the reagents such as compounds LXXXI and LXXXII in Appendix D may act as homobifunctional reagents cross-linking two sulfhydryl groups if the thiol group is in excess. On the other hand, in its absence, other nucleophiles will be cross-linked. In general, however, these reagents are used to cross-link proteins with known amino acid side chains in a sequence of reaction steps. Proteins that are known to contain free thiols will react first with the compound. After modification, the free end of the bifunctional reagent reacts with another protein with desired amino acid side chains. For example, to use 4-chloroactylphenylmaleimide (LXXXI in Appendix D) as a heterobifunctional cross-linker, the maleimido end is first reacted with a protein containing a thiol group. The alkyl halide end is then allowed to react with an amino group in another protein.[26,29]

There are several compounds in this category that are cleavable (e.g., I through XIV, L through LII, LXII, LXIII, LXVI, LXVII, LXXXIII, LXXXV, LXXXVII, LXXXVIII, and XCV through XCVIII in Appendix D). Obviously, compounds that result in the formation of disulfide bonds, for example, N-succinimidyl-3-(2-pyridyldithio)propionate (SPDP) (compound I in Appendix D), are cleavable in the presence of a free thiol.[2,21] The sulfone-containing compounds such as N-succinimidyl-3-(2-bromo-3-oxobutane-1-sulfonyl)propionate (LXII in Appendix D) are cleavable by dithionite. Compounds that lead to maleylation of amino groups such as N-[4-[(2,5-dioxo-3-furyl)methylsulfanyl]phenyl]-6-(2,5-dioxopyrrol-1-yl)hexanamide (LXXXIII in Appendix D) are cleavable under mildly acidic conditions (see Figure 4.17).[31]

Among all these heterobifunctional compounds, few are particularly worth noting. 2-Hydroxy-4-(N-maleimido)benzoylazide (HMB, compound LXXVII in Appendix D) is photosensitive but undergoes nucleophilic substitution in the dark. This compound contains a phenolic ring and is therefore iodionatable by various iodinating agents (see Chapter 4), providing the possibility of introducing radioisotope ^{125}I for various applications.[25,26]

There is a group of compounds referred to as equilibrium transfer alkylating cross-linkers (ETAC) described by Mitra and Lawton and others.[33,34] These compounds (CI through CXV in Appendix D) contain a good leaving group and an electron-withdrawing group in resonance with a double bond. In the first described compounds, 2-(p-nitrophenyl)allyl-4-nitro-3-carboxyphenylsulfide and 2-(p-nitrophenyl)allyl-4-trimethylammonium iodide (CI and CII in Appendix D), p-nitrophenyl served as an electron-withdrawing group and 5-thio-2-nitrobenzoate and ammonium iodide, respectively, served as leaving groups. As shown in Figure 6.2, protein nucleophilic residues undergo Michael addition at the double bond, eliminating the leaving group to generate a new resonant double bond. A second Michael addition is then possible with another nucleophile on the same or different protein, resulting in a cross-linkage. Alternatively, during the second Michael reaction process, the first added nucleophile may be eliminated reforming a double bond. A nucleophile may then undergo similar Michael addition and the process continues. The reagent can be transferred from the initial site of protein attachment to other groups until the most thermodynamically stable bond is formed. With the protein side chains of lysine, tyrosine, glutamic, and aspartic acids, the reagent will undergo Michael addition and Michael elimination indefinitely, until a thermodynamically stable bond is formed. A more stable bond is formed when thiol addition is encountered to form the thioether bridge. Since the publication of the first ETAC, several compounds have been synthesized (CIII through CXV in Appendix D).[34] Not only can these compounds be used to cross-link different groups heterobifunctionally or homobifunctionally, they have also been used to introduce various reporter groups to reduced immunoglobulins.[47]

The amino- and sulfhydryl-directed heterobifunctional reagents have been extensively used to cross-link various proteins. SPDP is probably the most popular of all these reagents. It has been used to conjugate antifibrin antibody to urokinase and tissue plasminogen activator,[48,49] in the preparation of bispecific antibodies and immunotoxins,[50,51] and in the study of stomatal and fenestral diaphragm and bacterial cell surface layer.[52,53] Several SPDP analogs such as MSPDP, LC-SPDP, and SMPT have been

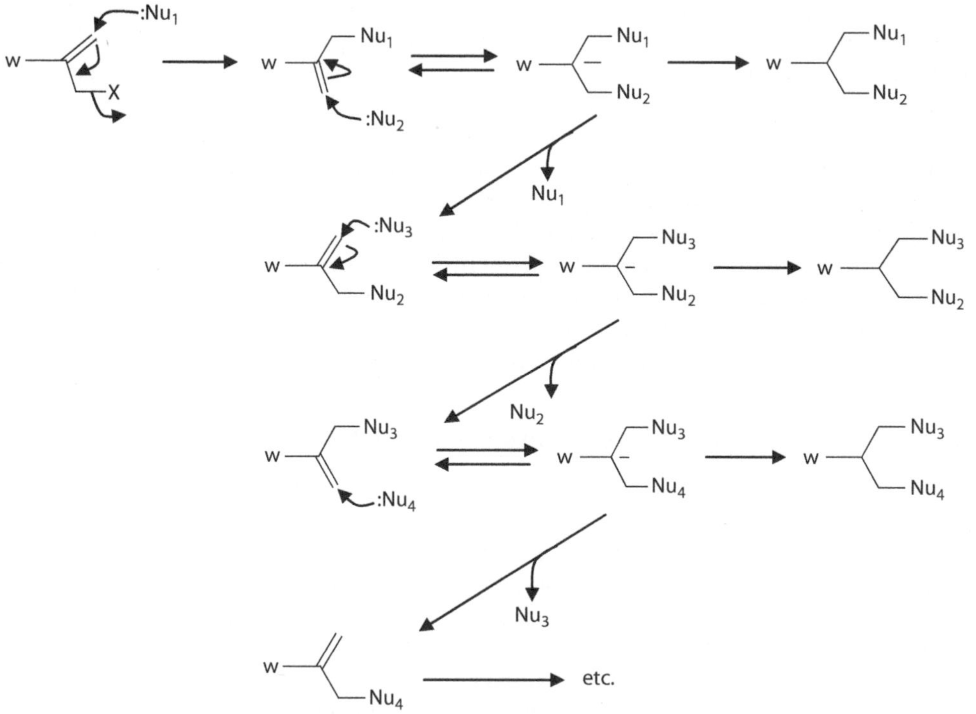

FIGURE 6.2 Mechanism of reaction of equilibrium transfer alkylating cross-linkers (ETAC). W designates an electron withdrawing group, X a good leaving group, and Nu a nucleophile. (Adapted from Libertore, F. A. et al., *Bioconjug. Chem.*, 1, 36, 1990.)

synthesized and made more soluble by adding a sulfonyl group to the succinimide ring (II through VII in Appendix D). *N*-Hydroxysuccinimidyl-3-methyl-3-(acetylthio)butanoate (SAMBA) (XV in Appendix D) is a special SPDP analog.[54] To cross-link proteins, it first reacts with an amino group. The modified protein is then treated with hydroxylamine to remove the acetyl protecting group to expose a free thiol, which then reacts with an activated protein disulfide to complete the cross-linking. The reaction procedure is similar for iminothiolane (CXVI in Appendix D), which reacts with an amino group and exposes a free thiol as discussed in Chapter 2. It has been used to immobilize avidin onto gelatin nanoparticales.[55] Goff and Carroll[56] have synthesized a series of iminothiolane analogs (CXVII through CXXII in Appendix D) and studied their effect on preparation of immunoconjugates.[54] They found that 5-methyl-2-iminothiolane (M2IT) (CXVII in Appendix D) has optimal properties for the preparation of disulfide cross-linked immunoconjugates with enhanced disulfide bond stability.

The mercurial compounds, 3-(acetoxymercurio)-5-nitrosalicylaldehyde and 3-(chloromercurio)-5-nitrosalicylaldehyde (CXXIII and CXXIV in Appendix D), are truly sulfhydryl specific. They have been used to study the interactions between cobratoxin and acetylcholine receptor through cross-linking of the reduced sulfhydryl group of AcChR and lysine 23 of cobratoxin.[57] Many other compounds have also been frequently used in protein structure and interaction,[58] in the preparation of enzyme–immunoglobulin conjugates,[59–62] and the stabilization of microtubules.[63]

6.2.2 Cross-Linkers Directed toward Carboxyl and Either Sulfhydryl or Amino Groups

The carboxyl group–selective agents used in these heterobifunctional compounds are shown in Table 6.1.II. As discussed in Chapter 5, compounds containing a diazoacetyl group are photosensitive. In the dark, however, they are reactive toward the carboxyl group at acidic pHs (see Chapter 3).[64] Therefore, in addition to being photoaffinity labels, these cross-linking reagents (Appendix D.B) are potential cross-linkers for carboxyl groups.[65–68] The reactive group at the opposite end of the molecule determines whether the cross-linker will react with sulfhydryl or amino group. For example, if the reagent contains a disulfide bond such as p-(2′-pyridyldithio)benzyldiazoacetate (PPD, Appendix D.B.CXXVII), it will be thiol specific because the disulfide bond can undergo thiol–disulfide interchange with sulfhydryl groups on the protein. Compounds containing an alkylating or acylating group will react with any nucleophile. For instance, in the presence of sulfhydryl groups, the compound, 1-diazoacetyl-1-bromo-2-phenyle-thane (Appendix D.B.CXXVIII), will be thiol directing and has been shown to react with a cysteine at the active site of pepsin.[66] In the absence of sulfhydryl groups, amino groups and other nucleophiles will react. However, for acylating groups such as that in p-nitrophenyl diazoacetate (Appendix D.B.CXXIX), more stable products will form with amino groups. These compounds have been used to photolabel the active site of chymotrypsin[67] and cross-link calmodulin and adenylate cyclase.[68]

6.2.3   CARBONYL- AND AMINO- OR SULFHYDRYL-GROUP–DIRECTED CROSS-LINKERS

As discussed in Chapter 3, carbonyl groups form Schiff bases with amino groups. Thus, the het-erobifunctional cross-linking reagents that have been used in literature contains an amino group, either free or as a hydrazide (see Table 6.1.III and IV). Only one carbonyl and amino cross-linker has been studied, N-succinimidyl-4-hydrazidoterephthalate (SHTH, Appendix D.E.CXLVIII). It has been used to cross-link the oxidized carbohydrate moiety in Fc region of polyclonal antibodies and lysine residues on the microtubule surface in a kinesin-based transport system.[69]

    A few compounds are designed to cross-link carbonyl and sulfhydryl groups. Maleimido hydra-zides (Appendix D.D.CXLIV through CXLVII) contain a maleimide and a hydrazide moiety. The other reagents (Appendix D.D.CXLII and CXLIII) contain an alkoxylamino group with a disul-fide bond. Through thiol–disulfide interchange, protein sulfhydryls will form new disulfide bonds with these compounds.[70] The alkoxylamino moiety of these compounds reacts readily with ketones and aldehydes to produce stable alkoxime as shown in Figure 6.3. Thus, dialdehydes derived from the carbohydrate moiety of glycoproteins on periodate oxidation will react with the alkoxylamino group or hydrazide. These compounds have also been used to cross-link adriamycin and thiolated antibody and the preparation of immunoconjugates.[70,71]

6.2.4   MISCELLANEOUS HETEROBIFUNCTIONAL CROSS-LINKERS WITH UNDEFINED SPECIFICITY

In addition to the above categories of heterobifunctional cross-linkers, there are other heterobifunc-tional reagents that cross-link various nucleophiles. The Cyssor reagent, 2-methyl-N^1-benzenesulfonyl-N^4-bromoacetylquinonediimide (Appendix D.F.CXLIX), has been found to cross-link antibodies at pH 8.0.[72] This molecule is essentially an alkylating agent. Nucleophiles attack both the quinone ring and the bromoacetyl group as shown in Figure 6.4. Although the nucleophiles in this particular reac-tion have not been identified, it is speculated that carboxyl groups of aspartate and glutamate and the indolyl ring of tryptophan may serve as nucleophiles in addition to sulfhydryls and amino groups.

    Compounds that contain an aldehyde group such as N-hydroxysuccinimidyl-p-formylbenzoate (HFB) and methyl-4-(6-formyl-3-azidophenoxy)butyrimidate (FAPOB) (Appendix D.F.CL and CLI) will form

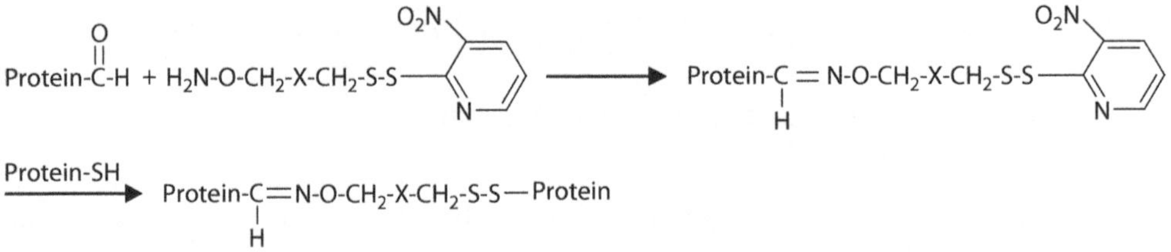

FIGURE 6.3   Cross-linking carbonyl and sulfhydryl groups with aminooxypyridyldithio derivatives.

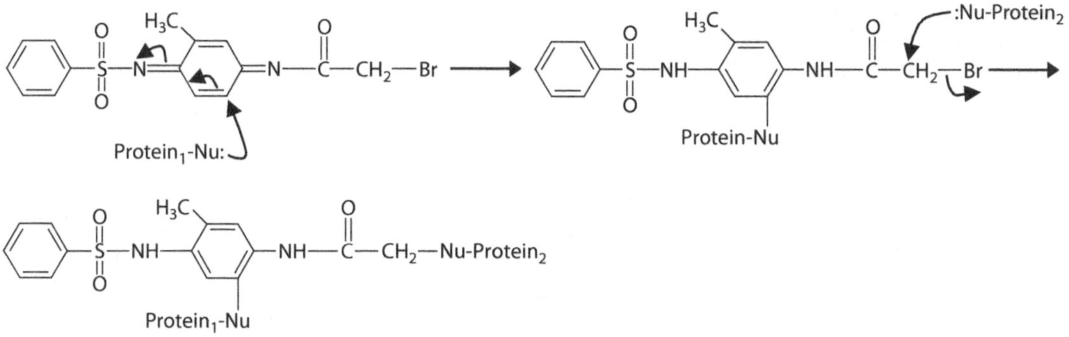

FIGURE 6.4 Speculated mechanism of cross-linking reaction with 2-methyl-N^1-benzenesulfonyl-N^4-bromo-acetylquinonediimide.

Schiff base with amino groups.[72–76] The succinimidyl of HFB and the imidoester of FAPOB are also reactive toward the ammo group, making these two compounds function like a homobifunctional reagent. FAPOB, for example, cross-links lysine 51 and lysine 29 of the ribosomal protein L7/L12.[74,75] FAPOB, containing an phenylazido group, is also photosensitive and may be used as a photochemical agent. In acrolein (Appendix D.F.CLIII), the aldehyde group may form a Schiff base with an amino group. Its double bond is subject to Michael addition with different nucleophiles and has been shown to cross-link collagen.[77] Ishii et al.[78] used model peptides and mass spectrometric techniques to show that cross-linking took place between amino groups and the side chain of histidine in the peptide. Compounds that contain either trimethoxysilyl or triethoxysilyl moiety such as 3-glycidyloxypropyltrimethoxy silane and N-(3-triethoxysilylpropyl)-4-(isothiocyanatomethyl)cyclohexane-1-carboxamide (TPICC) (compounds CXXXIV and CXXXVI in Appendix D) are capable of reacting with hydroxyl groups of silica on glass surfaces. The other end of the molecules containing reactive epoxide or isothiocyanates can react with various nucleophiles, thus immobilizing the molecules as shown in Figure 6.5. These compounds have been used for the preparation of bioconjugates and immobilization of biomolecules such as oligonucleotides, peptides, and proteins on the glass surface.[79,80]

Affinity labeling is a specially designed agent that will bind specifically to a desired location of a biomolecule, usually the active site. An example of an affinity bifunctional cross-linker is

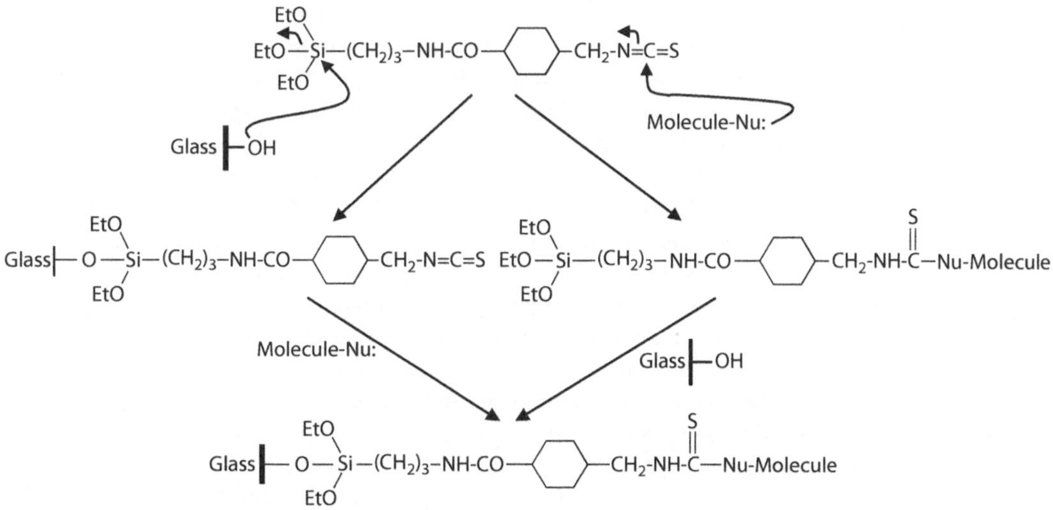

FIGURE 6.5 Reaction mechanism of immobilization of macromolecules on to glass surface using N-(3-triethoxysilylpropyl)-4-(isothiocyanatomethyl)cyclohexane-1-carboxamide. (Adapted from Misra, A. et al., *Bioorg. Med. Chem. Lett.*, 18, 5217, 2008.)

1-(4-methoxyphenyl)-3-acetamido-4-methoxyazetidin-2-one (Appendix D.F.CLVI), which is an analog of β-lactam.[81] It inhibits the class A β-lactamase from *Bacillus cereus* 569/H as an active site-directed inhibitor and cross-links ser 70 and lys 234 of the enzyme. However, the mechanism of reaction is unknown.

6.3 PROTEIN-PHOTOSENSITIVE HETEROBIFUNCTIONAL CROSS-LINKING REAGENTS

Photosensitive heterobifunctional cross-linkers represent by for the largest portion of the heterobifunctional reagents. For a comprehensive list of these compounds, please see Appendix E. Because these functionalities are inert until they are photolyzed, these reagents are first linked to the protein in the dark through a group-directed agent as shown in Table 6.2. The labeled protein is then irradiated to activate the photosensitive group, which reacts indiscriminatively with its environment as discussed in Chapter 4.

The photosensitive cross-linkers are generally classified according to the active species they produce, for example, the nitrenes and carbenes.[82] These reactive species are generated from various groups. As discussed in Section 4.5, the most widely used photoreactive groups are azides, diazo moiety, benzophenone, diazonium salts, and diazirines. Only a few carbene-generating diazo reagents are used for cross-linking,[65,67,68,82–87] probably because of the ability of carbenes to undergo a variety of reactions including the very efficient reaction with water.[82] In addition, the parent diazoacetyl compounds are generally unstable, particularly at low pH and are reactive toward nucleophiles including carboxyl groups.[64] However, 3-phenyl-3-(trifluoromethyl) diazirine (TPD) is gaining popularity and has been incorporated into photoaffinity labels and cross-linkers.[88,89] The reasons are the unexpected stability of the TPD three-membered ring and its ability to be photoreactivated with light over 350 nm to generate carbenes, which can rapidly form cross-links to biomolecules with short photoirradiation times.

Azido derivatives constitute the majority of the photoactivatable cross-linking agents (compounds I through XXXVIII, L through LXI, LXXI through XCIV, CXVI, CXVII, and CXXI through CXXXIX in Appendix E). Three types of azides have been synthesized: the aryl, alkyl, and acyl azides. Alkylazides, however, are not used in cross-linking for several reasons. First, they have absorption maxima in the UV region in which irradiation may damage proteins, nucleic acids, and other components. Second, the alkylnitrene intermediates readily undergo rearrangement to form inactive imines. Last, alkylazides are reactive and may undergo nucleophilic displacement reactions. For the same reasons, acylazides are generally used as acylating agents rather than photoaffinity labels.[90] Only arylazides have been extensively used in photoactivatable cross-linkers.[90–138] Aryl azides have a low activation energy and can be photolyzed in the long UV region.[21,82,139] The presence of electron-withdrawing substituents such as nitro and hydroxyl groups further increases the wavelength of absorption into the 300 nm region.[64] Arylnitrenes have a half-life on the order of 10^{-2}–10^{-4} s,[140,141] and, therefore, the cross-linking reaction is expected to be terminated within a short time. Arylazides are susceptible to reduction to amino groups. They are not stable in the presence of thiols. The half-life of arylazides is 5–15 min in 10 mM dithiothreitol (pH 8.0) and over 24 h in 50 mM mercaptoethanol (pH 8.0).[142]

The benzophenone derivatives (LXII through LXV, CXI, CXII, and CXL through CXLIV in Appendix E) constitute yet another class of photoaffinity labels.[143–145] These compounds, as shown in Chapter 4, can form covalent adducts on irradiation with nearby amino acid residues leading to cross-linking. Unlike the azides, which are irreversibly photolyzed in most cases, the excited triplet state of benzophenones may be resistant to reaction with water and may revert back to the starting material if no photoreaction takes place. Since benzophenones can be reexcited, their cross-linking efficiency can, in principle, reach 100%.[143,144] These compounds have been used to study virus-induced proteins,[146] ribosomal proteins,[147] troponin and tropomyosin,[148] actin,[149] thin filament proteins,[150] and chymotrypsin.[151]

Another class of photosensitive reagents is nitrophenyl ether (Appendix E.B.CXIV and CXV).[152] These compounds react quantitatively with amines at slightly alkaline conditions (pH 8) on irradiation with 366 nm light. The reaction involves the transfer of nitrophenyl group from the alcohol to the amine as shown in Figure 6.6 for the cross-linking reaction of *N*-(maleimidomethyl)-2-(*O*-methoxy-*p*-nitrophenoxy)carboamidopropane. With this compound, the 2-methoxy-4-nitrophenyl ether is

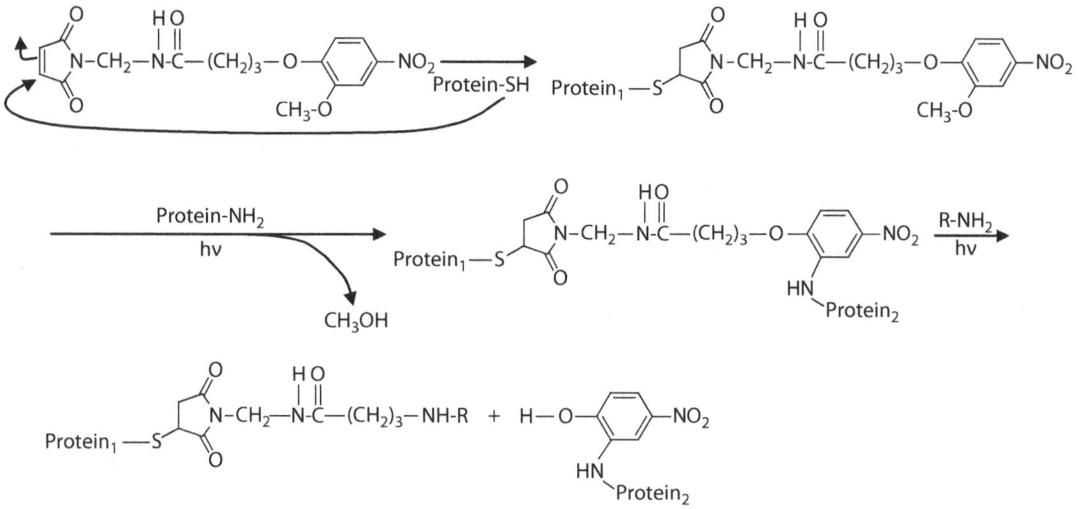

FIGURE 6.6 Photochemical cross-linking of protein sulfhydryl and amino groups effected by a nitrophenyl ether and the cleavage of the cross-linked product.

attached to the protein through the Michael addition reaction of a thiol group at the maleimide ring. Such reaction was demonstrated to occur at the γ-cysteine F9 of human fetal hemoglobin. On irradiation, the reagent yielded γ-γ-cross-linked hemoglobin.[152] Like other photoreagents, the nitrophenyl ethers are stable in the dark, but unlike other reagents, they are stable even upon irradiation in the absence of a nucleophile. Irradiation excites the compound to a triplet state with an extremely short lifetime of 10^{-7} to 10^{-9} s. The chemical reaction will have to occur during that time to prevent the reagent from wandering. Nonproductive deactivation regenerates the starting compound providing a relatively high yield as benzophenones.

One of the most important applications of these photoactivatable bifunctional reagents is in the identification of receptors. The photosensitive agent is first anchored onto the protein in the dark according to the group specificity of the reagent as will be discussed below. The labeled polypeptide ligand is then allowed to bind to its specific receptors. On photolysis, cross-linking will occur with molecules directly interacting with or adjacent to the derivatized ligand. This technique has been used to identify binding sites for Con A,[1,2] vasoactive intestinal polypeptide receptor,[133] epidermal growth factor,[130] insulin receptor,[153] fibronectin,[154] bungarotoxin,[100] choriogonadotropin,[96] calmodulin-binding protein,[155] interleukin-3 receptor,[108] glucagon receptor,[156] nerve growth factor,[91,92] and parathyroid hormone receptor,[157] to mention a few.

6.3.1 AMINO GROUP–ANCHORED PHOTOSENSITIVE REAGENTS

The functional groups used to react with amino groups in these photosensitive reagents are shown in Table 6.2.I. They contain such classical amino group–selective functionalities as N-hydroxysuccinimidyl ester, imidoester, aryl halide, isothiocyanates, acyl chloride, and p-nitrophenyl ester. The photosensitive components are made up of arylazide, benzophenone, diazoacetate, and diazirine. A comprehensive list of amino group–anchored photosensitive heterobifunctional reagents can be found in Appendix E.A. The majority of photophores are arylazides (Appendix E.A.I through XXXVIII and L through LXI), followed by diazirines (Appendix E.A.XXXIX through XLIX). These diazirines are relatively new additions and have many applications. For example, Bochkariov and Kogon[158] coupled N-hydroxysuccinimide and 3-(3-(3-(trifluoromethyl)-diazirin-3-yl)phenyl)-2,3-dihydroxypropionic acid with N,N'-dicyclohexylcarbodiimide to form compound LXVI of Appendix E, which was successfully linked to the amino acid end group of Phe-tRNA. The labeled tRNA was bound to ribosomes and photolyzed, which caused cross-linking to the 23S RNA. The cis-diol bond enabled the cross-linked product to be cleaved by periodate.

Besides arylazides and diazirines, there are only few reagents containing benzophenone (Appendix E.A.LXII through LXV) and diazo (Appendix E.A.LXVII through LXX) photophores.

A few of these compounds such as *N*-succinimidyl-4-azidosalicylate (Appendix E.A.III) contain the phenol ring and are directly iodinatable with reagents such as chloramine T. This arrangement provides a convenient way of introduction of the radioisotope ^{125}I. Some compounds are cleavable. Those containing the disulfide bond will be cleavable with excess mercaptans (Appendix E.A.XX through XXVIII, XLIII, XLIV, and LIV through LVI). The azo derivatives (Appendix E.A.XXIX through XXXII) are cleavable by dithionite.

These photoactivatable cross-linkers are mostly used for the identification of cell surface receptors.[1,139] For example, sulfosuccinimidyl-2-(*p*-azidosalicylamido)-1,3′-dithiopropionate (SASD, Appendix E.A.XXVII) was used to identify cell surface receptors,[159] and *N*-5-azido-2-nitrobenzoyloxysuccinimide (ANB-NOS, Appendix E.A.VIII) was used to identify the ligand-binding sites on integrin $\alpha4\beta1$.[160] In addition, some of these compounds have been applied to study transferrin-binding sites and ion channels.[99,161]

6.3.2 SULFHYDRYL GROUP–ANCHORED PHOTOACTIVATABLE REAGENTS

As shown in Table 6.2.II, reagents with maleimido groups, disulfides, alkyl halides, and thiol ethers undergo nucleophilic reaction with the thiolate ion as the most preferable agent. In the absence of the thiol group, however, other nucleophiles will react with these groups and, therefore, constitute the major side reaction, particularly at high pHs. For example, 2-nitro-4-azidophenylsulfenyl chloride (2,4-NAPSCl) (Appendix E.B.LXXVIII) was synthesized to react with sulfhydryl groups for the modification of tryptophan residues on photolysis, but it reacts with various nucleophiles.[119] Proteins lacking thiol groups may be first thiolated with 2-iminothiolane and then react with the cross-linking reagent.[162] Compounds that can undergo disulfide exchange with thiols are truly sulfhydryl-specific (Appendix E.B.LXXXI through LXXXIII, LXXXV through XCIII, CI, CIII, CV through CIX). The cleavable amino group reagents with disulfide bonds mentioned in Section 6.2.1 are therefore also thiol reagents. In this case, the reagent would behave as a heterobifunctional reagent if it reacts with the amino group first. If it reacts with a protein sulfhydryl group first, the disulfide exchange may result in its acting both as a cross-linker for reacting with an amino group or as a photosensitive cross-linker. In either case, the cross-linked products formed are cleavable in the presence of an excess mercaptan. The seleno ester, 3-(4-azido-2-nitrobenzoylseleno) propionic acid (ANBSP, Appendix E.B.XCIV), also selectively reacts with free sulfhydryl group.[125] Upon nucleophilic substitution by thiols, thiol esters are formed. The liberated selenol readily forms diselenides, providing a very favorable equilibrium for the reaction. The protein thiol–ester can be cleaved with excess free thiols or amines, thus offering a means of identifying the labeled amino acids of proteins after photolysis. The labeled amino acids will be within a span of 7 Å from the cysteine residue anchor.

There are fewer published thiol-anchored photoactivatable agents than amino group–anchored agents (Appendix E.B). Again, most of the photophores are arylazides (Appendix E.B.LXXI through XCIV). There are few benzophones (Appendix E.B.CXI and CXII) and two nitrophenyl ethers (Appendix E.B.CXVI and CXV) with few diazoacetate (Appendix E.B.CVII to CX). The second largest group of reagents are the diazirines (Appendix E.B.XCV through CVI).

In addition to being used for identifying receptors, these thiol-anchored reagents have been used to study protein interactions in troponin,[121,122] α-tropomyosin,[123] rhodospin,[163] F-plasmids,[164] cytochrome c,[165] and protein–nucleic acid interactions such as HIV reverse transcriptase (RT) where both BATDHP (Appendix E.B.C, a diazirine) and APTP (Appendix E.B.LXXX, an azide) were used to label mutant RT with specific cys locations.[166] These modified RTs were allowed to bind to a dsDNA template primer. Upon irradiation with mild UV light, the photoactivatable groups rapidly and nonspecifically reacts with nearby DNA to form protein–DNA cross-links. The benzophenone derivative, benzophenone-4-maleimide (Appendix E.B.CXI), has been used to study conformational changes in myosin subfragment 1.[167] Fluorinated compounds such as TFPAM-3 (Appendix E.B.LXXII) have been synthesized for mass spectral studies.[168]

6.3.3 Guanidinyl Group–Anchored Photoactivatable Reagents

There is only one published arginine-specific photoactivatable cross-linker, *p*-azidophenylglyoxal (Appendix E.C.CXXIV), which contains a vicinal dicarbonyl that is specific toward the guanidinyl group of arginine (Table 6.2.III). Ngo et al.[127] have used this reagent to inhibit LDH, lysozyme, alcohol dehydrogenase as an arginine-specific reagent. Politz et al.[169] have reacted the reagent with guanosine and cross-linked RNA to proteins in 30S ribosomal subunits after photolysis.

6.3.4 Carboxyl-, Carboxamide-, and Carbonyl-Group–Anchored Photoactivatable Reagents

As shown in Table 6.2.IV, photoactivatable reagents containing a free alkylamine are considered reactive toward carboxyl groups, carbonyl groups, and γ-carboxamide moiety of glutamine. In the presence of a carbodiimide, condensation occurs between the carboxyl group of a protein and the amino group of the photosensitive reagent. After labeling, the protein may be photolyzed to activate the photophor.[130] In the presence of transglutaminase, the amines are introduced covalently to the γ-carboxamide group of peptide-bound glutamine residues.[170–172] These compounds (Appendix E.D) have been incorporated into substance P, glucagon, and casein in this manner.[128] *N*-(azido-2-nitrophenyl)putrescine (ANP, Appendix E.D.CXXVII) was covalently bound to Gln-41 of rabbit skeletal muscle actin by a bacterial transglutaminase-mediated reaction[173] and cross-linked Cys-374 residues of two adjacent actin protomers. The cross-linked actin dimer was used to study its crystal structure.[174]

Carbonyls of aldehydes and ketones are normally not found in proteins. However, they can be derived from periodate oxidation of carbohydrates of glycoproteins.[175] These carbonyls can react with hydrazides and amines at pH 5–7 to form Schiff bases, which may be stabilized by borohydride. The reaction with hydrazides is faster than with amines, making them useful for site specific cross-linking. Aldehydes react with hydrazides in the formation of a hydrazone bond. Watkins et al.[176] have used *p*-azidobenzoyl hydrazide (ABH, Appendix E.D.CXXXIV) to immobilize human decay accelerating factor onto a cardiovascular bypass circuit. The reaction is particularly useful for antibodies in which the carbohydrate is located in the Fc region away from the binding sites.

6.3.5 Photoaffinity-Labeling Reagents

Affinity labels are reagents specifically designed to bind with high affinity to a protein molecule. They are usually analogs of substrates or inhibitors. The ATP/adenosine photoactivatable affinity labels (CXXXVI through CXLIV in Appendix E), such as 3′-arylazido-β-alanine-δ-azido-ATP (diN$_3$ATP), 5′(*p*-fluorosulfonylbenzoyl)-8-azidoadenosine (FSBAzA), adenosine 5′-triphosphate-γ-benzophenone (5-BzATP), and 3′-*O*-(4-benzoyl)benzoyl-adenosine 5′-triphosphate (3-BzATP) are expected to bind to ATP-binding proteins. Other photolabile nucleoside derivatives can be found in a review by Blencowe and Hayes.[89] diN$_3$ATP (Appendix E.E.CXXXIX) has been shown to bind to the ATP-binding site of bacterial F$_1$ATPases and actually serves as a substrate for the enzyme.[131] On photolysis, the 8-azidoadenosine moiety labels the adenine binding site, which is located on the β-subunit. The other photoactivatable moiety, azidophenyl, interacts with the neighboring polypeptide, which is the α-subunit. Thus, diN$_3$ATP cross-links the α- and β-subunits of F$_1$ATPase on photoactivation.

FSBAzA (Appendix E.E.CXXXVI) has been shown to bind to the adenine nucleotide binding site of glutamate dehydrogenase.[132] The electrophilic fluorosulfonyl moiety is capable of reacting with amino acid side chain nucleophiles at the binding site. This nucleophilic substitution reaction takes place in the dark and anchors the photoaffinity label to the protein nucleotide binding site. On photolysis, the azido group is activated and reacts with the neighboring amino acids, making a cross-link between the nucleophile- and the adenine-binding residues.

The benzophenone-attached ATP at either 3′ or 5′-γ-phosphate position (3-BzATP and 5-BzATP) have high affinity for ATP-binding sites. 3-BzATP was used to covalently label and identify the

ATP-binding site of the skeletal muscle ryanodine receptor,[177] and 5-BzATP was used to study the ATP-binding domain of ribulose-1,5-bisphosphate carboxylase/oxygenase.[178]

Photolabile carbohydrate derivatives shown in Appendix E.E.CXLV through E.CXLVII represent only a few carbohydrate photoactivatable affinity labels. More examples for photoaffinity ganglioside, phospho- and sphingolipid, phosphoramidite, and galactosylceramide probes can be found in the reviews by Vodovozova[143] and Blencowe and Hayes.[89] These affinity labels have been used study glucose transport system, carbohydrate binding proteins, membrane structure, receptors, etc. This area of research is far-reaching and is beyond the scope of this book. The compounds listed in Appendix E.E are true photosensitive cross-linkers. 9-AAz-NeuAc (Appendix E.E.CXLV), a sialic acid analog, was metabolically incorporated into glycoproteins. Photoactivation revealed *in situ* interaction of the glycoprotein with co-receptor CD22.[179] Similarly, Tanaka and Kohler[180] incorporated two diazirine labels, Ac5-5-SiaDAz (Appendix E.E.CXLVI) and Ac4-ManNDAz (Appendix E.E.CXLVII), into cellular glycoproteins in the form of sialic acid in a K20 cell line. They were able to use the compounds to capture carbohydrate-mediated interaction with CD22.

Two photoaffinity analogs of retinal (Appendix E.E.CXLVIII and E.E.CXLIX) have been synthesized to study rhodopsin.[181,182] Both have a free aldehyde group that forms a Schiff base with the amino group of Lys-296 of helix G of rhodopsin. However, the photoactivatable group is different. In *o*-dimethyl-*p*-trifluoromethyldiazirine phenyl retinal (Appendix E.E.CXLIX), diazirine is the photophore, which when photolyzed at 365 nm with rhodopsin covalently cross-linked predominantly to helices C or F.[182] When the analog reconstituted rhodopsin in rod outer segments was photolyzed, cross-linking was predominantly to helix C. In 3-diazo-4-oxo-10,13-ethano-11-*cis*-retinal (Appendix E.E.CXLVIII), the diazo moiety is the photophore that cross-linked exclusively to Trp-265/Leu-266 in helix F when bound to opsin.[181] Both compounds provided insight into the structure of rhodopsin.

6.4   NONCOVALENT IMMUNOGLOBULIN CROSS-LINKING SYSTEM

The high affinity of antibodies for antigens makes it possible to use immunoglobulins as cross-linking reagents. Antibodies of different specificities have been cross-linked to yield heterobifunctional agents. Bode et al.[49] have prepared a bispecific antibody by cross-linking antifibrin antibody and 2-iminothiolane-modified antitissue plasminogen activator (tPA) with *N*-succinimidyl-3-(2-pyridyldithio)propionate. Such a bispecific antibody recognizes both fibrin and tPA and is able to conjugate these components with an apparent dissociation constant of 10^{-9}–10^{-10} M. This application has extended immunoglobulins from homobifunctional cross-linkers to heterobifunctional cross-linkers. Besides direct cross-linking of two different antibodies to form bispecific antibodies, this chemical manipulation can be extended to involve the dissociation of the two different antibodies and reassociation of the two-half molecules.[183] With the advent of monoclonal technology, this approach proves to be particularly useful for developing bispecific (Fab′)2 antibodies.[183,184] Group-specific homobifunctional and heterobifunctions reagents can be used to cross-link the molecules. In addition to chemically cross-linking two antibody molecules or antibody fragments as demonstrated by Bode et al.,[49] bispecific antibodies can be created by fusion of two different cell lines to form a quadroma or trioma. Fusion of two established hybridomas generates a quadroma,[185] whereas fusion of one established hybridoma with lymphocytes derived from a mouse immunized with a second antigen generates trioma.[186] The produced bispecific antibodies are purified from the media of the cell cultures.[183] Bispecific antibodies can also be made by recombinant DNA-based approaches. This genetic engineering technique overcomes many of the shortcomings of chemical conjugation and cell fusion since homogeneous antibodies can be produced.[187] There are numerous methods for producing monoclonal-bispecific antibodies.[183,184,188] Single-chain bispecific monoclonal molecules have been made by combining two single chain Fv fragments using a polypeptide linker.[189] Bostrom et al.[190] described a new "two-in-one" designer antibody concept in which the same binding site on an antibody is engineered to recognize two different antigens, both with high affinity. A novel "knobs into holes" method was presented by Carter et al.,[191,192] where the knobs were created by replacing small amino acid side chains at the interface between C_H3 domains with larger ones, whereas the holes were generated by replacing large side chains with smaller ones. Another method for dimerization of monoclonal antibodies fragments is leucine zipper.[193,194] Holliger et al.[195] developed a diabody method to generate bispecific monoclonal antibody fragments. This approach is reviewed by Kipriyanov.[196] Lu and Zhu[197] described a recombinant method for the construction and production of a novel IgG-like bispecific antibody molecule, using the variable domains of two fully human antibodies as the building blocks. Glycosylation of bispecific diabody was investigated by Kim et al.[198]

There are many applications of bispecific antibodies.[183,184] They have been used in immunohisto-chemistry and enzyme immunoassays.[199] Milstein and Cuello[200] first developed an antisomatostatin/antiperoxidase bifunctional antibody for immunohistochemistry use. Bispecific monoclonal antibodies (bsMAb) directed against an enzyme (e.g., HRPO, alkaline phosphatase, and α-galactosidase) and a second antigen (e.g., tumor-specific antigen, peptide, or hormone) have been developed for use in enzyme immunoassay or immunohistochemistry.[201–204] Bispecific antibodies have also been used in radioimaging and radioimmunotherapy.[205] These bsMAb were designed to deliver radio-isotopes such as ^{99m}Tc, ^{90}Y, ^{67}Ga, and ^{111}In to a tumor quickly and specifically for imaging and radioimmunotherapy.[206–208] Cornelissen et al.[209] used ^{111}In-labeled bispecific immunoconjugates that specifically bound to EGFR and p27(Kip1) to probe intranuclear proteins in breast cancer cells. Radioimmunotherapy has also been used to treat solid tumors.[210] Kraeber-Bodéré et al.[211] described a bispecific monoclonal antibody recognizing carcinoembryonic antigen and the dipeptide hapten di-diethylenetriamine pentaacetic acid (DTPA)-indium-tyrosine-lysine labeled with ^{131}I for treatment of medullary thyroid carcinoma. The preparation is now under clinical trial. The antibody therapeutics are probably one of the important applications of bsMAb.[212,213] This approach may replace combination therapies of a mixture of monospecific antibodies.[214] These bsMAb are capable of activating and targeting the cellular immune defense system to kill tumor cells or other pathogens. Such bsMAB-mediated cancer immunotherapeutic strategies have been applied to antibodies of antiepidermal growth factor receptor for breast cancer,[215] anti-sialyl Lewis(a) for colon cancer,[216] antiovarian cancer,[217] antirenal cell carcinoma,[218] anti–small cell lung carcinomas,[219] anti-CD19 for B lymphoma,[220] anti-CD13 for acute myeloid leukemia,[221] and anti-tenascin for gliomas.[222] A new class of bsMAb called "bispecific T-cell engager" (BiTE antibodies), which are bispecific for a surface target antigen on cancer cells and for CD3 on T cells, have been developed.[223,224] Nagorsen et al.[225] constructed a single-chain bispecific antibody with specificity for CD19 for treatment of patients with CD19-expressing hematological malignancies. These BiTE antibodies are able to control tumor growth and survival in cancer patients. Immunotoxins are another tool for cancer therapy. The bsMAb allows a therapeutic agent such as a drug, toxin, enzyme, DNA, radionuclide, etc., to be placed on one arm of the antibody while allowing the other arm to specifically target the disease site.[226] Frankel and Woo[227] have prepared a bispecific immunotoxin in *Escherichia coli* with the first 389 amino acid residues of diphtheria toxin. These immunotoxins have been tested in prostate carcinoma, lung carcinoma, glioblastoma, pancreatic carcinoma cells, and antigen positive human cancer cell lines such as breast, ovary, colon, lung, prostate, and squamous cells. In a similar token, bsMAb are effective in delivery of drugs like methotrexate,[228] saporin,[229] doxorubicin,[230] and vinca alkaloids,[231] as well as helping liposomal drug delivery.[232] BsMAb can also be used to activate a prodrug *in vivo* to cause localized targeted cytotoxicity. De Sutter and Fiers[233] prepared a bsMAb against a tumor marker, human placental alkaline phosphatase, and *E. coli* α-lactamase as the prodrug activating group. Activation of cephalosporin-based anticancer prodrugs at the tumor site was achieved with this bsMAb. The plethora of applications of bsMAb as unique macromolecular heterobifunctional cross-linkers is expanding with promising prospects for clinical use.

A new approach with bifunctional antibodies is the development of chemically programmed antibodies (cpAbs). Gavrilyuk et al.[234,235] used the aldolase antibody 38C2 to bind a target of biological interest via amide bond formation between the low pK_a catalytic site lysine $H93_{1,2}$ and β-lactam-equipped targeting-modules, such as cyclic-RGD peptide-linked LHRH or small-molecule integrin

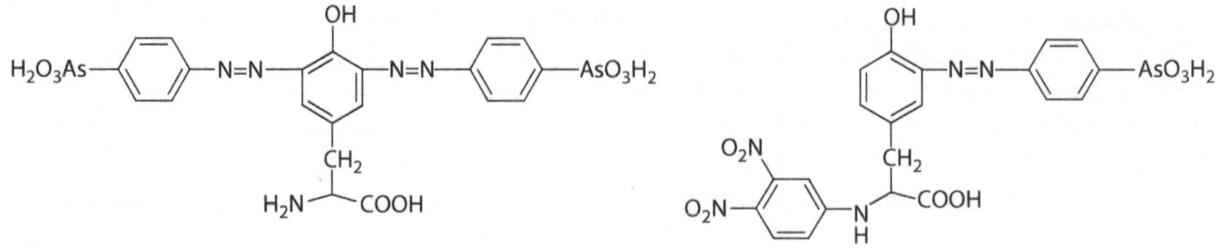

FIGURE 6.7   Chemical structure of bifunctional antigens: Bis-RAT, L-tyrosine-bis(p-azobenzenearsenate); DNP-RAT, dinitrophenyl-1-tyrosine-p-azobenzenearsenate.

inhibitor SCS-873 conjugated to LHRH. The result is a bifunctional and bivalent RDG/LHRH and SCS-783/LHRH antibody. The cpAbs are bound specifically to the LHRH receptors expressed on human ovarian cancer cells. This approach could provide efficient and economical high-valency therapeutic antibodies that target specific receptors.

Compounds that contain two antigenic determinants will cross-link antibody molecules. Simple molecules such as L-tyrosine-bis(p-azobenzenearsenate) (Bis-RAT) and dinitrophenyl-1-tyrosine-p-azobenzenearsenate (DNP-RAT) (Figure 6.7) have been shown to cross-link anti-RAT antibodies, and anti-RAT and anti-DNP antibodies, respectively.[236] Other examples of multiepitopic molecules are hapten-conjugated antigens such as dinitrophenyl-labeled bovine serum albumin.[237] In addition to proteins, polysaccharides have also been used as carriers. Ficoll, a polymer of fructose, has been covalently bonded with dinitrophenyl and phosphorylcholine.[238,239] Between 30 and 35 haptens can be conjugated per molecule of ficoll. Such molecules are able to cross-link immunoglobulins to form large immune complexes used for nephropathy studies.[240]

6.5   HETEROBIFUNCTIONAL NUCLEIC ACID CROSS-LINKING REAGENTS

Like homobifunctional DNA cross-linking agents, many of the reagents discussed above for protein cross-linking also cross-link DNAs and RNAs, as well as protein and nucleic acids. For example, 3-[3-(bromoacetylamino)phenyl]-3-(trifluoromethyl)diazirine (BAPTD, see Appendix E.B.XCIX), which is a sulfhydryl group–anchored photoactivable diazirine cross-linker, has been used to study tRNA and rRNA intereactions.[241] BAPTD was first reacted with thio group on tRNA^{Arg1} from *E. coli*, which has a 2-thiocytidine residue at position 32 in the anticodon loop. The labeled tRNA was then bound to ribosomes under a variety of conditions specific for binding to the A, P, or E sites. On photolysis, it is found that tRNA^{Arg1} cross-linked to 16S rRNA. With the tRNA bound to various sites, different cross-links at the rRNA were found revealing the overall topography of the decoding region of the 30S ribosomal subunit. Another reagent, acrolein, which cross-links an amino group and another nucleophile, also reacts with DNAs. In fact, a series of α,β-unsaturated aldehydes such as acrolein, crotonaldehyde, and 4-hydroxynonenal (4-HNE) as shown in Table 6.3.I.A cross-link interstrand DNAs between guanines in the neighboring C·G and G·C base pairs located in 5'-CpG-3' sequences.[242] The reaction mechanism is believed to involve Michael addition of the enals to dG N2-amine to give N2-(3-oxopropyl)-dG adducts, which can react with a protein to form protein–DNA conjugates or with another DNA guanine to from interstrand cross-links as shown in Figure 6.8. The interstrand linkage exists as an equilibrium mixture of carbinolamine, imine, and pyrimidopurinone species. In addition to guanine, other nucleobases may react.[242]

Other compounds that not only cross-links protein amino and sulfhydryl groups but also DNAs are epihalohydrins (Table 6.3.I.B). The epihalohydrins including both epichlorohydrin (ECH) and epibromohydrin (EBH) form interstrand cross-links between distal deoxyguanosine residues at 5'-GGC and 5'-GC sites.[46] EBH is a more efficient cross-linker than ECH. The optimal pH for cross-linking is pH 5.0 for ECH and pH 7.0 for EBH. The cross-linking reaction proceeds via a two step process. The formation of monoadducts could occur either by loss of the halide or by attacking

TABLE 6.3

Nucleic Acid and Nucleic Acid–Protein Heterobifunctional Cross-Linking Reagents

I. Alkylating Cross-Linkers

A. α,β-Unsaturated aldehydes

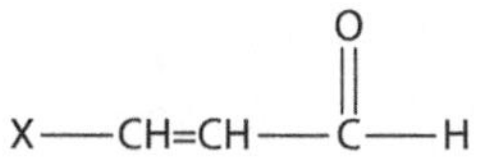

1. X=H: Acrolein
2. X=CH$_3$: Crotonaldehyde
3. X=C$_5$H$_{11}$: 4-Hydroxynonenal (4-HNE)

B. Epihalohydrin

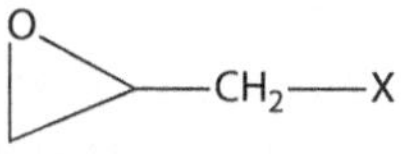

1. X=Cl: Epichlorohydrin (ECH)
2. X=Br: Epibromohydrin (EBH)

C. N'-chloroethylaziridine

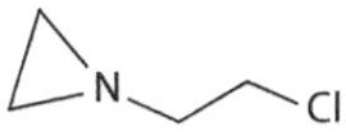

D. Mitomycin C

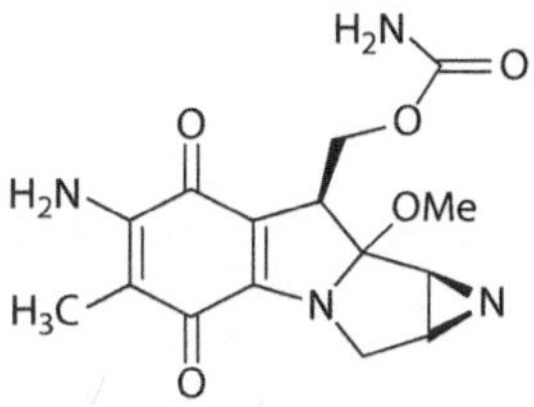

E. Mitomycin C pyrrole conjugates

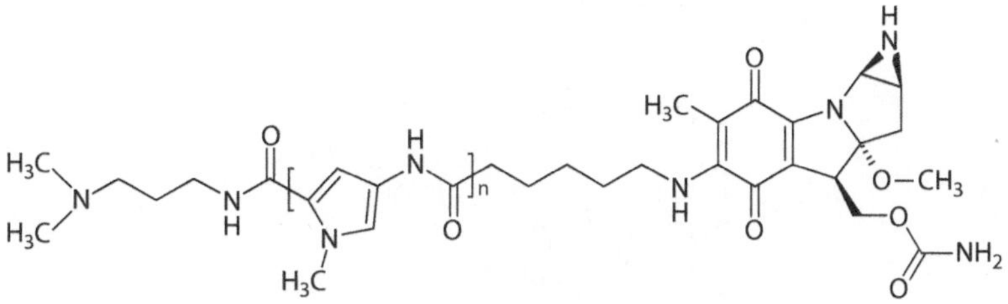

1. n = 1: Mitomycin C monopyrrole
2. n = 2: Mitomycin C bispyrrole
3. n = 3: Mitomycin C tripyrrole

F. N-Methylmitomycin A

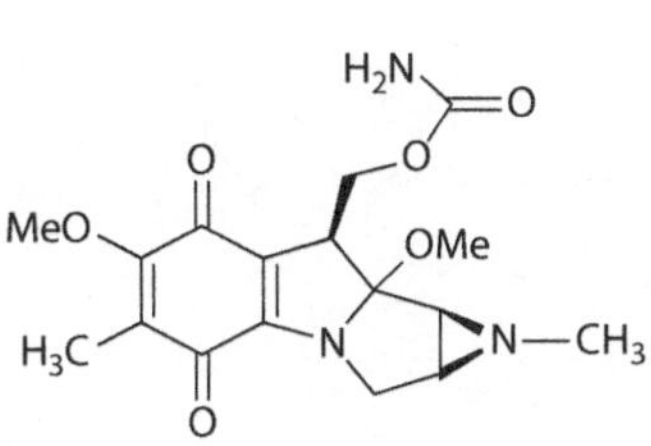

(*continued*)

TABLE 6.3 (Continued)

Nucleic Acid and Nucleic Acid–Protein Heterobifunctional Cross-Linking Reagents

G. Azinomycins

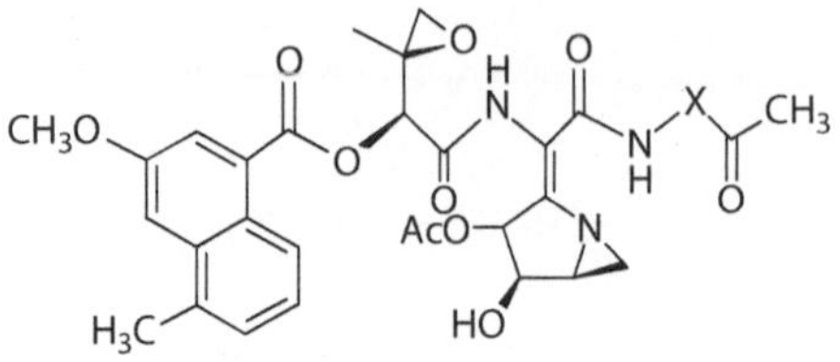

1. X=CH$_2$: Azinomycin A
2. X= C=CH-OH: Azinomycin B
3. X= C=CH-OMe: 4-Methyl azinomycin B

H. Epoxy aziridine azinomycin analog: [(1S)-2-[[(1E)-1-[(1S,5R)-1-azabicyclo[3.1.0]hexan-2-ylidene]-2-ethoxy-2-oxo-ethyl]amino]-1-[(2S)-2-methyloxiran-2-yl]-2-oxo-ethyl]-3-methoxy-5-methyl-naphthalene-1-carboxylate

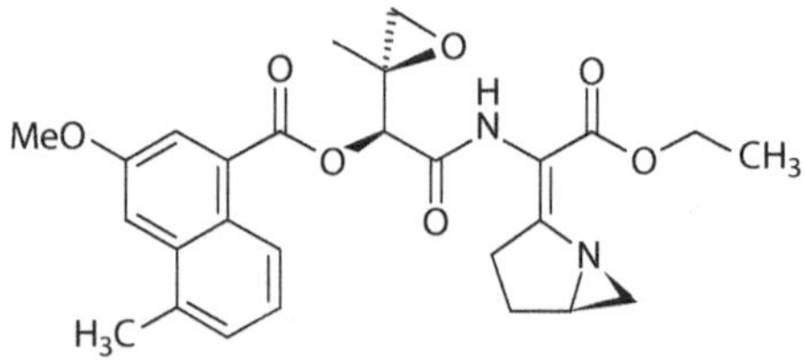

I. Epoxy mustard azinomycin analog: (S)-2-({[2-(3-Chloropiperidin-1-yl)ethyl]amino}-1-[(2S)-2-methyloxirane-2-yl])-2-oxoethyl-3-methoxy-5-methyl-1-naphthoate

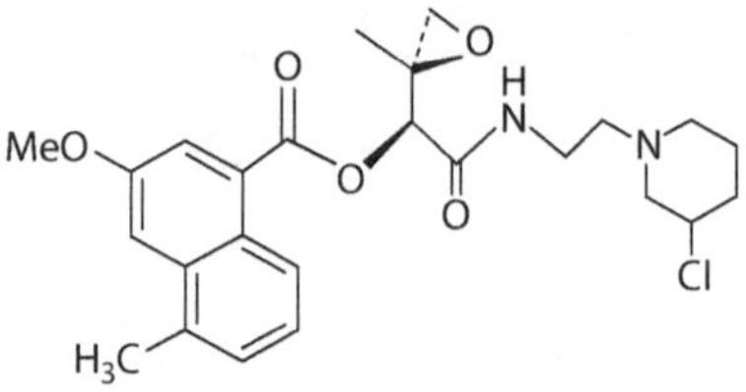

J. Adriamycin

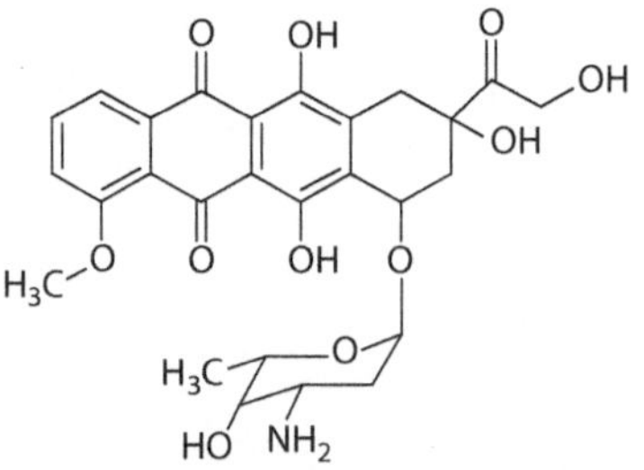

K. Deaminoadriamycin derivatives

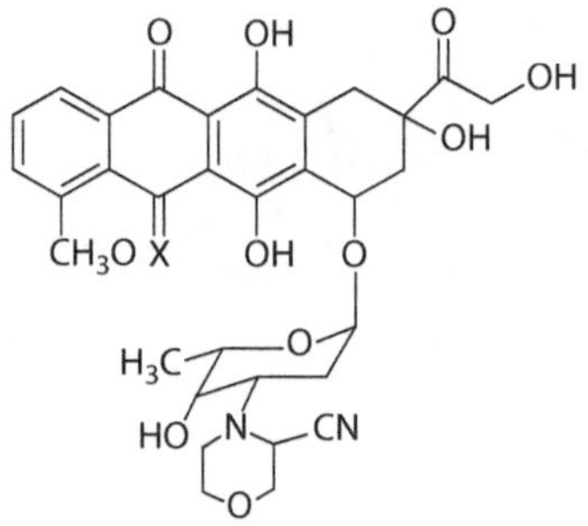

TABLE 6.3 (Continued)
Nucleic Acid and Nucleic Acid–Protein Heterobifunctional Cross-Linking Reagents

    1. X=O: 3′-(3-Cyano-4-morpholinyl)-3′-deaminoadriamycin (CMA)

    2. X=NH: 5-Imino-3′-(3-cyano-4-morpholinyl)-3′-deaminoadriamycin (ICMA)

L. Epoxide PBD: (11aS)-8-(2,3-Epoxypropoxy)-7-methoxy-1,2,3,11a-tetrahydro-5H-pyrrolo[2,1-c][1,4]
    benzodiazepin-5-one

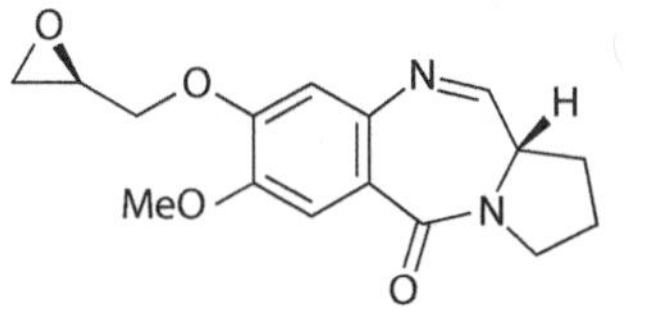

M. CPI-PBD

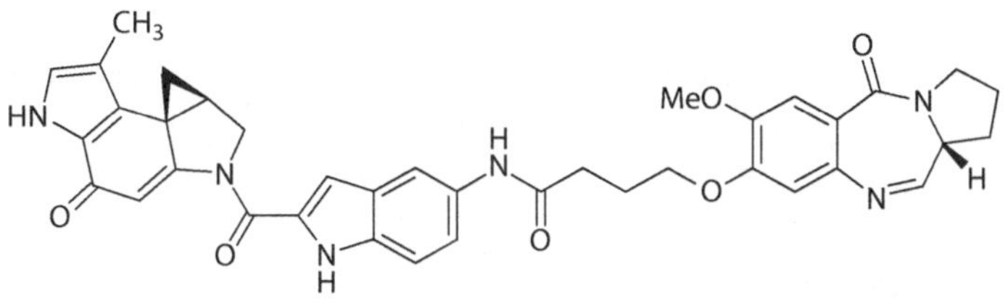

N. Achiral *seco*-amino-CBI-PBD

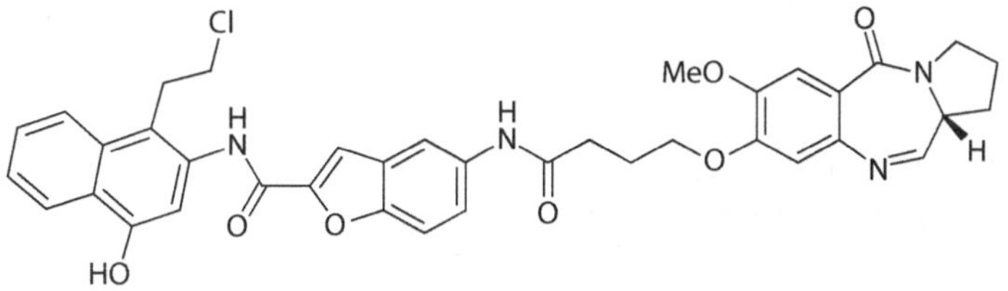

O. Achiral *seco*-CI-PBD

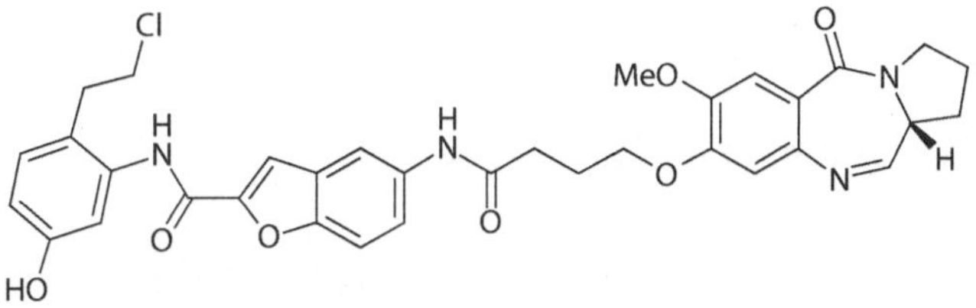

II. Photoactivatable Cross-Linkers
    A. Psoralens

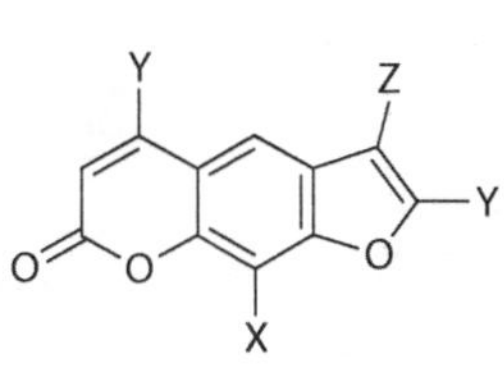

    1. X=Y=Z=H: Psoralen

    2. X=OMe, Y=Z=H: 8-Methoxypsoralen

    3. X=Y=CH₃, Z=H: 4,5′,8-Trimethylpsoralen

    4. X=Y=CH₃, Z=CH₂-NH₂: 4′-Aminomethyl-4,5′,8-trimethylpsoralen, 4′-Aminomethyl-trioxsalen (AMT)

    5. X=Y=CH₃, Z=CH₂-OH: 4′-Hydroxymethyl-4,5′,8-trimethylpsoralen

(*continued*)

TABLE 6.3 (Continued)

Nucleic Acid and Nucleic Acid–Protein Heterobifunctional Cross-Linking Reagents

B. Photoreactive Ru(II) complexes

    1. [Ru(tpy)(dppz-COOH)(CH₃CN)]²⁺ where tpy = 2,2′:6,2″-terpyridine; dppz = dipyrido[3,2-*a*:2′,3′-*c*]-phenazine

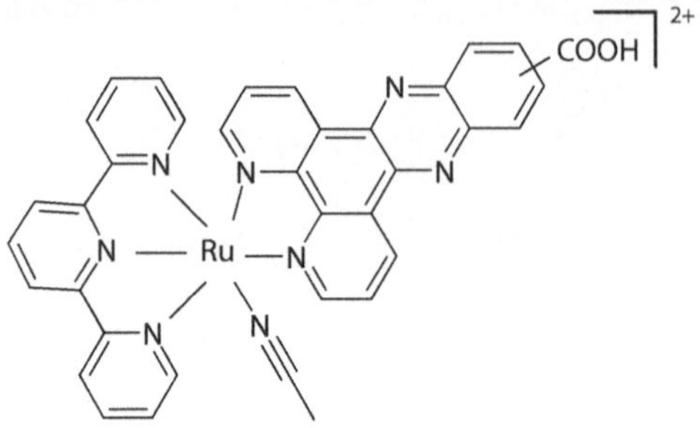

    2. [Ru(TAP)₂dip]²⁺ where TAP = 1,4,5,8-tetraazaphenanthrene; dip = 4,7-diphenyl-1,10-phenanthroline

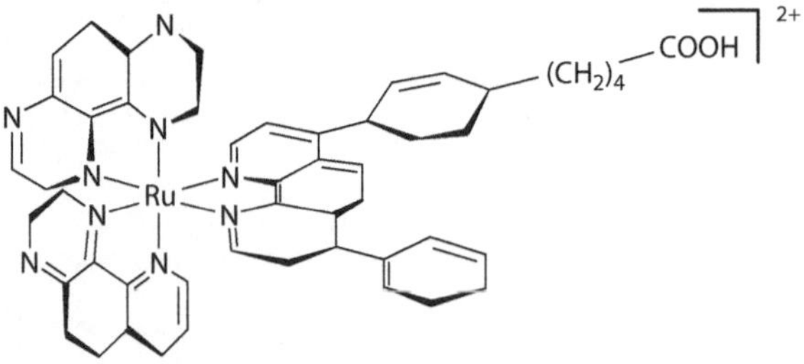

III. Photoaffinity Labeling Cross-Linkers

    A. Amino acid analog: *p*-Benzoyl-L-phenylalaninoe (pBpa):

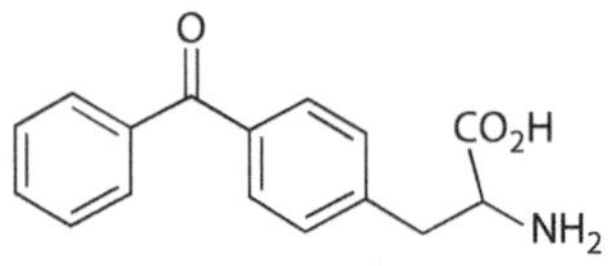

    B. Nucleic acid analogs

        1. Deoxyuridine triphosphate (dUTP) derivatives—dTTP analogs

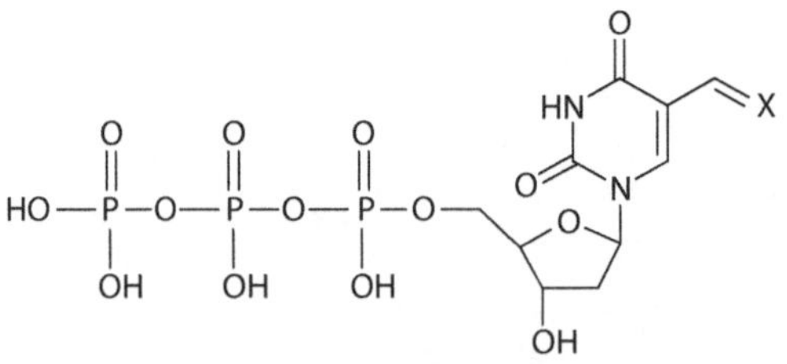

    a. X =

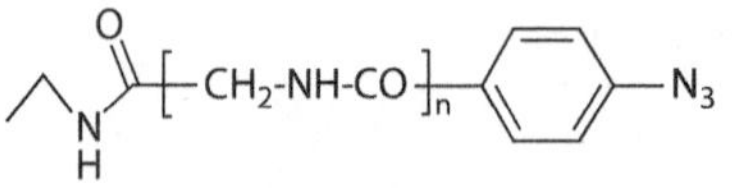

        i. n = 0: 5-[*N*-(4-Azidobenzoyl)-3-aminoallyl]-2′-deoxyuridine-5′-triphosphate (AB-dUTP)

        ii. n = 1: 5-[*N*-(4-Azidobenzoylglycyl)-3-aminoallyl]-2′-deoxyuridine-5′-triphosphate (ABG-dUTP)

        iii. n = 2: 5-[*N*-(4-Azidobenzoyldiglycyl)-3-aminoallyl]-2′-deoxyuridine-5′-triphosphate (ABG2-dUTP)

        iv. n = 3: 5-[*N*-(4-Azidobenzoyltriglycyl)-3-aminoallyl]-2′-deoxyuridine-5′-triphosphate (ABG3-dUTP)

    b. X = : 5-[*N*-(4-(4-Azidophenyl)butyrl)-3-aminoallyl]-deoxyuridine
                              triphosphate (APB-dUTP)

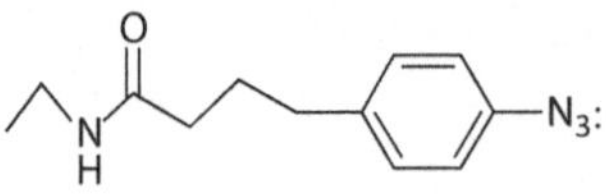

TABLE 6.3 (Continued)

Nucleic Acid and Nucleic Acid–Protein Heterobifunctional Cross-Linking Reagents

c. X= : 5-[N-(4-Azido-2,3,5,6-tetrafluorobenzoyl)-3-amino-propenyl-1]-2′-deoxyuridine-5′-triphosphate (FAB-4-dUTP)

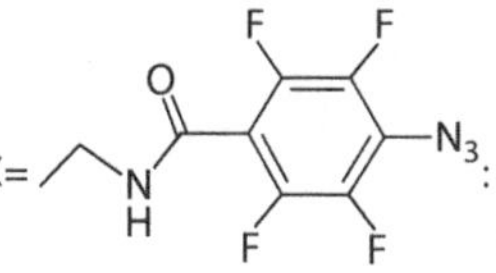

d. X= : 5-[N-[[(2,3,5,6-Tetrafluoro-4-azidobenzoyl)- butanoyl]-amino]-trans-3-aminopropenyl-1]-2′-deoxyuridine-5′-triphosphate (FAB-9-dUTP)

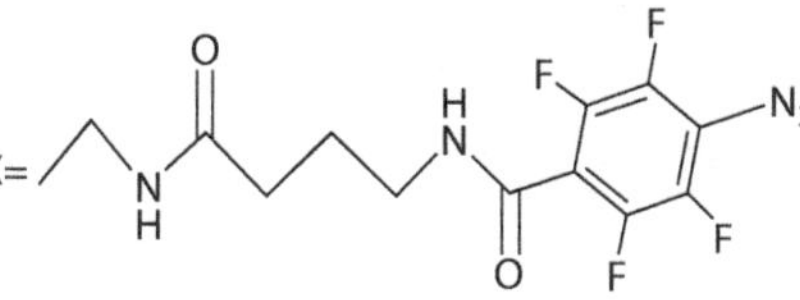

e. X= : 5-[N-[N-(4-Azido-2,5-difluoro-3-chloropyridine-6-yl)-3-aminopropionyl]-trans-3-aminopropenyl-1]-2′-deoxyuridine-5′-triphosphate (FAP-dUTP)

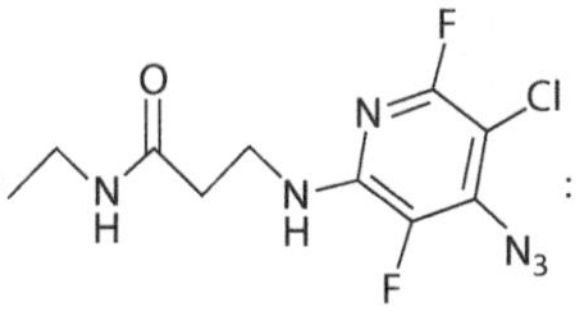

f. X= : 5-[N-(2-Nitro-5-azidobenzoyl)-trans-3-aminopropenyl-1]-2′-deoxyuridine-5′-triphosphate (NAB-4-dUTP)

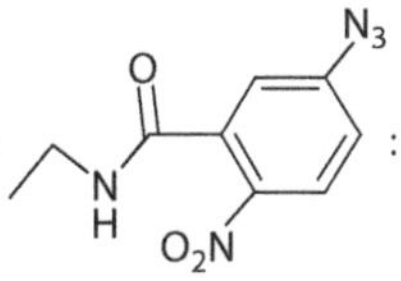

g. X=

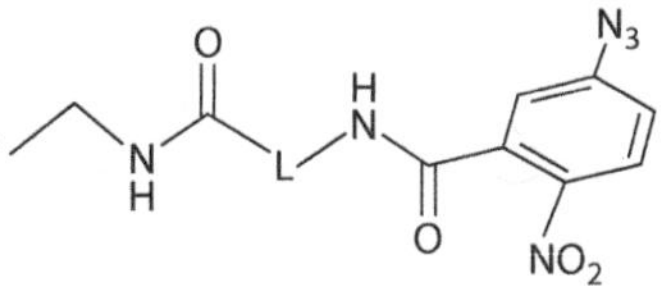

   i. L=CH$_2$: 5-[N-(N′-(2-Nitro-5-azidobenzoyl)-glycyl)-trans-3-aminopropenyl-1]-2′-deoxyuridine-5′-triphosphate (NAB-7-dUTP).

   ii. L=CH$_2$–CH$_2$: 5-[N-(N′-(2-Nitro-5-azidobenzoyl)-3-aminopropeonyl)-trans-3-aminopropenyl-1]-2′-deoxyuridine-5′-triphosphate (NAB-8-dUTP)

   iii. L=CH$_2$–CH$_2$–CH$_2$: 5-[N-(N′-(2-Nitro-5-azidobenzoyl)-4-aminobutyryl)-trans-3-aminopropenyl-1]-2′-deoxyuridine-5′-triphosphate (NAB-9-dUTP)

   iv. L=CH$_2$–CH$_2$–CH$_2$–CH$_2$: 5-[N-(N′-(2-Nitro-5-azidobenzoyl)-5-aminopentanoyl)-trans-3-aminopropenyl-1]-2′-deoxyuridine-5′-triphosphate (NAB-10-dUTP0029)

   v. L=CH$_2$–CH$_2$–CH$_2$–CH$_2$–CH$_2$: 5-[N-(N′-(2-Nitro-5-azido-benzoyl)-6-aminohexanoyl)-trans-3-aminopropenyl-1]-2′-deoxyuridine-5′-triphosphate (NAB-11-dUTP)

   vi. L=CH$_2$–CH$_2$–CH$_2$–CH$_2$–CH$_2$–CH$_2$: 5-[N-(N′-(2-Nitro-5-azidobenzoyl)-7-aminoheptanoyl)-trans-3-aminopropenyl-1]-2′-deoxyuridine-5′-triphosphate (NAB-12-dUTP)

   vii. L=CH$_2$–CH$_2$–CH$_2$–CH$_2$–CH$_2$–CH$_2$–CH$_2$: 5-[N-(N′-(2-Nitro-5-azidobenzoyl)-8-aminooctanoyl)-trans-3-amino-propenyl-1]-2′-deoxyuridine-5′-triphosphate (NAB-13-dUTP)

h. X = : 5-[N-(4-(3-Trifluoromethyldiazirine)benzoyl)-3-aminopropenyl-1]-2′-deoxyuridine-5′-triphosphate (TDB-5-dUTP)

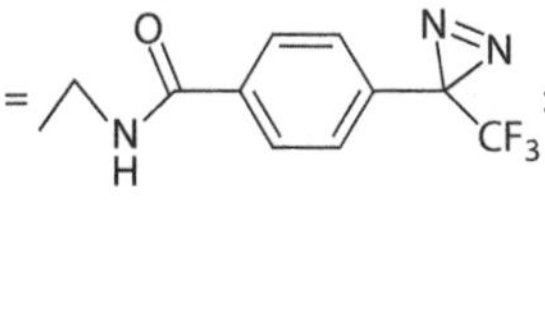

(continued)

TABLE 6.3 (Continued)
Nucleic Acid and Nucleic Acid–Protein Heterobifunctional Cross-Linking Reagents

i. X = 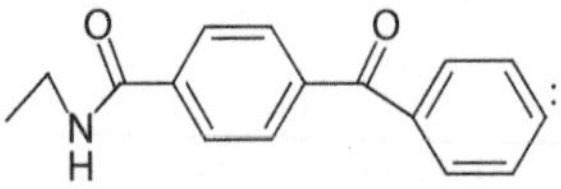 : 5-[N-(4-Benzoylbenzoyl)-3-aminoallyl]-2′-deoxyuridine-5′-triphosphate (BP-dUTP)

j. X = 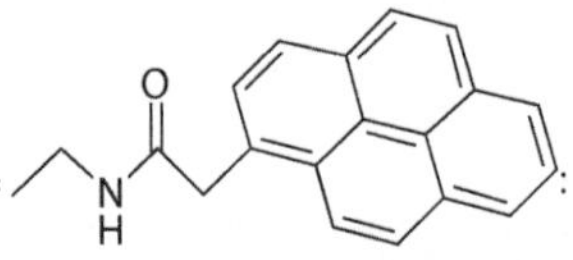 : 5-[N-(4-(1-Pyrenyl)-ethylcarbonyl)-amino-*trans*- propenyl-1]-2′-deoxyuridine-5′-triphosphate (Pyr-6-dUTP)

k. X = 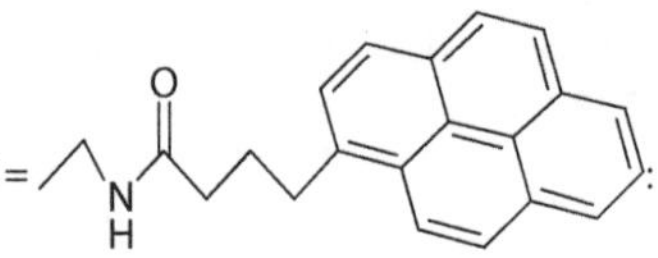 : 5-[N-(4-(1-Pyrenyl)-butylcarbonyl)-amino-*trans*-propenyl-1]-2′-deoxyuridine-5′-triphosphate (Pyr-8-dUTP)

2. Other uridine derivatives

   a. 5-N-[N-(2-Nitro-5-azidobenzoyl)aminomethyl]-2′-deoxyuridine-5′-triphosphate (NAB-2-dUTP)

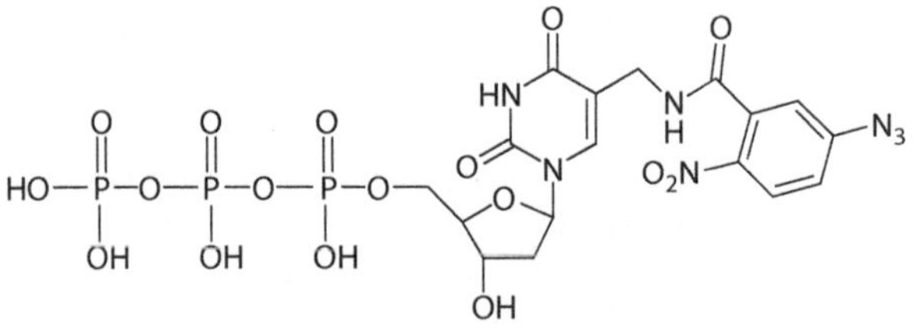

   b. 2′,3′-Deoxy-E-5-(4-(3-(trifluoromethyl)diazirine-3-yl)styryl)uridine triphosphate (TDS-dUTP)

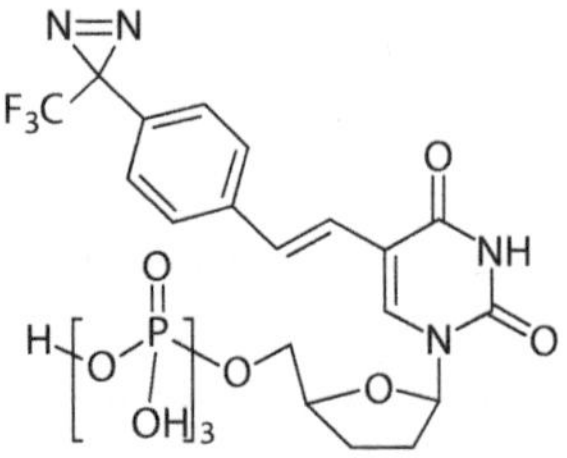

   c. 1-(4-Hydroxy-5-hydroxymethyl-tetrahydro-furan-2-yl)-5-{5-[4-(3- trifluoromethyl-3H-diazirin-3-yl)-benzyloxy]-pent-1-inyl}-1H-pyrimidin-2,4-dione

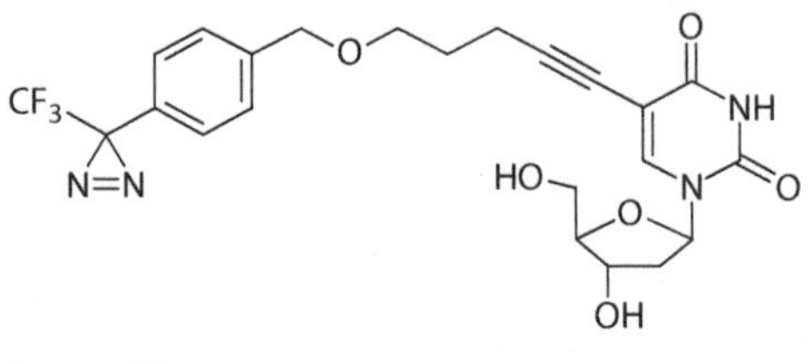

   d. 5-(3H-Diazirin-3-yl)-2′-deoxyuridine

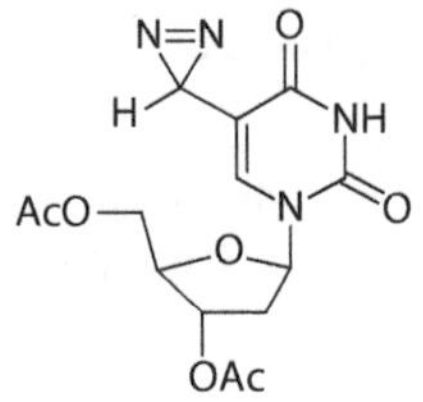

TABLE 6.3 (Continued)
Nucleic Acid and Nucleic Acid–Protein Heterobifunctional Cross-Linking Reagents

e. 5-[4-[3-(Trifluoromethyl)-3*H*-diazirin-3-yl]phenyl]-2′-deoxyuridine (Tfmdp-uD)

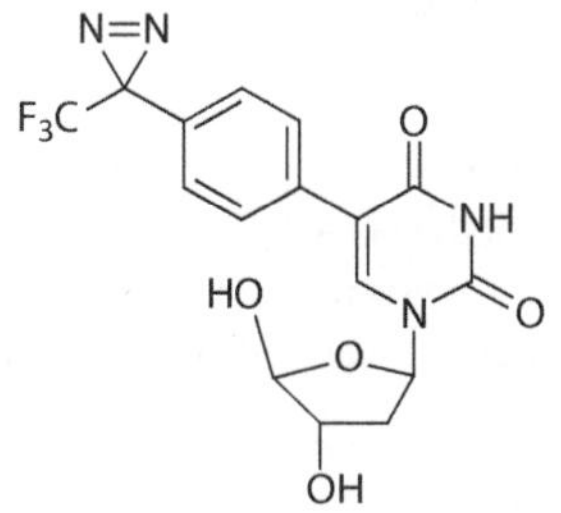

f. 2′-Deoxy-5-[4-[3-(trifluoromethyl)-3*H*-diazirin-3-yl]styryl]-uridine (TDS-dU)

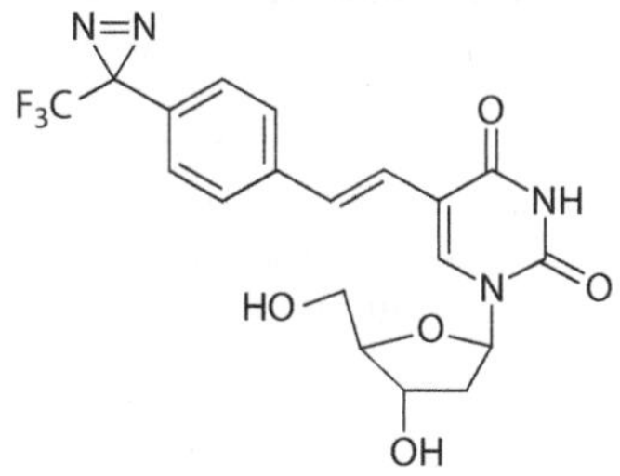

g. 4-Thio-2′-deoxyuridine-5′-triphosphate (4S-dUTP)

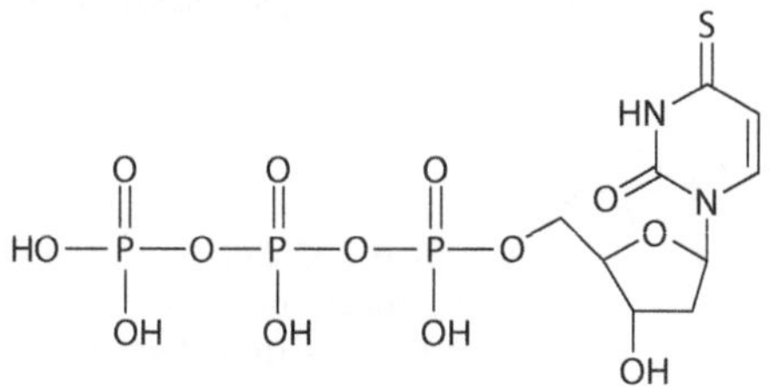

3. 4-Thiodeoxythymidine 5′-triphosphate (4S-dTTP)

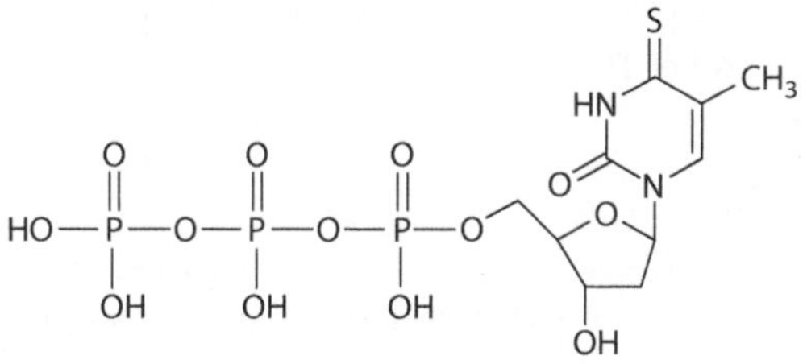

4. Deoxycytidine triphosphate (dCTP) analogs

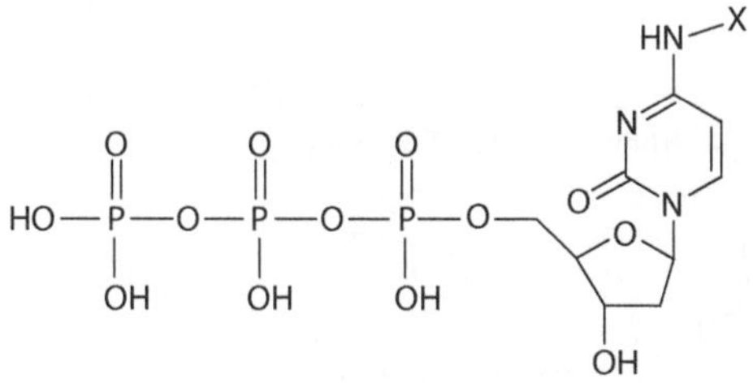

(*continued*)

TABLE 6.3 (Continued)
Nucleic Acid and Nucleic Acid–Protein Heterobifunctional Cross-Linking Reagents

a. X =

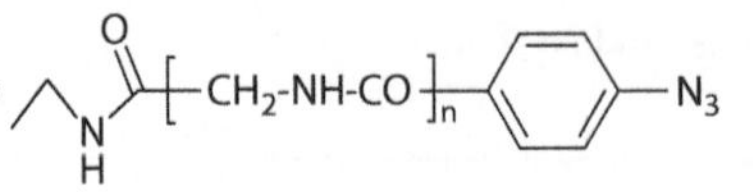

    i. n = 0: *exo-N-[N-*(4-Azidobenzoyl)aminoethyl]-2'-deoxycytidine-5'-triphosphate (AB-dCTP)
    ii. n = 1: *exo-N-[N-*(4-Azidobenzoylglycyl)aminoethyl]-2'-deoxycytidine-5'-triphosphate (ABG-dCTP)
    iii. n = 2: *exo-N-[N-*(4-Azidobenzoyldiglycyl)aminoethyl]-2'-deoxycytidine-5'-triphosphate (ABG2-dCTP)
    iv. n = 3: *exo-N-[N-*(4-Azidobenzoyltriglycyl)aminoethyl]-2'-deoxycytidine-5'-triphosphate (ABG-dCTP)

b. X =

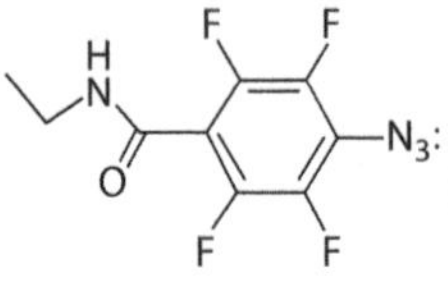

exo-N-[N-(4-Azido-2,3,5,6-tetrafluorobenzoyl)-aminoethyl]-2'-deoxycytidine-5'-triphosphate (FAB-dCTP)

c. X =

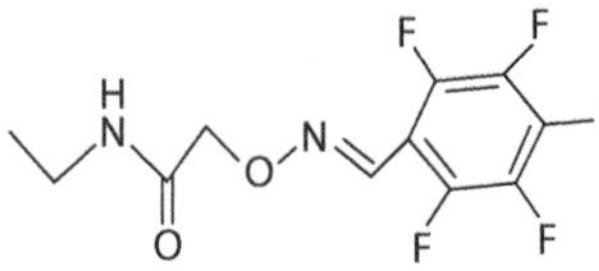

exo-N-[2-(4-Azido-2,3,5,6-tetrafluorobenzylidene-aminooxymethylcarbamoyl)ethyl]-2'-deoxycytidine-5'-triphosphate (FABC-dCTP)

d. X = O

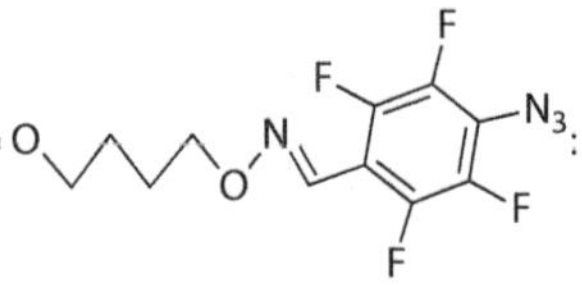

exo-N-[(4-Azido-2,3,5,6-tetrafluorobenzylidene-aminooxy)butyloxy]-2'-deoxycytidine-5'-triphosphate (FABO-dCTP)

e. X =

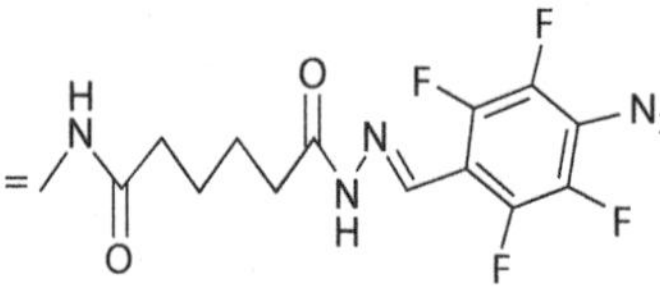

exo-N-[4-(4-Azido-2,3,5,6-tetrafluorobenzylidene-hydrazinocarbonyl)butylcarbamoyl]-2'-deoxycytidine 5'-triphosphate (FABG-dCTP)

f. X =

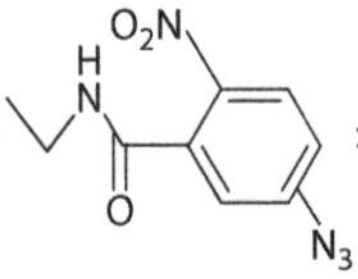

exo-N-[N-(2-Nitro-5-azidobenzoyl)aminoethyl]-2'-deoxycytidine-5'-triphosphate (NAB-dCTP)

g. X =

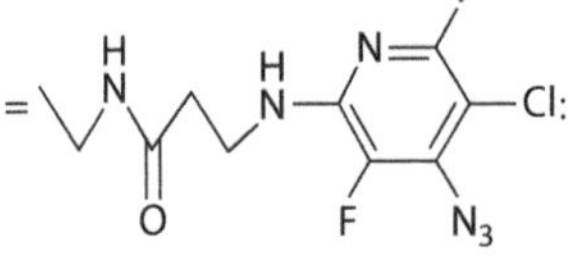

exo-N-{2-[*N-*(4-Azido-2,5-difluoro-3-chloropyridin-6-yl)-3-aminopropionyl]aminoethyl}-2'-deoxycytidine-5'-triphosphate (FAP-7-dCTP)

h. X =

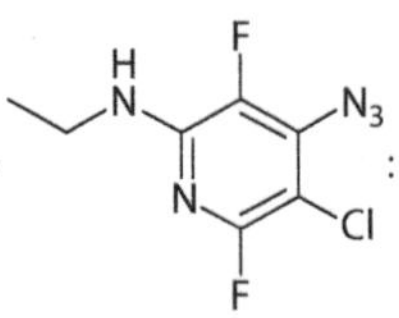

exo-N-[2-(4-Azido-2,5-difluoro-3-chloropyridin-6-yl)aminoethyl]-2'-deoxycytidine 5'-triphosphate (FAP-3-dCTP)

i. X =

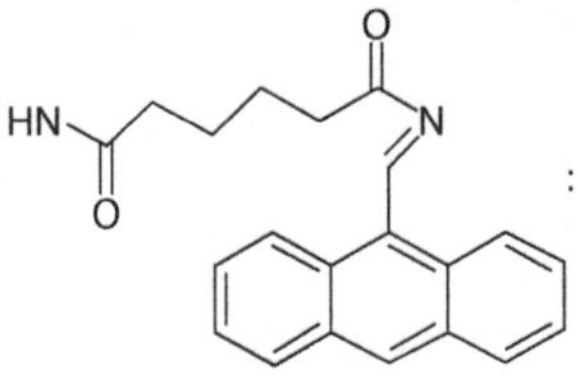

exo-N-[4-(9-Anthracenylhydrazinocarbonyl)butyl-carbamoyl]-2'-deoxycytidine-5'-triphosphate (Anthr-dCTP)

TABLE 6.3 (Continued)
Nucleic Acid and Nucleic Acid–Protein Heterobifunctional Cross-Linking Reagents

j. X = [structure with N=N diazirine and CF₃ group] *exo*-*N*-[*N*-(4-(3-Trifluoromethyldiazirine)benzoyl)-aminoethyl]-2′-
deoxyuridine-5′-triphosphate (TDB-4-dCTP)

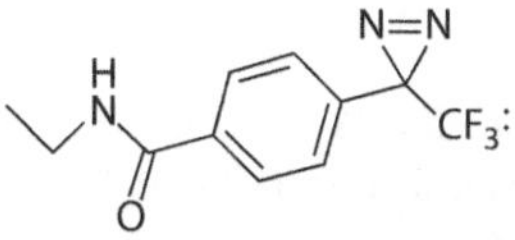

5. 5-Methylcytidine analog: *N*-[(4-((3-Trifluoromethyl)-3*H*-diazirin-3-yl)benzoyl)-aminopropyl]-5-methylcytidine
(TDB-dC)

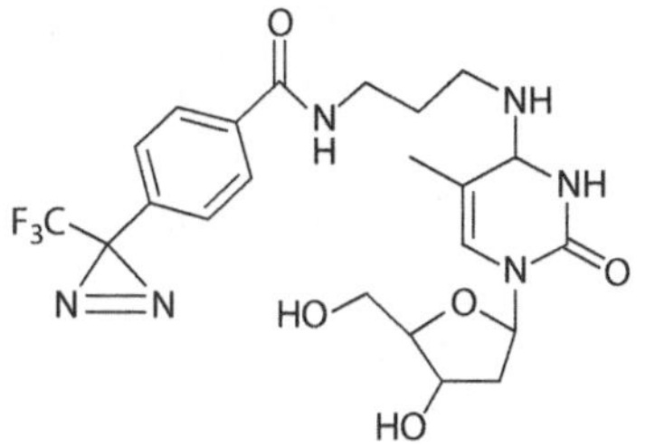

6. Deoxyadenosine triphosphate (dATP) analogs

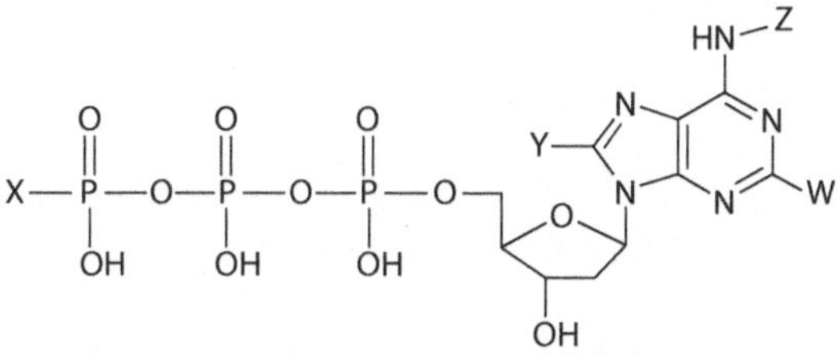

a. X = OH, Y = N₃, Z = W = H: 8-Azido-2′-deoxyadenosine-5′-triphosphate (8-Azido-dATP)

b. X = OH, Y = W = H, Z = [structure with N₃] *N*⁶-[4-Azidobenzoyl-(2-aminoethyl)]-2′-deoxyadenosine-
5′-triphosphate (AB-dATP)

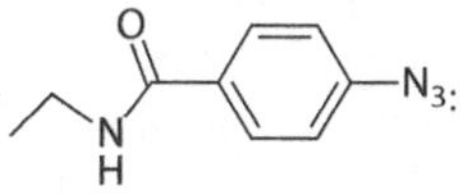

c. X = OH, Y = W = H, Z = [structure with N=N diazirine and CF₃] : *N*⁶-[4-[3-(Trifluoromethyl)diazirin-3-ly]benzoyl-(2-
aminoethyl)]-2′-deoxyadenosine-5′triphosphate
(DB-dATP)

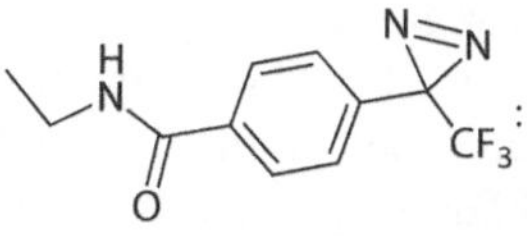

d. X = [structure with N₃, NO₂] , Y = Z = W = H: *N*-[2-(5-Azido-2-nitrobenzoyl)amino-ethyl]amide of
ATP (NAB-ATP)

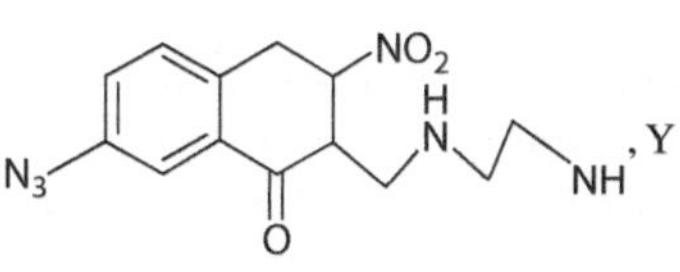

e. X = N₃ [tetrafluorobenzoyl structure] NH, Y = Z = W = H: *N*-[2-(4-Azido-2,3,5,6-tetrafluorobenzoyl)-aminoethyl]-
amide of ATP (FAB-ATP)

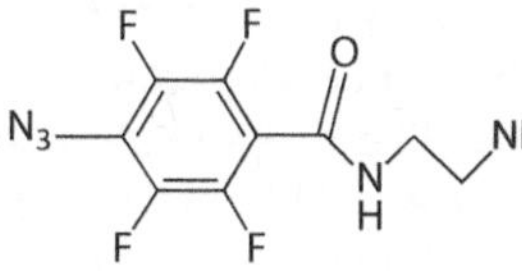

(continued)

TABLE 6.3 (Continued)
Nucleic Acid and Nucleic Acid–Protein Heterobifunctional Cross-Linking Reagents

f. X = N$_3$ ... NH, Y = Z = W = H: N-[2-(4-Azido-2,3,5,6-tetrafluoro-benzylideneaminooxymethylcarbamyl)ethyl]-amide of ATP (FABO-ATP)

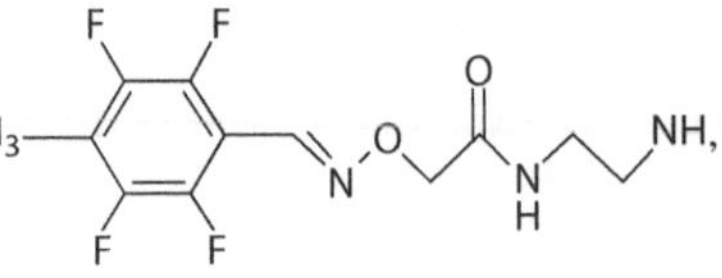

g. X = Y = Z = H, W = S ... N$_3$: 2-[(4-Azidophenacyl)thio]-2′-deoxy-adenosine 5′-triphosphate (AP-ATP)

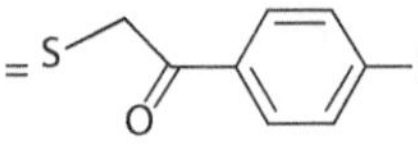

7. Thiopurines

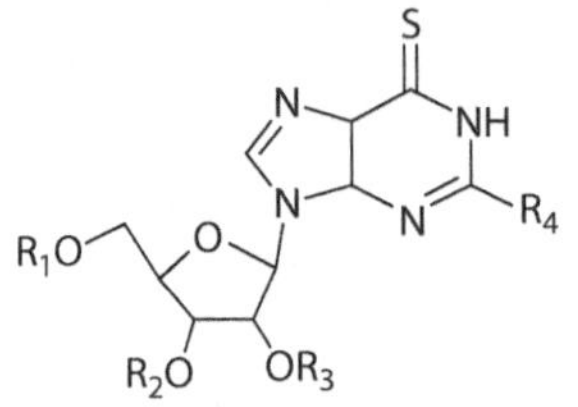

a. R1 = R2 = R3 = H, R4 = NHCO(p-C$_6$H$_4$)C(N$_2$)CF$_3$: 2-[N-(4-(3-Trifluoromethyldiazirine)benzoyl) amino]-6-Thioadenosine

b. R1 = CO(p-C$_6$H$_4$)C(N$_2$)CF$_3$, R2 = R3 = H, R4 = NH$_2$: 6-Thio-5′-(4-(3-trifluoromethyldiazirine)benzoyl) oxyadenosine

c. R1 = R3 = R4 = H, R2 = CO(p-C$_6$H$_4$)C(N$_2$)CF$_3$: 6-Thio-3′-(4-(3-trifluoromethyldiazirine)benzoyl)oxyadenosine

d. R1 = R2 = R4 = H, R3 = CO(p-C$_6$H$_4$)C(N$_2$)CF$_3$: 6-Thio-2′-(4-(3-trifluoromethyldiazirine)benzoyl) oxyadenosine

the epoxide. Nucleophilic addition to ECH has been reported to occur initially at the epoxide ring while EBH may react initially through the loss of the bromide because it is a better leaving group. The reaction of the monoadducts with another guanine base completes the cross-linking.

There are many reagents specifically designed for studying nucleic acids. These nucleic acid cross-linkers are presented in Table 6.3. N'-chloroethylaziridinium chloride (Table 6.3.I.C) is seemingly a heterobifunctional reagent. It was synthesized by Anderson et al.[243] as a proof that aziridinium ion is an intermediate in the reaction of 2-chloroethylamines with DNA. It is an efficient DNA cross-linker, alkylating primarily guanine-N7. The reaction was discussed in Chapter 5 and was shown in Figure 5.25.

Other heterobifunctional nucleic acid cross-linkers containing aziridinium are shown in Table 6.3.I.D through J. Mitomycin C (MC, Table 6.3.I.D) is a natural antibiotic that forms DNA interstrand and intrastrand cross-links with guanine residues of 5′-CG-3′ and 5′-GG-3′ sequences in the minor groove of DNA, respectively.[244] The reaction requires reductive activation of MC by sodium dithionite or metabolic enzymes such as diaphorase.[245] As shown in Figure 6.9, the monoadduct formed with a guanine base contains a structure of hydroxymethylpyrrole, which reacts with another guanine base to complete the cross-linking process. The proposed involvement of hydroxymethylpyrrole is supported by the demonstration that a series of bis(hydroxymethyl)pyrrole compounds such as 2,3-bis(hydroxymethyl)pyrrole as shown in Table 6.4 are able to react with guanine bases. When the bis(hydroxymethyl)pyrrole is attached to DNA minor groove binding agent as distamycin derivative (Table 6.4.II), the yield of DNA cross-linking increased 1000-fold. Therefore, to increase its reactivity, MC have been linked to pyrrole as shown in Table 6.3.I.E.[244] These mitomycin C-pyrrole compounds have been shown to cross-link pBR322 DNA. The monopyrrole conjugate (Table 6.3.I.E.1) is more active than MC while the di- and tri-pyrrole conjugates are less so. However, no adducts were formed with calf thymus DNA or short oligonucleotides.

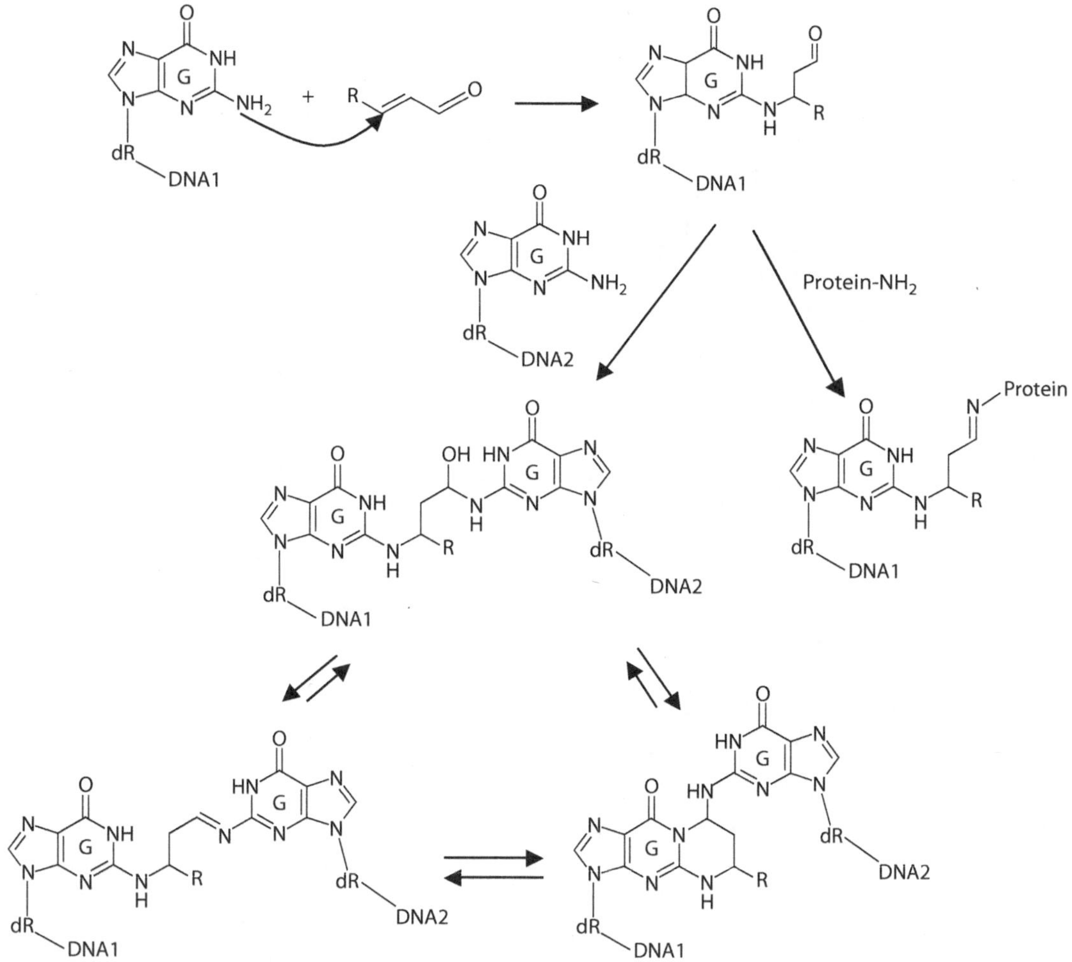

FIGURE 6.8 Speculated reaction mechanism of α,β-unsaturated aldehydes in cross-linking nucleic acids and nucleic acid-protein. The interstrand DNA cross-link exists as an equilibrium mixture of carbinolamine, imine, and pyrimidopurinone.

N-Methylmitomycin A (Table 6.3.I.F), similar to MC, forms an interstrand cross-link by alkylating the N2-exocyclic amino groups of guanine in a 5′-CG-3′ sequence.[246]

Another class of antitumor antibiotics that has the ability to cross-link duplex DNA are the azinomycins (Table 6.3.I.G), which are isolated from *Streptomyces griseofuscus* S42227.[247] Azinomycin B (Table 6.3.I.G.2) was shown to cross-link N7 of guanine of one strand and another purine residue (A or G) on the complementary DNA strand in the major groove of DNA.[247,248] The reaction is pH dependent, with more rapid cross-linking at lower pHs. It is widely accepted that the epoxide and aziridine functionalities are responsible for interstrand cross-linking of DNA as shown in Figure 6.10.[249,250] The preferential base is a purine, adenine, or guanine, on complementary strands with a sequence such as GNC and GNT where N is any base.[249] Alkylation occurs via N7 of both purine bases. Support for such a reaction mechanism comes from the synthesis of various azinomycin analogs such as epoxy-aziridine (Table 6.3.I.H)[250] and epoxy-mustard analogs (Table 6.3.I.I)[251] as well as analogs that are not capable of forming cross-links.[249–252]

In addition to azinomycin, an anthracycline antibiotic, adriamycin (Table 6.3.I.J), also induces DNA interstrand cross-links *in vitro* and in drug-treated cells in culture. The cross-links appear to form at 5′-GC-3′ sequences.[253] Despite intensive research effort, the mode of action of adriamycin

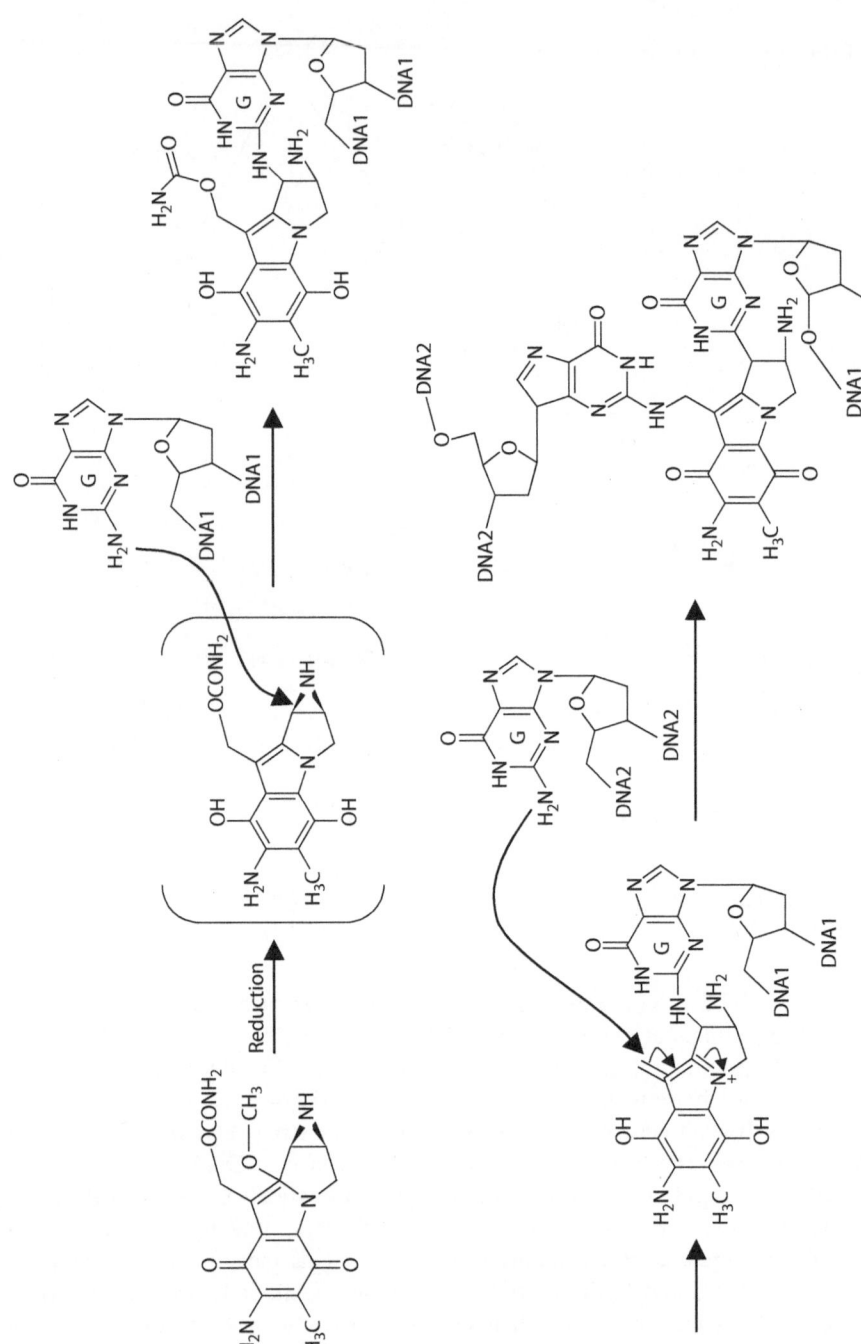

FIGURE 6.9   Speculated cross-linking mechanism of mitomycin C.

TABLE 6.4

DNA Cross-Linking Reagents Containing Bis(Hydroxymethyl)Pyrrole

I. 2,3-Bis(hydroxymethyl)pyrrole

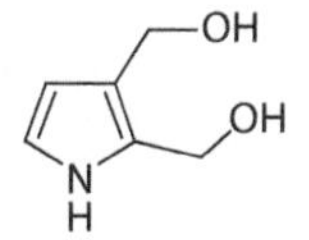

II. 2,3-Bis(hydroxymethyl)pyrrole-distamycin

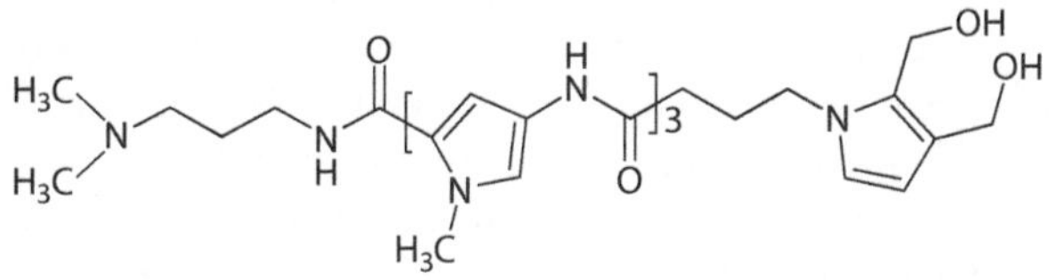

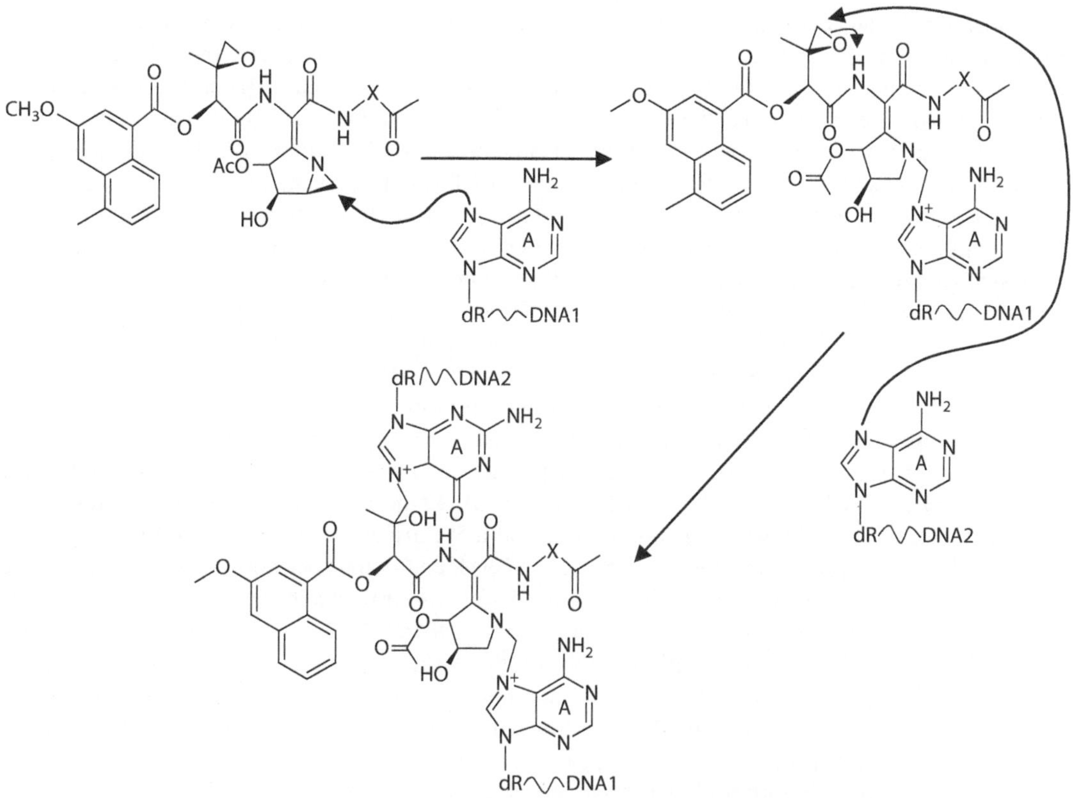

FIGURE 6.10 Speculated mechanism of DNA interstrand cross-linking by azinomycin.

still remains unclear. The adriamycin-induced interstrand cross-links appear to be unstable, with increasing lability at higher pHs, particularly at pH 12. The adducts are also heat labile. More stable cross-links are formed with deaminoadriamycin compounds, CMA and ICMA (Table 6.3.I.K).[254] Moreover, CMA (Table 6.3.I.K.1) is 100–1500-fold more cytotoxic than adriamycin *in vivo* and *in vitro*. CMA can act as an alkylating agent producing DNA–DNA cross-links in a number of tumor cell lines and in isolated DNA with a preference for G–C bases. ICMA (Table 6.3.I.K.2)

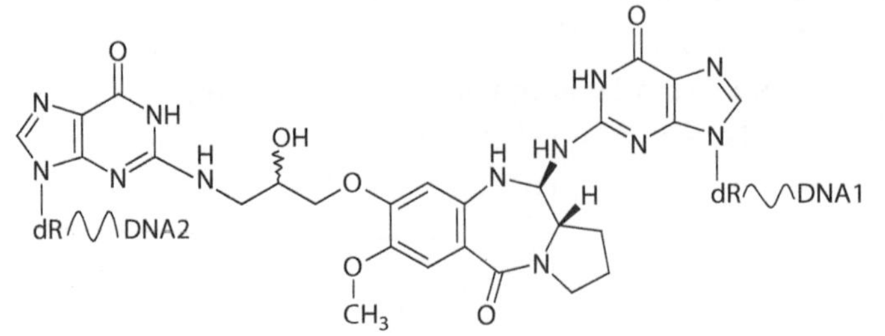

FIGURE 6.11   Structure of epoxide-PBD interstrand cross-linked DNA.

also cross-link DNA, but its cytotoxic and cross-linking activities are less than those observed with CMA. Although the exact reaction mechanism is not understood, it is speculated that the cyanide group and the quinine ring is involved in the cross-linking activity.[254]

As discussed in Chapter 5, a series of pyrrolobenzodiazepine (PBD) dimers were shown to cross-link DNA as homobifunctional reagents. Wilson et al.[255] have synthesized a heterobifunctional analog with an epoxide moiety at the C8 position (Table 6.3.I.L). The epoxide PBD forms DNA interstrand cross-links efficiently with the alkylating units bonded to the guanine C_2–NH_2. The cross-links involve either two adjacent guanines on opposite strands or an analogous sites separated by one spacing base pair as in (G/C)(C/G) or (G/C)N(C/G), respectively, where N is any other base pair. The cross-linked product is shown in Figure 6.11. Another heterobifunctional PBD cross-linker contains cyclopropylpyrolo[e]indolone (CPI) (Table 6.3.I.M).[256,257] It is found to exhibit potent cytotoxicity against human cancer cells grown in culture such as breast MCF-7, colon SW480, and lung A549, and it showed selectivity for producing interstrand cross-links at a 5′-T[A]AATTG-3′ site, alkylating at the adenine and guanine bases as underlined. The cross-linking reaction involves both the cyclopropyl and the benzodiazepine moieties as discussed in Section 5.11.3. Seco-compounds have also been synthesized such as achiral *seco*-amino-CBI (cyclopropanebenz[e]indoline)-PBD (Table 6.3.I.N) and achiral *seco*-CI (cyclopropylindolone)-PBD (Table 6.3.I.O).[258] These compounds demonstrated enhanced cytotoxicity over the monomer counterparts against the growth of P815 murine mastocytoma cells in culture. They also covalently react with adenine-N3 positions within the minor groove at AT-rich sequences and produce DNA interstrand cross-links. During the cross-linking reaction, the chloropropyl moiety is activated to cyclopropyl as shown in Chapter 5.XI.D. At a low concentration of 1 µM, *seco*-amino-CBI-PBD exhibited a clear and selective alkylation at A(3233) of the 5′-AAA(3233)TTTTC[G]C-3′ sequence, which is potentially an interstrand cross-link site due to the proximity between the alkylated adenine residue and a 5′-PurineGPurine-3′ site in the complementary strand. However, at higher concentrations, e.g., 10 µM, additional alkylation sites become evident.

Photoactivatable reagents have also been used to cross-link nucleic acids. Psoralens (Table 6.3.II.A), for example, efficiently intercalate into helical DNA or RNA with high binding affinity and upon irradiation with long wavelength ultraviolet light at 366 nm form covalent crosslinks between pyrimidines of opposite strands.[246,259] They preferentially form cross-link between thymines in the 5′-TA-3′ and 5′-AT-3′ sequences in DNA. Photoreaction takes place by cycloaddition between the furan or pyrone ring of psoralens and the adjacent thymine, leading to the formation of an interstrand cross-link as shown in Figure 6.12. The cross-link is chemically very stable. NMR studies of interstrand cross-linked products of eight base pair DNA duplexes with 4′-hydroxymethy-4,5′,8-trimethylpsoralen (Table 6.3.II.A.5) or 4′-aminomethyl-4,5′,8-trimethylpsoralen (Table 6.3.II.A.4) showed that, although there is considerable local distortion at the site of the cross-link, the duplex retains its B-form conformation three base pairs from the cross-link site. The cross-links did not induce any significant bending of the helix.[260] The psoralen cross-linking technique has successfully been used for the study of chromatin structure with trimethylpsoralen (Table 6.3.II.A.3) being the most commonly used psoralen.[259] Lustig et al.[261] described a novel "RNA walk" method to investigate the RNA–RNA interactions between a small RNA and its target. The method is based on UV-induced

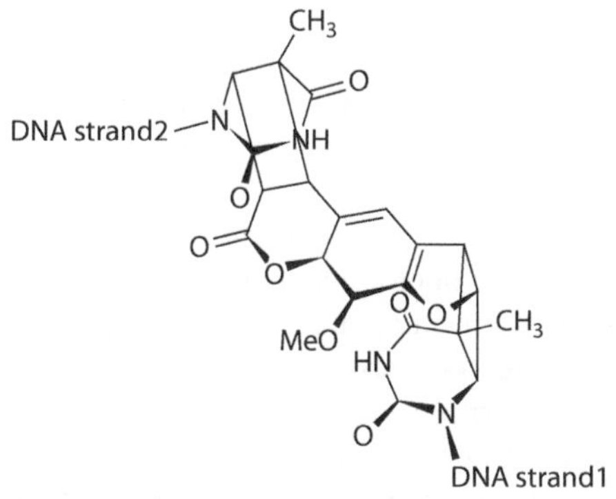

FIGURE 6.12 Structure of psoralen interstrand cross-linked DNA.

4′-aminomethyl-4,5′,8-trimethylpsoralen (AMT) (Table 6.3.II.A.4) cross-linking *in vivo* followed by affinity selection of the hybrid molecules and mapping the intermolecular adducts by RT–PCR or real-time PCR. The AMT molecule intercalates within the double-helical structure and upon irradiation at 365 nm covalently cross-links pyrimidine bases. The covalent linkage can be reversed by irradiation at 254 nm. Using this method, four contact sites between sRNA-85 (a special tRNA-like molecule) and rRNA were identified in trypanosome.

Photoreactive Ru(II) complexes such as that shown in Table 6.3.II.B have been used to study DNA.[262,263] These Ru[2+] complexes are tethered to oligodeoxynucleotides (ODNs) through the carboxyl group either at the 5′ or 3′ end or at specific locations of the ODNs. In the case of [Ru(tpy)(dppz)(CH₃CN)][2+]-ODN conjugate (for the Ru[2+] complex, see Table 6.3.II.B.1), photolysis gives the reactive aqua derivative, [Ru(tpy)(dppz)(H2O)][2+]-ODN, which reacts with the target complementary DNA strand in the duplex to give DNA interstrand cross-linked photoproducts.[262] For [Ru(TAP)₂dip][2+]-ODN conjugate (for the Ru[2+] complex, see Table 6.3.II.B.2), the Ru(II) complex containing π-deficient TAP ligand will be excited upon illumination. The excited state of the complex will then abstract an electron from guanine, the most reductive base of DNA, to form a covalent bond. Thus, an interstrand DNA cross-link is obtained. The complex has been used to determine the positions of the guanine bases in the duplex ODN that are able to form a photo-adduct with the tethered complex.[263]

Photoaffinity labeling techniques are valuable tools of proteomics to elucidate structural and functional aspects of protein–DNA and DNA–DNA interactions.[143,264] Photoreactive DNAs have been successfully used to study enzymes and protein factors of DNA replication and repair in reconstituted systems and cellular/nuclear extracts. They can also be applied for identification of proteins, including unknown ones, which interact with specific DNA intermediates in cellular/nuclear extracts. Photoreactive groups are introduced at specific locations in a protein or at desired points in DNA (base or phosphate) either at the ends or in inner positions of the DNA chain by combination of enzymatic and chemical synthesis. Photoaffinity labels that have been used for nucleic acid investigations are shown in Table 6.3.III. As can be seen a wide range of base-substituted dNTP analogs containing photoreactive groups of different photoreactivity and spacers of various lengths has been synthesized. A widely used amino acid photosensitive analog, *p*-benzoyl-phenylalanine (pBpa, Table 6.3.III.A), has been incorporated into proteins, particularly membrane receptor proteins, to study protein–protein interactions.[265–268] Recently, it has been used to study protein–DNA interactions. In this case, it was incorporated into position Lys26 of the *E. coli* catabolite activator protein (CAP), which regulates a number of catabolite-sensitive operons in *E. coli*.[268] On UV irradiation of the DNA/CAP-K26pBpa complex, pBpa formed a covalent complex with a DNA fragment containing the consensus operator sequence. Since pBpa can be incorporated into any DNA-binding protein in *E. coli*, yeast, or mammalian cells with minimal perturbation of protein structure, this method may be useful for investigating DNA– or RNA–protein interactions.

A number of photosensitive nucleosides and nucleotides have been synthesized (Table 6.3.III.B).[89,143,264,269–283] The photosensitive moieties include aryl azide, perfluorinated aryl azide, benzophenone, and trifluromethyldiazirine.[272] Of the nucleotide triphosphate (NTP) labels, the most abundant reagents are dUTP analogs. Since the photoactive moieties are attached to the 5-position of the pyrimidines ring, these dUTP derivatives are also dTTP analogs (Table 6.3.III.B.1). The other major analogs are dCTP, followed by dATP. Several methods have been developed to site-specifically incorporate photoreactive nucleotide analogs into DNA for the purpose of identifying the proteins and their domains that are in contact with particular regions of DNA.[269] These dNTP analogs have been shown to be effective substrates of viral, bacterial, and eukaryotic DNA polymerases, which can be used to incorporate the photoactive dNTP into definite positions of DNA molecules.[264] Thus, intermediates of DNA replication or DNA repair containing photoactive nucleic acids may be synthesized *in situ* for cross-linking studies of protein subunits interacting with certain positions of DNA. The various nucleosides such as trifluoromethyldiazirine-uridine (Table 6.3.III.B.2.c through f) have also been incorporated into DNA oligomers by a standard phosphoramidite chemical procedure through solid-phase DNA synthesis methodology.[275–277] Since the photosensitive dNTP compounds are substrates of DNA polymerase, these polymerases are subject of most of the photoaffinity cross-linking studies. Tate et al.[272] used four different photoreactive groups attached to DNA, an aryl azide (AB-dUTP, Table 6.3.III.B.1.a.i), benzophenone (BP-dUTP, Table 6.3.III.B.1.i), perfluorinated aryl azide (FAB-5-dUTP, Table 6.3.III.B.1.c), and diazirine (TDB-5-dUMP, Table 6.3.III.B.1.h) to cross-link yeast RNA polymerase III transcription complexes (Pol III). Photocross-linking with diazirine revealed contacts of Pol III with DNA that are not detected by other photoactive groups. Thus, different photoreactive group may reveal different interaction sites. Further studies of rat DNA polymerase b with radioactive uridine styryl diazirine, TDS-dU (Table 6.3.III.B.2.f), was carried out by Yamaguchi et al.[277] Persinger and Bartholomew[269] used azidobenzoyl and aziridinebenzoyl dUTP and dCTP analogs (AB-dUTP, ABG$_{1-3}$-dUTP, Table 6.3.III.B.1.a.i through iv; TBD-5-dUTP, Table 6.3.III.B.1.h; AB-dCTP, ABG$_{1-3}$-dCTP, Table 6.3 III.B.4.a.i through iv; TBD-4-dCTP, Table 6.3.III.B.4.J) as well as 4-thiodeoxythymidine triphosphate (4S-dTTP, Table 6.3.III.B.3) to probe the RNA polymerase III transcription complex and ATP-dependent chromatin remodeling complexes SWI/SNF and ISW2. Protein–DNA cross-links were obtained and analyzed. The 4-thiodeoxythymidine nucleotide, which can be photolyzed below 300 nm, has also been used for zero-distance cross-linking of protein to DNA and may be useful for probing very close protein–DNA contacts.[278] Another reagent considered to be zero-length photocross-linker is 5-(3*H*-Diazirin-3-yl)-2′-deoxyuridine (Table 6.3.III.B.2.d).[279]

The photosensitive group, 2-nitro-5-azidobenzoyl (NAB), connected by different methylene lengths to the fifth position of uracil (NAB-[2, 4, 7–13]-dUTP, Table 6.3.III.B.2.a, III.B.1.f, III.B.1.g.i through vii), was used to study interactions of DNA polymerase β (pol β) and subunits of the human heterotrimeric replicative protein A (RPA).[273,274] The DNA pol β efficiently incorporated these analogs into synthetic primer-template substrates in place of TTP. When the reaction mixtures were irradiated with monochromatic UV light (315 nm) in the presence of RPA, a heterotrimer consisting of three subunits, p70, p32, and p14, the photoreactive primers cross-linked directly with p70 and p32, but not p14. These data were taken to mean that RPA and pol β formed a complex on primer-template substrates. Lebedeva et al.[280] also used 4-azidotetrafluorobenzoyl group linked via spacers of varying length to 5-position of uridine ring to study DNA pol β (FAB-4-dUTP and FAB-9-dUTP, Table 6.3.III.B.1.c and d). The reaction mixtures were UV-irradiated in the absence or presence of a dTTP analog containing a pyrene moiety, Pyr-6-dUTP (Table 6.3.III.B.1.j) or Pyr-8-dUTP (Table 6.3.III.B.1.k). The most efficient cross-linking of pol β was observed in the case of photoreactive DNA primer, carrying the FAB-4-dUMP moiety at the 3′-end, and Pyr-6-dUTP as a sensitizer.

Photosensitive dATP analogs (Table 6.3.III.B.6) have also been used to study DNA polymerase. Moore et al.[281] used 4-azidophenacylthio dATP (AP-ATP, Table 6.3.III.B.6.f) to study DNA polymerase I Klenow fragment. Zofall and Bartholomew[282] used two dATP analogs, AB-dATP and DB-dATP (Table 6.3.III.B.6.b and c), to investigate DNA polymerase I. Khodyreva and Lavrik[264] have synthesized a series of dATP analogs, NAB-ATP, FAB-ATP, and FABO-ATP (Table 6.3. III.B.6.d through f), to study multisubunit factors of RPA and RFC (clamp loader).

In addition to polymerases, related enzymes such as HIV-1 reverse transcriptase (RT) has been studied with photosensitive styryl diazirine (TDS-dUTP, Table 6.3.III.B.2.b) and 5-methylcytidine (TDB-dC, Table 6.3.III.B.5).[89] TDS-dUTP was incorporated into the 3′ terminus of the primer strand of RT. Upon photolysis, the 3′ terminus of the primer strand was cross-linked to a 66 kDa subunit of the RT enzyme. Human apurine/apyrimidine endonuclease 1 (APE1) was investigated with dCTP photoactivable analogs FABC-dCTP, FABO-dCTP, FABG-dCTP, and FAP-7-dCTP (Table 6.3.III.B.4.c through e and g).[143,283,284] These dCTP analogs were incorporated at the 3′ or 5′-end of the single-strand nicked DNA. The usefulness of these photosensitive probes and the interactions between the proteins involved in the excision repair of nucleotides and DNA-duplexes containing bulky photophores were studied. Base excision repair (BER) and nucleotide excision repair (NER) systems were investigated using photocross-linker nucleotides that induce neither a destabilization nor a structural change of the DNA duplex such as diazirine alkyne deoxyuridine (Table 6.3.III.B.2.c).[275] The compound was incorporated into DNA oligonucleotides and the photolysis experiments showed selective binding of the formamidopyrimidine-DNA glycosylase (Fpg protein) to lesion-containing DNA. Taranenko et al.[276] used Tfmdp-dU (Table 6.3.III.B.2.e) to identify specific contacts between nucleobases of DNA and amino acids of E. coli Fpg protein. DNA duplexes containing Tfmdp-dU residues near the 7-hydro-8-oxoguanosine (oxoG) lesion were synthesized. The Fpg protein was found to bind specifically and tightly to the modified DNA duplexes to form cross-linked products after photolysis. Tfmdp-dU was also used to characterize regions of EcoRII DNA methyltransferase (M.EcoRII) involved in DNA binding and to investigate the DNA double helix conformational changes that took place during methylation. DNA photocross-linking results agreed with the flipping of the target cytosine out of the DNA double helix for catalysis.[285] Petruseva et al.[286] used 48-mer duplexes containing bulky substituted pyrimidine nucleotides in internal positions (Pyr-8-dUTP, FAP-dUTP, FAP-dCTP, Anthr-dCTP, and 4S-dUTP, Table 6.3.III.B.1.k, III.B.1.e, III.B.4.g, III.B.4.i, and III.B.2.g, respectively) to study NER from HeLa cell extracts. UV irradiation of reaction mixture containing the duplex and extract proteins resulted in formation of sets of covalent DNA–protein adducts. To study the active centre of thiopurine-S-methyl transferase, Blencowe and Hayes[89] prepared a series of photolabile thiopurines (Table 6.3.III.B.7), which were shown to be substrates for the transferase enzyme.

As evident from the above discussion, the photoaffinity reagents used are quite extensive. Many studies have revealed interesting data on protein–DNA and DNA–DNA interactions. The accomplishment of the genome project gave a major thrust for investigations in this area and further development of photoaffinity reagents is foreseeable.

REFERENCES

1. Ji, T. H., The application of chemical crosslinking for studies on cell, membranes and the identification of surface reporters, *Biochim. Bjophys. Acta*, 559, 39, 1979.
2. Ji, T. H., Crosslinking of lectins and receptors in membranes with hetero-bifunctional crosslinking reagents, in *Membranes and Neoplasia: New Approaches and Strategies*, Marchesi, V. T. (ed.), Alan R. Liss, New York, 1976, p. 171.
3. Carlsson, J., Drevin, H., and Axen, R., Protein thiolation and reversible protein–protein conjugation, N-succinimidyl 3-(2-pyridyldithio)propionate, a new heterobifunctional reagent, *Biochem. J.*, 173, 723, 1978.
4. Kitagawa, T., Shimozono, T., Aikawa, T., Yoshida, T., and Nishimura, H., Preparation and characterization of hetero-bifunctional cross-linking reagents for protein modifications, *Chem. Pharm. Bull.*, 29, 1130, 1981.
5. Keller, O. and Rudinger, J., Preparation and some properties of maleimido acids and malcoyl derivatives of peptides, *Helv. Chim. Acta*, 58, 531, 1975.
6. Yoshitake, S., Yamada. Y., Ishikawa, E., and Masseyeff, R., Conjugation of glucose oxidase from *Aspergillus niger* and rabbit antibodies using N-hydroxysuccinimide ester of N-(4-carboxycyclohexylmethyl)maleimide, *Eur. J. Biochem.*, 101, 395, 1979.

7. Staros, J. V. and Anjanejulu, P. S., Membrane-impermeable cross-linking reagents, *Methods Enzymol.*, 172, 609, 1989.

8. Kitagawa, T. and Aikawa, T., Enzyme coupled immunoassay of insulin using a novel coupling reagent, *J. Biochem.*, 79, 233, 1976.

9. Srinivasachar, K. and Neville, D. M., Jr., New protein cross-linking reagents that are cleaved by mild acid, *Biochemistry*, 28, 2501, 1989.

10. Weltman, J. K., Johnson, S. A., Langevin, J., and Riester, E. F., N-succinimidyl-(4-iodoacetyl)aminobenzoate: A new heterobifunctional crosslinker, *Biotechniques*, 1, 148, 1983.

11. Rector, E. S., Schwenk, R. J., Tse, K. S., and Sehon, A. H., A method for the preparation of protein–protein conjugates of predetermined composition, *J. lmmunol. Methods*, 24, 321, 1978.

12. Cuatrecasas, P., Wilchek, M., and Anfinsen, C. B., Affinity labeling of the active site of staphylococcal nuclease. Reactions with bromacetylated substrate analogs, *J. Biol. Chem.*, 244, 4316, 1969.

13. Fasold, H., Baumert, H., and Fink, G., Comparison of hydrophobic and strongly hydrophilic cleavable crosslinking reagents in intermolecular bond formation in aggregates of proteins or protein-RNA, in *Protein Crosslinking: Biochemical and Molecular Aspects*, Friedman, M., (ed.), Plenum Press, New York, 1976, p. 207.

14. Fink, G., Fasold, H., and Rammel, W., Reagents suitable for the crosslinking of nucleic acids to proteins, *Anal. Biochem.*, 108, 394, 1980.

15. McKenzie, J. A., Raison, R. L., and Rivett, D. E., Development of a bifunctional crosslinking agent with potential for the preparation of immunotoxins, *J. Protein Chem.*, 7, 581, 1988.

16. Kriwaczek, V. M., Bonnafous, J. C., Mueller, M., and Schwyzer, R., Tobacco mosaic virus as a carrier for small molecules. II. Cooperative affinity labeling of membrane vesicles with a TMV angiotensin conjugate, *Helv. Chim. Acta*, 61, 1241, 1978.

17. Bjorn, M. J., Groetsema, G., and Scalapino, L., Antibody-*Pseudomonas* exotoxin A conjugates cytotoxic to human breast cancer cells *in vitro*, *Cancer Res.*, 46, 3262, 1986.

18. Lorand, L., Brannen, W. T., and Rule, N. G., Thrombin-catalyzed hydrolysis of p-nitrophenyl esters, *Arch. Biochem. Biophys.*, 96, 147, 1962.

19. Kriwaczek, V. M., Eberle, A. N., Mueller, M., and Schwyzer, R., Tobacco mosaic virus as a carrier for small molecules. I. The preparation and characterization of a TMV/α-melanotropin conjugate, *Helv. Chim. Acta*, 61, 1232, 1978.

20. Fujii, N., Hayashi, Y., Katakura, S., Akaji, K., Yajima, H., Inouye, A., and Segawa, T., Studies on peptides. CXXVIII. Application of new heterobifunctional crosslinking reagents for the preparation of neurokinin (A and B)-BSA (bovine serum albumin) conjugates, *Int. J. Pept. Protein Res.*, 26, 121, 1985.

21. Peters, K. and Richards, F. M., Chemical cross-linking: Reagents and problems in studies of membrane Structure, *Annu. Rev. Biochem.*, 46, 523, 1977.

22. King, T. P., Li, Y., and Kochoumian, L., Preparation of protein conjugates via intermolecular disulfide bond formation, *Biochemistry*, 17, 1499, 1978.

23. Olomucki, M. and Diopoh, J., New protein reagents. I. Ethyl chloroacetimidate. Its properties and its reaction with ribonuclease, *Biochim. Biophys. Acta*, 263, 312, 1972.

24. Diopoh, J. and Olomucki, M., Ethyl bromoacetimidate, a NH_2-specific heterobifunctional reagent, *Hoppe-Seylers Z. Physiol. Chem.*, 360, 1257, 1979.

25. Trommer, W. E., Kolkenbrock., H., and Pfleiderer, G., Synthesis and properties of a new selective bifunctional cross-linking reagent, *Hoppe-Seylers Z. Physiol. Chem.*, 356, 1455, 1975.

26. Trommer, W. E., Friebel, K., Kiltz, H.-H., and Koldenbrock, H.-J., Synthesis and application of new bifunctional reagents, in *Protein Cross-Linking: Biochemical and Molecular Aspects*, Friedman, M. (ed.), Plenum Press, New York, 1976, Chapter 10.

27. Hermentin, P., Doenges, R., Gronski, P., Bosslet, K., Kraemer, H. P., Hoffmann, D., Zilag, H. et al., Attachment of rhodosaminyl anthracyclinone-type anthracyclines to the hinge region of monoclonal antibodies, *Bioconjug. Chem.*, 1, 100, 1990.

28. McClure, D. E., Arison, B. H., and Baldwin, J. J., Mode of nucleophilic addition to epichlorohydrin and related species: Chiral aryloxymethyloxiranes, *J. Am. Chem. Soc.*, 101, 3666, 1979.

29. Trommer, W. E. and Hendrick, M., The formation of maleimides by a new mild cyclization procedure, *Synthesis*, 484, 1973.

30. Simon, S. R. and Konigsberg, W. H., Chemical modification of hemoglobins: A study of conformation restraint by internal bridging, *Proc. Natl. Acad. Sci. U. S. A.*, 56, 749, 1966.

31. Blättler, W. A., Kuenzi, B. S., Lambert, J. M., and Senter, P. D., New heterobifunctional protein crosslinking reagent that forms an acid-labile link, *Biochemistry*, 24, 1517, 1985.

32. Bäumert, H. G. and Fasold, H., Cross-linking techniques, *Methods Enzymol.*, 172, 584, 1989.
33. Mitra, S. and Lawton, R. G., Reagent for cross-linking of proteins by equilibrium transfer alkylation, *J. Am. Chem. Soc.*, 101, 3097, 1979.
34. Liberatore, F. A., Comeau, R. D., Mckearin, J. M., Pearson, D. A., Belonga III, B. Q., Brocchini, S. J., Kath, J., Phillips, T., Oswell, K., and Lawton, R. G., Site-directed chemical modification and cross-linking of monoclonal antibody using equilibrium transfer alkylating cross-link reagents, *Bioconjug. Chem.*, 1, 36, 1990.
35. Uto, I., Ishimatsu, T., Hirayama, H., Ueda, S., Tsuruta, J., and Kambara, T., Determination of urinary Tamm-Horsfall protein by ELISA using a maleimide method for enzyme-antibody conjugation, *J. Immunol. Methods*, 138, 87, 1001, 1991.
36. Bieniarz, C., Husain, M., Barnes, G., King, C. A., and Welch, C. J., Extended length heterobifunctional coupling agents for protein conjugations, *Bioconjug. Chem.*, 7, 88, 1996.
37. Chrisey, L. A., Lee G. U., and O'Ferrall, C. E., Covalent attachment of synthetic DNA to self-assembled monolayer films, *Nucleic Acids Res.*, 24, 3031, 1996.
38. Kuijpers, W. H., Bos, E. S., Kaspersen, F. M., Veeneman, G. H., and van Boeckel, C. A., Specific recognition of antibody-oligonucleotide conjugates by radiolabeled antisense nucleotides: A novel approach for two-step radioimmunotherapy of cancer, *Bioconjug. Chem.*, 4, 94, 1993.
39. Brinkley, M. A., A survey of methods for preparing protein conjugates with dyes, haptens and crosslinking reagents, *Bioconjug. Chem.*, 3, 2, 1992.
40. Mattson, G., Conklin, E., Desai, S., Nielander, G., Savage, M. D., and Morgensen, S., A practical approach to crosslinking, *Mol. Biol. Rep.*, 17, 167, 1993.
41. Partis, M. D., Griffiths, D. G., Roberts, G. C., and Beechey, R. B., Crosslinking of proteins by ω-maleimido alkanoyl N-hydroxysuccinimide esters, *J. Protein Chem.*, 2, 263, 1983.
42. Yoshitake, S., Imagawa, M., Ishikawa, E., Niitsu, Y., Urushizaki, I., Nishiura, M., Kanazawa, R. et al., Mild and efficient conjugation of rabbit Fab and horseradish peroxidase using a maleimide compound and its use for enzyme immunoassay, *J. Biochem.*, 92, 1413, 1982.
43. Cumber, A. J., Forrester, J. A., Foxwell, B. M., Ross, W. C., and Thorpe, P. E., Preparation of antibody–toxin conjugates, *Methods Enzymol.*, 112, 207, 1985.
44. Itoh, Y., Cai, K., and Khorana, H. G., Mapping of contact sites in complex formation between light-activated rhodopsin and transducin by covalent crosslinking: Use of a chemically preactivated reagent, *Proc. Natl. Acad. Sci. U. S. A.*, 98, 4883, 2001.
45. Wang, D., Li, Q., Hudson, W., Berven, E., Uckun, F., and Kersey, J. H., Generation and characterization of an anti-CD19 single-chain Fv immunotoxin composed of C-terminal disulfide-linked dgRTA, *Bioconjug. Chem.*, 8, 878, 1997.
46. Romano, K. P., Newman, A. G., Zahran, R. W., and Millard, J. T., DNA interstrand cross-linking by epichlorohydrin, *Chem. Res. Toxicol.*, 20, 832, 2007.
47. Del Rosario, R. B., Wahl, R. L., Brocchini, S. J., Lawton, R. G., and Smith, R. H., Sulfhydryl site-specific cross-linking and labeling of monoclonal antibodies by a fluorescent equilibrium transfer alkylation cross-link reagent, *Bioconjug. Chem.*, 1, 51, 1990.
48. Haber, E., Quertermous, T., Matsueda, G. R., and Runge, M. S., Innovative approaches to plasminogen activator therapy, *Science*, 243, 51, 1989.
49. Bode, C., Runge, M. S., Branscomb, E. E., Newell, J. B., Matsueda, G. R, and Harber, E., Antibody-directed fibrinogen: An antibody specific for both fibrin and tissue plasminogen activator, *J. Biol. Chem.*, 264, 944, 1989.
50. Wawrzyaczak, E. J. and Thorpe, P. E., Methods for preparing immunotoxins: Effect of the linkage on activity and stability, in *Immunoconjugates, Antibody Conjugates in Radioimaging and Therapy of Cancer*, Vogel, C.-W. (ed.), Oxford University Press, New York, 1987, p. 28.
51. Marsh, J. W., Srinivasachar, K., and Neville, D. M., Jr., Antibody–toxin conjugation, in *Immunotoxins*, Frankel, A. E. (ed.), Kluwer Academic Publishers, Boston, MA, 1988, p. 213.
52 Stan, R. V., Multiple PV1 dimers reside in the same stomatal or fenestral diaphragm, *Am. J. Physiol. Heart Circ. Physiol.*, 286, H1347, 2004.
53. Mader, C., Huber, C., Moll, D., Sleytr, U. B., and Sára, M., Interaction of the crystalline bacterial cell surface layer protein SbsB and the secondary cell wall polymer of *Geobacillus stearothermophilus* PV72 assessed by real-time surface plasmon resonance biosensor technology, *J. Bacteriol.*, 186, 1758, 2004.
54. Carroll, S. F., Bernhard, S. L., Goff, D. A., Bauer, R. J., Leach, W., and Kung, A. H. C., Enhanced stability *in vitro* and *in vivo* of immunoconjugates prepared with 5-methyl-2-imino thiolane, *Bioconjug. Chem.*, 5, 248, 1994.

55. Balthasar, S., Michaelis, K., Dinauer, N., von Briesen, H., Kreuter, J., and Langer, K., Preparation and characterisation of antibody modified gelatin nanoparticles as drug carrier system for uptake in lymphocytes, *Biomaterials*, 26, 2723, 2005.

56. Goff, D. A. and Carroll, S. F., Substituted 2-iminothiolanes: Reagents for the preparation of disulfide cross-linked conjugates with increased stability, *Bioconjug. Chem.*, 1, 381, 1990.

57. Wohlfeil, E. R. and Hudson, R. A., Synthesis and characterization of a heterobifunctional mercurial cross-linking agent: Incorporation into cobratoxin and interaction with the nicotinic acetylcholine receptor, *Biochemistry*, 30, 7231, 1991.

58. Ruppert, C., Markart, P., Schmidt, R., Grimminger, F., Seeger, W., Lehr, C.-M., and Günther, A., Chemical crosslinking of urokinase to pulmonary surfactant protein B for targeting alveolar fibrin, *Thromb. Haemost.*, 89, 53, 2003.

59. Hashida, S., Imagawa, M., Inoue, S., Ruan, K.-H., and Ishikawa, E., More useful maleimide compounds for the conjugation of Fab' to horseradish peroxidase through thiol groups in the hinge, *J. Appl. Biochem.*, 6, 56, 1984.

60. Yoshitake, S., Hamaguchi, Y., and Ishikawa, E., Efficient conjugation of rabbit Fab' with β-D-galactosidase from *Escherichia coli*, *Scand. J. Immunol.*, 10, 81, 1979.

61. Ishikawa, E., Imagawa, M., Hashida, S., Yoshitake, S., Hamaguchi, Y., and Ueno, T., Enzyme-labeling of antibodies and their fragments for enzyme immunoassay and immunohistochemical staining, *J. Immunoassay*, 4, 209, 1983.

62. Ishikawa, E., Hashida, S., Kohno, T., and Tanaka, K., Methods for enzyme-labeling of antigens, antibodies and their fragments, in *Nonisotopic Immunoassays*, Ngo, T. T. (ed.), Plenum Press, New York, 1988, p. 27.

63. Boal, A. K., Tellez, H., Rivera, S. B., Miller, N. E., Bachand, G. D., and Bunker, B. C., The stability and functionality of chemically crosslinked microtubules, *Small*, 2, 793, 2006.

64. Bayley, H. and Knowles, J. R., Photoaffinity labeling, *Methods Enzymol.*, 46, 69, 1977.

65. Henkin, J., Photolabeling reagent for thiol enzymes. Studies on rabbit muscle creatine kinase, *J. Biol. Chem.*, 252, 4293, 1977.

66. Husain, S. S., Ferguson, J. B., and Fruton, J. S., Bifunctional inhibitors of pepsin, *Proc. Natl. Acad. Sci. U. S. A.*, 68, 2765, 1971.

67. Shafer, J., Baronowsky, D., Laursen, R., Finn, F., and Westheimer, F. H., Products from the photolysis of diazoacetyl chymotrypsin, *J. Biol. Chem.*, 241, 421, 1966.

68. Harrison, J. K., Lawton, R. G., and Gnegy, M. E., Development of novel photoreactive calmodulin derivative. Cross-linking purified adenylate cyclase from bovine brain, *Biochemistry*, 28, 6023, 1989.

69. Carroll-Portillo, A., Bachand, M., and Bachand, G. D., Directed attachment of antibodies to kinesin-powered molecular shuttles, *Biotechnol. Bioeng.*, 104, 1182, 2009.

70. Webb, R. R., II, and Kancko, E., Synthesis of 1-(aminooxy)-4-[(3-nitro-2-pyridyl)dithio)butane hydrochloride and 1-(aminooxy)-4-[(3-nitro-2-pyridyl)dithio]but-2-ene. Novel heterobifunctional cross-linking reagents, *Bioconjug. Chem.*, 1, 96, 1990.

71. Willner, D., Trail, P. A., Hofstead, S. J., King, H. D., Lasch, S. J., Braslawsky, G. R., Greenfield, R. S., Kaneko, T., and Firestone, R. A., (6-Maleimidocaproyl)hydrazone of doxorubicin—A new derivative for the preparation of immunoconjugates of doxorubicin, *Bioconjug. Chem.*, 4, 521,1993.

72. Liberatore, F. A., Comeau, R. D., and Lawton, R. G., Heterobifunctional cross-linking of a monoclonal antibody with 2-methyl-N'-benzenesulfonyl-N⁴-bromoacetylquinonediimide, *Biochem. Biophys. Res. Commun.*, 158, 640, 1989.

73. Kraehenbuhl, J. P., Galardy, R. E., and Jamieson, J. D., Preparation and characterization of an immuno-electron microscope tracer consisting of hemeoctapeptide coupled to Fab, *J. Exp. Med.*, 139, 208, 1974.

74. Maassen, J. A., Schop, E. N., and Moller, W., Structural analysis of ribosomal proteins L7/L12 by the heterobifunctional cross-linker: 4-(6-Formyl-3-azidophenoxy)butyrimidate, *Biochemistry*, 20, 1020, 1981.

75. Maassen, J. A., Cross-linking of ribosomal proteins by 4-(6-formyl-3-azido phenoxy)butyrimidate. A heterobifunctional cleavable cross-linker, *Biochemistry*, 18, 1288, 1979.

76. Maassen, J. A. and Terhorst, C., Identification of a cell-surface protein involved in the binding site of sindbis virus on human lymphobastoic cell lines using a heterobifunctional cross-linker, *Eur. J. Biochem.*, 115, 153, 1981.

77. Cater, C. W., The evaluation of aldehydes and other difunctional compounds as crosslinking agents for collagen, *J. Soc. Leather Trades Chem.*, 47, 259, 1963.

78. Ishii, T., Yamada, T., Mori, T., Kumazawa, S., Uchida, K., and Nakayama, T., Characterization of acrolein-induced protein cross-links, *Free Radical Res.*, 41, 1253, 2007.

79. Sethi, D., Kumar, A., Gupta, K. C., and Kumar, P., A facile method for the construction of oligonucleotide microarrays, *Bioconjug. Chem.*, 19, 2136, 2008.

80. Misra, A., Shahid, M., and Dwivedi, P., N-(3-triethoxysilylpropyl)-4-(isothiocyanatomethyl)-cyclohexane-1-carboxamide (TPICC): A heterobifunctional reagent for immobilization of biomolecules on glass surface, *Bioorg. Med. Chem. Lett.*, 18, 5217, 2008.

81. Ahluwalia, R., Day, R. A., and Nauss, J., A bifunctional monocyclic β-lactam cross-links across the active site of β-lactamase, *Biochem. Biophys. Res. Commun.*, 206, 577, 1995.

82. Gilchrist, T. L. and Rees, C. W., *Carbenes, Nitrenes and Arynes (Studies in Modern Chemistry)*, Nelson Publisher, London, 1969, p. 131.

83. Chowdhry, V., Vaughan, R., and Westheimer, F. H., 2-Diazo-3,3,3-trifluoropropionyl chlorides: Reagent for photoaffinity labeling, *Proc. Natl. Acad. Sci. U. S. A.*, 73, 1406, 1976.

84. Pascual, A., Casanova, J. S., and Herbert, H., Photoaffinity labeling of thyroid hormone nuclear receptors in intact cells, *J. Biol. Chem.*, 257, 9640, 1982.

85. Gupta, C. M., Radhadrishnan, R., Gerber, G. E., Olsen, W. L., Quay, S. C., and Khorana, H. G., Intermolecular cross-linking of fatty acylchains in phospholipids: use of photoactivable carbene precursors, *Proc. Natl. Acad. Sci. U. S. A.*, 76, 2595, 1979.

86. Takagaki, Y., Gupta, C. M., and Khorana, H. G., Thiols and the diazo group in photoaffinity labels, *Biochem. Biophys. Res. Commun.*, 95, 589, 1980.

87. Chowdhry, V. and Westheimer, F. H., Photoaffinity labeling of biological systems, *Annu. Rev. Biochem.*, 48, 293, 1979.

88. Hashimoto, M. and Hatanaka, Y., Recent progress in diazirine-based photoaffinity labeling, *Eur. J. Org. Chem.*, 15, 2513, 2008.

89. Blencowe, A. and Hayes, W., Development and application of diazirines in biological and synthetic macromolecular systems, *Soft Matter*, 1, 178, 2005.

90. Lutter, L. C., Ortanderl, F., and Fasold, H., The use of new series of cleavable protein-crosslinking on the *Escherichia coli* ribosomes, *FEBS Lett.*, 48, 288, 1974.

91. Grob, P. M., Berlot, C. H., and Bothwill, M. A., Affinity labeling and partial purification of nerve growth factor receptors from rat pheochromocytoma and human melanoma cells, *Proc. Natl. Acad. Sci. U. S. A.*, 80, 6819, 1983.

92. Puma, P., Buxser, S. E., Watson, L., Kellcher, D. J., and Johnson, G. L., Purification of the receptor for nerve growth factor from A875 melanoma cells by affinity chromatography, *J. Biol. Chem.*, 258, 3370, 1983.

93. Borst, D. W. and Sayare, M., Photoactivated crosslinking of protein to hepatic membrane binding sites, *Biochem. Biophys. Res. Commun.*, 105, 194, 1982.

94. Ji, T. H. and Ji, I., Macromolecular photoaffinity labeling with radioactive photoactivable heterobifunctional reagent, *Anal. Biochem.*, 121, 286, 1982.

95. Lewis, R. V., Roberts, M. F., Dennis, E. A., and Allison, W. S., Photoactivated heterobifunctional cross-linking reagents which demonstrate the aggregation state of phospholipase A$_2$, *Biochemistry*, 16, 5650, 1977.

96. Ji, I. and Ji, T. H., Both α and β subunits of human choriogonadotropin photoaffinity label the hormone receptor, *Proc. Natl. Acad. Sci. U. S. A.*, 78, 5465, 1981.

97. Guire, P., Fliger, D., and Hodgson, J., Photochemical coupling of enzymes to mammalian cells, *Pharmacol. Res. Commun.*, 9, 131, 1977.

98. Ballmer-Hofer, K., Schlup, V., Burn, P., and Burger, M. M., Isolation of *in situ* crosslinked ligand receptor complexes using an anticrosslinker specific antibody, *Anal. Biochem.*, 126, 246, 1982.

99. Schmidt, R. R. and Betz, H., Cross-linking of β-bungarotoxin to chick brain membranes. Identification of subunits of a putative voltage-gated K$^+$ channel, *Biochemistry*, 28, 8346, 1989.

100. Witzemann, V., Muchmore, D., and Raftery, M. A., Affinity directed cross-linking of membrane-bound acetylcholine receptor polypeptides with photolabile α-bungarotoxin derivatives, *Biochemistry*, 18, 5511, 1979.

101. Vanin, E. F. and Ji, T. H., Synthesis and application of cleavable photoactivable heterobifunctional reagents, *Biochemistry*, 20, 6754, 1981.

102. Schwartz, I. and Offengand, J., *E. coli* tRNA[Phe] modified at the 3-(3-amino-3-carboxypropyl)uridine with a photoaffinity label is fully functional for aminoacylation and for ribosomal interaction, *Biochim. Biophys. Acta*, 697, 330, 1982.

103. Baenziger, J. U. and Fiete, D., Photoactivatable glycopeptide reagents for site-specific labeling of lectins, *J. Biol. Chem.*, 257, 4421, 1982.

104. Jaffe, C. L., Lis, H., and Sharon, N., New cleavable photoreactive heterobifunctional cross-linking reagents for studying membrane organization, *Biochemistry*, 19, 4423, 1980.

105. Denny, J. B. and Blobel, G., Iodine-125-labeled crosslinking reagent that is hydrophilic, photoactivatable and cleavable through an azo linkage, *Proc. Natl. Acad. Sci. U. S. A.*, 81, 5286, 1984.

106. Schwartz, M. A., Das, O. P., and Hynes, R. O., A new radioactive cross-linking reagent for studying the interactions of proteins, *J. Biol. Chem.*, 257, 2343, 1982.

107. Wollenweber, H. and Morrison, D. C., Synthesis and biochemical characterization of a photoactivatable, iodinatable, cleavable bacterial lipopolysaccharide derivative, *J. Biol. Chem.*, 260, 15068, 1985.

108. Sorensen, P., Farber, N. M., and Krystal, G., Identification of the interleukin-3 receptor using an iodinatable, cleavable, photoreactive crosslinking agent, *J. Biol. Chem.*, 261, 9094, 1986.

109. Imai, N., Kometani, T., Crocker, P. J., Bowdan, J. B., Demir, A., Dwyer, L. D., Mann, D. M., Vanaman, T. C., and Watt, D. S., Photoaffinity heterobifunctional cross-linking reagents based on N-(azidobenzoyl) tyrosines, *Bioconjug. Chem.*, 1, 138, 1990.

110. Imai, N., Dwyer, L. D., Komentani, T., Ji, T., Vanaman, T. C., and Watt, D. S., Photoaffinity heterobifunctional cross-linking reagents based on azide-substituted salicylates, *Bioconjug. Chem.*, 1, 144, 1990.

111. Hinds, T. R. and Andreasen, T. J., Photochemical cross-linking of azidocalmodulin to the $(Ca^{2+} Mg^{2+})$ ATPase of the erythrocyte membrane, *J. Biol. Chem.*, 256, 7877, 1981.

112. Moreland, R. B., Smith, P. K., Fujimoto, E. K., and Dockter, M. E., Synthesis and characterization of N-(4-azidophenylthio)-phthalimide, *Analyt. Biochem.*, 121, 321, 1982.

113. Harris, R. and Findlay, J. B., Investigation of the organization of the major proteins in bovine myelin membranes. Use of chemical probes and bifunctional crosslinking reagents, *Biochim. Biophys. Acta*, 732, 75, 1983.

114. Fleet, G. W. J., Knowles, J. R., and Porter, R. R., The antibody binding site. Labelling of specific antibody against the photo-precursor of an aryl nitrene, *Biochem. J.*, 128, 499, 1972.

115. Wilson, D. F., Miyata, Y., Erecinska, M., and Vanderkooi, J. A., An aryl azide suitable for photoaffinity labeling of amine groups in proteins, *Arch. Biochem. Biophys.*, 171, 104, 1975.

116. Sigrist, H., Allegrini, P. R., Kempf, C., Schnippering, C., and Zahler, P., 5-Isothiocyanato-1-naphthalene azide and p-azido-phenylisothiocyanate, *Eur. J. Biochem.*, 125, 197, 1982.

117. Hixson, S. H. and Hixson, S. S., p-Azidophenylacyl bromide, a versatile photolabile bifunctional reagent: Reaction with glyceraldehyde-3-phosphate dehydrogenae, *Biochemistry*, 14, 4251, 1975.

118. Rudnick, G., Kaback, H. R., and Weil, R., Photoinactivation of the β-galactoside transport system in *Escherichia coli* membrane vesicles with an impermeant azidophenylgalactoside, *J. Biol. Chem.*, 250, 6847, 1975.

119. Demoliou, C. D. and Epand, R. M., Synthesis and characterization of a heterobifunctional photoaffinity reagent for modification of tryptophan residues and its application to the preparation of a photoreactive glucagon derivative, *Biochemistry*, 19, 4539, 1980.

120. Huang, C.-K. and Richards, F. M., Reaction of a lipid-soluble, unsymmetrical, cleavable, cross-linking reagent with muscle aldolase and erythrocyte membrane proteins, *J. Biol. Chem.*, 252, 5514, 1977.

121. Chong, P. C. S. and Hodges, R. S., A new heterobifunctional cross-linking reagent for the study of biological interaction between proteins. I. Design, synthesis and characterization, *J. Biol. Chem.*, 256, 5064, 1981.

122. Chong, P. C. S. and Hodges, R. S., A new heterobifunctional cross-linking reagent for the study of biological interaction between proteins. II. Application to the troponin C-troponin I interaction, *J. Biol. Chem.*, 256, 5071, 1981.

123. Chong, P. C. S. and Hodges, R. S., Photochemical cross-linking between rabbit skeleton troponin and alpha-tropomyosin, *J. Biol. Chem.*, 257, 9152, 1982.

124. Ohananasekaran, N., Wessling-Resnick, M., Kelleher, D. J., Johnson, G. L., and Ruoho, A. E., Mapping of the carboxyl terminus within the tertiary structure of transducin's alpha subunit using the heterobifunctional cross-linking reagent, ^{125}I-N-(3-iodo-4-azidophenylpropionamido-S-(2-thiopyridyl)cysteine, *J. Biol. Chem.*, 261, 17942, 1988.

125. Friebel, K., Huth, H., Jany, K. D., and Trummer, W. E., Semireversible cross-linking: Synthesis and application of a novel heterobifunctional reagent, *Hoppe-Seylers Z. Physiol. Chem.*, 362, 421, 1981.

126. Vanin, E. F., Burkhard, S. J., and Kaiser, I. I., p-Azidophenylglyoxal. A heterobifunctional photo-sensitive reagent, *FEBS Lett.*, 124, 89, 1981.

127. Ngo, T. T., Yam, C. F., Lenhoff, H. M., and Ivy, J., p-Azidophenylglyoxal: A heterobifunctional photoactivatable cross-linking reagent selective for arginine residue, *J. Biol. Chem.*, 256, 11313, 1981.

128. Gorman, J. J. and Folk, J. E., Transglutaminase amine substrates for photochemical labeling and cleavable cross-linking of proteins, *J. Biol. Chem.*, 255, 1175, 1980.

129. Drafler, F. L. and Marinetti, G. V., Synthesis of a photoaffinity probe for the β-adrenergic receptor, *Biochem. Biophys. Res. Commun.*, 79, 1, 1977.

130. Das, M. and Fox, F., Chemical cross-linking in biology, *Annu. Rev. Biophys. Bioeng.*, 8, 165, 1979.

131. Schäfer, H.-J., Divalent azido-ATP analog for photoaffinity cross-linking of F₁ subunits, *Methods Enzymol.*, 126, 649, 1986.

132. Dombroski, K. E. and Colman, R. F., 5′-(p-Fluorosulfonylbenzoyl)-8-azidoadenosine. A new bifunctional affinity label for nucleotide binding sites in proteins, *Arch. Biochem. Biophys.*, 275, 302, 1989.

133. Wood, C. L. and O'Dorisio, M. S., Covalent crosslinking of vasoactive intestinal polypeptide to its receptors on intact human lymphoblasts, *J. Biol. Chem.*, 260, 1243, 1985.

134. McMahan, S. A. and Burgess, R. R., Use of aryl azide cross-linkers to investigate protein–protein interactions: An optimization of important conditions as applied to *Escherichia coli* RNA polymerase and localization of a sigma 70-alpha cross-link to the C-terminal region of alpha, *Biochemistry*, 33, 12094, 1994.

135. van der Horst, G. T., Mancini, G. M., Brossmer, R., Rose, U., and Verheijen, F. W., Photoaffinity labeling of a bacterial sialidase with an aryl azide derivative of sialic acid, *J. Biol. Chem.*, 265, 10801, 1990.

136. Tiberi, M., Nash, S. R., Bertrand, L., Lefkowitz, R. J., and Caron, M. G., Differential regulation of dopamine D1A receptor responsiveness by various G protein-coupled receptor kinases, *J. Biol. Chem.*, 271, 3771, 1996.

137. Chattopadhyay, A., James, H. L., and Fair, D. S., Molecular recognition sites on factor Xa which participate in the prothrombinase complex, *J. Biol. Chem.*, 267, 12323, 1992.

138. Smith, J. W., Ruggeri, Z. M., Kunicki, T. J., and Cheresh, D. A., Interaction of integrins alpha v beta 3 and glycoprotein IIb-IIIa with fibrinogen. Differential peptide recognition accounts for distinct binding sites, *J. Biol. Chem.*, 265, 12267, 1990.

139. Knowles, J. R., Photogenerated reagents for biological receptor site labeling, *Acc. Chem. Res.*, 5, 155, 1972.

140. DeGraff, B. A., Gillespie, D. W., and Sundberg, R. J., Phenyl nitrene. A flash photolytic investigation of the reaction with secondary amines, *J. Am. Chem. Soc.*, 96, 7491, 1974.

141. Reiser, A., Willets, F. W., Terry, G. C., Williams, V., and Morley, R., Photolysis of aromatic azides. IV. Lifetimes of aromatic nitrenes and absolute rates of some of their reactions, *Trans. Faraday Soc.*, 64, 3265, 1968.

142. Staros, J. V., Bayley, H., Standring, D. N., and Knowles, J. R., Reduction of aryl azides by thiols: Implication for the use of photoaffinity reagents, *Biochem. Biophys. Res. Commun.*, 80, 568, 1978.

143. Vodovozova, E. L., Photoaffinity labeling and its application in structural biology, *Biochemistry (Moscow)*, 72, 1, 2007.

144. Dorman, G. and Prestwich, G. D., Benzophenone photophores in biochemistry, *Biochemistry*, 33, 5661, 1994.

145. Weber, P. J. and Beck-Sickinger, A. G., Comparison of the photochemical behavior of four different photoactivatable probes, *J. Pept. Res.*, 49, 375, 1997.

146. Harnish, D. G., Leung, W.-C., and Rawls, W. E., Characterization of polypeptides immunoprecipitable from Pichinde virus-infected BHK-21 cells, *J. Virol.*, 38, 840, 1981.

147. Odom, O. W., Deng, H.-Y., Subramanian, A. R., and Hardesty, B., Relaxation time, interthiol distance, and mechanism of action of ribosomal protein S1(1), *Arch. Biochem. Biophys.*, 230, 178, 1984.

148. Tao, T., Scheiner, C. J., and Lamkin, M., Site-specific photo-cross-linking studies on interactions between troponin and tropomyosin and between subunits of troponin, *Biochemistry*, 25, 7633, 1986.

149. Tao, T., Lamkin, M., and Scheiner, C. J., The conformation of the C-terminal region of actin: A site-specific photocrosslinking study using benzophenone-4-maleimide, *Arch. Biochem. Biophys.*, 240, 627, 1985.

150. Luo, Y., Wu, J. L., Li, B., Langsetmo, K., Gergely, J., and Tao, T., Photocrosslinking of benzophenone-labeled single cysteine troponin I mutants to other thin filament proteins, *J. Mol. Biol.*, 296, 899, 2000.

151. Campbell, P. and Gioannini, T. L., The use of benzophenone as a photoaffinity label. Labeling in p-benzoylphenylacetyl chymotrypsin at unit efficiency, *Photochem. Photobiol.*, 29, 883, 1979.

152. Jelenc, P. C., Cantor, C. R., and Simon, S. R., High yield photoreagents for protein crosslinking and affinity labeling, *Proc. Natl. Acad. Sci. U. S. A.*, 75, 3564, 1978.

153. Yip, C. C., Yeung, C. W. T., and Moule, M., Photoaffinity labeling of insulin receptor of rat adipocyte plasma membrane, *J. Biol. Chem.*, 253, 1743, 1978.

154. Perkins, M. E., Ji, T. H., and Hynes, R. O., Cross-linking of fibronectin to sulfated proteoglycan at the cell surface, *Cell*, 16, 941, 1979.

155. Andreasen, T. J., Keller, C. H., LaPorte, D. C., Edelman, A. M., and Storm, D. R., Preparation of azido-calmodulin: A photoaffinity labeling for calmodulin binding proteins, *Proc. Natl. Acad. Sci. U. S. A.*, 78, 2782, 1981.

156. Johnson, G. L., MacAndrew, Jr., V, I., and Pilch, P. F., Identification of glucagon receptor in rat liver membranes by photoaffinity crosslinking, *Proc. Natl. Acad. Sci. U. S. A.*, 78, 875, 1981.

157. Coltrera, M. D., Potts, J. T., and Rosenblatt, M., Identification of a renal receptor for parathyroid hormone by photoaffinity radiolabeling using a synthetic analogue, *J. Biol. Chem.*, 256, 10555, 1981.

158. Bochkariov, D. E. and Kogon, A. A., Application of 3-[3-(3-(trifluoromethyl)diazirin-3-yl)phenyl]-2,3-dihydroxypropionic acid, carbene-generating, cleavable cross-linking reagent for photoaffinity labeling, *Anal. Biochem.*, 204, 90, 1992.

159. Shephard, E., De Beer, F. C., von Holt, E., and Hapgood, J. P., The use of sulfosuccinimidyl-2-(p-azidosalicylamido)-1,3'-dithiopropionate as a crosslinking reagent to identify cell surface receptors, *Anal. Biochem.*, 168, 306, 1988.

160. Chen, L. L., Lobb, R. R., Cuervo, J. H., Lin, K.-C., Adams, S. P., and Pepinsky, R. B., Identification of ligand binding sites on integrin α4β1 through chemical cross-linking, *Biochemistry*, 37, 8743, 1998.

161. Steiner, M., Identification of the binding site for transferrin in human reticulocytes, *Biochem. Biophys. Res. Commun.*, 94, 861, 1980.

162. Erecinsska, M., A new photoaffinity labeled derivative of mitochondrial cytochrome-C, *Biochem. Biophys. Res. Commun.*, 76, 495, 1977.

163. Cai, K., Itoh, Y., and Khorana, H. G., Mapping of contact sites in complex formation between transducin and light-activated rhodopsin by covalent crosslinking: Use of a photoactivatable reagent, *Proc. Natl. Acad. Sci. U. S. A.*, 98, 4877, 2001.

164. Ghetu, A. F., Arthur, D. C., Kerppola, T. K., and Glover, J. N. M., Probing FinO–FinP RNA interactions by site-directed protein–RNA crosslinking and gelFRET, *RNA*, 8, 816, 2002.

165. Kiehm, D. J. and Ji, T. H., Photochemical cross-linking of cell membranes, *J. Biol. Chem.*, 252, 8524, 1977.

166. Peletskaya, E. N., Boyer, P. L., Kogon, A. A., Clark, P., Kroth, H., Sayer, J. M., Jerina, D. M., and Hughes, S., Cross-linking of the fingers subdomain of human immunodeficiency virus type 1 reverse transcriptase to template-primer, *J. Virol.*, 75, 9435, 2001.

167. Rajasekharan, K. N., Mayadevi, M., and Burke, M., Studies of ligand-induced conformational perturbations in myosin subfragment 1. An examination of the environment about the SH2 and SH1 thiols using a photoprobe, *J. Biol. Chem.*, 264, 10810, 1989.

168. Spanggord, R. J. and Beal, P. A., Site-specific modification and RNA crosslinking of the RNA-binding domain of PKR, *Nucleic Acid Res.*, 28, 1899, 2000.

169. Politz, S. M., Noller, H. F., and McWhirter, P. D., (4-Azido-pbenyl)glyoxal, a novel heterobifunctional reagent: RNA–protein cross-linking in *E. coli* ribosomes, *Biochemistry*, 20, 372, 1981.

170. Folk, J. E. and Chung, S. I., Molecular and catalytic properties of transglutaminase, *Adv. Enzymol.*, 38, 109, 1973.

171. Folk. J. E. and Finlayson, J. S., The ε-(γ-glutamyl)lysine crosslink and the catalytic role of transglutaminase, *Adv. Protein Chem.*, 31, 1, 1977.

172. Folk, J. E., Transglutaminase, *Annu. Rev. Biochem.*, 49, 517, 1980.

173. Hegyi, G., Mák, M., Kim, E., Elzinga, M., Muhlrad, A., and Reisler, E., Intrastrand cross-linked actin between Gln-41 and Cys-374. I. Mapping of sites cross-linked in F-actin by N-(4-azido-2-nitrophenyl) putrescine, *Biochemistry*, 37, 17784, 1998.

174. Kudryashov, D. S., Sawaya, M. R., Adisetiyo, H., Norcross, T., Hegyi, G., Reisler, E., and Yeates, T. O., The crystal structure of a cross-linked actin dimer suggests a detailed molecular interface in F-actin, *Proc. Natl. Acad. Sci. U. S. A.*, 102, 13105, 2005.

175. Morehead, H. W., Talmadge, K. W., O'Shannessy, D. J., and Siebert, C. J., Optimization of oxidation of glycoproteins: An assay for predicting coupling to hydrazide chromatographic supports, *J. Chromatogr.*, 587, 171, 1991.

176. Watkins, N. J., Braidley, P., Bray, C. J., Savill, C. M., and White, D. J., Coating of human decay accelerating factor (hDAF) onto medical devices to improve biocompatibility, *Immunopharmacology*, 38, 111, 1997.

177. Zarka, A. and Shoshan-Barmatz, V., Characterization and photoaffinity labeling of the ATP binding site of the ryanodine receptor from skeletal muscle, *Eur. J. Biochem.*, 213, 147, 1993.

178. Salvucci, M. E., Rajagopalan, K., Sievert, G., Haley, B. E., and Watt, D. S., Photoaffinity labeling of ribulose-1,5-bisphosphate carboxylase/oxygenase activase with ATP gamma-benzophenone. Identification of the ATP gamma-phosphate binding domain, *J. Biol. Chem.*, 268, 14239, 1993.

179. Han, S., Collins, B. E., Bengtson, P., and Paulson, J. C., Homomultimeric complexes of CD22 in B cells revealed by protein-glycan cross-linking, *Nat. Chem. Biol.*, 1, 93, 2005.

180. Tanaka, Y. and Kohler, J. J., Photoactivatable crosslinking sugars for capturing glycoprotein interactions, *J. Am. Chem. Soc.*, 130, 3278, 2008.

181. Nakanishi, K., Zhang, H., Lerro, K. A., Takekuma, S., Yamamoto, T., Lien, T. H., Sastry, L. et al., Photoaffinity labeling of rhodopsin and bacteriorhodopsin, *Biophys. Chem.*, 56, 13, 1995.

182. Nakayama, T. A. and Khorana, H. G., Orientation of retinal in bovine rhodopsin determined by crosslinking using a photoactivatable analog of 11-cis-retinal, *J. Biol. Chem.*, 265, 15762, 1990.

183. Segal, D. M. and Bast, B. J., Production of bispecific antibodies, *Curr. Protoc. Immunol.*, 2001, Chapter 2, Unit 2.13.

184. Cao, Y. and Suresh, M. R., Bispecific antibodies as novel bioconjugates, *Bioconjug. Chem.*, 9, 635, 1998.

185. Suresh, M. R., Cuello, A. C., and Milstein, C., Bispecific monoclonal antibodies from hybrid hybridomas, *Methods Enzymol.*, 121, 210,1986.

186. Nolan, O. and Kennedy O. R., Bifunctional antibodies: Concept, production and applications, *Biochim. Biophys. Acta*, 1040, 1, 1990.

187. Pluckthun, A. and Pack, P., New protein engineering approaches to multivalent and bispecific antibody fragments, *Immunotechnology*, 3, 83, 1997.

188. Tomlinson, I. and Holliger, P., Methods for generating multivalent and bispecific antibody fragments, *Methods Enzymol.*, 326, 461, 2000.

189. Schmidt, M., Hynes, N. E., Groner, B., and Wels, W., A bivalent single-chain antibody-toxin specific for ErbB-2 and the EGF receptor, *Int. J. Cancer*, 65, 538, 1996.

190. Bostrom, J., Yu, S. F., Kan, D., Appleton, B. A., Lee, C. V., Billeci, K., Man, W. et al., Variants of the antibody herceptin that interact with HER2 and VEGF at the antigen binding site, *Science*, 323, 1610, 2009.

191. Ridgway, J. B., Presta, L. G., and Carter, P., Knobs-into-holes engineering of antibody CH3 domains for heavy chain heterodimerization, *Protein Eng.*, 9, 617, 1996.

192. Merchant, A. M., Zhu, Z. P., Yuan, J. Q., Goddard, A., Adams, C. W., Presta, L. G., and Carter, P., An efficient route to human bispecific IgG, *Nat. Biotechnol.*, 16, 677, 1997.

193. Kostelny, S. A., Cole, M. S., and Tso, J. Y., Formation of a bispecific antibody by the use of leucine zippers, *J. Immunol.*, 148, 1547, 1992.

194. De Kruif, J. and Logtenberg, T., Leucine zipper dimerized bivalent and bispecific scFv antibodies from a semisynthetic antibody phage display library, *J. Biol. Chem.*, 271, 7630, 1996.

195. Holliger, P., Prospero, T., and Winter, G., "Diabodies": Small bivalent and bispecific antibody fragments, *Proc. Natl. Acad. Sci. U. S. A.*, 90, 6444, 1993.

196. Kipriyanov, S. M., Generation of bispecific and tandem diabodies, *Methods Mol. Biol.*, 562, 177, 2009.

197. Lu, D. and Zhu, Z., Construction and production of an IgG-like tetravalent bispecific antibody for enhanced therapeutic efficacy, *Methods Mol. Biol.*, 525, 377, 2009.

198. Kim, W. D., Tokunaga, M., Ozaki, H., Ishibashi, T., Honda, K., Kajiura, H., Fujiyama, K., Asano, R., Kumagai, I., Omasa, T., and Ohtake, H., Glycosylation pattern of humanized IgG-like bispecific antibody produced by recombinant CHO cells, *Appl. Microbiol. Biotechnol.*, 85, 535, 2010.

199. Wood, W. G., Immunoassays & co.: Past, present, future?—A review and outlook from personal experience and involvement over the past 35 years, *Clin. Lab.*, 54, 423, 2008.

200. Milstein, C. and Cuello, A. C., Hybrid hybridomas and their use in immunohistochemistry, *Nature*, 305, 537, 1983.

201. Cao, Y., Christian, S., and Suresh, M. R., Development of a bispecific monoclonal antibody as a universal immunoprobe for detecting biotinylated macromolecules, *J. Immunol. Methods*, 220, 85, 1998.

202. Khaw, B. A., Rammohan, R., and Abu-Taha, A., Bispecific enzyme-linked signal-enhanced immunoassay with subattomole sensitivity, *Assay Drug Dev. Technol.*, 3, 319, 2005.

203. Liu, F., Guttikonda, S., and Suresh, M. R., Bispecific monoclonal antibodies against a viral and an enzyme: Utilities in ultrasensitive virus ELISA and phage display technology, *J. Immunol. Methods*, 274, 115, 2003.

204. Kreutz, F. T. and Suresh, M. R., Novel bispecific immunoprobe for rapid and sensitive detection of prostatespecific antigen, *Clin. Chem.*, 43, 649, 1997.

205. Goldenberg, D. M., Rossi, E. A., Sharkey, R. M., McBride, W. J., and Chang, C. H., Multifunctional antibodies by the Dock-and-Lock method for improved cancer imaging and therapy by pretargeting, *J. Nucleic Med.*, 49, 158, 2008.

206. Chatal, J. F., Faivre-Chauvet, A., Bardies, M., Peltier, P., Gautherot, E., and Barbet, J., Bifunctional antibodies for radioimmunotherapy, *Hybridoma*, 14, 125, 1995.

207. Schuhmacher, J., Klivenyi, G., Matys, R., Stadler, M., Regiert, T., Hauser, H., Doll, J., Maier-Borst, W., and Zoller, M., Multistep tumor targeting in nude mice using bispecific antibodies and a gallium chelate suitable for immunoscintigraphy with positron emission tomography, *Cancer Res.*, 55, 115, 1995.

208. Dillehay, L. E., Mayer, R., Zhang, Y. G., Shao, Y., Song, S. Y., Mackensen, D. G., and Williams, J. R., Prediction of tumor response to experimental radioimmunotherapy with ^{90}Y in nude mice, *Int. J. Radiat. Oncol. Biol. Phys.*, 33, 417, 1995.

209. Cornelissen, B., Kersemans, V., McLarty, K., Tran, L., and Reilly, R. M., ^{111}In-labeled immunoconjugates (ICs) bispecific for the epidermal growth factor receptor (EGFR) and cyclin-dependent kinase inhibitor, p27Kip1, *Cancer Biother. Radiopharm.*, 24, 163, 2009.

210. Reilly, R. M., Radioimmunotherapy of solid tumors: The promise of pretargeting strategies using bispecific antibodies and radiolabeled haptens, *J. Nucleic Med.*, 47, 196, 2006.

211. Kraeber-Bodéré, F., Rousseau, C., Bodet-Milin, C., Ferrer, L., Faivre-Chauvet, A., Campion, L., Vuillez, J. P. et al., Targeting, toxicity, and efficacy of 2-step, pretargeted radioimmunotherapy using a chimeric bispecific antibody and 131I-labeled bivalent hapten in a phase I optimization clinical trial, *J. Nucleic Med.*, 47, 247, 2006.

212. Chames, P. and Baty, D., Bispecific antibodies for cancer therapy, *Curr. Opin. Drug Discov. Devel.*, 12, 276, 2009.

213. Nieri, P., Donadio, E., Rossi, S., Adinolfi, B., and Podestà, A., Antibodies for therapeutic uses and the evolution of biotechniques, *Curr. Med. Chem.*, 16, 753, 2009.

214. Parren, P. W. and Burton, D. R., Immunology. Two-in-one designer antibodies, *Science*, 323, 1567, 2009.

215. Knuth, A., Bernhard, H., Jager, E., Wolfel, T., Karbach, J., Jaggle, C., Strittmatter, W., and Meyer zum Buschenfelde, K. H., Induction of tumour cell lysis by a bispecific antibody recognising epidermal growth factor receptor (EGFR) and CD3, *Eur. J. Cancer*, 30A, 1103, 1994.

216. Ohta, S., Tsukamoto, H., Watanabe, K., Makino, K., Kuge, S., Hanai, N., Habu, S., and Nishimura, T., Tumor associated glycoantigen, sialyl Lewis(a) as a target for bispecific antibody-directed adoptive tumor immunotherapy, *Immunol. Lett.*, 44, 35, 1995.

217. Canevari, S., Stoter, G., Arienti, F., Bolis, G., Colnaghi, M. I., Di Re, E. M., Eggermont, A. M. et al., Regression of advanced ovarian carcinoma by intraperitoneal treatment with autologous T lymphocytes retargeted by a bispecific monoclonal antibody, *J. Natl. Cancer Inst.*, 87, 1463, 1995.

218. Van Dijk, J., Zegveld, S. T., Fleuren, G. J., and Warnaar, S. O., Localization of monoclonal antibody G250 and bispecific monoclonal antibody CD3/G250 in human renal cell carcinoma xenografts: Relative effects of size and affinity, *Int. J. Cancer*, 48, 738, 1991.

219. Azuma, A., Yagita, H., Matsuda, H., Okumura, K., and Niitani, H., Induction of intercellular adhesion molecule 1 on small cell lung carcinoma cell lines by gamma-interferon enhances spontaneous and bispecific anti-CD3 × antitumor antibody-directed lymphokine activated killer cell cytotoxicity, *Cancer Res.*, 52, 4890, 1992.

220. De Gast, G. C., Van Houten, A. A., Haagen, I. A., Klein, S., De Weger, R. A., Van Dijk, A., Phillips, J., Clark, M., and Bast, B. J., Clinical experience with CD3 × CD19 bispecific antibodies in patients with B cell malignancies, *J. Hematother.*, 4, 433, 1995.

221. Kaneko, T., Fusauchi, Y., Kakui, Y., Masuda, M., Akahoshi, M., Teramura, M., Motoji, T., Okumura, K., Mizoguchi, H., and Oshimi, K., A bispecific antibody enhances cytokine-induced killer-mediated cytolysis of autologous acute myeloid leukemia cells, *Blood*, 81, 1333, 1993.

222. Davico Bonino, L., De Monte, L. B., Spagnoli, G. C., Vola, R., Mariani, M., Barone, D., Moro, A. M. et al., Bispecific monoclonal antibody anti-CD3 × anti-tenascin: An immunotherapeutic agent for human glioma [published erratum appears in *Int. J. Cancer*, 62, 364, 1995], *Int. J. Cancer*, 61, 509, 1995.

223. Baeuerle, P. A. and Reinhardt, C., Bispecific T-cell engaging antibodies for cancer therapy, *Cancer Res.*, 69, 4941, 2009.

224. Baeuerle, P. A., Kufer, P., and Bargou, R., BiTE: Teaching antibodies to engage T-cells for cancer therapy, *Curr. Opin. Mol. Ther.*, 11, 22, 2009.

225. Nagorsen, D., Bargou, R., Ruttinger, D., Kufer, P., Baeuerle, P. A., and Zugmaier, G., Immunotherapy of lymphoma and leukemia with T-cell engaging BiTE antibody blinatumomab, *Leuk. Lymphoma*, 50, 886, 2009.

226. Cao, Y. and Lam, L., Bispecific antibody conjugates in therapeutics, *Adv. Drug Deliv. Rev.*, 55, 171, 2003.

227. Frankel, A. E. and Woo, J. H., Bispecific immunotoxins, *Leuk. Res.*, 33, 1173, 2009.

228. Affleck, K. and Embleton, M. J., Monoclonal antibody targeting of methotrexate (MTX) against MTX-resistant tumour cell lines, *Br. J. Cancer*, 65, 838, 1992.

229. French, R. R., Hamblin, T. J., Bell, A. J., Tutt, A. L., and Glennie, M. J., Treatment of B-cell lymphomas with combination of bispecific antibodies and saporin, *Lancet*, 346, 223, 1995.

230. Reddy, V. S. and Ford, C. H., Production of hybrids secreting bispecific antibodies recognising CEA and doxorubicin, *Anticancer Res.*, 13, 2077, 1993.

231. Corvalan, J. R., Smith, W., Gore, V. A., and Brandon, D. R., Specific in vitro and in vivo drug localisation to tumour cells using a hybrid-hybrid monoclonal antibody recognising both carcinoembryonic antigen (CEA) and vinca alkaloids, *Cancer Immunol. Immunother.*, 24, 133, 1987.

232. Cao, Y. and Suresh, M. R., Bispecific MAb aided liposomal drug delivery, *J. Drug Target*, 8, 257, 2000.

233. De Sutter, K. and Fiers, W., A bifunctional murine: Human chimeric antibody with one antigen-binding arm replaced by bacterial beta-lactamase, *Mol. Immunol.*, 31, 261, 1994.

234. Gavrilyuk, J. I., Wuellner, U., Salahuddin, S., Goswami, R. K., Sinha, S. C., and Barbas, C. F., 3rd., An efficient chemical approach to bispecific antibodies and antibodies of high valency, *Bioorg. Med. Chem. Lett.*, 19, 3716, 2009.

235. Gavrilyuk, J. I., Wuellner, U., and Barbas, C. F., 3rd., Beta-lactam-based approach for the chemical programming of aldolase antibody 38C2, *Bioorg. Med. Chem. Lett.*, 19, 1421, 2009.

236. Nitecki, D. E., Woods, V., and Goodman, J. W., Crosslinking of antibody molecules by bifunctional antigens, in *Protein Cross-Linking. Biochemical and Molecular Aspects*, Friedman, M. (ed.), Plenum Press, New York, 1976, p. 139.

237. Rifai, A., Experimental models for IgA-associated nephritis, *Kidney Int.*, 31, 1, 1987.

238. Plotz, P. H. and Rifai, A., Stable, soluble, model immune complexes made with a versatile multivalent affinity-labeling antigen, *Biochemistry*, 21, 301, 1982.

239. Rifai, A. and Wong, S. S., Preparation of phosphorylcholine-conjugate antigens, *J. Immunol. Methods*, 94, 25, 1986.

240. Chen, A., Wong, S. S., and Rifai, A., Glomerular immune deposits in experimental IgA nephropathy: A continuum of circulating and in situ formed immune complexes, *Am. J. Pathol.*, 130, 216, 1988.

241. Döring, T., Mitchell, P., Osswald, M., Bochkariov, D., and Brimacombe, R., The decoding region of 16S RNA; a cross-linking study of the ribosomal A, P and E sites using tRNA derivatized at position 32 in the anticodon loop, *EMBO J.*, 13, 2677, 1994.

242. Stone, M. P., Cho, Y.-J., Huang, H., Kim, H.-Y., Kozekov, I. D., Kozekova, A., Wang, H. et al., Interstrand DNA cross-links induced by α,β-unsaturated aldehydes derived from lipid peroxidation and environmental sources, *Acc. Chem. Res.*, 41, 793, 2008.

243. Anderson, F., O'Hare, C., Hartley, J., and Robins, D., Synthesis of new homochiral bispyrrolidines as potential DNA cross-linking antitumour agents, *Anti-Cancer Drug Des.*, 15, 119, 2000.

244. Paz, M. M., Das, A., and Tomasz, M., Mitomycin C linked to DNA minor groove binding agents: Synthesis, reductive activation, DNA binding and cross-linking properties and in vitro antitumor activity, *Bioorg. Med. Chem.*, 7, 2713, 1999.

245. Siegel, D., Beall, H., Senekowitsch, C., Kasai, M., Arai, H., Gibson, N. W., and Ross, D., Bioreductive activation of mitomycin C by DT-diaphorase, *Biochemistry*, 31, 7879, 1992.

246. Noll, D. M., Mason, T. M., and Miller, P. S., Formation and repair of interstrand cross-links in DNA, *Chem Rev.*, 106, 277, 2006.

247. Alcaro, S. and Coleman, R. S., A molecular model for DNA cross-linking by the antitumor agent azinomycin B, *J. Med. Chem.*, 43, 2783, 2000.

248. Fujiwara, T., Saito, I., and Sugiyama, H., Highly efficient DNA interstrand crosslinking induced by an antitumor antibiotic, carzinophilin, *Tetrahedron Lett.*, 40, 315, 1999.

249. Hodgkinson, T. J. and Shipman, M., Chemical systhesis and mode of action of the azinomycins, *Tetrahedron*, 57, 4467, 2001.

250. Hartley, J. A., Hazrati, A., Kelland, L. R., Khanim, R., Shipman, M., Suzenet, F., and Walker, L. F., A synthetic azinomycin analogue with demonstrated DNA cross-linking activity: Insights into the mechanism of action of this class of antitumor agent, *Angew. Chem. Int. Ed.*, 2009, 3467, 2000.

251. Casely-Hayford, M. A., Pors, K., James, C. H., Patterson, L. H., Hartley, J. A., and Searcey, M., Design and synthesis of a DNA-crosslinking azinomycin analogue, *Org. Biomol. Chem.*, 3, 3585, 2005.

252. David-Cordonnier, M.-H., Casely-Hayford, M., Kouach, M., Briand, G., Patterson, L. H., Bailly, C., and Searcey, M., Stereoselectivity, sequence specificity and mechanism of action of the azinomycin epoxide, *ChemBioChem*, 7, 1658, 2006.

253. van Rosmalen, A., Cullinane, C., Cutts, S. M., and Phillips, D. R., Stability of adriamycin-induced DNA adducts and interstrand crosslinks, *Nucleic Acids Res.*, 23, 42, 1995.

254. Jesson, M. I., Johnston, J. B., Robotham, E., and Begleiter, A., Characterization of the DNA–DNA cross-linking activity of 3′-(3-Cyano-4-morpholinyl)-3′-deaminoadriamycin, *Cancer Res.*, 49, 7031, 1989.

255. Wilson, S. C., Howard, P. W., Forrow, S. M., Hartley, J. A., Adams, L. J., Jenkins, T. C., Kelland, L. R., and Thurston, D. E., Design, synthesis, and evaluation of a novel sequence-selective epoxide-containing DNA cross-linking agent based on the pyrrolo[2,1-*c*][1,4]benzodiazepine system, *J. Med. Chem.*, 42, 4028, 1999.

256. Zhou, Q., Duan, W., Simmons, D., Shayo, Y., Raymond, M. A., Dorr, R. T., and Hurley, L. H., Design and synthesis of a novel DNA-DNA interstrand adenine-guanine cross-linking agent, *J. Am. Chem. Soc.*, 123, 4865, 2001.

257. Tercel, M., Stribbling, S. M., Sheppard, H., Siim, B. G., Wu, K., Pullen, S. M., Botting, K. J., Wilson, W. R., and Denny, W. A., Unsymmetrical DNA cross-linking agents: Combination of the CBI and PBD pharmacophores, *J. Med. Chem.*, 46, 2132, 2003.

258. Purnell, B., Sato, A., O'Kelley, A., Price, C., Summerville, K., Hudson, S., O'Hare, C. et al., DNA inter-strand crosslinking agents: Synthesis, DNA interactions, and cytotoxicity of dimeric achiral seco-amino-CBI and conjugates of achiral seco-amino-CBI with pyrrolobenzodiazepine (PBD), *Bioorg. Med. Chem. Lett.*, 16, 5677, 2006.

259. Wellinger, R. E. and Sogo, J. M., In vivo mapping of nucleosomes using psoralen–DNA crosslinking and primer extension, *Nucleic Acid Res.*, 26, 1544, 1998.

260. Hwang, G.-S., Kim, J.-K., and Choi, B.-S., The solution structure of a psoralen cross-linked DNA duplex by NMR and relaxation matrix refinement, *Biochem. Biophys. Res. Commun.*, 219, 191, 1996.

261. Lustig, Y., Wachtel, C., Safro, M., Liu, L., and Michaeli, S., 'RNA walk' a novel approach to study RNA-RNA interactions between a small RNA and its target, *Nucleic Acids Res.*, 38, e5, 2010.

262. Ossipov, D., Gohil, S., and Chattopadhyaya, J., Synthesis of the DNA-[Ru(tpy)(dppz)(CH3CN)]²⁺ conjugates and their photo cross-linking studies with the complementary DNA strand, *J. Am. Chem. Soc.*, 124, 13416, 2002.

263. Lentzen, O., Defrancq, E., Constant, J.-F., Schumm, S., Garcia-Fresnadillo, D., Moucheron, C., Dumy, P., and Kirsch-De Mesmaeker, A., Determination of DNA guanine sites forming photo-adducts with Ru(II)-labeled oligonucleotides: DNA polymerase inhibition by the resulting photo-crosslinking, *J. Biol. Inorg. Chem.*, 9, 100, 2004.

264. Khodyreva, S. N. and Lavrik, O. I., Photoaffinity labeling technique for studying DNA replication and DNA repair, *Curr. Med. Chem.*, 12, 641, 2005.

265. Okuda, S. and Tokuda, H., Model of mouth-to-mouth transfer of bacterial lipoproteins through inner membrane LolC, periplasmic LolA, and outer membrane LolB, *Proc. Natl. Acad. Sci. U. S. A.*, 106, 5877, 2009.

266. Wittelsberger, A., Mierke, D. F., and Rosenblatt, M., Mapping ligand-receptor interfaces: Approaching the resolution limit of benzophenone-based photoaffinity scanning, *Chem. Biol. Drug Des.*, 71, 380, 2008.

267. Ye, S., Köhrer, C., Huber, T., Kazmi, M., Sachdev, P., Yan, E. C., Bhagat, A., RajBhandary, U. L., and Sakmar, T. P., Site-specific incorporation of keto amino acids into functional G protein-coupled receptors using unnatural amino acid mutagenesis, *J. Biol. Chem.*, 283, 1525, 2008.

268. Lee, H. S., Dimla, R. D., and Schultz, P. G., Protein–DNA photo-crosslinking with a genetically encoded benzophenone-containing amino acid, *Bioorg. Med. Chem. Lett.*, 19, 5222, 2009.

269. Persinger, J. and Bartholomew, B., Incorporation of photosensitive DNA into DNA for x-linking: Site-directed DNA crosslinking of large multisubunit protein-DNA complexes, *Methods Mol. Biol.*, 543, 453, 2009.

270. Meffert, R., Dose, K., Rathgeber, G., and Schêfer, H. J., Ultraviolet crosslinking of DNA protein complexes via 8-azidoadenine, *Methods Mol. Biol.*, 543, 389, 2009.

271. Kimoto, M., Endo, M., Mitsui, T., Okuni, T., Hirao, I., and Yokoyama, S., Site-specific incorporation of a photo-crosslinking component into RNA by T7 transcription mediated by unnatural base pairs, *Chem. Biol.*, 11, 47, 2004.

272. Tate, J. J., Persinger, J., and Bartholomew, B., Survey of four different photoreactive moieties for DNA photoaffinity labeling of yeast RNA polymerase III transcription complexes, *Nucleic Acids Res.*, 26, 1421, 1998.

273. Kolpashchikov, D. M., Ivanova, T. M., Boghachev, V. S., Nasheuer, H.-P., Weisshart, K., Favre, A., Pestryakov, P. E., and Lavrik, O. I., Synthesis of base-substituted dUTP analogues carrying a photoreactive group and their application to study human replication protein A, *Bioconjug. Chem.*, 11, 445, 2000.

274. Kolpashchikov, D. M., Khodyreva, S. N., Khlimankov, D. Y., Wold, M. S., Favre, A., and Lavrik, O. I., Polarity of human replication protein A binding to DNA, *Nucl. Acids Res.*, 29, 373, 2001.

275. Winnacker, M., Breeger, S., Strasser, R., and Carell, T., Novel diazirine-containing DNA photoaffinity probes for the investigation of DNA–protein-interactions, *ChemBioChem.*, 10, 109, 2009.

276. Taranenko, M., Rykhlevskaya, A., Mtchedlidze, M., Laval, J., and Kuznetsova, S., Photochemical cross-linking of *Escherichia coli* Fpg protein to DNA duplexes containing phenyl(trifluoromethyl)diazirine groups, *Eur. J. Biochem.*, 270, 2945, 2003.

277. Yamaguchi, T., Suyama, K., Narita, K., Kohgo, S., Tomikawa, A., and Saneyoshi, M., Synthesis and evaluation of oligodeoxyribonucleotides containing an aryl(trifluoromethyl)diazirine moiety as the cross-linking probe: Photoaffinity labeling of mammalian DNA polymerase beta, *Nucleic Acid Res.*, 25, 2352, 2997.

278. Bartholomew, B., Braun, B. R., Kassavetis, G. A., and Geiduschek, E. P., Probing close DNA contacts of RNA polymerase III transcription complexes with the photoactive nucleoside 4-thiodeoxythymidine, *J. Biol. Chem.*, 269, 18090, 1994.

279. Taranenko, M., Mtchedlidze, M., Sumbatyan, N., and Korshunova, G., A zero-length diazirine photoactive nucleoside, *Nucleos. Nucleot. Nucleic Acids*, 22, 715, 2003.

280. Lebedeva, N. A., Kolpashchikov, D. M., Rechkunova, N. I., Khodyreva, S. N., and Lavrik, O. I., A binary system of photoreagents for high-efficiency labeling of DNA polymerases, *Biochem. Biophys. Res. Commun.*, 287, 530, 2001.

281. Moore, B. M., II, Jalluri, R. K., and Doughty, M. B., DNA polymerase photoprobe 2-[(4-azidophenacyl) thio]-2′-deoxyadenosine 5′-triphosphate labels an *Escherichia coli* DNA polymerase I Klenow fragment substrate binding site, *Biochemistry*, 35, 11642, 1996.

282. Zofall, M. and Bartholomew, B., Two novel dATP analogs for DNA photoaffinity labeling, *Nucleic Acids Res.*, 28, 4382, 2000.

283. Dyrkheeva, N. S., Khodyreva, S. N., Sukhanova, M. V., Safronov, I. V., Dezhurov, S.V., and Lavrik, O. I., 3′–5′ exonuclease activity of human apurinic/apyrimidinic endonuclease 1 towards DNAs containing dNMP and their modified analogs at the 3 end of single strand DNA break, *Biochemistry (Moscow)*, 71, 200, 2006.

284. Maltseva, E. A., Rechkunova, N. I., Petruseva, I. O., Silnikov, V. N., Vermeulen, W., and Lavrik, O. I., Interaction of nucleotide excision repair factors RPA and XPA with DNA containing bulky photoreactive groups imitating damages, *Biochemistry (Moscow)*, 71, 270, 2006.

285. Koudan, E. V., Subach, O. M., Korshunova, G. A., Romanova, E. A., Eritja, R., and Gromova, E. S., DNA duplexes containing photoactive derivatives of 2′-deoxyuridine as photocrosslinking probes for EcoRII DNA methyltransferase-substrate interaction, *J. Biomol. Struct. Dyn.*, 20, 421, 2002.

286. Petruseva, I. O., Tikhanovich, I. S., Maltseva, E. A., Safronov, I. V., and Lavrik, O. I., Photoactivated DNA analogs of substrates of the nucleotide excision repair system and their interaction with proteins of NER-competent HeLa cell extract, *Biochemistry (Moscow)*, 74, 491, 2009.

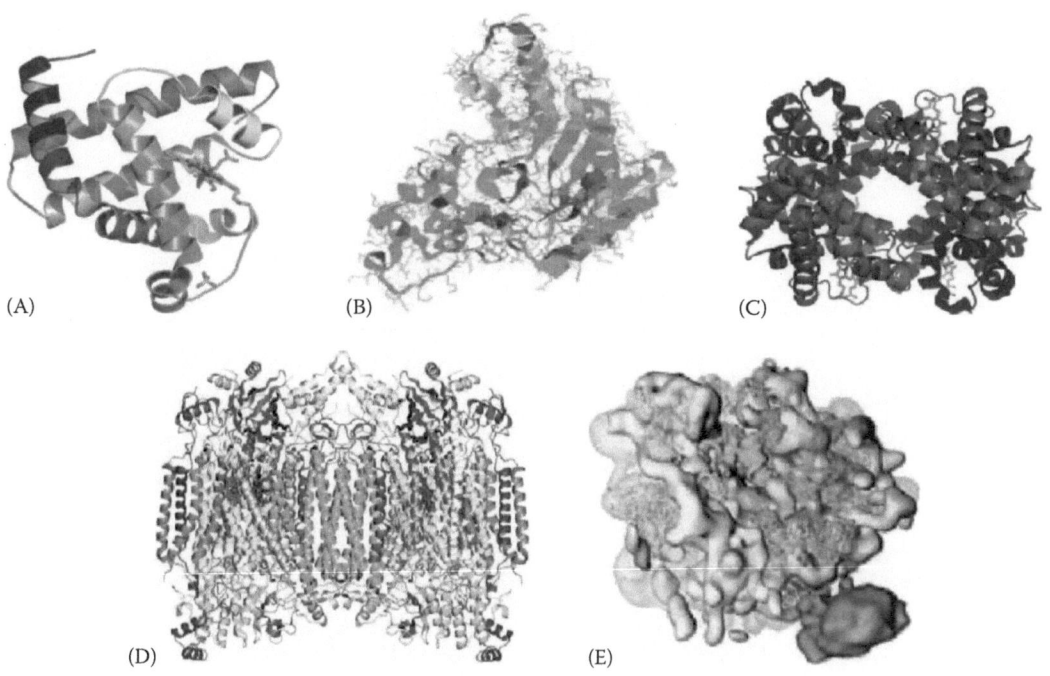

(A)

(B)

(C)

(D)

(E)

FIGURE 1.1 Examples of different molecular structures of proteins. (A) Myoglobin molecule. (After Phillips, S. E. V. *J. Mol. Biol.*, 142, 531, 1980.) (B) Dimeric creatine kinase. (After Shen, Y. Q. et al., *Acta Crystallogr. D Biol. Crystallogr.*, 57, 1196, 2001.) (C) Tetrameric hemoglobin. (After Paoli, M. et al., *J. Mol. Biol.*, 256, 775, 1996.) (D) Bovine cytochrome C oxidase with 2 copies of 13 different components. (From Shinzawa-Itoh, K. et al., *EMBO J.*, 26, 1713, 2007. With permission.) (E) Yeast 80S ribosome of multicomponent proteins and RNA. (Reprinted from *Cell*, 107, Beckmann, R. et al., Architecture of the protein-conducting channel associated with the translating 80S ribosome, 361, Copyright 2001, with permission from Elsevier.)

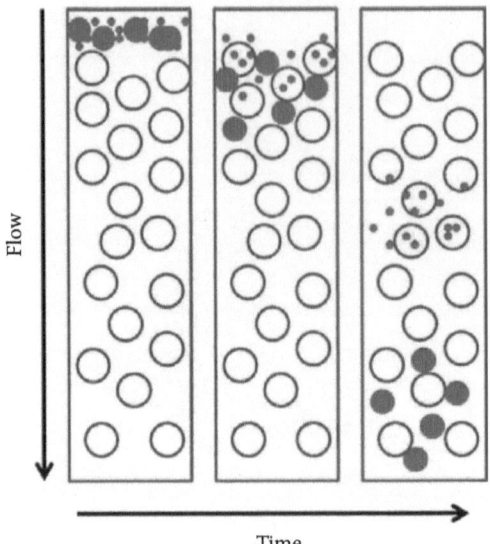

FIGURE 10.1 Depiction of a SEC experiment. As molecules flow down the column, the larger ones move fastest since they cannot enter the resin matrix.

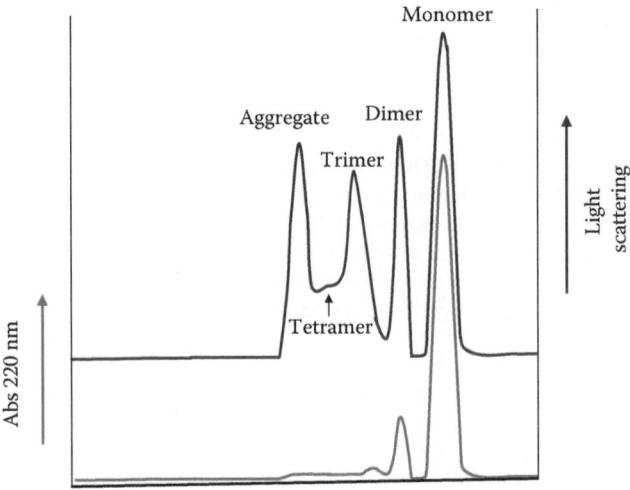

FIGURE 10.4 Depiction of a SEC experiment on a commercial preparation of BSA that has developed cross-linked products. The output from the column was monitored by absorbance at 220 nm as well as by light scattering.

7 Multifunctional Cross-Linking Reagents

7.1 INTRODUCTION

Multifunctional cross-linkers contain more than two functional groups that will react with proteins and/or nucleic acids. The most common ones are trifunctional reagents, as an extension of bifunctional cross-linkers. The multifunctional groups could be all identical as in homobifunctional reagents (Chapter 5), all different as in heterobifunctional (Chapter 6), or mixed homo- and heterotypes. The functional groups will react with amino acid side chains according to their selectivity or specificity as discussed in earlier chapters. Because most of the functional groups are not specific toward a special target, almost all of the multifunctional reagents will react with different side chains of amino acids or nucleobases and can be considered heterofunctional even in the case of homomultifunctional reagents. Thus, multifunctional agents will anchor at multiple sites of complex proteins, cross-linking multiple components or locales. For trifunctional reagents, three target sites will be linked together. The most important usage of these multifunctional cross-linkers is for the investigations of protein structures and protein–protein interactions, particularly for multisubunit complex protein aggregates. Many of the reagents are used to prepare immunotoxins and antibody conjugates. There are also important applications in tissue engineering, cross-linking polymeric protein complexes, cells, and tissues for medical purposes. There are relatively few published multifunctional reagents and a few trifunctional reagents are used as bifunctional cross-linkers. The latter are mentioned in earlier chapters, but will be discussed below as trifunctional compounds. All other new multifunctional cross-linkers that appear in the literature are listed in Table 7.1 and will be further elaborated upon below.

7.2 TRIFUNCTIONAL CROSS-LINKERS

From the discussion of bifunctional cross-linkers (Chapters 5 and 6), it seems logical to speculate that compounds with more than two functional groups could be useful in some cases. There are few compounds that contain three functional groups such as *N*-succinimidyl-2,3-dibromopropionate and chlorambucil-*N*-hydroxysuccinimide ester (see compounds LXIV and LXV in Appendix D). These reagents contain one succinimide ester and two alkyl halides and therefore have the potential of cross-linking three sites in a protein or proteins. However, to our knowledge, they have only been used as heterobifunctional cross-linkers.[1] It is uncertain if they could behave as trifunctional agents. Of particular interest is methyl-4-(6-formyl-3-azidophenoxy)-butyrimidate (FAPOB) (compound CLI in Appendix D), which has three reactive groups.[2] The imidate ester and the formyl moieties can react with amino groups, and as such may function as heterobifunctional reagents in the dark. The phenylazide moiety is photosensitive and constitutes the third reactive group. Maassen et al.[3,4] have used this compound both as a heterobifunctional cross-linker and as a trifunctional agent to study ribosomal protein–protein interactions. As a bifunctional reagent, the compound cross-linked the lysine residues of dimeric ribosomal protein L7/L12 using the imidate and aldehyde groups. The aldehyde group may also be used as a chromophore due to its absorbance at 325 nm. On reduction or Schiff base formation, the absorption band is diminished.[3] As a trifunctional agent, elongation factor G was cross-linked to 50S ribosomal proteins after photolysis.[3]

TABLE 7.1
Multifunctional Cross-Linking Reagents

I. Trifunctional Cross-Linkers

  A. Tri-maleimides

    1. Tris(2-maleimidoethyl)amine (TMEA)

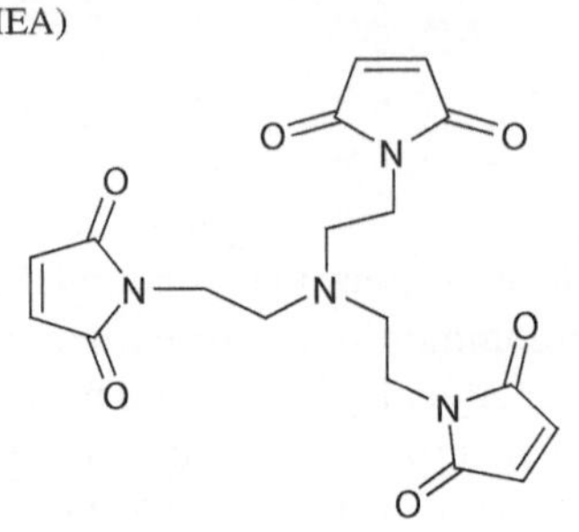

    2. *N*1,*N*3,*N*5-Tris[2-(2,5-dioxopyrrol-1-yl)ethyl]benzene-1,3,5-tricarboxamide (TMBA)

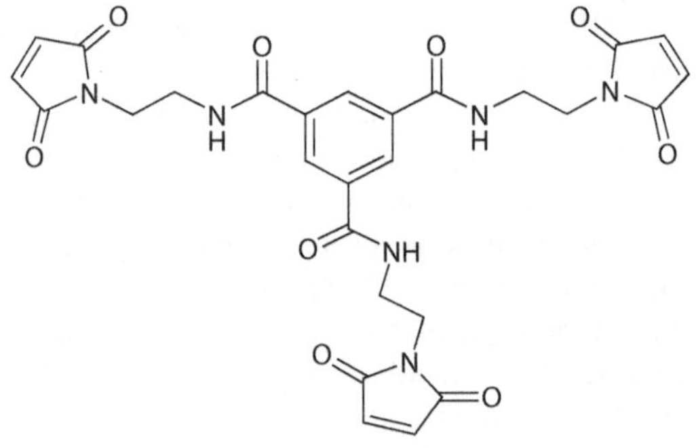

  B. Tri-aziridines

    1. 2,4,6-Tri(ethyleneimino)-s-triazine (TEM)

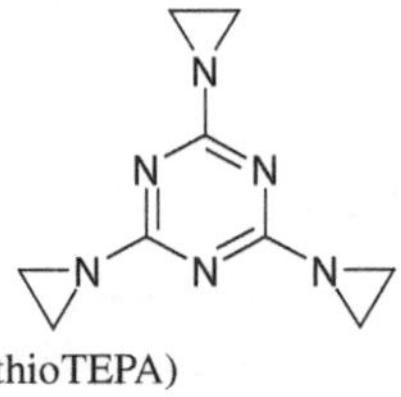

    2. *N*,*N*′,*N*″-Triethylenethiophosphoramide (thioTEPA)

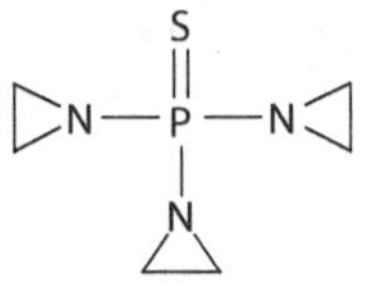

    3. *N*,*N*′,*N*″-Triethylenephosphoramide (TEPA)

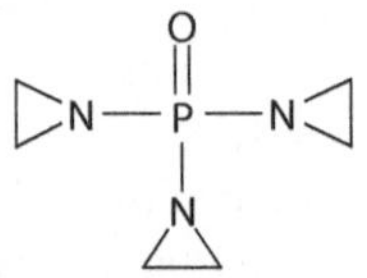

    4. Tri{1-(2-methylaziridenyl)}phosphine oxide

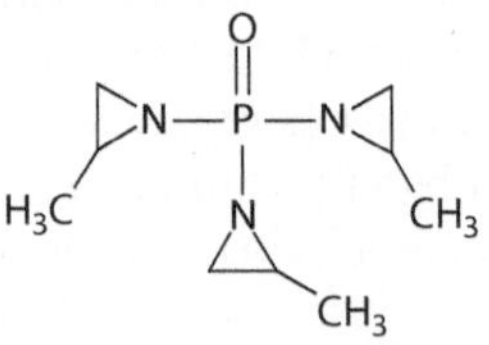

TABLE 7.1 (Continued)
Multifunctional Cross-Linking Reagents

C. Glycerol triglycidyl ether: 1,2,3-(tris(2,3-epoxypropoxy))propanol

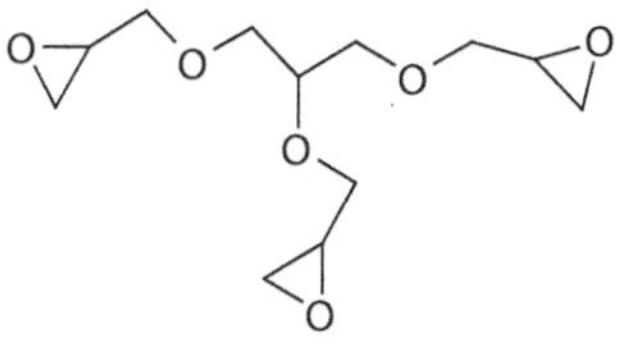

D. Tris-succinimidyl aminotriacetate (TSAT)

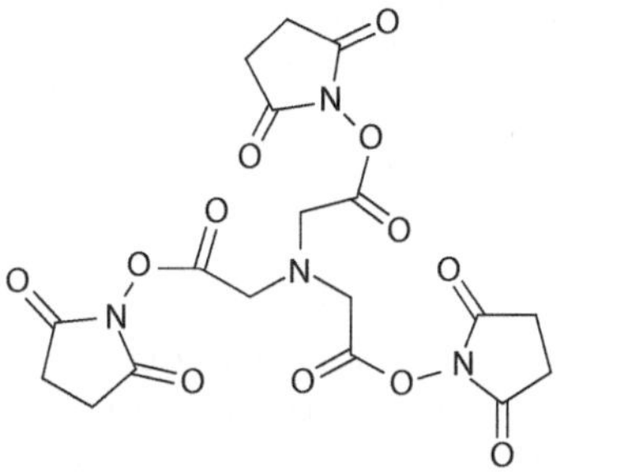

E. β-[Tris(hydroxymethyl)phosphinol]propionic acid (THPP)

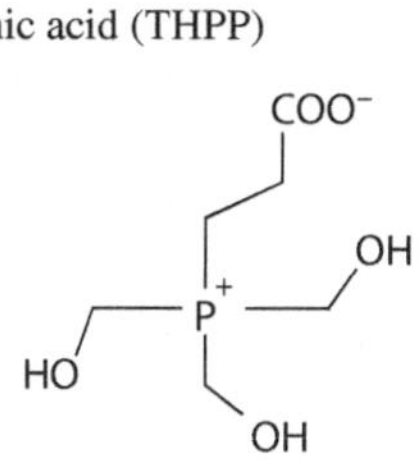

F. Trimesoyl tris(methyl phosphate) (TMMP)

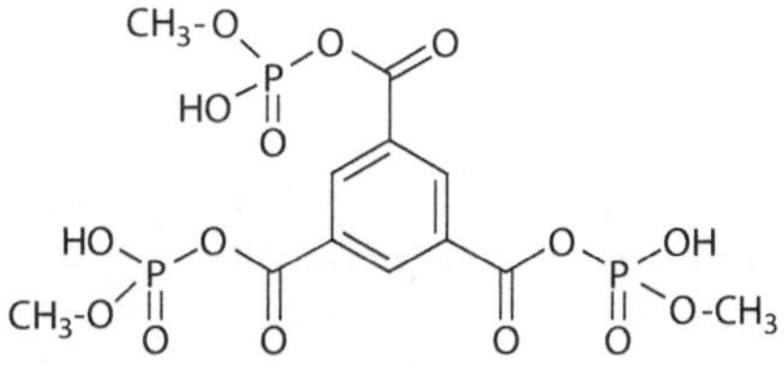

G. Trimesoyl tris(3,5-dibromosalicylate) (TTDS)

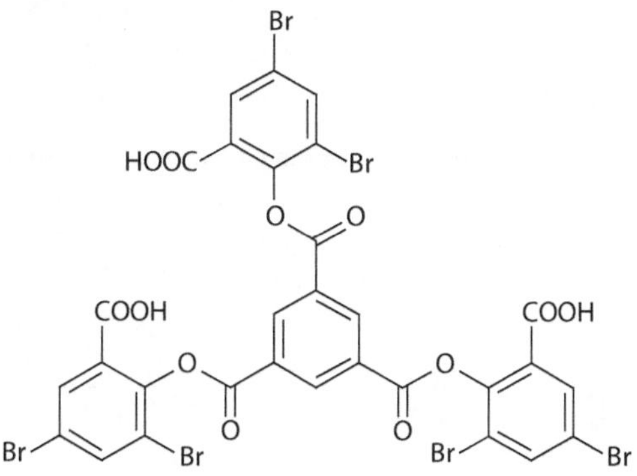

(continued)

TABLE 7.1 (Continued)
Multifunctional Cross-Linking Reagents

H. Tri(2-chloroethyl)amine (TCEA)

I. Triazamacrocyclic nitrogen mustards

1. x = y = z = 0: 1,4,7-Tris(2-chloroethyl)-1,4,7-triazacyclononane trihydrochloride
2. x = y = z = 1: 1,5,9-Tris(2-chloroethyl)-1,5,9-triazacyclododecane trihydrochloride
3. x = y = 1, z = 2: 1,5,9-Tris(2-chloroethyl)-1,5,9-triazacyclotridecane trihydrochloride
4. x = y = 1, z = 3: 1,5,9-Tris(2-chloroethyl)-1,5,9-triazacyclotetradecane trihydrochloride
5. x = 1, y = z = 2: 1,5,10-Tris(2-chloroethyl)-1,5,10-triazacyclotetradecane trihydrochloride
6. x = 1, y = 2, z = 3: 1,5,10-Tris(2-chloroethyl)-1,5,10-triazacyclopentadecane trihydrochloride
7. x = 2, y = z = 3: 1,6,12-Tris(2-chloroethyl)-1,6,12-triazacycloheptadecane trihydrochloride
8. x = 2, y = z = 4: 1,6,13-Tris(2-chloroethyl)-1,6,13-triazacyclononadecane trihydrochloride

J. N-succinimidyl 4-[{N,N-bis(2-chloroethyl)}amino]phenylbutyrate, chlorambucil-N-hydroxysuccinimide ester (CB-NHS)

K. N-succinimidyl-[1-methyl-3-[1-methyl-3-[1-methyl-3-[4-bis(2-chloroethyl) aminophenylamido]-pyrazole-5-carboxamido]pyrazole-5-carboxamido] pyrazole-5-carboxylate (BAMP-NHS)

L. 4-(2-Chloroethyl)-3-[[5-(4-bis-(2-chloroethyl)-amino)benzamido]indole-2-carboxamido]phenol (*seco*-CI-BAM)

TABLE 7.1 (Continued)
Multifunctional Cross-Linking Reagents

M. Methoxyaziridinyl benzoic acid mustard

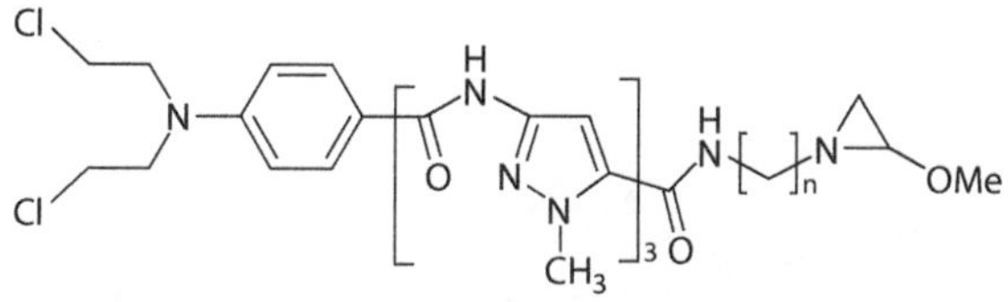

1. n = 2: *N*-(2-(Methoxyaziridinyl)ethyl)-1-methyl-4-[1-methyl-4-(1-methyl-4-{*p-N,N*-bis(2-chloroethyl)-benzamido}imidazole-2-carboxamido)imidazole-2-carboxamido]imidazole-2-carboxamide
2. n = 3: *N*-(3-(Methoxyaziridinyl)propyl)-1-methyl-4-[1-methyl-4-(1-methyl-4-[*p-N,N*-bis-(2-chloroethyl)aminobenzamido]imidazole-2-carboxamido)-imidazole-2-carboxamido]imidazole-2-carboxamide
3. n = 4: *N*-(4-(Methoxyaziridinyl)butyl)-1-methyl-4-[1-methyl-4-(1-methyl-4-[*p-N,N*-bis-(2-chloroethyl)aminobenzamido]imidazole-2-carboxamido)imidazole-2-carboxamido]imidazole-2-carboxamide

N. 4,4′-Bis(*N*-maleimido)benzophenone (BMBP)

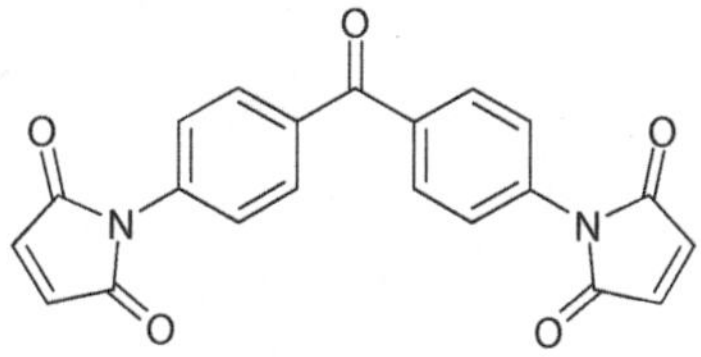

O. Patulin (PAT), 4-hydroxy-4*H*-furo[3,2c]pyran-2(6*H*)-one [PAT1999.pdf]

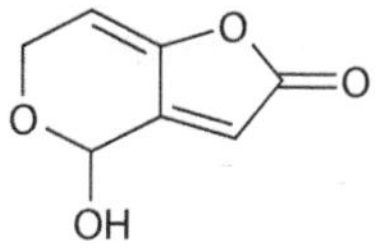

P. Biotin containing cross-linkers
1. Sulfosuccinimidyl-2-[6-(biotinamido)-2-(p-azidobenzamido)-hexanoamido]ethyl-1,3′-dithiopropionate (sulfo-SBED)

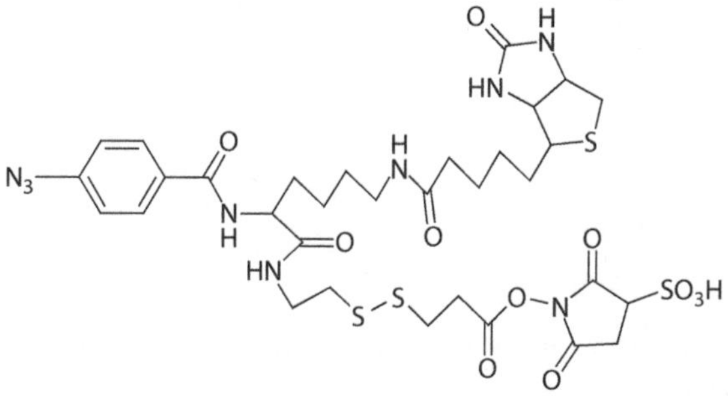

2. 4-Azido-*N*-[5-[5-(2-oxo-1,3,3a,4,6,6a-hexahydrothieno[3,4-d]imidazol-6-yl)pentanoylamino]-1-[2-(2-pyridyldisulfanyl)ethylcarbamoyl]pentyl] benzamide (APB)

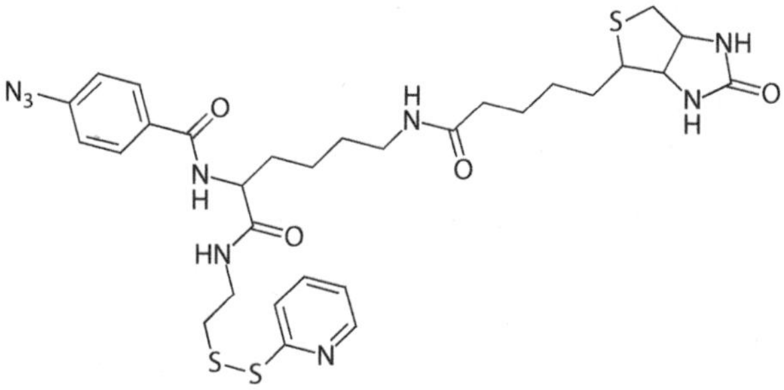

(*continued*)

TABLE 7.1 (Continued)
Multifunctional Cross-Linking Reagents

3. 2-[N2-(4-Azido-2,3,5,6-tetrafluorobenzoyl)-N6-(6-biotinamidocaproyl)-L-lysinyl]ethylmethanethiosulfonate
(Mts-Aff-Biotin)

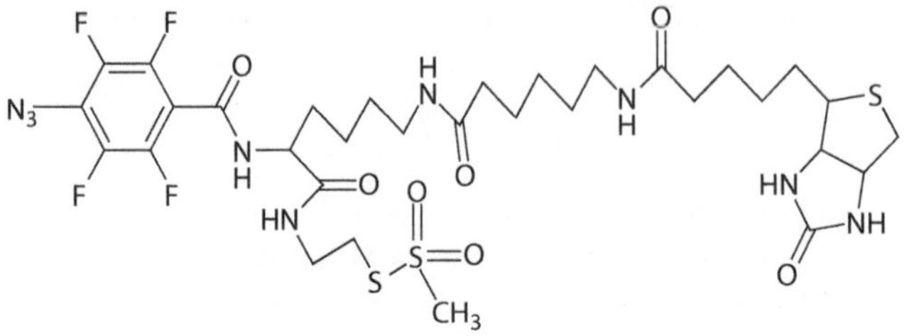

4. 2-{N2-[N6-(4-Azido-2,3,5,6-tetrafluorobenzoyl-6-aminocaproyl)-N6-(6-biotinamidocaproyl)-L-lysinylamido]}
ethylmethanethiosulfonate (Mts-Aff-LC-Biotin)

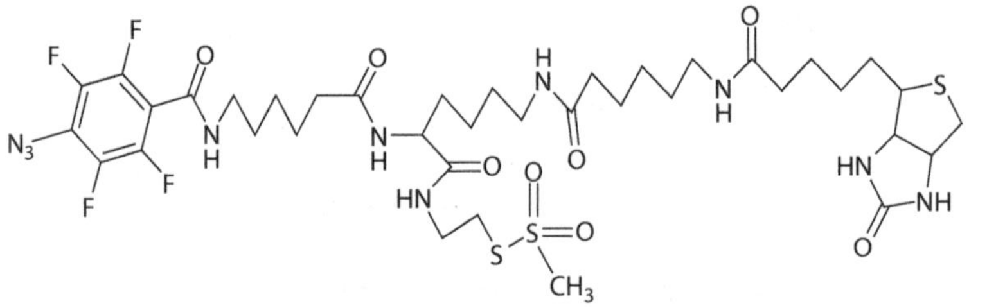

5. 4-Azido-2-nitrophenylbiocytin-4-nitrophenyl ester (ABNP)

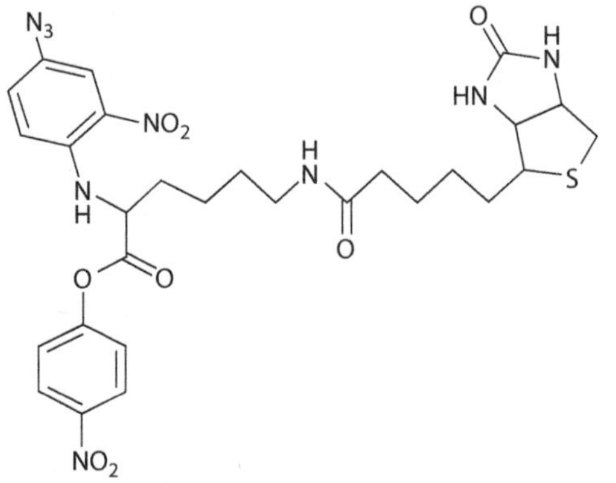

6. 2-[N^α-Benzoylbenzoicamido-N^6-(6-biotinamidocaproyl)-L-lysinylamido]ethyl methanethiosulfonate (MTS-BP-Bio)

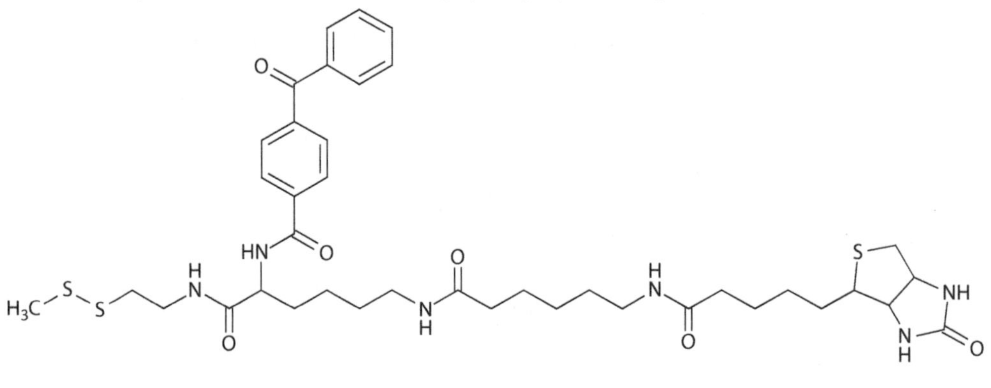

TABLE 7.1 (Continued)
Multifunctional Cross-Linking Reagents

7. (3-Methyl-2,5-dioxo-pyrrolidin-1-yl)4-[[1-[[(4-benzoylbenzoyl)amino] methyl]-2-oxo-2-[[2-oxo-2-[2-[2-[2-[5-(2-oxo-1,3,3a,4,6,6a-hexahydrothieno[3,4-d]imidazol-6-yl)pentanoylamino]ethoxy]ethoxy] ethylamino]ethyl]amino]ethyl]amino]-4-oxo-butanoate (BP-NHS- PEGBIO)

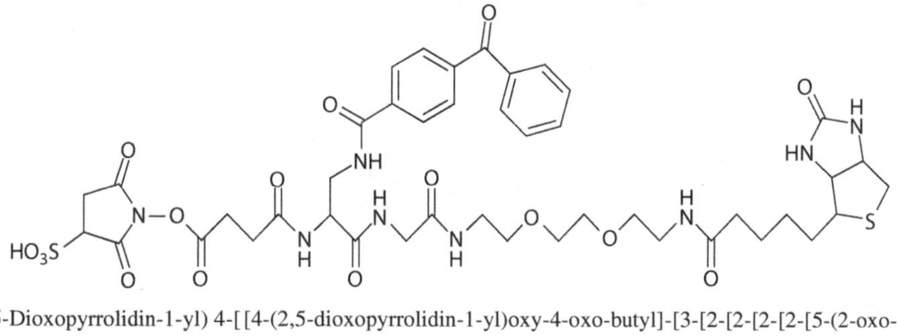

8. (2,5-Dioxopyrrolidin-1-yl) 4-[[4-(2,5-dioxopyrrolidin-1-yl)oxy-4-oxo-butyl]-[3-[2-[2-[2-[2-[5-(2-oxo-1,3,3a,4,6,6a-hexahydrothieno[3,4-d]imidazol-4-yl)pentanoylamino]ethoxy]ethoxy]ethoxy]ethoxy] propanoyl] amino]butanoate (PEG-Bio-diNHS)

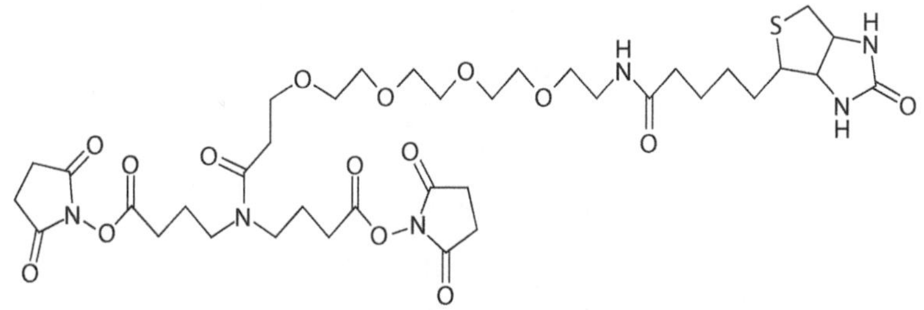

9. 1-[2-[[2-[[4-(2,5-Dioxo-3-sulfonato-pyrrolidin-1-yl)oxy-4-oxo-butanoyl]amino]-3-[[4-oxo-4-[2-[2-[2-[5-(2-oxo-1,3,3a,4,6,6a-hexahydrothieno[3,4-d]imidazol-6-yl)pentanoylamino]ethoxy]ethoxy] ethylamino]butanoyl]amino] propanoyl]amino]acetyl]oxy-2,5-dioxo-pyrrolidine-3-sulfonate (SulfoNHS-PEGBio)

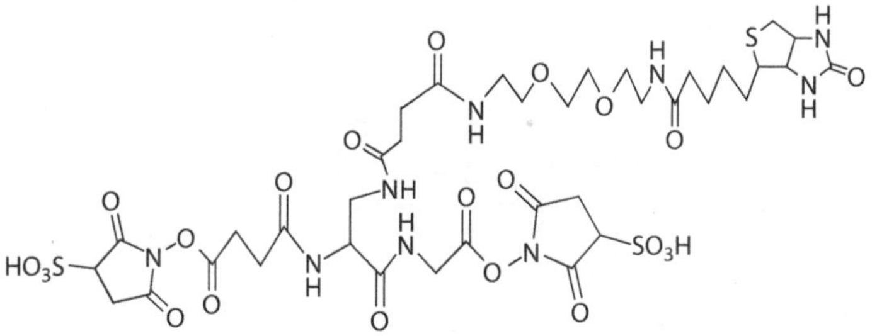

10. 1-[2-[[2-[2-[[2-[[4-(2,5-Dioxo-3-sulfonato-pyrrolidin-1-yl)oxy-4-oxo-butanoyl]amino]acetyl]amino]-3-[4-oxo-4-[2-[2-[2-[5-(2-oxo-1,3,3a,4,6,6a-hexahydrothieno[3,4-d]imidazol-6-yl)pentanoyl amino]ethoxy]ethoxy]ethylamino]butanoyl]oxy-propanoyl]amino] acetyl]oxy-2,5-dioxo-pyrrolidine-3-sulfonate (SulfoNHS-ES-PEGBio)

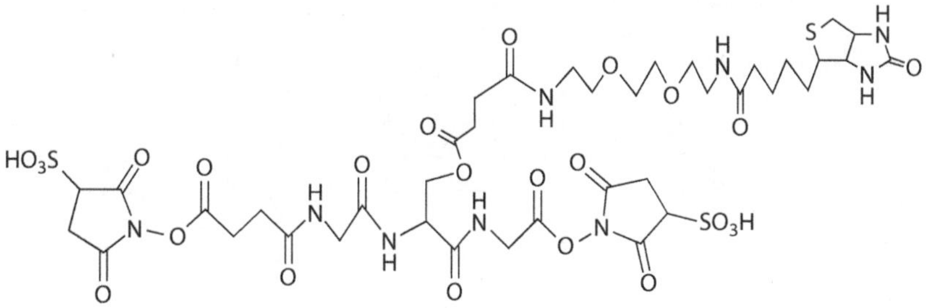

(continued)

TABLE 7.1 (Continued)
Multifunctional Cross-Linking Reagents

11. 1-[3-[2-[2-[3-[[4-[[2-[[2-(2,5-Dioxo-3-sulfonato-pyrrolidin-1-yl)oxy-2-oxo-ethyl]amino]-2-oxo-1-[[[4-oxo-4-[2-[2-[2-[5-(2-oxo-1,3,3a,4,6,6a-hexahydrothieno[3,4-d]imidazol-6-yl)pentanoylamino]ethoxy] ethoxy] ethylamino]butanoyl]amino]methyl]ethyl]amino]-4-oxo-butanoyl]amino]propoxy]ethoxy]ethoxy] propylcarbamoyloxy]-2,5-dioxo-pyrrolidine-3-sulfonate (PEGNHS-Bio)

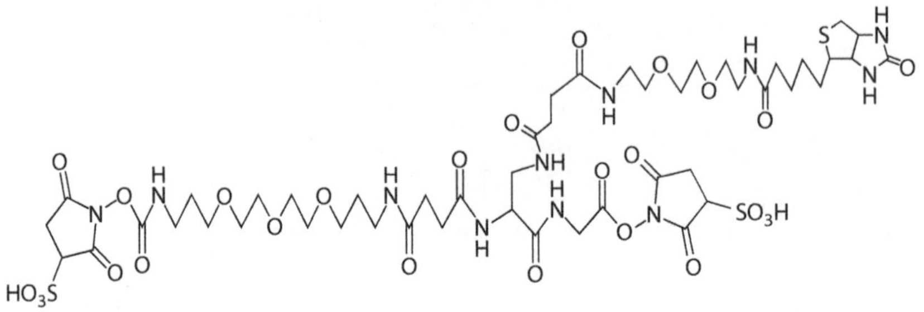

12. 1-[2-[[2-[3-(2,5-Dioxopyrrol-1-yl)propanoylamino]-3-[[4-oxo-4-[2-[2-[2- [5-(2-oxo-1,3,3a,4,6,6a-hexahydrothieno[3,4-d]imidazol-6-yl) pentanoylamino]ethoxy]ethoxy]ethylamino]butanoyl]amino] propanoyl] amino]acetyl]oxy-2,5-dioxo-pyrrolidine-3-sulfonate (Sulfo-NHS-M-PEGBio)

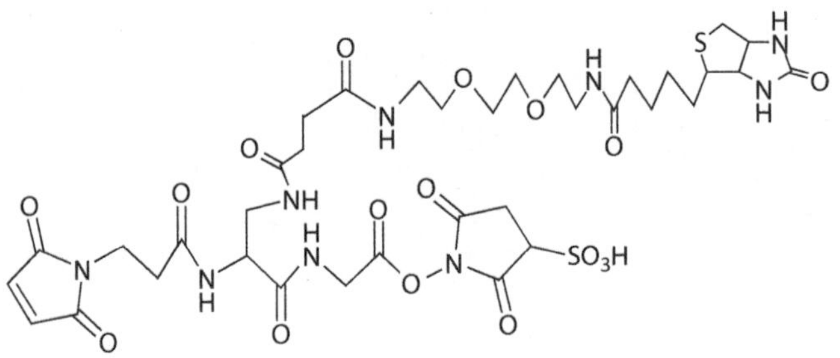

II. Tetrafunctional Cross-Linkers

A. N-[4-[[3,5-Bis(2,5-dioxopyrrol-1-yl)benzoyl]amino]phenyl]-3,5-bis(2,5-dioxopyrrol-1-yl)benzamide (Tetra-MBA)

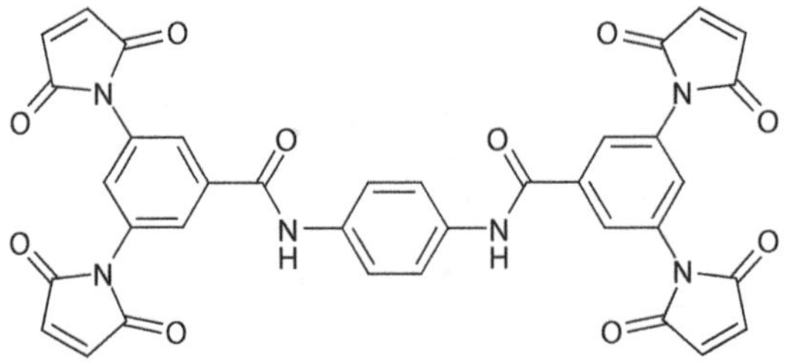

B. N,N'-5,5'-Bis[bis(3,5-dibromosalicyl)isophthalyl]terephthalamide (DBIT)

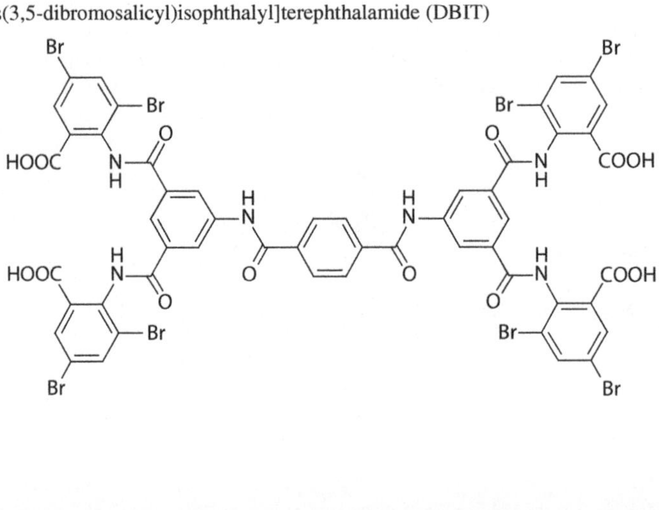

TABLE 7.1 (Continued)
Multifunctional Cross-Linking Reagents

C. 3,5,3′,5′-Biphenyltetracarbonyl tetrakis(3,5-dibromosalicylate) (BTDS)

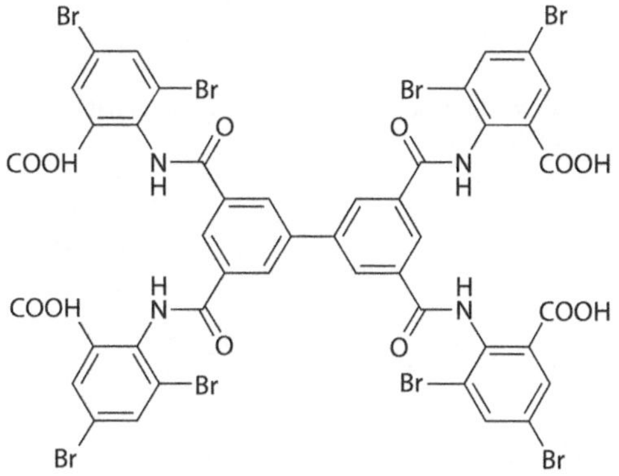

D. 1,2-Bis{2-[3,5-bis(3,5-dibromosalicyloxycarbonyl)phenoxy]ethoxy}ethane (DPEE)

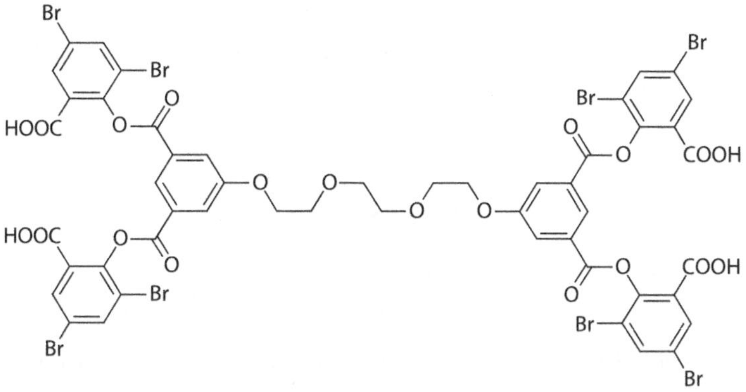

E. Tetrakis-dibromosalicylates

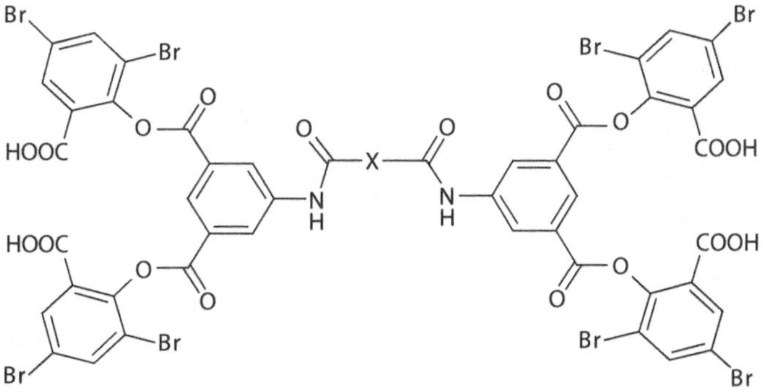

1. X=-Ø-SO₂-Ø-: *N,N′*-Bis[bis(3,5-dibromosalicyl)isophthalyl]-5,5′-[sulfonyl bis(1,4-phenylenecarbonylimino)] bis-(1,3-benzenedicarboxylate) (DBSS)

2. X=-Ø-C(CF₃)₂-Ø-: *N,N′*-Bis[bis(3,5-dibromosalicyl)isophthalyl]-2,2′-[hexafluoropropanyl bis(1,4-phenylenecarbonylimino)]bis-(1,3-benzenedicarboxylate) (DBSH)

3. X=-Ø-O-Ø-: *N,N′*-Bis[bis(3,5-dibromosalicyl)isophthalyl]-5,5′-[oxybis(1,4- phenylenecarbonylimino)] bis-(1,3-benzenedicarboxylic acid) (DBSO)

4. X= 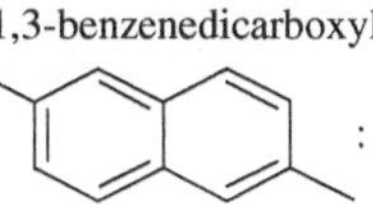 : *N,N′*-Bis[bis(3,5-dibromosalicyl)isophthalyl]-2,6-naphthalenedicarboxylate (DBSN)

(continued)

TABLE 7.1 (Continued)
Multifunctional Cross-Linking Reagents

5. X= 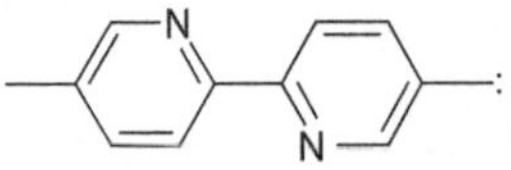 : *N,N'*-Bis[bis(3,5-dibromosalicyl)isophthalyl]-2,2'-bipyridinyl-5,5'-dicarboxylate
(DBSBP)

6. X= 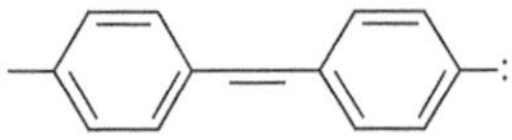 : *N,N'*-Bis[bis(3,5-dibromosalicyl)isophthalyl]-trans-stilbene-4,4'-dicarboxylate
(DBSSD)

F. Tetrakis acyl phosphate esters

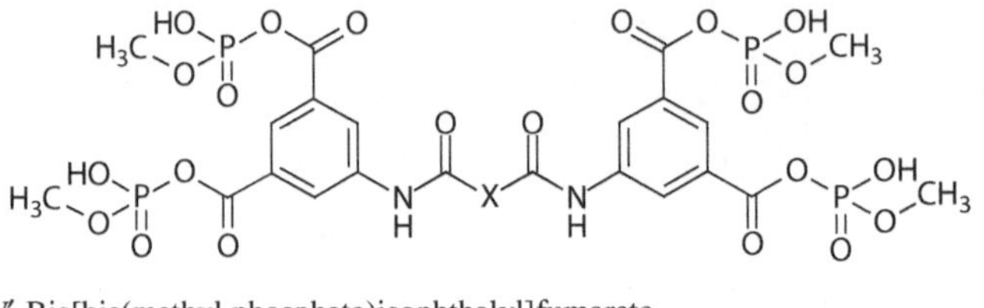

1. X= 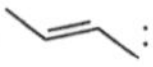 : *N,N'*-Bis[bis(methyl phosphate)isophthalyl]fumarate

2. X= 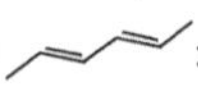 : *N,N'*-Bis[bis(methyl phosphate)isophthalyl] *trans, trans*muconate

3. X= 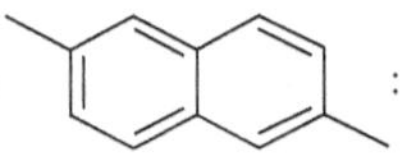 : *N,N'*-Bis[bis(methyl phosphate)isophthalyl]-2,6-naphthalenedicarboxylate

G. Tetraazamacrocyclic nitrogen mustards

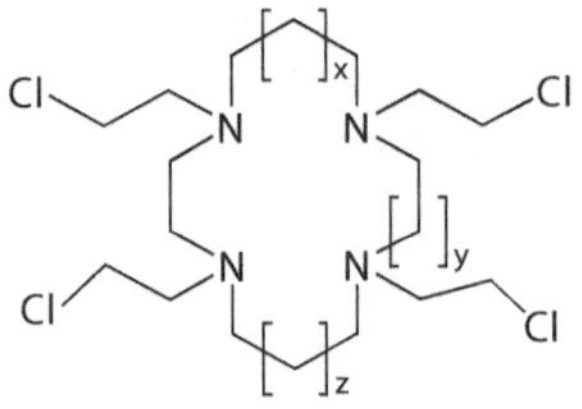

1. x = z = 0, y = 1: 1,4,8,10-Tetra(2-chloroethyl)-1,4,8,10-tetra-azacyclododecane tetrahydrochloride
2. x = y = z = 1: 1,4,8,12-Tetra(2-chloroethyl)-1,4,8,12-tetra-azacyclotetradecane tetrahydrochloride
3. x = z = 1, y = 2: 1,4,8,12-Tetra(2-chloroethyl)-1,4,8,12-tetra-azacyclopentadecane tetrahydrochloride

H. Bis-benzimidazole nitrogen mustards

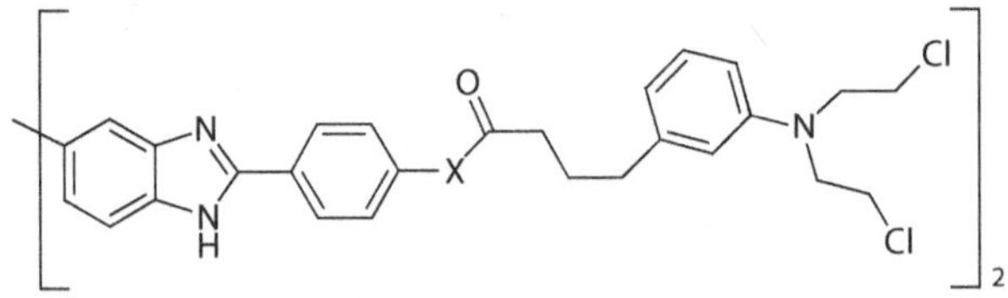

1. x=O: 2,2-Bis{4'-[4''-(*p-N,N*-(chloroethyl)aminophenyl)butanoyl(oxy)]-phenyl}-5,5-bi-1*H*-benzimidazole
2. x=NH: 2,2-Bis{4'-[4''-(*p-N,N*-(chloroethyl)aminophenyl)butanamido]-phenyl}-5,5-bi-1*H*-benzimidazole

I. 5-[[4-[Bis(2-chloroethyl)amino]benzoyl]amino]-*N*-[5-[[5-[[4-[bis(2-chloroethyl) amino]phenyl]carbamoyl]-1-methyl-pyrazol-3-yl]carbamoyl]-1-methyl- pyrazol-3-yl]-2-methyl-pyrazole-3-carboxamide (BAMP)

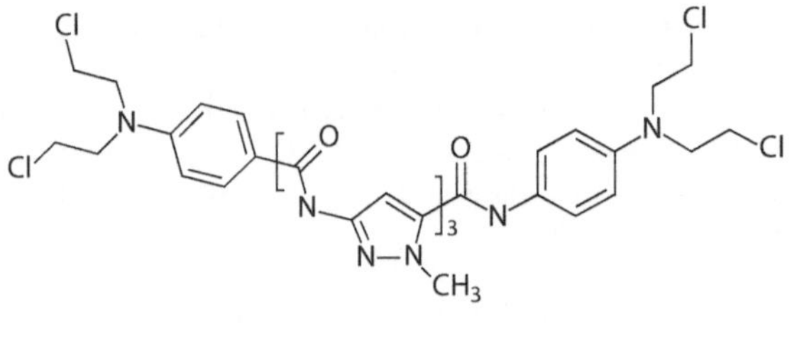

TABLE 7.1 (Continued)
Multifunctional Cross-Linking Reagents

III. Hexafunctional Cross-Linkers

A. Hexaazamacrocyclic nitrogen mustard: 1,5,9,12,17,20-Hexa(2-chloroethyl)-1,5,9,12,17,20-hexaazacyclodocosane hexahydrochloride

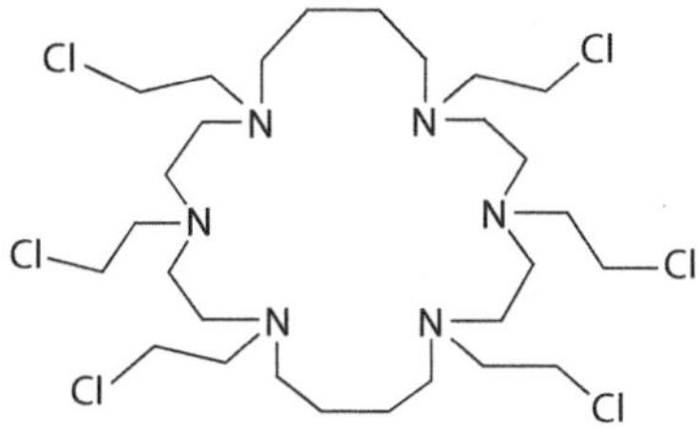

IV. Multifunctional Cross-Linkers

A. Diazonium precursors

1. Poly(4,4′-diaminodiphenylamine-3,3′-dicarboxylic acid)

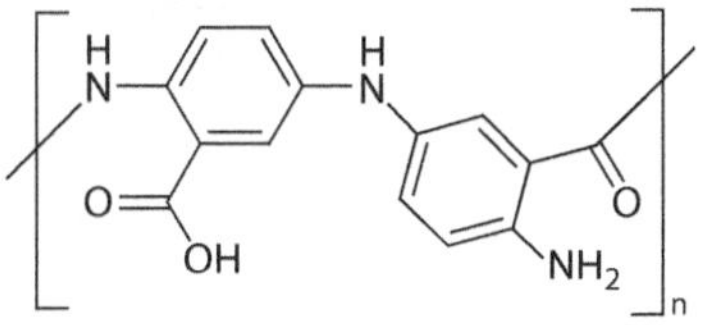

2. Poly(p-amino-D,L-phenylalanyl-L-leucine)

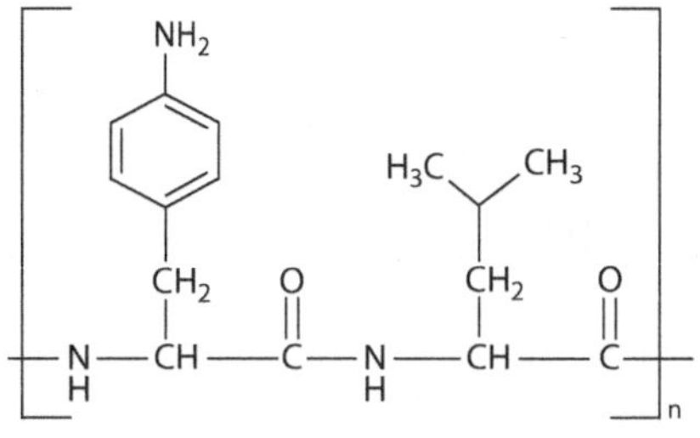

B. Poly(propyleneimine) dendrimers

1. N,N,N′,N′-tetrakis(3-aminopropyl)-1,4-butanediamine (DAB-Am-4)

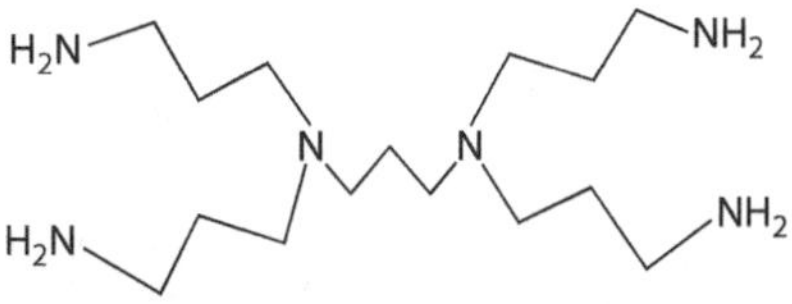

2. 4,17-Bis-(3-aminopropyl)-8,13-bis-[3-[bis-(3-aminopropyl)-amino]-propyl]-4,8,13,17-tetraazaeicosan-1,20-diamin, 4,17-Bis(3-aminopropyl)-8,13-bis[3-[bis(3-aminopropyl)amino]propyl]-4,8,13,17-tetraazaeicosane-1,20-diamine (DAB-AM-8)

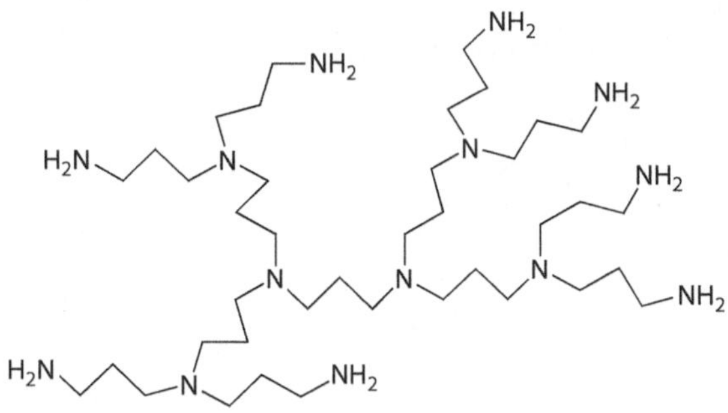

(*continued*)

TABLE 7.1 (Continued)
Multifunctional Cross-Linking Reagents

3. Polypropyleneimine hexadecaamine dendrimer (DAB-Am-16)

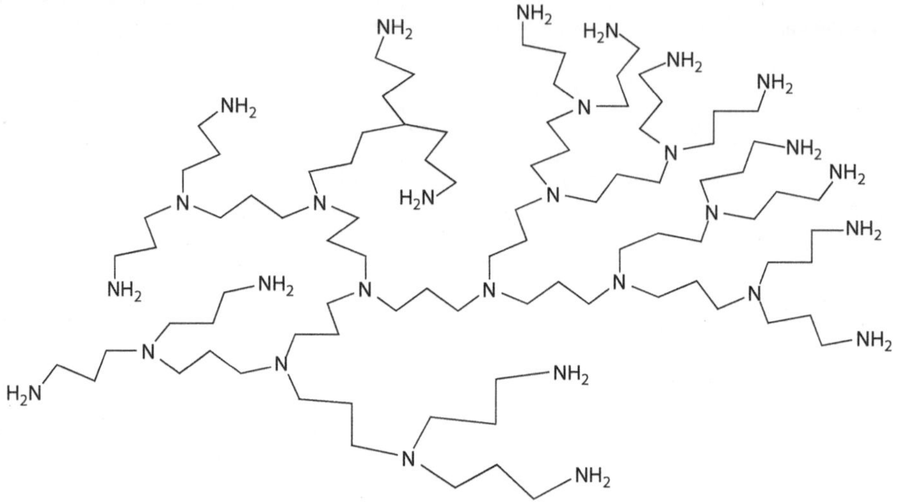

C. Poly(amidoamine) dendrimers
   1. PAMAM dendrimer, generation 1 (G1-PAMAM)

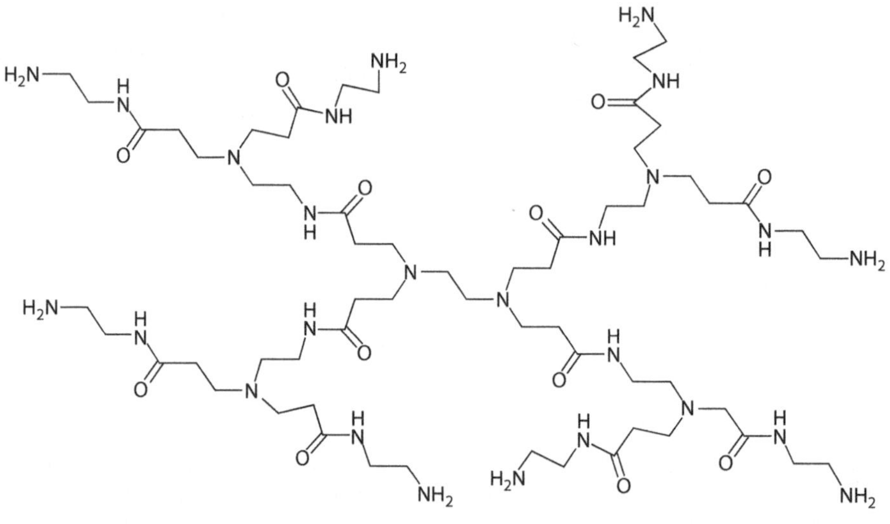

D. Azido terminated dendrimers
   1. Tetraazido terminated dendrimer, first generation, $(N_3)_4$-[G-1]

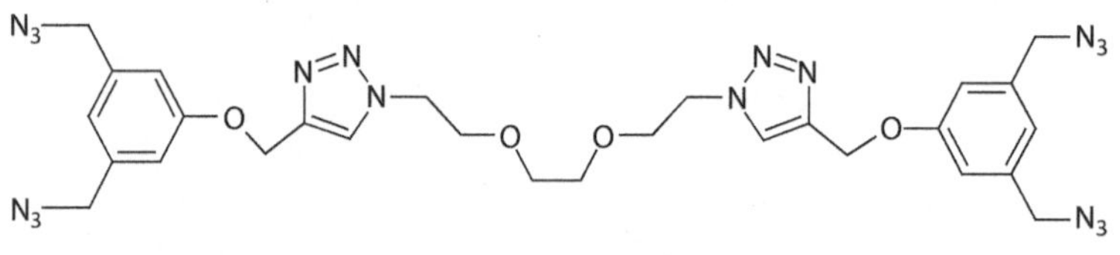

TABLE 7.1 (Continued)

Multifunctional Cross-Linking Reagents

2. Octaazido terminated dendrimer, second generation, $(N_3)_8$-[G-2]

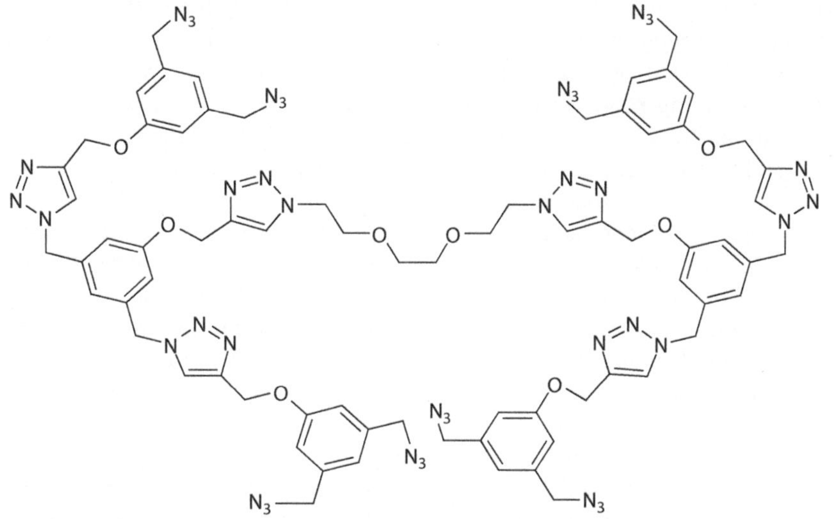

3. Hexadecaazido terminated dendrimer, third generation, $(N_3)_{16}$-[G-3]

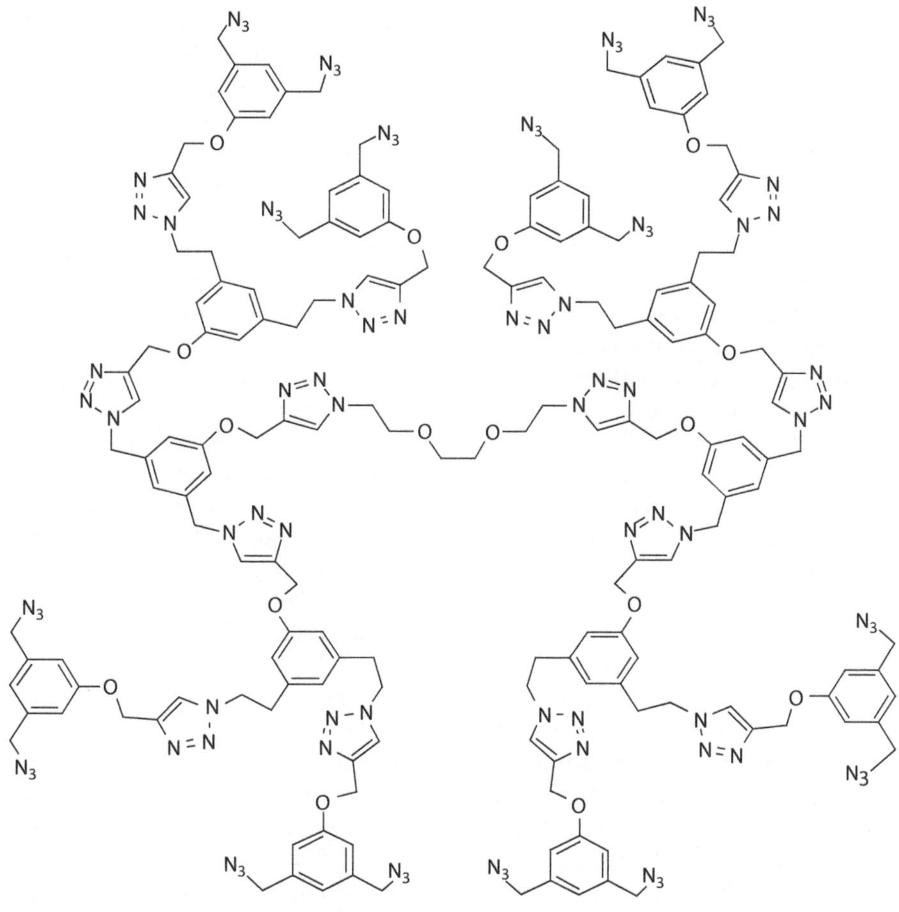

In addition to these bivalent bi- and trifunctional agents, there are compounds that contain three identical functional groups and are truly trifunctional. As shown in Table 7.1.I.A through N, the functional groups could be maleimide, aziridine, epoxyethane, succinimide, activated ester, nitrogen mustard, and benzophenol. Tris(2-maleimidoethyl)amine (TMEA, Table 7.1.I.A.1) is a trifunctional maleimide reagent directed toward sulfhydryl groups. Studdert and Parkinson[5] used TMEA to investigate the underlying molecular mechanisms of bacterial chemoreceptor signaling. When bacterial RP3098 or UU1581 cells carrying Tsr-S366C and/or Tar-S364C were treated with TMEA, three-subunit cross-linking products were obtained. Further analysis showed that TMEA cross-linked cysteines at position 366 of both Tsr-S366C and Tar-S364C replacement mutants, forming trimers of Tsr and Tsr-Tar mix-trimers. These results revealed that most of the cell's receptor molecules were organized in higher-order groups.

Another tri-maleimide compound is TMBA (Table 7.1.I.A.2). Schott et al.[6] used this and a tetra-maleimide (tetra-MBA, Table 7.1.II.A) to prepare multivalent antibody forms, tribody and tetra-body, respectively. Homogeneous Fab' was prepared from monoclonal antibody IgG F(ab')$_2$, which recognizes the tumor-associated antigen TAG-72, by mild reduction followed by selective reoxidation of interchain disulfide bonds, leaving a single hinge region sulfhydryl group. The tri- and tetra-maleimide compounds react with the cysteine sulfhydryl groups, cross-linking three and four Fab' to form F(ab')$_3$ (tribody) and F(ab')$_4$ (tetrabody), respectively.

As mentioned in Chapter 5, aziridines are reactive entities. Compounds with three aziridine rings are shown in Table 7.1.I.B.1 through 4 and are potential trifunctional cross-linkers. While 2,4,6-tri(ethyleneimino)-s-triazine (TEM) and tri{1-(2-methylaziridenyl)}phosphine oxide (Table 7.1.I.B.1 and B.4) are cytotoxic carcinogens, ThioTEPA and TEPA (Table 7.1.I.B.2 and I.B.3) have been used as antitumor therapeutics.[7-11] Experiments have demonstrated their cytotoxic properties. Although the aziridines are alkylating agents, their reaction mechanism with cellular components has not been illucidated. It is not clear whether these compounds actually exhibited their trifunctional activity. Nevertheless, Massen et al.[10] and Cohen et al.[11] showed that alkylation of DNA by thioTEPA and TEPA did occur *in vivo* and *in vitro*. ThioTEPA produced interstrand DNA cross-links, but it was not shown whether the cross-links occur trifunctionally. Similarly, the detailed cross-linking reaction for glycerol triglycidyl ether (Table 7.1.I.C) is not clear. Although it has been used to cross-link collagen, it is not known whether the three epoxide moieties react together to alkylate the amino acid side chains as a trifunctional cross-linker.[12]

Tris-succinimidyl aminotriacetate (TSAT, Table 7.1.I.D) contains three *N*-hydroxysuccimidyl esters, which are reactive to amino groups. Trabbic-Carlson et al.[13] have used this trifunctional cross-linker to cross-link elastin-like polypeptides (ELPs) consisting of Val-Pro-Gly-X-Gly repeats, where X is a lysine residue every 7 or 17 pentapeptides (otherwise X is valine). High-molecular-weight and high lysine content ELPs formed hydrogels upon TSAT intermolecular cross-linking. These hydrogels have a low-temperature gel structure that is nearly completely elastic. ELP hydrogels can also be formed from lysine-containing ELPs with β-[tris(hydroxymethyl)phosphino]propionic acid (THPP, Table 7.1.I.E), which is a trifunctional cross-linker.[14] Hydroxymethylphosphines react with primary and secondary amines of amino acids via a Mannich-type condensation reaction as shown in Figure 7.1.[15] With secondary amines, three of the amines are cross-linked. With primary amines, in addition to the formation of trifunctional cross-linked product, only two of the primary amines may be cross-linked.

The other cross-linkers containing three functional groups are trimesoyl tris(methyl phosphate) (TMMP, Table 7.1.I.F) and trimesoyl tris(3,5-dibromosalicylate) (TTDS, Table 7.1.I.G). Johnson et al.[16] used these compounds to cross-link hemoglobin subunits. The β-subunit processes two free amino groups at the interface between the two β-chains that are normally involved in binding allosteric effectors such as 2,3-bisphosphoglycerate. The ε-amino group of β82Lys and α-amino group of β19Val react with TTDS and TMMP, replacing 3,5-dibromosalicylate and methylphosphate, respectively, as shown in Figure 7.2. TTDS preferentially cross-links β82Lys-β82'Lys, and TMMP preferentially cross-links the ε-amino group of β82Lys with the α-amino group of β19Val.

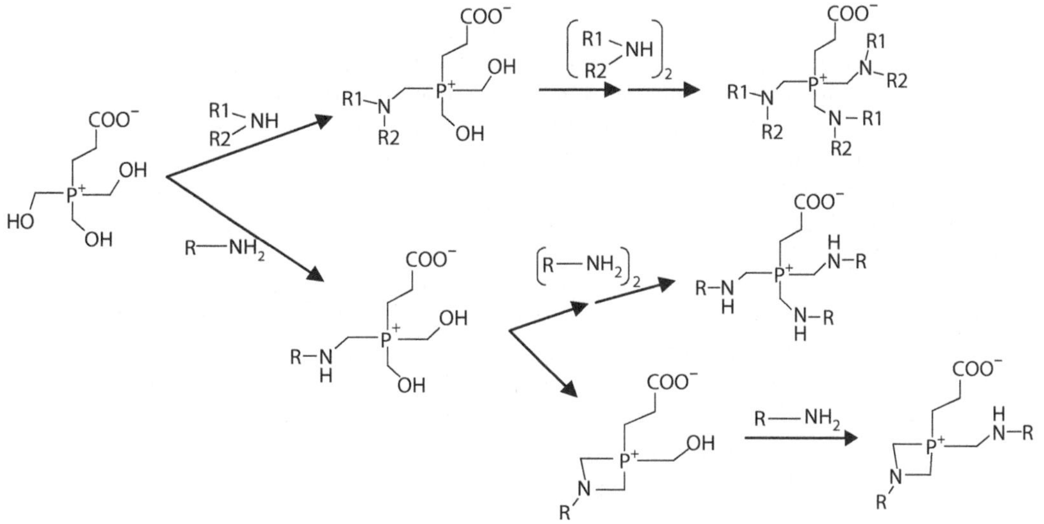

FIGURE 7.1 A Mannich-type condensation reaction of THPP with primary and secondary amines as a trifunctional cross-linker. Reaction with primary amines also produces a bifunctional product.

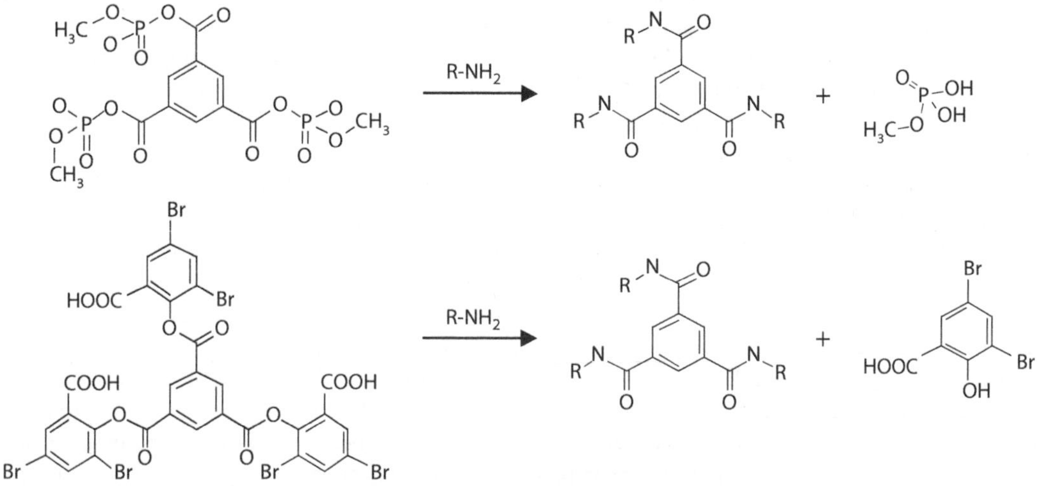

FIGURE 7.2 Reaction of amino groups with TTDS and TMMP replacing 3,5-dibromosalicylate and methylphosphate, respectively.

It is possible that three amino groups are cross-linked. For example, TMMP cross-linked β19Val, β82Lys, and β82′Lys giving rise to inter-subunit, but intramolecular cross-linked hemoglobin.[16]

The nitrogen mustard, TCEA (Table 7.1.I.H), containing three chloroethylamine groups with a cross-linking span of 5 Å, is reactive toward sulfhydryl and amino groups. Hiratsuka[17] used the compound to investigate protein–protein contact interactions of myosin subfragment 1 (S-1), which contains three domains, the 20, 26, and 50 kDa domains. Cross-linking experiments showed that TCEA first reacted with the highly reactive sulfhydryl group SH1 (Cys-707) on the 20 kDa domain, and then with a residue on the 50 kDa domain to generate the 74 kDa product. The 74 kDa product was further cross-linked to the 26 kDa domain in the presence of Mg-ATP, generating the 104 kDa product. It was interpreted that binding of the nucleotide induced conformational changes, which caused the 26 kDa domain to become close to the 74 kDa cross-linked product. Thus, TCEA successfully cross-linked the three domains of S-1.

Polyazamacrocyclic nitrogen mustards that have been studied as cross-linking agents include tri-azamacrocyclic (Table 7.1.I.I), tetra-azamacrocyclic (Table 7.1.II.G), and hexa-azamacrocyclic (Table 7.1.III.A) nitrogen mustards. The chloroethylamine moieties of these compounds are alkylating agents via aziridiniumion formation as in TCEA. Studies carried out by Parker et al.[18] showed that they were powerful DNA alkylating agents and were cytotoxic toward the human chronic myeloid leukemia cell line K562. Interstrand DNA cross-linking was produced in the presence of these polyazamacrocyclic nitrogen mustards. It was speculated that these compounds function as bifunctional reagents cross-linking N7-guanine of different DNA strands. However, multifunctional alkylation of DNA was not ruled out.

In addition to these mustard-containing cross-linkers, there are aniline mustards that are coupled to other functional groups to form trifunctional reagents (Table 7.1.I.J through M). The functional groups include *N*-hydroxysuccinimidyl ester as in *N*-succinimidyl 4-[{*N,N*-bis(2-chloroethyl)} amino]phenylbutyrate (CB-NHS, Table 7.1.I.J) and BAMP-NHS (Table 7.1.I.K), chloroethylphenol as in *seco*-CI-BAM (Table 7.1.I.L), and methoxyaziridine as in tallimustine analogues shown in Table 7.1.I.M. These functional groups would react with nucleophiles such as amino and sulfhydryl groups. The two 2-chloroethylamino moieties of aniline mustard form activated aziridinium ions, which alkylates potential amino groups as discussed earlier. Thus, three point reactions are possible. For CB-NHS, McKenzie et al.[1] demonstrated its cross-linking reactivity with benzylamine as a model system for the preparation of immunotoxins. Unfortunately, it was not tested with proteins or nucleic acids. BAMP-NHS was an intermediate compound in the synthesis of novel benzoic acid mustard (BAM) derivatives of distamycin A bearing one or more pyrazole moieties.[19] Although it was not tested for cross-linking ability, the compound has three reactive functional groups and is potentially a trifunctional cross-linker. *Seco*-CI-BAM is an achiral *seco*-CI analogue of duocarmycin and has been shown to have cytotoxicity toward human K562 (chronic myeloid leukemia cells).[20] As discussed in Chapter 5, BAM forms aziridinium ions, which alkylate DNA primarily at the guanine-N7 position to form an N7-alkylated guanine derivative. In the *seco*-CI-BAM compound, in addition to the reactive BAM moiety, the chloroethylphenol in the reagent is also potentially reactive. It could lose HCl to produce the putative spiro[2,5]cyclopropanecyclohexadienone alkylating agent, which could react with an adenine-N3 group in AT-rich DNA sequences as shown in Figure 7.3. Together with BAM, the reaction could potentially cross-link guanine and adenine of double-strand DNA. Whether three bases are cross-linked still needs to be studied. Another similar case is methoxyaziridinyl BAM (Table 7.1.I.M.1 through 3), which contains a methoxyaziridine instead of chloroethylphenol.[21] These compounds were found to be cytotoxic to human chronic myeloid leukemia K562 cells, which decreases with increasing C-terminus methylene chain length. They cross-link naked DNA more efficiently than corresponding compounds at equivalent doses without a methoxyaziridinyl moiety. Their preferential alkylation site is a purine (Pu), either guanine or adenine, of the sequence 5′-TTTTGPu-3′. The interstrand DNA cross-link ability increases

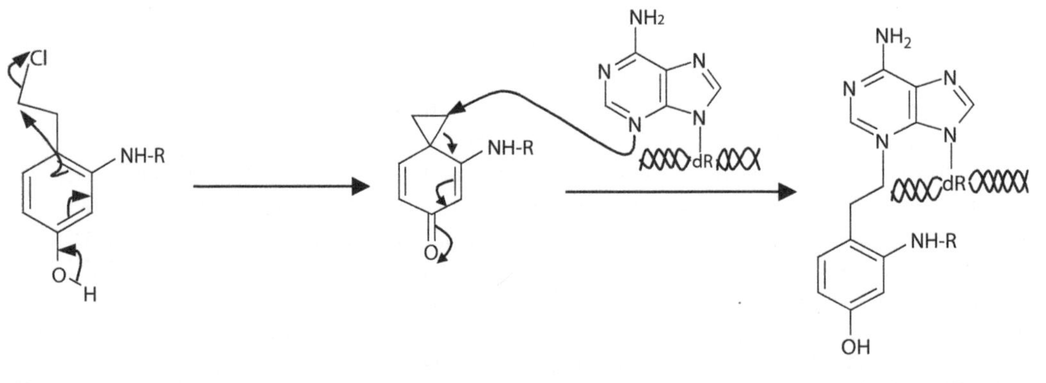

FIGURE 7.3 Reaction mechanism of chloroethylphenol moiety of *seco*-CI-BAM with adenine of DNA.

with increasing linker length. Presumably the methoxyaziridinyl moiety alkylates one of the Pu bases and the BAM alkylates another. It is not clear whether both of the chloroethylamino groups of BAM react with the bases giving rise to trifunctional cross-linking.

4,4′-Bis(*N*-maleimido)benzophenone (BMBP, Table 7.1.I.N) is a trifunctional photoactivatable cross-linker. The two maleimido moieties are thiol specific and the benzophenone can be photo-lyzed at 350–360 nm to generate reactive diradical.[22] Rajasekharan et al.[23] employed this reagent to monitor the conformational changes of myosin S-1 in the presence of either MgADP or MgPP. The maleimide groups of BMBP reacted with the two thiols, SH1 and SH2, in the 21 kDa segment of the heavy chain. On photolysis, the benzophenone bridge efficiently and specifically cross-linked the 21 kDa segment and the 50 kDa segment when MgADP is present. With MgPP, low efficiency cross-linking occurs between the bridged thiols and either the 27 kDa N-terminal or the 50 kDa segments of the heavy chain. These results indicate conformational changes of S-1 on binding adenosine nucleotide. MgADP causes the 50 kDa segment to move close to the bridged thiols. Without the adenosine moiety, the binding of MgPP leaves the protein in a flexible state such that residues in both the 27 kDa and the 50 kDa segment can move within the cross-linking span of the activated benzophenone triplet.

Patulin (PAT, Table 7.1.I.O), a mytotoxin isolated from various *Penicillium*, *Aspergillus*, and *Byssochlamys* molds, seemingly does not contain a functional group, but its electrophilic prop-erty enables it to act as a homobifunctional, a heterobifunctional, as well as multifunctional cross-linking agent. PAT-induced intermolecular protein–protein cross-links are demonstrated for bovine serum albumin, hen egg lysozyme, tubulin, and other species.[24] Presumably the thiol group of cysteine is preferred for PAT-mediated cross-link reactions. The initial reaction of a thiol group appears to activate PAT for the subsequent reactions with other nucleophiles as shown in Figure 7.4. Apparently, the side chains of lysine, histidine, and α-amino groups also exhibited reactivity to effect cross-linking. It is possible that the amino groups of nucleic acid bases can also react. Unfortunately, PAT has not been employed up to the present time as a multifunctional cross-linker to study complex biological systems.

A series of trifunctional cross-linkers containing biotin (Table 7.1.I.P) bind non-covalently with the biotin-binding proteins, avidin and streptavidin.[25–31] In addition to biotin, the other functional groups include amine-reactive *N*-hydroxysuccinimide and nitrophenyl ester, sulfhydryl-reactive maleimide, and methanethiosulfonate, photochemically reactive phenylazide and benzophenone, base-labile ester cleavage site, and disulfide bonds. A combination of these functional groups gener-ates various trifunctional cross-linkers directed for different side chains of amino acids. Many of these reagents are commercially available such as sulfo-SBED, Mts-Aff-Biotin, Mts-Aff-LC-Biotin, and MTS-BP-Bio (Table 7.1.I.P.1, 3, 4, and 6). These cross-linkers have been used to study protein complexes, for example, Sinz et al.[27] used the commercially available biotinylated cross-linking

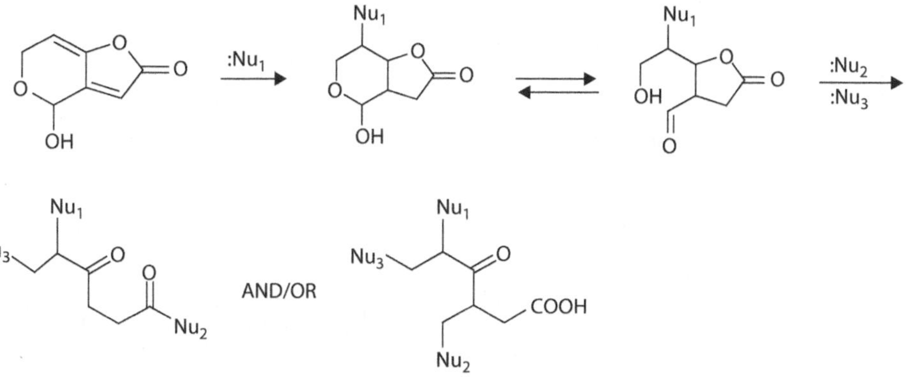

FIGURE 7.4 Proposed mechanism of reaction of PAT in cross-linking nucleophiles of proteins. (After Fliege, R. and Metzler, M., *Chem. Biol. Interact.*, 123, 85, 1999.)

reagent sulfo-SBED (Table 7.1.I.P.1) to cross-link the Ca^{2+}-dependent complex between calmodulin and its target peptide M13; Ahrends et al.[28] also used commercial MTS-BP-Bio (Table 7.1.I.P.6) to cross-link mismatch repair protein MutL and the strand discrimination endonuclease MutH; Wedekind et al.[29] synthesized ABNP (Table 7.1.I.P.5) to specifically cross-link insulin and its receptor; Fujii et al.[30] designed PEG-Bio-diNHS (Table 7.1.I.P.7) to study protein–protein interactions in ubiquitin; Ishmael et al.,[31] used APB (Table 7.1.I.P.2) to map protein–protein interactions in the bacteriophage T4 DNA polymerase holoenzyme; and Trester-Zedlitz et al.[25] prepared five flexible reagents (Table 7.1.I.P.8 through 12) to cross-link heterodimeric protein complex negative cofactor 2 (NC2) that functions as a negative regulator of basal transcription by blocking the assembly of the preinitiation complex. In all cases, after the cross-linking reaction is complete, the cross-linked products are isolated using avidin or streptavidin, either before or after protease digestion. Matrix-assisted laser desorption/ionization time-of-flight mass spectrometry (MALDI-TOFMS) and one-dimensional gel electrophoresis have been employed to check for the extent of cross-linking product formation.[25] The isolated cross-linked peptides are analyzed by MALDI-TOFMS, nano-high-performance liquid chromatography (HPLC)/nano-electrospray ionization Fourier transform ion cyclotron resonance mass spectrometry (ESI-FTICRMS), and tandem MS as will be discussed in Chapter 10.[25–28,30–32] These procedures have provided valuable information on protein structures and protein–protein interactions. For instance, Trester-Zedlitz et al.[25] have illustrated that the use of biotinylated trifunctional cross-linkers provided insight into the intermolecular interaction between the N-terminal regions of NC2α and β.

7.3   TETRAFUNCTIONAL CROSS-LINKERS

Tetrafunctional cross-linkers contain four functional groups. As mentioned above, tetra-MBA (Table 7.1.II.A), which has four thio-active maleimide functional groups has been used to cross-link four Fab′ IgG fragments to form tetrabodies.[6] Other tetrafunctional cross-linkers contain four 3,5-dibromosalicylate or methylphosphate groups (Table 7.1.II.B through F). Kluger et al.[33–38] used these series of reagents to cross-link hemoglobin to form hemoglobin bis-tetramer. As discussed above on the trifunctional cross-linkers, TTDS and TMMP (Table 7.1.I.G and F), the reagents cross-linked any of the available amino groups of β82Lys, β′82Lys, β19Val, and β′19Val of two hemoglobin tetramers as shown in Figure 7.5. The reaction is highly selective and efficient as shown by BTDS (Table 7.1.II.C).[35] Characterization of hemoglobin bis-tetramer by circular dichroism showed that the secondary structure of the globin chains is maintained while the microenvironment of the

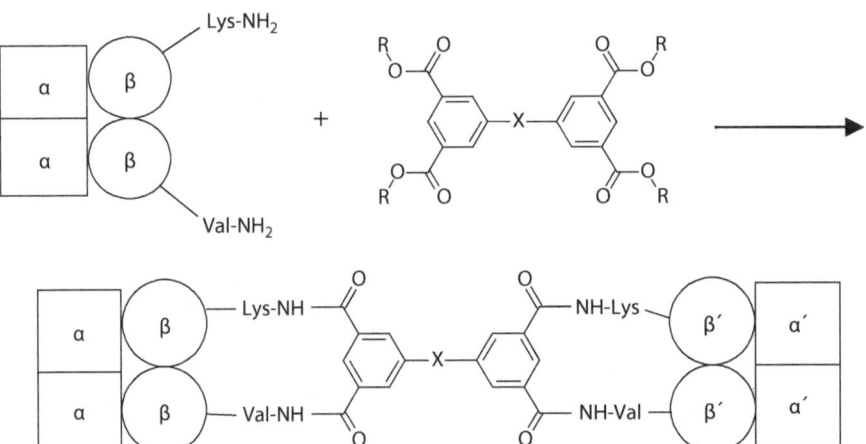

FIGURE 7.5   Cross-linking of hemoglobin with tetrafunctional cross-linkers (R being 3,5-dibromosalicylate or methylphosphate and X is spacer arm) to give hemoglobin bis-tetramer.

hemes is altered.[34] Tetrakis-dibromosalicylate cross-linkers contain both bridging spacer arms of different flexibility. In some cases, the cross-linked bis-tetramer has a decreased affinity for oxygen and very low cooperativity in oxygen binding such as that produced by DBIT (Table 7.1.II.B).[33] The compounds DBSS, DBSH, DBSO (Table 7.1.II.E.1 through 3) yielded bis-tetramers that permit tosional movement and have oxygen affinity and cooperativity similar to native hemoglobin. However, compounds DBSN, DBSBP, and DBSSD (Table 7.1.II.E.4 through 6) yielded products with decrease cooperativity.[34] Further conjugation of four polyethylene glycol chains to the DBSS cross-linked bis-tetramer at each of the β-Cys-93 greatly increased its nitrite reductase activity while retaining cooperativity.[38] Cross-linking with tetrakis acyl phosphate esters (Table 7.1.II.F.1 through 3) yielded tetramers, which bind oxygen cooperatively but with Hill coefficients lower than that of the native protein and with high average affinities. The bis-tetramers with longer connections between tetramers exhibit higher Hill coefficients.[37] In addition to bis-tetramers, the reaction of DPEE (Table 7.1.II.D) with deoxy hemoglobin gave intramolecular cross-links between the subunits, including β-β, α-β, and α-α products.[36]

Tetra-nitrogen mustards are represented by those shown in Table 7.1.II.G through I. The Tetraazamacrocyclic nitrogen mustards (Table 7.1.II.G.1 through 3) were discussed above.[18] The bis-benzimidazole nitrogen mustards (Table 7.1.II.H.1 and 2) are hybrid molecules comprising bis-benzimidazoles in ester and amide combination with the N-mustard chlorambucil. Chlorambucil has a preference for cross-linking via N7 of guanine in the DNA major groove, but can also bind to phosphate groups and to purine N3 in the minor groove, Le Sann et al.[39] conjugated bis-benzimidazoles to chlorambucil with a spacer distance of 10 base pairs providing one arm of each alkylating group positioned to interact covalently with N3 sites of purines. The amide hybrid (Table 7.1.II.H.2) was found to be around 12-fold more cytotoxic than the ester hybrid (Table 7.1.II.H.1). However, it was not determined whether the four chloroethylamino groups reacted with the DNA bases. Another hybrid is BAMP (Table 7.1.II.I), which is a BAM derivative of distamycin A bearing three pyrazole rings.[20] Distamycin A has a preference for AT sequence and alkylates the 3′-adenine-N3 atom located in the sequence 5′-TTTTGA-3′. However, BAMP was found to be ineffective against L1210 leukemia cells. But its reactivity with nucleic acids has not been investigated.

7.4 MULTIFUNCTIONAL CROSS-LINKERS

Multifunctional cross-linkers contain numerous functional groups with the potential to react with other moieties in proteins, nucleic acids, and other functionalities. As mentioned above, the hexaazamacrocyclic nitrogen mustard (Table 7.1.III.A) was designed as an antitumor agent, but its reactivity with DNA was not investigated.[18] Other potential multifunctional reagents are poly-diazonium precursors (Table 7.1.IV.A). The polyphenyl amines such as poly(4,4′-diaminodiphenylamine-3,3′-dicarboxylic acid) (Table 7.1.IV.A.1) and poly(p-amino-D,L-phenylalanyl-L-leucine) (Table 7.1.IV.A.2), as well as mixed poly-condensates of 4,4′-diaminodiphenylmethane-3,3′-dicarboxylic acid and 4-nitro-2-aminobenzoic acid can be activated to poly-diazonium salts with sodium nitrite in acidic conditions.[40] The resulting diazonium salts readily react with the amino acid side chains of tyrosine and histidine by electrophilic substitution reactions. These poly-diazoniums have been used to prepare antigens from insoluble proteins and enzymes, including bovine serum albumin, ribonuclease, trypsin, urease, and human chorionic gonadotropin.[41–43]

Dendrimers form another class of potential multifunctional cross-linking agents.[44–47] Several generations of dendrimers have been developed and some are commercially available. With each new generation, the functional terminal multiplies. For example, polypropylenimine tetramine dendrimer (DAB-Am-4, Table 7.1.IV.B.1) is the first generation propyleneimine-based dendrimer with four amine terminal arms. The second generation, DAB-Am-8 (Table 7.1.IV.B.2) expands the number of terminal arms to 8, and the third generation, DAB-Am-16 (Table 7.1.IV.B.3), to 16 amine terminal arms.[46] These multifunctional dendrimers were used to cross-link collagen in the presence of water-soluble carbodiimide, 1-ethyl-3-(3-dimethyl aminopropyl) carbodiimide (EDC).[46] Activation

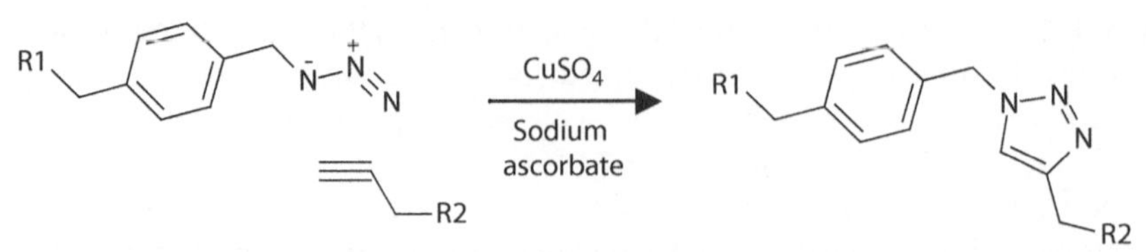

FIGURE 7.6 Click reaction cross-linking azide terminating dendrimers with alkynyl group–containing compounds.

of carboxylic acid groups of glutamine and aspartic acid residues in collagen by EDC enables the amine functional groups of the dendrimers to form ester bonds resulting in highly cross-linked collagens. Another series of dendrimers with amine terminal arms based on poly(amidoamine) (G1-PAMAM, Table 7.1.IV.C.1) have also been used to cross-link collagen.[44] Several generations from generation one (G1-PAMAM) to generation ten (G10-PAMAN) are available commercially. There are other chemical-based dendrimers, such as phosphorous-hydrazide dendrimers, poly(glycerol-succinic acid) dendrimers, and hydrazide-terminated ethylene diamine-based dendrimers. These reagents have been used to cross-link collagen, cells, bone, and cartilage as well as other areas of tissue engineering.[44] An interesting class of dendrimers contain azide terminating groups. Three generations with divergently grown azido groups, (N3)4-[G-1], (N3)8-[G-2], and (N3)16-[G-3] (Table 7.1.IV.D.1 through 3) have been studied.[47] These compounds react with alkynyl groups to form covalent cross-links via Click reactions as shown in Figure 7.6.[48] Joralemon et al.[47] have used the Click chemistry to cross-link azide terminating dendrimers to produce nanoparticles. Iha et al.[49] have also used Click chemistry to cross-link dendrimers with alkynyl groups to azide-containing compounds to form various soft materials.

In addition to the reagents discussed above, there are many commercially available multifunctional cross-linkers that have not been published. Mixed *N*-hydroxysuccinimidyl-maleimido trifunctional reagents are available from Molecular Biosciences, Inc., Boulder, Colorado (http://www.molbio.com/). Pierce Biotechnology, Inc., Rockford, IL (www.piercenet.com) also carries various dendrimers and other multifunctional cross-linkers.

7.5 NONCOVALENT CROSS-LINKERS

Noncovalent cross-linkers are agents, which bind to proteins and nucleic acids without forming a covalent bond. However, the binding affinities are so tight that they are essentially irreversible. There are many multifunctional noncovalent agents, which will be discussed in the following text.

7.5.1 AVIDIN AND STREPTAVIDIN

Avidin found in hen egg white is a glycoprotein with a molecular weight of 67,000 Da. It has four extraordinarily high-affinity binding sites for biotin and a dissociation constant of approximately 10^{-15} M, The avidin-biotin interaction has been exploited in immunoassays, in labeling techniques, and in the purification of macromolecules as discussed above for biotin-containing trifunctional cross-linkers (Table 7.1.I.P).[50] Streptavidin is similar to avidin except that it is free of carbohydrates and is isolated from *Streptomyces*. Streptavidin has a lower isoelectric point than avidin, namely, around 5.[51,52] A number of studies have shown that it exhibits reduced nonspecific binding, compared to avidin.[53–55] Since avidin and streptavidin are tetravalent, they can cross-link biotin-containing substrates. In most immunochemical applications, biotin is artificially incorporated into various macromolecules. Avidin is then used to cross-link these biotinylated molecules. In the avidin–biotin complex (ABC) method of immunoassay, both the immunoglobulin and the indicator enzyme are biotinylated and avidin is used to cross-link these species as shown in Figure 7.7A.[56] Because of the multiple binding sites on avidin, more biotinylated enzyme can be bound to increase the intensity of

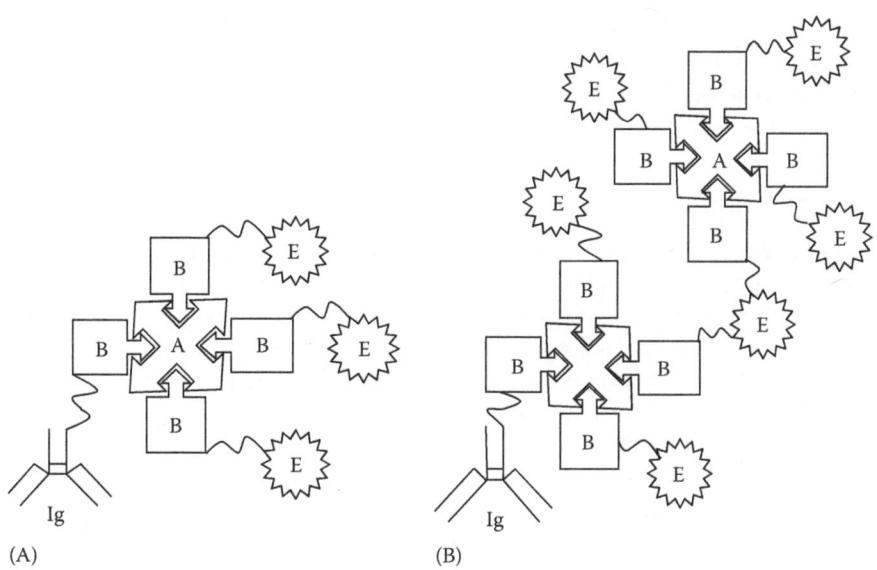

FIGURE 7.7 Tetrafunctional cross-linking of avidin immunoassays. (A) ABC; (B) BRAB. In the figure, B represents biotin, A represents avidin, E represents enzyme, and Ig represents immunoglobulin.

the substrate color development. A similar cross-linking complex is seen in the bridged avidin–biotin complex (BRAB) system (Figure 7.7B). In addition to immunoassays, the avidin–biotin technology has been used in gene probes, protein blotting, and immunohistochemistry.[57,58]

7.5.2 LECTINS

Lectins are sugar-binding proteins that agglutinate cells and precipitate carbohydrates or glyco-conjugates. They are multivalent, possessing at least two sugar-binding sites, which enable them to agglutinate animal and plant cells and/or to precipitate polysaccharides, glycoproteins, peptidoglycans, teichoic acids, glycolipids, etc. Each lectin binds specifically to a certain sugar sequence in oligosaccharides and glycoconjugates. There are many extensive reviews, symposia proceedings as well as books that have been written on the properties, functions, and applications of lectins.[59–67] It is beyond the scope of this book to review all the literature. The interested reader is urged to consult the references listed. In this section, we will summarize the relevant literature related to the application of lectins in cross-linking.

There must be hundreds, if not thousands, of different lectins that have been detected in and isolated from various plants and animals, bacteria, and viruses. The most celebrated example is probably concanavalin A (Con A) purified from jack bean. Con A at pH 7 is composed of four subunits of $M_r = 26,500\,\text{Da}$.[68,69] It is a metalloprotein, each subunit of which contains one Ca^{++} and one Mn^{++}, which are required for carbohydrate-binding activity.[70] Con A can bind α-mannosyl, α-glucosyl, and α-N-acetylglucosaminyl groups and can precipitate branched polysaccharides containing these sugar units as nonreducing termini.[71,72] Mannose, in its α-anomeric form, is the monosaccharide most complementary to the Con A sugar binding site.[73–84] Being tetravalent, Con A can cross-link and precipitate glycogen, yeast mannan, and various glycoproteins in a manner resembling antibodies. Such interactions have been used for structural analysis and isolation and resolution of glycoproteins.[75]

Like Con A, many lectins have been used for the detection, fractionation, isolation, and purification of glycoproteins, which do not necessarily employ their multivalent properties. Their cross-linking nature, however, is the basis of the Ouchterlony double diffusion and the affinoimmunoelectrophoresis technique in which the lectin is used in place of antibodies for the detection of glycoproteins.[76,77] Other applications include blood typing[78,79] and the identification of microorganisms.[80–83] It seems that the carbohydrate specificity and the multivalent nature of lectins can be further employed in cross-linking of components for enzyme immunoassays, for the study of immune complexes, and for colloidal gold staining techniques.[84]

7.5.3 MULTIFUNCTIONAL ANTIBODIES

In addition to bispecific antibodies discussed in Chapter 6 as heterobifunctional cross-linkers, trimeric and tetrameric antibodies with two or three specificities have been constructed using cross-linking reagents. Tutt et al.[85,86] constructed a series of trispecific F(ab')$_3$ derivatives by selective coupling of three mAb Fab' fragments primarily via their hinge-region sulfhydryl with thio-specific N,N'-(1,2-phenylene)bismaleimide (oPMD, see Appendix B.L). These Fab' fragments contain specificities for CD2, CD3, CD4, CD5, CD8, and CD37. The trispecific F(ab')$_3$ antibodies were shown to have more potent redirecting effector cell tumor cytotoxicity than corresponding dimeric antibodies. As mentioned above, Schott et al.[6] used trismaleimide and tetramaleimide (TMBA and tetra-MBA, Table 7.1.I.A.2 and II.A, respectively) to cross-link cysteine sulfhydryl groups present on the hinge region of Fab' to form F(ab')$_3$ tribody and F(ab')$_4$ tetrabody. Using Fab' subunits from a monoclonal antibody against tumor-associated antigen TAG-72, derived from murine CC49 IgG, Schott et al.[6] showed that the multifunctional antibodies retained antigen-binding ability. Biodistribution studies in tumor-bearing nude mice showed F(ab')$_4$ accumulated mainly in the liver, whereas the biodistribution behavior of F(ab')$_3$ in both Balb/c and nude mice was intermediate between that of IgG and F(ab')$_2$ for all organs studied. The progression of genetic engineering has enabled Li et al.[87] to develop two tetravalent antibodies (TetraMcAb), respectively, derived from the anti-CD20 mAbs C2B8 and 2F2. The tetraMcAbs were shown not only to be as effective in mediating complement-dependent cytotoxicity and antibody-dependent cellular cytotoxicity against B-lymphoma cells as native divalent antibodies, but also have superior antiproliferative and apoptosis-inducing activity. In a similar vein, Mertens et al.[88] applied molecular manifold modeling to create multispecific antibodies of intermediate molecular size by fusing one Fab and two single-chain variable fragments (scFv) antibody fragments into a trifunctional molecule (Tribody). Tribodies can be made trivalent, bispecific for one target, or trispecific. These multispecific antibodies have many uses, including targeting more than one receptor, or more than one epitope on target receptors, in a pretargeting strategy where the antibody reagent is first bound to the tumor cells followed by a small and fast clearing radiolabeled reagent. Cuesta et al.[89] also constructed a new class of multivalent antibodies, termed "trimerbodies," by fusion of the N-terminal association subdomain of collagen XVIII NC1, responsible for the noncovalent trimerization of collagen XVIII alpha chains, to the C-terminus of a scFv antibody through a 21 amino acid artificial flexible linker. Trimerbodies with specificity for human carcinoembryonic antigen (CEA) and an angiogenesis-associated laminin epitope were assembled and expressed as soluble secreted proteins in human HEK293 cells. Binding affinity studies with anti-CEA trimerbodies showed efficient tumor targeting of colorectal carcinomas in mice and with anti-laminin trimerbodies showed excellent tumor localization in several cancer types, including fibrosarcomas and carcinomas. Similarly, Lu and Zhu[90] have developed a recombinant method for the construction and production of a novel IgG-like tetravalent antibodies, using the variable domains of two fully human antibodies as the building blocks. Another new technology, named the dock and lock (DNL) method, was developed by Chang et al.[91] for making a myriad of bioactive molecules with multivalency, multifunctionality, and defined composition. It uses the dimerization and docking domain (DDD) of cyclic adenosine monophosphate-dependent protein kinase and the anchoring domain (AD) of A-kinase anchoring proteins as linkers for specifically docking a DDD-containing immunoglobulin module with an AD-containing antibody module. The resulting complex is covalently locked with disulfide bonds. Rossi et al.[92] used the DNL method to generate six different pairs of hexavalent anti-CD20/22 antibodies, designated as 22-20 and 20-22, which consist of an IgG linked to four Fab fragments. The authors showed that 22-20 and 20-22 have enhanced anti-lymphoma activity *in vitro* and comparable efficacy *in vivo*, indicating multiple advantages of hexavalent anti-CD20/22 antibodies over the individual parental antibodies.

With the advent of various multifunctional antibodies, Hirschhaeuser et al.[93] established a long-term coculture of human multicellular tumor spheroids with peripheral blood mononuclear cells to test the anticancer effects of these therapeutic molecules. Such new approaches may assist further development of multivalent, multifunctional antibodies.

REFERENCES

1. McKenzie, J. A., Raison, R. L., and Rivett, D. E., Development of a bifunctional crosslinking agent with potential for the preparation of immunotoxins, *J. Protein Chem.*, 7, 581, 1988.
2. Maassen, J. A., Cross-linking of ribosomal proteins by 4-(6-formyl-3-3-azidophenoxy)butyrimidate, a heterobifunctional, cleavable cross-linker, *Biochemistry*, 18, 1288, 1979.
3. Maassen, J. A., Schop, E. N., and Moller, W., Structural analysis of ribosomal proteins L7/L12 by the heterobifunctional cross-linker: 4-(6-Formyl-3-azidophenoxy)butyrimidate, *Biochemistry*, 20, 1020, 1981.
4. Maassen, J. A. and Möller, W., Photochemical cross-linking of elongation factor G to 70-S ribosomes from *Escherichia coli* by 4-(6-formyl-3-azidophenoxy)butyrimidate, *Eur. J. Biochem.*, 115, 279, 1981.
5. Studdert, C. A. and Parkinson, J. S., Crosslinking snapshots of bacterial chemoreceptor squads, *Proc. Natl. Acad. Sci., U. S. A.*, 101, 2117, 2004.
6. Schott, M. E., Frazier, K. A., Pollock, D. K., and Verbanact, K. M., Preparation, characterization, and in vivo biodistribution properties of synthetically cross-linked multivalent antitumor antibody fragments, *Bioconjug. Chem.*, 4, 153, 1993.
7. Ross, W. C., The chemistry of cytotoxic alkylating agents, *Adv. Cancer Res.*, 1, 397, 1853.
8. Alexander, P., The reactions of carcinogens with macromolecules, *Adv. Cancer Res.*, 2, 1, 1954.
9. Thiersch, J. B., Effect of 2, 4, 6, tri-amino-s-triazine (TR), 2, 4, 6 tris (ethyleneimino)-s-triazine (TEM) and N, N′, N″-triethylenephosphoramide (TEPA) on rat litter in utero, *Proc. Soc. Exp. Biol. Med.*, 94, 36, 1957.
10. Maanen, M. J., Smeets, C. J., and Beijnen, J. H., Chemistry, pharmacology and pharmacokinetics of N,N′,N″-triethylenethiophosphoramide (ThioTEPA), *Cancer Treat. Rev.*, 26, 257, 2000.
11. Cohen, N. A., Egorin, M. J., Snyder, S. W., Ashar, B., Wietharn, B. E., Pan, S. S., Ross, D. D., and Hilton, J., Interaction of N,N′,N″-triethylenethiophosphoramide and N,N′,N″-triethylenephosphoramide with cellular DNA, *Cancer Res.*, 51, 4360, 1991.
12. Tang, A. and Yue, Y., Crosslinkage of collagen by polyglycidyl ethers, *ASAIO J.*, 41, 72, 1995.
13. Trabbic-Carlson, K., Setton, L. A., and Chilkoti, A., Swelling and mechanical behaviors of chemically cross-linked hydrogels of elastin-like polypeptides, *Biomacromolecules*, 4, 572, 2003.
14. Lim, D. W., Nettles, D. L., Setton, L. A., and Chilkoti, A., Rapid crosslinking of elastin-like polypeptides with hydroxymethylphosphines in aqueous solution, *Biomacromolecules*, 8, 1463, 2007.
15. Henderson, W., Olsen, G. M., and Bonnington, L. S., Immobilised phosphines incorporating the chiral biopolymers chitosan and chitin, *J. Chem. Soc. Chem. Commun.*, 1863, 1994.
16. Johnson, M. B., Adamson, J. G., and Mauk, A. G., Functional comparison of specifically cross-linked hemoglobins biased toward the R and T states, *Biophys. J.*, 75, 3078, 1998.
17. Hiratsuka, T., Cross-linking of three heavy-chain domains of myosin adenosine triphosphatase with a trifunctional alkylating reagent, *Biochemistry*, 27, 4110, 1988.
18. Parker, L. L., Anderson, F. M., O'Hare, C. C., Lacy, S. M., Bingham, J. P., Robins, D. J., and Hartley, J. A., Synthesis of novel DNA cross-linking antitumour agents based on polyazamacrocycles, *Bioorg. Med. Chem.*, 13, 2389, 2005.
19. Baraldi, P. G., Cozzi, P., Geroni, C., Mongelli, N., Romagnolia, R., and Spalluto, G., Novel benzoyl nitrogen mustard derivatives of pyrazole analogues of distamycin A: Synthesis and antileukemic activity, *Bioorg. Med. Chem.*, 7, 251, 1999.
20. Kupchinsky, S., Centioni, S., Howard, T., Trzupek, J., Roller, S., Carnahan, V., Townes, H. et al., A novel class of achiral seco-analogs of CC-1065 and the duocarmycins: Design, synthesis, DNA binding, and anticancer properties, *Bioorg. Med. Chem.*, 12, 6221, 2004.
21. Brooks, N., Lee, M., Wright, S. R., Woo, S., Centioni, S., and Hartley, J. A., Synthesis, DNA binding, cytotoxicity and sequence specificity of a series of imidazole-containing analogs of the benzoic acid mustard distamycin derivative tallimustine containing an alkylating group at the C-terminus, *Anticancer Drug Des.*, 12, 591, 1997.

22. Dormtin, G. and Prestwich, G. D., Benzophenone photophores in biochemistry, *Biochemistry*, 33, 5661, 1994.
23. Rajasekharan, K. N., Sivaramakrishnan, M., and Burke, M., Proximity and ligand-induced movement of interdomain residues in myosin subfragment 1 containing trapped MgADP and MgPPi probed by multi-functional cross-linking, *J. Biol. Chem.*, 262, 11207, 1987.
24. Fliege, R. and Metzler, M., The mycotoxin patulin induces intra- and intermolecular protein crosslinks in vitro involving cysteine, lysine, and histidine side chains, and α-amino groups, *Chem. Biol. Interact.*, 123, 85, 1999.
25. Trester-Zedlitz, M., Kamada, K., Burley, S. K., Fenyo, D., Chait, B. T., and Muir, T. W., A modular cross-linking approach for exploring protein interactions, *J. Am. Chem. Soc.*, 125, 2416, 2003.
26. Sinz, A., Chemical cross-linking and mass spectrometry to map three-dimensional protein structures and protein–protein interactions, *Mass Spectrom. Rev.*, 25, 663, 2006.
27. Sinz, A., Kalkhof, S., and Ihling, C., Mapping protein interfaces by a trifunctional cross-linker combined with MALDI-TOF and ESI-FTICR mass spectrometry, *J. Am. Soc. Mass Spectrom.*, 16, 1921, 2005.
28. Ahrends, R., Kosinski, J., Kirsch, D., Manelyte, L., Giron-Monzon, L., Hummerich, L., Oliver Schulz, O., Spengler, B., and Friedhoff, P., Identifying an interaction site between MutH and the C-terminal domain of MutL by crosslinking, affinity purification, chemical coding and mass spectrometry, *Nucleic Acids Res.*, 34, 3169, 2006.
29. Wedekind, F., Baer-Pontzen, K., Bala-Mohan, S., Choli, D., Zahn, H., and Brandenburg, D., Hormone binding site of the insulin receptor: Analysis using photoaffinity-mediated avidin complexing, *Biol. Chem. Hoppe Seyler.*, 370, 251, 1989.
30. Fujii, N., Jacobsen, R. B., Wood, N. L., Schoeniger, J. S., and Guy, R. K., A novel protein crosslinking reagent for the determination of moderate resolution protein structures by mass spectrometry (MS3-D), *Bioorg. Med. Chem. Lett.*, 14, 427, 2004.
31. Ishmael, F. T., Alley, S. C., and Benkovic, S. J., Identification and mapping of protein–protein interactions between gp32 and gp59 by cross-linking, *J. Biol. Chem.*, 276, 25236, 2001.
32. Lee, Y. J., Mass spectrometric analysis of cross-linking sites for the structure of proteins and protein complexes, *Mol. BioSyst.*, 4, 816, 2008.
33. Kluger, R., Lock-O'Brien, J., and Teytelboym, A., Connecting proteins by design. Cross-linked bis-hemoglobin, *J. Am. Chem. Soc.*, 121, 6780, 1999.
34. Hu, D. and Kluger, R., Functional cross-linked hemoglobin bis-tetramers: Geometry and cooperativity, *Biochemistry*, 47, 12551, 2008.
35. Paal, K., Jones, R. T., and Kluger, R., A site-specific tetrafunctional reagent for protein modification: Cross-linked hemoglobin with two sites for further reaction, *J. Am. Chem. Soc.*, 118, 10380, 1996.
36. Kluger, R., Paal, K., and Adamson, J. G., An ether-linked tetrafunctional acylating reagent and its cross-linking reactions with hemoglobin, *Can. J. Chem.*, 77, 271, 1999.
37. Gourianov, N. and Kluger, R., Cross-linked bis-hemoglobins: Connections and oxygen binding, *J. Am. Chem. Soc.*, 125, 10885, 2003.
38. Lui, R. E. and Kluger, R., Enhancing nitrite reductase activity of modified hemoglobin: Bis-tetramers and their PEGylated derivatives, *Biochemistry*, 48, 11912, 2009.
39. Le Sann, C., Baron, A., Mann, J., van den Berg, H., Gunaratnam, M., and Neidle, S., New mustard-linked 2-aryl-bis-benzimidazoles with anti-proliferative activity, *Org. Biomol. Chem.*, 4, 1305, 2006.
40. Means, G. E. and Feeney, R. E., *Chemical Modification of Proteins*, Holden-Day, San Francisco, CA, 1971.
41. Anderer, F. A. and Schlumberger, H. D., Antigenic properties of proteins cross-linked by multidiazonium compounds, *Immunochemistry*, 6, 1, 1969.
42. Bar-Eli, A. and Katchalski, E., Preparation and properties of water-insoluble derivatives of trypsin, *J. Biol. Chem.*, 238, 1690, 1963.
43. Riesel, E. and Katchalski, E., Preparation and properties of water-insoluble derivatives of urease, *J. Biol. Chem.*, 239, 1521, 1964.
44. Joshi, N. and Grinstaff, M., Applications of dendrimers in tissue engineering, *Curr. Topics Med. Chem.*, 8, 1225, 2008.
45. Duan, X. and Sheardown, H., Crosslinking of collagen with dendrimers, *J. Biomed. Mater. Res. A*, 75, 510, 2005.
46. Duan, X. and Sheardown, H., Dendrimer crosslinked collagen as a corneal tissue engineering scaffold: Mechanical properties and corneal epithelial cell interactions, *Biomaterials*, 27, 4608, 2006.
47. Joralemon, M. J., O'Reilly, R. K., Hawker, C. J., and Wooley, K. L., Shell click-crosslinked (SCC) nanoparticles: A new methodology for synthesis and orthogonal functionalization, *J. Am. Chem. Soc.*, 127, 16892, 2005.

48. O'Reilly, R. K., Hawker, C. J., and Wooley, K. L., Cross-linked block copolymer micelles: Functional nanostructures of great potential and versatility, *Chem. Soc. Rev.*, 35, 1068, 2006.

49. Iha, R. K., Wooley, K. L., Nystrom, A. M., Burke, D. J., Kade, M. J., and Hawker, C. J., Applications of orthogonal "Click" chemistries in the synthesis of functional soft materials, *Chem. Rev.*, 109, 5620, 2009.

50. Bayer, E. A. and Wilchek, M., The use of avidin–biotin complex, *Methods Biochem. Anal.*, 26, 1, 1980.

51. Chaiet, L. and Wolf, F. J., The properties of streptavidin, a biotin-binding protein produced by *Streptomycetes*, *Arch. Biochem. Biophys.*, 106, 1, 1964.

52. Weber. P. C., Ohlendorf, D. H., Wendoloski, J. J., and Salemme, F. R., Structural origins of high affinity biotin binding to streptavidin, *Science (London)*, 243, 85, 1989.

53. Hacuptle, M. T., Aubert, M. L., Djiane, J., and Krachenbuhl, J. P., Binding sites for lactogenic and somatogenic hormones from rabbit mammary gland and liver, *J. Biol. Chem.*, 258, 305, 1983.

54. Gardner, L., Nonradioactive DNA labeling. Detection of specific DNA and RNA sequences on nitrocellulose and in situ hybridizations, *Biotechniques*, 1, 39, 1983.

55. Hofmann, K., Wood, S. W., Brinton, C. C., Montibeller, J. A., and Finn, F. M., Iminobiotin affinity columns and their application to retrieval of streptavidin, *Proc. Natl. Acad. Sci. U. S. A.*, 77, 4666, 1980.

56. Bratthauer, G. L., The avidin–biotin complex (ABC) method and other avidin–biotin binding methods, *Methods Mol. Biol.*, 588, 257, 2010.

57. Wilchek, M., and Bayer, E. A. (eds.), Avidin–biotin technology, in *Methods in Enzymology*, Vol. 184, Academic Press, Orlando, FL, 1990, pp. 5–672.

58. Mabuchi, Y., Yamoto, M., Minami, S., and Umesaki, N., Immunohistochemical localization of inhibin and activin subunits, activin receptors and Smads in ovarian endometriosis, *Int. J. Mol. Med.*, 25, 17, 2010.

59. Goldstein, I. J., My favorite enzyme: For love of lectins, *IUBMB Life*, 62, 247, 2010.

60. Sharon, N. and Lis, H., *Lectins*, Chapman & Hall, New York, 1989.

61. Bog-Hansen, T. C. (ed.), *Lectins: Biology, Biochemistry, Clinical Biochemistry*, Vols. 1–5, De Gruyter, New York, 1981–1986.

62. Lis, H. and Sharon, N., Lectins. Properties and applications to the study of complex carbohydrates in solution and on cell surfaces, in *Biology of Carbohydrates*, Ginsburg, V. and Robins, P. W. (eds.), Vol. 2, Wiley, New York, 1984, p. 1–85.

63. Etzler, M. E., Plant lectins: Molecular and biological aspects, *Annu. Rev. Plant Physiol.*, 36, 209, 1985.

64. Liener, I. E., Sharon, N., and Goldstein, I. J. (eds.), *The Lectins: Properties, Functions and Applications in Biology and Medicine*, Academic Press, New York, 1986.

65. Lis, H. and Sharon, N., Lectin as molecules and as tools, *Annu. Rev. Biochem.*, 55, 35, 1986.

66. Osawa, T. and Tsuji, T., Fractionation and structural assessment of oligosaccharides and glycopeptides by use of immobilized lectins, *Annu. Rev. Biochem.*, 21, 1987.

67. Rüdiger, H. and Gabius, H. J., Plant lectins: Occurrence, biochemistry, functions and applications, *Glycoconj. J.*, 18, 589, 2001.

68. Agawal, B. B. L. and Goldstein, I. J., Protein–carbohydrate interactions. VI. Isolation of concanavalin A by specific adsorption on cross-linked dextran gels, *Biochim. Biophys. Acta*, 147, 262, 1967.

69. McKenzie, G. H., Sawyer, W. H., and Nichol, L. W., The molecular weight and stability of concanavalin A, *Biochim. Biophys. Acta*, 263, 283, 1972.

70. Agawal, B. B. L. and Goldstein, I. J., Protein–carbohydrate interactions. XV. The role of bivalent cations in concanavalin A-polysaccharide interaction, *Can. J. Biochem.*, 46, 1147, 1968.

71. Goldstein, I. J., Studies on the combining sites of concanavalin A, *Adv. Exp. Med. Biol.*, 55, 35, 1974.

72. Goldstein, I. J. and Hayes, C. E., The lectins. Carbohydrate-binding proteins of plants and animals, *Adv. Carbohydr. Chem. Biochem.*, 35, 127, 1978.

73. So, L. L. and Goldstein, I. J., Protein–carbohydrate interaction. XXI. Interaction of concanavalin A with d-fructans, *Carbohydr. Res.*, 10, 231, 1969.

74. Poretz, R. D. and Goldstein, I. J., An examination of the topography of the saccharide binding sites of concanavalin A and of the forces involved in complexation, *Biochemistry*, 9, 2890, 1970.

75. Dulaney, J. T., Binding interactions of glycoproteins with lectins, *Mol. Cell. Biochem.*, 21, 43, 1978.

76. Goidstein, I. J., Use of concanavalin A for structural studies, *Methods Carbohydr. Chem.*, 6, 106, 1972.

77. Bog-Hansen, T. C., Affinity electrophoresis with lectins for characterization of glycoproteins, in *Proceedings of 3rd International Symposium on Affinity Chromatography and Molecular Interaction*, Strasbourg, France, June 26–29, 1979, Egly, J. M. (ed.), INSERM Symposium Series, 1979, p. 399.

78. Watkins, W. M., Yates, A. D., and Greeawell, P., Blood group antigens and the enzymes involved in their synthesis. Past and present, *Biochem. Soc. Trans.*, 9, 186, 1981.

79. Judd, W. J., The role of lectins in blood group serology, *CRC Crit. Rev. Clin. Lab. Sci.*, 12, 171, 1980.

80. Yajko, D. M., Chu, A., and Hadley, W. K., Rapid confirmatory identification of *Neisseria gonorrhoeae* with lectins and chromatogenic substrates, *J. Clin. Microbiol.*, 19, 380, 1984.
81. Raychowdhury, M. K., Goswami, R., and Chakrabarti, P., Agglutination of some bacterial cells by concanavalin A, *Indian J. Exp. Biol.*, 20, 748, 1982.
82. DeLucca, A. J., II., Lectins grouping of *Bacillus thuringiensis* serovars, *Can. J. Microbiol.*, 30, 1100, 1984.
83. Graham, K., Keller, K., Ezzell, J., and Doyle, R., Enzyme-linked lectinosorbent assay (ELlA) for detecting *Bacillus anthracis*, *Eur. J. Clin. Microbiol.*, 3, 210, 1984.
84. Roth, J., Application of lectin-gold complexes for electron microscopic localization of glycoconjugation on thin sections, *J. Histochem. Cytochem.*, 31, 987, 1983.
85. Tutt, A., Stevenson, G. T., and Glennie, M. J., Trispecific F(ab')$_3$ derivatives that use cooperative signaling via the TCR/CD3 complex and CD2 to activate and redirect resting cytotoxic T cells, *J. Immunol.*, 147, 60, 1991.
86. Tutt, A., Greenman, J., Stevenson, G. T., and Glennie, M. J., Bispecific F(ab'gamma)$_3$ antibody derivatives for redirecting unprimed cytotoxic T cells, *Eur. J. Immunol.*, 21, 1351, 1991.
87. Li, B., Shi, S., Qian, W., Zhao, L., Zhang, D., Hou, S., Zheng, L., Dai, J., Zhao, J., Wang, H., and Guo, Y., Development of novel tetravalent anti-CD20 antibodies with potent antitumor activity, *Cancer Res.*, 68, 2400, 2008.
88. Mertens, N., Filip Devos, F., Leoen, J., Van Deynse, E., Willems, A., Schoonooghe, S., Burvenich, I. et al., New strategies in polypeptide and antibody synthesis: An overview, *Cancer Biother. Radiopharm.*, 19, 99, 2004.
89. Cuesta, A. M., Sánchez-Martín, D., Sanz, L., Bonet, J., Compte, M., Kremer, L., Blanco, F. J., Oliva, B., and Alvarez-Vallina, L., In vivo tumor targeting and imaging with engineered trivalent antibody fragments containing collagen-derived sequences, *PLoS One*, 4, e5381, 2009.
90. Lu, D. and Zhu, Z., Construction and production of an IgG-like tetravalent bispecific antibody for enhanced therapeutic efficacy, *Methods Mol. Biol.*, 525, 377, 2009.
91. Chang, C.-H., Rossi, E. A., and Goldenberg, D. M., The dock and lock method: A novel platform technology for building multivalent, multifunctional structures of defined composition with retained bioactivity, *Clin. Cancer Res.*, 13, 5586S, 2007.
92. Rossi, E. A., Goldenberg, D. M., Cardillo, T. M., Stein, R., and Chang, C.-H., Hexavalent bispecific antibodies represent a new class of anticancer therapeutics: 1. Properties of anti-CD20/CD22 antibodies in lymphoma, *Blood*, 113, 6461, 2009.
93. Hirschhaeuser, F., Leidig, T., Rodday, B., Lindemann, C., and Mueller-Klieser, W., Test system for trifunctional antibodies in 3D MCTS culture, *J. Biomol. Screen*, 14, 980, 2009.

8 Monofunctional and Zero-Length Cross-Linking Reagents

8.1 INTRODUCTION

From a logical point of view, cross-linking reagents would contain two functional groups, which react with target components, thus linking them together. The two functional groups in the cross-linker could be identical as in homobifunctional reagents (Chapter 5), or different as in heterobifunctional reagents (Chapter 6). However, there are some compounds that do not belong to either of these two classes, but are still able to effect cross-linking of two proteins or other macromolecules. These compounds are generally simple and readily available and may be classified into two types: the monofunctional cross-linkers and the zero-length cross-linkers. Although monofunctional cross-linkers have one reactive group, part of the molecule is incorporated in the final cross-linked product. On the other hand, zero-length reagents induce direct joining of two intrinsic chemical groups of proteins or other molecules without introduction of any extrinsic material. During the cross-linking reaction, atoms are eliminated from the reactants, thus shortening the distance between the two linked moieties. This approach contrasts with other cross-linking reactions in which a spacer is always incorporated between the two cross-linked groups. For this reason, intramolecular cross-linking can only occur between very closely juxtaposed residues in a macromolecular complex. Reagents that catalyze the formation of disulfide bonds, for example, are zero-length cross-linkers. Other reagents condense carboxyl and primary amino groups to form amide bonds, hydroxyl and carboxyl groups to form esters, thiol and carboxyl groups to form thioesters, and so on. Many of these reagents simply act as activating agents converting one of the components, for example, carboxyl groups, into a reactive species. Examples of these reactions are multitudinous in organic synthesis wherein two molecules are condensed to form a new compound. Carboxylic acids, for instance, are activated to acyl chlorides for the synthesis of esters or amides. In the synthesis of polypeptides, the carboxyl group is activated by carbodiimide to a reactive intermediate that the amino group attacks. However, many of the activation reactions are too harsh to be used with native proteins. They either disrupt the three-dimensional structure of the macromolecule or cause extensive modification of the biologically active component. Only mild reagents that do not cause denaturation are useful for the cross-linking. This chapter will focus on these compounds. Other reagents that are not suitable for cross-linking proteins, but which have been used to couple macromolecules to solid supports, will be discussed in a later chapter.

There are many applications of these monofunctional and zero-length reagents. They have been used as specific probes of both the static and dynamic aspects of macromolecular structure, providing information about macromolecular contacts when the actual sites of cross-linking are determined. Thus, the exact contacts between histones in nucleosomes, protein and RNA in ribosomes, and RNA polymerase and DNA have been determined. They have also been used to probe macromolecular conformational changes upon binding of other molecules. These applications will be illustrated below for individual cross-linking reagents.

8.2 MONOFUNCTIONAL CROSS-LINKING REAGENTS

The commonly used monofunctional cross-linkers are listed in Table 8.1. These reagents are described below in detail.

8.2.1 IMIDOESTERS

Imidoesters readily react with primary amines eliminating an alcohol to form amidines (Figure 4.7). Acetimidates are simple monofunctional imidoesters that can cross-link proteins. The mechanism of such a reaction is discussed in detail by Peters and Richards[1] and is shown in Figure 8.1. As shown in the figure, primary amines attack imidates nucleophilically to produce an intermediate that breaks down to an amidine at high pHs or to a new imidate at low pH values. Both products can further react with another amino group, thus cross-linking the two components. The cross-linking span is only about 3 Å. Ethyl acetimidate (EA) (Table 8.1.I.B) has been used to cross-link the subunits of Na⁺, K⁺-ATPase resulting a dimeric enzyme with different phosphorylation and dephosphorylation properties.[2] EA also cross-linked tubulin with the formations of dimers, trimers, tetramers, pentamers, and hexamers.[3] Methyl acetimidate (MA, Table 8.1.I.A) has been shown to cross-link erythrocyte membrane proteins[4] and used with isethionyl acetimidate (IAI, Table 8.1.I.C) to study human peripheral lymphocytes in culture.[5] Its antisickling effect was investigating by Chao et al.[6] who treated sickle cell erythrocytes with varying concentrations of methyl acetimidate and found intermolecular cross-linking of hemoglobin to produce high-molecular-weight species. EA and IAI have been used to study the gating currents in voltage-clamped single frog nerve fibers.[7] EA can penetrate a membrane, whereas IAI is membrane impermeant, leading to the observation that IAI was almost without effect, whereas EA caused a considerable reduction of the gating currents.

8.2.2 FORMALDEHYDE

Formaldehyde (Table 8.1.II), capable of reacting bifunctionally, is the simplest of all cross-linking reagents.[8–13] In addition to its monomeric form, formaldehyde can also exist as a stable cyclic trimer (trioxane) and as the polymer paraformaldehyde, which can have 8–100 formaldehyde units. It reacts with water to form formaldehyde hydrate, $H_2C(OH)_2$. Formalin is a 37% solution of formaldehyde and water. "100%" formalin consists of a saturated solution of formaldehyde (roughly 40% by mass) in water. In formalin, it exists as a series of low-molecular-weight polymers.[14] In dilute solutions, it reverts to the monomeric form.[10] Formaldehyde is a good electrophile and can undergo electrophilic addition reactions. It cross-links amino groups in proteins or DNA/RNA through a methylene ($-CH_2-$) linkage, thus yielding protein–protein, protein–DNA, and protein–RNA cross-linked products. Cross-linking reaction involves the attack of an amino group to form a quaternary ammonium salt, which loses a molecule of water to produce an immonium cation (Figure 8.2). This strongly electrophilic cation then reacts with a number of nucleophiles producing a methylene-bridged cross-link.[11] In addition to amines, it reacts with sulfhydryl, phenolic, imidazolyl, indolyl, and guanidinyl groups.[15] Methylene bridges between lysine and tyrosine have been isolated from formaldehyde-treated tetanus and diphtheria toxins.[12] The cross-link distance span is only approximately 2 Å (although commercially available formaldehyde may be polymerized, and, thus, the actual cross-linking distance is unknown).[16] The reaction can be reversed by heating the product to rehydrate the methylene bridge and release formaldehyde. Reversible formaldehyde cross-linking is frequently utilized for analyzing protein–DNA interactions *in vivo*, for example, chromatin structure.[17,18] Han et al.[19] have used the method to map the RNA-binding regions of helicase-like domain of Brome mosaic virus with the help of mass spectrometry. Several peptidyl regions were identified. Hall and Struhl[20] used formaldehyde to study transcriptional activator proteins and found that the VP16 activation domain directly interacts with TATA-binding protein, TFIIB, and the SAGA histone acetylase complex *in vivo*. In conjunction with homobifunctional cross-linkers, such as

---

TABLE 8.1
Monofunctional Cross-Linking Reagents

I. Acetimidates

A. Methyl acetimidate.HCl (MA)

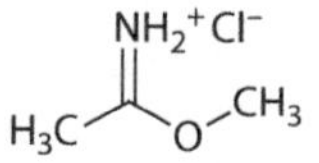

B. Ethyl acetimidate.HCl (EA)

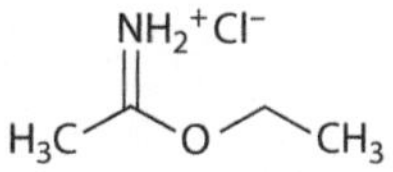

C. Isethionyl acetimidate (IAI)

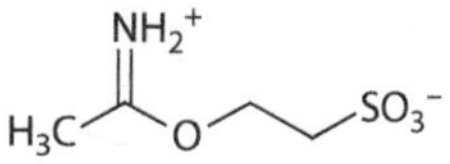

II. Formaldehyde

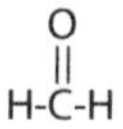

III. Chloroformates

A. Ethylchloroformate

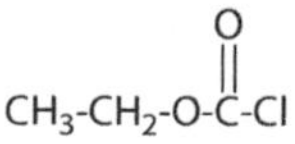

B. *p*-Nitrophenylchloroformate

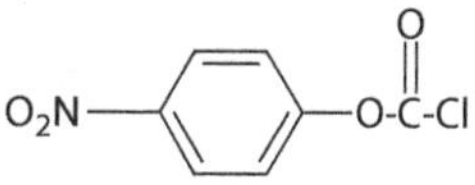

IV. Mercuric Ion

Hg^{2+}

V. Functional Group Modifying Reagents

A. Maleic anhydride

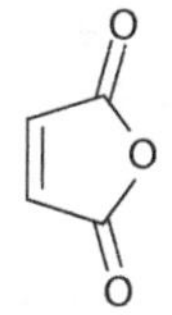

B. Succinic anhydride

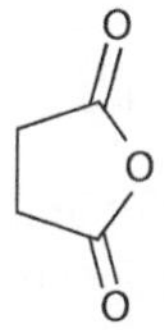

C. 2-Iminothiolane

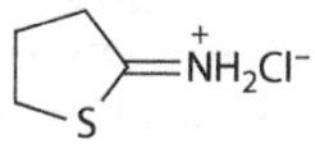

D. Maleimidocarboxylic acids

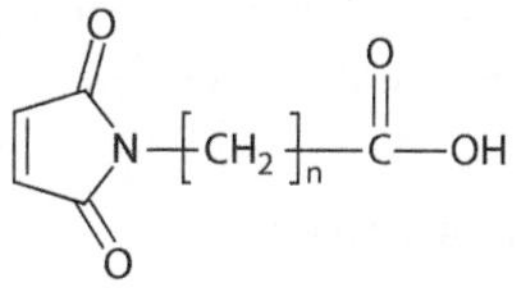

1. n = 2: *N*-β-Maleimidopropionic acid
2. n = 5: *N*-ε-Maleimidocaproic acid
3. n = 10: *N*-κ-Maleimidoundecanonic acid

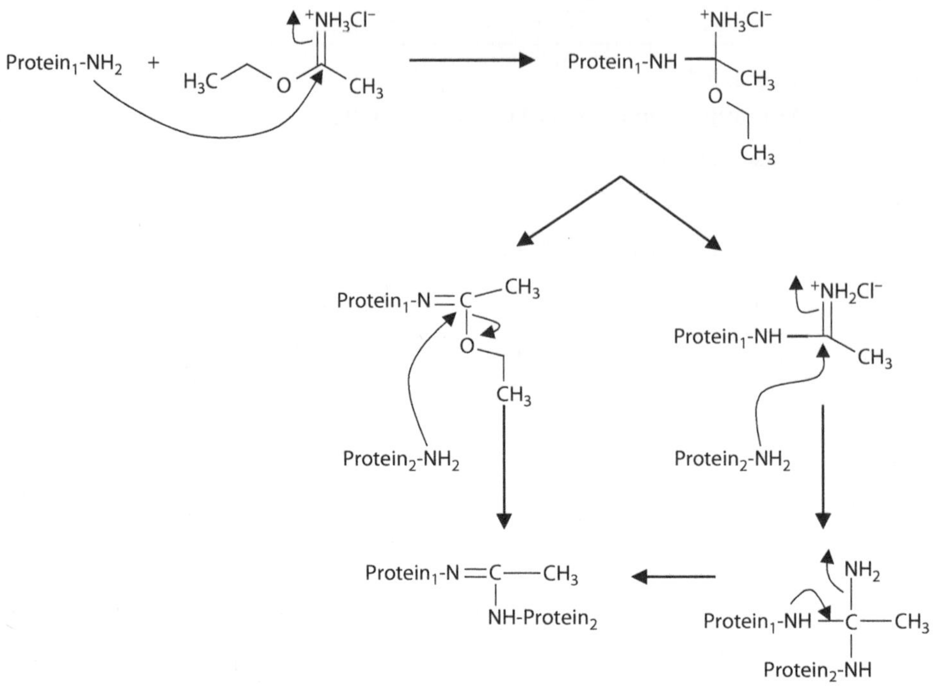

FIGURE 8.1　Speculated cross-linking reaction mechanism of ethyl acetimidate.

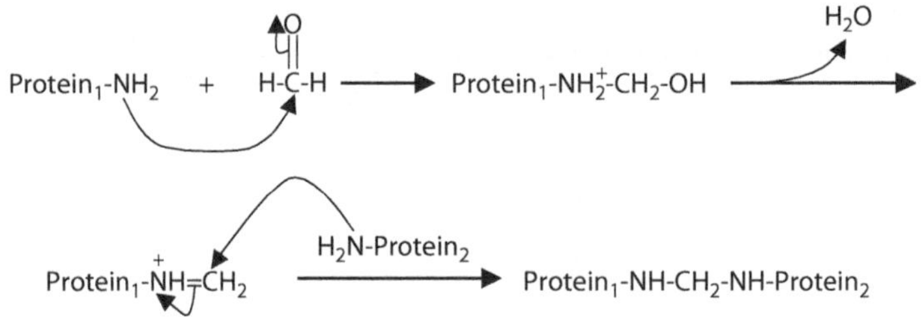

FIGURE 8.2　Cross-linking reaction mechanism of formaldehyde.

dimethyl adipimidate (DMA, compound V in Appendix A) and dimethyl 3,3'-dithiobispropionimidate (DTBP, compound XVII in Appendix A), Zeng et al.[21] used formaldehye to study protein–DNA cross-links by chromatin immunoprecipitation. In this regard, it has been used to examine proteins directly bound to DNA, such as transcription factors and histones, and their interactions.[22–25] The tissue fixation properties of formaldehyde have been applied to crosslink collagen-based constructs for vascular grafts and may be useful for other areas of tissue engineering[26] as well as protein and DNA analysis using microarrays.[27]

8.2.3　CHLOROFORMATES

Ethylchloroformate and *p*-nitrophenylchloroformate (Table 8.1.III) contain two good leaving groups on the same carbonyl carbon, chloride and ethanol. They are monofunctional cross-linkers when the nucleophiles such as amino groups replace the leaving groups sequentially as shown in Figure 8.3. Other chloroformates with a good leaving group would probably undergo similar reactions. The final cross-linked product incorporates a carbonyl group from the chloroformate. These reagents may also function as a zero length cross-linker, which will be discussed in detail below. Avrameas and

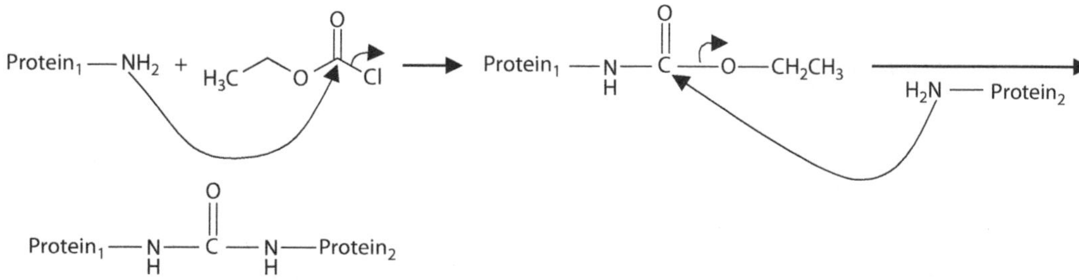

FIGURE 8.3 Speculated cross-linking reaction mechanism of ethylchloroformate.

Ternynck[28] employed ethylchloroformate (Table 8.1.III.A) to polymerize a series of proteins including various antigen and antibodies, including human serum albumin (HSA) and anti-HSA. These proteins were probably cross-linked through amino groups and carboxyl groups, with ethylchloroformate functioning both as a monofunctional and zero-length cross-linker. Ethylchloroformate was also used to polymerize rabbit antiserum against human placental alkaline phosphatase for radioimmunoassays.[29]

8.2.4 Mercuric Ion

Mercuric ion (Table 8.1.IV) reacts reversibly with two sulfhydryl groups, forming a thio-mercuric bridge, $-S-Hg-S-$, thus cross-linking them.[30–33] The first thiol group reacts very quickly followed by a slower reaction with the second group. The reaction of mercuric ion with free thiols is very specific. Cross-linking of two thiol-containing components occurs when the ratio Hg^{2+} to $-SH$ is 0.5 or less. Intramolecular cross-linking has been prepared with reduced papain[31] and pancreatic ribonuclease.[32] Vas and Csanády[34] also showed that the two fast-reacting thiol groups of pig muscle 3-phosphoglycerate kinase can be cross-linked by mercuric chloride. Intermolecular cross-links have been shown with mercaptalbumin[30,33] and *Escherichia coli* DNA polymerase.[35] Because mercuric ion is small and able to penetrate membranes, Soskine et al.[36] were able to cross-link cysteine sulfhydryl groups in the transmembrane domains of the EmrE multidrug transporter with mercuric chloride whereas other thiol-specific reagents were unable to give cross-linked products. Similarly, Hastrup et al.[37] cross-linked endogenous cysteines in transmembrane segments (TMs) of dopamine transporter (DAT) to form dimeric, trimeric, and tetrameric species, suggesting the interaction of various TMs of DAT. Mercuric ion cross-linking of cysteines in fourth transmembrane segment (TM4) in dopamine D2 receptor also provided information that conformational changes at the TM4 dimer interface is part of the receptor activation mechanism.[38]

8.2.5 Functional Group–Modifying Reagents

In Chapter 2, Section 2.3.2.2, the reactions for the conversion of protein amino acid functional groups into other functional groups are discussed. Any reagent used for this purpose is potentially a monofunctional cross-linker. For example, in succinylation and maleylation of a protein to convert an amino group to a carboxylic acid, the reagents used, succinic anhydride and maleic anhydride (Table 8.1.V.A and B), respectively, may be considered cross-linkers if the carboxyl group produced condenses with another protein's amino group, using carbodiimide as a coupling agent to form an amide bond (see below). Obviously, this is a two-step reaction. In thiolation of amino groups, 2-iminothiolane (Table 8.1.V.C) would be a good example. As discussed in Chapter 2, Section 2.3.2.2.2.3, 2-iminothiolane or Traut's reagent has been used for thiolation of proteins converting amino groups to sulfhydryl groups. Used as a cross-linker,[39] the first step in the cross-linking process is amidination of amino groups of proteins to introduce free thiols. The second step consists of the formation of disulfide bonds between the incorporated thiols in the presence of an oxidizing agent such as hydrogen peroxide as shown in Figure 8.4. In this manner, this monofunctional

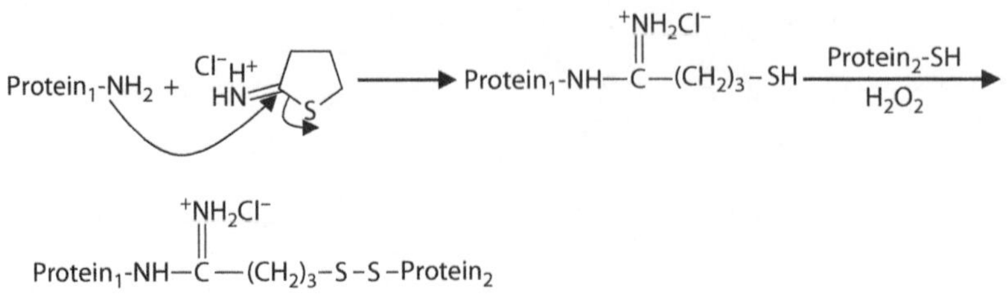

FIGURE 8.4   Reaction sequence of iminothiolane cross-linking of two proteins.

reagent has been used to form cross-links between progesterone-receptor subunits,[40] ribosomal proteins,[39,41,42] and ribosome proteins and RNA.[43]

Another series of group modifying reagents are the maleimidocarboxylic acids (Table 8.1.V.D), which contain one maleimide functional group directed toward sulfhydryl groups. A large homologous series has been synthesized.[44,45] The ones listed in Table 8.1.V.D.1 to 3 are available commercially. On reaction, the sulfhydryl group is converted to the carboxyl group, which can then form amide bonds with amino groups in the presence of a carbodiimide coupling agent as mentioned above.

8.3   ZERO-LENGTH CROSS-LINKING REAGENTS

As mentioned in Section 8.1, zero-length cross-linkers induce a covalent link directly between groups without introducing any bridge atoms. As a result, the cross-linked components are slightly closer together than before conjugation. Before condensation, the reagents activate one of the groups so that the other group can reactive. These reagents are listed in Table 8.2 and will be discussed in detail below.

8.3.1   CARBOXYL GROUP–ACTIVATING REAGENTS

There are many reagents that cross-link carboxyl and amino groups to form amide bonds. For cross-linking of proteins, the most commonly used agents are carbodiimides (Table 8.2.I.A),[46–48] isoxazolium salts (Table 8.2.I.B),[48–52] ethylchloroformate (Table 8.2.I.C),[46,53] diethylpyrocarbonate (Table 8.2.I.F),[54] and carbonyldiimidazole (Table 8.2.I.D).[55,56] A feature common to all these reagents in the mechanism of action is the initial activation of the carboxyl group. The formation of an amide bond or an ester bond facilitated by these reagents proceeds in two steps. In the first step, the reagent forms a highly reactive adduct with the carboxyl group. During the subsequent reaction, nucleophilic attack at the activated species eliminates the activating moiety, resulting in the formation of a bond that does not involve the incorporation of the cross-linking agent. While the amino group is implied as the nucleophile in these reactions, it should be borne in mind that other amino acid nucleophilic side chains of serine, histidine, tyrosine, arginine, and cysteine residues may also be involved in these reactions. In addition to cross-linking of proteins, these compounds have also been used to cross-link nucleic acids and immobilize proteins to solid supports (see Chapter 14). In nonaqueous solutions, many other reagents have been used to activate carboxyl groups. Under this condition, these reagents are used in most cases for peptide synthesis. The reaction conditions are usually too harsh for protein cross-linking. Some of the reagents are used for coupling of proteins to solid matrices, which will be discussed in Chapter 14.

8.3.1.1   Carbodiimides

The basic reaction mechanism of carbodiimides (Table 8.2.I.A) in the modification of carboxyl group has been discussed in Chapter 3, Section 3.4.2. A more detailed reaction mechanism for cross-linking of proteins is depicted in Figure 8.5. In the reaction sequence, the carboxyl group of a protein is first activated by the carbodiimide to form an O-acylisourea intermediate. If an amino group from a second protein reacts with the activated intermediate, the two proteins will be cross-linked through an amide bond without incorporation of any atoms from the carbodiimide. Intramolecular cross-linking may also occur if the amino nucleophile is from the same protein. Without productive

TABLE 8.2
Zero-Length Cross-Linking Reagents

I. Carboxyl Group–Activating Reagents

A. Carbodiimides

1. Dihexylcarbodiimide (DCC)

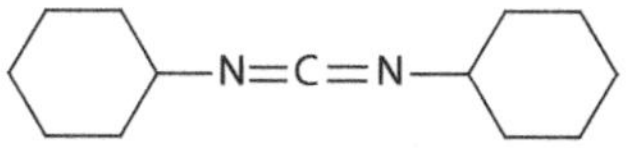

2. 1-Ethyl-3-(3-dimethylaminopropyl)carbodiimide hydrochloride (EDC)

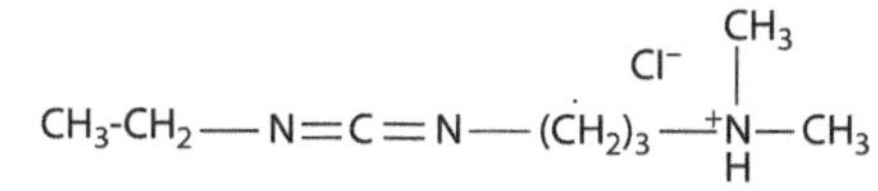

3. 1-Ethyl-3-(4-azonia-4,4-dimethylpentyl)carbodiimide iodide (EAC); 1-ethyl-3-(3-dimethylaminopropyl) carbodiimide methiodide

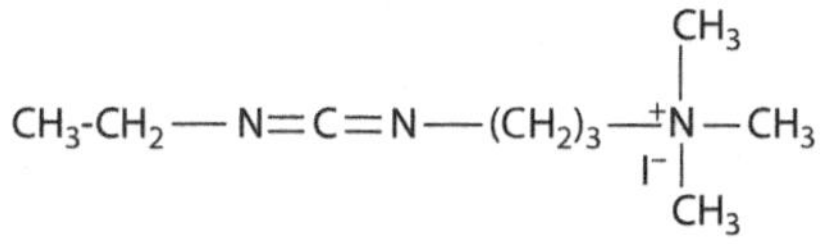

4. 1-Cyclohexyl-3-(2-morpholinoethyl)cardodiimide metho-*p*-toluenesulfonate (CMC); *N*-cyclohexyl-*N*′- [β-(*N*-methylmorpholino)ethyl]carbodiimide *p*-toluene sulfonate

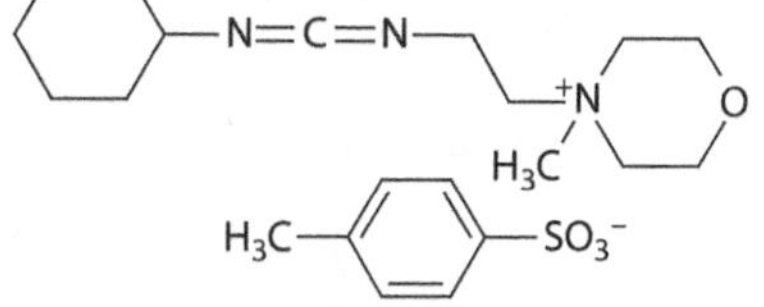

5. *N*-Benzyl-*N*′-3-dimethylaminopropylcarbodiimide hydrochloride; 1-benzyl-3-(3-dimethylaminopropyl) carbodiimide hydrochloride (BDC)

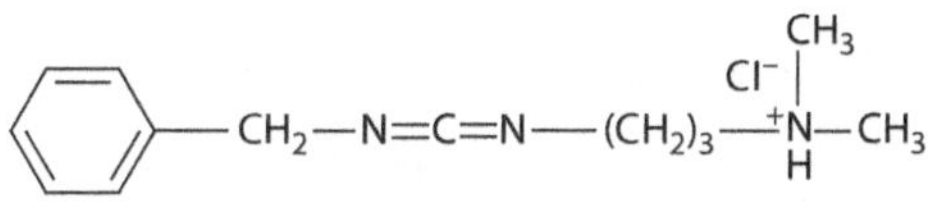

B. Isoxazolium compounds

1. *N*-Ethyl-5-phenylisoxazolium-3′-sulfonate (Woodward's reagent K)

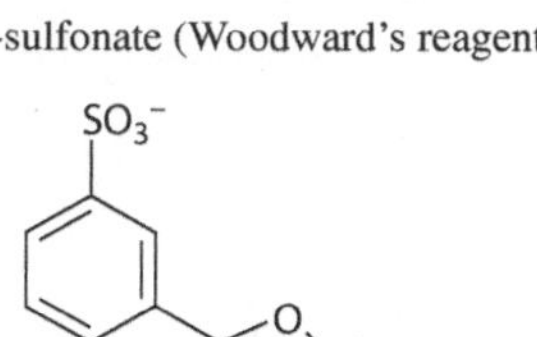

2. *N*-Ethylbenzisoxazolium tetrafluoroborate

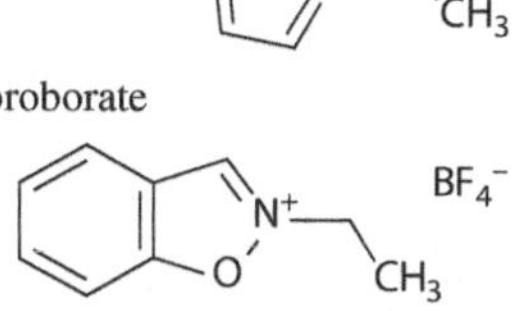

C. Chloroformates

1. Ethylchloroformate

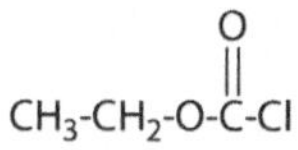

(*continued*)

TABLE 8.2 (Continued)
Zero-Length Cross-Linking Reagents

2. Isobutylchloroformate

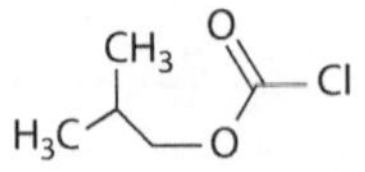

3. *p*-Nitrophenylchloroformate

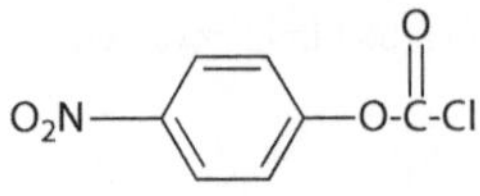

4. 2,4,5-Trichlorophenylchloroformates

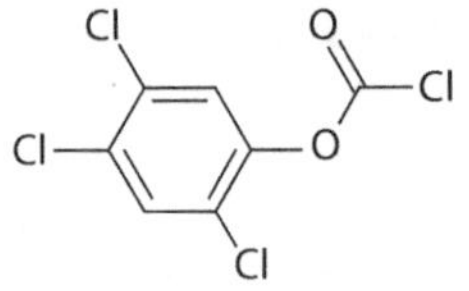

D. 1,1-Carbonyldiimidazole (CDI)

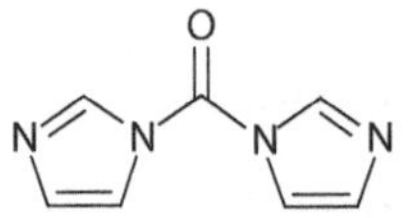

E. Dihydroquinolines
   1. *N*-(Ethoxycarbonyl)-2-ethoxy-1,2-dihydroquinoline (EEDQ)

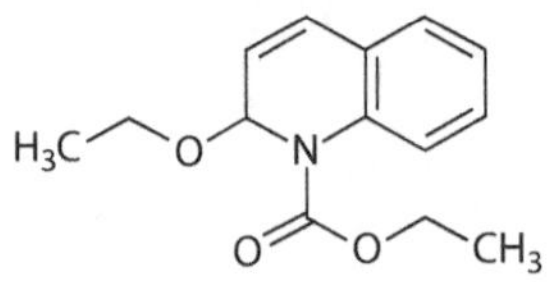

   2. *N*-(Isobutoxycarbonyl)-2-isobutoxy-1,2-dihydroquinoline (IIDQ)

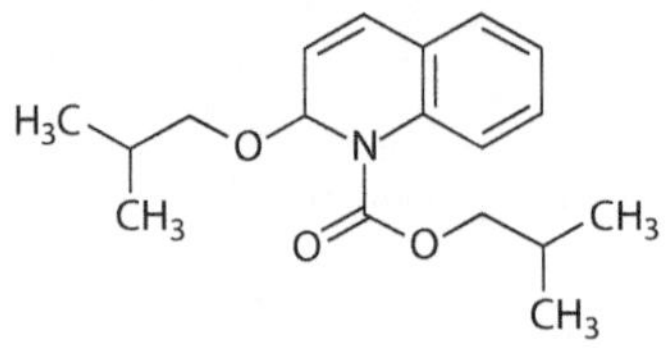

F. Diethylpyrocarbonate (DEPC); diethyl dicarbonate

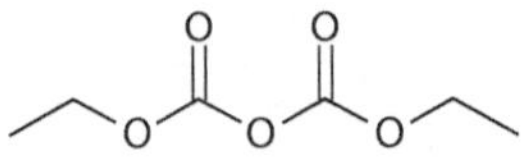

II. Disulfide-Forming Reagents
   A. Cupric di(1,10-phenanthroline) (CuP)

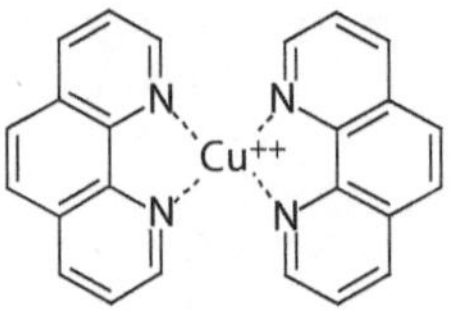

TABLE 8.2 (Continued)
Zero-Length Cross-Linking Reagents

B. Disulfide compounds
1. 5,5'-Dithiobis(2-nitrobenzoic acid) (DTNB), Ellman's reagent

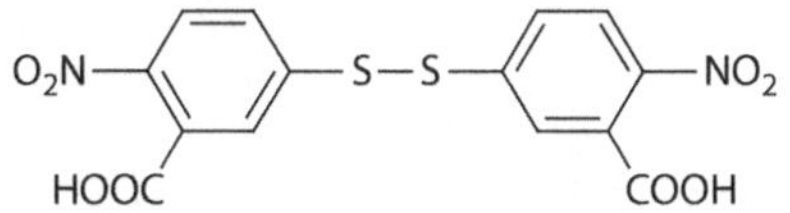

2. 4,4'-Dithiodipyridine (DTP)

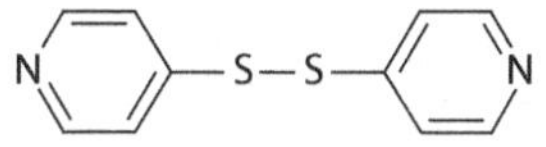

3. 2,2'-Dithiodipyridine

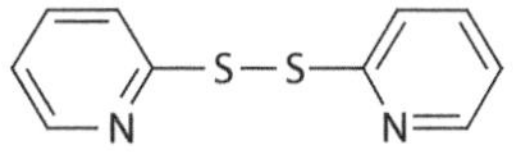

C. Aromatic sulfonyl chlorides
1. 3-Nitro-2-pyridylsulfenyl chloride

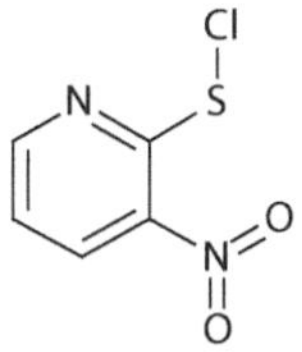

2. 2,4-Dinitrophenylsulfenyl chloride

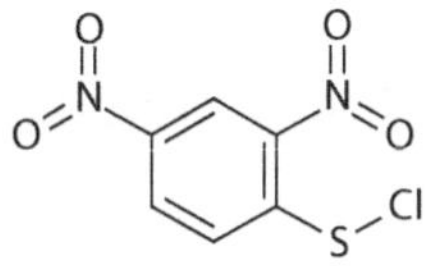

III. Oxidative Cross-Linking Reagents

A. Ni(II).NH$_2$-Gly-Gly-His complex

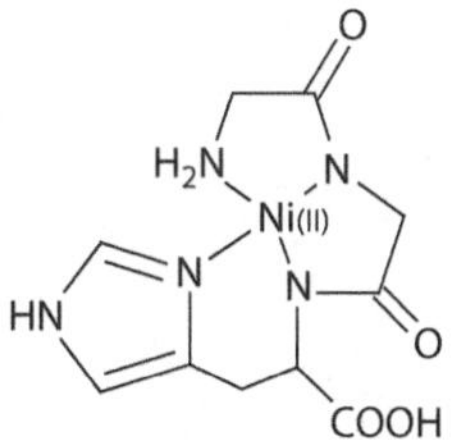

B. Tris-(2,2'-bipyridyl)ruthenium (II) dication, (Ru(bipy)$_3^{2+}$)

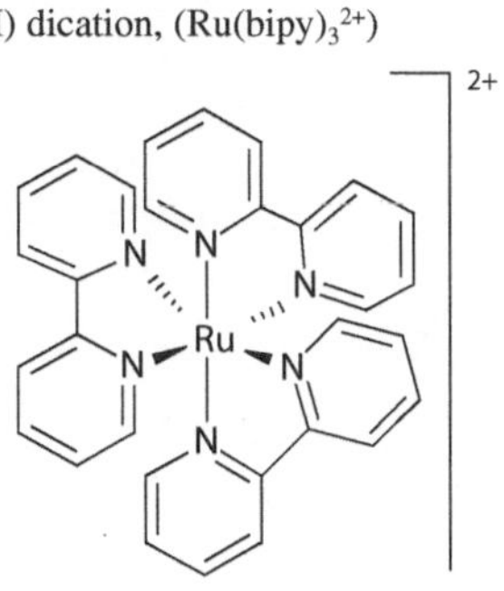

(continued)

TABLE 8.2 (Continued)
Zero-Length Cross-Linking Reagents

C. Palladium(II)-tetra(*p*-sulfonatophenyl)porphyrin

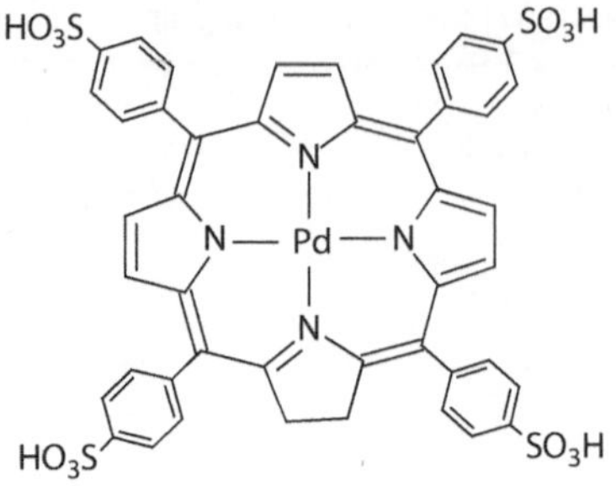

D. Chloro-manganese(III)-tetra(*p*-sulfonatophenyl)porphyrin

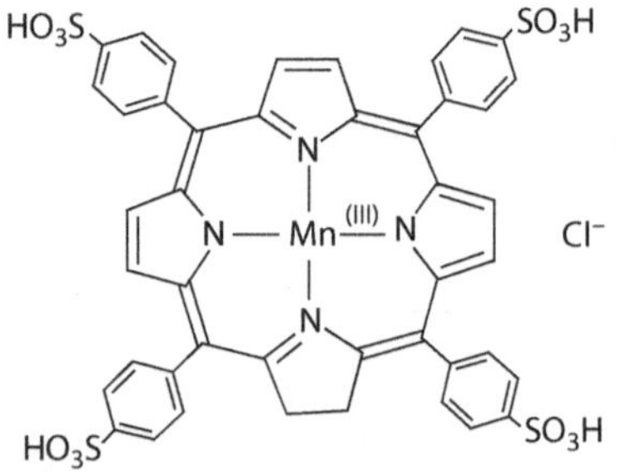

IV. Miscellaneous Reagents

A. Tetranitromethane (TNM)

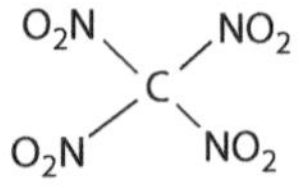

B. Potassium nitrosyl disulfonate (NDS)

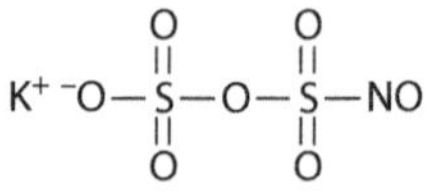

C. Bisulfite, hydrogen sulfite HSO_3^-

V. Enzymes as Zero-Length Cross-Linkers

A. Glutaminase

B. Tyrosinase

C. Peroxidase

D. Xanthine oxidase

E. Laccase

F. Lysyl oxidase

G. Superoxide dismutase

H. Catalase

VI. Radiation as Zero-Length Cross-Linker

A. Ultraviolet radiation

1. UV lamp

2. UV laser

B. Gamma ray irradiation

C. Electron beam irradiation

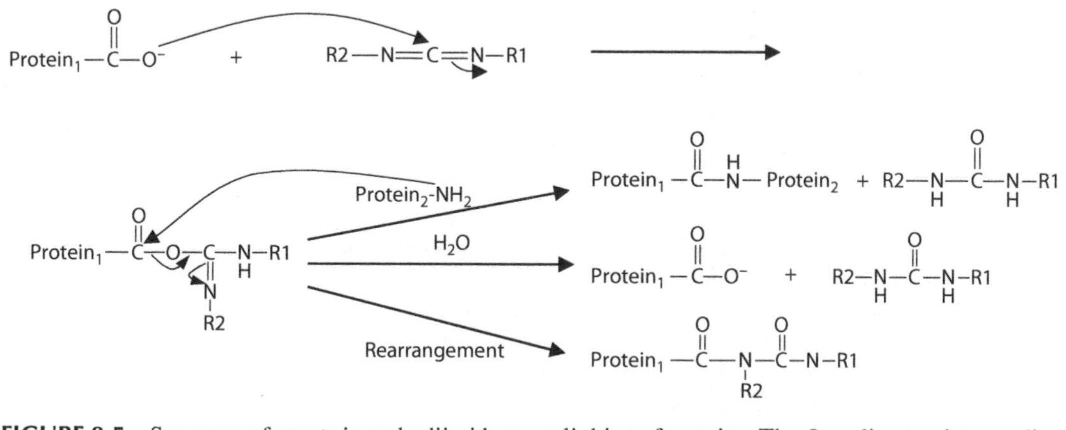

FIGURE 8.5 Sequence of events in carbodiimide cross-linking of proteins. The *O*-acylisourea intermediate may react with an amino group of a second protein to form a cross-linked adduct (top), hydrolyze to regenerate the original protein carboxyl group (middle), or rearrange to form a stable *N*-acylurea (bottom).

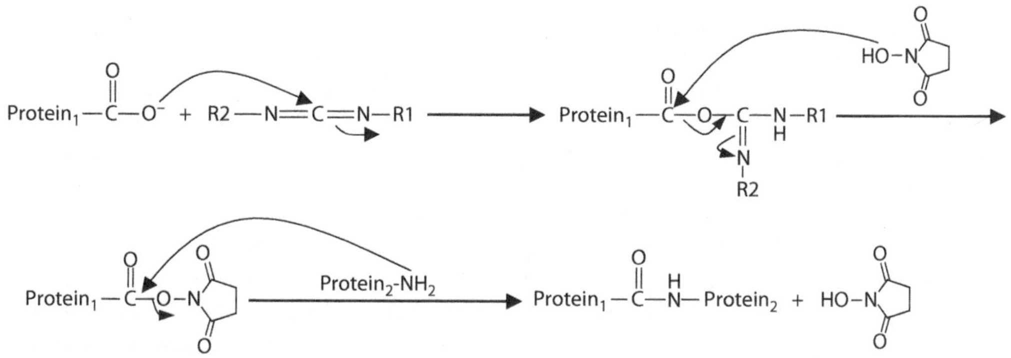

FIGURE 8.6 Carbodiimide and *N*-hydroxysuccinimide two-step cross-linking of proteins. The *O*-acylisourea intermediate reacts with *N*-hydroxysuccinimide to form succinimidyl ester, which can be isolated and react with an amino group of a second protein to form cross-linked product.

nucleophilic reaction, the *O*-acylisourea may undergo hydrolysis to regenerate the free carboxyl group, or undergo an intramolecular O to N-acyl shift to form a more stable *N*-acylurea.[47]

Another procedure for activating protein carboxyl group with carbodiimides is converting it into a succinimidyl ester in the presence of *N*-hydroxysuccinimide as shown in Figure 8.6.[57] This activated ester can be isolated and reacted with an amino group of another protein to form cross-linked amide bond. The advantage of this two-step method is reduction of cross-links among several proteins of a multicomponent complex. This approach has been employed to cross-link troponin C and troponin I,[58] as well as other biological materials.[59–62]

A large number of carbodiimides are available commercially. In addition to EDC and CMC (Table 8.2.I.A.2 and 4), other cardodiimides, commonly used in protein cross-linking because of their commercial availability, are listed in Table 8.2.I.A.1 to 5. The water-soluble reagents are usually the preferred choices. These compounds have different stabilities in aqueous solutions. EDC (Table 8.2.I.A.2), for instance, has a half-life of 37h at pH 7 whereas EAC (Table 8.2.I.A.3) is about tenfold less stable. Carbodiimides are not absolutely specific toward protein carboxyl groups and may react with other nucleophiles like thiol, imidazol, phenolic, and hydroxyl groups; hence, phosphate and phosphate-containing reagents, as well as hydroxylamine and other amine derivatives, increase the rate of loss of carbodiimides, in some cases dramatically.[63] As the amino acid residues in a protein may be affected by pH, different pH conditions may yield adducts with different properties. For example, at different pHs, EDC yields cross-linked collagen with different thermomechanical, tensile, and shear mechanical properties.[64]

An extensive application of these carbodiimides can be found in the literature. For example, when DCC (Table 8.2.I.A.1) was incubated with calmodulin and myosin light chain kinase, a cross-linked product was produced only in the presence of Ca^{2+}, indicating that the association of calmodulin was sufficient to stimulate kinase activity.[65] DCC is also used to conjugate enzymes and antibodies for immunohistology use.[66] When ubiquitin was treated with EDC, three cross-links were found.[67] Together with amine-selective homobifunctional reagents, EDC was used to investigate protein–protein associations within *Chlamydomonas* flagellar dynein, which is a highly complex enzyme containing up to 15 different protein components and consisting of several distinct domains.[68] Such an approach revealed new interactions between the dynein arms and other components of the Flagellar axoneme. While EDC is the most commonly used reagent, this and other soluble carbodiimides have been used to cross-link heavy meromyosin to F1 actin and between heavy meromyosin beads,[69] F-actin and myosin subfragment,[70] troponin C and troponin I,[71] F1-ATPase,[72] F1F0 ATP synthase,[73] ribosomal proteins and RNA,[74] voltage-gated K+ channel subunits,[75] cytochrome *c* and cytochrome *b*5,[76,77] factor VIIIa,[78] *E. coli* Fpg protein,[79] and photosystem I polypepetides.[80,81] EDC was also found to induced interchain cross-linking within double-stranded DNA.[82] In addition to the study of protein–protein and protein–nucleic acid interactions, the carbodiimides have been applied to protein engineering. For example, EDC has been used to cross-link collagen to improve mechanical and biological properties for vascular grafts.[26] Collagen has also been cross-linked to hyaluronic acid to form cross-linked type II collagen/hyaluronate–chondroitin-6-sulfate scaffolds for tissue engineering as a delivery vehicle,[60,62] to polyamidoamine dendrimer to improve the biostability and structural integrity,[83] and to fibroin to form fibroin/collagen hydrogels for tissue engineering.[84] Other macromolecules such as poly(ethylene glycol) were covalently attached to ferritin surfaces,[85] and proteins were conjugated to carbon nanotubes for various biological applications.[86]

8.3.1.2 Isoxazolium Compounds

N-ethyl-5-phenylisoxazolium-3′-sulfonate (Table 8.2.I.B.1) was first synthesized by Woodward et al.[87,88] and is thus known as Woodward's reagent K (WRK). Under alkaline conditions at room temperature, the reagent is converted to a reactive ketoketenimine, which reacts with the carboxylate anion to form an enol ester intermediate.[88,89] This reactive intermediate then reacts with an amino group to form the cross-linked product as shown in Figure 8.7.[90]

WRK is not stable in aqueous medium as it is hydrolyzed rapidly above pH 3.0. The rate constant for the conversion of the isoxazolium salt to the ketoketenimine at pH 5.8 is 0.44 min and the subsequent reaction step may occur at the same rate.[91] Nucleophilic side chains of proteins other than carboxylate such as sulfhydryl, imidazole, and amino groups could easily react with the ketoketenimine intermediate to form relatively stable enamine adducts.[92] Thus, although the reagent is used mostly as a carboxyl group–modifying agent, it was used to inactivate enzymes through the reaction with protein residues such as histidines.[93–95] As a cross-linking reagent,[48] it has also been used to prepare polymeric α-chymotrypsin,[96] and conjugate hemin and IgG,[97] and to label glutathione S-transferase with bilirubin,[98] to give but a few examples.

A related compound, *N*-ethylbenzisoxazolium tetrafluoroborate (Table 8.2.I.B.2), has also been shown to condense carboxyl and amino groups to form amide bonds. As in the case of WRK, ethylbenzisoxazolium decomposes to ethylbenzoketoketenimine, which reacts with various nucleophiles including the carboxylate ion.[50] The initial product of the reaction with the carboxyl group rearranges very rapidly to form an enol ester, which can acylate an amino group to form an amide bond. Although the reagent was used in peptide synthesis,[87] its application in protein cross-linking has not been extensive due to the availability of WRK. This compound was used to cross-link synthetic polypeptides,[99] angiotensin, and polylysine.[100]

8.3.1.3 Ethylchloroformate

As discussed above, ethylchloroformate and *p*-nitrophenylchloroformate (Table 8.2.I.C.1 and 3) function as homobifunctional cross-linkers but can react as zero-length cross-linking agents.

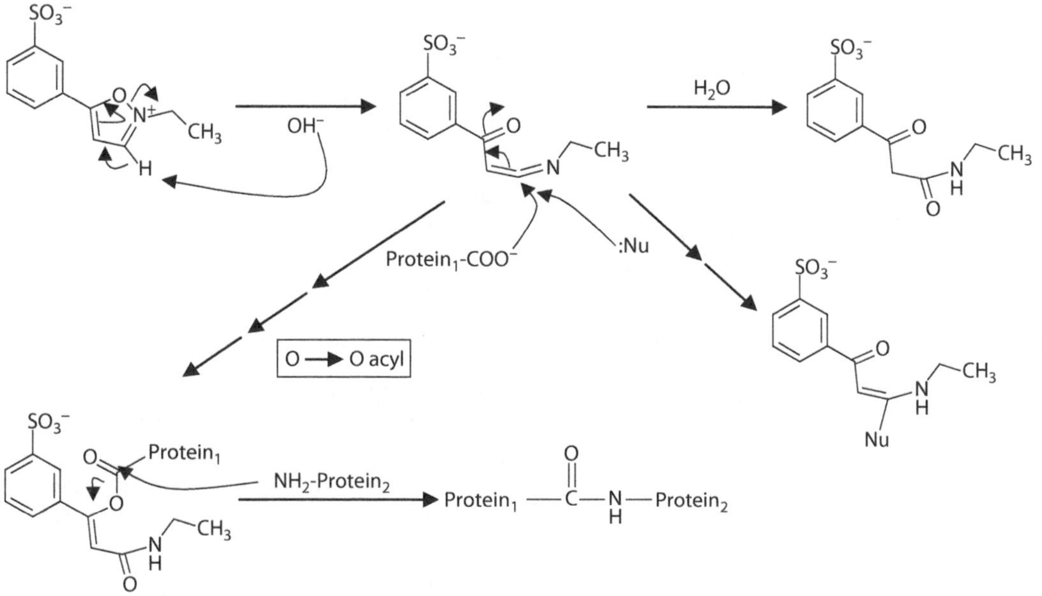

FIGURE 8.7 Speculated reaction mechanism of Woodward's reagent K. At slightly alkaline pH, Woodward' reagent K forms a ketoketenimine intermediate. The ketoketenimine can undergo hydrolysis, react with a nucleophile (:Nu) to form a stable enamine, or react with a protein carboxyl group to form an enol ester, which reacts with an amino group of a second protein to form a cross-linked product.

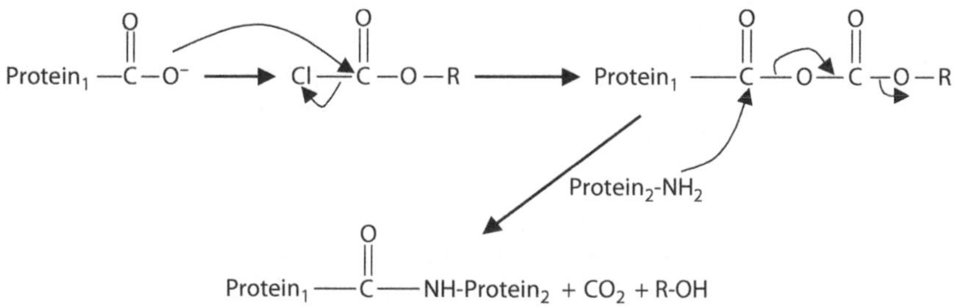

FIGURE 8.8 Cross-linking reaction mechanism of chloroformates as a zero-length agent.

When the carboxyl group attacks the reagent first instead of the amino group, a mixed acid anhydride is formed as shown in Figure 8.8. This active anhydride intermediate then transfers the carboxyl group to a nucleophile, such as an amino group to form an amide bond. Because of the reactivity of the anhydride, hydrolysis is the major competing reaction.

While these reagents have been used to prepare insoluble protein polymers[28] and the coupling of fluorescent or paramagnetic probes to gangliosides,[101] they have also been used for the immobilization of proteins.[102] In a similar manner, isobutyl chloroformate (Table 8.2.I.C.2) has been used for the synthesis of various conjugates.[103–106] Other chloroformates such as 2,4,5-trichlorophenyl chloroformate (Table 8.2.I.C.3) are also potential zero-length cross-linkers.[107]

8.3.1.4 Carbodiimidazole

1,1-Carbonyldiimidazole (CDI, Table 8.2.I.D) is another carboxyl group activating zero-length cross-linker. When the carboxyl group attacks the carbonyl carbon of CDI, a mixed anhydride is formed as shown in Figure 8.9. Subsequent reaction with an amino group will lead to an amide bond formation without the incorporation of any atoms from CDI.

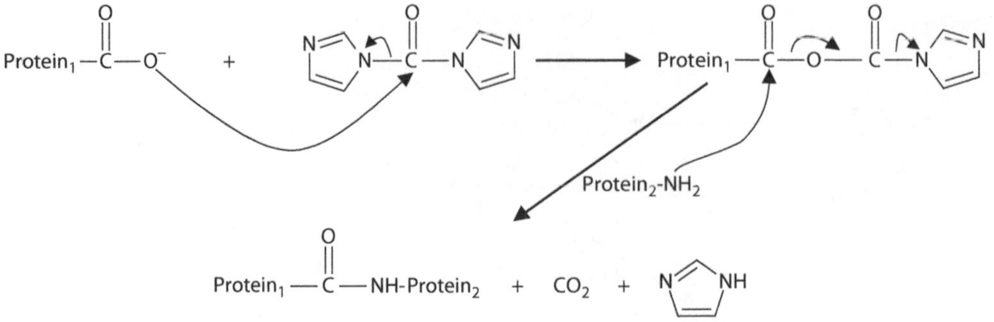

FIGURE 8.9 Cross-linking reaction of carbonyldiimidazole as a zero-length reagent.

CDI has been used to synthesize affinity probes by coupling two organic compounds[108] and for the formation of oligoglutamate.[109] It has also been used to cross-link collagen in the presence and absence of other molecules[56,110] and horseradish peroxidase.[111] The major application of CDI is in the immobilization of proteins and other macromolecules onto solid supports. For example, watermelon α-galactosidase was immobilized on chitin and a synthetic anion-exchange (Amberlite IRA-938) support with CDI first activating the solid matrix followed by addition of the enzyme.[112] CDI is also used in coupling oligonucleotides to cellulose acetate membrane,[113] poly(ethylene glycol) to ferritin,[114] and immobilization of peptides onto polyethylene film.[115]

8.3.1.5 *N*-Alkoxycarbonyl-2-Alkoxy-1,2-Dihydroquinolines

N-(ethoxycarbonyl)-2-ethoxy-1,2-dihydroquinoline (EEDQ, Table 8.2.I.E.1) was developed by Belleau et al.[116] as a carboxyl group reagent for the study of adrenergic receptors. This and another reagent, *N*-(isobutoxycarbonyl)-2-isobutoxy-1,2-dihydroquinoline (IIDQ, Table 8.2.I.E.2), also function as zero-length cross-linkers by activating the carboxyl group through a mixed anhydride as shown in Figure 8.10. The carboxyl group first forms an acyldihydroquinoline, which rearranges to a mixed anhydride. Reaction of the anhydride with an amino group leads to the formation of an amide bond.

While EEDQ is commonly used to access the carboxyl group function at the active centers of enzymes, it has also been used for chemical cross-linking of the rigor complex between F-actin and the skeletal myosin S-1.[117] The covalent attachment of F-actin to the S-1 heavy chain induced an elevated Mg^{2+}-ATPase activity. The spatial relations between the various regions of the S-1 molecule and the nucleotide- and actin-induced intramolecular movements were also studied by EEDQ.[118] In addition, the zero-length cross-linker was used to investigate molecular interactions between alpha-synuclein and lipids.[119] Lipids such as phosphatidic acid, phosphatidylinositol,

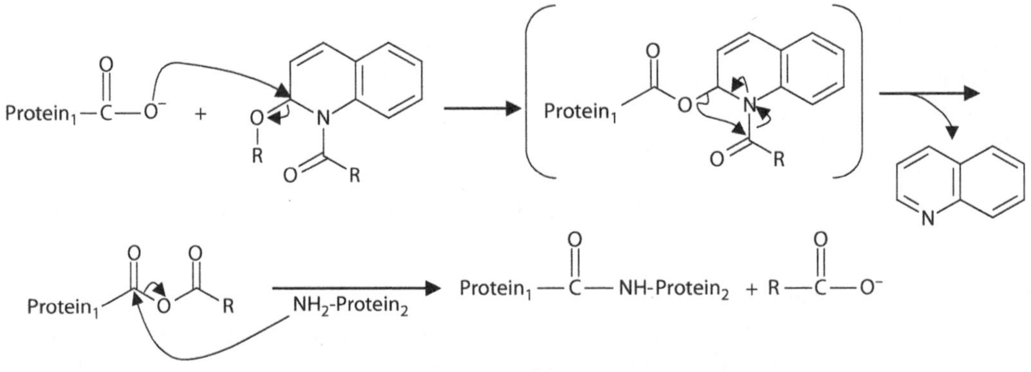

FIGURE 8.10 Cross-linking reaction of *N*-alkoxycarbonyl-2-alkoxy-1,2-dihydroquinoline as a zero-length reagent.

phosphatidylserine, phosphatidylethanolamine, and even arachidonic acid induced protein self-oligomerization. The reagent was also used for immobilization of enzymes such as horseradish peroxidase to solid support by first activating the matrix.[120]

8.3.1.6 Diethylpyrocarbonate

Diethylpyrocarbonate (DEPC, Table 8.2.I.F) is widely used as a protein-modifying reagent, which preferentially reacts with histidines more than with cysteine, lysine, tryptophan, and tyrosine residues and thus inactivates enzymatic activities in which the imidazole ring is involved.[121] However, cross-linking of multimeric protein structures upon DEPC treatment has been observed. Rexin et al.[122] demonstrated cross-linking of glucocorticoid receptors to form high-molecular-weight moieties. While it is not known which amino acid groups are the targets for reaction with DEPC, it has been proposed that a putative N-carboxyanhydride existed as an intermediate that might undergo transamidation with carboxyl groups to yield amide cross-links.[54] It is likely that the carboxyl group of glutamate or aspartate may react with the reagent first to form a mixed anhydride, which leads to an amide formation with an amino group of lysine, a reaction mechanism similar to the other carboxyl group–activating reagents mentioned above.

8.3.2 REAGENTS FOR DISULFIDE FORMATION

Proteins with sulfhydryl groups can be cross-linked by the simple expedient of oxidation of the free sulfhydryl groups to form disulfide bonds. Such linkages may occur intermolecularly or intramolecularly. Various oxidizing agents have been used for this purpose. These oxidizing agents are not incorporated into the disulfide bonds and are therefore zero-length cross-linkers. Although agents such as air,[123,124] iodide,[125] and hydrogen peroxide[124] have been demonstrated to be useful in forming disulfide bonds, cupric bis(1,10-phenanthroline) (CuP, Table 8.2.II.A) is a more commonly used reagent.[1] It has been used to study membrane proteins where micromolar concentrations suffice to complete the reaction in few minutes. With bovine erythrocytes, the membrane proteins were cross-linked where the cross-linking of Band 3 protein and of spectrin increased with increasing erythrocyte age.[126] With skeletal muscle sarcoplasmic reticulum vesicles, the passive permeability to Ca^{2+} ions was drastically increased upon addition of CuP due to protein cross-linking induced by oxidation of SH groups to disulfide bridges.[127] CuP has also been applied to investigate protein structures. For example, in rat liver cytochrome c oxidase, seven polypeptides (I–III, Va, Vb, VIIb and VIII) were cross-linked with each other and with itself, indicating the occurrence of free –SH groups in these polypeptides.[128] In the investigation of the structure of E. coli ATP synthase complex, a cysteine was site-directed into the c subunit. Oxidation with CuP revealed cross-linking of the c subunit to both the gamma and epsilon subunits showing the proximity of these subcomponents.[129] CuP also induced cross-linking of the alpha and beta subunits of the pyridine nucleotide transhydrogenase of E. coli by catalyzing the formation of interchain disulfide bonds.[130] Cross-linked dimers of alpha$_2$, alpha-beta and beta$_2$, and the trimer alpha$_2$-beta were obtained. A small amount of tetramer, probably alpha$_2$-beta$_2$, was also formed.

Another group of disulfide-forming reagents are disulfide compounds and arylsulfenyl chlorides.[131] Representatives of these compounds are shown in Table 8.2.II.B and C, respectively. There are many disulfide compounds available, but Ellman's reagent and dithiodipyridines (Table 8.2.II.B) are the most commonly used reagents. As mentioned in Chapter 3, Section 3.2.4, these compounds react with protein sulfhydryl groups to form activated protein mixed disulfides through a disulfide exchange reaction.[131] A second disulfide exchange reaction with another sulfhydryl group completes the cross-linking reaction as shown in Figure 8.11. As can be seen, this is a two-step reaction. For sulfonyl chlorides, the first step is not a disulfide exchange reaction, but simply a replacement of the chloride with a protein thiol group. The advantage of using Ellman's reagent is that the reaction can be followed spectrophotometrically, since the thionitrobenzoate anion released absorbs light at 412 nm. Another advantage is that free sulfhydryl groups can be introduced into proteins as

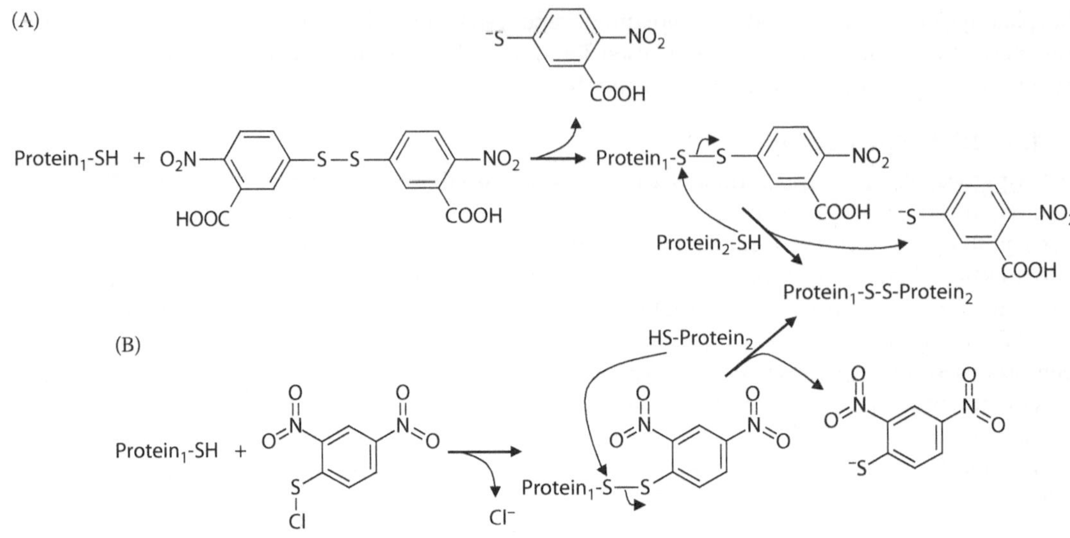

(A)

(B)

FIGURE 8.11 Two-step reaction procedure for cross-linking of proteins with free sulfhydryl groups through a disulfide bond using: (A) Ellman's reagent and (B) 2,4-dinitrophenylsulfenyl chloride.

discussed in Chapter 2. The disulfide cross-linking procedures have been used to introduce small molecules such as glutathione[131] into proteins for the creation of polymer hydrogels using thiolated poly(methacrylic acid),[132] to crosslink troponin-C and troponin-I,[133] for the inactivation and cross-linking of porcine heart thiolase I,[134] the formation of a disulfide link between two monomers of human brain GABA transaminase,[135] and for other applications.

8.3.3 OXIDATION CROSS-LINKING REAGENTS

Oxidative cross-linking of proteins is a methodology that uses low-valent metal complexes to cross-link proteins.[136] These low-valent metal complexes are elevated, through chemical or photochemical oxidation, to higher-valent complexes, which create protein-centered radicals that lead to cross-linked products. The reaction is rapid and with high efficiency in many cases. The metal complex cross-linking reagents are not incorporated into the product, but merely initiate direct bond formation between two residues at the protein–protein interface. Examples of lower-valent metal complexes are Gly-Gly-His–Ni^{2+} (Table 8.2.III.A),[137] tris-(2,2′-bipyridyl)ruthenium(II) dication (Ru(bipy)$_3$$^{2+}$, Table 8.2.III.B),[138,139] and palladium(II) porphyrin (Table 8.2.III.C).[139] In the presence of a peracid or magnesium mono-perphthalate, the peptide–metal complex, Gly-Gly-His–Ni^{2+}, mediates the cross-linking of closely associated proteins though a reactive intermediate thought to be a high-valent Ni(III) species, which is rapidly reduced by a tyrosine residue. The resultant tyrosyl radical then couples with another tyrosine, leading to the formation of a stable bityrosyl adduct.[137] The reaction mechanism is similar to that of tris-(2,2′-bipyridyl)ruthenium(II) dication and palladium(II) porphyrins. When these lower-valent species are subjected to photochemical oxidation by visible light in the presence of electron acceptors such as ammonium persulfate (APS), they give rise to oxidized metals. These oxidative reactions are thought to involve the generation of tyrosine, tryptophan, or cysteine radicals on peptides or proteins through one-electron oxidation of accessible tyrosine, tryptophan, or cysteine residues by the oxidized metal, followed by loss of a proton. These tyrosine, tryptophan, or cysteine radicals react with the same residues located on the interacting protein partner as shown in Figure 8.12 for tyrosine to form cross-linked adducts.[136] Because of the various radicals that can form on the protein, there are a variety of pathways by which covalent adducts could form. The system shows very little light-independent oxidation of protein residues and significant perturbation of the protein prior to the brief photolysis period does not occur.

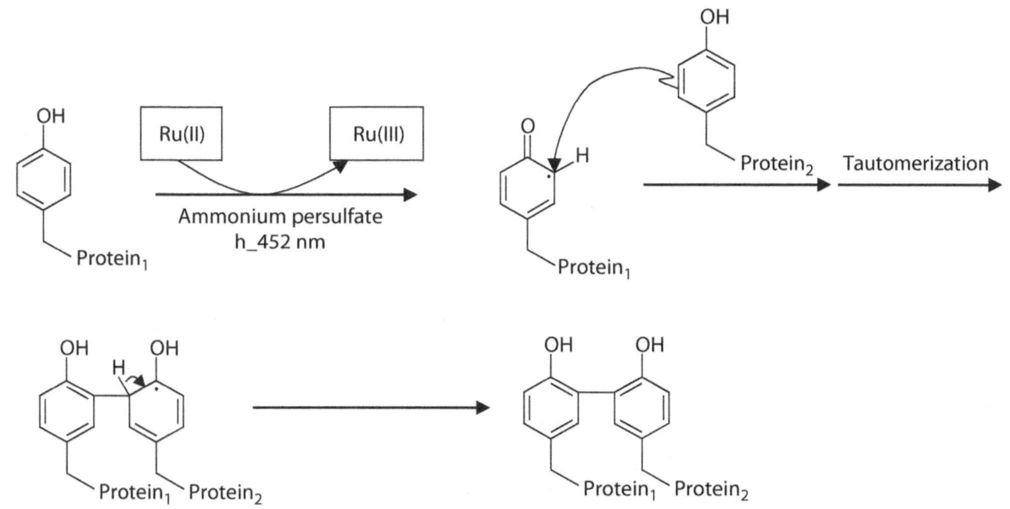

FIGURE 8.12 Speculated oxidative cross-linking reaction of $Ru(II)(bpy)_3^{2+}$ in the presence of ammonium persulfate and light in forming di-tyrosine cross-links of proteins.

It can be employed to analyze protein–protein interactions in crude cell extracts. For example, the acidic activation domain of the yeast Gal4 transcription factor and the Gal80 repressor were cross-linked in the presence of $Ru(bipy)_3^{2+}$ and APS when the sample was photolyzed for 0.5 s.[139] A similar result was obtained for 20S proteasome core complex, where many of the 14 different proteins in this 28-protein complex were specifically cross-linked to their nearest neighbors.[138] Oxidative cross-linking was also obtained with G-protein-coupled receptors in which intermolecular homo- or heterodimerization was detected.[136]

In addition to the Ni(II), Ru(II), and Pd(II) complexes, manganese or iron porphyrins also mediate cross-linking of closely associated proteins in the presence of oxidants such as peroxide and peracids.[140] When the dimeric protein glutathione-S-transferase was treated with chloromanganese(III)-tetra(p-sulfonatophenyl)porphyrin (Table 8.2.III.D) in the presence of the oxidant $KHSO_5$, new cross-linked species were formed. Similarly, TATA-binding protein, which is known to form stable dimers, was cross-linked to yield a mixture of oligomers, and the phage T4 UvsY protein was cross-linked forming large covalent multimers. The data obtained demonstrate that such oxidation cross-linking can be generally applied to biological systems and extends the traditional bifunctional cross-linking reagents.

8.3.4 CARBOHYDRATE ACTIVATION REAGENTS

Reductive alkylation between aldehydes derived from carbohydrates and amino groups is also a zero-length cross-linking process. The Schiff base formation does not incorporate any extrinsic atoms. Thus, reagents that oxidize vicinal diols of carbohydrates to dialdehydes may be regarded as zero-length cross-linkers in the coupling of proteins to carbohydrates or glycoproteins. The most commonly used mild oxidation agent is sodium periodate as discussed in Chapter 2, Section 2.3.2.4. Reagents used for stabilization of the Schiff base are also discussed in that chapter. Periodate oxidation has been applied to cellulose dispite its crystalline nature, which formed dialdehyde groups in the longitudinally spaced, bandlike domains.[141] In addition, it has been used to couple nucleosides to proteins. Each of the dialdehyde groups of a periodate-oxidized nucleoside is able to couple to lysine residues on different protein molecules through Schiff bases, thereby cross-linking different protein molecules.[142] Periodate oxidation followed by reductive alkylation has been used to prepare immunoconjugates of horseradish peroxidase for immunoassays,[143] to glycosylate penicillin G acylase with mannan,[144] and to form ferritin–avidin conjugates for electron microscopic cytochemistry.[145]

8.3.5 ENZYMES AS ZERO-LENGTH CROSS-LINKERS

In addition to chemical cross-linking, enzymes have been found to catalyze protein cross-linking. As catalysts, they lower the activation energy of the reaction. The end products are direct linkage between the substrates. There are numerous enzymes that have been found to induce protein cross-linking. The important ones are discussed below.

8.3.5.1 Transglutaminase

Transglutaminases (EC 2.3.2.13) are calcium-dependent enzymes that catalyze the incorporation of various exogenous amines into proteins through an acyl transfer reaction between γ-carboxamides of a peptide-bound glutamyl residue and various amine substrates.[146–150] Transglutaminase is ubiquitous with the tissue form being the most widespread member of this family. The enzyme has been isolated from liver, platelet, hair follicle, prostate, epidermis, umbilical vein endothelial cells, erythrocytes, and plants.[150] It belongs to a superfamily of cysteine proteases possessing the catalytic triad of Cys-His-Asp/Asn. The mechanism of action involves two transamination reactions as summarized in Figure 8.13.[151] In the first step, the active site thiol residue reacts with a glutamyl residue in the protein substrate, releasing ammonia and activating the glutamine acyl moiety. In the following step, the active thioester undergoes an acyl transfer to a primary amine of a second substrate completing the coupling reaction.[150] If the amino group is from a lysine residue in a protein, an amide bond is formed, yielding a protein–protein cross-linked product as shown in Figure 8.13. The substrate specificity for the glutamine is high and only protein-bound glutamines are cross-linked such as those in fibrin,[152] fibrinogen and other blood components,[153] whey protein,[154] gelatin,[155] type II collagen,[156] and numerous others.[157] On the other band, the substrate specificity for the amine donor is much less stringent and transglutaminase has been used for attachment of a number of primary amine-containing labels and probes into proteins such as cleavable diamine, N,N'-bis(β-aminoethyl)tartramide,[158] and putrescine (1,4-diaminobutane)–pectin conjugate, which in combination with soy flour proteins form edible films.[159] Transglutaminase as a zero-length protein cross-linker have been applied to medicine, food processing, and tissue enginerring.[160] Gelatin-transglutaminase gel, for example, has been used for cell delivery,[155] and cross-linked porous hydroxyapatite/collagen composites for bone tissue regeneration.[161]

8.3.5.2 Tyrosinase

Tyrosinases (EC 1.14.18.1) are copper-containing enzymes with flexible binuclear copper catalytic sites.[162] They are widely distributed in nature and have been isolated from bacteria, fungi, plants, insects, and from higher animals.[163–165] Their biological function is related to the biosynthesis of

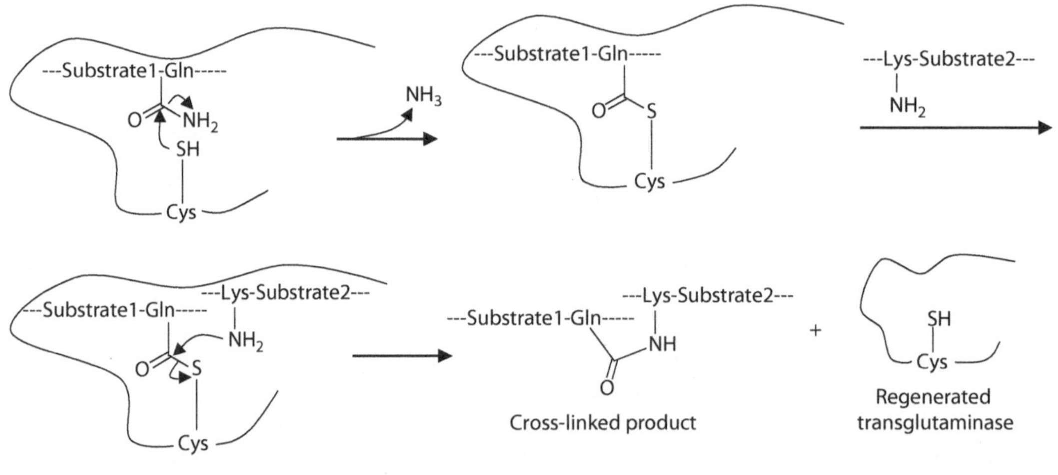

FIGURE 8.13 Catalytic reaction mechanism of transglutaminase.

melanin pigments where monophenolic tyrosine is oxidized via diphenolic L-dihydroxyphenyl ala-nine to dopaquinone, which is very reactive and can further polymerize to produce mixed melanins.[166] Tyrosinase-catalyzed oxidation and cross-linking of tyrosine-containing peptides takes place when the tyrosine residue is located in the N-terminus or in the middle of the peptide.[167] The accessible tyro-sine residues of proteins are oxidized into reactive o-quinone moieties. These quinone residues can undergo nonenzymatic reactions with available nucleophiles such as the nucleophilic amino groups of carbohydrates. They have been shown to induce cross-linking of whey,[168] wheat,[169] wool,[170] bovine serum albumin, and beta-casein proteins.[164] Tyrosinase-formed quinones have also been exploited for covalent grafting of chitosan to silk proteins[171] and to gelatin.[172] Activated tyrosine residues of wool protein fibers are accessed for cross-linking with different materials like collagen, elastin, and gelatin for hygienically sensitive applications.[173] In the food industry, enzymatic protein cross-linking con-tributes to gel formation as well as to other properties of the food matrix.[174] The protein modification can be utilized to stabilize structure in food products, to stabilize foams and emulsions, to mask off taste, flavor, or color.[175,176] Therefore, tyrosinase enzymes have great potential for improving food tex-ture, as protein-based fat replacers, as flavor enhancers, and to make health-functional carbohydrates.

8.3.5.3 Peroxidases

Peroxidases are a complex superfamily of heme-containing enzymes (EC 1.11.1.x), which utilize hydrogen peroxide to oxidize a wide variety of organic and inorganic compounds. Physiologically, their antibacterial and antioxidant properties provide a means of defense.[177] They are ubiquitous and have been isolated from various sources from animals to plants.[178] Industrially, they have many applications, for example, the treatment of industrial wastes, such as phenols, which are oxidized to form less toxic compounds.[179] The most important enzyme for chemical applications is horse-radish peroxidase whose biochemical characteristics have been studied extensively.[180] Stahmann[181] reported that horseradish peroxidase induced cross-linking of various proteins *in vitro* including cytochrome C, bovine serum albumin, catalase, ovalbumin, α-lactoglobulin, and pepsin. In the presence of hydrogen peroxide and a hydrogen donor such as benzidine, *p*-anisidine, *o*-phenylene diamine, or pyrogallol, the enzyme oxidatively deaminated some lysyl residues of the protein to form lysyl aldehyde. The aldehyde presumably forms a Schiff base with another amino group pro-ducing a cross-linked product upon reduction. However, other mechanism involving free radical chemistry may also occur.[182] There are many applications of horseradish peroxidase, for example, gelling of alginate possessing phenolic hydroxyl groups,[183] preparation of injectable chitosan-based hydrogels,[184] and formation of hydrogels were from tyramine-substituted hyaluronan.[185]

8.3.5.4 Xanthine Oxidase and Others

Xanthine oxidase (EC 1.1.3.22) is an enzyme that catalyzes the oxidation of hypoxanthine to xan-thine and can further catalyze the oxidation of xanthine to uric acid. It is a form of xanthine oxido-reductase and can be derived from xanthine dehydrogenase.[186] The enzyme contains two FAD, two molybdenum atoms, and eight iron atoms. The molybdenum atoms are contained as molybdopterin cofactors and are the active sites of the enzyme. The iron atoms are part of ferredoxin iron–sulfur clusters and participate in electron transfer reactions.[187] During the catalytic reaction, reactive oxy-gen species, both superoxide anion and hydrogen peroxide are produced. Girotti et al.[188] found that in the presence of xanthine and ferric ion, xanthine oxidase causes protein cross-linking of isolated erythrocyte membranes and that spectrin is more susceptible to cross-linking than the other poly-peptides. Thiol disulfide bonds as well as nonreducible bonds are generated. The mechanism of the cross-linking reaction is not known, although it is speculated that hydrogen peroxide is involved in the formation of disulfide bonds.

There are other oxidase enzymes that are involved in cross-linking of proteins. Laccases (EC 1.10.3.2), which are found in many plants, fungi, and microorganisms and which play a role in the formation of lignin by promoting the oxidative coupling of lignols, are able to induce intermo-lecular cross-links in β-lactoglobulin and α-casein.[189] The mechanism is unknown, but the bound copper in a trinuclear cluster is probably involved in performing a one-electron oxidation process.[190] Lysyl oxidase, also known as protein-lysine 6-oxidase (EC 1.4.3.13), is an extracellular copper enzyme that is essential for stabilization of extracellular matrixes, specifically the enzymatic cross-linking of collagen and elastin.[191] The lysine or hydroxylysine residues in the nonhelical portions of the molecules are oxidized to aldehydes. These aldehydes are highly reactive, and undergo spontaneous chemical reactions with lysine or hydroxylysine and other residues in neighboring molecules to form intermolecular bonds. These reactions result in cross-linking collagen and elastin, which is essential for stabilization of collagen fibrils and for the integrity and elasticity of mature elastin.[192] Other enzymes such as superoxide dismutase and catalase are found to cross-link hemoglobin.[193] All these enzymes generate reactive oxygen species that preferentially react with Trp, His, Tyr, Met, and Cys side chains. Reaction of these residues with each other may gives rise to cross-linked products.[194] However, whether these oxidative enzyme systems can be applied for chemical cross-linking of isolated proteins awaits to be determined.

8.3.6   RADIATION AS ZERO-LENGTH CROSS-LINKER

By means of incorporating photosensitive homobifunctional or heterobifunctional reagents into biological macromolecules, ultraviolet (UV) light has been used to cross-link proteins, protein and nucleic acid, and nucleic acids as discussed in Chapters 5 and 6. However, UV light has been shown to induce the formation of protein–protein, protein–nucleic acid, and nucleic acid–nucleic acid adducts directly without the use of any external material. Direct coupling of these molecules qualifies UV as zero-length cross-linker. Application of photo cross-linking of proteins is limited, since radiation often causes photodegradation of proteins.[195] Absorption of light occurs through either the peptide backbone or by the amino acid side chains of Trp, Tyr, Phe, and cystine. Excitation of these moieties to a higher energy states leads to the formation of various reactive species such as peroxy radical and radical cation. These intermediates can cause degradation, aggregation, and conformational changes of the protein structure as well as induce cross-linking.[194] Callaway et al.[196] showed that irradiation of isolated nuclei or of a complex of histones 2A (H2A) and 2B (H2B) with ultraviolet light cross-linking the histones. Analysis revealed that the cross-linkage resulted from a covalent bond between tyrosine-40 of H2B and proline-26 of H2A.[197] Irradiation of living cells with UV light of wavelength near 260 nm also produces covalent bonds between nucleic acids and proteins. Using UV crosslinking, Bohnsack et al.[198] identified multiple binding sites for the helicase Prp43 in the 18S and 25S rRNA regions of pre-rRNAs. The interactions of DNA and histones, RNA Pol II, and Q50 homeodomain proteins have been studied.[199–201] With the development of UV laser as UV light source, the number of photons required for cross-linking can be delivered in nano- or pico- or femto-second intervals. This extremely short irradiation time achieves higher cross-linking efficiency than with a conventional UV lamp and is well suitable for kinetic studies. As the protein–DNA interactions are dynamic *in vivo*, the rapid freezing of the interaction at a particular step during the assembly of protein–DNA complex can be studied with UV laser cross-linking.[202] While static site-specific protein–DNA photocross-linking permits identification of protein–DNA interactions within multiprotein–DNA complexes, kinetic site-specific protein–DNA photocross-linking involving rapid-quench-flow mixing and pulsed-laser irradiation permits elucidation of pathways and kinetics of formation of protein–DNA interactions within multiprotein–DNA complexes.[203] For example, irradiation of a single UV laser pulse of 5 ns, 50 mJ on yeast crude extract allows detectable complex formation of specific protein and DNA.[204] Zhang et al.[205] have successfully detected DNA-binding of specific transcription factor, heat shock factor 1, *in vivo* by UV laser irradiation of living *Arabidopsis* suspension culture cells for 60 s, while conventional UV light failed. The UV laser technique has been successfully applied on the study of DNA replication, transcription, chromatin structure, and genome-wide location of DNA-binding proteins.[205]

In Chapter 2, the photochemistry of nucleic acid bases was discussed, specifically the production of photoactivated thymine dimers. Thus, nucleic acid polymers can be cross-linked by UV radiation.

However, such UV irradiation may also cause DNA and RNA strand breaks. This unfavorable side reaction can be reduced by using low-intensity radiation. In this manner, tRNAs have been cross-linked directly to 16S RNA at the P site of *E. coli* ribosomes.[206] The linkage probably consists of a cyclobutane dimer formed between suitable oriented pyrimidines bases in each RNA.[207] Further analysis of cross-linking between tRNA(Phe) to 16S rRNA in the presence of either poly(U) or the mRNA analogue indicated a limited structural change in the small subunit around C967 and C1400 during tRNA P-site binding.[208] Similarly, long-range cross-links are induced by UV light in 16S rRNA of *E. coli*, *Bacillus subtilis*, and *Thermus aquaticus*. Many of these cross-links corresponded exactly to each other suggesting that the tertiary interactions are highly conserved but different in some regions.[209]

With the new development of laser technology, Shapkina et al.[210] have used UV pulsed lasers to determine the yields of RNA–RNA crosslinks in 16S rRNA and found that the pattern of cross-linking was similar to that observed with low intensity irradiation but with four additional long-range cross-links not previously seen in *E. coli* ribosomes. This technique may prove to be useful for future nucleic acid cross-linking studies.

In addition to the cross-linking processes mentioned above, UV irradiation has been applied to immobilize single-stranded capture oligonucleotides on silanated (amino and epoxy) glass surfaces.[211] Comparison to UV cross-linking of the same DNA oligonucleotides on unmodified glass surfaces showed that single-stranded DNA molecules do not require a special modification to immobilize them by UV cross-linking on epoxy- or amino-modified glass surfaces. This methodology of immobilizing biomolecules on solid supports is useful for the preparation of microarray chips, which will be discussed in Chapter 14. In a similar vein of research, electron beams have been used to cross-link biphenyl self-assembled monolayers to form "polymer carpets," which have potential applications as microsensors.[212] Electron beam irradiation have also been applied to induce cross-links in collagen membrane and gelatin gels.[213,214] Terao et al.[214] used an electron beam from a 3 MeV accelerator to cross-link gelatin to obtain a gelatin hydrogel without any added reagent. Because electron beam irradiation penetrates deeper into the gel medium than UV light, the hydrogel obtained may be more suitable for medical use. Similar results were also obtained when gamma ray generated from a Cobalt-60 source at a dose rate of 10 kGy/h used employed.[214] These experiments indicated that further application of irradiation to cross-link biomolecules as a zero-length cross-linker may be worth investigation.

8.3.7 MISCELLANEOUS REAGENTS

There are few reagents that seem to function as zero-length cross-linkers, but their mechanism is not fully understood and the chemical moieties involved in the reaction are only speculated upon and not definitely elucidated. These reagents are discussed below.

8.3.7.1 Tetranitromethane

The best known application of tetranitromethane (TNM, Table 8.2.IV.A) is nitration of tyrosine residues in proteins.[215] Tyrosine nitrosylation may also occur by a radical-mediated reaction.[216] However, the reaction mechanism of protein cross-linking is not clear. It is thought to involve two tyrosine residues. Various mechanisms have been proposed, but none is totally satisfactory. Bruice et al.[217] proposed that the reaction involves initial charge transfer complex formation between tetranitromethane and a phenoxide ion, generating a highly reactive phenoxide radical and nitroradical. These intermediates would then form coupled and nitrated adducts. In support of this mechanism, Williams and Lowe[218] as well as Aeschbach et al.[219] isolated tyrosine dimer in cross-linked proteins in which tyrosine units were linked through biphenyl bonds. However, such biphenyl bond formation is not consistent with the observation that the protein dimers can be cleaved with ammonium bicarbonate.[197]

TNM has been used to study protein interactions. Since it can penetrate hydrophobic clusters, the binding sites of histones H4 and H2B for each other have been localized through cross-linking to the carboxyl terminal regions of the two histones.[220] For membrane studies, cross-linking of lipids to apoproteins and of apoproteins to each other was detected.[221] TNB cross-linking documented the self-association of ADP-ribosyltransferase protein.[222] It has also been found to cross-linking other protein such as collagen, globulin, and carboxypeptidase A.[223]

8.3.7.2   Potassium Nitrosyl Disulfonate

Potassium nitrosyl disulfonate (NDS, Table 8.2.IV.B) has been shown to form cross-links in wool, silk, casein, insulin, and collagen.[224,225] The reaction mechanism is not understood. It is speculated that tyrosine and tryptophan residues are involved in the cross-linking process. Polis et al.[226] found that interaction of NDS with proteins such as albumin induced stable free radicals in the proteins and that the presence of tyrosine or tryptophan residues in the protein was essential for the free-radical formation. However, it is not known whether NDS cross-linking involves free radicals.

8.3.7.3   Bisulfite

Sodium bisulfite (Table 8.2.IV.C) induces protein–DNA cross-links at neutral pH and ambient temperature between protein amino groups and cytosines of DNA as a zero-length cross-linker.[197] The reaction probably involves initial activation of cytosine by addition across the 5–6 double bond, which in turn facilitates transamination at carbon 4 to yield a covalent link with amino groups. The products of the reaction are N4-substituted cytosine. The amino group can be derived from free amines, α- and ε-amino groups of lysine, free or bound in proteins, and polylysine.[227] Thus, cytosines in nucleic acids can be converted into N4-aminocytosine by treatment with a mixture of hydrazine and bisulfite. The hydrazino group thus formed can be linked to a sulfhydryl group in proteins by the use of bromopyruvate as a linker. In this way, RNA has been cross-linked with protein in the 30 S ribosomal subunit of *E. coli.*[228] The reagent has been used in the determination of the methylation site of DNA-methyltransferase NlaX from *Neisseria lactamica*, which was established to be the inner cytosine in the double-stranded pentanucleotide recognition sequence 5'-CCNGG-3' (where N = any nucleoside).[229] It has also been applied in the cross-linking virus RNA and viral coat protein.[227,230]

REFERENCES

1. Peters, K. and Richards, F. M., Chemical cross-linking: Reagents and problems in studies of membrane structure, *Annu. Rev. Biochem.*, 46, 523, 1977.
2. Sweadner, K. J., Crosslinking and modification of Na,K-ATPase by ethyl acetimidate, *Biochem. Biophys. Res. Commun.*, 78, 962, 1977.
3. Galella, G. and Smith, D. B., The cross-linking of tubulin with imidoesters, *Can. J. Biochem.*, 60, 71, 1982.
4. Shaw, A. B. and Marinetti, G. V., Cross linking of erythrocyte membrane proteins and phospholipids by chemical probes, *Membrane Biochem.*, 3, 1, 1980.
5. Richardson, V. B., Littlefield, L. G., and Colyer, S. P., Cytogenetic evaluations in human lymphocytes exposed to methyl acetimidate, a lysine-specific protein crosslinking agent, *Mutat. Res.*, 180, 121, 1987.
6. Chao, T. L., Berenfeld, M. R., Gelbart, T., and Gabuzda, T. G., The effects of oxygen affinity and gelation of hemoglobin S crosslinked by reaction with methyl acetimidate, *Hemoglobin*, 5, 47, 1981.
7. Drews, G. and Rack, M., Modification of sodium and gating currents by amino group specific cross-linking and monofunctional reagents, *Biophys. J.*, 54, 383, 1988.
8. Fraenkel-Conrat, H. and Olcott, H. S., Reaction of formaldehyde with proteins. VI. Cross-linking of amino groups with phenol, imidazole, or indole groups, *J. Biol. Chem.*, 174, 827, 1948.
9. Fraenkel-Conrat, H. and Olcott, H. S., Crosslinking between amino and primary amide or guanidyl groups, *J. Am. Chem. Soc.*, 70, 2673, 1948.
10. French, D. and Edsall, J. T., Reactions of formaldehyde with amino acids and proteins, *Adv. Prot. Chem.*, 2, 277, 1945.
11. Ji, T. H., Bifunctional reagents, *Methods Enzymol.*, 91, 580, 1983.

12. Blass, J., Bizzini, B., and Raynaud, M., Mechanism of detoxication by formol, *Compt. Rend.*, 261, 1448, 1965.

13. Hopwood, D., Comparison of the crosslinking abilities of glutaraldehyde, formaldehyde, and α-hydroxyadipaldehyde with bovine serum albumin and casein, *Histochemie*, 17, 1.51, 1969.

14. Means, G. E. and Feeney, R. E., *Chemical Modification of Proteins*, Holden-Day, San Francisco, CA, 1971.

15. Bernard, M., Kersten, G. F. A., Hoogerhout, P., Brugghe, H. F., Timmermans, H. A. M., Jong, A. D., Meiring, H. et al., Identification of formaldehyde-induced modifications in proteins, *J. Biol. Chem.*, 279, 6235, 2004.

16. Orlando, V., Strutt, H., and Paro, R., Analysis of chromatin structure by *in vivo* formaldehyde cross-linking, *Methods*, 11, 205, 1997.

17. Perez-Romero, P. and Imperiale, M. J., Assaying protein–DNA interactions *in vivo* and *in vitro* using chromatin immunoprecipitation and electrophoretic mobility shift assays, *Methods Mol. Med.*, 131, 123, 2007.

18. Kim, Y.-C., Russell, W. K., Ranjith-Kumar, C. T., Thomson, M., Russell, D. H., and Kao, C. C., Functional analysis of RNA binding by the Hepatitis C virus RNA-dependent RNA, Polymerase, *J. Biol. Chem.*, 280, 38011, 2005.

19. Han, Y. T., Hsu, Y. H., Lo, C. W., and Meng, M., Identification and functional characterization of regions that can be crosslinked to RNA in the helicase-like domain of BaMV replicase, *Virology*, 389, 34, 2009.

20. Hall, D. B. and Struhl, K., The VP16 activation domain interacts with multiple transcriptional components as determined by protein–protein cross-linking *in vivo*, *J. Biol. Chem.*, 277, 46043, 2002.

21. Zeng, P.-Y., Vakoc, C. R., Chen, Z.-C., Blobel, G. A., and Berger, S. L., *In vivo* dual cross-linking for identification of indirect DNA-associated proteins by chromatin immunoprecipitation, *Biotechniques*, 41, 694, 2006.

22. Evans, E., Sugawara, N., Haber, J. E., and Alani, E., The *Saccharomyces cerevisiae* Msh2 mismatch repair protein localizes to recombination *intermediates in vivo*, *Mol. Cell.*, 5, 789, 2000.

23. Kuo, M. H. and Allis, C. D., In vivo cross-linking and immunoprecipitation for studying dynamic protein: DNA associations in a chromatin environment, *Methods*, 19, 425, 1999.

24. Lo, A. W., Craig, J. M., Saffery, R., Kalitsis, P., Irvine, D. V., Earle, E., Magliano, D. J., and Choo, K. H., A 330 kb CENP-A binding domain and altered replication timing at a human neocentromere, *EMBO J.*, 20, 2087, 2001.

25. Orlando, V., Mapping chromosomal proteins in vivo by formaldehyde-crosslinked-chromatin immunoprecipitation, *Trends Biochem. Sci.*, 25, 99, 2000.

26. Madhavan, K., Belchenko, D., Motta, A., and Tan, W., Evaluation of composition and crosslinking effects on collagen-based composite constructs, *Acta Biomater.*, 6, 1413, 2010.

27. Robyr, D., Kurdistani, S. K., and Grunstein, M., Analysis of genome-wide histone acetylation state and enzyme binding using DNA microarrays, *Methods Enzymol.*, 376, 289, 2004.

28. Avrameas, S. and Ternynck, T., Biologically active water-insoluble protein polymers. 1. Their use for isolation of antigens and antibodies, *J. Biol. Chem.*, 242, 1651, 1967.

29. Chang, C. H., Raam, S., Angellis, D., Doellgast, G., and Fishman, W. H., A simple radioimmunoassay of human placental alkaline phosphatase (Regan isoenzyme) using specific antibody polymers, *Cancer Res.*, 35, 1706, 1975.

30. Hughes, W. L., Albumin fraction isolated from human plasma as a crystalline mercuric salt, *J. Am. Chem. Soc.*, 69, 1836, 1947.

31. Arnon, R. and Shapira, E., Crystalline papain derivative containing an intramolecular mercury bridge, *J. Biol. Chem.*, 244, 1033, 1969.

32. Sperling, R., Burstein, Y., and Steinberg, I. Z., Selective reduction and mercuration of cystine IV-V in bovine pancreatic ribonuclease, *Biochemistry*, 8, 3810, 1969.

33. Edsall, J. T., Maybury, R. H., Simpson, R. B., and Straessle, R., Dimerization of serum mercaptalbumin in the presence of mercurials. II. Studies with a bifunctional organic mercurial, *J. Am. Chem. Soc.*, 76, 3131, 1954.

34. Vas, M. and Csanády, G., The two fast-reacting thiols of 3-phosphoglycerate kinase are structurally juxtaposed. Chemical modification with bifunctional reagents, *Eur. J. Biochem.*, 163, 365, 1987.

35. Jovin, T. M., Englund, P. T., and Kornberg, A., Enzymatic synthesis of deoxyribonucleic acid, *J. Biol. Chem.*, 244, 3009, 1969.

36. Soskine, M., Steiner-Mordoch, S., and Schuldiner, S., Crosslinking of membrane-embedded cysteines reveals contact points in the EmrE oligomer, *Proc. Natl. Acad. Sci. U. S. A.*, 99, 12043, 2002.

37. Hastrup, H., Sen, N., and Javitch, J. A., The human dopamine transporter forms a tetramer in the plasma membrane: Cross-linking of a cysteine in the fourth transmembrane segment is sensitive to cocaine analogs, *J. Biol. Chem.*, 278, 45045, 2003.

38. Guo, W., Shi, L., Filizola, M., Weinstein, H., and Javitch, J. A., Crosstalk in G protein-coupled receptors: Changes at the transmembrane homodimer interface determine activation, *Proc. Natl. Acad. Sci. U. S. A.*, 102, 17495, 2005.

39. Traut, R. R., Bollena, A., Sun, T. T., Hershey, J. W. B., Sundberg, J., and Pierce, L. R., Methyl 4-mercaptobutyrimidate as a cleavable cross-linking reagent and its application to the *Escherichia coli* 30S ribosome, *Biochemistry*, 12, 3266, 1973.

40. Birnbaumer, M. E., Schrader, W. T., and O'Malley, B. W., Chemical cross-linking of chick oviduct progesterone-receptor subunit by using a reversible bifunctional cross-linking reagent, *Biochem. J.*, 181, 201, 1979.

41. Lambert, J. M., Boileau, G., Cover, J. A., and Traut, R. R., Cross-links between ribosomal proteins of 30S subunits in 70S tight couples and in 30S subunits, *Biochemistry*, 22, 3913, 1983.

42. Walleczek, J., Redi, B., Stoffler-Meiliche, M., and Stoffler, G., Protein–protein cross-linking of the 50S ribosomal subunit of *Escherichia coli* using 2-imiothiolane: Identification of cross-links by immunoblotting techniques, *J. Biol. Chem.*, 264, 4231, 1989.

43. Gulle, H., Hoppe, E., Osswald, M., Greuer, B., Brimacombe, R., and Stoffler, G., RNA-protein cross-linking in *Escherichia coli* 50S ribosomal subunits. Determination of sites on 23S RNA that are cross-linked to proteins L2, L4, L24 and L27 by treatment with 2-iminothiolane, *Nucleic Acids Res.*, 16, 815, 1988.

44. Rich, D. H., Gesellchen, P. D., Tong, D., Cheung, A., and Buckner, C. K., Alkylating derivatives of amino acids and peptides. Synthesis of N-maleoylamino acids, [1-(N-maleoylglycyl)cysteinyl]oxytocin. Effects on vasopressin-stimulated water loss from isolated toad bladder, *J. Med. Chem.*, 18, 1004, 1975.

45. Griffiths, D. G., Partis, M. D., Sharp, R. N., and Beechey, R. B., *N*-Polymethylenecarboxymaleimides—A new class of probes for membrane sulphydryl groups, *FEBS Lett.*, 134, 261, 1981.

46. Carraway, K. L. and Koshland, D. E., Jr., Cardodiimide modification of proteins, *Methods Enzymol.*, 25, 616, 1972.

47. Kurzer, F. and Douraghi-Zadeh, K., Advances in the chemistry of carbodiimides, *Chem. Rev.*, 67, 107, 1967.

48. Bodanszky, M., The myth of coupling reagents, *Pept. Res.*, 5, 134, 1992.

49. Pikuleva, I. A., Lapko, A. G., and Chashchin, V. L., Functional reconstitution of cytochrome P-450scc with hemin activated with Woodward's reagent K. Formation of a hemeprotein cross-link, *J. Biol. Chem.*, 267, 1438, 1992.

50. Kemp, D. S. and Woodward, R. B., N-Ethylbenzisoxazolium cation. I. Preparation and reactions with nucleophilic species, *Tetrahedron*, 21, 3019, 1965.

51. Kooistra, C. and Sluyterman, L. A., Isosteric conversion of protein carboxyl groups into carboxamide groups. II. Application to lysozyme, *Int. J. Pept. Protein Res.*, 29, 357, 1987.

52. Kooistra, C. and Sluyterman, L. A., Isosteric and non-isosteric modification of carboxyl groups of papain, *Biochim. Biophys. Acta*, 997, 115, 1989.

53. Patramani, I., Katsiri, K., Pistevou, E., Kalogerakos, T., Pawlatos, M., and Evangelopoulos, A. E., Glutamic-aspartic transaminase-antitransaminase interaction: A method for antienzyme purification, *Eur. J. Biochem.*, 11, 28, 1969.

54. Wolf, B., Lesnaw, J. A., and Reichmann, M. E., Mechanism of the irreversible inaction of bovine pancreatic ribonuclease by diethylpyrocarbonate. General reaction of diethylpyrocarbonate with proteins, *Eur. J. Biochem.*, 13, 519, 1970.

55. Korshun, V. A., Stetsenko, D. A., and Gait, M. J., Uridine 2′-carbamates: Facile tools for oligonucleotide 2′-functionalization, *Curr. Protoc. Nucleic Acid Chem.*, Chapter 4, Unit 4.21, 2004.

56. Osborne, C. S., Barbenel, J. C., Smith, D., Savakis, M., and Grant, M. H., Investigation into the tensile properties of collagen/chondroitin-6-sulphate gels: The effect of crosslinking agents and diamines, *Med. Biol. Eng. Comput.*, 36, 129, 1998.

57. Grabarek, Z. and Gergely, J., Zero-length crosslinking procedure with the use of active esters, *Anal. Biochem.*, 185, 131, 1990.

58. Kobayashi, T., Grabarek, Z., Gergely, J., and Collins, J. H., Extensive interactions between troponins C and I. Zero-length cross-linking of troponin I and acetylated troponin C, *Biochemistry*, 34, 10946, 1995.

59. Marzec, E. and Pietrucha, K., The effect of different methods of cross-linking of collagen on its dielectric properties, *Biophys. Chem.*, 132, 89, 2008.

60. Cao, H. and Xu, S. Y., EDC/NHS-crosslinked type II collagen-chondroitin sulfate scaffold: Characterization and in vitro evaluation, *J. Mater. Sci. Mater. Med.*, 19, 567, 2008.

61. Yue, T. W., Chien, W. C., Tseng, S. J., and Tang, S. C., EDC/NHS-mediated heparinization of small intestinal submucosa for recombinant adeno-associated virus serotype 2 binding and transduction, *Biomaterials*, 28, 2350, 2007.

62. Li, C. Q., Huang, B., Luo, G., Zhang, C. Z., Zhuang, Y., and Zhou, Y., Construction of collagen II/hyaluronate/chondroitin-6-sulfate tri-copolymer scaffold for nucleus pulposus tissue engineering and preliminary analysis of its physico-chemical properties and biocompatibility, *J. Mater. Sci. Mater. Med.*, 21, 741, 2010.

63. Gilles, M. A., Hudson, A. Q., and Borders, Jr., C. L., Stability of water-soluble carbodiimides in aqueous solution, *Anal. Biochem.*, 184, 244, 1990.

64. Gratzer, P. F. and Lee, J. M., Control of pH alters the type of cross-linking produced by 1-ethyl-3-(3-dimethylaminopropyl)-carbodiimide (EDC) treatment of acellular matrix vascular grafts, *J. Biomed. Mater. Res.*, 58, 172, 2001.

65. Zot, H. G. and Puett, D., An enzymatically active cross-linked complex of calmodulin and rabbit skeletal muscle myosin light chain kinase, *J. Biol. Chem.*, 264, 15552, 1989.

66. Clyne, D. H., Norris, S. H., Modesto, R. R., Pesce, A. J., and Polak, V. E., Antibody enzyme conjugates. The preparation of intermolecular conjugates of horseradish peroxidase and antibody and their use in immunohistology of ranal cortex, *J. Histochem. Cytochem.*, 21, 233, 1973.

67. Novak, P. and Kruppa, G. H., Intra-molecular cross-linking of acidic residues for protein structure studies, *Eur. J. Mass Spectrom. (Chichester, Eng.)*, 14, 355, 2008.

68. Benashski, S. E. and King, S. M., Investigation of protein–protein interactions within flagellar dynein using homobifunctional and zero-length crosslinking reagents, *Methods*, 22, 365, 2000.

69. Onishi, H., Maita, T., Matsuda, G., and Fujiwara, K., Carbodiimide-catalyzed cross-linking sites in the heads of gizzard heavy meromyosin attached to F-actin, *Biochemistry*, 28, 1905, 1989.

70. Andreeva, A. L., Andreev, O. A., and Borejdo, J., Structure of the 265-kilodalton complex formed upon EDC cross-linking of subfragment 1 to F-actin, *Biochemistry*, 32, 13956, 1993.

71. Leszyk, J., Grabarek, Z., Gergely, J., and Collins, J. H., Characterization of zero-length cross-links between rabbit skeletal muscle troponin C and troponin I: Evidence for direct interaction between the inhibitory region of troponin I and the NH2-terminal, regulatory domain of troponin C, *Biochemistry*, 29, 299, 1990.

72. Dallmann, H. G., Flynn, T. G., and Dunn, S. D., Determination of the 1-ethyl-3-[(3-dimethylamino) propyl]-carbodiimide-induced cross-link between the beta and epsilon subunits of *Escherichia coli* F1-ATPase, *J. Biol. Chem.*, 267, 18953, 1992.

73. Gardner, J. L. and Cain, B. D., Amino acid substitutions in the a subunit affect the epsilon subunit of F1F0 ATP synthase from *Escherichia coli*, *Arch. Biochem. Biophys.*, 361, 302, 1999.

74. Petridou, B., Guerin, M. F., and Hayes, F., Protein–RNA crosslinking in the subunits of the cytoplasmic ribosome of *Tetrahymena thermophila*, *Biochimie*, 71, 667, 1989.

75. Schmidt, R. R. and Betz, H., Cross-linking of beta-bungarotoxin to chick brain membranes. Identification of subunits of a putative voltage-gated K+ channel, *Biochemistry*, 28, 8346, 1989.

76. Mauk, M. R. and Mauk, A. G., Crosslinking of cytochrome c and cytochrome b5 with a water-soluble carbodiimide. Reaction conditions, product analysis and critique of the technique, *Eur. J. Biochem.*, 186, 473, 1989.

77. Zhou, J. S., Brothers, H. M., 2nd, Neddersen, J. P., Peerey, L. M., Cotton, T. M., and Kostić, N. M., Metalloprotein complexes for the study of electron-transfer reactions. Characterization of diprotein complexes obtained by covalent cross-linking of cytochrome c and plastocyanin with a carbodiimide, *Bioconjug. Chem.*, 3, 382, 1992.

78. Takeyama, M., Nogami, K., Saenko, E. L., Nishiya, K., Ogiwara, K., and Shima, M., Identification of a protein S-interactive site within the A2 domain of the factor VIII heavy chain, *Thromb. Haemost.*, 102, 645, 2009.

79. Kuznetsova, S., Rykhlevskaya, A., Taranenko, M., Sidorkina, O., Oretskaya, T., and Laval, J., Use of crosslinking for revealing the DNA phosphate groups forming specific contacts with the *E. coli* Fpg protein, *Biochimie*, 85, 511, 2003.

80. Armbrust, T. S., Odom, W. R., and Guikema, J. A., Structural analysis of photosystem I polypeptides using chemical crosslinking, *J. Exp. Zool.*, 269, 205, 1994.

81. Armbrust, T. S., Chitnis, P. R., and Cuikema, J. A., Organization of photosystem I polypeptides examined by chemical cross-linking, *Plant Physiol.*, 111, 1307, 1996.

82. Moshnikova, A. B., Afanasyev, V. N., Proussakova, O. V., Chernyshov, S., Gogvadze, V., and Beletsky, I. P., Cytotoxic activity of 1-ethyl-3-(3-dimethylaminopropyl)-carbodiimide is underlain by DNA inter-chain cross-linking, *Cell. Mol. Life Sci.*, 63, 229, 2006.

83. Zhong, S. and Yung, L. Y., Enhanced biological stability of collagen with incorporation of PAMAM dendrimer, *J. Biomed. Mater. Res. A.*, 91, 114, 2009.

84. Lv, Q., Hu, K., Feng, Q., and Cui, F., Fibroin/collagen hybrid hydrogels with crosslinking method: Preparation, properties, and cytocompatibility, *J. Biomed. Mater. Res. A.*, 84, 198, 2008.

85. Sengonul, M., Ruzicka, J., Attygalle, A. B., and Libera, M., Surface modification of protein nanocontainers and their self-directing character in polymer blends, *Polymer (Guildf)*, 48, 3632, 2007.

86. Gao, Y. and Kyratzis, I., Covalent immobilization of proteins on carbon nanotubes using the cross-linker 1-ethyl-3-(3-dimethylaminopropyl)carbodiimide—A critical assessment, *Bioconjug. Chem.*, 19, 1945, 2008.

87. Woodward, R. B., Olofson, R. A., and Mayer, H., A new synthesis of peptides, *J. Am. Chem. Soc.*, 83, 1010, 1961.

88. Woodward, R. B. and Olofson, R. A., The reaction of isoxazolium salts with base, *J. Am. Chem. Soc.*, 83, 1007, 1961.

89. Kosters, H. A. and de Jongh, H. H., Spectrophotometric tool for the determination of the total carboxylate content in proteins; molar extinction coefficient of the enol ester from Woodward's reagent K reacted with protein carboxylates, *Anal. Chem.*, 75, 2512, 2003.

90. Sinha, U. and Brewer, J. M., A spectrophotometric method for quantitation of carboxyl group modification of proteins using Woodward's reagent K, *Anal. Biochem.*, 151, 327, 1985.

91. Dunn, B. M. and Chaiken, I. M., Quantitative affinity chromatography. Determination of binding constants by elution with competitive inhibitors, *Proc. Natl. Acad. Sci. U. S. A.*, 71, 2382, 1974.

92. Llamas, K., Owens, M., Blakeley, L., Zerner, B., N-ethyl-5-phenylisoxazolium-3′-sulfonate (Woodward's reagent K) as a reagent for nucleophilic side chains of proteins, *J. Am. Chem. Soc.*, 108, 5543, 1986.

93. Lai, L. S., Chang, P. C., and Chang, C. T., Isolation and characterization of superoxide dismutase from wheat seedlings, *J. Agric. Food Chem.*, 56, 8121, 2008.

94. Paoli, P., Fiaschi, T., Cirri, P., Camici, G., Manao, G., Cappugi, G., Raugei, G., Moneti, G., and Ramponi, G., Mechanism of acylphosphatase inactivation by Woodward's reagent K, *Biochem. J.*, 328, 855, 1997.

95. Carvajal, N., Uribe, E., López, V., and Salas, M., Inactivation of human liver arginase by Woodward's reagent K: Evidence for reaction with His141, *Protein J.*, 23, 179, 2004.

96. Patel, R. P. and Price, S., Derivatives of proteins. I. Polymerization of α-chymotrypsin by use of N-ethyl-5-phenylisoxazolium-3′-sulfonate, *Biopolymers*, 5, 583, 1967.

97. Pikuleva, I. A. and Turko, I. V., A new method of preparing hemin conjugates with rabbit IgG, *Bioorg. Khim.*, 15, 1480, 1989.

98. Boyer, T. D., Covalent labeling of the nonsubstrate ligand-binding site of glutathione S-transferases with bilirubin-Woodward's reagent K, *J. Biol. Chem.*, 261, 5363, 1986.

99. Satre, M., Lunardi, J., Dianoux, A. C., Dupuis, A., Issartel, J. P., Klein, G., Pougeois, R., and Vignais, P. V., Modifiers of F1-ATPases and associated peptides, *Methods Enzymol.*, 126, 712, 1986.

100. Goodfriend, T., Fasman, G., Kemp, D., and Levine, L., Immunochemical studies of angiotensin, *Immunochemistry*, 3, 223, 1966.

101. Acquotti, D., Sonnino, S., Masserini, M., Casella, L., Fronza, G., and Tettamanti, G., A new chemical procedure for the preparation of gangliosides carrying fluorescent or paramagnetic probes on the lipid moiety, *Chem. Phys. Lipids*, 40, 71, 1986.

102. Aleixo, J. A., Swaminathan, B., Minnich, S. A., and Wallshein, V. A., Enzyme immunoassay: Binding of *Salmonella* antigens to activated microtiter plates, *J. Immunoassay*, 6, 391, 1985.

103. Samokhin, G. P. and Filimonov, I. N., Coupling of peptides to protein carriers by mixed anhydride procedure, *Anal. Biochem.*, 145, 311, 1985.

104. Ghosh, M. K., Kildsig, D. O., and Mitra, A. K., Preparation and characterization of methotrexate-immunoglobulin conjugates, *Drug Des. Deliv.*, 4, 13, 1989.

105. Bain, B. M., Harrison, G., Jenkins, K. D., Pateman, A. J., and Shenoy, E. V., A sensitive radioimmunoassay, incorporating solid-phase extraction, for fluticasone 17-propionate in plasma, *J. Pharm. Biomed. Anal.*, 11, 557, 1993.

106. Kuklev, D. V. and Smith, W. L., A procedure for preparing oxazolines of highly unsaturated fatty acids to determine double bond positions by mass spectrometry, *J. Lipid Res.*, 44, 1060, 2003.

107. Veronese, F. M., Largajolli, R., Boccù, E., Benassi, C. A., and Schiavon, O., Surface modification of proteins. Activation of monomethoxy-polyethylene glycols by phenylchloroformates and modification of ribonuclease and superoxide dismutase, *Appl. Biochem. Biotechnol.*, 11, 141, 1985.

108. Chang, S. I. and Hammes, G. G., Interaction of spin-labeled nicotinamide adenine dinucleotide phosphate with chicken liver fatty acid synthase, *Biochemistry*, 25, 4661, 1986.

109. Chu, B. C. and Orgel, L. E., Inhibition of oligo(glutamine) precipitation by glutamine-containing peptides, *Biochem. Biophys. Res. Commun.*, 283, 351, 2001.

110. Osborne, C. S., Reid, W. H., and Grant, M. H., Investigation into the biological stability of collagen/chondroitin-6-sulphate gels and their contraction by fibroblasts and keratinocytes: The effect of cross-linking agents and diamines, *Biomaterials*, 20, 283, 1999.

111. Bartling, G. J., Chattopadhyay, S. K., Barker, C. W., Forrester, L. J., and Brown, H. D., Preparation and properties of horseradish peroxidase cross-linked in nonaqueous media, *Int. J. Pept. Protein Res.*, 6, 287, 1974.

112. Onal, S. and Telefoncu, A., Comparison of chitin and Amberlite IRA-938 for alpha-galactosidase immobilization, *Artif. Cells Blood Substit. Immobil. Biotechnol.*, 31, 19, 2003.

113. Okutucu, B. and Telefoncu, A., Covalent attachment of oligonucleotides to cellulose acetate membranes, *Artif. Cells Blood Substit. Immobil. Biotechnol.*, 32, 599, 2004.

114. Kumashiro, Y., Ikezoe, Y., Tamada, K., and Hara, M., Dynamic interfacial properties of poly(ethylene glycol)-modified ferritin at the solid/liquid interface, *J. Phys. Chem. B*, 112, 8291, 2008.

115. Shenoy, N. R., Bailey, J. M., and Shively, J. E., Carboxylic acid-modified polyethylene: A novel support for the covalent immobilization of polypeptides for C-terminal sequencing, *Protein Sci.*, 1, 58, 1992.

116. Belleau, B., DiTullio, V., and Godin, D., The mechanism of irreversible adrenergic blockade by N-carbethoxydihydroquinolines—Model studies with typical serine hydrolases, *Biochem. Pharmacol.*, 18, 1039, 1969.

117. Bertrand, R., Chaussepied, P., Kassab, R., Boyer, M., Roustan, C., and Benyamin, Y., Cross-linking of the skeletal myosin subfragment 1 heavy chain to the N-terminal actin segment of residues 40-113, *Biochemistry*, 27, 5728, 1988.

118. Blotnick, E. and Muhlrad, A., Effect of nucleotides and actin on the intramolecular cross-linking of myosin subfragment-1, *Biochemistry*, 33, 6867, 1994.

119. Lee, E. N., Lee, S. Y., Lee, D., Kim, J., and Paik, S. R., Lipid interaction of alpha-synuclein during the metal-catalyzed oxidation in the presence of Cu^{2+} and H_2O_2, *J. Neurochem.*, 84, 1128, 2003.

120. Bartling, G. J., Chattopadhyay, S. K., Barker, C. W., and Brown, H. D., Preparation and properties of matrix-supported horseradish peroxidase, *Can. J. Biochem.*, 53, 868, 1975.

121. Wu, S. N. and Chang, H. D., Diethyl pyrocarbonate, a histidine-modifying agent, directly stimulates activity of ATP-sensitive potassium channels in pituitary GH(3) cells, *Biochem. Pharmacol.*, 71, 615, 2006.

122. Rexin, M., Busch, W., and Gehring, U., Chemical cross-linking of heteromeric glucocorticoid receptors, *Biochemistry*, 27, 5593, 1988.

123. Kluger, R. and Li, X., Efficient chemical introduction of a disulfide cross-link and conjugation site into human hemoglobin at beta-lysine-82 utilizing a bifunctional aminoacyl phosphate, *Bioconjug. Chem.*, 8, 921, 1997.

124. Shu, X. Z., Liu, Y., Palumbo, F., and Prestwich, G. D., Disulfide-crosslinked hyaluronan-gelatin hydrogel films: A covalent mimic of the extracellular matrix for in vitro cell growth, *Biomaterials*, 24, 3825, 2003.

125. Wan, F. Y., Wang, Y. N., and Zhang, G. J., The influence of oxidation of membrane thiol groups on lysosomal proton permeability, *Biochem. J.*, 360(Pt 2), 355, 2001.

126. Gaczyñska, M. and Bartosz, G., Crosslinking of membrane proteins during erythrocyte ageing, *Int. J. Biochem.*, 8, 377, 1986.

127. Chiesi, M., Cross-linking agents induce rapid calcium release from skeletal muscle sarcoplasmic reticulum, *Biochemistry*, 23, 3899, 1984.

128. Jarausch, J. and Kadenbach, B., Structure of the cytochrome c oxidase complex of rat liver. 1. Studies on nearest-neighbour relationship of polypeptides with cross-linking reagents, *Eur. J. Biochem.*, 146, 211, 1985.

129. Watts, S. D., Zhang, Y., Fillingame, R. H., and Capaldi, R. A., The gamma subunit in the *Escherichia coli* ATP synthase complex (ECF1F0) extends through the stalk and contacts the c subunits of the F0 part, *FEBS Lett.*, 368, 235, 1995.

130. Hou, C., Potier, M., and Bragg, P. D., Crosslinking and radiation inactivation analysis of the subunit structure of the pyridine nucleotide transhydrogenase of *Escherichia coli*, *Biochim. Biophys. Acta*, 1018, 61, 1990.

131. Faulstich, H. and Heintz, D., Reversible introduction of thiol compounds into proteins by use of activated mixed disulfides, *Methods Enzymol.*, 251, 357, 1995.

132. Chong, S. F., Chandrawati, R., Städler, B., Park, J., Cho, J., Wang, Y., Jia, Z. et al., Stabilization of polymer-hydrogel capsules via thiol-disulfide exchange, *Small*, 5, 2601, 2009.

133. Park, H. S., Gong, B. J., and Tao, T., A disulfide crosslink between Cys98 of troponin-C and Cys133 of troponin-I abolishes the activity of rabbit skeletal troponin, *Biophys. J.*, 66, 2062, 1994.

134. Izbicka-Dimitrijević, E. and Gilbert, H. F., Multiple oxidation products of sulfhydryl groups near the active site of thiolase I from porcine heart, *Biochemistry*, 23, 4318, 1984.

135. Yoon, C. S., Kim, D. W., Jang, S. H., Lee, B. R., Choi, H. S., Choi, S. H., Kim, S. Y. et al., Cysteine-321 of human brain GABA transaminase is involved in intersubunit cross-linking, *Mol. Cells*, 18, 214, 2004.

136. Kodadek, T., Duroux-Richard, I., and Bonnafous, J.-C., Techniques: Oxidative cross-linking as an emergent tool for the analysis of receptor-mediated signalling events, *Trends Pharmacol. Sci.*, 26, 210, 2005.

137. Brown, K. C., Yang, S. H., and Kodadek, T., Highly specific oxidative cross-linking of proteins mediated by a nickel-peptide complex, *Biochemistry*, 34, 4733, 1995.

138. Denison, C. and Kodadek, T., Toward a general chemical method for rapidly mapping multi-protein complexes, *J. Proteome Res.*, 3, 417, 2004.

139. Fancy, D. A., Denison, C., Kim, K., Xie, Y., Holdeman, T., Amini, F., and Kodadek, T., Scope, limitations and mechanistic aspects of the photo-induced cross-linking of proteins by water-soluble metal complexes, *Chem. Biol.*, 7, 697, 2000.

140. Brown, K. C. and Kodadek, T., Protein cross-linking mediated by metal ion complexes, *Met. Ions Biol. Syst.*, 38, 351, 2001.

141. Kim, U. J., Kuga, S., Wada, M., Okano, T., and Kondo, T., Periodate oxidation of crystalline cellulose, *Biomacromolecules*, 1, 488, 2000.

142. Senapathy, P., Ali, M. A., and Jacob, M. T., Mechanism of coupling periodate-oxidized nucleosides to proteins, *FEBS Lett.*, 190, 337, 1985.

143. Wilson, M. B. and Nakane, P. K., Recent developments in the periodate method of conjugating horseradish peroxidase (HRPO) to antibodies, in *Immunofluorescence and Related Staining Techniques*, Kanpp, W., Holuber, K., and Wilck, G. (eds.), Elsevier/North Biomedical Press, Amsterdam, the Netherlands, 1978, p. 215.

144. Masárová, J., Mislovicová, D., Gemeiner, P., and Michalková, E., Stability enhancement of *Escherichia coli* penicillin G acylase by glycosylation with yeast mannan, *Biotechnol. Appl. Biochem.*, 34(Pt 2), 127, 2001.

145. Bayer, E. A., Skutelsky, E., Wynne, D., and Wilchek, M., Preparation of ferritin–avidin conjugates by reductive alkylation for use in election microscopy cytochemistry, *J. Histochem.*, 24, 033, 1976.

146. Folk, J. E., Transglutaminases, *Annu. Rev. Biochem.*, 49, 517, 1980.

147. Griffin, M., Casadio, R., and Bergamini, C. M., Transglutaminases: Nature's biological glues, *Biochem. J.*, 368, 377, 2002.

148. Lorand, L. and Graham, R. M., Transglutaminases: Crosslinking enzymes with pleiotropic functions, *Nat. Rev. Mol. Cell Biol.*, 4, 140, 2003.

149. Beninati, S., Bergamini, C. M., and Piacentini, M., An overview of the first 50 years of transglutaminase research, *Amino Acids*, 36, 591, 2009.

150. Serafini-Fracassini, D. and Del Duca, S., Transglutaminases: Widespread cross-linking enzymes in plants, *Ann. Bot.*, 102, 145, 2008.

151. Nemes, Z., Petrovski, G., Csosz, E., and Fésüs, L., Structure–function relationships of transglutaminases— A contemporary view, *Prog. Exp. Tumor Res.*, 38, 19, 2005.

152. Akpalo, E. and Larreta-Garde, V., Increase of fibrin gel elasticity by enzymes: A kinetic approach, *Acta Biomater.*, 6, 396, 2010.

153. Lorand, L., Crosslinks in blood: Transglutaminase and beyond, *FASEB J.*, 21, 1627, 2007.

154. Agyare, K. K. and Damodaran, S., pH-Stability and thermal properties of microbial transglutaminase-treated whey protein isolate, *J. Agric. Food Chem.*, 58, 1946, 2010.

155. Kuwahara, K., Yang, Z., Slack, G. C., Nimni, M., and Han, B., Cell delivery using an injectable and adhesive transglutaminase gelatin gel, *Tissue Eng. Part C Methods*, 16, 609, 2010.

156. Jubeck, B., Muth, E., Gohr, C. M., and Rosenthal, A. K., Type II collagen levels correlate with mineralization by articular cartilage vesicles, *Arthritis Rheum.*, 60, 2741, 2009.

157. Hitomi, K., Kitamura, M., and Sugimura, Y., Preferred substrate sequences for transglutaminase 2: Screening using a phage-displayed peptide library, *Amino Acids*, 36, 619, 2009.

158. Gorman, J. J. and Folk, J. E., Transglutaminase amine substrates for photochemical labeling and cleavable cross-linking of proteins, *J. Biol. Chem.*, 255, 1175, 1980.

159. Di Pierro, P., Mariniello, L., Sorrentino, A., Villalonga, R., Chico, B., and Porta, R., Putrescine–polysaccharide conjugates as transglutaminase substrates and their possible use in producing crosslinked films, *Amino Acids*, 38, 669, 2010.

160. Santos, M. and Torné, J. M., Recent patents on transglutaminase production and applications: A brief review, *Recent Pat. Biotechnol.*, 3, 166, 2009.
161. Ciardelli, G., Gentile, P., Chiono, V., Mattioli-Belmonte, M., Vozzi, G., Barbani, N., and Giusti, P., Enzymatically crosslinked porous composite matrices for bone tissue regeneration, *J. Biomed. Mater. Res. A*, 92, 137, 2010.
162. Matoba, Y., Kumagai, T., Yamamoto, A., Yoshitsu, H., and Sugiyama, M., Crystallographic evidence the dinuclear copper center of tyrosinase is flexible during catalysis, *J. Biol. Chem.*, 281, 8981, 2006.
163. Halaouli, S., Asther, M., Sigoillot, J.-C., Hamdi, M., and Lomascolo, A., Fungal tyrosinases: New prospects in molecular characteristics, bioengineering and biotechnological applications, *J. Appl. Microbiol.*, 100, 219, 2006.
164. Mattinen, M. L., Lantto, R., Selinheimo, E., Kruus, K., and Buchert, J., Oxidation of peptides and proteins by *Trichoderma reesei* and *Agaricus bisporus* tyrosinases, *J. Biotechnol.*, 133, 395, 2008.
165. Selinheimo, E., Lampila, P., Mattinen, M. L., and Buchert, J., Formation of protein–oligosaccharide conjugates by laccase and tyrosinase, *J. Agric. Food Chem.*, 56, 3118, 2008.
166. Marumo, K. and Waite, J. H., Optimization of hydroxylation of tyrosine and tyrosine containing peptides by mushroom tyrosinase, *Biochim. Biophys. Acta*, 872, 98, 1986.
167. Jee, J.-G., Park, S.-J., and Kim, H.-J., Tyrosinase-induced cross-linking of tyrosine-containing peptides investigated by matrix-assisted laser desorption/ionization time-of-flight mass spectrometry, *Rapid. Commun. Mass Spectrom.*, 14, 1563, 2000.
168. Thalmann, C. R. and Lotzbeyer, T., Enzymatic cross-linking of proteins with tyrosinase, *Eur. Food Res. Technol.*, 214, 276, 2002.
169. Selinheimo, E., Autio, K., Kruus, K., and Buchert, J., Elucidating the mechanism of laccase and tyrosinase in wheat bread making, *J. Agric. Food Chem.*, 55, 6357, 2007.
170. Lantto, R., Heine, E., Freddi, G., Lappalainen, A., Miettinen-Oinonen, A., Niku-Paavola, M.-L., and Buchert, J., Enzymatic modification of wool with tyrosinase and peroxidase, *J. Textile Inst.*, 96, 109, 2005.
171. Anghileri, A., Lantto, R., Kruus, K., Arosio, C., and Freddi, G., Tyrosinase-catalyzed grafting of sericin peptides onto chitosan and production of protein–polysaccharide bioconjugates, *J. Biotechnol.*, 127, 508, 2007.
172. Chen, T., Embree, H. D., Wu, L. Q., and Payne, G. F., In vitro protein–polysaccharide conjugation: Tyrosinase-catalyzed conjugation of gelatin and chitosan, *Biopolymers*, 64, 292, 2002.
173. Jus, S., Kokol, V., and Guebitz, G. M., Tyrosinase-catalysed coating of wool fibres with different protein-based biomaterials, *J. Biomater. Sci. Polym. Ed.*, 20, 253, 2009.
174. Buchert, J., Selinheimo, E., Kruus, K., Mattinen, M.-L., Lantto, R., and Autio, K., Cross-linking enzymes in food processing, in *Novel Enzyme Technology for Food Applications*, Rastall, R. (ed.), Woodhead Publishing Ltd., Cambridge, U.K., 2007, pp. 101–126.
175. Semo, E., Kesselman, E., Danino, D., and Livney, Y. D., Casein micelle as a natural nano-capsular vehicle for nutraceuticals, *Food Hydrocolloids*, 21, 936, 2007.
176. Conde, J. M. and Patino, J. M. R., Phospholipids and hydrolysates from a sunflower protein isolate adsorbed films at the air–water interface, *Food Hydrocolloid*, 21, 212, 2007.
177. van der Veen, B. S., de Winther, M. P., Heeringa, P., Augusto, O., Chen, J. W., Davies, M., Ma, X. L., Malle, E., Pignatelli, P., and Rudolph, T., Myeloperoxidase: Molecular mechanisms of action and their relevance to human health and disease, *Antioxid. Redox. Signal.*, 11, 2899, 2009.
178. Cosio, C. and Dunand, C., Specific functions of individual class III peroxidase genes, *J. Exp. Bot.*, 60, 391, 2009.
179. Alvarado, B. and Torres, E., Recent patents in the use of peroxidases, *Recent Pat. Biotechnol.*, 3, 88, 2009.
180. Veitch, N. C., Horseradish peroxidase: A modern view of a classic enzyme, *Phytochemistry*, 65, 249, 2004.
181. Stahmann, M. A., Cross-linking of protein by peroxidase, *Adv. Exp. Med. Biol.*, 86B, 285, 1977.
182. Wong, J. L. and Wessel, G. M., Free-radical crosslinking of specific proteins alters the function of the egg extracellular matrix at fertilization, *Development*, 135, 431, 2008.
183. Sakai, S., Hirose, K., Moriyama, K., and Kawakami, K., Control of cellular adhesiveness in an alginate-based hydrogel by varying peroxidase and H_2O_2 concentrations during gelation, *Acta Biomater.*, 6, 1446, 2010.
184. Jin, R., Moreira Teixeira, L. S., Dijkstra, P. J., Karperien, M., van Blitterswijk, C. A., Zhong, Z. Y., and Feijen, J., Injectable chitosan-based hydrogels for cartilage tissue engineering, *Biomaterials*, 30, 2544, 2009.

185. Darr, A. and Calabro, A., Synthesis and characterization of tyramine-based hyaluronan hydrogels, *J. Mater. Sci. Mater. Med.*, 20, 33, 2009.
186. Nishino, T., Okamoto, K., Eger, B. T., Pai, E. F., and Nishino, T., Mammalian xanthine oxidoreductase— Mechanism of transition from xanthine dehydrogenase to xanthine oxidase, *FEBS J.*, 275, 3278, 2008.
187. Hille, R., Structure and function of xanthine oxidoreductase, *Eur. J. Inorg. Chem.*, 10, 1905, 2006.
188. Girotti, A. W., Thomas, J. P., and Jordan, J. E., Xanthine oxidase-catalyzed crosslinking of cell membrane proteins, *Arch. Biochem. Biophys.*, 251, 639, 1986.
189. Steffensen, C. L., Andersen, M. L., Degn, P. E., and Nielsen, J. H., Cross-linking proteins by laccase-catalyzed oxidation: Importance relative to other modifications, *J. Agric. Food Chem.*, 56, 12002, 2008.
190. Solomon, E. I., Sundaram, U. M., and Machonkin, T. E., Multicopper oxidases and oxygenases, *Chem. Rev.*, 96, 2563, 1996.
191. Csiszar, K., Lysyl oxidases: A novel multifunctional amine oxidase family, *Prog. Nucleic Acid Res. Mol. Biol.*, 70, 1, 2001.
192. Robins, S. P., Biochemistry and functional significance of collagen cross-linking, *Biochem. Soc. Trans.*, 35(Pt 5), 849, 2007.
193. D'Agnillo, F. and Chang, T. M., Absence of hemoprotein-associated free radical events following oxidant challenge of crosslinked hemoglobin-superoxide dismutase catalase, *Free Radic. Biol. Med.*, 24, 906, 1998.
194. Davies, M. J., Reactive species formed on proteins exposed to singlet oxygen, *Photochem. Photobiol. Sci.*, 3, 17, 2004.
195. Kerwin, B. A. and Remmele, R. L. Jr., Protect from light: Photodegradation and protein biologics, *J. Pharm. Sci.*, 96, 1468, 2007.
196. Callaway, J. E., DeLange, R. J., and Martinson, H. G., Contact site of histones 2A and 2B in chromatin and in solution, *Biochemistry*, 24, 2686, 1985.
197. Kunkel, G. R., Mehrabian, M., and Martinson, H. G., Contact-site cross-linking agents, *Mol. Cell. Biochem.*, 34, 3, 1981.
198. Bohnsack, M. T., Martin, R., Granneman, S., Ruprecht, M., Schleiff, E., and Tollervey, D., Prp43 bound at different sites on the pre-rRNA performs distinct functions in ribosome synthesis, *Mol. Cell*, 36, 583, 2009.
199. Dimitrov, S. T., Stefanovsky, V. Y., Karagyozovl, L., Angelov, D., and Pashev, I. G., The enhancers and promoters of the *Xenopus laevis* ribosomal spacer are associated with histones upon active transcription of the ribosomal genes, *Nucleic Acids Res.*, 18, 6393, 1990.
200. Solano, P. J., Mugat, B., Martin, D., Girard, F., Huibant, J. M., Ferraz, C., Jacq, B., Demaille, J., and Maschat, F., Genome-wide identification of in vivo Drosophila engrailed-binding DNA fragments and related target genes, *Development*, 130, 1243, 2003.
201. Gilmour, D. S. and Lis, J. T., Protein–DNA cross-linking reveals dramatic variation in RNA polymerase II density on different histone repeats of *Drosophila melanogaster*, *Mol. Cell. Biol.*, 7, 3341, 1987.
202. Hager, G. L., Elbi, C., and Becker, M., Protein dynamics in the nuclear compartment, *Curr. Opin. Genet. Dev.*, 12, 137, 2002.
203. Naryshkin, N., Druzhinin, S., Revyakin, A., Kim, Y., Mekler, V., and Ebright, R. H., Static and kinetic site-specific protein–DNA photocrosslinking: Analysis of bacterial transcription initiation complexes, *Methods Mol. Biol.*, 543, 403, 2009.
204. Ho, D. T., Sauve, D. M., and Roberge, M., Detection and isolation of DNA-binding proteins using single-pulse ultraviolet laser crosslinking, *Anal. Biochem.*, 218, 248, 1994.
205. Zhang, L., Zhang, K., Prändl, R., and Schöffl, F., Detecting DNA-binding of proteins in vivo by UV-crosslinking and immunoprecipitation, *Biochem. Biophys. Res. Commun.*, 322, 705, 2004.
206. Ehresmann, C., Ehresmann, B., Millon, R., Ebel, J. P., Nurse, K., and Ofengand, J., Cross-linking of the anticodon of *Escherichia coli* and *Bacillus subtilis* acetylvalyl-tRNA to the ribosomal P site. Characterization of a unique site in both *E. coli* 16S and yeast 18S ribosomal RNA, *Biochemistry*, 23, 429, 1984.
207. Ofengand, J. and Liou, R., Evidence for pyrimidine–pyrimidine cyclobutane dimer formation in the covalent cross-linking between transfer ribonucleic acid and 16S ribonucleic acid at the ribosomal P site, *Biochemistry*, 19, 4814, 1980.
208. Noah, J. W., Shapkina, T. G., Nanda, K., Huggins, W., and Wollenzien, P., Conformational change in the 16S rRNA in the *Escherichia coli* 70S ribosome induced by P/P- and P/E-site tRNAPhe binding, *Biochemistry*, 42, 14386, 2003.
209. Noah, J. W., Shapkina, T., and Wollenzien, P., UV-induced crosslinks in the 16S rRNAs of *Escherichia coli*, *Bacillus subtilis* and *Thermus aquaticus* and their implications for ribosome structure and photochemistry, *Nucleic Acids Res.*, 28, 3785, 2000.

210. Shapkina, T., Lappi, S., Franzen, S., and Wollenzien, P., Efficiency and pattern of UV pulse laser-induced RNA-RNA cross-linking in the ribosome, *Nucleic Acids Res.*, 32, 1518, 2004.

211. Schüler, T., Nykytenko, A., Csaki, A., Möller, R., Fritzsche, W., and Popp, J., UV cross-linking of unmodified DNA on glass surfaces, *Anal. Bioanal. Chem.*, 395, 1097, 2009.

212. Amin, I., Steenackers, M., Zhang, N., Beyer, A., Zhang, X., Pirzer, T., Hugel, T., Jordan, R., and Gölzhäuser, A., Polymer carpets, *Small*, 6, 1623–30, 2010.

213. Jiang, B., Wu, Z., Zhao, H., Tang, F., Lu, J., Wei, Q., and Zhang, X., Electron beam irradiation modification of collagen membrane, *Biomaterials*, 27, 15–23, 2006.

214. Terao, K., Nagasawa, N., Nishida, H., Furusawa, K., Mori, Y., Yoshii, F., and Dobashi, T., Reagent-free crosslinking of aqueous gelatin: Manufacture and characteristics of gelatin gels irradiated with gamma-ray and electron beam, *J. Biomater. Sci. Polym. Ed.*, 14, 1197, 2003.

215. Abello, N., Kerstjens, H. A., Postma, D. S., and Bischoff, R., Protein tyrosine nitration: Selectivity, physicochemical and biological consequences, denitration and proteomics methods for the identification of tyrosine-nitrated proteins, *J. Proteome Res.*, 8, 3222, 2009.

216. Lee, S. J., Lee, J. R., Kim, Y. H., Park, Y. S., Park, S. I., Park, H. S., and Kim, K. P., Investigation of tyrosine nitration and nitrosylation of angiotensin II and bovine serum albumin with electrospray ionization mass spectrometry, *Rapid Commun. Mass Spectrom.*, 21, 2797, 2007.

217. Bruice, T. C., Gregory, M. J., and Walters, S. L., Reactions of tetranitromethane. I. Kinetics and mechanism of nitration of phenols by tetranitromethane, *J. Am. Chem. Soc.*, 90, 1612, 1968.

218. Williams, J. and Lowe, J. M., The cross-linking of tyrosine by treatment with tetranitromethane, *Biochem. J.*, 121, 203, 1971.

219. Aeschbach, R., Amadò, R., and Neukom, H., Formation of dityrosine cross-links in proteins by oxidation of tyrosine residues, *Biochim. Biophys. Acta*, 439, 292, 1976.

220. Martinson, H. G., True, R., Lau, C. K., and Mehrabian, M., Histone–histone interactions within chromatin. Preliminary location of multiple contact sites between histones 2A, 2B, and 4, *Biochemistry*, 18, 1075, 1979.

221. Chacko, G. K., Mahlberg, F. H., and Johnson, W. J., Cross-linking of apoproteins in high density lipoprotein by dimethylsuberimidate inhibits specific lipoprotein binding to membranes, *J. Lipid Res.*, 29, 319, 1988.

222. Bauer, P. I., Buki, K. G., Hakam, A., and Kun, E., Macromolecular association of ADP-ribosyltransferase and its correlation with enzymic activity, *Biochem. J.*, 270, 17, 1990.

223. Doyle, R. J., Bello, J., and Robolt, O. A., Probable protein cross-linking with tetranitromethane, *Biochim. Biophys. Acta*, 160, 274, 1968.

224. Earland, C. and Stell, J. P. C., Formation of new cross-linkages in proteins by oxidation of tyrosine residues with potassium nitrosyldisulfonate, *Polymer*, 7, 549, 1966.

225. Consden, R. and Kirrane, J. A., Cross-linking in collagen by potassium nitrolyldisulfonate, *Nat. (London)*, 218, 957, 1968.

226. Polis, B. D., Wyeth, J., Goldstein, L., and Graedon, J., Stable free-radical forms of plasma proteins or simpler related structures which induce brain excitatory effects, *Proc. Natl. Acad. Sci. U. S. A.*, 64, 755, 1969.

227. Shapiro, R. and Gazit, A., Crosslinking of nucleic acids and proteins by bisulfite, *Adv. Exp. Med. Biol.*, 86A, 633, 1977.

228. Nitta, N., Kuge, O., Yui, S., Tsugawa, A., Negishi, K., and Hayatsu, H., A new reaction useful for chemical cross-linking between nucleic acids and proteins, *FEBS Lett.*, 166, 194, 1984.

229. Kubareva, E. A., Walter, J., Karyagina, A. S., Vorob'eva, O. V., Lau, P. C., and Trautner, T., Determination of methylation site of DNA-methyltransferase NlaX by a hybrid method, *Biotechniques*, 33, 526, 2002.

230. Ehresmann, B., Briand, J. P., Reinbolt, J., and Witz, J., Identification of binding sites of turnip yellow mosaic virus protein and RNA by crosslinks induced in situ, *Eur. J. Biochem.*, 108, 123, 1980.

9 General Approaches for Chemical Cross-Linking

9.1 INTRODUCTION

Considering the diversity of cross-linking reagents as well as their varying degrees of selectivity and specificity and their different chemical reaction mechanisms, and also the vast number of bio-molecular systems in which they may be used, the technical procedures of conjugation and cross-linking are clearly highly diverse. The reaction protocols depend on which reagents are used and for what purpose.[1] In Chapter 4, we discussed the choice of these reagents based on their chemistry. In that regard, it is obvious that the right reagent is necessary for the right chemical reaction. The other criterion of choice of these reagents is based on the biological systems to be studied and the exact purpose.[2,3] For analysis of subunit organizations in a stable multimeric complex system, for example, various homobifunctional, heterobifunctional, or multifunctional reagents can be used. In this case, analysis of subunit interactions is relatively straightforward. Typically, a purified complex is incubated with one or several different cross-linkers to form covalent bonds with amino acid side chains, such as the ε-amino groups of lysines, the sulfhydryl groups of cysteines, or the carboxyl side chains of aspartate and glutamate. For investigations of weak or transient subunit interactions, where the entities associate only under certain physiological conditions such as ligand–receptor interactions, or transient association of subunits for regulatory or enzymatic purposes, a different set of cross-linkers would be used. These may include photoactivatable cross-linkers, cleavable reagents, or oxidation cross-linking reagents. For detection of conformational changes of proteins, different reagents would be employed. Furthermore, the biological systems to be studied also dictate which cross-linkers should be used. Different reagents would be used for studying nucleic acids, protein–nucleic acid interactions, membrane-bound systems, glycoproteins, etc. The choice of reagent would be based on the specific properties of the target macromolecules. Each of these approaches requires a different reaction protocol. It is therefore not possible to give a standard description of reagents and reaction conditions for general applications. However, a broad classification of reaction types may be useful for understanding the applications of these reagents. It is possible to generalize the reaction procedures for common functional groups. One must be aware, though, that in some cases different protocols may give rise to different products or ratios of products, and undesired side reactions may occur. One should always be open to the possibility that conditions may be varied in order to achieve the desired results. Thus, if a procedure is not established, trial and error may be necessary to attain the cross-linking objectives. In the following sections, established reaction procedures are provided as a guide for well-known cross-linkers.

9.2 CLASSIFICATION OF CROSS-LINKING PROCEDURES

9.2.1 ONE-STEP CROSS-LINKING REACTIONS

The simplest cross-linking procedure would be the addition of a cross-linker to a solution of target macromolecules. This approach is most useful for stable multisubunit complexes where the subunits do not dissociate under the conditions of the study. In general, the purpose of this approach is to map the architectural organization of the subunits in the complex, as in nearest neighbor analysis.

This procedure has many applications. For example, in the study of ribosomal protein topography, Hultin[4] added sulfhydryl- and amino-group–directed bifunctional cross-linkers directly to a suspension of ribosomes as described in the text box below. Similarly, Baskin and Yang[5] added dithiobis(succinimidylproprionate) directly to a suspension of microsomes for studying the protein topography of rat liver microsomes. However, such one-step procedures are not desirable for nonassociating molecules because side reactions cross-linking the same molecules may occur, giving rise to homopolymeric adducts with relatively few heterodimeric cross-links. This scenario would be the case for the preparation of immunoconjugates where the proteins are free in solution.[6] IgG, for example, has a higher reactivity with 4,4′-difluoro-3,3′-dinitrodiphenyl sulfone than does horseradish peroxidase. This reagent thus preferentially reacts with IgG, giving rise to homopolymerization. The yield of conjugation between peroxidase and IgG is very low in the one-step procedure. Modesto and Pesce[7] have shown that in some instances the rate of reagent addition can influence the conjugation yield. In general, slow addition of reagent over time, as opposed to addition of the entire reagent at one time, leads to increased yields of coupled proteins. This slow addition protocol is particularly appropriate in the case of large protein assemblies such as those found in membrane preparations. Rapid addition procedures are less desirable since they may lead to the production of both homopolymers and heteropolymers, particularly for nonreacting molecules. In this case, a two-step reaction procedure would be a better choice.

CROSS-LINKING OF PROTEIN CONTACT SITES IN MAMMALIAN RIBOSOMES BY A ONE-STEP PROCEDURE[4]

Ribosomes (40–60 ODU/mL) were suspended in 20 mM triethanolamine/HCl buffer (pH 6.8) containing 150 mM sucrose, 75 mM KCl, and 5 mM MgCl$_2$. The cross-linking reagent (e.g., (2,5-dioxopyrrolidin-1-yl)-2-[2-(2,5-dioxopyrrol-1-yl)-4-hydroxy-phenyl]azobenzoate), preserved at −20°C as a 10 mM solution in dimethylsulfoxide, was added with stirring to a concentration of 0.1–0.4 mM. After incubation for 10 min at 35°C, the suspension was mixed with 10 volumes of 75 mM KCl, 5 mM MgCl$_2$, and 5 mM mercaptoethanol to abolish remaining maleimide functions. The diluted suspension was centrifuged and the pellet preserved at −20°C for analysis.

9.2.2 Two-Step Cross-Linking Reactions

In this procedure, one of the components to be conjugated is first reacted with the cross-linker. The modified molecule is then isolated or the unreacted reagent removed prior to addition of the second component. This approach takes advantage of the differential reactivities of the functional groups in heterobifunctional reagents as well as the differential selectivity of homobifunctional reagents toward the molecules to be coupled.

Practically all heterobifunctional cross-linkers, particularly photosensitive reagents, are used according to a two-step procedure.[3] With N-hydroxysuccinimidyl (NHS)-ester-maleimido hetrobifunctional reagents, cross-linking is initiated with the NHS-ester reaction first to minimize hydrolysis of the NHS-ester in aqueous systems. The sulfhydryl then reacts with the maleimide group in the second step since it is significantly more resistant to hydrolysis than the NHS-ester moiety. For example, in the preparation of β-galactosidase–IgG conjugate with m-maleimidobenzoyl-N-hydroxysuccinimde (MBS), IgG (which does not have a free thiol group) is first labeled with the reagent through an amino group reaction with the NHS-ester. After removal of excess reagent, by either dialysis or gel filtration chromatography, β-galactosidase (which contains free thiol groups) is added to react with the maleimide moiety of labeled IgG, resulting in the desired immunoconjugate product.[6] This two-step procedure is described in the text box. Many other immunoconjugates are prepared in a similar way.

A notable example of a homobifunctional reagent that shows differential reactivities usable in a two-step coupling reaction is toluene-2,4-diisocyanate. The para-isocyante group is much more reactive than that at the ortho-position owing to steric hindrance of the latter due to the methyl group.[8] As a first step in the reaction, the protein is mixed with the reagent at near 0°C where modification of the protein will take place with the para-isocyanate group. After this reaction, the second protein to be cross-linked is added and the temperature is raised to 37°C, the temperature at which the ortho-isocyanate group will react and cross-link the proteins.

CONJUGATION OF β-GALACTOSIDASE AND IGG BY A TWO-STEP PROCEDURE[6]

Step 1: Purified donkey antisheep IgG antibodies were dissolved in 1.5 mL of 0.1 M phosphate buffer, pH 7.0, containing 50 mM NaCl to give an optical density of approximately 1.4 at 280 nm. A 15 μL aliquot of dioxan containing 0.32 mg of MBS was added to the antibody solution, mixed, and maintained at 30°C for 1 h. The solution was applied to a Sephadex G-25 column (30 × 0.9 cm), equilibrated with 10 mM phosphate buffer, pH 7.0, containing 10 mM MgCl$_2$ and 50 mM NaCl, and eluted with the same buffer.

Step 2: A total of 3 mL of eluant having an optical density of 0.70 (equivalent to a total protein content of 1.5 mg) was pooled. One milliliter of phosphate buffer containing 1.5 mg of β-galactosidase was immediately mixed with the antibody eluted from the column and maintained at 30°C for 1 h. The reaction was terminated by the addition of 1 M mercaptoethanol to give a final concentration of 10 mM mercaptoethanol.

Similar differences in reactivity are seen in 1,5-difluoro-2,4-dinitrobenzene, bis(4-fluoro-3-niro) sulfone, and 2,4-dichloro-6-methoxy-*s*-triazine probably due to electronic effects after nucleophilic replacement. For 2,4-dichloro-6-methoxy-*s*-triazine, coupling occurs with tyrosine residues at pH 7; however, the second chloro-group will only react with a tyrosine residue of another protein under alkaline conditions.[9] Similarly, under acid conditions, only one of the diazo groups of bis-diazotized *o*-dianisidine is reactive. For coupling of ferritin to rabbit gamma-globulin, the first step of the reaction is carried out at pH 5 and the second at pH 9.4.[10]

Differential reactivity of a homobifunctional reagent toward different proteins has also been used in two-step reactions. Glutaraldehyde, for example, reacts with γ-immunoglobulins much faster than with horseradish peroxidase. Reaction of horseradish peroxidase with an excess of glutaraldehyde constitutes the first step of the reaction. After removal of excess reagent, the immunoglobulin is added to generate the enzyme–immunoglobulin conjugate. Self-coupling of horseradish peroxidase is minimal due to the lack of available reactive groups.[11] Ferritin has also been coupled to γ-immunoglobulins under similar conditions.[12]

Another example of a two-step procedure is the cross-linking reaction with carbodiimides in the presence NHS. Carbodiimides react with a carboxyl group to form an active *o*-acylisourea intermediate, which is unstable in aqueous solutions and is therefore not useful for a two-step conjugation procedure. However, the intermediate can be stabilized using NHS, converting the protein carboxyls into succinimidyl esters that react with amino groups.[13] A reaction of such a procedure in cross-linking troponin C (TnC) and troponin T (TnT) using 1-ethyl-3-(3-dimethylaminopropyl)carbodiimide (EDC) is described in the text box below. The advantage of this two-step procedure over a one-step zero-length cross-linking is that only one component of the complex is exposed to the cross-linker, which reduces the formation of cross-links among several proteins of a multicomponent complex. Furthermore, cross-links can be formed even in the presence of other reagents, such as dithiothreitol and EDTA, which would interfere with direct cross-linking with EDC.

A TWO-STEP PROCEDURE CROSS-LINKING TNC AND TNT USING EDC[13]

TnC (1 mg/mL) was incubated for 15 min at 25°C in a solution containing 0.1 M Mes (2-(N-morpholino)ethanesulfonic acid) buffer, pH 6.0, containing 0.5 M NaCl, 2 mM EDC, and 5 mM NHS. The reaction was terminated by addition of 2-mercaptoethanol to a final concentration 20 mM and the solution was further incubated at the same temperature for a few minutes. The solution was passed through a Sephadex G-25 column (0.6 × 5 cm) equilibrated with 0.5 M NaCl and 0.1 M Mes to remove low-molecular-weight components by centrifugation. The effluent was combined with TnT in a 1:1 molar ratio in 0.5 M NaCl, 0.1 M Mes and further incubated for 2 h. TnT can be added directly to the quenched solution without gel filtration and incubated for 2 h. The cross-linked TnC–TnT adduct was analyzed by polyacrylamide gel electrophoresis.

9.2.3 Three-Step Cross-Linking Reactions

These procedures involve an additional step for the preparation of the proteins to be coupled, for instance, the introduction of a thiol group as discussed in Chapter 2. Figure 9.1 shows an example of a three-step coupling process for coupling IgG to albumin. In the first step, IgG is labeled with pyrrole-α-acyl azide. The second protein, albumin, is reacted with the cross-linker, bis-diazotized p-phenylenediamine, under acidic conditions. After isolation, the pyrrole-modified IgG and the diazo-albumin are mixed and allowed to react at pH 6. Specific reaction occurs between the diazo group and the pyrrole ring as shown in Figure 9.1. This method reduces the side reactions by adjusting the coupling conditions. At acidic pHs, the first diazotized group is reactive whereas the second group is reactive at higher pH.[14]

Many immunoconjugates and immunotoxins are prepared by this three-step process. These reactions will be further discussed in Chapter 12.

9.2.4 Multistep Cross-Linking Reactions

More sophisticated cross-linking reactions involve multistep procedures. The process usually involves several protein preparation steps before the final coupling reaction. Various multistep cross-linking schemes are possible, depending on the chemical reagents and biological systems used. A simple example is the use of N-succinimidyl-3-(2-pyridyldithio)propionate (SPDP) to prepare immunotoxins.[15] Both immunoglobulin and toxin are separately reacted with SPDP. One of the labeled proteins is then reduced to generate a free thiol group and then purified. The two modified

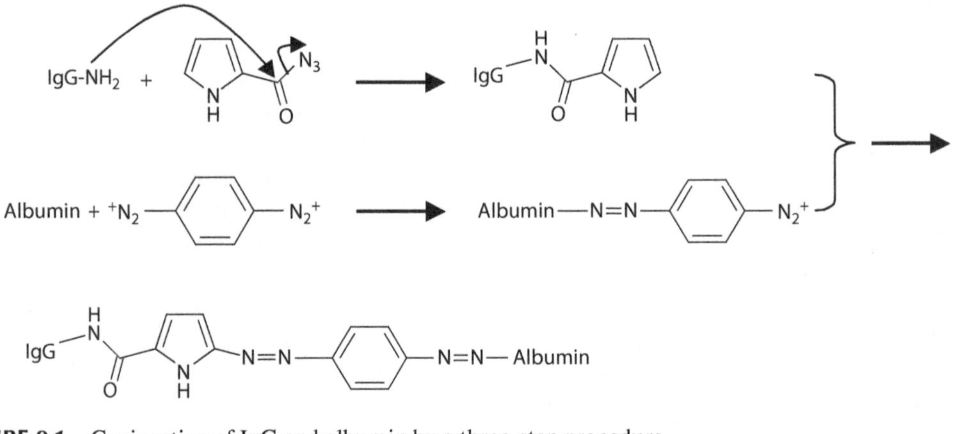

FIGURE 9.1 Conjugation of IgG and albumin by a three-step procedure.

proteins are then mixed to allow cross-linking to proceed, which provides an adduct with a cleavable disulfide bond. Using two different reagents, Rector et al.[16] described a cross-linking procedure for conjugating ricin and immunoglobulin as shown in the text box below. An iodoacetyl group was introduced into the toxin with N-hydroxysuccinimidyl iodoacetate and the immunoglobulin was thiolated with SPDP. Both modified proteins were isolated and mixed together to form a conjugate with the structure: IgG-NH-CO-CH$_2$CH$_2$-S-CH-CO-NH-ricin.

Cross-linking effected by masked or disguised cross-linkers proceeds through several steps. Disguised or masked reagents are compounds that can be easily converted to heterobifunctional reagents in a simple reaction. These reagents are usually disguised as monofunctional agents. During the procedure, extra steps are necessary to generate the functional group needed for cross-linking process. There are relatively few published reports on these cross-linkers. Figure 9.2 shows the reactions of three of these compounds. 3-Amino-4-methoxyphenylvinyl sulfone (Figure 9.2A) has been used to couple the enzymes catalase, trypsin, chymotrypsin, and ribonuclease to cellulose.[17] The first step involves a Michael addition reaction of a hydroxyl group to the vinyl sulfone. Treatment of the label with sodium nitrite in acid converts the amino group on the benzene ring to a diazonium compound, which will react with tyrosine residues of the proteins to complete the cross-linking process.

A MULTISTEP CONJUGATION OF RICIN AND IMMUNOGLUBULIN[16]

Step 1: A solution of the N-hydroxysuccinimidyl iodoacetic acid (0.3 mg) in 200 µL dried dimethylformamide was added to 1.55 mL solution of ricin (20 mg) in borate buffer. After stirring for 30 min at room temperature, the reaction mixture was applied to a column (22 × 1.6 cm) of Sephadex G-25 pre-equilibrated with phosphate buffer to isolate iodoacetylated ricin in about 12 mL of eluate. The number of iodoacetyl groups introduced into each molecule of ricin was 1.5, as determined spectrophotometrically after reacting the modified protein with 3-carboxy-4-nitrothiophenol.

Step 2: To a solution of immunoglobulin (43.6 mg) in borate buffer (4.3 mL) was added a solution of SPDP reagent (0.218 mg) in dry dimethylformamide (70 µL). After stirring at room temperature for 30 min, the mixture was applied to a column (20 × 1.6 cm) of Sephadex G-25 equilibrated with acetate buffer. Elution with the same buffer isolated the substituted protein in 11 mL. The eluate (10 mL) was concentrated to 3 mL in an Amicon ultrafiltration cell with a PMIO membrane.

Step 3: To the concentrate was added dithiothreitol (22 mg) in acetate buffer (0.5 mL). After stirring for 30 min at room temperature the mixture was applied to a column (20 × 1.6 cm) of Sephadex G-25 that had been equilibrated in nitrogen-flushed phosphate buffer.

Step 4: The protein solution, removed by elution in the same buffer, was added directly into the iodoacetylated ricin solution prepared above. The mixture was then concentrated to 5 mL by ultrafiltration. After stirring for 18 h at room temperature, the mixture was treated with N-ethylmaleimide (1 mg) dissolved in dimethylformamide (100 µL). An hour later the solution was applied to a column (82 × 3.2 cm) of Sephadex G-200 equilibrated with borate buffer and eluted with the same buffer solution to isolate the conjugated product.

Similarly, ethyl N-(carbamoylcyanomethyl)acetimidate (Figure 9.2B), under mild conditions, reacts with protein amino groups and cyclizes to form an aminoimidazoyl derivative. The resulting imidazole-derivatized amino group can be diazotized and then reacts with the tyrosyl residue of another protein to form a cross-linked product. By this method, antigens have been coupled to antihuman group O erythrocyte IgG.[18]

Another masked reagent, N-(2,2-dimethoxyethyl)-5-(hydrazidecarbonyl)pentanamide (Figure 9.2C), has been used to couple glycopeptides to proteins.[19] The reagent is first converted to acylazide

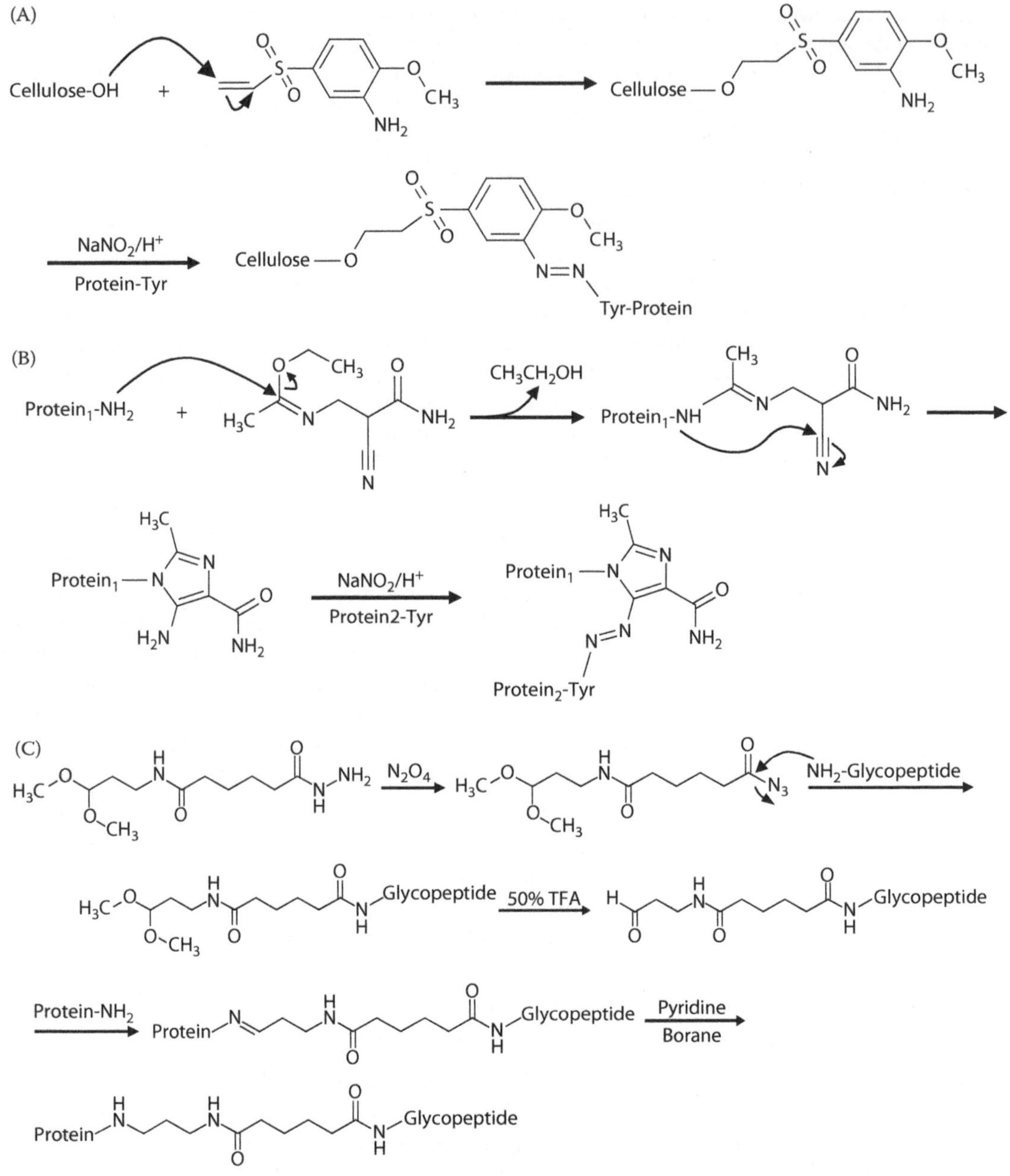

FIGURE 9.2   Cross-linking reactions of disguised reagents. (A) 3-Amino-4-methoxyphenylvinyl sulfone; (B) Ethyl *N*-(carbamoylcyanomethyl)acetimidate; (C) *N*-(2,2-dimethoxyethyl)-5-(hydrazide carbonyl) pentanamide.

on treatment with dinitrogen tetraoxide. After reaction with an amino-containing compound, the blocked methyl acetal is removed with 50% trifluoroacetic acid to generate an aldehyde group. Cross-linking is achieved between the aldehyde and a protein amino group by reductive alkylation.

The examples above demonstrated the versatility of the multistep process. Even more complex procedures involve applications to protein–oligonucleotide and DNA–DNA cross-linking. Ghosh et al.[20] described two methods for the synthesis of oligonucleotide–enzyme conjugates. 5′-Thiolated oligonucleotide was synthesized in two steps using cystamine and EDC followed by dithiothreitol.

In the first approach, heterobifunctional *N*-succinimidyl-6-maleimidohexanoate (SMH) was used to modify an amino group of calf intestine alkaline phosphatase and then coupled to 5′-thiolated oligonucleotide to give a 1:1 conjugate. In the second strategy, homobifunctional *N,N′*-(1,2-phenylene) bismaleimide was used to activate the thiol group of the 5′-thiolated oligonucleotide. In the mean time, horseradish peroxidase or beta-galactosidase was thiolated with 2-iminothiolane and then reacted with the modified oligonucleotide to give an enzyme–oligonucleotide conjugate. Another complex reaction is shown by Tona and Haner[21] in interstrand cross-linking of DNA. In this case, 1,3-butadiene-containing building blocks were integrated into DNA oligomers such that they are in opposite positions to each other in the double helix. Cross-linking between the diene-modified duplexes was achieved by bifunctional dienophiles such as homobifunctional bismaleimide, *N,N′*-ethylenedimaleimide through a double Diels–Alder reaction. The reaction resulted in clean cross-linking of the two DNA strands.

9.3 GENERAL CONDITIONS FOR CROSS-LINKING

Optimal cross-linking conditions are highly dependent on the type of reagent used and the particular system under investigation.[3,22] For example, reaction times may range from minutes to hours and the reagent concentration varies with its relative reactivity as well as the stability of each reagent. Some reagents are readily hydrolyzed and consequently may have to be used in excess. Cross-linking can generally be carried out in buffers such as phosphate-buffered saline (PBS) or isotonic phosphate. Since reagent type dictates the reaction conditions, the following discussion represents generalized parameters for the most frequently used reagents only. For specific applications, the reader is referred to the specific literature.

9.3.1 CHOICE OF REACTION MEDIUM

Conjugations should be carried out in a well-buffered system at a pH that maintains the integrity of the biological macromolecules and is optimal for the reaction. The ionic strength should be in the range of 25–100 mM in most cases. For thiol and α-amino groups, modification at pH 7.0–7.5 in phosphate buffers is ideal. More basic ε-amine of lysine requires more alkaline pH, in the range of 8.0–9.5, where carbonate/bicarbonate or borate buffers are satisfactory. The optimal pH also depends on the reagents used. Reactions with NHS esters are best carried out in pH 8.2 bicarbonate buffer, and isothiocyanates at pH 9.0–9.5, provided by carbonate or borate buffers. It should be noted that for these reactions, the buffer should not contain any free amino groups such as in Tris buffer.

For reagents that are poorly soluble in water or highly reactive with water, a water-miscible cosolvent must be employed to dissolve the reagent before adding to the conjugation medium. Some cosolvents are methanol, ethanol, 2-propanol, 2-methoxyethanol, dioxane, dimethylformamide (DMF), and dimethylsulfoxide (DMSO). The most versatile of these are DMF and DMSO because they are miscible with water in all proportions, are inert to many reactive reagents, and are compatible with most aqueous protein solutions even at up to 30% v/v ratios. For sulfonyl chlorides, DMF is the solvent of choice since they react with DMSO. The cosolvent should be carefully dried and stored over a drying agent to prevent competing hydrolysis of the reactive reagent.

9.3.2 CHOICE OF REACTION TEMPERATURE AND TIME

As a general rule of thumb, conjugation reactions should be done at or below room temperature since most reactions are rapid. Also, low temperatures tend to increase selectivity of the reaction, resulting in fewer side reactions. To avoid overreaction, a high temperature should be accompanied with a short reaction time and vice versa. It should be noted that the higher the temperature, the faster will be the reaction. A convenient procedure is to add the reagent to a gently stirred buffered solution of the macromolecules to be cross-linked in an ice-bath and then allow the bath to warm up to room temperature over a period of about 2 h, although many published procedures specify overnight reaction times. Generally, 1–2 h is sufficient for conjugation reactions to go to completion. However, depending on the reagent, longer reaction times may be necessary. The more reactive the reagent, the shorter the recommended reaction time.

9.3.3 CHOICE OF REACTANT CONCENTRATIONS

The degree of intermolecular and intramolecular cross-linking depends on the concentrations of the macromolecules to be cross-linked. The higher the concentration, the higher will be the degree of intermolecular cross-linking. For nonassociating proteins, concentrations above 10 μM are strongly recommended with an optimum in the range of 50–100 μM. For associated macromolecular complexes, a dilute solution will decrease intercomplex cross-linking. The concentration of the cross-linker is also important. The higher the concentration of reagent, the faster will be the reaction and the more cross-linking will take place. The degree of cross-linking is generally limited by the ratio of the reagent to the biological substance. The choice of molar ratio depends on the available reactive moieties of the macromolecules, such as reactive amino acid side chains of the proteins. Low to moderate ratios will decrease the side reactions. If there is no prior knowledge on the reaction, trial and error may be necessary to get the desired products. The cross-linking reagent should be added dropwise as slowly as possible to a slowly stirred solution. Stirring can be done with a magnetic stir-bar at a slow speed to avoid denaturation of proteins and other biological molecules.

9.4 CROSS-LINKING PROTOCOLS FOR COMMONLY USED REAGENTS

In the study of protein–protein interactions by chemical cross-linking, the selection of a reagent is critical. For each reagent to achieve a cross-linking goal, different reaction protocols are to be established. The following sections describe some general reaction conditions for the different classes of cross-linkers and provide examples of protocols for some common reagents in each class. Many other reaction procedures can be found in Niemeyer's edited bioconjugation protocols.[1] Specific protocols may be found in publications of commercial companies that provide the cross-linking reagent.

9.4.1 EXAMPLES FOR ZERO-LENGTH CROSS-LINKER

Carbodiimides are probably the most commonly used zero-length cross-linkers. Cross-linking is usually performed at a pH between 4.5 and 5 where the reaction rate is rapid, requiring only a few minutes for many applications. Carbodiimides are subject to hydrolysis and should be stored desiccated. 4-Morpholine ethanesulfonic acid (MES) is a good carbodiimide reaction buffer. Tris, glycine, acetate, and phosphate buffers can react with the cross-linkers and should be avoided. Below are protocols for general applications. Other examples are discussed above.

9.4.1.1 Cross-Linking a Peptide and a Protein Using EDC

  a. Prepare the protein at 10 mg/mL in 0.1 M MES buffer, pH 5. (The protein should be free of other buffers such as Tris or interfering substances such as thios, amines, acetate, DTT.)
  b. Prepare peptide (10 mg/mL) in the same MES buffer.
  c. Prepare EDC at 10 mg/mL in distilled water (should be used immediately).
  d. Mix 2 mg peptide and 2 mg protein (ratio may be optimize depending on the desired coupled ratio).
  e. Add 0.5–1 mg EDC to the mixture (0.05–0.4 mg for each mg of total protein, usually 0.5 mg EDC for 1 mg BSA) with stirring.
  f. Incubate for 2–3 h at room temperature.
  g. Desalt by dialysis or gel filtration.

9.4.1.2 Cross-Linking of Porcine Luteinizing Hormone with EDC to Study a and b Subunit Interactions[23]

a. Dissolve hormone in deionized water to a concentration of 0.2 mg/mL.

b. Adjust pH to 4.75 with HCl.

c. Dissolve EDC in deionized water and add to the hormone solution to a final concentration of 0.02 M.

d. Gently stir at room temperature for 1.5 h, keeping the pH constant.

e. Terminate the reaction by dialyzing the reaction mixture against 0.001 N HCl in the cold and lyophilize for analysis.

9.4.2 EXAMPLES FOR HOMOBIFUNCTIONAL REAGENTS

9.4.2.1 Bis-Imidoesters

Imidates are generally readily soluble in aqueous solutions and are hydrolyzed rapidly with a pH-dependent half-life ranging from several minutes up to half an hour. Below pH 8.5, the half-life for ethyl acetimidate is 2–5 min. The rate of hydrolysis increases substantially at higher pH values.[24,25] Up to a 100-fold excess of reagent, at a concentration range of 0.1–10 mM, is required for complete reaction. To circumvent the degradation problem, incremental additions of reagents may be used. Imidoesters react over a wide pH range from 7 to 10[24–27] and temperatures from 0°C to 40°C.[24,25,28] Alkaline pH increases the rate of the reaction of imidates with amines to form amidines.[24,25,29,30] Imidoester conjugation is usually performed between pH 8.5 and 9. The reaction rate decreases several fold as the temperature drops from 39°C to 25°C and again from 25°C to near 0°C.[24] At or below zero, amidination occurs at considerably slower rates and requires longer reaction times of several hours to overnight.[28,31,32] The product carries a positive charge at physiological pH, as does the primary amine it replaces and therefore does not affect the overall charge of the protein. The final product should be kept at neutral to acidic pH to retard hydrolysis.

9.4.2.1.1 Cross-Linking of the Subunits of HIV-1 Reverse Transcriptase with Dimethylsuberimidate (DMS)[33]

1. DMS was prepared immediately before use by dissolving 10 mg in 180 μL of ice-cold triethanolamine (TEA)-HCl (0.15 M, pH 8.2). The pH of the DMS solution was readjusted to 8.2 by addition of 20 μL of 1 M NaOH.

2. Cross-linking was initiated by the addition of 2 μL DMS (final concentration 10 mg/mL) to 8 μL of reaction mixture at 4°C. The final reaction mixture (45 mM TEA buffer, pH 8.2) contained 110 nM heterodimeric HIV-I reverse transcriptase, 650 nM of polyuridylic acid, 10 mM MgCl, 100 mM NaCl, and 1 mM DTT.

3. Incubation was performed for 30 min at 37°C.

4. The cross-linking reaction was stopped by adding an equal volume of 1 M glycine.

9.4.2.2 Bis-N-Hydroxysuccinimide (NHS) Esters

The NHS cross-linkers are more stable in solution than their imidate counterparts and are typically more reactive at neutral pH. The half-life of hydrolysis of NHS esters is approximately 10 min at pH 8.6 and 4°C,[34] 1 h at pH 8.0 and 25°C,[35] and 4–5 h at pH 7 at 0°C.[36] NHS esters are more stable when dissolved in dry organic solvents. In absolute ethanol at 23°C, the NHS ester retains 80% activity after 20 days.[37] The imidazole group of histidine effectively accelerates the rate of hydrolysis of the NHS groups in solution. The reaction product of NHS esters with histidine is unstable and hydrolyzes very rapidly. Therefore, when histidine side chains are present, higher concentration of NHS cross-linkers are required to achieve a given degree of conjugation. The extent of hydrolysis of NHS esters in aqueous solutions may be determined by measuring the increase in absorbance at 260 nm. The molar extinction coefficient for the NHS group is 8.2×10^3 M^{-1} cm^{-1} at 260 nm, pH 9.0.[37] The optimal pH for the cross-linking reaction is pH 9. At pH 7, the reactivity is only half of that at pH 9. Temperature has little effect on the reaction, allowing it to react efficiently even near freezing. NHS esters have been used in the concentration range 0.05–9 mM with reaction time from minutes to hours depending on the conditions used.[36,38,39] Buffers that contain primary amines, such as Tris, or reducing agents should be not be used. Most common buffers are phosphate, bicarbonate/carbonate, HEPES, and borate at concentrations between 50 and 200 mM. The NHS ester reaction with amines is typically performed between pH 7.0 and 9.0 at room temperature for 30 min to 2 h. Reaction times at 4°C should be increased fourfold over room temperature incubation times to give similar results.[22] NHS cross-linkers are usually used in 2- to 50-fold molar excess to protein. The concentration of the cross-linker can vary from 0.1 to 10 mM. Insoluble NHS esters can be first dissolved in water-miscible organic solvents such as DMSO and DMF. The final organic solvent can be up to 10% final volume in aqueous reaction. When only the surface of a cell or membrane is to be modified, it is best to use water-soluble reagents since they will not permeate the membrane. It is recommended that the protein concentration be kept above 10 μM (50–100 μM) because more dilute protein solutions result in excessive hydrolysis.

9.4.2.2.1 Cross-Linking of Proteins with Ethylene Glycol Bis(Succinimidylsuccinate) (EGS)[40]

1. The protein of interest has to be taken up in a buffer free of amines, such as HEPES. The pH of the buffer should be between 7 and 9.
2. Prepare a fresh stock solution of EGS (50 mM) in DMSO. If the protein cannot tolerate DMSO, choose a different cosolvent or use water soluble sulfo-EGS.
3. Use 10 μg of protein sample per reaction and add EGS to a final concentration of 0.01–5 mM.
4. Incubate the reaction for 20 min at room temperature or on ice for 1 h.
5. Add 50 mM Tris pH 8.0 buffer to quench the reaction; incubate for 5 min at room temperature.

9.4.2.2.2 Cross-Linking ATP Synthase Complex with Dithiobis(Succinimidyl Propionate) (DSP)[41]

1. Prepare protein sample (1.5 mg/mL) in 50 mM triethanolamine HCl buffer, pH 8.0, containing 0.25 M sucrose.
2. Prepare 0.15 mM DSP stock solution in methanol:acetone (1:1).
3. Incubate protein sample with 1% by volume of DSP at 0°C for 30 min.
4. Quench the cross-linking reaction by addition of lysine to 5 mM final concentration. If preservation of enzymatic activity is not essential, the reaction can be quenched with 125 mM Tris-HCl buffer, pH6.8, containing 4% SDS, 4 mM EDTA, and 20% glycerol for gel electrophoresis analysis.
5. Cross-linked proteins can be cleaved by incubating with 80 mM 2-mercaptoethanol or 10 mM DTT.

9.4.2.3 Bis-Maleimido Reagents

The maleimide group is selective for sulfhydryl groups when the pH of the reaction mixture is kept between 6.5 and 7.5,[42] with an optimum pH for the reaction near 7.0. At pH 7, the rate of the reaction of maleimides with sulfhydryls is 100-fold faster than with amines. Above this pH range, the reaction rate with primary amines becomes more significant. Above pH 8.0, hydrolysis of maleimides to nonreactive maleamic acid can occur.[43] Reducing agents should be excluded from buffers, since they will quench the reactivity. β-Mercaptoethanol, dithiothreitol, mercaptoethylamine, and other thiol compounds must be removed prior to the cross-linking reaction.

9.4.2.3.1   *Cross-Linking of Cysteine-Containing Mutant of Mitochondrial
             F1Fo ATP Synthase Complex with 1,6-Bis-Maleimidohexane (BMH)
             or N,N'-o-Phenylene-Dimaleimide (OPD)*[44]

1. Stock solutions (100 mM) of BMH and OPD in dimethylformamide were stored at −20°C.
2. Mitochondria with cysteine-containing mutant of F1Fo ATP synthase complex were prepared and suspended in the isolation buffer (0.6 M mannitol, 2 mM EGTA, 10 mM Tris-maleate, pH 6.8).
3. These prepared mitochondria were washed twice with 0.6 M mannitol, 2 mM EGTA, 50 mM HEPES, pH 7.0, and suspended in the same buffer at a protein concentration of 5 mg/mL.
4. This suspension was incubated with 300 µM of either OPD or BMH for 1 h at room temperature.
5. Reactions were stopped by the addition of 25 mM of 2-mercaptoethanol.

9.4.2.4   Bis-α-Haloacetyl Reagents

Haloacetyl compounds are directed toward the sulfhydryl but also react with other functional groups. The most common reagents are α-iodoacetyl derivatives. Selectivity for sulfhydryl groups is achievable by using only a slight excess of γ-haloacetyl groups over the number of free sulfhydryls and by keeping the pH of the reaction mixture between 7.5 and 8.5, with an optimum specificity at pH 8.3. If there are no free sulfhydryls present, or a gross excess of haloacetyl group is used, the haloacetyl group can react with other amino acids such as imidazoles at pH 6.9–7.0, although the reaction is slow. Extraneous reducing agents should be excluded from buffers for α-haloacetyl reactions. Iodoacetamides commonly used to modify thiols will react with amines of proteins if the pH is in the range 9.0–9.5.[45]

9.4.2.4.1   *Cross-Linking of Aldolase Subunits with
             N,N-Bis(α-Iodoacetyl)-2,2'-Dithiobis(Ethylamine) (BIDBE)*[46]

1. Prepare 7.15 mg/mL aldolase solution in 0.05 M Tris-HCl buffer, pH 8.0.
2. Prepare a 34.1 mM solution of BIDBE, freshly dissolved in dimethylsulfoxide.
3. 0.25 mL of aldolase was mixed with 10 µL of BIDBE. The final concentration of BIDBE is 1.31 mM and the molar ratio of BIDBE to aldolase sulfhydryls is 0.95:1.
4. Incubate the mixture for 3.5 h at 25°C in the dark.
5. At the end of the incubation, the sample is dialyzed for 16 h at 1°C in the dark against 0.059 M Tris-phosphate buffer, pH 7.0, containing 20% sucrose.
6. The sample is then analyzed by gel electrophoresis.
7. The cross-linked protein can be cleaved by 1% β-mercaptoethanol.

9.4.3   EXAMPLES FOR HETEROBIFUNCTIONAL REAGENTS

Heterobifunctional reagents are diverse compounds as discussed in Chapter 6. The reaction conditions described above for specific reactive moieties also apply to the heterobifunctional cross-linkers if they contain such a reactive group. In general, the most reactive or unstable moiety end of the bifunctional agent is reacted first. For example, with NHS-ester-disulfide cross-linkers such as *N*-succinimidyl-3-(2-pyridyldithio)propionate (SPDP) reaction with an amine is carried out first to react with the NHS ester. The pyridyl–disulfide derivative obtained can either be reduced with reducing agents such as β-mercaptoethanol to generate a free thiol or reacted directly with protein or peptide with a free sulfhydryl for coupling via disulfide exchange. The pyridine-2-thione released can be quantified at 343 nm (molar extinction coefficient at 343 nm = 8.08×10^3 M^{-1} cm^{-1}).[47,48] For *N*-succinimidyl *S*-acetylthioacetate (SATA) after reaction with an amine, excess and hydrolyzed

SATA is typically removed by gel filtration or other similar techniques. The activated protein is usually quite stable and can be stored for long periods. The blocked sulfhydryl group on the derivatized molecule can be deprotected by incubation with 50 mM hydroxylamine hydrochloride for 2 h at pH 7.5. It is not necessary to remove the hydroxylamine from the deprotected derivative prior to reaction with maleimide cross-linkers. Similarly, for NHS-ester-maleimide compounds, reaction with NHS ester is first carried out between pH 6.5 and 7.5 to prevent hydrolysis and reaction of the maleimide group with amines. If it is necessary to perform the maleimide reaction prior to coupling with the NHS ester, a buffer with pH below 7.0 should be used and the reaction times should be kept to a minimum. This procedure will be most successful if there are no primary amines available and if a sulfhydryl is readily accessible. An example is sulfosuccinimidyl-4-(N-maleimidomethyl) cyclohexane-1-carboxylate (sulfo-SMCC), which is very stable in solution exhibiting essentially no degradation after 6 h at 30°C in pH 7 buffer. A SMCC activated protein is stable and often can be stored several months prior to the second stage of coupling.[47]

9.4.3.1   Conjugation of Human Serum Albumin (HSA) and Monoclonal Antibody (mAb) with SPDP[49]

a. One mL of mAb Dal K20 (10 mg/mL) in PBS (0.01 M phosphate in 0.15 M sodium chloride, pH 7.2) was derivatized with fivefold molar excess of SPDP for 30 min. SPDP was dissolved in a small volume of DMF, which did not exceed 20% of the volume of the HSA solution. The solutions were then desalted into buffer B (0.1 M sodium phosphate, pH 7.2, containing 1 mM EDTA (ethylenediaminetetraacetic acid)). The number of pyridyldithio groups incorporated into mAb Dal K20 was determined in the presence of 0.1 M DTT, using $E = 8.08 \times 10^3$ M^{-1} cm^{-1} at 343 nm for released pyridine-2-thione.

b. To HSA at 10 mg/mL in PBS was added a fivefold molar excess of SPDP with stirring at room temperature. The SPDP had been dissolved in an amount of DMF that did not exceed 20% of the volume of the HSA solution. After 30 min, the solution was desalted into 0.1 M acetate buffer, pH 4.5, containing 0.1 M NaCl, and DTT added to give a concentration of 10 mM. After 20 min at room temperature, the DTT-treated mixture was desalted into buffer B. The number of pyridyldithio groups incorporated into HSA was determined.

c. mAb Dal K20 (3.6 mg/mL, 3.0 mL) derivatized with SPDP in buffer B was mixed with HSA-SPDP-SH in buffer B at a 4:1 molar ratio of HSA over mAb Dal K20 in a final volume of 6.2 mL. All reaction mixtures were stirred briefly and left at room temperature for 16 h at which time thiols were blocked by the addition of a 20-fold molar excess of N-ethylmaleimide over mAb Dal K20.

d. The mAb K20-HSA conjugate was purified by gel filtration chromatography on Bio-Gel P 300 (2.5 cm × 90 cm) equilibrated with PBS.

9.4.3.2   Cross-Linking of Demineralized Bone Matrix (DBM) and Monoclonal Antibody with Sulfo-SMCC[50]

a. Traut's Reagent (5 mg/mL) was dissolved in PBS with 4 mM EDTA (pH = 8).

b. DBM (5 mm × 5 mm × 2 mm) was soaked in Traut's Reagent for 3 h at room temperature.

c. In a separate reaction, 5 μg monoclonal antibody was diluted in PBS containing 4 mM EDTA (pH = 7.2), and sulfo-SMCC (25 μg/mL) was reacted with antibodies at room temperature for 30 min.

d. The DBM treated by Traut's Reagent was washed by PBS for several times, and incubated with antibodies treated by Sulfo-SMCC for 2 h at room temperature.

e. The antibody-conjugated DBM was then washed with PBS for several times and incubated with 5% (W/V) glycine and 5% (W/V) bovine serum albumin for 3 h to block remaining reactive groups on DBM.

9.4.4 EXAMPLES FOR HETEROBIFUNCTIONAL PHOTOSENSITIVE REAGENTS

Photosensitive reagents are, of course, sensitive to light and they are particularly sensitive to ultraviolet radiation, that is, wavelengths less than 300 nm. Consequently, they must usually be handled in the dark or under conditions of dim or red light. For example, under normal white fluorescent lamp illumination, some photoreactive reagents may have half-lives of only a few hours or less. Azides are also sensitive to sulfhydryl reducing agents such as dithiothreitol, 2-mercaptoethanol, and glutathione, since they can be reduced to the corresponding amine.[51,52] This reduction is pH dependent. In 10 mM dithiothreitol, various azides are found to have half-lives of 5–15 min at pH 8. At pH 10, this rate is increased 12-fold.[52] With 50 mM glutathione or 2-mercaptoethanol at pH 8, the azides were reduced 60%–70% and 10%–20%, respectively, in 24 h. Therefore, if reducing conditions are required during cross-linking, the use of 2-mercaptopethanaol is recommended.

Aryl azides are generally insoluble in aqueous buffers and may be solubilized with the aid of water-miscible organic solvents such as acetone, methanol, ethanol, dioxane, dimethylformamide, pyridine, acetonitrile and dimethylsulfoxide.[39,53–56] The final concentration of organic solvent may be as high as 20%. Alternatively, a fine powder of the reagent may be added to the reaction mixture, although the rate of reaction may be reduced.

The photolysis of photosensitive cross-linkers and their subsequent chemical reactions are temperature independent. A common method to photoactivate azides is to irradiate with a short-wavelength UV lamp (typically 254 nm). Arylazides are photolyzed at wavelengths between 250 and 460 nm forming a reactive aryl nitrene. The half-time of photolysis is usually on the order of 10–50 s with the sample positioned close (e.g., 1 cm) to the lamp.[57] An alternative method is flash photolysis using electronic flash units.[32,58,59] A bright camera flash works well with the nitro and hydroxyl-substituted aryl azides. Unsubstituted aryl azides require UV light or numerous flashes. In a typical experiment, less than 10 flashes are sufficient to photolyze reagents associated with proteins.[59] With molecules that contain chemical moieties that are photosensitive or prone to photo-oxidation, longer wavelength illumination may be necessary, for example, by illumination through glass filters which are particularly useful for activating aryl azides with nitro substituents.[60] In this case, irradiation must be lengthened to many minutes or even hours, depending on the molar absorptivity of the aryl azides at the longest wavelengths.[39,61,62] It should be kept in mind that the yield resulting from a photoreactive cross-linker is inherently low. Yields of less than 10% should be considered acceptable.[47]

9.4.4.1 Cross-Linking of Proteins with the Photoreagent N-(4-Azido-2,3,5, 6-Tetrafluorobenzyl)-3-Maleimidylpropionamide (TFPAM-3)[56]

9.4.4.1.1 Labeling

1. Pass protein through a BioSpin 6 centrifuge column equilibrated with 50 mM Mops buffer (pH 7.0), containing 0.5 mM EDTA, and 10% glycerol to remove thiols and primary amines present in storage buffer.
2. Prepare a stock solution of 4 mM TFPAM-3 in DMSO.
3. Mix 40 µM of a protein with 250 µM TFPAM-3 at room temperature for 1 h with gentle stirring.
4. Quench the reaction with 1 mM cysteine for 30 min at room temperature with gentle stirring.
5. Remove excess TFPAM-3 by passing the sample through a column equilibrated with 25 mM Hepes (pH 7.5), containing 60 mM potassium acetate, 6 mM magnesium acetate, and 10% glycerol (conjugating buffer). Proteins can be stored in this buffer at 20°C.

9.4.4.1.2 Photo-Conjugation

1. Mix 1 µM of TFPAM-3 labeled protein with 1 µM of a protein of interest in conjugating buffer.
2. Irradiate the sample for 45 min at room temperature at a distance of about 2 cm from the UV lamp (6W UV lamp, model UVL-56, Blak-Ray lamp).
3. Analyze conjugated products.

9.4.4.2 Cross-Linking UvsY Hexamer Protein Complex with the Photo-Reagent Ruthenium(II) Tris-Bipyridyl Dichloride (Ru(II)bpy$_3$Cl$_2$)[63]

a. Free UvsY protein from DTT and β-mercaptoethanol as they can be oxidized in the reaction.

b. Mix proteins (0.01–20 μM in 15 mM sodium phosphate (pH 7.5) containing 150 mM NaCl) with 0.125 mM Ru(II)bpy$_3$Cl$_2$.

c. Place the solution in a 1.7 mL Eppendorf tube parallel to the beam of light at a distance of 50 cm from a 150-W xenon arc lamp, or at a distance of 5 cm for the high-intensity standard flashlight.

d. Add ammonium persulfate to 2.5 mM concentration just before irradiation.

e. Irradiate sample for 0.5 s with xenon arc lamp (the light should be filtered first through 10 cm of distilled water and then through a 380–2500 nm cut-on filter. Exposure time can be controlled by shining light through the shutters of a single lens reflex camera with the lens and back cover removed) or for 5–30 s for flashlight.

f. Quench the reaction with approximately 10 μL of 0.2 M Tris, 2.88 M β-mercaptoethanol.

g. Analyze the cross-linked products.

9.5 CROSS-LINKING PROTOCOLS BASED ON BIOLOGICAL SYSTEMS

The above discussion on reaction protocols is based on the chemical properties of the cross-linking reagents. While understanding the chemistry of the reagents is important, in most cases, we would be most interested in the biological system under study. Based on the aim in studying the biological macromolecules, a specific cross-linker will be chosen. For example, to study a biological transmembrane system, a reagent that can penetrate the membrane would be desired. Thus, more hydrophobic and less hydrophilic cross-linkers should be chosen. The reverse would be true for studying aqueous macromolecules. The following will demonstrate the principles in choosing the right reagents for some general biological systems to be investigated.

9.5.1 Soluble Macromolecules

Water-soluble macromolecules are those entities that exist freely in aqueous solutions. Most of the water-soluble proteins have a shell of hydrophilic amino acids such as serine, aspartic acid, and lysine that can form hydrogen bonds with surrounding water molecules. The core of the soluble protein may contain hydrophobic amino acids such as leucine and tryptophan. Some macromolecules tend to associate into higher aggregates when present in high concentrations, depending on their dissociation constants. On the other hand, some macromolecules exist in stable multisubunit entities. Under certain circumstances, different molecules may associate to form complexes, such as ligand–receptor interactions. An intriguing example is the interaction between α-lactalbumin and galactosyltransferase. It is only in the presence of glucose and UDP-galactose that the two proteins associate to form lactose synthase.[64] Thus, for studying protein–protein interactions in such systems, appropriate conditions must be chosen for the interaction to occur. Since there are so many different possible variations of soluble proteins to be studied, only a few examples are chosen to illustrate the principles involved in the cross-linking process. In general, for cross-linking of soluble macromolecules, water-soluble reagents would be used. As discussed in Chapter 4, compounds with ether-oxygen, hydroxyl groups, ester and amide bonds, formal charges, and sulfonation have increased solubility in aqueous solutions. These reagents should be considered first when designing a cross-linking experiment.

9.5.1.1 Cross-Linking Nonassociated Proteins

Intramolecular cross-linking of nonassociated proteins can be carried out with a one-step procedure. As discussed above in Section 9.2.1, the cross-linker can be added directly to a solution of target proteins. Since intermolecular cross-linking would also occur, dilute solutions should be used to avoid close encounter of the macromolecules. For intermolecular conjugation of two proteins, such as the preparations of immunotoxins or enzyme-antibody conjugates, a two-step procedure is desired as discussed above under Section 9.2.2, with examples presented therein. Three-step and multistep procedures are also applicable as shown in Sections 9.2.3 and 9.2.4.

9.5.1.2 Cross-Linking Multisubunit Complexes

Similar to single subunit proteins, dilute solutions of strongly associated, stable multisubunit complexes should be used to avoid intermolecular cross-linking as discussed in Section 9.3.3. In most cases, a one-step procedure can be used for intracomplex subunit cross-linking as showed above in Section 9.2.1 for ribosome protein complexes. Various reagents, either homofunctional or heterofunctional, may be used depending on the aims of the experiments. These are illustrated in Section 9.4 for reverse transcriptase (Section 9.4.2.1.1), ATP synthase (Sections 9.4.2.1.2 and 9.4.2.3.1), aldolase (Section 9.4.2.4.1), and UvsY protein hexamer (Section 9.4.4.2).

Investigation of protein–protein interactions of readily dissociable protein complexes is more challenging. Some proteins interact with several other proteins transiently for regulatory or enzymatic purposes. In these cases, a condition that causes the formation of the complex must be achieved during the cross-linking experiment, either by inclusion of substrates or cofactors that induce the interaction or by increasing the concentrations of the interacting subunits. The most commonly used reagents are photoactivatable cross-linkers, although oxidative cross-linking such as hexahistidine-mediated cross-linking methodology has been evolving.[65,66] Any of the photoactivatable reagents listed in Appendix E may be used. In the first step, the reagent is covalently attached to lysine or cysteine side chains of the prey protein. The derivatized prey protein is then incubated with an interacting protein and exposed to UV light to induce cross-linking between the prey protein and the interacting proteins. For label transfer experiments, cleavable reagents, particularly those containing disulfide bonds, are used. After photoactivation, the reagent is cleaved, thereby transferring the label to the interacting proteins. The transferred label enables the identification or isolation of the interacting proteins.

For hexahistidine-cross-linking, a hexahistidine tag or a NH2-Gly-Gly-His tag is first genetically introduced into the prey protein.[66–68] On incubation of the tagged prey protein with other interacting protein in the presence of nickel, fast and efficient cross-linking of proteins is achieved when oxidized by peracids like magnesium monoperoxyphthalic acid or $KHSO_5$. The cross-linked adducts can be identified immunologically. Klein et al.[69] have used this method to study the transcriptional activation domain–coactivator complexes.

Advances in genetic cloning have facilitated the development of a new methodology of cross-linking to study protein–protein interaction, particularly in living cells. Photoactivatable amino acid analogues such as p-benzoyl-L-phenylalanine (pBpa), p-azido-L-phenylalanine, and 4′-[3-(trifluoromethyl)-3H-diazirin-3-yl]-L-phenylalanine are site-specifically incorporated into interested proteins by means of heterologous amber suppressor tRNA/aminoacyl-tRNA synthetase pairs that recognize the unnatural amino acids.[70] After association of the modified protein with target macromolecules, the complex is photoactivated to initiate cross-linking. Lee et al.[71] have used pBpa to study the interaction of *Escherichia coli* catabolite activator protein (CAP) with DNA. pBpa was genetically incorporated into CAP in bacteria in response to an amber nonsense codon using an orthogonal tRNA/aminoacyl-tRNA synthetase pair. On binding with DNA and after UV irradiation, SDS–PAGE analysis showed that the mutant CAP containing pBpa formed a covalent complex with a DNA fragment containing the consensus operator sequence.

9.5.2 MEMBRANE-BOUND PROTEINS

Membrane-bound proteins present a special challenge for studying protein–protein interactions. These proteins consist of transmembrane helices that are hydrophobic in nature and an extramembraneous region exposed to the aqueous media. The choice of cross-linking reagent depends on objectives of the experiment. Generally, the exposed section of the protein is of interest, such as the membrane surface receptors. In this case, the use of membrane-impermeable cross-linkers will ensure cell-surface specific cross-linking.[35] Compounds containing formal charges such as imidates, hydroxyl and amino groups, polyethylene glycol, and sulfonyl groups such as sulfo-NHS-esters are water-soluble, membrane-impermeable, and nonreactive with inner-membrane proteins. Thus, bis-imidoesters, sulfo-NHS-esters, and sulfonated photoreactive cross-linkers are good choices. Cross-linkers with spacer arms formed from polyethylene glycol provide the added benefit of transferring their hydrophilic spacer to the cross-linked complex, decreasing the potential for aggregation and precipitation of the cross-linked adduct. For these reagents, reaction time and concentration are less critical. If water-insoluble cross-linkers are used, the amount of reagent and reaction must be well controlled to reduce membrane penetration and reaction with inner membrane proteins. Numerous examples of studies on membrane surface proteins exist in the literature.[66,72,73] For example, Patzke et al.[74] investigated the structural organization of the coxsackievirus–adenovirus receptor (CAR) using homobifunctional bis(sulfosuccinimidyl) suberate (BS[3]) cross-linker. CAR is a type I transmembrane protein composed of two Ig domains, a membrane distal D1 and a membrane proximal D2. Cross-linking was started by the addition of BS[3] to extracellular CAR domains to a final concentration of 1 mM and incubated on ice for 1–2 h. Protein concentrations were chosen such that one of the two putative binding partners was used in up to 20-fold molar excess. The reaction was quenched with 50 mM Tris-HCl with subsequent heating. The cross-linked product was analyzed by SDS–PAGE and Western blot. The results showed the homophilic and heterophilic binding activities of D1 and D2. Western blot analysis identified monomeric, homodimeric, trimeric, and tetrameric complexes of CAR-D1D2. An additional cross-linked species, with an apparent mass of 40 kDa, indicates that the D1 monomer also binds D2.

To study the inner membrane proteins, reagents of greater hydrophobicity for membrane penetration are required. NHS-esters and photoactivatable phenyl compounds are useful. Although imidates are water-soluble, they can still penetrate membranes and may be used under certain circumstances. Water-insoluble dicyclohexylcarbodiimide can also provide valuable information. Various cross-linkers with differing spacer arm lengths can be used to determine the distance between molecules located in the membrane. Successful cross-linking with shorter cross-linkers is a strong indication that the molecules are in close approximation and may be interacting. Failure to obtain cross-linking with short cross-linkers, while obtaining conjugation with reagents with longer spacer arms, would indicate that the molecules are located in the same region of the membrane but not interacting. Wu et al.[75] investigated the position of β4 transmembrane helices in the BK potassium channel by disulfide cross-linking. BK channels are composed of α-subunits and four types of β-subunits. The locations of the two β4 transmembrane (TM) helices, TM1 and TM2, relative to the seven αTM helices, S0–S6, were analyzed from the extent of disulfide bond formation between cysteines substituted in the extracellular flanks of these TM helices. Disulfide cross-linking was effected by oxidation of reduced sulfhydryl groups with 4,4′-(azodicarbonyl)-bis-[1,1-dimethylpiperazinium, diiodide]. From the highly cross-linked cysteine pairs, the authors inferred that β4 TM2 is close to αS0 and that β4 TM1 is close to both αS1 and S2.

9.5.3 NUCLEIC ACIDS AND NUCLEIC ACID–PROTEIN COMPLEXES

Nucleic acid cross-linking, either DNA–DNA, RNA–RNA, or DNA–RNA coupling, can be achieved chemically or photochemically. Nucleic acid–specific homobifunctional, heterobifunctional, and photosensitive cross-linkers are described in Chapters 5 and 6. Many of these reagents are a result of the search for tumor-therapeutic drugs and may be used for chemical studies of nucleic acid interactions. In some cases, the cross-linked nucleic acids are a result of carcinogenic oxidative or UV-induced damages. Since thymine is photosensitive, it can be activated by UV-irradiation to form pyrimidine-pyrimidine cross-links.[76] Such a mechanism can be used to prepare cross-linked DNA simply by exposing the nucleic acid to 250–270 nm UV light. Photocross-linking can also be effected in the presence of metal complexes and other photosensitive compounds as described in Chapters 5 and 6. In addition to chemical and photochemical approaches, disulfide cross-links have been engineered into DNA and RNA.[77,78] This method of cross-linking has been used to probe solution structures, to monitor dynamic motion and thermodynamics, and to study the process of tertiary structure folding and function of DNA and RNA.

Of more interest in the study of nucleic acids are their interactions with proteins. Numerous studies on protein regulation of gene express, DNA damage repair, chromosome structure, and general nucleic acid–protein interactions have been published.[79–82] Reagents described in previous chapters, especially in Chapters 5 and 6, have been used for these studies. In addition, photo-cross-linking of DNA and protein have also been explored. For example, Neher et al.[83] used the method to study the interaction of Xeroderma pigmentosum group C (XPC)-Rad23B complex and cisplatin-damaged DNA strand. XPC–Rad23B is a protein complex involved in the recognition of damaged bulky DNA adducts and initiates the global genomic nucleotide excision repair pathway. When a mixture of XPC–RAD23B and cisplatin damaged DNA on ice was photolyzed by UV irradiation using General Electric-15 Watt bulbs, which emitted a wavelength of 254 nm, cross-linking between the protein complex and DNA was achieved. Analysis of the adducts by SDS–PAGE revealed that the XPC–Rad23B complex makes direct contact with the cisplatin-damaged DNA strand. Using denaturation and immunoprecipitation analysis, it was found that the XPC subunit was shown to directly bind with the damaged DNA, while the Rad23B–DNA interaction was largely indirect via its interaction with XPC. The power of photo-cross-linking may be realized from these experiments.

9.6 CONDITIONS FOR CLEAVAGE OF CROSS-LINKED COMPLEXES

Cleavage of the cross-linkers requires specific conditions depending on the type of bonding in the reagents. A general approach is given below for some of the common linkages.

9.6.1 DISULFIDE LINKAGES

Cleavage of disulfide bonds can be easily achieved by incubating with sulfhydryl compounds such as β-mercaptoethanol, dithiothreitol, or dithioerythritol at concentrations of about 10–100 mM, between pH 7 and 9 at 25°C–37°C for 10–30 min. Occasionally, concentrations of reducing reagents up to 0.4 M may be used.[84] Common buffers such as Tris and phosphate, as well as detergents such as Triton X-100 and sodium dodecylsulfate (SDS) do not interfere with the cleavage reaction. Disulfide bonds can also be conveniently cleaved during electrophoresis by addition of a reducing agent to the electrophoresis buffer.

9.6.2 GLYCOL BONDS

Vincinal glycol bonds can be cleaved by 15 mM sodium periodate, pH 7–7.5, for 4–5 h at 25°C.[85] Buffers such as triethanolamine and phosphate, as well as SDS, do not interfere. But Tris cannot be used since it reacts with sodium periodate.

9.6.3 AZO BONDS

Azo linkages can be cleaved by reduction with 0.1 M sodium dithionite in 0.15 M NaCl, buffered at pH 8 with 0.1 M NaHCO$_3$ for 25 min.[86] Disulfide reducing agents do not interfere with this process.

9.6.4 SULFONE LINKAGES

Sulfone bonds can be easily hydrolyzed in 100 mM sodium phosphate adjusted to pH 11–12 with Tris, 6 M urea, 0.1% SDS, and 2 mM dithiothreitol for 2 h at 37°C.[87] The presence of dithiothreitol is not absolutely necessary and the denaturants may not be needed.

9.6.5 Ester and Thioester Bonds

Theoretically esters and thioesters can be hydrolyzed under both acidic and alkaline conditions. They are most conveniently cleaved by 1 M hydroxylamine, in 50 mM Tris, pH 7.5–8.5, 25 mM CaCl$_2$ and 1 mM benzamidine for 3–6 h 25°C–37°C.[88]

9.6.6 Acetals, Ketals, and Orthoesters

These acetal, ketal, and orthoester bonds are acid labile and base stable.[89] Cross-linked products are stable at basic conditions and can be manipulated at pH 8–9.[90] They are approximately 100 times more stable at pH 7.4 and 1000 times more stable at pH of 8.4. They can be cleaved at pH 5.4 in minutes to hours, orthoesters being the fastest, acetals the slowest, and ketals intermediate.

9.7 REACTION COMPLICATIONS

9.7.1 General Considerations

Cross-linking of two different proteins with a bifunctional reagent can give rise to intramolecular and intermolecular products. Intermolecularly, a range of products including the desired 1:1-conjugate and the undesired polyconjugates and polymers of each of the reactant proteins can occur. Other possible products include aggregates of the various newly formed dimers, oligomers, and polymers. These side reactions may lead to the loss of catalytic or immunologic activities of the original reactants. Intrachain cross-linking may give rise to a variety of complications, including change of structure, which may lead to different mobility on polyacrylamide gel electrophoresis. To address some of these problems, cleavable reagents are particularly helpful. Upon cleavage, the molecule should return to its original molecular state, including electrophoretic mobility. In most of the applications of cross-linking reagents a parallel experiment using the analogous monofunctional reagent should be carried out. These experiments may elucidate whether chemical modification itself is the cause of the problem, such as loss of the biological activity or immunogenicity. If such is the case a different reagent may be chosen. Alternatively, protection of enzyme-active sites or immunological activities may be carried out with inhibitors,[91] substrates,[92] or antigens.[93,94]

While sulfhydryl cleavable reagents have many advantages, there are also several disadvantages. Major advantages include (1) rapid cleavage of the disulfide bond under mild conditions, (2) quantitative completion of the cleavage reaction, (3) ability to be cleaved both before and after electrophoresis, and (4) the specificity of the reduction reaction. The disadvantages of the use of these reagents are the following: (1) they are susceptible to disulfide exchange with the possibility of linking noninteracting molecules; (2) their use precludes the application of reducing agents for the isolation of cross-linked complexes; and (3) they cannot be used in a system that is sensitive to oxidation and which would normally be kept under reducing conditions. Disulfide exchange usually involves the presence of free sulfhydryl groups that must be present in significant excess over disulfides.[95] This reaction can be decreased by lowering the pH of the reaction below the pK$_a$ of sulfhydryl groups.[96,97]

For other cleavable reagents, there are also advantages and disadvantages. The major disadvantages with the use of glycol reagents include (1) the reduced rate of cleavage relative to that achieved with disulfides, (2) the difficulty in achieving complete cleavage, (3) the lack of specificity of the cleavage reaction, namely, the carbohydrate portions of glycoproteins can also be disrupted, and (4) the oxidative side reaction of the carbohydrates may lead to potential formation of Schiff bases with the protein amino groups. A cross-linking reagent will have to be carefully chosen to suit the particular system under investigation to reduce the complications to a minimum.

9.7.2 IMMUNOGENICITY

In the application of cross-linking reagents to prepare antigen-carrier conjugates, the major concern is the effect of the reagent on the immunogenicity of the antigen. Peeters et al.[98] have systematically investigated the problem with four cross-linking reagents. In a model system using angiotensin and tetanus toxoid, it was found that cross-linking did not affect the immunogenicity of either the protein or the carrier in inducing antibody production. Antibodies were also induced against the cross-linking reagent. However, flexible nonaromatic linkers of succinimidyl 6-(N-maleimido)-n-hexanoate and succinimidyl 3-(2-pyridyldithio)propionate showed the least immunogenicity. It seems reasonable to assume that cross-linking does not affect the antigenicity or immunogenicity of an antigen and that flexible cross-linkers are the best choices for this application.

9.7.3 STABILITY

The stability of immunoconjugates and immunotoxins is of paramount importance in the application of these conjugated proteins. In general, immobilization of proteins tends to increase their stability toward both mechanical and thermal denaturation.[99] The same observation is also reported for cross-linked proteins.[100–103] In fact, higher activity of β-galactosidase was obtained after cross-linking with glutaraldehyde and dimethyladipimidate. Glutaraldehyde cross-linking of enzyme crystals and polyethylene glycol modification of enzyme surface amino groups also enhance biocatalyst stability.

Some sense of the stability of a conjugate may be obtained from consideration of the Arrhenius equation (Equation 9.1):

$$\ln k = \frac{A - E_a}{RT} \tag{9.1}$$

When the natural logarithm of the rate constant of inactivation, k, is plotted against the reciprocal of the absolute temperature, T, the slope of the line gives the value of E_a/R where R is the gas constant. The intercept is then equal to A, the Arrhenius constant. The magnitude of E_a, the activation energy of the process, provides information about the stability of the conjugate toward denaturation. The larger this value, the more stable the conjugate, since more energy is required for the inactivation. For horseradish peroxidase, peroxidase–IgG conjugate, and peroxidase–jacalin conjugate cross-linked with gluteraldehyde, the values of E_a are shown in Table 9.1.[104] These values indicate that conjugated horseradish peroxidase is more stable than the free enzyme.

From the Arrhenius plot, it is also possible to predict the half-life of a conjugate at a certain temperature. The rate constant of inactivation at that temperature is determined by extrapolation of the plot. The relationship of the half-life to the rate constant is given by the following equation:

$$T_{1/2} = \frac{0.693}{k} \tag{9.2}$$

TABLE 9.1
Heat of Inactivation of Horseradish Peroxidase Conjugates

Conjugates	E_a (kcal/mol)
Horseradish peroxidase (free)	35
Horseradish peroxidase-IgG	51
Horseradish peroxidase-Jacalin	43

For horseradish peroxidase at the conditions where the rate constants are determined, the half-life at 4°C is found to be 21 years, This mechanism may be useful for predicting the stability of an immunoconjugate or immunotoxin at a particular temperature.

REFERENCES

1. Niemeyer, C. M. (eds.), Bioconjugations protocols: Strategies and methods, in *Methods in Molecular Biology*, Humana Press, Totowa, NJ, 2004, p. 283.
2. Trakselis, M. A., Alley, S. C., and Ishmael, F. T., Identification and mapping of protein–protein interactions by a combination of cross-linking, cleavage, and proteomics, *Bioconjug. Chem.*, 16, 741, 2005.
3. Kluger, R. and Alagic, A., Chemical cross-linking and protein–protein interactions—A review with illustrative protocols, *Bioorg. Chem.*, 32, 451, 2004.
4. Hultin, T., A class of cleavable heterobifunctional reagents for thiol-directed high-efficiency protein crosslinking: Synthesis and application to the analysis of protein contact sites in mammalian ribosomes, *Anal. Biochem.*, 155, 262, 1986.
5. Baskin, L. S. and Yang, C. S., Cross-linking studies of the protein topography of rat liver microsomes, *Biochim. Biophys. Acta*, 684, 263, 1982.
6. O'Sullivan, M. J., Gnemmi, E., Morris, D., Chieregatti, G., Simmonds, A. D., Simmons, M., Bridges, J. W., and Marks, V., Comparison of two methods of preparing enzyme–antibody conjugates: Application of these conjugates for enzyme immunoassay, *Anal. Biochem.*, 100, 100, 1979.
7. Modesto, R. R. and Pesce, A. J., The reaction of 4,4'-difluoro-3,3'-dinitro-diphenyl sulfone with gamma-globulin and horseradish peroxidase, *Biochim. Biophys. Acta*, 229, 384, 1971.
8. Schick, A. F. and Singer, S. J., On the formation of covalent linkages between two protein molecules, *J. Biol. Chem.*, 236, 2477, 1961.
9. Agarwal, K. L., Grudzinski, S., Kenner, G. W., Rogers, N. H., Sheppard, R. C., and McGuigan, J. E., Immunochemical differentiation between gastrin and related peptide hormones through a novel conjugation of peptides to proteins, *Experientia*, 27, 514, 1971.
10. Borek, F., A new two-stage method for cross-linking proteins, *Nature (London)*, 191, 1293, 1961.
11. Nakane, P. K., Sri Ram, J., and Pierce, G. B., Enzyme-labeled antibodies: Preparation and application for the localization of antigens, *J. Histochem. Cytochem.*, 14, 789, 1966.
12. Otto, H., Takamiya, H., and Vogt, A., Two-stage method for crosslinking antibody globulin to ferritin by glutaraldehyde. Comparison between the one-stage and the two-stage method, *J. Immunol. Methods*, 3, 137, 1973.
13. Grabarek, Z. and Gergely, J., Zero-length crosslinking procedure with the use of active esters, *Anal. Biochem.*, 185, 131, 1990.
14. Howard, A. N. and Wild, F., A two-stage method of cross-linking proteins suitable for use in serological techniques, *Br. J. Exp. Pathol.*, 38, 640, 1957.
15. Cumber, A. J., Forrester, J. A., Foxwell, B. M. J., Ross, W. C. J., and Thorpe, P. E., Preparation of antibody–toxin conjugate, *Methods Enzymol.*, 112, 207, 1985.
16. Rector, E. S., Schwenk, R. J., Tse, K. S., and Sehon, A. H., A method for the preparation of protein–protein conjugates of predetermined composition, *J. Immunol. Methods*, 24, 321, 1978.
17. Kennedy, J. H., Kricka, L. J., and Wilding, P., Protein–protein coupling reactions and the application of protein conjugates, *Clin. Chim. Acta*, 70, 1, 1976.
18. Wilson, D. V. and Devey, M., A new coupling procedure for red cell-linked antigen antiglobulin reaction, *Int. Arch. Allergy Appl. Immunol.*, 44, 77, 1973.
19. Lee, R. T., Wong, T.-C., Lee, R., Yue, L., and Lee, Y. C., Efficient coupling of glycopeptides to proteins with a heterobifunctional reagent, *Biochemistry*, 28, 1856, 1989.
20. Ghosh, S. S., Kao, P. M., McCue, A. W., and Chappelle, H. L., Use of maleimide-thiol coupling chemistry for efficient syntheses of oligonucleotide–enzyme conjugate hybridization probes, *Bioconjug. Chem.*, 1, 71, 1990.
21. Tona, R. and Haner, R., Crosslinking of diene-modified DNA with bis-maleimides, *Mol. Biosyst.*, 1, 93, 2005.
22. Brinkley, M., A brief survey of methods for preparing protein conjugates with dyes, haptens and cross-linking reagents, *Bioconjug. Chem.*, 3, 2, 1992.
23. van Dijk, S. and Ward, D. N., Chemical cross-linking of porcine luteinizing hormone: Location of the cross-link and consequences for stability and biological activity, *Endocrinology*, 132, 534, 1993.

24. Hunter, M. J. and Ludwig, M. L., The reaction of imidoesters with proteins and related small molecules, *J. Am. Chem. Soc.*, 84, 3491, 1962.

25. Browne, D. T. and Kent, S. B. H., Formation of nonamidine products in the reaction of primary amines with imido esters, *Biochem. Biophys. Res. Commun.*, 67, 126, 1975.

26. Liu, S. C., Fairbanks, G., and Palek, J., Spontaneous reversible protein cross-linking in the human erythrocyte membrane. Temperature and pH dependence, *Biochemistry*, 16, 4066, 1977.

27. Ji, T. H., The application of chemical crosslinking for studies of cell membrane and the identification of surface reporters, *Biochim. Biophys. Acta*, 559, 39, 1979.

28. Haller, I. and Henning, U., Cell envelope and shape of *Escherichia coli* K12. Crosslinking with dimethyl imidoesters of the whole cell wall, *Proc. Natl. Acad. Sci. U. S. A.*, 71, 2018, 1974.

29. Peters, K. and Richards, R. M., Chemical cross-linking. Reagents and problems in studies of membrane structure, *Annu. Rev. Biochem.*, 46, 523, 1977.

30. Hand, E. S. and Jencks, W. P., Mechanism of the reaction of imido esters with amines, *J. Am. Chem. Soc.*, 84, 3505, 1962.

31. Dutton, A., Adams, M., and Singer, S. J., Bifunctional imido-esters as cross-linking reagents, *Biochem. Biophys. Res. Commun.*, 23, 730, 1966.

32. Carpenter, F. H. and Harrington, K. T., Intermolecular cross-linking of monomeric proteins and cross-linking of oligomeric proteins as a probe of quaternary structure, *J. Biol. Chem.*, 247, 5580, 1972.

33. Debyser, Z. and De Clercq, E., Chemical crosslinking of the subunits of HIV-1 reverse transcriptase, *Protein Sci.*, 5, 278, 1996.

34. Cuatrecasas, P. and Parikh, I., Absorbents for affinity chromatography. Use of N-hydroxysuccinimide ester of agarose, *Biochemistry*, 11, 2291, 1972.

35. Staros, J. V., Membrane-impermeant cross-linking reagents: Probes of the structure and dynamics of membrane proteins, *Acc. Chem. Res.*, 21, 435, 1988.

36. Lomant, A. J. and Fairbanks, G., Chemical probes of extended biological structures. Synthesis and properties of the cleavable protein cross-linking reagent [^{35}]dithiobis(succinimidylpropionate), *J. Mol. Biol.*, 104, 243, 1976.

37. Carlsson, J., Drevin, H., and Axén, R., Protein thiolation and reversible protein–protein conjugation. N-succinimidyl 3-(2-pyridyldithio)propionate, a new heterobifunctional reagent, *Biochem. J.*, 173, 723, 1978.

38. Smith, R. J., Capaldi, R. A., Muchmore, D., and Dahlquist, F., Cross-linking of ubiquinone cytochrome c reductase (complex III) with periodate-cleavable bifunctional reagent, *Biochemistry*, 17, 3719, 1978.

39. Lewis, R. V., Roberts, M. F., Dennis, E. A., and Allison, W. S., Photoactivated heterobifunctional cross-linking reagents which demonstrate the aggregation state of phospholipase A$_2$, *Biochemistry*, 16, 5650, 1977.

40. Scianimanico, S., Schoehn, G., Timmins, J., Ruigrok, R. H. W., Klenk, H.-D., and Weissenhorn, W., Membrane association induces a conformational change in the Ebola virus matrix protein, *EMBO J.*, 19, 6732, 2000.

41. Joshi, S. and Burrows, R. J., ATP synthase complex from bovine heart mitochondria. Subunit arrangement as revealed by nearest neighbor analysis and susceptibility to trypsin, *J. Biol. Chem.*, 265, 14518, 1990.

42. Smyth, D. G., Blumenfeld, O. O., and Konigsberg, W., Reactions of N-ethylmaleimide with peptides and amino acids, *Biochem. J.*, 91, 589, 1964.

43. Ishi, Y. and Lehrer, S. S., Effects of the state of the succinimido-ring on the fluorescence and structural properties of pyrene maleimide labeled alpha-tropomyosin, *Biophys. J.*, 50, 75, 1986.

44. Fronzes, R., Chaignepain, S., Bathany, K., Giraud, M. F., Arselin, G., Schmitter, J. M., Dautant, A., Velours, J., and Brèthes, D., Topological and functional study of subunit h of the F1Fo ATP synthase complex in yeast *Saccharomyces cerevisiae*, *Biochemistry*, 42, 12038, 2003.

45. Gurd, F. R. N., Carboxymethylation, *Methods Enzymol.*, 11, 532, 1967.

46. Ludueña, R. F., Roach, M. C., Trcka, P. P., and Weintraub, S., N,N-Bis(alpha-iodoacetyl)-2,2'-dithiobis(ethylamine), a reversible crosslinking reagent for protein sulfhydryl groups, *Anal. Biochem.*, 117, 76, 1981.

47. Mattson, G., Conklin, E., Desai, S., Nielander, G., Savage, M. D., and Morgensen, S., A practical approach to crosslinking, *Mol. Biol. Rep.*, 17, 167, 1993.

48. Na, D. H., Woo, B. H., and Lee, K. C., Quantitative analysis of derivatized proteins prepared with pyridyl disulfide-containing cross-linkers by high-performance liquid chromatography, *Bioconjug. Chem.*, 10, 306, 1999.

49. Kondejewski, L. H., Kralovec, J. A., Blair, A. H., and Ghose, T., Synthesis and characterization of carbo-hydrate-linked murine monoclonal antibody K20-human serum albumin conjugates, *Bioconjug. Chem.*, 5, 602, 1994.

50. Zhao, Y., Zhang, J., Wang, X., Chen, B., Xiao, Z., Shi, C., Wei, Z., Hou, X., Wang, Q., and Dai, J., The osteogenic effect of bone morphogenetic protein-2 on the collagen scaffold conjugated with antibodies, *J. Control Release*, 141, 30, 2010.

51. Cartwright, I. L., Hutchinson, D. W., and Armstrong, V. W., The reaction between thiols and 8-azido-adenosine derivatives, *Nucleic Acids Res.*, 3, 2331, 1996.

52. Staros, J. V., Bayley, H., Standring, D. N., and Knowles, J. R., Reduction of aryl azides by thiols: Implications for the use of photoaffinity reagents, *Biochem. Biophys. Res. Commun.*, 80, 568, 1978.

53. Spanggord, R. J. and Beal, P. A., Site-specific modification and RNA crosslinking of the RNA-binding domain of PKR, *Nucleic Acids Res.*, 28, 1899, 2000.

54. Hixson, S. H. and Hixson, S. S., p-Azidophenacyl bromide, a versatile photolabile bifunctional reagent. Reaction with glyceraldehyde-3-phosphate dehydrogenase, *Biochemistry*, 14, 4251, 1975.

55. Henkin, J., Photolabeling reagent for thiol enzymes. Studies on rabbit muscle creatine kinase, *J. Biol. Chem.*, 252, 4293, 1977.

56. Pietroni, P., Young, M. C., Latham, G. J., and von Hippel, P. H., Structural analyses of gp45 sliding clamp interactions during assembly of the bacteriophage T4 DNA polymerase holoenzyme. I. Conformational changes within the gp44/62-gp45-ATP complex during clamp loading, *J. Biol. Chem.*, 272, 31666, 1997.

57. Ji, T. H., A novel approach to the identification of surface receptors. The use of photosensitive hetero-bifunctional cross-linking reagent, *J. Biol. Chem.*, 252, 1566, 1977.

58. Kiehm, D. J. and Ji, T. H., Photochemical cross-linking of cell membranes. A test for natural and random collisional cross-links by millisecond cross-linking, *J. Biol. Chem.*, 252, 8524, 1977.

59. Middaugh, C. R. and Ji, T. H., A photochemical crosslinking study of the subunit structure of membrane-associated spectrin, *Eur. J. Biochem.*, 110, 587, 1980.

60. Miyakawa, T., Takemoto, L. J., and Fox, C. F., Membrane permeability of bifunctional, amino site-specific, cross-linking reagents, *J. Supramol. Struct.*, 8, 303, 1978.

61. Mas, M. T., Wang, J. K., and Hargrave, P. A., Topography of rhodopsin in rod outer segment disk mem-branes. Photochemical labeling with N-(4-azido-2-nitrophenyl)-2-aminoethanesulfonate, *Biochemistry*, 19, 684, 1980.

62. Staros, J. V. and Richards, F. M., Photochemical labeling of the surface proteins of human erythrocytes, *Biochemistry*, 13, 2720, 1974.

63. Fancy, D. A. and Kodadek, T., Chemistry for the analysis of protein–protein interactions: Rapid and effi-cient cross-linking triggered by long wavelength light, *Proc. Natl. Acad. Sci. U. S. A.*, 96, 6020, 1999.

64. Qasba, P. K., Ramakrishnan, B., and Boeggeman, E., Structure and function of beta-1,4-galactosyltrans-ferase, *Curr. Drug Targets*, 9, 292, 2008.

65. Melcher, K., New chemical crosslinking methods for the identification of transient protein–protein inter-actions with multiprotein complexes, *Curr. Prot. Pept. Sci.*, 5, 287, 2004.

66. Kodadek, T., Duroux-Richard, I., and Bonnafous, J.-C., Techniques: Oxidative cross-linking as an emer-gent tool for the analysis of receptor-mediated signalling events, *Trends Pharmacol. Sci.*, 26, 210, 2005.

67. Fancy, D. A., Melcher, K., Johnston, S. A., and Kodadek, T., New chemistry for the study of multiprotein complexes: The six-histidine tag as a receptor for a protein crosslinking reagent, *Chem. Biol.*, 3, 551, 1996.

68. Brown, K. C., Yu, Z., Burlingame, A. L., and Craik, C. S., Determining protein–protein interactions by oxidative cross-linking of a glycine–glycine–histidine fusion protein, *Biochemistry*, 37, 4397, 1998.

69. Klein, J., Nolden, M., Sanders, S. L., Kirchner, J., Weil, P. A., and Melcher, K., Use of a genetically introduced cross-linker to identify interaction sites of acidic activators within native transcription factor IID and SAGA, *J. Biol. Chem.*, 278, 6779, 2003.

70. Tippmann, E. M., Liu, W., Summerer, D., Mack, A. V., and Schultz, P. G., A genetically encoded diazirine photocrosslinker in *Escherichia coli*, *ChemBioChem*, 8, 2210, 2007.

71. Lee, H. S., Dimla, R. D., and Schultz, P. G., Protein–DNA photo-crosslinking with a genetically encoded benzophenone-containing amino acid, *Bioorg. Med. Chem. Lett.*, 19, 5222, 2009.

72. Vodovozova, E. L., Photoaffinity labeling and its application in structural biology, *Biochemistry (Moscow)*, 72, 1, 2007.

73. Deller, M. C. and Yvonne, J. E., Cell surface receptors, *Curr. Opin. Struct. Biol.*, 10, 213, 2000.

74. Patzke, C., Max, K. E., Behlke, J., Schreiber, J., Schmidt, H., Dorner, A. A., Kröger, S. et al., The cox-sackievirus–adenovirus receptor reveals complex homophilic and heterophilic interactions on neural cells, *J. Neurosci.*, 30, 2897, 2010.

75. Wu, R. S., Chudasama, N., Zakharov, S. I., Doshi, D., Motoike, H., Liu, G., Yao, Y. et al., Location of the β4 transmembrane helices in the BK potassium channel, *J. Neurosci.*, 29, 8321, 2009.

76. Begley, T. P., Thymine dimer photochemistry: A mechanistic perspective, in *Comprehensive Natural Product Chemistry*, Vol. 5, Poulter, C. D. (ed.), Elsevier, Amsterdam, the Netherlands, 1999, p. 371.

77. Glick, G. D., Engineering terminal disulfide bonds into DNA, *Curr. Protoc. Nucleic Acid Chem.*, Chapter 5, Unit 5.7, 2003.

78. Maglott, E. J. and Glick, G. D., Engineering disulfide cross-links in RNA via air oxidation, *Curr. Protoc. Nucleic Acid Chem.*, Chapter 5, Unit 5.4, 2001.

79. Christie, R. J., Nishiyama, N., and Kataoka, K., Delivering the code: Polyplex carriers for deoxyribonucleic acid and ribonucleic acid interference therapies, *Endocrinology*, 151, 466, 2010.

80. Noll, D. M., Mason, T. M., and Miller, P. S., Formation and repair of interstrand cross-links in DNA, *Chem. Rev.*, 106, 277, 2006.

81. Vassetzky, Y., Gavrilov, A., Eivazova, E., Priozhkova, I., Lipinski, M., and Razin, S., Chromosome conformation capture (from 3C to 5C) and its ChIP-based modification, *Methods Mol. Biol.*, 567, 171, 2009.

82. Duca, M., Vekhoff, P., Oussedik, K., Halby, L., and Arimondo, P. B., The triple helix: 50 years later, the outcome, *Nucleic Acids Res.*, 36, 5123, 2008.

83. Neher, T. M., Rechkunova, N. I., Lavrik, O. I., and Turchi, J. J., Photo-cross-linking of XPC-Rad23B to cisplatin-damaged DNA reveals contacts with both strands of the DNA duplex and spans the DNA adduct, *Biochemistry*, 49, 669, 2010.

84. Ruoho, A., Bartlett, P. A., Dutton, A., and Singer, S. J., A disulfide-bridge bifunctional imidoester as a reversible cross-linking reagent, *Biochem. Biophys. Res. Commun.*, 63, 417, 1975.

85. Lutter, L. C., Ortanderl, F., and Fasold, H., The use of a new series of cleavable protein-crosslinkers on the *Escherichia coli* ribosome, *FEBS Lett.*, 48, 288, 1974.

86. Jaffe, C. L., Lis, H., and Sharon, N., New cleavable photoreactive heterobifunctional cross-linking reagents for studying membrane organization, *Biochemistry*, 19, 4423, 1980.

87. Zarling, D. A., Watson, A., and Bach, F. H., Mapping of lymphocyte surface polypeptide antigens by chemical cross-linking with BSOCOES, *J. Immunol.*, 124, 913, 1980.

88. Abdella, P. M., Smith, P. K., and Royer, G. P., A new cleavable reagent for cross-linking and reversible immobilization of proteins, *Biochem. Biophys. Res. Commun.*, 87, 734, 1979.

89. Cordes, E. H. and Bull, H. G., Mechanism and catalysis for hydrolysis of acetals, ketals, and ortho esters, *Chem. Rev.*, 74, 581, 1974.

90. Srinivasachar, K. and Neville, D. M. Jr., New protein cross-linking reagents that are cleaved by mild acid, *Biochemistry*, 28, 2501, 1989.

91. Jansen, E. F., Tomimatsu, Y., and Olson, A. C., Cross-linking of α-chymotrypsin and other proteins by reaction with glutaraldehyde, *Arch. Biochem. Biophys.*, 144, 394, 1971.

92. Nicolson, G. L. and Singer, S. J., The distribution and asymmetry of mammalian cell surface saccharides utilizing ferritin-conjugated plant agglutinins as specific saccharide stains, *J. Cell Biol.*, 60, 236, 1974.

93. Mannik, M. and Downey, W., Studies on the conjugation of horseradish peroxidase to Fab fragments, *J. Immunol. Methods*, 3, 233, 1973.

94. Kraehenbuhl, J. P. and Jamieson, J. D., Solid-phase conjugation of ferritin to Fab-fragments of immunoglobulin G for use in antigen localization on thin sections, *Proc. Natl. Acad. Sci. U. S. A.*, 69, 1771, 1972.

95. Liu, T. Y., The role of sulfur in proteins, in *The Proteins*, Vol. 3, 3rd edn., Neurath, H. and Hill, R. L. (eds.), Academic Press, New York, 1977, p. 239.

96. Fava, A., Iliceto, A., and Camera, E., Kinetics of the thiol-disulfide exchange, *J. Am. Chem. Soc.*, 79, 833, 1957.

97. Eldjarn, L. and Pihl, A., The equilibrium constants and oxidation–reduction potentials of some thiol-disulfide systems, *J. Am. Chem. Soc.*, 79, 4589, 1957.

98. Peeters, J. M., Hazendonk, T. G., Beuvery, E. C., and Tesser, G. I., Comparison of four bifunctional reagents for coupling peptides to proteins and the effect of the three moieties on the immunogenicity of the conjugates, *J. Immunol. Methods*, 120, 133, 1989.

99. Demers, A. G. and Wong, S. S., Increased stability of galactosyltransferase on immobilization, *J. Appl. Biochem.*, 7, 122, 1985.

100. Khare, S. K. and Gupta, M. N., A crosslinked preparation of *E. coli* beta-D-galactosidase, *Appl. Biochem. Biotechnol.*, 16, 1, 1987.

101. Wong, S. S. and Wong, L. J., Chemical crosslinking and the stabilization of proteins and enzymes, *Enzyme Microb. Technol.*, 14, 866, 1992.
102. Govardhan, C. P., Crosslinking of enzymes for improved stability and performance, *Curr. Opin. Biotechnol.*, 10, 331, 1999.
103. DeSantis, G. and Jones, J. B., Chemical modification of enzymes for enhanced functionality, *Curr. Opin. Biotechnol.*, 10, 324, 1999.
104. Wong, S. S., Losiewicz, M., and Wong, L.-J. C., An overview of protein chemical cross-linking: Implication for protein stabilization, in *Biocatalyst Design for Stability and Specificity*, Himmel, M. E. and Georgiou, G. (eds.), ACS Publisher, Washington, DC, Chapter 22, 1993.

10 Analysis of Cross-Linked Products

10.1 INTRODUCTION

After cross-linking has been accomplished, the next step is usually purification and characterization of the product(s) of the reaction. The target product must first be isolated from excess or unreacted reagent. In many cases, simple dialysis may suffice to remove unreacted reagent from the reaction solution. Typically, the solution is placed in a dialysis bag made of a semipermeable material usually based on cellulose. The dialysis tubing has pores that will allow smaller molecules to pass through while retaining larger species, that is, macromolecules. Some type of chromatography, for example, size-exclusion chromatography (SEC) (discussed below), may also be used to either remove excess reagent or isolate and characterize the cross-linked product. The isolated cross-linked protein may then be further characterized by biochemical or biophysical techniques. In the following sections, various analytical methodologies and examples of their application to protein cross-linking will be described. Once the product has been purified, it may be subjected to many different types of studies including spectroscopic (e.g., fluorescence, NMR, EPR, and Raman), immunochemical, biochemical, and enzymatic, and numerous examples of these type of studies have been given throughout this book. In this chapter, however, we shall focus on methods to purify and characterize the cross-linked product.

10.2 TECHNIQUES

10.2.1 SIZE-EXCLUSION CHROMATOGRAPHY

Perhaps the most common analytical method used to separate unused cross-linking reagents from reacted products as well as to separate different reaction products is SEC, sometimes known as gel-filtration chromatography, when aqueous phases are used or gel-permeation chromatography in the case of organic solvents. The method, originally developed in the 1950s, allows for the separation of molecules based on their molecular size. SEC utilizes a stationary phase material, typically a polymer, which is subjected to varying extents of cross-linking to create pores, which allow molecules below a certain size to enter the polymer matrix while excluding larger molecules (Figure 10.1). The large excluded molecules elute from the gel matrix in the so-called void volume, before the smaller size molecules. Stationary phase materials are available with different pore sizes, and, hence, different classes of SEC products should be used depending on the size of the target molecule. One of the most popular SEC matrix is Sephadex, a trade name for a cross-linked dextran, in bead form, originally produced by the Pharmacia company (now a part of GE Healthcare). The name Sephadex is an acronym from Separation Phamacia Dextran. Sephadex resins are listed according to their separation-size range, for example, Sephadex G-100 is useful for the separation of globular proteins in the range of approximately 4,000–150,000 Da, whereas G-25 is useful for the range of 1,000–5,000 Da. The resin is typically packed inside a column, and the solution containing the cross-linked material is loaded onto the column, which may be run using gravity or via a pump system such as fast protein liquid chromatography (FPLC) or high-performance liquid chromatography (HPLC). Each column must be calibrated with molecular weight standards covering the

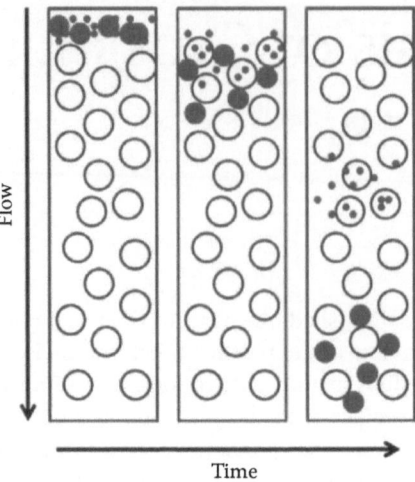

FIGURE 10.1 **(See color insert.)** Depiction of a SEC experiment. As molecules flow down the column, the larger ones move fastest since they cannot enter the resin matrix.

range of interest; for example, a typical mix of molecular weight protein standards may include thyroglobulin (67 kDa), IgG (156 kDa), BSA (66 kDa), ovalbumin (43 kDa), peroxidase (40 kDa), myoglobin (17 kDa), and cytochrome c (12.4 kDa). In the case of HPLC columns, resins that can mechanically withstand the elevated pressures are required. Such resins are often composed of silica-based polymers, for example, the TSK-GEL SW-type packings, from Tosoh Bioscience LLC, comprising rigid spherical silica gel particles chemically derivatized with diol-containing ligands. This particular product line includes three pore sizes: 125 Å pore size for the analysis of small proteins and peptides, 250 Å pore size for most protein samples, and 450 Å pore size for very large proteins and nucleic acids. An example of a standardization curve is depicted in Figure 10.2.

SEC separation depends approximately upon differences in the hydrodynamic volumes of the molecules, which in turn are related to the molecular weights as well as the molecular shapes. We note that SEC cannot only be used to separate unreacted reagent from target molecules but may cannot also be used to separate a limited range of intermolecularly linked target molecules. For example, if the cross-linking results in dimers, trimers, et cetera of a target protein the proper gel filtration

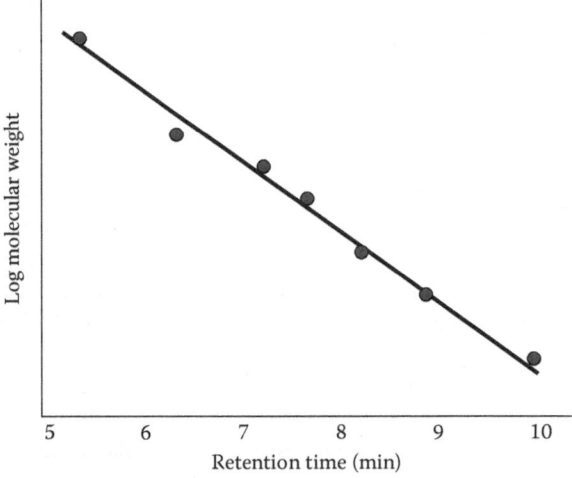

FIGURE 10.2 A generalized plot of the log of the molecular weight versus retention time for a SEC experiment such as that depicted in Figure 10.1.

resin may be able to separate some or all of these products, which may then be subjected to more detailed characterization. Numerous examples of the use of SEC are given throughout this book.

10.2.2 ELECTROPHORESIS

Electrophoresis refers to the movement of particles in an electric field. The most common electrophoresis technique used in the characterization of cross-linked molecules is gel electrophoresis in which the stationary phase is a cross-linked gel polymer, typically based on acrylamide, which has different effective pore sizes achieved via cross-linking of the acrylamide into a three-dimensional polyacrylamide matrix. Agarose is also sometimes used as a gel matrix. The pore size within a gel matrix depends upon the concentration of both the acrylamide solution and the cross-linking solution. A common cross-linker is N,N'-methylenebisacrylamide (Bis). The gel polymerization process is usually initiated by the addition of N,N,N',N'-tetramethylethylenediamine (TEMED) and ammonium persulfate. Pore sizes decrease with increasing percentages of cross-linker. Nowadays, precast gels can be purchased with pore sizes to match the particular application. For sodium dodecyl sulfate polyacrylamide gel electrophoresis (SDS–PAGE) (see below), a standard percentage of cross-linker to monomer is 37.5:1 (2.6%) while for denaturing DNA/RNA, electrophoresis ratios of 19:1 (5%) are common while native DNA/RNA gels are typically 29:1 (3.3%). Most often, the gel matrix is in the shape of a relatively thin rectangular sheet. Typically, two chambers are utilized, each containing different buffer solutions, namely, the anode buffer and the cathode buffer, with the cathode buffer covering the gel in the upper negative electrode chamber, and the anode buffer covering the gel in the lower positive electrode chamber. An electric field is applied across two buffer solutions. The sample is loaded onto one end of the gel, typically the cathode end, and the molecules move within the gel matrix toward the anode at a rate, which depends upon their charge-to-mass ratio. After the electrophoresis is run for sufficient time to allow separation of the target materials [often judged by running the gel until the fast moving dye bromophenol blue (3′,3″,5′,5″-tetra-bromophenolsulfonphthalein) is near the bottom of the gel], the bands corresponding to the different cross-linked products can be visualized by staining the gel with an appropriate dye. For proteins, common dyes are Coomassie Blue (note there are different forms such as R-250 and G-250) or the more sensitive Silver stain. Ethidium bromide or the newer SYBR® dye (this so-called safe dye is less carcinogenic than ethidium bromide but also much more expensive) is often used to visualize DNA. SDS–PAGE is commonly used to separate proteins based on their molecular weight. In this method, the protein solution is first mixed with SDS and heated to at least 60°C. SDS is an anionic detergent, which binds to the proteins and helps denature them while conferring a negative charge to each protein in proportion to the number of SDS molecules bound, which in turn is proportional to the size or molecular mass of the protein. When the gel is run, a mixture of molecular weight standards is often used to determine the approximate position of different size products in the gel matrix. If cross-linking leads to oligomeric products such a dimers, trimers, and tetramers., then the gel lane containing the cross-linked material should show a "ladder" of proteins (Figure 10.3). One notes that the lower molecular weight species run farther into the gel than the larger molecules. The amount of protein in each band can be quantified using a gel scanner or densitometer. In 1970, Ulrich K. Laemmli published a paper[1] refining the SDS–PAGE method, and this paper has become one of the most cited papers in science.

10.2.3 LIGHT SCATTERING

Light scattering may also be used in connection with column chromatography (e.g., SEC) to determine the extent of intermolecular cross-linking in a protein mixture. For example, Figure 10.4 depicts the light scattering trace and one might observe from a solution of cross-linked BSA

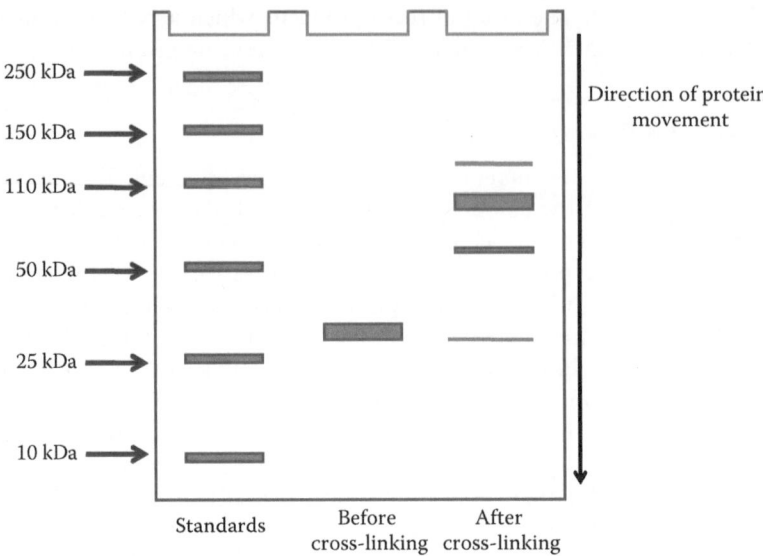

FIGURE 10.3 An idealized SDS–PAGE experiment showing molecular weight standards (left lane), a 30 kDa protein before cross-linking (middle lane), and the product after cross-linking (right lane).

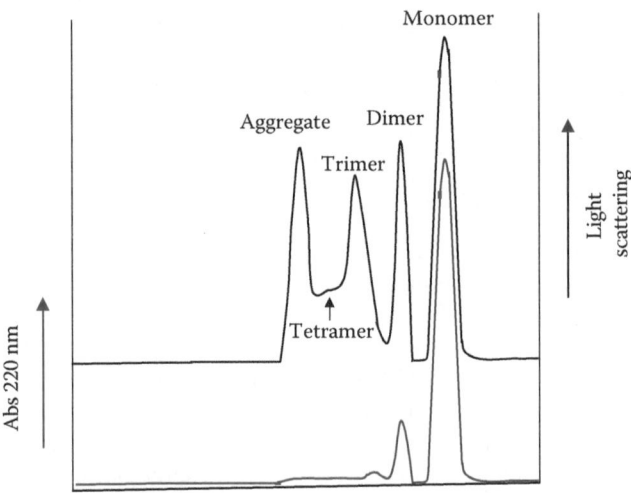

FIGURE 10.4 **(See color insert.)** Depiction of a SEC experiment on a commercial preparation of BSA that has developed cross-linked products. The output from the column was monitored by absorbance at 220 nm as well as by light scattering.

molecules eluting from a SEC column (adapted from Varian Application Note SI-02008). We note that commercially available lyophilized preparations of albumins typically have 10%–15% of cross-linked proteins due to rearrangement of disulfide bonds, and, for some purposes, the monomeric species may have to be purified from higher oligomers. Also shown in Figure 10.4 is the absorbance trace one might observe at 220 nm. The amount of protein in each band is approximated by the 220 nm absorbance trace while the light scattering trace emphasizes the larger oligomers disproportionately to their concentrations. This disproportionate weighting is due to the fact that the scattered light signal is proportional to the square of the molecular weight of the particle—at least in the range of Rayleigh scatter. Light scattering is preferably used in conjunction with a SEC chromatography system, that is, in direct connection with the column output, since flow through the column will eliminate particles, such as dust or bubbles, which can result in significant scattering.

Modern multiangle light scattering (MALS) instruments (e.g., from Wyatt Instruments), which monitor the extent of scattered light as a function of the angle from the incident beam, can provide accurate information on the molecular mass of the eluting products.

10.2.4 Mass Spectrometry

During the last two decades, the use of various forms of mass spectrometry (MS) in conjunction with chemical cross-linking has increased tremendously. The combination of MS, cross-linking, and bioinformatics has proved to be an exceptionally powerful approach.[2–6] The application of MS to protein cross-linking has served to determine which protein residues have been modified with the aim of either ascertaining precisely which regions of a protein or which regions of two proteins are cross-linked or, more ambitiously, to provide structural information. Although several approaches may be used to ionize the target molecules, the most common methods, at least for our purposes, are the matrix-assisted laser desorption ionization and the electrospray ionization (ESI) techniques. With modern time-of-flight-MS instrumentation, even small mass differences resulting from different isotopes (e.g., hydrogen compared to deuterium or C^{13} compared to C^{12}) can be resolved. In fact, when interpreting MS data on proteins and peptides, the isotope abundance of relevant atoms, such as carbon, nitrogen, and oxygen, must be taken into account. An excellent source of information relevant to mass spectrometry of proteins and other biomolecules, including isotope abundance data and mass calculators for amino acids and peptides, can be found at http://ionsource.com. Direct coupling of the output of a liquid chromatography (LC) system to the MS, an approach known as LC–MS, is a common method. Tandem MS, also known as MS/MS, is another popular method in which two MS steps are carried out in tandem. These steps can be carried out using either spatial or temporal separation. One increasingly popular type of instrument in protein studies is the Fourier transform ion cyclotron resonance mass spectrometry (FTICR-MS), which permits exceptionally high-mass resolution accuracy.

Two general MS approaches to the study of proteins are possible, namely, the "top-down" or "bottom-up" approaches.[7] In the bottom-up approach, the protein or, in our case, the cross-linked protein products would be first digested with proteases, and the resulting peptides would be subjected to MS analysis. In the top-down approach, the intact protein would be subjected directly to MS analysis without protease treatment. A recent variant in which proteins are subjected to partial protease treatment and larger peptides are analyzed is termed the "middle-down" approach.[8] The bottom-up approach is presently the most common methodology utilized although as new MS instrumentation and techniques become available, and this bias may change. The growing interest in proteomics continues to provide the impetuous to develop new and more powerful MS approaches to analyze proteins.

The role of bioinformatics in analysis of cross-linking of proteins and nucleic acids has become extremely important in the last decade, in no small part due to the growing interest in proteomics. In particular, the bioinformatics approach coupled to MS methodologies has been particularly significant. For example, Rinner et al.[9] have described a method to identify cross-linked peptides from complex samples using isotopically labeled cross-linkers attached to proteins and large protein sequences using a novel search engine called xQuest. Rappsilber recently reviewed these type of approaches and the role of bioinformatics and MS in analysis of cross-linked products.[2] The details of the various approaches to analysis of cross-linked proteins including the use of isotopically labeled cross-linkers and/or proteins are beyond the scope of this chapter.

Clearly, readers desiring a detailed discussion of the many MS approaches being used for investigations on cross-linked biomolecules should refer to the primary literature. Numerous examples of applications of MS to cross-linking studies are given throughout this book. Additional relevant references are also given here.[10–22] As should be evident from this brief overview, however, mass spectrometry, in all its various implementations, is an extremely powerful tool for the characterization of cross-linking in biomolecules and in proteomics in general.

REFERENCES

1. Laemmli, U. K., Cleavage of structural proteins during the assembly of the head of bacteriophage T4, *Nature*, 227, 680, 1970.
2. Rappsilber, J., The beginning of a beautiful friendship: Cross-linking/mass spectrometry and modelling of proteins and multi-protein complexes, *J. Struct. Biol.*, 173, 530, 2011.
3. Siuzdak, G., *Mass Spectrometry for Biotechnology*, Elsevier, New York, 1996.
4. Cole, R. B. (ed.), *Electrospray and MALDI Mass Spectrometry: Fundamentals, Instrumentation, Practicalities, and Biological Applications*, John Wiley & Sons, Inc., New York, 2010.
5. Chhabil, D., *Principles and Practice of Biological Mass Spectrometry*, John Wiley & Sons, Inc., New York, 2001.
6. Walla, P. J., *Modern Biophysical Chemistry*, Wiley-VCH, Weinheim, Germany, 2009.
7. Bogdanov, B. and Smith, R. D., Proteomics by FTICR mass spectrometry: Top down and bottom up, *Mass Spectrom. Rev.*, 24, 168, 2005.
8. Kelleher, N. L. and Siuti, N., Decoding protein modifications using top-down mass spectrometry, *Nat. Methods*, 4, 817, 2007.
9. Rinner, O., Seebacher, J., Walzthoeni, T., Mueller, L. N., Beck, M., Schmidt, A., Mueller, M., and Aebersold, R., Identification of cross-linked peptides from large sequence databases, *Nat. Methods*, 5, 315, 2008.
10. Singh, P., Panchaud, A., and Goodlett, D. R., Chemical cross-linking and mass spectrometry as a low-resolution protein structure determination technique, *Anal. Chem.*, 82, 2636, 2010.
11. Huang, B. X. and Kim, H. Y., Probing Akt-inhibitor interaction by chemical cross-linking and mass spectrometry, *J. Am. Soc. Mass. Spectrom.*, 20, 1504, 2009.
12. Ishii, T., Yamada, T., Mori, T., Kumazawa, S., Uchida, K., and Nakayama, T., Characterization of acrolein-induced protein cross-links, *Free Radical Res.*, 41, 1253, 2007.
13. Ihling, C., Schmidt, A., Kalkhof, S., Schulz, D. M., Stingl, C., Mechtler, K., Haack, M., Beck-Sickinger, A. G., Cooper, D. M., and Sinz, A., Isotope-labeled cross-linkers and Fourier transform ion cyclotron resonance mass spectrometry for structural analysis of a protein/peptide complex, *J. Am. Soc. Mass Spectrom.*, 17, 1100–1113, 2006.
14. Müller, D. R., Schindler, P., Towbin, H., Wirth, U., Voshol, H., Hoving, S., and Steinmetz, M. O., Isotope-tagged cross-linking reagents. A new tool in mass spectrometric protein interaction analysis, *Anal. Chem.*, 73, 1927–1934, 2001.
15. Seebacher, J., Mallick, P., Zhang, N., Eddes, J. S., Aebersold, R., and Gelb, M. H., Protein cross-linking analysis using mass spectrometry, isotope-coded cross-linkers, and integrated computational data processing, *J. Proteome Res.*, 5, 2270, 2006.
16. Leitner, A., Walzthoeni, T., Kahraman, A., Herzog, F., Rinner, O., Beck, M., and Aebersold, R., Probing native protein structures by chemical cross-linking, mass spectrometry, and bioinformatics, *Mol. Cell Proteomics*, 9, 1634, 2010.
17. Brewis, I. A. and Brennan, P., Proteomics technologies for the global identification and quantification of proteins, *Adv. Protein Chem. Struct. Biol.*, 80, 1, 2010.
18. Collins, C. J., Schilling, B., Young, M., Dollinger, G., and Guy, R. K., Isotopically labeled crosslinking reagents: Resolution of mass degeneracy in the identification of crosslinked peptides, *Bioorg. Med. Chem. Lett.*, 13, 4023, 2003.
19. Müller, M. Q., Zeiser, J. J., Dreiocker, F., Pich, A., Schäfer, M., and Sinz, A., A universal matrix-assisted laser desorption/ionization cleavable cross-linker for protein structure analysis, *Rapid Commun. Mass Spectrom.*, 15, 155, 2011.
20. Müller, M. Q., Dreiocker, F., Ihling, C. H., Schäfer, M., and Sinz, A., Cleavable cross-linker for protein structure analysis: Reliable identification of cross-linking products by tandem MS, *Anal. Chem.*, 82, 6958, 2010.
21. Sinz, A., Investigation of protein–protein interactions in living cells by chemical crosslinking and mass spectrometry, *Anal. Bioanal. Chem.*, 397, 3433, 2010.
22. Sinz, A., Chemical cross-linking and mass spectrometry to map three-dimensional protein structures and protein-protein interactions, *Mass Spectrom. Rev.*, 25, 663, 2006.

11 Applications of Chemical Cross-Linking to the Study of Biological Macromolecules

11.1 INTRODUCTION

Elucidation of the location of intra- and interchain chemical cross-links in proteins has helped unravel the folding process of polypeptide chains. The most common naturally occurring inter- and intramolecular linkage is the disulfide bond. Other examples include transglutaminase-catalyzed formation of amide bonds between γ-carboximide groups of glutamines and the ε-amino groups of lysines in fibrin[1] and the extremely complex cross-linking in collagen.[2,3] Determination of the locations of these cross-linkages has contributed to our understanding of the tertiary and quaternary structures of proteins. It follows, therefore, that the introduction of stable covalent linkages between amino acid residues in the native states of proteins should provide additional means for the study of interresidue distances, the relationship between various protein domains, the conformational states of a protein and the interactions of polypeptide chains in solution.

The application of bifunctional reagents to proteins will result in intramolecular as well as inter-molecular cross-linking (see Chapter 1). Each reaction mode provides different ways of gathering information, and the reaction condition may be adjusted to favor the yield of one product over the other. Intramolecular cross-linking can be enhanced by low protein concentrations (<0.1 mg/mL) and high protein to reagent ratios. The opposite conditions will increase the yield of intermolecular linkages. From the product of intramolecular cross-linking, determination of the location of each cross-link will give specific information on the three-dimensional (3D) folding of the polypeptide. Studies of the physical and biological properties of protein derivatives with different extents of cross-linking should also reveal the effect of covalent bonds in stabilizing the tertiary structure. In order to draw meaningful conclusions from cross-linking studies, the effect of such cross-linking with respect to the catalytic and biological function must be assessed. The protein should be studied with the corresponding monofunctional analogs. That is, if the cross-linking reagent is x-R-R-x or x-R-R-y, where x and y are the reactive moieties, the cross-linked product should be compared with the reaction product of x-R and/or R-y. For example, the effect of methyl acetimidate may be compared with that of dimethyl suberimidate.[4]

Intermolecular cross-linking provides two different types of products. Cross-linking between identical protein molecules yields homopolymers. Intermolecular linking between different proteins yields heteropolymers. These derivatives provide excellent means for the study of protein–protein interactions in multisubunit protein systems and could provide practical use of stable active insoluble proteins.[5]

A large volume of work on the application of cross-linking reagents to the study of biological macromolecules has been published. This chapter will explore some aspects of the information obtainable from such studies.

11.2 DETERMINATION OF TERTIARY STRUCTURES OF PROTEINS

The 3D protein structure is of paramount importance to its function. The conformational structure provides a framework of the active sites, ligand-binding pockets, and interaction domains. Although there are many proteins that exist as a single independent unit, there are many unit proteins that carry out their function as part of large complexes, which are composed of either similar or different subunits. The interplay of the subunits in the complexes is the basis of cellular function. Therefore, 3D structure determination is one of the key steps in understanding protein action. In cases where the structural analysis is directly applicable, such as x-ray crystallography and nuclear magnetic resonance (NMR) spectroscopy, the 3D structure of a protein gives insights into stable interactions within a protein complex. However, in many cases, these physical techniques are not applicable to map the 3D protein structure. Also, in the case of x-ray structure determination, the protein structure is subject to crystal packing forces, which may result in some differences from the protein structure free in solution. An alternative method is chemical cross-linking. By intramolecular cross-linking, the distance between the amino acid residues involved can be measured, and, by the use of a series of cross-links, aspects of the 3D structure may be elucidated. To measure the molecular distance, we need to understand how the cross-linkers can be used as molecular rulers.

11.2.1 MOLECULAR DISTANCES OF CROSS-LINKING REAGENTS

For the measurement of distances between reactive amino acid residues, cross-linking reagents of various chain lengths containing different numbers of carbon atoms have been synthesized.[6] These homologous series include the imidoester,[7] N-maleimido,[8,9] azido acyl,[10] N-succinimidyl,[11] and thiosulfonate[12] functional groups. The molecular chain lengths of these reagents are listed in these published articles. In addition to the analogous series, different cross-linkers of different functionalities and spacer arms can be grouped together for investigating molecular distances. An example is shown in Table 11.1.[149]

As an estimate of the maximum span of these molecules, the following calculations may be noted. Because of the hybridization of the carbon atom, the length of the carbon chain is not the sum of the individual bond distances. Taking the C–C bond length to be 1.54 Å and a bond angle of 109.5°, one can calculate that the distance between alternate carbons is about 2.52 Å. This arrangement translates into 1.26 Å for the "projected" C–C bond (Figure 11.1). For the thioether bond, the C–S projected distance is 1.44 Å. Since the C–O bond is shorter (1.43 Å), the projected C–O distance is only 1.17 Å. On the other hand, the S–S bond length is the longest, being 1.89 Å, and the projected distance is estimated to be 1.50 Å as shown in Figure 11.1. It should be noted that these values are calculated from methylene carbons. Any substituents on the carbon that affect its bond angle or bond length will affect the value of the projected distance. For aromatic compounds, the distance between two groups bonded to the meta-positions is about 2.8 Å, for ortho-positions about 4.8 Å and for the para-positions about 5.3 Å. From these values, the distance span of any cross-linked species can be estimated. For example, bis[2-(succinimidooxycarbonyloxy)ethyl]sulfone (BSOCOES, see compound LVI in Appendix A) can be estimated to have a maximum span of 10.4 Å between the reactive carbons. The distance between two cross-linked groups will be about 13 Å, since a bond is formed between the reactive carbon of the reagent and a reactive group in a protein. Thus, the maximum distance between two coupled groups is that of the span of the reagent plus two bond lengths.

As the C–C single bond is freely rotatable, the parent molecule can assume various conformations. Many of these conformational states will have spans much less than the estimated maximum distance. It is therefore possible for a cross-linker to join two groups within the calculated distance but not beyond.

TABLE 11.1
Molecular Distance of Some Amine-Reactive, Homobifunctional Cross-Linking Reagents

Reagent	Cross-Linked Distance (Å)
Disulfosuccinimidyl tartarate (sulfo-DST) 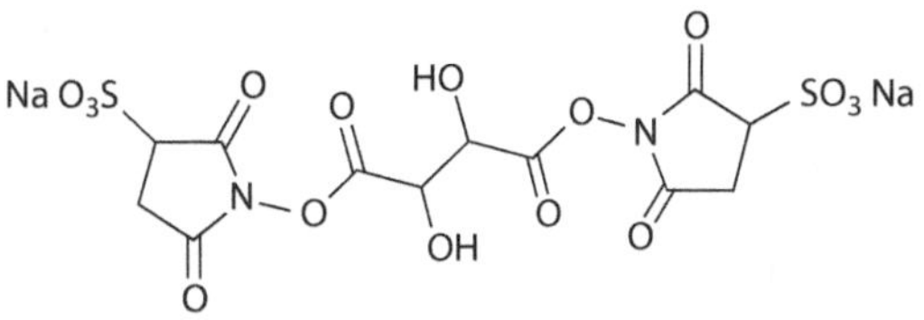	6.4

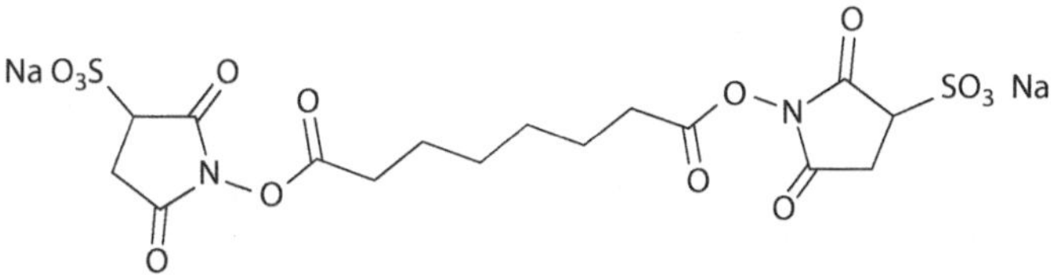 Bis(sulfosuccinimidyl) suberate (BS³)	11.4

Ethylene glycolbis(sulfosuccinimidyl succinate) (Sulfo-EGS) 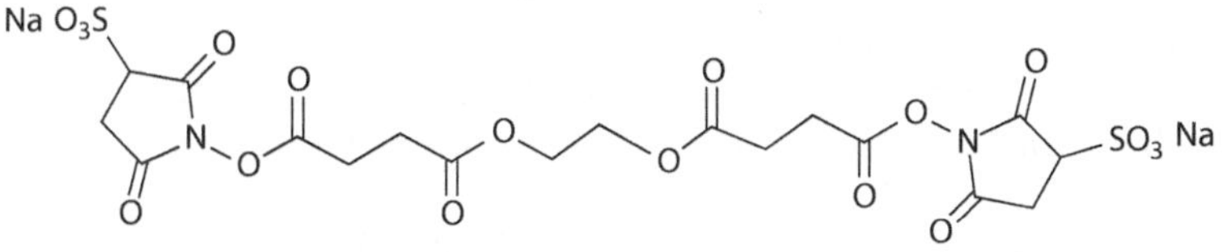	16.1

Dimethyl adipimidate 2HCl (DMA) 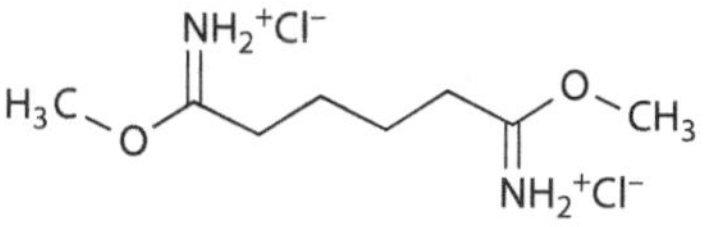	8.6

Dimethyl suberimidate 2HCl (DMS) 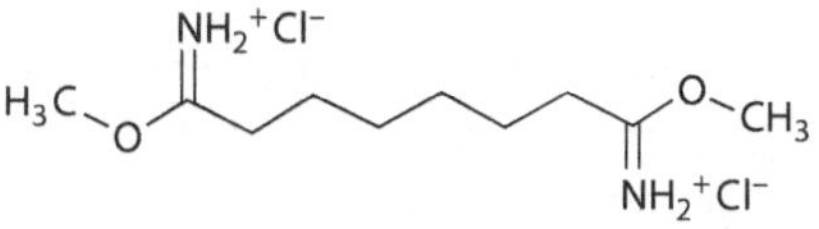	11.0

Source: Dihazi, G.H. and Sinz, A., *Rapid Commun. Mass Spectrom.*, 17, 2005, 2003.

11.2.2 EXAMPLES OF INTERRESIDUE DISTANCE MEASUREMENTS

The strategic approach for measuring the distance between reactive amino acid residues in a protein is to cross-link the groups with a homologous series of cross-linkers of different lengths. Only cross-linkers with molecular spans equal to or greater than the interresidue distance will react. Not only must the flexibility of the cross-linking reagents be considered, the ability of the reactive amino acid residues to move must also be taken into account. Many studies that focus on measuring the interresidue distances have been published. The following are some examples. Husain and Lowe[13] have found that 1,3-dibromoacetone cross-linked the active-site cysteine and histidine in ficin and stem-bromelain. From the molecular structure of the reagent, the authors concluded that the active-site residues were about 5 Å apart.

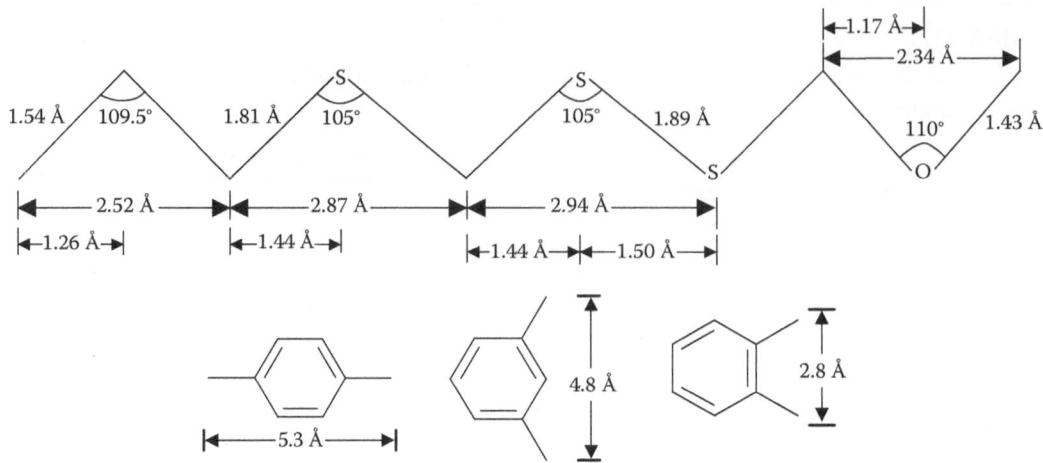

FIGURE 11.1 Projected molecular distances between different atoms.

In a study of tryptophan synthase, Heilmann and Holzner[8] isolated and modified one of the nonidentical subunits with N-succinimidyl-3-(2-pyridyldithio)propionate (SPDP) to generate a sulfhydryl group. After the reaction, the subunits were mixed to form a tetrameric complex. A series of homologous bis-maleimide reagents were used to cross-link the subunits. Only bis-N-maleimido-1,6-hexane and bis-N-maleido-1,8-octane were found to be effective cross-linkers. These compounds have a maximum span of 13.9 and 16.4 Å, respectively. Together with the incorporated portion of SPDP, the distance between the reactive groups in the nonidentical subunits was estimated to be at least 10–20 Å.

A series of bis-imidates was used by Rümbeli et al.[14] to cross-link phycobiliprotein complexes. The highest yield was obtained with dimethyl pimelimidate, which has a maximum cross-linking distance of 10 Å. From the x-ray crystallographic data of phycocyanin, the authors were able to identify the cross-linked amino acid residues.

11.2.3 Examples of Applications to 3D Protein Structure Determination

X-ray crystallography and NMR spectroscopy are the classical means to determine the tertiary structure of proteins. However, due to difficulties in obtaining suitable crystals (in the case of crystallography) and the size limits accessible (NMR), an emerging approach is to derive a set of amino acid pairs using chemical cross-linkers and then assemble the protein structure by computational manipulation.[15] This approach is particularly powerful when partial structural information is available, for example, the structures of analogous proteins. When reagents covalently connect two functional groups in a protein, the location of the created cross-links reveals the distance between the two groups. With a set of structurally defined interactions obtained by covalently connecting pairs of functional groups within a protein or a protein complex, one can deduce the 3D structure of the protein or a protein complex. Recently, cross-linked peptides have been analyzed with mass spectrometry (MS), and various mass spectrometric techniques have been explored.[15–19] Thus, chemical cross-linking with mass spectrometric analysis provides an important means to elucidate the 3D structures of proteins; this approach is discussed in Chapter 10. The following provides some examples to illustrate this advancement in 3D protein analysis.

HDL apolipoprotein (apo)A-I is a 28 kDa protein containing a 43-residue N-terminal globular region, followed by a series of 10 repeating units that comprise approximately 80% of the protein. Eight of these repeats contain 22 amino acids while two contain 11 amino acids, most of which are separated by helix-breaking amino acids such as proline. The structure of lipid-bound apoA-I is unknown, although the x-ray crystal structure of full-length lipid-free apoA-I was reported by

Ajees et al.[20] The lipid-bound conformation of apoA-I was studied with chemical cross-linking combined with mass spectrometric identification of cross-linked peptides.[21–25] Using the amine reactive homobifunctional cross-linker dithiobis(succinimidylpropionate) (DSP) at various concentrations, Thomas et al.[22] identified five intra- and three intermolecular cross-links. Sequence analysis with tandem MS/MS showed that the N-terminal end of lipid-bound apoA-I folded back upon itself maximizing interactions with the C-terminal hydrophobic domain. The authors proposed a belt buckle 3D model for the lipid-bound apoA-I.[23,24] The results were generally consistent with those of Silva et al.[25] who used an amine reactive cross-linker, bis(sulfosuccinimidyl)suberate (BS³), with an arm length similar to that of DSP. Thus, chemical cross-linking with MS analysis has provided additional structural information on apoA-I folding that compliments previous x-ray approaches.

Studies on the 3D structure of bovine serum albumin (BSA) provide another example of chemical cross-linking with MS analysis. While the crystal structure of human serum albumin (HSA) has been characterized, the 3D structure of BSA, which shares 76% sequence homology with human serum albumin (HSA), has not been determined. To probe the tertiary structure of BSA, Huang et al.[26] reacted BSA (which contains 60 lysines) with lysine-specific cross-linkers, BS³, disuccinimidyl suberate (DSS), and disuccinimidyl glutarate (DSG). While both BS³ and DSS have spacer arm of 11.4 Å, the former is water soluble whereas the latter is not and therefore would probe different regions of the molecule. The spacer arm length in DSG is only 7.7 Å. After the cross-linking reaction, BSA was digested with trypsin in either O-16 or O-18 water, and the cross-linked peptides were analyzed by tandem LC/MS/MS with the assistance of O-isotope labeling. The results showed that modification with DSS generated 17 cross-linked lysine pairs, that with DS³ generated 13 pairs and DSG generated 12 pairs. Twelve pairs of lysine residues were separated within 20 Å (distance from C_α to C_α of cross-linked lysines), while 5 pairs were spaced between 20 and 24 Å. From these distance constrains, and using the available structure information for HSA, a 3D structure of BSA was proposed.

An additional example of 3D protein determination using cross-linkers is the case of the prion protein (PrP)^{Sc} structure.[27] PrP^{Sc} is one of the two conformations of a PrP associated with a group of fatal neurodegenerative diseases, the transmissible spongiform encephalopathies. The other conformation is PrP^C whose structure has been characterized by NMR and which can be converted to PrP^{Sc} in the presence of preformed PrP^{Sc}. PrP^{Sc} and PrP^C differ with respect to secondary, tertiary, and quaternary structure, but not the amino acid sequence. The 3D structure of PrP^{Sc} is largely unknown. Onisko et al.[27] investigated the PrP^{Sc} structure using chemical cross-linking and MS. BS³ was used to react with PrP 27–30 isolated from brains of terminally ill Syrian Hamsters. The reaction resulted in cross-linked monomer, dimer, trimer, and higher-order multimers of PrP 27–30. Analysis of trypsin digested cross-linked monomer and dimer by matrix-assisted laser desorption ionization time-of-flight (MALDI-TOF) and nanobore LC electrospray quadrupole/quadrupole/time-of-flight (nano-LC-ESI-QqTOF) MS/MS showed five intermolecular cross-linked species in the dimer and four tryptic peptides in the intramolecular linked monomer when reacted with BS³. Further analysis showed that BS³ reacted preferentially with Gly90. A cross-link involving two Gly90 amino termini was found in cross-linked PrP 27–30 dimers. From the BS³ distance constrains analysis of the cross-linking data together with structural information obtained by electron crystallography as well as computer analysis, the authors proposed that PrP 27–30 monomers were stacked vertically along the fiber axis.

11.3   DETERMINATION OF QUATERNARY STRUCTURES OF PROTEINS

While tertiary structure refers to 3D structure of a single protein subunit, quaternary structure refers to the organization of subunits of a multiprotein complex. As mentioned earlier, many proteins interact to form multisubunit complexes. In order to understand their biological function, it is necessary to characterize interactions and spatial relationships within large assemblies, which in themselves may be transient both in time and in composition. The following is a summary on how cross-linking reagents can provide such needed information. Melcher and Chen[28] have provided protocols for cross-linking and analysis of large multiprotein complexes. Trakselis et al.[29] also presented a general approach for the identification and mapping of protein–protein interactions of multiprotein complexes.

11.3.1 NEAREST NEIGHBOR ANALYSIS

The principle behind the nearest neighbor analysis is the fact that proteins are cross-linked only when they are within the reaction distance of the cross-linking reagent. In a protein complex where the subunits associate to form a distinct organization, only those subunits situated next to each other will be linked together. By analyzing which proteins are coupled, a topographical model of their location in the complex may be deduced. This powerful premise of probing the protein–protein interacting sites has been applied to investigate the arrangement of the subunits in multiprotein complexes. The following examples will serve to illustrate the usefulness of this method.

11.3.2 EXAMPLES OF DETERMINATION OF GEOMETRIC ARRANGEMENTS OF SUBUNITS WITHIN A MULTIPROTEIN COMPLEX

11.3.2.1 Subunit Arrangement in Hexameric Protein Oligomers

The simplest application of chemical cross-linking to study subunit arrangements in complex proteins probably started in the 1970s, when Carpenter and Harrington[30] showed that the six polypeptide chains of beef lens leucine aminopeptidase were arranged according to D_3 symmetry in two layers as a trimer-of-dimers, and Hucho and Janda[31] demonstrated that the six polypeptide chains of beef liver glutamate dehydrogenase (GDH) were arranged according to D_3 symmetry in two layers of dimers-of-trimers. The analyses of the cross-linked peptides were based on sodium dodecyl sulfate (SDS) gel electrophoresis. Recently, Azem et al.[32] published a theoretical approach for the analysis of cross-linking patterns derived for protein homohexamers. The equations derived were used for the analysis of glutardialdehyde cross-linked protein hexamers: beef liver GDH, jack bean urease, hemocyanin from the spiny lobster *Panulirus pencillatus* (PpHc), and *Escherichia coli* glutamate decarboxylase (GDC). The published data of cross-linked hexameric *E. coli* rho by dimethyl suberimidate were also analyzed. By computer best fit analysis, the authors concluded that the 3D arrangement of the homosubunits in GDH and urease was a two-layered eclipsed dimer-of-trimers structure, whereas that for PpHc and GDC was a two-layered staggered trimer-of-dimers. Analysis of cross-linking data of *E. coli* rho showed a dihedral symmetry, a one-layered ring arrangement of subunits with a trimer-of-dimers description, which was in complete agreement with the x-ray crystal structure of the hexameric molecule.

11.3.2.2 Three-Dimensional Arrangement of F_1-Adedosine Triphosphatase Subunits

Perhaps the best model to demonstrate the use of cross-linking reagents in elucidation of the quaternary structure of a protein complex is F_1-adedosine triphosphatase (F_1-ATPase). F_1-ATPase is part of the energy-transduction system of mitochondria, chloroplasts, chromatophores, and bacteria and can be released from membranes as a Ca^{2+}, Mg^{2+}-dependent ATPase. The solubilized F_1-ATPase consists of five different subunits, α, β, γ, δ, and ε. The stoichiometry of the subunits in the complex is $\alpha_3\beta_3\gamma\delta\varepsilon$. Chemical cross-linking has been used to study the arrangement of these subunits. Using DSP, methyl-4-mercaptobutyrimidate, dimethyl-3,3′-dithiobispropionimidate, disuccinimidyl tartarate, and cupric 1,10-phenanthrolinate, Bragg and Hou[33] have obtained the following cross-linked subunit dimers of F_1-ATPase from *E. coli*: $\alpha\alpha$, $\beta\beta$, $\alpha\beta$, $\beta\gamma$, $\beta\delta$, $\beta\varepsilon$, and $\gamma\varepsilon$. In another study, the same investigators have used 2.2′- and 3,3′-dithiobis(succinimidyl propionate), 3,3′-dithiobis(sulfosuccinimidyl propionate), disuccinimidyl tartarate, dimethyl adipimidate, 1-ethyl-3-[3-(dimethylamino)propyl]carbodiimide (EDC), and 1,2:3,4-diepoxybutane to study the subunits and

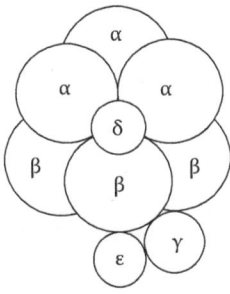

FIGURE 11.2   3D arrangement of F_1-ATPase subunits.

found cross-linked dimers of αα, αβ, βγ, αδ, βε, and γε.[34] The presence of trimer ααδ was also noted. These findings were consistent with a model wherein three α and three β subunits stacked in two planes as a triangular antiprism shown in Figure 11.2. A similar conclusion was reached by Joshi and Wang[35] who first used 1-fluoro-2,4-dinitrobenzene to modify the F1-ATPase from bovine heart mitochondria. After reduction with sodium hydrosulfite, the modified complex was cross-linked with EDC. A cross-linked ββ dimer was detected. The proposed model is similar to that obtained by x-ray crystallography.[36]

Chemical cross-linking has also been used to study the structure of F_1-ATPase from another source, *Micrococcus lysodeikticus*. Muñoz et al.[37] have used dimethyl suberimidate and DSP to explore the nearest neighbor relationship of the subunits and found that the γ subunit cross-linked with itself as well as with other subunits except β. The α subunit also cross-linked with itself and with other subunits. In addition, cross-linked $δ_2$ was also detected. From these data, the authors proposed a subunit stoichiometry of $α_3β_3γ_2δ_2$. A spatial organization of three sets of α subunit stacked on another three sets of β subunit was derived. The other subunits were arranged around this core. This spatial organization is similar to that proposed by Bragg and Hou[33] as shown in Figure 11.2.

Photoaffinity cross-linking has also been employed to investigate the relative location of the active site of F_1-ATPase from *Micrococcus luteus*. The bifunctional photosensitive substrate analog, 3′-arylazido-β-alanine-8-azido ATP (diN_3TP), was found to bind to the active site,[38,39] which was located in the β subunit. On photoactivation, diN_3ATP cross-linked the β and α subunits. It was concluded that the catalytic nucleotide-binding sites were located on the β subunits very close to the α subunits.

11.3.2.3   Three-Dimensional Structure of the RNA Polymerase II–TFIIF Complex

The polymerase II (Pol II) transcription initiation complex is composed of Pol II and the basal transcription factors (TFs) IIB, -D, -E, -F, and -H. Pol II is a 12-subunit, 513 kDa complex, while yeast TFIIF is composed of the essential subunits Tfg1 and Tfg2, and the nonessential subunit Tfg3. TFIIF is required for stable preinitiation complex formation. In the yeast *Saccharomyces cerevisiae*, about half of Pol II is bound by TFIIF. This complex comprises 15 polypeptides and has a total molecular weight of 670 kDa. Although the crystal structure of the 12-subunit Pol II is known,[40] structural information on the Pol II–TFIIF complex remains incomplete. Electron microscopy of Pol II complex with endogenous TFIIF and recombinant Tfg2 at 18 Å resolution suggests that Tfg2 extends along the polymerase cleft and Tfg1 binds around the Rpb4/7 subcomplex. However, site-specific radical generating probing placed TFIIF on the other side of the cleft near Rpb2.[41] In order to understand the 3D architecture of the Pol II initiation complex, Chen et al.[42] used protein cross-linking coupled to MS. First, free Pol II was reacted with amino group–specific BS,[3] which has a maximal length spacer of 11.4 Å capable of linking lysine $C_α$ with a distance of 27.4 Å. One hundred and forty-six linked peptide pairs were identified. Analysis of the cross-linked data base on BS[3] distance constraints showed that the structural features of Pol II are similar to the crystallographic data. With the verification that chemical cross-linking coupled to MS is applicable as a tool for the structural analysis of large multiprotein complexes, the authors cross-linked the Pol II–TFIIF complex and identified by MS 402 linkage sites of which 220 fell within TFIIF and 182 between Pol II and TFIIF. Of these, 253 were interprotein and 149 intraprotein links. Distance constraint analysis of the detailed map of cross-links between Pol II and TFIIF, together with previous crystallographic data and computer molecular modeling, revealed that the N-terminal regions of TFIIF subunits Tfg1 and Tfg2 form a dimerization domain, which anchors TFIIF on the Pol II lobe near the location of downstream DNA in initiation and elongation complexes. The linkers between the dimerization domain and the C-terminal domains in Tfg1 and Tfg2 are located in the jaws and the protrusion, respectively. The authors also speculated that the C-terminal winged helix domains of Tfg1 and Tfg2 are mobile, but that the Tfg2 winged helix domain can reside at the Pol II protrusion near upstream DNA in the initiation complex. The resulting 3D architecture of the Pol II–TFIIF complex provides insights into the function of TFIIF during transcription.

11.3.2.4 Three-Dimensional Structure of the Ribosome

Proteins within the large and small ribosomal subunits from *E. coli* have been studied for many years with chemical cross-linking reagents to determine their 3D arrangements.[43–48] In most reports, 2-iminothiolane was used as the cross-linking reagent. Other reagents such as *p*-phenylenebis(maleimide), bis-imidoesters, both cleavable and noncleavable, tartryl-containing acyl azides, and tetranitromethane have also been used. For the 30S subunit, Lambert et al.[47] have compiled all cross-linked protein pairs. A total of 33 pairs were documented. These cross-linking data were compared to the 3D arrangement of the ribosomal proteins derived from immunoelectron microscopy and neutron scattering.[48,49] Although there were cross-links that were incompatible with the derived model, many of the chemical data were in agreement as illustrated in Figure 11.3. In this figure, the arrangement of the proteins in 30S ribosomal subunit based on the neutron map is presented.[49] The cross-linked protein pairs that are compatible with this model are indicated by the double-headed arrows.

For the 50S ribosomal subunit, similar results have been derived. Traut et al.[46,48] have collected all the cross-linking data revealing 81 different cross-linked protein pairs. Stöffler-Meilicke et al.[50–52] further used immunoblotting technique to analyze protein–protein cross-links using cross-linkers of various lengths, dimethylsuberimidate, 2-iminothiolane, diepoxybutane, *o*- and

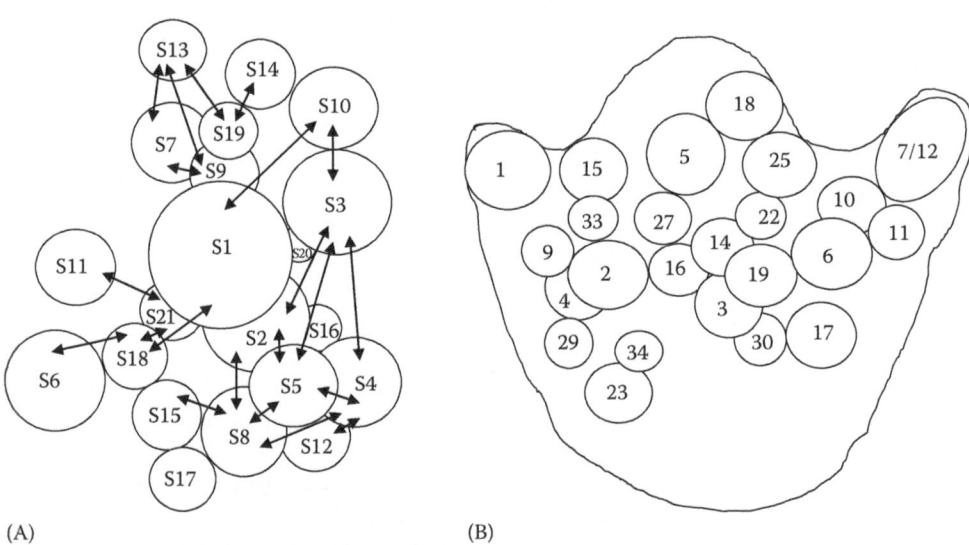

(A)   (B)

FIGURE 11.3 Cross-linking of ribosomal proteins: (A) 30S ribosomal subunit: the proteins cross-linked are indicated by two-headed arrows; (B) 50S subunit: location of the proteins cross-linked in the subunit based on immunoelectron microscopy structure.

p-phenylenedimaleimide, and DSP. Different reagents provided different cross-linked products due to their spacer arm distance. Based on the 3D model derived from immunoelectron microscopy, the authors were able to identify 29 of the 33 proteins of the 50S ribosomal subunit as shown in Figure 11.3.

Not only have the proteins in the 30S and 50S ribosomal subunits been cross-linked, RNA–protein coupling has also been obtained. Cross-linking with various cross-linkers has defined the 3D organization of the *E. coli* 16S and 23S RNA molecules in the respective 30S and 50S subunits.[53] In the 30S ribosomal subunit, proteins S4, S5, S7, S8, S11, S12, and S18 were efficiently cross-linked to domains 1 and 2 of 16S RNA using EDC.[54] With various chain lengths of photoactivatable aryl azides tethered to oligonucleotides complementary to 16S rRNA nucleotides 517–527, Wang et al.[55] were able to cross-link the probe to 16S rRNA and the surrounding peptides. Analysis of these cross-links enabled the authors to elucidate the 16S rRNA-binding region in the 30S ribosomal subunit. The structure of 16S was also studied by Nanda and Wollenzien[56] using RNA incorporated with 4-thiouridine. On UV irradiation, a limited number of RNA–RNA cross-links in nine regions were obtained. From the analysis of these cross-links, the authors speculated that the organization of the RNA in the 30S subunit. RNA–protein cross-links were introduced into *E. coli* 50S ribosomal subunits by treatment with 2-iminothiolane followed by mild ultraviolet irradiation. Protein L2 cross-linked an oligonucleotide within positions 1818–1823 in the 23S RNA, protein L4 within positions 320–325, protein L24 within positions 99–107, and protein L27 within positions 2320–2323.[57] Mild ultraviolet irradiation has shown that cross-links could be precisely localized between positions 1782 and 2608–2609, 1940 and 2554, 1941–1942 and 1964–1965, 1955 and 2552–2553, 2145–2146 and 2202, 2518–2519 and 2544–2545, and between positions 2790–2791 and 2892–2895 in the 23S RNA sequence.[58] Brimacombe et al.[59] have further analyzed a large number of intra-RNA and RNA–protein cross-links within the 23S RNA in 50S ribosomal subunit. These data have provided information on the three-dimensional relationship between the RNA and the protein arrangements in the ribosomal subunit.[53]

The 3D structural data of the ribosome derived from chemical cross-linking were compared with the x-ray crystallographic structures when they became available.[60–62] The agreement between the two methods is striking. The gross mismatches are regarded by Ivanov and Mears[63] as a result of large-scale dynamics of selected ribosomal elements.

In addition to ribosomes from *E. coli*, 40S and 60S ribosomal subunits isolated from rat liver and other mammalian sources have also been studied with cross-linking agents. Ogata et al.[64,65] used dimethyl suberimidate and dimethyl 3,3′-dithiobispropionimidate to study rat liver ribosome. Seven protein pairs were obtained from the 40S subunit and 14 pairs were obtained from the 60S subunit. Tolan et al.[66] used the reversible protein cross-linking reagent, 2-iminothiolane, to cross-link 40S ribosomal subunits and initiation factor 3 from rabbit reticulocytes. Proteins at or near the binding site for initiation factor 3 were identified. Takahashi et al.[67] used 1,3-butadiene diepoxide to study the interaction between mRNA and proteins in the 40S initiation complex. Three ribosomal proteins, S6, S8, and S23/S24, together with small amounts of S3/S3a, S27, and S30 were identified and thought to constitute the ribosomal-binding site for mRNA in the 40S initiation complex. In addition to the initiation complex, various interactions with ribosomes have also been investigated with cross-linking reagents. Gu et al.[68] used photoactivatable cross-linker *p*-azidophenacyl bromide to study the interaction of 4.5S RNA in the signal recognition particle (SRP) with the 30S ribosomal subunit. The cross-linker attached to positions 1 and 54 of 4.5S RNA was cross-linked to ribosomal protein S7, S2, S18, and S21. However, the significance of these interactions is unknown. Nechifor and Wilson[69] investigated the binding of translation factor EF-G to the bacterial ribosome using 4-azidophenacyl bromide and DSG cross-linkers together with immunoblotting and mass spectroscopy analysis. Three ribosomal proteins, L7/L12, L6, and S12, were found specifically cross-linked to EF-G at three different domains (G′, 3 and 5). The identification of EF-G in close proximity to functionally important ribosomal proteins suggested that these proteins might regulate GTP hydrolysis and translocation. These studies have demonstrated the usefulness of cross-linking reagents.

11.3.2.5   Organization of Contractile Protein Systems

Another complex cellular organization in which the application of cross-linking reagents has provided valuable information is the muscle contractile system. This contractile system is made up of myofibrils, which are composed of thin and thick filaments. Thin filaments are primarily made up of three types of proteins: actin, tropomyosin (Tm), and troponin (Tn) in a ratio of 7:1:1. *In vitro*, actin molecules exist in two states: globular G-actin and fibrous F-actin. Troponin consists of a complex of three separate proteins: troponin-T (TnT, tropomyosin binding), troponin-C (TnC, Ca^{2++} binding), and troponin-I (TnI, inhibitory). The basic structure of the thin filament consists of two strands of F-actin polymers intertwined in the conformation of a double helix; tropomyosin overlays the actin, with a single tropomyosin molecule spanning 7 actin subunits, and is fixed in this position by troponin. Thick filaments consist primarily of myosin, which has two globular heads, possessing the ATPase activity. Treatment of myosin with trypsin gives two components: light meromyosin, which contains the tail part of the native molecule, and heavy meromyosin, the globular end of the molecule. Muscle contraction occurs via a cyclic interaction between myosin heads and actin. The contraction of striated muscle is regulated by Ca^{++} via the thin filament proteins Tn and Tm. Interaction of these various protein components under different conditions has been extensively investigated using cross-linking reagents.

Leszyk et al.[70] have used EDC to cross-link TnC and TnI of the thin filament. TnC was first activated with EDC and *N*-hydroxysuccinimide and then mixed with TnI. Analysis of the cross-linked peptides revealed that the regulatory Ca^{++}-binding site II in the N-terminal domain of TnC is closely associated with the inhibitory region of TnI. The interaction between the inhibitory segment of TnI and an α-helical segment of TnC adjacent to Ca^{++}-binding site III was studied using 4-maleimidobenzophenone.[71] Luo et al.[72] studied the association of TnI and actin by using TnI mutants with 4-maleimidobenzophenone photo-cross-linking probes attached to genetically engineered cysteine residues in each of the two actin-binding regions. The modified TnI was reconstituted with native rabbit skeletal TnC, TnT, Tm, and actin to form the thin filament complex. After irradiation, the cross-linked peptides were isolated and analyzed with MALDI-TOF MS. It was demonstrated that TnI residues 133 and 104, which are near or in the critical segment, are close to Met47 of actin in the complete thin filament. Based on this information, the authors proposed a mechanistic action of the thin filament. In a similar manner, the interaction between Tm and TnI was investigated. Mudalige et al.[73] performed photochemical cross-linking experiments using Tm labeled with 4-maleimidobenzophenone at position 146 or 174. After reconstitution with actin and troponin to form thin filament followed by photolysis, both Tm146 and Tm174 were found cross-linked to TnI, TnT, and TnC. However, the Tm146-TnI cross-link was only formed in the absence of Ca^{++}. These data indicate that a part of TnI interacts near the Tm residue 146 in a Ca^{++}-dependent manner consistent with the early proposed model for thin filament regulation.

The structural organization of the thick filament myosin was investigated by several groups. Chantler and Bower[74] used 4,4′-dimaleimidylstilbene-2,2′-disulfonate to study the relative locations of the myosin light chains on the two myosin heads. *Mercenaria* myosin and scallop pure hybrid myosin whose native regulatory light chains are replaced by sulfhydryl-modified (Cys-50) regulatory light chains from *Mercenaria* were reacted with the cross-linker. The regulatory light chains of either *Mercenaria* or scallop hybrid myosin were cross-linked irrespective of the presence or absence of calcium and/or ATP. This result indicates that two light-chain sulfhydryl groups (Cys-50 of the regulatory light-chain on either head of hybrid myosin) can come within 18 Å of each other, the span of the reacted cross-linker. Since the dimaleimidylstilbene cross-linker acquired fluorescence upon reaction with sulfhydryl groups, the product was used for Förster resonance energy transfer (FRET) measurements. The analysis suggested that the distance between the two myosin heads was >50 Å.[74,75] Another study using *Mercenaria* regulatory light chain that possesses a sulfhydryl group closest to the N-terminus, at position 55, has shown that the thiol reagent, 5,5′-dithiobis(2-nitrobenzoic acid) (DTNB) cross-linked the two Cys-55 sites on either myosin head.[76] Since DTNB has a span of 2 Å, the result indicated that the light chains have a finite probability of coming within 2 Å of each other. Taken together, these experiments demonstrated considerable flexibility of the two myosin heads with respect to each other. The head portion of myosin, subfragment 1 (S1) is a key component of the machinery that converts the chemical energy of ATP to mechanical energy. The heavy chains of S1 can be cleaved with proteases into three distinct fragments with apparent molecular masses of 50, 25, and 20 kDa. Of particular interest is the helix containing the reactive sulfhydryls, Cys707 (SH1) and Cys697 (SH2) located in the C-terminal 20 kDa region. Lu et al.[77] used the photo-activatable reagent, benzophenone-4-iodoacetamide (BPIA) to study the spatial relationship of these various domains. One of the two reactive thiols, SH1, was modified with the reagent. Photolysis revealed a cross-link between the N-terminal 25 kDa fragment and the C-terminal 20 kDa fragment. In the presence of Mg^{++}-ATP, an additional cross-link between the C-terminal 20 kDa fragment and the central 50 kDa fragment was obtained. Further analysis revealed that the segment located about 12–16 kDa from the N-terminus of the heavy chain can be cross-linked to SH1 via BPIA independent of the presence of a nucleotide, whereas the segment located 57–60 kDa from the N-terminus can be cross-linked to SH1 only in the presence of a Mg^{++}-nucleotide.[78] Amino acid sequence analysis of the cross-linked 57–60 kDa segment showed that Glu-88 was the major site, and Asp-89 and Met-92 were the minor sites involved in cross-linking with SH1 via BPIA.[79] Since BPIA spans less than 1 nm, these results show that the two regions located in the 25 and 50 kDa domains of S1 are also part of the head structure about 12–14 nm from the head/rod junction.

Cross-linking reagents have also been applied to investigate the interactions between the thin and thick filaments of the contractile apparatus. Arata[80] cross-linked myosin S1 to F-actin in the presence and absence of nucleotides using EDC. FRET analysis suggested that there was an increase in distance between SH1 thiol of S1 and Cys-374 of actin in the presence of a nucleotide. The same zero-length cross-linker was used to join F-actin and heavy meromyosin.[81] Two cross-linked products were obtained. One was formed between F-actin and the C-terminal 68 kDa fragment of heavy meromyosin. The other was a cross-linked dimer of the N-terminal 24 kDa fragment of heavy meromyosin whose amino acids cross-linked were identified as Glu-168 on one head and Lys-65 of the other in chicken gizzard.[82] These data showed that the two heads of heavy meromyosin were in contact with each other as well as with actin. Pliszka et al.[83] further showed that in the lysine-rich sequence of loop 2 comprising residues 626–647 (residues 636–642: Lys-Lys-Gly-Gly-Lys-Lys-Lys), Lys-636 and Lys-637 can interact with negatively charged residues of actin other than those located at its N-terminal part, most probably with residues Glu-99 and Glu-100. The cross-links between myosin loop 2 and F-actin are formed independent of the presence or absence of MgADP.

The interaction of F-actin and S1 from cardiac muscle was also investigated with EDC. Andreev and Borejdo[84] showed that the heavy chain of cardiac S1 can be cross-linked to one or two actin protomers in the actin filament depending on the molar ratio of S1:actin. When the ratio was high, S1 bound only to a single actin and heavy-chain S1 was cross-linked to actin only through a primary site located on 20 kDa proteolytic fragment. On the other hand, when the ratio was low, S1 bound to two actins, and the heavy chain of S1 was cross-linked both through a primary and a secondary site located on 50 kDa proteolytic fragment. Similar observations were made by Bonafé and Chaussepied.[85] The light chain of S1 formed a complex with actin only when F-actin was nonsaturated with S1. Thus, the production of cross-linked complexes of the heavy and light chain with actin decreased with an increase in the molar ratio of S1:actin. While the significance of these observations is not defined, Miyanishi et al.[86] found that cardiac myosin heads induced actin filaments to form bundles, which reached saturation at the 1:1 S1:actin ratio. EDC cross-linked the essential light chain of cardiac myosin to the actin molecule in the bundle as well as to actin, S1 heavy chain, and essential light chain. However, the former was not formed in high salt concentrations. From this study, the authors argued that the cardiac myosin heads bind to two actin molecules: one actin molecule at the conventional actin-binding region and the other at the essential light-chain-binding region. Such an interaction may affect the ATPase activity of the myosin head.

It seems that EDC has been the most popular reagent for the study of interactions of myosin and actin. Yet other reagents may be useful. Even though the contractile apparatus is already highly complex, it still interacts with other regulatory proteins such as titin and cofilin.[87] The 3D arrangement of these proteins may be further investigated by the chemical cross-linking technique.

11.4   DETERMINATION OF PROTEIN–PROTEIN INTERACTIONS

The protein–protein interactions between subunits in a protein complex may be so strong that the aggregate is not dissociable under usual conditions. As we have seen in the examples cited above, the association of the subunits has a defined unique arrangement such that the quaternary structure of the complex is maintained. There are many proteins, however, that interact with association constants that would allow them to associate or dissociate under certain physiological conditions. The formation and dissociation of protein complexes plays an important role in numerous biological processes such as signal transduction, transcription, apoptosis, protein degradation, and localization. In these regulatory processes, the proteins interact with each other to elicit their effects. Depending on the system, the dissociation constants vary dramatically with ligand–receptor interactions having among the highest affinities. On the other hand, the association may be so weak that the interaction may be transient, occurring only under specific conditions. Some of these interactions may be difficult to detect. In this section, the use of cross-linking techniques to study protein–protein interactions of these dissociable proteins is explored. Kluger and Alagic[88] have provided some protocols for studying these interactions with chemical cross-linking. Melcher[89] also presented chemical cross-linking methods to map direct protein contacts in transient interactions with multimeric complexes.

11.4.1   Examples of Determinations of Protein–Protein Interactions of Soluble Proteins

There are many examples of protein–protein interactions in soluble proteins. An interesting case of a change in substrate specificity is exemplified by galactosyltransferase upon interaction with α-lactalbumin. The enzyme functions to transfer the galactose unit of UDP-galactose to a receptor. In the absence of α-lactalbumin, the receptor is *N*-acetylglucosamine or any carbohydrate with terminal *N*-acetylglucosamine. However, in the presence of α-lactalbumin, the two proteins associate to form lactose synthase where the galactose acceptor is glucose with the formation of lactose as a product. Thus, the association and dissociation of α-lactalbumin with galactosyltransferase determine the substrate specificity of the enzyme.[90] The interaction between galactosyltransferase and α-lactalbumin in solution was detected by the cross-linking agent, dimethyl pimelimidate.[91] Cross-linked complex was obtained in the presence of substrates or inhibitors of galactosyltransferase. Thus, *N*-acetylglucosamine, or a combination of *N*-acetylglucosamine, Mn++, and UDP, or Mn++ and UDP-galactose, facilitated the formation of the cross-linked species. The isolation of a cross-linked complex provided direct evidence for the association of these two proteins.

Another example of a protein–protein interaction is provided by calmodulin (CaM). CaM is a regulatory protein, which mediates a wide variety of calcium-dependent cellular processes.[92] Over 30 proteins are regulated by CaM through a calcium-dependent interaction. It is a small 148 amino acids acidic protein with four calcium-binding sites, and it is found ubiquitously in animals, plants, fungi, and protozoa. Photolabile derivatives of CaM have been synthesized and used to label CaM-binding proteins. For example, Harrison et al.[93] have prepared photosensitive derivatives of CaM using the heterobifunctional cross-linking agent, *p*-nitrophenyl 3-diazopyruvate. Interaction of CaM with adenylate kinase was demonstrated by the formation of a cross-link product on photolysis. This process was Ca++ dependent.

Other CaM–enzyme complexes have also been detected by cross-linking. SPDP was used to cross-link CaM and cyclic nucleotide phosphodiesterase in the presence of Ca++. CaM was first modified with the reagent through an amino group. The 3-(2-pyridyldithio)propionyl-substituted CaM was then mixed with cyclic nucleotide phosphodiesterase. In the presence of Ca^{++} an enzymatically active cross-linked complex was obtained.[94] Zot and Puett[95] have used N,N'-dicyclohexyl carbodiimide to study the interaction between bovine testes CaM and rabbit skeletal muscle myosin light-chain kinase. A cross-linked complex between the proteins was isolated in the presence of Ca^{++}. This complex was active without exogenous Ca^{++} or CaM.

Recently, MS has been used to study CaM interaction with target molecules. Schulz et al.[96] have employed chemical cross-linking in combination with nano-HPLC/nano-ESI-FTICR (electrospray ionization Fourier transform ion cyclotron resonance MS) MS and computational structure determination methodologies derived from NMR spectroscopy to study the CaM–melittin complex. Melittin, a small 26-residue peptide, is the principal component of honeybee (*Apis mellifera*) venom and is known to interact with a number of proteins, including myosin light chains and calsequestrin. The CaM–melittin complex was cross-linked with cross-linkers of different spacer lengths such as EDC/sulfo-NHS (*N*-hydroxysulfosuccinimide), BS^3, and disulfosuccinimidyl tartarate (sulfo-DST). The cross-linked products were analyzed with ESI-FTICR MS from which the interacting regions within the calcium-dependent CaM–melittin complex were derived. Using the ambiguous distance restraints from the chemical cross-linking data in combination with recently developed computational methods, the authors generated low-resolution 3D structure models of the CaM–melittin complex. Similar approach was applied to map the interface of the calcium-dependent complex between CaM and a peptide derived from the C-terminal region of adenylyl cyclase 8,[97] a 26-amino acid peptide derived from the skeletal muscle myosin light chain kinase,[98] as well as a peptide from Munc13 proteins, which are essential regulators of synaptic vesicles.[99] All these studies provided insight into the interactions of CaM with target proteins.

Numerous other protein–protein interactions have been studied with chemical cross-linking methodology. An additional example is the determination of self-association of ADP-ribosyltransferase in solution. It was speculated that the enzyme existed in different forms. Bauer et al.[100] cross-linked the enzyme with glutaraldehyde, dimethyl pimelimidate, dimethyl suberimidate, dimethyl 3,3′-dithiobisproprionimidate, and tetranitromethane and found that it self-associates to form dimers.

In cases where protein–protein interactions are transient or of such low affinity as to prevent isolation and are dependent on the specific cellular condition and environment in which they occur, *in vivo* chemical cross-linking can be used to stabilize these interactions through covalent-bond formation before isolation. The most commonly used cross-linker for this purpose is formaldehyde. Guerrero et al.[101] used formaldehyde to cross-link protein complexes in yeast *in vivo* and identified the full composition of the 26S proteasome complex, including two known ubiquitin receptors, Rad23 and Dsk2. Quantitative MS analysis also identified 64 potential proteasome-interacting proteins, 42 of which were novel interactions. This method, termed quantitative analysis of tandem affinity-purified *in vivo* cross-linked protein complexes (QTAX), uses *in vivo* cross-linking in combination with stable isotope labeling of amino acids in cell culture, affinity purification, and LC–MS/MS analysis and is a powerful integrated approach that could be used for the study of other protein complexes *in vivo*.[102]

11.4.2 EXAMPLES OF PROTEIN–PROTEIN INTERACTIONS OF MEMBRANE-BOUND PROTEINS: LIGAND–RECEPTOR INTERACTIONS

Chemical cross-linking procedures used to study protein–protein interactions of membrane-bound proteins are similar to those used for soluble proteins. Usually, cross-linking can be achieved by simply mixing a cross-linker with a membrane preparation. The cross-linked products are then isolated and analyzed as for soluble proteins. There are two types of protein–protein interactions in membranes. The interactions can occur between membrane-bound proteins and also between membrane-bound proteins and soluble proteins.

11.4.2.1   Interactions between Membrane-Bound Proteins

Leduc et al.[103] studied interactions of membrane proteins and lipoproteins with murein of *Escherichia* cells. The cells were mixed with DSP at a reagent to protein ratio of 2:1. After incubation for 30 min at room temperature, the cell envelopes obtained after lysis were mixed in 8% SDS at 100°C to isolate murein in the presence of 2-mercaptoethanol. The murein-modified products were solubilized and analyzed on SDS–polyacrylamide gel electrophoresis (SDS–PAGE). Nine murein-associated peptides were identified, five of which were lipoproteins. The results suggested that membrane lipoproteins may play a significant role in structural integration of the murein.

Interactions between membrane protein aggregates are exemplified by the bacterial chemoreceptors, which are predominantly methyl-accepting chemotaxis proteins (MCPs). There are five MCP-like receptors with different detection specificities in *E. coli*. The most abundant types are the serine receptor (Tsr) and the aspartate receptor (Tar). These MCPs are integral membrane proteins. Natural MCP exists in the homodimer form, but the crystal structure revealed a trimer-of-dimers arrangement. Studdert and Parkinson[104] used the trifunctional cysteine-targeted reagent, tris-(2-maleimidoethyl)amide (TMEA), to target the cysteine residues in the Tar and Tsr receptors to obtain *in vivo* snapshots of trimer composition in the receptor population. Cell suspensions were treated with 50 μM TMEA for 20 s at 30°C. Reactions were quenched by the addition of 10 mM *N*-ethylmaleimide. Cells were pelleted and then lyzed by boiling. The lysate proteins were analyzed by SDS–PAGE. The results showed that TMEA-cross-linked products contained a variety of two- and three-MCP dimer subunits indicating that one molecule of TMEA was able to react with three MCP dimers. In addition to pure products with only one type of receptor subunit, there were mixed products containing different receptor subunits. The mixed trimer species were derived from the interaction of mixed Tar and Tsr receptor dimers. These results were in agreement with the team-signaling model for bacterial chemoreceptors that receptor dimers of different detection specificities form mixed trimers of dimers.

11.4.2.2   Interactions between Membrane-Bound Proteins and Soluble Proteins

Cross-linking can also detect the interaction of soluble proteins with membrane bound proteins. The interaction of calcium-activated neutral proteinase (CANP) with erythrocyte membrane was demonstrated by cross-linking with 4,4′-dithiobisphenylazide (DTBPA).[105] Erythrocytes were reacted with DTBPA in the dark for 30 min on ice and then irradiated with 365 nm light. After irradiation, the cells were lyzed and membrane proteins were solubilized with 1% Triton X-100. CANP-linked proteins were selectively isolated by immunoprecipitation with polyclonal anti-CANP antibodies and analyzed by SDS–PAGE. In the calcium-free incubation medium, CANP was cross-linked to cytosolic soluble proteins, such as hemoglobin. With calcium ion in the incubation medium, CANP was found cross-linked to membrane skeletal proteins such as spectrin, band 3, 4.1, 4.2, and 6 proteins. These data demonstrated the effect of calcium on the dynamic interaction of CANP and the membrane proteins.

Binding of a protein ligand to a membrane receptor is the first step in eliciting a series of regulatory reactions. Photoactivatable heterobifunctional agents are usually used to identify a receptor.[106] Shephard et al.[107] used sulfosuccinimidyl-2-(*p*-azidosalicylamido)-1,3′-dithiopropionate (SASD) to identify phytohemagglutinin cell surface receptors. The reagent can be iodinated with ^{125}I using iodogen to facilitate the identification of cross-linked products. In this study, lectin phytohemagglutinin was first labeled with the radio-iodinated SASD through a displacement reaction of the sulfosuccinimidyl group. Blood mononuclear cells were then incubated with the labeled phytohemagglutinin at 4°C for 30 min, after which the unbound ligand was removed by washing the cells. The ligand-bound cells on ice were irradiated to activate the photosensitive group. Membrane proteins were solubilized with Triton X-100 and separated on SDS–PAGE under reducing conditions. Several major bands arising specifically from photo-cross-linking were identified. Cofano et al.[108] used *N*-succinimidyl-4-azidobenzoate (HSAB) to identify the receptor for murine interferon-γ. Thymoma cell line, EL-4, was mixed with radio-iodinated interferon-γ (^{125}I-IFN-γ) for 16 h at 4°C.

After incubation, the cells were washed to remove free ^{125}I-IFN-γ. Different concentrations of HSAB were added and incubated in the dark for 1 h, after which the cells were irradiated at 366 nm for 10 min. The cell membranes were then solubilized and analyzed on SDS–PAGE. At low HSAB concentrations (<50 μM), the results showed only a single band of about 95 kDa. At high concentrations of HSAB (500 μM), two bands of 95 and 110 kDa were detected. These results were interpreted as indicating that a receptor of about 80 kDa existed and that ^{125}I-IFN-γ could bind as a monomer (16 kDa) or as a dimer (32 kDa) to the receptor.

Not all identifications of receptors used photosensitive cross-linkers. Lin et al.[109] used ethylene glycolbis(succinimidylsuccinate) (EGS) or sulfo-EGS to identify the receptor of astakine 1, an invertebrate cytokine isolated from a freshwater crayfish. Isolated hematopoietic tissue cells (Hpt cells) were incubated with recombinant astakine 1 on ice for 1 h and then washed to remove unbound ligand. EGS or sulfo-EGS, which were dissolved in DMSO or water, respectively, were added to the washed cells. After incubation for 30 min at 4°C with gentle agitation, the unreacted cross-linking reagents were blocked with 1 M Tris, pH 7.4. After incubating for 15 min on ice, the cells were harvested and solubilized with Triton X-100. The cross-linked proteins containing astakine was isolated with antiastakine antibody and analyzed by SDS–PAGE. A major band corresponding to 64 kDa was detected as the putative receptor–ligand complex. Mass spectrometric analysis identified the complex-containing F1ATP synthase β-subunit. Further molecular cloning of F1ATP synthase β-subunit confirmed that the receptor for astakine 1 on hematopoietic tissue (Hpt) cells is the β-subunit of F1ATP.

11.5 DETECTION OF PROTEIN CONFORMATIONAL CHANGES

Conformational changes that occur during protein–protein interactions are important phenomena in the regulation of biological processes. These conformational changes essentially transduce information to enable the desired biological effect to take place. Conformational changes of proteins on binding of ligands or effectors can be detected by cross-linking reagents. Cross-linking experiments are valuable tools to monitor conformational changes in functional biological complexes under physiological conditions. The structural changes that occur in myosin S1 upon binding Mg^{++}-ATP and of troponin and CaM induced by Ca^{++} have already been discussed. Since the contractile apparatus converts chemical energy of ATP to mechanical energy, one would expect that conformational changes of these contractile proteins occur upon ATP hydrolysis. As ATP is hydrolyzed, different transitional states are created in the thin and thick filaments. Further evidence of changes in conformation is afforded by cross-linking of the two sulfhydryl groups SH1 and SH2 of the myosin S1, which are shown by crystal structure to be 19 Å apart. Wells et al.[110] used rigid heterobifunctional sulfhydryl cross-linkers (*N,N'-p*-phenylene dimaleimide, *N,N'-o*-phenylene dimaleimide,1,5-difluoro-2,4-dinitrobenzene, 4,4'-difluoro-3,3'-dinitrophenylsulfone, and naphthalene-1,5-dimaleimide) that span from 5 to 14 Å to react with the two sulfhydryl groups. All the reagents cross-linked the protein, indicating the abilities of the thiol groups to assume different positions. In addition, the thiol group cross-linked myosin was able to trap MgADP and other nucleotides stably at the active site revealing a different structural state when the thiols are cross-linked.[111] Further studies with five dimaleimide reagents (1,1'-(methylenedi-4,1-phenylene)bismaleimide, naphthalene-1,5-dimaleimide, *N,N'*-1,3-phenylenedimaleimide, *N,N'*-1,2-phenylenedimaleimide, and *N,N'*-1,4-phenylenedimaleimide) with spans ranging from 5 to 15 Å reacting with the thiol groups indicated helix melting or increased flexibility in the presence of nucleotides.[9] Kinetic effects of nucleotides on the rates of cross-linking reactions were measured. In the absence of nucleotides, the rates of the SH1–SH2 cross-linking were slow and reagent insensitive. In the presence of MgADP, MgATP, and MgATPγS, the rates of SH1 and SH2 cross-linking were increased about 2–7-fold for the shortest reagent (5–8 Å). Rate accelerations were much greater for the longer reagents (9–15 Å): 40–50-fold for MgADP, 25–40-fold for MgATP, and 80–270-fold for MgATPγS. These findings were interpreted in terms of nucleotide-induced shifts in the equilibria among conformational states of the SH1–SH2 helix.[112,113] Similar observations were obtained by Kliche et al.[114] who used reagents of the type $RSS(CH_2)_nSSR$, with R = 3-carboxy-4-nitrophenyl and n = 3, 6, 7, 8, 9, 10, and 12, spanning distances from 9 to 20 Å. The cross-linking rates for n = 3 or 6 were less than for n = 7 and n = 8. However, the cross-linking rates for n = 3, 6, and 7 in the absence of MgADP were about 15 times lower than in the presence of MgADP, suggesting a change in the structure of the SH2 region that depends on nucleotide binding. Blotnick and Muhlrad[115] also found that cross-linking of myosin S1 was influenced by actin and nucleotides. Using N-(ethoxycarbonyl)-2-ethoxy-1,2-dihydroquinoline (EEDQ), EDC, phenylenediglyoxal (PDG), and glutaraldehyde to cross-link myosin S1 trypsin fragments, the authors found the formation of 50/20 kDa, 27/20 kDa, 27/50 kDa, and 20 kDa/light-chain cross-linked products with all cross-linkers. MgATP or MgADP promoted the formation of the 20/50 kDa cross-linked products when EEDQ was used as a cross-linker. With PDG as a cross-linker, MgATP also affected the cross-link formation between the 20 kDa fragment and the light chains, whereas it had no influence on the formation of other products. On the other hand, actin reduced cross-linking of various S1 domains. The authors concluded that both nucleotides and actin induce intramolecular movements in S1.

Ribosomes are another dynamic structural complex, which involves mechanical movement during protein biosynthesis. Similar to the contractile system, one would expect conformational changes to take place during the biological process. In comparing the cross-linking data and x-ray crystallographic data, Ivanov and Mears[63] demonstrate that the cross-linking method can be applied to study ribosomal dynamics, including large-scale functional movements. Through statistical analysis of over one hundred cases of *E. coli* ribosomal cross-links, the researchers were able to define three well-separated groups of motions in the ribosome: (1) mean magnitude of 10 Å; (2) most abundant, centered at 20 Å and of wide dispersion; and (3) sparsely populated, with large distances up to 95 Å. The largest distance (88 Å) belongs to the L7/12 stalk region, and a distance of 69 Å is related to the L1-protuberance region. This movement of L7/L12 is consistent with the flexibility and motional properties of L7/L12 free in solution, which had been demonstrated by steady-state and time-resolved fluorescence.[116] These elements are shown to adopt different positions in crystallographic or electron micrographic structures. Ermolenko et al.[117] used FRET to monitor changes in the relative orientation of the bacterial ribosomal subunits in different complexes trapped at intermediate stages of translocation in solution. These FRET studies provided evidence for the coupling of ribosomal subunit rotation to translocation.

Related to the conformational changes of ribosomal structures is the movement of nascent protein during the translocation process of protein biosynthesis. Pitonzo et al.[118] studied the movement of newly synthesized transmembrane segments (TMs) along the ribosome exit tunnel and into the Sec61 translocon prior to insertion into the endoplasmic reticulum membrane. A photo-cross-linking approach was used to examine the co-translational proximity of a native TM to Sec61α during synthesis of TM7 and TM8 from the cystic fibrosis transmembrane conductance regulator. N^ε-ANB-lys-tRNA$_{CUA}$ was prepared by reacting 5-azido-2-nitrobenzoyl (ANB) hydroxysuccinimide ester with lys-tRNA$_{CUA}$ and added to translation reactions at a final concentration of 1 μM. The translation products were irradiated on ice with a collimated beam of UV light from an Oriel 500 W mercury arc lamp. Cross-linked membranes were isolated, and the membrane fractions containing targeted ribosome-nascent chain complexes were solubilized and analyzed by SDS–PAGE. The results showed that nascent polypeptides can form stable interactions within the translocon that inhibit TM release from Sec61α even after cleavage of the peptidyl-tRNA bond and that these interactions are dependent on nascent chain structure and involve not only the presence but also the precise orientation of an acidic side chain group in the hydrophobic TM segment. The results also show that the kinetics of TM release from Sec61α can be regulated in an ATP-dependent manner. The authors contributed these findings to indicate that the ribosome and translocon cooperatively play an active role in the recognition and integration of TMs into the endoplasmic reticulum membrane.

As mentioned under Section 11.4.1, Ca++ binding causes conformation changes in CaM and Troponin C. Kareva et al.[119] provided additional evidence using cross-linking agent, 1,3-difluoro-4,6-dinitrobenzene. Troponin C was first modified with the reagent at the Cys-98 residue. In the presence of Ca⁺⁺, the second reactive group of the reagent reacted with Lys-90. In the absence of Ca⁺⁺, two additional cross-links were obtained, reaction with Tyr-l09 and Lys-136. Thus, Ca^{2++} binding caused a change in the distance between the residues indicating a conformational change has taken place.

There are many other examples on the detection of conformational changes by chemical cross-linking. The insulin receptor is a disulfide-linked heterotetramer with two α subunits and two β subunits. Insulin binds to the α subunits, which induces autophosphorylation by the β subunits where the tyrosine kinase activity is located. The conformational states of the insulin receptor were demonstrated by cross-linking with 4,4'-dithiostilbene-2,2'-disulfonate. Binding of insulin to the α-subunit of purified insulin receptor increased the rate of cross-linking, especially the formation of β–β dimers.[120] These results provide evidence of conformational changes following insulin binding, leading to the activation of autophosphorylation.

Glycogen phosphorylase is the key enzyme of glycogen breakdown. The structure of both phosphorylase a and phosphorylase b has been solved to better than 2.3 Å resolution. The binding sites for the various substrates and effectors have been located in the 3D structure. Structural changes of the highly regulated glycogen phosphorylase system are revealed by cross-linking with a homologous series of bis-imidoesters, malonic, adipic, pimelic, suberic, and dodecanedioic diimidates, with different lengths ranging from 3.7 to 14.5 Å. In the presence of various activators, inhibitors, and substrates, glycogen phosphorylase b exhibited different cross-linked patterns.[121] The effect of these ligands on the conformation of the enzyme corrected with biological activity of the enzyme. Similar results were obtained with phosphorylase a.[122] These data reflect structural changes on binding of effectors and were used to construct a model of allostericity for the enzyme system.

Another system is protein kinase B (Akt), which is a serine/threonine kinase involved in the regulation of cell survival including apoptosis, proliferation, and diabetes. It has been well established that the Akt-membrane interaction is a crucial step for the activation of the enzyme. The interaction with membrane results in conformational changes in Akt, which are supported by x-ray crystallography, NMR, in-cell fluorescence lifetime imaging microscopy, and chemical cross-linking and MS. Huang and Kim[123] used DSS cross-linker and MS approaches to probe the mechanism of Akt interactions with Akt inhibitors. Akt was incubated with various inhibitors followed by additional liposomes for 40 min at 30°C. The mixture was incubated with a 50 M excess of freshly prepared DSS in DMSO at room temperature for 10 min. The cross-linking reaction was quenched by adding 1 M Tris–HCl (pH 7.4) to a final concentration of 50 mM. The sample was digested with trypsin in 16O/18O containing water. The peptides were isolated and analyzed by MS. It was found that Akt-membrane interaction induced an open interdomain conformation. By quantitative comparisons of Akt conformational changes using two interdomain cross-linked peptides, the authors were able to construct distinctively different molecular mechanisms by which the inhibitors function.

Like soluble proteins, conformational changes of membrane proteins can be detected by chemical cross-linking techniques. Numerous studies on structural changes of various membrane components have been reported. For example, band 3 protein of erythrocyte membrane was found to undergo structural changes on reacting with pyridoxal 5'-phosphate (PLP).[124] Cross-linking of intact erythrocyte with BS³ revealed a mixture of band 3 protein dimers and tetramers. Treatment of the red blood cells with PLP prior to conjugation provided exclusive tetrameric band 3 protein. The results indicated that PLP induced a conformational change in the band 3 protein. In another study, the band 3 protein anion antiport was found similarly affected by the transport inhibitor, 4,4'-dinitrostilbene-2,2'-disulfonate.[125] In the presence of the inhibitor, only cross-linked tetrameric product was obtained. In addition, two quaternary conformational states of the band 3 dimer, modulated by ligands of the stilbenedisulfonate site, have been identified. Photo-cross-linking with 4,4'-diazido-2,2'-stilbenedisulfonate also demonstrated pH dependence of changes in quaternary structure/conformational state of band 3 subunits.[126] Lowering the pH promoted intersubunit cross-linking and no photo-cross-linking was observed between pH 6.0 and 8.0, where band 3 subunits are known to exist as stable dimers and tetramers. It was speculated that proton binding to band 3 either altered the conformation at the interface between subunits of preexisting tetramers or promoted self-association of stable dimers to a "novel" tetrameric conformational state. A model was proposed wherein band 3 transport-site ligands allosterically modulated the global conformation of a tetrameric porter between two quaternary structures.

11.6 DETERMINATION OF NUCLEIC ACID INTERACTION AND NUCLEIC ACID–PROTEIN INTERACTION

An example of nucleic acid interaction is the RNA–RNA interaction in ribosomes. Huggins et al.[127] reviewed the UVB (280–320 nm) and UVC (220–280 nm) light-induced intra- and intermolecular RNA cross-links, and the UVA (320–400 nm) light photo-cross-linked RNA containing 4-thiouridine (s⁴U). The cross-linked products were analyzed and compared to the x-ray structure. The majority of the UVB/C and UVA-s⁴U-induced cross-links are located in four regions in the 30S subunit: within or at the ends of rRNA helix 34, in the tRNA P-site, in the distal end of helix 28, and in the helix 19/helix 27 region. These regions are implicated in different aspects of tRNA accommodation, translocation, and in the termination reaction. The frequencies and patterns of photo-cross-link formation in the 16S rRNA are connected to its tertiary structure. These data are consistent with transient conformational changes during cross-linking reaction in which the arrangements needed for photochemical reaction are attained during the electronic excitation times.

Lustig et al.[128] developed a "RNA walk" method to investigate the RNA–RNA interactions between a small RNA and its target. The method is based on UV-induced 4′-aminomethyl-trioxsalen hydrochloride (AMT) cross-linking *in vivo* followed by affinity selection of the cross-linked molecules and mapping the intermolecular adducts by reverse transcription polymerase chain reaction (RT–PCR) or real-time PCR. The method was used to study the interaction between a special tRNA-like molecule (sRNA-85) that is part of the trypanosome SRP complex and the ribosome. *Leptomonas collosoma* cells were incubated with AMT on ice and irradiated using a UV lamp at 365 nm for 60 min. After cross-linking, the cells were deproteinized by digestion with proteinase K and RNA was isolated. The cross-linked species are mapped by using RT-CR and primers spanning the region of interest, identifying domains that cannot be copied efficiently due to the presence of cross-linked adducts. For reversal of the cross-linking, the RNA was irradiated at 254 nm for 10 min. Four contact sites between sRNA-85 and rRNA were identified. The finding that sRNA-85 interacts with helix 95 of rRNA (SRL), an interaction analogous to the mammalian Alu domain led the authors to suggest molecular mimicry between sRNA-85 and the mammalian Alu domain. A contact site was also found with the peptidyl-transferase center (PTC). The authors speculated that these contact sites with the SRL and the PTC may suggest that this novel tRNA can transiently block translation.

Photo-cross-linking with AMT has also been used *in vivo* in trypanosomes to identify interactions between the spliced leader RNA and other small RNAs.[129]

There are numerous examples of detecting protein–nucleic acid interactions with chemical cross-linking. Recently, Bich et al.[130] used nondenaturing nano-electrospray (nanoESI) MS and MALDI MS to study the nucleic acid-nuclear hormone receptor cross-links. Retinoids exert their pleiotropic effects through two transcriptional regulators, the retinoic acid (RA) nuclear receptors (RARα, β, and γ isotypes), and the retinoid X nuclear receptors (RXRα, β, and γ isotypes), which form functional RAR/RXR heterodimer. The heterodimers preferentially bind DNA on two sites of a direct repeat (DR) configuration, separated by one to five nucleic acids, called DR1–DR5, respectively. In the cross-linking reaction, RAR and the RAR·RXR with 9-cis RA ligand were incubated with DR5 for 15 min before adding the cross-linker, which includes disuccinimidyl tartrate, iodoacetic acid *N*-hydroxysuccinimidyl ester, DSP, and octanedioic acid di-*N*-hydroxysuccinimidyl ester. After incubation for 2 h at room temperature, the cross-linked products were analyzed by MS. A complex between RAR·RXR and the dsDR5 was detected, so was RAR and DR5 complex indicating that complex formation between RAR and the DR motive of the DNA does not require the presence of

RXR. The data obtained on these RARRXR/DNA and RAR-RAR/DNA complexes showed that the RAR binds the single stranded DR5, and the RAR dimer binds both single and double-stranded DR5. The authors speculated that these results suggested a gene-regulatory site on DNA, which can induce quaternary structural changes in a TF such as RAR.

Other examples of protein–nucleic acid interactions have been mentioned earlier in the chapter as well as in earlier chapters. Additional procedures for detecting protein–nucleic acid interactions by chemical cross-linking techniques have been reviewed.[129–133]

11.7 EFFECTS OF CROSS-LINKING ON STRUCTURAL STABILITY AND BIOLOGICAL ACTIVITY

11.7.1 INCREASED STRUCTURAL STABILITY AND ACTIVITY

Cross-linking bas been investigated as a means for stabilizing protein structures.[134–138] Glyceraldehyde-3-phosphate dehydrogenase and asparaginase were cross-linked with dicarbox-ylic acids in the presence of water-soluble carbodiimide as well as a series of homologous bisimi-doesters.[135,139] The cross-linked enzymes were shown to have increased thermal stability as well as increased stability against other denaturing factors such as urea, guanidinium chloride, detergents, and different solvents. The stability of lactate dehydrogenase cross-linked with glutaraldehyde was also reported to be enhanced.[140] Cross-linking was shown not to affect the pH-dependent inactiva-tion, but the effect on thermal inactivation and guanidinium denaturation was greatly reduced. Tatsumoto et al.[141] have also demonstrated that intramolecular cross-linking of amyloglucosidase with various bifunctional reagents more than doubled the half-life of the enzyme at 65°C.

It seems in general that intramolecular cross-linking confers increased thermal and chemical stability of the structure of proteins. This technology is crucial for their use as industrial catalysts. For example, extracellular invertase of *S. cerevisiae* was stabilized against thermal denaturation by intermolecular and intramolecular cross-linking of the surface nucleophilic functional groups with diisocyanate homobifunctional reagents, particularly 1,4-diisocyanatobutane.[142] Molecular engi-neering by cross-linking reduced the first-order thermal denaturation constant at 60°C fourfold. In addition to enzymes, other biomaterials such as collagen, silk, and fiber networks are stabilized by chemical cross-linking. Walton et al.[143] showed that different Type I collagen preparations cross-linked with EDC/NHS have increased mechanical strength and resistance to enzymatic degrada-tion. In the study, Type I collagen air-dried films were prepared from acid solubilized tropocollagen and pepsin-digested atelocollagen in both afibrillar and reconstituted fibrillar forms were treated with EDC/HNS for 3 h. The cross-linked samples were washed with deionized water for 1 h and then left to air dry. Using differential scanning calorimetry and enzymatic stability assays, the authors showed that cross-linking with EDC/NHS significantly increases the hydrothermal stability of all samples. These cross-linked products also have increased stability toward collagenase.

The use of cross-linking reagents to stabilize membrane proteins is exemplified by the superox-ide-generating respiratory burst oxidase system. This NADPH-oxidase of neutrophil plasma mem-brane is highly unstable, thus hampering a detailed study of its components. However, Tamura et al.[144] have used EDC to enhance its stability. Cross-linking increased its half-life from 2 min to over 20 min at 37°C. Its stability toward high salt and detergent was also increased. Cross-linking also stabilized the V_{max} and K_m of the enzyme, which were noted to increase on storage in its non-cross-linked state. The increased stability paves the way to further study of this enzyme system.

11.7.2 CONFORMATION LOCK

While conformational changes of proteins lead to different cross-linking patterns, cross-linking will in turn prevent structural changes from taking place. Thus, a specific conformation of a protein may be locked in by intramolecular cross-linking. Cross-linked hemoglobin provides an example.[145,146]

Cross-linking of human deoxy-hemoglobin between α_1Lys^{99} and α_2Lys^{99} residues by bis(3,5-dibromosalicyl)fumarate resulted in a cross-linked dimer, which demonstrated lower oxygen affinity but which maintained cooperative oxygen binding. The cooperativity was slightly reduced, and all heterotropic effects were diminished. These results were attributable to the locking of the hemoglobin molecule into a particular conformation, which the authors attribute as the so-called T-state. An argument was also presented that the reduced oxygen affinity arose from smaller binding constants for other conformation present, attributed as both T- and R-states.

When hemoglobin S was cross-linked with dimethyl adipimidate, there was an increase in oxygen affinity in either the presence or absence of 2,3-diphosphoglycerate, a slight decrease in Bohr effect and cooperativity, and a small but significant destabilization of the conformation of deoxyhemoglobin.[147] More importantly, the modification of both α and β^s subunits increased the solubility of cross-linked hemoglobin S. This antisickling effect possibly was a consequence of locking the molecule into a specific conformation.

Another example of conformation lock by cross-linking reagents is presented by cross-linking α_2-macroglobulin with cis-dichlorodiammine platinum(II) (cis-DDP), which caused extensive intersubunit cross-links.[148] Treatment of native α_2-macroglobulin with protease results in cleavage of the molecule and subsequently in a conformational change in the inhibitor leading to the generation of four thiol groups as well as exposure of receptor recognition sites. Treatment of cross-linked α_2-macroglobulin with trypsin leads to complete subunit cleavage; however, no conformational change, receptor recognition site exposure, or the appearance of thiol groups is detectable. These results demonstrate that cross-linking of α_2-macroglobulin by cis-DDP locks the molecule in the native or slow conformation.

REFERENCES

1. Furlan, M., Structure of fibrinogen and fibrin, in *Fibrinogen, Fibrin Stabilisation and Fibrinolysis: Clinical, Biochemical and Laboratory Aspects*, Francis, J. L. (ed.), Ellis Horwood, Chichester, U.K., 1988, p. 16.
2. Eyre, D. R., Paz, M. A., and Gallop, P. M., Cross-linking in collagen and elastin, *Annu. Rev. Biochem.*, 53, 717, 1984.
3. Mimni, M. E. (ed.), *Collagen*, Vol. I, CRC Press, Boca Raton, FL, 1988.
4. Monneron, A. and d'Alayer, J., Effects of imido-esters on membrane-bound adenylate cyclase, *FEBS Lett.*, 122, 241, 1980.
5. Kennedy, J. F. and Scabral, J. M. S., Immobilized enzymes, in *Solid Phase Biochemistry: Analytical, and Synthetic Aspects*, Scouten, W. H. (ed.), Wiley-Interscience, New York, 1983, p. 253.
6. Green, N. S., Reisler, E., and Houk, K. N., Quantitative evaluation of the lengths of homobifunctional protein cross-linking reagents used as molecular rulers, *Protein Sci.*, 10, 1293, 2001.
7. Ji, T. H., Cross-linking of glycolipids in erythrocyte ghost membrane, *J. Biol. Chem.*, 249, 7841, 1974.
8. Heilmann, H.-D. and Holzner, M., The spatial organization of the active sites of the bifunctional oligomeric enzyme tryptophan synthase: Cross-linking by a novel method, *Biochem. Biophys. Res. Commun.*, 99, 1146, 1981.
9. Nitao, L. K. and Reisler, E., Probing the conformational states of the SH1–SH2 helix in myosin: A cross-linking approach, *Biochemistry*, 37, 16704, 1998.
10. Lutter, L. C., Ordanderl, F., and Fasold, H., The use of a new series of cleavable protein-crosslinkers on the *Escherichia coli* ribosome, *FEBS Lett.*, 48, 288, 1974.
11. Hill, M., Bechet, J.-J., and d'Albis, A., Disuccinimidyl esters as bifunctional crosslinking reagents for proteins, *FEBS Lett.*, 102, 282, 1979.
12. Blosbam, D. P. and Sharma, R. P., The development of S,S'-polymethylene bis(methanesulfonates) as reversible cross-linking reagents for thiol groups and their use to form stable catalytically active cross-linked dimers within glyceraldehyde 3-phosphate dehydrogenase, *Biochem. J.*, 181, 355, 1979.
13. Husain, S. S. and Lowe, G., Evidence for histidine in the active sites of ficin and stem-bromelain, *Biochem. J.*, 110, 53, 1968.
14. Rümbeli, R., Wirth, M., and Zuber, H., Crosslinking of phycobiliproteins from cyanobacterium *Mastigocladus laminosus* with bis-imidates: Localization of an intersubunit and an intrasubunit crosslink in C-phycocyanin, *Biol. Chem. H.-S.*, 368, 1179, 1987.

15. Singh, P., Panchaud, A., and Goodlett, D. R., Chemical cross-linking and mass spectrometry as a low-resolution protein structure determination technique, *Anal. Chem.*, 82, 2636, 2010.

16. Sinz, A., Chemical cross-linking and mass spectrometry to map three-dimensional protein structures and protein-protein interactions, *Mass Spectrom. Rev.*, 25, 663, 2006.

17. Mouradov, D., King, G., Ross, I. L., Forwood, J. K., Hume, D. A., Sinz, A., Martin, J. L., Kobe, B., and Huber, T., Protein structure determination using a combination of cross-linking, mass spectrometry, and molecular modeling, *Methods Mol. Biol.*, 426, 459, 2008.

18. Jin, L. Y., Mass spectrometric analysis of cross-linking sites for the structure of proteins and protein complexes, *Mol. Biosyst.*, 4, 816, 2008.

19. Back, J. W., de Jong, L., Muijsers, A. O., and de Koster, C. G., Chemical cross-linking and mass spectrometry for protein structural modeling, *J. Mol. Biol.*, 331, 303, 2003.

20. Ajees, A. A., Anantharamaiah, G. M., Mishra, V. K., Hussain, M. M., and Murthy, H. M., Crystal structure of human apolipoprotein A-I: Insights into its protective effect against cardiovascular diseases, *Proc. Natl. Acad. Sci. U. S. A.*, 103, 2126, 2006.

21. Thomas, M. J., Bhat, S., and Sorci-Thomas, M. G., Three-dimensional models of HDL apoA-I: Implications for its assembly and function, *J. Lipid Res.*, 49, 1875, 2008.

22. Thomas, M. J., Bhat, S., and Sorci-Thomas, M. G., The use of chemical cross-linking and mass spectrometry to elucidate the tertiary conformation of lipid-bound apolipoprotein A-I, *Curr. Opin. Lipidol.*, 17, 214, 2006.

23. Bhat, S., Sorci-Thomas, M. G., Alexander, E. T., Samuel, M. P., and Thomas, M. J., Intermolecular contact between globular N-terminal fold and C-terminal domain of ApoA-I stabilizes its lipid-bound conformation: Studies employing chemical cross-linking and mass spectrometry, *J. Biol. Chem.*, 280, 33015, 2005.

24. Bhat, S., Sorci-Thomas, M. G., Tuladhar, R., Samuel, M. P., and Thomas, M. J., Conformational adaptation of apolipoprotein A-I to discretely sized phospholipid complexes, *Biochemistry*, 46, 7811, 2007.

25. Silva, R. A., Hilliard, G. M., Li, L., Segrest, J. P., and Davidson, W. S., A mass spectrometric determination of the conformation of dimeric apolipoprotein A-I in discoidal high density lipoproteins, *Biochemistry*, 44, 8600, 2005.

26. Huang, B. X., Kim, H. Y., and Dass, C., Probing three-dimensional structure of bovine serum albumin by chemical cross-linking and mass spectrometry, *J. Am. Soc. Mass Spectrom.*, 15, 1237, 2004.

27. Onisko, B., Fernandez, E. G., Freire, M. L., Schwarz, A., Baier, M., Camina, F., Garcia, J. R., Villamarin, S. R.-S., and Requena, J. R., Probing PrP^{Sc} structure using chemical cross-linking and mass spectrometry: Evidence of the proximity of Gly90 amino termini in the Pr27–30 aggregate, *Biochemistry*, 44, 10100, 2005.

28. Melcher, K. and Chen, H.-T., Identification and analysis of multiprotein complexes through chemical crosslinking, *Curr. Protoc. Cell Biol.*, 17, 1, 2006.

29. Trakselis, M. A., Alley, S. C., and Ishmael, F. T., Identification and mapping of protein-protein interactions by a combination of cross-linking, cleavage, and proteomics, *Bioconjug. Chem.*, 16, 741, 2005.

30. Carpenter, F. H. and Harrington, K. T., Intermolecular cross-linking of monomeric proteins and cross-linking of oligomeric proteins as a probe of quaternary structure. Application to leucine aminopeptidase (bovine lens), *J. Biol. Chem.*, 247, 5580, 1972.

31. Hucho, F. and Janda, M., Investigation of the quaternary structure of beef liver glutamate dehydrogenase with bifunctional reagents, *Biochem. Biophys. Res. Commun.*, 57, 1080, 1974.

32. Azem, A., Tsfadia, Y., Hajouj, O., Shaked, I., and Daniel, E., Cross-linking with bifunctional reagents and its application to the study of the molecular symmetry and the arrangement of subunits in hexameric protein oligomers, *Biochim. Biophys. Acta*, 1804, 768, 2010.

33. Bragg, P. D. and Hou, C., A cross-linking study of the Ca^{2+}, Mg^{2+}-activated adenosine triphosphatase of *Escherichia coli*, *Eur. J. Biochem.*, 106, 495, 1980.

34. Bragg, P. D. and Hou, C., Chemical crosslinking of α-subunits in the F1-adeosine triphosphatase of *Escherichia coli*, *Arch. Biochem. Biophys.*, 244, 361, 1986.

35. Joshi, V. K. and Wang, J. B., Cross-linking study of the quaternary fine structure of mitochondrial F_1-ATPase, *J. Biol. Chem.*, 262, 15721, 1987.

36. Okuno, D., Fujisawa, R., Iino, R., Hirono-Hara, Y., Imamura, H., and Noji, H., Correlation between the conformational states of F1-ATPase as determined from its crystal structure and single-molecule rotation, *Proc. Natl. Acad. Sci. U. S. A.*, 105, 20722, 2008.

37. Muñoz, E., Palacios, P., Marquet, A., and Andreu, J. M., Substructure of F1-ATPase (BF1 factor) from *Micrococcus lysodeikticus*. A cross-linking study with diimido esters, *Mol. Cell. Biochem.*, 33, 3, 1980.

38. Schäfer, H. J., Scheurich, P., Rathgeber, G., and Dose, K., Fluorescent photoaffinity labeling of F1 ATPase from *Micrococcus luteus* with 8-azido-1,N6-etheno-adenosine 5′-triphosphate, *Anal. Biochem.*, 104, 106, 1980.

39. Schäfer, H. J. and Dose, K., Photoaffinity cross-linking of the coupling factor 1 from *Micrococcus luteus* by 3′-arylazido-8-azido-ATP, *J. Biol. Chem.*, 259, 15301, 1984.

40. Armache, K. J., Mitterweger, S., Meinhart, A., and Cramer, P., Structures of complete RNA polymerase II and its subcomplex, Rpb4/7, *J. Biol. Chem.*, 280, 7131, 2005.

41. Chen, H.-T., Warfield, L., and Hahn, S., The positions of TFIIF and TFIIE in the RNA polymerase II transcription initiation complex, *Nat. Struct. Mol. Biol.*, 8, 696, 2007.

42. Chen, Z. A., Jawhari, A., Fischer, L., Buchen, C., Tahir, S., Kamenski, T., Rasmussen, M. et al., Architecture of the RNA polymerase II-TFIIF complex revealed by cross-linking and mass spectrometry, *EMBO J.*, 29, 717, 2010.

43. Bickle, T. A., Hershey, J. W., and Traut, R. R., Spatial arrangement of ribosomal proteins: Reaction of the *Escherichia coli* 30S subunit with bis-imidoesters, *Proc. Natl. Acad. Sci. U. S. A.*, 69, 1327, 1972.

44. Lutter, L. C., Bode, U., Kurland, C. G., and Stöffler, G., Ribosomal protein neighborhoods. 3. Cooperativity of assembly, *Mol. Gen. Genet.*, 129, 167, 1974.

45. Traut, R. R., Lambert, J. M., Boileau, G., and Kenny, J. W., Protein topography of *Escherichia coli* ribosomal subunits as inferred from protein crosslinking, in *Ribosomes: Structure, Function and Genetics*, Chambliss, G., Craven, G. R., Davies, J., Davis, K., Kahan, L., and Normura, M. (eds.), University Park Press, Baltimore, MD, 1980, p. 89.

46. Traut, R. R., Tewari, D. S., Sommer, A., Gavino, G. R., Olson, H. M., and Glitz, D. G., Protein topography of ribosomal functional domains: Effect of monoclonal antibodies to different epitopes in *Escherichia coli* protein L7/L12 on ribosome function and structure, in *Structure, Function and Genetics of Ribosomes*, Hardesty, B. and Kramer, G. (eds.), Springer-Verlag, New York, 1986, p. 286.

47. Lambert, J. M., Boileau, G., Cover, J. A., and Traut, R. R., Cross-links between ribosomal proteins of 30S subunits in 70S tight couples and in 30S subunits, *Biochemistry*, 22, 3913, 1983.

48. Traut, R. R., Lambert, J. M., and Kenny, J. W., Ribosomal protein L7/L12 cross-links to proteins in separate regions of the 50 S ribosomal subunit of *Escherichia coli*, *J. Biol. Chem.*, 258, 14592, 1983.

49. Capel, M. S., Kjeldgaard, M., Engelman, D. M., and Moore, P. B., Positions of S2, S13, S16, S17, S19 and S21 in the 30 S ribosomal subunit of *Escherichia coli*, *J. Mol. Biol.*, 200, 65, 1988.

50. Redl, B., Walleczek, J., Stöffler-Meilicke, M., and Stöffler, G., Immunoblotting analysis of protein–protein crosslinks within the 50S ribosomal subunit of *Escherichia coli*. A study using dimethylsuberimidate as crosslinking reagent, *Eur. J. Biochem.*, 181, 351, 1989.

51. Walleczek, J., Redl, B., Stöffler-Meilicke, M., and Stöffler, G., Protein–protein cross-linking of the 50 S ribosomal subunit of *Escherichia coli* using 2-iminothiolane. Identification of cross-links by immunoblotting techniques, *J. Biol. Chem.*, 264, 4231, 1989.

52. Walleczek, J., Martin, T., Redl, B., Stöffler-Meilicke, M., and Stöffler, G., Comparative cross-linking study on the 50S ribosomal subunit from *Escherichia coli*, *Biochemistry*, 28, 4099, 1989.

53. Brimacombe, R., RNA–protein interactions in the *Escherichia coli* ribosome, *Biochimie*, 73, 927, 1991.

54. Chiaruttini, C., Milet, M., Hayes, D. H., and Expert-Bezancon, A., Crosslinking of ribosomal proteins S4, S5, S7, S8, S11, S12 and S18 to domains 1 and 2 of 16S rRNA in the *Escherichia coli* 30S particle, *Biochimie*, 71, 839, 1989.

55. Wang, R., Alexander, R. W., VanLoock, M., Vladimirov, S., Bukhtiyarov, Y., Harvey, S. C., and Cooperman, B. S., Three-dimensional placement of the conserved 530 loop of 16 S rRNA and of its neighboring components in the 30 S subunit, *J. Mol. Biol.*, 286, 521, 1999.

56. Nanda, K. and Wollenzien, P., Pattern of 4-thiouridine-induced cross-linking in 16S ribosomal RNA in the *Escherichia coli* 30S subunit, *Biochemistry*, 43, 8923, 2004.

57. Gulle, H., Hoppe, E., Osswald, M., Greuer, B., Brimacombe, R., and Stöffler, G., RNA–protein cross-linking in *Escherichia coli* 50S ribosomal subunits; determination of sites on 23S RNA that are cross-linked to proteins L2, L4, L24 and L27 by treatment with 2-iminothiolane, *Nucleic Acids Res.*, 16, 815, 1988.

58. Mitchell, P., Osswald, M., Schueler, D., and Brimacombe, R., Selective isolation and detailed analysis of intra-RNA cross-links induced in the large ribosomal subunit of *E. coli*: A model for the tertiary structure of the tRNA binding domain in 23S RNA, *Nucleic Acids Res.*, 18, 4325, 1990.

59. Brimacombe, R., Gornicki, P., Greuer, B., Mitchell, P., Osswald, M., Rinke-Appel, J., Schüler, D., and Stade, K., The three-dimensional structure and function of *Escherichia coli* ribosomal RNA, as studied by cross-linking techniques, *Biochim. Biophys. Acta*, 1050, 8, 1990.

60. Yusupov, M. M., Yusupova, G. Z., Baucom, A., Lieberman, K., Earnest, T. N., Cate, J. H., and Noller, H. F., Crystal structure of the ribosome at 5.5 A resolution, *Science*, 292, 883, 2001.

61. Sergiev, P. V., Dontsova, O. A., and Bogdanov, A. A., Chemical methods for the structural study of the ribosome: Judgment day, *Mol. Biol.*, 35, 472, 2001.

62. Whirl-Carrillo, M., Gabashvili, I. S., Bada, M., Banatao, D. R., and Altman, R. B., Mining biochemical information: Lessons taught by the ribosome, *RNA*, 8, 279, 2002.

63. Ivanov, V. I. and Mears, J. A., Using cross-links to study ribosomal dynamics, *J. Biomol. Struct. Dyn.*, 21, 691, 2004.

64. Terao, K., Uchiumi, T., Kobayashi, Y., and Ogata, K., Identification of neighbouring protein pairs in the rat liver 40-S ribosomal subunits cross-linked with dimethyl suberimidate, *Biochim. Biophys. Acta*, 621, 72, 1980.

65. Uchiumi, T., Terao, K., and Ogata, K., Identification of neighboring protein pairs in rat liver 60S ribosomal subunits cross-linked with dimethyl suberimidate or dimethyl 3,3′-dithiobispropionimidate, *J. Biochem.*, 88, 1033, 1980.

66. Tolan, D. R., Hershey, J. W., and Traut, R. T., Crosslinking of eukaryotic initiation factor eIF3 to the 40S ribosomal subunit from rabbit reticulocytes, *Biochimie*, 65, 427, 1983.

67. Takahashi, Y., Mitsuma, T., Hirayama, S., and Odani, S., Identification of the ribosomal proteins present in the vicinity of globin mRNA in the 40S initiation complex, *J. Biochem.*, 132, 705, 2002.

68. Gu, S. Q., Jöckel, J., Beinker, P., Warnecke, J., Semenkov, Y. P., Rodnina, M. V., and Wintermeyer, W., Conformation of 4.5S RNA in the signal recognition particle and on the 30S ribosomal subunit, *RNA*, 11, 1374, 2005.

69. Nechifor, R. and Wilson, K. S., Crosslinking of translation factor EF-G to proteins of the bacterial ribosome before and after translocation, *J. Mol. Biol.*, 368, 1412, 2007.

70. Leszyk, J., Grabarek, Z., Gergely, J., and Collins, J. H., Characterization of zero-length cross-links between rabbit skeletal muscle troponin C and troponin I: Evidence for direct interaction between the inhibitory region of troponin I and the NH2-terminal, regulatory domain of troponin C, *Biochemistry*, 29, 299, 1990.

71. Leszyk, J., Collins, J. H., Leavis, P. C., and Tao, T., Cross-linking of rabbit skeletal muscle troponin subunits: Labeling of cysteine-98 of troponin C with 4-maleimidobenzophenone and analysis of products formed in the binary complex with troponin T and the ternary complex with troponins I and T, *Biochemistry*, 27, 6983, 1988.

72. Luo, Y., Leszyk, J., Li, B., Li, Z., Gergely, J., and Tao, T., Troponin-I interacts with the Met47 region of skeletal muscle actin. Implications for the mechanism of thin filament regulation by calcium, *J. Mol. Biol.*, 316, 429, 2002.

73. Mudalige, W. A., Tao, T. C., and Lehrer, S. S., Ca2+-dependent photocrosslinking of tropomyosin residue 146 to residues 157–163 in the C-terminal domain of troponin I in reconstituted skeletal muscle thin filaments, *J. Mol. Biol.*, 389, 575, 2009.

74. Chantler, P. D. and Bower, S. M., Cross-linking between translationally equivalent sites on the two heads of myosin. Relationship to energy transfer results between the same pair of sites, *J. Biol. Chem.*, 263, 938, 1988.

75. Chantler, P. D., Tao, T., and Stafford, W. F., III, On the relationship between distance information derived from cross-linking and from resonance energy transfer, with specific reference to sites located on myosin heads, *Biophys. J.*, 59, 1242, 1991.

76. Bower, S. M., Wang, Y., and Chantler, P. D., Regulatory light-chain Cys-55 sites on the two heads of myosin can come within 2A of each other, *FEBS Lett.*, 310, 132, 1992.

77. Lu, R. C., Moo, L., and Wong, A. G., Both the 25-kDa and 50-kDa domains in myosin subfragment 1 are close to the reactive thiols, *Proc. Natl. Acad. Sci. U. S. A.*, 83, 6392, 1986.

78. Sutoh, K. and Lu, R. C., Identification of two segments, separated by approximately 45 kilodaltons, of the myosin subfragment 1 heavy chain that can be cross-linked to the SH-1 thiol, *Biochemistry*, 26, 4511, 1987.

79. Lu, R. C. and Wong, A., Glutamic acid-88 is close to SH-1 in the tertiary structure of myosin subfragment 1, *Biochemistry*, 28, 4826, 1989.

80. Arata, T., Structure of the actin-myosin complex produced by crosslinking in the presence of ATP, *J. Mol. Biol.*, 191, 107, 1986.

81. Onishi, H., Maita, T., Matsuda, G., and Fujiwara, K., Evidence for the association between two myosin heads in rigor acto-smooth muscle heavy meromyosin, *Biochemistry*, 28, 1898, 1989.

82. Onishi, H., Maita, T., Matsuda, G., and Fujiwara, K., Lys-65 and Glu-168 are the residues for carbodiimide-catalyzed cross-linking between the two heads of rigor smooth muscle heavy meromyosin, *J. Biol. Chem.*, 265, 19362, 1990.

83. Pliszka, B., Martin, B. M., and Karczewska, E., Ionic interaction of myosin loop 2 with residues located beyond the N-terminal part of actin probed by chemical cross-linking, *Biochim. Biophys. Acta*, 1784, 285, 2008.

84. Andreev, O. A. and Borejdo, J., Interaction of the heavy and light chains of cardiac myosin subfragment-1 with F-actin, *Circ. Res.*, 81, 688, 1997.

85. Bonafé, N. and Chaussepied, P., A single myosin head can be cross-linked to the N termini of two adjacent actin monomers, *Biophys. J.*, 68, 35S, 1995.

86. Miyanishi, T., Ishikawa, T., Hayashibara, T., Maita, T., and Wakabayashi, T., The two actin-binding regions on the myosin heads of cardiac muscle, *Biochemistry*, 41, 5429, 2002.

87. Grintsevich, E. E., Benchaar, S. A., Warshaviak, D., Boontheung, P., Halgand, F., Whitelegge, J. P., Faull, K. F. et al., Mapping the cofilin binding site on yeast G-actin by chemical cross-linking, *J. Mol. Biol.*, 377, 395, 2008.

88. Kluger, R. and Alagic, A., Chemical cross-linking and protein-protein interactions: A review with illustrative protocols, *Bioorg. Chem.*, 32, 451, 2004.

89. Melcher, K., New chemical crosslinking methods for the identification of transient protein–protein interactions with multiprotein complexes, *Curr. Protein Pept. Sci.*, 5, 287, 2004.

90. Lambright, D. G., Lee, T. K., and Wong, S. S., Association-dissociation modulation of enzyme activity: The case of lactose synthase, *Biochemistry*, 24, 910, 1985.

91. Brew, K., Shaper, J. H., Olsen, K. W., Trayer, I. P., and Hill, R. L., Cross-linking of the components of lactose synthetase with dimethylpimelimidate, *J. Biol. Chem.*, 250, 1434, 1975.

92. Valeyev, N. V., Heslop-Harrison, P., Postlethwaite, I., Kotov, N. V., and Bates, D. G., Multiple calcium binding sites make calmodulin multifunctional, *Mol. Biosyst.*, 4, 66, 2008.

93. Harrison, J. K., Lawton, R. G., and Gnegy, M. E., Development of a novel photoreactive calmodulin derivative: Cross-linking of purified adenylate cyclase from bovine brain, *Biochemistry*, 28, 6023, 1989.

94. Kincaid, R. L., Preparation of an enzymatically active cross-linked complex between brain cyclic nucleotide phosphodiesterase and 3-(2-pyridyldithio)propionyl-substituted calmodulin, *Biochemistry*, 23, 1143, 1984.

95. Zot, H. G. and Puett, D., An enzymatically active cross-linked complex of calmodulin and rabbit skeletal muscle myosin light chain kinase, *J. Biol. Chem.*, 264, 15552, 1989.

96. Schulz, D. M., Ihling, C., Clore, G. M., and Sinz, A., Mapping the topology and determination of a low-resolution three-dimensional structure of the calmodulin-melittin complex by chemical cross-linking and high-resolution FTICRMS: Direct demonstration of multiple binding modes, *Biochemistry*, 43, 4703, 2004.

97. Schmidt, A., Kalkhof, S., Ihling, C., Cooper, D. M., and Sinz, A., Mapping protein interfaces by chemical cross-linking and Fourier transform ion cyclotron resonance mass spectrometry: Application to a calmodulin/adenylyl cyclase 8 peptide complex, *Eur. J. Mass Spectrom.*, 11, 525, 2005.

98. Kalkhof, S., Ihling, C., Mechtler, K., and Sinz, A., Chemical cross-linking and high performance Fourier transform ion cyclotron resonance mass spectrometry for protein interaction analysis: Application to a calmodulin/target peptide complex, *Anal. Chem.*, 77, 495, 2005.

99. Dimova, K., Kalkhof, S., Pottratz, I., Ihling, C., Rodriguez-Castaneda, F., Liepold, T., Griesinger, C., Brose, N., Sinz, A., and Jahn, O., Structural insights into the calmodulin-Munc13 interaction obtained by cross-linking and mass spectrometry, *Biochemistry*, 48, 5908, 2009.

100. Bauer, P. I., Buki, K. G., Hakam, A., and Kun, E., Macromolecular association of ADP-ribosyltransferase and its correlation with enzymic activity, *Biochem. J.*, 270, 17, 1990.

101. Guerrero, C., Tagwerker, C., Kaiser, P., and Huang, L., An integrated mass spectrometry-based proteomic approach: Quantitative analysis of tandem affinity-purified in vivo cross-linked protein complexes (QTAX) to decipher the 26S proteasome interacting network, *Mol. Cell. Proteomics*, 5, 366, 2006.

102. Vasilescu, J. and Figeys, D., Mapping protein–protein interactions by mass spectrometry, *Curr. Opin. Biotechnol.*, 17, 394, 2006.

103. Leduc, M., Joseleau-Petit, D., and Rothfield, L. I., Interactions of membrane lipoproteins with the murein sacculus of *Escherichia coli* as shown by chemical crosslinking studies of intact cells, *FEMS Microbiol. Lett.*, 51, 11, 1989.

104. Studdert, C. A. and Parkinson, J. S., Insights into the organization and dynamics of bacterial chemoreceptor clusters through in vivo crosslinking studies, *Proc. Natl. Acad. Sci. U. S. A.*, 102, 15623, 2005.

105. Sakai, K., Hayashi, M., Kawashima, S., and Akanuma, H., Calcium-induced localization of calcium-activated neutral proteinase on plasma membranes, *Biochim. Biophys. Acta*, 985, 51, 1989.

106. Ji, T. H., A novel approach to the identification of surface receptors. The use of photosensitive hetero-bifunctional cross-linking reagent, *J. Biol. Chem.*, 252, 1566, 1977.

107. Shephard, E. G., de Beer, F. C., von Holt, C., and Hapgood, J. P., The use of sulfosuccinimidyl-2-(p-azidosalicylamido)-1,3′-dithiopropionate as a crosslinking reagent to identify cell surface receptors, *Anal. Biochem.*, 168, 306, 1988.

108. Cofano, F., Landolfo, S., Appella, E., and Ullrich, S. J., Analysis of murine interferon-gamma binding to its receptor on intact cells and solubilized membranes. Identification of an 80 kDa receptor, *FEBS Lett.*, 242, 233, 1989.

109. Lin, X., Kim, Y. A., Lee, B. L., Söderhäll, K., and Söderhäll, I., Identification and properties of a receptor for the invertebrate cytokine astakine, involved in hematopoiesis, *Exp. Cell Res.*, 315, 1171, 2009.

110. Wells, J. A., Knoeber, C., Sheldon, M. C., Werber, M. M., and Yount, R. G., Cross-linking of myosin subfragment 1. Nucleotide-enhanced modification by a variety of bifunctional reagents, *J. Biol. Chem.*, 255, 11135, 1980.

111. Dalbey, R. E., Wells, J. A., and Yount, R. G., Trapping of transition metal-nucleotide complexes in myosin subfragment 1 by cross-linking thiols; divalent transition metal probes of the active site, *Biochemistry*, 22, 490, 1983.

112. Nitao, L. K., Yeates, T. O., and Reisler, E., Conformational dynamics of the SH1–SH2 helix in the transition states of myosin subfragment-1, *Biophys. J.*, 83, 2733, 2002.

113. Nitao, L. K., Loo, R. R., O'Neall-Hennessey, E., Loo, J. A., Szent-Györgyi, A. G., and Reisler, E., Conformation and dynamics of the SH1–SH2 helix in scallop myosin, *Biochemistry*, 42, 7663, 2003.

114. Kliche, W., Pfannstiel, J., Tiepold, M., Stoeva, S., and Faulstich, H., Thiol-specific cross-linkers of variable length reveal a similar separation of SH1 and SH2 in myosin subfragment 1 in the presence and absence of MgADP, *Biochemistry*, 38, 10307, 1999.

115. Blotnick, E. and Muhlrad, A., Effect of nucleotides and actin on the intramolecular cross-linking of myosin subfragment-1, *Biochemistry*, 33, 6867, 1994.

116. Hamman, B. D., Oleinikov, A. V., Jokhadze, G. G., Traut, R. R., and Jameson, D. M., Rotational and conformational dynamics of *Escherichia coli* ribosomal protein L7/L12, *Biochemistry*, 35, 16672, 1996.

117. Ermolenko, D. N., Majumdar, Z. K., Hickerson, R. P., Spiegel, P. C., Clegg, R. M., and Noller, H. F., Observation of intersubunit movement of the ribosome in solution using FRET, *J. Mol. Biol.*, 370, 530, 2007.

118. Pitonzo, D., Yang, Z., Matsumura, Y., Johnson, A. E., and Skach, W. R., Sequence-specific retention and regulated integration of a nascent membrane protein by the endoplasmic reticulum Sec61 translocon, *Mol. Biol. Cell*, 20, 685, 2009.

119. Kareva, V. V., Dobrovol'sky, A. B., Baratova, L. A., Friedrich, P., and Gusev, N. B., Ca2+-induced structural change in the Ca^{2+}/Mg^{2+} domain of troponin C detected by crosslinking, *Biochim. Biophys. Acta*, 869, 322, 1986.

120. Schenker, E. and Kohanski, R. A., Conformational states of the insulin receptor, *Biochem. Biophys. Res. Commun.*, 157, 140, 1988.

121. Hajdu, J., Dombrádi, V., Bot, G., and Friedrich, P., Structural changes in glycogen phosphorylase as revealed by cross-linking with bifunctional diimidates: Phosphorylase b, *Biochemistry*, 18, 4037, 1979.

122. Dombrádi, V., Hajdu, J., Bot, G., and Friedrich, P., Structural changes in glycogen phosphorylase as revealed by cross-linking with bifunctional diimidates: Phospho-dephospho hybrid and phosphorylase a, *Biochemistry*, 19, 2295, 1980.

123. Huang, B. X. and Kim, H. Y., Probing Akt-inhibitor interaction by chemical cross-linking and mass spectrometry, *J. Am. Soc. Mass Spectrom.*, 20, 1504, 2009.

124. Salhany, J. M. and Sloan, R. L., Partial covalent labeling with pyridoxal 5′-phosphate induces bis(sulfosuccinimidyl)suberate crosslinking of band 3 protein tetramers in intact human red blood cells, *Biochem. Biophys. Res. Commun.*, 156, 1215, 1998.

125. Salhany, J. M., Allosteric effects in stilbenedisulfonate binding to band 3 protein (AE1), *Cell. Mol. Biol. (Noisy-le-grand)*, 42, 1065, 1996.

126. Salhany, J. M., Cordes, K. S., and Sloan, R. L., Band 3 (AE1, SLC4A1)-mediated transport of stilbenedisulfonates. III: Role of solute and protein structure in proton-activated stilbenedisulfonate influx, *Blood Cells Mol. Dis.*, 37, 155, 2006.

127. Huggins, W., Ghosh, S. K., Nanda, K., and Wollenzien, P., Internucleotide movements during formation of 16S rRNA-rRNA photocrosslinks and their connection to the 30S subunit conformational dynamics, *J. Mol. Biol.*, 354, 358, 2005.

128. Lustig, Y., Wachtel, C., Safro, M., Liu, L., and Michaeli, S., 'RNA walk' a novel approach to study RNA–RNA interactions between a small RNA and its target, *Nucleic Acids Res.*, 38, e5, 2010.

129. Watkins, K. P., Dungan, J. M., and Agabian, N., Identification of a small RNA that interacts with the 5′ splice site of the *Trypanosoma brucei* spliced leader RNA *in vivo*, *Cell*, T6, 171, 1994.

130. Bich, C., Bovet, C., Rochel, N., Peluso-Iltis, C., Panagiotidis, A., Nazabal, A., Moras, D., and Zenobi, R., Detection of nucleic acid-nuclear hormone receptor complexes with mass spectrometry, *J. Am. Soc. Mass Spectrom.*, 21, 635, 2010.

131. Kumar, R. and Lown, J. W., Recent developments in novel pyrrolo[2,1-c][1,4]benzodiazepine conjugates: Synthesis and biological evaluation, *Mini Rev. Med. Chem.*, 3, 323, 2003.

132. Hecht, A., Strahl-Bolsinger, S., and Grunstein, M., Mapping DNA interaction sites of chromosomal proteins. Crosslinking studies in yeast, *Methods Mol. Biol.*, 119, 469, 1999.

133. Jiang, D., Jarrett, H. W., and Haskins, W. E., Methods for proteomic analysis of transcription factors, *J. Chromatogr. A*, 1216, 6881, 2009.

134. Torchilin, V. P., Maksimenko, A. V., Smirnov, V. N., Berezin, I. V., and Martinek, K., Principles of enzyme stabilization. V. The possibility of enzyme selfstabilization under the action of potentially reversible intramolecular cross-linkages of different length, *Biochim. Biophys. Acta*, 568, 1, 1979.

135. Torchilin, V. P. and Trubetskoy, V. S., Stabilization of subunit enzymes by intramolecular crosslinking with bifunctional reagents, *Ann. N. Y. Acad. Sci.*, 434, 27, 1984.

136. Trubetskoy, V. S. and Torchilin, V. P., Artificial and natural thermostabilization of subunit enzymes. Do they have similar mechanism? *Int. J. Biochem.*, 17, 661, 1985.

137. Govardhan, C. P., Crosslinking of enzymes for improved stability and performance, *Curr. Opin. Biotechnol.*, 10, 331, 1999.

138. Martinek, K. and Torchilin, V. P., Stabilization of enzymes by intramolecular cross-linking using bifunctional reagents, *Methods Enzymol.*, 137, 615, 1988.

139. Torchilin, V. P., Trubetskoy, V. S., Omelyanenko, V. G., and Martinek, K., Stabilization of subunit enzymes by intersubunit cross-linking with bifunctional reagents: Studies with glyceraldehyde-3-phosphate dehydrogenase, *J. Mol. Catal.*, 19, 291, 1983.

140. Gottschalk, N. and Jaenicke, R., Chemically crosslinked lactate dehydrogenase: Stability and reconstitution after glutaraldehyde fixation, *Biotechnol. Appl. Biochem.*, 9, 389, 1987.

141. Tatsumoto, K., Oh, K. K., Baker, J. O., and Himmel, M. E., Enhanced stability of glucoamylase through chemical crosslinking, *Appl. Biochem. Biotechnol.*, 20, 293, 1989.

142. Tananchai, P. and Chisti, Y., Stabilization of invertase by molecular engineering, *Biotechnol. Prog.*, 26, 111, 2010.

143. Walton, R. S., Brand, D. D., and Czernuszka, J. T., Influence of telopeptides, fibrils and crosslinking on physicochemical properties of type I collagen films, *J. Mater. Sci. Mater. Med.*, 21, 451, 2010.

144. Tamura, M., Tamura, T., Burnham, D. N., Uhlinger, D. J., and Lambeth, J. D., Stabilization of the superoxide-generating respiratory burst oxidase of human neutrophil plasma membrane by crosslinking with 1-ethyl-3-(3-dimethylaminopropyl) carbodiimide, *Arch. Biochem. Biophys.*, 275, 23, 1989.

145. Vandegriff, K. D., Medina, F., Marini, M. A., and Winslow, R. M., Equilibrium oxygen binding to human hemoglobin cross-linked between the alpha chains by bis(3,5-dibromosalicyl) fumarate, *J. Biol. Chem.*, 264, 17824, 1989.

146. Vandegriff, K. D., Le Tellier, Y. C., Winslow, R. M., Rohlfs, R. J., and Olson, J. S., Determination of the rate and equilibrium constants for oxygen and carbon monoxide binding to R-state human hemoglobin cross-linked between the alpha subunits at lysine 99, *J. Biol. Chem.*, 266, 17049, 1991.

147. Pennathur-Das, R., Heath, R. H., Mentzer, W. C., and Lubin, B. H., Modification of hemoglobin S with dimethyl adipimidate. Contribution of individual reacted subunits to changes in properties, *Biochim. Biophys. Acta*, 704, 389, 1982.

148. Roche, P. A., Jensen, P. E., and Pizzo, S. V., Intersubunit cross-linking by cis-dichlorodiammineplatinum(II) stabilizes an alpha 2-macroglobulin "nascent" state: Evidence that thiol ester bond cleavage correlates with receptor recognition site exposure, *Biochemistry*, 27, 759, 1988.

149. Dihazi, G. H. and Sinz, A., Mapping low-resolution three-dimensional protein structures using chemical cross-linking and Fourier transform ion-cyclotron resonance mass spectrometry, *Rapid Commun. Mass Spectrom.*, 17, 2005, 2003.

12 Applications of Chemical Conjugation in the Preparation of Immunoconjugates and Immunogens

12.1 INTRODUCTION

As mentioned in Chapter 1, chemical conjugation is a process to cross-link unrelated molecules that normally do not have any affinity for each other. In this chapter, we will consider covalent bonding between immunoglobulins and reporter groups (usually an enzyme) to form immunoconjugates for immunoassays and haptens and proteins to form hapten-carrier conjugates for immunization to produce hapten-specific antibodies. In the next chapter, we will consider covalent binding between toxic molecules and antibodies or other proteins to form immunotoxins and tissue-directed conjugates for therapeutic applications. As may be realized, chemical conjugation has widespread applications in biotechnology. From the preparation of immunogens to cell-targeted cancer drugs, from immunoassays to purification of macromolecules, chemical cross-linking reagents are used to prepare the necessary components. Examples are provided in this chapter to illustrate how these cross-linkers are employed in the area of chemical analysis involving immunoglobulins.

12.2 PREPARATION OF IMMUNOCONJUGATES

Since the introduction of radioimmunoassay (RIA) by Yalow and Berson[1] in 1959, immunological methods have been most widely used for the quantification and detection of a wide variety of compounds. These biochemical tests rely on the affinity of antibodies to bind to specific molecular structures (antigens). They have become the most prevalent technology in diagnostics, from home pregnancy testing kits to AIDS testing.[2] While radioisotopes are sensitive reporter groups for RIA, other groups have also been developed. These include enzymes, fluorescent, chemiluminescent, chromophoric, and spin probes. In all these cases, the reporter group is covalently attached to either the antibody or the antigen (analyte). In these applications, chemical cross-linking reagents are particularly needed to prepare the enzyme–antibody conjugates for enzyme immunoassay (EIA).

EIA uses enzyme activity to determine the concentration of analytes. There are many types of EIAs, which have been treated in various textbooks and monographs.[2–7] In general, they can be divided into homogeneous (or separation free) and heterogeneous (or separation required) assays. The heterogeneous EIA is also known as the enzyme-linked immunosorbent assay or ELISA.[8,9] It encompasses competitive-binding assays and immunoenzymometric or sandwich assays. The classical competitive EIA is analogous to the traditional RIA. Either enzyme–antigen or enzyme–antibody conjugates are employed to measure the concentration of analytes. In the sandwich or immunoenzymometric assays, enzyme-coupled antibodies are used. The analyte is sandwiched between two antibodies, and the enzyme activity of the isolated sandwich complex is proportional to the concentration of the analyte detected. Several variations of the immunometric assays have been devised. These include the use of enzyme-coupled species-specific antibodies,[10] enzyme-labeled protein A,[11] and the affinity column-mediated immunometric assay.[12]

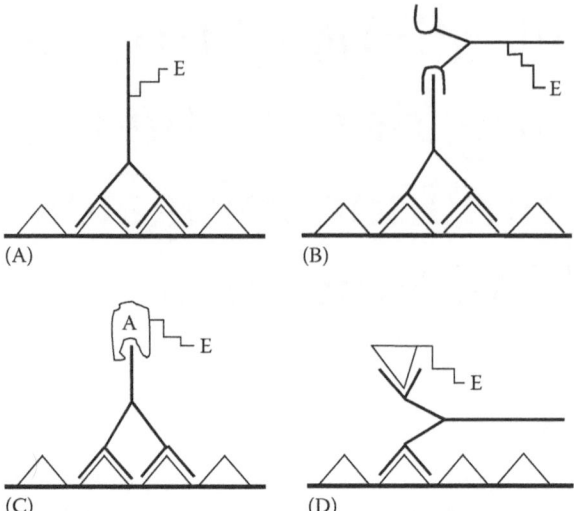

FIGURE 12.1   Immunoperoxidase staining methods. (A) Direct staining procedure: peroxidase–antibody conjugate is used. (B) Indirect method: a peroxidase–second antibody conjugate is used. (C) Two-stage protein A–peroxidase method: peroxidase–protein A conjugate is used. (D) Labeled antigen method: peroxidase–antigen conjugate is used.

Homogeneous EIAs do not require separation of labeled antigen–antibody complexes. They depend on a change in enzyme-specific activity when enzyme–antigen conjugates bind to the antibody. There are many different versions of the homogeneous EIA. The best-known example is the enzyme-multiplied immunoassay technique (EMIT).[3] A decrease in enzyme activity is seen when an antibody is bound to the enzyme–antigen conjugate. In the presence of the antigen analyte, such complex formation is prevented. Thus, the amount of enzyme activity is proportional to the concentration of analyte.

The EIA requires an enzyme-conjugated antigen or antibody, and these conjugates are also used in a variety of histochemical and cytochemical studies.[13] For example, immunostaining, using horseradish peroxidase (HRP)-labeled antibody (immunoperoxidase), has become a prevalent technique. Different methods, including direct and indirect immunoperoxidase procedures, two-stage protein A–peroxidase, and antigen-labeled methods have been devised as shown in Figure 12.1. The immunoperoxidase methods are also applicable to electron microscopy. It is obvious that enzyme–antigen and enzyme–antibody conjugates are of paramount importance in immunoassays. The ability to produce active and staple conjugates is critical to such analytical techniques. This section will present the various versatile methods for the cross-linking of enzymes to antibodies and antigens.

In addition to covalent cross-linking methods, biotin–avidin/streptavidin interactions have been used to amplify EIA.[14,15] As discussed in Chapter 4, the system is based on the principle that avidin possesses four binding sites and can act as a bridge between two different biotinylated proteins. Similarly, lectins, which possess two or more active sites, have been used to amplify immunoassays, as alluded to in Chapter 4.[16,17]

12.2.1   Components of Enzyme Immunoconjugates

12.2.1.1   Enzymes

Theoretically, any enzyme can be used as a label in EIA. However, certain properties are more desirable than others. These include the following:

   a. High substrate turnover rate, that is, high specific activity with low K_m
   b. High stability, that is, long shelf life
   c. An easy, cheap, nontoxic, and sensitive assay procedure

d. Reactive groups for coupling to other molecules

e. Easily purified

f. Lack of enzyme activity in test fluids

Over 25 different enzymes have been used as labels in EIA. Some of the characteristics of these enzymes are listed in Table 12.1.[18–55] It should be noted that the information given in the table is dependent on the source of the enzyme. The kinetic parameters may vary for enzymes isolated from different species. Interested readers should consult the literature for the detailed enzymatic

TABLE 12.1
Examples of Enzymes Used in EIA

Enzyme	Specific Activity (U/mg)	K_m (mM)	Mol. Wt. (kDa)	pH_{max}	References
Acetylcholine esterase (EC 3.1.1.7)	11,000	0.09 (acetylcholine)	54	7–8	13,18
Adenosine deaminase (EC 3.5.4.4)	437	0.04 (adenosine)	110	7–9	19,20
Alkaline phosphatase (EC 3.1.3.1)	350	0.2 (p-nitrophenyl phosphate)	100	8–10	21,22 23,24
α-Amylase (EC 3.2.1.1)	2,240	1 g/mL (starch)	97	5–9	25,26
β-Amylase (EC 3.2.1.2)	1,640	0.07 (amylose)	152	4–6	27
Catalase (EC 1.11.1.6)	9,000,000	1,100 (H_2O_2)	250	6–8	28
Carbonic anhydrase (EC 4.2.1.1)	30,000	2.8 (CO_2)	30	7–8	29
β-Galactosidase (EC 3.2.1.23)	340	1 (o-nitrophenyl-β-D-galactopyranoside)	540	6–8	30,31 32,33
β-Glucosidase (EC 3.2.1.21)	38	0.08 (p-nitrophenyl-β-D-glucopyranoside)	300	6–7	32,34
Glucose oxidase (EC 1.1.3.4)	80	33 (glucose)	186	4–7	32,35 36,43
G6PDH (EC 1.1.1.49)	700	0.02 (glucose-6-phosphate)	128	7–8	32,37 43
Glucoamylase (EC 3.2.1.3)	37	0.03 (amylose)	48	4–6	38
Hexokinase (EC 2.7.1.1)	800	0.1 (glucose)	99	6–8	39
HRP (EC 1.11.1.7)	4,500	0.2 (H_2O_2)	40	5–7	22,23 36,40
Inorganic phosphatase (EC 3.6.1.1)	1,433	0.05 (pyrophosphate)	63	6–8	41,42
Invertase: (EC 3.2.1.26)	3,000	9.1 (glucose)	270	4–7	43
Δ5,3-Ketosteroid isomerase (EC 5.3.3.1)	88,000	0.3 (Δ5-androstene-3,17-dione)	40	6–8	44
Lactate dehydrogenase (EC 1.1.1.27)	11,500	0.8 (pyruvate)	140	6–8	45
Luciferase (EC 1.13.12.7)	28,000	0.2 (ATP) 0.02 (luciferin)	100	6–8	46,47
Lysozyme (EC 3.2.1.17)	2×10^{-5}	4 (p-nitrophenyl-β-D-chitobioside)	14.4	4–6	48,49
Malate dehydrogenase (EC 1.1.1.37)	1,000	0.3 (malate) 0.1 (NAD^+)	70	8–10	10,50
Penicillinase (EC 3.5.2.6)	5,400	0.05 (benzylpenicillin)	23	7	51,52
Phospholipase C (EC 3.1.4.3)	15	0.1 (phosphoinositol)	85	7–8	53
Pyruvate kinase (EC 2.7.1.40)	445	0.07 (PEP)	237	6–8	54
Urease (EC 3.5.1.5)	2,150	11 (urea)	483	6–8	55

properties of these proteins. By far, the most widely employed enzymes are HRP, alkaline phosphatase (ALP), glucose oxidase (GO), glucose-6-phosphate dehydrogenase (G6PDH), and β-galactosidase (GS). HRP is cheap and readily available in fairly pure form. There are many substrates available. In addition, it has 10%–15% of carbohydrate, which can be used for conjugation. GS has 20 free thiol groups that provide a useful functionality for coupling reactions. It is gaining popularity because of the availability of fluorogenic substrates. G6PDH has been used in the EMIT-type assays. Other enzymes have also been popular because of the reasons cited above. Since it is impossible to provide all the coupling procedures for all the enzymes, only the most commonly used enzymes are described to illustrate the versatility of the conjugation methodology.

12.2.1.2 Antibodies and Their Fragments

The basic unit of an immunoglobulin (Ig) molecule consists of two identical light chains and two identical heavy chains. These chains are held together by disulfide bonds as well as noncovalent interactions. The light chain contains one variable domain and one constant domain, whereas the heavy chain contains a variable domain and three separate constant domains. There are five major classes of Ig's: IgA, IgG, IgD, IgM, and IgE. Each class has different molecular complexities. Among these, IgG is the most abundant, constituting 8–16 mg/mL in human serum. It is therefore the most widely used antibody in immunoassays. IgG is easily purified from serum or ascites fluid by fractionation with sodium sulfate or ammonium sulfate (35%–45%) followed by DEAE ion exchange chromatography[56,57] or affinity chromatography on a protein A column,[56,58] protein G,[59] or recombinant protein A/G-Sepharose.[56,60] IgG can be further fractionated into its subclasses. For example, mouse IgG can be fractionated into IgG_1, IgG_{2a}, IgG_{2b}, and IgG_3 by various affinity chromatography.[56–61] Further information regarding procedures for the purification of immunoglobulins can be obtained from the literature.[62–64]

Fragmentation of IgG into antigen-binding (Fab) and effector-activating (Fc) fragments can be achieved by enzymatic cleavage of the hinge-region between constant domains and two of the heavy chain. Treatment of IgG with papain will produce two Fab and one Fc fragments,[65,66] whereas treatment with pepsin generates only $F(ab')_2$ fragment as shown in Figure 12.2.[66,67] $F(ab')_2$ can be further reduced to Fab' in the presence of 2-mercaptoethylamine.[68,69] Enzyme–Fab' conjugates are more useful than the enzyme–IgG conjugate in both immunohistochemical staining of tissue sections and EIAs.[70] They also give a lower nonspecific binding and a higher sensitivity in solid-phase EIA[71] and more readily penetrate into tissue sections and provide a lower background staining.

Since Fab and $F(ab')_2$ fragments retain their antigen-binding capability, the use of these fragments reduces nonspecific binding due to the removal of Fc. In addition, the generation of free sulfhydryl group in Fab' fragment facilitates some of the coupling reactions.

12.2.2 Introduction of Thiol Groups into Immunoglobulins

Although both enzymes and antibodies contain various functional groups, the most reactive functionality is the thiol. The generation of free sulfhydryl groups has been achieved by two different methods. The first involves reductive cleavage of the native cystine residues in the protein with reagents such as dithiothreitol (DTT), and the second involves chemical introduction of thiol groups (thiolation). Reductive cleavage of disulfides can be employed to functionalize antibodies lacking free thiol groups. This approach is feasible because reduction conditions can be kept mild enough not to significantly alter the functional features of Igs.[72] The concentration of reducing agents needed varies with the antibody, and the optimum levels have to be individually determined. Thiol groups in IgG can be generated by the reduction of disulfide bonds in the hinge region. This process is

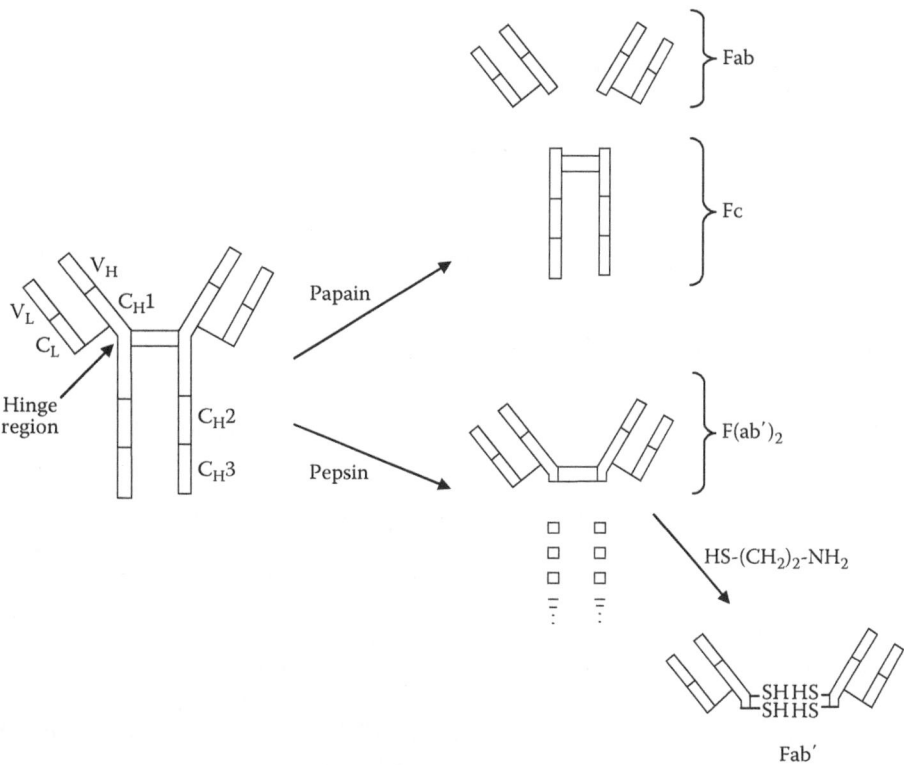

FIGURE 12.2 Fragmentation of IgG immunoglobulin by papain and pepsin.

easily achieved by incubating the sample at 37°C for 2 h with 10 mM 2-mercaptoethylamine in 0.1 M sodium phosphate buffer, pH 6.0, containing 5 mM EDTA.[73] One can also reduce F(ab')₂ fragments to introduce a free thiol group into Fab'. Reduced Fab' from rabbit contains one thiol group per molecule. Fab' from other animal sources may contain up to three sulfhydryl groups. The content of thiols can be measured by adding 0.2 mM 4,4'-dithiodipyridine to the sample. After incubation at room temperature for 20 min, the concentration of thiol may be obtained from the absorbance at 324 nm using a molar absorptivity of 19,800 mol/cm.[74]

Alternatively, thiol groups can be introduced into IgG and F(ab')₂ by mercaptosuccinylation. Succinylation is first carried out by incubating IgG or F(ab')₂ in 0.1 M sodium phosphate buffer, pH 6.5, with 100-fold molar excess of S-acetylmercaptosuccinic anhydride (previously dissolved in N,N-dimethylformamide) at room temperature for 90 min. Hydroxylamine is then added to attain a final concentration of 0.2 M. Free thiol will be obtained on incubation at 30°C for 5 min (see Chapter 2). The content of thiol groups can be measured with dithiodipyridine as mentioned above. Thiol groups are fairly stable in the presence of EDTA.[75] Other chemical methods for the introduction of thiol groups using N-succinimidyl 3-(2-pyridyldithio)propionate (SPDP), 2-iminothiolane, sulfosuccinimidyl 6-(3'-[2-pyridyldithio]propionamido)hexanoate (Sulo-LC-SPDP),[76] and other compounds have been discussed in Chapter 2. Examples illustrating the involvement of these thiols in the preparation of immunoconjugates will be shown below.

Although various reagents have been used to cross-link the functional groups present in proteins, the most favorable cross-linkers are those directed toward the thiol and amino groups. Theoretically, any heterobifunctional reagents in this category can be used. However, there are certain preferences toward different enzymes and Ig fragments. Detailed experimental methods for the preparation of some of these conjugates are described below.

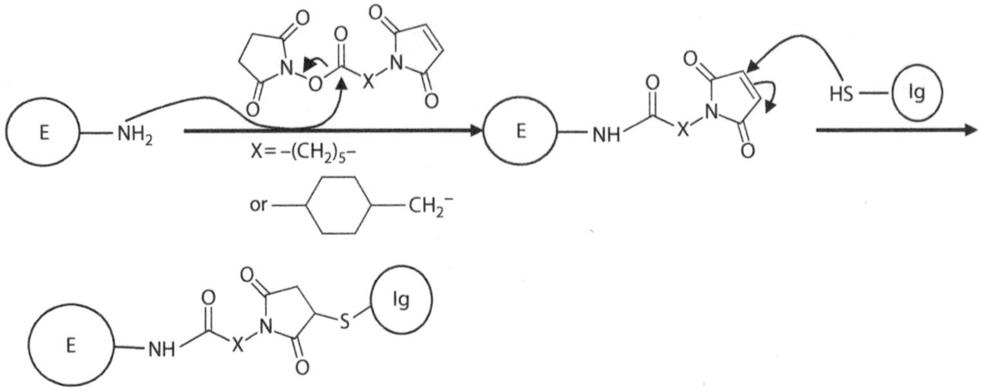

FIGURE 12.3   Coupling of HRP (E) to thiol containing immunoglobulin (Ig) with SMCC or SMH. G6PDH and alkaline phosphatase may be similarly conjugated.

12.2.3   PREPARATION OF HORSERADISH PEROXIDASE IMMUNOCONJUGATES

12.2.3.1   Conjugation with Amino- and Thiol-Directed Cross-Linkers

Among the heterobifunctional reagents selective toward amino and thiol groups, N-succinimidyl 4-(N-maleimidomethyl)cyclohexane-1-carboxylate (SMCC) and N-succinimidyl 6-maleimidohexanoate (SMH) have been most favorably used to cross-link the amino group of horseradish peroxidase (HRP) with the thiol group of Fab′, reduced IgG, or thiolated F(ab′)$_2$. The strategy of the coupling reaction is shown in Figure 12.3. In short, HRP (6 mg/mL) in 0.1 M sodium phosphate buffer, pH 7.0, is first reacted with 0.3–0.7 mg of cross-linker dissolved in N,N-dimethylformamide (final total organic solvent 10%). The reaction is incubated for 0.5–1.0 h at 30°C. Labeled HRP can be isolated by gel filtration on Sephadex G-25 (equilibrated with 0.1 M sodium phosphate, pH 6.0) by gravity or centrifugation.[77] The number of maleimide groups introduced into the enzyme can be estimated by first reacting with excess 2-mercaptoethylamine followed by back titration with 4,4′-dithiodipyridine, as discussed above.[74]

Coupling of labeled HRP with Fab′, reduced IgG, or thiolated IgG and F(ab′)$_2$ can be achieved by incubating the components (1:1 mol/mol; 0:01–0.15 mM) in 0.1 M sodium phosphate buffer, pH 6.0, containing 2.5 mM EDTA, at 4°C for 20 h or at 30°C for 1 h. The conjugated product may be separated by gel filtration. Since the incorporation of cross-linker into HRP is limited and the number of thiols in immunoglobulins is small, this method affords 1:1 conjugates.

12.2.3.2   Conjugation through Disulfide Formation

HRP and IgG or its fragments can be conjugated through a disulfide bond.[70] The reaction of SPDP with HRP will activate the amino groups through labeling with pyridyl disulfide, which can react further with free thiol groups of Fab′ or thiolated IgG and F(ab′)$_2$ (Figure 12.4A). Introduction of pyridyl disulfide groups into HRP can be achieved by incubating HRP (about 6 mg/mL) with fivefold molar excess of SPDP (previously dissolved in ethanol) in 0.1 M sodium phosphate buffer, pH 7.5, for 30 min at 25°C. After the reaction, pyridyl disulfide-labeled HRP may be isolated by gel filtration through a Sephadex G-25 column. Coupling of the labeled enzyme to Fab′ can be achieved by mixing equimolar concentrations of the components in 0.1 M sodium phosphate buffer, pH 6, containing 5 mM EDTA for 2.5 h at 30°C. The conjugate may be purified by gel filtration.

Alternatively, HRP can be thiolated with S-acetylmercaptosuccinic anhydride to generate a free thiol, which can react with dithiopyridine introduced into Fab′ or thiolated IgG (Figure 12.4B). Thiolation of HRP can be carried out by reacting HRP (8 mg/mL) in 0.1 M sodium phosphate buffer, pH 7.5, with 200-fold excess S-acetylmercaptosuccinic anhydride (previously dissolved in N,N-dimethylformamide). After 30 min at 30°C, hydroxylamine (1:5 v/v) is added to hydrolyze the thioester.

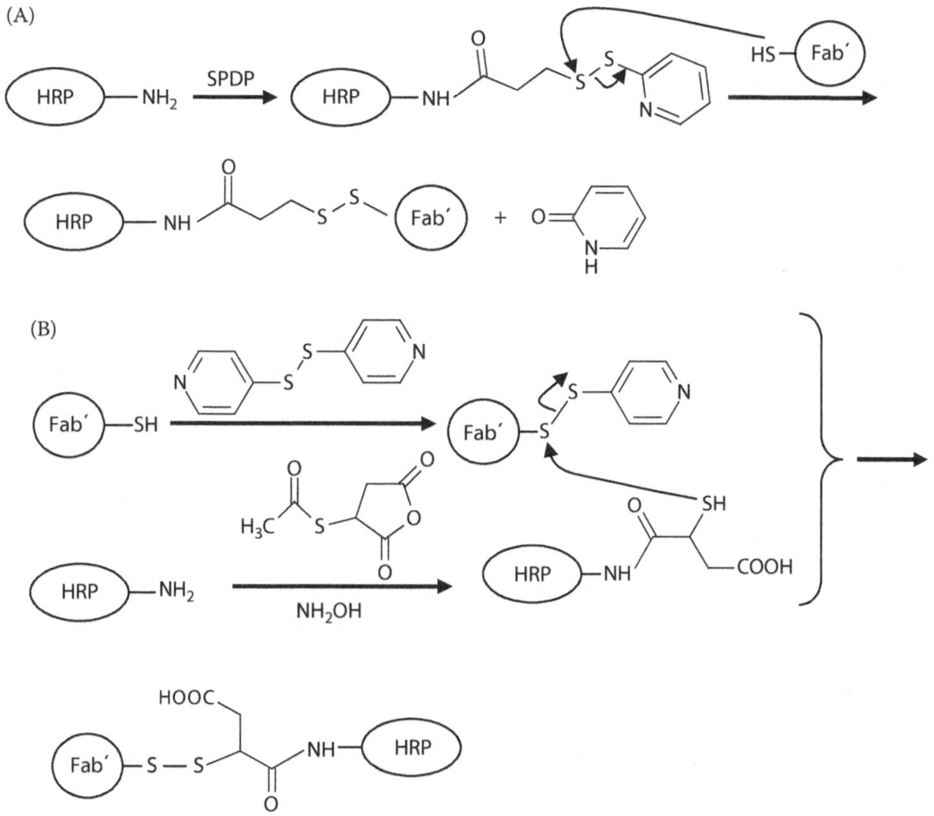

FIGURE 12.4 Coupling of horseradish peroxidase (HRP) and Fab′ through disulfide bond. (A) HRP is activated by SPDP and then reacts with Fab′; (B) Fab′ is activated with dithiodipyridine and then reacts with thiolated HRP.

Thiolated HRP can be purified by gel filtration in the presence of 5 mM EDTA. To prepare for conjugation, Fab′ is activated with 100-fold excess of 4,4′-dithiodipyridine for 15 min and isolated by gel filtration. The two components are cross-linked on mixing, similar to the procedure described above. Before purification by gel filtration, N-ethylmaleimide is added to terminate the reaction.

12.2.3.3 Conjugation with Glutaraldehyde

Glutaraldehyde readily reacts with amino groups of enzymes, antigens, and antibodies under mild conditions to form stable conjugates. However, the reaction of glutaraldehyde with HRP is much slower than with antibodies. The products of direct one-step coupling of HRP and antibodies are generally heterogeneous, high molecular-weight aggregates with less than 10% recovery of enzyme activity.[78] Thus, a two-step procedure is adapted.[79] First, HRP (50 mg/mL) is modified with 0.2% glutaraldehyde in 0.1 M phosphate buffer, pH 6.8, at ambient temperature for 18 h. Excess reagent and peroxidase polymers are then removed by gel filtration. The isolated glutaraldehyde-activated HRP (about 10 mg/mL) is then mixed with IgG, F(ab′)₂, or Fab′ (pretreated with N-ethylmaleimide to block the free thiol group) in a 1:0.2 molar ratio in 0.1 M sodium carbonate buffer, pH 9.5, containing 0.15 M NaCl. After 24 h at 4°C, L-lysine (15 mM) is added to terminate the reaction, and the mixture is incubated for another 2 h. Peroxidase–antibody conjugate can be isolated on Ultragel AcA 44. The conjugate is composed mostly of 1:1 component.[79] However, only a small proportion of HRP is recovered in the conjugate. Some dimerization of glutaraldehyde-activated HRP has been reported.[80]

12.2.3.4 Conjugation Using Periodate Oxidation

Since HRP is a glycoprotein, oxidation of its carbohydrate moiety with sodium periodate will result in the formation of aldehyde groups, which will form Schiff bases with amino groups, as discussed in Chapter 2. The enzyme has a few amino groups, but self-coupling can be prevented. When IgG is added, stable conjugates can be obtained after reduction with sodium borohydride or other reducing agents (see Chapter 2).[81] Basically, HRP (4 gm/mL) is oxidized with 16 mM NaIO$_4$ at room temperature for 10–20 min. At the end of that incubation period, ethylene glycol is added to stop the reaction. Excess reagents are removed by either gel filtration with 1 mM sodium acetate buffer, pH 4.4, or dialysis against the same buffer overnight at °C. Oxidized peroxidase is then incubated with IgG, F(ab')$_2$, or Fab' (pretreated with *N*-ethylmaleimide) in a 1:1 molar ratio in 0.1 M sodium carbonate buffer, pH 9.5, containing 0.15 M NaCl, at 25°C for 2 h. The pH is adjusted, if necessary, with 0.2 M sodium bicarbonate. The Schiff base formed is reduced by adding freshly prepared sodium borohydride (final concentration about 0.2 mg/mL) and incubated at 4°C for 2 h. Peroxidase–Ig conjugate can be isolated by gel filtration on a column of Ultragel AcA 44.

Tijssen and Kurstak[82] designed a simplified version of the periodate method. In this procedure, HRP (1 mg/mL) is oxidized with 4–8 mM sodium periodate in 0.1 M sodium carbonate in a sealed tube. After 2 h at ambient temperature, excess IgG (1:7 mg/mg) is added followed by dry Sephadex (one sixth the combined weight of peroxidase and IgG) to destroy unreacted periodate. This mixture is incubated for 3 h at ambient temperature. Finally, the proteins are eluted from the Sephadex and reduced with sodium borohydride in two additions at 30 min intervals (total 0:5 mg/mL). Peroxidase–IgG conjugate is purified by first precipitation with 50% ammonium sulfate and then on a Concanavalin A affinity column from which peroxidase–IgG can be eluted with 0.05 M α-methyl-D-mannopyranoside.

To prevent self-coupling, the reactive amino groups of HRP can be first irreversibly blocked by alkylation with dinitrofluorobenzene (DNFB).[83] After activation by periodate, the amino group blocked HRP is then coupled to IgG, F(ab')$_2$, or Fab', as discussed above. With this procedure, a maximum of 5–6 mol of HRP could be bound per mole of IgG.

12.2.3.5 Zero-Length Conjugation *In Vacuo*

Simons et al.[84] developed a new method of *in vacuo* conjugation of HRP and IgG. Basically, 5 mg of HRP and 2 mg IgG were dissolved in 2 mL of distilled water. The solution was dialyzed in a 10 kDa MWCO dialysis membrane against distilled water for four changes of 1 L each over 5 h to remove glycerol, sodium azide, and storage buffers present in commercial antibody preparations. The HRP and IgG mixture was then placed in a glass tube (13 × 100 mm), pH was adjusted to 7.5 with 1 N NaOH, and the sample was lyophilized. Once the proteins were dry, the glass sample tube was narrowed, sealed under vacuum (<50 mTorr), and incubated in an oven at 85°C for 96 h. Following the reaction, the vacuum was broken, and the dried sample was reconstituted in 1 mL of 0.1 M Tris-HCl, 0.15 M NaCl at pH 7.5. The covalent cross-linking observed between HRP and IgG presumably is caused by condensation of a protonated amino group with a negatively charged carboxyl group to form an amide bond. This condensation reaction is driven by the removal of water under vacuum. Cross-linking is optimal when the proteins are lyophilized at neutral to slightly alkaline pH, between pH 7 and 9.0, in which amino and carboxyl groups are expected to be in their positively and negatively charged states, respectively. *In vacuo* cross-linked proteins generally retain their biological activity when returned to an aqueous environment.

12.2.3.6 Conjugation with Miscellaneous Cross-Linkers

Many other cross-linking reagents have been used to couple HRP to antibodies. However, most of these either give low yield or cause considerable loss of activity. For example, the use of *p,p'*-difluoro-*m,m'*-dinitrophenyl sulfone gives less than 1% yield.[85] Other reagents include cyanuric chloride,[86] toluene diisocyanate,[87,88] water-soluble carbodiimides,[88] and *p*-benzoquinone.[89]

12.2.4 PREPARATION OF ALKALINE PHOSPHATASE IMMUNOCONJUGATES

12.2.4.1 Conjugation with Amino- and Thiol-Directed Reagents

Like HRP, alkaline phosphatase (ALP) can be coupled to antibodies by SMCC or SMH, according to the scheme shown in Figure 12.3. ALP from calf intestine (4 mg/mL) is first labeled with the cross-linker (1:25) in 50 mM sodium borate buffer, pH 7.6, containing 1 mM MgCl$_2$ and 0.1 mM ZnCl$_2$ at 30°C for 30 min. The labeled enzyme is then isolated by gel filtration. About four to six cross-linkers can be introduced per mol of enzyme, but up to 40%–50% of activity may be lost. Fivefold excess of Fab' [or reduced IgG, mercaptosuccinimidyl IgG, or F (ab')] is then incubated with labeled ALP (2 mg/mL) at 4°C for 20 h. The reaction is terminated by incubating with 0.2 mM mercaptoethylamine for 20 min at room temperature. ALP immunoconjugates are isolated by gel filtration on Ultragel AcA 34. A significant proportion of Ig remains unconjugated. To correct this situation, excess labeled ALP may be used. Complete conversion will avoid the difficulty in separating free IgG from the conjugate.

Jeanson et al.[90] have used SPDP to conjugate ALP and IgG. The reaction scheme is shown in Figure 12.5. The principle involves reaction of both ALP and IgG separately with excess SPDP for 30 min at room temperature. Excess reagent and its N-hydroxysuccinimide products are removed by dialysis. DTT (25 mM) is added to the labeled ALP to expose the sulfhydryl group. After 20 min, excess DTT is removed by gel filtration on Sephadex G-25. Modified IgG and ALP are mixed and incubated for 20 h. The conjugates can be isolated by gel filtration. The authors claimed that this method yielded better performing conjugates.

12.2.4.2 Conjugation with Glutaraldehyde

Both one-step and two-step procedures of coupling ALP to antibodies have been used. In the one-step method, the enzyme and IgG (2:1 wt/wt) or Fab' (4:1 wt/wt) are mixed with 0.2% (v/v) glutaraldehyde in 0.1 M phosphate buffer, pH 6.8, for 2 h at ambient temperature.[90,91] L-Lysine (0.1 M) is used to block the reactive sites. However, the conjugates are heterogeneous aggregates, and the recovery of enzyme activity is low.[92] For the two-step procedure, ALP (2.5 mg/mL) in 50 mM potassium phosphate buffer, pH 7.2, is first treated with 0.2% glutaraldehyde for 50 min at 24°C and then mixed with IgG (1 mg/mL). The mixture is incubated for 75 min at 24°C before dialysis.

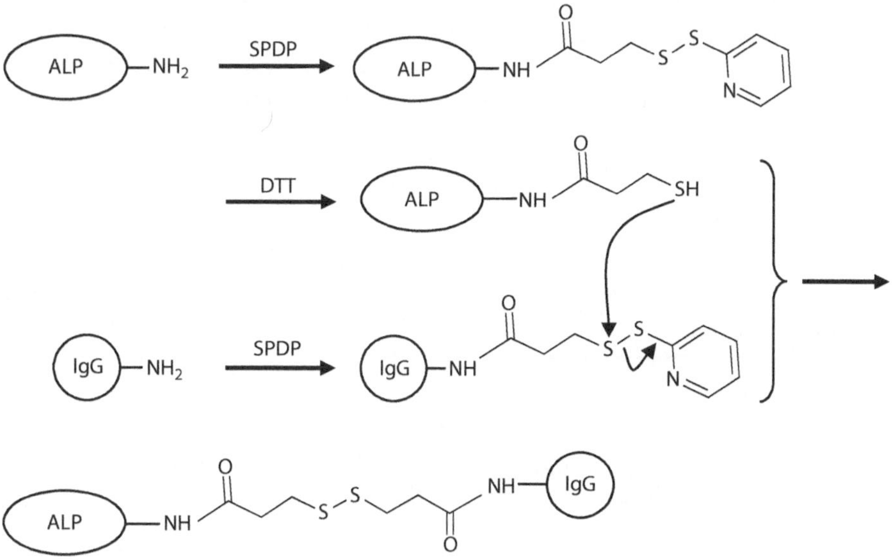

FIGURE 12.5 Conjugation of alkaline phosphatase (ALP) and IgG with SPDP.

12.2.4.3 Conjugation with Periodate Oxidation

The procedure for coupling ALP to IgG with sodium periodate is similar to that for HRP.[93] ALP is first treated with DNFB (0.1%) for 2h at 23°C to block its amino groups. Sodium periodate (20 mM) is then added. After 6h, ethylene glycol is added to terminate the reaction. Free reagents are removed by dialysis. The oxidized ALP is then mixed with IgG and incubated for 24h at 4°C. Monoethanoamine is then added, and the conjugate is dialyzed to get rid of free agents.

12.2.4.4 Zero-Length Conjugation

ALP is conjugated to IgG *in vacuo* using the similar procedure described above for HRP–IgG cross-linking with the following changes:[84] 2 mg of ALP and 2 mg IgG were mixed, and the final volume was adjusted to 2 mL with distilled water. The solution was dialyzed against water in 50 kDa MWCO dialysis tubing.

12.2.5   PREPARATION OF α-D-GALACTOSIDASE IMMUNOCONJUGATES

12.2.5.1 Conjugation with Amino- and Thiol-Directed Reagents

In addition to SMCC, sulfo-SMCC,[94] and SMH, other amino- and thiol-group directed reagents such as maleimidobenzoyl-*N*-hydroxysuccinimide ester (MBS) have been used to cross-link GS to IgG. These reagents are of particular value when one of the proteins involved has no free thiol groups, for example, IgG. Although maleimide groups attached to benzene rings are labile at neutral pH,[95] MBS has been successfully applied to GS at pH 6. The principle of the reaction is shown in Figure 12.6. Since GS contains free thiol groups, IgG is first modified with MBS. Basically, IgG (1 mg/mL) in 0.1 M phosphate buffer, pH 6.0, containing 0.05 M sodium chloride, is first reacted with 20 mg/mL MBS (dissolved in dioxin, final organic solvent 10%) at 25°C for 1 h. The labeled IgG is isolated by gel filtration on Sephadex G-25, equilibrated with 0.1 M phosphate buffer, pH 6.0, containing 20 mM magnesium chloride and 0.05 M sodium chloride, and incubated with equal weight of GS. After 1 h at 30°C, the reaction is stopped by adding a reducing agent such as 10 mM 2-mercaptoethanol.[96] The conjugate may be purified by gel filtration on Ultragel AcA 34.

GS-antibody conjugates are also prepared using *N*-succinimidyl 4-maleimidobutyrate (GMBS).[96] The reaction is similar to MBS except that GMBS is used.

12.2.5.2 Conjugation with Thiol Group–Directed Dimaleimides

Coupling the thiol groups of GS and Fab′, reduced or thiolated IgG, or F(ab′)₂ can be achieved with dimaleimide bifunctional reagents such as *N,N′-o*-phenylenedimaleimide or *N,N′*-oxydimethylenedimaleimide.[97] Either the enzyme or the immunoglobulin can be labeled first according

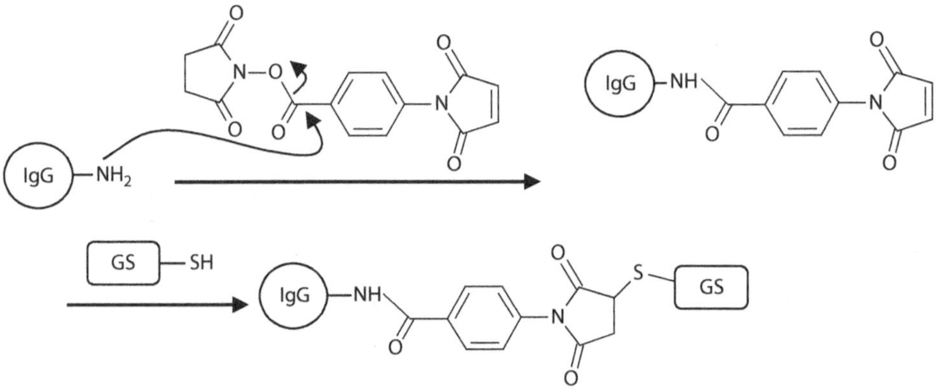

FIGURE 12.6   Conjugation of α-D-galactosidase (GS) and IgG with MBS.

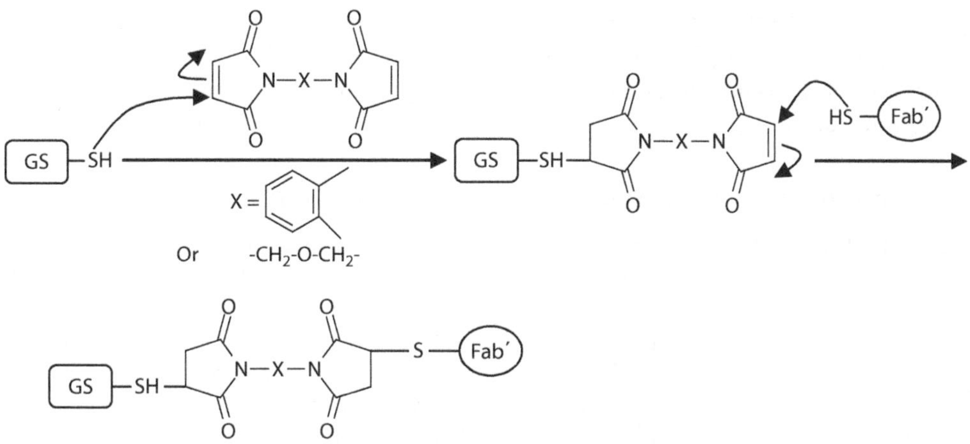

FIGURE 12.7 Conjugation of the β-D-galactosidase (GS) and Fab′ with a dimaleimide, N,N'-o-phenylene dimaleimide, or N,N'-oxydimethylene dimaleimide.

to the scheme depicted in Figure 12.7. For example, the enzyme (5 mg/mL) is reacted with 0.5 mg/mL cross-linker (previously dissolved in N,N'-dimethylformamide) in 0.1 M sodium phosphate buffer, pH 6.0. After incubation at 30°C for 20 min, the labeled enzyme is isolated by gel filtration. In general, about 13–16 maleimide groups may be introduced per GS molecule with less than 10% loss of enzyme activity. The labeled enzyme (1 mg/mL) is incubated with Fab′, reduced or thiolated IgG, or $F(ab')_2$ (about 1:4 mol/mol) in 0.1 M sodium phosphate buffer, pH 6.0, containing 2.5 mM EDTA, at 4°C for about 20 h. GS-antibody conjugates may be isolated by gel filtration on Sepharose 6B. The use of excess Ig to completely convert GS to the conjugate is important, since the enzyme has a molecular weight of 540,000 Da and cannot be easily separated from the conjugate by gel filtration.

12.2.5.3 Conjugation with Phenolate and Thiol Group–Directed Reagent

Fujiwara et al.[97] used N-[β-(4-diazopbenyl)ethyl]maleimide (DPEM) to conjugate IgG and GS. The diazo functional group will react selectively with tyrosine and histidine residues and the maleimide group with thiol groups, as shown in Figure 12.8. DPEM is derived from N-[β-(4-aminophenyl) ethyl]maleimide with sodium nitrite and acetic acid. Basically, the diazotized compound reacts with IgG through the diazo functional group at pH 9 for 20 min at 20°C. The labeled antibody is isolated by gel filtration and then mixed with GS. Conjugation is complete after 40 min at 30°C or can be terminated with mercaptoethanol.

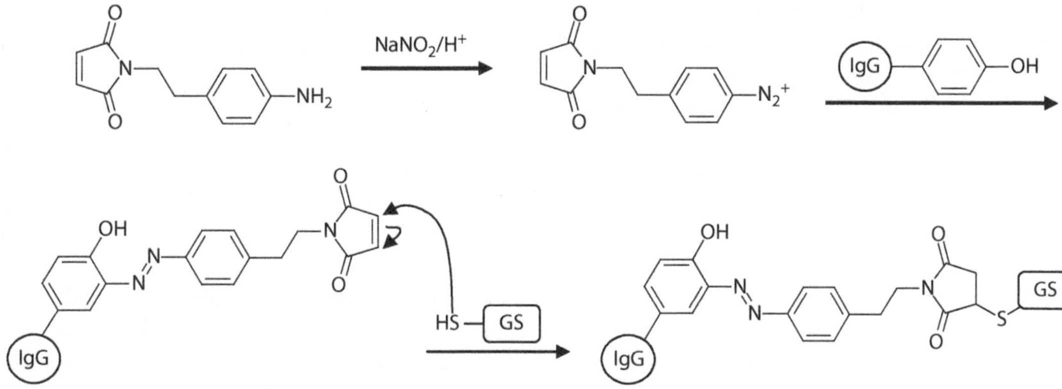

FIGURE 12.8 Conjugation of IgG and GS with DPEM.

12.2.5.4 Conjugation with Glutaraldehyde

The one-step procedure of glutaraldehyde coupling has been applied to conjugate GS to antibodies.[98] However, due to the heterogeneous cross-linking nature of the procedure, the use of this method is limited.

12.2.6 PREPARATION OF GLUCOSE-6-PHOSPHATE DEHYDROGENASE IMMUNOCONJUGATES

Similar to HRP and ALP, SMCC and SMH have been used to couple G6PDH to Ig, according to the same scheme shown in Figure 12.3. The enzyme (1 mg/mL) reacts with the cross-linker (1:2000 mol/mol) in 0.2 M sodium phosphate buffer, pH 7.0, containing 4.8 mM glucose-6-phosphate and 2 mM NAD. After 30 min at 30°C, the labeled enzyme is isolated by gel filtration. An average of about five cross-linkers is introduced per mole of enzyme, but the enzyme may lose 50%–70% of its activity.[71] The labeled enzyme (0.3–0.5 mg/mL) is then incubated with Fab′ (1:10) in 0.1 M sodium phosphate buffer, pH 6.0, containing 5 mM EDTA at 4°C for 20 h. The reaction is terminated by adding 1.5 mM 2-mercaptoethylamine, and the conjugate can be isolated on Ultragel AcA 34.

12.2.7 PREPARATION OF GLUCOSE OXIDASE IMMUNOCONJUGATES

12.2.7.1 Coupling with N-Ethoxycarbonyl-2-Ethoxy-1,2-Dihydroquinoline

GO can be coupled to IgG through the activation of its carboxyl groups with N-ethoxycarbonyl-2-ethoxy-1,2-dihydroquinoline (EEDQ) according to Guesdon.[99] The amino groups of GO are first biotinylated (or blocked with DNFB) by incubating with D-biotinyl-N-hydroxysuccinimide ester (dissolved in N,N′-dimethylformamide; 1:400 mol/mol, organic solvent 10%) in 0.1 M NaHCO$_3$ containing 0.15 M NaCl, for 1 h at ambient temperature. After dialysis against 0.1 M KH$_2$PO$_4$ containing 0.15 M NaCl, 10 mg EEDQ in 0.05 mL dimethylformamide is added and incubated at ambient temperature for 1 h. The labeled enzyme is separated by gel filtration on Sephadex G-25 and mixed with 1 mg IgG in potassium phosphate saline buffer, pH 8.0, at 4°C for 24 h. The immunoconjugate is then dialyzed or purified.

12.2.7.2 Coupling with Amino- and Thiol-Directed Reagents

SMCC has been used to couple GO to Fab′, reduced IgG, or thiolated antibody as described for HRP. The enzyme is first treated with SMCC to introduce maleimide groups, which react with the thiol group of Ig. Because of the number of reactive amino groups and the high molecular weight of GO (153,000 Da), excess antibody is used to convert all maleimide-labeled GO to the conjugate. Conjugated and unconjugated GO are difficult to separate on gel filtration.[70] In place of SMCC, MBS, SMH, and SPDP may be used to prepare GO-Ig.

12.2.7.3 Coupling with Other Cross-Linkers

GO immunoconjugates have also been prepared by the one-step procedure with glutaraldehyde.[78,100] However, the recovery of the enzyme is low with extensive polymerization of IgG. Cyanuric chloride has also been used to couple IgG to GO, but the yield of the conjugate is low, and loss of enzyme activity is often noted.[101] Periodate oxidation of the polysaccharide moiety of GO to aldehydes for cross-linking with amino groups of Ig is also applicable.[102]

12.2.8 PREPARATION OF OTHER ENZYME IMMUNOCONJUGATES

Various methods have been employed to prepare other enzyme immunoconjugates. Glutaraldehyde is probably the most widely used reagent.[103] Acetylcholinesterase,[104] inorganic pyrophosphatase,[42] lactoperoxidase,[105] lactate dehydrogenase,[45] glucoamylase,[106] ribonuclease,[107] and urease[55] immunoconjugates have been prepared with this reagent. Periodate oxidation is applicable to cross-link glucoamylase and IgG. Succinate bis(N-hydroxysuccinimide ester) is used to couple microperoxidase to IgG. Water-soluble carbodiimide, 1-cyclohexyl-3-(2-morpholinyl-4-ethyl) carbodiimide, has been used to link phospholipase C and IgG.[108] Phospholipase C immunoconjugate has also been prepared with SMCC.[109] The same reagent was also used to conjugate Fab' and α-amylase.[26] It should be realized that a diversity of reagents can be used to prepare the immunoconjugates.

12.2.9 PREPARATION OF NONENZYME PROTEIN IMMUNOCONJUGATES

Many proteins, other than enzymes, have been conjugated to Ig for use in various forms of immunoassays. BSA immunoconjugate, for example, has been used to detect an antigen, and an enzyme-labeled anti-BSA antibody is employed to reveal the BSA-conjugated antibody.[110] Similarly, ferritin can be used to label antibodies for antigen detection, and the ferritin–antibody–antigen complex can be quantified with an antiferritin antibody.[111] On the other hand, phycobiliprotein immunoconjugates can be used as fluorescent tracers for immunoassays.[112–116]

BSA-IgG immunoconjugate can be prepared by a one-step reaction by mixing IgG (2 mg) and BSA (5 mg) with glutaraldehyde (0.1%) in 0.1 M phosphate buffer, pH 6.8, for 3 h at room temperature. Glycine (0.1 M) is added to terminate the reaction.[99] Ferritin–antibody conjugates have been prepared by various cross-linkers. These include p,p'-difluoro-m,m'-dinitrophenyl sulfone,[117] glutaraldehyde,[111,118,119] and toluene diisocyanate.[120]

Phycobiliproteins from a variety of algae and cryptomonads have been coupled to antibodies or its fragments by SPDP,[113,115] SMCC,[114,116] and succinimidyl-4-(p-maleimidophenyl)butyrate (SMPB).[116] Free thiol groups may be introduced into both IgG and phycobiliproteins with iminothiolane,[115] SPDP,[113] or by mild reduction of endogenous disulfides.[116] These proteins may be cross-linked through a disulfide bond similar to the scheme shown in Figure 12.4.

12.2.10 COUPLING ENZYMES TO PROTEINS OTHER THAN ANTIBODIES

12.2.10.1 Examples of Conjugations of Enzymes and Biotin-Binding Proteins

As mentioned in Chapter 7, the avidin–biotin system provides another dimension of EIA. Labeling of enzymes or other proteins with biotin can be easily achieved by incubation with biotinyl-N-hydroxysuccinimide ester, which is readily available commercially. The conjugation of avidin with enzymes can be attained by different cross-linkers. Thiolation of avidin can be carried out with either N-succinimidyl-S-acetylacetate [SATA; 5 mg avidin in 0.5 mL, pH 7.5 phosphate buffer, and 7 μL SATA (25 mg/mL in DMSO)[121] are incubated for 90 min at 30°C] or 20-fold excess of S-acetylmercaptosuccinic anhydride dissolved in N,N-dimethylformamide (final organic solvent is less than 10%) in 0.2 M sodium phosphate buffer, pH 7.0, and incubated for 30 min at 30°C. Free thiol groups are generated by adjusting to 0.2 M hydroxylamine-HCl and 5 mM EDTA at pH 7.0. An average of 1.6 thiol groups are introduced per molecule of avidin.

Coupling the thiolated avidin to HRP can be carried out by any amino- and thiol-group–directed cross-linkers, as described above. Avidin has also been coupled to HRP using glutaraldehyde.[122]

12.2.10.2 Examples of Conjugation of Enzymes and Other Proteins

Other proteins that have been coupled to enzymes include protein A and ferritin. Both conjugates have been used as a powerful tool in cytochemistry and immunochemistry. Coupling of ALP to protein A can be achieved by either the one- or two-step glutaraldehyde procedure.[123] In the one-step procedure, ALP (5 mg/mL) and protein A (5 mg/mL) in phosphate-buffered saline (PBS), pH 7.2, is added glutaraldehyde to a final concentration of 0.2%–0.5%. After incubation at room temperature for 2–3 h, the reaction is stopped by dialysis. In the two-step procedure, ALP (5 mg/mL) in PBS is first treated with 1% glutaraldehyde for 16 h. After the removal of excess glutaraldehyde, protein A (5 mg/mL) is added and incubated for 16 h. The conjugate is fractionated by gel filtration. A HRP-ferritin conjugate can be prepared by periodate oxidation of the carbohydrate moiety of the enzyme, as described earlier.[124]

12.2.10.3 Examples of Conjugation of Enzymes and Antigens

Enzyme–antigen conjugates are used in EIAs, particularly in the homogeneous immunoassays. Coupling of small organic antigens to enzymes is generally achieved by introducing active functional groups into the antigen through organic synthesis. Functional groups such as acyl chloride, reactive esters, and anhydrides are frequently employed. Occasionally, bifunctional cross-linkers may be used. Bifunctional reagents are particularly useful for antigens of larger molecular weight where organic synthetic modification is difficult. These antigens can be coupled to enzymes in the same way as IgG. For example, SMCC is used to couple vitamin B_{12} to GS enzyme-donor fragment.[125] A maleimido group is introduced into B_{12} by reacting with SMCC. Conjugation is completed by the reaction between the thiol group of the GS enzyme and the B_{12} maleimide. The same reagent is used to couple acetylcholinesterase to substance P and atriopeptide. DPEM is used to conjugate neurotensin to GS. Neurotensin is first modified with DPEM and then coupled GS.[126] GS has also been conjugated to insulin by the dimaleimide cross-linkers.[127] Sulfhydryl groups can be introduced into insulin by mercaptosuccinylation, as described earlier. Other reagents, such as difluoronitrobenzene, have been used to couple succininyl-cAMP to acetylcholinesterase, and dicyclohexylcarbodiimide for conjugation of succinyl-GMP and acetylcholinesterase.[128] Gentamicin is coupled to catalase by water-soluble carbodiimide, 1-ethyl-3-(3-dimethylamino-propyl) carbodiimide.[28] Many other antigens have been coupled to enzymes and other proteins using various cross-linkers.

12.3   PREPARATION OF IMMUNOGENS

Immunogens are entities that induce the immune response for the preparation of immunoglobulins. Low-molecular-weight molecules (haptens), such as peptides (MW 1–3000 Da), are often not sufficiently immunogenic to elicit an immune response alone. These compounds can be made immunogenic by conjugation to a suitable carrier. A wide range of carriers are available, including proteins and synthetic polymeric carriers. Examples of carrier proteins are BSA, keyhole limpet hemocyanin (KLH), ovalbumin (OVA), thyroglobulin (THY), synthetic multiple antigenic peptides,[129] purified protein derivative,[130] and DendriGraft poly-lysine DGL-G3 (third generation).[131] Depending on the nature of the carriers and the heptans, there are many conjugation methods available for the preparation of immunogens. The following paragraph will provide some examples of the methodology.

12.3.1   Examples of Conjugation of Hapten to Albumin

BSA is a commonly used hapten carrier. It has been used to conjugate to small nonimmunogenic compounds to form immunogens. Considering the three-dimensional structure of BSA, 26 ε-NH2 groups of the total 59 lysines residues in BSA are present on the protein surface and therefore available for cross-linking to haptens. Zhang et al.[132–134] have coupled various insecticides to BSA to raise antibodies against these immunogens. For example, various carboxylic acid derivatives of organophosphorus insecticide fenthion with different spacer-arms are attached to amino groups of BSA with the carbodiimide method.[132] Over 20 haptens are attached per mole of BSA. Polyclonal antisera are raised against these immunogens. The results show that heterology in the hapten spacer-arm could achieve a remarkable improvement in the quantity, sensitivity, and/or specificity of antibody. Likewise, N-methylcarbamate insecticide metolcarb carboxylate haptens are conjugated to BSA with carbodiimide, and polyclonal antisera are raised against the immunogen.[133] In the same manner, pyrethroid insecticide is conjugated to BSA, and polyclonal antibodies are raised against the immunogen.[134]

The carbodiimide/NHS method is also used to couple other carboxylate derivatives of haptens to BSA. Examples of such compounds are cortisol-3-O-carboxymethyl-oxime,[135] cortiso-21-hemisuccinate,[135] herbicide acetochlor,[136] C(1)-carboxymethyl oxime (CMO) derivative of AFB(1),[51] fungicide pentachloronitrobenzene,[137] and monoacetylmorphine[138] to BSA. Generally, this procedure involves activation of hapten (e.g., 50 μM) with carbodiimide and NHS (e.g., 75 μM EDC and 75 μM sulfo-NHS) for 1 h at room temperature. The activated hapten (3–30.0 µmol) is added to BSA (e.g., 10 mg; 0.15 µmol) to prepare the immunogen. The conjugate is separated from free hapten by gel filtration. Antibodies are successfully raised to these immunogens.

Isobutylchloroformate, a zero-length cross-linker, has been used to couple the systemic herbicide 2,4-dichlorophenoxyacetic acid (2,4-D) to BSA.[139] Briefly, isobutylchloroformate (10 µL) and *N*-triethylamine (10 µL) are added to 2,4-D (5.0 mg) dissolved in 1 mL dioxane and stirred at room temperature for 2 h. H_2O (2 mL) containing BSA (5.0 mg) is added to this solution, and the reaction is stirred at room temperature for 16 h. The reaction mixture is dialyzed against H_2O five times and freeze dried. The method is also used to couple carboxylic acid haptens of glycitein to BSA.[140]

A homobifunctional cross-linker, disuccinimidyl tartarate (DST), has been used to couple 3,4-dihydroisocoumarin moiety to the BSA carrier protein.[1,141] Basically, BSA (33.5 mg, 0.5 µmol) is dissolved in 5 mL of 0.1 M $NaHCO_3$ (pH 8.5) and cooled to 4°C. Fifty- to hundred-fold excess of dihydroisocoumarin derivative, (3S)-3-[(1′S)-*t*-butyloxycarbonylamino-3′-methylbutyl]-8-hydroxy-3,4-dihydroisocoumarin hydrobromide, in 0.5 mL DMF is added followed by DST (25 µmol–50 mol in 0.2 mL dry DMF). After stirring at 4°C for 4 h, the mixture is dialyzed against distilled water and lyophilized.

A solid polystyrene-supported 2-isobutoxy-1-isobutoxycarbonyl-1,2-dihydroquinoline (PS-IIDQ) is used as an efficient cross-linking reagent for amide bond formation to conjugate phytanic acid derivative to BSA for immunogen preparation.[142] In the reaction, PS-IIDQ functions as a zero-length cross-linker. The coupling reaction mechanism is shown in Figure 12.9. Basically, PS-IIDQ (1.87 mmol/g, 66.6 mg, and 124.6 µmol) is added to a 1 mL solution of the phytanic acid derivative (dissolved in DMF, 50 µmol) and BSA (dissolved in borate buffer, pH 9.0, and 10 mg/mL) in a protein to hapten molar ratio up to 1:109. The solution was stirred overnight at room temperature. The coupling agent is removed by centrifugation, and the immunogen is purified by gel filtration.

In conjugation of carbohydrates to carrier proteins, reductive amination (also termed reductive alkylation) is used. For monosaccharides, there is no need for periodate activation to oxidize vicinal hydroxyl groups to aldehydes. The potential aldehyde in the monosaccharides directly forms Schiff bases with amino group of proteins, which can be stabilized upon reduction. This method of reductive alkylation has been applied to couple mannose to BSA.[143] In this process, BSA (68 mg),

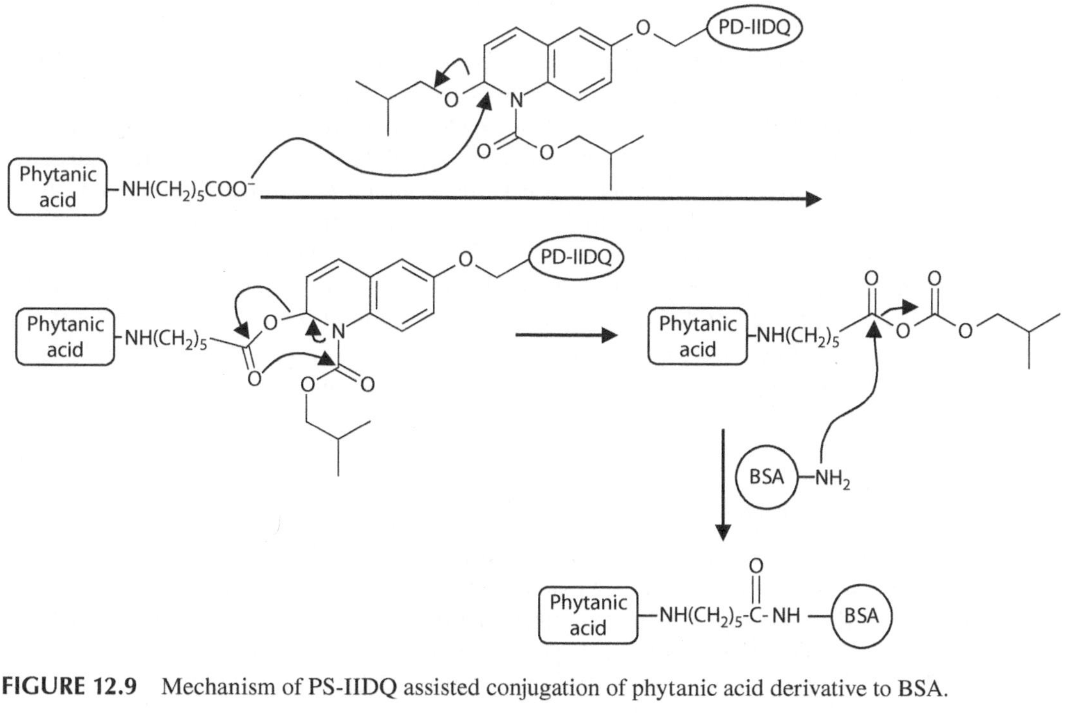

FIGURE 12.9 Mechanism of PS-IIDQ assisted conjugation of phytanic acid derivative to BSA.

D-mannose (100 mg), and sodium cyanoborohydride (100 mg) were dissolved in 5.0 mL of 0.2 M borate buffer (pH 8.0) and incubated at 37°C for 1–2 h. The solution is then dialyzed against PBS at 4°C. The method is also applicable to prepare D-xylose-BSA.[144]

Another method for preparing BSA immunogens involves maleimide activation of BSA. Chemical modification of ε-amino groups of BSA with SMCC or MPS (succinimidyl-3-*N*-maleimidopropionic acid) heterobifunctional cross-linkers result in the formation of maleimido BSA, which may contain up to 15 maleimide groups per BSA molecule. The maleimide group will react with small haptens containing a free thiol group. Such activated maleimide BSA is now commercially available and can be easily used. For example, the bacterial iron chelator, vulnibactin, is modified to contain a free thiol group, 1-(2,3-dihydroxybenzoyl)-5,9-bis[[(4S,5R)-2-(2,3-dihydroxyphenyl)-4,5-dihydro-5-methyl-4-oxazolyl]carbonyl]-14-(3-mercaptopropanoyl)-1,5,9,14-tetraazatetradecan, which is linked to maleimide BSA.[145]

As can be seen, various methods have been used to conjugate haptens to BSA carrier protein to form immunogens. Some of the haptens can be coupled with different cross-linkers. The choice of which procedure to use is up to the investigator.

12.3.2 Examples of Conjugation of Hapten to Keyhole Limpet Hemocyanin

Keyhole limpet hemocyanin (KLH), a large molecular weight protein (MW 2–3 × 10⁶ Da) isolated from the mollusk *Megathura crenulata*, is often used as a carrier protein due to its highly immunogenic properties and the large number of lysine residues available for modification. Like BSA, small molecules such as peptides can be coupled to the carrier proteins using a wide range of cross-linking reagents such as the water-soluble carbodiimide (EDC), bifunctional cross-linkers such as glutaraldehyde, bis-succinimidyl esters of dicarboxylic acids (DSS and DST), or heterobifunctional cross-linkers such as *N*-hydroxysuccinimidyl esters of maleimido-alkyl-carboxylate derivatives. Commercially available maleimide KLH is also available, which may contain up to 80 maleimide groups per KLH molecule. The reaction of the maleimide group and the sulfhydryl group proceeds rapidly and selectively under mild coupling conditions (pH 6.5–7.5) to yield a stable, covalently linked immunogen. Essentially, all the coupling reactions with BSA are applicable to KLH. A few examples are shown below.

Similarly to BSA, carbodiimide/NHS system is also used to couple hapten carboxylic acids to KLH. For example, organophosphorus insecticide is modified to contain carboxyl group and activated with carbodiimide/NHS. The activated insecticide is then coupled to KLH to form immunogen.[146] Other examples of haptens that have been linked to KLH using this method include carboxylic acid derivative of organophosphorus insecticide acephate, *O,S*-dimethyl acetylphosphoramidothioate,[147] 2,4-dinitroaniline and 2,6-dinitroaniline,[148] and fungicide fenarimol.[149]

Also, isobutyl chloroformate has been used to activate haptens for conjugation to KLH as demonstrated by Ramon-Azcon[150] for the preparation of miticide bromopropylate derivative, 2,2-bis(4-bromophenyl)-*N*-2-hydroxyacetamide-butanoic acid, and KLH immunogen. Basically, the hapten (15 μmol) is reacted with tributylamine (4 μL, 16.5 μmol) and (3 μL, 18 μmol) in anhydrous dimethylformamide (DMF; 160 μL). The activated hapten is then added dropwise to a solution of KLH in 0.2 M borate buffer (1.8 mL).

12.3.3 Examples of Hapten Conjugation to Other Carriers

Many haptens that have been coupled to BSA have also been conjugated to OVA. For example, *N*-methylcarbamate insecticide derivatives are conjugated to OVA with carbodiimide,[133] and the bacterial iron chelator with a free thiol modified vulnibactin is coupled to OVA activated with SMCC.[145]

Likewise, carboxylic acid derivatives of glycitein are coupled to BSA as well as THY using the zero-length cross-linker, isobutyl chloroformate.[140] Briefly, tributylamine (17.3 μL, 0.072 mmol), isobutyl chloroformate (4.7 μL, 0.036 mmol), and hapten (0.036 mmol) are added to 1 mL of DMF at 4°C and stirred on an ice bath for 30 min. The resulting activated hapten solution in DMF is added dropwise to the protein solution (THY: 48 mg, 0.072 μmol) in 5 mL of borate buffer (borate-boric buffer 0.2 M, pH 8.7). The mixture is stirred for 6 h at room temperature and dialyzed against phosphate buffer saline (PBS; 0.01 M, pH 7.4, 0.9% NaCl) and against distilled water. The hapten-protein conjugates are lyophilized. Also, glycine-modified pyrethroid insecticide permethrin is coupled to BSA and TYR using water-soluble carbodiimide.[151]

In addition to natural existing proteins as hapten carriers, Romestand et al.[131] have studied the third generation of dendrigraft poly-L-lysine (DGL-G3) as a hapten carrier. The core is a linear poly-L-lysine with an average of eight monomers. DGL-G3 has an MW of 22 kDa and presents a mean number of 123 lysine residues on its surface. The amino groups can be coupled to haptens such as fluorescein or histamine to form immunogens; however, the antigenicity is low. But the conjugate will improve the production of specific hapten-directed antibodies following a second immunization with DGL-G3 conjugate after a primary immunization using a BSA conjugate.

12.4 CHARACTERIZATION OF CONJUGATION METHODS

Since functional groups are present in proteins, whether an enzyme or an Ig, it is evident that specific coupling producing an exclusive homogeneous enzyme–protein conjugate is almost impossible. In fact, all conjugation reactions produce a mixture of heterogeneous immunoconjugates, such as enzyme–enzyme, enzyme–Ig, and Ig–Ig conjugates, in addition to multiconjugates. Theoretically, a two-step procedure provides better control of the reaction. Homogeneous conjugates or conjugates of limited heterogeneity may be expected.

Because of the limited number of amino groups (two to three) on HRP available for reaction, HRP immunoconjugates prepared by amino and thiol-specific reagents largely contain a one-to-one ratio.[76,95] Similar conjugates are obtained using glutaraldehyde. However, the periodate oxidation method yielded polymers as well. Blockage of amino groups with DNFB before coupling greatly improves the result. Immunoconjugates of other enzymes are generally heterogeneous. GO, for example, has many reactive amino groups. The resulting conjugates consist of molecules with various numbers of Ig. Although the enzymes such as GS and ALP are not polymerized with various cross-linkers, they are polymerized using glutaraldehyde.[76] When this reaction occurs, the yield is always low. Monomeric conjugates are best obtained with other methods.[152]

Various immunoconjugates prepared by different methods have been compared.[76,78,95,153] However, because of the diversity of the conditions used, it is difficult to access the validity of these comparisons. For example, Beyzavi et al.[153] found the periodate method to be most effective for preparing HRP–IgG conjugate, particularly at low pH, and the glutaraldehyde method to be more effective than periodate for conjugating ALP to antibodies. Ishikawa et al.,[91] on the other hand, reported polymerization of HRP with the periodate method. The results of conjugation are procedure dependent. The choice of a suitable reaction condition is important. In general, the methods described here provide relatively good results.

The stability of immunoconjugates has been studied.[91,154] Enzyme immunoconjugates have been reported to remain active after 3–4 years when stored at 4°C in the presence of 1 mg/mL bovine serum albumin.[91] Montoya and Castell[154] estimated the half-life of the whole HRP–IgG activity to be 9 years at 4°C. Cross-linking generally increases the life expectancy of enzymes.

REFERENCES

1. Yalow, R. S. and Berson, S. A., Assay of plasma insulin in human subjects by immunological methods, *Nature*, 184, 1648, 1959.
2. Wild, D., *The Immunoassay Handbook*, 3rd edn., Elsevier, Oxford, U.K., 2005.
3. Diamandis, E. P. and Christopoulos, T. K., *Immunoassay*, Academic Press, San Diego, CA, 1996.
4. Law, B. (ed.), *Immunoassay: A Practical Guide*, Taylor & Francis Ltd., Bristol, PA, 1996.

5. Gosling, J. P., *Immunoassays: A Practical Approach*, Oxford University Press, New York, 2000.

6. Deshpande, S. S., *Enzyme Immunoassays: From Concept to Product Development*, Chapman & Hall, New York, 1996.

7. Ishikawa, E., Ultrasensitive and rapid enzyme immunoassay, in *Laboratory Techniques in Biochemistry and Molecular Biology*, Vol. 27, Pillai, S. and van der Vliet, P. C. (eds.), Elsevier, Amsterdam, the Netherlands, 1999.

8. Hornbeck, P., Enzyme-linked immunosorbent assays, *Curr. Protoc. Immunol.*, Chapter 2, Unit 2.1, 2001.

9. Crowther, J. R., Stages in ELISA, *Methods Mol. Biol.*, 516, 43, 2009.

10. Voller, A., Bartlett, A., and Bidwell, D. E., Enzyme immunoassay with special reference to ELISA techniques, *J. Clin. Pathol.*, 31, 507, 1978.

11. Schuurs, A. H. W. and van Weemen, B. K., Enzyme immunoassay: A powerful analytical tool, *J. Immunoassay*, 1, 229, 1980.

12. Freytag, J. W., Affinity column mediated immunoenzymometric assays, in *Enzyme-Mediated Immunoassay*, Ngo, T. T. and Lenhoff, H. M. (eds.), Plenum Press, New York, 1985, p. 277.

13. Kurita, R., Arai, K., Nakamoto, K., Kato, D., and Niwa, O., Development of electrogenerated chemiluminescence-based enzyme linked immunosorbent assay for sub-pM detection, *Anal. Chem.*, 82, 1692, 2010.

14. Scalia, G., Halonen, P. E., Condorelli, F., Mattila, M. L., and Hierholzer, J. C., Comparison of monoclonal biotin–avidin enzyme immunoassay and monoclonal time-resolved fluoroimmunoassay in detection of respiratory virus antigens, *Clin. Diagn. Virol.*, 3, 351, 1995.

15. Meyer, H. H., Eisele, K., and Osaso, J., A biotin–streptavidin amplified enzyme immunoassay for 13,14-dihydro-15-keto-PGF2 alpha, *Prostaglandins*, 38, 375, 1989.

16. Suzuki, Y., Aoyagi, Y., Muramatsu, M., Igarashi, K., Saito, A., Oguro, M., Isemura, M., and Asakura, H., A lectin-based monoclonal enzyme immunoassay to distinguish fucosylated and non-fucosylated alphafetoprotein molecular variants, *Ann. Clin. Biochem.*, 27 (Pt 2), 121, 1990.

17. Cullina, M. J. and Greally, J. F., A novel lectin-based enzyme-linked immunosorbent assay for the measurement of IgA1 in serum and secretory IgA1 in secretions, *Clin. Chim. Acta*, 216, 23, 1993.

18. Finley, P. R., Williams, R. J., and Lichti, D. A., Evaluation of a new homogeneous enzyme inhibitor immunoassay of serum thyroxine with use of a bichromatic analyzer, *Clin. Chem.*, 26, 1723, 1980.

19. Brown, D. V. and Meyerhoff, M. E., Potentiometric enzyme channeling immunosensor for proteins, *Biosens. Bioelectron.*, 6, 615, 1991.

20. Gebauer, C. R. and Rechnitz, G. A., Deaminating enzyme labels for potentiometric enzyme immunoassay, *Anal. Biochem.*, 124, 338, 1982.

21. Yin, Z., Liu, Y., Jiang, L. P., and Zhu, J. J., Electrochemical immunosensor of tumor necrosis factor alpha based on alkaline phosphatase functionalized nanospheres, *Biosens. Bioelectron.*, 26, 1890, 2011.

22. Bratthauer, G. L., Overview of antigen detection through enzymatic activity, *Methods Mol. Biol.*, 588, 231, 2010.

23. Fanjul-Bolado, P., González-García, M. B., and Costa-García, A., Voltammetric determination of alkaline phosphatase and horseradish peroxidase activity using 3-indoxyl phosphate as substrate: Application to enzyme immunoassay, *Talanta*, 64, 452, 2004.

24. Kaw, C. H., Hefle, S. L., and Taylor, S. L., Sandwich enzyme-linked immunosorbent assay (ELISA) for the detection of lupine residues in foods, *J. Food Sci.*, 73, T135, 2008.

25. Nanda, S., Muralidhar, K., and Kar, S. K., Thermostable alpha-amylase conjugated antibodies as probes for immunodetection in ELISA, *J. Immunoassay Immunochem.*, 23, 327, 2002.

26. Shimura, T., Nakamura, T., Kawakami, A., Haga, M., and Kato, Y., A new type of enzyme immunosensor using antigen-bound membrane and multivalent antibody (Fab′-alpha-amylase conjugate), *Chem. Pharm. Bull. (Tokyo)*, 34, 5020, 1986.

27. Oellerich, M., Enzyme immunoassays in clinical chemistry: Present status and trends, *J. Clin. Chem. Clin. Biochem.*, 18, 197, 1980.

28. Mattiasson, B., Svensson, K., Borrebaeck, C., Jonsson, S., and Kronvall, G., Non-equilibrium enzyme immunoassay of gentamicin, *Clin. Chem.*, 24, 1770, 1978.

29. O'Sullivan, M. J., Bridges, J. W., and Marks, V., Enzyme immunoassay: A review, *Ann. Clin. Biochem.*, 16, 221, 1979.

30. Szucs, J., Pretsch, E., and Gyurcsányi, R. E., Potentiometric enzyme immunoassay using miniaturized anion-selective electrodes for detection, *Analyst*, 134, 1601, 2009.

31. Saita, T., Tokunaga, A., Egoshi, M., Tokushima, H., and Fujito, H., Quantification of cibenzoline by enzyme-linked immunosorbent assay, *Yakugaku Zasshi*, 127, 1007, 2007.

32. Kopetzki, E., Lehnert, K., and Buckel, P., Enzymes in diagnostics: Achievements and possibilities of recombinant DNA technology, *Clin. Chem.*, 40, 688, 1994.

33. Ko, F. H. and Monbouquette, H. G., Photometric and electrochemical enzyme-multiplied assay techniques using beta-galactosidase as reporter enzyme, *Biotechnol. Prog.*, 22, 860, 2006.

34. Arakawa, H., Maeda, M., and Tsuji, A., Chemiluminescent assay of various enzymes using indoxyl derivatives as substrate and its applications to enzyme immunoassay and DNA probe assay, *Anal. Biochem.*, 199, 238, 1991.

35. Piao, Y., Lee, D., Lee, J., Hyeon, T., Kim, J., and Kim, H. S., Multiplexed immunoassay using the stabilized enzymes in mesoporous silica, *Biosens. Bioelectron.*, 25, 906, 2009.

36. Krämer, P. M., Weber, C. M., Forster, S., Rauch, P., and Kremmer, E., Analysis of DDT isomers with enzyme-linked immunosorbent assay and optical immunosensor based on rat monoclonal antibodies as biological recognition elements, *J. AOAC Int.*, 93, 44, 2010.

37. Kim, B., Park, E. Y., Lee, Y. T., Lee, J. H., and Lee, S. H., Development of homogeneous enzyme immunoassay for the organophosphorus insecticide fenthion, *J. Microbiol. Biotechnol.*, 17, 1002, 2007.

38. Tateishi, K., Yamamoto, H., Ogihara, T., and Hayashi, C., Enzyme immunoassay of serum testosterone, *Steroids*, 30, 25, 1977.

39. Litman, D. J., Hanlon, T. M., and Ullman, E. F., Enzyme channeling immunoassay: A new homogeneous enzyme immunoassay technique, *Anal. Biochem.*, 106, 223, 1980.

40. Dong, T., Sun, J., Liu, B., Zhang, Y., Song, Y., and Wang, S., Development of a sensitivity-improved immunoassay for the determination of carbaryl in food samples, *J. Sci. Food Agric.*, 90, 1106, 2010.

41. Peuravuori, H. and Korpela, T., Pyrophosphatase-based enzyme-linked immunosorbent assay of total IgE in serum, *Clin. Chem.*, 39, 846, 1993.

42. Baykov, A. A., Kasho, V. N., and Avaeva, S. M., Inorganic pyrophosphatase as a label in heterogeneous enzyme immunoassay, *Anal. Biochem.*, 171, 271, 1988.

43. Tsuji, A., Maeda, M., Arakawa, H., Shimizu, S., Ikegami, T., Sudo, Y., Hosoda, H., and Nambara, T., Fluorescence and chemiluminescence enzyme immunoassays of 17 alpha-hydroxyprogesterone in dried blood spotted on filter paper, *J. Steroid Biochem.*, 27, 33, 1987.

44. Terouanne, B., Nicolas, J. C., Descomps, B., and Crastes de Paulet, A., Coupling of delta 5,3-ketosteroid isomerase to human placental lactogen with intermolecular disulfide bond formation. Use of this conjugate for a sensitive enzyme immunoassay, *J. Immunol. Methods*, 35, 267, 1980.

45. Casu, A. and Avrameas, S., Conjugation of lactic-dehydrogenase with proteins by the use of glutaraldehyde: Detection of antigens in the immunocompetent cells by means of the conjugates, *Ital. J. Biochem.*, 18, 166, 1969.

46. Wu, C., Irie, S., Yamamoto, S., and Ohmiya, Y., A bioluminescent enzyme immunoassay for prostaglandin E(2) using *Cypridina luciferase*, *Luminescence*, 24, 131, 2009.

47. Wu, C., Kawasaki, K., Ogawa, Y., Yoshida, Y., Ohgiya, S., and Ohmiya, Y., Preparation of biotinylated cypridina luciferase and its use in bioluminescent enzyme immunoassay, *Anal. Chem.*, 79, 1634, 2007.

48. Samanta, A. K. and Ali, E., Homogeneous enzyme immunoassay of estriol using lysozyme, *Indian J. Med. Res.*, 87, 615, 1988.

49. Dhar, T. K., Samanta, A. K., and Ali, E., Homogeneous enzyme immunoassay of estradiol using estradiol-3-O-carboxymethyl ether as hapten, *Steroids*, 51, 519, 1988.

50. Rodgers, R., Crowl, C. P., Eimstad, W. M., Hu, M. W., Kam, J. K., Ronald, R. C., Rowley, G. L., and Ullman, E. F., Homogeneous enzyme immunoassay for cannabinoids in urine, *Clin. Chem.*, 24, 95, 1978.

51. Paknejad, M., Javad Rasaee, M., Mohammadnejad, J., Pouramir, M., Rajabibazl, M., and Kakhki, M., Development and characterization of enzyme-linked immunosorbent assay for aflatoxin B1 measurement in urine sample using penicillinase as label, *J. Toxicol. Sci.*, 33, 565, 2008.

52. Venkatesa Perumal, S., Umapathi, V., Ambwani, T., and Lakhchaura, B. D., A competitive dipstick enzyme immunoassay for diagnosis of early pregnancy in bovine, *Reprod. Domest. Anim.*, 43, 744, 2008.

53. Kim, C. K. and Lim, S. J., Liposome immunoassays using phospholipase C or alkaline phosphatase, *Methods Enzymol.*, 373, 260, 2003.

54. Fromell, K., Hulting, G., Ilichev, A., Larsson, A., and Caldwell, K. D., Particulate platform for bioluminescent immunosensing, *Anal. Chem.*, 79, 8601, 2007.

55. Chandler, H. M., Cox, J. C., Healey, K., MacGregor, A., Premier, R. R., and Hurrell, J. G., An investigation of the use of urease-antibody conjugates in enzyme immunoassays, *J. Immunol. Methods*, 53, 187, 1982.

56. Goding, J. W., Methods useful in antibody purification, in *Monoclonal Antibodies: Principles and Practice*, 2nd edn., Academic Press, New York, 1986, p. 1011.

57. Menozzi, F. D., Vanderpoorten, P., Dejaiffe, C., and Miller, A. O. A., One-step purification of mouse monoclonal antibodies by mass ion exchange chromatography on zetaprep, *J. Immunol. Methods*, 99, 229, 1987.

58. Lindmark, R., Thoren-Tolling, K., and Sjoquist, J., Binding of immunoglobulins to protein A and immunoglobulin levels in mammalian sera, *J. Immunol. Methods*, 62, 1, 1983.

59. Björck, L. and Kronvall, G., Purification and some properties of streptococcal protein G, a novel IgG binding reagent, *J. Immunol.*, 133, 969, 1984.

60. Fassina, G., Ruvo, M., Palombo, G., Verdoliva, A., and Marino, M., Novel ligands for the affinity-chromatographic purification of antibodies, *J. Biochem. Biophys. Methods*, 49, 481, 2001.

61. Eliasson, M., Andersson, R., Olsson, A., Wigzell, H., and Uhlén M., Differential IgG-binding characteristics of staphylococcal protein A, streptococcal protein G, and a chimeric protein AG, *J. Immunol.*, 142, 575, 1989.

62. Andrew, S. M. and Titus, J. A., Purification of immunoglobulin G, *Curr. Protoc. Immunol.*, Chapter 2, Unit 2.7, 2001.

63. Andrew, S. M., Titus, J. A., Coico, R., and Amin, A., Purification of immunoglobulin M and immunoglobulin D, *Curr. Protoc. Immunol.*, Chapter 2, Unit 2.9, 2001.

64. Pack, T. D., Purification of human IgA, *Curr. Protoc. Immunol.*, Chapter 2, Unit 2.10B, 2001.

65. Zhao, Y., Gutshall, L., Jiang, H., Baker, A., Beil, E., Obmolova, G., Carton, J., Taudte, S., and Amegadzie, B., Two routes for production and purification of Fab fragments in biopharmaceutical discovery research: Papain digestion of mAb and transient expression in mammalian cells, *Protein Expr. Purif.*, 67, 182, 2009.

66. Andrew, S. M. and Titus, J. A., Fragmentation of immunoglobulin G, *Curr. Protoc. Cell Biol.*, Chapter 16, Unit 16.4, 2003.

67. Jones, R. G. and Landon, J., Enhanced pepsin digestion: A novel process for purifying antibody F(ab′)(2) fragments in high yield from serum, *J. Immunol. Methods*, 263, 57, 2002.

68. Rousseaux, J., Rousseaux-Prévost, R., and Bazin, H., Optimal conditions for the preparation of Fab and F(ab′)2 fragments from monoclonal IgG of different rat IgG subclasses, *J. Immunol. Methods*, 64, 141, 1983.

69. Rousseaux, J., Rousseaux-Prevost, R., and Bazin, H., Optimal conditions for the preparation of proteolytic fragments from monoclonal IgG of different rat IgG subclasses, *Methods Enzymol.*, 121, 663, 1986.

70. Ishikawa, E., Imagawa, M., Hashida, S., Yoshitake, S., Hamaguchi, Y., and Ueno, T., Enzyme-labeling of antibodies and their fragments for enzyme immunoassay and immunohistochemical staining, *J. Immunoassay*, 4, 209, 1983.

71. Ngo, T. T. (ed.), *Nonisotopic Immunoassay*, Plenum Press, New York, 1988.

72. Hong, J., Lee, A., Han, H., and Kim, J., Structural characterization of immunoglobulin G using time-dependent disulfide bond reduction, *Anal. Biochem.*, 384, 368, 2009.

73. Palmer, J. L. and Nisonoff, A., Dissociation of rabbit gamma-globulin into half molecules after reduction of one labile disulfide bond, *Biochemistry*, 3, 863, 1964.

74. Grassetti, D. R. and Murray, J. F., Jr., Determination of sulfhydryl groups with 2,2′- or 4,4′-dithiopyridine, *Arch. Biochem. Biophys.*, 103, 1132, 1967.

75. Yoshitake, S., Hamaguchi, Y., and Ishikawa, E., Efficient conjugation of rabbit Fab′ with β-D-galactosidase from *Escherichia coli*, *Scand. J. Immunol.*, 10, 81, 1979.

76. Cherkaoui, S., Bettinger, T., Hauwel, M., Navetat, S., Allémann, E., and Schneider, M., Tracking of antibody reduction fragments by capillary gel electrophoresis during the coupling to microparticles surface, *J. Pharm. Biomed. Anal.*, 53, 172, 2010.

77. Penefsky, H. S., A centrifuged-column procedure for the measurement of ligand binding by beef heart F1, *Methods Enzymol.*, 56, 527, 1979.

78. Boorsma, D. M. and Kalsbeek, G. L., A comparative study of horseradish peroxidase conjugates prepared with a one-step and a two-step method, *J. Histochem. Cytochem.*, 23, 200, 1975.

79. Avrameas, S. and Ternynck, T., Peroxidase labelled antibody and Fab conjugates with enhanced intracellular penetration, *Immunochemistry*, 8, 1175, 1971.

80. Boorsma, D. M. and Streefkerk, J. G., Peroxidase-conjugate chromatography isolation of conjugates prepared with glutaraldehyde or periodate using polyacrylamide-agarose gel, *J. Histochem. Cytochem.*, 24, 481, 1976.

81. Boorsma, D. M. and Streefkerk, J. G., Periodate or glutaraldehyde for preparing peroxidase conjugates? *J. Immunol. Methods*, 30, 245, 1979.

82. Tijssen, P. and Kurstak, E., Highly efficient and simple methods for the preparation of peroxidase and active peroxidase–antibody conjugates for enzyme immunoassays, *Anal. Biochem.*, 136, 451, 1984.

83. Nakane, P. K., Recent progress in the peroxidase-labeled antibody method, *Ann. N. Y. Acad. Sci.*, 254, 203, 1975.

84. Simons, B., Kaplan, H., and Hefford, M. A., Novel cross-linked enzyme–antibody conjugates for Western blot and ELISA, *J. Immunol. Methods*, 315, 88, 2006.

85. Modesto, R. R. and Pesce, A. J., The reaction of 4,4'-difluoro-3,3'-dinitro-diphenyl sulfone with gamma-globulin and horseradish peroxidase, *Biochim. Biophys. Acta*, 229, 384, 1971.
86. Abuknesha, R. A., Luk, C. Y., Griffith, H. H. M., Maragkou, A., and Iakovaki, D., Efficient labelling of antibodies with horseradish peroxidase using cyanuric chloride, *J. Immunol. Methods*, 306, 211, 2005.
87. Modesto, R. R. and Pesce, A. J., Use of tolylene diisocyanate for the preparation of a peroxidase-labelled antibody conjugate. Quantitation of the amount of diisocyanate bound, *Biochim. Biophys. Acta*, 295, 283, 1973.
88. Clyne, D. H., Norris, S. H., Modesto, R. R., Pesce, A. J., and Pollak, V. E., Antibody enzyme conjugates: The preparation of intermolecular conjugates of horseradish peroxidase and antibody and their use in immunohistology of renal cortex, *J. Histochem. Cytochem.*, 21, 233, 1973.
89. Ternynck, T. and Avrameas, S., Conjugation of p-benzoquinone treated enzymes with antibodies and Fab fragments, *Immunochemistry*, 14, 767, 1977.
90. Jeanson, A., Cloes, J. M., Bouchet, M., and Rentier, B., Preparation of reproducible alkaline phosphatase-antibody conjugates for enzyme immunoassay using a heterobifunctional linking agent, *Anal. Biochem.*, 172, 392, 1988.
91. Ishikawa, E., Hashida, S., Kohno, T., and Ranaka, K., Methods for enzyme-labeling of antigens, antibodies and their fragments, in *Nonisotopic Immunoassay*, Ngo, T. T. (ed.), Plenum Press, New York, 1988, p. 27.
92. Ford, D. J., Radin, R., and Pesce, A. J., Characterization of glutaraldehyde coupled alkaline phosphatase-antibody and lactoperoxidase–antibody conjugates, *Immunochemistry*, 15, 237, 1978.
93. Falini, B. and Taylor, C. R., New developments in immunoperoxidase techniques and their application, *Arch. Pathol. Lab. Med.*, 107, 105, 1983.
94. Liu, Z., Gurlo, T., and von Grafenstein, H., Cell-ELISA using beta-galactosidase conjugated antibodies, *J. Immunol. Methods*, 234, 153, 2000.
95. Hashida, S., Imagawa, M., Inoue, S., Ruan, K. H., and Ishikawa, E., More useful maleimide compounds for the conjugation of Fab' to horseradish peroxidase through thiol groups in the hinge, *J. Appl. Biochem.*, 6, 56, 1984.
96. Kurstak, E., *Enzyme Immunodiagnosis*, Academic Press, Orlando, FL, 1986, p. 20.
97. Fujiwara, K., Saita, T., and Kitagawa, T., The use of N-[beta-(4-diazophenyl)ethyl]maleimide as a coupling agent in the preparation of enzyme–antibody conjugates, *J. Immunol. Methods*, 110, 47, 1988.
98. Deelder, A. M. and de Water, R., A comparative study on the preparation of immunoglobulin–galactosidase conjugates, *J. Histochem. Cytochem.*, 29, 1273, 1981.
99. Guesdon, J.-L., Amplification systems for enzyme immunoassay, in *Nonisotopic Immunoassay*, Ngo, T. T. (ed.), Plenum Press, New York, 1988, p. 85.
100. Avrameas, S., Coupling of enzymes to proteins with glutaraldehyde. Use of the conjugates for the detection of antigens and antibodies, *Immunochemistry*, 6, 43, 1969.
101. Engvall, E. and Perlmann, P., Enzyme-linked immunosorbent assay (ELISA). Quantitative assay of immunoglobulin G, *Immunochemistry*, 8, 871, 1971.
102. Muzykantov, V. R., Sakharov, D. V., Sinitsyn, V. V., Domogatsky, S. P., Goncharov, N. V., and Danilov, S. M., Specific killing of human endothelial cells by antibody-conjugated glucose oxidase, *Anal. Biochem.*, 169, 383, 1988.
103. Avrameas, S., Ternynck, T., and Guesdon, J.-L., Coupling of enzyme to antibodies and antigens, in *Quantitative Enzyme Immunoassay*, Engvall, E. and Pesce, A. J. (eds.), Blackwell Scientific, Oxford, U.K., 1978, p. 7.
104. Van der Waart, M. and Schuurs, A. H. M. W., Towards the development of a radioenzyme-immunoassay (REIA), *J. Anal. Chem.*, 279, 142, 1976.
105. Pene, J., Rousseau, V., and Stanislawski, M., In-vitro cytolysis of myeloma tumor cells with glucose oxidase and lactoperoxidase antibody conjugates, *Biochem. Int.*, 13, 233, 1986.
106. Ishikawa, E., Enzyme immunoassay of insulin by fluorimetry of the insulin-glucoamylase complex, *J. Biochem.*, 73, 1319, 1973.
107. Herrmann, J. E. and Morse, S. A., Conjugation of enzymes to anti-poliovirus globulin: Effect of enzyme molecular weight on virus neutralization capacity, *Immunochemistry*, 11, 79, 1974.
108. Wei, R. and Riebe, S., Preparation of a phospholipase C-antihuman IgG conjugate, and inhibition of its enzymatic activity by human IgG, *Clin. Chem.*, 23, 1386, 1977.
109. Lal, R. B., Brown, E. M., Seligmann, B. E., Edison, L. J., and Chused, T. M., Selective elimination of lymphocyte subpopulations by monoclonal antibody-enzyme conjugates, *J. Immunol. Methods*, 79, 307, 1985.

110. Guesdon, J. L., Jouanne, C., and Avrameas, S., An amplification system using BSA-antibody conjugate for sensitive enzyme immunoassay, *J. Immunol. Methods*, 58, 133, 1983.

111. Rauterberg, E. W., Schieck, C., Kreft, H., and Römer, W., Optimal conditions for the preparation of ferritin-labeled antibodies defined by binding to their antigen in an ELISA, *Immunobiology*, 166, 439, 1984.

112. Shapiro, H. M., Glazer, A. N., Christenson, L., Williams, J. M., and Strom, T. B., Immunofluorescence measurement in a flow cytometer using low-power helium-neon laser excitation, *Cytometry*, 4, 276, 1983.

113. Kronick, M. N. and Grossman, P. D., Immunoassay techniques with fluorescent phycobiliprotein conjugates, *Clin. Chem.*, 29, 1582, 1983.

114. Triantafilou, K., Triantafilou, M., and Wilson, K. M., Phycobiliprotein-Fab conjugates as probes for single particle fluorescence imaging, *Cytometry*, 41, 226, 2000.

115. Oi, V. T., Glazer, A. N., and Stryer, L., Fluorescent phycobiliprotein conjugates for analyses of cells and molecules, *J. Cell Biol.*, 93, 981, 1982.

116. Hardy, R. R., Purification and coupling of fluorescent proteins for use in flow cytometry, in *Handbook of Experimental Immunology*, 4th edn., Vol. 1, Weir, D. M. (ed.), Blackwell Scientific, Edinburgh, U.K., 1986, Chapter 13.

117. Tawde, S. S. and Ram, J. S., Conjugation of antibody to ferritin by means of p,p'-difluoro-m, m'-dinitrodiphenylsulphone, *Arch. Biochem. Biophys.*, 97, 429, 1962.

118. Rauterberg, E. W. and Schieck, C., Ferritin-labeling of antibodies by glutaraldehyde. Comparison of conjugates prepared at different antibody: Ferritin: glutaraldehyde ratios, *Immunobiology*, 159, 307, 1981.

119. Rudick, R. A., Bloechl, E. K., and Knutson, D. W., Preparation of monoclonal antibody-ferritin conjugates of high specific activity, *Histochemistry*, 80, 269, 1984.

120. Singer, S. J. and Schick, A. F., The properties of specific stains for electron microscopy prepared by the conjugation of antibody molecules with ferritin, *J. Biophys. Biochem. Cytol.*, 9, 519, 1961.

121. van Gijlswijk, R. P., van Gijlswijk-Janssen, D. J., Raap, A. K., Daha, M. R., and Tanke, H. J., Enzyme-labelled antibody-avidin conjugates: New flexible and sensitive immunochemical reagents, *J. Immunol. Methods*, 189, 117, 1996.

122. Boorsma, D. M., Van Bommel, J., and Vanden Heuvel, J., Avidin-HRP conjugates in biotin–avidin immunoenzyme cytochemistry, *Histochemistry*, 84, 333, 1986.

123. Engvall, E., Preparation of enzyme-labelled staphylococcal protein A and its use for detection of antibodies, in *Quantitative Enzyme Immunoassay*, Engvall, E. and Pesce, A. J. (eds.), Blackwell Scientific, Oxford, U.K., 1978, p. 25.

124. Denisov, V. N. and Metelitsa, D. I., Catalytic and immunochemical properties of ferritin conjugates with horseradish peroxidase, *Biokhimiia*, 52, 1248, 1987.

125. Khanna, P. L., Dworschack, R. T., Manning, W. B., and Harris, J. D., A new homogeneous enzyme immunoassay using recombinant enzyme fragments, *Clin. Chim. Acta*, 185, 231, 1989.

126. Fujiwara, K. and Saita, T., The use of N-[beta-(4-diazophenyl)ethyl]maleimide as a heterobifunctional agent in developing enzyme immunoassay for neurotensin, *Anal. Biochem.*, 161, 157, 1987.

127. Kato, K., Hamaguchi, Y., Fukui, H., and Ishikawa, E., Enzyme-linked immunoassay. I. Novel method for synthesis of the insulin-beta-D-galactosidase conjugate and its applicability for insulin assay, *J. Biochem.*, 78, 235, 1975.

128. Pradelles, P., Grassi, J., Chabardes, D., and Guiso, N., Enzyme immunoassays of adenosine cyclic 3′,5′-monophosphate and guanosine cyclic 3′,5′-monophosphate using acetylcholinesterase, *Anal. Chem.*, 61, 447, 1989.

129. Butz, S., Rawer, S., Rapp, W., and Birsner, U., Immunization and affinity purification of antibodies using resin-immobilized lysine-branched synthetic peptides, *Pept. Res.*, 7, 20, 1994.

130. De Silva, B. S., Egodage, K. L., and Wilson, G. S., Purified protein derivative (PPD) as an immunogen carrier elicits high antigen specificity to haptens, *Bioconjug. Chem.*, 10, 496, 1999.

131. Romestand, B., Rolland, J. L., Commeyras, A., Coussot, G., Desvignes, I., Pascal, R., and Vandenabeele-Trambouze, O., Dendrigraft poly-L-lysine: A non-immunogenic synthetic carrier for antibody production, *Biomacromolecules*, 11, 1169, 2010.

132. Zhang, Q., Wang, L., Ahn, K. C., Sun, Q., Hu, B., Wang, J., and Liu, F., Hapten heterology for a specific and sensitive indirect enzyme-linked immunosorbent assay for organophosphorus insecticide fenthion, *Anal. Chim. Acta*, 596, 303, 2007.

133. Zhang, Q., Wu, Y., Wang, L., Hu, B., Li, P., and Liu, F., Effect of hapten structures on specific and sensitive enzyme-linked immunosorbent assays for N-methylcarbamate insecticide metolcarb, *Anal. Chim. Acta*, 625, 87, 2008.

134. Zhang, Q., Zhang, W., Wang, X., and Li, P., Immunoassay development for the class-specific assay for types I and II pyrethroid insecticides in water samples, *Molecules*, 15, 164, 2010.
135. Basu, A., Nara, S., Chaube, S. K., Rangari, K., Kariya, K. P., and Shrivastav, T. G., The influence of spacer-containing enzyme conjugate on the sensitivity and specificity of enzyme immunoassays for hapten, *Clin. Chim. Acta*, 366, 287, 2006.
136. Chen, Y., Wang, X., Wang, J., and Tang, S., Preparation of acetochlor antibody and its application on immunoaffinity chromatography cleanup for residue determination in peanuts, *J. Agric. Food Chem.*, 57, 7640, 2009.
137. Xu, T., Shao, X. L., Li, Q. X., Keum, Y. S., Jing, H. Y., Sheng, W., and Li, J., Development of an enzyme-linked immunosorbent assay for the detection of pentachloronitrobenzene residues in environmental samples, *J. Agric. Food Chem.*, 55, 3764, 2007.
138. Gandhi, S., Sharma, P., Capalash, N., Verma, R. S., and Suri, C. R., Group-selective antibodies based fluorescence immunoassay for monitoring opiate drugs, *Anal. Bioanal. Chem.*, 392, 215, 2008.
139. Tanaka, H., Yan, S., Miura, N., and Shoyama, Y., Preparation of anti-2,4-dichlorophenol and 2,4-dichlorophenoxyacetic acid monoclonal antibodies, *Cytotechnology*, 42, 101, 2003.
140. Shinkaruk, S., Lamothe, V., Schmitter, J. M., Fructus, A., Sauvant, P., Vergne, S., Degueil, M., Babin, P., Bennetau, B., and Bennetau-Pelissero, C., Synthesis of haptens and conjugates for ELISA of glycitein: Development and validation of an immunological test, *J. Agric. Food Chem.*, 56, 6809, 2008.
141. Shinkaruk, S., Bennetau, B., Babin, P., Schmitter, J. M., Lamothe, V., Bennetau-Pelissero, C., and Urdaci, M. C., Original preparation of conjugates for antibody production against Amicoumacin-related antimicrobial agents, *Bioorg. Med. Chem.*, 16, 9383, 2008.
142. Sathe, M., Derveni, M., Allen, M., and Cullen, D. C., Use of polystyrene-supported 2-isobutoxy-1-isobutoxycarbonyl-1,2-dihydroquinoline for the preparation of a hapten-protein conjugate for antibody development, *Bioorg. Med. Chem. Lett.*, 20, 1792, 2010.
143. Hegde, V. L. and Venkatesh, Y. P., Generation of antibodies specific to D-mannitol, a unique haptenic allergen, using reductively aminated D-mannose-bovine serum albumin conjugate as the immunogen, *Immunobiology*, 212, 119, 2007.
144. Sreenath, K. and Venkatesh, Y. P., Reductively aminated D-xylose-albumin conjugate as the immunogen for generation of IgG and IgE antibodies specific to D-xylitol, a haptenic allergen, *Bioconjug. Chem.*, 18, 1995, 2007.
145. Bergeron, R. J., Bharti, N., Singh, S., McManis, J. S., Wiegand, J., and Green, L. G., Vibriobactin antibodies: A vaccine strategy, *J. Med. Chem.*, 52, 3801, 2009.
146. Shim, J. Y., Kim, Y. A., Lee, E. H., Lee, Y. T., and Lee, H. S., Development of enzyme-linked immunosorbent assays for the organophosphorus insecticide EPN, *J. Agric. Food Chem.*, 56, 11551, 2008.
147. Lee, J. K., Ahn, K. C., Stoutamire, D. W., Gee, S. J., and Hammock, B. D., Development of an enzyme-linked immunosorbent assay for the detection of the organophosphorus insecticide acephate, *J. Agric. Food Chem.*, 51, 3695, 2003.
148. Krämer, P. M., Forster, S., and Kremmer, E., Enzyme-linked immunosorbent assays for the sensitive analysis of 2,4-dinitroaniline and 2,6-dinitroaniline in water and soil, *Anal. Bioanal. Chem.*, 391, 1821, 2008.
149. Lee, J. K., Park, S. H., Lee, E. Y., Kim. Y. J., and Kyung, K. S., Development of an enzyme-linked immunosorbent assay for the detection of the fungicide fenarimol, *J. Agric. Food Chem.*, 52, 7206, 2004.
150. Ramón-Azcón, J., Sánchez-Baeza, F., Sanvicens, N., and Marco, M.-P., Development of an enzyme-linked immunosorbent assay for determination of the miticide bromopropylate, *J. Agric. Food Chem.*, 57, 375, 2009.
151. Ahn, K. C., Watanabe, T., Gee, S. J., and Hammock, B. D., Hapten and antibody production for a sensitive immunoassay determining a human urinary metabolite of the pyrethroid insecticide permethrin, *J. Agric. Food Chem.*, 52, 4583, 2004.
152. Imagawa, M., Hashida, S., Ishikawa, E., and Freytag, J. W., Preparation of a monomeric 2,4-dinitrophenyl Fab'-beta-D-galactosidase conjugate for immunoenzymometric assay, *J. Biochem.*, 96, 1727, 1984.
153. Beyzavi, K., Hampton, S., Kwasowski, P., Fickling, S., Marks, V., and Clift, R., Comparison of horseradish peroxidase and alkaline phosphatase-labelled antibodies in enzyme immunoassays, *Ann. Clin. Biochem.*, 24 (Pt 2), 145, 1987.
154. Montoya, A. and Castell, J. V., Long-term storage of peroxidase-labelled immunoglobulins for use in enzyme immunoassay, *J. Immunol. Methods*, 99, 13, 1987.

13 Application of Chemical Conjugation for the Preparation of Immunotoxins and Other Drug Conjugates for Targeting Therapeutics

13.1 INTRODUCTION

Drug conjugates developed as a result of systemic pharmacotherapy are target-specific cytotoxic agents.[1] The underlying concept involves coupling a therapeutic agent to a delivery molecule with specificity for a defined target-cell population. Antibodies with high affinity for antigens are natural choices for targeting agents.[2–5] With the availability of high-affinity monoclonal antibodies (mAbs) and their fragments, the prospects of antibody-targeting therapeutics have become promising. Toxic substances that have been conjugated to mAbs include protein toxins, low-molecular-weight drugs, biological response modifiers, and radionuclides.[6] Antibody–toxin conjugates are frequently termed immunotoxins, whereas immunoconjugates consisting of antibodies and low-molecular-weight drugs such as methotrexate and adriamycin are called chemoimmunoconjugates. Immunomodulators contain biological response modifiers that are known to have regulatory functions such as lymphokines, growth factors, and complement-activating cobra venom factor (CVF). Radioimmunoconjugates consist of radioactive isotopes, which may be used as therapeutics to kill cells with radiation or used for imaging.

In addition to antibodies, other molecules that have specific receptors or binding sites on target cells have also been used as targeting agents. These include transferrin, α_2-macroglobulin, epidermal growth factor (EGF), and hormones. When hormones are used as the targeting agents, the term hormonotoxin is frequently used. Toxins have also been conjugated to antigens to selectively kill antigen-responsive B cells.[7]

The coupling of targeting agents with toxic moieties is most commonly performed with heterobifunctional cross-linking reagents. Earlier attempts to conjugate toxins and antibodies by homobifunctional reagents generated nonspecific cross-linking products. More specific and efficient cross-linking techniques have been developed.[8] This chapter will review the procedures that have been used to prepare immunotoxins and other cytotoxic drug conjugates. Theoretically, a cross-linking agent used to couple one toxin is applicable to other toxins. In fact, such is the case for most of the toxins described.

13.2 TARGETING AGENTS AND TOXINS

13.2.1 CHOICE OF TARGETING AGENTS

In almost all cases, the cytotoxic chemotherapeutic agents, such as paclitaxel, cisplatin, and doxorubicin (DOX), cannot distinguish cancer cells from normal cells. Consequently, in the development or choice of tumor-specific delivery systems for anticancer agents, recognizing the intrinsic differences between normal and tumor cells is the most important success factor for efficacious cancer chemotherapy. While new drug delivery systems continue to evolve, there are many choices of targeting agents as listed below. Understanding the specificity of these agents facilitates the development of tumor-directed drug conjugates.

13.2.1.1   Antibodies

As mentioned above, the toxin carriers in immunotoxins are antibodies.[5,9] One of the most important aspects of targeting therapy is the specificity of the carrier agents toward the target.[8] Among all antibodies, mAbs raised against specific markers on the surface of tumor cells are the most highly selective. Although these mAbs may have therapeutic value on their own, their cytotoxicities are greatly enhanced by conjugating them to highly cytotoxic drugs, which may be too toxic to be used alone. Thus, to achieve the full potential of immunotoxins, the antibody should be carefully selected such that it binds selectively to the target tissue with high affinity and has little cross-reactivity with healthy tissues. Ideally, the antibodies should be raised against those antigens that are highly expressed on the cell surface for maximum therapeutic potential. To avoid immunogenicity of these antibodies, nonimmunogenic humanized forms of antibodies should be used. These considerations are very important as demonstrated by the fact that a KS1/4 antibody–methotrexate conjugate for nonsmall cell lung cancer[10] and an antibody–DOX conjugate, BR96-Dox, for gastric adenocarcinoma[11] and metastatic breast cancer[12] were all found to lack therapeutic benefit. On the other hand, the antibody-maytansinoid conjugates seemed to provide superior antitumor activity.[8] Also, chimeric antitransferrin receptor antibodies in which the constant region of the antibody is substituted by a human-constant-region-inhibited proliferation and directly induced apoptosis in hematopoietic-derived cell lines.[13]

13.2.1.2   Other Naturally Occurring Molecules

In addition to antibodies, many other naturally occurring biopolymers and substances have been used as toxin carriers. These include vitamins, fatty acids, carbohydrates, transferrin, lectin, inulin, and regulatory and signaling molecules. Chen et al.[14] have developed a mechanism-based tumor-targeting drug delivery system based on tumor-specific vitamin-receptor-mediated endocytosis. In this case, biotin (a vitamin) is linked to cytotoxic agents such as taxoid through a disulfide linkage attached to a phenyl group. Such a disulfide bond can be cleaved upon endocytosis by endogenous thiols, for example, glutathione (GSH) and thioredoxin, to generate the desirable thiophenolate or sulfhydrylphenyl species and the free toxin. The synthesis involves pyridine disulfide and N-hydroxysuccinimide. Other essential vitamins that have overexpressed receptors on cancer cell surface have also been used. These include folic acid,[15] vitamin B12,[16] and riboflavin.[17] Unfortunately, in most cases, the synthesis of the linker between the vitamin derivatives and the toxin is complicated involving many steps.

Another naturally occurring substance used as a drug carrier based on endocytosis is transferrin.[13] Transferrins are iron-binding proteins. The human transferrin consists of two domains, homologous to each other with 679 amino acids. Iron-bound transferrin is internalized after binding to transferrin-specific receptors by receptor-mediated endocytosis. Since the transferrin receptor is expressed in inflammation and in proliferating malignant cells, transferrin may be used as a drug carrier for therapy. Several drugs have been conjugated to transferrin and tested for efficacy against cancer cells. These drugs include adriamycin, cisplatin, chlorambucil, daunorubicin, protein synthesis inhibitors, plant-derived toxins, and many others.[18] Adriamycin–transferrin conjugate has been shown to be toxic to human leukemia, erythroleukemia, colorectal carcinoma, breast adenocarcinoma, mesothelioma, liver carcinoma, and cervical adenocarcinoma cell lines.[19] Cisplatin-transferrin inhibits the growth of the human epidermoid carcinoma cell line.[20] The drugs chlorambucil and daunorubicin conjugated with transferrin show increased cytotoxicity in human breast cancer cell and small cell lung carcinoma compared with unconjugated drug.[21,22] The effect of other drug-transferrin conjugates has been reported.[18] Many methods have been used to prepare the transferring-drug conjugates such as Schiff base formation,[18] acid-sensitive maleimide derivatives, and glutaraldehyde.[18–22]

Polyunsaturated fatty acids (PUFAs) have also been used as tumor-specific molecules. Representative naturally occurring PUFAs include linolenic acid (LNA), linoleic acid (LA), arachidonic acid (AA), eicosapentaenoic acid (EPA), and docosahexaenoic acid (DHA). Kuznetsova et al.[23] prepared the conjugates of DHA, LNA, and LA with second-generation taxoids and studied their efficacy *in vivo* against ovarian and colon tumor xenografts. The coupling procedure for esterification uses N,N'-dicyclohexylcarbodiimide (DCC) as a coupling agent, methylene chloride as a solvent system, and dimethylaminopyridine (DMAP) as a catalyst. The authors demonstrated that two of the PUFA-taxoid conjugates show total regression of drug-resistant and drug-sensitive tumors in animal models.

Hyaluronic acid (HA, also called hyaluronan) is a polymer of disaccharides, composed of D-glucuronic acid and D-N-acetylglucosamine, linked together via alternating β-1,4 and β-1,3 glycosidic bonds. HA can contain up to 25,000 disaccharide repeats in length with molecular weights ranging from 5,000 to 20,000,000 Da. It is distributed widely throughout connective, epithelial, and neural tissues. There are three main groups of cell receptors for HA: CD44, RHAMM (receptor for HA-mediated motility) and ICAM-1 (intracellular adhesion molecule-1). Because of these receptors, HAs have been employed as tumor-specific modules to construct tumor-targeting drug conjugates.[24] Akima et al.[25] coupled mitomycin C and epirubicin to HA by carbodiimide chemistry and found that the former adduct was selectively taken up by, and was toxic to, a lung carcinoma xenograft. Luo and Prestwich[26] covalently attached taxol with N-hydroxysuccinimide to HA modified with adipic. The HA-taxol conjugates were shown to have selective toxicity toward the human breast, colon, and ovarian cancer cell lines that overexpress HA receptors. Saravanakumar et al.[27] modified HA with an amine-terminated hydrotropic N,N-diethylnicotinamide (DENA) oligomer to synthesize hydrotropic HA (HydroHA) derivatives. Paclitaxel (PTX), a highly hydrophobic chemotherapeutic agent, was coupled to a HydroHA using carbodiimide chemistry. The HydroHA-PTX conjugates were selectively taken up by the cancer cell line with overexpressing CD44. Luo et al.[28] also prepared a modified HA with N-(2-hydroxypropyl)methacrylamide (HPMA) copolymer and conjugated to DOX. The HPMA-HA-DOX conjugate was found to have higher cytotoxicity against human breast cancer, ovarian cancer, and colon cancer.

Lectins are sugar-binding proteins that are highly specific for the structures of sugar moieties. Lectin wheat germ agglutinin (WGA), for example, binds to N-acetyl-D-glucosamine and sialic acid of carbohydrates. It has a high binding rate to intestinal cell lines of human origin, human colonocytes, and prostate cancer cells. Moreover, it is also taken up into the cytoplasm of enterocyte-like Caco-2 cells.[29] Gabor et al.[30] conjugated fluorescein-bovine serum albumin (F-BSA) to lectin WGA using a homobifunctional cross-linker, divinyl sulfone, to react with the amino groups of both proteins. The F-BSA-WGA conjugate were bound specifically to Caco-2 cells and exhibited uptake into the cells. The authors concluded that WGA-mediated drug delivery is a promising strategy. With another lectin, concanavalin A (Con-A), which binds specifically to mannose, Anande et al.[31] prepared mucoadhesive microspheres of diloxanide furoate (DF) for the effective treatment of amoebiasis. The carboxyl groups of Eudragit microspheres of DF were linked to Con-A using ethyl-3,3-(dimethylaminopropyl) carbodiimide (EDC) and N-hydroxysuccinimide. The conjugate increased the mucoadhesiveness and provided controlled release of DF in simulated GI fluids.

Heparin is a highly sulfated glycosaminoglycan that has been conjugated to deoxycholic acid (DOCA) for drug delivery.[32] Park et al.[32,33] prepared heparin-DOCA (HD) conjugates by covalently bonding N-(2-aminoethyl)deoxycholylamide to heparin via amide formation with EDC. The HD conjugates were loaded with DOX through heparin nanoparticle formation. These DOX-loaded heparin nanoparticles displayed a sustained drug release pattern and enhanced therapeutic effect against squamous cell carcinoma and B16F10 melanoma. Wang et al.[34] also prepared various heparin conjugates by activating heparin to a mixed anhydride intermediate to which paclitaxel and amino acid-paclitaxel derivatives have been linked. These conjugates were shown to arrest MCF-7 cells in the G2/M phase of cell cycle.

Gellan gum, also known commercially as Phytagel or Gelrite, is a water-soluble polysaccharide produced by bacterium, *Sphingomonas elodea*. The repeating unit of the polymer is a tetrasaccharide, which consists of two residues of D-glucose along with one residue of L-rhamnose and one residue of D-glucuronic acid connected through an (α1 → 3) glycosidic bond. Krauland et al.[35] conjugated L-cysteine to deacetylated gellan gum (DGG) by a carbodiimide. The DGG-cysteine conjugate was capable of forming inter- and/or intramolecular disulfide bonds in aqueous solution. The authors suggested that the conjugated polymer represents a promising novel excipient for various drug delivery systems.

13.2.1.3 Synthetic Peptides and Nucleotides

Many peptides, polypeptides, and proteins have been shown to posses the ability to traverse biological membranes. These matrices have been successfully used for the intracellular delivery of many therapeutic agents including small molecules, proteins, peptides, oligonucleotides, plasmids, and nanoparticles.[36] Unbiased biopanning of phage-displayed peptide libraries has generated a suite of cancer targeting peptidic ligands.[37] Among these, cell-penetrating peptides, which are 9–35 mer cationic and/or amphipathic peptides, can be linked to a variety of anticancer therapeutics, making them an efficient, effective, and nontoxic mechanism for drug delivery.[38]

Specifically, receptors for certain peptide hormones are expressed in a relatively high concentration on a variety of cancer cells. These peptide hormones can serve as carriers for a local delivery of cytotoxic agents or radiopharmaceuticals to the tumors. The most widely investigated of these peptide hormones is the hypothalamic hormone, somatostatin.[39] The short plasma half-life of the native forms of somatostatin prompted an avid search for more stable and more potent synthetic analogs as demonstrated by the successful clinical use of radiolabeled somatostatin analog, Octreoscan, for the detection and treatment of some somatostatin receptor-positive tumors. In recent years, a series of other cytotoxic peptide hormone conjugates based on derivatives of hypothalamic hormones such as luteinizing hormone-releasing hormone (LHRH) and the brain-gut hormone bombesin were prepared. A derivative of DOX, 2-pyrrolino-DOX, which is 500–1000 times more active than its parent compound was coupled to somatostatin octapeptide, to LHRH analogue, [D-Lys6]LHRH, and to bombesin-like peptide. These conjugates were investigated for their effectiveness in various cancers including ovarian and breast cancers,[39] hepatocellular carcinoma,[40] and prostate cancer.[41]

Oligonucleotides have also been used as targeting agent for drug delivery. Of particular importance are the aptamers, which are short single-stranded nucleic acids with a defined three-dimensional shape that allows them to interact with high affinity with a target molecule.[42,43] They have been used to target distinct cell subtypes and tissues. The most established and best characterized aptamer in this regard is the prostate-specific membrane antigen (PSMA)-binding nucleic acid molecule A10. This aptamer has been conjugated with cytotoxic molecules, thereby allowing their selective delivery to cancer cells. For example, Farokzhad et al.[44] have attached chemotherapeutics, such as docetaxel, to the 5′-amino end of the aptamer by an NHS/EDC approach and showed the efficacy of such an aptamer-drug conjugate in an *in vivo* xenograft rat tumor model system. The same group also directly complexed aptamer A10 with DOX and demonstrated that treatment of prostate cancer cells with the complex resulted in a significant reduction of tumor-cell proliferation.[45] Chu et al.[46] also conjugated the aptamer to the toxin gelonin, a ribosome-inactivating protein (RIP), with *N*-succinimidyl-3-(2-pyridylodithio)propionate (SPDP). The aptamer conjugate not only promoted uptake of gelonin into target cells but also decreased the toxicity of gelonin in nontarget cells.

In addition to chemotherapeutics, siRNA molecules have also been coupled to aptamer A10 either directly by nucleotidic extensions or indirectly through the assembly of tetrameric streptavidin–biotin complexes.[47,48] Several other aptamers have also been used as aptamer-based siRNA-delivering agents.[49]

13.2.1.4 Synthetic Polymers

There are many synthetic polymers that have been devised as toxin carriers. Different molecules were designed for different purposes.[50–52] For example, cationic biopolymers such as poly (L-lysine), linear and branched polyethylenimine (LPEI), branched polyethylenimine (brPEI), and dendrimers including polyamidoamines (PAMAMs) have been constructed to bind nucleic acids via electrostatic interactions with the negatively charged phosphate backbone.[50] Those commonly used for dermal drug delivery include hydrocolloids, alginates, hydrogels, polyurethane, and poly(lactic-co-glycolic acid).[51] Polymers conjugated to drugs with either affinity-based targeting moieties or cleavage mechanisms have been constructed in soluble and micellar form. The types of synthetic polymers used include hydrophilic and nonhydrolytically degradable materials such as poly(ethylene glycol) (PEG), poly(vinyl alcohol) (PVA), and poly(acrylamide) (PAAm). Hydrophobic polymers, such as poly(n-butyl acrylate), as well as hydrophobic and hydrolytically susceptible materials such as poly-(α-esters) are also widely employed. Amphiphilic block polymers such as (PEG-b-PPO-b-PEG) and thermally sensitive polymers such as poly(N-isopropylacrylamide) (pNIPAAM) have also been widely employed.[51,52] Anticancer drugs such as DOX have been linked to amphiphilic block copolymers of HPMA with a functional monomer 2-(2-pyridyldisulfide)ethylmethacrylate (PDSM) by a thiol-maleimide reaction.[53] HPMA copolymer-DOX conjugates have also been prepared using a polymerizable derivative of DOX (N-methacryloylglycylphenylalanylleucylglycyl DOX) and a crosslinking agent, N(2),N(5)-bis(N-methacryloylglycylphenylalanyl-leucylglycyl)ornithine methyl ester.[54] Polysaccharides, proteins, or peptides have been conjugated to synthetic polymers to impart desired bioactivity. Current techniques for conjugation include aldehyde, carbodiimide, epoxide, hydrazide, active ester, radical, and addition reactions. Examples of polysaccharides linked to synthetic polymers are chitin, dextran, HA, chondroitin sulfate, and heparin.[52]

Many of the synthetic polymers are hydrogels like poly(2-hydroxyethyl methacrylate) (pHEMA). Hydrogels are polymeric networks of hydrophilic polymeric chains, which absorb and retain a large amount of water. They have found widespread applications in different technological areas including matrices for cell encapsulation and devices for the controlled release of drugs and proteins. While some hydrogels are prepared by chemical polymerization of hydrophilic monomers such as 2-hydroxyethyl methacrylate (HEMA) or derivatized dextran, others are made by cross-linking hydrophilic polymers.[55] Hennink and van Nostrum[56] have reviewed various cross-linking procedures. Hydrophilic polymers with functional groups of OH, COOH, and NH_2 are used for the formation of hydrogels. Cross-linkers such as glutaraldehyde, 1,6-hexamethylenediisocyanate, divinylsulfone, 1,6-hexanedibromide, carbodiimides, and transglutaminase, to mention a few, have been used. The diversity of hydrogels therefore provides another dimension for drug delivery.

Thiolation of polymers provides another means to increase the amendable properties of these polymers for therapeutic drug delivery.[57] Thiolated polymers, known as thiomers, are mucoadhesive polymers, which display thiol-bearing side chains. Thiomers are generated by the immobilization of thiol-bearing ligands to mucoadhesive polymeric excipients. For example, Shahnaz et al.[58] covalently attached L-cysteine to carboxymethyl dextran (CMD) by a carbodiimide. The resulting CMD-cysteine conjugate displayed 273 ± 20 µmol thiol groups per gram of polymer. Other thiomers such as cationic chitosan-cysteine, chitosan-thiobutylamidine as well as chitosan-thioglycolic acid and anionic thiomers poly(acrylic acid)-cysteine, poly(acrylic acid)-cysteamine, carboxy-methylcellulose-cysteine, and alginate-cysteine have also been generated.[59–61] Based on thiol/disulfide exchange reactions and/or a simple oxidation process between such polymers and cysteine-rich subdomains of mucus glycoproteins, the mucoadhesive properties of these polymers are improved up to 130-fold. The free thiols on the thiomers also provide a means for the attachment of cytotoxic agents. Additionally, some thiomers exhibit improved inhibitory properties toward peptidases. Thus, thiomers are promising novel polymers for drug delivery.

13.2.2 Choice of Toxins

The most commonly used toxins in the preparation of immunotoxins are diphtheria toxin (DT),[62,63] anthrax toxin,[63] ricin,[8,63,64] abrin,[64] pokeweed antiviral protein (PAP),[63,65] CVF,[66] *Pseudomonas* exotoxin (PE),[8,63,67] gelonin,[63] saporin,[63,68] and other RIPs.[68] These proteins have been purified and characterized.[68–70] Some of these toxins such as ricin, abrin, modeccin, viscumin, and volkensin have two dissimilar A and B polypeptides attached through a disulfide bond and are referred to as true toxins.[8,64,67,69] Only the A chain contains the enzymatic activity that is cytotoxic, and the B chain is for receptor binding that usually involves a carbohydrate. Other toxins such as PAP and gelonin contain a single polypeptide that have similar or identical enzymatic activity to the A chains of the true toxins. These hemitoxins can generate a true toxin when covalently bound to the B chain of ricin.[71] DT, on the other hand, contains a single polypeptide, but can be cleaved with trypsin to generate fragment A on reduction. There are therefore two ways of preparing immunotoxins. The first is to link the whole intact toxin to the antibody or its fragment. The second is to couple the antibody to the isolated A chain or the single-polypeptide toxin. The method dictates the choice of cross-linker to be used. Since the A and B subunits are separated during the action of cell killing, intramolecular cross-linking of true toxins must be avoided. Similarly, the linkage of an antibody-A chain conjugate must be reducible, as in an intact toxin, to be cytotoxic. Another constraint on the choice of coupling agent is the retention of cytotoxicity of the toxin and the antigen-binding capacity of the antibody. Thus, each conjugate should be tested for both cytotoxicity and antigen-binding activity. Limited modification of lysine residues in abrin, for example, reduces its galactose-binding affinity, but the toxicity of the A chain is not affected.[72] On the other hand, limited modification of lysine residues of RIP from *Momordica charantio* reduces its ability to inhibit protein synthesis.[73] Furthermore, the cross-linking should be stable under *in vivo* conditions. The biological stability of immunotoxins *in vivo* might be the major limitation of its efficacy. Most of the active conjugates have been prepared with a disulfide link between the effector and the antibody. The disulfide bond may be disrupted by disulfide exchange with GSH or other serum borne or cellular factors. Generally, sulfosuccinimidyl-6-[(-methyl-(-(2-pyridyldithio)toluamido]hexanoate (S-LC-SMPT) and sulfosuccinimidyl-6-[3-(2-pyridyldithio)-propionamido]hexanoate (S-LC-SPDP) cross-linked conjugates show good disulfide stability.[74]

The choice of therapeutic agents depends on the sensitivity of the tumor toward the drug.[8] For example, ovarian and breast cancers are sensitive to tubulin agents, while lymphomas are sensitive to DNA-interacting agents. Thus, it is important to understand the mechanisms of action of the drugs, so that the right choice can be made. The following show a few examples of drugs with different mechanisms of actions.

Maytansinoids are members of the ansamycin class of natural products. They possess strong antitumor activity as potent antimitotic agents. The mechanism of action entails interference with the formation of microtubules through the inhibition of the polymerization of the microtubule protein, tubulin.[75] The antitumor activity of maytansine was extensively evaluated in human clinical trials, but the results were not favorable.[76] Chari's laboratory synthesized several disulfide-containing maytansinoids and conjugated them to appropriately modified specific antibodies by disulfide exchange reaction.[8] The results of these conjugates against colorectal, pancreatic, and gastric cancer cells seem promising.

The dolastatins, originally isolated from the sea hare, *Dolabella auricularia*, have a unique pentapeptide structure and are highly cytotoxic.[76] They share a common mechanism of action with maytansine and cause cell death by inhibiting tubulin polymerization. Auristatins are synthetic analogues. Auristatin E and F have been linked via a peptidase-labile linker to mAbs. Conjugates with CD30-directed mAb have been tested for the treatment of Hodgkin Lymphoma.[77] After binding to CD30 on the cell surface, the antibody drug conjugate is internalized and traffics to lysosomes, where the peptide linker is selectively cleaved and monomethyl auristatin E is released, resulting in cell cycle arrest and apoptosis through tubulin binding. Auristatin E has also been conjugated to an antiglycoprotein nonmetastatic melanoma protein B.[78] The conjugate is found to have therapeutic potential in the treatment of melanoma and breast cancer.

Another family of compounds that acts on microtubules is taxane diterpenoids. The natural product paclitaxel and its semisynthetic analog, docetaxel, are members of the family and are two of the most active anticancer agents used against ovarian and breast cancers. Photoaffinity labeling using photoreactive radiolabeled taxoids has disclosed the drug-binding domain of tubulin and P-glycoprotein. Presumably taxoids bind to a putative binding site located in one of the two different types of pores in the B-lattice of the microtubules and are later transported to the luminal site.[79] Miller et al.[80] have synthesized a series of highly potent taxoids with improved toxicity and solubility in aqueous systems and disulfide bonds for linkage to specific antibodies. These antibody-taxane conjugates are very cytotoxic.

CC-1065, duocarmycin A, and duocarmycin SA are naturally occurring antitumor antibiotics. These agents have been shown to exert their biological effects through a sequence-selective alkylation of DNA.[81] The mechanism of action involves binding of the antibiotics to selected AT rich regions in the minor groove followed by alkylation of N3 of adenine flanked by two 5' A or T bases with a preference for this three-base sequence: 5'-AAA > 5'-TTA > 5'-TAA >5'-ATA. The cytotoxicity of these compounds is 1000-fold more than other DNA-interacting agents, such as DOX and *cis*-platin. Chari et al.,[8,82] modified a highly potent synthetic analogue, adozelesin, with phosphate to make the molecule more water soluble and with a disulfide group to conjugate to an antibody that binds to CD19 antigen expressed in B-cell lymphoma. The conjugate is far superior to clinically used anticancer drugs in the treatment of lymphoma.

Another group of antitumor antibiotics that acts on DNA is the calicheamicins, which are produced from cultures of *Micromonospora echinospora*.[83] Structurally, they possess a bicyclic core containing an enediyne bridge along with the allylic trisulfide and a set of carbohydrates. The bicyclic core seems to serve as the key moiety to cause the DNA double-strand breaks, causing cell death, and the sugar moiety is believed to anchor the whole molecule to the DNA minor groove via appropriate recognition processes. It is reported that the sugar moiety plays a key role in sequence specificity and tight DNA binding. Calicheamicin γ1 displayed selectivity for pyrimidine. Calicheamicins are potent at subpicomolar concentrations *in vitro*. To link them to antibodies via acid-labile bonds, an hydrazide functionality has been introduced into calicheamicin γ1. A humanized anti-CD33 antibody-calicheamicin conjugate has been approved for the treatment of acute myeloid leukemia.[8,84] Another conjugate of calicheamicin to an antibody directed against the CD22 antigen expressed in B-cell lymphomas has been reported.[85]

Specific therapeutics has also been conjugated to targeting agents other than antibodies. Intraalveolar fibrin formation is a hallmark of acute inflammatory and chronic interstitial lung diseases, such as the acute respiratory distress syndrome (ARDS) of the adult and idiopathic pulmonary fibrosis (IPF). Polymerization of fibrin in the presence of pulmonary surfactant results in far-reaching incorporation of the hydrophobic surfactant compounds into the growing fibrin matrix, with the loss of surface activity, altered fibrin structure, and reduced susceptibility of the clot to fibrinolysis. For specific targeting of such alveolar fibrin, Ruppert et al.[86] designed a hybrid molecule consisting of the catalytic domain of urokinase (B-chain) and the hydrophobic surfactant protein B (SP-B), termed SPUC. The urokinase B-chain was chemically coupled to SP-B using a heterobifunctional crosslinker, sulfosuccinimidyl-4-(*p*-maleidophenyl)butyrate (S-SMPB). The incorporation of SP-B offers SPUC compartmentalized fibrinolysis at the alveolar level and improvement of gas exchange and lung compliance under conditions of acute lung injury and fibrotic lung disease.

13.3 PREPARATION OF THERAPEUTIC CONJUGATES

13.3.1 CHOICE OF CROSS-LINKING REAGENTS

The amino side chains of lysines, the carboxyl groups of aspartic and glutamic acids, the thiol group of cysteine, and the carbohydrate moiety of antibodies, toxins, and other drugs have been used to prepare drug conjugates. However, the side chains of histidines, methionine, arginine, serine, threonine, tryptophan, and tyrosine have not been fully utilized. In earlier studies, immunotoxins were prepared using homobifunctional reagents such as glutaraldehyde, diethylmalonimidate hydrochloride, and difluorodinitrophenylsulfone that cross-linked the free amino groups found in abundance in protein species. As discussed in earlier chapters, such cross-linking reagents produce a mixture of heterogeneous conjugates with extensive polymerization of both antibody and toxin. In addition, intramolecular cross-linking between toxin subunits generally destroys their cytotoxicity. The development of two-step coupling procedures using heterobifunctional reagents such as *m*-maleimidobenzoyl-*N*-hydroxysuccinimide ester (MBS), and SPDP has resulted in more efficient use of reactants, increased yields, and more desirable structural features in synthesized antibody–toxin conjugates. These reagents generally first involve generating a free thiol group on one protein followed by reacting selectively with a functionality on a second protein. The art of this type of conjugation will be illustrated below in detail. However, the choice of which cross-linker to use depends on the purpose of the conjugate to be prepared. For example, a design of cross-linking between antibody and maytansinoid (MD1) can circumvent multidrug resistance. Because preferred substrates for the multidrug transporter MDR1 are hydrophobic compounds, Kovtun et al.[87] conjugated antibody to MD1 with a hydrophilic heterobifunctional cross-linker, succinimidyl-[(*N*-maleimidopropionamido)tetraethyleneglycol] ester (NHS-PEG4Mal). The metabolite, lysine-PEG4Mal-DM1, released from the conjugate inside the cell and thereby avoided MDR1-mediated efflux. The increased retention of the metabolite of PEG4Mal-linked conjugate correlated with the greater antimitotic and cytotoxic potency of the conjugate.

The drug-antibody conjugate has to be designed in a manner that ensures its stability during circulation in the bloodstream but allows for rapid release of the cytotoxic agent in its fully active form inside the tumor cells. Furthermore, the drug and its carrier must remain intact during storage to allow formulations for convenient intravenous administration. Several types of cleavable cross-linkers have been evaluated, notably acid- and peptidase-labile reagents. However, disulfide cross-linkers are a better choice for the reasons that they are stable at physiological pH and that intracellular levels of reduced GSH and other reducing agents can cleave the disulfide bond and release the drug inside the cell. On the other hand, levels of reduced GSH in blood are very low, thus preserving the disulfide bond while the conjugate is in circulation. In consideration of the superiority of disulfide-containing cross-linkers, various ways of generating disulfide cross-links will be considered below.

13.3.2 CONJUGATION THROUGH DISULFIDE BOND

The most common amino acid side chains of toxins and immunoglobulins that have been used to link the two molecules together are the amino and sulfhydryl groups. Amino groups are reactive, abundant, and, in most cases, expendable. That is, a limited number of amino groups can be modified without diminishing the biological activity of the protein. Because the linkage with a cleavable bond between the toxin and antibody is of critical importance in determining the cytotoxicity of the conjugate,[1,5,8,9] cross-linkers with disulfide bonds are the preferred choices. A free thiol of the protein or toxin can be used to form reducible disulfide bonds. In cases where a free thiol is not available, such as in native antibodies, free sulfhydryl groups can be generated by reductive cleavage of native cystine residues with thiol reagents (e.g., dithiothreitol [DTT]) or by thiolation of the amino groups using SPDP, 2-iminothiolane, or methyl-3-mercaptopropionimidate as described in an earlier chapter. The conversion of an amino group into a sulfhydryl group has been discussed in Chapter 2. The following sections illustrate various possible ways for the preparation of drug conjugates.

13.3.2.1 Coupling with *N*-Succinimidyl-3-(2-Pyridylodithio)propionate

13.3.2.1.1 Preparation of Immunotoxins

N-succinimidyl-3-(2-pyridyldithio)propionate (SPDP) is probably the most widely used cross-linker for preparing antibody–toxin conjugates. Not only has it been used to cross-link between ammo and thiol groups but also has also been used to introduce a free thiol in immunoglobulins or other molecules that do not have a free sulfhydryl group. The reaction scheme is shown in Figure 13.1A. First, the ammo groups of both an immunoglobulin and a toxin are reacted with the reagent to form a carboxamide bond. One of the 2-pyridyldisulfide groups introduced, for example, that on the antibody, is then reduced by DTT to generate a free thiol. The introduction of sulfhydryl groups into immunoglobulins is preferable over introduction into the toxin in order to avoid the exposure of the latter to reducing conditions, which could result in the separation of its A and B subunits of binary toxins, such as ricin. Cross-linking reactions will take place on mixing between the two modified components where the antibody reduced thiol will nucleophilically displace pyridine-2-thione from the 3-(2-pyridyldithio)propionyl group incorporated into the toxin. The resulting conjugate contains a disulfide bond derived from two molecules of SPDP. The yield of the 1:1 antibody:toxin conjugate is usually about 20%–40%. A disadvantage of SPDP conjugation is the reported instability of the disulfide bond, which is prone to *in vivo* cleavage and exchange reactions.[88]

This method has been used to prepare immunotoxins of ricin,[89] momordin,[90] gelonin,[90] PAP,[91] saporin,[92] human lymphoblastoid interferon-α,[93,94] and CVF.[95] The same approach has also been used to prepare F(ab′)$_2$ conjugates with saporin,[96,97] ricin,[98] PAP,[97] gelonin,[97] and CVF.[99]

Instead of modifying both components with SPDP, one of the components can be thiolated with 2-iminothiolane as shown in Figure 13.1B. It is reported that 5-methyl-2-iminothiolane is a better thiolating reagent because it enhances the stability of the resulting immunoconjugate due to steric hindrance of the methyl group.[100] Usually, the toxin is thiolated and antibodies modified with SPDP.[97] On mixing the two components, the thiolated toxin will be conjugated to the antibody by disulfide exchange as discussed above. Bolognesi et al.[97] have prepared gelonin, saporin, PAP, momordin, and dianthin immunotoxins this way. Gelonin immunotoxins have also been prepared by other groups using this method.[101,102]

When a free thiol is present in the toxin, such as in ricin A, these toxins can directly react with SPDP-modified antibodies to yield immunotoxins as in Figure 13.1C. A disulfide exchange reaction between the free thiol of toxin and 3(2-pyridyldithio)propionyl on the antibody will lead to a new disulfide bond linking the two components. A free thiol on a toxin may be produced on the reduction of a disulfide bond in toxins such as ricin, abrin, and other ribosome-inactivating proteins, which contain two subunits, A and B, linked by a disulfide bond.[64] This approach has been used to link antibodies to A chains of ricin,[70,103,104] abrin,[70,103,105] recombinant ricin,[106] and DT.[70] Usually, a yield of up to 50% of the 1:1 conjugates is attainable.

13.3.2.1.2 Preparation of Other Toxin Conjugates

In addition to immunoglobulins, other targeting molecules have also been conjugated to various toxins using SPDP. Very specific and potent-conjugated cytotoxins have been prepared from ligands such as transferrin, α$_2$-macroglobulin, EGF, and other molecules for receptor-mediated endocytosis.[107,108] Transferrin was conjugated to DT via a disulfide linkage following derivatization of both components using SPDP and reduction of SPDP-modified DT according to Figure 13.1A.[109] Ovine luteinizing hormone (oLH)-gelonin homonotaxin was also prepared this way; gelonin was first modified with SPDP and reduced with DTT to generate gelonin-SH. The free thiol then reacts with SPDP-modified oLH to generate the homonotoxin.[110] Similarly, Stirpe et al.[111] coupled gelonin-SH to SPDP-modified Con-A to give a Con-A-gelonin conjugate that was moderately cytotoxic to HeLa cells.

Gelonin-oLH disulfide-linked hormonotoxin is also prepared by thiolation of oLH with 2-iminothiolane and then reacted with SPDP-activated gelonin according to Figure 13.1B.[112] Similarly, a transferrin–saporin conjugate is obtained by reacting 2-iminothiolane-thiolated saporin with SPDP-activated transferrin.[113] Instead of 2-iminothiolane, EDC is used to introduce a free thiol group into methotrexate with cystine dimethyl ester or cystamine followed by hydrolysis and reduction. The free thiol containing methotrexate reacts with SPDP-activated mAbs to form methotrexate immunoconjugates.[114]

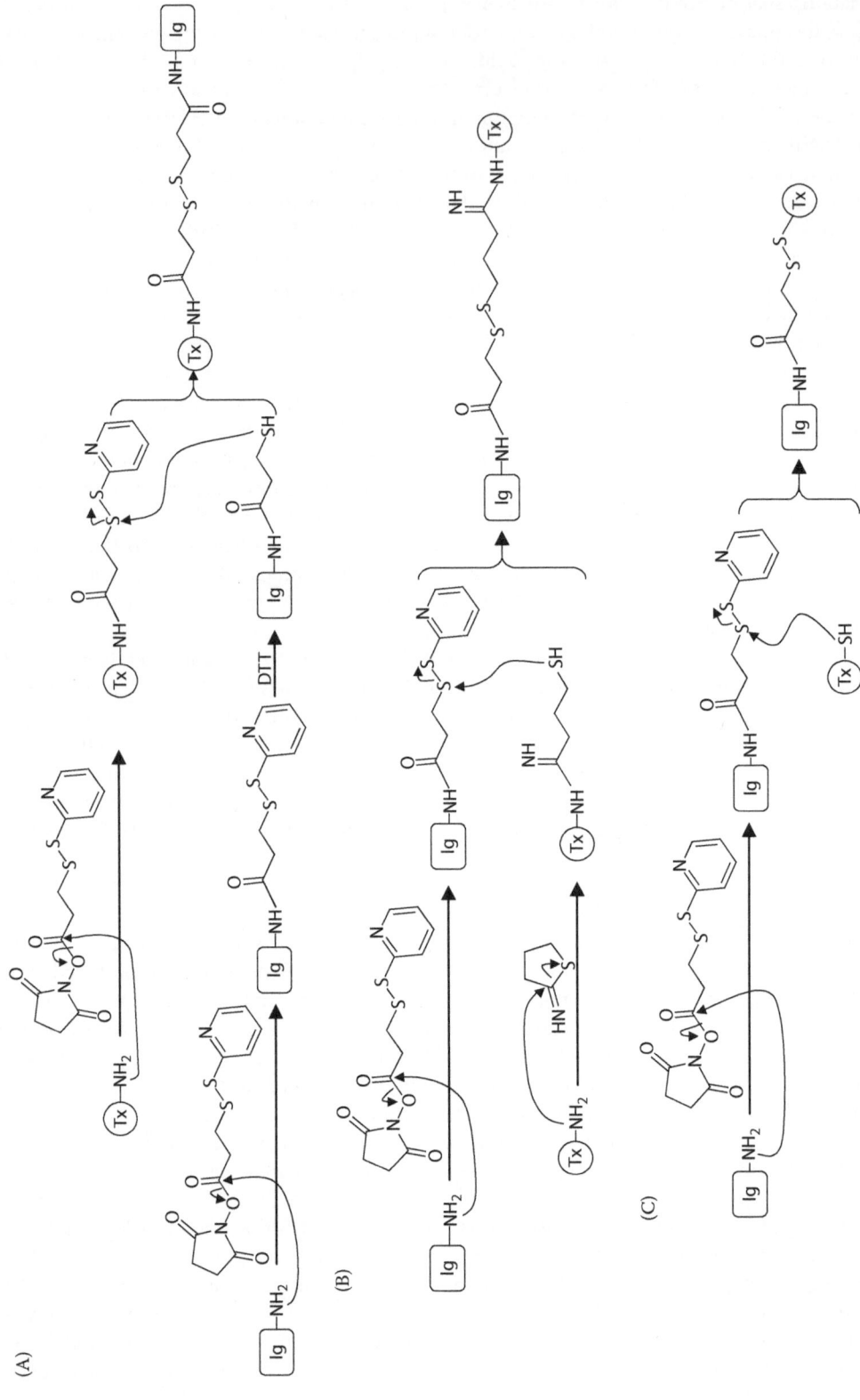

FIGURE 13.1 Conjugation with SPDP: (A) conjugation with two SPDP molecules; (B) conjugation with SPDP and iminothiolane; (C) conjugation with SPDP and free thiol.

As mentioned above, toxins with free thiols can be directly conjugated to SPDP-activated targeting molecules. Thus, ricin A chain is linked to SPDP-activated transferrin to form transferrin–ricin A chain conjugate.[115] It also reacts with α_2-macroglobulin derivatized with SPDP.[116] The A chains of ricin and DT have conjugated to SPDP pyridyldithiopropionate-derivatized fetuin,[117] asialofetuin,[117] EGF,[117,118] and human chorionic gonadotropin.[119,120] Likewise, asialoorosomucoid has been disulfide-linked to DT A chain.[121]

13.3.2.2 Coupling with Other Disulfide Generating Agents

13.3.2.2.1 Use of Pyridyldisulfide

In addition to SPDP, introduction of pyridyldisulfide into proteins can also be achieved by using 3-(2-pyridyldithio)propionate and water-soluble carbodiimide such as EDC according to the reaction shown in Figure 13.2A.[122] Immunoglobulins, transferrin, and other proteins can be activated by this procedure. Free thiols of toxins will react with the pyridyldisulfide through disulfide exchange to form conjugates. Using this procedure, Bourrie et al.[123] prepared antibody-ricin

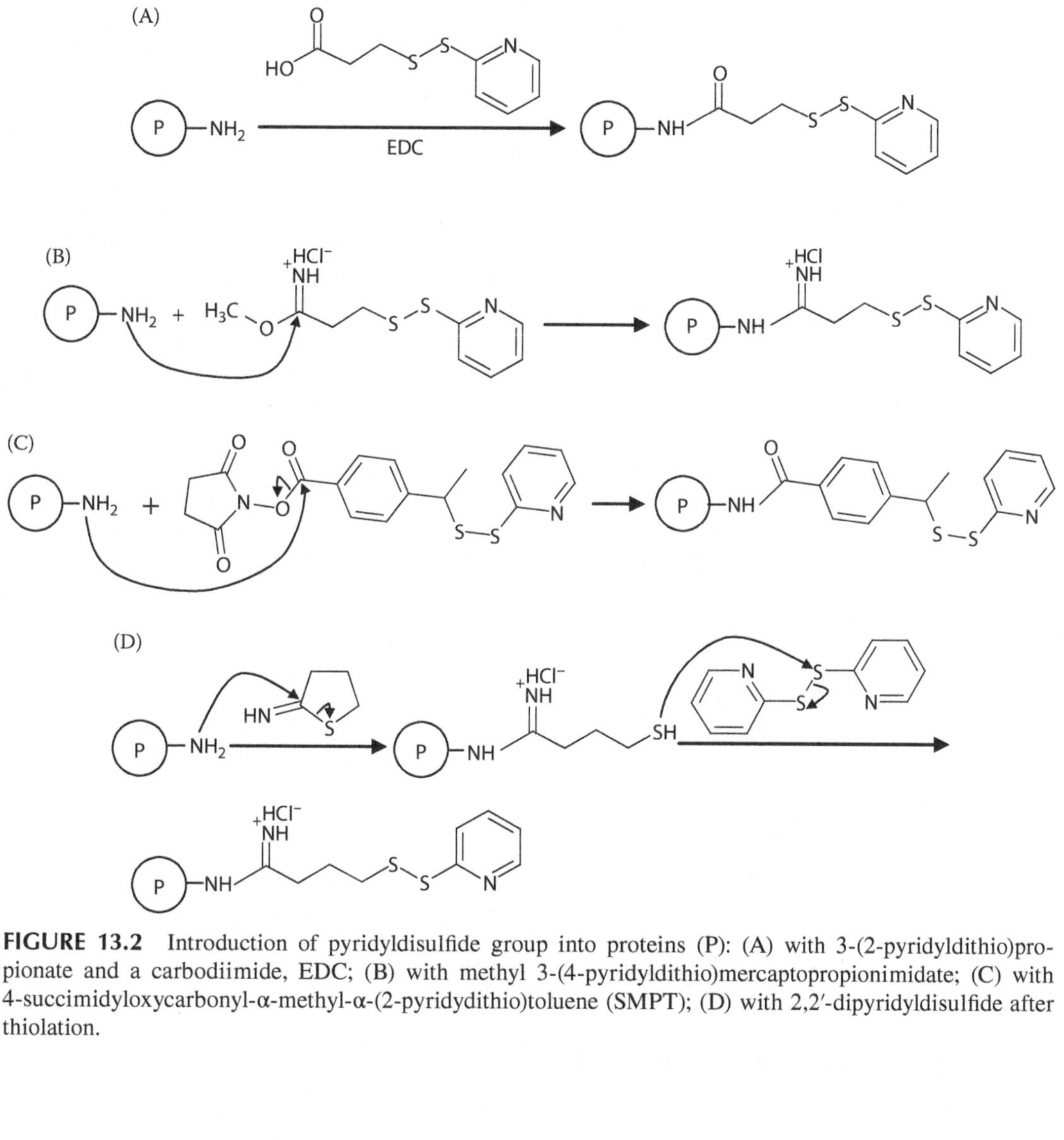

FIGURE 13.2 Introduction of pyridyldisulfide group into proteins (P): (A) with 3-(2-pyridyldithio)propionate and a carbodiimide, EDC; (B) with methyl 3-(4-pyridyldithio)mercaptopropionimidate; (C) with 4-succimidyloxycarbonyl-α-methyl-α-(2-pyridydithio)toluene (SMPT); (D) with 2,2′-dipyridyldisulfide after thiolation.

A-chain immunotoxins. In the case of transferrin, three to four dithiopyridyl substituents were linked to amino groups on transferrin in this way. Reacting this activated protein with the free sulfhydryl of isolated ricin A chain resulted in the formation of disulfide-linked transferrin–ricin A-chain conjugate.[115]

A direct method of incorporation of 4-pyridyldisulfide can be achieved by using methyl 3-(4′-dithiopyridyl)mercaptopropionimidate ester as shown in Figure 13.2B.[124] Amidination occurs at the free amino groups of proteins, which can undergo disulfide exchange with other thiolated compounds. Unfortunately, the literature does not at this time present any application of this reagent for the preparation of toxin conjugates.

Another cross-linker containing the pyridyldisulfide group that can be bonded to amino groups of proteins is 4-succimidyloxycarbonyl-α-methyl-α-(2-pyridydithio)toluene (SMPT) as shown in Figure 13.2C.[125] Kreitman and Pastan[126] derivatized antibody with SMPT and prepared a series of immunotoxins of ricin, abrin, saporin, gelonin, and PAP. Recombinant gelonin and recombinant ricin A immunotoxins have also been prepared using SMPT.[101,127] Using SMPT, the disulfide bond between the antibody and toxin is in a hindered state, which results in a prolonged half-life for deglycosylated ricin A chain immunotoxins and hence an improved therapeutic index.[125]

Free sulfhydryl groups may be directly converted to pyridyldisulfide by 2,2′-dipyridyldisulfide. The reaction involves a disulfide interchange, displacing 2-thiopyridine as shown in Figure 13.2D. Free sulfhydryl groups may be introduced into the proteins by 2-iminothiolane as described above.

13.3.2.2.2 Use of Cystamine

Cystamine incorporated into proteins provides another means of conjugation through disulfide interchange with a free thiol. The labeling of proteins with cystamine is generally carried out with a water-soluble carbodiimide (e.g., EDC) to cross-link with a carboxyl group as shown in Figure 13.3. Several conjugates are prepared with cystamine. For example, in the preparation of EGF–DT conjugate, the carboxyl groups of EGF are modified with cystamine. This cystaminyl derivative was mixed with reduced DT fragment A (DTA) to produce a disulfide-coupled EGF–DTA conjugate.[128] Similarly, Miskimins and Shimizu[129] prepared an insulin-DTA conjugation by first derivatizing the carboxyl groups of porcine insulin with cystamine using EDC. An excess of cystaminyl-insulin was mixed with DTA to allow disulfide interchange, and the product was then isolated by gel filtration. In a similar manner, Oeltmann and Forbes[130] linked anti-Thy-l.2 antibody to DTA through a disulfide bond by reacting the cystaminyl-antibody with the thiol of DTA. In a slightly modified version, two DT polypeptides, CRM26 and CRM45, were disulfide-linked to thyrotropin-releasing hormone (TRH, L-pyroglutamyl-L-histidyl-L-proline amide).[131] The histidyl imidazole of TRH was first modified with iodoacetylcystamine, which was derived from iodoacetic acid and cystamine using EDC condensation as shown in Figure 13.4. The acetylcystaminylated TRH was then mixed with reduced CRM peptides to produce the conjugate.

13.3.2.2.3 Use of S-Sulfonate

Another method of forming disulfides linkage is the reaction through S-sulfonate. S-sulfonate can be introduced into proteins by reacting a free thiol with Na_2SO_3/NaS_4O_6. Thorpe et al.[125]

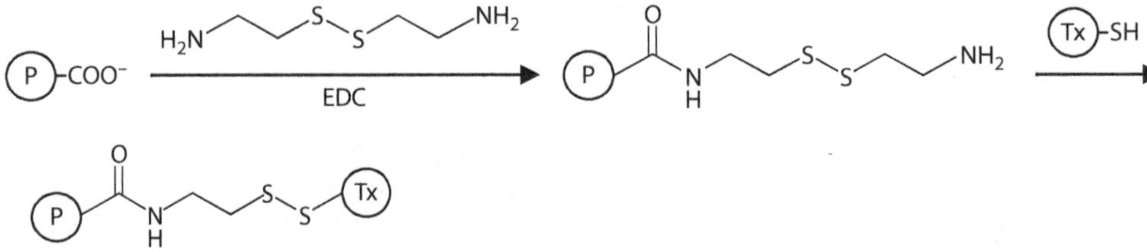

FIGURE 13.3 Conjugation of protein and toxin using cystamine.

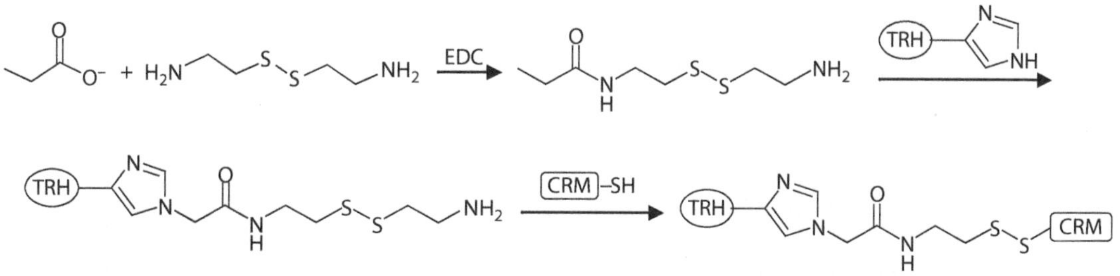

FIGURE 13.4 Reaction of TRH with iodoacetylcystamine and subsequent conjugation with CRM polypeptides.

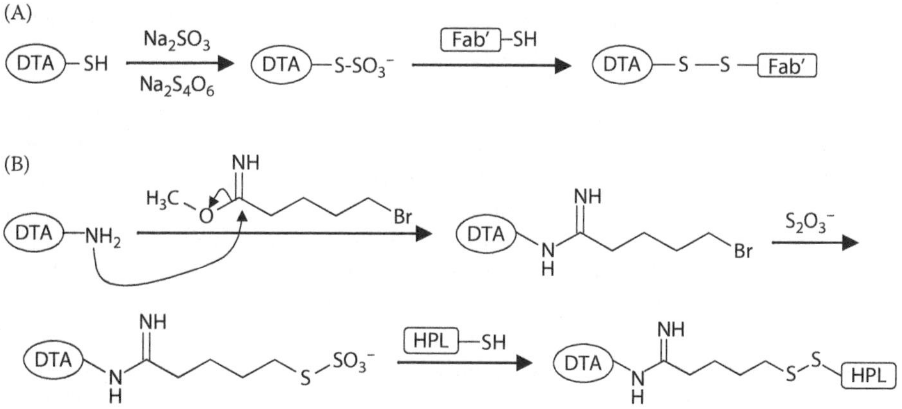

FIGURE 13.5 Conjugation using thio-sulfonate: (A) coupling DTA and Fab′; (B) coupling DAT and human placental lactogen (HPL) using methyl-5-bromovalerimidate.

synthesized two compounds, sodium S-4-succinimidyloxycarbonyl benzyl thiosulfate (SBT) and sodium S-4-succinimidyloxycarbonyl-α-methyl benzyl thiosulfate (SMBT), which react with amino groups in proteins to introduce benzyl S-sulfonate. The S-sulfonate group can also be introduced by the reaction of an amino group with methyl-5-bromovalerimidate followed by reaction with sodium thiosulfate.[132] The sulfite ion is easily replaced by an attacking free thiol. Masuho et al.[133] introduced the S-sulfonate into DTA and cross-linked it to the monovalent Fab′ fragment of a polyclonal antibody directed against L1210 cells as shown in Figure 13.5A. Chang et al.[132] introduced 5-S-sulfomercaptovaleramidinate into DTA and coupled the activated toxin with human placental lactogen through the S-sulfonate (Figure 13.5B). Ricin A-chain was also disulfide cross-linked to the β subunit of hCG utilizing the same methyl-5-bromovalerimidate method.[134]

13.3.2.2.4 Use of Ellman's Reagent

Ellman's reagent, 5,5′-dithiobis(2-nitrobenzoic acid) (DTNB), can be used to conjugate two free thiols as a zero-length cross-linker by two disulfide interchange reactions.[135] One thiol is first activated with DTNB to form mixed disulfide to which the second thiol reacts as shown in Figure 13.6.[9] For example, reduced Fab′ fragment is coupled to ricin A-chain by first reacting with excess Ellman's reagent. After dialysis to remove excess reagent, the activated Fab′ fragment is then incubated with ricin A-chain to generate the desired immunotoxins. If one or both of the components to be coupled is devoid of a free sulfhydryl group, they can be thiolated with 2-iminothiolane (Traut's reagent). For example, pseudomonas toxin (PE) and antibodies were thiolated with 2-iminothiolane. Two moles of thiol groups were introduced per mole of thiolated toxin, which was activated with DTNB.

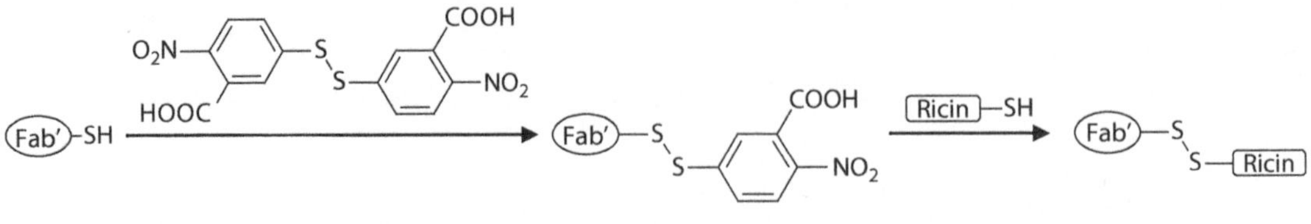

FIGURE 13.6 Coupling reduced Fab' and ricin A-chain with Ellman's reagent, DTNB.

Conjugates were obtained on mixing the activated toxin with the thiolated antibodies.[136,137] PE was similarly linked to EGF.[138] Ricin A-chain immunotoxins were also prepared by thiolating the antibody with either iminothiolane[139] or methyl 3-mercaptopropionimidate[140] followed by reaction with DTNB. The disulfide-activated antibodies were conjugated with ricin A-chain by disulfide interchange. In the preparation of IL-2-gelonin conjugates, McIntyre et al.[141] thiolated both components with 2-iminothiolane, and the thiolated gelonin was activated with DTNB. Thiolated IL-2 then reacted with the activated gelonin to achieve the conjugate.

As mentioned earlier, SMBT can be used to introduce *S*-thiosulfate groups. Although the *S*-thiosulfate protein derivatives can be used directly to conjugate other free thiols by disulfide interchange, Thorpe et al.[125] reduced the antibody-*S*-thiosulfate with DTT and then activated the free sulfhydryl group by Ellman's reagent to form an activated mixed disulfide. The activated disulfide then reacted with free thiols of ricin A-chain to form a ricin A-chain immunotoxin. Anti-Thy-1.1-abrin A-chain conjugate was also prepared this way.[125,142] The presence of a methyl group next to the disulfide bond seems to stabilize the bond as in SMBT mentioned above.[125,143] Arpicco et al.[143] further provided evidence that methyl groups linked to the α-carbon adjacent to the dithio bond (hindered disulfide bond) increased stability of the immunotoxins. They reacted antibodies and gelonin with either ethyl *S*-acetyl-3-mercaptobutyrothioimidate (M-AMPT) or 3-(4-carboxamidophenyldithio)propionthioimidate (CDPT). The M-AMPT-modified antibody was treated with hydroxylamine and activated with DTNB followed by conjugation with M-AMPT-gelonin in the presence of hydroxyamine. CDPT-antibody was similarly conjugated to M-AMPT-derivatized gelonin. The results showed that the dimethyl-substituted disulfide bond exhibited a substantially greater *in vivo* stability than the mono-methyl-substituted conjugate.[144]

13.3.3 CONJUGATION THROUGH THIOETHER LINKAGE

Covalent cross-linking of toxins and proteins through a thioether bond can be achieved by reacting free thiols with maleimides or alkyl halides introduced into the components. The thioether linkage is stable under reducing conditions. Because it is probable that the toxic moiety of the toxin has to be released from the conjugate in order to diffuse into the cells to elicit cytotoxicity effects, the conjugates made without disulfide linkages are usually inactive unless the whole toxin molecules are used. In this case, the cytotoxic A-chain of RIP can still be released *in vivo*.

13.3.3.1 Use of Iodoacetyl Compounds

Iodoacetyl groups can be introduced into toxins by using *N*-hydroxysuccinimidyl iodoacetate or *N*-succinimidyl(4-iodoacetyl)aminobenzoate. The alkyl iodide incorporated can react with thiols introduced into immunoglobulins to produce immunotoxins according to the reaction shown in Figure 13.7A. This procedure has been used to couple CVF,[145] ricin,[70,146] abrin,[70] and DT[70] to several antibodies. The yield of 1:1 antibody:toxin conjugate is usually 20%–40%. The linkage of ricin to antibodies by this method produces impairment of the galactose-binding sites of the B-chain of the toxin in the majority of conjugates formed, reducing the tendency of the conjugates to bind to and kill cells nonspecifically.[146] McIntyre et al.[141] have constructed

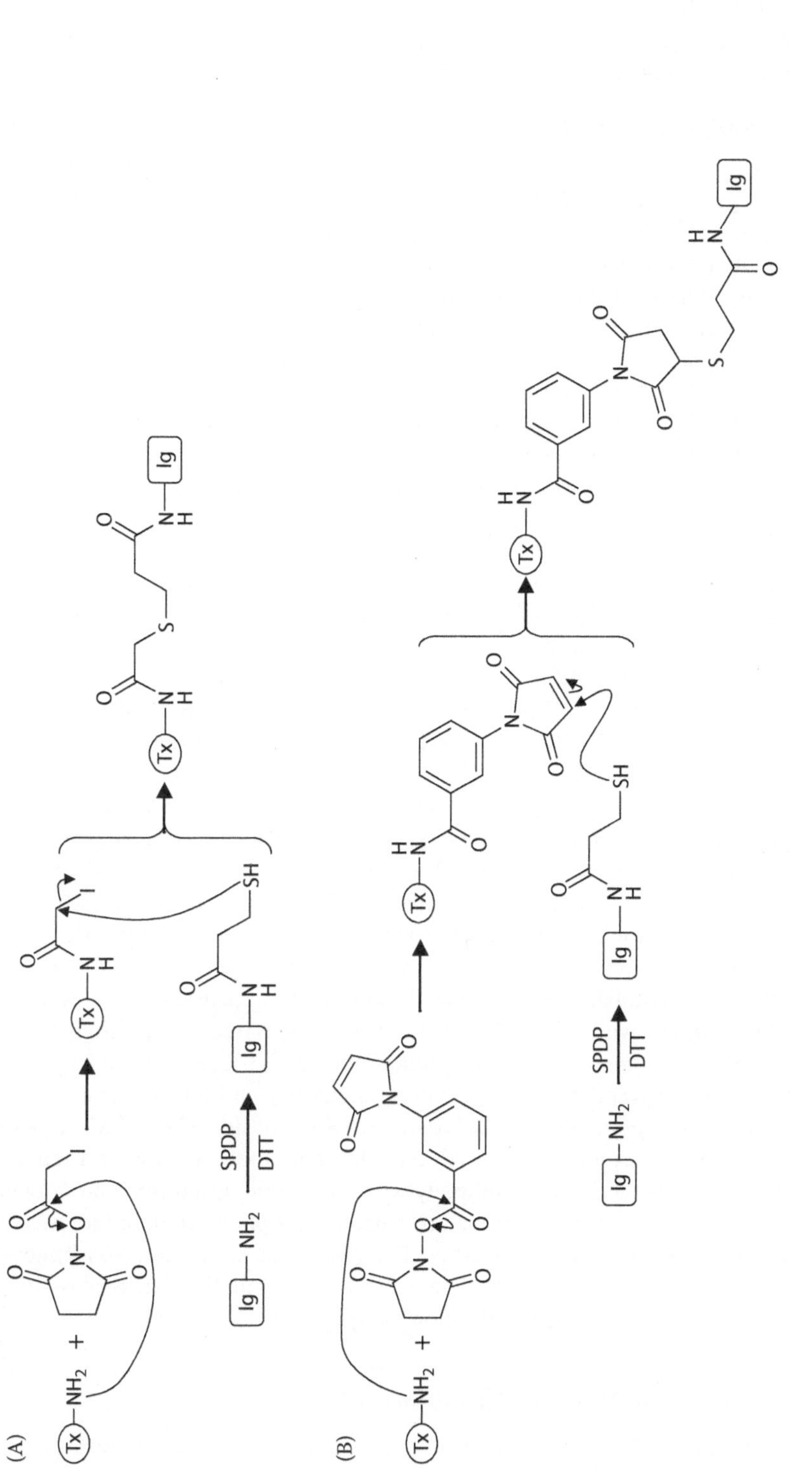

FIGURE 13.7 Coupling of toxin to antibody through a thioether linkage: (A) using iodoacetyl group; (B) using maleimide containing cross-linker, for example, MBS.

IL-2-gelonin conjugate by introducing iodoacetamido groups into IL-2 using *N*-succinimidyl iodoacetate. The conjugate was obtained on reacting the iodoacetamido IL-2 with sulfhydryl groups introduced into gelonin using 2-minothiolane.

13.3.3.2   Use of Amino and Thiol Directed Cross-Linkers

Another method of forming conjugates with the thioether linkage is the reaction of a free thiol with the maleimide group as shown in Figure 13.7B. Maleimido groups can be introduced into proteins containing no free thiols with maleimido-containing cross-linkers. For example, MBS has been used to conjugate antibodies and ricin to prepare ricin immunotoxins.[147] The two-step reaction procedure is generally employed. The amino groups of ricin (true toxin) first react with MBS. After the removal of excess reagent, the derivatized ricin is mixed with reduced or thiolated antibody for the sulfhydryls to react with the maleimido group on ricin to form a thioether bond. Purification of the conjugate is simplified by the presence of the free galactose-binding site on the B chain of intact ricin.[148] Similarly, tetanus toxoid thiolated with SPDP (reduced with DTT after reaction) has been coupled to whole ricin molecule derivatized with MBS.[149] For reduced recombinant ricin A-chain, the antibody reacts with MBS first to provide a maleimide group for the free thiol of ricin to react.[139] Other conjugates such as asialoorosomucoid-diphtheria A conjugate and CVF-Ig immunotoxin are also prepared in this way.[145]

Analogs of MBS have also been used. Sulfo-MBS (*m*-maleimidobenzoyl sulfosuccinimide ester) reportedly increased the yield of antibody T101-ricin conjugate two- to fourfold as compared to MBS. This result is probably due to the increased water solubility of the reagent. Another analog, succinimidyl-4-(*p*-maleimidophenyl)butyrate (SMPB) and its sulfo-analog, sulfosuccinimidyl-4-(*p*-maleimidophenyl)butyrate (sulfo-SMPB) provide an extended chain length and also give four- to sevenfold greater yield than with MBS. Sulfo-SMPB has been used to prepare thioether conjugates of PE-immunotoxin.[136]

Other amino and thiol-directed heterobifunctional reagents that have been used to prepare thioether-linked immunotoxins include succinimidyl 4-(*N*-maleimidomethyl)cyclohexane-1-carboxylate (SMCC), *N*-succinimidyl 4-maleimidobutyrate (GMBS, γ-*N*-maleimidobutyryl *N*-hydroxysuccinimide ester), and (2-nitro-4-sulfonic acid-phenyl)-6-maleimidocaproate (maleimido-6-aminocaproyl-4-hydroxy-3-nitrobenzene sulfonic acid ester [MAHNSA]) (see Chapter 6 for structures and reaction mechanisms). Various conjugates have been constructed with these reagents. Gelonin and PAP were thiolated with 2-iminothiolane and conjugated to antibodies reacted previously with SMCC.[150] Conjugates of D-Lys⁶-GnRH with PAP (GnRH-PAP) were prepared by reacting 2-iminothiolane thiolated D-Lys⁶-GnRH (i.e., SH-GnRH) with maleimidobutyryl-PAP obtained by reacting PAP with sulfo-GMBS.[151] For the preparation of CVF immunotoxins, antibodies were first thiolated with 2-iminothiolane before conjugation with SMCC or MAHNSA-modified CVF.[145] MAHNSA was used to synthesize thioether-linked conjugate of mAb and PE. The antibody was first reacted with the reagent. After purification, the derivatized antibody was mixed with iminothiolane-thiolated PE.[152]

Myers et al.[153] compared the use of various aromatic and aliphatic maleimide cross-linkers for the preparation of anti-CD5-ricin immunotoxins. The results showed that cross-linkers with aromatic moiety such as sulfo-SMPB gave higher yield. However, aliphatic GMBS cross-linker yielded the most toxic immunotoxin in cell-free translation assays.

13.3.4   Conjugation with Activated Chlorambucil

N-succinimidyl chlorambucil (4-[bis(2-chloroethyl)amino]-benzene butanoic acid *N*-hydroxysuccinimide ester) contains functional groups that can cross-link amino groups of proteins. The compound is synthesized by condensation of chlorambucil and *N*-hydroxysuccinimide with dicyclohexyl carbodiimide. The activated ester first reacts with amino groups in immunoglobulins at 4°C at which the *N*-chloroethylamine groups are relative inert. After isolation, the substituted immunoglobulin is mixed at pH 9 with excess toxin at ambient temperature to promote the reaction

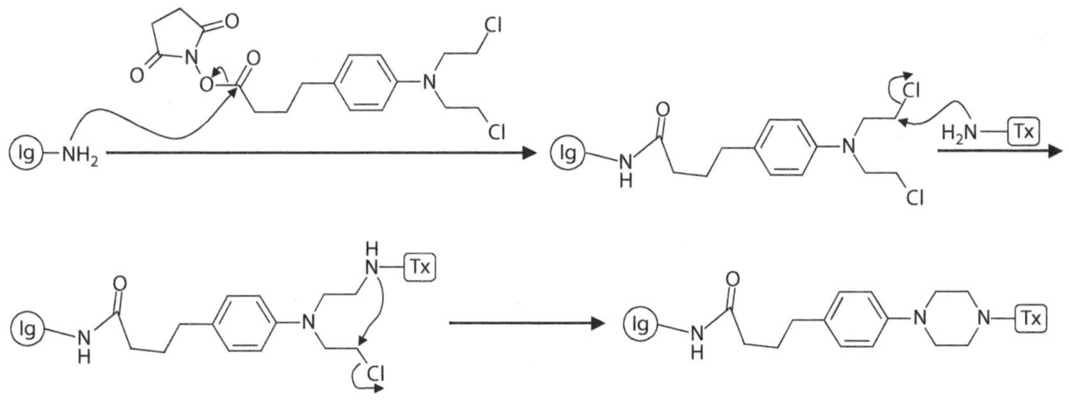

FIGURE 13.8 Conjugation of antibodies and toxins with *N*-hydroxysuccinimidyl chlorambucil.

between amino side chains of the toxin and the bis-2-chloroethylamine group to form a piperazine ring as shown in Figure 13.8.[70,72,154] This method has been used to prepare conjugates of DT,[70,72] abrin,[70,154,155] ricin,[70] and RIP gelonin[70,72] to various antibodies. Six to eight molecules of chlorambucil can be introduced per molecule of immunoglobulin. The overall yield of 1:1 antibody:toxin conjugate is about 5%–10% based on the amount of antibody used. Some antibody–antibody polymers are also formed. According to Edwards et al.[155] disulfide-linked abrin was 10 times more potent than that linked through chlorambucil.

13.3.5 Conjugation with Acid-Labile Cross-Linkers

A maleimide derivative of 2-methylmaleic anhydride has been used to generate an antibody-gelonin conjugate with an acid-labile bond. The principle of the reactions is illustrated in Figure 13.9A.[156] The reaction of the amino group of gelonin with 2-methylmaleic anhydride gives a substituted maleyl derivative whose carboxamide bond is susceptible to hydrolysis under mildly acidic conditions (pH 4–5). The derivative is then coupled to an antibody thiolated with 2-iminothiolane.

Another compound that generates an acid-cleavable amide linkage on reaction is 4-(iodoacetylamino)-3,4,5,6-tetrahydrophthalic anhydride. It has been used to conjugate interleukin 2 (IL-2) to gelonin.[157] The reaction is shown in Figure 13.9B. Although the modification reduced the activity significantly, toxins are released in their native and fully active form by a mild acid treatment.

McIntyre et al.[141] prepared 4-(iodoacetamido)-1-cyclohexene-1,2-dicarboxylic acid anhydride and used it to modify gelonin. The iodoacetamido-modified gelonin was reacted with IL-2, which was previously thiolated with 2-iminothiolane to give acid labile IL-2-gelonin conjugate similar to that shown in Figure 13.9B except that the reaction with gelonin was carried out first.

Cross-linkers containing hydrazone functionalities are also acid-labile. Doronina et al.[158] prepared conjugates of mAbs and the hydrazone of 5-benzoylvaleric acid-AE (auristatin E) ester (AEVB). Because AE is totally synthetic, acid-labile linkers were formed at the C terminus of AE by condensing maleimidocaproyl hydrazide with AEVB ketoester. The mAb were reduced and then alkylated with the maleimido-containing AEVB drug derivative as shown in Figure 13.9C, forming conjugates with about eight drugs attached per mAb.

Carbohydrate conjugates (see below for other carbohydrate conjugates) containing hydrazone bonds are capable of releasing active drug by hydrolysis of the hydrazone bond in the lysosomes where the pH is low. Hamann et al.[159–161] conjugated a series mAb to calicheamicins, a highly potent antitumor antibiotics of the enediyne family. The reactions involved oxidation of the carbohydrate moiety of the mAb with sodium periodate to form carbohydrate-derived aldehydes. Conjugates were

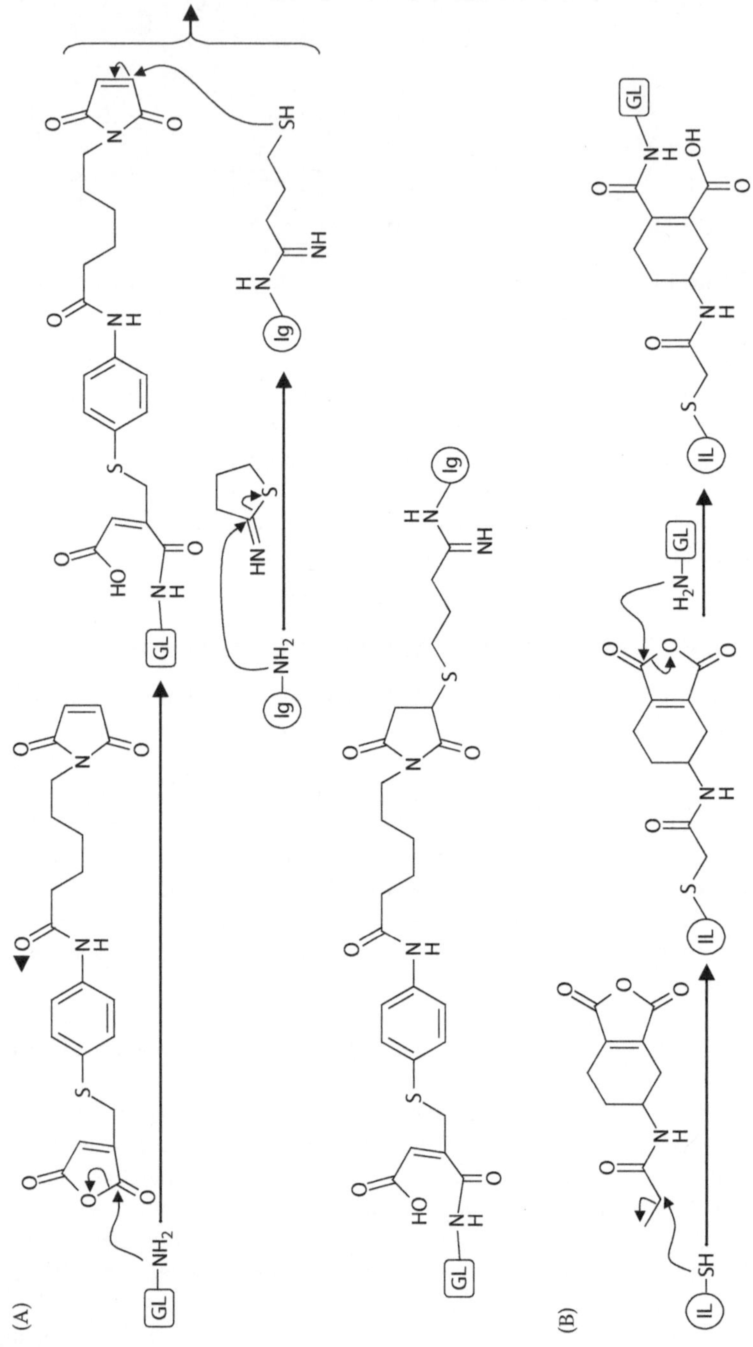
(A)
(B)

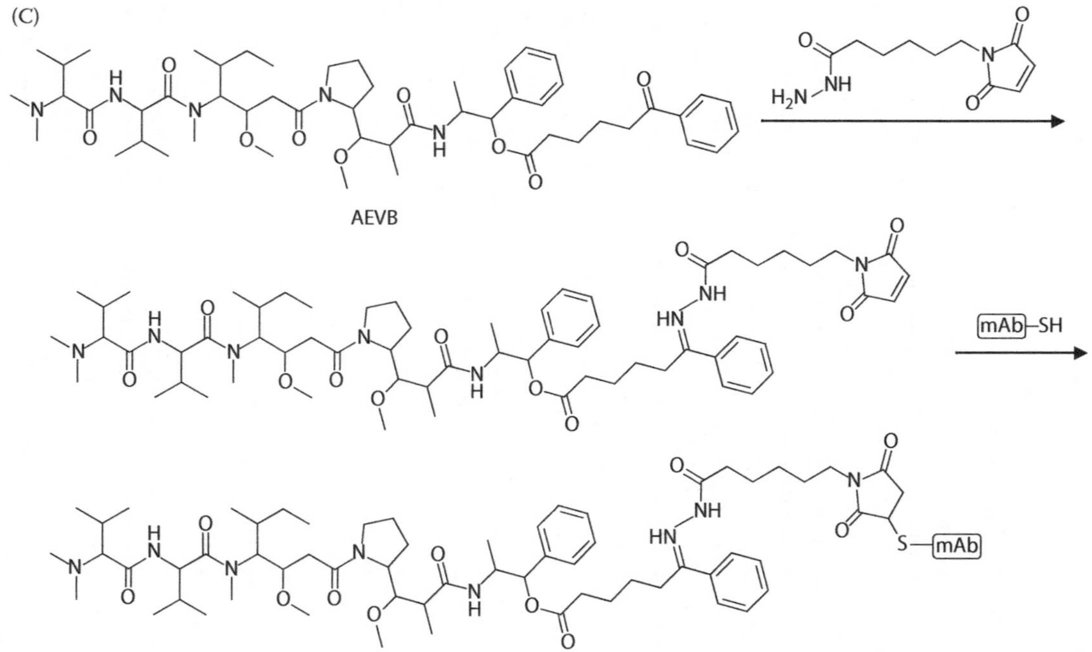

FIGURE 13.9 Conjugation with acid-labile bonds: (A) using methylmaleic anhydride derivative, *N*-[4-[(2,5-dioxo-3-furyl)methylsulfanyl]phenyl]-6-(2,5-dioxopyrrol-1-yl)hexanamide to couple GL and antibody (Ig); (B) using tetrahydrophthalic anhydride to link gelonin and interleukin 2 (IL); (C) using maleimidocaproyl hydrazide to link 5-benzoylvaleric acid-auristatin E ester (AEVB) and mAb.

obtained by attaching hydrazide derivatives of calicheamicins to the oxidized carbohydrates with the formation of acid-labile hydrazone bond. Zara et al.[162] used hydrazide containing heterobifunctional cross-linkers, S-(2-thiopyridyl)-L-cysteine hydrazide (TPCH) and S-(2-thiopyridyl)mercaptopropionohydrazide (TPMPH) to conjugate an antibody with ricin A-chain and barley toxin. Human monoclonal IgM antibody 16–88 was treated with periodate to generate dialdehydes, which reacted with the hydrazide moiety of TPCH and TPMPH. Ricin A-chain with its free sulfhydryl group was then coupled to the antibody by reacting with the dithiopyridyl moiety. For barley toxin, it was thiolated with SPDP and DTT before displacing thiopyridine from the TPCH and TPMPH-modified antibody to form the antibody–barley toxin conjugate.

13.3.6   CONJUGATION WITH PHOTOCLEAVABLE CROSS-LINKERS

Senter et al.[163] have prepared two photolabile heterobifunctional cross-linking reagents for the conjugation of antibody to PAP. One of these, [4-nitro-3-(1-chlorocarbonyloxyethyl)phenyl]methyl-S-acetylthioic acid ester, containing a protected sulfhydryl group, is first reacted with the amino group of PAP. The sulfhydryl group is then deprotected with hydroxylamine and reacted with an antibody that has been modified with a maleimide group with SMCC as shown in Figure 13.10A. The second compound, [4-nitro-3-(1-chlorocarbonyloxyethyl)phenyl]methyl-3-(2-pyridyldithiopropionic acid) ester, is similarly reacted as shown in Figure 13.10B. Antibody-PAP conjugates were predominantly in the ratio of 1:1. The benzylic carbon–oxygen bond is photosensitive. On irradiation at 365 nm, fully active PAP is released after spontaneous liberation of CO_2.

Other photocleavable conjugates have also been constructed. Goldmacher et al.[164] conjugated PAP with mAb 5E9 directed against the human transferrin-receptor. PAP first reacted with the chloroformate of S-[4-nitro-3-(1-hydroxyethyl)phenyl]methyl thioacetate followed by hydroxylamine to generate a free sulfhydryl group. The modified PAP was then mixed with 5E9 antibody, which had been functionalized with SMCC to introduce an average of 1.5 maleimido groups per antibody molecule. Reaction of PAP-thiol with the maleimide group provided the PAP-5E9 mAb conjugate. To conjugate PAP to ricin B-chain, PAP was modified with a chloroformate derivative of 1-[5-(N-maleimidomethyl)-2-nitrophenyl]ethanol. The free thiol of ricin B-chain then reacted with the maleimido group of PAP to generate PAP-ricin B-chain conjugate. When irradiated with 350 nm light, the linker of these conjugates undergoes photolytic degradation, resulting in the release of native toxin that is fully functional.

13.3.7   COUPLING THROUGH CARBOHYDRATE RESIDUES

13.3.7.1   Use of Intrinsic Carbohydrate Moieties

The carbohydrate moieties of antibodies and toxins are potentially valuable sites for cross-linking. As mentioned in Section 13.3.5, periodate oxidized carbohydrate moieties of antibodies can form hydrazone bonds with hydrazide-containing molecules. In addition to antibodies, many toxins such as ricin, CVF, and abrin as well as targeting agents including α_2-macroglobulin are glycoproteins. Olsnes and Pihl[165] have prepared an abrin immunotoxin using the oligosaccharide side chain of the toxin. The amino groups in abrin were first blocked with formaldehyde and $NaBH_4$ before reacting with periodate. After oxidation, the oligosaccharides of the abrin B chain generated aldehyde groups, which form Schiff bases with free amino groups on the antibody. Conjugates formed were stabilized with $NaBH_4$. Intact ricin was also conjugated to the acetylcholine receptor after periodate oxidation of the carbohydrate moiety.[166]

Similar procedures have been used to conjugate daunomycin and ouabain to melanotropin and thyrotropin.[167] The sugar residues in these drugs were first oxidized by periodate to generate the aldehydes, which formed Schiff bases with the amino groups of peptide hormones. The Schiff bases were reduced with $NaBH_4$.

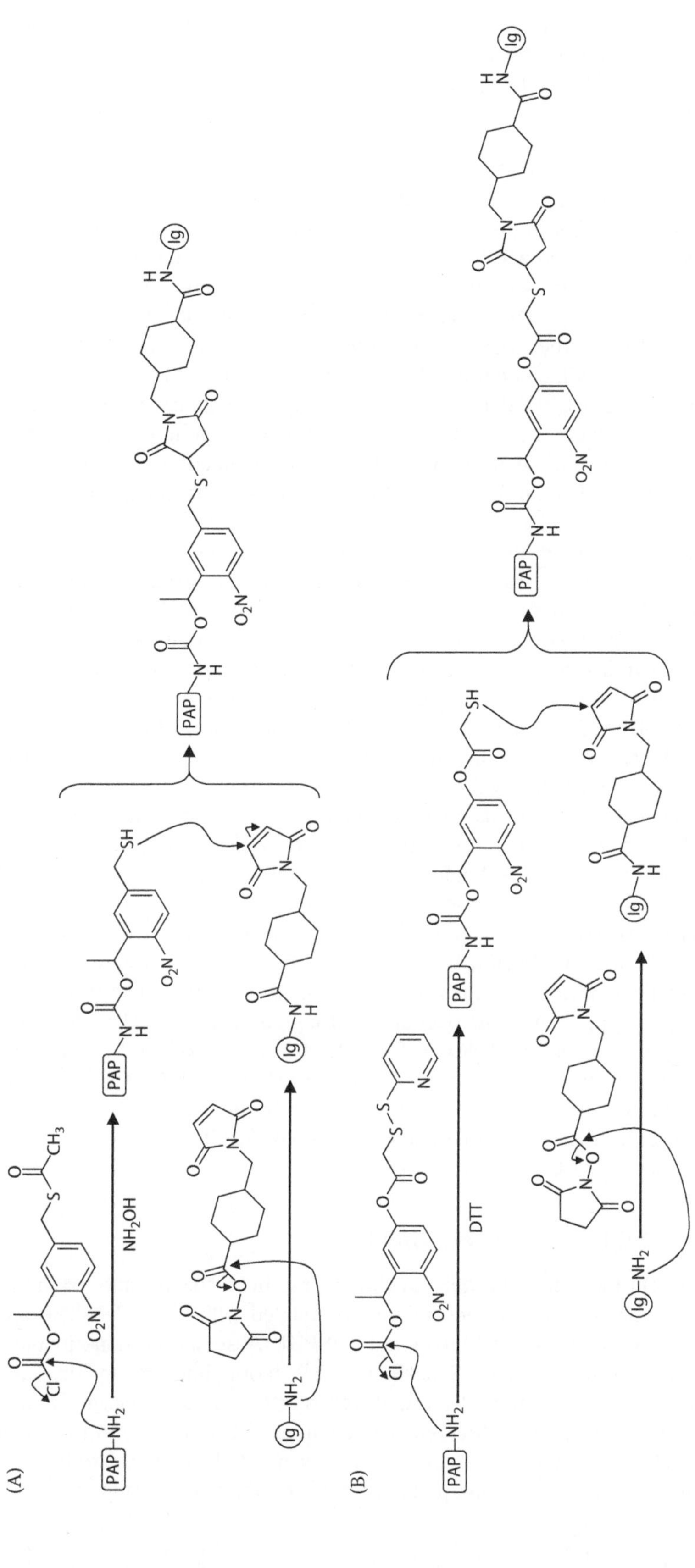

FIGURE 13.10 Conjugating antibody and PAP with photocleavable cross-linkers: (A) use of [4-nitro-3-(1-chlorocarbonyloxyethyl)phenyl]methyl-S-acetylthioic acid ester; (B) use of [4-nitro-3-(1-chlorocarbonyloxyethyl)phenyl]methyl-3-(2-pyridyldithiopropionic acid) ester.

In a different approach, Susuki et al.[168] activated the glycoside moiety of antibodies before the binding of drugs using cyanogen bromide, which reacts with vicinal diols to form a cyclic reactive imidocarbonate. In this way, mitomycin C, daunomycin, and adriamycin have been linked to anti-α-fetoprotein antibodies.

13.3.7.2   Use of Polysaccharide Spacers

Carbohydrates, such as dextran, have been used as a spacer in linking drugs and antibodies. Hurwitz et al.[169–172] prepared a series of antibody-drug/toxin conjugates via a dextran spacer. Dextrans of 10,000 and 40,000 Da were oxidized by sodium periodate. The oxidized dextran was mixed with toxin/drug overnight and finally with the mAbs for 24 h at the same temperature. The conjugates were isolated by gel filtration. Drugs such as cytosine arabinoside,[169] DOX,[170] daunomycin,[171] and adriamycin[172] and antibodies were bound through their amino groups to the aldehyde functions generated from the oxidized dextran. The Schiff bases thus formed were reduced by sodium borohydride or cyanoborohydride, depending on the drug's sensitivity to the reducing agent. The arabinoside–dextran conjugate was completely reduced, daunomycin–dextran conjugate was partially reduced, and adriamycin–dextran conjugate was not reduced at all, but they were stable in phosphate-buffered saline. The antibody bound to periodate-oxidized dextran was stable without reduction.[173]

Another derivative of dextran that has been employed is carboxymethyldextran (CM-dextran), which is obtained by reacting dextran with monochloroacetic acid. Condensation of hydrazine with CM-dextran effected by a carbodiimide results in the formation of CM-dextran-hydrazide, which reacts with the carbonyl group of daunomycinone moiety of either daunomycin or adriamycin.[174,175] The antibody is cross-linked to the CM-dextran-drug conjugate with glutaraldehyde to yield drug-CM-dextran-antibody conjugate. The same CM-dextran-hydrazide is used for the binding of 5-fluorouridine derivatives. Periodate oxidation of the ribose moiety of these compounds gave two aldehyde groups, which in turn reacted with CM-dextran-hydrazide. The antibody was then cross-linked to the spacer by glutaraldehyde.[173]

CM-dextran itself has been used as a carrier because of its mucoadhesive properties by forming covalent bonds with the mucus layer, and attempts have been made to improve its properties by thiolation as mentioned earlier.[58] Analogues of camptothecin (CPT), particularly T-2513, with antitumor activity by inhibiting topoisomerase I was conjugated to CM-dextran through a triglycine spacer.[176] Triglycine was linked to CM-dextran and T-2513 through amide bond formation caused by water soluble carbodiimide. Paclitaxel, an antimicrotubule agent, was bound via its hydroxyl group to carboxymethyldextran by means of a gly-gly-phe-gly linker. The linker was introduced into the 29- or 7-hydroxyl group of the paclitaxel through an ester bond.[177] CM-dextran has been further polyhydroxylated to form CM-dextran polyalcohol (CM-dex-PA) as a water-soluble carrier, which has structural flexibility and similarity to polyethylene glycol. Effects of peptide spacers for the design of CM-dex-PA-peptide spacer-drug conjugates are systematically studied.[178] To that end, DOX hydrochloride (DXR) was linked to CM-Dex-PA via Gly-Gly-Phe-Gly (GGFG) peptidyl spacer to form (CM-Dex-PA)-GGFG-DXR conjugate.[179]

13.3.8   CONJUGATION USING AVIDIN–BIOTIN LINKAGE

The strong noncovalent interaction between avidin and biotin has been exploited in the preparation of immunotoxins.[180] For example, antibodies are biotinylated with biotinyl-N-hydroxysuccinimide ester, and toxin ricin A-chain is coupled to avidin by SPDP. Association of biotin and avidin forms the immunotoxin. In a similar manner, Schechter et al.[181] biotinylated monoclonal antibody (mAb 108) against the extracellular domain of the EGF receptor on human epidermoid carcinoma (KB) cells and prepared a CM-dextran-avidin conjugate complexed with cis-diamminedichloroplatinum (II) (cis-Pt). The KB cells were first treated with biotinylated mAb 108 followed with cis-Pt-CM-dextran-avidin. The treatment was potentially more effective in suppressing the growth of established KB tumor xenografts or in inhibiting the development of lung metastases in nude mice than free MAb 108, free drug, or MAb 108 followed by drug. In a slightly different fashion, Schechter et al.[182] biotinylated both mAb 108 and ricin. Streptavidin was then complexed with biotinylated ricin, which was targeted to KB cells via the biotinylated monoclonal antibody 108.

An avidin–biotin system has also been designed for pretargeting immunotherapy. Moro et al.[183] described a three-step antibody/avidin-targeting approach to increase the local concentration and the persistence of biotinylated human tumor necrosis factor α (bio-TNF) on a mouse tumor, which expressed tumor-associated antigen, RMA-Thy 1.1. Anti-Thy 1.1 mAb 19E12 was biotinylated with sulfo-NHS-LC-biotin (bio-19E12) and human TNF with D-biotinyl-6-aminocaproic acid N-hydroxysuccinimide. In pretargeting, RMA-Thy 1.1 cells were treated with bio-19E12 and avidin. Bio-TNF was then added to bind to the avidin. The approach increased the amount of bio-TNF bound to the cell.

13.3.9 Conjugation Using Enzymes

Enzymes have been used to prepare polymer–drug, polymer–protein, and protein–drug conjugates. Of particular importance are the transglutaminases. Buxman[184] used epidermal transglutaminase to cross-link isoniazid and hydralazine to serum and cell nuclear proteins. These drug–albumin and drug–histone conjugates were particularly effective in eliciting drug-specific antibodies.

Besheer et al.[185] have proved that transglutaminase is useful for the preparation of polymer–protein conjugates for therapeutic proteins. Hydroxyethyl starch (HES) was investigated as a substitute for the popular PEG. HES is a semisynthetic, water soluble, biodegradable polymer. When it was modified with N-carbobenzyloxy glutaminyl glycine (Z-QG) and hexamethylene diamine (HMDA), the modified HES acted as acyl donor and acyl acceptor, respectively, in the microbial transglutaminase-catalyzed coupling reaction. The microbial transglutaminase was able to link HES-Z-GQ with monodansyl cadaverine as an acyl donor and HES-HMDA with dimethylcasein as acyl acceptor. The successful conjugation of proteins to HES by transglutaminase provides a competitive method to the well-established PEGylation.

13.3.10 Conjugation Using Solid-Phase Procedures

Solid-phase conjugation of agents to carriers has the advantage that the products are easier to purify than in the case of solution conjugation. Using solid supports, the unreacted reagents may be removed by simple washing. The use of a solid support may also help to avoid problems arising from insoluble reactants and opens a way to automate the conjugation procedure. The solid-phase synthesis of oligonucleotides conjugated with lipids, polyamines, peptides, carbohydrates, PEG, and some small molecules useful for delivery and targeting of potential nucleic acid therapeutics has been reviewed.[186]

Immunoconjugates have also been prepared by solid-phase methodology. Moroney et al.[187] prepared a lactose-blocked ricin–mAb conjugate through a series of reactions involving an affinity column. Lactose was introduced into a free sulfhydryl group with cystamine by reductive alkylation followed by reduction with DTT. The lactose derivative was bound to pyridyldithio-activated polyacrylamide beads through the disulfide interchange reaction. The ligand was activated with the bifunctional cross-linking reagent, 2,4-dichloro-6-methoxytriazine. On binding to the lactose moiety of the affinity beads, ricin was covalently bound to the cross-linker. The lactose ricin complex was released from the solid support by DTT. The free sulfhydryl group of the complex was finally reacted with the maleimide group introduced into mAb by SMCC. The conjugate was purified on a series of affinity chromatography and was 10-fold less toxic than the native ricin. Similarly, Dosio et al.[188] prepared immunotoxins using ribosome-inactivating protein anchored on an affinity gel derivatized with triazinic dye. The adsorbed toxins were activated with 2-imino-thiolane and then conjugated to mAb activated by SPDP.

13.3.11 CONJUGATION WITH GLUTARALDEHYDE AND CARBODIIMIDES

One-step direct coupling of toxins or drugs to antibodies and other proteins can be carried out with either glutaraldehyde or water-soluble carbodiimide, particularly EDC and 1-cyclohexyl-3-[2-morpholinyl-4-ethyl]carbodiimide (CMC). The popularity of these cross-linker in conjugate preparation may be realized from the citations mentioned throughout this book. The following are some additional examples of the use of these reagents. DOX,[189] and daunorubicin[190] were coupled to mAbs by glutaraldehyde. DOX,[189] and daunomycin[191] were also conjugated to mAb with EDC and/or CMC. Similarly, mitomycin C (MMC) was linked to glutarylated bovine serum albumin (G-BSA) in the presence of EDC.[192] MMC was liberated from MMC-G-BSA when suspended in a phosphate buffer. The conjugate was employed as a sustained drug release delivery system. EDC and CMC were used to conjugate amatoxins and phallotoxins to various proteins and poly(amino acids) by incubating the components at room temperature for 24 h.[193] Likewise, methotrexate was coupled to IgG with EDC.[194] Clearly these cross-linkers are useful in some instances, but their applications in the preparation of immunoconjugates remain limited.

REFERENCES

1. Pastan, I., Hassan, R., FitzGerald, D. J., and Kreitman, R. J., Immunotoxin treatment of cancer, *Annu. Rev. Med.*, 58, 221, 2007.
2. Shapira, S., Lisiansky, V., Arber, N., and Kraus, S., Targeted immunotherapy for colorectal cancer: Monoclonal antibodies and immunotoxins, *Expert Opin. Investig. Drugs*, 19, Suppl. 1, S67, 2010.
3. Nelson, A. L., Antibody fragments: Hope and hype, *MAbs*, 2, 77, 2010.
4. Teicher, B. A., Antibody-drug conjugate targets, *Curr. Cancer Drug Targets*, 9, 982, 2009.
5. Hall, W. A. (ed.), Immunotoxin methods and protocols, in *Methods in Molecular Biology*, Vol. 166, Humana Press Inc., Totowa, NJ, 2001.
6. Goldenberg, D. M. and Sharkey, R. M., Novel radiolabeled antibody conjugates, *Oncogene*, 26, 3734, 2007.
7. Tanemura, M., Ogawa, H., Yin, D. P., Chen, Z. C., DiSesa, V. J., and Galili, U., Elimination of anti-Gal B cells by alpha-Gal ricin1, *Transplantation*, 73, 1859, 2002.
8. Chari, R. V., Targeted cancer therapy: Conferring specificity to cytotoxic drugs, *Acc. Chem. Res.*, 41, 98, 2008.
9. Ghetie, V. and Vitetta, E. S., Chemical construction of immunotoxins, *Mol. Biotechnol.*, 18, 251, 2001.
10. Elias, D. J., Hirschowitz, L., Kline, L. E., Kroener, J. F., Dillman, R. O., Walker, L. E., Robb, J. A., and Timms, R. M., Phase I clinical comparative study of monoclonal antibody KS1/4 and KS1/4-methotrexate immunoconjugate in patients with non-small cell lung carcinoma, *Cancer Res.*, 50, 4154, 1990.
11. Ajani, J. A., Kelsen, D. P., and Haller, D. A., Multi-institutional phase II study of BMS-182248-01 (BR96-doxorubicin conjugate) administered every 21 days in patients with advanced gastric adenocarcinoma, *Cancer J.*, 6, 78, 2000.
12. Tolcher, A. W., Sugarman, S., Gelman, K. A., Cohen, R., Saleh, M., Isaacs, C., Young, L., Healey, D., Onetto, N., and Slichenmeyer, W., Randomized phase II study of BR96–doxorubicin conjugate in patients with metastatic breast cancer, *J. Clin. Oncol.*, 17, 478, 1999.
13. Macedo, M. F. and de Sousa, M., Transferrin and the transferrin receptor: Of magic bullets and other concerns, *Inflamm. Allergy Drug Targets*, 7, 41, 2008.
14. Chen, S., Zhao, X., Chen, J., Chen, J., Kuznetsova, L., Wong, S. S., and Ojima, I., Mechanism-based tumor-targeting drug delivery system. Validation of efficient vitamin receptor-mediated endocytosis and drug release, *Bioconjug. Chem.*, 21, 979, 2010.
15. Leamon, C. P., Folate-targeted drug strategies for the treatment of cancer, *Curr. Opin. Investig. Drugs*, 9, 1277, 2008.
16. Gupta, Y., Kohli, D. V., and Jain, S. K., Vitamin B12-mediated transport: A potential tool for tumor targeting of antineoplastic drugs and imaging agents, *Crit. Rev. Ther. Drug Carrier Syst.*, 25, 347, 2008.
17. Tomei, S., Yuasa, H., Inoue, K., and Watanabe, J., Transport functions of riboflavin carriers in the rat small intestine and colon: Site difference and effects of tricyclic-type drugs, *Drug Deliv.*, 8, 119, 2001.
18. Daniels, T. R., Delgado, T., Helguera, G., and Penichet, M. L., The transferrin receptor part II: Targeted delivery of therapeutic agents into cancer cells, *Clin. Immunol.*, 121, 159, 2006.

19. Singh, M., Atwal, H., and Micetich, R., Transferrin directed delivery of adriamycin to human cells, *Anticancer Res.*, 18, 1423, 1998.

20. Hoshino, T., Misaki, M., Yamamoto, M., Shimizu, H., Ogawa, Y., and Toguchi, H., In vitro cytotoxicities and in vivo distribution of transferrin-platinum(II) complex, *J. Pharm. Sci.*, 84, 216, 1995.

21. Beyer, U., Roth, T., Schumacher, P., Maier, G., Unold, A., Frahm, A. W., Fiebig, H. H., Unger, C., and Kratz, F., Synthesis and in vitro efficacy of transferrin conjugates of the anticancer drug chlorambucil, *J. Med. Chem.*, 41, 2701, 1998.

22. Kratz, F., Beyer, U., Roth, T., Tarasova, N., Collery, P., Lechenault, F., Cazabat, A., Schumacher, P., Unger, C., and Falken, U., Transferrin conjugates of doxorubicin: Synthesis, characterization, cellular uptake, and in vitro efficacy, *J. Pharm. Sci.*, 87, 338, 1998.

23. Kuznetsova, L., Chen, J., Sun, L., Wu, X., Pepe, A., Veith, J. M., Pera, P., Bernacki, R. J., and Ojima, I., Syntheses and evaluation of novel fatty acid-second-generation taxoid conjugates as promising anticancer agents, *Bioorg. Med. Chem. Lett.*, 16, 974, 2006.

24. Ziegler, J., Hyaluronan seeps into cancer treatment trials, *J. Natl. Cancer Inst.*, 88, 397, 1996.

25. Akima, K., Ito, H., Iwata, Y., Matsuo, K., Watari, N., Yanagi, M., Hagi, H. et al., Evaluation of antitumor activities of hyaluronate binding antitumor drugs: Synthesis, characterization and antitumor activity, *J. Drug Target.*, 4, 1, 1996.

26. Luo, Y. and Prestwich, G. D., Synthesis and selective cytotoxicity of a hyaluronic acid-antitumor bioconjugate, *Bioconjug. Chem.*, 10, 755, 1999.

27. Saravanakumar, G., Choi, K. Y., Yoon, H. Y., Kim, K., Park, J. H., Kwon, I. C., and Park, K., Hydrotropic hyaluronic acid conjugates: Synthesis, characterization, and implications as a carrier of paclitaxel, *Int. J. Pharm.*, 394,154, 2010.

28. Luo, Y., Bernshaw, N. J., Lu, Z.-R., Kopecek, J., and Prestwich, G. D., Targeted delivery of doxorubicin by HPMA copolymer-hyaluronan bioconjugates, *Pharm. Res.*, 19, 396, 2002.

29. Gabor, F., Bogner, E., Weissenboeck, A., and Wirth, M., The lectin-cell interaction and its implications to intestinal lectin-mediated drug delivery, *Adv. Drug Deliv. Rev.*, 56, 459, 2004.

30. Gabor, F., Schwarzbauer, A., and Wirth, M., Lectin-mediated drug delivery: Binding and uptake of BSA-WGA conjugates using the Caco-2 model, *Int. J. Pharm.*, 237, 227, 2002.

31. Anande, N. M., Jain, S. K., and Jain, N. K., Con-A conjugated mucoadhesive microspheres for the colonic delivery of diloxanide furoate, *Int. J. Pharm.*, 359, 182, 2008.

32. Park, K., Lee, G. Y., Kim, Y. S., Yu, M., Park, R. W., Kim, I. S., Kim, S. Y., and Byun, Y., Heparin-deoxycholic acid chemical conjugate as an anticancer drug carrier and its antitumor activity, *J. Control. Release*, 114, 300, 2006.

33. Park, K., Lee, G. Y., Park, R. W., Kim, I. S., Kim, S. Y., and Byun, Y., Combination therapy of heparin-deoxycholic acid conjugate and doxorubicin against squamous cell carcinoma and B16F10 melanoma, *Pharm. Res.*, 25, 268, 2008.

34. Wang, Y., Xin, D., Liu, K., and Xiang, J., Heparin-paclitaxel conjugates using mixed anhydride as intermediate: Synthesis, influence of polymer structure on drug release, anticoagulant activity and in vitro efficiency, *Pharm. Res.*, 26, 785, 2009.

35. Krauland, A. H., Leitner, V. M., and Bernkop-Schnürch, A., Improvement in the in situ gelling properties of deacetylated gellan gum by the immobilization of thiol groups, *J. Pharm. Sci.*, 92, 1234, 2003.

36. Ezzat, K., El Andaloussi, S., Abdo, R., and Langel, U., Peptide-based matrices as drug delivery vehicles, *Curr. Pharm. Des.*, 16, 1167, 2010.

37. Brown, K. C., Peptidic tumor targeting agents: The road from phage display peptide selections to clinical applications, *Curr. Pharm. Des.*, 16, 1040, 2010.

38. Bitler, B. G. and Schroeder, J. A., Anti-cancer therapies that utilize cell penetrating peptides, *Recent Pat. Anticancer Drug Discov.*, 5, 99, 2010.

39. Nagy, A. and Schally, A. V., Targeting cytotoxic conjugates of somatostatin, luteinizing hormone-releasing hormone and bombesin to cancers expressing their receptors: A "smarter" chemotherapy, *Curr. Pharm.*, 11, 1167, 2005.

40. Szepeshazi, K., Schally, A. V., Treszl, A., Seitz, S., and Halmos, G., Therapy of experimental hepatic cancers with cytotoxic peptide analogs targeted to receptors for luteinizing hormone-releasing hormone, somatostatin or bombesin, *Anticancer Drugs*, 19, 349, 2008.

41. Stangelberger, A., Schally, A. V., Nagy, A., Szepeshazi, K., Kanashiro, C. A., and Halmos, G., Inhibition of human experimental prostate cancers by a targeted cytotoxic luteinizing hormone-releasing hormone analog AN-207, *Prostate*, 66, 200, 2006.

42. Mayer, G., The chemical biology of aptamers, *Angew. Chem. Int. Ed. Engl.*, 48, 2672, 2009.

43. Barbas, A. S. and White, R. R., The development and testing of aptamers for cancer, *Curr. Opin. Investig. Drugs*, 10, 572, 2009.

44. Farokhzad, O. C., Jon, S., Khademhosseini, A., Tran, T. N., Lavan, D. A., and Langer, R., Nanoparticle-aptamer bioconjugates: A new approach for targeting prostate cancer cells, *Cancer Res.*, 64, 7668, 2004.

45. Bagalkot, V., Farokhzad, O. C., Langer, R., and Jon, S., An aptamer-doxorubicin physical conjugate as a novel targeted drug-delivery platform, *Angew. Chem. Int. Ed. Engl.*, 45, 8149, 2006.

46. Chu, T. C., Marks, J. W., 3rd, Lavery, L. A., Faulkner, S., Rosenblum, M. G., Ellington, A. D., and Levy, M., Aptamer: Toxin conjugates that specifically target prostate tumor cells, *Cancer Res.*, 66, 5989, 2006.

47. Cerchia, L., Giangrande, P. H., McNamara, J. O., and de Franciscis, V., Cell-specific aptamers for targeted therapies, *Methods Mol. Biol.*, 535, 59, 2009.

48. Chu, T. C., Twu, K. Y., Ellington, A. D., and Levy, M., Aptamer mediated siRNA delivery, *Nucleic Acids Res.*, 34, e73, 2006.

49. Zhou, J. and Rossi, J. J., Aptamer-targeted cell-specific RNA interference, *Silence*, 1, 4, 2010.

50. Edinger, D. and Wagner, E., Bioresponsive polymers for the delivery of therapeutic nucleic acids, *Wiley Interdiscip. Rev. Nanomed. Nanobiotechnol.*, 3, 33, 2011.

51. Basavaraj, K. H., Johnsy, G., Navya, M. A., Rashmi, R., and Siddaramaiah, Biopolymers as transdermal drug delivery systems in dermatology therapy, *Crit. Rev. Ther. Drug Carrier Syst.*, 27, 155, 2010.

52. Baldwin, A. D. and Kiick, K. L., Polysaccharide-modified synthetic polymeric biomaterials, *Biopolymers*, 94, 128, 2010.

53. Jia, Z., Wong, L., Davis, T. P., and Bulmus, V., One-pot conversion of RAFT-generated multifunctional block copolymers of HPMA to doxorubicin conjugated acid- and reductant-sensitive crosslinked micelles, *Biomacromolecules*, 9, 3106, 2008.

54. Shiah, J. G., Dvorák, M., Kopecková, P., Sun, Y., Peterson, C. M., and Kopecek, J., Biodistribution and antitumour efficacy of long-circulating N-(2-hydroxypropyl)methacrylamide copolymer-doxorubicin conjugates in nude mice, *Eur. J. Cancer*, 37, 131, 2001.

55. Gehrke, S. H., Synthesis and properties of hydrogels used for drug delivery, *Drugs Pharm. Sci.*, 102, 473, 2000.

56. Hennink, W. E. and van Nostrum, C. F., Novel crosslinking methods to design hydrogels, *Adv. Drug Deliv. Rev.*, 54, 13, 2002.

57. Bernkop-Schnürch, A., Krauland, A. H., Leitner, V. M., and Palmberger, T., Thiomers: Potential excipients for non-invasive peptide delivery systems, *Eur. J. Pharm. Biopharm.*, 58, 253, 2004.

58. Shahnaz, G., Perera, G., Sakloetsakun, D., Rahmat, D., and Bernkop-Schnürch, A., Synthesis, characterization, mucoadhesion and biocompatibility of thiolated carboxymethyl dextran-cysteine conjugate, *J. Control. Release*, 144, 32, 2010.

59. Albrecht, K. and Bernkop-Schnürch, A., Thiomers: Forms, functions and applications to nanomedicine, *Nanomed. (Lond.)*, 2, 41, 2007.

60. Bernkop-Schnürch, A., Clausen, A. E., and Hnatyszyn, M., Thiolated polymers: Synthesis and in vitro evaluation of polymer-cysteamine conjugates, *Int. J. Pharm.*, 226, 185, 2001.

61. Marschütz, M. K. and Bernkop-Schnürch, A., Thiolated polymers: Self-crosslinking properties of thiolated 450 kDa poly(acrylic acid) and their influence on mucoadhesion, *Eur. J. Pharm. Sci.*, 15, 387, 2002.

62. Frankel, A. E., Powell, B. L., and Lilly, M. B., Diphtheria toxin conjugate therapy of cancer, *Cancer Chemother. Biol. Response Modif.*, 20, 301, 2002.

63. Mathew, M. and Verma, R. S., Humanized immunotoxins: A new generation of immunotoxins for targeted cancer therapy, *Cancer Sci.*, 100, 1359, 2009.

64. Olsnes, S., The history of ricin, abrin and related toxins, *Toxicon*, 44, 361, 2004.

65. Parikh, B. A. and Tumer, N. E., Antiviral activity of ribosome inactivating proteins in medicine, *Mini Rev. Med. Chem.*, 4, 523, 2004.

66. Vogel, C. W. and Fritzinger, D. C., Humanized cobra venom factor: Experimental therapeutics for targeted complement activation and complement depletion, *Curr. Pharm. Des.*, 3, 2916, 2007.

67. Wolf, P. and Elsässer-Beile, U., *Pseudomonas* exotoxin A: From virulence factor to anti-cancer agent, *Int. J. Med. Microbiol.*, 299, 161, 2009.

68. Ng, T. B., Wong, J. H., and Wang, H., Recent progress in research on ribosome inactivating proteins, *Curr. Protein Pept. Sci.*, 11, 37, 2010.

69. Jansen, F. K., Bourrie, B., Casellas, P., Dussossoy, D., Gros, O., Vic, P., Vidal, H., and Gros, P., Toxin selection and modification: Utilization of the A chain of ricin, *Cancer Treat. Res.*, 37, 97, 1988.

70. Cumber, A. J., Forrester, J. A., Foxwell, B. M., Ross, W. C., and Thorpe, P. E., Preparation of antibody–toxin conjugates, *Methods Enzymol.*, 112, 207, 1985.

71. Houston, L. L., Ramakrishnan, S., and Hermodson, M. A., Seasonal variations in different forms of poke-weed antiviral protein, a potent inactivator of ribosomes, *J. Biol. Chem.*, 258, 9601, 1983.

72. Thorpe, P. E. and Ross, W. C., The preparation and cytotoxic properties of antibody–toxin conjugates, *Immunol. Rev.*, 62, 119, 1982.

73. Blakey, D. C., Wawrzynczak, E. J., Stirpe, F., and Thorpe, P. E., Anti-tumour activity of a panel of anti-Thy 1.1 immunotoxins made with different ribosome-inactivating proteins, in *Membrane-Mediated Cytotoxicity*, Bonavida, B. and Collier, R. J. (eds.), *UCLA Symposium on Molecular and Cellular Biology*, New Series, Alan R. Liss, New York, Vol. 45, 1986, p. 195.

74. Woo, B. H., Lee, J. T., Park, M. O., Lee, K. R., Han, J. W., Park, E. S., Yoo, S. D., and Lee, K. C., Stability and cytotoxicity of Fab-ricin A immunotoxins prepared with water soluble long chain heterobifunctional crosslinking agents., *Arch. Pharm. Res.*, 22, 459, 1999.

75. Issell, B. F. and Crooke, S. T., Maytansine, *Cancer Treat. Rev.*, 5, 199, 1978.

76. Banerjee, S., Wang, Z., Mohammad, M., Sarkar, F. H., and Mohammad, R. M., Efficacy of selected natural products as therapeutic agents against cancer, *J. Nat. Prod.*, 71, 492, 2008.

77. Foyil, K. V. and Bartlett, N. L., Anti-CD30 Antibodies for Hodgkin Lymphoma, *Curr. Hematol. Malig. Rep.*, 5, 140, 2010.

78. Naumovski, L. and Junutula, J. R., Glembatumumab vedotin, a conjugate of an anti-glycoprotein non-metastatic melanoma protein B mAb and monomethyl auristatin E for the treatment of melanoma and breast cancer, *Curr. Opin. Mol. Ther.*, 12, 248, 2010.

79. Díaz, J. F., Barasoain, I., Souto, A. A., Amat-Guerri, F., and Andreu, J. M., Macromolecular accessibility of fluorescent taxoids bound at a paclitaxel binding site in the microtubule surface, *J. Biol. Chem.*, 280, 3928, 2005.

80. Miller, M. L., Roller, E. E., Zhao, R. Y., Leece, B. A., Ab, O., Baloglu, E., Goldmacher, V. S., and Chari, R. V. J., Synthesis of taxoids with improved cytotoxicity and solubility for use in tumor-specific delivery, *J. Med. Chem.*, 47, 4802, 2004.

81. Boger, D. L. and Johnson, D. S., CC-1065 and the duocarmycins: Unraveling the keys to a new class of naturally derived DNA alkylating agents, *Proc. Natl. Acad. Sci. U. S. A.*, 92, 3642, 1995.

82. Chari, R. V. J., Jackel, K. A., Bourret, L. A., Derr, S. M., Tadayoni, B. M., Mattocks, K. M., Shah, S. A., Liu, C., Blättler, W. A., and Goldmacher, V. S., Enhancement of the selectivity and antitumor efficacy of a CC-1065 analog through immunoconjugate formation, *Cancer Res.*, 55, 4079, 1995.

83. Lee, S. H., Disulfide and multisulfide antitumor agents and their modes of action, *Arch. Pharm. Res.*, 32, 299, 2009.

84. Bross, P. F., Beitz, J., Chen, G., Chen, X. H., Duffy, E., Kieffer, L., Roy, S. et al., Approval summary: Gemtuzumab ozogamicin in relapsed acute myeloid leukemia, *Clin. Cancer Res.*, 7, 1490, 2001.

85. DiJoseph, J. F., Armellino, D. C., Boghaert, E. R., Khandke, K., Dougher, M. M., Sridharan, L., Kunz, A. et al., Antibody-targeted chemotherapy with CMC-544: A CD22-targeted immunoconjugate of calicheamicin for the treatment of B-lymphoid malignancies, *Blood*, 103, 1807, 2004.

86. Ruppert, C., Markart, P., Schmidt, R., Grimminger, F., Seeger, W., Lehr, C. M., and Günther, A., Chemical crosslinking of urokinase to pulmonary surfactant protein B for targeting alveolar fibrin, *Thromb. Haemost.*, 89, 53, 2003.

87. Kovtun, Y. V., Audette, C. A., Mayo, M. F., Jones, G. E., Doherty, H., Maloney, E. K., Erickson, H. K. et al., Antibody-maytansinoid conjugates designed to bypass multidrug resistance, *Cancer Res.*, 70, 2528, 2010.

88. Jansen, F. K., Blythman, H. E., Carrière, D., Casellas, P., Gros, O., Gros, P., Laurent, J. C. et al., Immunotoxins: Hybrid molecules combining high specificity and potent cytotoxicity, *Immunol. Rev.*, 62, 185, 1982.

89. Houston, L. L. and Nowinski, R. C., Cell-specific cytotoxicity expressed by a conjugate of ricin and murine monoclonal antibody directed against thy 1.1 antigen, *Cancer Res.*, 41, 3913, 1981.

90. Cumber, A. J., Henry, R. V., Parnell, G. D., and Wawrzynczak, E. J., Purification of immunotoxins containing the ribosome-inactivating proteins gelonin and momordin using high performance liquid immunoaffinity chromatography compared with blue sepharose CL-6B affinity chromatography, *J. Immunol. Methods*, 135, 15, 1990.

91. Ramakrishnan, S. and Houston, L. L., Comparison of the selective cytotoxic effects of immunotoxins containing ricin A chain or pokeweed antiviral protein and anti-Thy 1.1 monoclonal antibodies, *Cancer Res.*, 44, 201, 1984.

92. Flavell, D. J., Boehm, D. A., Noss, A., Warnes, S. L., and Flavell, S. U., Therapy of human T-cell acute lymphoblastic leukaemia with a combination of anti-CD7 and anti-CD38-SAPORIN immunotoxins is significantly better than therapy with each individual immunotoxins, *Br. J. Cancer*, 84, 571, 2001.

93. Flannery, G. R., Pelham, J. M., Gray, J. D., and Baldwin, R. W., Immunomodulation: NK cells activated by interferon-conjugated monoclonal antibody against human osteosarcoma, *Eur. J. Cancer Clin. Oncol.*, 20, 791, 1984.

94. Pelham, J. M., Gray, J. D., Flannery, G. R., Pimm. M. V., and Baldwin, R. W., Interferon-alpha conjugation to human osteogenic sarcoma monoclonal antibody 791T/36, *Cancer Immunol. Immunother.*, 15, 210, 1983.

95. Petrella, E. C., Wilkie, S. D., Smith, C. A., Morgan, A. C., Jr., and Vogel, C. W., Antibody conjugates with cobra venom factor. Synthesis and biochemical characterization, *J. Immunol. Methods*, 104, 159, 1987.

96. Flavell, D. J., Warnes, S., Noss, A., and Flavell, S. U., Host-mediated antibody-dependent cellular cytotoxicity contributes to the in vivo therapeutic efficacy of an anti-CD7-saporin immunotoxin in a severe combined immunodeficient mouse model of human T-cell acute lymphoblastic leukemia, *Cancer Res.*, 58, 5787, 1988.

97. Bolognesi, A., Tazzari, P. L., Tassi, C., Gromo, G., Gobbi, M., and Stirpe, F., A comparison of anti-lymphocyte immunotoxins containing different ribosome-inactivating proteins and antibodies, *Clin. Exp. Immunol.*, 89, 341, 1992.

98. Vallera, D. A., Taylor, P. A., Panoskaltsis-Mortari, A., and Blazar, B. R., Therapy for ongoing graft-versus-host disease induced across the major or minor histocompatibility barrier in mice with anti-CD3F(ab')2-ricin toxin A chain immunotoxins, *Blood*, 86, 4367, 1995.

99. Juhl, H., Petrella, E. C., Cheung, N. K., Bredehorst, R., and Vogel, C. W., Complement killing of human neuroblastoma cells: A cytotoxic monoclonal antibody and its F(ab')2-cobra venom factor conjugate are equally cytotoxic, *Mol. Immunol.*, 27, 957, 1990.

100. Carroll, S. F., Bernhard, S. L., Goff, D. A., Bauer, R. J., Leach, W., and Kung, A. H., Enhanced stability in vitro and in vivo of immunoconjugates prepared with 5-methyl-2-iminothiolane, *Bioconjug. Chem.*, 5, 248, 1994.

101. Rosenblum, M. G., Marks, J. W., and Cheung, L. H., Comparative cytotoxicity and pharmacokinetics of antimelanoma immunotoxins containing either natural or recombinant gelonin, *Cancer Chemother. Pharmacol.*, 44, 343, 1999.

102. Duzkale, H., Pagliaro, L. C., Rosenblum, M. G., Varan, A., Liu, B., Reuben, J., Wierda, W. G. et al., Bone marrow purging studies in acute myelogenous leukemia using the recombinant anti-CD33 immunotoxin HuM195/rGel, *Biol. Blood Marrow Transplant.*, 9, 364, 2003.

103. Wawrzynczak, E. J., Cumber, A. J., Henry, R. V., May, J., Newell, D. R., Parnell, G. D., Worrell, N. R., and Forrester, J. A., Pharmacokinetics in the rat of a panel of immunotoxins made with abrin A chain, ricin A chain, gelonin, and momordin, *Cancer Res.*, 50, 7519, 1990.

104. Zhou, X. X., Ji, F., Zhao, J. L., Cheng, L. F., and Xu, C. F., Anti-cancer activity of anti-p185(HER-2) ricin A chain immunotoxin on gastric cancer cells, *J. Gastroenterol. Hepatol.*, 25, 1266, 2010.

105. Tsai, L. C., Chen, Y. L., Lee, C., Chen, H. M., Chang, Z. N., Hung, M. W., Chao, P. L., and Lin, J. Y., Growth suppression of human colorectal carcinoma in nude mice by monoclonal antibody C27-abrin A chain conjugate, *Dis. Colon Rectum.*, 38, 1067, 1995.

106. Weiner, L. M., O'Dwyer, J., Kitson, J., Comis, R. L., Frankel, A. E., Bauer, R. J., Konrad, M. S., and Groves, E. S., Phase I evaluation of an anti-breast carcinoma monoclonal antibody 260F9-recombinant ricin A chain immunoconjugate, *Cancer Res.*, 49, 4062, 1989.

107. Saito, A., Sato, H., Iino, N., and Takeda, T., Molecular mechanisms of receptor-mediated endocytosis in the renal proximal tubular epithelium, *J. Biomed. Biotechnol.*, 2010, 403272, 2010.

108. Grant, B. D. and Donaldson, J. G., Pathways and mechanisms of endocytic recycling, *Nat. Rev. Mol. Cell Biol.*, 10, 597, 2009.

109. O'Keefe, D. O. and Draper, R. K., Characterization of a transferrin-diphtheria toxin conjugate, *J. Biol. Chem.*, 260, 932, 1985.

110. Singh, V., Mavila, A. K., and Kar, S. K., Comparison of the cytotoxic effect of hormonotoxins prepared with the use of heterobifunctional cross-linking agents N-succinimidyl 3-(2-pyridyldithio)propionate and N-succinimidyl 6-[3-(2-pyridyldithio)propionamido]hexanoate, *Bioconjug. Chem.*, 4, 473, 1993.

111. Stirpe, F., Olsnes, S., and Pihl, A., Gelonin, a new inhibitor of protein synthesis, nontoxic to intact cells. Isolation, characterization, and preparation of cytotoxic complexes with concanavalin A, *J. Biol. Chem.*, 255, 6947, 1980.

112. Singh, V. and Curtiss, R., 3rd, Hormonotoxins: The role of positive charge of lysine residue on the immunological, biological and cytotoxic properties of ovine lutropin-S-S-gelonin conjugates, *Mol. Cell Biochem.*, 130, 91, 1994.

113. Ippoliti, R., Lendaro, E., D'Agostino, I., Fiani, M. L., Guidarini, D., Vestri, S., Benedetti, P. A., and Brunori, M., A chimeric saporin-transferrin conjugate compared to ricin toxin: Role of the carrier in intracellular transport and toxicity, *FASEB J.*, 9, 1220, 1995.

114. Umemoto, N., Kato, Y., and Hara, T., Cytotoxicities of two disulfide-bond-linked conjugates of methotrexate with monoclonal anti-MM46 antibody, *Cancer Immunol. Immunother.*, 28, 9, 1989.

115. Raso, V. and Basala, M., A highly cytotoxic human transferrin–ricin A chain conjugate used to select receptor-modified cells, *J. Biol. Chem.*, 259, 1143, 1984.

116. Martin, H. B. and Houston, L. L., Arming alpha 2-macroglobulin with ricin A chain forms a cytotoxic conjugate that inhibits protein synthesis and kills human fibroblasts, *Biochim. Biophys. Acta*, 762, 128, 1983.

117. Simpson, D. L., Cawley, D. B., and Herschman, H. R., Killing of cultured hepatocytes by conjugates of asialofetuin and EGF linked to the A chains of ricin or diphtheria toxin, *Cell*, 29, 469, 1982.

118. Shimizu, N., Shimizu, Y., and Miskimins, W. K., EGF-ricin A conjugates: Kinetic profiles of cytotoxic effects and resistant cell variants, *Cell Struct. Funct.*, 9, 203, 1984.

119. Sakai, A., Sakakibara, R., and Ishiguro, M., Human chorionic gonadotropin-ricin A chain hybrid protein: A hormone analog for the study of signal transduction, *J. Biochem.*, 105, 275, 1989.

120. Oeltmann, T. N. and Wiley, R. G., Preparation of diphtheria toxin fragment A coupled to hormone, *Methods Enzymol.*, 165, 204, 1988.

121. Chang, T. M. and Kullberg, D. W., Studies of the mechanism of cell intoxication by diphtheria toxin fragment A-asialoorosomucoid hybrid toxins. Evidence for utilization of an alternative receptor-mediated transport pathway, *J. Biol. Chem.*, 257, 12563, 1982.

122. Gros, O., Gros, P., Jansen, F. K., and Vidal, H., Biochemical aspects of immunotoxin preparation, *J. Immunol. Methods*, 81, 283, 1985.

123. Bourrie, B. J., Casellas, P., Blythman, H. E., and Jansen, F. K., Study of the plasma clearance of antibody—Ricin-A-chain immunotoxins. Evidence for specific recognition sites on the A chain that mediate rapid clearance of the immunotoxins, *Eur. J. Biochem.*, 155, 1, 1986.

124. King, T. P., Li, Y., and Kochoumian, L., Preparation of protein conjugates via intermolecular disulfide bond formation, *Biochemistry*, 17, 1499, 1978.

125. Thorpe, P. E., Wallace, P. M., Knowles, P. P., Relf, M. G., Brown, A. N., Watson, G. J., Knyba, R. E., Wawrzynczak, E. J., and Blakey, D. C., New coupling agents for the synthesis of immunotoxins containing a hindered disulfide bond with improved stability in vivo, *Cancer Res.*, 47, 5924, 1987.

126. Kreitman, R. J. and Pastan, I., Immunotoxins for targeted cancer therapy, *Adv. Drug Deliv. Rev.*, 31, 53, 1998.

127. Van Horssen, P. J., Preijers, F. W., Van Oosterhout, Y. V., and De Witte, T., Highly potent CD22-recombinant ricin A results in complete cure of disseminated malignant B-cell xenografts in SCID mice but fails to cure solid xenografts in nude mice, *Int. J. Cancer*, 68, 378, 1996.

128. Shimizu, N., Miskimins, W. K., and Shimizu, Y., A cytotoxic epidermal growth factor cross-linked to diphtheria toxin A-fragment, *FEBS Lett.*, 118, 274, 1980.

129. Miskimins, W. K. and Shimizu, N., Synthesis of a cytotoxic insulin cross-linked to diphtheria toxin fragment A capable of recognizing insulin receptors, *Biochem. Biophys. Res. Commun.*, 91, 143, 1979.

130. Oeltmann, T. N. and Forbes, J. T., Inhibition of mouse spleen cell function by diphtheria toxin fragment A coupled to anti-mouse Thy-1.2 and by ricin A chain coupled to anti-mouse IgM, *Arch. Biochem. Biophys.*, 209, 362, 1981.

131. Bacha, P., Murphy, J. R., and Reichlin, S., Thyrotropin-releasing hormone-diphtheria toxin-related polypeptide conjugates. Potential role of the hydrophobic domain in toxin entry, *J. Biol. Chem.*, 258, 1565, 1983.

132. Chang, T. M. and Neville, D. M., Jr., Artificial hybrid protein containing a toxic protein fragment and a cell membrane receptor-binding moiety in a disulfide conjugate. I. Synthesis of diphtheria toxin fragment A-S-S-human placental lactogen with methyl-5-bromovalerimidate, *J. Biol. Chem.*, 252, 1505, 1977.

133. Masuho, Y., Hara, T., and Noguchi, T., Preparation of a hybrid of fragment Fab' of antibody and fragment A of diphtheria toxin and its cytotoxicity, *Biochem. Biophys. Res. Commun.*, 90, 320, 1079.

134. Oeltmann, T. N. and Heath, E. C., A hybrid protein containing the toxic subunit of ricin and the cell-specific subunit of human chorionic gonadotropin. I. Synthesis and characterization, *J. Biol. Chem.*, 254, 1022, 1979.

135. Ellman, G. L., Tissue sulfhydryl groups, *Arch. Biochem. Biophys.*, 82, 70, 1959.

136. FitzGerald, D. J., Willingham, M. C., and Pastan, I., *Pseudomonas* exotoxin—Immunotoxin, in *Immunotoxins*, Frankel, A. E. (ed.), Kluwer Academic Publishers, Boston, MA, 1988, p. 161.

137. FitzGerald, D. J., Construction of immunotoxins using *Pseudomonas* exotoxin A, *Methods Enzymol.*, 151, 139, 1987.

138. FitzGerald, D. J., Padmanabhan, R., Pastan, I., and Willingham, M. C., Adenovirus-induced release of epidermal growth factor and pseudomonas toxin into the cytosol of KB cells during receptor-mediated endocytosis, *Cell*, 32, 607, 1983.

139. Ramakrishnan, S., Bjorn, M. J., and Houston, L. L., Recombinant ricin A chain conjugated to monoclonal antibodies: Improved tumor cell inhibition in the presence of lysosomotropic compounds, *Cancer Res.*, 49, 613, 1989.

140. Miyazaki, H., Beppu, M., Terao, T., and Osawa, T., Preparation of antibody (IgG)-ricin A-chain conjugate and its biologic activity, *Gann*, 71, 766, 1980.

141. McIntyre, G. D., Scott, C. F., Jr., Ritz, J., Blättler, W. A., and Lambert, J. M., Preparation and characterization of interleukin-2-gelonin conjugates made using different cross-linking reagents, *Bioconjug. Chem.*, 5, 88, 1994.

142. Wawrzynczak, E. J. and Thorpe, P. E., Methods for preparing immunotoxins: Effect of the linkage on activity and stability, in *Immunoconjugates, Antibody Conjugates in Radioimaging and Therapy of Cancer*, Vogel, E.-W. (ed.), Oxford University Press, New York, 1987, p. 28.

143. Arpicco, S., Dosio, F., Brusa, P., Crosasso, P., and Cattel, L., New coupling reagents for the preparation of disulfide cross-linked conjugates with increased stability, *Bioconjug. Chem.*, 8, 327, 1997.

144. Dosio, F., Arpicco, S., Adobati, E., Canevari, S., Brusa, P., De Santis, R., Parente, D. et al., Role of cross-linking agents in determining the biochemical and pharmacokinetic properties of Mgr6-clavin immunotoxins, *Bioconjug. Chem.*, 9, 372, 1998.

145. Vogel, C.-W., Antibody conjugate without inherent toxicity: The targeting of cobra venom factor and other biological response modifiers, in *Immunoconjugates: Antibody Conjugates in Radioimaging and Therapy of Cancer*, Vogel, C.-W. (ed.), Oxford University Press, New York, 1987, p. 170.

146. Thorpe, P. E., Ross, W. C., Brown, A. N., Myers, C. D., Cumber, A. J., Foxwell, B. M., and Forrester, J. T., Blockade of the galactose-binding sites of ricin by its linkage to antibody. Specific cytotoxic effects of the conjugates, *Eur. J. Biochem.*, 140, 63, 1984.

147. Youle, R. J. and Neville, D. M., Jr., Anti-Thy 1.2 monoclonal antibody linked to ricin is a potent cell-type-specific toxin, *Proc. Natl. Acad. Sci. U. S. A.*, 77, 5483, 1980.

148. Vallera, D. A. and Myers, D. E., Immunotoxins containing ricin, in *Immunotoxins*, Frankel, A. E. (ed.), Kluwer Academic Publishers, Boston, MA, 1988, p. 141.

149. Volkman, D. J., Ahmad, A., Fauci, A. S., and Neville, D. M., Jr., Selective abrogation of antigen-specific human B cell responses by antigen-ricin conjugates, *J. Exp. Med.*, 156, 634, 1982.

150. Lambert, J. M., Senter, P. D., Yau-Young, A., Blättler, W. A., and Goldmacher, V. S., Purified immunotoxins that are reactive with human lymphoid cells. Monoclonal antibodies conjugated to the ribosome-inactivating proteins gelonin and the pokeweed antiviral proteins, *J. Biol. Chem.*, 260, 12035, 1985.

151. Yang, W. H., Wieczorck, M., Allen, M. C., and Nett, T. M., Cytotoxic activity of gonadotropin-releasing hormone (GnRH)-pokeweed antiviral protein conjugates in cell lines expressing GnRH receptors, *Endocrinology*, 144, 1456, 2003.

152. Bjorn, M. J., Groetsema, G., and Scalapino, L., Antibody-*Pseudomonas* exotoxin A conjugates cytotoxic to human breast cancer cells in vitro, *Cancer Res.*, 46, 3262, 1986.

153. Myers, D. E., Uckun, F. M., Swaim, S. E., and Vallera, D. A., The effects of aromatic and aliphatic maleimide crosslinkers on anti-CD5 ricin immunotoxins, *J. Immunol. Methods*, 121, 129, 1989.

154. Edwards, D. C., Smith, A., Ross, W. C., Cumber, A. J., Thorpe, P. E., and Davies, A. J., The effect of abrin, anti-lymphocyte globulin and their conjugates on the immune response of mice to sheep red blood cells, *Experientia*, 37, 256, 1981.

155. Edwards, D. C., Ross, W. C., Cumber, A. J., McIntosh, D., Smith, A., Thorpe, P. E., Brown, A., Williams, R. H., and Davies, A. J., A comparison of the in vitro and in vivo activities of conjugates of anti-mouse lymphocyte globulin and abrin, *Biochim. Biophys. Acta*, 717, 272, 1982.

156. Blättler, W. A., Kuenzi, B. S., Lambert, J. M., and Senter, P. D., New heterobifunctional protein crosslinking reagent that forms an acid-labile link, *Biochemistry*, 24, 1517, 1985.

157. Lambert, J. M., Blättler, W. A., McIntyre, G. D., Goldmacher, V. S., and Scott, C. F., Jr., Immunotoxins containing single chain ribosome-inactivating proteins, in *Immunotoxins*, Frankel, A. E. (ed.), Kluwer Academic Publishers, Boston, MA, 1988, p. 175.

158. Doronina, S. O., Toki, B. E., Torgov, M. Y., Mendelsohn, B. A., Cerveny, C. G., Chace, D. F., DeBlanc, R. L. et al., Development of potent monoclonal antibody auristatin conjugates for cancer therapy, *Nat. Biotechnol.*, 21, 778, 2003.

159. Hinman, L. M., Hamann, P. R., Wallace, R., Menendez, A. T., Durr, F. E., and Upeslacis, J., Preparation and characterization of monoclonal antibody conjugates of the calicheamicins: A novel and potent family of antitumor antibiotics, *Cancer Res.*, 53, 3336, 1993.

160. Hamann, P. R., Hinman, L. M., Beyer, C. F., Lindh, D., Upeslacis, J., Flowers, D. A., and Bernstein, I., An anti-CD33 antibody calicheamicin conjugate for treatment of acute myeloid leukemia. Choice of linker, *Bioconjug. Chem.*, 13, 40, 2002.

161. Hamann, P. R., Hinman, L. M., Beyer, C. F., Greenberger, L. M., Lin, C., Lindh, D., Menendez, A. T., Wallace, R., Durr, F. E., and Upeslacis, J., An anti-MUC1 antibody-calicheamicin conjugate for treatment of solid tumors. Choice of linker and overcoming drug resistance, *Bioconjug. Chem.*, 16, 346, 2005.

162. Zara, J. J., Wood, R. D., Boon, P., Kim, C. H., Pomato, N., Bredehorst, R., and Vogel, C. W., A carbohydrate-directed heterobifunctional cross-linking reagent for the synthesis of immunoconjugates, *Anal. Biochem.*, 194, 156, 1991.

163. Senter, P. D., Tansey, M. J., Lambert, J. M., and Blattler, W. A. Novel photocleavable protein crosslinking reagents and their use in the preparation of antibody–toxin conjugates, *Photochem. Photobiol.*, 42, 231, 1985.

164. Goldmacher, V. S., Senter, P. D., Lambert, J. M., and Blättler, W. A., Photoactivation of toxin conjugates, *Bioconjug. Chem.*, 3, 104, 1992.

165. Olsnes, S. and Pihl, A., Chimeric toxins, *Pharmacol. Ther.*, 15, 355, 1981.

166. Killen, J. A. and Lindstrom, J. M., Specific killing of lymphocytes that cause experimental autoimmune myasthenia gravis by ricin toxin-acetylcholine receptor conjugates, *J. Immunol.*, 133, 2549, 1984.

167. Varga, J. M., Hormone-drug conjugates, *Methods Enzymol.*, 112, 259, 1985.

168. Suzuki, T., Sato, E., Goto, K., Katsurada, Y., Unno, K., and Takahashi, T., The preparation of mitomycin C, adriamycin and daunomycin covalently bound to antibodies as improved cancer chemotherapeutic agents, *Chem. Pharm. Bull. (Tokyo)*, 29, 844, 1981.

169. Shouval, D., Adler, R., Wands, J. R., and Hurwitz, E., Conjugates between monoclonal antibodies to HBsAg and cytosine arabinoside, *J. Hepatol.*, 3, Suppl. 2, S87, 1986.

170. Aboud-Pirak, E., Hurwitz, E., Bellot, F., Schlessinger, J., and Sela, M., Inhibition of human tumor growth in nude mice by a conjugate of doxorubicin with monoclonal antibodies to epidermal growth factor receptor, *Proc. Natl. Acad. Sci. U. S. A.*, 86, 3778, 1989.

171. Hurwitz, E., Adler, R., Shouval, D., Takahashi, H., Wands, J. R., and Sela, M., Immunotargeting of daunomycin to localized and metastatic human colon adenocarcinoma in athymic mice, *Cancer Immunol. Immunother.*, 35, 186, 1992.

172. Adler, R., Hurwitz, E., Wands, J. R., Sela, M., and Shouval, D., Specific targeting of adriamycin conjugates with monoclonal antibodies to hepatoma associated antigens to intrahepatic tumors in athymic mice, *Hepatology*, 22, 1482, 1995.

173. Sela, M. and Hurwitz, E., Conjugates of antibodies with cytotoxic drugs, in *Immunoconjugates: Antibody Conjugates in Radioimaging and Therapy of Cancer*, Vogel, C.-W. (ed.), Oxford University Press, New York, 1987, p. 189.

174. Hurwitz, E., Wilchek, M., and Pitha, J., Soluble macromolecules as carriers for daunoribicin, *J. Appl. Biochem.*, 2, 25, 1980.

175. Hurwitz, E., Arnon, R., Sahar, E., and Danon, Y., A conjugate of adriamycin and monoclonal antibodies to Thy-1 antigen inhibits human neuroblastoma cells in vitro, *Ann. N. Y. Acad. Sci.*, 417, 125, 1983.

176. Okuno, S., Harada, M., Yano, T., Yano, S., Kiuchi, S., Tsuda, N., Sakamura, Y., Imai, J., Kawaguchi, T., and Tsujihara, K., Complete regression of xenografted human carcinomas by camptothecin analogue-carboxymethyl dextran conjugate (T-0128), *Cancer Res.*, 60, 2988, 2000.

177. Sugahara, S., Kajiki, M., Kuriyama, H., and Kobayashi, T. R., Carrier effects on antitumor activity and neurotoxicity of AZ10992, a paclitaxel-carboxymethyl dextran conjugate, in a mouse model, *Biol. Pharm. Bull.*, 31, 223, 2008.

178. Shiose, Y., Kuga, H., Ohki, H., Ikeda, M., Yamashita, F., and Hashida, M., Systematic research of peptide spacers controlling drug release from macromolecular prodrug system, carboxymethyldextran polyalcohol-peptide-drug conjugates, *Bioconjug. Chem.*, 20, 60, 2009.

179. Oyama, T., Kawamura, M., Abiko, T., Izumi, Y., Watanabe, M., Kumazawa, E., Kuga, H., Shiose, Y., and Kobayashi, K., Hyperthermia-enhanced tumor accumulation and antitumor efficacy of a doxorubicin-conjugate with a novel macromolecular carrier system in mice with non-small cell lung cancer, *Oncol. Rep.*, 17, 653, 2007.

180. Gruaz-Guyon, A., Raguin, O., and Barbet, J., Recent advances in pretargeted radioimmunotherapy, *Curr. Med. Chem.*, 12, 319, 2005.

181. Schechter, B., Arnon, R., Wilchek, M., Schlessinger, J., Hurwitz, E., Aboud-Pirak, E., and Sela, M., Indirect immunotargeting of cis-Pt to human epidermoid carcinoma KB using the avidin–biotin system, *Int. J. Cancer*, 48, 167, 1991.

182. Schechter, B., Arnon, R., and Wilchek, M., Cytotoxicity of streptavidin-blocked biotinyl-ricin is retrieved by in vitro immunotargeting via biotinyl monoclonal antibody, *Cancer Res.*, 52, 4448, 1992.

183. Moro, M., Pelagi, M., Fulci, G., Paganelli, G., Dellabona, P., Casorati, G., Siccardi, A. G., and Corti, A., Tumor cell targeting with antibody-avidin complexes and biotinylated tumor necrosis factor alpha, *Cancer Res.*, 57, 1922, 1997.

184. Buxman, M. M., The role of enzymatic coupling of drugs to proteins in induction of drug specific antibodies, *J. Invest. Dermatol.*, 73, 256, 1979.

185. Besheer, A., Hertel, T. C., Kressler, J., Mäder, K., and Pietzsch, M., Enzymatically catalyzed HES conjugation using microbial transglutaminase: Proof of feasibility, *J. Pharm. Sci.*, 98, 4420, 2009.

186. Lönnberg, H., Solid-phase synthesis of oligonucleotide conjugates useful for delivery and targeting of potential nucleic acid therapeutics, *Bioconjug. Chem.*, 20, 1065, 2009.

187. Moroney, S. E., D'Alarcao, L. J., Goldmacher, V. S., Lambert, J. M., and Blättler, W. A., Modification of the binding site(s) of lectins by an affinity column carrying an activated galactose-terminated ligand, *Biochemistry*, 26, 8390, 1987.

188. Dosio, F., Brusa, P., Delprino, L., Ceruti, M., Grosa, G., Cattel, L., Bolognesi, A., and Barbieri, L., A new 'solid phase' procedure to synthesize immunotoxins (antibody-ribosome inactivating protein conjugates), *Farmaco*, 48, 105, 1993.

189. Sinkule, J. A., Rosen, S. T., and Radosevich, J. A., Monoclonal antibody 44–3A6 doxorubicin immunoconjugates: Comparative in vitro anti-tumor efficacy of different conjugation methods, *Tumour Biol.*, 12, 198, 1991.

190. Thibeault, D. and Pagé, M., Coupling daunorubicin to monoclonal antialphafoetoprotein with a new activated derivative, *Int. J. Immunopharmacol.*, 12, 503, 1990.

191. Hudecz, F., Ross, H., Price, M. R., and Baldwin, R. W., Immunoconjugate design: A predictive approach for coupling of daunomycin to monoclonal antibodies, *Bioconjug. Chem.*, 1, 197, 1990.

192. Tanaka, T., Kaneo, Y., Miyashita, M., and Shiramoto, S., Properties of water-insoluble mitomycin C-albumin conjugate as a sustained release drug delivery system in mice inoculated with Sarcoma 180, *Biol. Pharm. Bull.*, 18, 1724, 1995.

193. Faulstich, H. and Fiume, L., Protein conjugates of fungal toxins, *Methods Enzymol.*, 112, 225, 1985.

194. Ghosh, M. K., Kildsig, D. O., and Mitra, A. K., Preparation and characterization of methotrexate-immunoglobulin conjugates, *Drug Des. Deliv.*, 4, 13, 1989.

14 Application of Chemical Conjugation to Solid-State Chemistry

14.1 INTRODUCTION

Insolubilized biomolecules such as proteins and nucleic acids have found increasing uses in medicine, clinical analysis, affinity chromatography, and synthetic chemistry applications.[1–10] Latex particles coated with antibodies, for example, have become standard in rapid diagnostic tests.[9] These immobilized proteins have also provided valuable information about basic protein–protein interactions.[11] In addition to enzymes, many different types of proteins, including antibodies, protein antigens, enzyme inhibitors, protein toxins, and peptide hormones, have been attached to insoluble carriers and have been shown to be biologically active. Quite often, immobilization of an enzyme actually results in an increased stability of the protein.[2–4]

Various approaches have been used to insolubilize proteins.[2–6] These methods involve either noncovalent adsorption, encapsulation, and entrapment, or covalent linkage. In this chapter, only covalent immobilization will be considered in detail, particularly the use of chemical cross-linkers. Readers interested in other methods of protein insolubilization are encouraged to consult the numerous books and reviews[3,12,13] that address this area. Covalent attachment of a protein to a solid support involves the formation of a covalent bond between a functional group on the protein and a reactive group on the surface of the solid phase. The methods of such covalent bonding are as varied as organic chemistry itself. During the last few decades, thousands of proteins have been immobilized, and hundreds of matrices have become available. In immobilization of proteins to solid supports, several factors must be considered in order to retain the optimum activity of the proteins. The proteins must be attached in such an orientation that their active sites or binding domains are accessible to the surrounding milieu and not buried in or blocked by the matrix or other components on the matrix surface. During the coupling reaction, the protein must be in its active state. Royer and Uy[14] have demonstrated, for example, that, in some cases, enzymes can be immobilized in an active conformation in the presence of their substrates. Another factor that must be considered in the covalent attachment of proteins is the possibility of chemically altering the protein in such a way that its reactivity is reduced. Since reactive groups on the proteins are involved in the chemical bonding, it is possible that groups associated with the active site or binding site of a protein could be involved in the reaction. In addition, chemical cross-linking could take part intramolecularly causing conformational changes in the protein, thereby reducing the attachment efficiency. New developments in the techniques of immobilization of proteins involving emulsification in the presence of cross-linkers, cross-linked enzyme crystals, cross-linked enzyme aggregates, and carrier-free enzyme particles may resolve some of these problems. In this chapter, the chemistry of protein insolubilization will be discussed.

14.2 FUNCTIONALITIES OF MATRICES

Commercial matrices used for attachment of proteins are made of different materials and contain different functional groups. In order to select the best method and reagent for coupling biomolecules, the type of functional groups on the matrix must be known. Probably the most prevalent and most widely used matrices are the polysaccharides, such as agarose, cellulose, and their derivatives. These materials contain vicinal hydroxyl groups, which can be oxidized with sodium periodate and activated by using epoxy reactants like glycidol and epichlorohydrin to generate reactive aldehyde glyoxal groups. One of the derivatives of interest is chitin, which consists of a repeating β-(1,4)-linked 2-acetamido-2-deoxy-D-glucose (N-acetylglucosamine) moiety, which is deacetylated to form chitosan with a linear polysaccharide composed of randomly distributed β-(1,4)-linked D-glucosamine (deacetylated) and N-acetylglucosamine (acetylated) units.[15] Chitosan has reactive amino and hydroxyl groups, which, after further chemical modification, can make covalent bonds with reactive groups of the proteins and other biochemicals. Due to the amine groups, chitosan is a cationic polyelectrolyte (pK$_a$ 6.5), which is insoluble in water above pH 6.5. Another natural substance that contains hydroxyls is silica.[16] Silica is usually used in the form of controlled porous glass, which can be modified to contain various groups.[17] Newer hydroxylic matrices are derived from polymeric materials such as polyacrylamide.[18]

There are hundreds of polymeric matrices that have been made of various polymers, copolymers, and terpolymers; and new ones are constantly being synthesized.[19] These materials may be in the form of membranes, balls, plates, beads, or slides. Many of these supports are commercially available in different sizes, forms, and colors. Fluorescent and magnetic particles are also available. The surface chemistries range from plain polystyrene (benzene) to a wide variety of different surface functional groups such as sulfonate, sulfate, carboxylate, primary amine, aromatic amine, amide, arylhydrazide, hydroxyl, aldehyde, vinylbenzyl chloride, and chloromethyl.[20,21] In addition, many activated matrices ready for protein immobilization are also available, such as CNBr-activated sepharose, bromoacetylcellulose, N-hydroxysuccinimide (NHS) ester-agarose, streptavidin bound magnetic microspheres, aromatic aldehyde magnetic microspheres, and others.

Many of these groups can be further converted to other functionalities. For example, the surface hydroxyl groups of glass beads (silica) can be converted to amino groups by silanization with 3-aminopropyltriethoxysilane or amine-containing alkoxysilanes. The amino groups thus generated can be further converted to carboxylic acids by succinyl chloride or succinic anhydride as shown in Figure 14.1A.[14] The resulting glass surfaces can be directly used for attachment of biomolecules; or they can be further derivatized with functional groups, such as thiol, maleimide, alkyne, NHS ester, epoxide, hydrazide, aminooxy, fluoroalkyl, and photoreactive groups.[17,22–24] During the process of derivatization, a spacer may also be introduced.

Matrix amide groups can be converted to amino groups by reaction with hydrazine or other alkyl diamines. The resulting amines can be further modified to aldehydes with glutaraldehyde or to carboxylic acids with succinic anhydride as shown in Figure 14.1B.

Carboxylic acids can be converted to amines by coupling with diamino alkanes such as 1,6-diaminohexane or 1,7-diaminoheptane in the presence of a water-soluble carbodiimide (e.g., EDC) as shown in Figure 14.1C. These reactions provide a spacer between the solid surface as well as the reactive group. In this regard, the carboxyl group can be further removed from the surface by coupling to 5-aminocaproic acid or other amino acids (Figure 14.1C). Many other manipulations of the matrix surface chemistry are possible. In fact, the matrices can be treated in many cases as organic chemicals and organic solvents may be used for derivatization or modification of the functional groups. The reactions for protein modification mentioned throughout this book are certainly applicable as well to solid surfaces.

Instead of modification of existing surface chemistry, another method is to create a coating of different functional groups on a solid surface.[25] For example, Burnham et al.[18] developed a method for creating surface-patterned, biofunctionalized hydrogels on glass or silicon, using polyacrylamide and the disulfide-containing polyacrylamide crosslinker, bis(acryloyl)cystamine. Upon treatment with a reducing agent, reactive sulfhydryl groups were created throughout

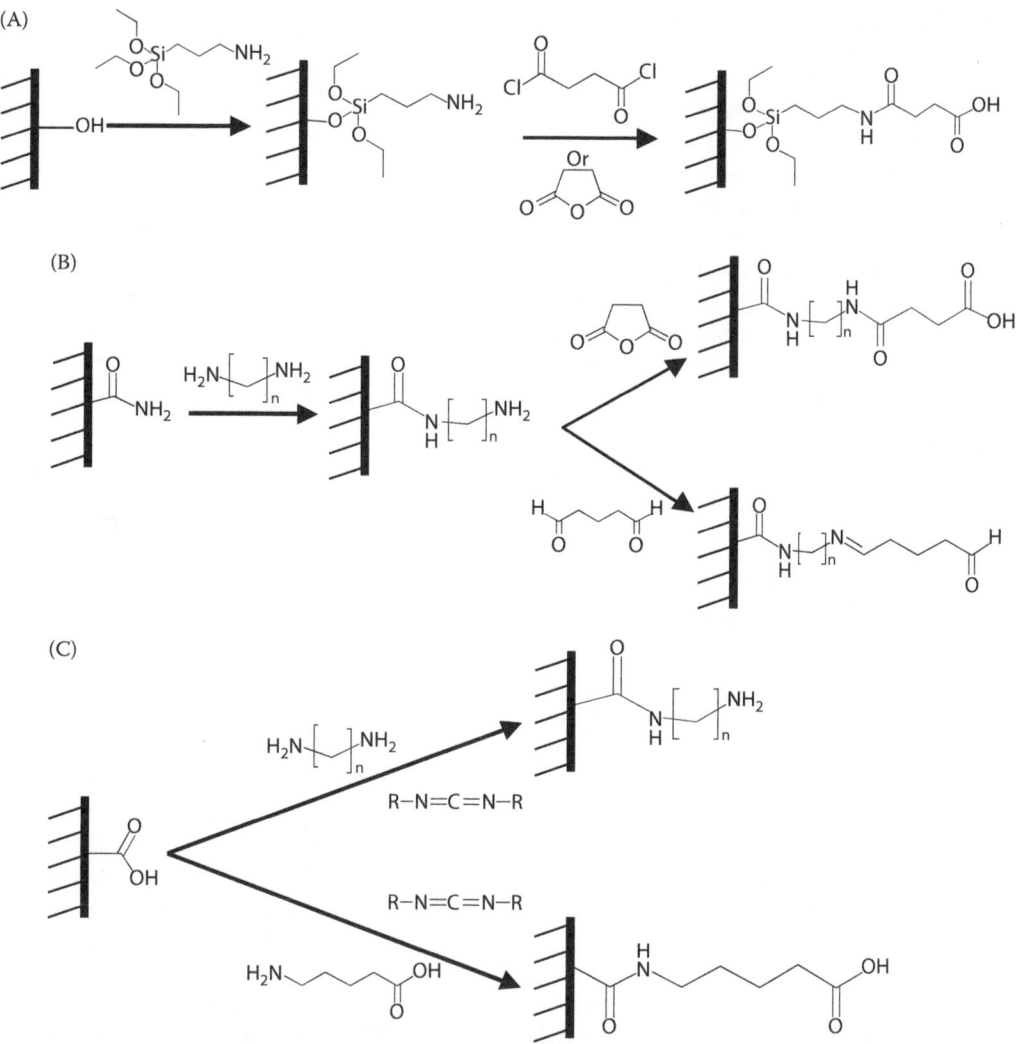

FIGURE 14.1 Modification of matrix functional groups: (A) conversion of matrix hydroxyl groups to carboxylic acids; (B) conversion of matrix amides to amines, carboxylic acids, and aldehydes; (C) conversion of matrix carboxylic acids to amines and the introduction of a spacer.

these hydrogels that could cross-link to desired biomolecules. Stillman and Tonkinson[26] also prepared a nitrocellulose-based polymer surface on slides that have higher binding capacity for DNA than the traditional poly-lysine-coated slides. Using a cross-linked monolayer prepared by electron-beam-induced cross-linking of 4'-nitro-1,1'-biphenyl-4-thiol 1,1'-biphenyl-4-thiol on silicon, Amin et al.[27] prepared a new class of polymer material, so-called polymer carpets, by self-initiated surface photopolymerization and photografting of styrene. Such polymer carpets may be modified for immobilization of biochemicals. In addition, metal oxides and nanocrystalline films of gold, titanium, nickel, and other metals have been deposited onto glasses and other solid surfaces for protein immobilization.[28–32] These metallic-coated surfaces are particularly useful for the construction of enzyme electrodes.[28,30,32] Nanoparticles of iron have also been prepared to immobilize lipase.[33]

14.3 PROTEIN IMMOBILIZATION BY MATRIX ACTIVATION

The choice of a chemical procedure for immobilization of a protein depends on both the protein and the surface functionality of the solid matrix. As discussed in Chapter 2, many of the amino acid side chains of proteins are available for reaction. One of the most common groups used for attachment is the amino group present at the N-terminus and the side chains of lysines. Carboxyl groups from the C-terminus, aspartic, and glutamic acids are also available for bonding as well as sulfhydryl group of cysteines, hydroxyl groups of serines and threonines, and phenyl moieties of phenylalanines and tyrosines. Because of the variety of reactive groups on the proteins, emphasis is placed on the functional groups on the solid surface. Thus, the method of immobilization is more or less dependent on the solid matrix chosen. Generally, the matrix is first activated by a specific chemical reaction depending on the surface chemistry. Proteins are then immobilized by spotting on the activated matrix. The following examples will demonstrate the kind of cross-linking that can be achieved through a two-step reaction.

14.3.1 ACTIVATION OF HYDROXYL GROUPS

The most commonly used method for protein immobilization involves the use of hydroxyl groups of cellulose and agarose derivatives. The vicinal diols can be activated by cyanogen bromide. Indeed, any polymeric matrices possessing cleavable vicinal hydroxy groups can be activated using this method. Agarose is a linear polymer consisting of D-galactose and 3,6-anhydro-D-galactose. Sepharose™ is a trademark for an agarose derivative commercially available. Sephadex is a cross-linked dextran originally prepared by Pharmacia (which was acquired by Amersham Biosciences, which in turn was acquired by GE Healthcare). Cyanogen bromide reacts with these materials to form a cyclic reactive imidocarbonate, which is susceptible to nucleophilic attack by amino groups present in the proteins to form N-substituted imidocarbonate as demonstrated in Figure 14.2.[16,34] The cyclic imidocarbonate may be converted to cyclic carbonate in aqueous solutions yielding N-substituted carbamates as the end product. A water-soluble, nonvolatile, and less toxic analog, 1-cyano-4-dimethylaminopyridinium tetrafluoroborate, developed by Kohn and Wilchek,[35] has also provided efficient coupling. Because of the versatility of the cyanogen bromide reaction, the activated matrix can be used for derivatization with diamines or dihydrazides. In practice, the reaction is quite simple. The carrier is first activated with cyanogen bromide for a brief period of time and then washed to remove excess reagent. It is then mixed with the protein for immobilization. The ease of the reaction and the mild conditions employed, together with the high yields of the immobilized proteins, account for the popularity of this method. In fact, cyanogen bromide-activated matrices are now commercially available for immediate use. This method has been used in catalytic chromatography[36] and to prepare immunosorbent media for various immunoaffinity systems.[37]

The vicinal diols of the carbohydrate derivatives can also be activated by oxidation with sodium periodate to aldehydes, which will form Schiff bases directly with amino groups of proteins. Alternatively, the oxidized dialdehyde can be condensed through reduction of the Schiff bases formed with 4,4'-diaminodiphenylmethane. The phenyl amine can then be diazotized and coupled to proteins (see below under Section 14.3.4).[34]

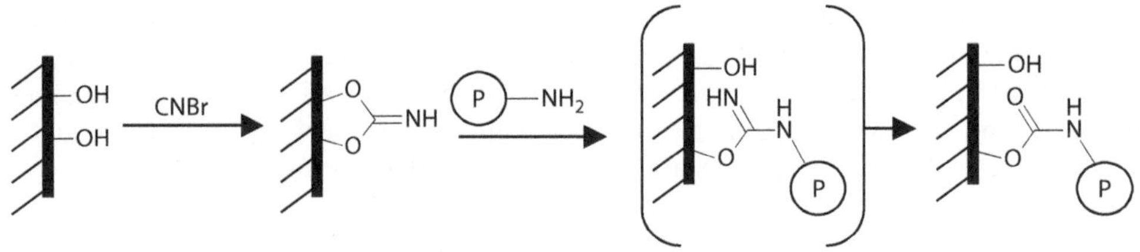

FIGURE 14.2 Immobilization of proteins with cyanogen bromide activation of vicinal diols.

14.3.2 ACTIVATION OF CARBOXYL GROUPS

Matrix carboxyl groups can be activated with *N*-hydroxybenzotriazole in the presence of a water-soluble carbodiimide.[38] Before coupling to proteins, the activated matrix is washed to remove excess carbodiimide. The reactive ester reacts very rapidly with amino groups of the ligand to form stable amide bonds as shown in Figure 14.3. Among the other nucleophiles, only sulfhydryl groups compete effectively with the amino group during the reaction.

p-Nitrophenol[39] and NHS[34,40,41] are also commonly used to form active esters with carboxylic acids in the presence of a carbodiimide. These active ester derivatives are stable when stored in dioxane. The coupling reaction is similar to that of *N*-hydroxybenzotriazole ester. Sam et al.[41] used EDC/NHS to activate porous silicon layers grafted with carboxyl-terminated alkyl chains.

Carboxylate functional groups can also be directly activated to the acyl chloride by reacting with thionyl chloride. The very reactive acyl chloride reacts with proteins at low temperatures and is used to immobilize alpha-amylase on poly(methyl methacrylate-acrylic acid) microspheres.[42] Thionyl chloride is also used to activate carboxyl groups of multiwalled carbon nanotubes for coupling poly(L-lactic acid).[43]

14.3.3 ACTIVATION OF ACYL HYDRAZIDE

As mentioned above, reaction of a hydrazide-modified surface with glutaraldehyde generates aldehyde groups, which can be used to couple to proteins by forming Schiff bases with the proteins' amino groups. Acyl hydrazides can also be converted to acyl azides with nitrous acid via the diazotization reaction.[44] The azide group will be replaced by amino groups on the proteins to form amide bonds as shown in Figure 14.4. Thus, by a series of reactions, surface hydroxyls can be activated to acyl azides as illustrated in Figure 14.4. First, the hydroxyl is carboxymethylated with chloroacetic acid, which is converted to an acyl hydrazide through an ester. The final activated acyl hydrazide then reacts with the proteins. Since matrix amide groups can also be converted to acyl hydrazide, they can be linked to proteins through this reaction sequence.

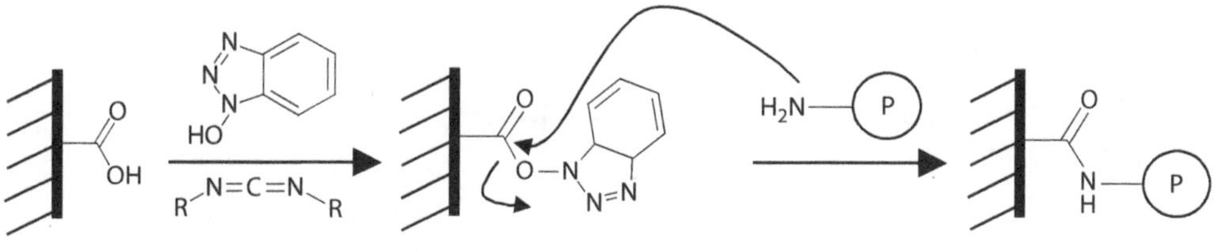

FIGURE 14.3 Immobilization of proteins to carboxylate matrix with carbodiimide and *N*-hydroxybenzotriazol.

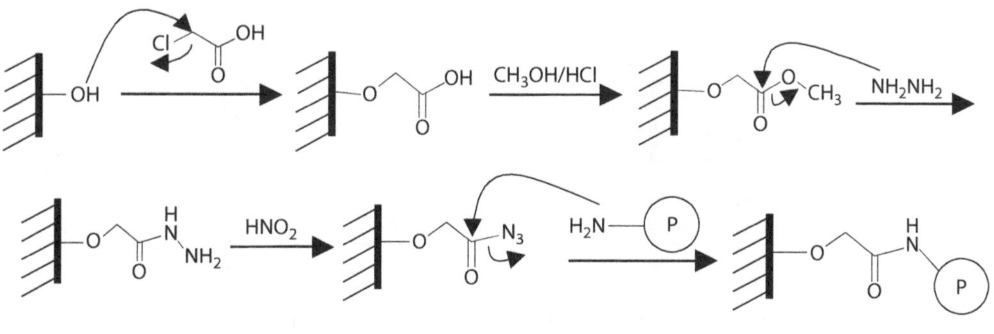

FIGURE 14.4 Coupling of proteins to hydroxyl matrix via diazotization of acyl hydrazide.

14.3.4 ACTIVATION OF AMINES

14.3.4.1 Use of Nitrous Acid

Aromatic amino groups from poly-*p*-aminostyrene can be diazotized with nitrous acid to form the corresponding diazonium derivative.[34] The diazonium group will react with phenolic, imidazole, and amino side chains of a protein to form a diazo bond (see Chapter 2). By this method, phenyl groups on polystyrene can be coupled to proteins through nitration and amination.[45] Reaction of the benzene ring on polystyrene with fuming red nitric acid results in nitrostyrene. The nitro benzene group is then converted to amino benzene by reduction with sodium dithionite and then diazotized for coupling with proteins. This sequence of steps is shown in Figure 14.5. Hydroxyl-containing carriers, such as cellulose, can also be coupled through the diazonium group after reaction with *p*-nitrobenzylchloride.[34,46]

14.3.4.2 Use of Phosgene and Thiophosgene

Aromatic amines can also be activated at alkaline pH with phosgene or thiophosgene to yield the corresponding isocyanate and isothiocyante, respectively, as shown in Figure 14.6.[34,47] These active derivatives react with free amino groups of proteins to form an amide or thioamide bond of the substituted urea or thiourea (see Chapter 2). The isothiocyanate derivative can also be obtained from aliphatic amines and acyl azides.

14.3.4.3 Use of Cyanogen Bromide

Like hydroxyl groups, amino group containing matrices can also be activated with cyanogen bromide to the corresponding cyanamides, which reacts with proteins amino groups to form guanidine linkages as shown in Figure 14.7.[48] It was claimed that amine matrices yielded more stable products than the hydroxyl group–based materials.

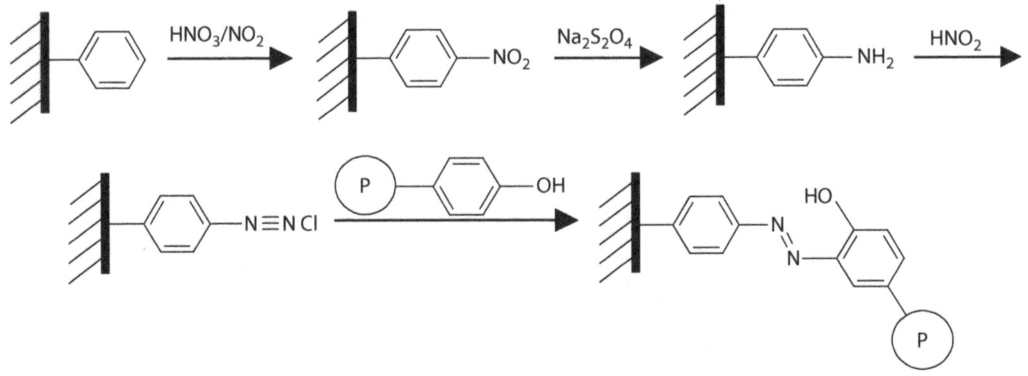

FIGURE 14.5 Coupling of proteins to aromatic amines via diazotization.

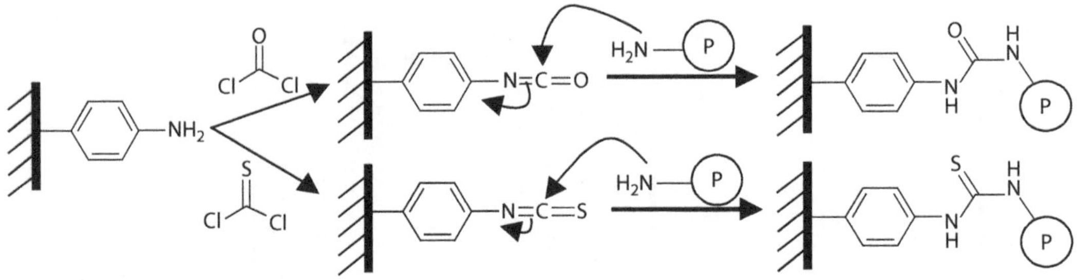

FIGURE 14.6 Coupling of proteins to aromatic amines through isocyanates and isothiocyanates.

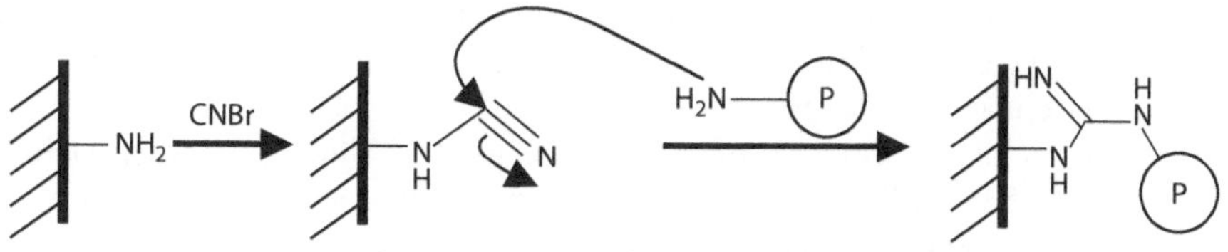

FIGURE 14.7 Coupling of proteins to cyanogen bromide-activated amine matrices.

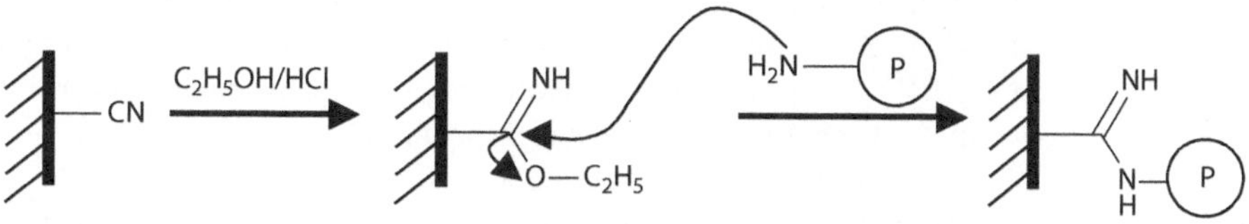

FIGURE 14.8 Coupling proteins to polyacrylonitrile after activation to imidoester.

14.3.5 ACTIVATION OF POLYACRYLONITRILE

Polyacrylonitrile can be activated with absolute ethanol and bubbling hydrogen chloride to an imidoester, which is readily attacked by amino groups of proteins at basic pH to yield amidine as shown in Figure 14.8.[49] There are several other methods to activate polyacrylonitrile. Jain et al.[50] reduced the pendant nitrile group of polyacrylonitrile with lithium aluminum hydride to an amine functionality, which was further activated by using glutaraldehyde for the covalent linking of immunoglobulins. The nitrile groups may also be transformed into the corresponding carboxyl group by strong acids or bases.[51,52] Alkaline hydrolysis introduces amide and carboxylic groups, which improve its hydrophilicity.[53] These group can be further chlorinated with thionyl chloride to form acyl chloride derivatives, which react with amino and hydroxyl groups of proteins to form amide and ester bonds, respectively.[54] Furthermore, Battistel et al.[55] have used an enzyme, nitrile hydratase, a member of the class of nitrile-converting enzymes, to selectively convert the pendant nitrile into the corresponding amides.

14.4 CROSS-LINKING REAGENTS COMMONLY USED FOR IMMOBILIZATION OF BIOMOLECULES

Theoretically, any of the cross-linking reagents discussed earlier in this book can be used to couple biomolecules to solid supports. The choice of reagents depends on the system to be studied. Suitable cross-linkers target the functionalities on the matrix and the functional groups of the biological molecules to be cross-linked. The use of these cross-linkers is, in a way, similar to the activation of the matrix as narrated in the earlier sections. In almost all cases, the matrix is first reacted with the cross-linker (an activation process), and the biological compound to be immobilized is then added to complete the conjugation reaction. The following sections illustrate the use of various types of cross-linking reagents to couple proteins to solid supports.

14.4.1 USE OF ZERO-LENGTH CROSS-LINKING REAGENTS

The commonly known zero-length cross-linkers that have been used for immobilization of biomolecules include carbodiimides, Woodward's reagent K,[56] chloroformates,[57] and carbonyldiimidazole.[58] While all of these compounds condense carboxyl and amino groups to form amide bonds, the latter two also activate hydroxyl groups. When hydroxyl groups are involved, a carbonyl moiety is incorporated. These compounds function as monobifunctional reagents as will be discussed below (see also Chapter 8).

Different carbodiimides, including water-soluble compounds described in Chapter 8, have been used for coupling amino groups of proteins to carboxyl group–containing solid matrices.[40] Among the chloroformates, ethyl chloroformate, p-nitrophenyl chloroformate, 2,4,5-trichlorophenyl chloroformate, and N-hydroxysuccinimide chloroformate have been used.[57,59] During the reaction, the carboxyl group first reacts with the cross-linker to form an activated species, which is attacked by the amino group nucleophile. Some reaction schemes for these reagents are shown in Chapter 8, Figure 8.5.

Other reagents that can be considered zero-length cross-linkers are sulfonyl chlorides and 2-fluoro-N-methylpyridinium tosylate.[60] The most commonly used sulfonyl chlorides are tresyl chloride (2,2,2-trifluoroethanesulfonyl chloride),[61] although p-toluene sulfonyl chloride (tosyl chloride) and colored sulfonyl chloride, 3,5-dinitro-4-dimethylaminobenzenesulfony chloride (diabsyl chloride), have also been used.[57,62] These reagents activate the primary hydroxyl groups into good leaving groups as shown in Figures 14.9A for tresyl chloride. The activated function is then displaced by a biological nucleophile, chiefly the amino side chain of lysine.

It should be pointed out that not only amino groups of proteins serve as nucleophiles; free sulfhydryl, if present, can potentially be coupled. The linking between a carboxyl group and an amino group forms an amide bond, whereas a thioester bond is formed when a sulfhydryl group is involved. Similarly, coupling primary alcohols using sulfonyl chloride and 2-fluoro-N-methylpyridinium salt affords a secondary amine with amino group and a thioether bond with free thiol, as shown in Figure 14.9.

Various different solid matrices have been activated by these zero-length cross-linkers, and many enzymes and proteins have been immobilized this way. Carboxyl groups containing supports such as carboxymethyl cellulose, acrylamide and acrylic acid copolymer, carboxymethyl Sephadex, BioGel, carboxymethyl agarose, and polyacrylic acid have been used. Polyhydroxylic matrices that have been activated by zero-length cross-linkers are agarose, glycerylpropyl-silica, cellulose, and hydroxyethyl methacrylate.

Thiol-containing matrices such as thiopropyl-sepharose can be activated by 2,2′-dipyridyldisulfide through the thiol-disulfide interchange reaction.[63,64] The protein is bonded via its free thiol through a second thiol-disulfide interchange with the liberation of 2-thiopyridine according to Figure 14.10. This method generates a disulfide bond between the protein and the solid support, which is stable under nonreducing conditions.[64] The enzyme can be released by low–molecular-weight thiol compounds. For proteins that do not contain a free thiol, it can be thiolated using various reagents such as N-acetylhomocysteine thiolactone as described in Chapter 2.

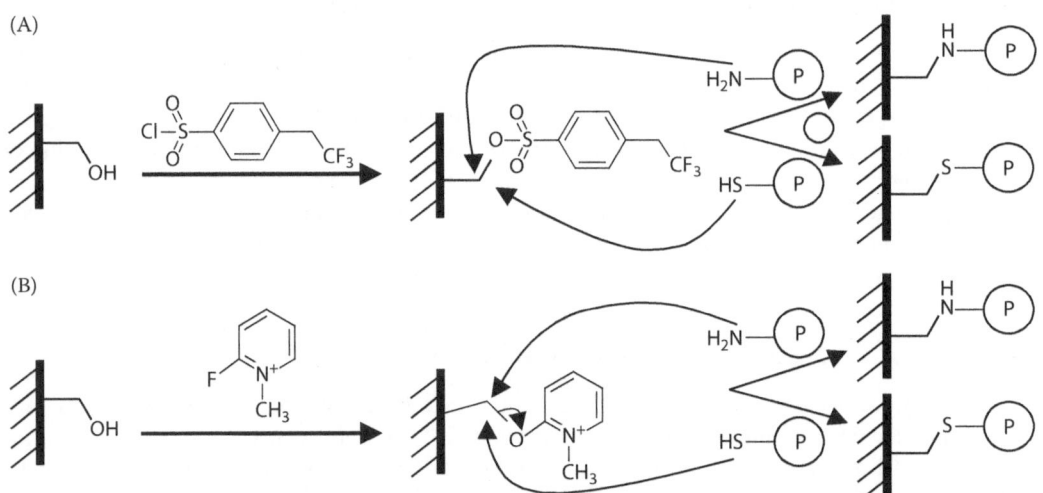

(A)

(B)

FIGURE 14.9 Coupling proteins to hydroxyl matrix with (A) tresyl chloride and (B) 2-fluoro-1-methylpyridinium salt.

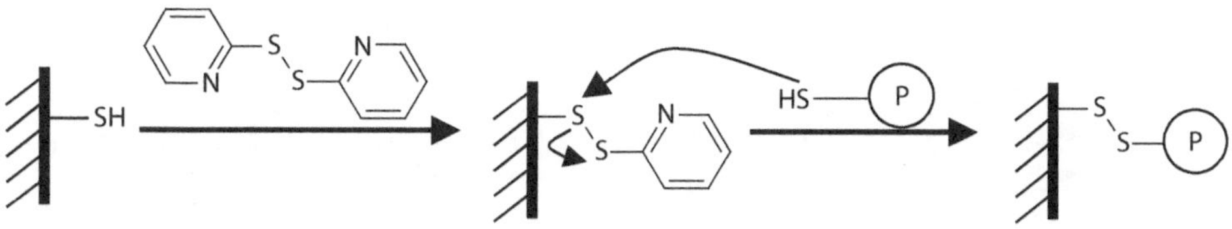

FIGURE 14.10 Coupling of proteins to thiol-containing matrices by thiol-disulfide interchange.

14.4.2 Use of Mono- and Homobifunctional Cross-Linkers

Almost any of the mono- and homobifunctional cross-linkers listed in Chapters 5 and 8 can be used one way or another to immobilize proteins. Depending on the matrix functional groups, different cross-linkers will have to be used. The following reactions demonstrate the use of these reagents.

14.4.2.1 Glutaraldehyde

Glutaraldehyde is the most-prevalent homobifunctional reagent used for the immobilization of bio-molecules. The reaction is dependent on pH, temperature, and ionic strength.[65] As presented in Chapter 5, the chemistry of cross-linking by glutaraldehyde is complex. There is no consensus on the main reactive species that participates in the cross-linking process because monomeric and polymeric forms are in equilibrium. Although the amino group is assumed to be the primary function to react, other functional groups, such as imidazole, thiol, and hydroxyl, have also been implicated.[66] Thus, glutaraldehyde has been used to couple proteins and other biomolecules to cellulosic materials,[67] silica gel,[68] polyacrylhydrazide,[69] nylon,[70] polyethyleneimine-treated magnetite,[71] and carbon nanotube fiber[72] as well as in the immobilization of glucose oxidase (GOD) for enzyme glucose electrode use and cross-linking chitosan/polyethylene glycol beads for drug delivery.[73] Both one-step and two-step cross-linking procedures have been used. For the two-step procedure, the matrix is first activated with glutaraldehyde, washed, and then coupled with the protein.

14.4.2.2 Chloroformates and Carbonyldiimidazole

A reaction similar to cyanogen bromide activation of vicinal hydroxyl groups is the formation of the cyclic carbonate derivative with 1,1′-carbonyldiimidazole, ethyl chloroformate, or other alkyl or aryl chloroformates (which are monofunctional agents, see Chapter 8) in anhydrous organic solvents. The activated cyclic carbonate will react with nucleophiles in biomolecules at pH 7–8 to form substituted carbonate bonds as shown in Figure 14.11.[74] When an amino group from a protein serves as the attacking nucleophile, a N-substituted carbamate bond is formed. A thiol carbonate bond results when thiol is involved. A carbonyl moiety from the cross-linker

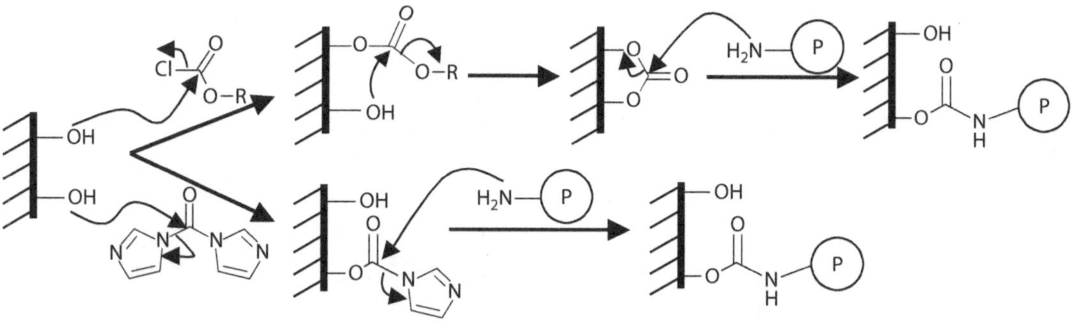

FIGURE 14.11 Immobilization of proteins on matrices containing vicinal diols with either chloroformates or 1,1′-carbonyldiimidazole.

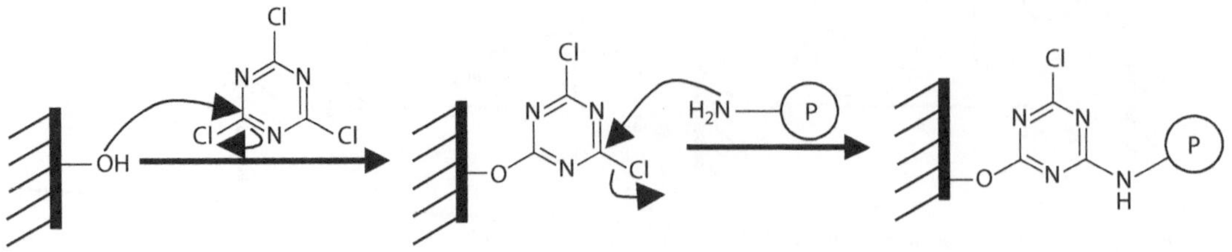

FIGURE 14.12 Immobilization of proteins with cyanuric chloride.

is incorporated into the bond. The carbamate and thiol carbonate bonds are not very stable and may be hydrolyzed under extreme pHs.

14.4.2.3 Heterocyclic Halides

s-Triazines (see Chapter 5), such as cyanuric chloride and some of its dichloro derivatives, 2-amino 4,6-dichloro-s-triazine, 2-carboxy-methylamino-4,6-dichloro-s-triazine, and 2-carboxymethoxy-4,6-dichloro-s-triazine, have been used to couple enzymes to hydroxyl and amino group containing matrices such as cellulose, agarose, and polyvinylalcohol through the activation of the hydroxyl group as shown in Figure 14.12.[75,76] The coupling process is an alkylation reaction involving the primary amino groups of proteins with the activated carbon of the s-triazine molecule.

14.4.2.4 Bisoxiranes

Introduction of epoxides onto matrix surfaces can be achieved by bisoxiranes, for example, 1,4-butanediol diglycidyl ether (1,4-bis(2,3-epoxypropoxy)butane). The reaction occurs readily at alkaline pH to yield derivatives containing a long-chain hydrophilic function with a reactive epoxide. This method is suitable for hydroxyl-containing supports, which form ether linkages through their hydroxyl groups.[77] Other matrices containing the amino and thiol groups can also be modified. de Souza et al.[78] immobilized ω-aminohexyl diamine onto bisoxirane-activated agarose gel for purification of IgG by negative chromatography. Immobilization occurs when nucleophilic amino group reacts with the epoxide as shown in Figure 14.13A. The bisoxiranes have also been used to couple enzymes to agarose solid supports.[64] Epoxides can be directly linked to hydroxyl group–containing matrices like silica by direct silianization with silane compounds containing an epoxide such as 3-glycidoxypropyltrimethoxysilane (which would be a heterofunctional agent).[79] Like the bisoxiranes, the epoxy function of the organomodified silica reacts with the amino function to immobilize the protein as shown in Figure 14.13B. Due to its high reactivity, other nucleophiles like hydroxyl or thiol moieties can also react with the epoxy group; hence, the selectivity to exclusively link specific moieties is low.

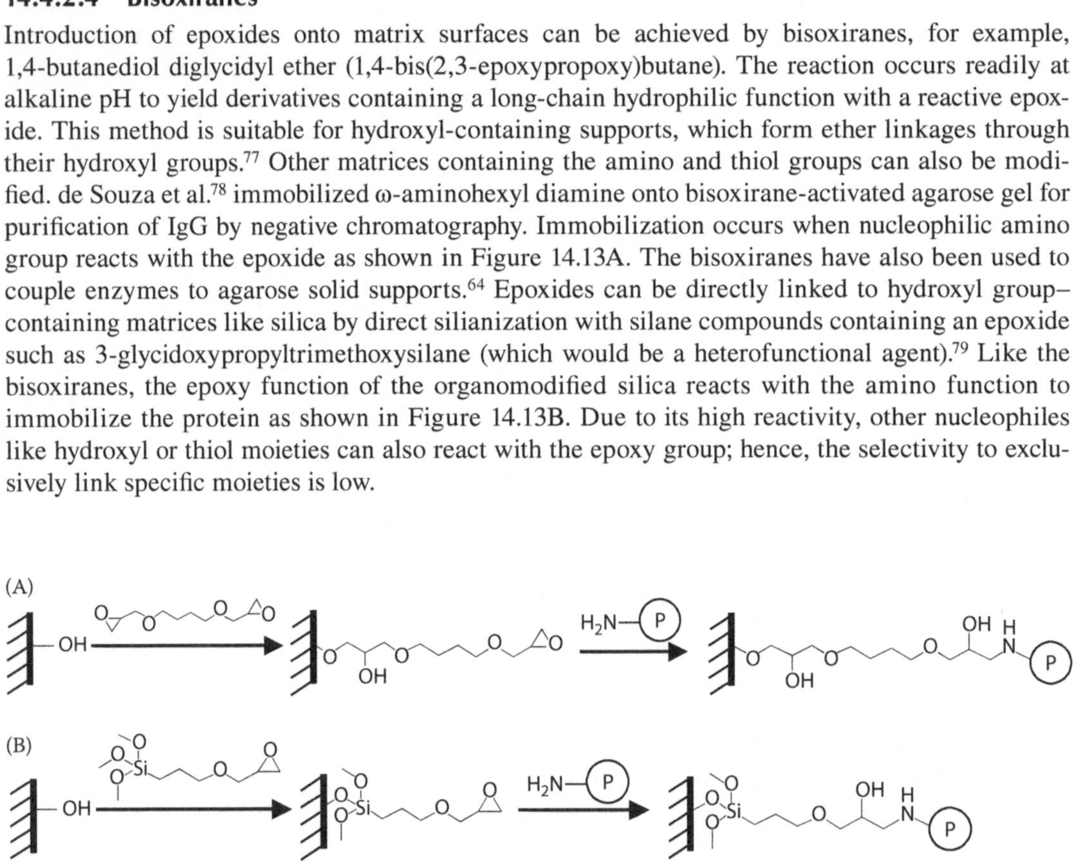

FIGURE 14.13 Immobilization of proteins with (A) bisoxirane: 1,4-butanediol diglycidyl ether; (B) 3-glycidoxypropyltrimethoxysilane.

14.4.2.5 Divinylsulfone

Divinylsulfone has been used to modify hydroxyl-containing matrices to vinylsulfone similar to its use with soluble proteins as discussed in Chapter 5.[64,77] The vinyl group is very reactive and reacts rapidly with nucleophiles of proteins, such as amines, alcohols, and phenols. However, the final linkage is unstable. The product with a hydroxyl function is unstable above pH 9 and that with an amino group is unstable above pH 8.[77] However, Rekuć et al.[80] showed that linkage to the amino anchor of derivatized cellulose gave increased stability of the immobilized laccase compared to other anchor groups.

14.4.2.6 Quinones

Proteins can be immobilized onto solid supports such as agarose by quinones, for example, p-benzoquinone.[64] The reaction is similar to protein cross-linking as discussed in Chapter 5. The matrix is first activated with the quinone. Proteins are then bound through their nucleophilic groups with high yields. The coupling reaction can occur in a broad range from pH 3 to 10. Undesirable side reactions may occur rendering a dirty color to the matrix. The reagent has also been used for the activation of other matrices including silica and polyacrylamide gels.[81,82]

14.4.2.7 Transition Metal Ions

Transition metal ions that form stable hexaqua complexes with water molecules can be used to activate certain polysaccharide matrices such as cellulose that contain vicinal diol groups.[83] These diols are amenable to chelation by transition metal ions. In addition, amino and carboxyl groups of solid matrices as in nylon can also be involved in the coordination chemistry.[84] In this process, the support, which may consist of materials as diverse as borosilicate, soda glass, filter paper, cellulose derivatives, and nylon 66, is first reacted with a transition metal salt, such as $Ti(IV)Cl_4$, $Co(II)Cl_2$, $Cu(II)Cl_2$, $Fe(II)Cl_2$, $Mn(II)Cl_2$, $Sn(IV)Cl_4$, $Sn(II)Cl_2$, $Cr(III)Cl_3$, $Zr(IV)Cl_4$, $V(III)Cl_3$, or $Fe(III)Cl_3$ for 24 h. After washing to remove unreacted metal salts, the immobilized metal ions further coordinate proteins through the carboxyl groups of C-terminus and acidic amino acids, the hydroxyl groups of tyrosyl, seryl, and threonyl residues, the free sulfhydryl groups of cysteines, or amino groups of the N-terminus and ε-amino groups of lysyl residues as shown in Figure 14.14.[85,86] Several enzymes, such as glucose dehydrogenase, α-amylase, glucoamylase, GOD, chymotrypsin, and urease, have been bonded to various solid supports by this method.[83–86] Relatively high retention and operational stability are demonstrated for some of these enzymes.

14.4.2.8 Other Homobifunctional Cross-Linkers

There are several other homobifunctional cross-linking reagents that have been used for immobilization of proteins such as toluene 2,4-diisocyanate and hexamethylene diisocyanate.[87,88] A two-step reaction procedure is usually followed. First, the solid surface is activated by the bifunctional reagent. The second step involves the reaction of the protein with the other end of the bifunctional cross-linker. The advantage of this two-step procedure is that it permits removal of the unreacted bifunctional reagent from the solid matrix before addition of the protein, thus preventing cross-linking of the protein in solution.

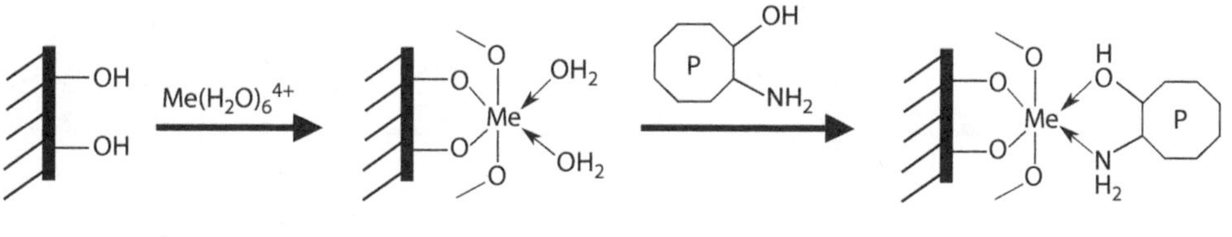

FIGURE 14.14 Immobilization of proteins with transition metal ions.

Bifunctional *N*-hydroxysuccinimide esters have been used to immobilize proteins to amino group-containing solid matrices. Ethylene glycol bis(succinimidyl succinate) (EGS) and dithiobis(succinimidyl propionate) (DSP) have been used to immobilize trypsin to hexamethylenediamine-sepharose CL-4B.[89] The immobilized protein can be released by treatment with hydroxylamine or thiol compounds, respectively, since these compounds contain cleavable bonds. A bis-sulfonated derivative of DSP, 3,3'-dithiobis(sulfosuccinimidyl propionate), which is more water soluble, has also been used.

14.4.3   USE OF HETEROBIFUNCTIONAL CROSS-LINKERS

Heterobifunctional cross-linkers have been increasingly used to immobilize biomolecules to solid carriers. A diverse array of reagents of different functionalities has been employed for this purpose. The following paragraphs discuss some representative examples.

14.4.3.1   Monohalogenacetyl Halide

Jagendorf et al.[90] used bromoacetyl bromide to immobilize serum globulins including antibodies to cellulose. The hydroxyl group of cellulose was first activated to form bromoacetyl cellulose, which alkylated a nucleophile from the protein, such as an amino group, to form conjugates as shown in Figure 14.15. Sato et al.[91] used the same approach to immobilize aminoacylase and found that, of the three halides, iodide conferred the best reactivity and stability to the enzyme.

14.4.3.2   Epichlorohydrin

Epichlorohydrin activates matrices containing nucleophiles such as amino or hydroxyl groups to an epoxide derivative as shown in Figure 14.16. This epoxide derivative reacts with nucleophilic groups of proteins in the order of thiol > amino > hydroxyl, although aromatic hydroxyl, guanidino, and imidazole groups also react. Bayramoglu et al.[92] used epichlorohydrin to both cross-link magnetic chitosan beads and to couple laccase under alkaline condition to the solid support. In the first step, the heterobifunctional reagent reacted with the amino and hydroxyl groups of chitosan to cross-link the polysaccharide chains. In the second step, laccase enzyme was coupled to the epoxide-activated chitosan. Chymotrypsin was also coupled to epichlorohydrin-activated chitosan, either in pure form or mixed with another biopolymer such as alginate, gelatin, or carrageenan. The immobilized enzyme was evaluated and compared to other immobilization methods.[93] Other enzymes that have been immobilized include lipase on fibrous membranes[94] and insoluble yeast beta-glucan,[95] peroxidase on chitosan,[96] catalase on starch-based polymers,[97] acetylcholinesterase, and choline oxidase co-immobilized on poly(2-hydroxyethyl methacrylate)

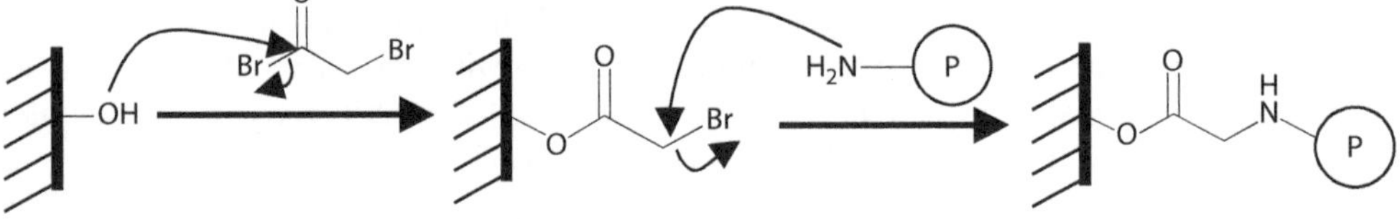

FIGURE 14.15   Immobilization of proteins with monohalogenacetyl halide, such as bromoacetyl bromide.

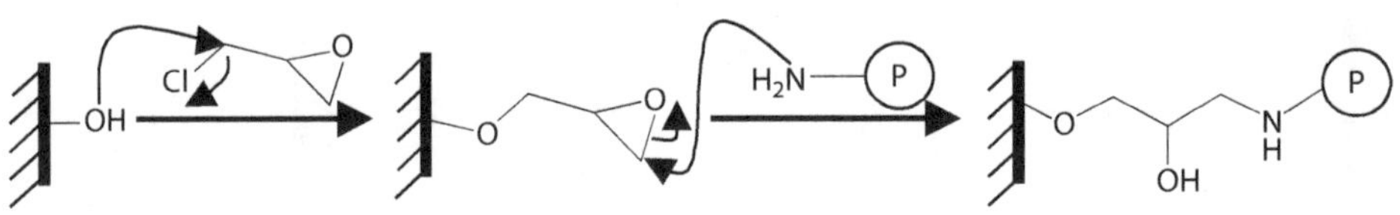

FIGURE 14.16   Immobilization of proteins with epichlorohydrin.

membranes,[98] α-amylase on poly(2-hydroxyethyl methacrylate) microspheres,[99] and GOD on poly(2-hydroxyethyl methacrylate) membranes.[100]

14.4.3.3 Amino- and Thiol-Group–Directed Reagents

Several heterobifunctional reagents of this type have been used to immobilize biomolecules. Compounds containing a succinimidyl group (amino group directed) and a maleimide group (thiol directed) are popular choices. The general reaction mechanism involves first reacting the cross-linker with the solid support. The activated matrices then react with the biomolecule to complete the immobilization process. For example, N-sulfosuccinimidyl-4-(N-maleimidomethyl) cyclohexane-1-carboxalate (SSMCC) was used to immobilize cysteine-terminated peptides on gold surface. SSMCC first reacted with an amine-modified gold surface to form an amide linkage. The solid-bound maleimide function was then linked to a sulfhydryl group of cysteine-terminated peptides through thioether bonds as shown in Figure 14.17.[101] Schlapak et al.[102] used another succinimidyl maleimido cross-linker, N-succinimidyl 4-(p-maleimidophenyl)butyrate, to immobilize thiol-modified DNA oligonucleotides to the terminal amino groups of the diamine-modified poly(ethylene glycol) (PEG) layer grafted onto silanized glass slides. The reaction of this type of cross-linkers can also occur with thiol groups on the solid support. For instance, Charles and Kusterbeck[103] used N-succinimidyl 4-maleimidobutyrate (GMBS) to immobilize antibodies onto a microcapillary surface. GMBS first activated the surface by reacting with thiol-terminated silane. The antibodies were then immobilized by reacting with the succinimidyl group. Bhatia et al.[104] have used mercaptom-ethyldimethylethyoxysilane or mercaptopropyltrimethyloxysilane to convert the hydroxyl groups of silica to thiol groups. N-γ-Maleimidobutyryloxy succinimide ester and N-succinimidyl 4-(p-maleidophenyl)butyrate, as well as other heterobifunctional reagents, N-succinimidyl-(4-iodoacetyl) amino benzoate and N-succinimidyl-3-(2-pyridyldiothio)propionate (SPDP), were used to conjugate immunoglobulins to the support.

As implied above, another group of commonly used heterobifunctional cross-linkers contain succinimidyl and pyridyldithio groups. The reagent, SPDP, whose N-hydroxysuccinimidyl ester reacts with amino groups and whose 2-pyridyldisulfide moiety reacts with free thiols, has been used to link calmodulin to thiol-sepharose 4B.[105] The immobilized protein can be cleaved with dithiothreitol. An analog with a longer flexible spacer arm, N-succinimidyl-6-[3'-(2-pyridyldithio) propionamido] hexanoate (LC-SPDP), was used to immobilize 5' end thiolated DNA to mica.[106] The thiolated DNA prepared by polymerase chain reaction reacted with the cross-linker to replace its 2-pyridyl disulphide group via sulfhydryl exchange. The modified DNA was deposited onto amino-silanized mica where the NHS-ester moiety of the cross-linker reacted with the primary amino group on the surface. Another analog, sulfosuccinimidyl-6-[3-(2-pyridyldithio)propionamido]hexanoate (sulfo-LC-SPDP), which is more water soluble, was used to immobilize antibodies and acetylcholine esterase onto a gold electrode surface in the piezoelectric quartz crystal.[107,108] The proteins were thiolated by reacting

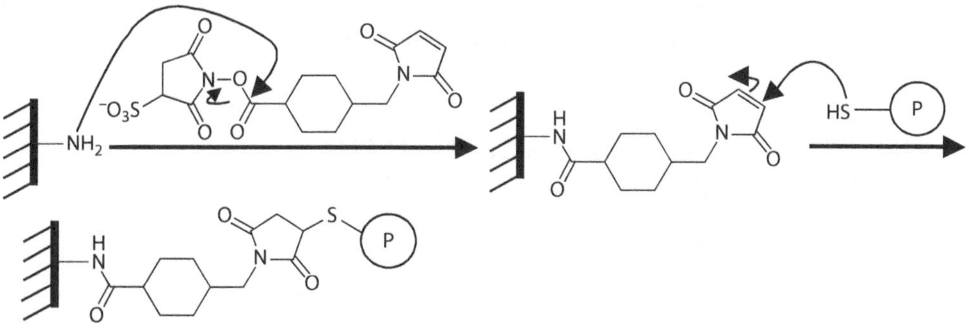

FIGURE 14.17 Immobilization of proteins with succinimidyl and maleimido group containing heterobifunctional reagents, such as SSMCC.

with sulfo-LC-SPDP followed by dithiothreitol reduction. The free thiol released on the proteins then formed a strong bond with gold on the electrode surface to immobilize the proteins.

Among other heterobifunctional cross-linkers, p-maleimidophenyl isocyanate (PMPI) was used to immobilize thiol-modified oligonucleotides onto Si surface coated with high-density PEG molecules.[109] The isocyanate moiety of PMPI first reacted with the –OH group of the surface to form a stable carbamate linkage. To complete the immobilization process, the thiol-modified oligonucleotides were then added to react with the maleimido group.

Photosensitive heterobifunctional reagents have also been used to immobilize biomolecules. Karakeçili et al.[110] used sulfosuccinimidyl-6-(4'-azido-2'-nitrophenyl-amino)hexanoate (sulfo-SANPAH) to immobilize epidermal growth factor (EGF) to chitosan membrane. EGF was first modified with the cross-linker though the reaction of an amino group with the sulfosuccinimidyl moiety. The immobilization to chitosan was achieved by UV irradiation. Yan et al.[111] synthesized a series of NHS perfluorophenyl azides (PFPAs) for immobilization of proteins to polymer surfaces. For example, N-succinimidyl-4-azido-2,3,5,6-tetrafluorobenzoate and N-succinimidyl-5-(4-azido-2,3,5,6-tetrafluorobenzamido)pentanoate were covalently bonded to polystyrene beads by photolysis at 254 nm. Horseradish peroxidase was immobilized on incubation with the NHS PFPAs-modified surface by nucleophilic replacement of the NHS group.

14.5 IMMOBILIZATION BY CROSS-LINKING THROUGH CARBOHYDRATE CHAINS

The immobilization of proteins to solid matrices via carbohydrate moieties was demonstrated by Royer.[112] The sugar moieties of the glycoprotein were first oxidized by sodium periodate to aldehydes, which form Schiff bases with either ethylenediamine or glycyltyrosine. Sodium borohydride was then used to stabilize the bonds. The derivatized glycoprotein was then immobilized to NHS ester-activated agarose or to diazotized arylamine supports as shown in Figure 14.18. Using this procedure, glucoamylase, peroxidase, GOD, and carboxypeptidase Y have been immobilized to solid supports.

Vicinal cis-hydroxyl groups of solid carriers such as agarose and many other polysaccharides are also susceptible to oxidation by periodate to yield aldehydes that can be used to insolubilize proteins by reductive amination.[113] Schiff bases are formed between the protein amino groups and the oxidized polysaccharide matrix. Subsequent reduction with sodium borohydride or pyridine borane stabilizes the bonds. The aldehydes can be further converted to N-alkyl amines by reacting with alkyl diamines or to hydrazides by reacting with dihydrazides such as adipic dihydrazide. The hydrazide-modified matrix is preferred over that of alkyl amine because the amino group is positively charged at neutral pH ($pK_a = 10$), whereas the hydrazide moiety is uncharged at pHs above 2.5 ($pK_a = 2.45$). These derivatives can then be cross-linked to biomolecules by different methods as mentioned above. The technique of reductive amination of Schiff bases formed between an aldehyde and an amine has been applied to the study of carbohydrate–protein and carbohydrate–carbohydrate interactions by surface plasmon resonance imaging and the construction of saccharide microarrays.[114,115]

14.6 EXAMPLES OF APPLICATIONS OF SOLID-PHASE IMMOBILIZATION CHEMISTRY

14.6.1 AFFINITY CHROMATOGRAPHY

Affinity chromatography is one of the most powerful techniques in the purification and isolation of biomolecules. The principle depends on the specific affinity between the target biomolecules and the entity immobilized on a solid support. For example, the affinity between enzymes and enzyme inhibitors, receptors and their ligands, and the specific binding of protein A and protein G

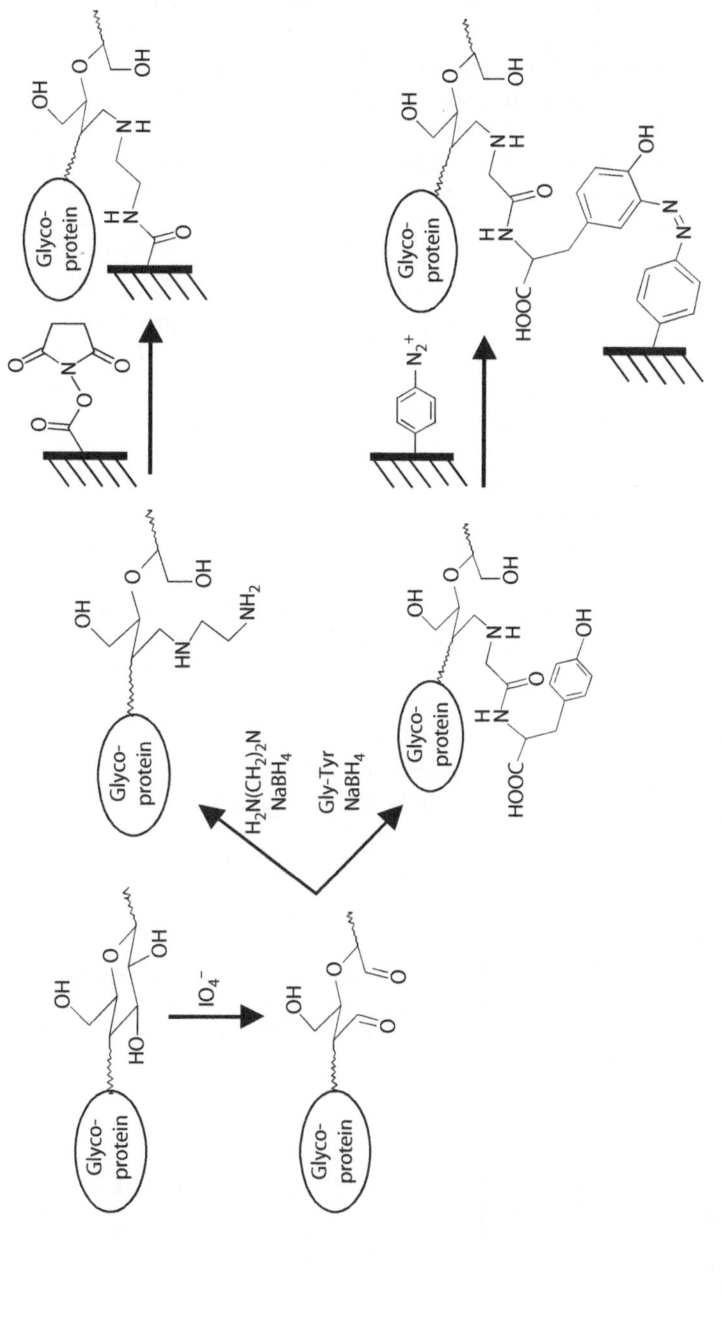

FIGURE 14.18 Immobilization of glycoproteins through oxidation and reductive amination of the carbohydrate moiety.

to the Fc region of antibodies have been utilized in affinity chromatography. Generally, a ligand is immobilized on a solid support, which is packed into a column. The selective affinity of an analyte for the ligand enables it to be isolated in pure form. For instance, protein A is used for the purification of IgG, IgM, IgA, and IgD. In some cases, the ligands themselves need to be purified, which may be difficult and costly. Their purity is an important factor in the success of selective isolation of analytes. In addition, their stability determines the life span of the absorbent material. Because of all these factors, synthetic ligands that combined the selectivity of the natural ligand with the high capacity, durability, and cost-effectiveness of the synthetic systems are being designed and utilized. Synthetic peptides with these properties have been employed in affinity chromatography.[116] Numerous functional groups on such synthetic peptides can be used for immobilization and various coupling approaches are applicable. The peptide is generally linked by using the carboxylic group from the C-terminal or the amino group from the N-terminal, although the side chains of any of the amino acids can also be used, taking the precaution that immobilization does not affect its binding capability or selectivity.

The selective interactions between antibodies and antigens are of great interest in affinity chromatography. When antibodies or antibody-related reagents are used as ligands for the purification of antigens, the process is referred to as immunoaffinity chromatography.[117] Immunoaffinity columns can serve as selective online precleanup steps for the isolation of a group of compounds, which are captured by immobilized antibodies.[118] There are numerous methods to immobilize antibodies such as those described above, which include reductive amination, epoxy chemistry, and matrice activation.[119] Zhang et al.[120] developed a novel solid support for antibody immobilization by using acrylamide as monomer, ethylene glycoldimethacrylate as cross-linker, and bulk polymerization as the synthetic method. The authors prepared a polymer in which the Cu(II) was embedded. The Cu(II)-embedded polymer displayed a strong binding with antibodies and was used as a novel solid support for antibody immobilization

When immobilized metal ions are used to isolate biomolecules, the version of affinity chromatography is referred to as immobilized-metal affinity chromatography (IMAC).[121] It is based on the known affinity of transition metal ions such as Zn^{2+}, Cu^{2+}, Ni^{2+}, and Co^{2+} to histidine and cysteine in aqueous solutions. The metal ions are immobilized onto a solid support to fractionate protein solutions. Several chelating ligands are used to fix the metal to solid supports like agarose and sepharose. The common ligands are iminodiacetic acid (IDA), N,N-bis[carboxymethyl]glycine (nitrilotriacetic acid [NTA]), and tris(carboxymethyl)ethylene diamine (TED) as shown in Figure 14.19.

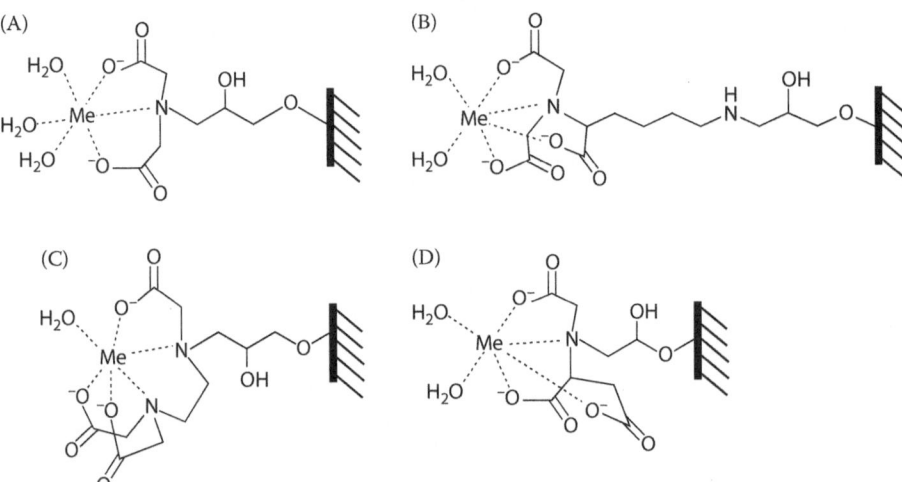

FIGURE 14.19 Structures of some chelating ligands used in immobilized-metal affinity chromatography: (A) IDA; (B) NTA; (C) TED; (D) N-carboxymethyl-2-carboxymethylglycine.

These chelating ligands are coupled to solid matrices though various cross-linking methodologies described earlier. For example, sepharose is activated using epichlorohydrin and IDA is attached by reacting its amino group with the epoxy group. Other ligand such as *N,N*-bis[carboxymethyl]lysine and carboxymethyl aspartate have also been developed.[121,122] However, many of the solid matrices with chelating ligands are commercially available. There are numerous applications of IMAC for the purification of proteins as well as immunoassays.[121] The choice of the metal ion immobilized on the IMAC ligand depends on the application. While trivalent cations such as Al^{3+}, Ga^{3+}, and Fe^{3+} and tetravalent Zr^{4+} are preferred for isolation of phosphoproteins and phosphopeptides, divalent Cu^{2+}, Ni^{2+}, Zn^{2+}, and Co^{2+} ions are used for purification of His-tagged proteins.

As discussed above, there are many matrices that have been used for affinity chromatography.[123] The natural polysaccharides, such as agarose, cellulose, and cross-linked dextran, possess a high content of hydroxyl groups available for activation and derivatization. Other organic polymers such as polyacrylamide, polyacrylate, and polyvinyl polymers have also been used. The inorganic polymer silica is very stable and can easily be derivatized to introduce functional ligands. All these solid matrices have their strengths and weaknesses. To improve their properties, a variety of protocols have been developed to modify the surface of these supports by either chemical modification or physical adsorption of polymers. Recently, monolithic materials have been developed.[124] Monoliths are continuous stationary phases that are made as a homogeneous column in a single piece. Organic monoliths are prepared by *in situ* polymerization of monomers, cross-linkers, porogens, and an initiator. Methacrylate and acrylamide-based polymers, poly(styrenedivinylbenzene), agarose, and cryogels are examples of organic monoliths. Inorganic monoliths (silica) are prepared by the sol–gel method or from bare silica particles. The monoliths can be molded into various shapes and forms. Ligands are coupled onto the monolithic stationary phase by various functional groups. Monolith with epoxide functionality can be used directly for immobilizations of ligands with amino groups. The ligands can be attached via different spacers. For example, ethylene diamine can be used to extend the distance from the epoxide moiety. Monoliths containing 1,2-diol groups can be oxidized with periodate to generate *in situ* aldehyde groups, which are used to attach the ligand via an amino group by reductive amination. In addition, monolith surface can be coated with various materials. A streptavidin-coated monolith is used to immobilize biotinylated DNA.

14.6.2 Biosensors

Biosensors are devices used for the detection and measurement of biological molecules. Although various forms of biosensors have been designed, they basically consist of two parts, a biologically derived part and a physical part referred to as the physicochemical transducer. The biological part is mostly based on enzymes. For example, biosensors based on acetylcholinesterase or butyrylcholinesterase principally monitor the activity of bound enzyme and change of its activity due to the presence of an analyte and are used for detection of organophosphorus and carbamate pesticides, nerve agents (e.g., sarin), and other natural toxins (e.g., aflatoxins).[125] The biorecognition component is coupled to electrochemical, optical, or piezoelectric transducers for measurement of the enzyme activity. There are many methods for the immobilization of cholinesterases onto membranes such as adsorption, entrapment, and chemical cross-linking. The chemical procedure uses bifunctional reagents with glutaraldehyde being the most widely used.

Another common biosensor is based on GOD.[126] Its use includes monitoring blood glucose for the control of diabetes, online glucose monitoring for fermentations, and analyzing glucose concentrations in soft drinks. A variety of supports have been used for immobilization of GOD such as cellulose, solid glass particles, porous glass particles, and nickel screens with porous glass and cellulose being the most popular supports. Nickel oxide screens can be silanized and GOD can be coupled by thiophosgene method. GOD can be cross-linked to solid supports with glutaraldehyde in the presence of polyethyleneimine. Immobilization of GOD in the presence of its substrates has been shown to protect the enzyme activity.

There are many other enzymes used in biosensors and some are genetically engineered to produce more stable structure, higher activity, broader dynamic range, and additional functional groups for immobilization.[127,128] Methylamine dehydrogenase, for example, has been engineered to improve sensitivity of a histamine senor.[129] Pyrroloquinoline quinone glucose dehydrogenase was engineered to expand the dynamic range of a biosensor.[130] A fusion enzyme between a P450 monooxygenase and a NADPH-cytochrome P450 oxidoreductase was genetically engineered to give higher enzyme activity than natural P450s.[131] Cysteine residues have been introduced into protein sequences for the immobilization of enzymes based on the dative binding between thiol groups and gold surfaces. For instance, six mutants of recombinant monomeric superoxide dismutase were engineered to contain one or two additional cysteine residues for the self-assembling on gold electrodes.[132] Enzymes have also been modified with histidine residues for their immobilization on specific metals. A hexa-His-tagged acetylcholinesterase was immobilized on a Ni-NTA screen-printed electrode as discussed for IMAC above.[133] The hexa-His-tagged AChE has also been conjugated to Ni-IDA-modified magnetic beads. Polyhistidine-tagged proteins have also been specifically immobilized on to Ni-NTA functionalized silicon nanowire for the preparation of biosensor field effect transistors.[134]

14.6.3 MICROARRAYS

Microarray technology has become a fundamental tool for fast detection and analysis of biological molecules. This rapidly growing field started with DNA microarrays for the analysis of nucleic acid sequence information. Major applications of this technology include studying gene expression profiles and the detection of single nucleotide polymorphisms (SNPs). This informative technology has expanded to other arrays such as protein/peptide microarrays, carbohydrate microarrays, and antibody microarrays. In all these microarray applications, an array biochip is produced, which involves immobilization of probe molecules onto a solid support. A basic technological advance is the methodology to immobilize all necessary probe molecules onto solid matrices. The following paragraphs will illustrate the importance of chemical cross-linking in the formulation of these microarray biochips.

14.6.3.1 DNA Microarrays

DNA microarrays and DNA chips are well developed and are commercially available for various applications.[135] DNA hybridization microarrays are generally fabricated on glass, silicon, or plastic surfaces. Other surface materials including nylon, nitrocellulose, polyacrylonitrile, polypropylene, polystyrene, teflon, gold, polypyrrole, polyurethane, polymethylmethacrylate (PMMA), and polyethylene glycol-grafted silica surfaces and PMMA have also been used.[135–138] The actual construction of microarrays involves the immobilization or *in situ* synthesis of DNA probes onto the specific test sites of the solid support or substrate material. Affymetrix has developed a photolithographic technique to carry out the parallel synthesis of large numbers of oligonucleotides on solid surfaces.[139] In this process, the glass is first modified with a silane reagent to provide hydroxyalkyl groups, which are extended with linker groups protected with special photolabile protecting moieties. The protecting groups are selectively removed on photolysis, allowing the sites to be coupled by standard phosphoramidite DNA synthesis protocols with an appropriate nucleoside phosphoramidite monomer, which is protected at their 5′ position with the photolabile 5′-(O-methyl-6-nitropiperonyloxycarbonyl) group. The 5′-terminal protecting groups are selectively removed from growing oligonucleotide chains by controlled exposure to light for reaction with another protected monomer. The cycles of photodeprotection and nucleotide addition are repeated numerous times to build any given array of oligonucleotide sequences This methodology has enabled the large-scale manufacture of arrays containing thousands of oligonucleotide probe sequences on glass slides. Polymeric photoresist films as the photoimageable component have also been developed for fabricating DNA arrays that utilize photolithographic methods.

Nanogen has developed an alternative DNA array technology using microelectronic chip devices to increase the DNA hybridization rate by concentrating probe/target DNA at the test site.[140] Arrays or chips are fabricated on silicon wafers, having a base structure of silicon with an insulating layer of silicon dioxide. Reversing the electric field on the test site produces an electronic stringency effect, which can greatly improve hybridization specificity and provide the ability to discriminate point mutation and SNPs.

Chemical cross-linking has been used to attach or immobilize DNA probes onto support materials.[141,142] The individual oligonucleotides are separately synthesized.[138] The synthetic nucleotides can be immobilized onto zirconylated surface based on coordination chemistry involving phosphoryl group of the oligonucleotides.[143] Lane et al.[144] found that poly(dG) spacer-bearing oligonucleotide probes immobilized on a zirconium phosphonate surface have a higher target capture rate than probes with either no spacer or a different polynucleotide (polyA, poly C, and polyT) spacer during hybridization with complementary targets. In a different approach, Wu et al.[145] were able to covalently immobilize unmodified oligonucleotides onto glass slides coated with acrylic acid-co-acrylamide copolymer after activation of the copolymer coating with EDC/NHS.

In most cases, the oligonucleotides are derivatized by nucleotide chemistry processes prior to immobilization. Various functional groups can be attached to the 3'- or 5'-end of the oligonucleotide such as mercaptoalkylation, thiophosphorylation, aminoalkylation, and silanization. The solid support media, for example, glass slide, nylon filter, or other surfaces, are activated with bifunctional or other reagents to generate special functionalities. For example, glass slides can be silanized with 3-glycidyloxypropyltrimethoxysilane to generate epoxy functions.[146] Aminated supports can be reacted with phenylene diisothiocyanate, disuccinimidyl carbonate or disuccinimidyl oxalate, dimethylsuberimidate, and succinimidyl 4-(maleimidophenyl) butyrate to generate isothiocyanate-, succinimidyl-, imidoester-, and maleimido-activated surface as shown in Figure 14.20.[147,148]

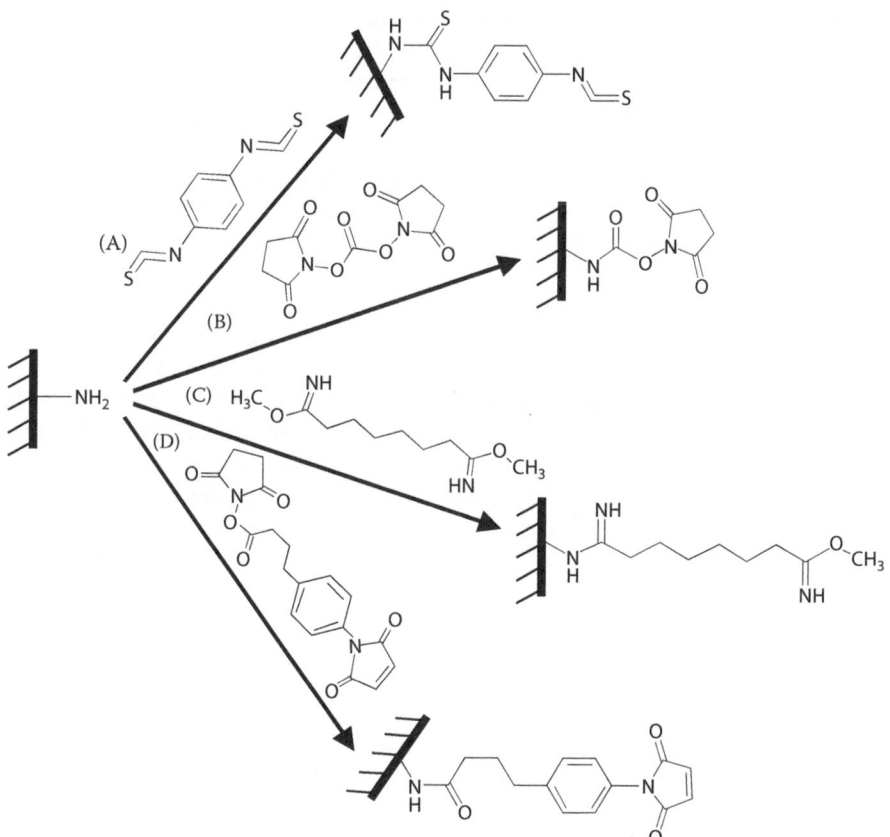

FIGURE 14.20   Some examples of microchip surface activation by bifunctional reagents for the immobilization of presynthesized DNA: (A) phenylene diisothiocyanate; (B) disuccinimidyl carbonate; (C) dimethylsuberimidate; and (D) succinimidyl 4-(maleimidophenyl) butyrate.

Other bifunctional cross-linkers such as *m*-maleimidobenzoyl-*N*-hydroxysuccinimide ester (MBS), succinimidyl 4-(*N*-maleimidomethyl) cyclohexane-1-carboxylate (SMCC), *N*-(γ-maleimidobutryloxy) succinimide ester (GMBS), *m*-maleimidopropionic acid-*N*-hydroxysuccinimide ester (MPS), and *N*-succinimidyl (4-iodoacetyl) aminobenzoate (SIAB) have also been used.[148,149] In addition, activation with NHS ester groups has been employed to activate glass surfaces.[6] Many of these activated support media are now available commercially. These reactive groups can be used to specifically immobilize oligonucleic acid with appropriate moieties as illustrated below.

Aminoalkylated oligonucleotides can react with amino-group–directed reactive functionalities introduced onto solid supports using the cross-linkers mentioned above. For instance, amine-modified nucleic acid can be immobilized onto epoxy silane-derivatized or isothiocyanate-coated glass slides.[146,150] The epoxylated surface can also immobilize oligonucleotides with electrophilic groups such as mercaptoalkyl-, aminooxyalkyl-, phosphoryl-, and thiophosphorylated oligonucleotides.[146,151–153] Amine-modified oligonucleotides can also be immobilized onto an aldehyde-containing surface by Schiff base formation, which can be stabilized with sodium cyanoborohydride (reductive amination).[154] A similar reaction can occur between aldehyde modified oligonucleotides and aminooxyalkylated glass surface where a stable oxime bond is formed, obviating the need for chemical reduction with sodium cyanoborohydride.[155] Aminooxyalkyl groups can be generated by UV treatment of 2-(2-nitrophenyl)propyloxycarbonyl group protected aminooxy silane. Another approach to immobilized amine modified nucleic acids is to react with tetrafluorophenyl (TFP)-activated self-assembled monolayers (SAMs) on gold-coated glass slides as shown in Figure 14.21.[156] TFP-SAMs are formed by coupling thiol-containing TFP to gold surfaces. This methodology can be extended to prepare NHS-, aldehyde-, and maleimide-activated SAMs, which can then be used to immobilized appropriately modified oligonucleotides.

Succinylated oligonucleotides can be immobilized onto aminophenyl- or aminopropyl-derivatized glass slides through amide bonds formation.[157] Disulfide-modified oligonucleotides can be immobilized onto a mercaptosilanized glass support by the thiol/disulfide exchange reaction.[158]

Thioalkylated oligonucleotides can be immobilized onto maleimido-activated surfaces[148] or directly on gold surfaces.[159] They can also be immobilized using a heterobifunctional reagents, *N*-(3-triethoxysilylpropyl)-4-(*N′*-maleimidylmethyl) cyclohexanamide (TPMC)[160] or *N*-(3-triethoxysilylpropyl)-6-(*N*-maleimido)-hexanamide (TPMH).[161] The triethoxysilyl functionality of the compounds has specificity toward a glass surface, whereas the maleimide functionality can react with the thiol-modified oligonucleotides to form a stable thioether linkage. Immobilization of DNA can be achieved by either reacting TPMC or TPMH with oligonucleotides to generate triethoxysilyl-oligonucleotide conjugates, which are then covalently attached via specific triethoxysilyl functionalities to an unmodified glass surface, or by first reacting the reagent with an unmodified glass surface to place maleimide functionalities on the surface, which are then used for immobilization of oligonucleotides via stable thioether linkages (Figure 14.22). Oligonucleotides can also be immobilized with *N*-(3-triethoxysilylpropyl)-4-(isothiocyanatomethyl) cyclohexane-1-carboxamide (TPICC)[162] or

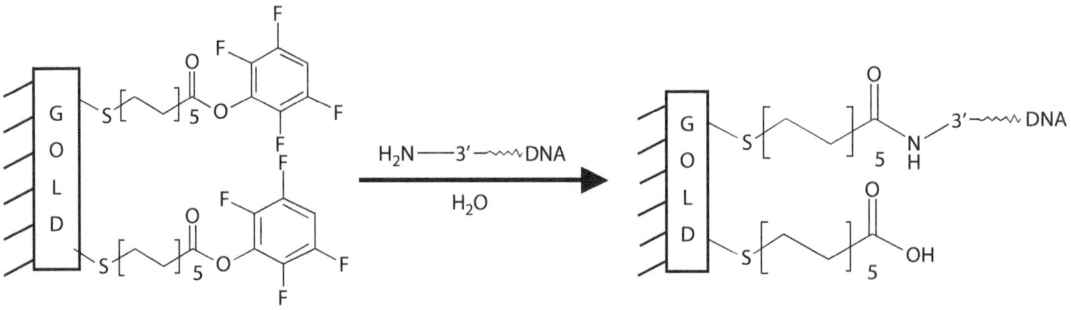

FIGURE 14.21 Coupling of amine-modified oligonucleotides to tetrafluorophenyl-activated self-assembled monolayers on gold-coated glass slides. Hydrolysis of unreacted esters completes the reaction.

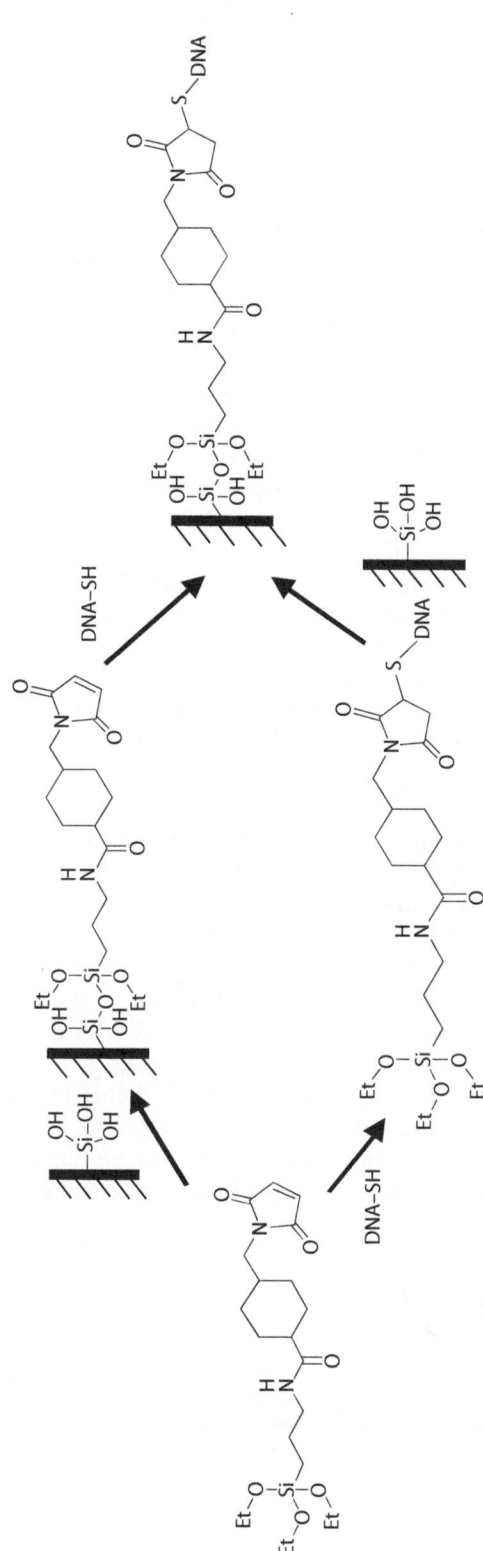

FIGURE 14.22 Coupling of thiol-containing oligonucleotides to glass slides using *N*-(3-triethoxysilylpropyl)-4-(*N'*-maleimidylmethyl) cyclohexanamide in two different approaches.

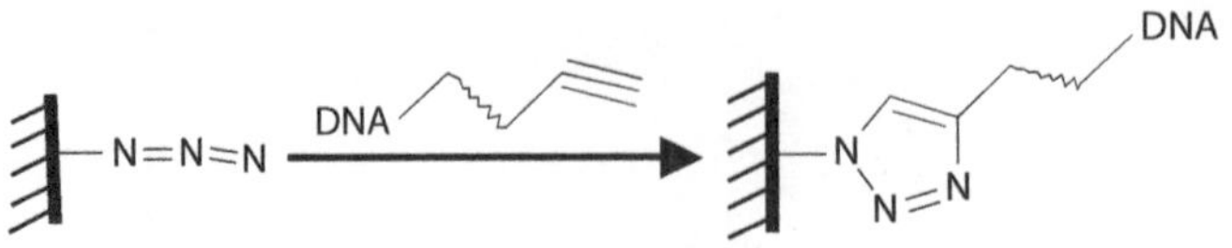

FIGURE 14.23 Coupling of acetylene-modified oligonucleotide to azide-terminated glass slide using click chemistry.

N-(2-trifluoroethanesulfonatoethyl)-N-(methyl)-triethoxysilylpropyl-3-amine (NTMTA)[163] in a similar manner. Again, the triethoxysilyl functionality of TPICC and NTMTA can react with a glass surface and the isothiocyanate functionality of TPICC, and the trifluoroethanesulfonate ester group can react with either aminoalkyl (R-NH$_2$) or mercaptoalkyl (R-SH) functionalities of modified oligonucleotides to form stable covalent bonds.

Many other approaches are also possible. Rozkiewicz et al.[164] used click chemistry to immobilize oligonucleotides. In this approach, acetylene-modified oligonucleotides form stable covalent bonds with a azide-terminated glass substrate as shown in Figure 14.23. Pack et al.[165,166] introduced oxa-nucleotide at the 3'- or 5'-end of the desired oligonucleotide sequences and reacted these oxa-nucleic acid with amine-functionalized glass slides under defined conditions as shown in Figure 14.24. The resulting microarrays were found to be highly efficient. Benters et al.[167] increased reactive functionalities on glass by attaching poly(amidoamine) (PAMAM) dendrimers with 64 amino groups in its outer sphere onto the aminated glass surface. The attached PAMAM dendrimers were subsequently activated with 1,4-phenylenediisothiocyanate (PDITC) or disuccinimidylglutarate (DSG) to generate a chemically reactive isothiocyanato- or succinimido-polymeric film, respectively. The resulting surface was used for the immobilization of amine-modified oligonucleotides. In a similar approach, Hong et al.[168] used a cone-shaped dendron instead of PAMAN to ensured proper spacing between immobilized biomolecules. Alternatively, glass slides can be coated with chitosan. The amine-rich polysaccharide is then activated with PDITC to generate reactive isothiocyanato functions, which can subsequently immobilize amine-modified oligonucleotides.[169] Abramov et al.[170] used nucleic acid analogs, hexitol nucleic acids (HNA), and altritol nucleic acids (ANA), which contain an anhydrohexitol sugar moiety, as probes in DNA microarray. HNA and ANA have high affinity toward DNA and RNA targets. The nucleic acid analog arrays are fabricated by immobilization of diene-modified oligonucleotides on maleimido-activated glass slides. The HNA/ANA arrays could become a useful tool for nucleic acid diagnostics. Another immobilization method involves the interaction of biotin and avidin. Lim et al.[171] developed a protocol for immobilizing biotinyl-oligonucleotides onto solid support coated with PAMAM dendrimers, which were peripherally modified with biotin and avidin. Breitenstein et al.[172] used the same avidin/biotin approach to immobilize biomolecules. A glass slide was first silanized with 3-amino-propyltriethoxysilane to provide the surface with reactive amino groups that could be biotinylated with sulfo-NHS-biotin. Neutravidin was spotted with an atomic force microscope (AFM) tip to

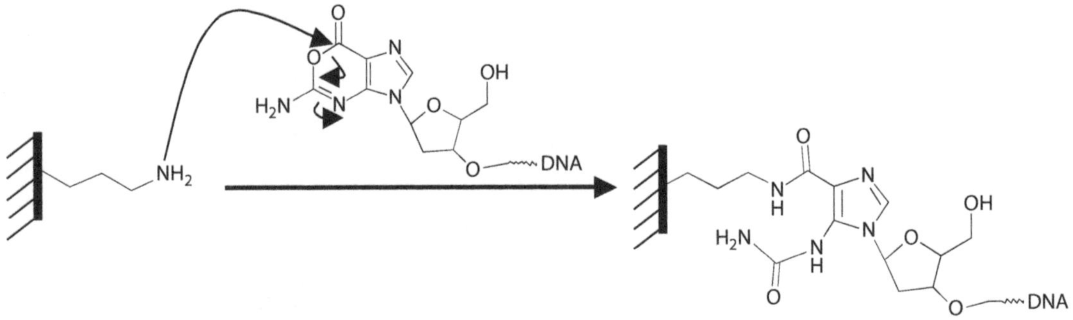

FIGURE 14.24 Coupling of oxa-nucleotide containing oligonucleotide to amine-terminated glass slides.

bind to the biotinylated surface. Biotinylated DNA was then added to bind to neutravidin. After washing, neutravidin was deposited a second time with the same AFM tip, and then a second biotinylated DNA was coupled by incubation. The authors claimed that the method can be used to deposit biological molecules that can be coupled to biotin.

In addition to these modified surfaces, unactivated microscope slides may be used with activated, silanized oligonucleotides. Silanized nucleic acids can be immobilized instantly onto glass surfaces. Kumar et al.,[173] for example, prepared silanized oligonucleotides by reacting 5′-thiolated oligonucleotides with either (3-mercaptopropyl)-trimethoxysilane followed by oxidation or 3-aminopropyl trimethoxysilane in the presence of SPDP. The trimethoxysilanized oligonucleotides can also be prepared by reacting aminoalkyl, mercaptoalkyl, and phosphorylated oligonucleotides with glycidoxypropyltriethoxysilane.[174] These silanized nucleic acids can be directly immobilized onto unmodified glass slides. Lim et al.[175] introduced distinct functional chemical groups into double-stranded DNA fragments of a random sequence using short synthetic DNA oligomers carrying desired terminal functional groups. When a thiol group is used, these thiol-modified DNA can be immobilized onto gold-coated or silane-modified solid substrates.

Photochemical reactions can also be used to immobilize oligonucleotides, although a number of side reactions can occur. Kumar et al.[176,177] synthesized two anthraquinone-based photosensitive heterobifunctional cross-linkers, N-(3-trifluoroethanesulfonyloxypropyl)anthraquinone-2-carboxamide (NTPAC)[176] and N-(iodoacetyl)-N'-(anthraquinon-2-oyl)-ethylenediamine (IAED).[177] The anthraquinone moiety is photosensitive and can be activated by irradiation (365 nm) to react with a variety of carbon-containing polymers. The trifluoroethanesulfonate ester group and the iodoacetyl group of the reagents react preferentially with aminoalkyl or mercaptoalkyl functions present in the biomolecules. Like TPMC and TPMH discussed above, there are two ways the reagents can immobilize aminoalkyl- and mercaptoalkyl-oligonucleotides (Figure 14.22). They can first react with the modified nucleic acid to form the oligonucleotide-anthraquinone conjugate and can then be photolyzed to covalently bind to solid support. They can also be first photolyzed to couple to the solid support followed by reaction with the modified oligonucleotides. Either method seems to give the same efficiency. Another anthraquinone photoreactive heterobifunctional reagent, 1-N-(maleimidohexanoyl)-6-N-(anthraquinon-2-oyl) hexanediamine (MHAHD), possessing an electrophilic maleimide group, has also been shown to efficiently immobilize thiolated oligonucleotides on a modified glass surface.[178]

14.6.3.2 Protein/Peptide Microarrays

Protein microarrays, which evolved after DNA microarrays, are particularly powerful for analyzing gene function, regulation, and a variety of other applications including medical diagnostics.[179,180] In protein microarrays, it is important that the proteins are retained on the solid substrate in an active state at high densities. Materials such as polystyrene, poly(vinylidene fluoride), nitrocellulose, glass microscope slides, and other matrices, to which proteins can be attached in high densities, have been used.[181] A variety of biochips has been designed, including 3D surface structures, nanowell, and plain glass biochips.[182] Since polyacrylamide gel packets and agarose thin films form highly porous and hydrophilic matrices, probe molecules can diffuse into the porous structure and can be immobilized by chemical cross-linking to the reactive group on the matrices. These 3D matrices can be formed on glass surfaces to increase the binding capacity of the planar surface.[25,182,183]

As in DNA microarrays, the challenge of constructing a protein/peptide biochip is the immobilization of proteins or peptides onto solid support surfaces. Proteins can be noncovalently adsorbed onto solid matrices via intermolecular forces, mainly ionic bonds, hydrophobic and polar interactions.[183] An example is the use of thin poly-L-lysine films on glass slides. Another example involves coordination chemistry where histidine residues of proteins physical bind to nickel or cobalt ions contained in a self-assembled monolayer at the surface. Chang and Chang[184] fabricated a nickel–cobalt alloy layer by electrodeposition as a protein biochip surface. His$_6$-tagged urate oxidase protein and penta-His biotin were successfully immobilized on the biochip. In general, proteins/peptides are covalently linked to solid matrices to achieve more specific binding and in such a way that the binding sites are exposed to the sample solution. Reactive functional groups are introduced to activate the solid biochip surfaces as discussed earlier in this chapter. Since most of the protein immobilization methodologies are presented in earlier sections, the following paragraphs will focus on demonstrating their specific use in fabricating protein/peptide microarray biochips. MacBeath and Schreiber[185] used glass slides containing aldehyde groups to print protein arrays. The aldehyde groups reacted with primary amines on the proteins to form Schiff base linkages, which were stabilized on reduction creating stable secondary amine linkages. To print peptides and very small proteins, the authors first attached a molecular layer of BSA to the surface of glass slides and then activated the BSA with N,N'-disuccinimidyl carbonate. The NHS-activated lysine, aspartate, and glutamate residues on the BSA reacted readily with surface amines on the printed proteins to form covalent urea or amide linkages. Because typical proteins display many lysines, with ε-amine moieties, on their surfaces as well as the α-amine at the NH_2-termini, they can attach to the slide in a variety of orientations, permitting different sides of the protein to interact with other proteins or small molecules in solution. In a different application, Funeriu et al.[186] used hydrogel-aldehyde functionalized glass slides to prepare enzyme microarray using a commercial microarrayer. The authors used the microarrayed enzymes to study low-molecular-weight fluorescent-affinity labels as activity probes.

The NHS chemistry, as mentioned above, is commonly used because the NHS ester can readily react with amine or hydrazide functionalities in substances of interest. There are many methods by which NHS-activated surfaces can be prepared. Recently, Park et al.[17] developed an efficient, single-step method for the derivatization of glass surfaces with NHS ester groups. The procedure involved reaction of NHS-ester functionalized dimethallylsilanes with glass surface silanols in the presence of acid catalyst trifluoromethanesulfonic acid. The NHS-activated glass surface was used to prepare microarrays containing various substances, such as proteins, peptides, sugars, and small molecules.

Bifunctional silane cross-linkers, which have one functional group that reacts with the hydroxyl group on glass surface and another to react with primary amine groups of proteins, can form a SAM. For example, (3-glycidyloxypropyl)trimethoxysilane (GOPS) activates the glass surface with an epoxide group, which reacts with primary amino groups of proteins forming a covalent bond. Fall et al.[187] immobilized allergen solutions on the GOPS-activated glass slides with a piezoelectric arrayer for screening of allergen-specific IgE in human serum. The researchers demonstrated that it was possible to distinguish between patients with and without elevated levels of allergen-specific IgE.

Proteins and peptide-containing free thiol groups are specifically able to be immobilized onto solid matrices that are activated by various thiol-directed cross-linkers as discussed in early sections of this chapter. Both homobifunctional and heterobifunctional reagents containing N-maleimido groups have been used. Disulfide-activated matrices, such as pyridyl disulfides, can react with proteins containing free thiols through disulfide exchange reactions leading to the formation of a new mixed disulfide. Vinyl sulfone–modified solid matrices provide another means to immobilize proteins with free thiols groups by a conjugate addition reaction. In addition, gold-coated glass surfaces can capture thiol-proteins through an Au-SH bond formation. Ressine et al.[29] used three-dimensional gold-coated silicon chips to immobilize laccase onto an aminothiophenol SAM. Proteins can be immobilized by Staudinger ligation between azides and appropriately substituted phosphanes.[188] In this approach, alkyl azides are introduced into small molecules, peptides, and proteins, which react with phosphane to form a chemically stable amide bond as shown in Figure 14.25. Köhn[188] used this method to prepare peptide and protein microarrays where azide-functionalized N-Ras proteins were immobilized onto phosphane-modified glass surfaces.

Photochemical reactions have also been employed to create protein/peptide, glycopeptides, and oligonucleotide microarrys.[189] Weinrich et al.[190] utilized the thiolene reaction, which is a photoinduced addition of thiols to terminal alkenes yielding a thioether linkage to couple various biomolecules to silicon wafers as shown in Figure 14.26. Olefin- or thiol-functionalized biomolecules were immobilized onto thiol-PAMAM- or olefin-PAMAM-functionalized glass slides, respectively, after

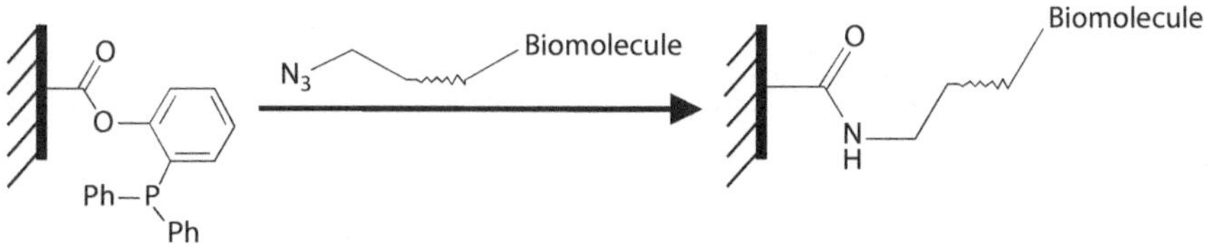

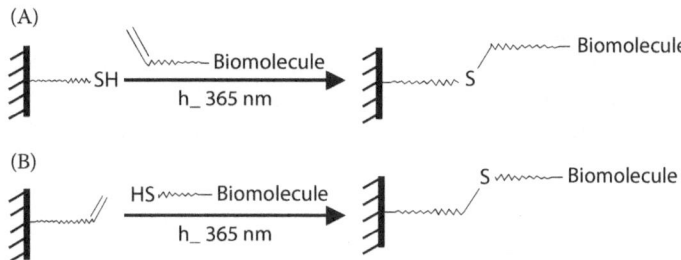

FIGURE 14.25 Immobilization of azide functionalized biomolecules by Staudinger ligation onto phosphane-modified glass surfaces.

FIGURE 14.26 Thiol-ene photoimmobilization of olefin- or thiol-functionalized biomolecules onto (A) thiol-PAMAM- or (B) olefin-PAMAM-functionalized glass slides, respectively.

exposure to UV light at 365 nm. For example, the allyl amides of streptavidin and biotin were synthesized and immobilized on thiol-modified solid surfaces.[190] Other photoreactive functional groups have been used to immobilize proteins onto solid matrices. These include diazirine, azidophenyl, benzophenone, dithiocarbamate, and camphorquinone as UV-reactive groups, and fluorescein, eosin, and Rose Bengal as visible light-sensitive groups. Miller et al.[191] used azidophenyl groups for the preparation of an antibody microarray. In this process, microscope slides were coated with poly-L-lysine, which was reacted with a second layer of N-hydroxysuccinimide-4-azidobenzoate (HSAB). Antibodies were deposited and cross-linked to the HSAB upon UV irradiation. Kanoh et al.[192] prepared small-molecule arrays on gold using surface-bound aryl diazirine. SAMs of alkanethiolates terminated with a phenyl diazirine group are formed on gold substrates. Small organic molecules are immobilized onto SAM via highly reactive carbene species generated from surface-bound aryl diazirine upon UV irradiation at 365 nm. Ito et al.[193] prepared photoreactive water-soluble polymers containing azidoaniline or azidobenzoic acid to prepare protein microarrays. These polymers include poly(acrylic acid), poly(vinyl alcohol), poly(phosphatidylcholine methacrylate), and PEG. To synthesize a protein microarray chip, the photoreactive polymer was mixed with proteins and spotted on a solid surface. The solution was dried and UV irradiated to immobilize the proteins. Moschallski et al.[25] also used photosensitive polymers to immobilize proteins onto unmodified plastic surface. A terpolymer based on water-soluble dimethylacrylamide with the photo-crosslinker methacryloyloxybenzophenone and methacrylic acid glycidylester (MAGE) was synthesized. MAGE introduced epoxide side groups to the polymer that could bind proteins through chemical reactions. The terpolymers were printed together with the protein onto the surface of a plastic substrate. On irradiation with UV-light at a wavelength of 254 nm, the polymer was cross-linked and immobilized the protein to the plastic surface.

14.6.3.3 Antibody Microarrays

An antibody microarray is a special version of a protein microarray. It is an analytical array in which antibodies or antibody mimic reagents that bind specific antigens are arrayed on solid supports.[194] In a way, it is essentially an immunoassay, viz a microarray immunoassay. Like protein microarrays, in general, solid-surface chemistry is an important aspect of a robust microarray platform, particularly with respect to immobilization of antibodies. A large repertoire of solid supports has been fabricated based on glass-, plastic-, or silicon-slides with structured surface modifications.[195] A myriad of surface chemistries is available commercially, some of which can be used to immobilize antibodies directly. Seurynck-Servoss et al.[196] evaluated these surface chemistries and found that glass slides coated with aldehyde silane, poly-L-lysine, or aminosilane, with or without activation with a cross-linker, produced superior results in the sandwich ELISA microarray analyses.

Immobilizations of antibodies and other antibody mimic reagents are accomplished in manners similar to other proteins. The approaches include physical adsorption and covalent binding via amino groups of lysines and arginines, sulfhydryl groups either in the hinge region or between light and heavy chains, and carbohydrate residues of the Fc region. A more specific method pertaining to antibodies is the coupling with proteins A or G. Since the active sites of antibodies should be accessible to analytes in immunoassays, care should be taken to immobilize the antibodies in the correct orientation. There are multiple choices of immobilization strategies.[197] The solid support could be coated with proteins A or G for direct coupling with antibodies.[198] Activated glass slides for antibody immobilization could be prepared with mercaptosilane followed with homo- or heterobifunctional cross-linkers such as maleimido-N-hydroxysuccinimide esters.[199,200] They can also be prepared with epoxy-silanization and amino-silanization with maleimido-succinimidyl cross-linker treatment.[200] Antibodies biotinylated via carbohydrate and amino groups can be immobilized on streptavidin-coated solid support.[201] Fab fragments can also be biotinylated. Chemical reduction of the disulfide bridges of antibodies will generate free sulfhydryl groups, which can react with thiol reactive surfaces. Thiol-derived antibodies can also be bound onto gold-coated surfaces.[202] Surfaces modified with PEG such as poly-L-lysine-grafted PEG copolymers give better signal intensities and less unspecific protein binding than silanized surfaces.[203] Alkane thiols that terminate in short PEG groups also reduces nonspecific adsorption of proteins.[204]

Monoclonal antibodies based on recombinant DNA technology,[205] engineered microbial proteins such as affibodies,[206] and short single-stranded nucleic acids with protein binding properties, known as aptamers[207] provide suitable probes for microarray immunoassay. Single-chain antibodies (scFv), which contain the variable light and heavy-chain regions from IgG, typically connected by a short linker, are significantly smaller and much less complex than IgG and can be expressed in either bacterial or yeast hosts. scFv against prostate-specific antigen was printed onto glass slides for immunoassays.[208] Kumada et al.[209] genetically fused scFv with polystyrene-binding peptides by recombinant *Escherichia coli* for direct and site-specific immobilization of scFv on polystyrene supports with high antigen-binding activity. They showed that high antigen-binding activities were comparable to, or greater than, that for a whole monoclonal antibody. Recombinant fusion antibodies and affibodies offer another strategy. For example, Zhu et al.[210] produced fusion proteins of glutathione-S-transferase with a His$_6$-tag and spotted them on functionalized glass slides. Paborsky et al.[122] demonstrated that different recombinant proteins, with the six His tag at either the N- or C-terminus, can bind to the Ni^{2+}-NTA plate and can be used in a modified enzyme-linked immunoabsorbent assay format to quantify protein concentrations and to determine the affinity of protein-ligand interactions. Lesaicherre et al.[211] proposed a new strategy for a site-directed attachment of fusion protein using biotinylated affinity tags on avidin-coated slides.

14.6.3.4 Carbohydrate Microarrays

Carbohydrates are vital components of glycoconjugates that are involved in interactions with proteins.[212] Carbohydrate–protein recognition forms the basis of carbohydrate microarrays, which are composed of diverse glycans, orderly and densely attached to a single-chip substrate for fast, quantitative, and simultaneous analyses of carbohydrate-based biomolecular interactions.[213] A variety of carbohydrate microarray formats are available that use different solid surfaces and immobilization methods. Unmodified sugars can be adsorbed onto nitrocellulose-coated or black polystyrene-coated surfaces.[214] However, this immobilization technique is nonspecific and size dependent. Another method for immobilizing unmodified sugars is by photoinduced covalent attachment to

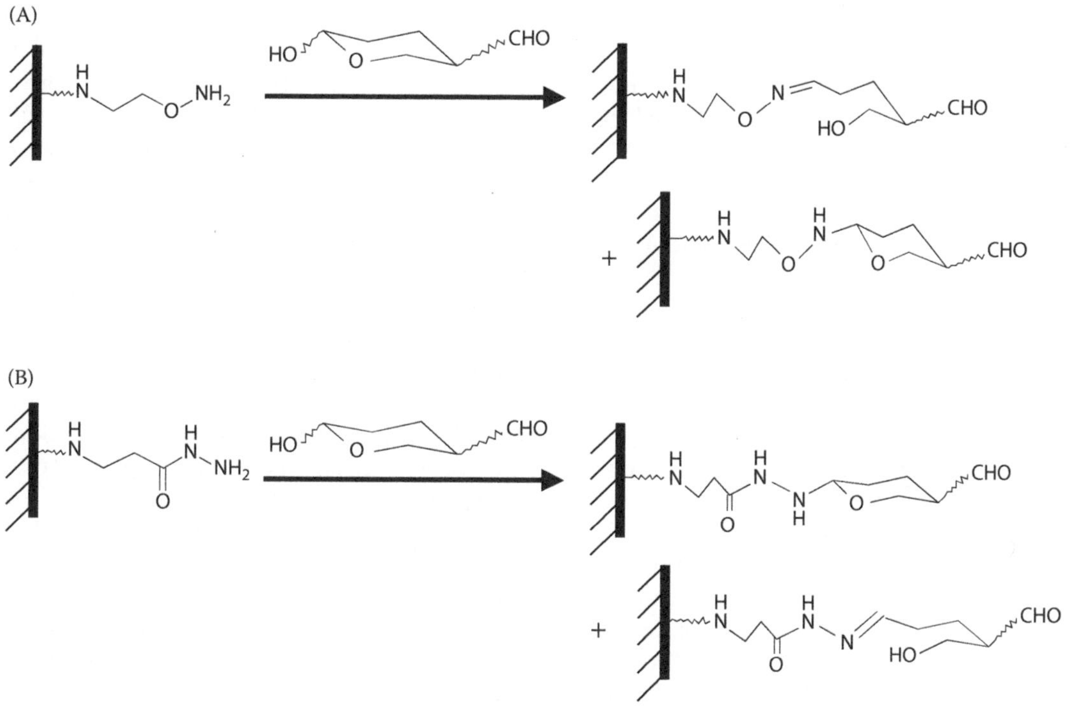

FIGURE 14.27 Immobilization of carbohydrates onto (A) aminooxy- and (B) hydrazide-coated glass slides.

surfaces that coated with photolabile groups such as aryltrifluoromethyl-diazirine.[215] On irradiation at 350 nm, aryltrifluoromethyl-diazirine groups form reactive carbenes, which eventually react with vicinal sugar molecules to form covalent bonds. Phthalimide-functionalized surfaces have also been used to photochemically immobilize underivatized glycans.[216] Lee and Shin[217] developed a direct, site-specific technique for the immobilization of unmodified carbohydrates, including simple carbohydrates, oligosaccharides, and polysaccharides, on hydrazide- and aminooxy-derivatized surfaces as shown in Figure 14.27. Carbohydrate microarrays containing a variety of glycans have been constructed by this method.[218]

The most general method involves site-specific and covalent immobilization of chemically modified carbohydrates to properly derivatize surfaces. Maleimide- or hydrazide-modified glycans can be immobilized on thiol- or epoxide-coated glass surfaces, respectively.[24,219] Sugars can be functionalized by one-step procedures for glycan microarray preparation. For example, free carbohydrates can be aminated with 2,6-diaminopyridine in the presence of sodium cyanoborohydride or with 2-aminomethyl *N,O*-hydroxyethyl bifunctional linkers.[220,221] These amine-appended sugars can be covalently immobilized onto the NHS ester-coated surfaces. Tateno et al.[222] took advantage of the primary amino groups of glycoproteins, neo-glycoproteins, and polyacrylamide to immobilize them on an epoxy-activated glass slide. Other method of immobilizing carbohydrates onto solid supports can be designed with bifunctional cross-linkers described earlier in this book.

14.6.4 INDUSTRIAL APPLICATIONS

In addition to the applications already discussed, other industrial applications involve immobilized enzymes. Numerous enzymes, for example, have found many commercial applications. Immobilization of enzymes enables reuse of the enzyme and continuous production of products in a suitable reactor. There are many methods for immobilizing enzymes.[12] The following illustrates a few examples. β-Fructofuranosidase is a well-known commercial enzyme for the production of fructooligosaccharides (FOSs) such as 1-kestose, nystose, and fructosyl-nystose. Kurakake et al.[223] immobilized β-fructofuranosidase from *Aspergillus oryzae KB* using an anion-exchange resin (WA-30; polystyrene with tertiary amine) and cross-linking with glutaraldehyde. The immobilized enzyme was packed in a tubular reactor for continuous production of FOSs from sucrose. These sugars are used as food ingredients. They have beneficial health effects in humans and improve the growth of bifidobacteria in intestinal flora. A related enzyme, invertase, is commercially important for the hydrolysis of sucrose, which yields glucose and fructose, known as invert syrup, widely used in food and beverage industries.[224] This enzyme is also used for the manufacture of artificial honey, as a plasticizing agent in the cosmetics, pharmaceutical, and paper industries and as an enzyme electrode for the detection of sucrose. Invertase has been immobilized on several hydrophilic as well as inorganic supports and polystyrene resins by various methods, including adsorption. Examples of covalent immobilization techniques include use of glutaraldehyde-activated aminoalkylsilylated magnetite, glutaraldehyde-activated aminopropyl silica, *O*-alkylated nylon tubes, by acyl-azide method, silane glutaraldehyde method, diazo method, and via its carbohydrate moieties. GOD is another important industrial enzyme.[126] It catalyzes the oxidation of β-D-glucose to gluconic acid by utilizing molecular oxygen as an electron acceptor with simultaneous production of hydrogen peroxide. Its commercial applications include glucose removal from dried eggs, improvement of color, flavor, and shelf life of food materials, oxygen removal from fruit juices, canned beverages, and mayonnaise to prevent rancidity, as a food preservative, and as a toothpaste ingredient. It has also been used in an automatic glucose assay and in biosensors for the detection and estimation of glucose in industrial solutions and in body fluids. GOD has been immobilized on both solid and porous glasses. It has been coupled by the thiophosgene method onto silanized nickel oxide screens. Its industrial applications will continue to expand.

In addition to immobilization of enzymes, immobilization of substrates to solid surfaces is becoming an increasingly important biotechnology.[10] This technique has been used for carbohydrate, peptide, oligonucleotide, and glycopeptide syntheses, microarray and biopolymer fabrication, design of antibiotics, and enzyme activity screening.

Advances in enzyme technology are still expanding. New applications of enzymes in industry are continuously being discovered, and new ways of immobilization are being developed. The use of chitosan, for example, as solid matrices increases thermal stability of immobilized enzyme and various immobilization techniques such as glutaraldehyde-chitosan, glyoxyl-chitosan, and epoxy-chitosan are being developed.[15] These and other developments will expand the use of cross-linking methodology in immobilization of enzymes for industrial use.

REFERENCES

1. Chang, T. M. S., Medical applications of immobilized proteins, enzymes, and cells, *Methods Enzymol.*, 137, 444, 1988.
2. Guilbault, G. G. and Mascini, M. (eds.), *Analytical Uses of Immobilized Biological Compounds for Detection, Medical and Industrial Uses*, Springer, New York, 1988.
3. Cao, L., *Carrier-Bound Immobilized Enzymes: Principles, Application and Design*, Wiley-VCH, New York, 2006.
4. Taylor, R. F. (ed.), *Protein Immobilization: Fundamentals and Applications*, Marcel Dekker Inc., New York, 1991.
5. Cass, A. E. G., Cass, T., and Ligler, F. S., *Immobilized Biomolecules in Analysis: A Practical Approach*, Oxford University Press, New York, 1998.
6. Guisan, J. M., *Immobilization of Enzymes and Cells*, 2nd edn., Humana Press, Totowa, NJ, 2006.
7. Zachariou, M. (ed.), *Affinity Chromatography: Methods and Protocols*, 2nd edn., Humana Press, Totowa, NJ, 2008.
8. Liang, J. F., Li, Y. T., and Yang, V. C., Biomedical application of immobilized enzymes, *J. Pharm. Sci.*, 89, 979, 2000.

9. Bangs, L. B., Particle-based tests and assays—Pitfalls and problems in preparation, *Am. Clin. Lab.*, 9, 16, 1990.

10. Laurent, N., Haddoub, R., and Flitsch, S. L., Enzyme catalysis on solid surfaces, *Trends Biotechnol.*, 26, 328–37, 2008.

11. Horng, W.-C., Yen, Y.-H., and Chang,Y.-C., A novel solid phase- and chemical crosslinking-based technology for determining protein localization in biological supramolecules, *Proteomics*, 8, 4642, 2008.

12. Brady, D. and Jordaan, J., Advances in enzyme immobilization, *Biotechnol. Lett.*, 31, 1639–1650, 2009.

13. Cao, L., Immobilised enzymes: Science or art? *Curr. Opin. Chem. Biol.*, 9, 217, 2005.

14. Royer, G. P. and Uy, R., Evidence for the induction of a conformational change of bovine trypsin by a specific substrate at pH 8.0, *J. Biol. Chem.*, 248, 2627, 1973.

15. Manrich, A., Komesu, A., Adriano, W. S., Tardioli, P. W., and Giordano, R. L., Immobilization and stabilization of xylanase by multipoint covalent attachment on agarose and on chitosan supports, *Appl. Biochem. Biotechnol.*, 161, 455, 2010.

16. Srere, P. A. and Uyeda, K., Functional groups on enzymes suitable for binding to matrices, *Methods Enzymol.*, 44, 11, 1976.

17. Park, S., Pai, J., Han, E.-H., Jun, C.-H., and Shin, I., One-step, acid-mediated method for modification of glass surfaces with N-hydroxysuccinimide esters and its application to the construction of microarrays for studies of biomolecular interactions, *Bioconjug. Chem.*, 21, 1246, 2010.

18. Burnham, M. R., Turner, J. N., Szarowski, D., and Martin, D. L., Biological functionalization and surface micropatterning of polyacrylamide hydrogels, *Biomaterials*, 27, 5883, 2006.

19. Ferruti, P., Ranucci, E., Sartore, L., Bignotti, F., Marchisio, M. A., Bianciardi, P., and Veronese, F. M., Recent results on functional polymers and macromonomers of interest as biomaterials or for biomaterial modification, *Biomaterials*, 15, 1235, 1994.

20. Izquierdo, M. P., Martín-Molina, A., Ramos, J., Rus, A., Borque, L., Forcada, J., and Galisteo-González, F., Amino, chloromethyl and acetal-functionalized latex particles for immunoassays: A comparative study, *J. Immunol. Methods*, 287, 159, 2004.

21 Sajeesh, S., Bouchemal, K., Sharma, C. P., and Vauthier, C., Surface-functionalized polymethacrylic acid based hydrogel microparticles for oral drug delivery, *Eur. J. Pharm. Biopharm.*, 74, 209, 2010.

22. Sun, X.-L., Stabler, C. L., Cazalis, C. S., and Chaikof, E. L., Carbohydrate and protein immobilization onto solid surfaces by sequential Diels-Alder and azide-alkyne cycloadditions, *Bioconjug. Chem.*, 17, 52, 2006.

23. Lee, M.-R. and Shin, I., Fabrication of chemical microarrays by efficient immobilization of hydrazide-linked substances on epoxide-coated glass surfaces, *Angew. Chem. Int. Ed.*, 44, 2881, 2005.

24. Park, S., Lee, M.-R., and Shin, I., Fabrication of carbohydrate chips and their use to probe protein–carbohydrate interactions, *Nat. Protoc.*, 2, 2747, 2007.

25. Moschallski, M., Baader, J., Prucker, O., and Rühe, J., Printed protein microarrays on unmodified plastic substrates, *Anal. Chim. Acta*, 671, 92, 2010.

26. Stillman, B. A. and Tonkinson, J. L., FAST slides: A novel surface for microarrays, *Biotechniques*, 29, 630, 2000.

27. Amin, I., Steenackers, M., Zhang, N., Beyer, A., Zhang, X., Pirzer, T., Hugel, T., Jordan, R., and Gölzhäuser, A., Polymer carpets, *Small*, 6, 1623, 2010.

28. Astuti, Y., Topoglidis, E., Cass, A. G., Durrant, J. R., Direct spectroelectrochemistry of peroxidases immobilised on mesoporous metal oxide electrodes: Towards reagentless hydrogen peroxide sensing, *Anal. Chim. Acta*, 648, 2, 2009.

29. Ressine, A., Vaz-Domínguez, C., Fernandez, V. M., De Lacey, A. L., Laurell, T., Ruzgas, T., and Shleev, S., Bioelectrochemical studies of azurin and laccase confined in three-dimensional chips based on gold-modified nano-/microstructured silicon, *Biosens. Bioelectron.*, 25, 1001, 2010.

30. Sousa, C. P., Polo, A. S., Torresi, R. M., de Torresi, S. I., and Alves, W. A., Chemical modification of a nanocrystalline TiO_2 film for efficient electric connection of glucose oxidase, *J. Colloid Interface Sci.*, 346, 442, 2010.

31. Prakasham, R. S., Devi, G. S., Rao, C. S., Sivakumar, V. S., Sathish, T., and Sarma, P. N., Nickel-impregnated silica nanoparticle synthesis and their evaluation for biocatalyst immobilization, *Appl. Biochem. Biotechnol.*, 160, 1888, 2010.

32. Crespilho, F. N,. Iost, R. M., Travain, S. A., Oliveira, O. N., Jr., and Zucolotto, V., Enzyme immobilization on Ag nanoparticles/polyaniline nanocomposites, *Biosens. Bioelectron.*, 24, 3073, 2009.

33. Wu, Y., Wang, Y., Luo, G., and Dai, Y., In situ preparation of magnetic Fe_3O_4-chitosan nanoparticles for lipase immobilization by cross-linking and oxidation in aqueous solution, *Bioresour. Technol.*, 100, 3459, 2009.

34. Weetall, H. H., Covalent coupling methods for inorganic support materials, *Methods Enzymol.*, 44, 134, 1976.

35. Kohn, J. and Wilchek, M., 1-Cyano-4-dimethylamino pyridinium tetrafluoroborate as a cyanylating agent for the covalent attachment of ligand to polysaccharide resins, *FEBS Lett.*, 154, 209, 1983.

36. Nagore, L. I., Mitra, S., Jiang, D., Jiang, S., Zhou, Y., Loranc, M., and Jarrett, H. W., Cyanogen bromide-activated coupling: DNA catalytic chromatography purification of EcoRI endonuclease, *Nat. Protoc.*, 1, 2909, 2006.

37. Desai, M. A., Immunoaffinity adsorption: Process-scale isolation of therapeutic-grade biochemicals, *J. Chem. Technol. Biotechnol.*, 48, 105, 1990.

38. Xiong, J., Zhu, H. F., Zhao, Y. J., Lan, Y. J., Jiang, J. W., Yang, J. J., and Zhang, S. F., Synthesis and anti-tumor activity of amino acid ester derivatives containing 5-fluorouracil, *Molecules*, 14, 3142, 2009.

39. Mier, W., Hoffend, J., Krämer, S., Schuhmacher, J., Hull, W. E., Eisenhut, M., and Haberkorn, U., Conjugation of DOTA using isolated phenolic active esters: The labeling and biodistribution of albumin as blood pool marker, *Bioconjug. Chem.*, 16, 237, 2005.

40. Fischer, M. J., Amine coupling through EDC/NHS: A practical approach, *Methods Mol. Biol.*, 627, 55, 2010.

41. Sam, S., Touahir, L., Salvador Andresa, J., Allongue, P., Chazalviel, J. N., Gouget-Laemmel, A. C., Henry de Villeneuve, C. et al., Semiquantitative study of the EDC/NHS activation of acid terminal groups at modified porous silicon surfaces, *Langmuir*, 26, 809, 2010.

42. Aksoy, S., Tumturk, H., and Hasirci, N., Stability of alpha-amylase immobilized on poly(methyl methacrylate-acrylic acid) microspheres, *J. Biotechnol.*, 60, 37, 1998.

43. Chen, G. X., Kim, H. S., Park, B. H., and Yoon, J. S., Controlled functionalization of multiwalled carbon nanotubes with various molecular-weight poly(L-lactic acid), *J. Phys. Chem. B*, 109, 22237, 2005.

44. Lee, R. T., Wong, T. C., Lee, R., Yue, L., and Lee, Y. C., Efficient coupling of glycopeptides to proteins with a heterobifunctional reagent, *Biochemistry*, 28, 1856, 1989.

45. Wood, W. G. and Gadow, A., Immobilisation of antibodies and antigens on macro solid phases—A comparison between adsorptive and covalent binding. A critical study of macro solid phases for use in immunoassay systems. Part I, *J. Clin. Chem. Clin. Biochem.*, 21, 789, 1983.

46. Campbell, D. H., Luescher, E., and Lerman, L. S., Immunologic adsorbents: I. Isolation of antibody by means of a cellulose-protein antigen, *Proc. Natl. Acad. Sci. U. S. A.*, 37, 575, 1951.

47. Zaborsky, O. R., *Immobilized Enzymes*, CRC Press, Boca Raton, FL, 1973.

48. Niwa, R., Kamada, H., Shitara, E., Horiuchi, J., Kibushi, N., and Kato, T., Synthesis of isomelamines and isocyanurates and their biological evaluation, *Chem. Pharm. Bull. (Tokyo)*, 44, 2314, 1996.

49. Biondi, P. A., Pace, M., Brenna, O., and Pietta, P. G., Coupling of enzymes to polyacrylonitrile, *Eur. J. Biochem.*, 61, 171, 1976.

50. Jain, S., Chattopadhyay, S., Jackeray, R., and Singh, H., Surface modification of polyacrylonitrile fiber for immobilization of antibodies and detection of analyte, *Anal. Chim. Acta*, 654, 103, 2009.

51. Geller, A. A., Yereshchenko, A. G., Zgibneva, Z. A., and Polovnikova, M. V., Transformations of polymers and copolymers of acrylonitrile during heterogeneous saponification, *J. Polym. Sci.*, 12, 2327, 1974.

52. Şanli, O., Homogeneous hydrolysis of polyacrylonitrile by potassium hydroxide, *Eur. Polym. J.*, 26, 9, 1990.

53. Liu, H. and Hsieh, Y.-L., Preparation of water-absorbing polyacrylonitrile nanofibrous membrane, *Macromol. Rapid Commun.*, 27, 142, 2006.

54. Jia, Z. and Shanyi Du, S., Grafting of casein onto polyacrylonitrile fiber for surface modification, *Fibers Polym.*, 7, 235, 2006.

55. Battistel, E., Morra, M., and Marinetti, M., Enzymatic surface modification of acrylonitrile fibers, *Appl. Surface Sci.*, 177, 32, 2001.

56. Vasil'eva, I. S., Morozova, O. V., Shumakovich, G. P., and Iaropolov, A. I., Synthesis of electroconductive polyaniline using immobilized laccase, *Prikl. Biokhim. Mikrobiol.*, 45, 33, 2009.

57. Luong, J. H. and Scouten, W. H., Affinity purification of natural ligands, *Curr. Protoc. Protein Sci.*, 9, Unit 9.3, 2008.

58. Gonçalves, R., Martins, M. C., Oliveira, M. J., Almeida-Porada, G., and Barbosa, M. A., Bioactivity of immobilized EGF on self-assembled monolayers: Optimization of the immobilization process, *J. Biomed. Mater. Res. A*, 94, 576, 2010.

59. Miron, T. and Wilchek, M., Immobilization of proteins and ligands using chlorocarbonates, *Methods Enzymol.*, 135, 84, 1987.

60. Narinesingh, D., Jaipersad, D., and Chang-Yen, I., Immobilization of linamarase and its use in the determination of bound cyanide in cassava using flow injection analysis, *Anal. Biochem.*, 172, 89, 1988.

61. Mei, J., Xu, J. R., Xiao, Y. X., Liao, X. Y., Qiu, G. F., and Feng, Y. Q., A novel covalent coupling method for coating of capillaries with liposomes in capillary electrophoresis, *Electrophoresis*, 29, 3825, 2008.

62. Scouten, W. H., van den Tweel, W., Kranenburg, H., and Dekker, M., Colored sulfonyl chloride as an activating agent for hydroxylic matrices, *Methods Enzymol.*, 135, 79, 1987.

63. Fedorova, N. V., Ksenofontov, A. L., Viryasov, M. B., Baratova, L. A., Timofeeva, T. A., and Zhirnov, O. P., Covalent chromatography of influenza virus membrane M1 protein on activated thiopropyl Sepharose-6B, *J. Chromatogr. B Biomed. Sci. Appl.*, 706, 83, 1998.

64. Porath, J. and Axén, R., Immobilization of enzymes to agar, agarose, and Sephadex supports, *Methods Enzymol.*, 44, 19, 1976.

65. Migneault, I., Dartiguenave, C., Bertrand, M. J., and Waldron, K. C., Glutaraldehyde: Behavior in aqueous solution, reaction with proteins, and application to enzyme crosslinking, *Biotechniques*, 37, 790, 2004.

66. Jayakrishnan, A. and Jameela, S. R., Glutaraldehyde as a fixative in bioprostheses and drug delivery matrices, *Biomaterials*, 17, 471, 1996.

67. Han, J., Salmieri, S., Le Tien, C., and Lacroix, M., Improvement of water barrier property of paperboard by coating application with biodegradable polymers, *J. Agric. Food Chem.*, 58, 3125, 2010.

68. Yang, G., Wu, J., Xu, G., and Yang, L., Comparative study of properties of immobilized lipase onto glutaraldehyde-activated amino-silica gel via different methods, *Colloids Surf. B Biointerfaces*, 78, 351, 2010.

69. Rapatz, E., Travnicek, A., Fellhofer, G., and Pittner, F., Studies on the immobilization of glucuronidase (Part 1). Covalent immobilization on various carriers (a comparison), *Appl. Biochem. Biotechnol.*, 19, 223, 1988.

70. Hendry, R. M. and Herrmann, J. E., Immobilization of antibodies on nylon for use in enzyme-linked immunoassay, *J. Immunol. Methods*, 35, 285, 1980.

71. Dekker, R. F., Immobilization of a lactase onto a magnetic support by covalent attachment to polyethyleneimine-glutaraldehyde-activated magnetite, *Appl. Biochem. Biotechnol.*, 22, 289, 1989.

72. Zhu, Z., Song, W., Burugapalli, K., Moussy, F., Li, Y. L., and Zhong, X. H., Nano-yarn carbon nanotube fiber based enzymatic glucose biosensor, *Nanotechnology*, 21, 165501, 2010.

73. Buranachai, T., Praphairaksit, N., and Muangsin, N., Chitosan/polyethylene glycol beads crosslinked with tripolyphosphate and glutaraldehyde for gastrointestinal drug delivery, *AAPS PharmSciTech.*, 11, 1128, 2010.

74. Wu, X. Q., Wu, C., and He, R. Q., Immobilized earthworm fibrinolytic enzyme III-1 with carbonyldiimidazole activated-agarose, *Protein Pept. Lett.*, 9, 75, 2002.

75. Kay, G. and Crook, E. M., Coupling of enzymes to cellulose using chloro-s-triazines, *Nature*, 216, 514, 1967.

76. Kartal, F. and Kilinç, A., Immobilization of pancreatic lipase on polyvinyl alcohol by cyanuric chloride, *Prep. Biochem. Biotechnol.*, 36, 139, 2006.

77. Porath, J., General methods and coupling procedures, *Methods Enzymol.*, 34, 13, 1974.

78. de Souza, M. C., Bresolin, I. T., and Bueno, S. M., Purification of human IgG by negative chromatography on omega-aminohexyl-agarose, *J. Chromatogr. B Anal. Technol. Biomed. Life Sci.*, 878, 557, 2010.

79. Jung, D., Streb, C., and Hartmann, M., Covalent anchoring of chloroperoxidase and glucose oxidase on the mesoporous molecular sieve SBA-15, *Int. J. Mol. Sci.*, 11, 762, 2010.

80. Rekuć, A., Kruczkiewicz, P., Jastrzembska, B., Liesiene, J., Peczyńska-Czoch, W., and Bryjak, J., Laccase immobilization on the tailored cellulose-based Granocel carriers, *Int. J. Biol. Macromol.*, 42, 208, 2008.

81. Wang, A., Wang, H., Zhu, S., Zhou, C., Du, Z., and Shen, S., An efficient immobilizing technique of penicillin acylase with combining mesocellular silica foams support and p-benzoquinone cross linker, *Bioprocess Biosyst. Eng.*, 31, 509, 2008.

82. Kálmán, M., Szajáni, B., and Boross, L., A novel polyacrylamide-type support prepared by p-benzoquinone activation, *Appl. Biochem. Biotechnol.*, 8, 515, 1983.

83. Kennedy, J. F. and Cabral, J. M., Immobilization of enzymes on transition metal-activated supports, *Methods Enzymol.*, 135, 117, 1987.

84. Bisse, E. and Vonderschmitt, D. J., Immobilization of glucose dehydrogenase by titanium tetrachloride, *FEBS Lett.*, 93, 102, 1978.

85. Moriya, K., Tanizawa, K., and Kanaoka, Y., Immobilized chymotrypsin by means of Schiff base copper(II) chelate, *Biochem. Biophys. Res. Commun.*, 162, 408, 1989.

86. Kennedy, J. F. and Kalogerakis, B., Immobilization of glucoamylase on gelatin by transition-metal chelation, *Biochimie*, 62, 549, 1980.

87. Kobayashi, H. and Ikada, Y., Covalent immobilization of proteins on to the surface of poly(vinyl alcohol) hydrogel, *Biomaterials*, 12, 747, 1991.

88. Saito, T. and Nagai, F., Immobilization of antibody to a plastic surface by toluene 2,4-diisocyanate and its application to radioimmunoassay, *Clin. Chim. Acta*, 133, 301, 1983.

89. Royer, G. P., Ikeda, S., and Aso, K., Cross-linking of reversibly immobilized enzymes, *FEBS Lett.*, 80, 89, 1977.

90. Jagendorf, A. T., Patchornik, A., and Sela, M., Use of antibody bound to modified cellulose as an immunospecific adsorbent of antigens, *Biochim. Biophys. Acta*, 78, 516, 1963.

91. Sato, T., Mori, T., Tosa, T., and Chibata, I., Studies on immobilized enzymes. IX. Preparation and properties of aminoacylase covalently attached to halogenoacetylcelluloses, *Arch. Biochem. Biophys.*, 147, 788, 1971.

92. Bayramoglu, G., Yilmaz, M., and Arica, M. Y., Preparation and characterization of epoxy-functionalized magnetic chitosan beads: Laccase immobilized for degradation of reactive dyes, *Bioproc. Biosyst. Eng.*, 33, 439, 2010.

93. Adriano, W. S., Mendonça, D. B., Rodrigues, D. S., Mammarella, E. J., and Giordano, R. L., Improving the properties of chitosan as support for the covalent multipoint immobilization of chymotrypsin, *Biomacromolecules*, 9, 2170, 2008.

94. Huang, X. J., Yu, A. G., and Xu, Z. K., Covalent immobilization of lipase from Candida rugosa onto poly(acrylonitrile-co-2-hydroxyethyl methacrylate) electrospun fibrous membranes for potential bioreactor application, *Bioresour. Technol.*, 99, 5459, 2008.

95. Vaidya, B. K. and Singhal, R. S., Use of insoluble yeast beta-glucan as a support for immobilization of Candida rugosa lipase, *Colloids Surf. B Biointerfaces*, 61, 101, 2008.

96. Zwirtes de Oliveira, I. R., Fernandes, S. C., and Vieira, I. C., Development of a biosensor based on gilo peroxidase immobilized on chitosan chemically crosslinked with epichlorohydrin for determination of rutin, *J. Pharm. Biomed. Anal.*, 41, 366, 2006.

97. Costa, S. A. and Reis, R. L., Immobilisation of catalase on the surface of biodegradable starch-based polymers as a way to change its surface characteristics, *J. Mater. Sci. Mater. Med.*, 15, 335, 2004.

98. Kok, F. N. and Hasirci, V., Determination of binary pesticide mixtures by an acetylcholinesterase-choline oxidase biosensor, *Biosens. Bioelectron.*, 19, 661, 2004.

99. Arica, M. Y., Hasirci, V., and Alaeddinoğlu, N. G., Covalent immobilization of alpha-amylase onto pHEMA microspheres: Preparation and application to fixed bed reactor, *Biomaterials*, 16, 761, 1995.

100. Arica, M. Y. and Hasirci, V., Bioreactor applications of glucose oxidase covalently bonded on pHEMA membranes, *Biomaterials*, 14, 803, 1993.

101. Inamori, K., Kyo, M., Matsukawa, K., Inoue, Y., Sonoda, T., Tatematsu, K., Tanizawa, K., Mori, T., and Katayama, Y., Optimal surface chemistry for peptide immobilization in on-chip phosphorylation analysis, *Anal. Chem.*, 80, 643, 2008.

102. Schlapak, R., Pammer, P., Armitage, D., Zhu, R., Hinterdorfer, P., Vaupel, M., Frühwirth, T., and Howorka, S., Glass surfaces grafted with high-density poly(ethylene glycol) as substrates for DNA oligonucleotide microarrays, *Langmuir*, 22, 277, 2006.

103. Charles, P. T. and Kusterbeck, A. W., Trace level detection of hexahydro-1,3,5-trinitro-1,3,5-triazine (RDX) by microimmunosensor, *Biosens. Bioelectron.*, 14, 387, 1999.

104. Bhatia, S. K., Shriver-Lake, L. C., Prior, K. J., Georger, J. H., Calvert, J. M., Bredehorst, R., and Ligler, F. S., Use of thiol-terminal silanes and heterobifunctional crosslinkers for immobilization of antibodies on silica surfaces, *Anal. Biochem.*, 178, 408, 1989.

105. Kincaid, R. L. and Vaughan, M., Affinity chromatography of brain cyclic nucleotide phosphodiesterase using 3-(2-pyridyldithio)propionyl-substituted calmodulin linked to thiol-sepharose, *Biochemistry*, 22, 826, 1983.

106. Gad, M., Machida, M., Mizutani, W., and Ishikawa, M., Method for orienting DNA molecules on mica surfaces in one direction for atomic force microscopy imaging, *J. Biomol. Struct. Dyn.*, 19, 471, 2001.

107. Park, I.-S., Kim, W. Y., and Kim, N., Operational characteristics of an antibody-immobilized QCM system detecting *Salmonella* spp., *Biosens. Bioelectron.*, 15, 167, 2000.

108. Kim, N., Park, I.-S., and Kim, D. K., High-sensitivity detection for model organophosphorus and carbamate pesticide with quartz crystal microbalance-precipitation sensor, *Biosens. Bioelectron.*, 22, 1593, 2007.

109. Cha, T. W., Boiadjiev, V., Lozano, J., Yang, H., and Zhu, X. Y., Immobilization of oligonucleotides on poly(ethylene glycol) brush-coated Si surfaces, *Anal. Biochem.*, 311, 27, 2002.

110. Karakeçili, A. G., Satriano, C., Gümüşderelioğlu, M., and Marletta, G., Enhancement of fibroblastic proliferation on chitosan surfaces by immobilized epidermal growth factor, *Acta Biomater.*, 4, 989, 2008.

111. Yan, M., Cai, S. X., Wybourne, M. N., and Keana, J. F., N-hydroxysuccinimide ester functionalized perfluorophenyl azides as novel photoactive heterobifunctional cross-linking reagents. The covalent immobilization of biomolecules to polymer surfaces, *Bioconjug. Chem.*, 5, 151, 1994.

112. Royer, G. P., Immobilization of glycoenzymes through carbohydrate chains, *Methods Enzymol.*, 135, 141, 1987.

113. Stults, N. L., Asta, L. M., and Lee, Y. C., Immobilization of proteins on oxidized crosslinked Sepharose preparations by reductive amination, *Anal. Biochem.*, 180, 114, 1989.

114. Kopitzki, S., Jensen, K. J., and Thiem, J., Synthesis of benzaldehyde-functionalized glycans: A novel approach towards glyco-SAMs as a tool for surface plasmon resonance studies, *Chemistry*, 16, 7017, 2010.

115. Suda, Y., Arano, A., Fukui, Y., Koshida, S., Wakao, M., Nishimura, T., Kusumoto, S., and Sobel, M., Immobilization and clustering of structurally defined oligosaccharides for sugar chips: An improved method for surface plasmon resonance analysis of protein-carbohydrate interactions, *Bioconjug. Chem.*, 17, 1125, 2006.

116. Tozzi, C., Anfossi, L., and Giraudi, G., Affinity chromatography techniques based on the immobilisation of peptides exhibiting specific binding activity, *J. Chromatogr. B Anal. Technol. Biomed. Life Sci.*, 797, 289, 2003.

117. Moser, A. C. and Hage, D. S., Immunoaffinity chromatography: An introduction to applications and recent developments, *Bioanalysis*, 2, 769, 2010.

118. Gallant, S. R., Koppaka, V., and Zecherle, N., Immunoaffinity chromatography, *Methods Mol. Biol.*, 421, 53, 2008.

119. Beyer, N. H., Hansen, M. Z., Schou, C., Højrup, P., and Heegaard, N. H. H., Optimization of antibody immobilization for on-line or off-line immunoaffinity chromatography, *J. Sep. Sci.*, 32, 1592, 2009.

120. Zhang, S., Wang, J., Li, D., Huang, J., Yang, H., and Deng, A., A novel antibody immobilization and its application in immunoaffinity chromatography, *Talanta*, 82, 704, 2010.

121. Block, H., Maertens, B., Spriestersbach, A., Brinker, N., Kubicek, J., Fabis, R., Labahn, J., and Schäfer, F., Immobilized-metal affinity chromatography (IMAC): A review, *Methods Enzymol.*, 463, 439, 2009.

122. Paborsky, L. R., Dunn, K. E., Gibbs, C. S., and Dougherty, J. P., A nickel chelate microtiter plate assay for six histidine-containing proteins, *Anal. Biochem.*, 234, 60, 1996.

123. Rhemrev-Boom, M. M., Yates, M., Rudolph, M., and Raedts, M., (Immuno)affinity chromatography: A versatile tool for fast and selective purification, concentration, isolation and analysis, *J. Pharm. Biomed. Anal.*, 24, 825, 2001.

124. Tetala, K. K. and van Beek, T. A., Bioaffinity chromatography on monolithic supports, *J. Sep. Sci.*, 33, 422, 2010.

125. Pohanka, M., Musilek, K., and Kuca, K., Progress of biosensors based on cholinesterase inhibition, *Curr. Med. Chem.*, 16, 1790, 2009.

126. Bankar, S. B., Bule, M. V., Singhal, R. S., and Ananthanarayan, L., Glucose oxidase—An overview, *Biotechnol. Adv.*, 27, 489, 2009.

127. Campàs, M., Prieto-Simón, B., and Marty, J. L., A review of the use of genetically engineered enzymes in electrochemical biosensors, *Semin. Cell Dev. Biol.*, 20, 3, 2009.

128. Caruana, D. J. and Howorka, S., Biosensors and biofuel cells with engineered proteins, *Mol. Biosyst.*, 6, 1548, 2010.

129. Bao, L., Sun, D., Tachikawa, H., and Davidson, V. L., Improved sensitivity of a histamine sensor using an engineered methylamine dehydrogenase, *Anal. Chem.*, 74, 1144–1148, 2002.

130. Yamazaki, T., Kojima K., and Sode, K., Extended-range glucose sensor employing engineered glucose dehydrogenases, *Anal. Chem.*, 72, 4689–4693, 2000.

131. Omura, T., Structural diversity of cytochrome P450 enzyme system, *J. Biochem.*, 147, 297, 2010.

132. Beissenhirtz, M. K., Scheller, F. W., Viezzoli, M. S., and Lisdat, F., Engineered superoxide dismutase monomers for superoxide biosensor applications, *Anal. Chem.*, 78, 928, 2006.

133. Andreescu, S. and Marty, J. L., Twenty years research in cholinesterase biosensors: From basic research to practical applications, *Biomol. Eng.*, 23, 1, 2006.

134. Liu, Y. C., Rieben, N., Iversen, L., Sørensen, B. S., Park, J., Nygård, J., and Martinez, K. L., Specific and reversible immobilization of histidine-tagged proteins on functionalized silicon nanowires, *Nanotechnology*, 21, 245, 2010.

135. Heller, M. J., DNA microarray technology: Devices, systems, and applications, *Annu. Rev. Biomed. Eng.*, 4, 129, 2002.

136. Vaijayanthi, B., Kumar, P., Ghosh, P. K., and Gupta, K. C., Recent advances in oligonucleotide synthesis and their applications, *Indian J. Biochem. Biophys.*, 40, 377, 2003.

137. Beaucage, S. L., Strategies in the preparation of DNA oligonucleotide arrays for diagnostic applications, *Curr. Med. Chem.*, 8, 1213, 2001.

138. Seliger, H., Hinz, M., and Happ, E., Arrays of immobilized oligonucleotides—Contributions to nucleic acids technology, *Curr. Pharm. Biotechnol.*, 4, 379, 2003.

139. McGall, G. H. and Fidanza, F. A., Photolithographic synthesis of arrays, *Methods Mol. Biol.*, 170, 71, 2001.

140. Tu, E., Forster, A., and Heller, M. J., Active microelectronic chip devices which utilize controlled electrophoretic fields for multiplex DNA hybridization and other genomic applications, *Electrophoresis*, 21, 157, 2000.

141. Lindroos, K., Liljedahl, U., Raitio, M., and Syvänen, A. C., Minisequencing on oligonucleotide microarrays: Comparison of immobilisation chemistries, *Nucleic Acids Res.*, 29, E69, 2001.

142. Sethi, D., Gandhi, R. P., Kumar. P., and Gupta, K. C., Chemical strategies for immobilization of oligonucleotides, *Biotechnol. J.*, 4, 1513, 2009.

143. Nonglaton, G., Benitez, I. O., Guisle, I., Pipelier, M., Léger, J., Dubreuil, D., Tellier, C., Talham, D. R., and Bujoli, B., New approach to oligonucleotide microarrays using zirconium phosphonate-modified surfaces, *J. Am.Chem. Soc.*, 126, 1497, 2004.

144. Lane, S. M., Monot, J., Petit, M., Tellier, C., Tellier, C., Bujoli, B., and Talham, D. R., Poly(dG) spacers lead to increased surface coverage of DNA probes: An XPS study of oligonucleotide binding to zirconium phosphonate modified surfaces, *Langmuir*, 24, 7394, 2008.

145. Wu, Q., Ma, W., Zhu, L., Guo, Q., Zhang, B., Kulageri, S. M., and Zheng, W., Oligo-microarray based on oligonucleotide immobilization on glass surface modified with activated acrylic acid-co-acrylamide copolymer, *J. Biomed. Mater. Res. B Appl. Biomater.*, 87, 67, 2008.

146. Mahajan, S., Kumar, P., and Gupta, K. C., Oligonucleotide microarrays: Immobilization of phosphorylated oligonucleotides on epoxylated surface, *Bioconjug. Chem.*, 17, 1184, 2006.

147. Beier, M. and Hoheisel, J. D., Versatile derivatisation of solid support media for covalent bonding on DNA-microchips, *Nucleic Acids Res.*, 27, 1970, 1999.

148. Chrisey, L. A., Lee, G. U., and O'Ferrall, C. E., Covalent attachment of synthetic DNA to self-assembled monolayer films, *Nucleic Acids Res.*, 24, 3031, 1996.

149. Wong, A. K. Y. and Krull, U. J., Surfaces for tuning of oligonucleotide biosensing selectivity based on surface-initiated atom transfer radical polymerization on glass and silicon surfaces, *Anal. Chim. Acta*, 639, 1, 2009.

150. Guo, Z., Guilfoyle, R. A., Thiel, A. J., Wang, R., and Smith, L. M., Direct fluorescence analysis of genetic polymorphisms by hybridization with oligonucleotide arrays on glass supports, *Nucleic Acids Res.*, 22, 5456, 1994.

151. Mahajan, S., Kumar, P., and Gupta, K. C., An efficient and versatile approach for the construction of oligonucleotide microarrays, *BioMed. Chem. Lett.*, 16, 5654, 2006.

152. Sethi, D., Patnaik, S., Kumar, A., Gandhi, R. P., Gupta, K. C., and Kumar, P., Polymer supported synthesis of aminooxyalkylated oligonucleotides, and some applications in the fabrication of microarrays, *Bioorg. Med. Chem.*, 17, 5442, 2009.

153. Mahajan, J., Sethi, D., Seth, S., Kumar, A., Kumar, P., and Gupta, K. C., Construction of oligonucleotide microarrays (biochips) via thioether linkage for the detection of bacterial meningitis, *Bioconjug. Chem.*, 20, 1703, 2009.

154. Razumovitch, J., Meier, W., and Vebert, C. A., A microcontact printing approach to the immobilization of oligonucleotide brushes, *Biophys. Chem.*, 139, 70, 2009.

155. Dendane, N., Hoang, A., Defrancq, E., Vinet, F., and Dumy, P., Use of gamma-aminopropyl-coated glass surface for the patterning of oligonucleotides through oxime bond formation, *BioMed. Chem. Lett.*, 18, 2540, 2008.

156. Lockett, M. R., Phillips, M. F., Jarecki, J. L., Peelen, D., and Smith, L. M., A tetrafluorophenylactivated ester self-assembled monolayer for the immobilization of amine-modified oligonucleotides, *Langmuir*, 24, 69, 2008.

157. Joos, B., Kuster, H., and Cone, R., Covalent attachment of hybridizable oligonucleotides to glass supports, *Anal. Biochem.*, 247, 96, 1997.

158. Rogers, Y. H., Jiang-Baucom, P., Huang, Z. J., Bogdanov, V., Anderson, S., and Boyce-Jacino, M. T., Immobilization of oligonucleotides onto a glass support via disulfide bonds: A method for preparation of DNA microarrays, *Anal. Biochem.*, 266, 23, 1999.

159. Dandy, D. S., Wu, P., and Grainger, D.W., Array feature size influences nucleic acid surface capture in DNA microarrays, *Proc. Natl. Acad. Sci. U. S. A.*, 104, 8223, 2007.

160. Misra, A. and Dwivedi, P., Immobilization of oligonucleotides on glass surface using an efficient hetero-bifunctional reagent through maleimide-thiol combination chemistry, *Anal, Biochem.*, 369, 248, 2007.

161. Choithani, J., Kumar, P., and Gupta, K. C., N-(3-Triethoxysilylpropyl)-6-(N-maleimido)-hexanamide: An efficient heterobifunctional reagent for the construction of oligonucleotide microarrays, *Anal. Biochem.*, 357, 240, 2006.

162. Misra, A., Shahid, M., and Dwivedi, P., N-(3-triethoxysilylpropyl)-4-(isothiocyanatomethyl)-cyclohexane-1-carboxamide (TPICC): A heterobifunctional reagent for immobilization of biomolecules on glass surface, *Bioorg. Med. Chem. Lett.*, 18, 5217, 2008.

163. Kumar, P., Choithani, J., and Gupta, K. C., Construction of oligonucleotide arrays on a glass surface using a heterobifunctional reagent, N-(2-trifluoroethanesulfonatoethyl)-N-(methyl)-triethoxysilylpropyl-3-amine (NTMTA), *Nucleic Acids Res.*, 32, e80, 2004.

164. Rozkiewicz, D. I., Gierlich, J., Burley, G. A., Gutsmiedl, K., Carell, T., Ravoo, B. J., and Reinhoudt, D. N., Transfer printing of DNA by click chemistry, *ChemBioChem*, 8, 1997, 2007.

165. Pack, S. P., Kamisetty, N. K., Nonogawa, M., Devarayapalli, K. C., Ohtani, K., Yamada, K., Yoshida, Y., Kodaki, T., and Makino, K., Direct immobilization of DNA oligomers onto the amine-functionalized glass surface for DNA microarray fabrication through the activation-free reaction of oxanine, *Nucleic Acids Res.*, 35, e110, 2007.

166. Pack, S. P. and Makino, K., Synthesis of 2'-deoxyoxanosine from 2'-deoxyguanosine, conversion to its phosphoramidite, and incorporation into oxanine-containing oligodeoxynucleotides, *Curr. Protoc. Nucleic Acid Chem.*, Chapter 4, Unit 4.39, 2010.

167. Benters, R., Niemeyer, C. M., Drutschmann, D., Blohm, D., and Wohrle, D., DNA microarrays with PAMAM dendritic linker systems, *Nucleic Acids Res.*, 30, e10, 2002.

168. Hong, B. J., Sunkara, V., and Park, J. W., DNA microarrays on nanoscale-controlled surface, *Nucleic Acids Res.*, 33, e106. 2005.

169. Consolandi, C., Severgnini, M., Castiglioni, B., Bordoni, R., Frosini, A., Battaglia, C., Bernardi, L. R., and Bellis, G. D., A structured chitosan-based platform for biomolecule attachment to solid surfaces: Application to DNA microarray reparation, *Bioconjug. Chem.*, 17, 371, 2006.

170. Abramov, M., Schepers, G., Aerschot, A. V., Hummelen, P. V., and Herdewin, P., HNA and ANA high-affinity arrays for detections of DNA and RNA single-base mismatches, *Biosens. Bioelectron.*, 23, 1728, 2008.

171. Lim, S. B., Kim, K.-W., Lee, C.-W., Kim, H.-S., Lee, C.-S., and Oh, M.-K., Improved DNA chip with poly(aminoamine) dendrimers peripherally modified with biotin and avidin, *Biotechnol. Bioproc. Eng.*, 13, 683, 2008.

172. Breitenstein, M., Hölzel, R., and Bier, F. F., Immobilization of different biomolecules by atomic force microscopy, *J. Nanobiotechnol.*, 8, 10, 2010.

173. Kumar, A., Larsson, O., Parodi, D., and Liang, Z., Silanized nucleic acids: A general platform for DNA immobilization, *Nucleic Acids Res.*, 28, E71, 2000.

174. Sethi, D., Kumar, A., Gupta, K. C., and Kumar, P., A facile method for the construction of oligonucleotide microarrays, *Bioconjug. Chem*, 19, 2136, 2008.

175. Lim, H. I., Oliver, P. M., Marzillier, J., and Vezenov, D. V., Heterobifunctional modification of DNA for conjugation to solid surfaces, *Anal. Bioanal. Chem.*, 397, 1861, 2010.

176. Kumar, P., Agarwal, S. K., and Gupta, K. C., N-(3-Trifluoroethanesulfonyloxypropyl)anthraquinone-2-carboxamide: A new heterobifunctional reagent for immobilization of biomolecules on a variety of polymer surfaces, *Bioconjug. Chem.*, 15, 7, 2004.

177. Patnaik, S., Swami, A., Sethi, D., Pathak, A., Garg, B. S., Gupta, K. C., and Kumar, P., N-(Iodoacetyl)-N'-(anthraquinon-2-oyl)-ethylenediamine (IAED): A new heterobifunctional reagent for the preparation of biochips, *Bioconjug. Chem.*, 18, 8, 2007.

178. Choithani, J., Sethi, D., Kumar, P., and Gupta, K. C., Preparation of oligonucleotide microarrays on modified glass using a photoreactive heterobifunctional reagent, 1-*N*-(maleimidohexanoyl)-6-*N*-(anthraquinon-2-oyl) hexanediamine (MHAHD), *Surf. Sci.*, 602, 2389, 2008.

179. Weinrich, D., Jonkheijm, P., Niemeyer, C. M., and Waldmann, H., Applications of protein biochips in biomedical and biotechnological research, *Angew. Chem. Int. Ed. Engl.*, 8, 7744, 2009.

180. Spisák, S. and Guttman, A., Biomedical applications of protein microarrays, *Curr. Med. Chem.*, 16, 2806, 2009.

181. Grainger, D. W., Greef, C. H., Gong, P., and Lochhead, M. J., Current microarray surface chemistries, *Methods Mol. Biol.*, 381, 37, 2007.

182. Zhu, H. and Snyder, M., Protein chip technology, *Curr. Opin. Chem. Biol.*, 7, 55, 2003.

183. Rusmini, F., Zhong, Z., and Feijen, J., Protein immobilization strategies for protein biochips, *Biomacromolecules*, 8, 1775, 2007.

184. Chang, Y. J. and Chang, C. H., Protein microarray chip with Ni-Co alloy coated surface, *Biosens. Bioelectron.*, 25, 1748, 2010.

185. MacBeath, G. and Schreiber, S. L., Printing proteins as microarrays for high-throughput function determination, *Science*, 289, 1760, 2000.

186. Funeriu, D. P., Eppinger, J., Denizot, L., Miyake, M., and Miyake, J., Enzyme family-specific and activity-based screening of chemical libraries using enzyme microarrays, *Nat. Biotechnol.*, 23, 622, 2005.

187. Fall, B. I., Eberlein-König, B., Behrendt, H., Niessner, R., Ring, J., and Weller, M. G., Microarrays for the screening of allergen-specific IgE in human serum, *Anal. Chem.*, 75, 556, 2003.

188. Köhn, M., Immobilization strategies for small molecule, peptide and protein microarrays, *J. Pept. Sci.*, 15, 393, 2009.

189. Ito, Y., Photoimmobilization for microarrays, *Biotechnol. Prog.*, 22, 924, 2006.

190. Weinrich, D., Köhn, M., Jonkheijm, P., Westerlind, U., Dehmelt, L., Engelkamp, H., Christianen, P. C. et al., Preparation of biomolecule microstructures and microarrays by thiol-ene photoimmobilization, *ChemBioChem*, 11, 235, 2010.

191. Miller, J. C., Zhou, H., Kwekei, J., Cavallo, R., Burke, J., Butler, E. D., Teh, B. S., and Haab, B. B., Antibody microarray profiling of human prostate cancer sera: Antibody screening and identification of potential markers, *Proteomics*, 3, 56, 2003.

192. Kanoh, N., Kyo, M., Inamori, K., Ando, A., Asami, A., Nakao, A., and Osada, H., SPR imaging of photo-cross-linked small-molecule arrays on gold, *Anal. Chem.*, 78, 2226, 2006.

193. Ito, Y., Nogawa, M., Takeda, M., and Shibuya, T., Photoreactive polyvinylalcohol for photoimmobilized microarray, *Biomaterials*, 26, 211, 2005.

194. Wingren, C. and Borrebaeck, C. A., Antibody-based microarrays, *Methods Mol. Biol.*, 509, 57, 2009.

195. Wingren, C. and Borrebaeck, C. A., Antibody microarrays: Current status and key technological advances, *OMICS*, 10, 411, 2006.

196. Seurynck-Servoss, S. L., White, A. M., Baird, C. L., Rodland, K. D., and Zangar, R. C., Evaluation of surface chemistries for antibody microarrays, *Anal. Biochem.*, 371, 105, 2007.

197. Kusnezow, W. and Hoheisel, J. D., Solid supports for microarray immunoassays, *J. Mol. Recognit.*, 16, 165, 2003.

198. Anderson, G. P., Jacoby, M. A., Ligler, F. S., and King, K. D., Effectiveness of protein A for antibody immobilization for a fiber optic biosensor, *Biosens. Bioelectron.*, 12, 329, 1997.

199. Shriver-Lake, L. C., Donner, B., Edelstein, R., Breslin, K., Bhatia, S. K., and Ligler, F. S., Antibody immobilization using heterobifunctional crosslinkers, *Biosens. Bioelectron.*, 12, 1101, 1997.

200. Kusnezow, W., Jacob, A., Walijew, A., Diehl, F., and Hoheisel, J. D., Antibody microarrays: An evaluation of production parameters, *Proteomics*, 3, 254, 2003.

201. Peluso, P., Wilson, D. S., Do, D., Tran, H., Venkatasubbaiah, M., Quincy, D., Heidecker, B. et al., Optimizing antibody immobilization strategies for the construction of protein microarrays, *Anal. Biochem.*, 312, 113, 2003.

202. Karyakin, A. A., Presnova, G. V., Rubtsova, M. Y., and Egorov, A. M., Oriented immobilization of antibodies onto the gold surfaces via their native thiol groups, *Anal. Chem.*, 72, 3805, 2000.

203. Huang, N.-P., Michel, R., Voros, J., Textor, M., Hofer, R., Rossi, A., Elbert, D. L., Hubbell, J. A., and Spencer, D., Poly(L-lysine)-γ-Poly (ethylene glycol) layers on metal oxide surfaces: Surfaceanalytical characterization and resistance to serum and fibrinogen adsorption, *Langmuir*, 17, 489, 2001.

204. Zhang, F., Kang, E. T., Neoh, K. G., Wang, P., and Tan, K. L., Modification of Si(100) surface by the grafting of poly(ethylene glycol) for reduction in protein adsorption and platelet adhesion, *J. Biomed. Mater. Res.*, 56, 324, 2001.

205. Siegel, D. L., Recombinant monoclonal antibody technology, *Transfus. Clin. Biol.*, 9, 15, 2002.

206. Nord, K., Gunneriusson, E., Ringdahl, J., Stahl, S., Uhlen, M., and Nygren, P. A., Binding proteins selected from combinatorial libraries of an alpha-helical bacterial receptor domain, *Nat. Biotechnol.*, 15, 772, 1997.

207. James, W., Nucleic acid and polypeptide aptamers: A powerful approach to ligand discovery, *Curr. Opin. Pharmac.*, 1, 540, 2001.

208. Seurynck-Servoss, S. L., Baird, C. L., Miller, K. D., Pefaur, N. B., Gonzalez, R. M., Apiyo, D. O., Engelmann, H. E. et al., Immobilization strategies for single-chain antibody microarrays, *Proteomics*, 8, 2199, 2008.

209. Kumada, Y., Hamasaki, K., Shiritani, Y., Nakagawa, A., Kuroki, D., Ohse, T., Choi, D. H., Katakura, Y., and Kishimoto, M., Direct immobilization of functional single-chain variable fragment antibodies (scFvs) onto a polystyrene plate by genetic fusion of a polystyrene-binding peptide (PS-tag), *Anal. Bioanal. Chem.*, 395, 759, 2009.
210. Zhu, H., Bilgin, M., Bangham, R., Hall, D., Casamayor, A., Bertone, P., Lan, N. et al., Global analysis of protein activities using proteome chips, *Science*, 293, 2101, 2001.
211. Lesaicherre, M. L., Lue, R. Y., Chen, G. Y., Zhu, Q., and Yao, S. Q., Inteinmediated biotinylation of proteins and its application in a protein microarray, *J. Am. Chem. Soc.*, 124, 8768, 2002.
212. Park, S., Lee, M.-R., and Shin, I., Chemical tools for functional studies of glycans, *Chem. Soc. Rev.*, 37, 1579, 2008.
213. Horlacher, T. and Seeberger, P. H., Carbohydrate arrays as tools for research and diagnostics, *Chem. Soc. Rev.*, 37, 1414, 2008.
214. Willats, W. G. T., Rasmussen, S. E., Kristensen, T., Mikkelsen, J. D., and Knox, J. P., Sugar-coated microarrays: A novel slide surface for the high-throughput analysis of glycans, *Proteomics*, 2, 1666, 2002.
215. Angeloni, S., Ridet, J. L., Kusy, N., Gao, H., Crevoisier, F., Guinchard, S., Kochhar, S., Sigrist, H., and Sprenger, N., Glycoprofiling with micro-arrays of glycoconjugates and lectins, *Glycobiology*, 15, 31, 2005.
216. Oyelaran, O. and Gildersleeve, J. C., Glycan arrays: Recent advances and future challenges, *Curr. Opin. Chem. Biol.*, 13, 406, 2009.
217. Lee, M.-R. and Shin, I., Facile preparation of carbohydrate microarrays by site-specific, covalent immobilization of unmodified carbohydrates on hydrazide-coated glass slides, *Org. Lett.*, 7, 4269, 2005.
218. Sungjin Park, S., Lee, M.-R., and Shin, I., Construction of carbohydrate microarrays by using one-step, direct immobilizations of diverse unmodified glycans on solid surfaces, *Bioconj. Chem.*, 20, 155, 2009.
219. Park, S., Lee, M.-R., Pyo, S.-J., and Shin, I., Carbohydrate chips for studying high-throughput carbohydrate–protein interactions, *J. Am. Chem. Soc.*, 126, 4812, 2004.
220. Xia, B., Kawar, Z. S., Ju, T., Alvarez, R. A., Sachdev, G. P., and Cummings, R. D., Versatile fluorescent derivatization of glycans for glycomic analysis, *Nat. Methods*, 2, 845, 2005.
221. Bohorov, O., Andersson-Sand, H., Hoffmann, J., and Blixt, O., Arraying glycomics: A novel bi-functional spacer for one-step microscale derivatization of free reducing glycans, *Glycobiology*, 16, 21C, 2006.
222. Tateno, H., Mori, A., Uchiyama, N., Yabe, R., Iwaki, J., Shikanai, T., Angata, T., Narimatsu, H., and Hirabayashi, J., Glycoconjugate microarray based on an evanescent-field fluorescence-assisted detection principle for investigation of glycan-binding proteins, *Glycobiology*, 18, 789, 2008.
223. Kurakake, M., Masumoto, R., Maguma, K., Kamata, A., Saito, E., Ukita, N., and Komaki, T., Production of fructooligosaccharides by β-fructofuranosidases from *Aspergillus oryzae* KB, *J. Agric. Food Chem.*, 58, 488, 2010.
224. Kotwal, S. M. and Shankar, V., Immobilized invertase, *Biotechnol. Adv.*, 27, 311, 2009.

Appendix A: Amino Group–Directed Homobifunctional Cross-Linkers

Name (Abbreviation)	Structure	References
A. Bisimidoesters (bisimidates)		
I		
Diethyl malonimidate 2HCl (DEM)	$CH_3CH_2-O-\underset{\overset{\|}{\underset{^{+}NH_2\,\bar{Cl}}{}}}{C}-CH_2-\underset{\overset{\|}{\underset{^{+}NH_2\,\bar{Cl}}{}}}{C}-O-CH_2CH_3$	1
II		
Dimethyl malonimidate 2HCl (DMM)	$H_3C-O-\underset{\overset{\|}{\underset{^{+}NH_2\,\bar{Cl}}{}}}{C}-CH_2-\underset{\overset{\|}{\underset{^{+}NH_2\,\bar{Cl}}{}}}{C}-O-CH_3$	2–4
III		
Dimethyl succinimidate 2HCl (DMSC)	$H_3C-O-\underset{\overset{\|}{\underset{^{+}NH_2\,\bar{Cl}}{}}}{C}-(CH_2)_2-\underset{\overset{\|}{\underset{^{+}NH_2\,\bar{Cl}}{}}}{C}-O-CH_3$	2–6
IV		
Dimethyl glutarimidate 2HCl (DMG)	$H_3C-O-\underset{\overset{\|}{\underset{^{+}NH_2\,\bar{Cl}}{}}}{C}-(CH_2)_3-\underset{\overset{\|}{\underset{^{+}NH_2\,\bar{Cl}}{}}}{C}-O-CH_3$	4,5
V		
Dimethyl adipimidate 2HCl (DMA)	$H_3C-O-\underset{\overset{\|}{\underset{^{+}NH_2\,\bar{Cl}}{}}}{C}-(CH_2)_4-\underset{\overset{\|}{\underset{^{+}NH_2\,\bar{Cl}}{}}}{C}-O-CH_3$	2–9
VI		
Dimethyl pimelimidate 2HCl (DMP)	$H_3C-O-\underset{\overset{\|}{\underset{^{+}NH_2\,\bar{Cl}}{}}}{C}-(CH_2)_5-\underset{\overset{\|}{\underset{^{+}NH_2\,\bar{Cl}}{}}}{C}-O-CH_3$	2,4–6,10
VII		
Dimethyl suberimidate 2HCl (DMS)	$H_3C-O-\underset{\overset{\|}{\underset{^{+}NH_2\,\bar{Cl}}{}}}{C}-(CH_2)_6-\underset{\overset{\|}{\underset{^{+}NH_2\,\bar{Cl}}{}}}{C}-O-CH_3$	2–6,10,11
VIII		
Dimethyl azelaimidate 2HCl (DMAZ)	$H_3C-O-\underset{\overset{\|}{\underset{^{+}NH_2\,\bar{Cl}}{}}}{C}-(CH_2)_7-\underset{\overset{\|}{\underset{^{+}NH_2\,\bar{Cl}}{}}}{C}-O-CH_3$	4

IX

Dimethyl sebacimidate 2HCl (DMSC)

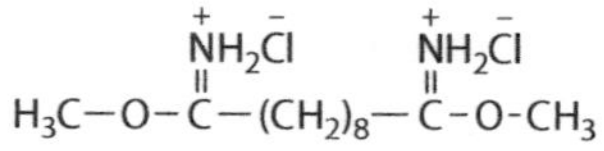

$$H_3C-O-\overset{\overset{+}{N}H_2\bar{C}l}{\underset{\|}{C}}-(CH_2)_8-\overset{\overset{+}{N}H_2\bar{C}l}{\underset{\|}{C}}-O-CH_3$$

4,6,10

X

Dimethyl dodecimidate 2HCl (DMDD)

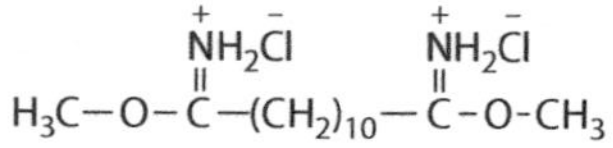

$$H_3C-O-\overset{\overset{+}{N}H_2\bar{C}l}{\underset{\|}{C}}-(CH_2)_{10}-\overset{\overset{+}{N}H_2\bar{C}l}{\underset{\|}{C}}-O-CH_3$$

6,10

XI

Dimethyl 3,3'-oxydipropion-imidate 2HCl (DODP)

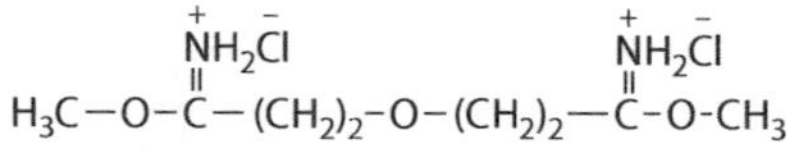

$$H_3C-O-\overset{\overset{+}{N}H_2\bar{C}l}{\underset{\|}{C}}-(CH_2)_2-O-(CH_2)_2-\overset{\overset{+}{N}H_2\bar{C}l}{\underset{\|}{C}}-O-CH_3$$

2,12,13

XII

Dimethyl 3,3'-(methylenedioxane)-dipropionimidate 2HCl (DMPD)

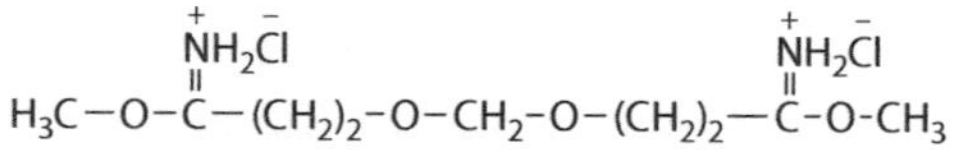

$$H_3C-O-\overset{\overset{+}{N}H_2\bar{C}l}{\underset{\|}{C}}-(CH_2)_2-O-CH_2-O-(CH_2)_2-\overset{\overset{+}{N}H_2\bar{C}l}{\underset{\|}{C}}-O-CH_3$$

2,12,13

XIII

Dimethyl 3,3'-(dimethylenedioxane)-dipropionimidate 2HCl (DDDP)

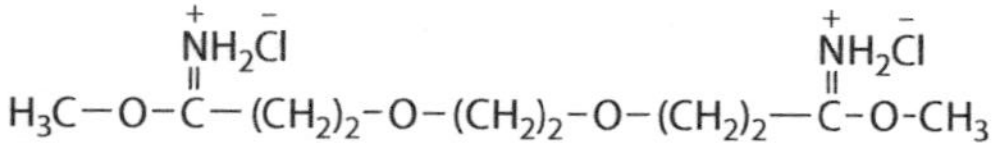

$$H_3C-O-\overset{\overset{+}{N}H_2\bar{C}l}{\underset{\|}{C}}-(CH_2)_2-O-(CH_2)_2-\overset{\overset{+}{N}H_2\bar{C}l}{\underset{\|}{C}}-O-CH_3$$

2,12,13

XIV

Dimethyl 3,3'-(tetramethylenedioxy)-dipropionimidate 2HCl (DTDP)

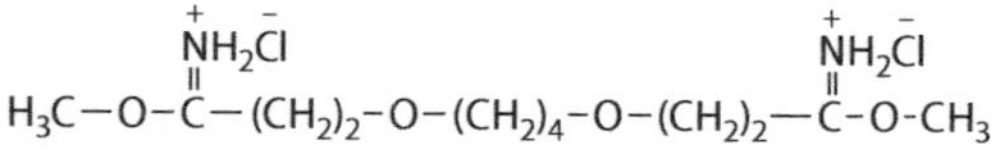

$$H_3C-O-\overset{\overset{+}{N}H_2\bar{C}l}{\underset{\|}{C}}-(CH_2)_2-O-(CH_2)_4-O-(CH_2)_2-\overset{\overset{+}{N}H_2\bar{C}l}{\underset{\|}{C}}-O-CH_3$$

2,12,13

XV

Dimethyl 3,3'-(diethyletherdioxy)-dipropionimidate 2HCl (DDDD)

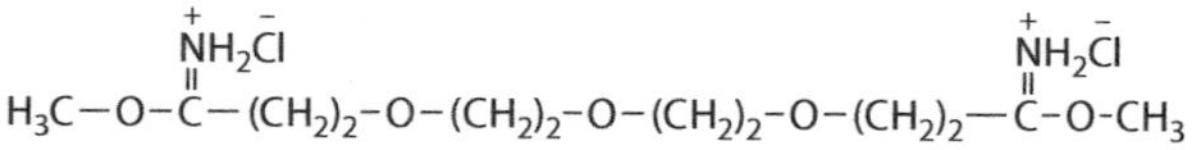

$$H_3C-O-\overset{\overset{+}{N}H_2\bar{C}l}{\underset{\|}{C}}-(CH_2)_2-O-(CH_2)_2-O-(CH_2)_2-\overset{\overset{+}{N}H_2\bar{C}l}{\underset{\|}{C}}-O-CH_3$$

12,13

(continued)

(Continued)

Name (Abbreviation)	Structure	References
XVI Diisoethionyl 3,3'-dithiobis-propionimidate 2HCl (DIDIT)	$HO_3S-(CH_2)_2-O-C(=\overset{+}{N}H_2\ \overset{-}{Cl})-(CH_2)_2-S-S-(CH_2)_2-C(=\overset{+}{N}H_2\ \overset{-}{Cl})-O-(CH_2)_2-SO_3H$	2,14
XVII Dimethyl 3,3'-dithiobispropionimidate 2HCl (DTBP)	$H_3C-O-C(=\overset{+}{N}H_2\ \overset{-}{Cl})-(CH_2)_2-S-S-(CH_2)_2-C(=\overset{+}{N}H_2\ \overset{-}{Cl})-O-CH_3$	2,6,7,15,16
XVIII Dimethyl 4,4'-dithiobisbutyroimidate 2HCl (DTBB)	$H_3C-O-C(=\overset{+}{N}H_2\ \overset{-}{Cl})-(CH_2)_3-S-S-(CH_2)_3-C(=\overset{+}{N}H_2\ \overset{-}{Cl})-O-CH_3$	2,16
XIX Dimethyl 5,5'-dithiobisvaleroimidate 2HCl (DTBV)	$H_3C-O-C(=\overset{+}{N}H_2\ \overset{-}{Cl})-(CH_2)_4-S-S-(CH_2)_4-C(=\overset{+}{N}H_2\ \overset{-}{Cl})-O-CH_3$	2,16
XX Dimethyl 7,7'-dithiobissenanthoimidate 2HCl (DTBE)	$H_3C-O-C(=\overset{+}{N}H_2\ \overset{-}{Cl})-(CH_2)_6-S-S-(CH_2)_6-C(=\overset{+}{N}H_2\ \overset{-}{Cl})-O-CH_3$	2,16
XXI Dimethyl 11,11'-dithiobisundecanoimidate 2HCl (DTBU)	$H_3C-O-C(=\overset{+}{N}H_2\ \overset{-}{Cl})-(CH_2)_{10}-S-S-(CH_2)_{10}-C(=\overset{+}{N}H_2\ \overset{-}{Cl})-O-CH_3$	16
XXII Dimethyl 3,3'-(dithiodimethylene-dioxy)bisimidate 2HCl (DDMB)	$H_3C-O-C(=\overset{+}{N}H_2\ \overset{-}{Cl})-(CH_2)_2-O-(CH_2)_2-S-S-(CH_2)_2-O-(CH_2)_2-C(=\overset{+}{N}H_2\ \overset{-}{Cl})-O-CH_3$	12,13

XXIII

Dimethyl 3,3'-(di-thiodimethylene-diamido)-bis-propionimidate 2HCl (DDDBP) — 12,13

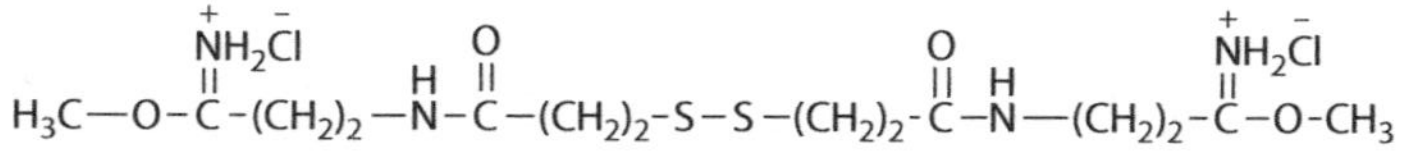

$$H_3C-O-\overset{\overset{+}{N}H_2\bar{Cl}}{\underset{\|}{C}}-(CH_2)_2-\overset{H}{\underset{\|}{N}}-\overset{O}{\underset{\|}{C}}-(CH_2)_2-S-S-(CH_2)_2-\overset{O}{\underset{\|}{C}}-\overset{H}{N}-(CH_2)_2-\overset{\overset{+}{N}H_2\bar{Cl}}{\underset{\|}{C}}-O-CH_3$$

XXIV

N,N'-Bis(2-carboxyimido-methyl)tartarimide dimethyl ester 2HCl (CMTD); Tartrydi(methyl-2-aminoacetimidate) 2HCl (TDAA) — 11

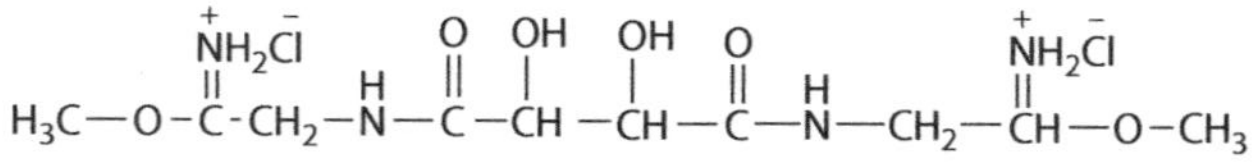

$$H_3C-O-\overset{\overset{+}{N}H_2\bar{Cl}}{\underset{\|}{C}}-CH_2-\overset{OH}{\underset{|}{C}H}-\overset{O}{\underset{\|}{C}}-\overset{H}{N}-CH_2-\overset{\overset{+}{N}H_2\bar{Cl}}{\underset{\|}{C}}-O-CH_3$$

XXV

N,N'-Bis(2-carboxyimido-ethyl)tartarimide dimethyl ester 2HCl (CETD); Tartrydi(methyl-2-amino-propionimidate) 2HCl — 6,11

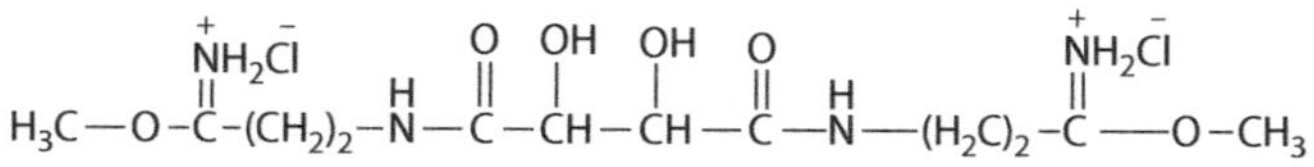

$$H_3C-O-\overset{\overset{+}{N}H_2\bar{Cl}}{\underset{\|}{C}}-(CH_2)_2-\overset{H}{\underset{\|}{N}}-\overset{O}{\underset{\|}{C}}-\overset{OH}{\underset{|}{C}H}-\overset{OH}{\underset{|}{C}H}-\overset{O}{\underset{\|}{C}}-\overset{H}{N}-(H_2C)_2-\overset{\overset{+}{N}H_2\bar{Cl}}{\underset{\|}{C}}-O-CH_3$$

XXVI

3,4,5,6-Tetrahydroxy-suberimidate 2HCl (THS) — 17

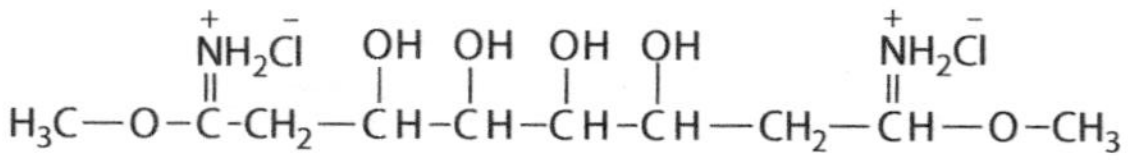

$$H_3C-O-\overset{\overset{+}{N}H_2\bar{Cl}}{\underset{\|}{C}}-CH_2-\overset{OH}{\underset{|}{C}H}-\overset{OH}{\underset{|}{C}H}-\overset{OH}{\underset{|}{C}H}-\overset{OH}{\underset{|}{C}H}-CH_2-\overset{\overset{+}{N}H_2\bar{Cl}}{\underset{\|}{C}}-O-CH_3$$

XXVII

Dimethyl 3,3'-(N-2,4-dinitrophenyl)-bispropionimidate 2HCl (DDPB) — 18

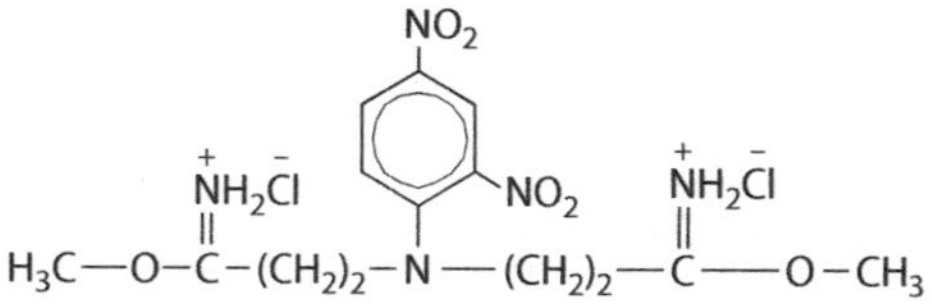

$$H_3C-O-\overset{\overset{+}{N}H_2\bar{Cl}}{\underset{\|}{C}}-(CH_2)_2-N-(CH_2)_2-\overset{\overset{+}{N}H_2\bar{Cl}}{\underset{\|}{C}}-O-CH_3$$

XXVIII

Dimethyl 3,3'-(N-2,4-dinitro-5-carboxyphenyl)-bispropionimidate 2HCl (DDCB) — 18

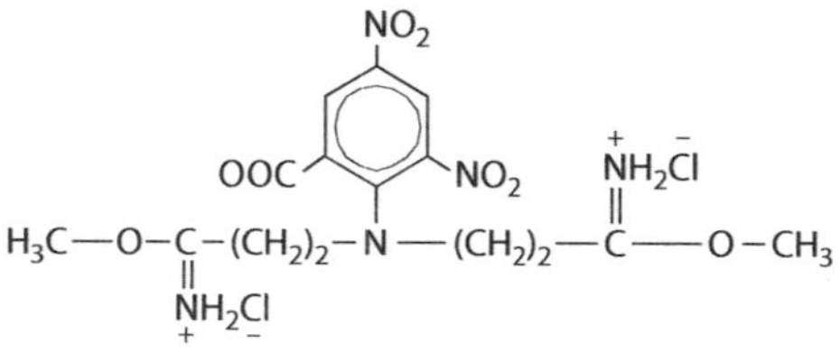

$$H_3C-O-\overset{\overset{+}{N}H_2\bar{Cl}}{\underset{\|}{C}}-(CH_2)_2-N-(CH_2)_2-\overset{\overset{+}{N}H_2\bar{Cl}}{\underset{\|}{C}}-O-CH_3$$

(continued)

(Continued)

Name (Abbreviation)	Structure	References

XXIX

Dimethyl 3,3'-[N-(5-(N,N-dimethylamino)napthyl) sulfonyl)bispropionimidate 2HCl (DDNBP)

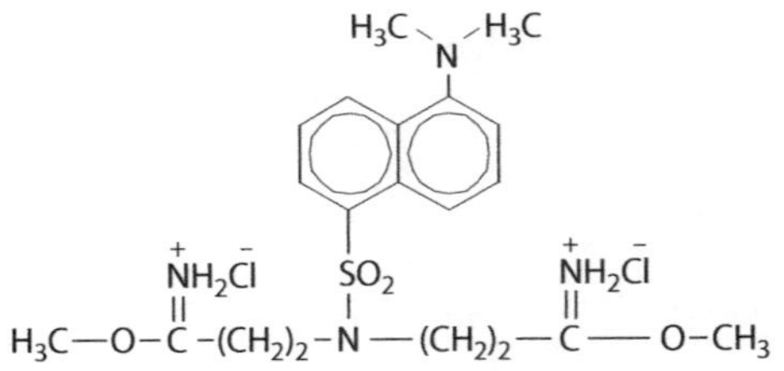

19

B. Bis-N-Succinimidyl Derivatives

XXX

Disuccinimidylcarbonate (DSC)

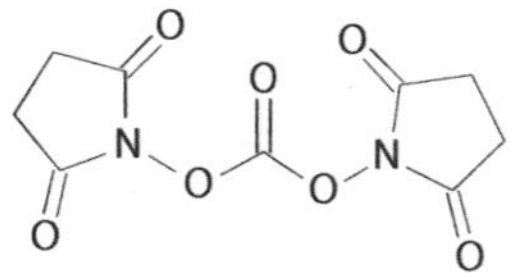

20

XXXI

Disuccinimidyloxalate (DSO)

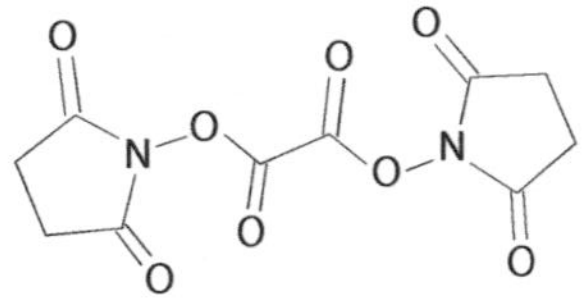

20

XXXII

Bis(N-succinimidyl)-2,2-dimethyl malonate (NHS-DMM)

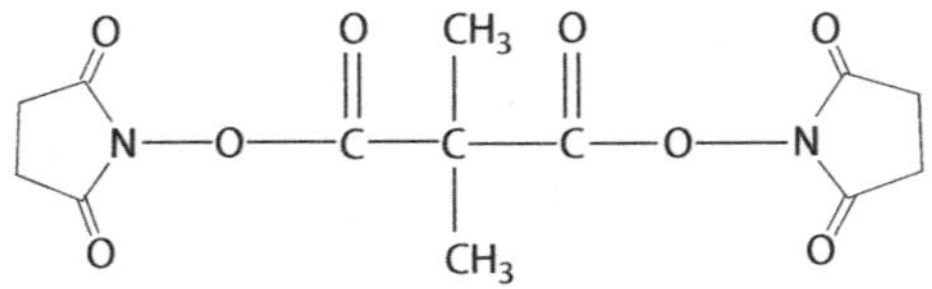

21

(continued)

XXXIII

Bis(*N*-hydroxysuccinimidyl)succinate (NHS-S);
disuccinimidyl succinate

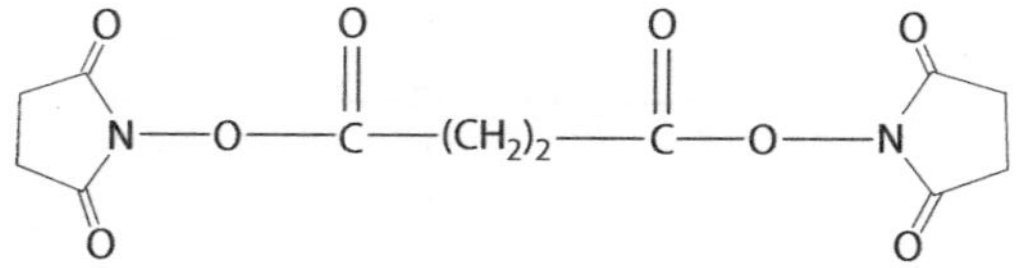

21

XXXIV

Bis(*N*-hydroxysuccinimidyl) glutarate (NHS-G);
disuccinimidyl glutarate (DSG)

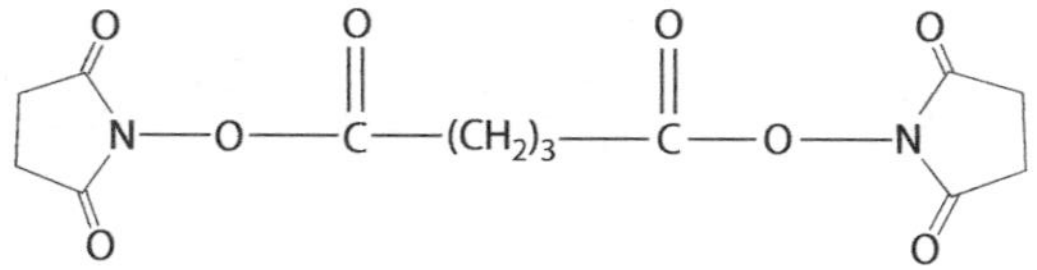

22

XXXV

Bis(sulfosuccinimidyl) glutarate (BSSG)

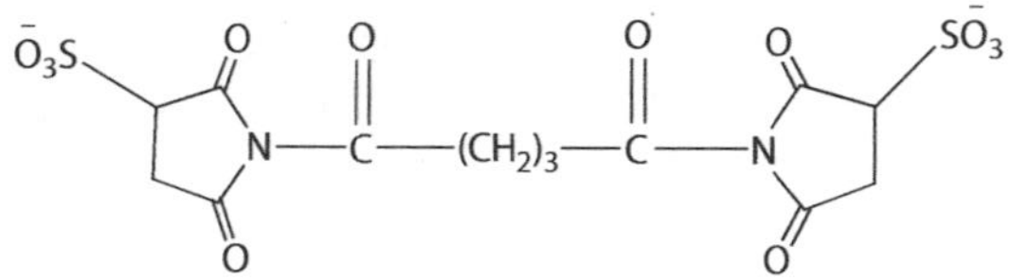

23,24

XXXVI

Tetradeuterated bis(sulfosuccinimidyl) glutarate
(BSSG-D4)

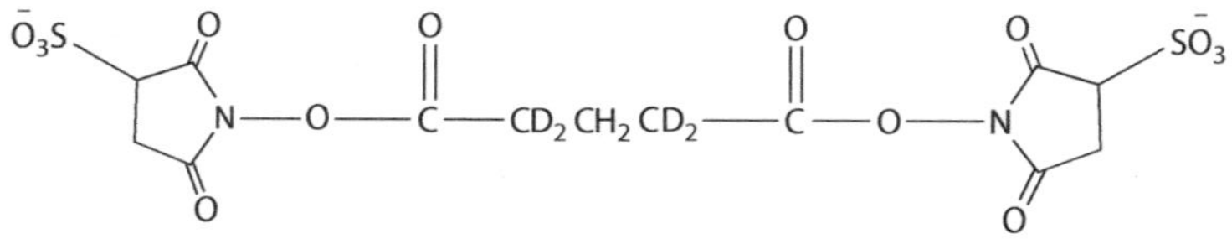

23,24

XXXVII

Bis(succinimidyl) adipate

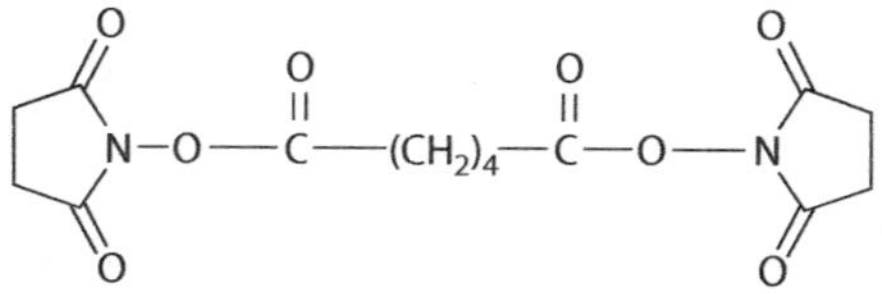

25

(Continued)

Name (Abbreviation)	Structure	References
XXXVIII		
N-Hydroxysuccinimidyl pimelate (NHS-SP); Disuccinimidyl pimelate (DSP)	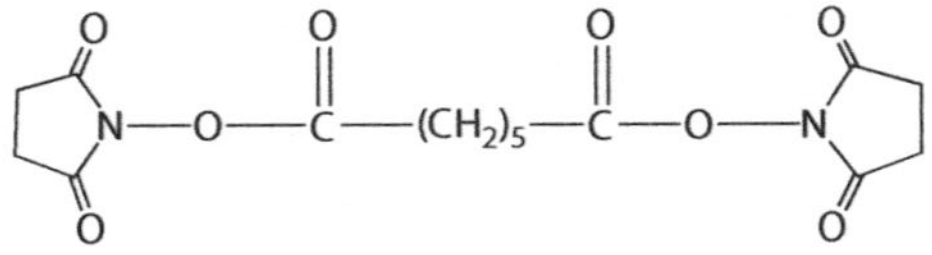	26
XXXIX		
Bis(sulfosuccinimidyl) pimelate (BSP)	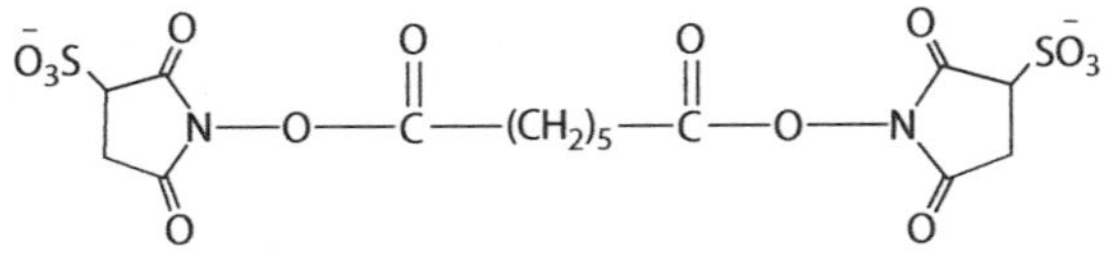	24
XL		
Tetradeuterated bis(sulfosuccinimidyl) pimelate (BSP-D4)	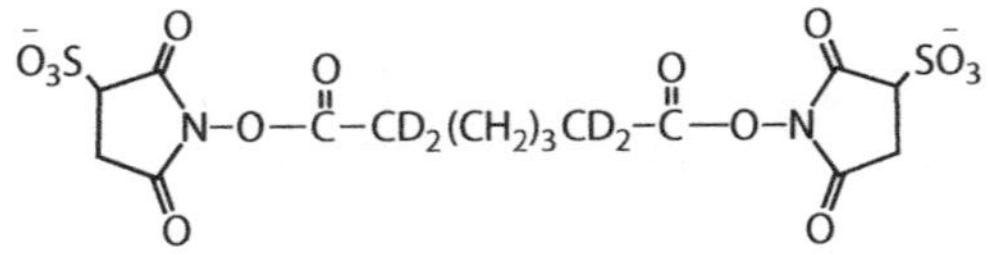	24
XLI		
N-Hydroxysuccinimidylsuberate (NHS-SA); disuccinimidyl suberate (DSS)	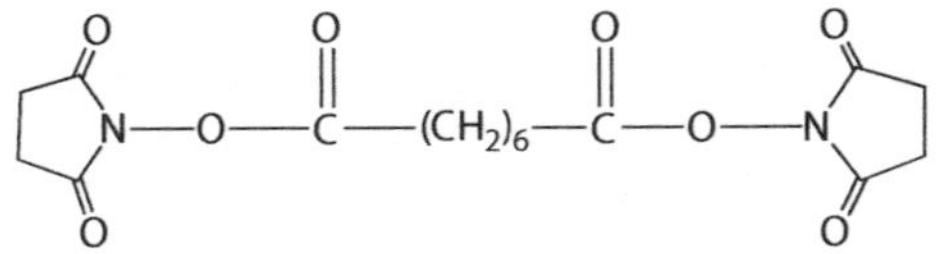	2,6,27
XLII		
Bis(sulfosuccinimidyl)-suberate (BSSS)	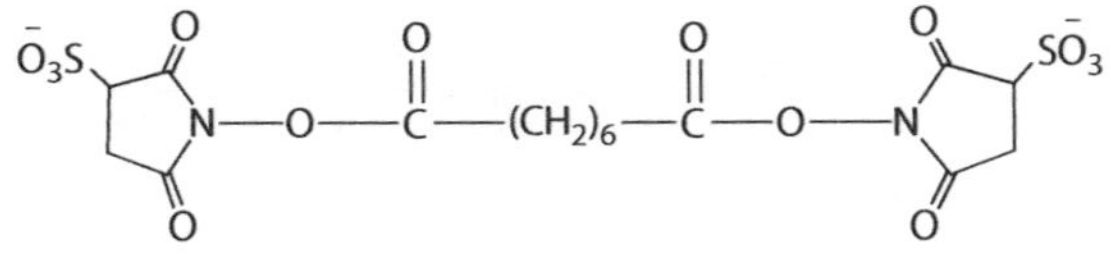	23,28

XLIII

Tetradeuterated bis(sulfosuccinimidyl)-suberate (BSSS-D4)

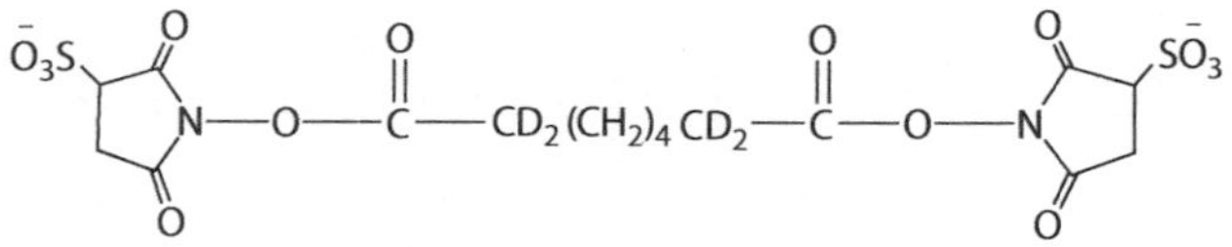

24

XLIV

Bis(succinimidyl) azelate

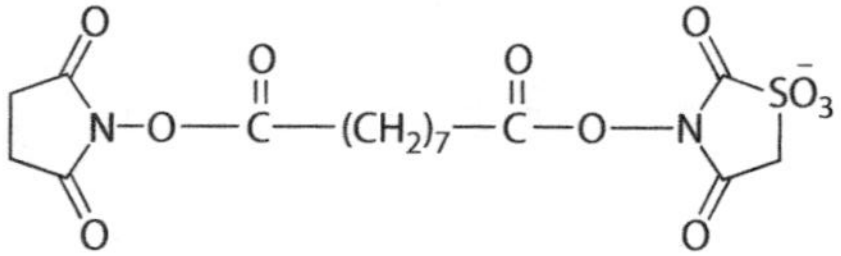

25

XLV

Bis(sulfosuccinimidyl) sebacate (BSS)

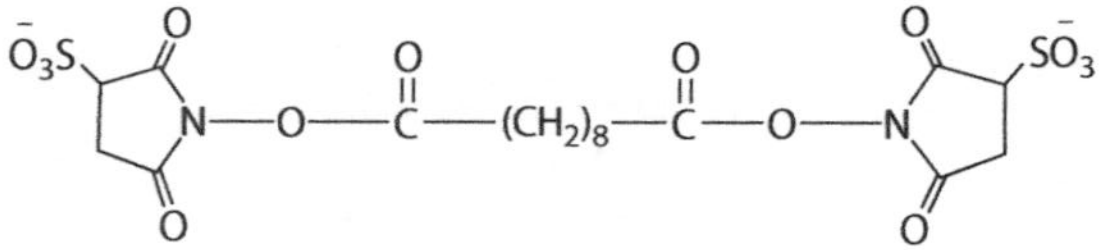

24

XLVI

Tetradeuterated bis(sulfosuccinimidyl) sebacate (BSS-D4)

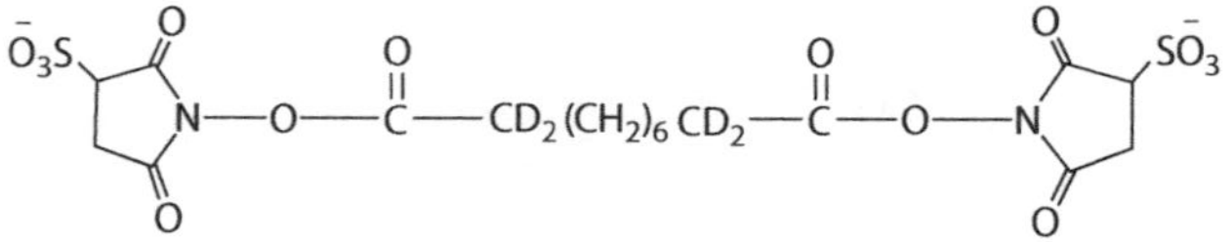

24

XLVII

(2,5-Dioxopyrrolidin-1-yl) 2-[4-[2-(2,5-dioxopyrrolidin-1-yl)oxy-2-oxo-ethoxy]but-2-ynoxy] acetate

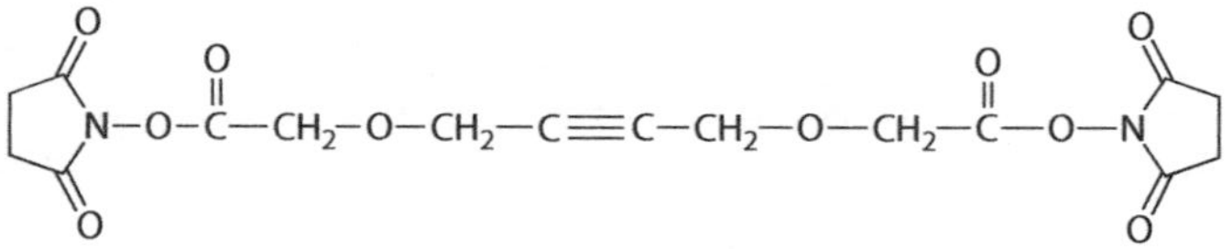

25

(continued)

(Continued)

Name (Abbreviation)	Structure	References
XLVIII (2,5-Dioxopyrrolidin-1-yl) 2-[2-[4-[2-(2,5-dioxopyrrolidin-1-yl)oxy-2-oxo-ethoxy]ethoxy] but-2-ynoxy]acetate		25
XLIX (2,5-Dioxopyrrolidin-1-yl) 2-[6-[2-(2,5-dioxopyrrolidin-1-yl)oxy-2-oxo-ethoxy]hexa-2,4-diynoxy] acetate		25
L Bis(succinimidyl) pentapolyethylene glycol (BS(PEG)5)		29
LI Bis(succinimidyl) nanopolyethylene glycol (BS(PEG)9)		29
LII *O,O′*-Bis[2-(*N*-succinimidyl-succinylamino)ethyl] polyethylene glycol (2,000, 3,000, and 10,000)		30

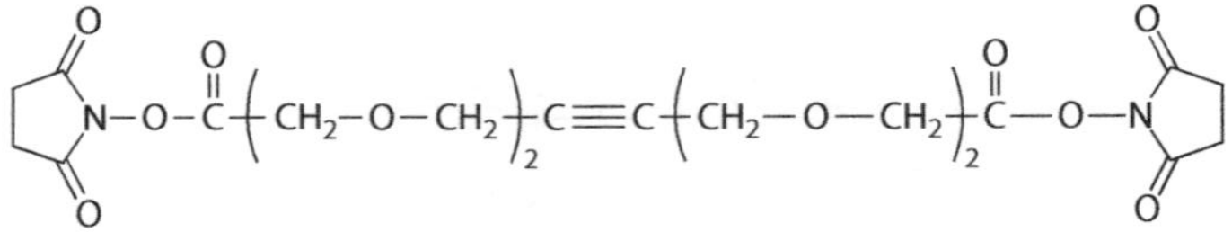

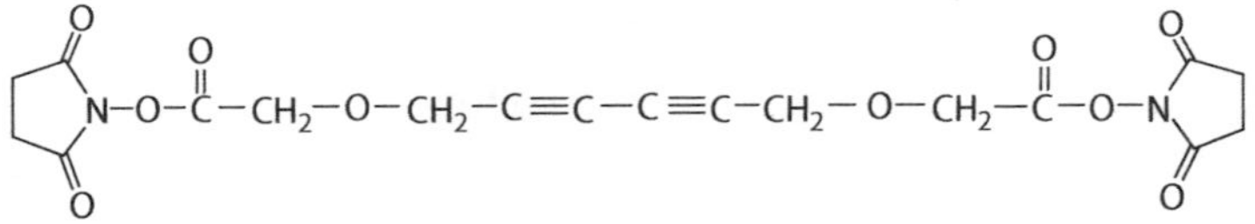

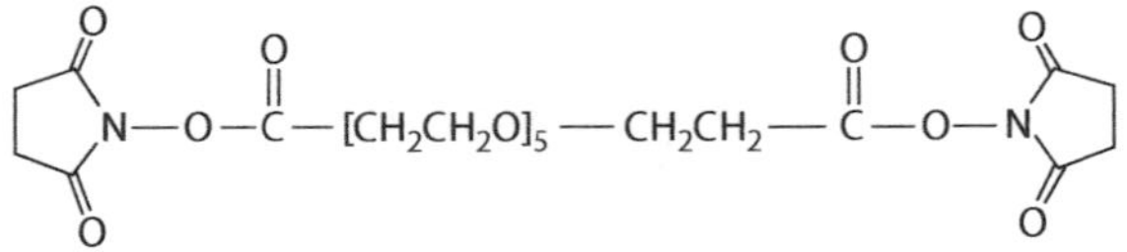

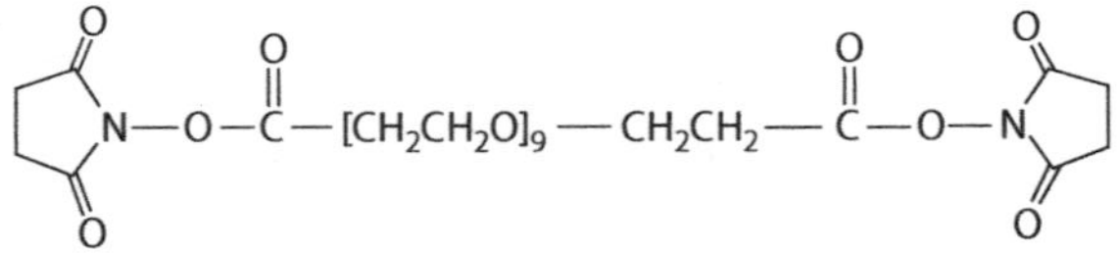

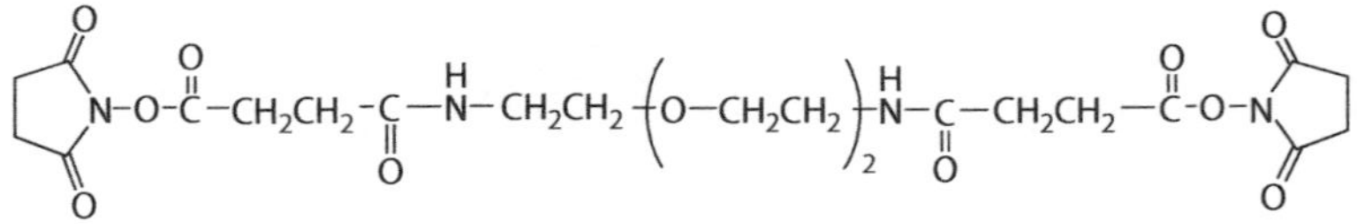

LIII

Ethyleneglycol bis-(succinimidylsuccinate) (EGS)

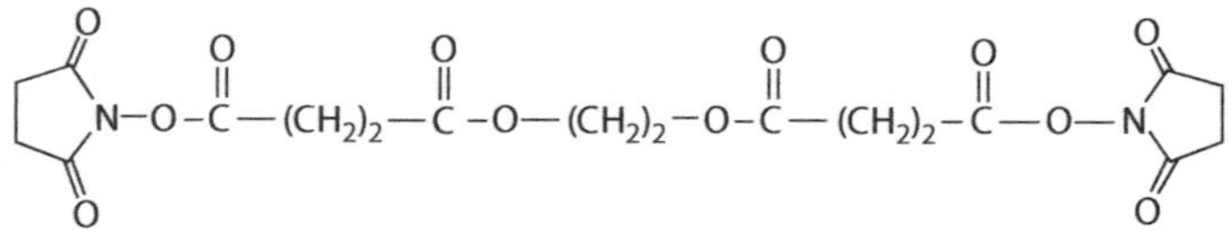

2,6,31

LIV

Ethyleneglycol bis-(sulfosuccinimidyl) succinate
(S-EGS)

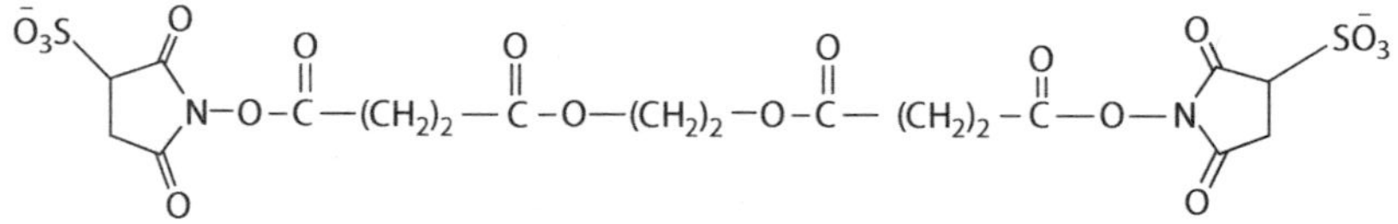

32

LV

Bis[2-(succinimido-oxycarbonyl)ethyl]sulfone (BSES)

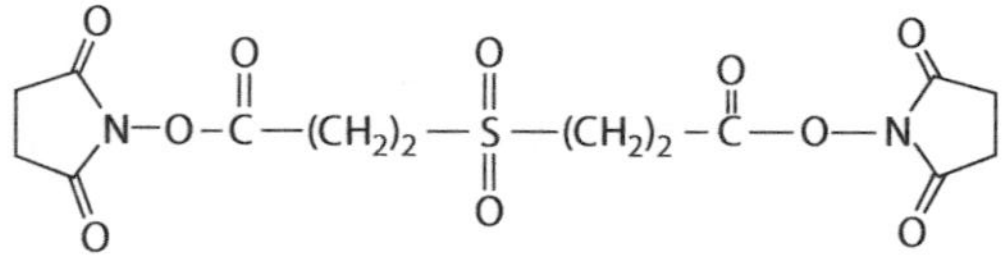

2

LVI

Bis[2-(succinimido-oxycarbonyloxy)ethyl]sulfone
(BSOCOES)

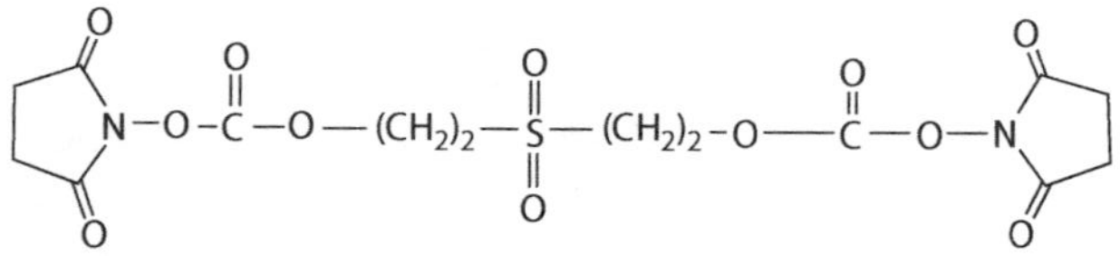

6,33

LVII

4,7,10,13,16,19,22,25,32,35,38,41,44,47,50,53-
Hexadecaoxa-28,29-dithia-hexapentacontanedioic
acid di-N-succinimidyl ester

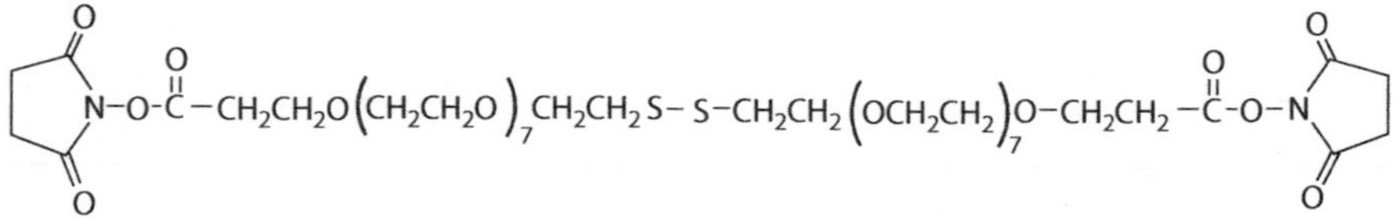

30

(continued)

(Continued)

Name (Abbreviation)	Structure	References
LVIII		
3,3′-Dithiobis(succinimidyl-propionate) (DTSP,DPS) [Lomant's reagent]	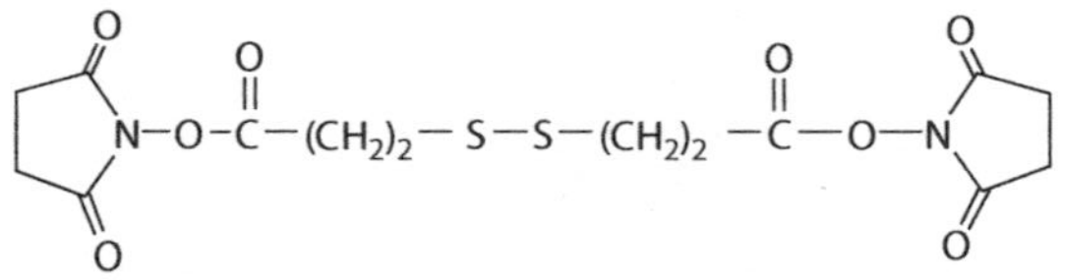	2,6,28,34
LIX		
3,3′-Dithiobis-(sulfosuccinimidyl-propionate) (DTSSP)	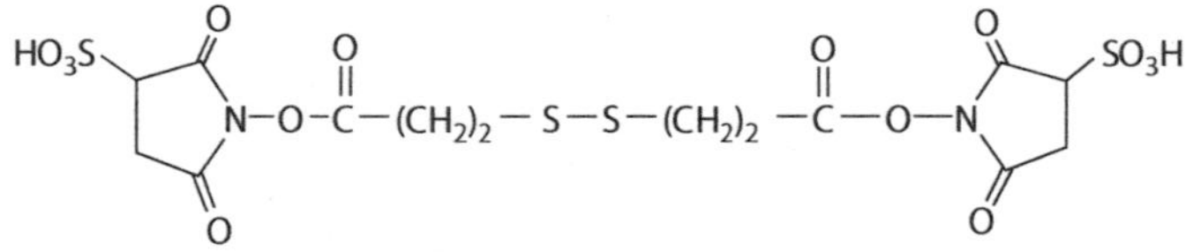	2,28,34
LX		
2,2′-Dithiobis(succinimidyl propionate) (2,2′-DSP)	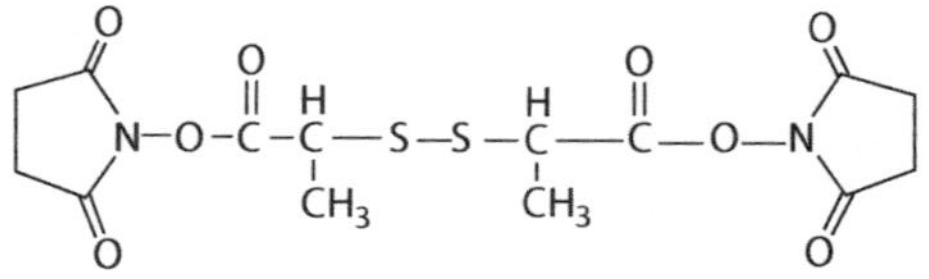	34
LXI		
Disuccinimidyl-(*N*,*N*′-diacetylhomocystine) (DDHC)	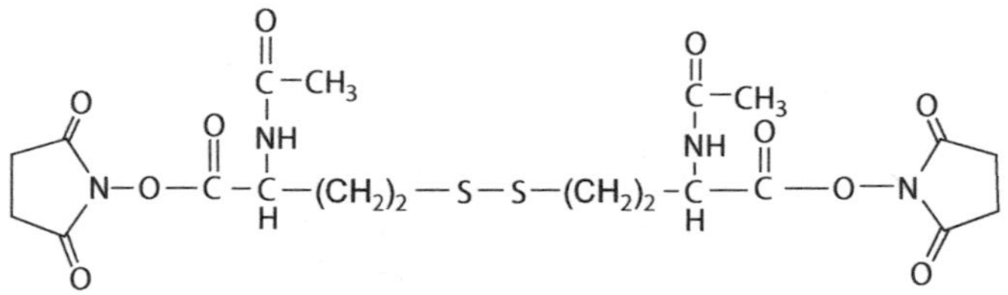	6

(continued)

LXII

Disuccinimidyl tartarate (DST)

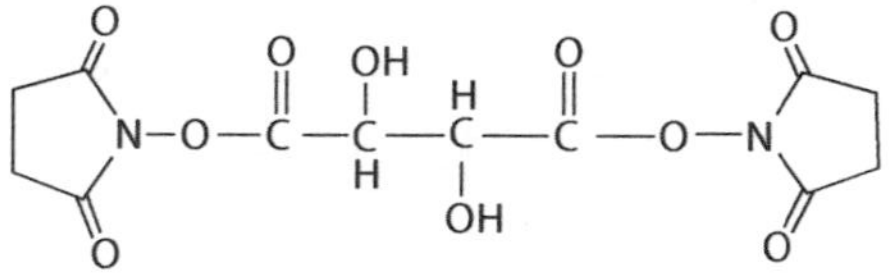

2,6,34,35

LXIII

Bis(sulfosuccinimidyl) tartarate (S-DST)

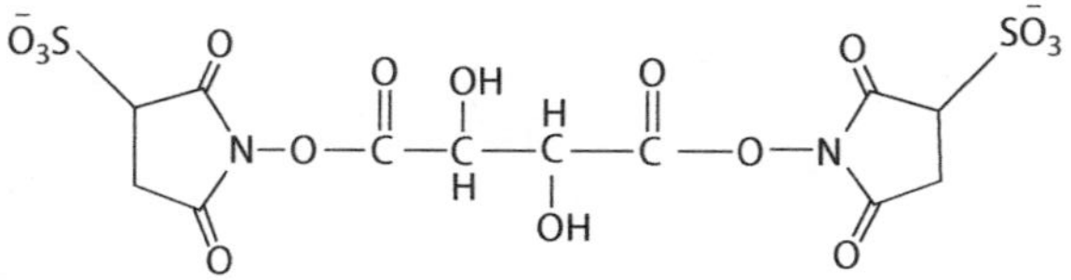

33

LXIV

N,N'-Bis(3-succinimidyloxy-carbonylpropyl) tartaramide (BSOPT)

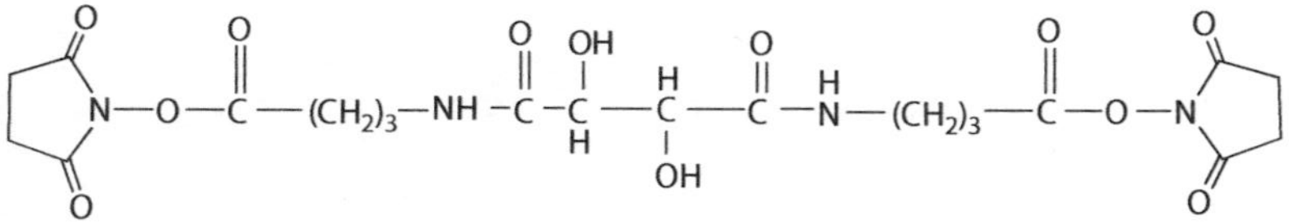

2,35

LXV

Succinimidyl selenodiacetate (DSSDA)

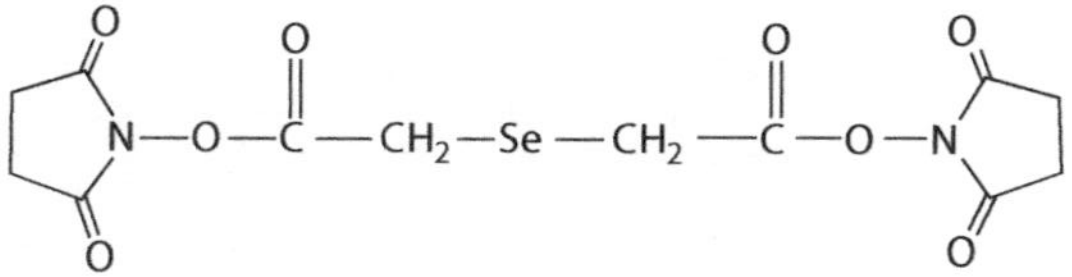

26

LXVI

Succinimidyl selenodipropionate (DSSDP)

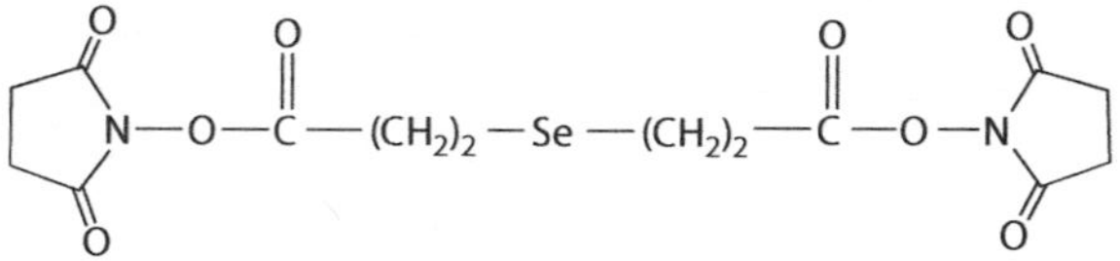

26

(Continued)

Name (Abbreviation)	Structure	References
LXVII Bis(*N*-succinimidyl)-*p*-carboxybenzoate (BSPCB)		21
LXVIII Bis(*N*-succinimidyl)-*m*-carboxybenzoate (BSMCB)		21
LXIX Biphenyl-4,4′-dicarboxylic acid bis(2,5-dioxopyrrolidin-1-yl) ester (B4DCA-NHS)		21
LXX Biphenyl-3,3′-dicarboxylic acid bis(2,5-dioxopyrrolidin-1-yl) ester (B3DCA-NHS)		21

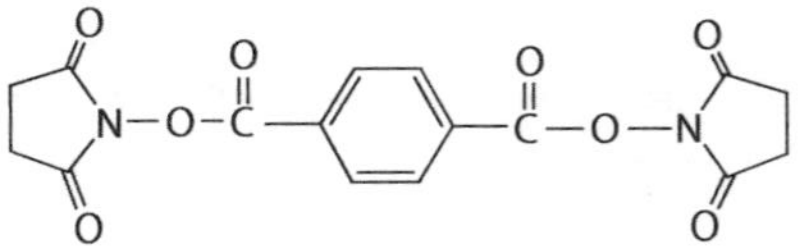

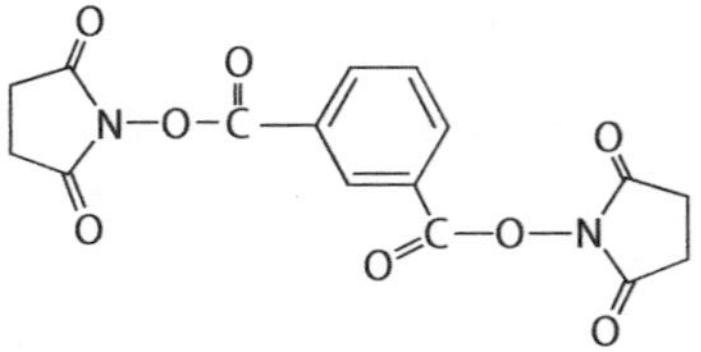

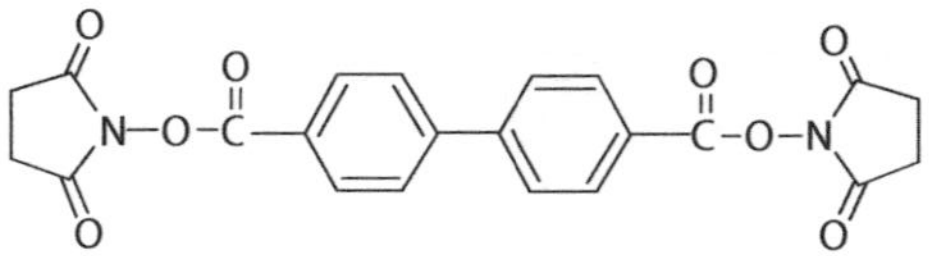

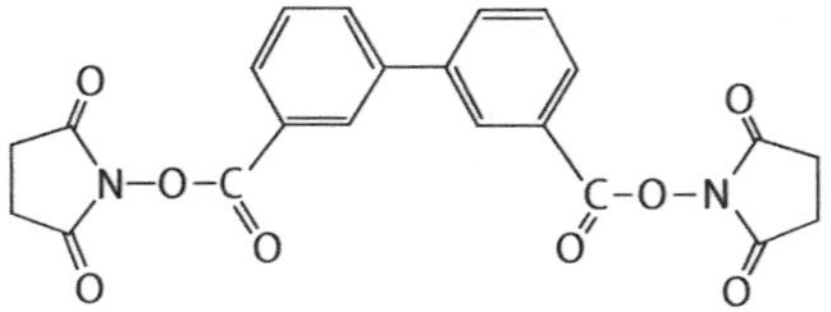

(*continued*)

LXXI

Biphenyl-2,2′-dicarboxylic acid bis(2,5-dioxopyrrolidin-1-yl) ester (B2DCA-NHS)

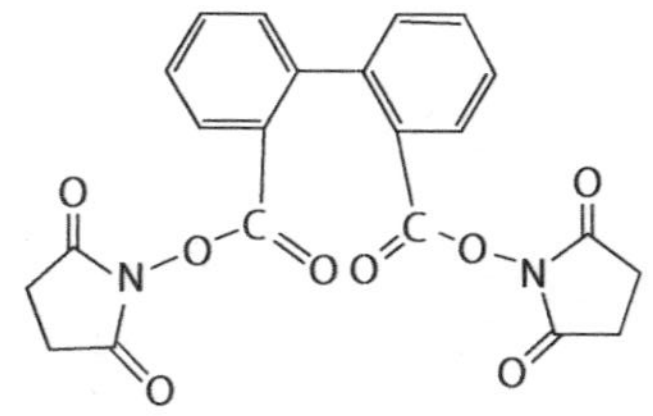

21

LXXII

Naphthalene-2,6-dicarboxylic acid bis(2,5-dioxopyrrolidin-1-yl) ester (NDCA-NHS)

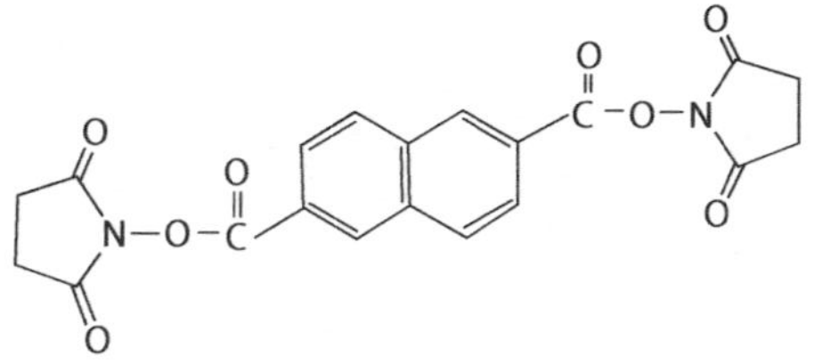

21

LXXIII

Phenanthrene-3,6-dicarboxylic acid bis(2,5-dioxopyrrolidin-1-yl) ester (PDCA-NHS)

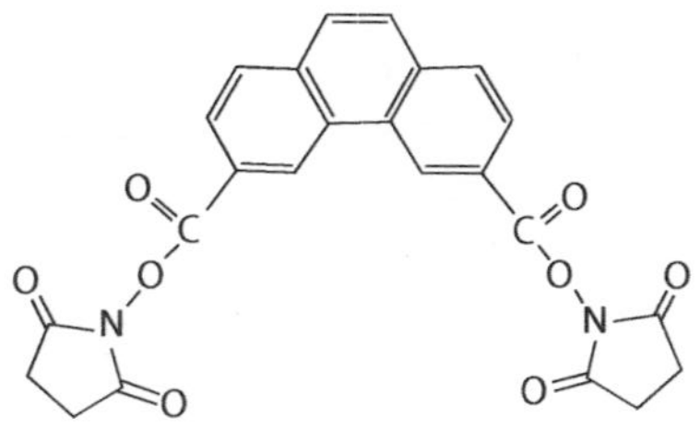

21

LXXIV

Fluorene-2,7-dicarboxylic acid bis(2,5-dioxopyrrolidin-1-yl) ester (FDCA-NHS)

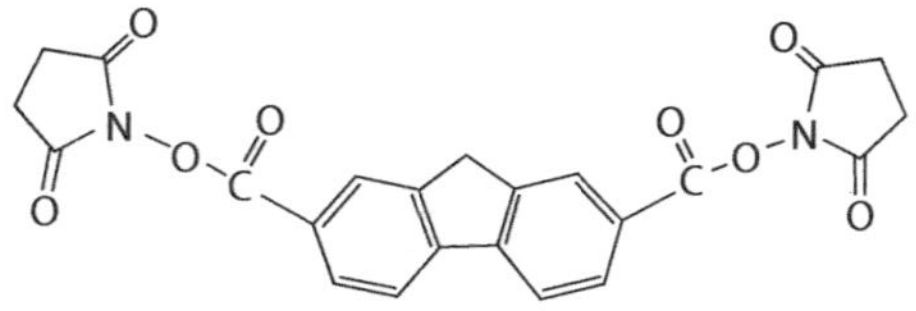

21

(Continued)

Name (Abbreviation)	Structure	References
LXXV Bimane bisthiopropionic acid *N*-succinimidyl ester (BiPS)		36
LXXVI Bimane bisthio-perdeuterated-propionic acid *N*-succinimidyl ester (BiPS-D8)		36
LXXVII Spin-labeled bis-(*N*-hydroxysuccinimide ester) n = 6: 3-Oxy-2,2-bis[6-((*N*-succinimidyloxy)carbonyl)hexanyl]-4,4-dimethyloxazolidine (NHS-HSL)		37,38

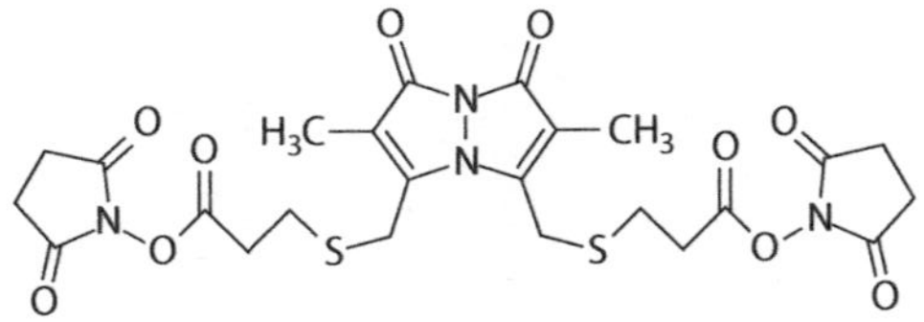

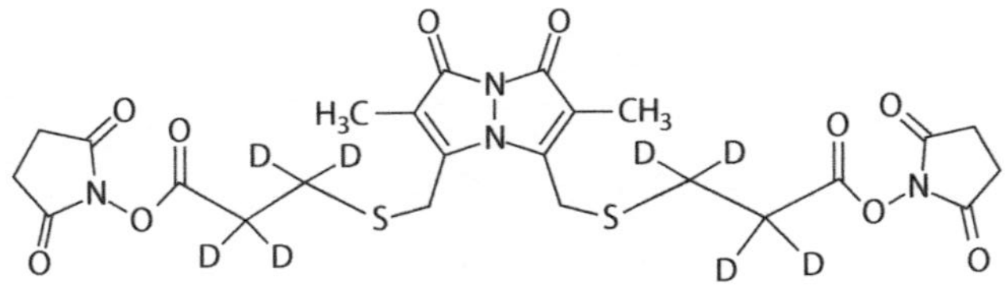

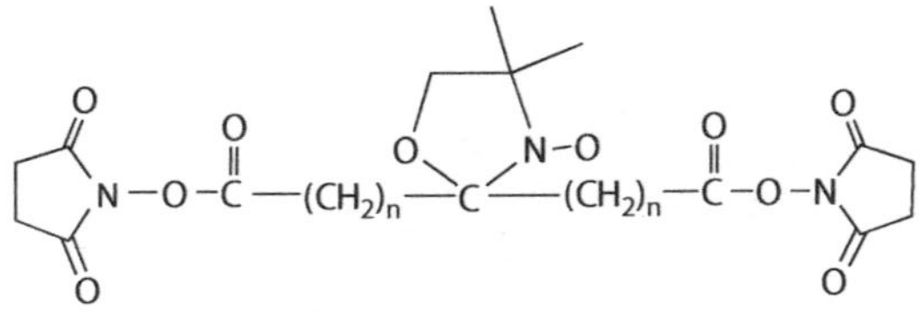

LXXVIII n = 8: 3-Oxy-2,2-bis[8-((*N*-succinimidyloxy)carbonyl) octanyl]-4,4-dimethyloxazolidine (NHS-OSL)		37
Other spin-labeled bis(*N*-hydroxysuccinimide esters)	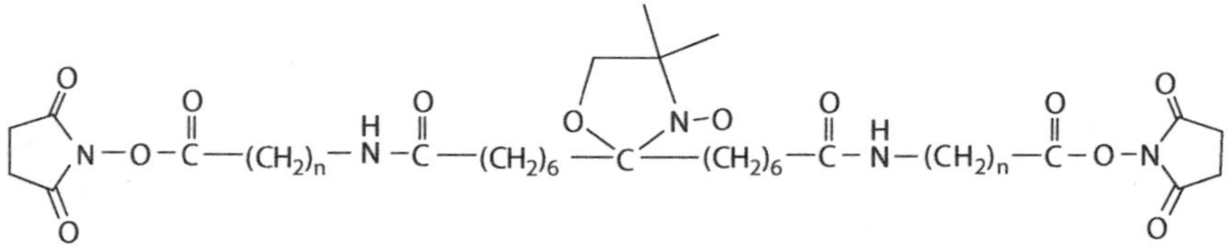	
LXXIX n = 3: 3-Oxy-2,2-bis[6-(((3-((*N*-succinimidyloxy) carbonyl)carbonyl)propyl)amino)-carbonyl)hexanyl]-4,4-dimethyloxazolidine (NHS-PPSL)		37
LXXX n = 5: 3-Oxy-2,2-bis[6-(((5-((*N*-succinimidyloxy) carbonyl) carbonyl)pentanyl)amino)-carbonyl) hexanyl]-4,4-Dimethyloxazolidine (NHS-PTSL)		37
LXXXI Bis(sulfo-*N*-succinimidyl)-doxyl-2-spiro-4′-pimelate (BSSDP)	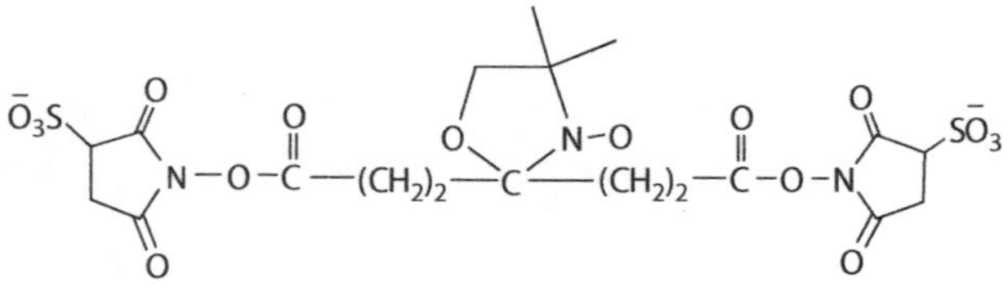	39
LXXXII Bis(sulfo-*N*-succinimidyl)-doxyl-2-spiro-5′-azelate(BSSDA)	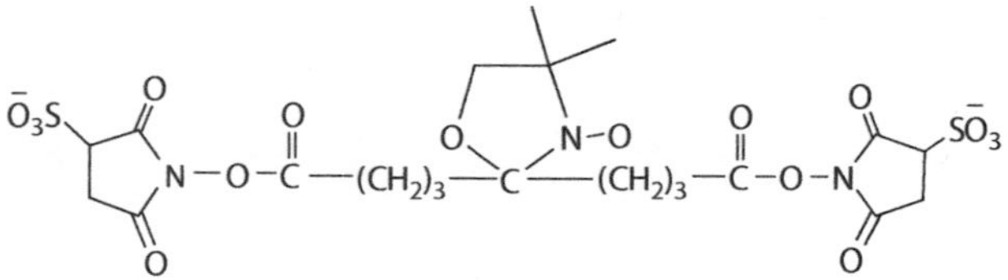	40

(*continued*)

(Continued)

Name (Abbreviation)	Structure	References
LXXXIII Dibenzoyloxysuccinimidyl ethyl phosphate (IRCX)		41
C. Bifunctional Aryl Halides **LXXXIV** 1,5-Difluoro-2,4-dinitrobenzene (FFDNB, DFNB, DFDNB)		2,6
LXXXV 1,5-Dichloro-2,4-dinitrobenzene (DCDNB)		42
LXXXVI 1,5-Dibromo-2,4-dinitrobenzene (DBDNB)		42
LXXXVII 2,4-Difluoro-5-nitrobenzene sulfonic acid		43

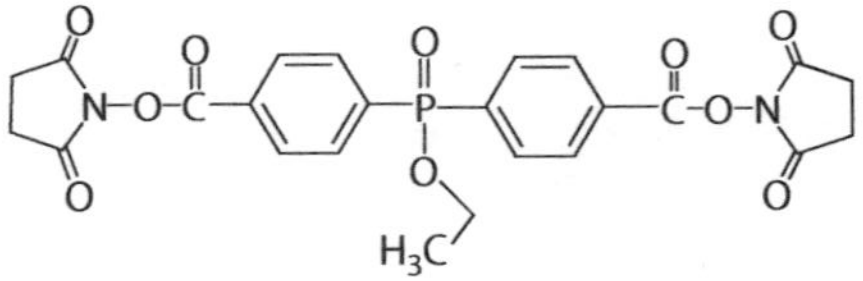

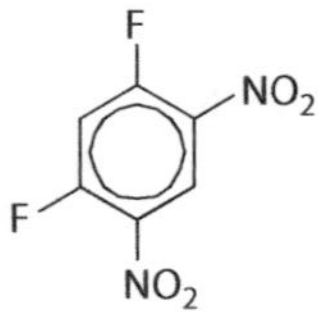

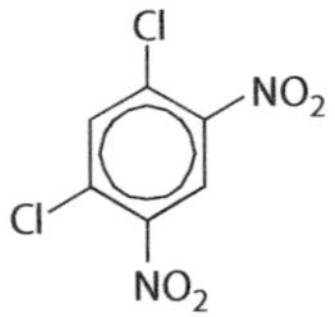

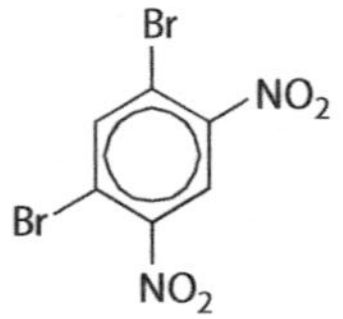

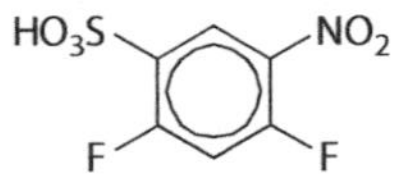

LXXXVIII

Bis(3,5-dibromosalicyl) fumerate (DBSF) — 44

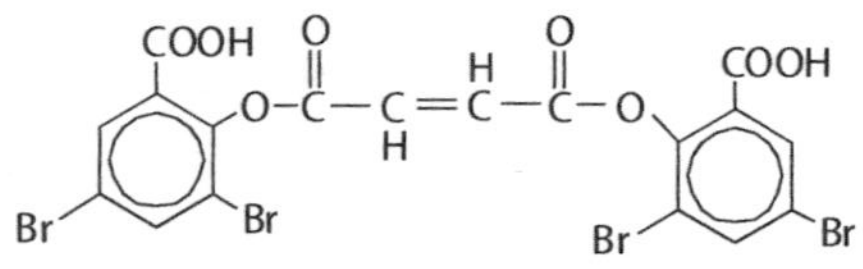

Other dibromosalicyl compounds

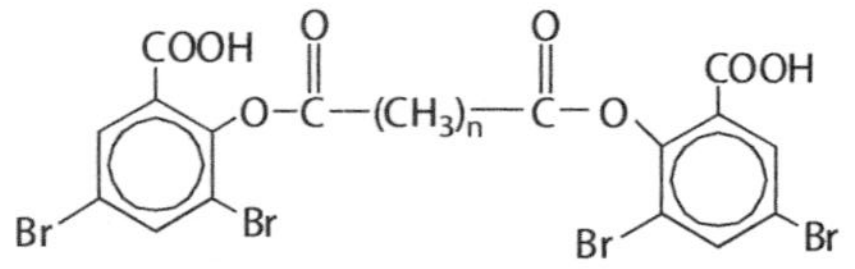

LXXXIX
n = 2: Bis(3,5-dibromosalicyl) succinate (DBSS) — 44,45

XC
n = 3: Bis(3,5-dibromosalicyl) gluterate (DBSG) — 45

XCI
n = 4: Bis(3,5-dibromosalicyl) adipate (DBSA) — 46

XCII
n = 5: Bis(3,5-dibromosalicyl) pimelate (DBSP) — 46

XCIII
n = 6: Bis(3,5-dibromosalicyl) suberate (DBSSU) — 46

XCIV
n = 7: Bis(3,5-dibromosalicyl) azelate (DBSAZ) — 46

XCV
n = 8: Bis(3,5-dibromosalicyl) sebacate (DBSSE) — 47

XCVI

4,4'-Difluoro-3,3'-dinitro-diphenylsulfone; bis(3-nitro-4-fluorophenyl)sulfone (BNFBS) — 2

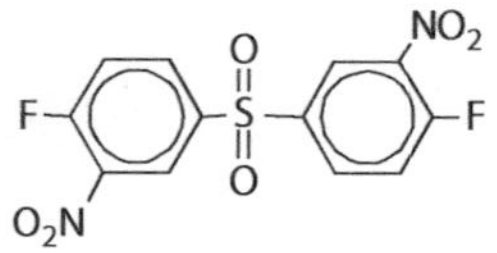

(continued)

(Continued)

Name (Abbreviation)	Structure	References
D. Diisocyanates and Diisothiocyanates		
XCVII		
1,6-Hexamethylene diisocyanate (HMDI); 1,6-diisocyantohexane	$O=C=N-(CH_2)_6-N=C=O$	48
XCVIII		
1,3-Dicyanatobenzene (o-DB)	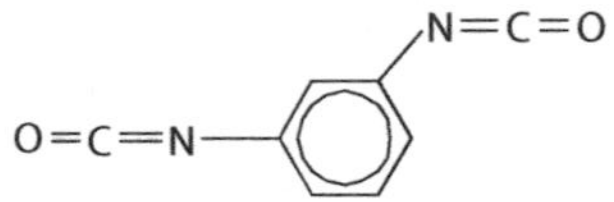	49
XCIX		
1,4-Dicyanatobenzene (p-DB)		49
C		48–50
Toluene-2,4-diisocyanate	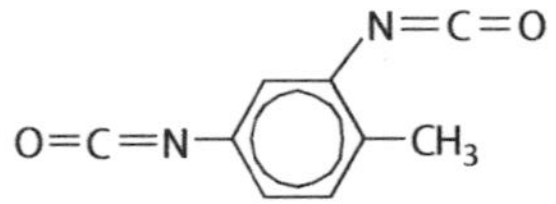	
CI		49
Toluene-2-isocyanate-4-isothiocyanate; 2-cyano-4-isothio-cyanatotoluene	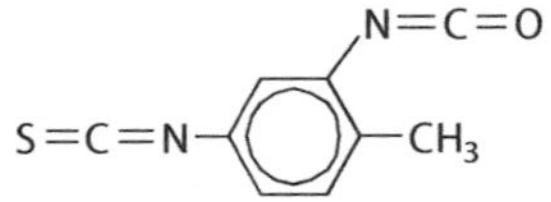	
CII		
Xylene diisocyanate	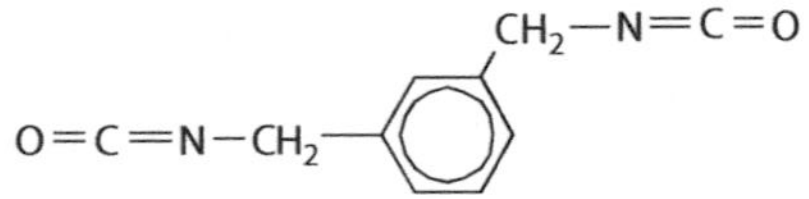	48,49
CIII		
Benzidine diisocyanate (BDI)	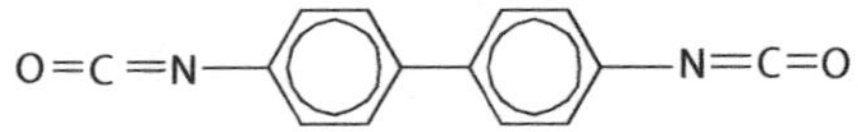	49,50

CIV

2,2′-Dimethoxybenzidine diisocyanate

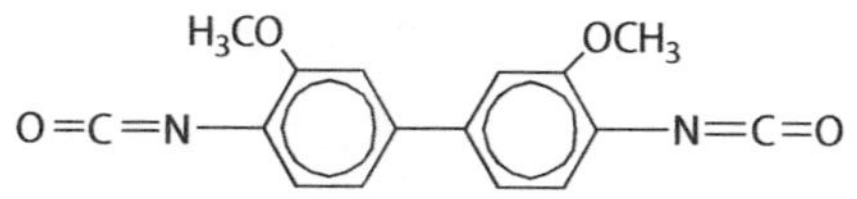

49

CV

Diphenylmethane-4,4′-diisocyanate

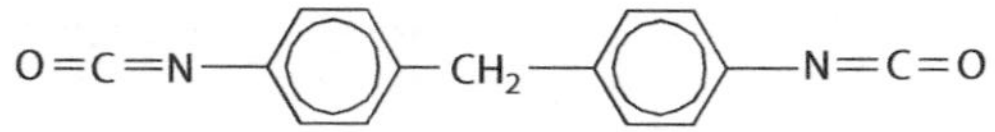

50

CVI

3-Methoxydiphenylmethane 4,4′-diisocyanate

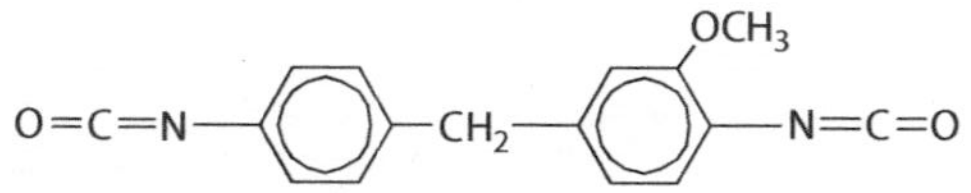

48,49

CVII

Dicyclohexylmethane 4,4′-diisocyanate

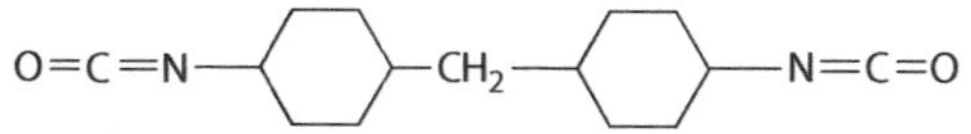

50

CVIII

Hexahydrobiphenyl-4,4′-diisocyanate

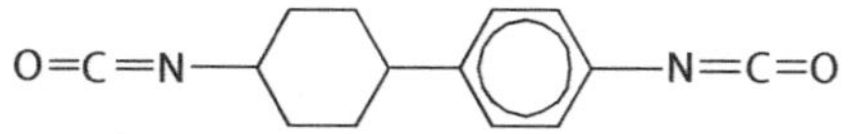

50

CIX

p-Diisocyanteoazobenzene

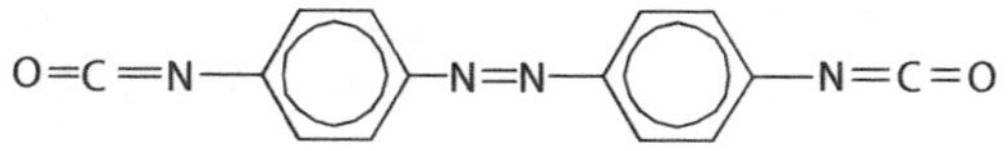

51

CX

2,2′-Dicarboxy-4,4′-azo-phenyldithioisocyanate

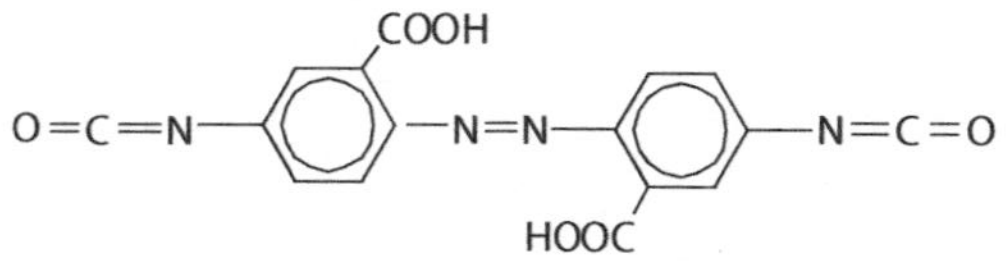

17,46,47

(continued)

(Continued)

Name (Abbreviation)	Structure	References
CXI		
2,2'-Dicarboxy-4,4'-azo-phenyldithioisocyanate	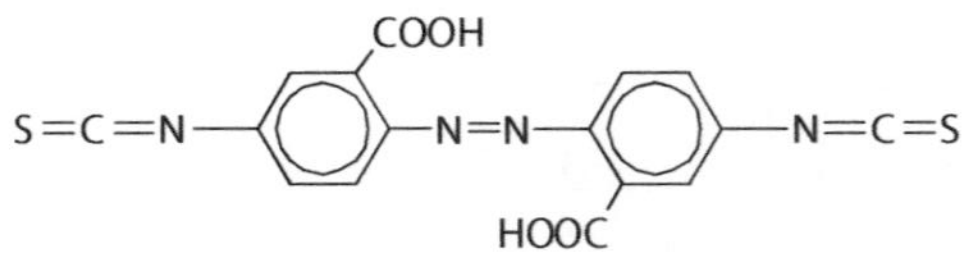	17,48,49
CXII		
2,2'-Dicarboxy-6,6'azo-phenyldiisocyanate	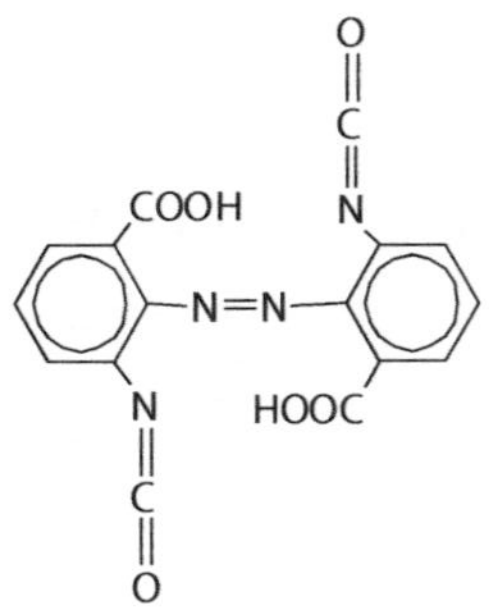	52
CXIII		
Bis-*p*-(2-carboxy-4'-azophenyl)isocyanate	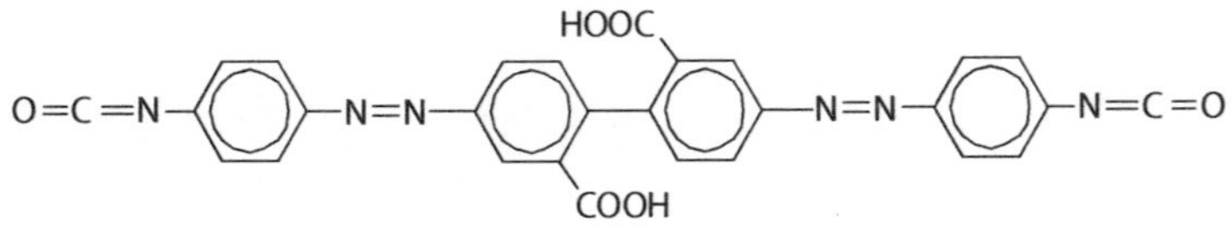	17,52
CXIV		
Bis-*p*-(2-carboxy-4'-azophenyl)isothiocyanate	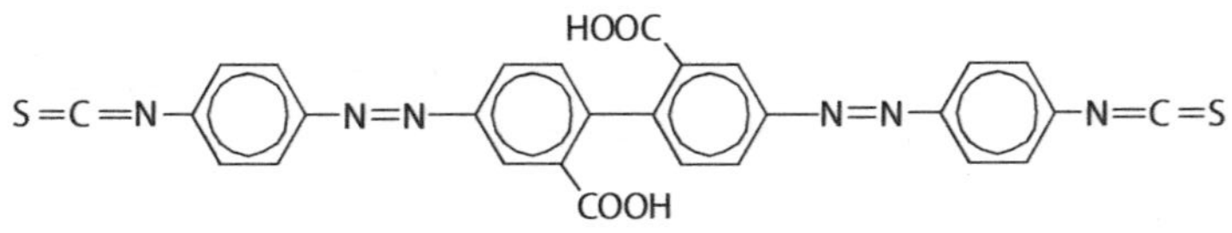	17,52
CXV		
p-Phenylene-diisothiocyanate; 1,4-phenylene diisothiocyanate		2,53

CXVI

Diphenyl-4,4'-Diisothiocyanto-2,2'-disulfonic acid 54

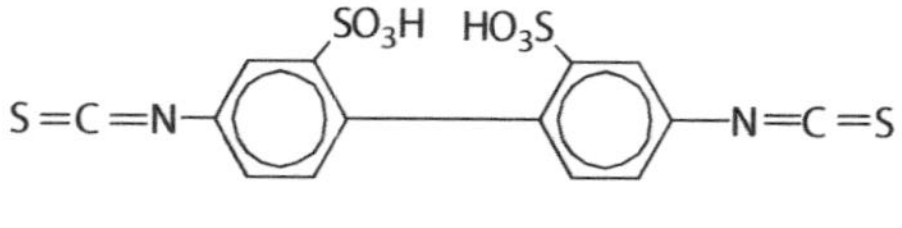

CXVII

4,4'-Diisothiocyanato-2,2'-disulfonic acid stilbene (DIDS) 2,55

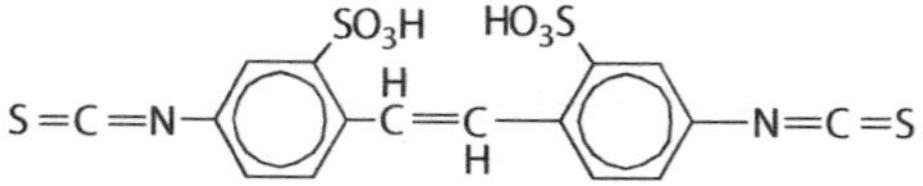

CXVIII

4,4'-Diisocyanato-dihydrostilbene-2,2' disulfonic acid 56

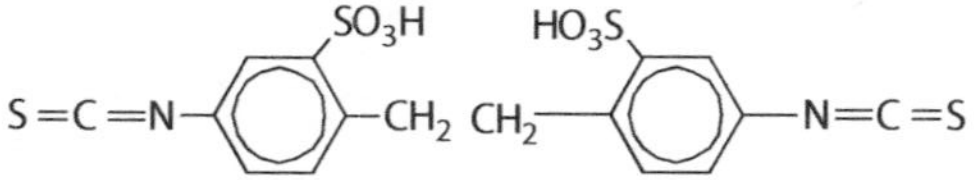

E. Bifunctional Sulfonyl Halides

CXIX

Phenol-2,4-disulfonyl chloride 57

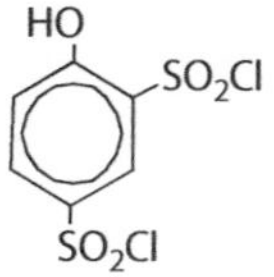

CXX

α-Naphthol-2,4-disulfonyl chloride 57

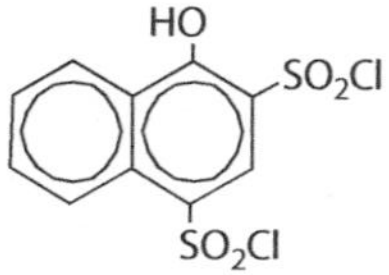

CXXI

Naphthalene-1,5-disulfonyl chloride 50

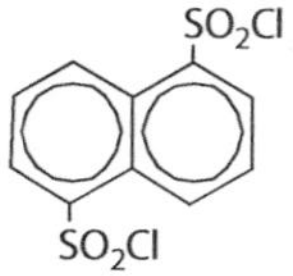

(continued)

(Continued)

Name (Abbreviation)	Structure	References
F. Bis-Nitrophenol Esters		
Bis-nitrophenol esters	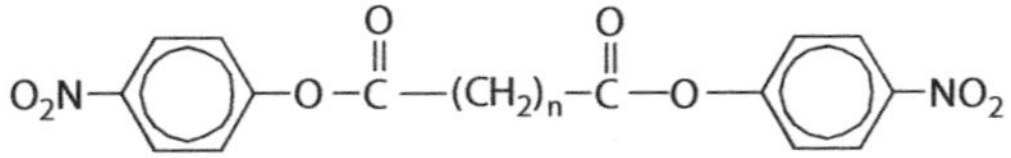	
CXXII		
n = 4: Bis-(p-nitrophenyl) adipate		58,59
CXXIII		
n = 5: Bis-(p-nitrophenyl) pimelate		59
CXXIV		
n = 8: Bis-(p-nitrophenyl) suberate		59
CXXV		
Carbonyl bis(L-methionine p-nitrophenyl ester) (CBMNPE)	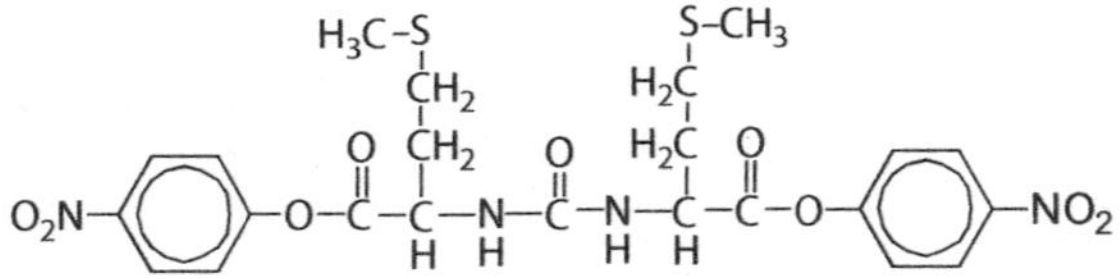	6,60
G. Bifunctional Acylazides		
CXXVI		
Tartryl diazide (TDA)	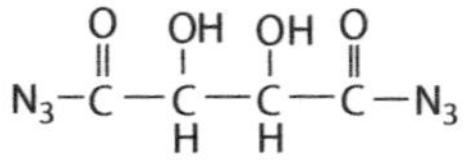	2,59
Other tartryl diazides	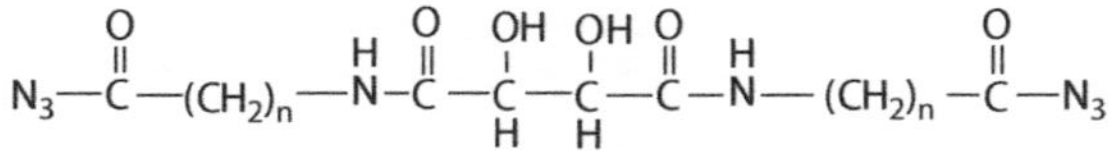	2,61
CXXVII		
n = 1: Tartryl di(glycylazide) (TDGA)		
CXXVIII		
n = 2: Tartryl di(β-alanylazide) (TDAA)		
CXXIX		
n = 3: Tartryl di(γ-aminobutyrylazide) (TDBA)		

CXXX

n = 4: Tartryl di(δ-aminovaleryazide) (TDVA)

CXXXI

n = 5: Tartryl di(ε-aminocaproylazide) (TDCA)

CXXXII

p-Bis-(ureido)azido-oligoprolylazo-benzene (PAPA)   62

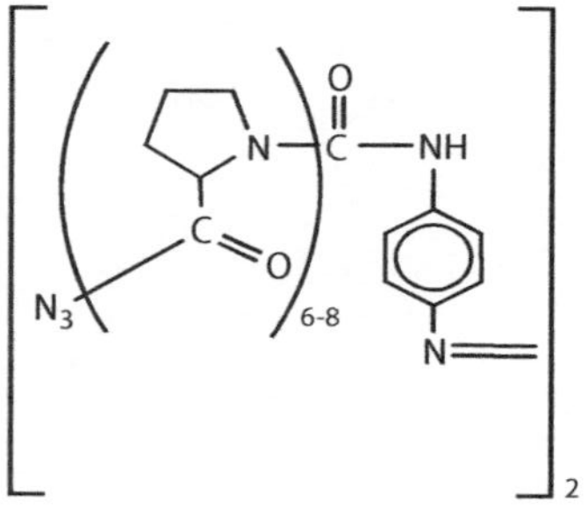

H. Dicarbonyl Compounds

CXXXIII

Glyoxal   63,64,71

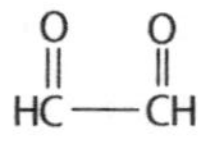

Other dialdehydes

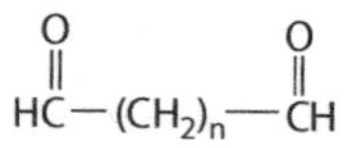

CXXXIV

n = 1: Malondialdehyde (MDA)   63,65

CXXXV

n = 2: Succinaldehyde   63

CXXXVI

n = 3: Glutaraldehyde   2,63,64,66,67

CXXXVII

n = 4: Adipaldehyde   62,66

(continued)

(Continued)

Name (Abbreviation)	Structure	References
CXXXVIII		
α-Hydroxyadipaldehyde	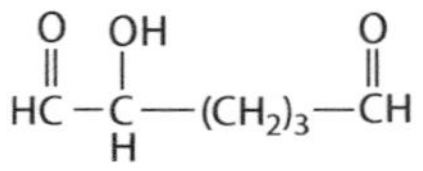 $HC-\overset{O}{\underset{H}{C}}-(CH_2)_3-\overset{O}{C}H$ (with OH on the second carbon)	67,68
CXXXIX		
3-Methylglutaraldehyde	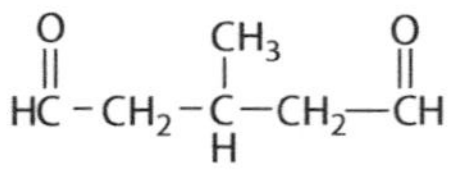 $HC-CH_2-\overset{CH_3}{\underset{H}{C}}-CH_2-\overset{O}{C}H$	67,68
CXL		
2-Methoxy-2,4-dimethyl-glutaraldehyde	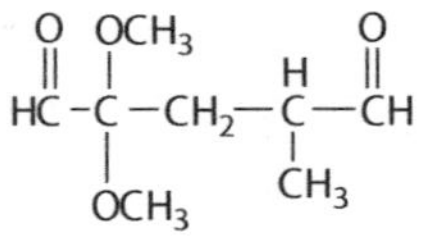 $HC-\overset{OCH_3}{\underset{CH_3}{C}}-CH_2-\overset{H}{\underset{OCH_3}{C}}-\overset{O}{\underset{CH_3}{C}}H$	67,68
CXLI		
o-Phthalaldehyde	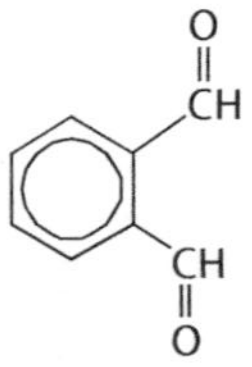	69
CXLII		70
P^1,P^2-bis(5'-pyridoxal)polyphosphates	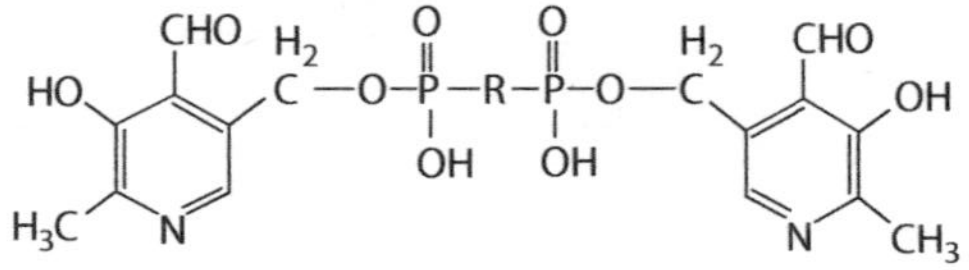	
R = O: P^1,P^2-bis(5'-pyridoxal)diphosphate (Bis-PLP)		

CXLIII

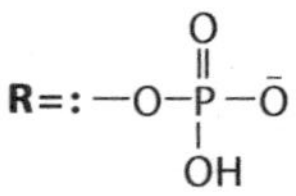

R=: —O—P—O
P¹,P³-bis(5′-pyridoxal)triphosphate

70

CXLIV

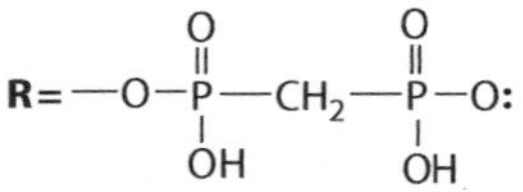

R= —O—P—CH₂—P—O:
Bis(5′-pyridoxalpyrophospho) methane

70

CXLV

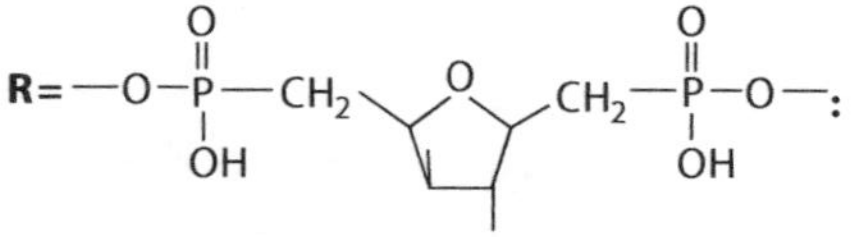

R= —O—P—CH₂
Bis(5′-pyridoxalpyrophospho)-1,6-fructose

70

CXLVI

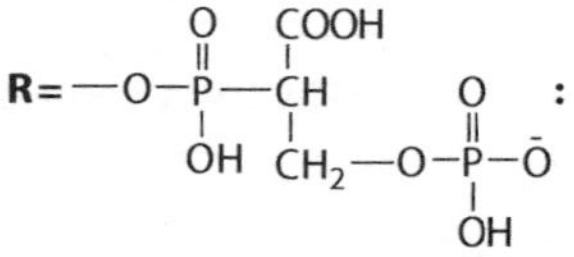

R= —O—P—CH
Bis(5′-pyridoxalpyrophospho)-2,3-glycerate

70

CXLVII

Methylglyoxal (MGO)

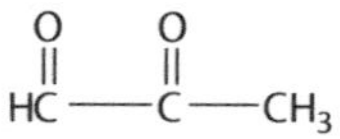

71,72

CXLVIII

Diacetyl 2,3-butadione

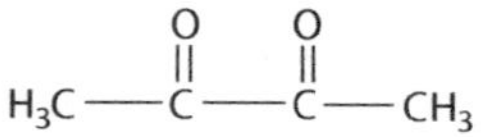

71

(continued)

(Continued)

Name (Abbreviation)	Structure	References
CXLIX		
2,5-Hexanedione	$H_3C-\overset{O}{\underset{\|}{C}}-CH_2CH_2-\overset{O}{\underset{\|}{C}}-CH_3$	73,74
CL		
3-(Trifluoromethyl)-2,5-hexanedione	$H_3C-\overset{O}{\underset{\|}{C}}-\overset{CF_3}{\underset{\|}{C}}HCH_2-\overset{O}{\underset{\|}{C}}-CH_3$	75
CLI		
3,4-Dimethyl-2,5-hexanedione	$H_3C-\overset{O}{\underset{\|}{C}}-\overset{CH_3}{\underset{\|}{C}}H-\overset{}{\underset{\underset{CH_3}{\|}}{C}}H-\overset{O}{\underset{\|}{C}}-CH_3$	72
CLII		
Cyclobutane-1,2-dione (CCBDO)	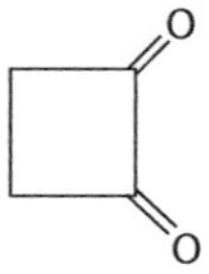	72
CLIII		
Cyclopentane-1,2-dione (CPDO)	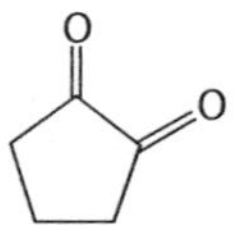	72
CLIV		
Cyclopentane-1,3-dione	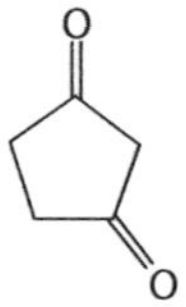	72

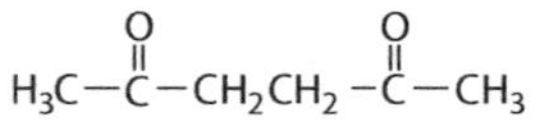

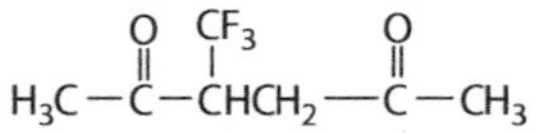

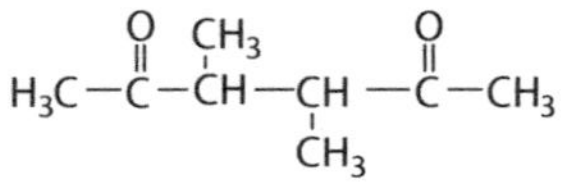

CLV

Cyclotene (CCT)

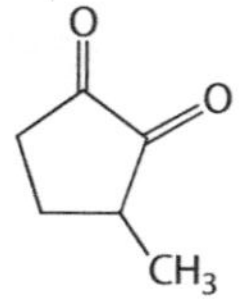

72

CLVI

3,5-Dimethylcyclopentane-1,2-dione (DMCPDO)

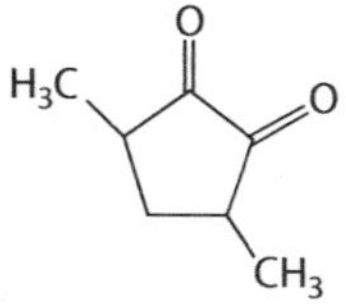

72

CLVII

Cyclohexane-1,2-dione (CHDO)

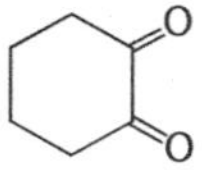

72

CLVIII

Cyclohexane-1,3-dione

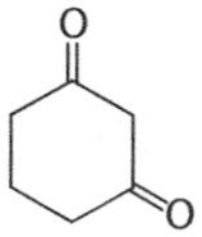

72

CLIX

Cyclohexane-1,4-dione

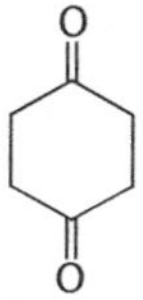

72

(continued)

(Continued)

Name (Abbreviation)	Structure	References
CLX		
Dehydroascorbic acid (DACA)	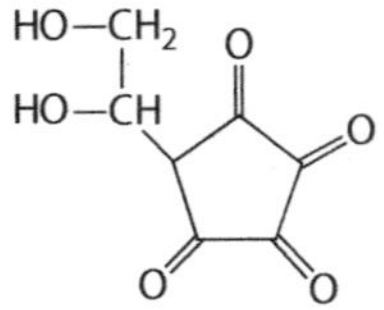	76
I. Other Amino Group–Directed Reagents		
CLXI		
Benzoquinone	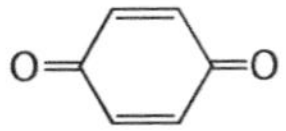	77
CLXII		
Erythreitobiscarbonate (EBC)	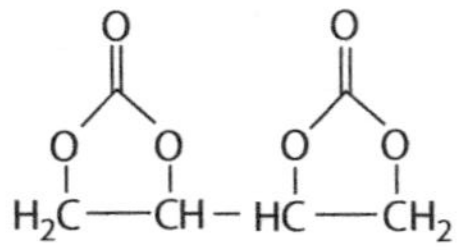	6
CLXIII		
Mucobromic acid	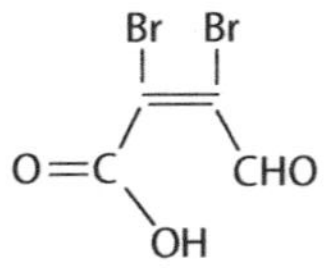	78

CLXIV

Mucochloric acid

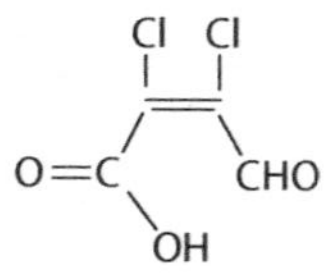

78

CLXV

Sebacoyl chloride

79

$$Cl\text{-}C\text{-}CH_2\text{-}CH_2\text{-}CH_2\text{-}CH_2\text{-}CH_2\text{-}CH_2\text{-}CH_2\text{-}CH_2\text{-}C\text{-}Cl$$

CLXVI

Genipin

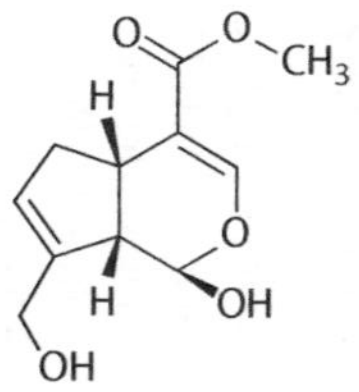

80

CLXVII

Homogentisic acid (HGA)
2,5-Dihydropenylacetic acid

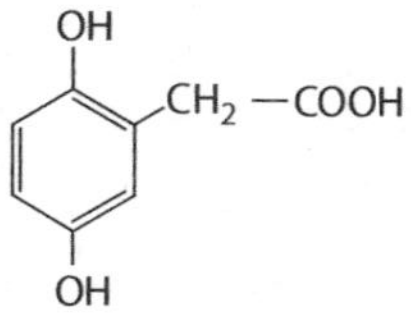

81

REFERENCES

1. Dutton, A., Adams, M., and Singer, S. J., Bifunctional imidoesters as cross-linking reagents, *Biochem. Biophys. Res. Commun.*, 23, 730–739, 1966.
2. Ji, T. H., Bifunctional reagents, *Methods Enzymol.*, 91, 580–609, 1983.
3. Ji, T. H., Cross-linking of glycolipids in erythrocyte ghost membrane, *J. Biol. Chem.*, 249, 7841–7847, 1974.
4. Collins, P. L. and Mottet, G., Homooligomerization of the hemagglutinin-neuraminidase glycoprotein of human parainfluenza virus type 3 occurs before the acquisition of correct intramolecular disulfide bonds and mature immunoreactivity, *J. Virol.*, 65, 2362–2371, 1991.
5. Packman, L. C. and Perham, R. N., Quaternary structure of the pyruvate dehydrogenase multienzyme complex of Bacillus stearothermophilus studied by a new reversible cross-linking procedure with bis(imidoesters), *Biochemistry*, 21, 5171–5175, 1982.
6. Han, K.-K., Richard, C., and Delacourte, A., Chemical cross-links of proteins by using bifunctional reagents, *Int. J. Biochem.*, 16, 129–252, 1984.
7. Siezen, R. J., Bindels, J. G., and Hoenders, H. J., The quaternary structure of bovine alpha-crystallin. Chemical crosslinking with bifunctional imido esters, *Eur. J. Biochem.*, 107, 243–249, 1980.
8. Hartman, F. C. and Wold, F., Bifunctional Reagents. Cross-linking of pancreatic ribonuclease with a diimido ester, *J. Am. Chem. Soc.*, 88, 3890–3891, 1966.
9. Niehaus, W. G., Jr. and Wold. F., Cross-linking of erythrocyte membranes with dimethyl adipimidate, *Biochim. Biophys. Acta*, 196,170–175, 1970.
10. Hucho, F., Müllner, H., and Sund, H., Investigation of the symmetry of oligomeric enzymes with bifunctional reagents, *Eur. J. Biochem.*, 59, 79–87, 1975.
11. Coggins, J. R., Hooper, E. A., and Perham, R. N., Use of dimethyl suberimidate and novel periodate-cleavable bis(imido esters) to study the quaternary structure of the pyruvate dehydrogenase multienzyme complex of *Escherichia coli*, *Biochemistry*, 15, 2527–2533, 1976.
12. Schramm, H. J. and Dülffer, T., Synthesis and application of cleavable and hydrophilic crosslinking reagents, *Adv. Exp. Med. Biol.*, 86A, 197–206, 1977.
13. Schramm, H. J. and Dülffer, T., Synthesis and application of cleavable and hydrophilic cross-linking reagents, in *Protein Cross-Linking: Biochemical and Molecular Aspects*, Friedman, M. (ed.), Plenum Press, New York, 1076, p. 197.
14. Staros, J. V., Morgan, D. G., and Appling, D. R., A membrane-impermeant, cleavable cross-linker. Dimers of human erythrocyte band 3 subunits cross-linked at the extracytoplasmic membrane face, *J. Biol. Chem.*, 256, 5890–5893, 1981.
15. Shivdasani, R. A. and Thomas, D. W., Molecular associations of IA antigens after T-B cell interactions. I. Identification of new molecular associations, *J. Immunol.*, 141, 1252–1260, 1988.
16. Aizawa, S., Kurimoto, F., and Yokono, O., Crosslinking studies with different length dithiobisalkylimidates. (I) Solubilized erythrocyte spectrin, *Biochem. Biophys. Res. Commun.*, 75, 870–878, 1977.
17. Fasold, H., Baumert, H., and Fink, G., Comparison of hydrophobic and strongly hydrophilic cleavable cross-linking reagents in intermolecular by formation in aggregates of proteins or protein-RNA, in *Protein Cross-Linking: Biochemical and Molecular Aspects*, Friedman, M. (ed.), Plenum Press, New York, 1977, p. 207.
18. Schramm, H. J., Synthese von farbigen Nitrilen, Dinitril und bifunktimellen Imidsoureestern [The use of imide acid esters for the chemical modification of proteins. I. Synthesis of colored nitriles, dinitriles and bifunctional imide acid esters], *Hoppe Seylers Z. Physiol. Chem.*, 348, 289–292, 1967.
19. Schramm, H. J., The synthesis of mono- and bifunctional nitriles and imidoesters carrying a fluorescent group, *Hoppe Seylers Z. Physiol. Chem.*, 356, 1375–1379, 1975.
20. Beier, M. and Hoheisel, J. D., Versatile derivatisation of solid support media for covalent bonding on DNA-microchips, *Nucl. Acids Res.*, 27, 1970–1977, 1999.
21. Fujimoto, K., Kajino, M., and Inouye, M., Development of a series of cross-linking agents that effectively stabilize α-helical structures in various short peptides, *Chem. Eur. J.*, 14, 857–863, 2008.
22. Mattson, G., Conklin, E., Desai, S., Nielander, G., Savage, M. D., and Morgensen, S., A practical approach to crosslinking, *Mol. Biol. Rep.*, 11, 167–183, 1993.
23. Ihling, C., Schmidt, A., Kalkhof, S., Schulz, D. M., Stingl, C., Mechtler, K., Haack, M., Beck-Sickinger, A. G., Cooper, D. M., and Sinz, A., Isotope-labeled cross-linkers and Fourier transform ion cyclotron resonance mass spectrometry for structural analysis of a protein/peptide complex, *J. Am. Soc. Mass Spectrom.*, 17, 1100–1113, 2006.

24. Müller, D. R., Schindler, P., Towbin, H., Wirth, U., Voshol, H., Hoving, S., and Steinmetz, M. O., Isotope-tagged cross-linking reagents. A new tool in mass spectrometric protein interaction analysis, *Anal. Chem.*, 73, 1927–1934, 2001.

25. Fujimoto, K., Oimoto, N., Katsuno, K., and Inouye, M., Effective stabilisation of alpha-helical structures in short peptides with acetylenic cross-linking agents, *Chem. Commun. (Camb).*, 7, 1280–1281, 2004.

26. Buchardt, O., Elsner, H. I., Nielsen, P. E., Petersen, L. C., and Suenson, E., Protein crosslinking reagents containing a selenoethylene linker are cleaved by mild oxidation, *Anal. Biochem.*, 158, 87–92, 1986.

27. Cox, G. W., Mathieson, B. J., Giardina, S. L., and Varesio, L., Characterization of IL-2 receptor expression and function on murine macrophages, *J. Immunol.*, 145, 1719–1726, 1990.

28. Staros, J. V., N-Hydroxysulfosuccinimide active esters: Bis(N-hydroxysulfosuccinimide) esters two carboxylic acids are hydrophilic, membrane-impermeable, protein cross-linkers, *Biochemistry*, 21, 3950, 1982.

29. *Pierce Applications Handbook and Catalog*, Pierce Biotechnology, Inc., Rockford, IL, 2006.

30. *Sigma-Aldrich Catalog*, Sigma-Aldrich Corporate, St. Louis, MO, 2010.

31. Abdella, P. M., Smith, P. K., and Royer, G. P., A new cleavable reagent for cross-linking and reversible immobilization of proteins, *Biochem. Biophys. Res. Commun.*, 87, 734–742, 1979.

32. Sinz, A., Chemical cross-linking and mass spectrometry for mapping three-dimensional structures of proteins and protein complexes, *J. Mass Spectrom.*, 38, 1225–1237, 2003.

33. Zarling, D. A., Watson, A., and Bach, F. H., Mapping of lymphocyte surface polypeptide antigens by chemical cross-linking with BSOCOES, *J. Immunol.*, 124, 913–920, 1980.

34. Bragg, P. D. and Hou, C., Chemical crosslinking of alpha subunits in the F1 adenosine triphosphatase of *Escherichia coli*, *Arch. Biochem. Biophys.*, 244, 361–372, 1986.

35. Smith, R. J., Capaldi, R. A., Muchmore, D., and Dahlquist, F., Cross-linking of ubiquinone cytochrome c reductase (complex III) with periodate-cleavable bifunctional reagents, *Biochemistry*, 17, 3719–3723, 1978.

36. Petrotchenko, E. V., Xiao, K., Cable, J., Chen, Y., Dokholyan, N. V., and Borchers, C. H., BiPS, a photocleavable, isotopically coded, fluorescent cross-linker for structural proteomics, *Mol. Cell Proteomics*, 8, 273–286, 2009.

37. Gaffney, B. J., Willingham, G. L., and Schepp, R. S., Synthesis and membrane interactions of spin-label bifunctional reagents, *Biochemistry*, 22, 881–892, 1983.

38. Willingham, G. L. and Gaffney, B. J., Reactions of spin-label cross-linking reagents with red blood cell proteins, *Biochemistry*, 22, 892–898, 1983.

39. Anjaneyulu, P. S., Beth, A. H., Sweetman, B. J., Faulkner, L. A., and Staros, J. V., Bis(sulfo-N-succinimidyl) [15N,2H16]doxyl-2-spiro-4′-pimelate, a stable isotope-substituted, membrane-impermeant bifunctional spin label for studies of the dynamics of membrane proteins: Application to the anion-exchange channel in intact human erythrocytes, *Biochemistry*, 27, 6844–6851, 1988.

40. Anjaneyulu, P. S., Beth, A. H., Cobb, C. E., Juliao, S. F., Sweetman, B. J., and Staros, J. V., Bis(sulfo-N-succinimidyl) doxyl-2-spiro-5′-azelate: Synthesis, characterization, and reaction with the anion-exchange channel in intact human erythrocytes, *Biochemistry*, 28, 6583–6590, 1989.

41. Gardner, M. W., Vasicek, L. A., Shabbir, S., Anslyn, E. V., and Brodbelt, J. S., Chromogenic cross-linker for the characterization of protein structure by infrared multiphoton dissociation mass spectrometry, *Anal. Chem.*, 80, 4807–4819, 2008.

42. Fraenkel-Conrat, H., The chemistry of proteins and peptides, *Annu. Rev. Biochem.*, 25, 91–330, 1956.

43. Kremer, U. M., Synthesis of a water-soluble protein cross-linking reagent, *Biol. Chem. Hoppe Seyler*, 371, 861–864, 1990.

44. Walder, J. A., Zaugg, R. H., Walder, R. Y., Steele, J. M., and Klotz, I. M., Diaspirins that cross-link beta chains of hemoglobin: Bis(3,5-dibromosalicyl) succinate and bis(3,5-dibromosalicyl) fumarate, *Biochemistry*, 18, 4265–4270, 1979.

45. Huang, H. and Olsen, K. W., Thermal stabilities of hemoglobins crosslinked with different length reagents, *Artif. Cells Blood Substit. Immobil. Biotechnol.*, 22, 719, 1994.

46. Bobofchak, K. M., Tarasov, E., and Olsen, K. W., Effect of cross-linker length on the stability of hemoglobin, *Biochim. Biophys. Acta*, 1784, 1410, 2008.

47. Zhang, Q. and Olsen, K. W., The modification of hemoglobin by a long crosslinking reagent: Bis(3,5-dibromosalicyl)sebacate, *Biochem. Biophys. Res. Commun.*, 203, 1463, 1994.

48. Ozawa, H., Bridging reagent for protein. I. The reaction of diisocyanates with lysine and enzyme proteins, *J. Biochem. (Tokyo)*, 62, 419, 1967.

49. Schick, A. F. and Singer, S. J., On the formation of covalent linkages between two protein molecules, *J. Biol. Chem.*, 236, 2477, 1961.

50. Borek, F. and Silverstein, A. M., Characterization and purification of ferritin-antibody globulin conjugates, *J. Immunol.*, 87, 555, 1961.

51. Wetz, K., Fasold, H., and Meyer, C., Synthesis of long, hydrophilic, protein cross-linking reagents, *Anal. Biochem.*, 58, 347, 1974.

52. Fasold, H. and Lusty, C. J., The application of azo dyes to identify reactive groups and determine distances in proteins, in *Seventh International Congress of Biochemistry Abstracts*, Vol. III, 1967, p. 1.

53. Hartman, F. C. and Wold, F., Bifunctional reagents. Cross-linking of pancreatic ribonuclease with a diimido ester, *J. Am. Chem. Soc.*, 88, 3890, 1966.

54. Manecke, G. and Gunzel, G., Darstellung eines wasserunlöslichen, aktiven papains (demonstration of a water insoluble, active papain), *Naturwissenschaften*, 54, 647–648, 1967.

55. Cabantchik, I. Z., Balshin, M., Breuer, W., and Rothstein, A., Pyridoxal phosphate. An anionic probe for protein amine groups exposed on the outer and inner surfaces of intact human red blood cells, *J. Biol. Chem.*, 250, 5130, 1975.

56. Macara, I. G. and Cantley, L. C., Mechanism of anion exchange across the red cell membrane by band 3: Interactions between stilbene sulfonate and NAP-taurine binding sites, *Biochemistry*, 20, 5695, 1981.

57. Herzig, D. J., Rees, A. W., and Day, R. A., Bifunctional reagents and protein structure determination. The reaction of phenolic disulfonyl chlorides with lysozyme, *Biopolymers*, 2, 349, 1964.

58. Brandenburg, D., Peptides. 87. Preparation of $N^{\alpha A1}$, $N^{\epsilon B29}$-adipoyl-insulin, an intramolecularly cross-linked derivative of beef insulin, *Hoppe-Seyler's Z. Physiol. Chem.*, 353, 869, 1972.

59. Plotz, P. H., Bivalent affinity labeling haptens in the formation of model immune complexes, *Methods Enzymol.*, 46, 505, 1977.

60. Busse, W. D. and Carpenter, F. H., Carbonyl bis(L-methionine p-nitrophenyl ester). A new reagent for the reversible intramolecular cross-linking of insulin, *J. Am. Chem. Soc.*, 96, 5947, 1974.

61. Lutter, L. C., Ortanderl, F., and Fasold, H., The use of a new series of cleavable protein-crosslinkers on the *Escherichia coli* ribosome, *FEBS Lett.*, 48, 288–292, 1974.

62. Wetz, K., Fasold, H., and Meyer, C., Synthesis of "long", hydrophilic, protein-cross-linking reagents, *Anal. Biochem.*, 58, 347–360, 1974.

63. Cater, C. W., The evaluation of aldehydes and other bifunctional compounds as cross-linking agents for collagen, *J. Soc. Leather Trade Chem.*, 47, 259, 1963.

64. Brooks, B. R. and Klamerth, O. L., Interaction of DNA with bifunctional aldehydes, *Eur. J. Biochem.*, 5, 178, 1968.

65. Kergonou, J. F., Marais, D., Lafite, C., Pennacino, I., and Ducousso, R., Immunological relevance of malonic dialdehyde. I. Preparation of Schiff's bases from lysozyme or polylysine reacted with malonic dialdehyde, *Biochimie*, 69, 1153–1159, 1987.

66. Richard, F. M. and Knowles, J. R., Glutaraldehyde as a protein cross-linking reagent, *J. Mol. Biol.*, 37, 231, 1968.

67. Hopwood, D., Comparison of the crosslinking abilities of glutaraldehyde, formaldehyde, and α-hydroxyadipaldehyde with bovine serum albumin and casein, *Histochemie*, 17, 1.51, 1969.

68. Fein, M. L. and Filachione, E. M., Tanning studies with aldehydes, *J. Am. Leather Chem. Assoc.*, 52, 17, 1957.

69. Yilmaz, S. and Ozer, I., Subunit level crosslinking of rabbit muscle pyruvate kinase by o-phthaldialdehyde, *Arch. Biochem. Biophys.*, 279, 32–36, 1990.

70. Benescb, R. E. and Kwong, S., *Bis*-pyridoxal polyphosphates: A new class of specific intramolecular crosslinking agents for hemoglobin, *Biochem. Biophys. Res. Commun.*, 156, 9, 1988.

71. Miller, A. G. and Gerrard, J. A., Assessment of protein function following cross-linking by alpha-dicarbonyls, *Ann. N. Y. Acad. Sci.*, 1043, 195–200, 2005.

72. Meade, S. J., Miller, A. G., and Gerrard, J. A., The role of dicarbonyl compounds in non-enzymatic crosslinking: A structure–activity study, *Bioorg. Med. Chem.*, 11, 853–862, 2003.

73. Sager, P. R., Cytoskeletal effects of acrylamide and 2,5-hexanedione: Selective aggregation of vimentin filaments, *Toxicol. Appl. Pharmacol.*, 97, 141, 1989.

74. Anthony, D. C., Boekelheide, K., Anderson, C. W., and Graham, D. G., The effect of 3,4-dimethyl substitution on the neurotoxicity of 2,5-hexanedione. II. Dimethyl substitution accelerates pyrrole formation and protein crosslinking, *Toxicol. Appl. Pharmacol.*, 71, 372–382, 1983.

75. Xu, G. and Sayre, L. M., Cross-linking of proteins by 3-(trifluoromethyl)-2,5-hexanedione. Model studies implicate an unexpected amine-dependent defluorinative substitution pathway competing with pyrrole formation, *J. Org. Chem.*, 67, 3007–3014, 2002.

76. Fayle, S. E., Gerrard, J. A., Simmons, L., Meade, S. J., Reid, E. A., and Johnston, A. C., Crosslinkage of proteins by dehydroascorbic acid and its degradation products, *Food Chem.*, 70, 193, 2000.

77. Ternynck, T. and Avrameas, S., Conjugation of p-benzoquinone treated enzymes with antibodies and Fab fragments, *Immunochemistry*, 14, 767, 1977.

78. Robinson, I. D., Role of crosslinking of gelatin in aqueous solutions, *J. Appl. Polymer Sci.*, 8, 1903, 1964.

79. Brady, D. and Jordaan, J., Advances in enzyme immobilization, *Biotechnol. Lett.*, 31, 1639–1650, 2009.

80. Sung, H. W., Chang, Y., Chiu, C. T., Chen, C. N., and Liang, H. C., Crosslinking characteristics and mechanical properties of a bovine pericardium fixed with a naturally occurring crosslinking agent, *J. Biomed. Mater. Res.*, 47, 116–126, 1999.

81. Milch, R. A., Viscometric hardening of gelatin sols in the presence of certain intermediary metabolites, *J. Surg. Res.*, 3, 254, 1963.

Appendix B: Sulfhydryl Group–Directed Homobifunctional Cross-Linkers

Name	Structure	References
A. Mercurial Reagents		
I		
Mercuric ion	Hg^{2+}	1
3,6-Bis(mercurymethyl) dioxane derivatives	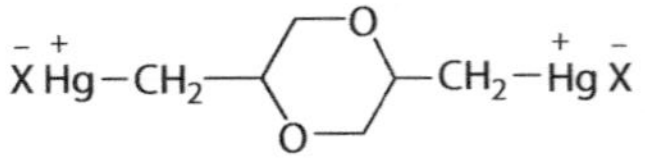	
II		2
X = CH₃-COO⁻ : 3,6-Bis (acetoxymercurimethyl) dioxane		
III		2
X = Cl⁻ : 3,6-Bis(chloromercurimethyl)dioxane		
IV		2
X = NO₂⁻ : 3,6-Bis(nitromercurimethyl)dioxane		
V		2
1,4-Bis(bromomercuri)butane		
B. Disulfide-Forming Reagents		
Alkylenebis(5,5′-dithio-2-nitrobenzoic acid) derivatives	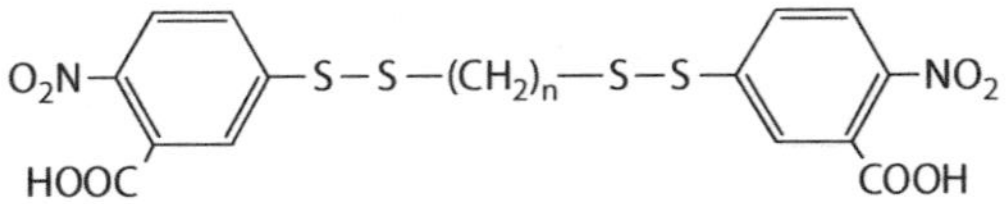	3,4
VI		
n = 3: Propylene-1, 3-bis(5,5′- dithio-2- nitrobenzoic acid)		
VII		4
n = 4: Butylene-1, 4-bis(5,5′- dithio-2- nitrobenzoic acid)		
VIII		3,4
n = 6: Hexylene-1, 6-bis(5,5′- dithio-2- nitrobenzoic acid)		

IX

n = 7: Heptylene-1,
7-bis (5,5′-dithio-2-nitrobenzoic acid)

X

n = 8: Octylene-1,8-bis (5,5′-dithio-2-nitrobenzoic acid)

XI

n = 9: Nonylene-1,9-bis
(5,5′-dithio-2-nitrobenzoic acid) (NBDN)

XII

n = 10: Dectylene-1,10-bis
(5,5′-dithio-2-nitrobenzoic acid)

XIII

n = 12: Dodectylene-1,12-bis
(5,5′-dithio-2-nitrobenzoic acid)

XIV

3-Oxy-2,2-bis[[((2-((3-carboxy-4-nitrophenyl) dithio)ethylamino)carbonyl] hexanyl]-4,4-dimethyloxazolidine

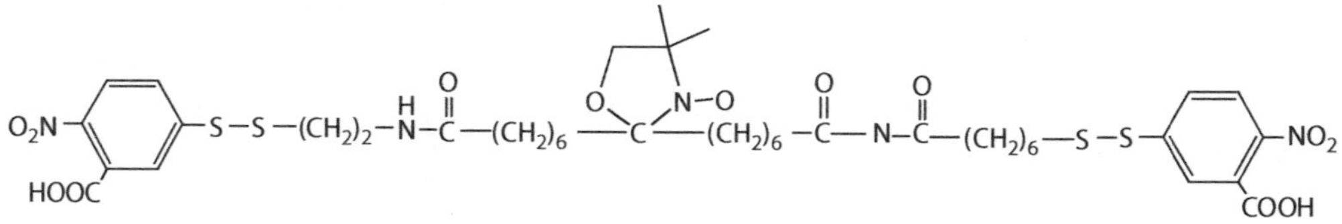

XV

(3S,4S)-1-hydroxy-2,2,5,5-tetramethyl-N3,N4-bis (4-methylsulfonylsulfanylbutyl) pyrrolidine-3,4-dicarboxamide

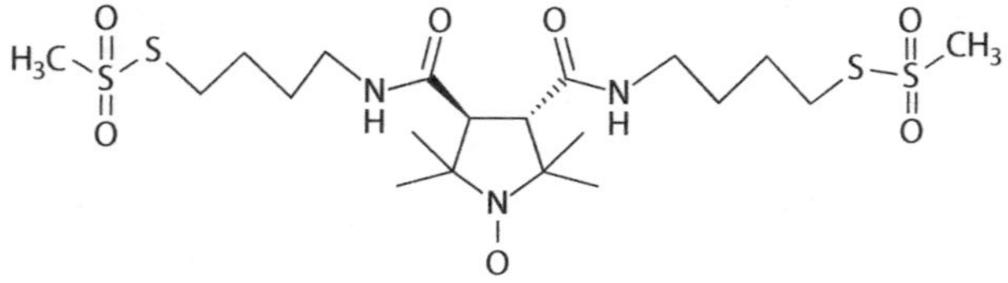

3,4

3,4

3–5

3

3

6

7

(continued)

(Continued)

Name	Structure	References
XVI		7
(3R,4R)-1-Hydroxy-2,2,5,5-tetramethyl-N3, N4-bis(4-methylsulfonylsulfanylbutyl) pyrrolidine-3,4-dicarboxamide	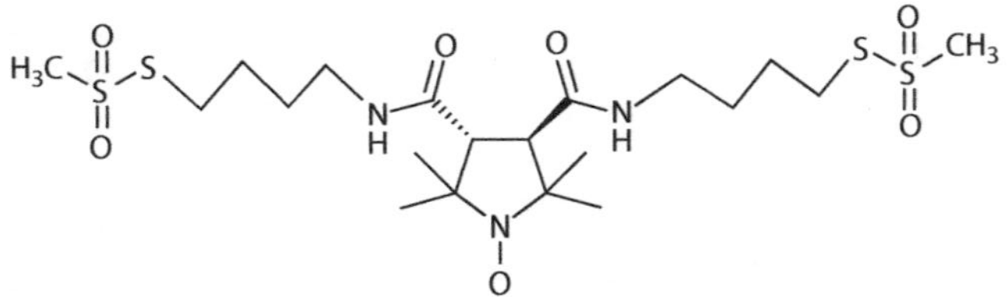	
Polymethylenebis-(methanethiosulfonate)	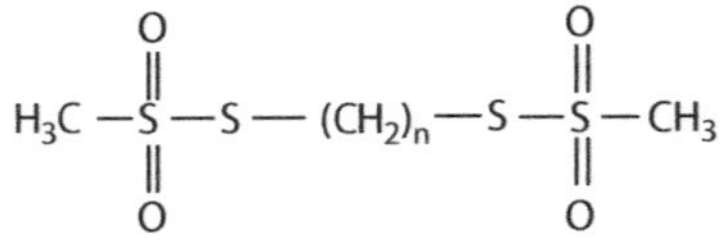	
XVII		8,9
n = 1: 1,1-Methanediyl bis(methanethiosulfonate)		
XVIII		8
n = 2: 1,2-Ethanediyl bis(methanethiosulfonate)		
XIX		8,9
n = 3: 1,3-Propanediyl bis(methanethiosulfonate)		
XX		8,9
n = 4: 1,4-Butanediyl bis(methanethiosulfonate)		
XXI		8,10
n = 5: 1,5-Pentanediyl bis(methanethiosulfonate)		
XXII		8–10
n = 6: 1,6-Hexanediyl bis(methanethiosulfonate)		
XXIII		10,11
n = 8: 1,8-Octanediyl bis(methanethiosulfonate)		

XXIV

n = 10: 1,10-Decanediyl bis(methanethiosulfonate)

XXV

n = 12: 1,12-Dodecanediyl
bis(methanethiosulfonate)

XXVI

3,6-Dioxaoctane-1,8-diyl
bis(methanethiosulfonate)

XXVII

3,6,9,12,15-Pentaoxa-heptadecane-1,17-diyl
bis(methane thiosulfonate)

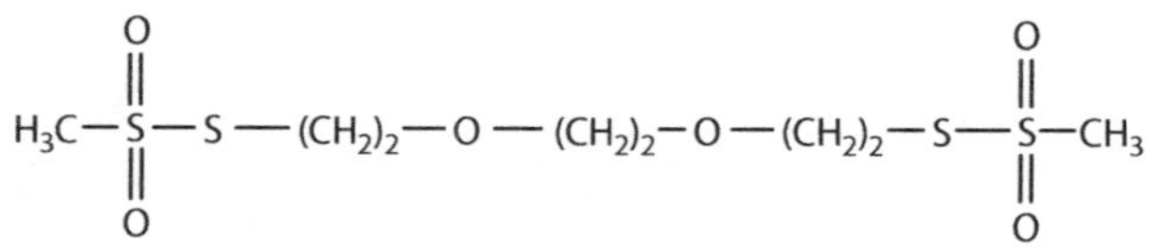

XXVIII

1,9-Bis(2-pyridyl)-1,2,8,9-tetrathia-5-oxanonane

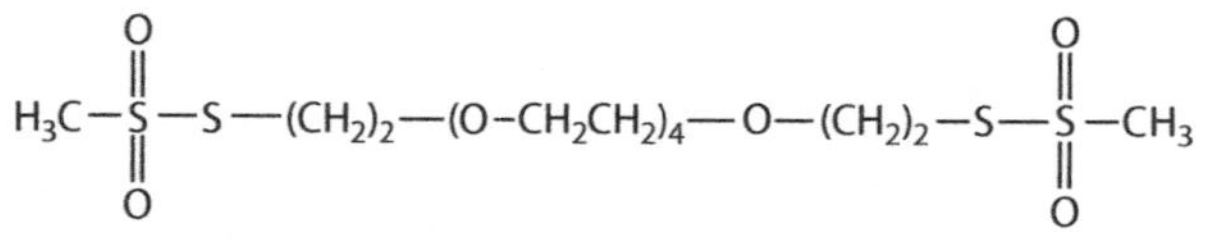

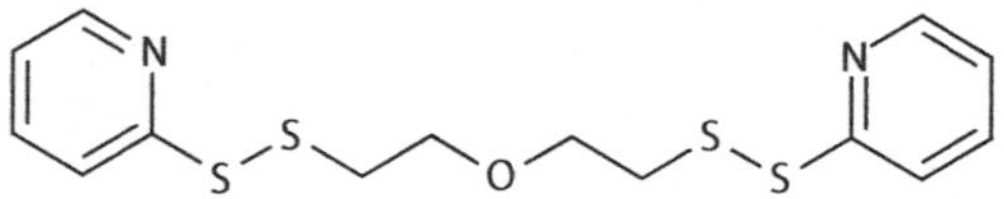

XXIX

1,12-Bis(2-pyridyl)-1,2,11,12-tetrathia-5,
8-dioxadodecane

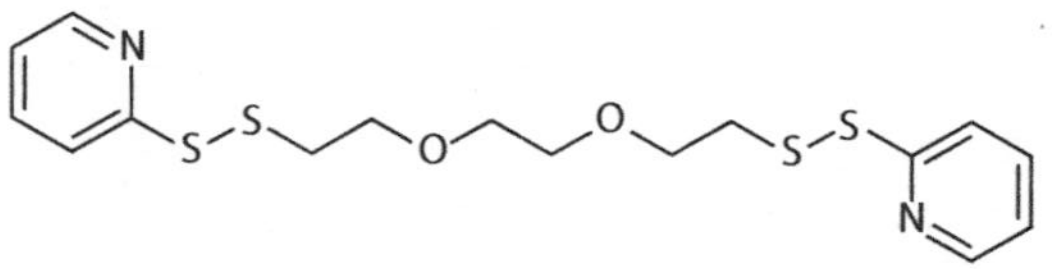

8,10

8,10

12

12

13

13

(continued)

Name	Structure	References
XXX		
1,4-Di[3′-(2′-pyridyldithio) propionamido]butane (DPDPB)	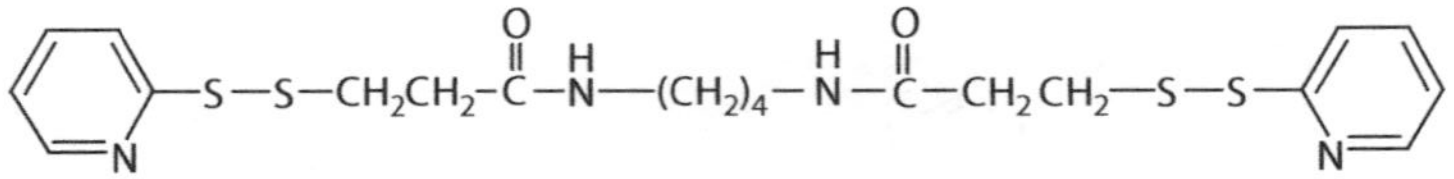	4
XXXI		
Crabescein	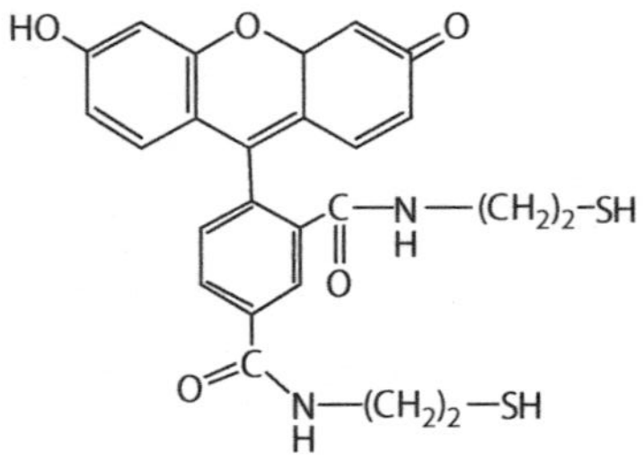	14
C. Bismaleimides		
Bis(N-maleimido)alkane; N,N′-alkylenebismaleimide	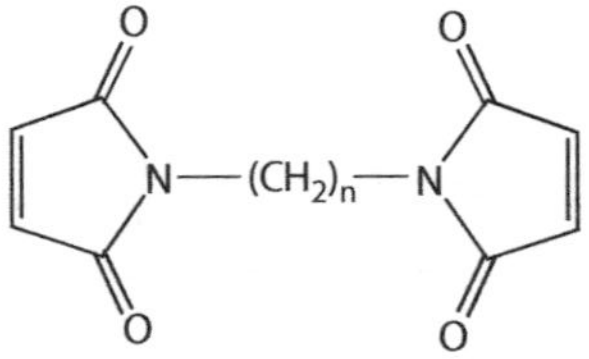	
XXXII n = 1: Bis(N-maleimido)methane (BMM); N,N′-methylenebismaleimide		4
XXXIII n = 2: 1,3-Bis(N-maleimido)ethane (BMOE); N,N′-ethylenebismaleimide		4

(Continued)

XXXIV
n = 4: 1,4-Bis(N-maleimido)butane (BMB);
N,N'-butanenebismaleimide   4

XXXV
n = 6: 1,6-Bis(N-maleimido)hexane (BMH);
N,N'-hexanenebismaleimide   4,15,16

XXXVI
n = 8: 1,8-Bis(N-maleimido)octane (BMO);
N,N'-octanenebismaleimide   16

XXXVII
n = 12: 1,12-Bis(N-maleimido)dodecane (BMD);
N,N'-dodecanenebismaleidme   16

XXXVIII
Bis(N-maleimidomethyl) ether   16,17

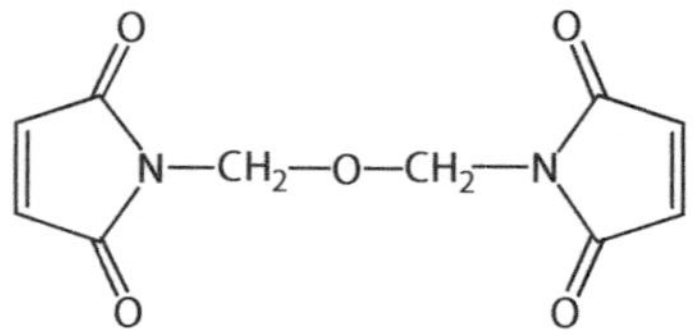

$N-CH_2-O-CH_2-N$ (bismaleimide structure)

XXXIX
1,8-Bismaleimido diethyleneglycol (BM(PEG)2)   4

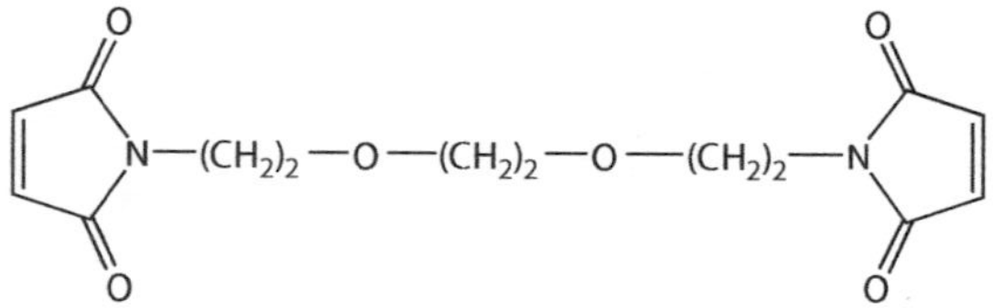

$N-(CH_2)_2-O-(CH_2)_2-O-(CH_2)_2-N$ (bismaleimide structure)

XL
1,11-Bismaleimido triethyleneglycol (BM(PEG)3)   4

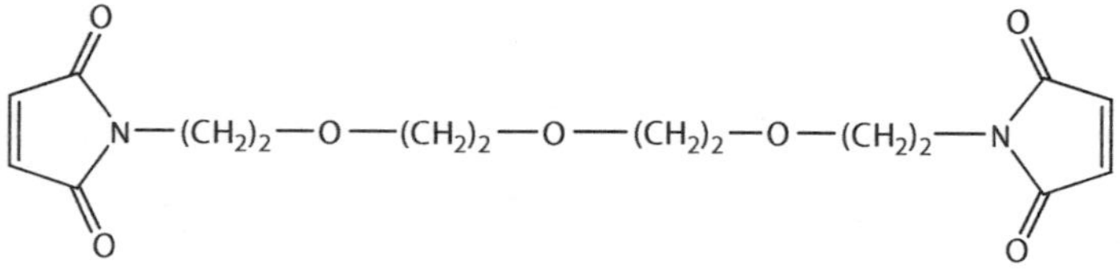

$N-(CH_2)_2-O-(CH_2)_2-O-(CH_2)_2-O-(CH_2)_2-N$ (bismaleimide structure)

(continued)

(Continued)

Name	Structure	References
XLI		
Maleimidomethyl-3-maleimido-propionate (MMP)	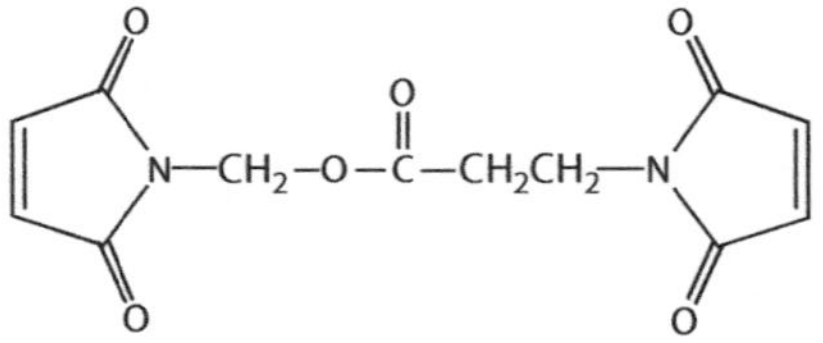	18
XLII		
2,2-Bis(maleimidomethoxy)-propane	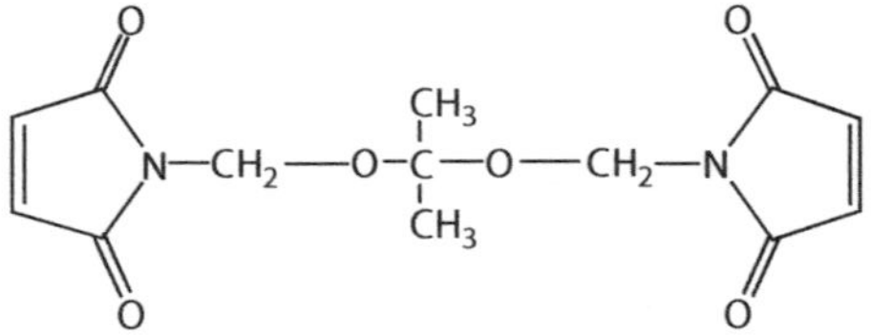	19
XLIII		
2,2-Bis(maleimidoethoxy)-propane	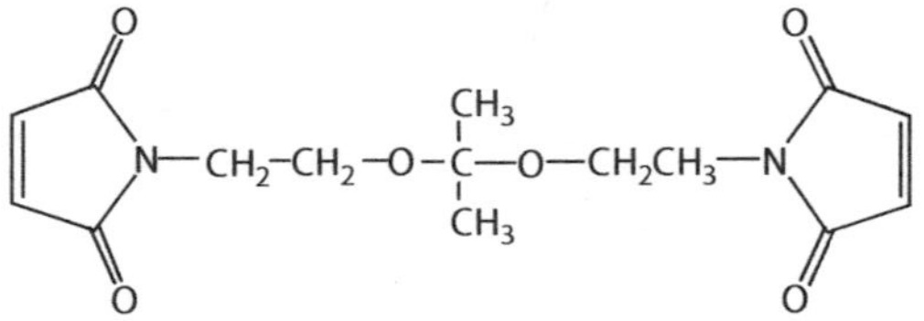	19
XLIV		
1,1′-[3,9-Diethyl-2,4,8,10-tetraoxaspiro[5,5] undecane-3,9-diyl] Bis(oxymethylene)]-bis-1*H*-pyrrole-2,5-dione	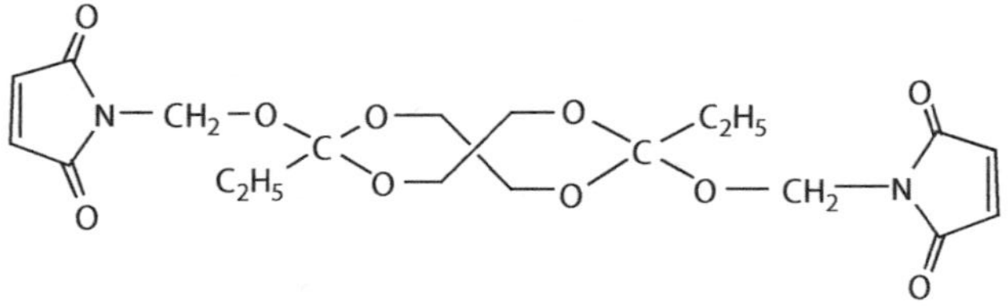	19

19

4

20

4

(continued)

XLV

1,1′-[3,9-Diethyl-2,4,8,10-tetraoxaspiro[5,5]
undecane-3,9-diyl]
bis(oxy-2,1-ethane-diyl-bis-1*H*-pyrrole-2,5-dione

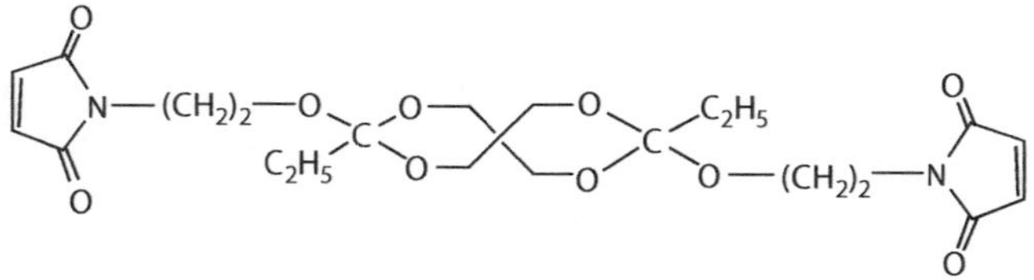

XLVI

1,4-Bismaleimidyl-2,3-dihydroxybutane (BMDB)

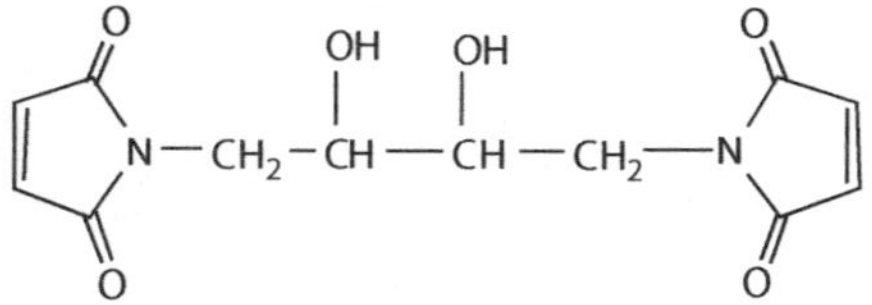

XVII

N,*N*′-bis(3-maleimidopropionyl) -2-hydroxy-1,3-
propanediamine (BMHP)

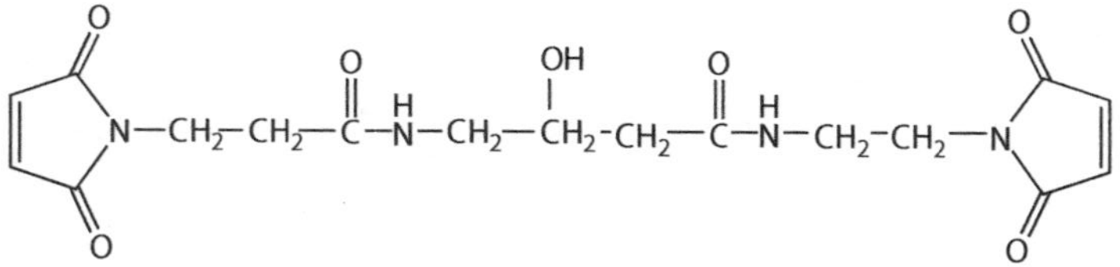

XLVIII

Dithiobismaleimido ethane (DTME)

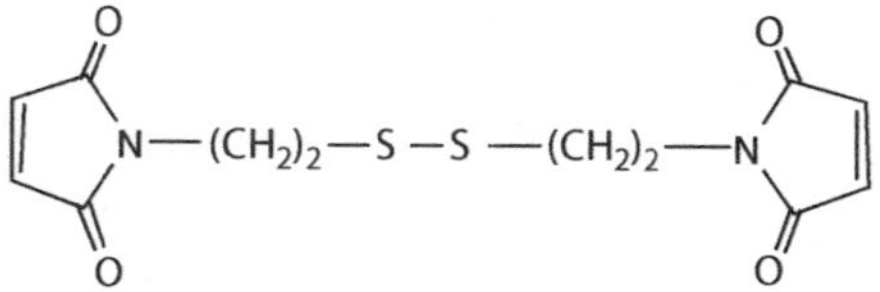

(Continued)

Name	Structure	References
XLIX N,N'-(1,3-Phenylene) bismaleimide (mPDM)		4,15
L N,N'-(1,2-Phenylene) bismaleimide (oPDM)		4,21
LI N,N'-(1,4-Phenylene) bismaleimide (pPDM)		4,21
LII Bis(N-maleimido)-4,4'-bibenzyl (BMB); N,N'-(methylene-1,4-phenylene)bismaleimide (MPBM)		15,16

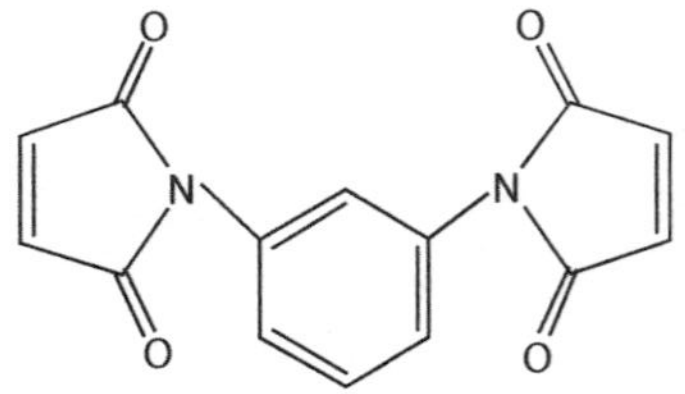

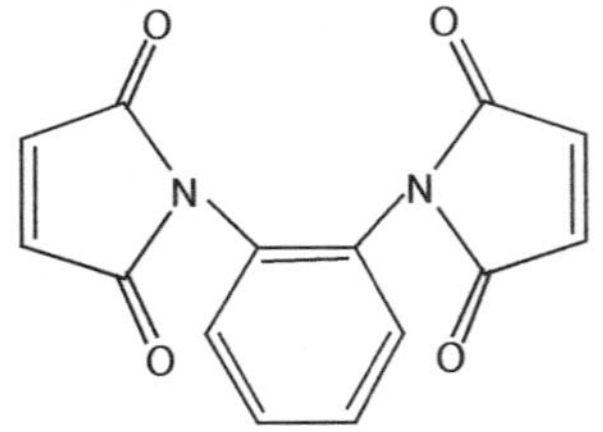

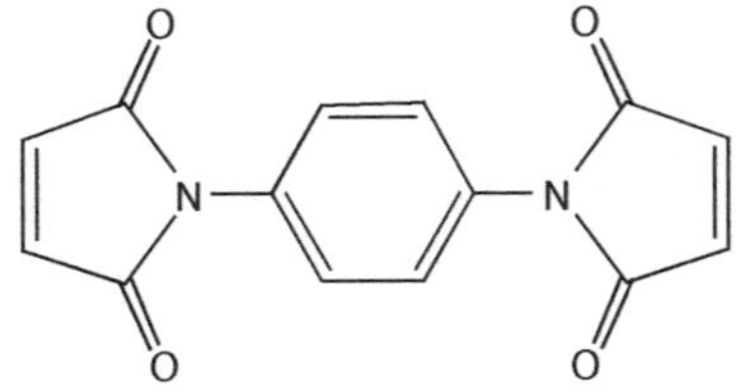

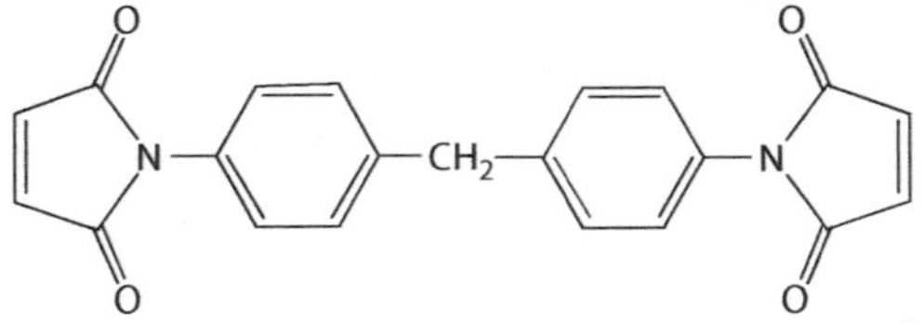

LIII

4,4′-Dimaleimidylstilbene

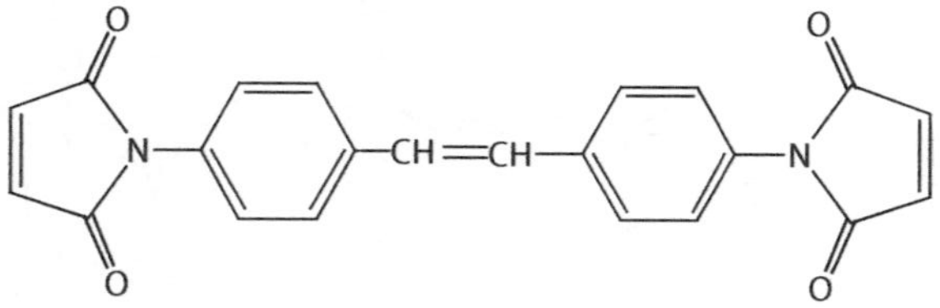

22

LIV

4,4′-Dimaleimidylstilbene-2,2′-disulfonic acid (DMSDS)

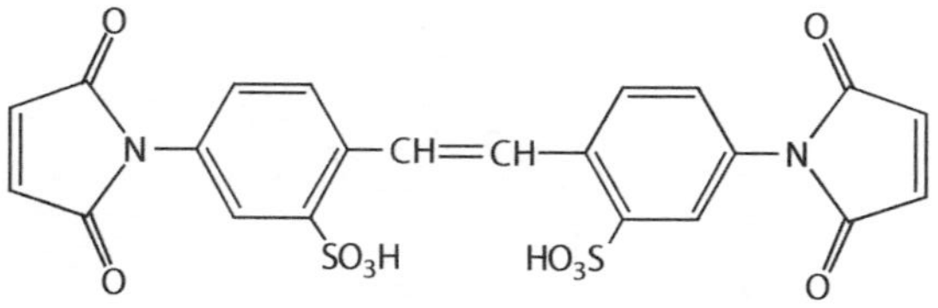

22

LV

4,4′-Azobenzene dimaleimide; azophenyl dimaleimide (APDM)

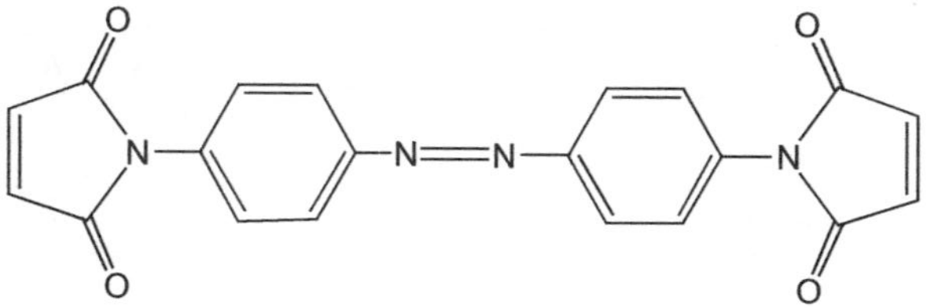

23,24

LVI

Naphthalene-1,5-dimaleimide (NDM)

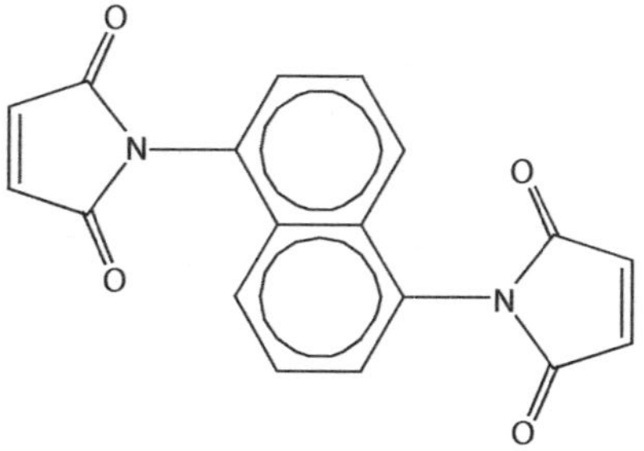

25,26

(continued)

(Continued)

Name	Structure	References
LVII		26
3,6-Dimaleimidyl chromen-2-one	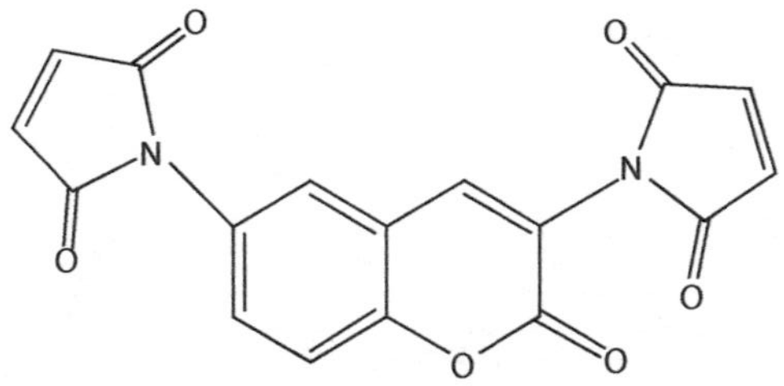	
LVIII		26
6-Maleimidyl-3-*p*-Maleimidylphenyl chromen-2-one	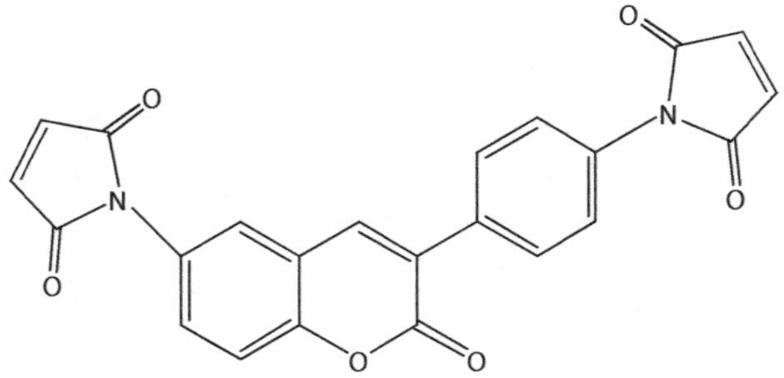	

26

27

(continued)

LIX

N-(3,6-Dimaleimidyl-1,8-naphthalyl)-L-aspartic acid dimethyl ester

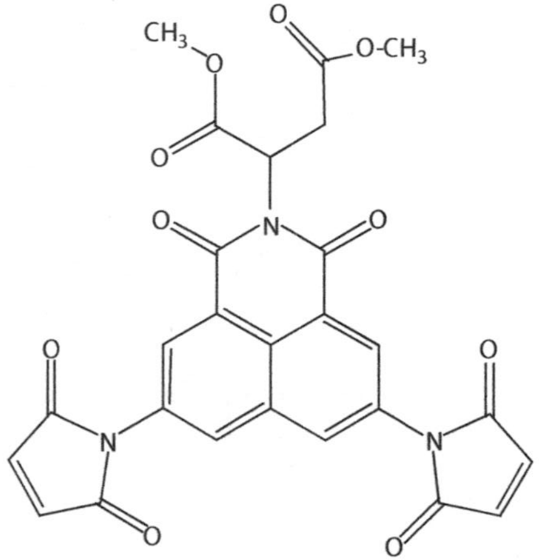

LX

N-[2-(1-(Dimethylamino)-5-naphthalene-sulfonyl) aminoethyl]-3,5-dimaleimidobenzamide

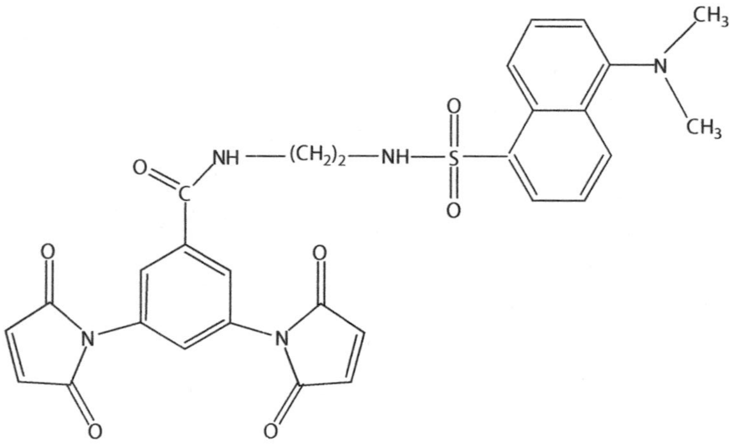

(Continued)

Name	Structure	References
LXI		27
N-[8-(1-(Dimethylamino)-5-naphthalenesulfonamide)-3,6-dioxaoctyl]-3,5-dimaleimidyl-1-benzamide	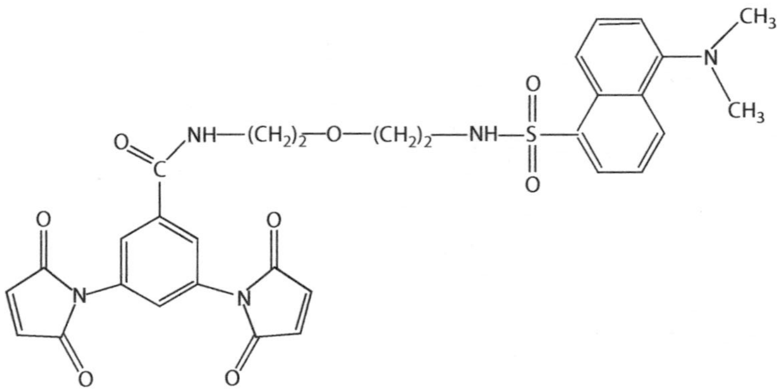	
D. Bis-Haloacetyl Derivatives		
LXII		
1,3-Dibromoacetone	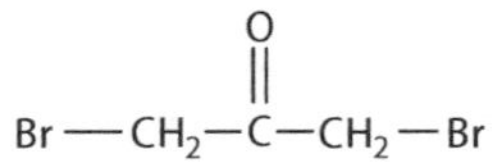	28
N,N′-Bis(iodoacetyl)polymethylenediamine	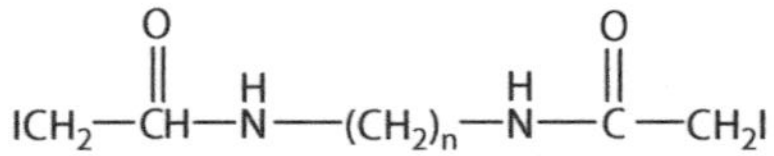	
LXIII		29
n = 2: *N,N′*-Bis(iodoacetyl)ethylenediamine; *N,N′*-ethylene-bis(iodoacetamide)		
LXIV		29
n = 6: *N,N′*-Bis(iodoacetyl)hexamethylenediamine; *N,N′*-hexamethylene-bis(iodoacetamide)		

LXV

n = 11: N,N'-bis(iodoacetyl)
undecamethylenediamine;
N,N'-undecamethylene-bis(iodoacetamide)

29

LXVI

N,N'-Bisbromoacetyl-1,2-diaminoethane

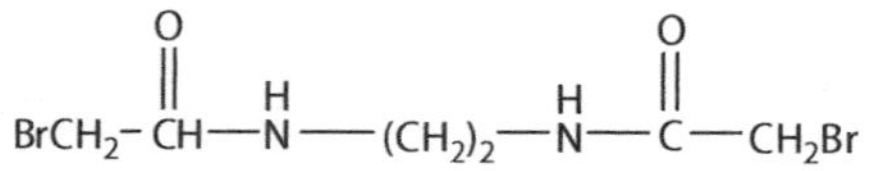

30

LXVII

N,N'-Bisbromoacetyl-hydrazine

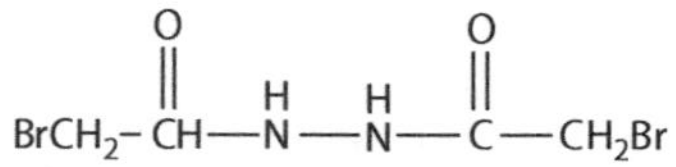

30

LXVIII

N,N'-Bisbromoacetyl-1,4-diaminobenzene

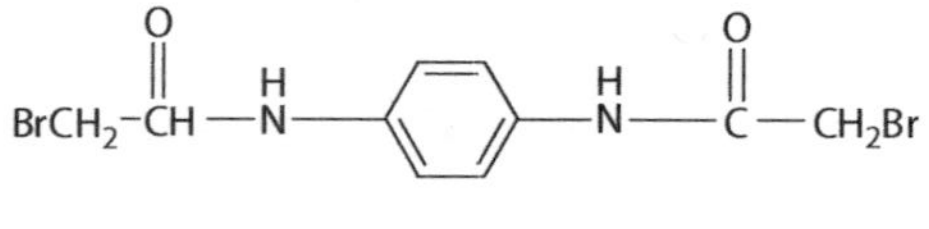

30

LXIX

N,N'-Bisbromoacetyl-1,2-diaminobenzene

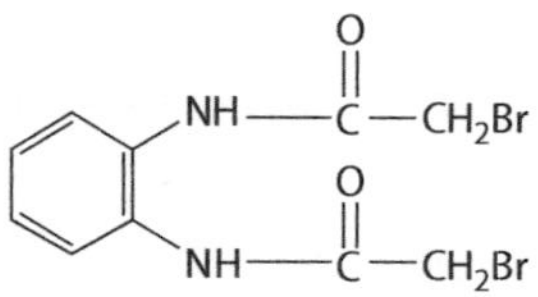

30

LXX

N,N'-Bisbromoacetyl-1,3-diaminobenzene

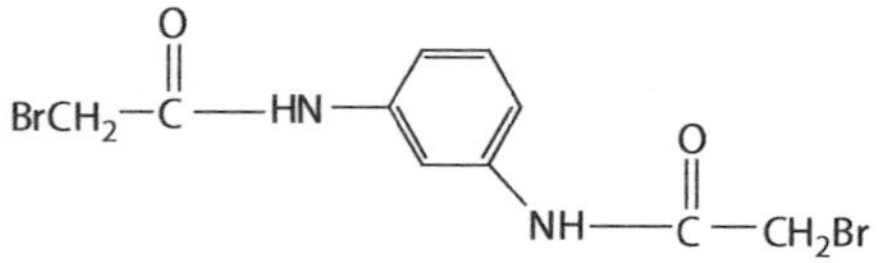

30

(continued)

(Continued)

Name	Structure	References
LXXI		
N,N′-Di(bromoacetyl)phenyl-hydrazine		31
LXXII		
1,2-Di(bromoacetyl)amino-3-phenylpropane		31
LXXIII		
γ-(2,4-Dinitrophenyl)-α-bromoacetyl-L-diamino-butyric acid bromoacetyl hydrazide (DIBAB)		32

LXXIV

Bis-[α-bromoacetyl-ε-(2,4-dinitrophenyl)-lysylproyline] ethylenediamine

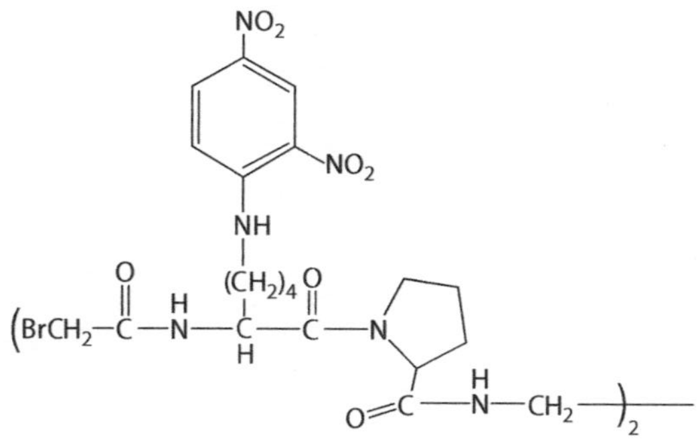

33

LXXV

2,2′-Dicarboxy-4,4′-diiodo-acetamidoazobenzene

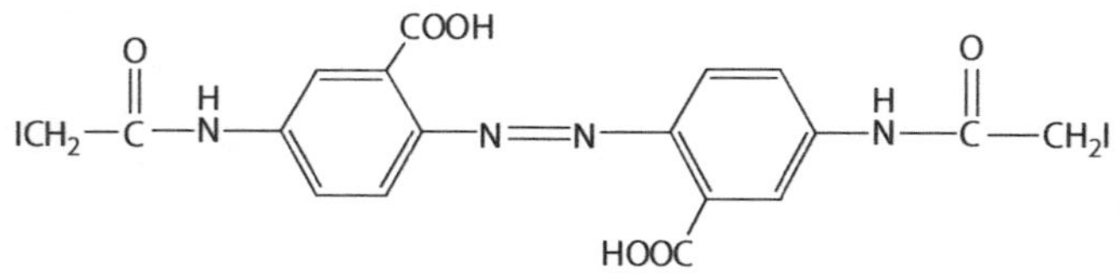

31,34

LXXVI

2,2′-Dicarboxy-4,4′-dibromo-acetamidoazobenzene

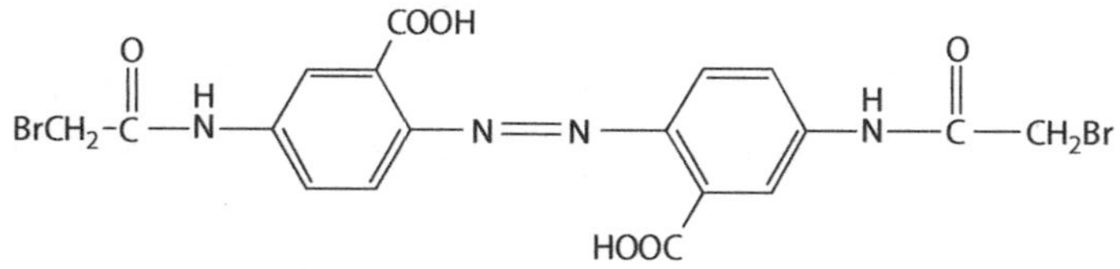

34

LXXVII

3,3′-Bis(sulfonato)-4,4′-bis (chloroacetamido) azobenzene (BSBCA)

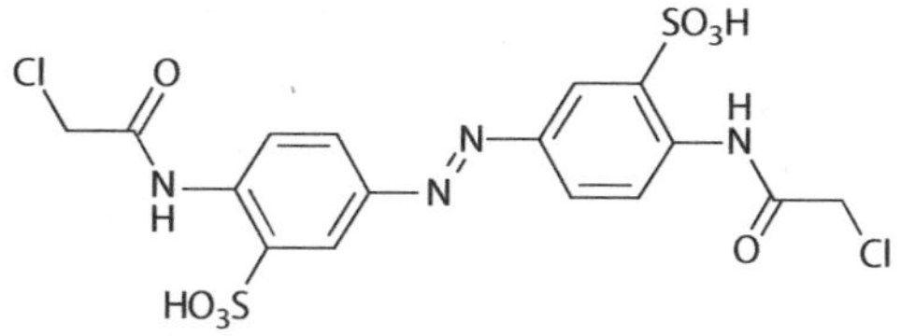

35

(continued)

(Continued)

Name	Structure	References
LXXVIII		
p-Bis(ureido)-(1-iodoacet-amido-2-ethylamino) oligoprolylazobenzene	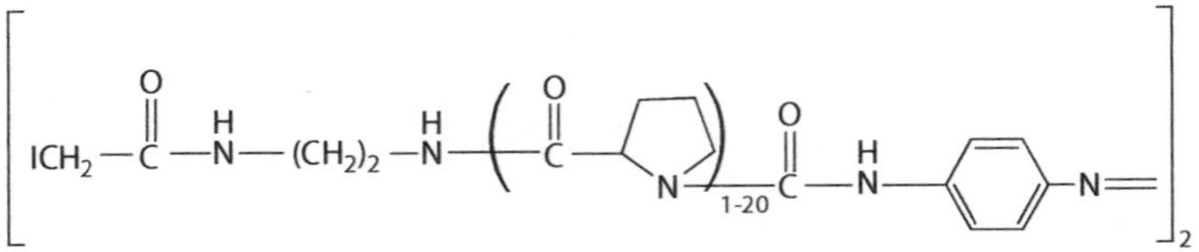	36
LXXIX		
N,N′-Bis(α-iodoacetyl)-2,2′-dithiobis(ethylamine) (DIDBE)	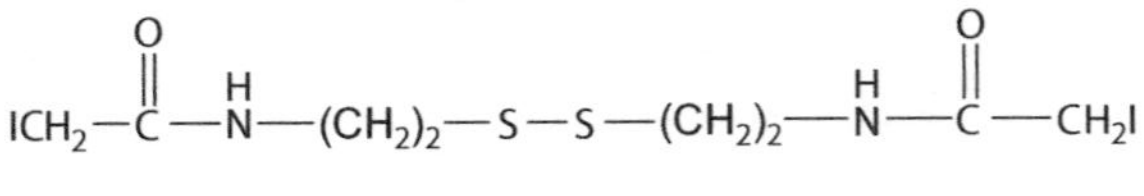	11,37
LXXX		
4,5′-Di[{(iodoacetyl)-amino}-methyl]fluorescein	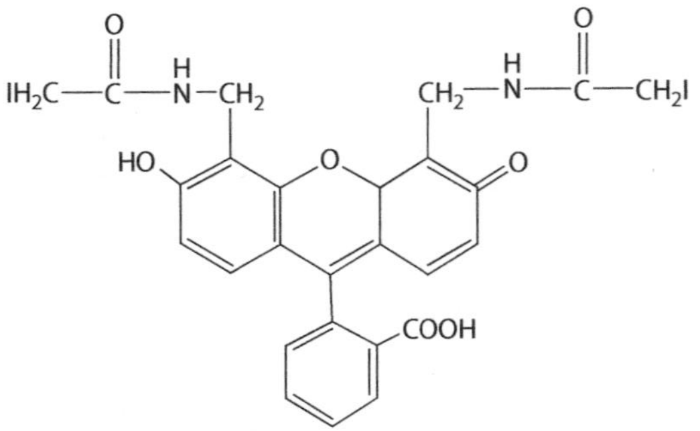	38
LXXXI		
3′,6′-Bis{*N*-[2-(iodoacetamido)ethyl]-*N*-methylamino}spiro[isobenzofuran-1(3*H*),9′-[9*H*] xanthen]-3-one	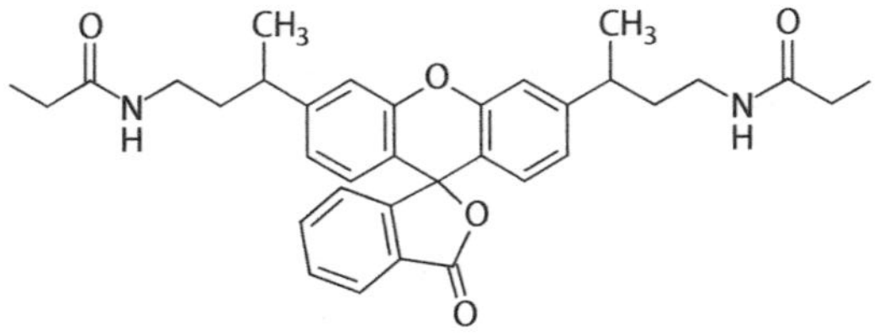	39

LXXXII

Bis((*N*-iodoacetyl)piperazinyl) sulfonerhodamine (BSR)

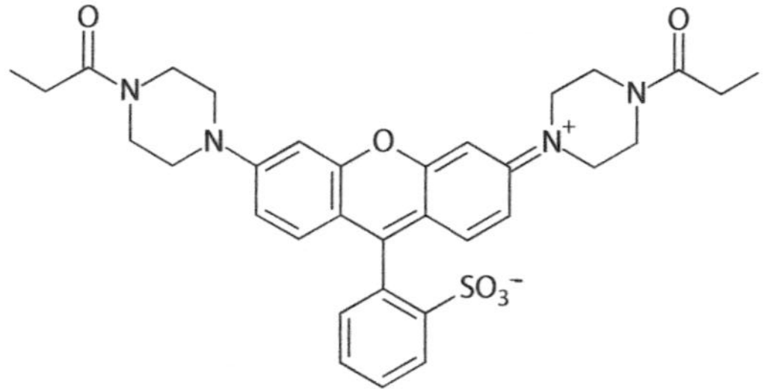

40

LXXXIII

4,4-Difluoro-3,5-di(iodoacetimidomethyl)-4-bora-3*a*,4*a*-*s*-diaza-*s*-indacene (BODIPY-FL)

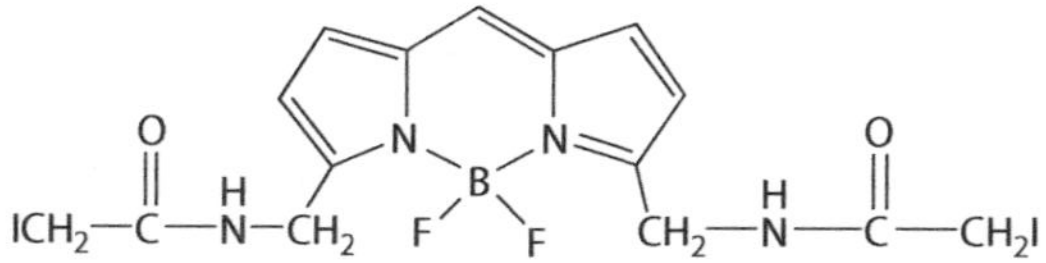

41

E. Di-Alkyl Halides

LXXXIV

α,α′-Dibromo-*p*-xylene sulfonic acid

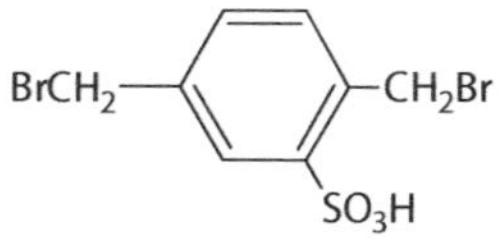

31,42

LXXXV

α,α′-Diiodo-*p*-xylene sulfonic acid

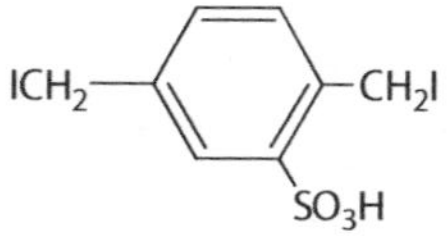

42

LXXXVI

Di(2-chloroethyl) sulfide

$Cl—CH_2CH_2—S—CH_2CH_2—Cl$

43,44

(continued)

(Continued)

Name	Structure	References
LXXXVII Di(2-chloroethyl) sulfone	$Cl-CH_2CH_2-\overset{\displaystyle O}{\underset{\displaystyle O}{S}}-CH_2CH_2-Cl$	44
LXXXVIII Di(2-chloroethyl)methylamine (BCEA); mechloroethamine; nitrogen mustard (NH_2)	$Cl-CH_2CH_2-\overset{\displaystyle CH_3}{N}-CH_2CH_2-Cl$	45,46
LXXXIX Phosphoramide	$O=\overset{\displaystyle NH_2}{\underset{\displaystyle OH}{P}}-N\overset{\displaystyle CH_2CH_2Cl}{\underset{\displaystyle CH_2CH_2Cl}{}}$	46
XC L-Phenylalanine mustard; melphalan	$HOOC-\overset{\displaystyle NH_2}{\underset{\displaystyle H}{C}}-CH_2-\bigcirc-N\overset{\displaystyle CH_2CH_2Cl}{\underset{\displaystyle CH_2CH_2Cl}{}}$	47
XCI N,N-Bis(β-bromoethyl)benzyl-amine	$\bigcirc-CH_2-N\overset{\displaystyle CH_2-}{}\underset{\displaystyle Br-CH_2CH_2-}{}-CH_2CH_2-Br$	31

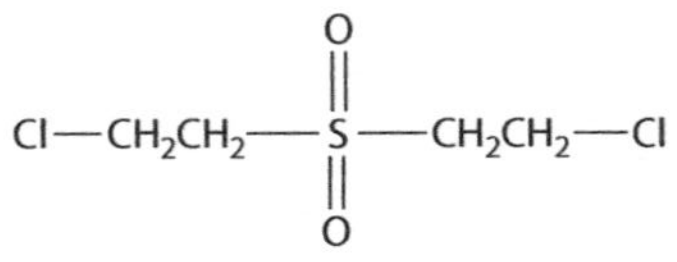

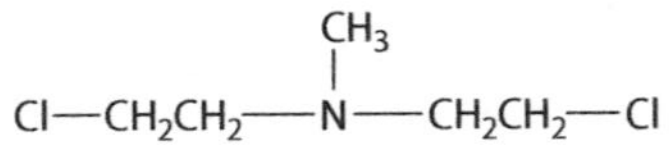

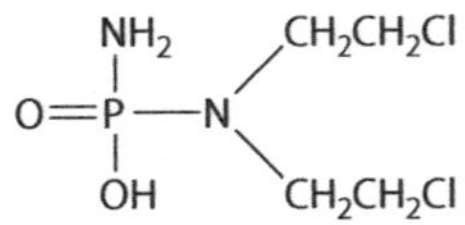

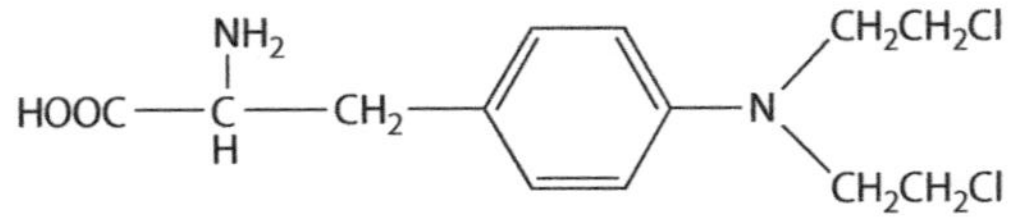

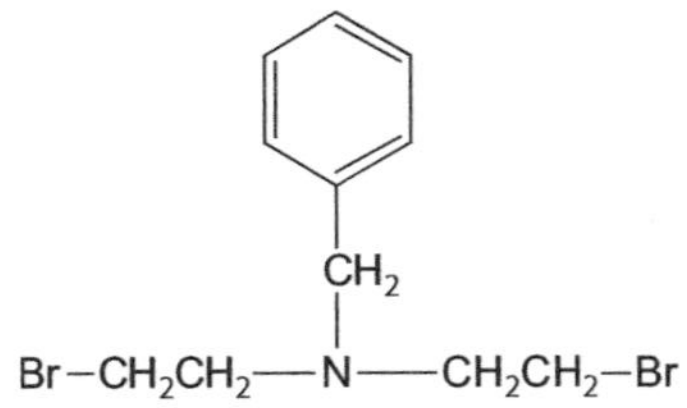

Benzoquinone mustards

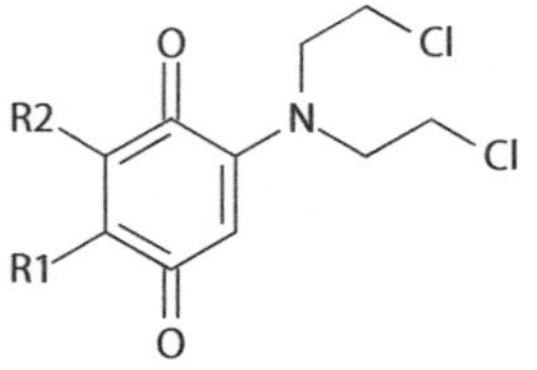

XCII

R1 = R2 = H: 2-[Di(chloroethyl)amino]-1,4-benzoquinone (BM)   47

XCIII

R1 = CH₃, R = H: 5-Methyl-2-[di(chloroethyl) amino]-1,4-benzoquinone (MeBM)   47

XCIV

R1 = OCH₃, R2 = H: 5-Methoxy-2-[di(chloroethyl)amino]-1, 4-benzoquinone (BMB)   47

XCV

R1 = Cl, R2 = H: 5-Chloro-2-[di(chloroethyl) amino]-1,4-benzoquinone (CBM)   47

XCVI

R1 = F, R2 = H: 5-Fluoro-2-[di(chloroethyl) amino]-1,4-benzoquinone (FBM)   47

XCVII

R1 = phenyl, R2 = H: 5-Phenyl-2-[di(chloroethyl) amino]-1,4-benzoquinone (PBM)   47

XCVIII

R1 = H, R2= CH₃; 6-Methyl-2-[di(chloroethyl) amino]-1,4-benzoquinone (m-MeBM)   47

(continued)

(Continued)

Name	Structure	References
XCIX		
R1 = H, R2 = phenyl: 6-Phenyl-2-[di(chloroethyl)amino]-1,4-benzoquinone (*m*PBM)		47
C		
R1 = H, R2 = C(CH₃)₃: 6-(trimethyl)methyl-2-[di(chloroethyl)amino]-1, 4-benzoquinone (*m*TBM)		47
CI		
Dibromobimane (BD); 4,6-bis(bromomethyl))-3,7-dimethyl-1, 5-diazabicyclo [3,3,0] octane-3,6-diene-2,8-dione	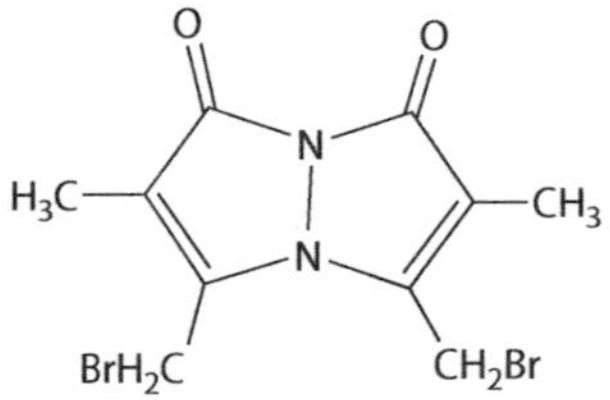	4,48
F. Dichloro-s-Triazines		
CII		
2,4,6-Trichlo-s-triazine; cyanuric chloride	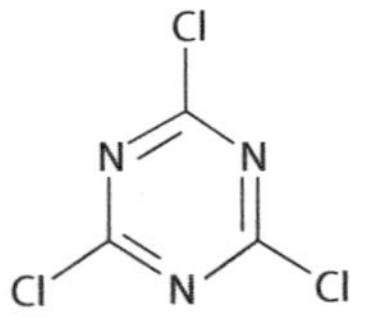	49
CIII		
2,4-Dichloro-6-methoxy-s-trizine	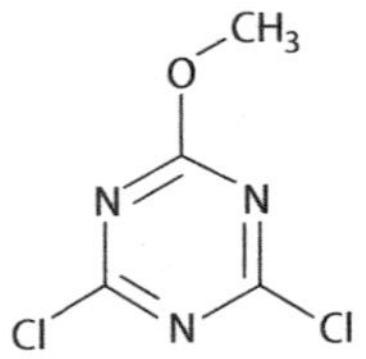	50

CIV

2-Carboxymethylamino-4,6-dichloro-s-triazine

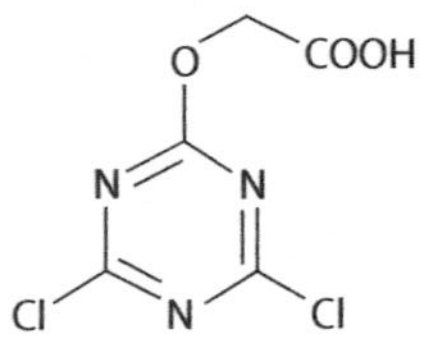

51

CV

2,4-Dichloro-6-(sulfonic acid)-s-triazine;
2,4-dichloro-s-triazin-2-yl-6-sulfonic acid

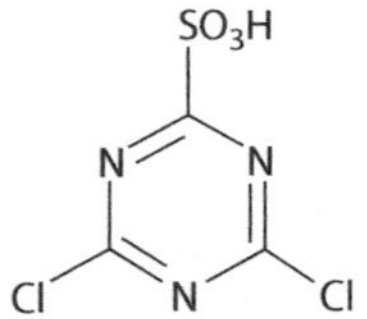

52,53

CVI

2,4-Dichloro-6-amino-s-triazine

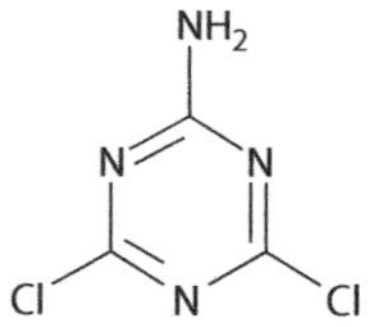

52,53

CVII

2-Carboxymethoxy-4,6-dichloros-triazine

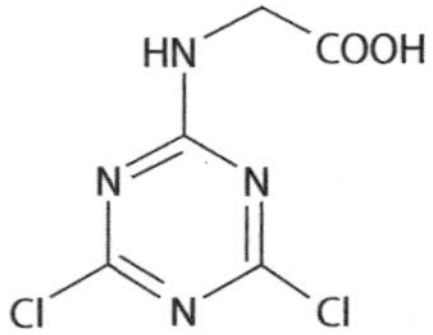

51

(continued)

(Continued)

Name	Structure	References
CVIII		
2,4-Dichloro-6-(3′-methyl-4-aminoanilino)-s-triazine	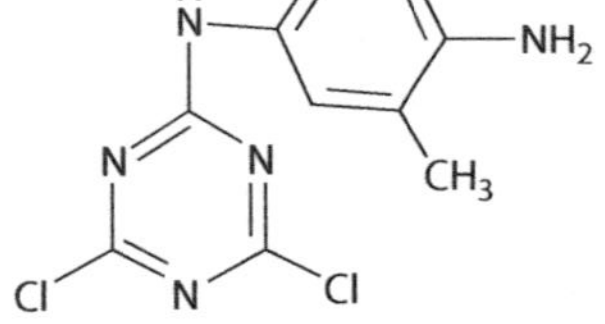	52,53
CIX		
2,4-Dichloro-6-(5′-sulfonic acid-naphthaleamino)-s-triazine; 5-[(4,6-dichloro-s-triazin-2-yl) amino] naphthalene-1-sulfonic acid	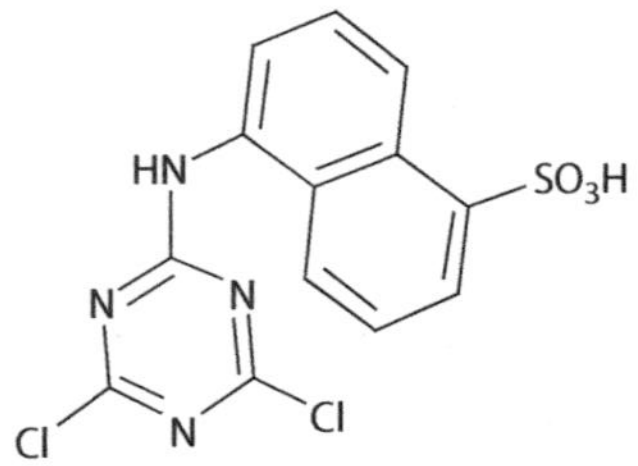	53
G. Aziridines		
CX		
N,N′-Ethyleneiminoyl-1,6-diaminohexane	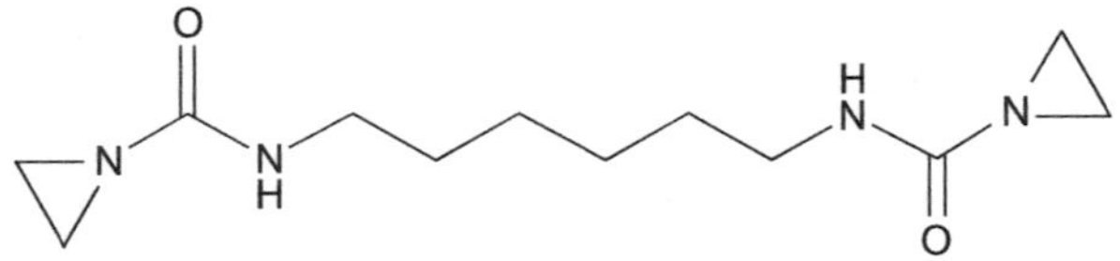	54
CXI		
2,5-Diaziridinyl-1,4-benzoquinone (oxidized-DZQ)	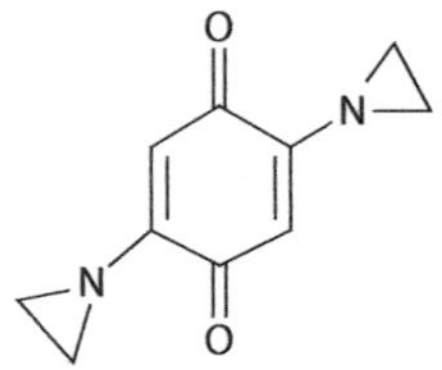	55

CXII

2,5-Diaziridinyl-1,4-dihydroxybenzene (reduced-DZQ)

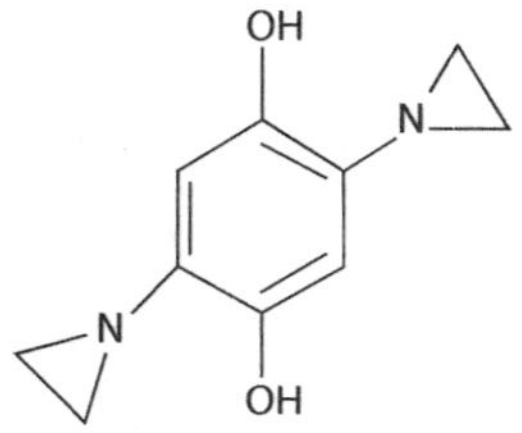

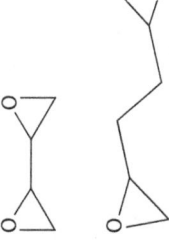

55

H. Bis-Epoxides

CXIII

1,2:3,4-Diepoxybutane

56–58

CXIV

1,2:5,6-Diepoxyhexane

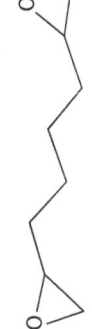

57

CXV

1,2:7,8-Diepoxyoctane

58

CXVI

4,6-Decadiyne-2,3:8,9-diepoxy-1,10-diol; repandiol

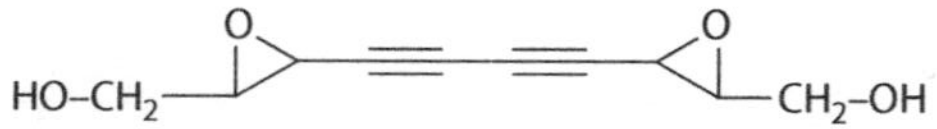

58

CXVII

Bis(2,3-epoxypropyl)ether

59

CXVIII

Ethylene glycol diglycidyl ether

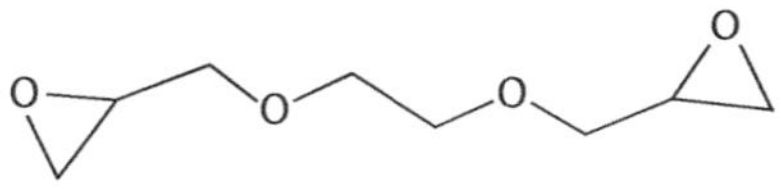

60

(continued)

(Continued)

Name	Structure	References
CXIX		
1,4-Butadioldiglycidoxyether		61
CXX		
Glycerol diglycidyl ether (GDGE)		62
CXXI		
3,4-Isopropylidene-1,2:5, 6-dianhydromannitol		57
I. Nitrosourea Derivatives		
Nitrosoureas		
CXXII		
R = CH₂CH₂Cl: 1,3-Bis(2-chloroethyl)-1-nitrosourea (BCNU)		63,64

CXXIII

R = [cyclohexane structure] —OH:

64

1,3-Bis(*trans*-4-hydroxyhexyl)-1-nitrosourea

J. Sulfone Derivatives

CXXIV

Divinyl sulfone

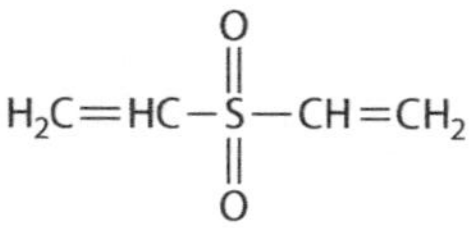

65

$$H_2C=HC-S-CH=CH_2$$

CXXV

1,6-Hexane-bis-vinylsulfone (HBVS)

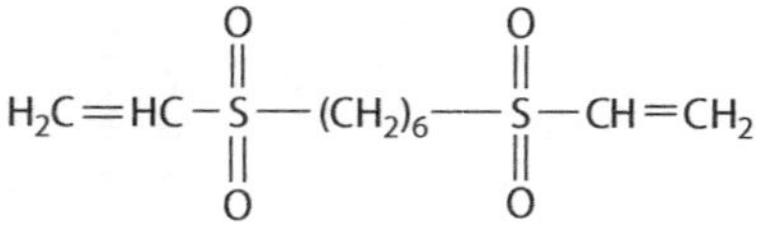

66

$$H_2C=HC-S-(CH_2)_6-S-CH=CH_2$$

CXXVI

Poly(2-methacryloyloxyethyl phosphorylcholine)-
bis-sulfone (PMPC-bis-sulfone)

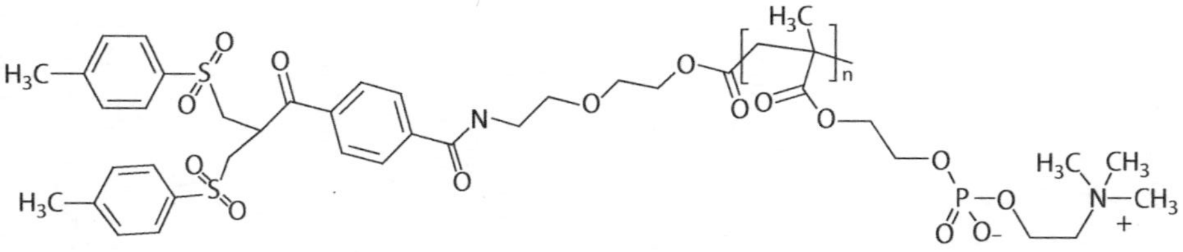

67

REFERENCES

1. Arnon, R. and Shapira, E., Crystalline papain derivative containing in intramolecular mercury bridge, *J. Biol. Chem.*, 244, 1033–1038, 1969.
2. Edsall, J. T., Maybury, R. H., Simpson, R. B., and Straessle, R., Dimerization of serum mercaptalbumin in the presence of mercurials. II. Studies with a bifunctional organic mercurial, *J. Am. Chem. Soc.*, 76, 3131, 1954.
3. Kliche, W., Pfannstiel, J., Tiepold, M., Stoeva, S., and Faulstich, H., Thiol-specific cross-linkers of variable length reveal a similar separation of SH1 and SH2 in myosin subfragment 1 in the presence and absence of MgADP, *Biochemistry*, 38, 10307–10317, 1999.
4. Green, N. S., Reisler, E., and Houk, K. N., Quantitative evaluation of the lengths of homobifunctional protein cross-linking reagents used as molecular rulers, *Protein Sci.*, 10, 1293–1304, 2001.
5. Faulstich, H., Heintz, D., and Drewes, G., Interchain and intrachain crosslinking of actin thiols by a bifunctional thiol reagent, *FEBS Lett.*, 302, 201–205, 1992.
6. Gaffney, B. J., Willingham, G. L., and Schopp, R. S., Synthesis and membrane interactions of spin-label bifunctional reagent, *Biochemistry*, 22, 881, 1983.
7. Chatani, S., Nakamura, M., Akahane, H., Kohyama, N., Taki, M., Arata, T., and Yamamoto, Y., Synthesis of C2-chiral bifunctionalised spin labels and their application to troponin C, *Chem. Commun. (Camb.)*, 14, 1880–1882, 2005.
8. Moore, K. J. and Fillingame, R. H., Structural interactions between transmembrane helices 4 and 5 of subunit a and the subunit c ring of *Escherichia coli* ATP synthase, *J. Biol. Chem.*, 283, 31726, 2008.
9. Shvetsov, A., Stamm, J. D., Phillips, M., Warshaviak, D., Altenbach, C., Rubenstein, P. A., Hideg, K., Hubbell, W. L., and Reisler, E., Conformational dynamics of loop 262–274 in G- and F-actin, *Biochemistry*, 45, 6541, 2006.
10. Bloxham, D. P. and Sharma, R. P., The development of S,S′-polymethylenebis(methanethiosulfonates) as reversible cross-linking reagent for thiol groups and their use to form stable catalytically active cross-linked dimers with glyceraldehyde-3-phosphate dehydrogenase, *Biochem. J.*, 181, 355, 1979.
11. Han, K.-K., Richard, C., and Delacourte, A., Chemical cross-links of proteins by using bifunctional reagents, *Int. J. Biochem.*, 16, 129, 1984.
12. Ermolova, N., Guan, L., and Kaback, H. R., Intermolecular thiol cross-linking via loops in the lactose permease of *Escherichia coli*, *Proc. Natl. Acad. Sci. U. S. A.*, 100, 10187, 2003.
13. Kalia, J. and Raines, R. T., 1,9-Bis(2-pyridyl)-1,2,8,9-tetrathia-5-oxanonane, *Molbank*, 2009, M642, 2009.
14. Packard, B., Edidin, M., and Komoriya, A., Site-directed labeling of a monoclonal antibody: Targeting to a disulfide bond, *Biochemistry*, 25, 3548, 1986.
15. Kovacic, P. and Hein, R. W., Cross-linking of polymers with dimaleimides, *J. Am. Chem. Soc.*, 81, 1187–1190, 1959.
16. Heilmann, H. D. and Holzner, M., The spatial organization of the active sites of the bifunctional oligomeric enzyme tryptophan synthase: Cross-linking by a novel method, *Biochem. Biophys. Res. Commun.*, 99, 1146–1152, 1981.
17. Simon, S. R. and Konigsberg, W. H., Chemical modification of hemoglobins: A study of conformation restraint by internal bridging, *Proc. Natl. Acad. Sci. U. S. A.*, 56, 749–756, 1966.
18. Sato, S. and Nakao, M., Cross-linking of intact erythrocyte membrane with a newly synthesized cleavable bifunctional reagent, *J. Biochem.*, 90, 1177–1185, 1981.
19. Srinivasachar, K. and Neville, D. M., Jr., New protein cross-linking reagents that are cleaved by mild acid, *Biochemistry*, 28, 2501–2509, 1989.
20. Boal, A. K., Tellez, H., Rivera, S. B., Miller, N. E., Bachand, G. D., and Bunker, B. C., The stability and functionality of chemically crosslinked microtubules, *Small*, 2, 793, 2006.
21. Chen, L. L., Rosa, J. J., Turner, S., and Pepinsky, R. B., Production of multimeric forms of CD4 through a sugar-based cross-linking strategy, *J. Biol. Chem.*, 266, 18237–18243, 1991.
22. Chantler, P. D. and Bower, S. M., Cross-linking between translationally equivalent sites on the two heads of myosin. Relationship to energy transfer results between the same pair of sites, *J. Biol. Chem.*, 263, 938–944, 1988.
23. Fasold, H., Groeschel-Stewart, U., and Turba, F., Azophenyl dimaleimides as splittable reagents forming peptide chains between cysteine residues, *Biochem. Z.* 337, 425–430, 1963.
24. Umeki, N., Yoshizawa, T., Sugimoto, Y., Mitsui, T., Wakabayashi, K., and Maruta, S., Incorporation of an azobenzene derivative into the energy transducing site of skeletal muscle myosin results in photo-induced conformational changes, *J. Biochem.*, 136, 839, 2004.

25. Wells, J. A., Knoeber, C., Sheldon, M. C., Werber, M. M., and Yount, R. G., Cross-linking of myosin subfragment 1. Nucleotide-enhanced modification by a variety of bifunctional reagents, *J. Biol. Chem.*, 255, 11135–11140, 1980.

26. Girouard, S., Houle, M. H., Grandbois, A., Keillor, J. W., and Michnick, S. W., Synthesis and characterization of dimaleimide fluorogens designed for specific labeling of proteins, *J. Am. Chem. Soc.*, 127, 559–566, 2005.

27. Guy, J., Caron, K., Dufresne, S., Michnick, S. W., Skene, W. G., and Keillor, J. W., Convergent preparation and photophysical characterization of dimaleimide dansyl fluorogens: Elucidation of the maleimide fluorescence quenching mechanism, *J. Am. Chem. Soc.*, 129, 11969–11977, 2007.

28. Husain, S. S. and Lowe, G., Evidence for histidine in the active sites of ficin and stem-bromelain, *Biochem. J.*, 110, 53–57, 1968.

29. Ozawa, H., Bridging reagent for protein. II. The reaction of N,N'-polymethylene-bis(iodoacetamide) with cysteine and rabbit muscle aldolase, *J. Biochem.*, 62, 531–536, 1967.

30. Coleman, R. S. and Pires, R. M., Covalent cross-linking of duplex DNA using 4-thio-2'-deoxyuridine as a readily modifiable platform for introduction of reactive functionality into oligonucleotides, *Nucleic Acids Res.*, 25, 4771–4777, 1997.

31. Wold, F., Bifunctional reagents, *Methods Enzymol.*, 11, 617–640, 1967.

32. Wilchek, M. and Givol, D., Affinity cross-linking of heavy and light chains, *Methods Enzymol.*, 46, 501–504, 1977.

33. Segal, D. M. and Hurwitz, E., Dimers and trimers of immunoglobulin G covalently cross-linked with a bivalent affinity label, *Biochemistry*, 15, 5253–5258, 1976.

34. Fasold, H., Groeschel-Stewart, U., and Turba, F., Synthese and reaktionen eines wasserlöslichen, spaltbarenreagens zur verknupfung frier SH-gruppen in proteinen, *Biochem. Z.*, 339, 487–490, 1964.

35. Burns, D. C., Zhang, F., and Woolley, G. A., Synthesis of 3,3'-bis(sulfonato)-4,4'-bis(chloroacetamido) azobenzene and cysteine cross-linking for photo-control of protein conformation and activity, *Nat. Protoc.*, 2, 251–8, 2007.

36. Wetz, K., Fasold, H., and Meyer, C., Synthesis of "long", hydrophilic, protein-cross-linking reagents, *Anal. Biochem.*, 58, 347–360, 1974.

37. Ludueña, R. F., Roach, M. C., Trcka, P. P., and Weintraub, S., N,N-Bis(alpha-iodoacetyl)-2,2'-dithiobis(ethylamine), a reversible crosslinking reagent for protein sulfhydryl groups, *Anal. Biochem.*, 117, 76–80, 1981.

38. Deshpande, S. S., Conjugation techniques, *Enzyme Immunoassays: From Concept to Product Development*, Chapman & Hall, New York, 1996, p. 89.

39. Corrie, J. E., Craik, J. S., and Munasinghe, V. R., A homobifunctional rhodamine for labeling proteins with defined orientations of a fluorophore, *Bioconjug. Chem.*, 9, 160–167, 1998.

40. Peterman, E. J., Sosa, H., Goldstein, L. S., and Moerner, W. E., Polarized fluorescence microscopy of individual and many kinesin motors bound to axonemal microtubules, *Biophys. J.*, 81, 2851–2863, 2001.

41. Ehrhardt, A. G., Kang, H. C., Tuft, R. A., Fay, F. S., and Ikebe, M., Labeling of calmodulin with a new bifunctional BODIPY for improved fluorophore stability in anisotropy measurements, *Biophys. J.*, 74(2 Pt 2), A380, Th-Pos200, 1998.

42. Hiremath, C. B. and Day, R. A., Introduction of covalent cross-linkages into lysozyme by reaction with α,α'-dibromo-p-xylenesulfonic acid, *J. Am. Chem. Soc.*, 86, 5027, 1964.

43. Philips, F. S., Recent contributions to the pharmacology of bis(2-haloethyl)amines and sulfides, *J. Pharmacol. Exp. Ther.*, 99, 281, 1950.

44. Berenblum, I. and Wormall, A., The immunological properties of proteins treated with betabeta'-dichlorodiethylsulphide (mustard gas) and betabeta'-dichlorodiethylsulphone, *Biochem. J.*, 33, 75–80, 1939.

45. Burnop, V. C., Francis, G. E., Richards, D. E., and Wormall, A., Studies with 15N-labelled nitrogen mustards: The combination of di-2-(chloroethyl)methylamine with proteins, *Biochem. J.*, 66, 504–515, 1957.

46. Struck, R. F., Davis, R. L., Jr., Berardini, M. D., and Loechler, E. L., DNA guanine-guanine crosslinking sequence specificity of isophosphoramide mustard, the alkylating metabolite of the clinical antitumor agent ifosfamide, *Cancer Chemother. Pharmacol.*, 45, 59–62, 2000.

47. Fourie, J., Guziec, F., Jr., Guziec, L., Monterrosa, C., Fiterman, D. J., and Begleiter, A., Structure–activity study with bioreductive benzoquinone alkylating agents: Effects on DT-diaphorase-mediated DNA crosslink and strand break formation in relation to mechanisms of cytotoxicity, *Cancer Chemother. Pharmacol.*, 53, 191–203, 2004.

48. Kim, J. S. and Raines, R. T., Dibromobimane as a fluorescent crosslinking reagent, *Anal. Biochem.*, 225, 174–176, 1995.

49. Avrameas, S., Taudou, B., and Chuilon, S., Glutaraldehyde, cyanuric chloride and tetrazotized O-dianisidine as coupling reagents in the passive hemagglutination test, *Immunochemistry*, 6, 67–76, 1969.

50. Agarwal, K. L., Grudzinski, S., Kenner, G. W., Rogers, N. H., Sheppard, R. C., and McGuigan, J. E., Immunochemical differentiation between gastrin and related peptide hormones through a novel conjugation of peptides to proteins, *Experientia*, 27, 514–515, 1971.

51. Kay, G. and Crook, E. M., Coupling of enzymes to cellulose using chloro-s-triazines, *Nature*, 216, 514–515, 1967.

52. Bowes, J. H. and Cater, C. W., Crosslinking of collagen, *J. Appl. Chem.*, 15, 296, 1965.

53. Cater, C. W., The efficiency of dialdehydes and other compounds as cross-linking agents for collagen, *J. Soc. Leather Trades Chem.*, 49, 455, 1965.

54. Alexander, P., The reactions of carcinogens with macromolecules, *Adv. Cancer. Res.*, 2, 1–72, 1954.

55. Shamma, T. and Haworth, I. S., Spermine inhibition of the 2,5-diaziridinyl-1,4-benzoquinone (DZQ) crosslinking reaction with DNA duplexes containing poly(purine)·poly(pyrimidine) tracts, *Nucleic Acids Res.*, 27, 2601–2609, 1999.

56. Sköld, S, E., Chemical crosslinking of elongation factor G to the 23S RNA in 70S ribosomes from *Escherichia coli*, *Nucleic Acids Res.*, 11, 4923–4932, 1983.

57. Fearnley, C. and Speakman, J. B., Cross-linkage formation in keratin, *Nature*, 166, 743–744, 1950.

58. Millard, J. T., Katz, J. L., Goda, J., Frederick, E. D., Pierce, S. E., Speed, T. J., and Thamattoor, D. M., DNA interstrand cross-linking by a mycotoxic diepoxide, *Biochimie*, 86, 419–423, 2004.

59. Kohn, K. W., Spears, C. L., and Doty, P., Inter-strand crosslinking of DNA by nitrogen mustard, *J. Mol. Biol.*, 19, 266–288, 1966.

60. Leach, J. B., Wolinsky, J. B., Stone, P. J., and Wong, J. Y., Crosslinked alpha-elastin biomaterials: Towards a processable elastin mimetic scaffold, *Acta Biomater.*, 1, 155–164, 2005.

61. Sundberg, L. and Porath, J., Preparation of adsorbents for biospecific affinity chromatography. Attachment of group-containing ligands to insoluble polymers by means of bifunctional oxiranes, *J. Chromatogr.*, 90, 87–98, 1974.

62. Tang, Z. and Yue, Y., Crosslinkage of collagen by polyglycidyl ethers, *ASAIO J.*, 41, 72–78, 1995.

63. Ali-Osman, F., Caughlan, J., and Gray, G. S., Decreased DNA interstrand cross-linking and cytotoxicity induced in human brain tumor cells by 1,3-bis(2-chloroethyl)-1-nitrosourea after in vitro reaction with glutathione, *Cancer Res.*, 49, 5954–5958, 1989.

64. Ali-Osman, F., Giblin, J., Berger, M., Murphy, M. J., Jr., and Rosenblum, M. L., Chemical structure of carbamoylating groups and their relationship to bone marrow toxicity and antiglioma activity of bifunctionally alkylating and carbamoylating nitrosoureas, *Cancer Res.*, 45, 4185–4191, 1985.

65. Houen, G. and Jensen, O. M., Conjugation to preactivated proteins using divinylsulfone and iodoacetic acid, *J. Immunol. Methods*, 181, 187–200, 1995.

66. Dickinson, B. C., Varadan, R., and Fushman, D., Effects of cyclization on conformational dynamics and binding properties of Lys48-linked di-ubiquitin, *Protein Sci.*, 16, 369–378, 2007.

67. Lewis, A., Tang, Y., Brocchini, S., Choi, J. W., and Godwin, A., Poly(2-methacryloyloxyethyl phosphorylcholine) for protein conjugation, *Bioconjug. Chem.*, 19, 2144–2155, 2008.

Appendix C: Phenolate- and Imidazolyl-Group–Directed Reagents: Bisdiazonium Precursors

Name	Structure	References
I		
p-Phenylenediamine		1
II		
Benzidine	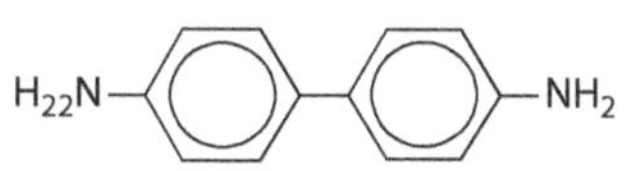	2–4
III		
3,3′-Dimethylbenzidine; o-tolidine	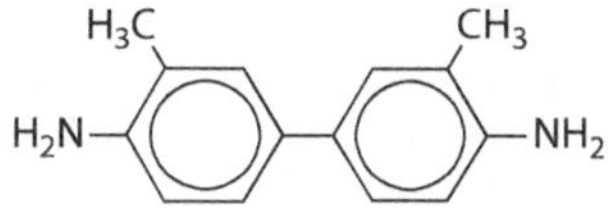	3
IV		
3,3′-Dimethoxybenzidine; o-dianisidine	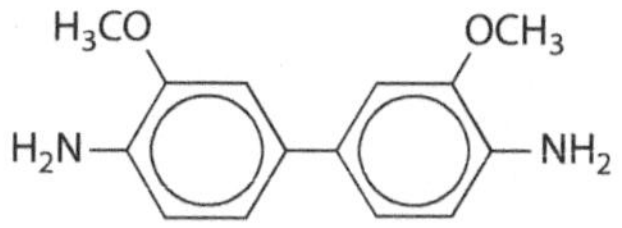	3,4
V		
Benzidine-3,3′-dicarboxylic acid	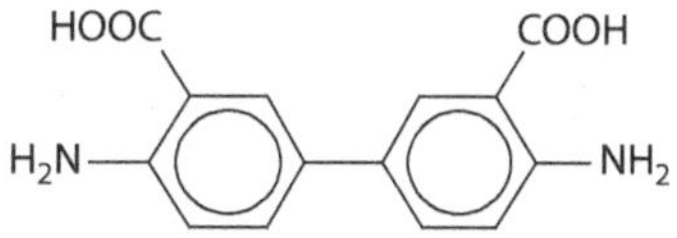	4
VI		
Benzidine-2,2′-disulfonic acid	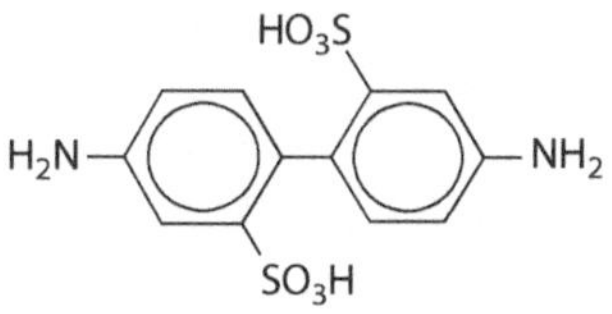	4
VII		
Benzidine-3,3′-disulfonic acid	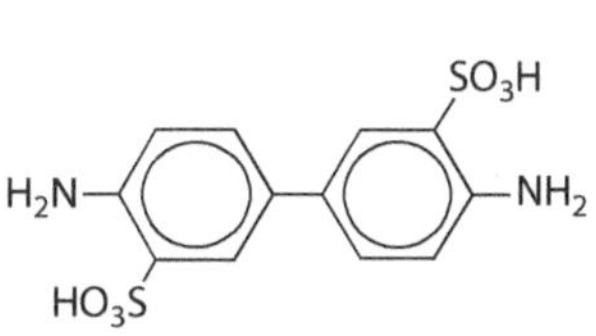	4

(*continued*)

(Continued)

Name	Structure	References
VIII		
4,4′-Diaminodiphenylamine	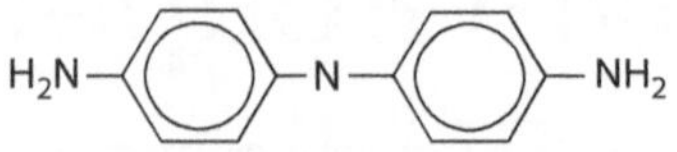 H_2N—⟨◯⟩—N—⟨◯⟩—NH_2	1
IX		
4,4′-Diaminodiphenylsulfide	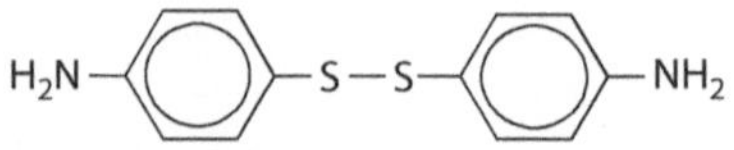 H_2N—⟨◯⟩—S—S—⟨◯⟩—NH_2	5,6
X		
2,2′-Dinitro-4,4′-diamino-diphenylsulfide	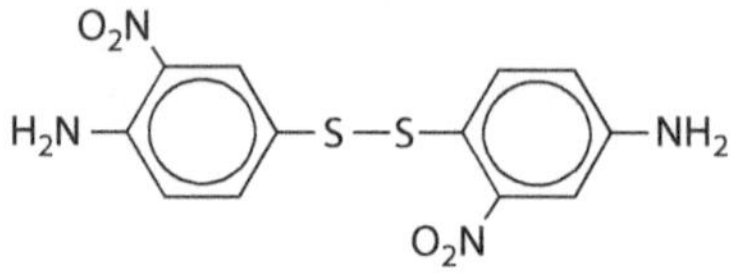 O_2N on ring; H_2N—⟨◯⟩—S—S—⟨◯⟩—NH_2 with O_2N	5–7

REFERENCES

1. Howard, A. N. and Wild, F., A two-stage method of cross-linking proteins suitable for use in serological techniques, *Br. J. Exp. Pathol.*, 38, 640–643, 1957.
2. DeCarvalho, S., Lewis, A. J., Rand, H. J., and Uhrick, J. R., Immunochromatographic partition of soluble antigens on columns of insoluble diazo-gamma-globulins, *Nature*, 204, 265–266, 1964.
3. Hermanson, G. T., Homobifunctional crosslinkers, *Bioconjugate Techniques*, 2nd edn., Elsevier, Burlington, MA, 2008, p. 272.
4. Goldman, R., Kedem, O., Silman, I. H., Caplan, S. R., and Katchalski, E., Papain-collodion membranes. I. Preparation and properties, *Biochemistry*, 7, 486–500, 1968.
5. Ji, T. H., Bifunctional reagents, *Methods Enzymol.*, 91, 580–609, 1983.
6. Wang, K. and Richards, F. M., Behavior of cleavable cross-linking reagents based on the disulfide group, *Israel J. Chem.*, 12, 375, 1974.
7. Demoliou, C. D. and Epand, R. M., Synthesis and characterization of a heterobifunctional photoaffinity reagent for modification of tryptophan residues and its application to the preparation of a photoreactive glucagon derivative, *Biochemistry*, 19, 4539–4546, 1980.

Appendix D: Group Selective Heterobifunctional Cross-Linkers

Name (Abbreviation)	Structure	References
A. Amino- and Sulfhydryl-Group–Directed Bifunctional Reagents		
I		
N-Succinimidyl-3-(2-pyridyldithio) proprionate (SPDP)	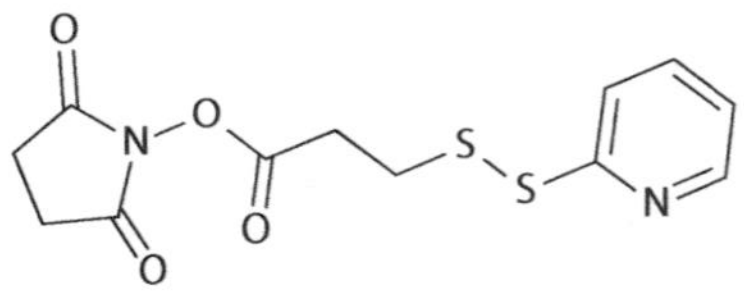	1–3
II		
N-Succinimidyl-3-(2-pyridyldithio) butyrate (MSPDP)	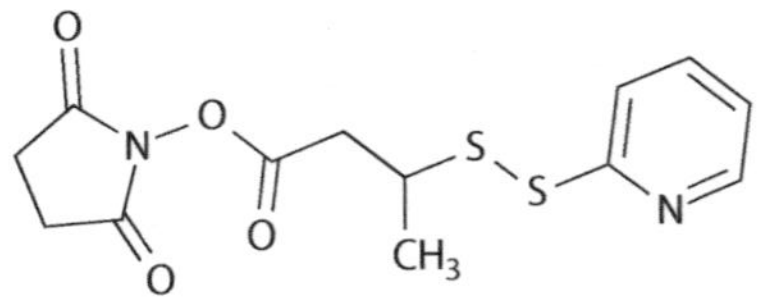	3
III		
N-Succinimidyl-6-(3′-[2-pyridyldithio]propionamide) hexanoate (LC-SPDP)	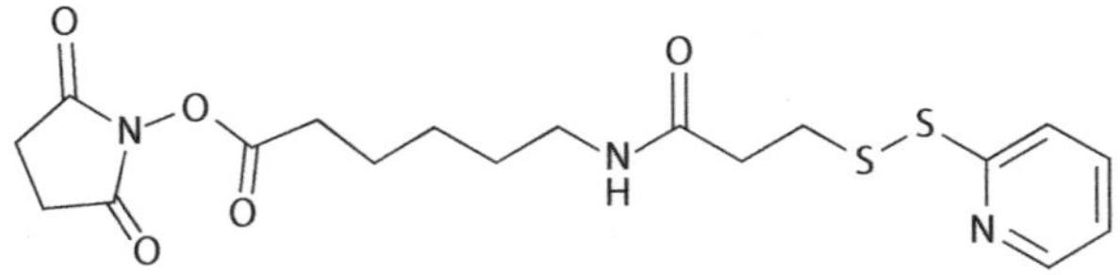	2
IV		
Sulfosuccinimidyl-6-(3′-[2-pyridyldithio]propionamide) hexanoate (Sulfo-LC-SPDP)	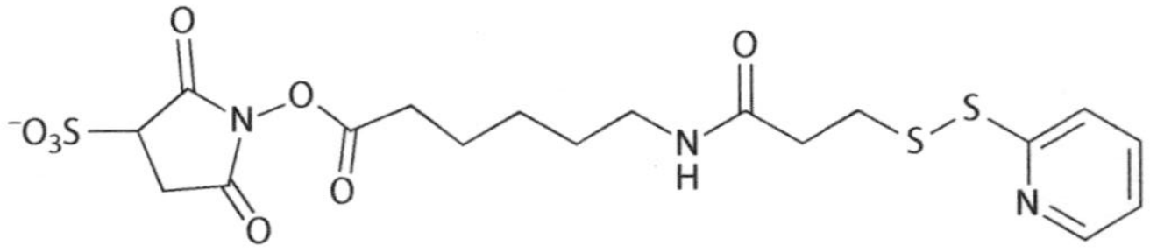	2
V		
3-(2-Pyridyl)dithiopropionyl poly(ethylene glycol) N-hydroxysuccinimide Ester (PDP–PEG–NHS)	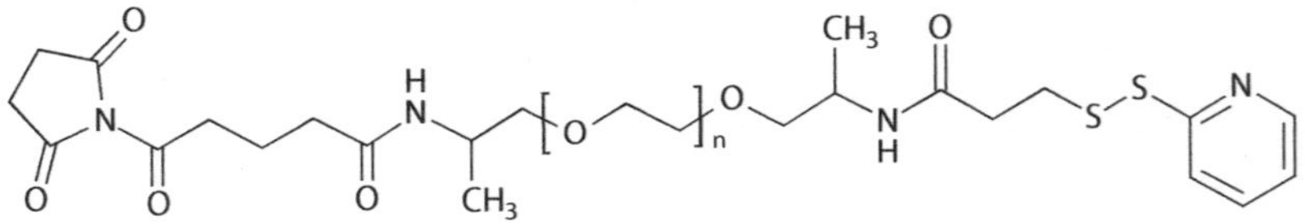	4

VI

4-[(N-Succinimidyloxy)carbonyl]-α-methyl-α-(2-pyridyldithio)
toluene (SMPT)

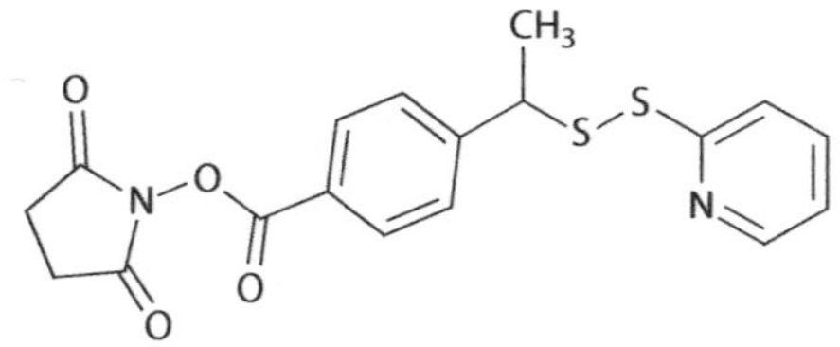

2,3,5

VII

Sulfo-4-Succinimidyloxycarbonyl-6-[α-methyl-α-(2-pyridylthio)
toluamido]hexanoate (sulfo-LC-SMPT)

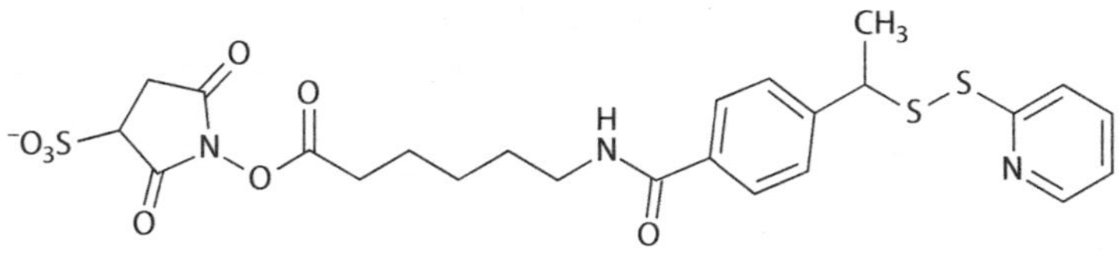

6

VIII

N-Succinimidyl-3-(4-carboxamido-phenyldithio)propionate
(SCDP)

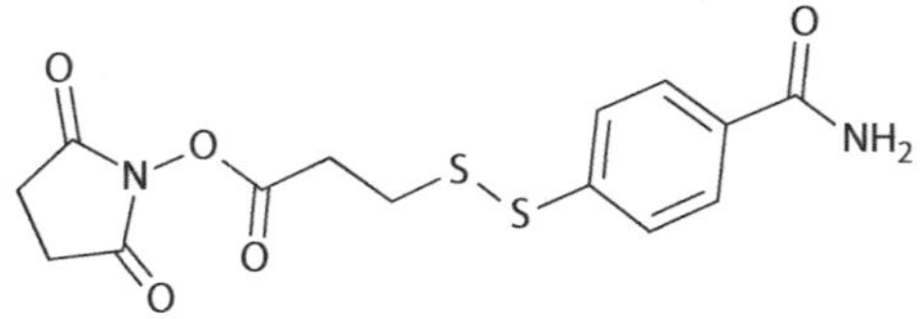

7

IX

N-Succinimidyl-S-acetyl-thioacetate (SATA)

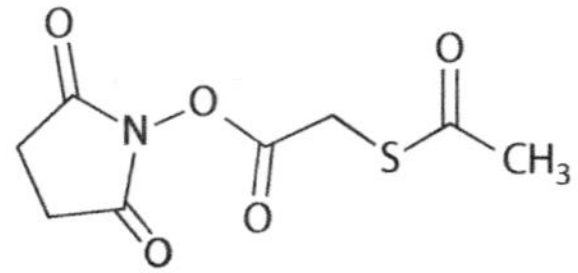

7

X

Sulfosuccinimidyl- N-[3-(acetylthio)-3-methyl-butyryl]-β-alanina
(sNHS-ATMBA)

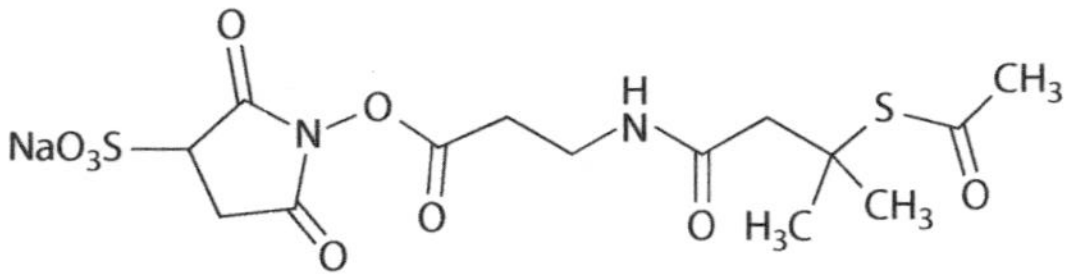

8

(continued)

Name (Abbreviation)	Structure	References
XI Sodium S-4-succinimidyl-oxycarbonyl benzyl thiosulfate (SBT)		5
XII Sodium S-4-succinimidyl-oxycarbonyl-α-methylbenzylthiosulfate (SMBT)		5
XIII Vinyl sulfone poly(ethylene glycol)-N-hydroxysuccinimide (VS–PEG–NHS)		4
XIV N-Succinimidyloxycarbonyl ethyl methanethiosulfonate (NHS-MTS)		9
XV N-Hydroxysuccinimidyl-3-methyl-3-(acetylthio)butanoate (SAMBA)		10

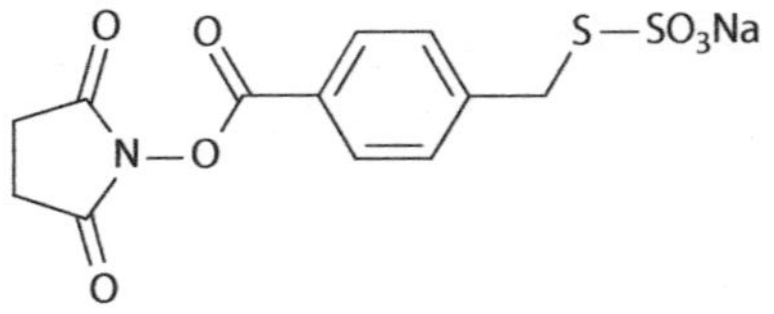

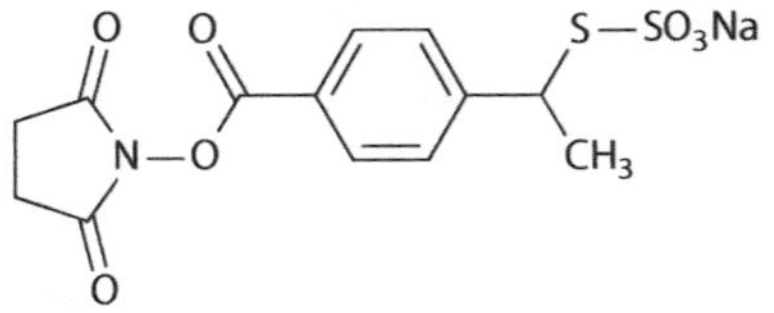

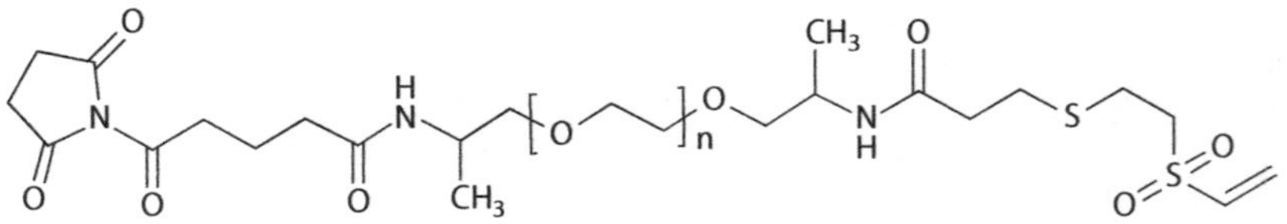

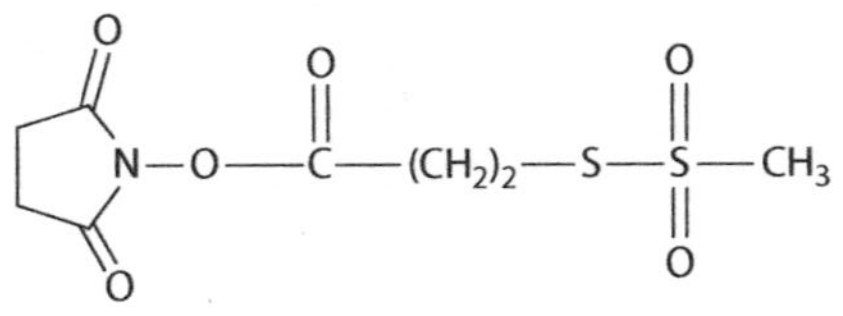

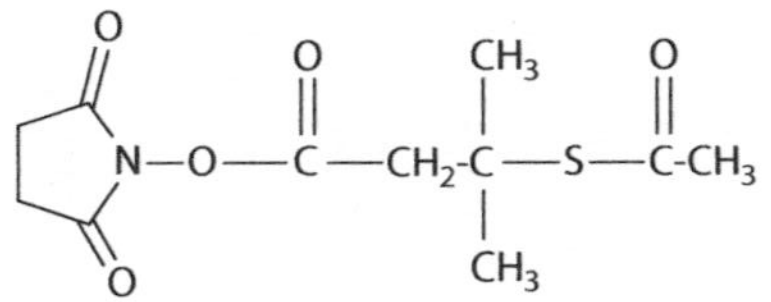

(Continued)

XVI

N-Succinimidyl maleimido-acetate;
N-(α-maleimidoacetoxy)-succinimide ester
(AMAS); α-maleimidoacetic
acid-N-hydroxysuccinimide (MANS)

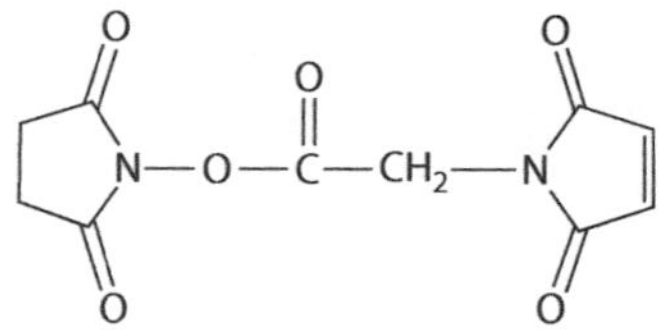

11,12

XVII

N-Succinimidyl 3-maleimido-proprionate (BMPS);
N-(β-maleimido-propyloxy)succinimide
ester; β-maleimidopropionic acid
N-hydroxysuccinimide (β-MHS)

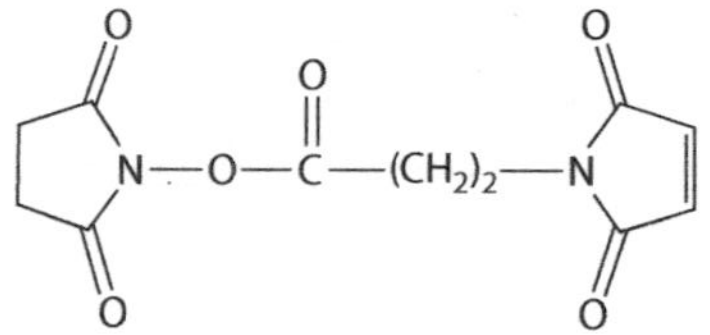

13

XVIII

N-Succinimidyl 4-maleimido-butyrate;
N-(γ-maleimidobutyryloxy) succinimide
ester (GMBS); maleimido-butyric
acid-N-hydroxysuccinimide (MBNS)

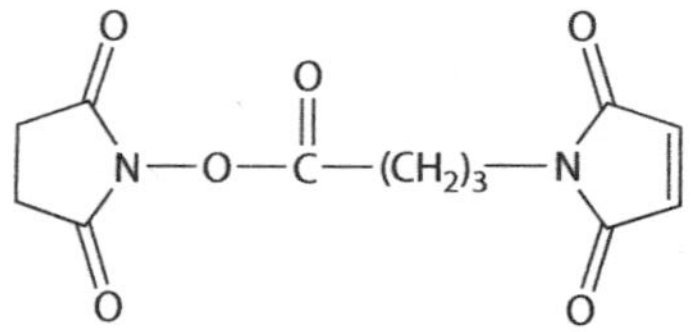

11,12

XIX

Sulfo-N-succinimidyl 4-maleimido-butyrate; N-(γ-maleimidobutyryloxy)sulfosuccinimide ester (Sulfo-GMBS)

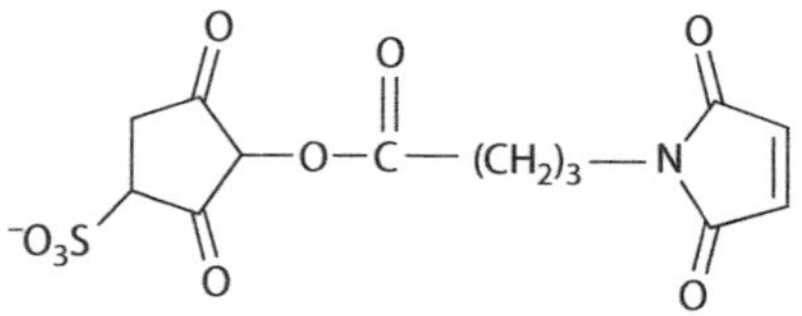

14

XX

N-Succinimidyl 6-maleimidocaproate;
N-Succinimidyl 6-maleimidohexanoate (SMH);
N-(ε-maleimidocaproyloxy) succinimide ester
(EMCS); ε-maleimido-caproic acid
N-hydroxysuccinimide (ε-MHS)

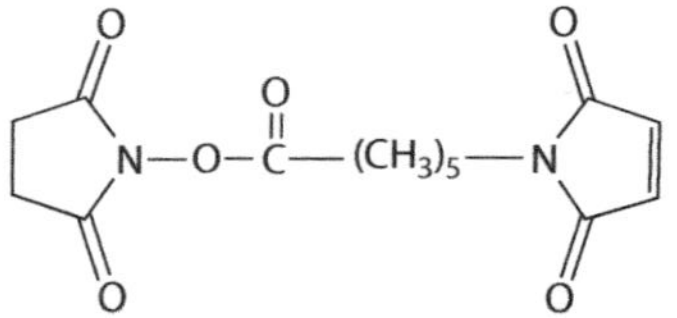

11

(continued)

(Continued)

Name (Abbreviation)	Structure	References
XXI Sulfo-*N*-succinimidyl 6-maleimido-caproate; *N*-(ε-maleimidocaproyl-*oxy*)sulfo-succinimide ester (Sulfo-EMCS)		15
XXII Sulfo-*N*-succinimidyl 11-maleimido-decanoate *N*-(κ-maleimido-undecanoyloxy)sulfosuccinimide ester (Sulfo-KMUS)		16
XXIII Succinimidyl-6-[(β-maleimido-propionamido)hexanoate] (SMPH)		17
XXIV *N*-Hydroxysuccinimide-poly(ethylene glycol)-maleimide with MW 2300 Da (NHS-PEG$_{2300}$-MAL)		12
XXV *N*-Hydroxysuccinimide-poly(ethylene glycol)-maleimide with MW 875 (NHS-PEG$_{875}$-MAL)		18

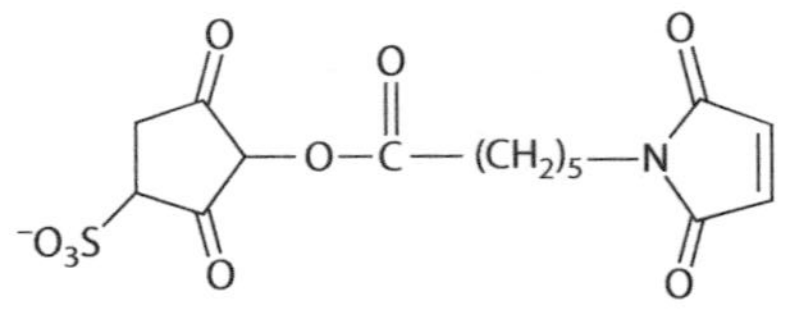

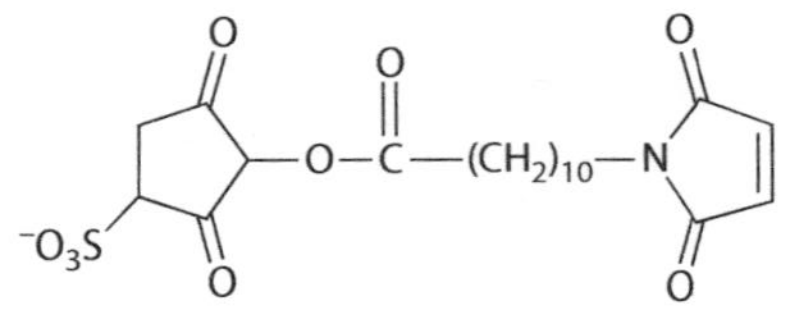

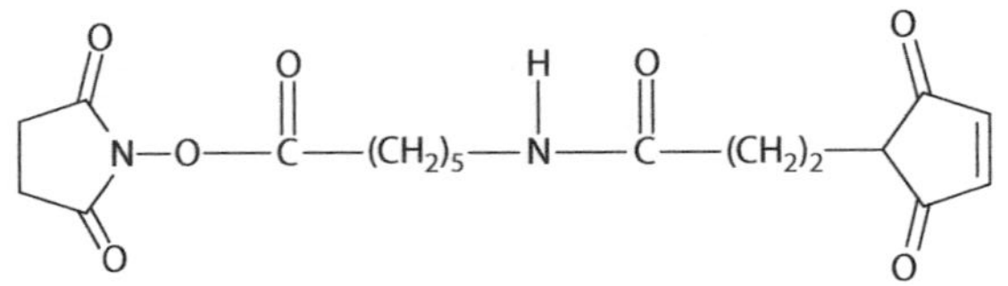

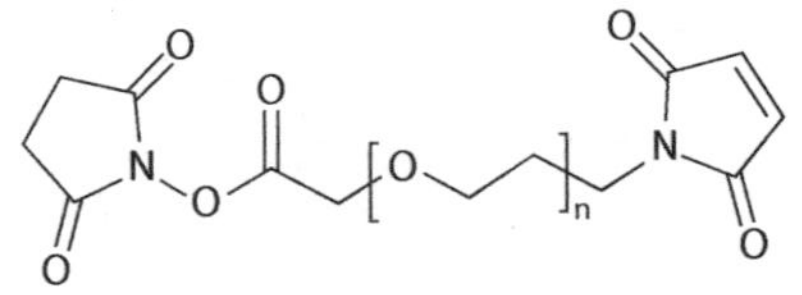

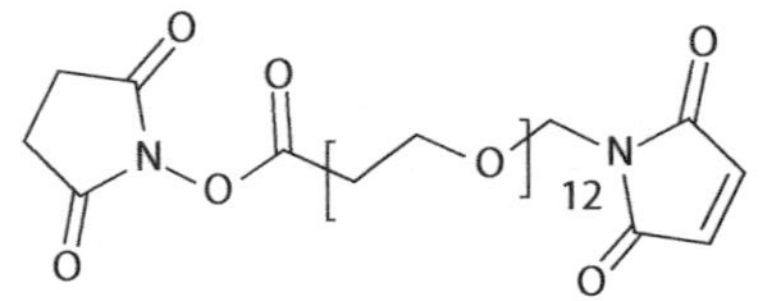

XXVI

N-Hydroxysuccinimide-poly(ethylene glycol)-maleimide with MW 3400 (NHS-PEG$_{3400}$-MAL)

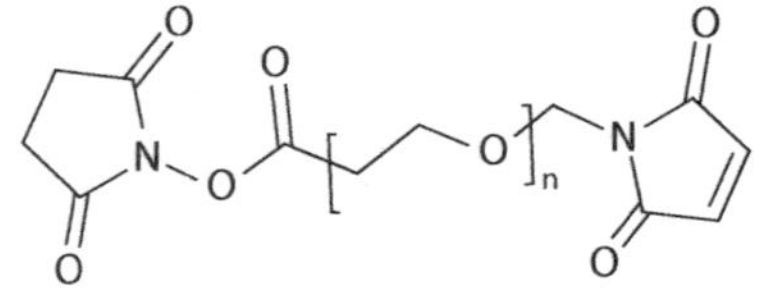

18

NHS-PEG-Maleimide cross-linkers

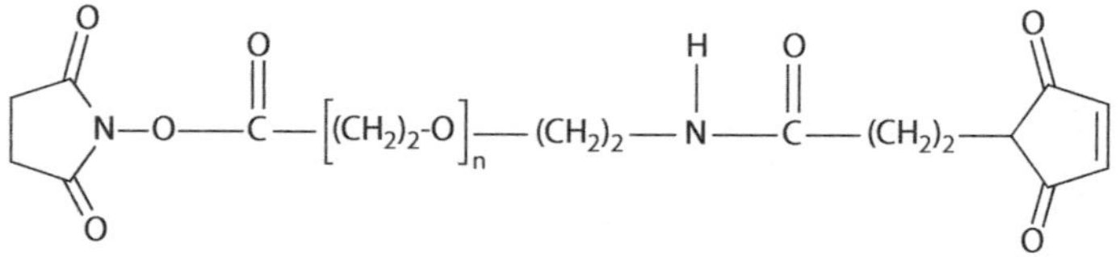

19

XXVII

n = 2: Succinimidyl-[(*N*-maleimidopropionamido)diethyleneglycol] ester (SM[PEG]$_2$)

XXVIII

n = 4: Succinimidyl-[(*N*-maleimidopropionamido)tetraethyleneglycol] ester (SM[PEG]$_4$)

XXIX

n = 6: Succinimidyl-[(*N*-maleimidopropionamido)hexaethyleneglycol] ester (SM[PEG]$_6$)

XXX

n = 8: Succinimidyl-[(*N*-maleimidopropionamido)octaethyleneglycol] ester (SM[PEG]$_8$)

XXXI

n = 12: Succinimidyl-[(*N*-maleimidopropionamido)dodecaethyleneglycol] ester (SM[PEG]$_{12}$)

XXXII

n = 24: Succinimidyl-[(*N*-maleimidopropionamido)tetracosaethyleneglycol] ester (SM[PEG]$_{24}$)

XXXIII

N-Succinimidyl 4-(*N*-maleimido-methyl)cyclohexane-1-carboxylate (SMCC)

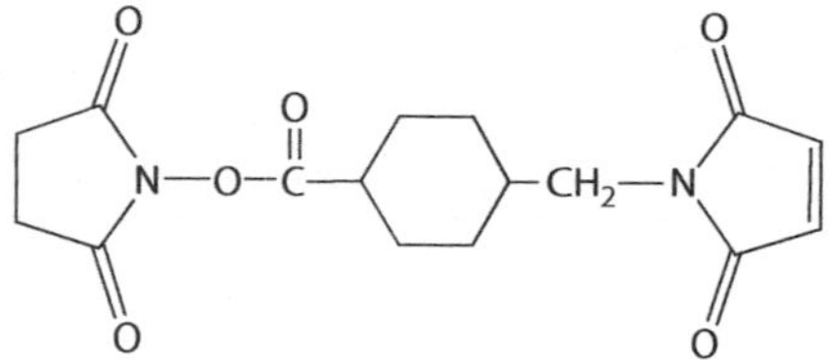

1,11

(continued)

(Continued)

Name (Abbreviation)	Structure	References
XXXIV *N*-Sulfosuccinimidyl 4-(*N*-maleimidomethyl) cyclohexane-1-carboxylate (Sulfo-SMCC)	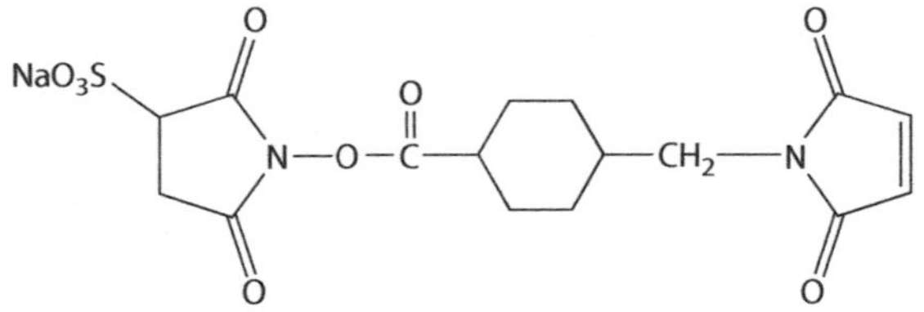	11,20
XXXV *O*-(*N*-Succinimidyl)-6-{[4-(*N*-maleimidomethyl)-cyclohexane-1-carbonyl]-amino}-hexanoate; succinimidyl-4-(*N*-maleimido-methyl)cyclohexane-1-carboxy-(6-amido-caproate) (LC-SMCC)	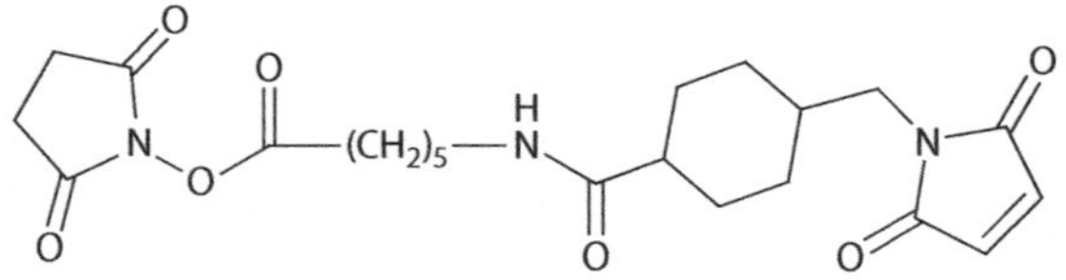	12,21
XXXVI 4-[(2,5-Dihydro-2,5-dioxo-1*H*-pyrrol-1-yl) methyl]-*N*-[6-[[6-[(2,5- dioxo-1-pyrrolidinyl)-oxy]-6-oxohexyl] amino]-6-oxo-hexyl] cyclohexanecarboxamide	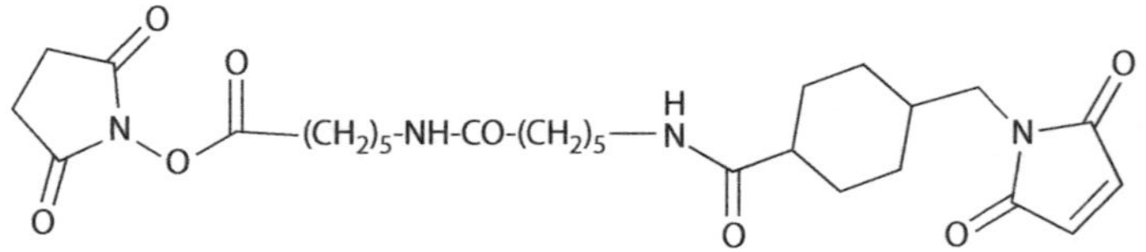	21
XXXVII 4-[(2,5-Dihydro-2,5-dioxo-1*H*-pyrrol-1-yl) methyl]-*N*-[6-[[6-[(2,5-dioxo-1-pyrrolidinyl)oxy]-6-oxo-hexyl]amino]-6-oxohexyl]-amino]-6-oxohexyl] cyclo-hexanecarboxamide	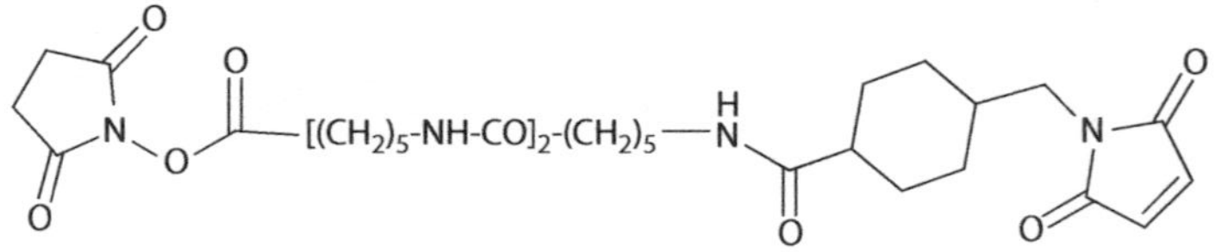	21

XXXVIII

N-Hydroxysuccinimide alkyl-poly(ethylene glycol)-alkylcyclo-
hexyl-maleimide (mal–PEG–NHS)

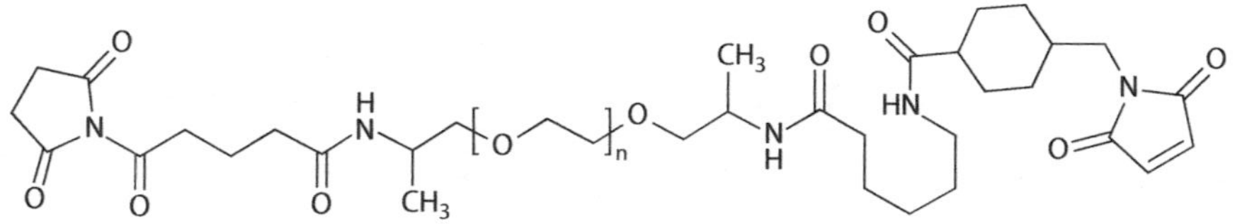

4

XXXIX

N-Succinimidyl 4-(*p*-maleimido-phenyl)butyrate (SMBP)

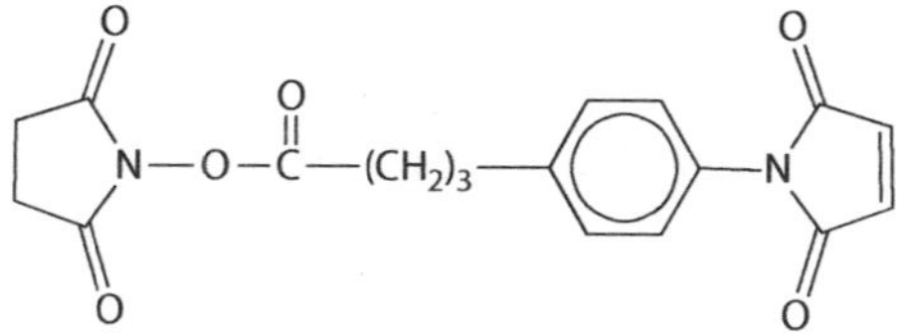

1,11,20

XL

N-Sulfosuccinimidyl 4-(*p*-maleimidophenyl) butyrate
(Sulfo-SMPB)

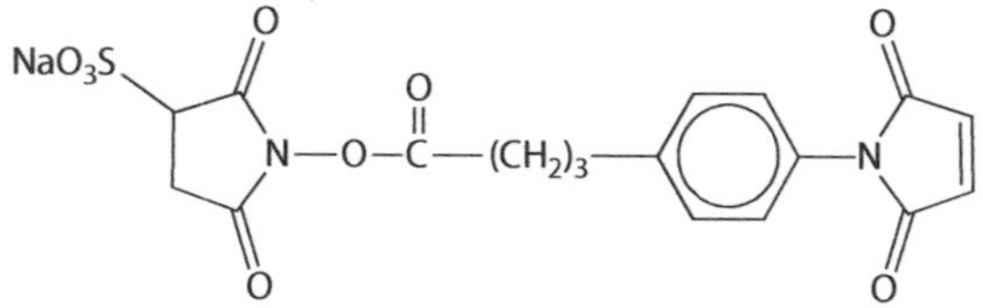

6,22

XLI

N-Succinimidyl *o*-maleimido-benzoate; *N*-(*o*-maleimidobenzoyl-
oxy)-succinimide (*o*-MBS)

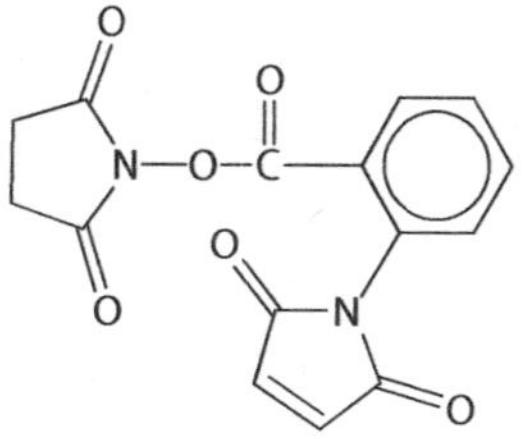

23

(continued)

(Continued)

Name (Abbreviation)	Structure	References
XLII *N*-Succinimidyl *m*-maleimido-benzoate; *N*-(*m*-Maleimidobenzoyl-oxy)-succinimide (MBS); *m*-maleimido-benzoyl *N*-hydroxysuccinimide ester		20,23,24
XLIII *N*-Sulfosuccinimidyl *m*-maleimido-benzoate; *m*-Maleimidobenzoyl-*N*-hydeoxysulfosuccinimide ester; *N*-(*m*-Maleimidobenzoyloxy)-succinimide (Sulfo-MBS)		25
XLIV *N*-Succinimidyl *p*-maleimido-benzoate; *N*-(*p*-maleimidobenzoyl-oxy)succinimide (*p*-MBS)		23
XLV *N*-Succinimidyl 4-maleimido-3-methoxybenzoate		23

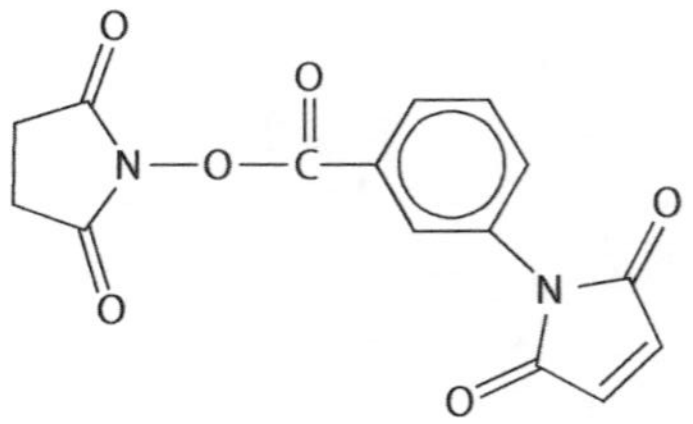

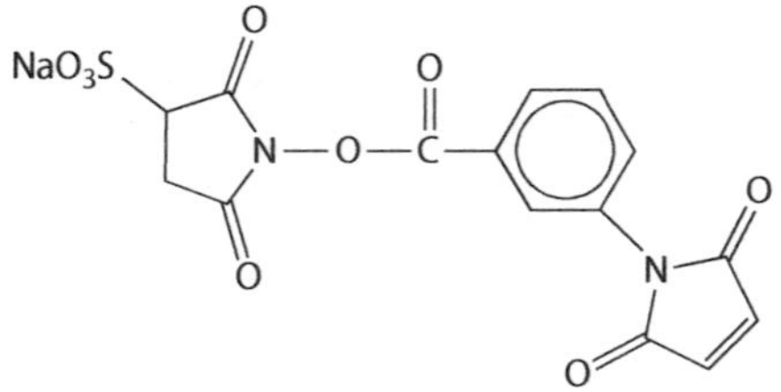

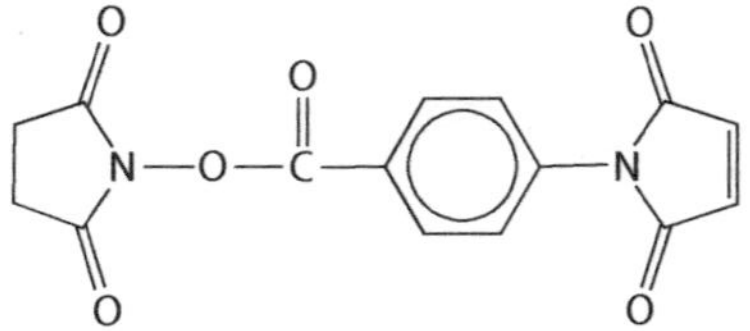

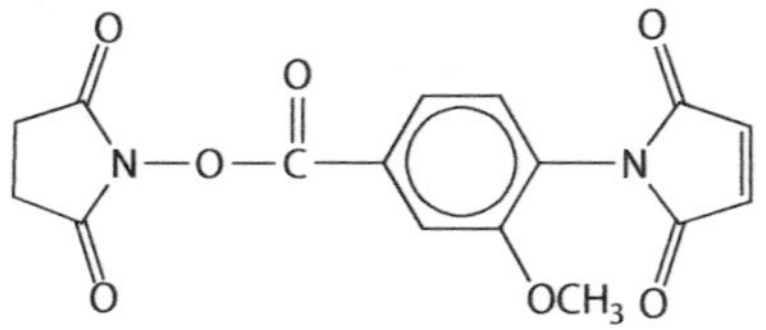

23

23

23

26

(continued)

XLVI

N-Succinimidyl-5-maleimido-2-methoxybenzoate

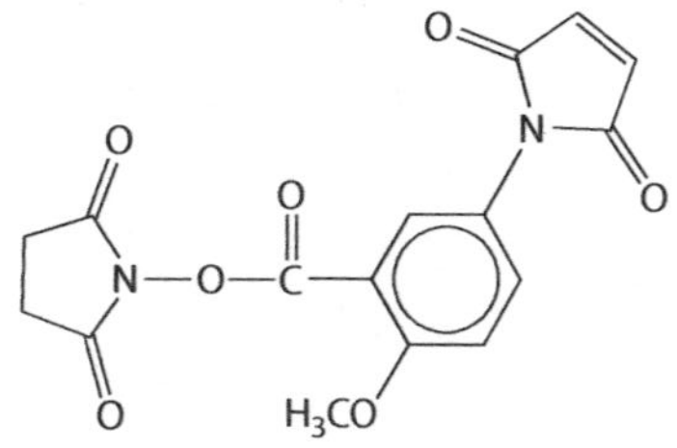

XLVII

N-Succinimidyl-3-maleimido-4-methoxybenzoate

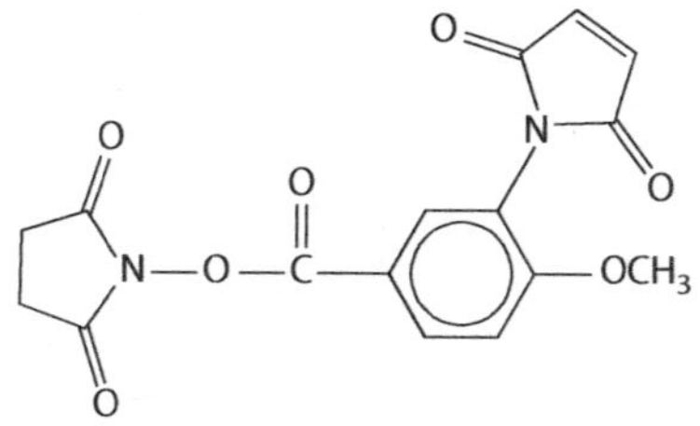

XLVIII

N-Succinimidyl 3-maleimido-4-(*N,N*-dimethyl)aminobenzoate

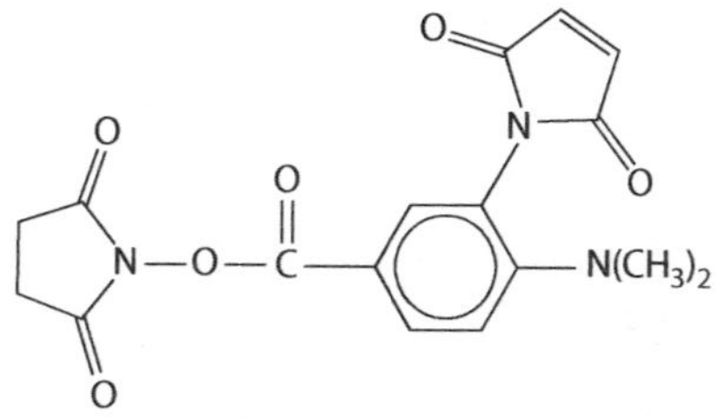

XLIX

Maleimidoethoxy {*p*-(*N*-succinimidylproprionato)-phenoxy}ethane

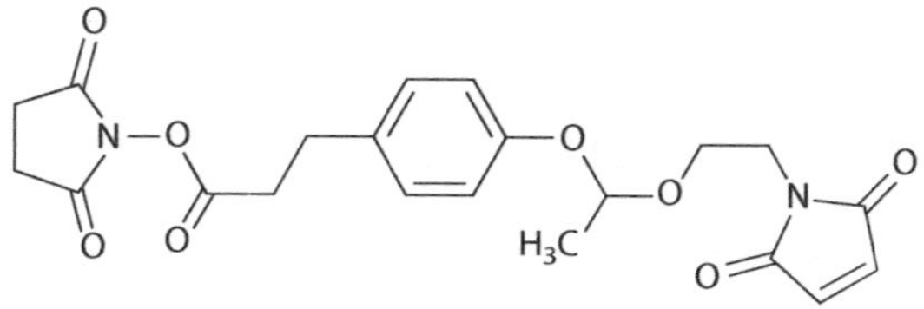

(Continued)

Name (Abbreviation)	Structure	References
L		
(2,5-Dioxopyrrolidin-1-yl) 2-[2-(2,5-dioxopyrrol-1-yl)-4-hydroxy-phenyl]azobenzoate	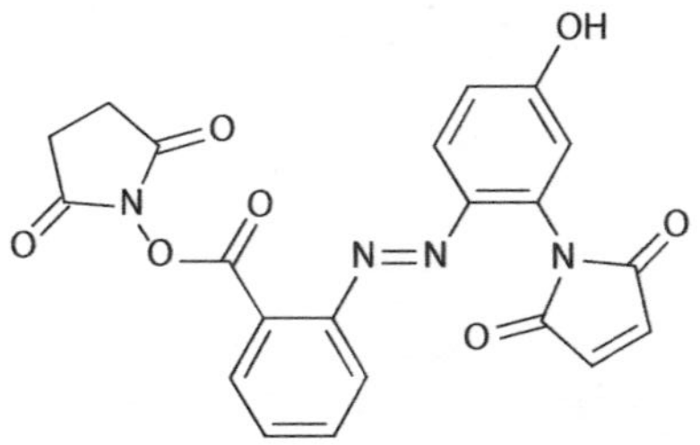	27
LI		
(2,5-Dioxopyrrolidin-1-yl) 2-[3-(2,5-dioxopyrrol-1-yl)-4-hydroxy-phenyl]azobenzoate	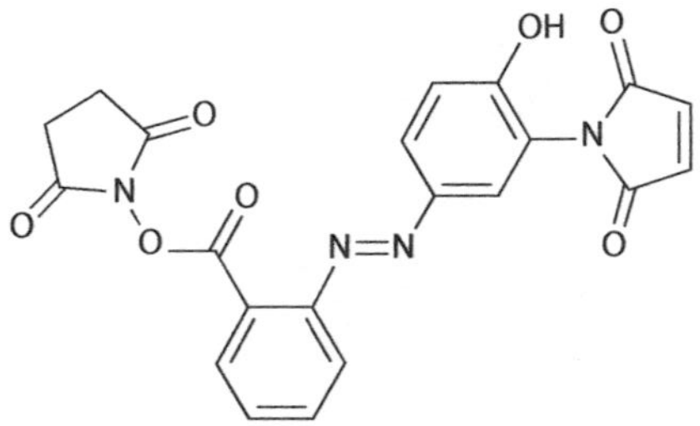	27
LII		
(2,5-Dioxopyrrolidin-1-yl) 4-[3-(2,5-dioxopyrrol-1-yl)-4-hydroxy-phenyl]azobenzoate	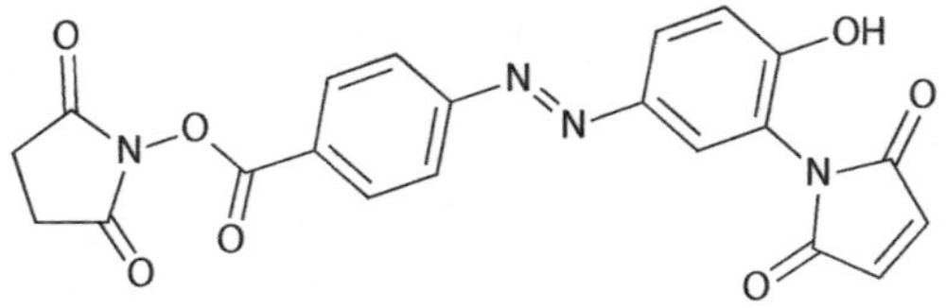	27
LIII		
N-Succinimidyl iodoacetate (SIA)	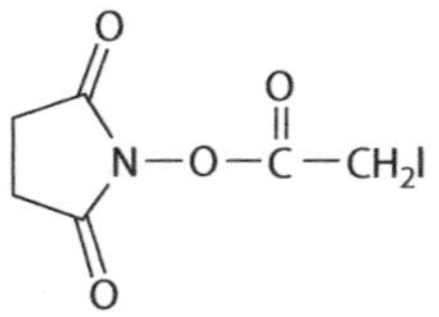	10,28

LIV

N-Succinimidyl-6-[(iodoacetyl)amino] hexanoate (SIAX)

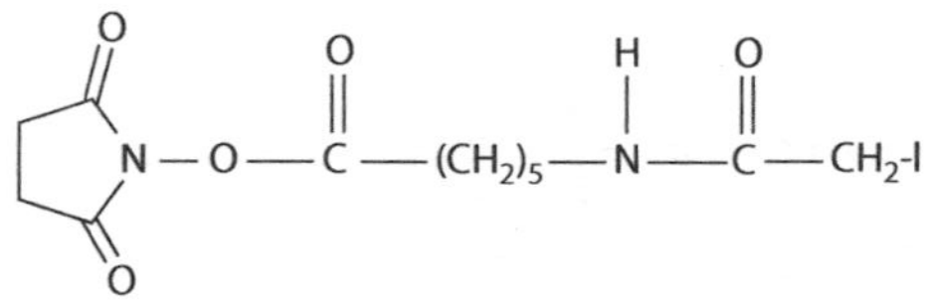

29

LV

N-Succinimidyl-6-(6-(((iodoacetyl)amino)hexanoyl)amino) hexanoate (SIAXX)

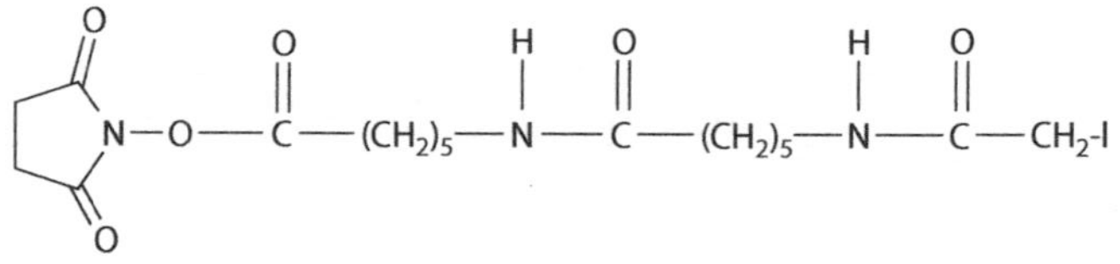

30

LVI

N-Succinimidyl-4-(((iodoacetyl)amino)methyl) cyclohexane-1-carboxylate (SIAC)

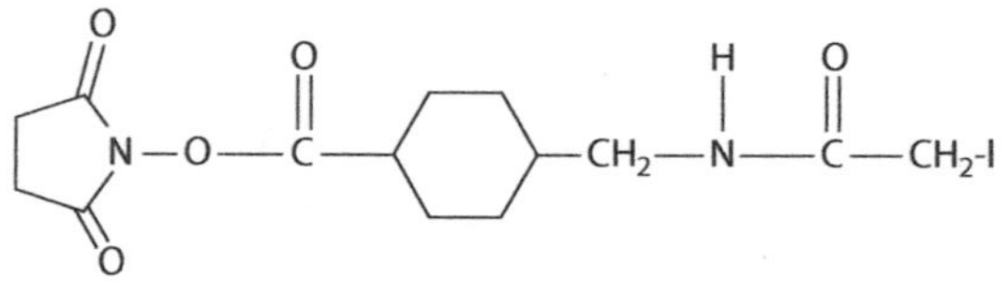

30

LVII

N-Succinimidyl-6-((((4-iodoacetyl)amino)methyl) cyclohexane-1-carbonylamino)hexanoate (SIACX)

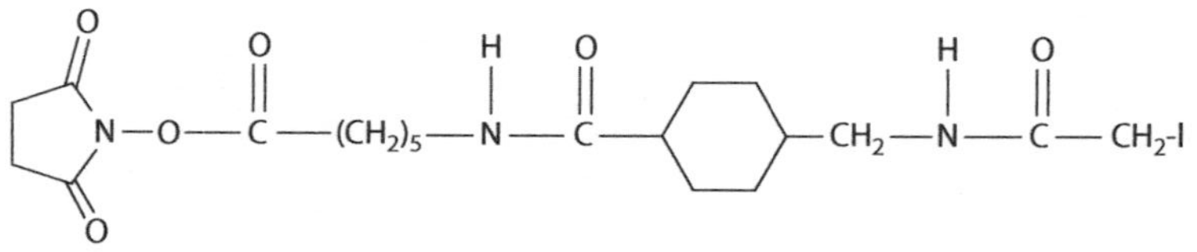

30

LVIII

N-Succinimidyl 4-[(*N*-iodo-acetyl)amino]benzoate (SIAB)

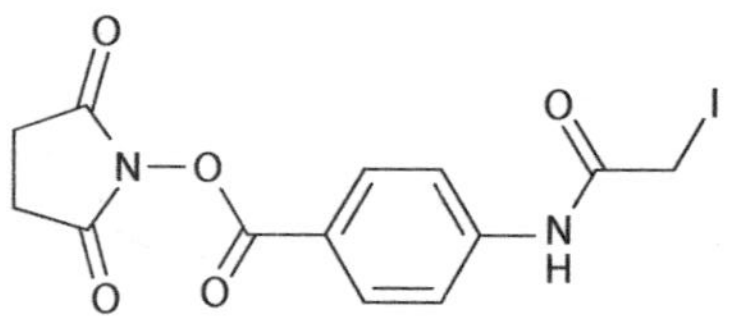

31

(continued)

(Continued)

Name (Abbreviation)	Structure	References
LIX *N*-Sulfosuccinimidyl 4-{(*N*-acetyl)amino)}benzoate (Sulfo-SIAB)		32
LX *N*-Succinimidyl bromoacetate		33
LXI *N*-Succinimidyl-3-(bromoacetamido) propionate (SBAP)		34
LXII *N*-Succinimidyl 3-(2-bromo-3-oxybutane-1-sulfonyl)proprionate		35

LXIII

N-Succinimidyl 3-(4-bromo-3-oxybutane-1-sulphonyl)
proprionate

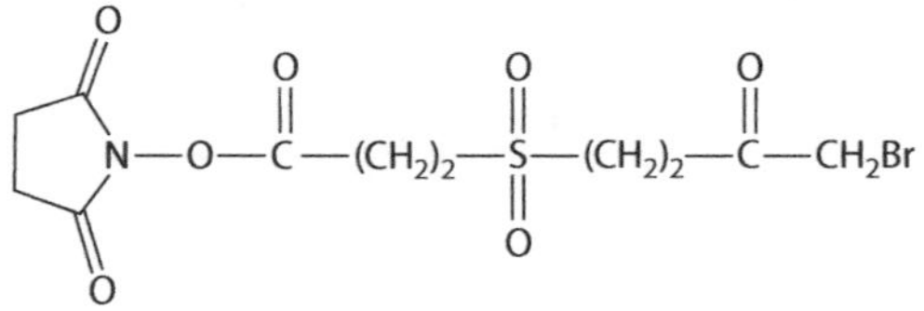

36

LXIV

N-Succinimidyl 2,3-dibromoproprionate (SDBP)

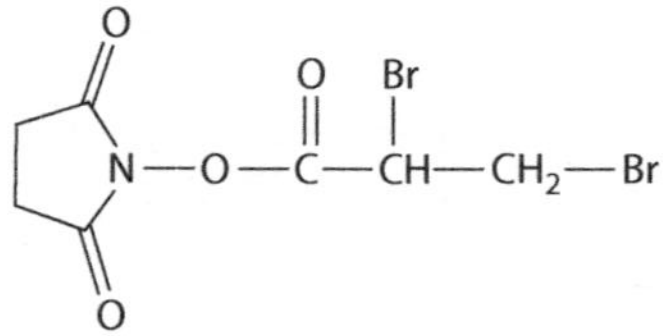

37

LXV

N-Succinimidyl 4-[(N,N-bis(2-chloroethyl)amino)phenyl]-
butyrate; Chlorambucil-N-hydroxysuccinimde ester

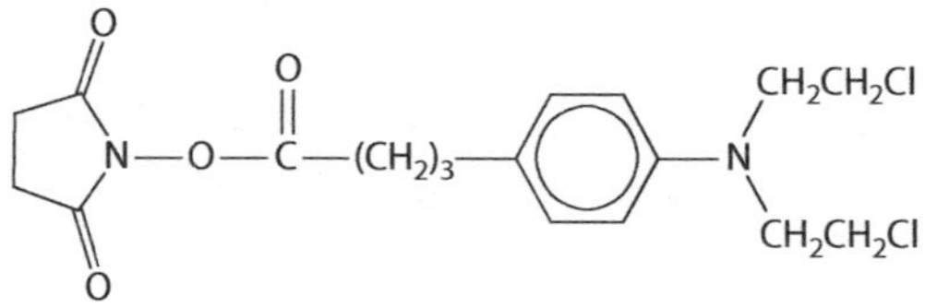

37

LXVI

p-Nitrophenyl 3-(2-bromo-3-oxybutane-1-sulfonyl) propionate

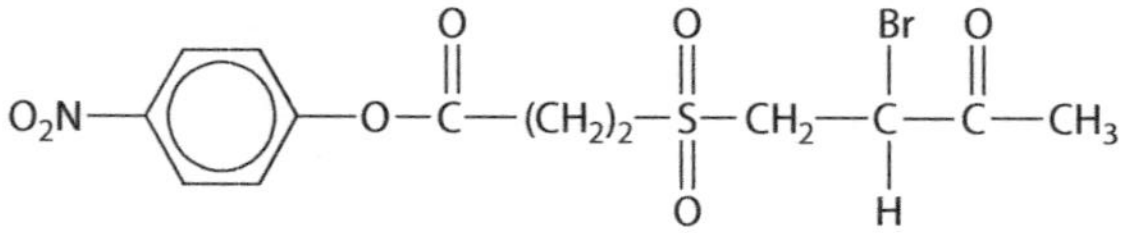

35

LXVII

p-Nitrophenyl 3-(4-bromo-3-oxybutane-1-sulfonyl) proprionate

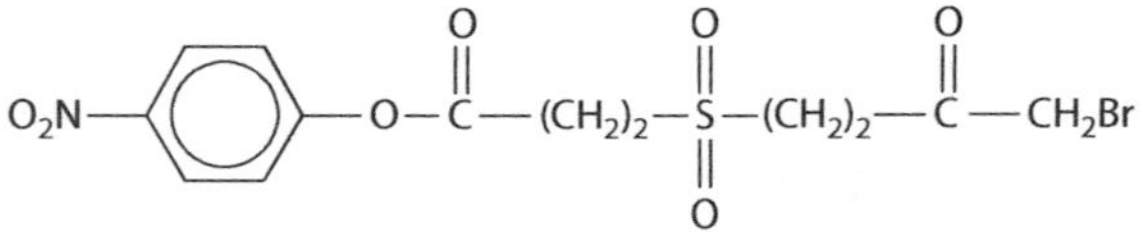

36

(continued)

(Continued)

Name (Abbreviation)	Structure	References
LXVIII		
p-Nitrophenyl iodoacetate		38
LXIX		
p-Nitrophenyl bromoacetate		38
LXX		
2,4-Dinitrophenyl-*p*-(β-nitrovinyl) benzoate		39
LXXI		
p-Nitrophenyl 6-maleimido caproate; 6-maleimido hexanoic acid 4-nitrophenyl ester (MHNp)		40
LXXII		
(2-Nitro-4-sulfonic acid phenyl)-6-maleimidocaproate (Mal-sac-HNSA)		41

40

40

42

43

44,45

(continued)

LXXIII

6-(*N*-Maleimido)hexanoic acid pentafluorophenyl ester (MHPf)

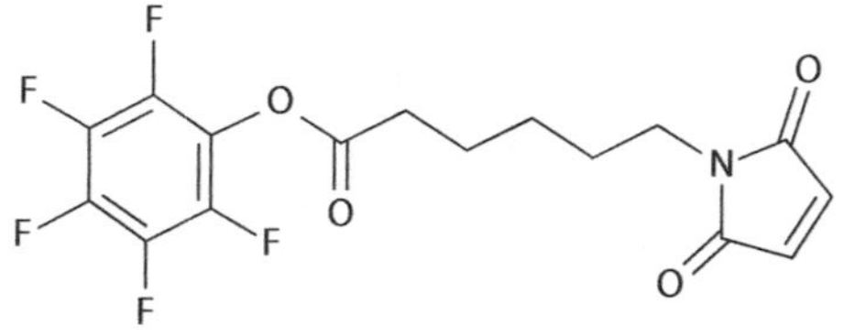

LXXIV

6-(*N*-Maleimido)hexanoic acid *N*-hydroxy-5-norbornene-endo-2,3-dicarboximide ester (MHNb)

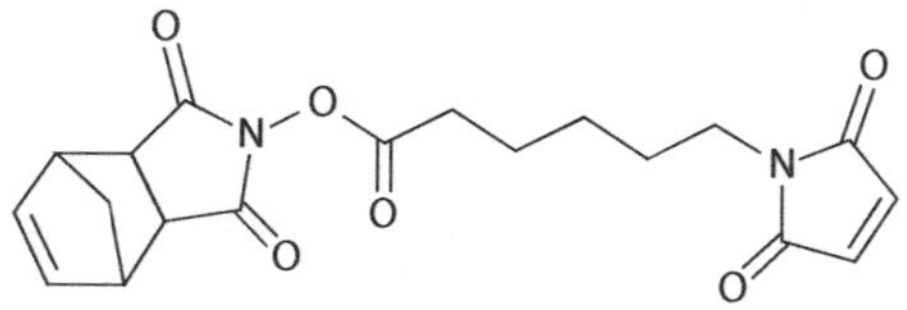

LXXV

N-(Maleimidomethyl)-2-(*p*-nitrophenoxy)carboxamidoethane

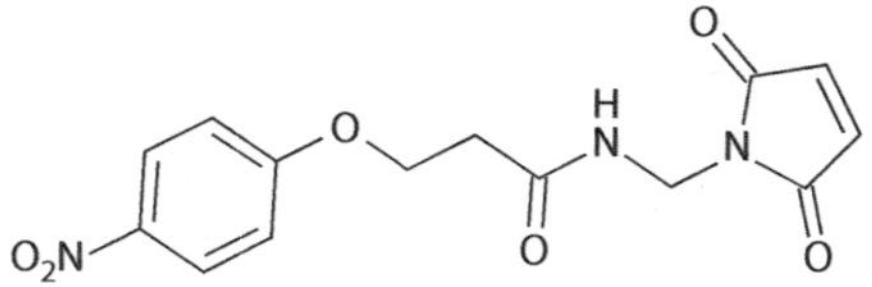

LXXVI

3-(*N*-Maleimido)propionyl chloride

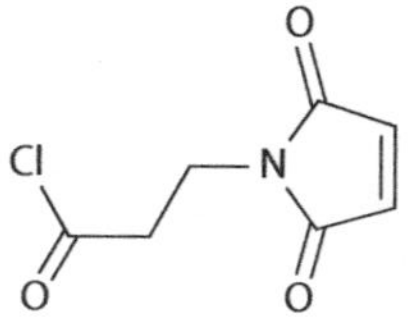

LXXVII

N-(4-Azidocarbonyl-3-hydroxy-phenyl) maleimide; 2-hydroxy-4-(*N*-maleimido) benzoylazide (HMB)

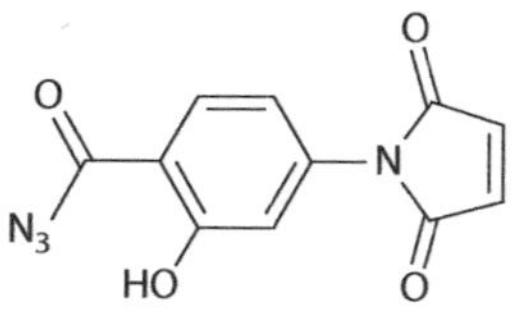

(Continued)

Name (Abbreviation)	Structure	References
LXXVIII		
4-(N-Maleimido) benzoyl chloride	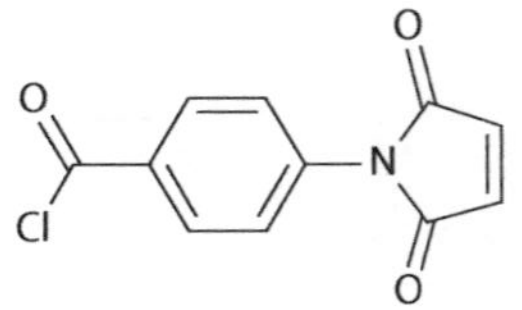	43
LXXIX		
2-Chloro-4-(N-maleimido) benzoyl chloride	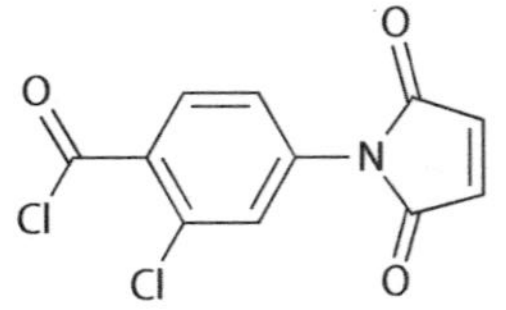	43
LXXX		
2-Acetoxy-4-(N-maleimido) benzoyl chloride	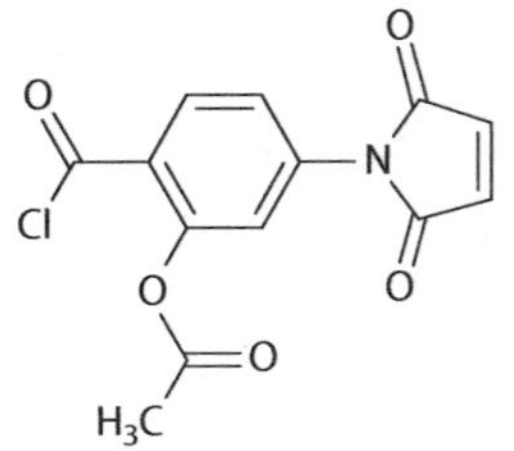	43
LXXXI		
4-Chloroacetylphenyl-N-maleimide	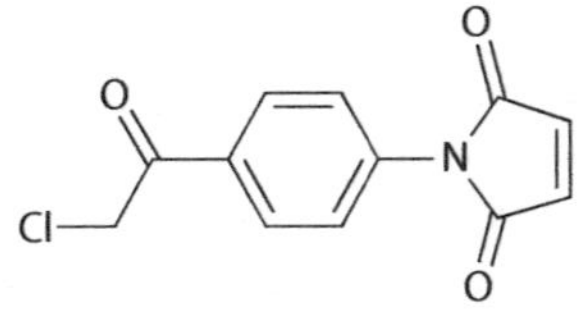	45

46

47

48

49

50

(*continued*)

LXXXII

N-α-(Bromoacetoxymethyl) maleimide

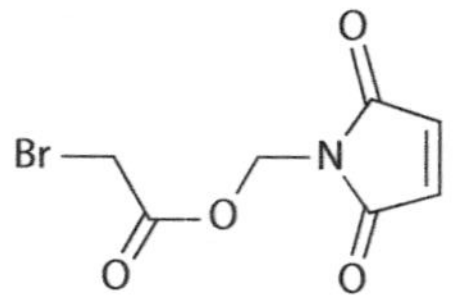

LXXXIII

2-[(4-(6-N-Maleimidocaproic acid)-aminophenyl)thiomethyl]maleic anhydride; N-[4-[(2,5-dioxo-3-furyl)methylsulfanyl]phenyl]-6-(2,5-dioxopyrrol-1-yl)hexanamide

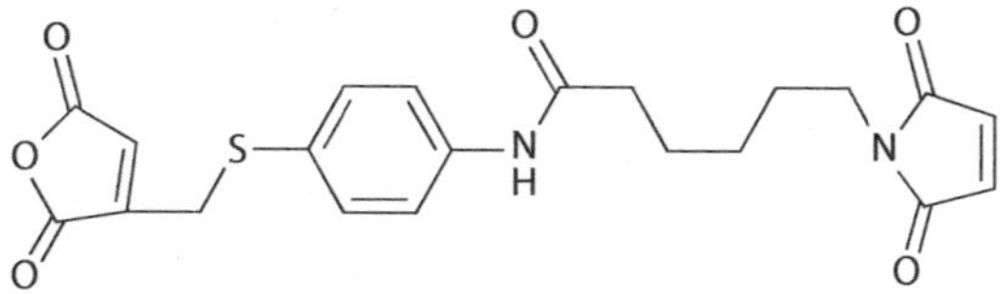

LXXXIV

1-[5-[(2,5-Dioxopyrrol-1-yl)methyl]-2-nitro-phenyl]ethyl carbonochloridate; 1-[5-(N-maleimidomethyl)-2-nitrophenyl] ethyl chloroformate

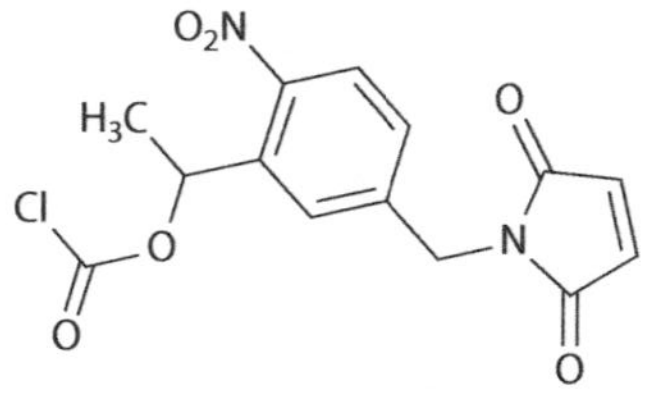

LXXXV

4-Nitro-3-(1-chlorocarbonyloxyethyl) phenyl] methyl-3-(2- pyridyldithio-propionic acid) ester

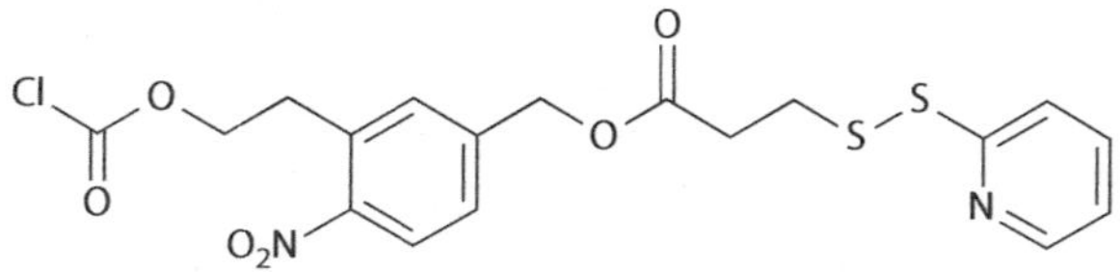

LXXXVI

4-(Iodoacetamido)-1-cyclohexene-1,2-dicarboxylic acid anhydride; 4-(iodo-acetylamino)-3,4,5,6-tetrahydrophthalic anhydride

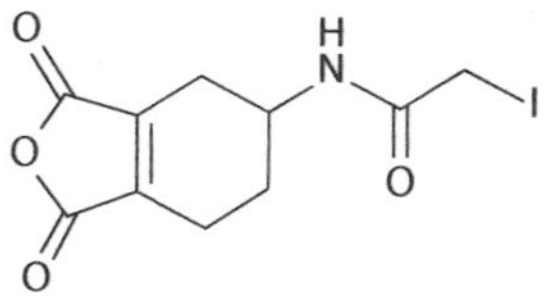

(Continued)

Name (Abbreviation)	Structure	References
LXXXVII		
N-(3-Fluoro-4,6-dinitrophenyl) cystamine	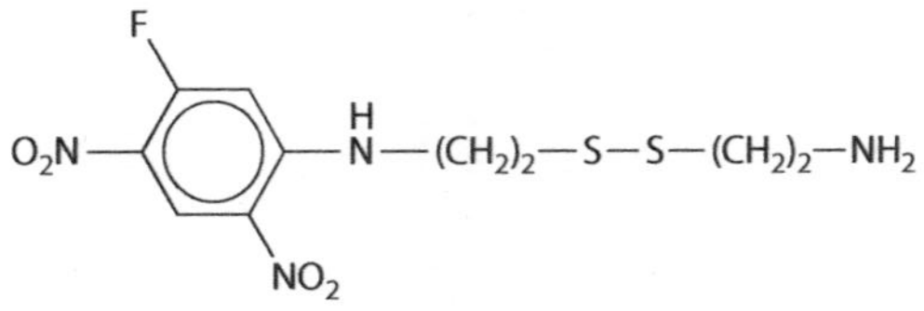 Aromatic ring with F, O_2N, NO_2 substituents; N—H—$(CH_2)_2$—S—S—$(CH_2)_2$—NH_2	42,51
LXXXVIII		
Methyl 3-(4-pyridyldithio)-propionimidate HCl; methyl 3-(4′-dithiopyridyl)mercapto-propionimidate ester HCl	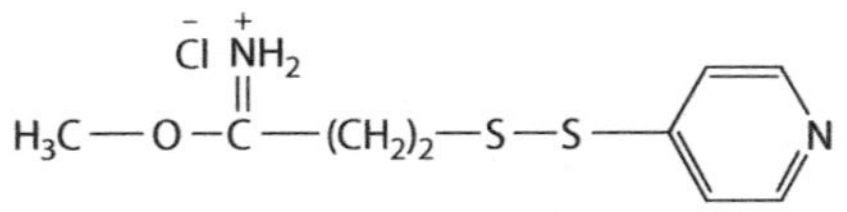 Pyridine ring—S—S—$(CH_2)_2$—C(=$\overset{+}{N}H_2$ $\bar{Cl}$)—O—CH_3	52
LXXXIX		
Ethyl iodoacetimidate HCl	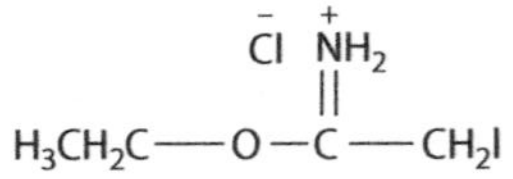 H_3CH_2C—O—C(=$\overset{+}{N}H_2$ $\bar{Cl}$)—CH_2I	53
XC		
Ethyl bromoacetimidate HCl	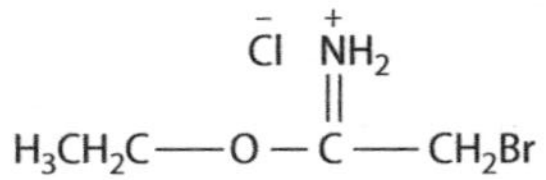 H_3CH_2C—O—C(=$\overset{+}{N}H_2$ $\bar{Cl}$)—CH_2Br	20,54
XCI		
Ethyl chloroacetimidate HCl	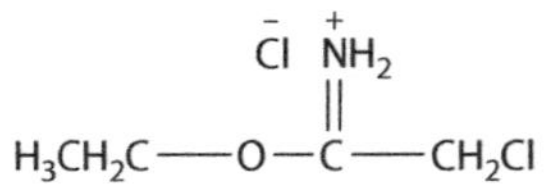 H_3CH_2C—O—C(=$\overset{+}{N}H_2$ $\bar{Cl}$)—CH_2Cl	53
XCII		
Ethyl *S*-acetylpropionthioimidate ester hydrochloride (AMPT)	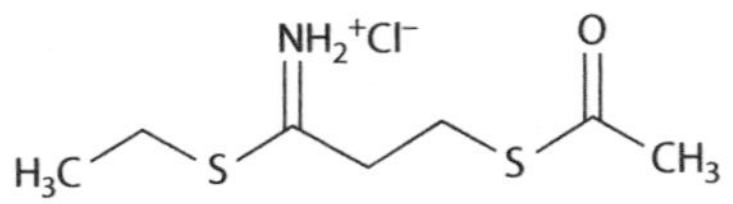 H_3C—S—C(=$\overset{+}{N}H_2\ Cl^-$)—CH_2CH_2—S—C(=O)—CH_3	7,10,55

XCIII

10,55

Ethyl S-acetyl-3-mercaptobutyro-thioimidate ester hydrochloride (M-AMPT)

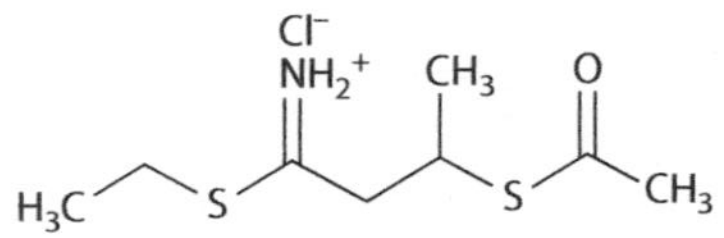

XCIV

55

Ethyl S-acetyl-3-mercapto-3-phenylpropionthioimidate ester hydrochloride (Ph-AMPT)

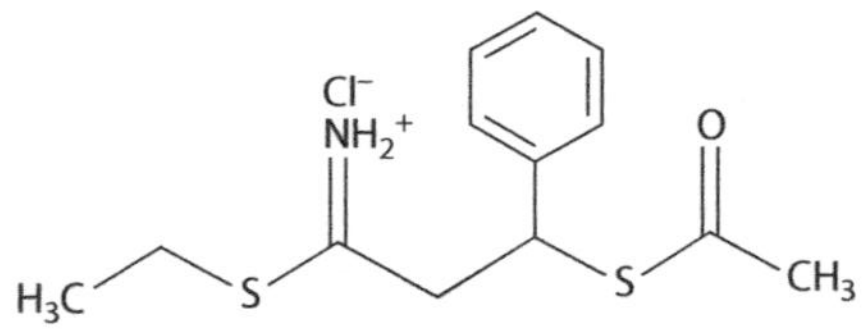

XCV

7,10,55

3-[(4-carboxamidophenyl)dithio]-propionthiomidate ester hydrochloride (CDPT)

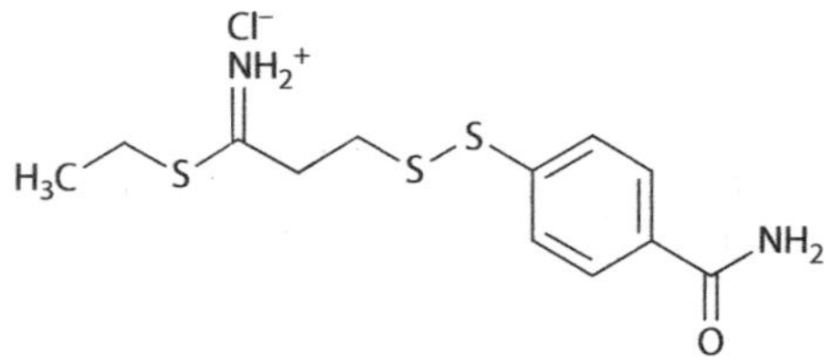

XCVI

55

Ethyl 3-[((4-carboxamidophenyl)-dithio]butyrothioimidate ester hydrochloride (M-CDPT)

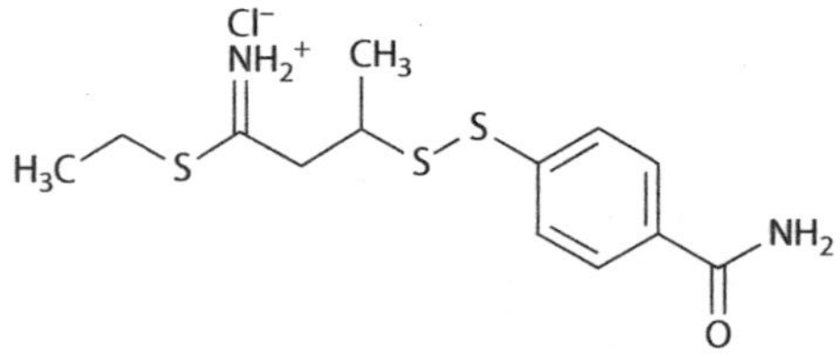

(continued)

(Continued)

Name (Abbreviation)	Structure	References
XCVII		
Ethyl 3-phenyl-3-[(4-carboxamido-phenyl)dithio] propionthioimidate ester hydrochloride (Ph-CDPT)	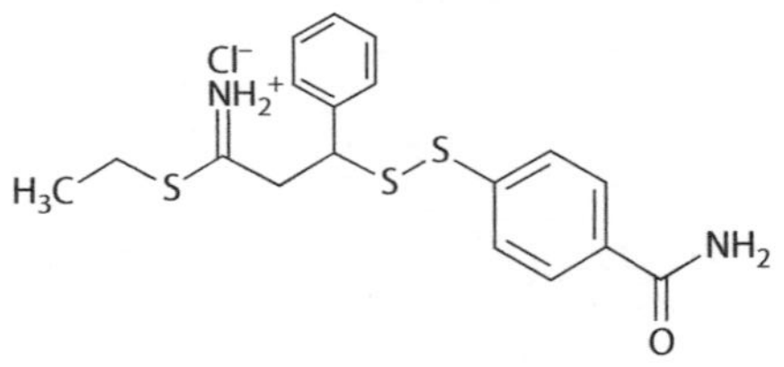	7
XCVIII		
Carboxymethyl 3-phenyl-3-[(4-carboxamidophenyl)dithio] propion-thioimidate ester hydrochloride (Ph-CDCT)	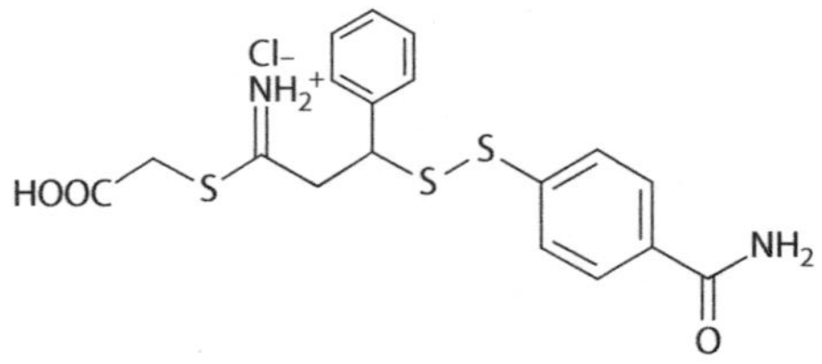	55
XCIX		
Epichorohydrin (ECH)	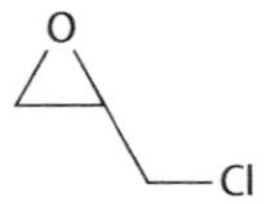	56
C		
Epibromohydrin (EBH)	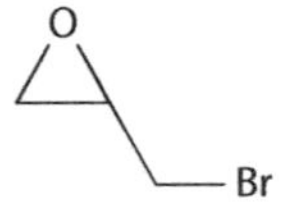	56
CI		
(2-p-Nitrophenyl)allyl-4-nitro-3-carboxyphenylsulfide (ETAC-II)	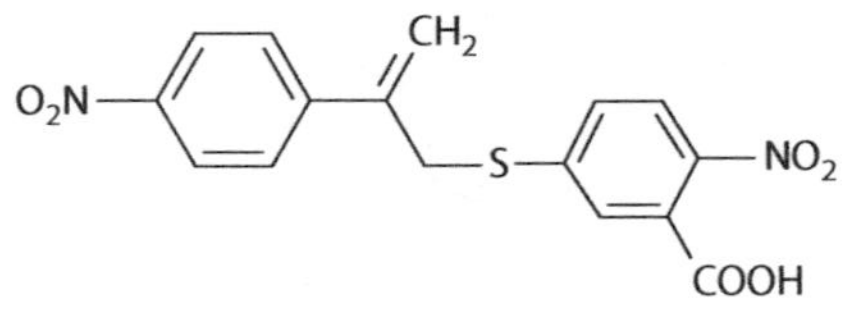	20,57

57

58

CII

2-(p-Nitrophenyl)allyl-4-trimethylammonium iodide

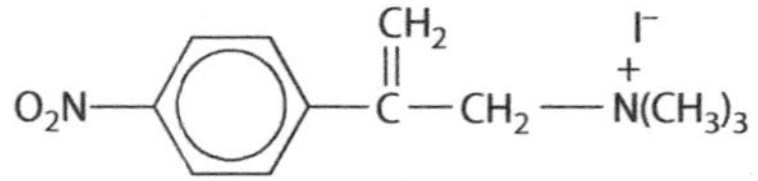

Acetophenone equilibrium transfer alkylation cross-link (ETAC) reagents

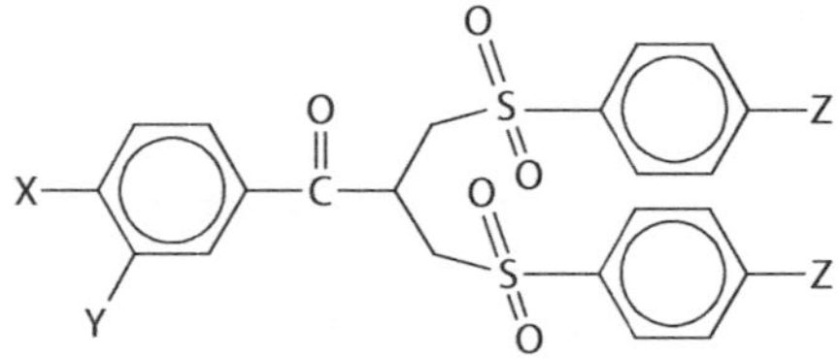

CIII. X=H, Y=H, Z=Cl: α,α-Bis[{(p-chlorophenyl)sulfonyl}methyl]acetophenone

CIV. X=Cl, Y=H, Z=Cl: α,α-Bis[{(p-chlorophenyl)sulfonyl}methyl]-p-chloroacetophenone

CV. X=NO$_2$, Y=H, Z=Cl: α,α-Bis[{(p-chlorophenyl)sulfonyl}methyl]-4-nitroacetophenone

CVI. X=NO$_2$, Y=H, Z=CH$_3$: α,α-Bis[{(p-tolylsulfonyl)methyl]-4-nitroacetophenone

CVII. X=H, Y= NO$_2$, Z=Cl: α,α-Bis[{(p-chlorophenyl)sulfonyl}methyl]-m-nitroacetophenone

CVIII. X=H, Y= NO$_2$, Z=CH$_3$: α,α-Bis{(p-tolylsulfonyl)methyl}-m-nitroacetophenone

CIX. X=COOH, Y= NO$_2$, Z=CH$_3$: 4-[2,2-Bis{(p-tolylsulfonyl)methyl}acetyl]benzoic acid

CX. X= 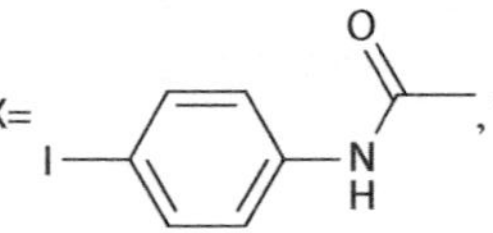 , Y=H, Z=CH$_3$; N-[4[2,2-{(p-Tolylsulfonyl)methyl}acetyl]benzoyl]-4-iodoaniline

CXI. X=NH$_2$, Y=H, Z=CH$_3$; α,α-Bis {(p-tolylsulfonyl)methyl}-p-aminoacetophenone

CXII. X=H$_3$C , Y=H, Z=CH$_3$; N-[{5-(Dimethylamino)naphthyl}sulfonyl]-α,α-bis{(p-tolylsulfonyl)methyl}-p-aminoacetophenone

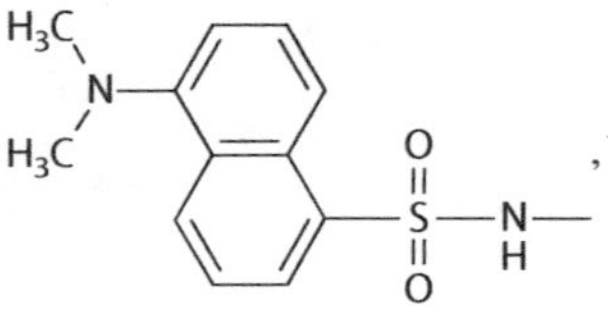

(continued)

(Continued)

Name (Abbreviation)	Structure	References

CXIII. X=HOOC , Y=H, Z=CH₃ ; N-[4-{2,2-Bis(p-tolylsulfonyl)methyl} acetyl]benzoyl-1-(p-aminobenzlyl) ethylenetriaminepentaacetic acid

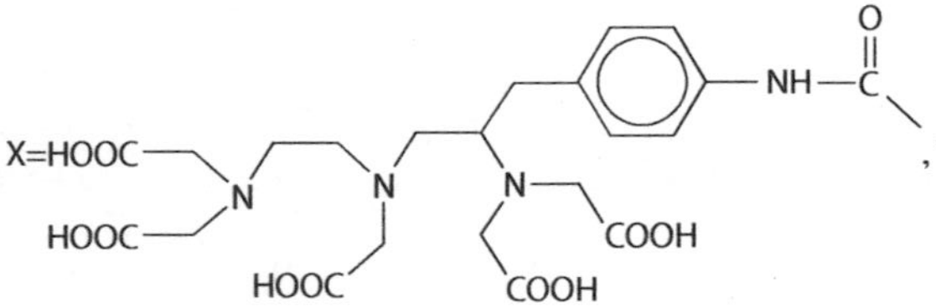

59

Aminoacetophenone equilibrium transfer alkylation cross-link (ETAC) reagents

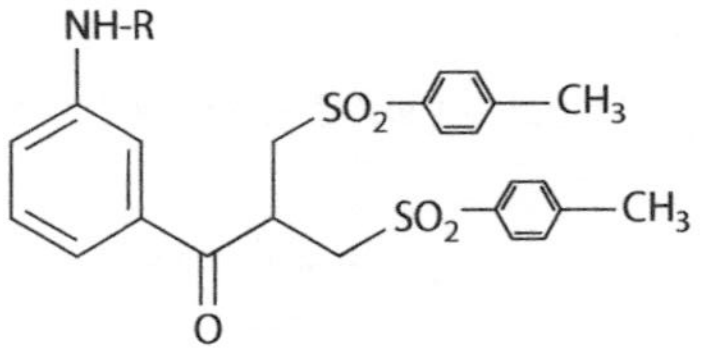

CXIV. R= H: α,α-Bis[(tolylsulfonyl)methyl]-m-aminoacetophenone

CXV. R = : Sulforhodamine B m-[α,α-bis[(ptolylsulfonyl)methyl]acetyl]anilide derivative

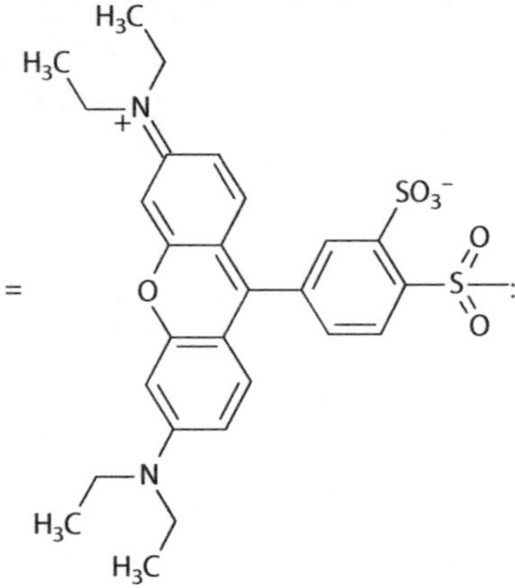

60

Iminothiolanes

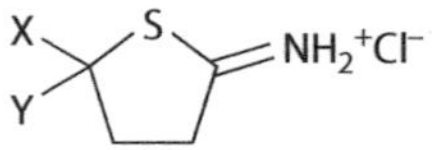

CXVI: X=Y=H: Iminothiolane HCl

CXVII. X=H, Y=H$_3$C-: 5-Methyl-2-iminothiolane HCl (M2IT.HCl)

CXVIII. X=H, Y=(CH$_3$)$_3$C-: 5-(1,1-Dimethylethyl)-2-iminothiolane HCl (TB2IT.HC)

CXIX. X=H, Y=Phenyl-: 5-Phenyl-2-iminothiolane HCl (Ph2IT.HCl)

CXX. X=Y= H$_3$C-: 5,5-Dimethyl-2-iminothiolane HCl (DM2IT.HCl)

CXXI

8-Imino-7-thiaspiro[5.4]-decane HCl (S2IT.HCl)

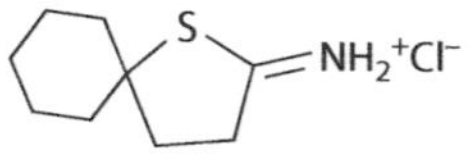

CXXII

t-8-Imino-7-thiabicyclo[4.3.0]-nonane hydrochloride (RZIT.HCl)

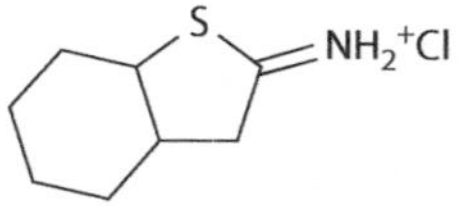

CXXIII

3-(Acetoxymercurio)-5-nitrosalicylaldehyde

CXXIV

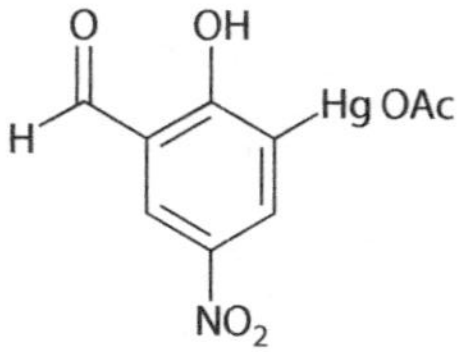

3-(Chloromercurio)-5-nitrosalicylaldehyde

B. Carboxyl- and Either Sulfhydryl- or Amino-Group–Directed Bifunctional Reagents

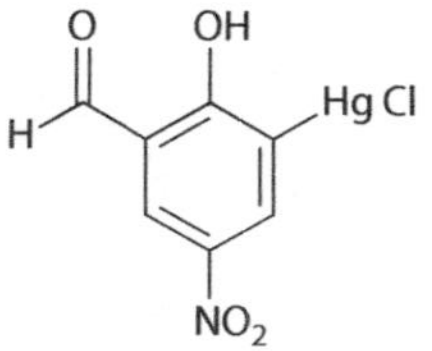

Pyridyldithiobenzyl diazoacetate

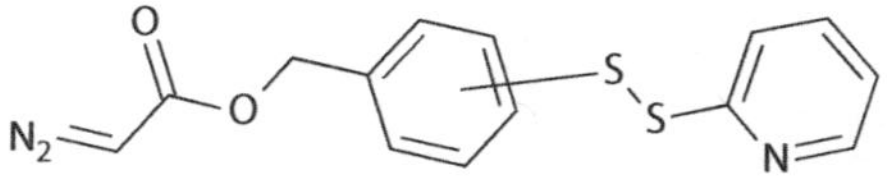

60

60

61

61

62

(continued)

(Continued)

Name (Abbreviation)	Structure	References
CXXV. *o*-(2′-Pyridyldithio)benzyldiazoacetate (OPD)		
CXXVI. *m*-(2′-Pyridyldithio)benzyldiazoacetate (MPD)		
CXXVII. *p*-(2′-Pyridyldithio)benzyldiazoacetate (PPD)		
CXXVIII		63
1-Diazoacetyl-1-bromo-2-phenylethane	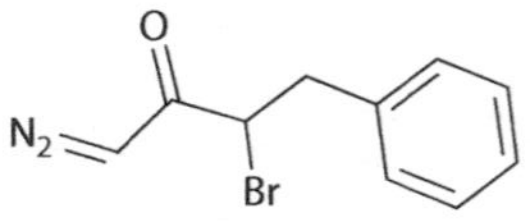	
CXXIX		64
p-Nitrophenyl diazoacetate	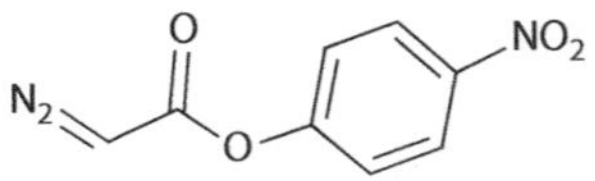	
CXXX		65
p-Nitrophenyl diazopyruvate	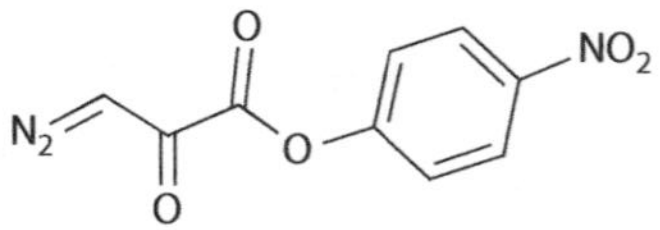	
C. Hydroxyl- and Either Sulfhydryl- or Amino-Group–Directed Bifunctional Reagents		
CXXXI		66
N-(*p*-Maleimidophenyl) isocyanates (PMPI)	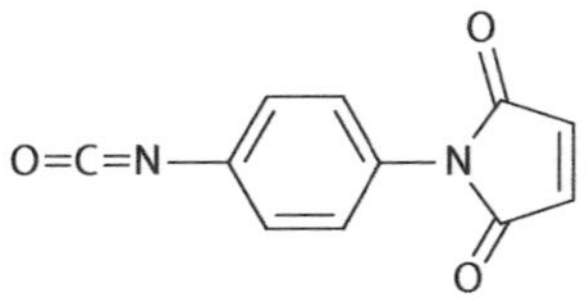	
CXXXII		67
N-(3-Triethoxysilylpropyl)-6-(*N*-maleimido)-hexanamide (TPMH)	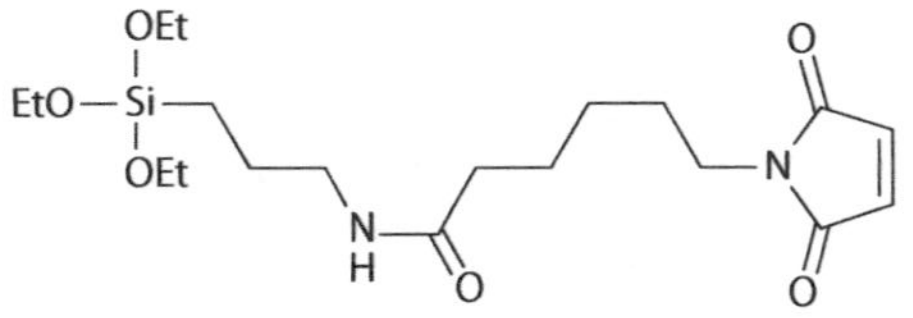	

68

69

70

71

72

73

(continued)

CXXXIII

N-(2-Trifluoroethanesulfonatoethyl)-N-(methyl)-triethoxysilylpropyl-3-amine (NTMTA)

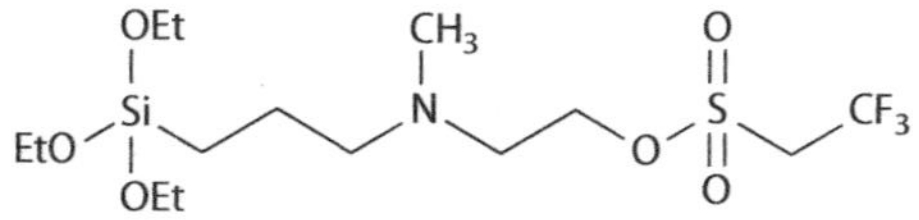

CXXXIV

3-Glycidyloxypropyl-trimethoxysilane

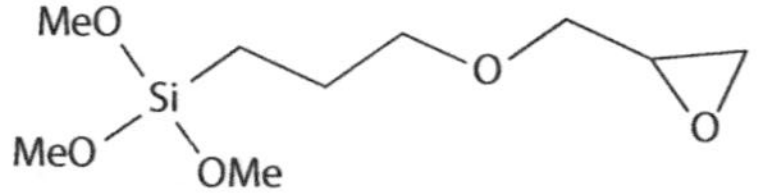

CXXXV

N-(3-Triethoxysilylpropyl)-4-(N'-maleimidylmethyl)cyclohexanamide (TPMC)

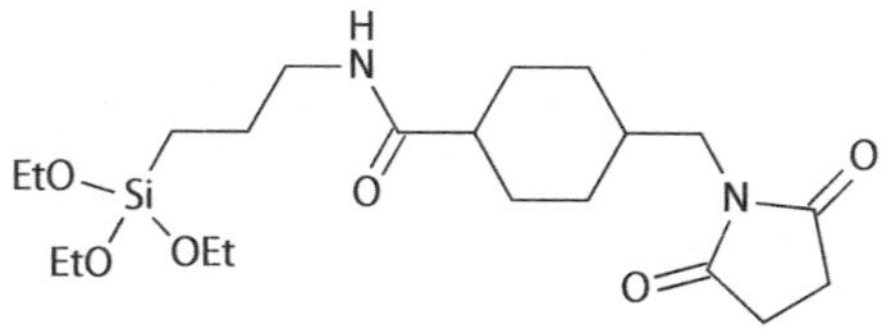

CXXXVI

N-(3-Triethoxysilylpropyl)-4-(isothiocyanatomethyl)cyclo-hexane-1-carboxamide (TPICC)

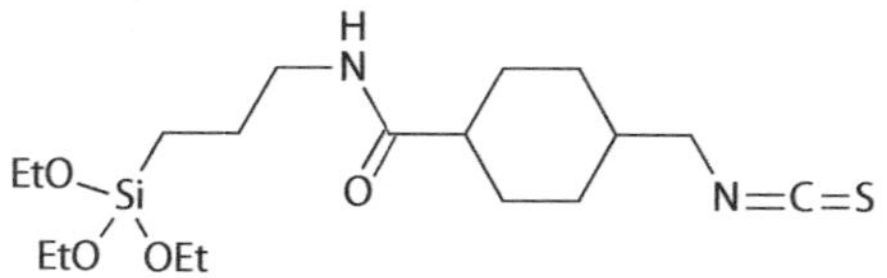

Haloacetyl halide

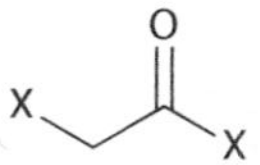

CXXXVII X = I: Iodoacetyl iodide
CXXXVIII X = Br: Bromoacetyl bromide
CXXXIX X = Cl: Chloroacetyl chloride

D. Carbonyl- and Sulfhydryl-Group–Directed Bifunctional Reagents
CXL

S-(2-Thiopyridyl)mercapto-propionohydrazide (TPMPH);
3-(2-pyridyldithio)propionyl hydrazide (PDPH)

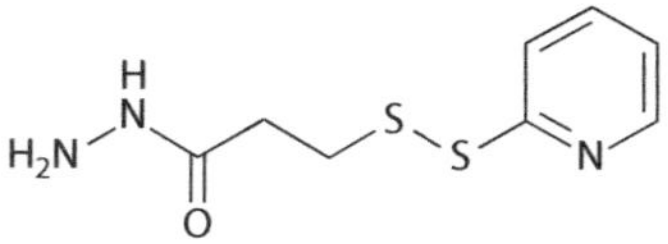

(Continued)

Name (Abbreviation)	Structure	References
CXLI		
S-(2-Thiopyridyl)-L-cysteine hydrazide (TPCH)	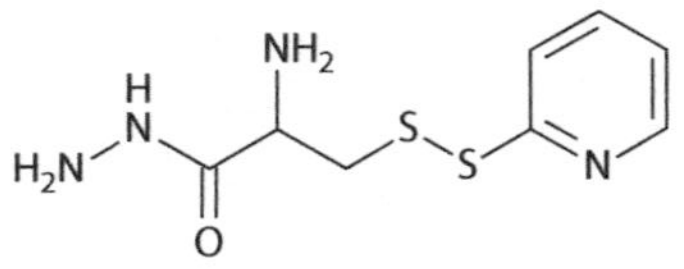	73
CXLII		
1-(Aminooxy)-4-[(3-nitro-2-pyridyl)dithio]butane	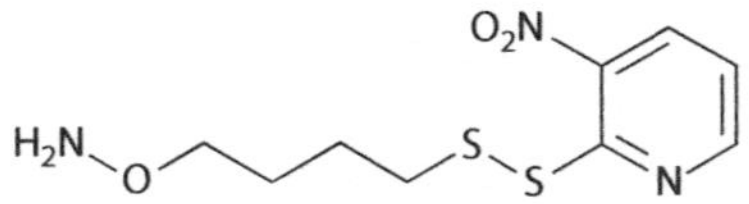	74
CXLIII		
1-(Aminooxy)-4-[(3-nitro-2-pyridyl)dithio]but-2-ene	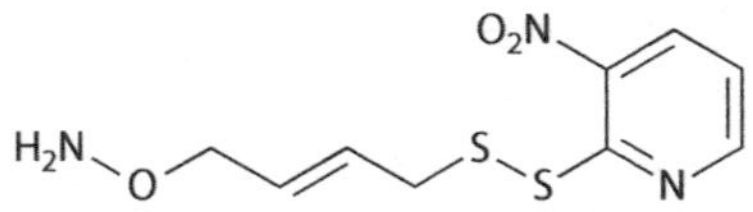	74
Maleimido alkylacid hydrazides	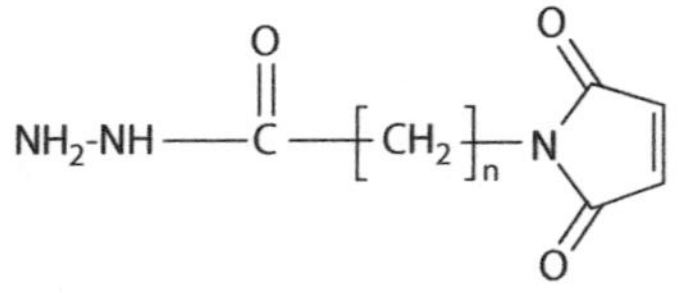	
CXLIV. n = 2: N-(β-Maleimidopropionic acid) hydrazide (BMPH); 3-(Maleimido)propionyl hydrazide (MPH)		75
CXLV. n = 5: N-(ε-Maleimidocaproic acid) hydrazide (EMCH); 6-(N-Maleimido)caproyl hydrazide		4,76
CXLVI. n = 10: N-(κ-Maleimidoundecanoic acid) hydrazide (KMUH)		77

CXLVII

78

4-(4-*N*-Maleimidophenyl) butyric acid hydrazide (MPBH)

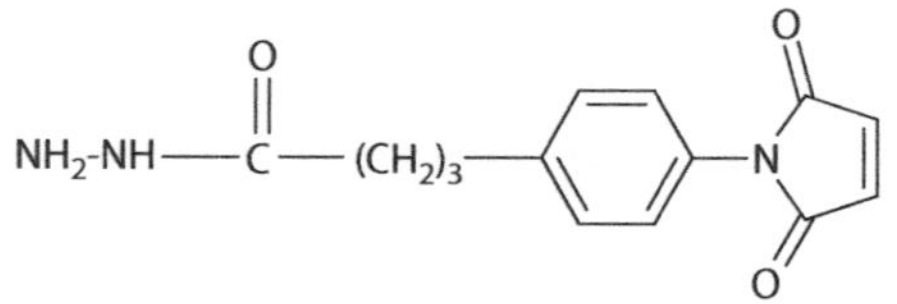

E. Carbonyl- and Amino-Group–Directed Bifunctional Reagents
CXLVIII

79

N-Succinimidyl-4-hydrazido-terephthalate (SHTH)

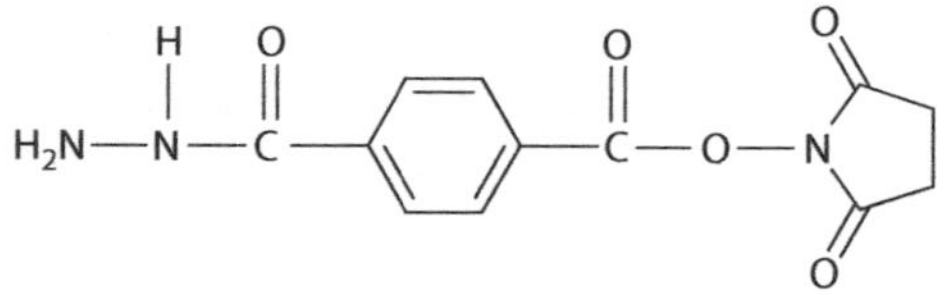

F. Miscellaneous Heterobifunctional Cross-Linkers
CXLIX

80

2-Methyl-*N*[1]-benzenesulfonyl-*N*[4]-bromoacetylquinonediimide

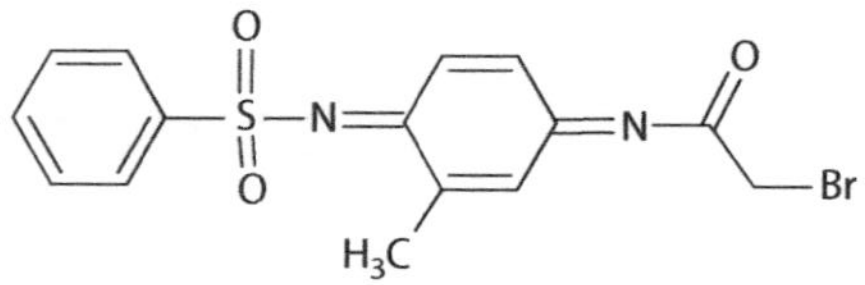

CL

81

N-Hydroxysuccinimidyl-*p*-formylbenzoate (HFB)

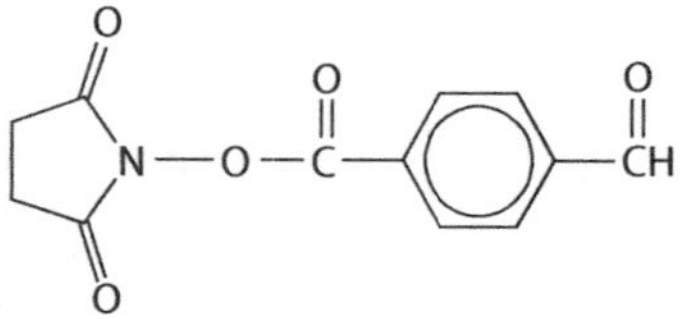

(continued)

(Continued)

Name (Abbreviation)	Structure	References
CLI Methyl-4-(6-formyl-3-azido-phenoxy)butyrimidate HCl (FAPOB)		82
CLII 1-(2-Chloroethyl)-3-cyclohexyl-1-nitrosourea (CCNU), Lomustine		83
CLIII Ethylene aldehyde; acrolein		84–86
CLIV Crotonaldehyde		84,85
CLV 4-Hydroxy-2-nonenal		86,87
CLVI 1-(4-Methoxyphenyl)-3-acetamido-4-methoxyazetidin-2-one		88

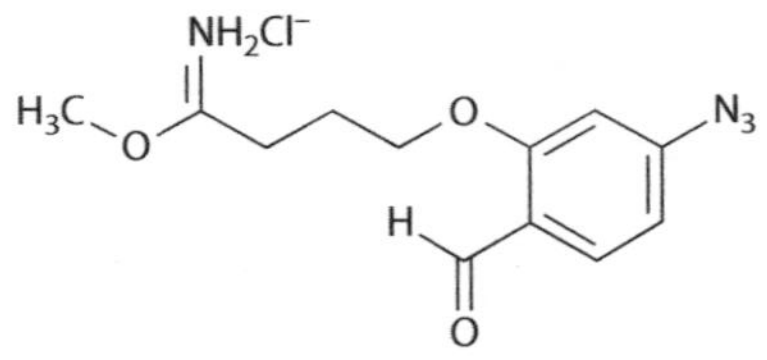

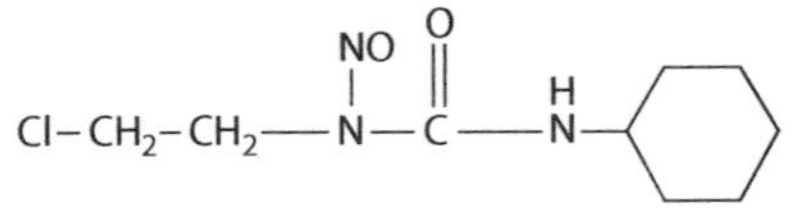

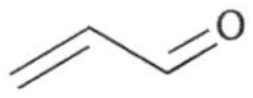

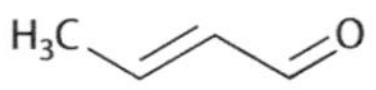

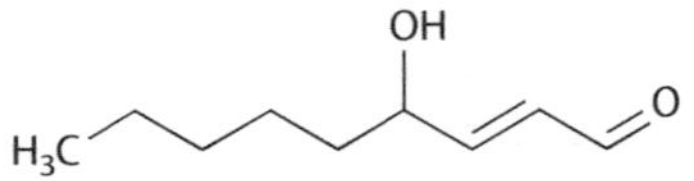

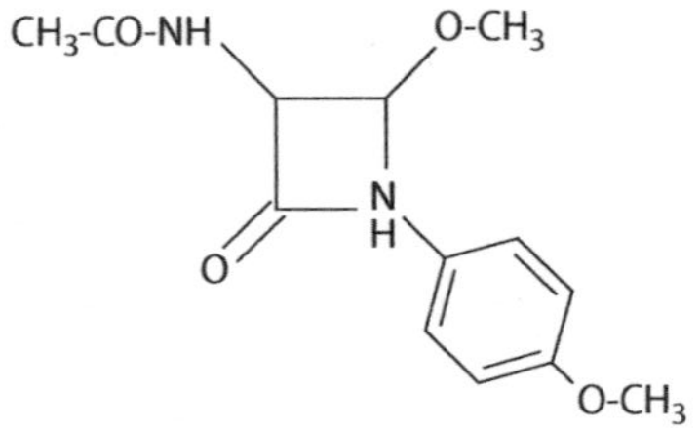

REFERENCES

1. Ji, T. H., Bifunctional reagents, *Methods Enzymol.,* 91, 580–609, 1983.
2. Carlsson, J., Drevin, H., and Axén, R., Protein thiolation and reversible protein-protein conjugation. N-Succinimidyl 3-(2-pyridyldithio)propionate, a new heterobifunctional reagent, *Biochem. J.,* 173, 723–737, 1978.
3. Carroll, S. F., Bernhard, S. L., Goff, D. A., Bauer, R. J., Leach, W., and Kung, A. H., Enhanced stability in vitro and in vivo of immunoconjugates prepared with 5-methyl-2-iminothiolane, *Bioconjug. Chem.,* 5, 248–256, 1994.
4. Riener, C. K., Kienberger, F., Hahn, C. D., Buchinger, G. M., Egwim, I. O. C., Haselgrübler, T., Ebner, A. et al., Heterobifunctional crosslinkers for tethering single ligand molecules to scanning probes, *Anal. Chim. Acta,* 497, 101, 2003.
5. Thorpe, P. E., Wallace, P. M., Knowles, P. P., Relf, M. G., Brown, A. N., Watson, G. J., Knyba, R. E., Wawrzynczak, E. J., and Blakey, D. C., New coupling agents for the synthesis of immunotoxins containing a hindered disulfide bond with improved stability in vivo, *Cancer Res.,* 47, 5924–5931, 1987.
6. Sanchez, E. F., Bush, L. R., Swenson, S., and Markland, F. S., Chimeric fibrolase: Covalent attachment of an RGD-like peptide to create a potentially more effective thrombolytic agent, *Thromb. Res.,* 87, 289–302, 1997.
7. Delprino, L., Giacomotti, M., Dosio, F., Brusa, P., Ceruti, M., Grosa, G., and Cattel, L., Toxin-targeted design for anticancer therapy. II: Preparation and biological comparison of different chemically linked gelonin–antibody conjugates, *J. Pharm. Sci.,* 82, 699–704, 1993.
8. Greenfield, L., Bloch, W., and Moreland, M., Thiol-containing cross-linking agent with enhanced steric hindrance, *Bioconjug. Chem.,* 1, 400–410, 1990.
9. Sampathkumar, S. G., Jones, M. B., and Yarema, K. J., Metabolic expression of thiol-derivatized sialic acids on the cell surface and their quantitative estimation by flow cytometry, *Nat. Protoc.,* 1, 1840–1851, 2006.
10. Dosio, F., Arpicco, S., Adobati, E., Canevari, S., Brusa, P., De Santis, R., Parente, D., Pignanelli, P., Negri, D. R,, Colnaghi, M. I., and Cattel, L., Role of cross-linking agents in determining the biochemical and pharmacokinetic properties of Mgr6-clavin immunotoxins, *Bioconjug. Chem.,* 9, 372–381, 1998.
11. Hashida, S., Imagawa, M., Inoue, S., Ruan, K. H., and Ishikawa, E., More useful maleimide compounds for the conjugation of Fab′ to horseradish peroxidase through thiol groups in the hinge, *J. Appl. Biochem.,* 6, 56–63, 1984.
12. Gittens, S. A., Kitov, P. I., Matyas, J. R., Löbenberg, R., and Uludağ, H., Impact of tether length on bone mineral affinity of protein-bisphosphonate conjugates, *Pharm. Res.,* 21, 608–616, 2004.
13. Heyse, S., Vogel, H., Sänger, M., and Sigrist, H., Covalent attachment of functionalized lipid bilayers to planar waveguides for measuring protein binding to biomimetic membranes, *Protein Sci.,* 4, 2532–2544, 1995.
14. Scholl, M., Sprössler, C., Denyer, M., Krause, M., Nakajima, K., Maelicke, A., Knoll, W., and Offenhäusser, A., Ordered networks of rat hippocampal neurons attached to silicon oxide surfaces, *J. Neurosci. Methods,* 104, 65–75, 2000.
15. Kominami, G., Agou, T., Kanda, A., and Ohno, M., Immunoenzymometric assay for recombinant methioninase in biological fluids, *J. Pharm. Biomed. Anal.,* 30, 733–738, 2002.
16. Ho, J. A. and Hsu, H. W., Procedures for preparing *Escherichia coli* O157:H7 immunoliposome and its application in liposome immunoassay, *Anal. Chem.,* 75, 4330–4334, 2003.
17. Storni, T., Ruedl, C., Schwarz, K., Schwendener, R. A., Renner, W. A., and Bachmann, M. F., Nonmethylated CG motifs packaged into virus-like particles induce protective cytotoxic T cell responses in the absence of systemic side effects, *J. Immunol.,* 172, 1777–1785, 2004.
18. Inamori, K., Kyo, M., Matsukawa, K., Inoue, Y., Sonoda, T., Tatematsu, K., Tanizawa, K., Mori, T., and Katayama, Y., Optimal surface chemistry for peptide immobilization in on-chip phosphorylation analysis, *Anal. Chem.,* 80, 643–650, 2008.
19. Hermanson, G. T., *Bioconjugate Techniques,* Academic Press, New York, 2008.
20. Han, K.-K., Richard, C., and Delacourte, A., Chemical cross-links of proteins by using bifunctional reagents, *Int. J. Biochem.,* 16, 129–252, 1984.
21. Bieniarz, C., Husain, M., Barnes, G., King, C. A., and Welch, C. J., Extended length heterobifunctional coupling agents for protein conjugations, *Bioconjug. Chem.,* 7, 88–95, 1996.
22. Goldoni, S., Owens, R. T., McQuillan, D. J., Shriver, Z., Sasisekharan, R., Birk, D. E., Campbell, S., and Iozzo, R. V., Biologically active decorin is a monomer in solution, *J. Biol. Chem.,* 279, 6606–6612, 2004.

23. Kitagawa, T., Shimozono, T., Aikawa, T., Yoshida, T., and Nishimura, H., Preparation and characterization of hetero-bifunctional crosslinking reagents for protein modifications, *Chem. Pharm. Bull.*, 29, 1130–1135, 1981.

24. Kitagawa, T. and Aikawa, T., Enzyme coupled immunoassay of insulin using a novel coupling reagent, *J. Biochem.*, 79, 233–236, 1976.

25. Myers, D. E., Uckun, F. M., Swaim, S. E., and Vallera, D. A., The effects of aromatic and aliphatic maleimide crosslinkers on anti-CD5 ricin immunotoxins, *J. Immunol. Methods*, 121, 129–142, 1989.

26. Srinivasachar, K. and Neville, D. M., Jr., New protein cross-linking reagents that are cleaved by mild acid, *Biochemistry*, 28, 2501–2509, 1989.

27. Hultin, T., A class of cleavable heterobifunctional reagents for thiol-directed high-efficiency protein crosslinking: Synthesis and application to the analysis of protein contact sites in mammalian ribosomes, *Anal. Biochem.*, 155, 262–269, 1986.

28. Rector, E. S., Schwenk, R. J., Tse, K. S., and Sehon, A. H., A method for the preparation of protein-protein conjugates of predetermined composition, *J. Immunol. Methods*, 24, 321–336, 1978.

29. Brinkley, M., A brief survey of methods for preparing protein conjugates with dyes, haptens, and crosslinking reagents, *Bioconjug. Chem.*, 3, 2–13, 1992.

30. Goddard, J. M. and Hotchkiss, J. H., Polymer surface modification for the attachment of bioactive compounds, *Prog. Polym. Sci.*, 32, 698–725, 2007.

31. Weng, G., Li, J., Dingus, J., Hildebrandt, J. D., Weinstein, H., and Iyengar, R., Gbeta subunit interacts with a peptide encoding region 956–982 of adenylyl cyclase 2. Cross-linking of the peptide to free Gbetagamma but not the heterotrimer, *J. Biol. Chem.*, 271, 26445–26448, 1996.

32. Hess, R., Rau, P., Schwab, M., Paetzold, S., Kuther, M., Obert, M., Agostini, H., Haessler, C., Braun, D. G., and Brandner, G., Covalent immunochemical membrane labeling of viable cells with K698–T708, a simian virus 40 tumor antigen-derived peptide, *Pept. Res.*, 7, 146–152, 1994.

33. Cuatrecasas, P., Wilchek, M., and Anfinsen, C. B., Affinity labeling of the active site of staphylococcal nuclease. Reactions with bromoacetylated substrate analogues, *J. Biol. Chem.*, 244, 4316–4329, 1969.

34. Inman, J. K., Highet, P. F., Kolodny, N., and Robey, F. A., Synthesis of N alpha-(tert-butoxycarbonyl)-N epsilon-[N-(bromoacetyl)-beta-alanyl]-L-lysine: Its use in peptide synthesis for placing a bromoacetyl cross-linking function at any desired sequence position, *Bioconjug. Chem.*, 2, 458–463, 1991.

35. Fasold, H., Baumert, H., and Fink, G., Comparison of hydrophobic and strongly hydrophilic cleavable crosslinking reagents in intermolecular bond formation in aggregates of proteins or protein-RNA, in *Protein Crosslinking: Biochemical and Molecular Aspects*, Plenum Press, New York, 1976, p. 207.

36. Fink, G., Fasold, H., Rommel, W., and Brimacombe, R., Reagents suitable for the crosslinking of nucleic acids to proteins, *Anal. Biochem.*, 108, 394–401, 1980.

37. McKenzie, J. A., Raison, R. L., and Rivett, D. E., Development of a bifunctional crosslinking agent with potential for the preparation of immunotoxins, *J. Protein Chem.*, 7, 581–592, 1988.

38. Hiratsuka, T., Nucleotide-induced change of the interaction between the 20- and 26-kilodalton heavy-chain segments of myosin adenosinetriphosphatase revealed by chemical cross-linking via the reactive thiol SH2, *Biochemistry*, 26, 3168–3173, 1987.

39. Fujii, N., Hayashi, Y., Katakura, S., Akaji, K., Yajima, H., Inouye, A., and Segawa, T., Studies on peptides. CXXVIII. Application of new heterobifunctional crosslinking reagents for the preparation of neurokinin (A and B)-BSA (bovine serum albumin) conjugates, *Int. J. Pept. Protein Res.*, 26, 121–129, 1985.

40. Kida, S., Maeda, M., Hojo, K., Eto, Y., Nakagawa, S., and Kawasaki, K., Studies on heterobifunctional cross-linking reagents, 6-maleimidohexanoic acid active esters, *Chem. Pharm. Bull. (Tokyo)*, 55, 685–687, 2007.

41. Aldwin, L. and Nitecki, D. E., A water-soluble, monitorable peptide and protein crosslinking agent, *Anal. Biochem.*, 164, 494–501, 1987.

42. Bäumert, H. G. and Fasold, H., Cross-linking techniques, *Methods Enzymol.*, 172, 584–609, 1989.

43. Hermentin, P., Doenges, R., Gronski, P., Bosslet, K., Kraemer, H. P., Hoffmann, D., Zilg, H. et al., Attachment of rhodosaminylanthracyclinone-type anthracyclines to the hinge region of monoclonal antibodies, *Bioconjug. Chem.*, 1, 100–107, 1990.

44. Trommer, W. E., Kolkenbrock, H., and Pfleiderer, G., Synthesis and properties of a new selective bifunctional cross-linking reagent, *Hoppe Seylers Z. Physiol. Chem.*, 356, 1455–1458, 1975.

45. Trommer, W. E., Friebel, K., Kiltz, H.-H., and Kolkenbrock, H.-J., Synthesis and application of new bifunctional reagents, in *Protein Crosslinking: Biochemical and Molecular Aspects*, Plenum Press, New York, 1976, Chapter 10.

46. Arndt, D. J., Simon, S. R., Maita, T., and Konigsberg, W., The characterization of chemically modified hemoglobins. 3. Reaction with various N-substituted maleimides, *J. Biol. Chem.*, 246, 2602–2608, 1971.

47. Walter, A., Blattler, W. A., Kuenzi, B. S., Lambert, J. M., and Senter, P. D., New heterobifunctional protein crosslinking reagent that forms an acid-labile link, *Biochemistry*, 24, 1517, 1985.

48. Goldmacher, V. S., Senter, P. D., Lambert, J. M., and Blättler, W. A., Photoactivation of toxin conjugates, *Bioconjug. Chem.*, 3, 104–107, 1992.

49. Senter, P. D., Tansey, M. J., Lambert, J. M., and Blattler, W. A. Novel photocleavable protein crosslinking reagents and their use in the preparation of antibody–toxin conjugates, *Photochem. Photobiol.*, 42, 231, 1985.

50. McIntyre, G. D., Scott, C. F., Jr., Ritz, J., Blättler, W. A., and Lambert, J. M., Preparation and characterization of interleukin-2-gelonin conjugates made using different cross-linking reagents, *Bioconjug. Chem.*, 5, 88–97, 1994.

51. Peters, K. and Richards, F. M., Chemical cross-linking: Reagents and problems in studies of membrane structure, *Annu. Rev. Biochem.*, 46, 523–551, 1977.

52. King, T. P., Li, Y., and Kochoumian, L., Preparation of protein conjugates via intermolecular disulfide bond formation, *Biochemistry*, 17, 1499–1506, 1978.

53. Olomucki, M. and Diopoh, J., New protein reagents. I. Ethyl chloroacetimidate, its properties and its reaction with ribonuclease, *Biochim. Biophys. Acta*, 263, 213–219, 1972.

54. Diopoh, J. and Olomucki, M., Ethyl bromoacetimidate, a NH2-specific heterobifunctional reagent. Model reactions with ribonuclease, *Hoppe Seylers Z. Physiol. Chem.*, 360, 1257–1262, 1979.

55. Arpicco, S., Dosio, F., Brusa, P., Crosasso, P., and Cattel, L., New coupling reagents for the preparation of disulfide cross-linked conjugates with increased stability, *Bioconjug. Chem.*, 8, 327–337, 1997.

56. Romano, K. P., Newman, A. G., Zahran, R. W., and Millard, J. T., DNA interstrand cross-linking by epichlorohydrin, *Chem. Res. Toxicol.*, 20, 832–838, 2007.

57. Mitra, S. and Lawton, R. G., Reagents for the crosslinking of proteins by equilibrium transfer alkylation, *J. Am. Chem. Soc.*, 101, 3097–3110, 1979.

58. Liberatore, F. A., Comeau, R. D., McKearin, J. M., Pearson, D. A., Belonga, B. Q., III, Brocchini, S. J., Kath, J., Phillips, T., Oswell, K., and Lawton, R. G., Site-directed chemical modification and cross-linking of a monoclonal antibody using equilibrium transfer alkylating cross-link reagents, *Bioconjug. Chem.*, 1, 36–50, 1990.

59. del Rosario, R. B., Wahl, R. L., Brocchini, S. J., Lawton, R. G., and Smith, R. H., Sulfhydryl site-specific cross-linking and labeling of monoclonal antibodies by a fluorescent equilibrium transfer alkylation cross-link reagent, *Bioconjug. Chem.*, 1, 51–59, 1990.

60. Goff, D. A. and Carroll, S. F., Substituted 2-iminothiolanes: Reagents for the preparation of disulfide cross-linked conjugates with increased stability. *Bioconjug. Chem.*, 1, 381–386, 1990.

61. Wohlfeil, E. R. and Hudson, R. A., Synthesis and characterization of a heterobifunctional mercurial cross-linking agent: Incorporation into cobratoxin and interaction with the nicotinic acetylcholine receptor, *Biochemistry*, 30, 7231–7241, 1991.

62. Paquatte, O., Fried, A., and Tu, S. C., Delineation of bacterial luciferase aldehyde site by bifunctional labeling reagents, *Arch. Biochem. Biophys.*, 264, 392–399, 1988.

63. Husain, S. S., Ferguson, J. B., and Fruton, J. S., Bifunctional inhibitors of pepsin, *Proc. Natl. Acad. Sci. U. S. A.*, 68, 2765–2768, 1971.

64. Shafer, J., Baronowsky, P., Laursen, R., Finn, F., and Westheimer, F. H., Products from the photolysis of diazoacetyl chymotrypsin, *J. Biol. Chem.*, 241, 421–427, 1966.

65. Harrison, J. K., Lawton, R. G., and Gnegy, M. E., Development of a novel photoreactive calmodulin derivative: Cross-linking of purified adenylate cyclase from bovine brain, *Biochemistry*, 28, 6023–6027, 1989.

66. Annunziato, M. E., Patel, U. S., Ranade, M., and Palumbo, P. S., p-Maleimidophenyl isocyanate: A novel heterobifunctional linker for hydroxyl to thiol coupling, *Bioconjug. Chem.*, 4, 212–218, 1993.

67. Choithani, J., Kumar, P., and Gupta, K. C., *N*-(3-Triethoxysilylpropyl)-6-(*N*-maleimido)-hexanamide: An efficient heterobifunctional reagent for the construction of oligonucleotide microarrays, *Anal. Biochem.*, 357, 240–248, 2006.

68. Kumar, P., Choithani, J., and Gupta, K. C., Construction of oligonucleotide arrays on a glass surface using a heterobifunctional reagent, *N*-(2-trifluoroethanesulfonatoethyl)-*N*-(methyl)-triethoxysilylpropyl-3-amine (NTMTA), *Nucleic Acids Res.*, 32, e80, 2004.

69. Jung, D., Streb, C., and Hartmann, M., Covalent anchoring of chloroperoxidase and glucose oxidase on the mesoporous molecular sieve SBA-15, *Int. J. Mol. Sci.*, 11, 762, 2010.

70. Misra, A. and Dwivedi, P., Immobilization of oligonucleotides on glass surface using an efficient heterobifunctional reagent through maleimide-thiol combination chemistry, *Anal. Biochem.*, 369, 248–255, 2007.

71. Misra, A., Shahid, M., and Dwivedi, P., N-(3-triethoxysilylpropyl)-4-(isothiocyanatomethyl)-cyclohexane-1-carboxamide (TPICC): A heterobifunctional reagent for immobilization of biomolecules on glass surface, *Bioorg. Med. Chem. Lett.*, 18, 5217–5221, 2008.

72. Sato, T., Mori, T., Tosa, T., and Chibata, I., Studies on immobilized enzymes. IX. Preparation and properties of aminoacylase covalently attached to halogenoacetylcelluloses, *Arch. Biochem. Biophys.*, 147, 788–796, 1971.

73. Zara, J. J., Wood, R. D., Boon, P., Kim, C. H., Pomato, N., Bredehorst, R., and Vogel, C. W., A carbohydrate-directed heterobifunctional cross-linking reagent for the synthesis of immunoconjugates, *Anal. Biochem.*, 194, 156–162, 1991.

74. Webb, R. R., II, and Kaneko, T., Synthesis of 1-(aminooxy)-4-[(3-nitro-2-pyridyl)dithio]butane and 1-(aminooxy)-4-[(3-nitro-2-pyridyl)dithio]but-2-ene, novel heterobifunctional cross-linking reagents, *Bioconjug. Chem.*, 1, 96–99, 1990.

75. Hölzl, M., Tinazli, A., Leitner, C., Hahn, C. D., Lackner, B., Tampé, R., and Gruber, H. J., Protein-resistant self-assembled monolayers on gold with latent aldehyde functions, *Langmuir*, 23, 5571–5577, 2007.

76. Trail, P. A., Willner, D., Lasch, S. J., Henderson, A. J., Hofstead, S., Casazza, A. M., Firestone, R. A., Hellström, I., and Hellström, K. E., Cure of xenografted human carcinomas by BR96-doxorubicin immunoconjugates, *Science*, 261, 212–215, 1993

77. Wu, G., Barth, R. F., Yang, W., Chatterjee, M., Tjarks, W., Ciesielski, M. J., and Fenstermaker, R. A., Site-specific conjugation of boron-containing dendrimers to anti-EGF receptor monoclonal antibody cetuximab (IMC-C225) and its evaluation as a potential delivery agent for neutron capture therapy, *Bioconjug. Chem.*, 15, 185–194, 2004.

78. Chamow, S. M., Kogan, T. P., Peers, D. H., Hastings, R. C., Byrn, R. A., and Ashkenazi, A., Conjugation of soluble CD4 without loss of biological activity via a novel carbohydrate-directed cross-linking reagent, *J. Biol. Chem.*, 267, 15916–15922, 1992.

79. Carroll-Portillo, A., Bachand, M., and Bachand, G. D., Directed attachment of antibodies to kinesin-powered molecular shuttles, *Biotechnol. Bioeng.*, 104, 1182–1188, 2009.

80. Liberatore, F. A., Comeau, R. D., and Lawton, R. G., Heterobifunctional cross-linking of a monoclonal antibody with 2-methyl-N1-benzenesulfonyl-N4-bromoacetylquinonediimide, *Biochem. Biophys. Res. Commun.*, 158, 640–645, 1989.

81. Kraehenbuhl, J. P., Galardy, R. E., and Jamieson, J. D., Preparation and characterization of an immuno-electron microscope tracer consisting of a heme-octapeptide coupled to fab, *J. Exp. Med.*, 139, 208–223, 1974.

82. Maassen, J. A., Cross-linking of ribosomal proteins by 4-(6-formyl-3–3-azidophenoxy)butyrimidate, a heterobifunctional, cleavable cross-linker, *Biochemistry*, 18, 1288–1292, 1979.

83. Schaefer, E. L., Morimoto, R. I., Theodorakis, N. G., and Seidenfeld, J., Chemical specificity for induction of stress response genes by DNA-damaging drugs in human adenocarcinoma cells, *Carcinogenesis*, 9, 1733–1738, 1988.

84. Kuykendall, J. R. and Bogdanffy, M. S., Efficiency of DNA-histone crosslinking induced by saturated and unsaturated aldehydes in vitro, *Mutat. Res.*, 283, 131–136, 1992.

85. Kuchenmeister, F., Schmezer, P., and Engelhardt, G., Genotoxic bifunctional aldehydes produce specific images in the comet assay, *Mutat. Res.*, 419, 69–78, 1998.

86. LoPachin, R. M., Gavin, T., Petersen, D. R., and Barber, D. S., Molecular mechanisms of 4-hydroxy-2-nonenal and acrolein toxicity: Nucleophilic targets and adduct formation, *Chem. Res. Toxicol.*, 22, 1499–1508, 2009.

87. Huang, H., Kozekov, I. D., Kozekova, A., Wang, H., Lloyd, R. S., Rizzo, C. J., and Stone, M. P., DNA cross-link induced by trans-4-hydroxynonenal, *Environ. Mol. Mutagen.*, 51, 625–634, 2010.

88. Ahluwalia, R., Day, R. A., and Nauss, J., A bifunctional monocyclic beta-lactam cross-links across the active site of beta-lactamase, *Biochem. Biophys. Res. Commun.*, 206, 577–583, 1995.

Appendix E: Photoactivatable Heterobifunctional Cross-Linking Reagents

Name (Abbreviation)	Structure	References

A. Amino Group–Anchored Photosensitive Reagents

I

N-Hydroxysuccinimidyl-4-azidobenzoate (NHS-ABA);
N-succinimidyl-4-azidobenzoate (HSAB)

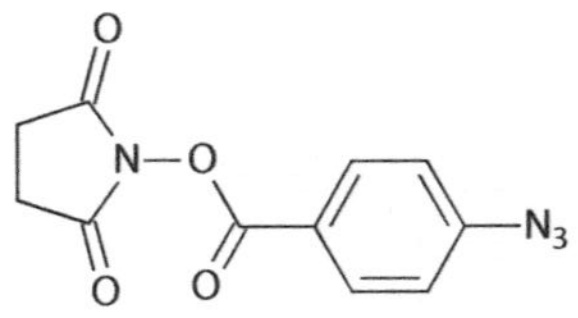

1–3

II

N-Sulfosuccinimidyl-4-azido-benzoate; N-hydroxysulfo-succinimidyl-4-azidobenzoate (Sulfo-HSAB)

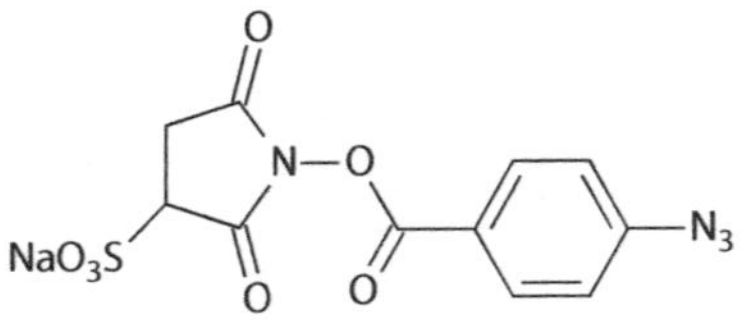

3

III

N-Hydroxysuccinimidyl-4-azidosalicylic acid; N-succinimidyl-4-azido-salicylate (NHS-ASA)

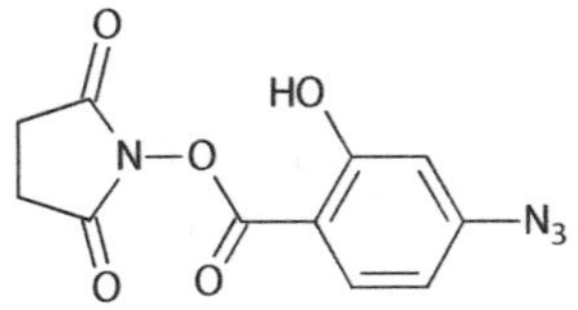

4

IV

N-Succinimidyl-N'-(4-azido-salicyl)-6-aminocaproate (NHS-ASC)

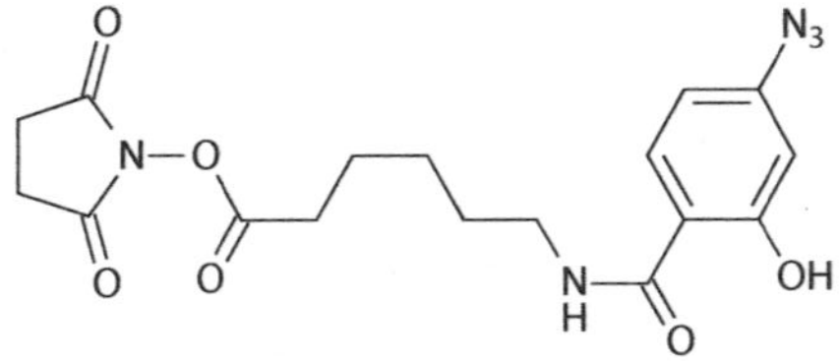

4

4

5

5

6

(continued)

V

N-Sulfosuccinimidyl-4-azidosalicylamidohexanoate (Sulfo-NHS-LC-ASA)

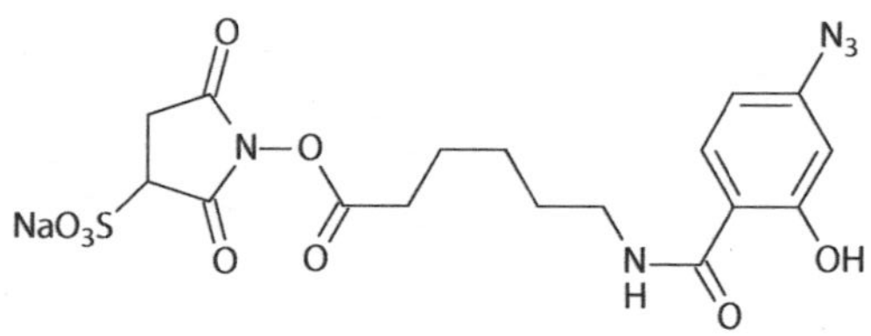

VI

1-[*N*-(2-Hydroxy-5-azidobenzoyl)-2-aminoethyl]-4-(*N*-hydroxy-succinimidyl)succinate (HAHS)

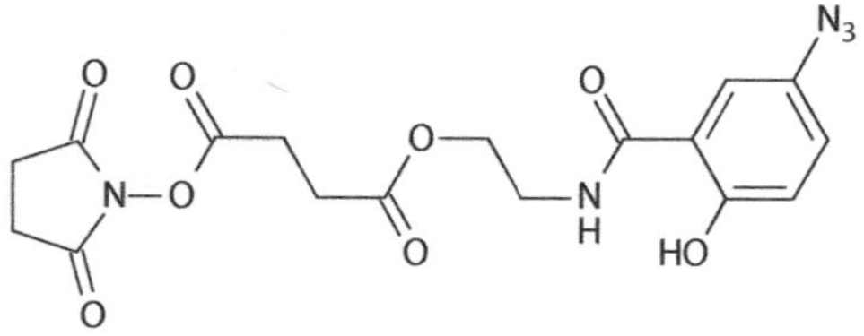

VII

1-[*N*-(2-Hydroxy-3-iodo-5-azidobenzoyl)-2-aminoethyl]-4-(*N*-hydroxy-succinimidyl)succinate (I-HAHS)

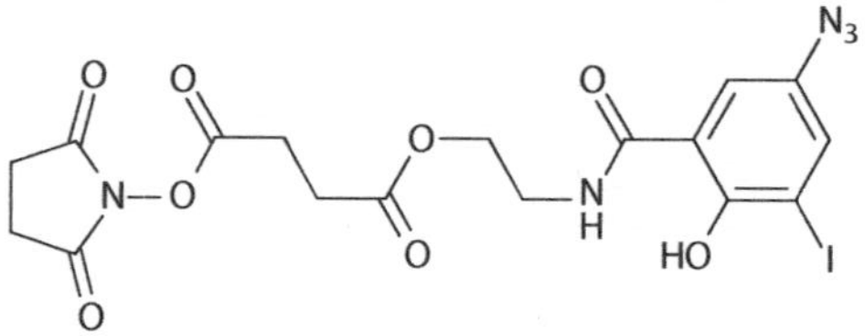

VIII

N-Succinimidyl-5-azido-2-nitrobenzoate (NHS-ANBA); *N*-5-azido-2-nitrobenzoyloxy-succinimide (ANB-NOS)

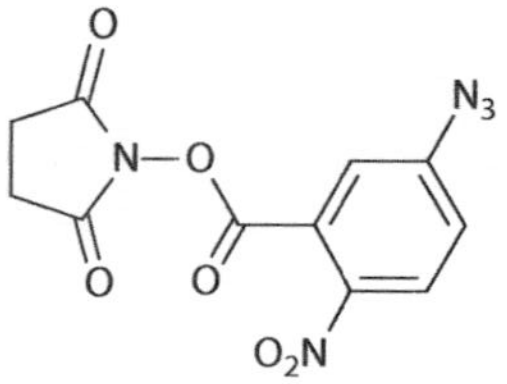

(Continued)

Name (Abbreviation)	Structure	References
IX		7
N-Succinimidyl-4-azidobenzoyl-glycinate (NHS-ABG)	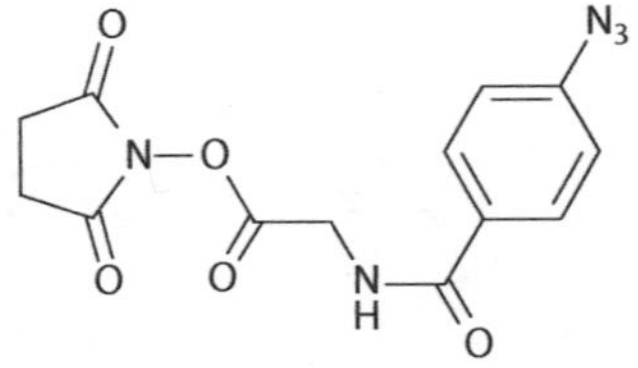	
X		7
N-Succinimidyl-4-azidobenzoyl-glycylglycinate (NHS-ABGG)	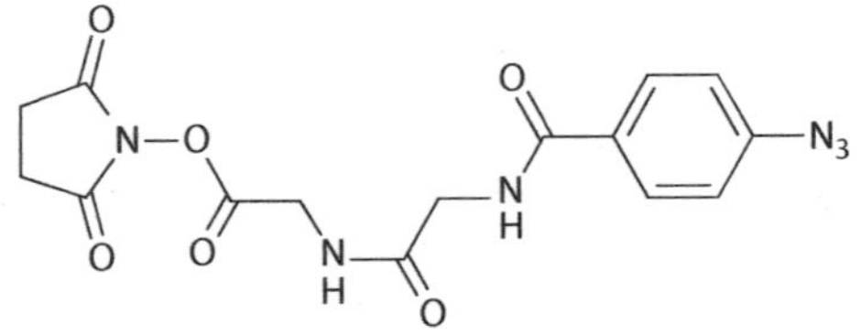	
XI		6
N-Succinimidyl-4-azidobenzoyl-glycyltyrosinate (NHS-ABGT)	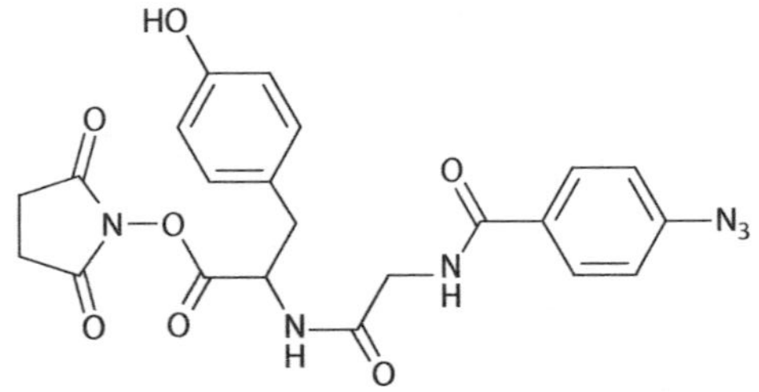	
XII		8
N-Succinimidyl-(4-azido-2-nitrophenyl)glycinate	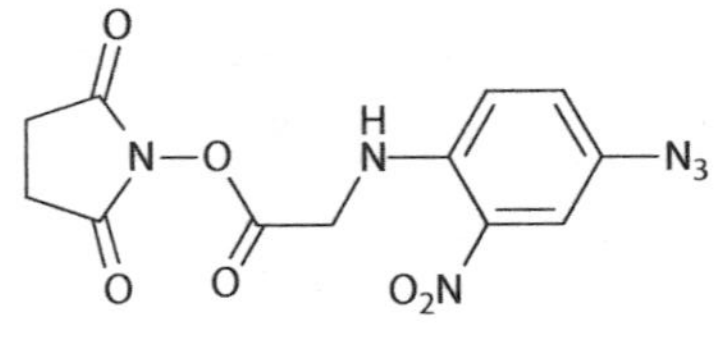	

XIII

N-Succinimidyl-(4-azido-2-nitrophenyl)-γ-aminobutyrate (NHS-ANAB)

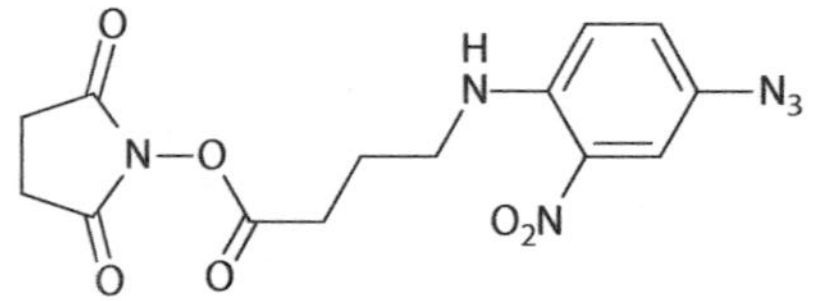

9

XIV

N-Succinimidyl-6-(4′-azido-2′-nitro-phenylamino)hexanoate (SANPAH, SANAH, Loman's reagent II)

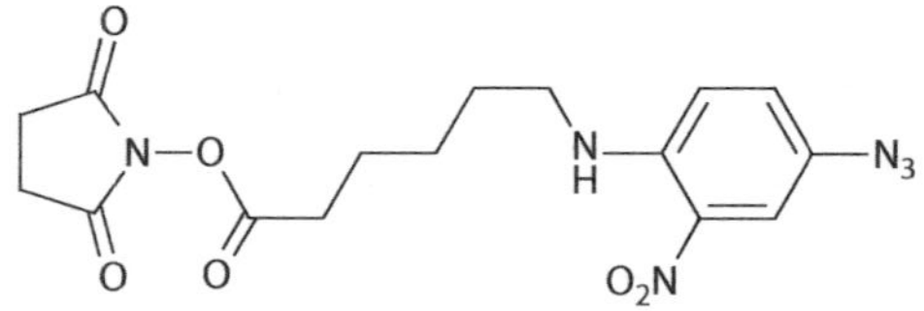

1,2

XV

N-Sulfosuccinimidyl-6-(4′-azido-2′-nitrophenylamino)hexanoate (Sulfo-SANPAH)

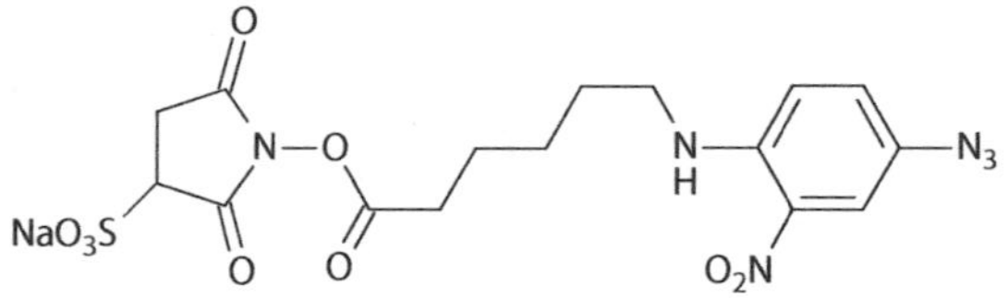

10

XVI

N-Sulfosuccinimidyl-*N*′-(4-azido-nitrophenyl)dodecanoate

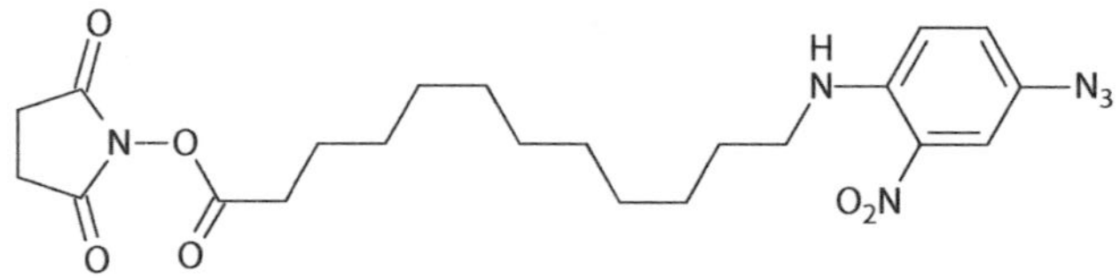

11

XVII

N-Succinimidyl 4-azido-2,3,5,6-tetrafluorobenzoate

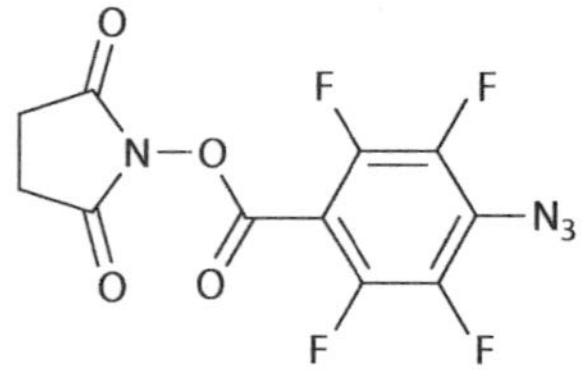

12

(continued)

(Continued)

Name (Abbreviation)	Structure	References
XVIII N-(4-Azido-2,3,5,6-tetrafluoro-benzoyl)glycine N-succinimidyl ester		12
XIX N-Succinimidyl 5-(4-Azido-2,3,5,6-tetrafluorobenzamido)pentanoate		12
XX N-Sulfosuccinimidyl(perfluoroazido-benzamide)ethyl-1,3′-dithiopropionate (Sulfo-SFAD)		13
XXI N-Sulfosuccinimidyl(N-methylamino-perfluoroazidobenzamido)-ethyl-1,3′-dithiopropionate		13

XXII

N-Succinimidyl-2-[(4-azidophenyl)-dithio]acetate (NHS-APDA)   14

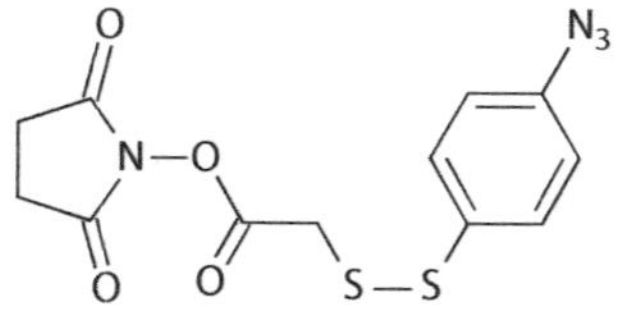

XXIII

N-Succinimidyl-3-[(4-azidophenyl)-dithio]propionate (SADP);
N-succinimidyl-(4-azidophenyl)-1,3'-dithiopropionate (NHS-APDP)   1,14

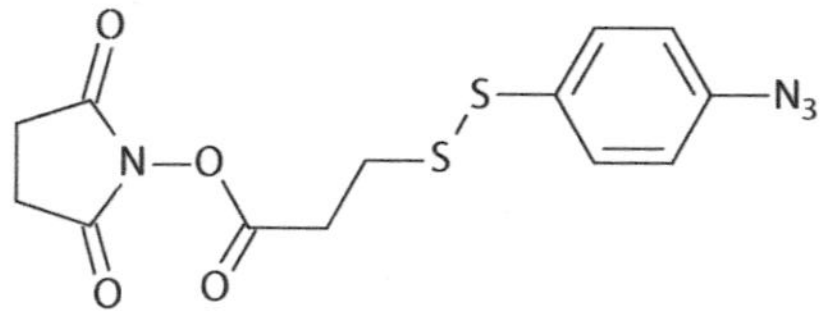

XXIV

N-Sulfosuccinimidyl-3-[(4-azidophenyl)-dithio]propionate
(Sulfo-SADP)   15

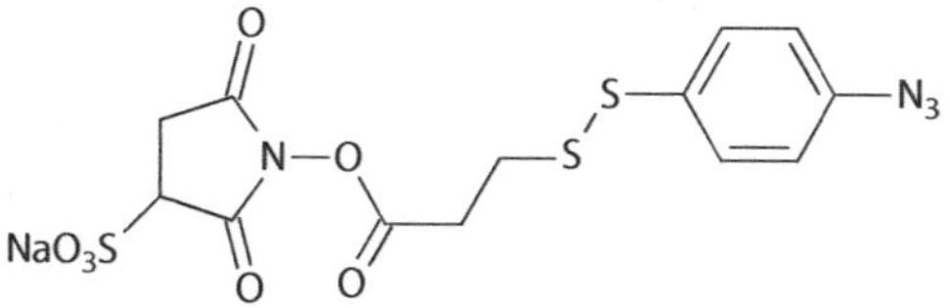

XXV

3-[(2-Nitro-4-azidophenyl)-2-aminoethyl-dithio]-*N*-
succinimidylpropionate (NAP-AEDSP); *N*-succinimidyl-3-[(2-nitro-
4-azidophenyl)-2-aminoethyldithio]propionate (SNAP)   1,16

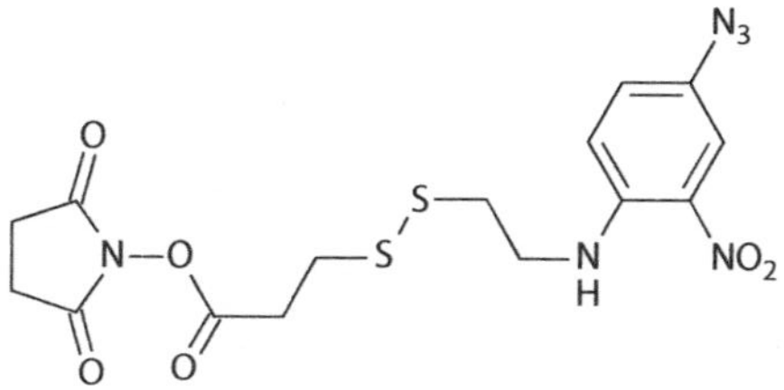

XXVI

N-Sulfosuccinimidyl-2-(*p*-azido-*o*-nitro-benzamido)ethyl-1,3'-dithio-
propionate (Sulfo-SAND)   6

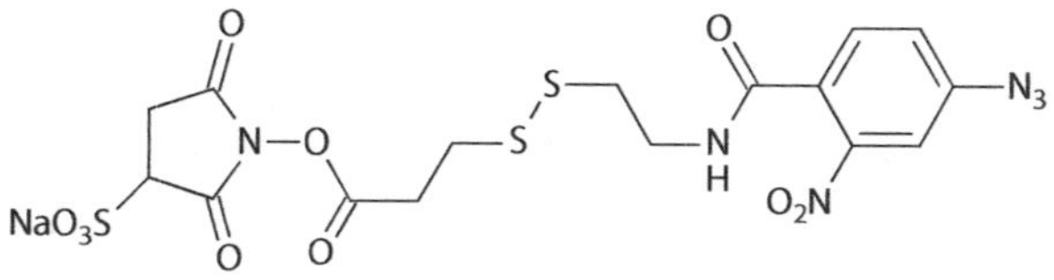

(continued)

(Continued)

Name (Abbreviation)	Structure	References
XXVII N-Sulfosuccinimidyl-2-(p-azido-salicylamido)ethyl-1,3′-dithiopropionate (SASD)		4,13
XXVIII N-Sulfosuccinimidyl-2-(7-azido-4-methylcoumarin-3-acetamido)ethyl-1,3′-dithiopropionate (Sulfo-SAED)		17
XXIX N-[4-(p-Azidophenylazo)benzoyl]-3-aminopropyl-N′-oxysuccinimide ester		18
XXX N-[4-(p-Azido-m-iodophenylazo)benzoyl]-3-aminopropyl-N′-oxysuccinimide ester		19

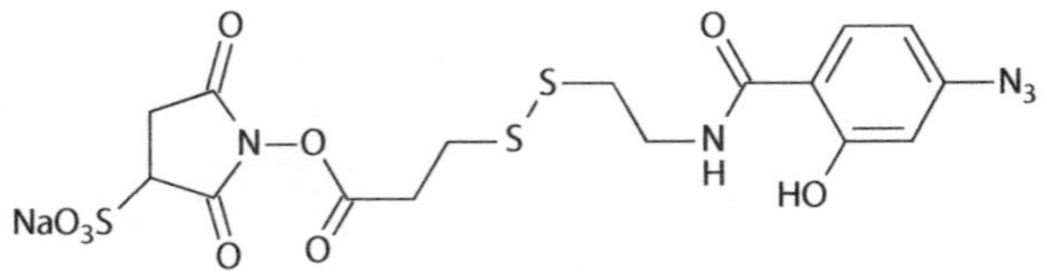

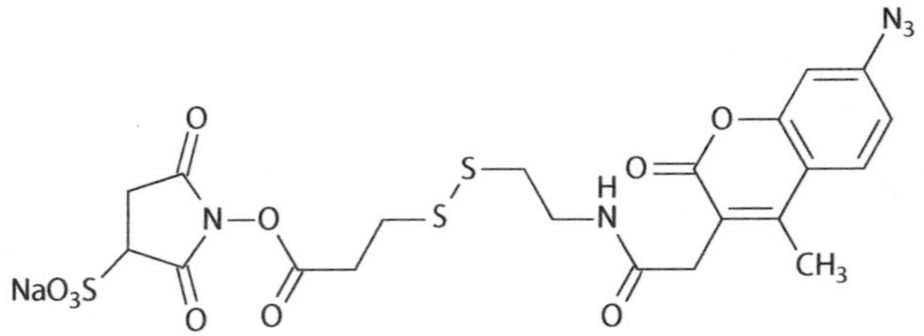

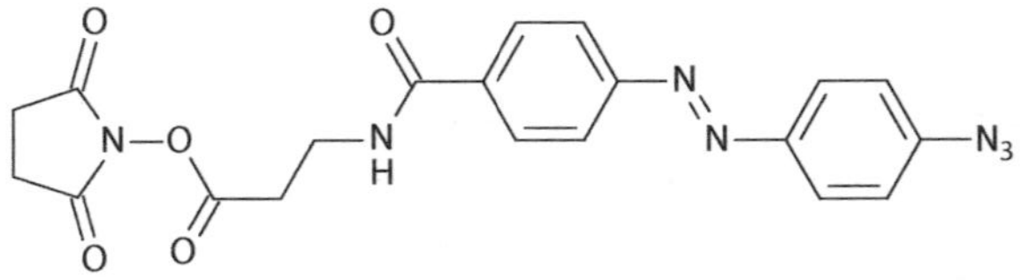

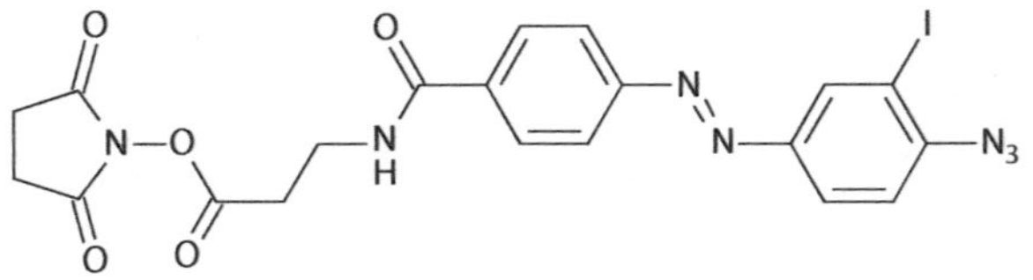

18

18

20

20

(continued)

XXXI

N-[4-(p-Azidophenylazo)benzoyl]-6-aminohexyl-N'-oxysuccinimide
ester

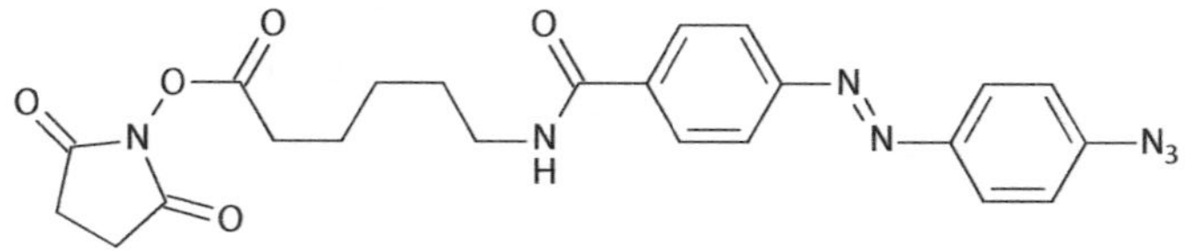

XXXII

N-[4-(p-Azidophenylazo)benzoyl]-11-aminoundecyl-N'-oxysuccinimide
ester

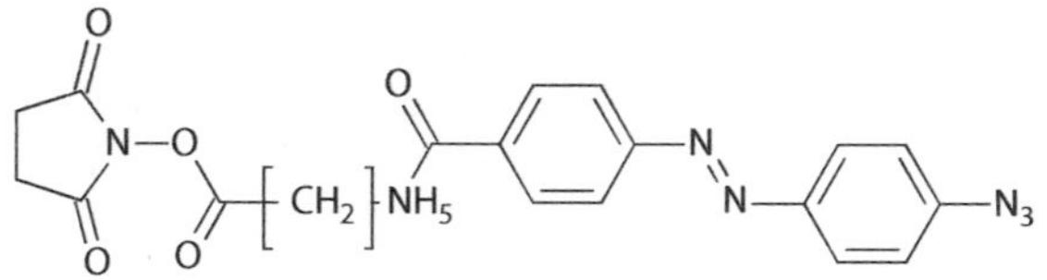

XXXIII

N-Succinimidyl-N-[N'-(4-azidobenzoyl)-tyrosyl]-β-alanate

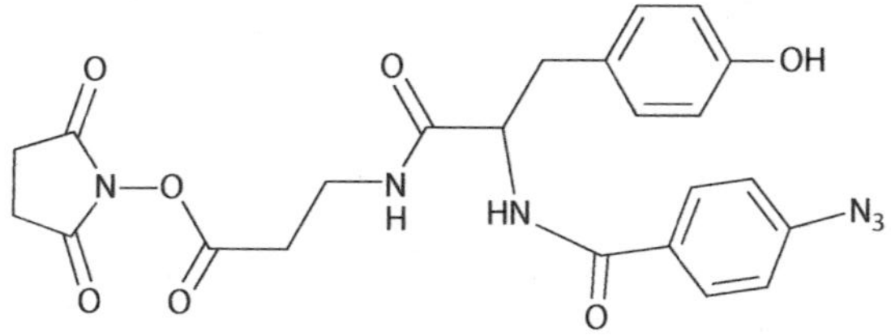

XXXIV

N-Succinimidyl-N-[N'-(3-azidobenzoyl)-tyrosyl]-β-alanate

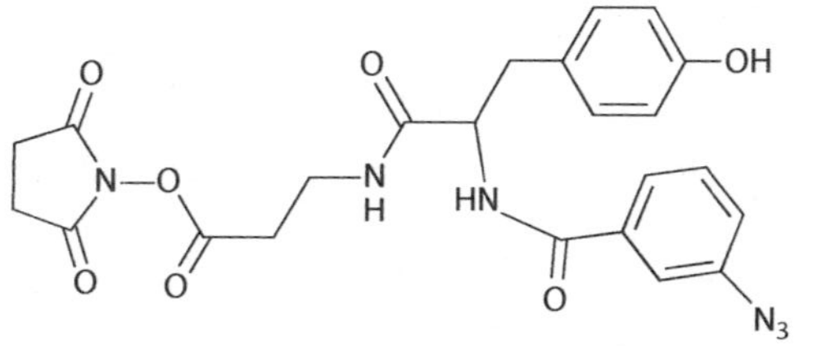

(Continued)

Name (Abbreviation)	Structure	References
XXXV		
N-Succinimidyl-*N*-[*N'*-(3-azido-5-nitro-benzoyl)tyrosyl]-β-alanate	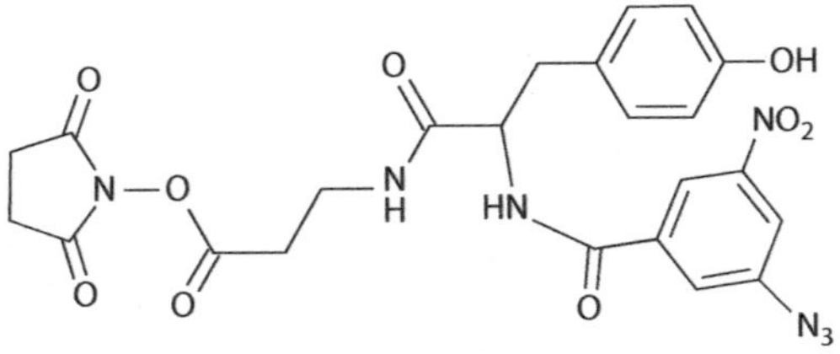	20
Salicylate azides	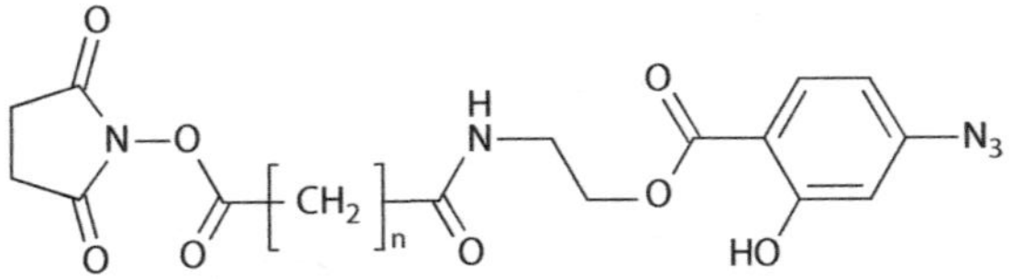	21
XXXVI. n = 2: *N*-Succinimidyl *N*-[2-{(4-azidosalicyloyl)oxy}ethyl] succinamate		
XXXVII. n = 4: *N*-Succinimidyl *N*-[2-{(4-azidosalicyloyl)oxy}ethyl] adipamate		
XXXVIII. n = 6: *N*-Succinimidyl *N*-[2-[(4-azidosalicyloyl)oxy]ethyl] suberamate		
XXXIX		
N-Succinimidyl-4,4′-azipentanoate (SDA, NHS-Diazirine)	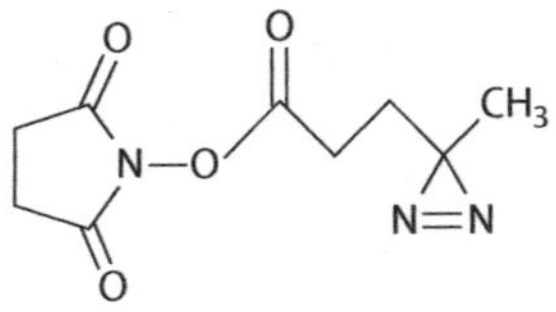	22,23
XL		
N-Sulfosuccinimidyl-4,4′-azipentanoate (Sulfo-SDA, Sulfo-NHS-Diazirine)	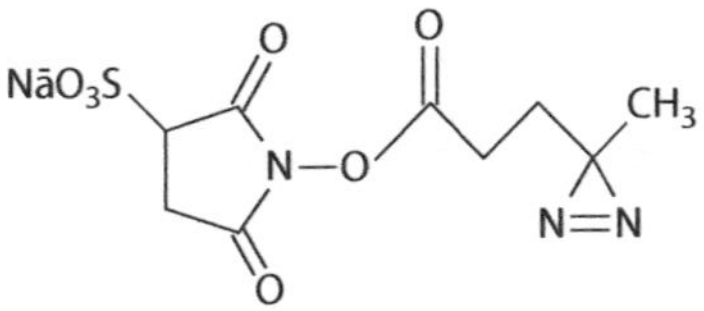	22,23

XLI

N-Succinimidyl-6-(4,4'-azipentanamido) hexanoate (LC-SDA, NHS-LC-Diazirine)

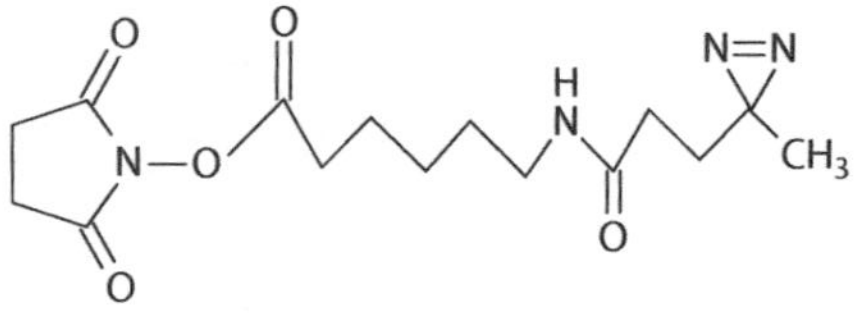

22

XLII

N-Sulfosuccinimidyl-6-(4,4'-azipentan-amido)hexanoate (Sulfo-LC-SDA, Sulfo-NHS-LC-Diazirine)

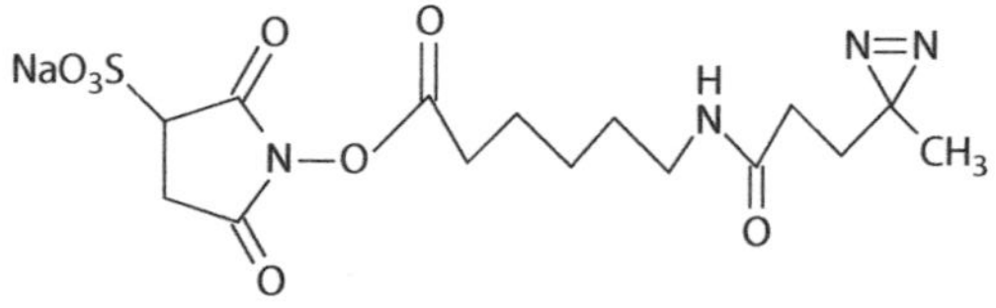

22,24

XLIII

N-Succinimidyl-2-([4,4'-azipentan-amido]-ethyl)-1,3'-dithiopropionate (SDAD, NHS-SS-Diazirine)

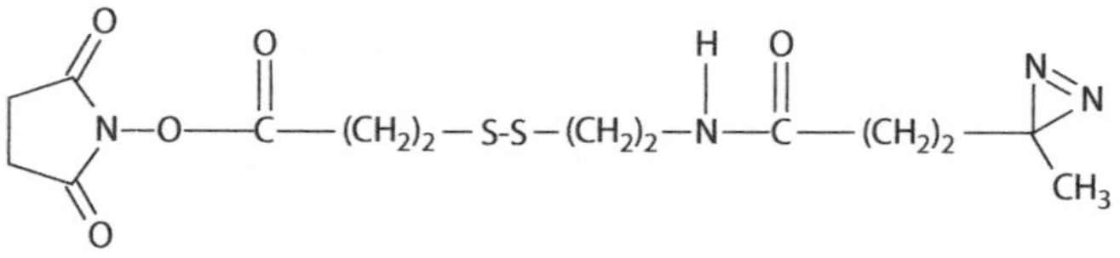

22,25

XLIV

N-sulfosuccinimidyl-2-([4,4'-azipentan-amido]ethyl)-1,3'-dithiopropionate (Sulfo-SDAD, Sulfo-NHS-SS-Diazirine)

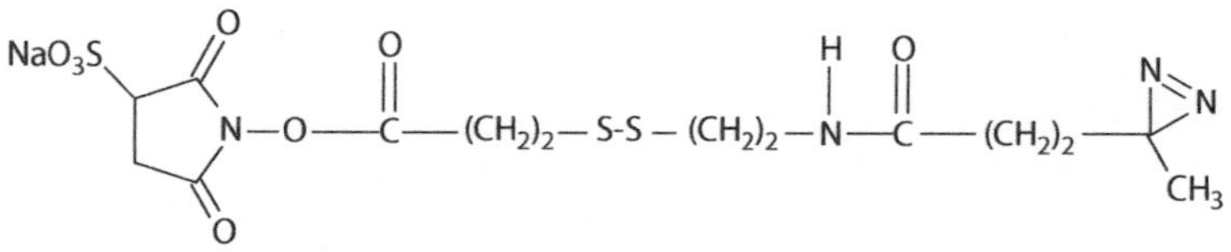

22,24

XLV

O-(4-(3-(Trifluoromethyl)diazirin-3-yl)-benzoyl)-N-hydroxyl succinimide; N-succinimidyl 4-(3-trifluoromethyldiazirino)benzoate (NHS-TDB)

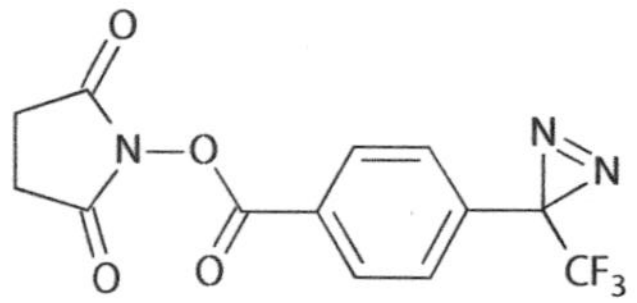

22,26

(continued)

(Continued)

Name (Abbreviation)	Structure	References
XLVI		
N-Succinimidyl-3-(3-(3-(trifluoromethyl)-diazirin-3-yl)phenyl)-2,3-dihydroxy-propionate (TDPDP-ONSu)	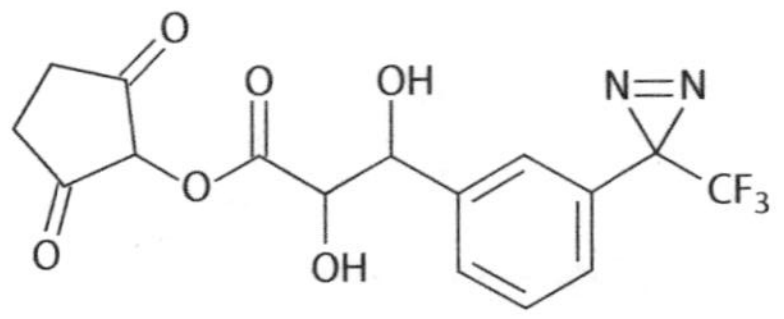	27
XLVII		
3-(Trifluoromethyl)-3-(*m*-isothiocyano-phenyl) diazirine	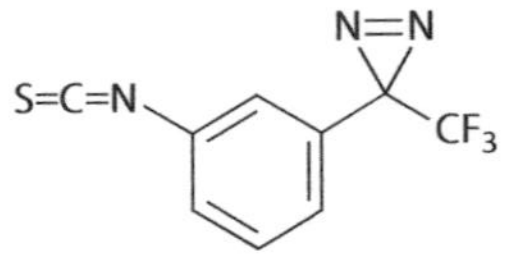	28
XLVIII		
3-(Trifluoromethyl)-3-(*p*-[125I]iodo-*m*-isothiocyanophenyl) diazirine	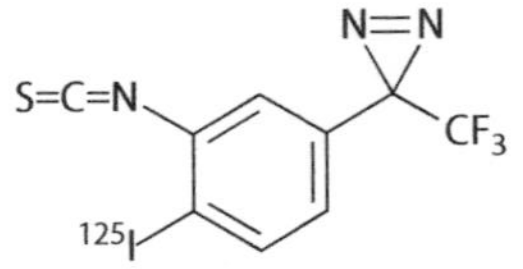	28
XLIX		
3-(Trifluoromethyl)-3-(*m*-nitro-*p*-iso-thiocyanomethylenephenyl) diazirine	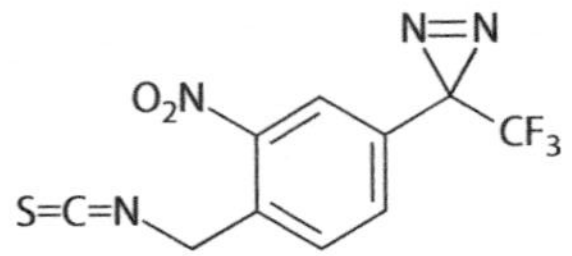	29
L		
Methyl-4-azidobenzoimidate HCl (MABI)	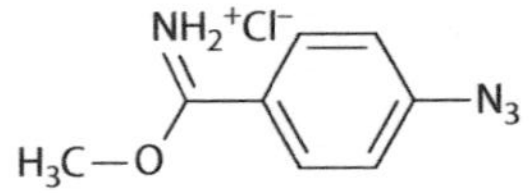	1,30,31

LI

Methyl-3-(4-azidophenyl)acetimidate HCL (MAPA; APAI)   31,32

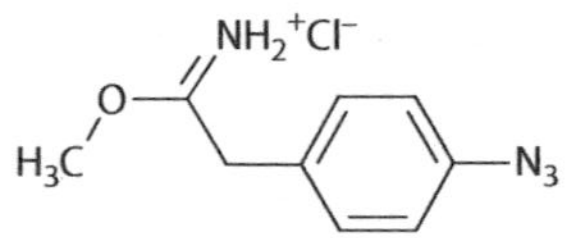

LII

Ethyl *N*-(5-azido-2-nitrobenzoyl) amino acetimidate HCl (ABNA)   6,31

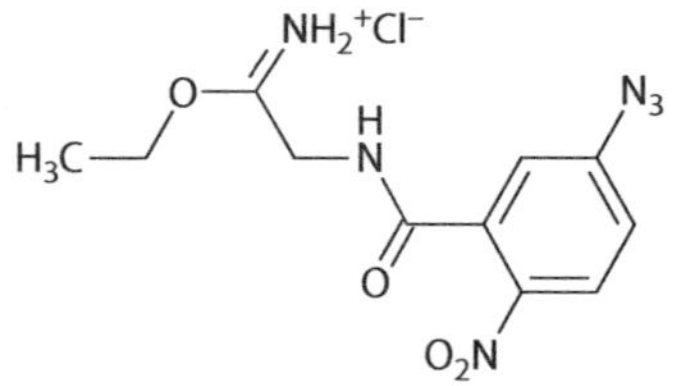

LIII

Methyl 4-(6-formyl-3-azidophenoxy)-butyrimidate HCl (FAPOB)   1,33

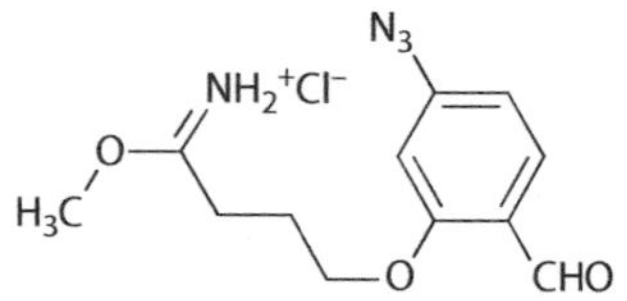

LIV

Methyl-[3-(4-azidophenyl)dithio]propionimidate HCl (MADP); methyl 3-(*p*-azidophenyl)dithio]propionimidate (PAPDIP)   14,31,34

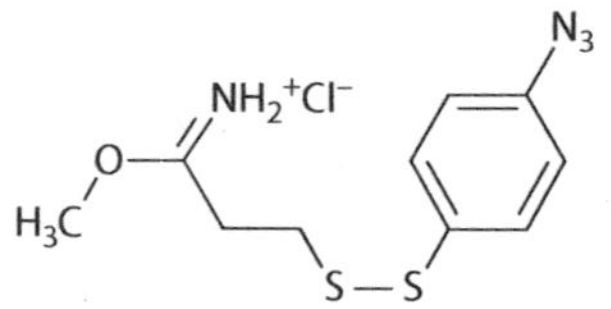

LV

Methyl-4-[(4-azidophenyl)dithio]-butyrimidate (MADB)   31,34

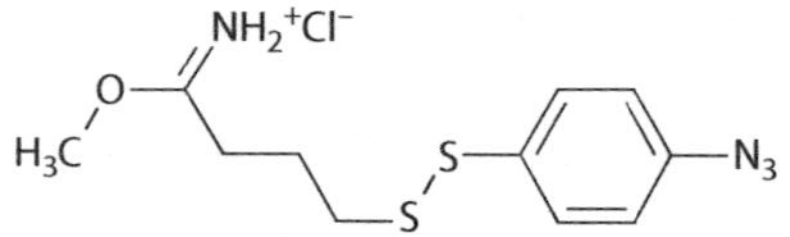

(continued)

(Continued)

Name (Abbreviation)	Structure	References
LVI		6,35
Ethyl-(4-azidophenyl)-1,4-dithio-butyrimidate HCl (EADB)	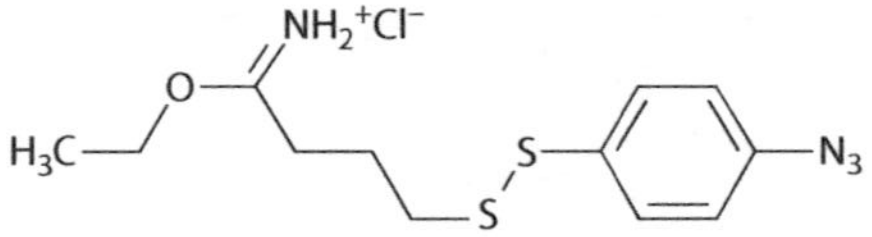	
LVII		36
4-Azidoiodobenzene	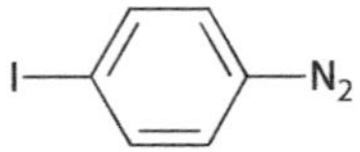	
LVIII		1,31,37
4-Fluoro-3-nitrophenylazide (FNA, FNPA)	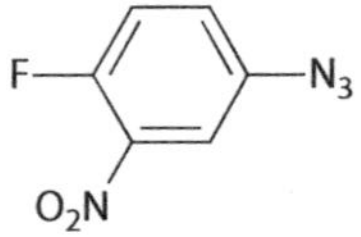	
LIX		31,38
2,4-Dinitro-5-fluorophenylazide (DNFA)	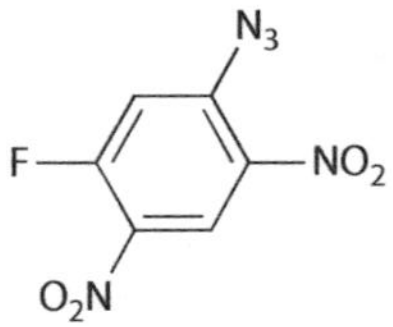	
LX		39
p-Azidophenylisothiocyanate	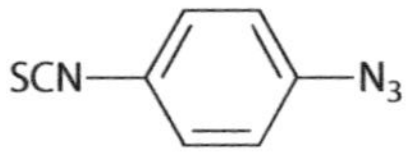	
LXI		39
1-Azido-5-naphthaleneisothiocyanate; 5-isothiocyanato-1-naphthalene azide	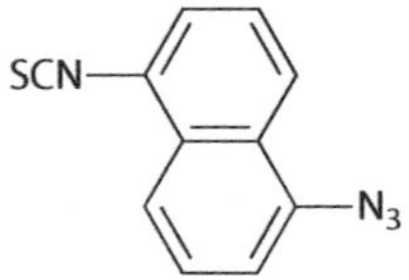	

LXII

Benzophenone-4-isothiocyanate

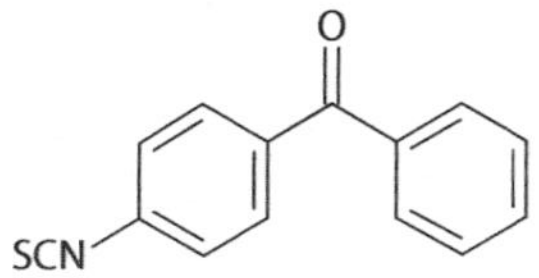

40

LXIII

Methyl-(4-benzophenone) acetimidate HCl

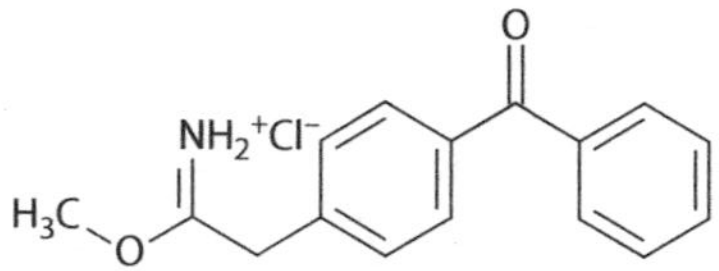

40,41

LXIV

p-Benzoyldihydrocinnamyl succinimido ester (BZDC-NHS);
N-succinimido-[2,3]-*p*-benzoyl dihydrocinnamate

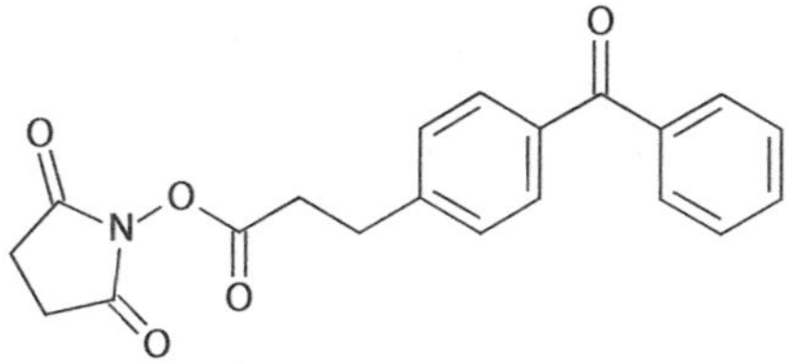

40

LXV

N-Succinimidyl-(*p*-benzoyl)cinnamate

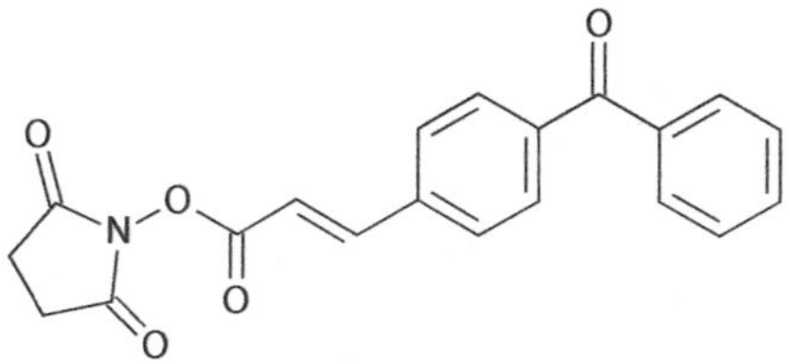

42

(continued)

(Continued)

Name (Abbreviation)	Structure	References
LXVI		
Succinimidyl-[4-(psoralen-8-yloxy)]butyrate (SPB)	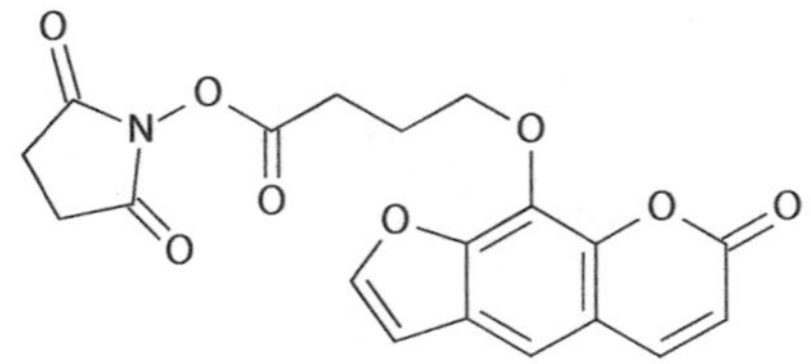	43
LXVII		
2-Diazo-3,3,3-trifluoropropionyl chloride (DTPC)	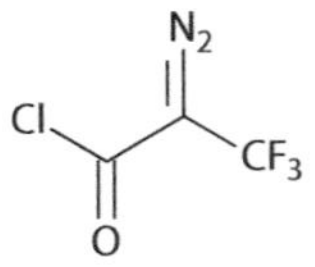	44
LXVIII		
p-Nitrophenyl-2-diazo-3,3,3-tri-fluoropropionate (NDTFP; PNP-DTP)	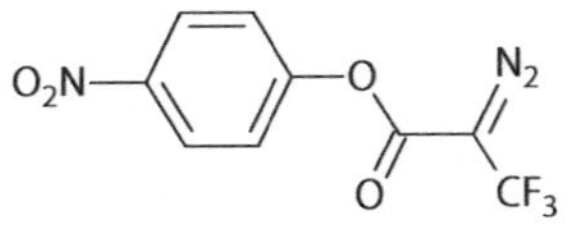	1,45
LXIX		
p-Nitrophenyldiazoacetate	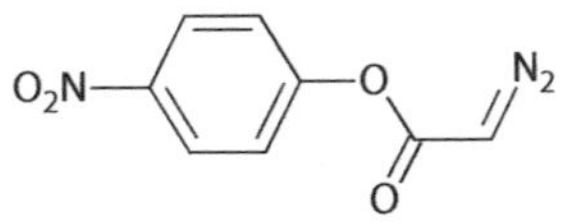	46
LXX		
p-Nitrophenyl-3-diazopyruvate (DAPpNP)	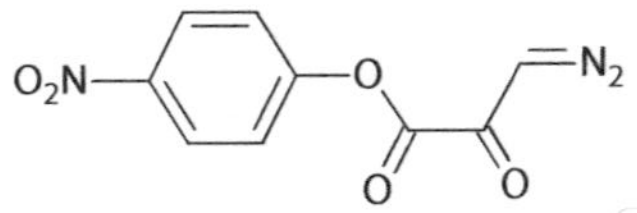	47

B. Sulfhydryl Group–Anchored Photoactivatable Reagents

LXXI

4-Azidophenylmaleimide (APM)

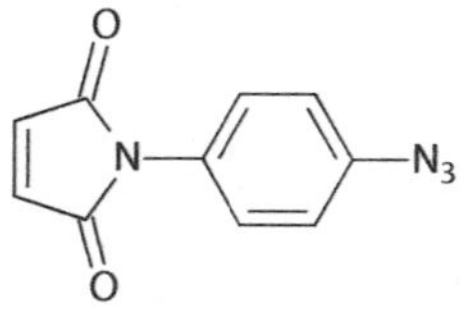

48

LXXII

N-(4-Azido-2,3,5,6-tetra-fluorobenzyl)maleimide propionamide (TFPAM-3)

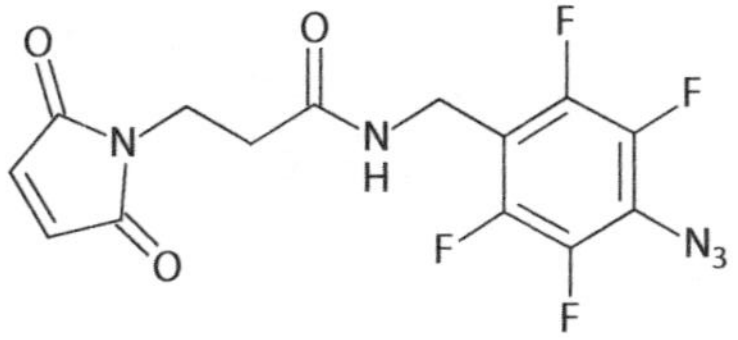

49

LXXIII

N-(4-Azido-2,3,5,6-tetrafluoro-benzyl)-6-maleimidohexanamide (TFPAM-6)

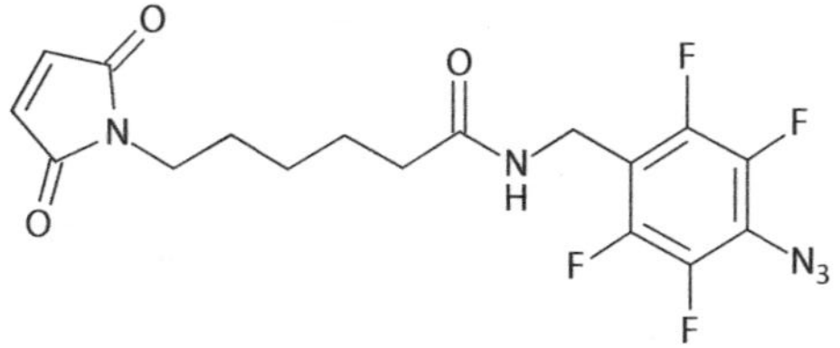

49

LXXIV

p-Azidophenacyl bromide (APB)

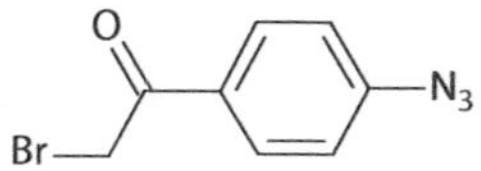

1,31,50

LXXV

4-(Bromoaminoethyl)-3-nitro-phenylazide (BANPA)

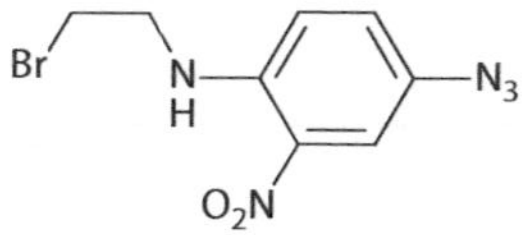

31

(continued)

(Continued)

Name (Abbreviation)	Structure	References
LXXVI		
1-(p-Azidosalicylamido)-4-(iodoacetamido)butane (ASIB)	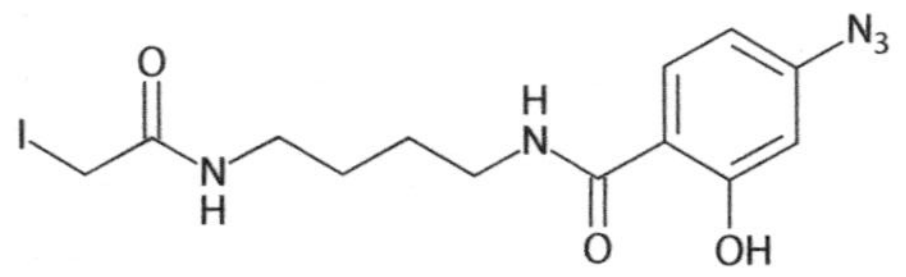	51
LXXVII		
4-Azidophenylsulfenyl chloride	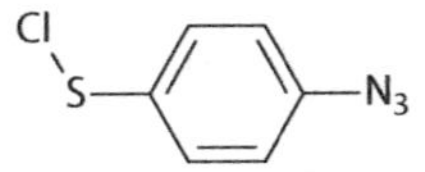	14,31
LXXVIII		
2-Nitro-4-azidophenylsulfenyl chloride (2,4-NAPS-Cl)	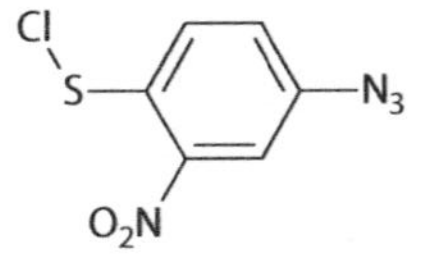	31,52
LXXIX		
2-Nitro-5-azidophenylsulfenyl chloride (2,5-NAPS-Cl)	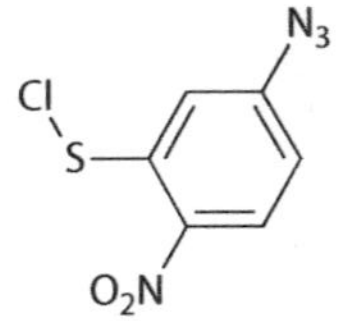	52
LXXX		
N-(4-Azidophenylthio)phthalimide (APTP)	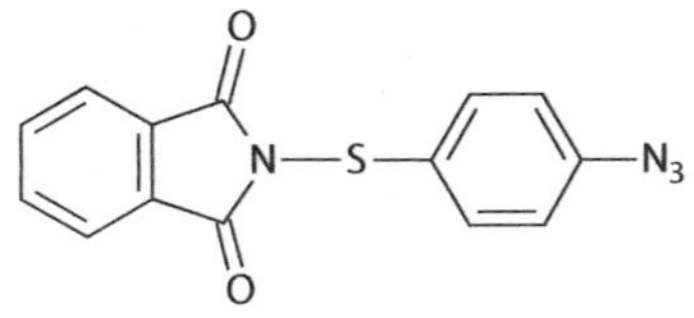	1,14,31,53

LXXXI

Di-*N*-(2-nitro-4-azidophenyl)-cystamine-*S,S*-dioxide (DNCO)

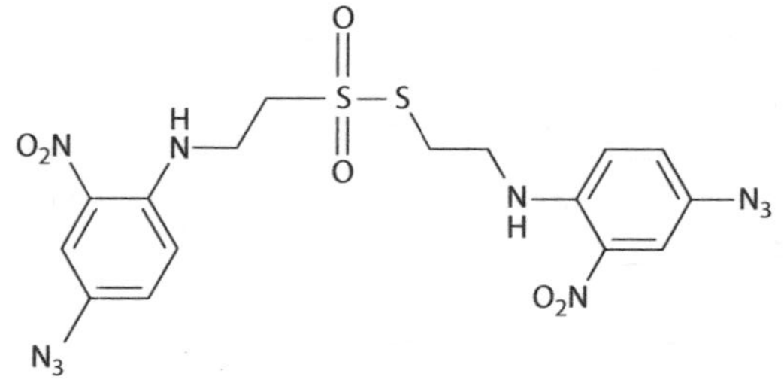

54

LXXXII

4,4'-Dithiobisphenylazide

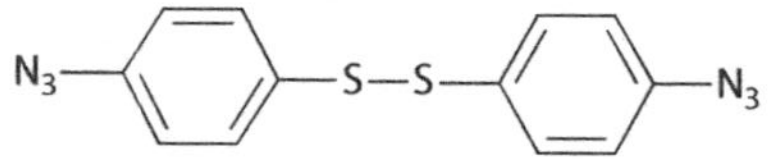

55

LXXXIII

Bis-[β-(4-azidosalicylamido)-ethyl]disulfide (BASED)

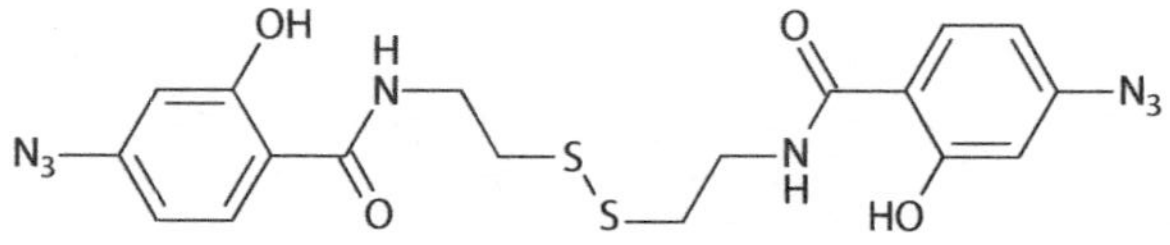

51

LXXXIV

Azidophenacylthiopyridine

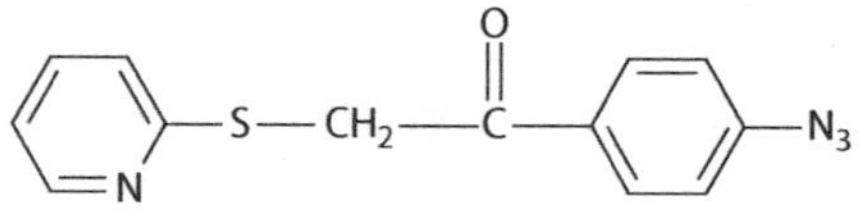

56

LXXXV

N-[(2-Pyridyldithio)ethyl]-4-azidosalicylamide (PEAS); *S*-[2-(4-azidosalicylamido)ethyl-thio]-2-thiopyridine (AET, ACT)

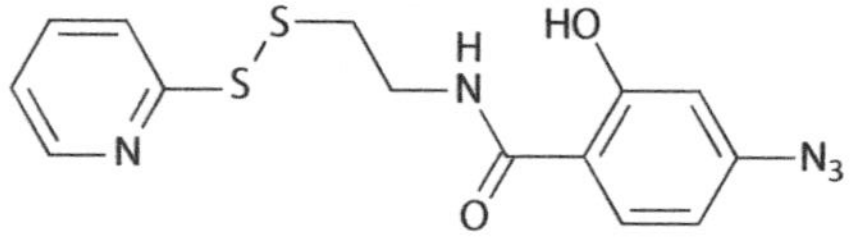

57

LXXXVI

S-[2-(3-Iodo-4-azidosalicylamido) ethylthio]-2-thiopyridine (I-AET)

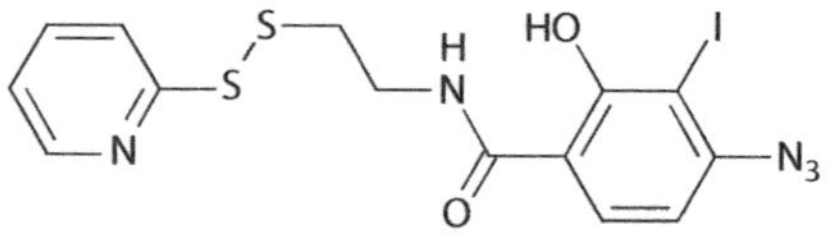

58

(continued)

(Continued)

Name (Abbreviation)	Structure	References
LXXXVII *S*-[*N*-(3-Iodo-4-azidosalicyl) cysteaminyl]-2-thiopyridine (I-ACT)		59
LXXXVIII *N*-(4-Azidobenzoyl-2-glycyl)-*S*-(2-thiopyridyl)cysteine (AGTC)		60
LXXXIX *N*-(3-Iodo-4-azidophenyl-propionamide-*S*-(2-thio-pyridyl)cysteine		61
XC *N*-[4-(*p*-Azidosalicylamido)-butyl]-3′-(2′-pyridylthio)propion-amide (APDP); *S*-[2-[*N*-[4-(4-azido-salicylamido) butyl]carbamoyl] ethyl-thio]-2-thiopyridine		62
XCI *N*-[4-(3-Iodo-4-azidosalicylamido)-butyl]-3-(2′-pyridyldithio) propionamide (I-APDP)		63

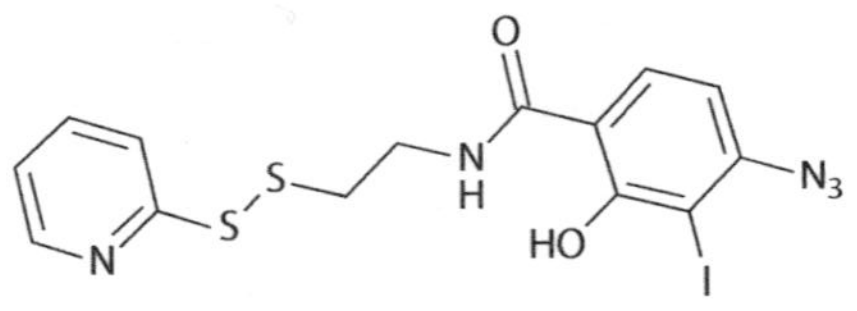

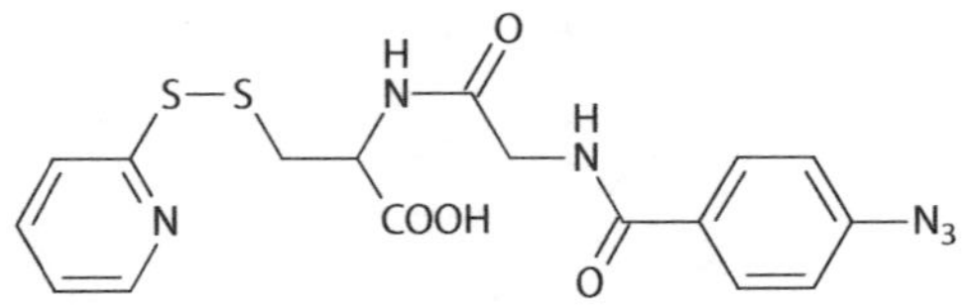

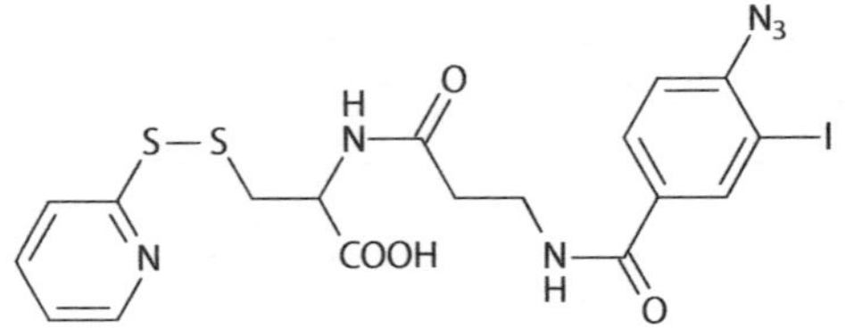

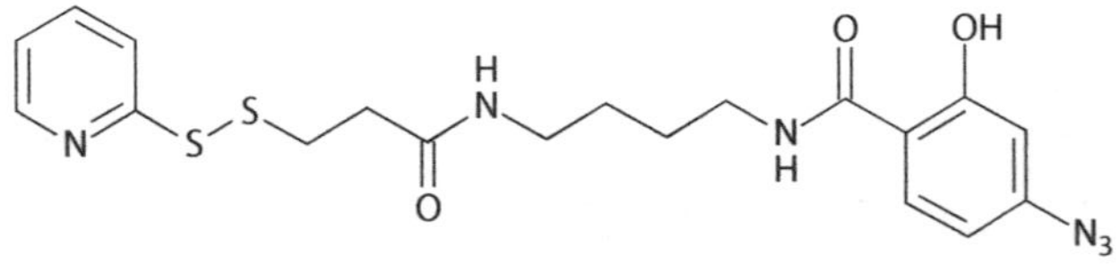

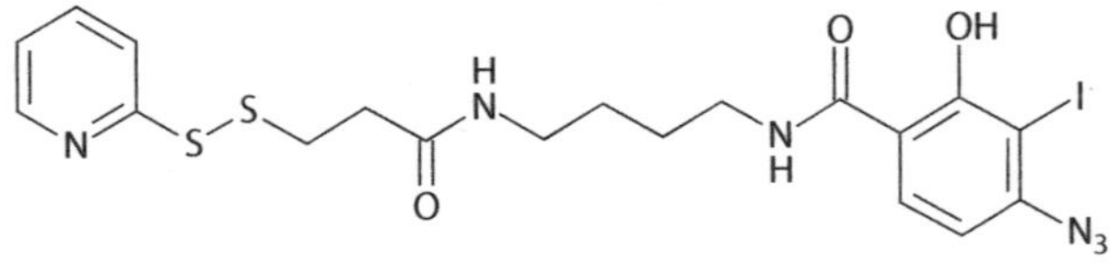

11

11

64

65

66

67

(continued)

XCII

2-[(4-Azido-2-nitrophenyl)-amino]
ethyl-3-carboxy-4-nitrophenyldisulfide

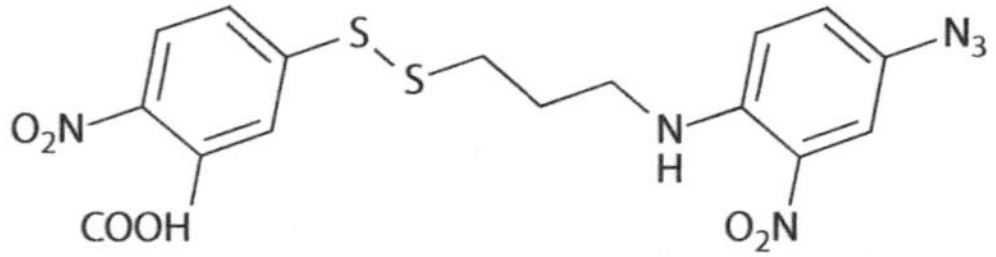

XCIII

2-[[[(4-Azido-2-nitrophenyl)amino] dodecanoyl]amino]ethyl-3-4-
nitrophenyl disulfidecarboxy-4-nitrophenyl disulfide

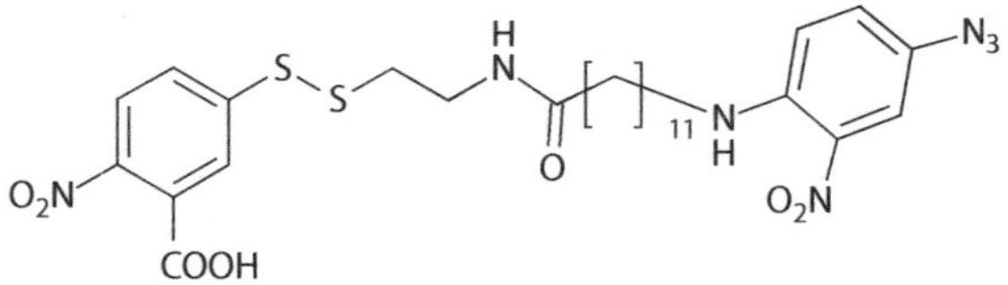

XCIV

3-(4-Azido-2-nitrobenzoylseleno)-propionic acid (ANBSP);
2-carboethylseleno-4-azido-2-nitrobenzoate

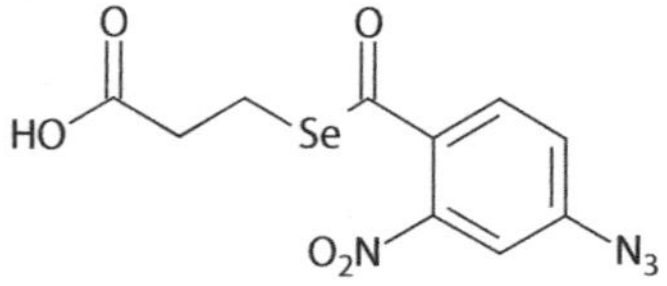

XCV

3-(Trifluoromethyl)-3-(*m*-iodophenyl)-diazirine (*I*-TID)

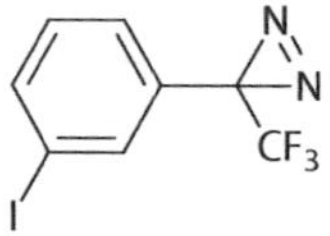

XCVI

3-[4-(Bromomethyl)phenyl]-3-(trifluoromethyl)-*3H*-diazirine

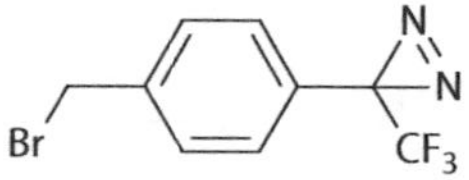

XCVII

3-(3-Bromomethyl-5-methylphenyl)-3-trifuloromethyl-*3H*-diazirine

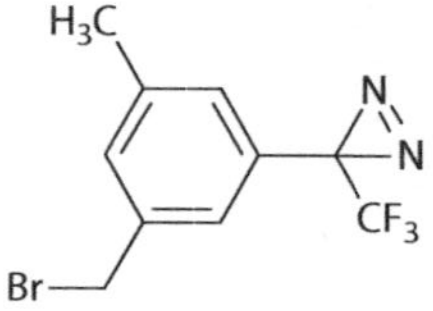

(Continued)

Name (Abbreviation)	Structure	References
XCVIII		
3-[3,5-Bis(bromomethyl)phenyl] 3-trifuloromethyl-3*H*-diazirine	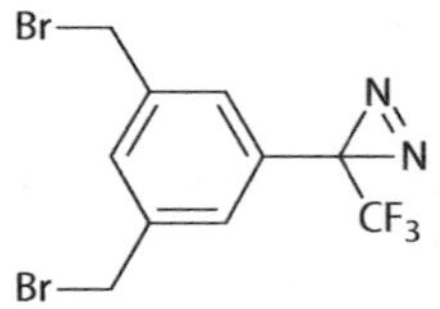	67
XCIX		
3-(3-(Bromoacetylamino)phenyl)-3-(trifluoromethyl)diazirine (BAPTD); Trifluoromethyl-(3-bromacetyl)-aminophenyl diazirine (TBAPD)	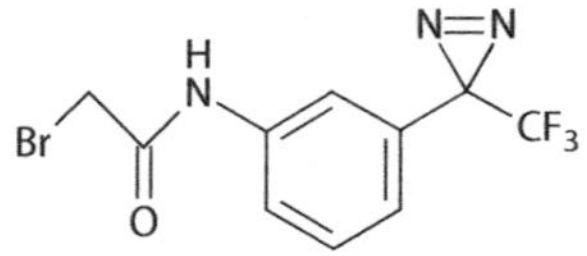	68
C		
N-Bromoacetyl-*N'*-{2,3-dihydroxy-3-[3-(3-(trifluoromethyl)diazirin-3-yl)-phenyl]propionyl}ethylenediamine (BATDHP)	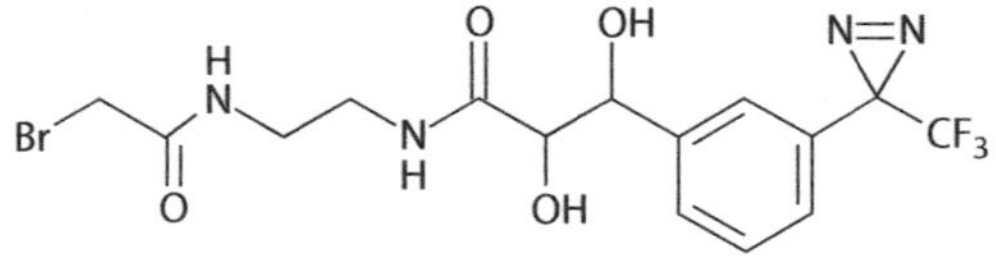	69
CI		
S-(2-Pyridyl)-*S'*-{*N*-[3-(3-trifluoro-methyl)diazirin-3-yl] benzoylamino-ethyl}disulfide (PTHBEDS)	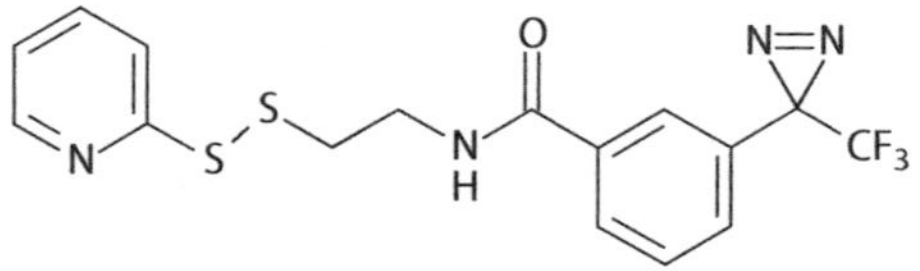	69
CII		
N-(3-(3-(Trifluoromethyl)diazirin-3-yl) phenyl)-4-maleimidobutyramide	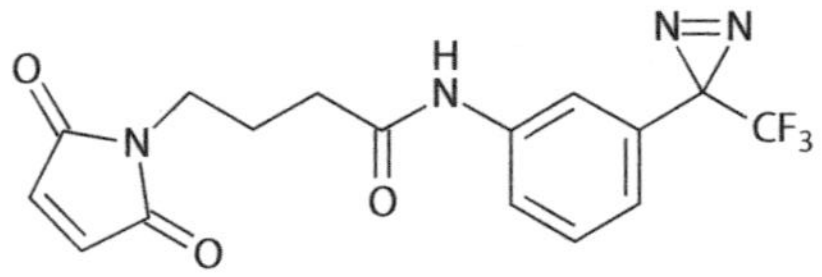	70

(continued)

CIII

3-(4-(((4-Nitro-3-carboxyphenyl)dithio)-*3H*-methyl)-phenyl)-3-(trifluoromethyl)-diazirine; 3-(4-(((4-nitro-3-carboxyphenyl) dithio) methyl-t)-phenyl)-3-(trifluoromethyl)-*3H*-diazirine (DTDA)

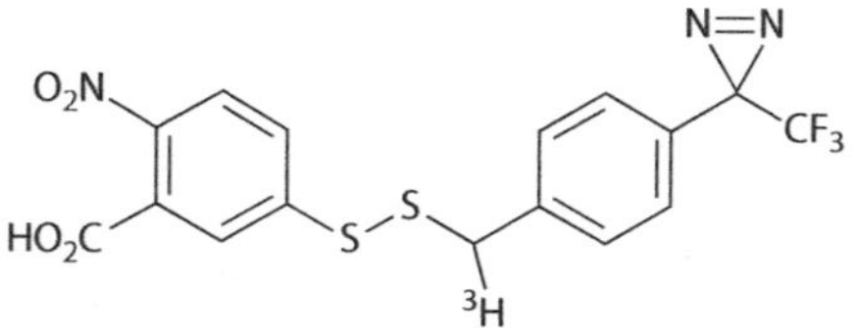

71

CIV

N-4-[3-(Trifluoromethyl)-*3H*-diazirin-3-yl]-benzyl-*N'*-cyclohexylcarbodiimide

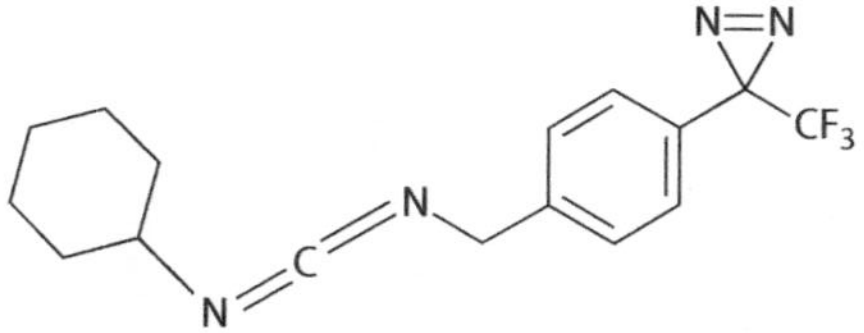

72

CV

1-[[3-[3-(Trifluoromethyl)-*3H*-diazirine-3-yl]-phenyl] methyldithio]-1-deoxy-α-D-mannopyranoside

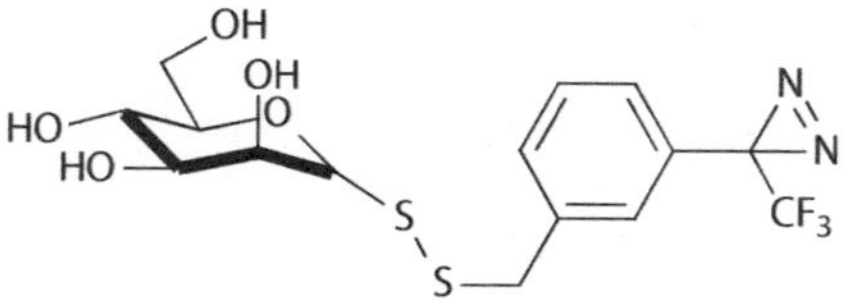

73

CVI

1-[[4-[3-(Trifluoromethyl)-*3H*-diazirine-3-yl]-phenyl] methyldithio]-1-deoxy-α-D-mannopyranose

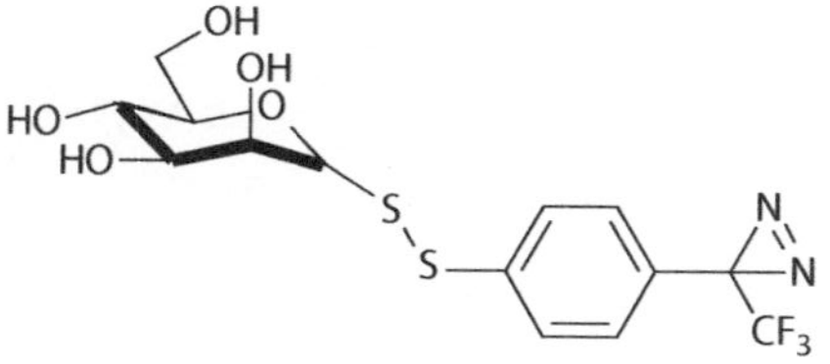

73

Pyridyldithiobenzyl diazoacetate

(Continued)

Name (Abbreviation)	Structure	References
CVII. *o*-(2′-Pyridyldithio)benzyldiazoacetate (OPD); pyridyl-2,2′-dithiobenzyldiazoacetate (PDD)		74,75
CVIII. *m*-(2′-Pyridyldithio)benzyldiazoacetate (MPD)		75
CIX. *p*-(2′-Pyridyldithio)benzyldiazoacetate (PPD)		75
CX		76
1-Diazoacetyl-1-bromo-2-phenylethane	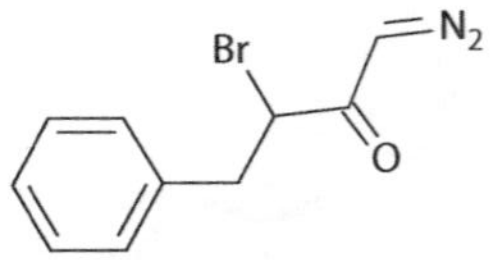	
CXI		40
Benzophenone-4-maleimide; 4-maleimidobenzo-phenone (MBP)	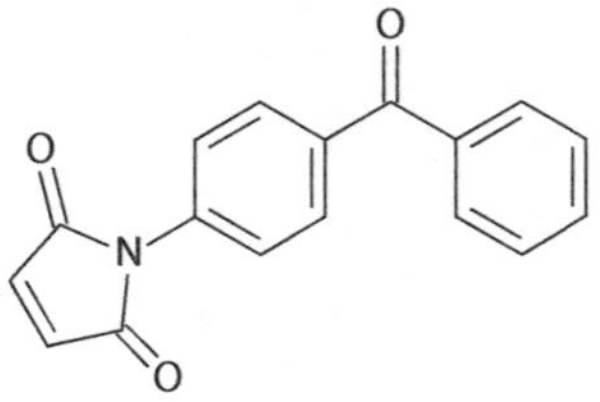	
CXII		40
4-(2-Iodoacetamido)benzo-phenon (BPIA)	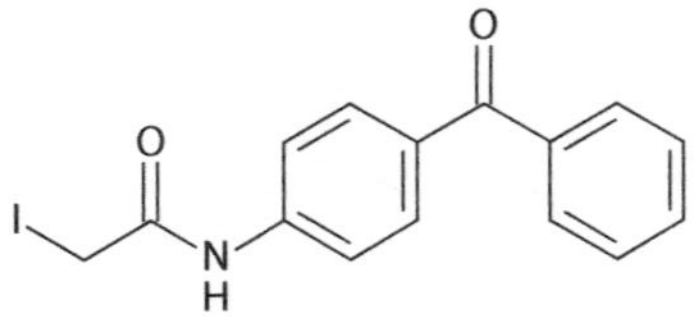	
CXIII		77
4-(*N*-Maleimidopropionamido)-4′-[(2-hydroxy-5-β-D-pyranosyl-phenyl)azo]benzophenone	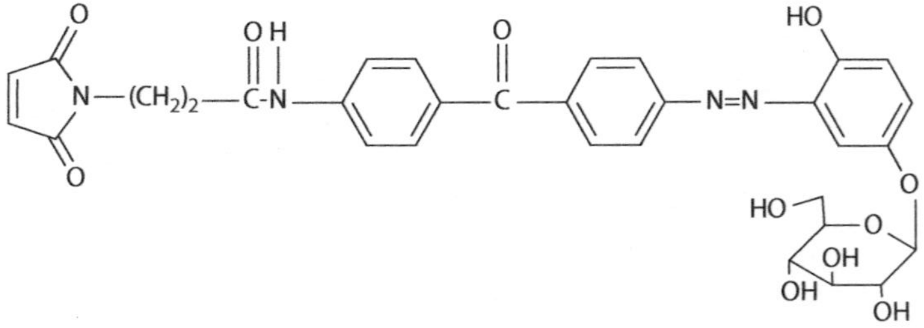	

Nitrophenyl ethers

CXIV. n = 3: *N*-(Maleimidomethyl)-2-(*o*-methoxy-*p*-nitrophenoxy) carboamidopropane

CXV. n = 6: *N*-(Maleimidomethyl)-2-(*o*-methoxy-*p*-nitrophenoxy) carboamidohexane

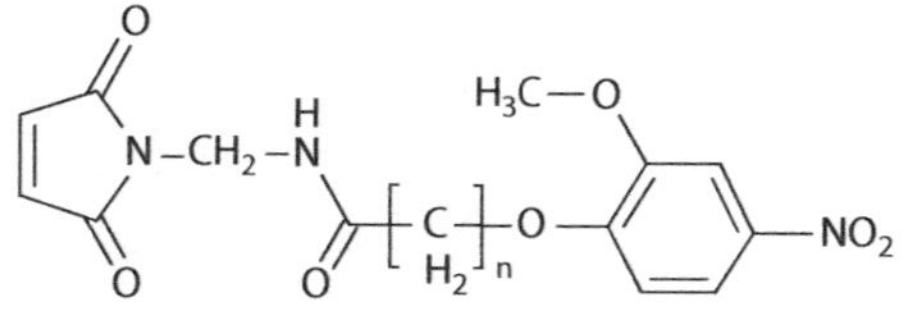

CXVI

N-(2-Methoxy-6-azidoacridin-9-yl)-*N'*-(4-zidobenzoyl)hexanediamine

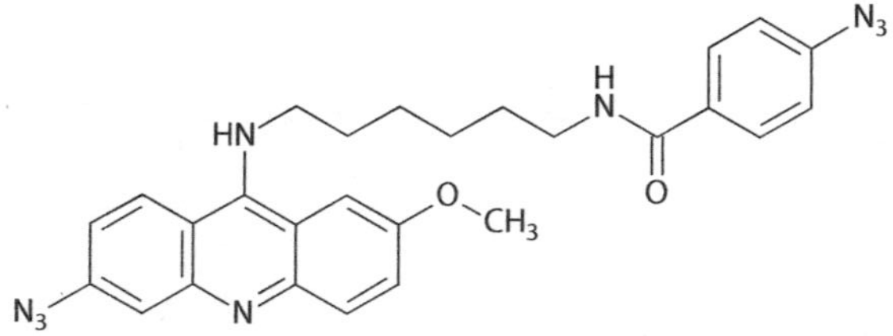

CXVII

N-(2-Methoxy-6-azidoacridin-9-yl)-*N'*-(4-azidobenzoyl)cystamine)

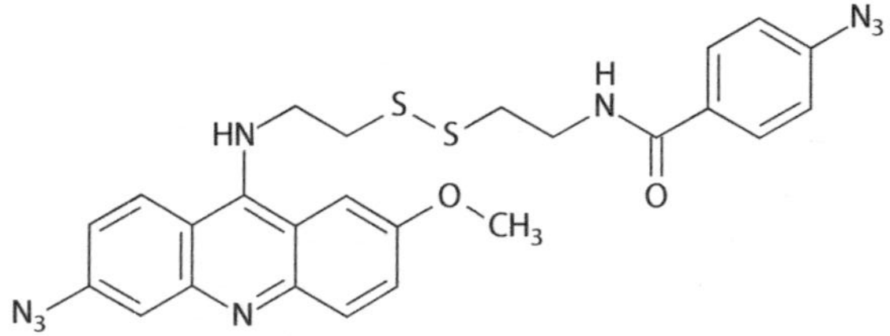

78

79

79

(*continued*)

(Continued)

Name (Abbreviation)	Structure	References

CXVIII

1-N-(Maleimidohexanoyl)-6-N-(anthraquinon-2-oyl) hexanediamine (MHAHD)

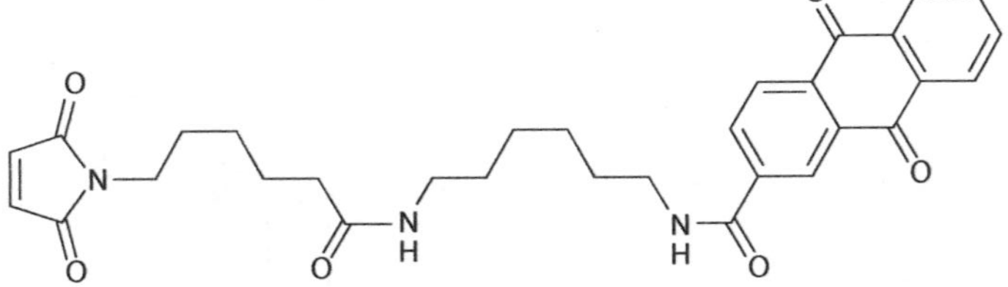

80

CXIX

N-(Iodoacetyl)-N″-(anthraquinon-2-oyl)-ethylenediamine (IAED)

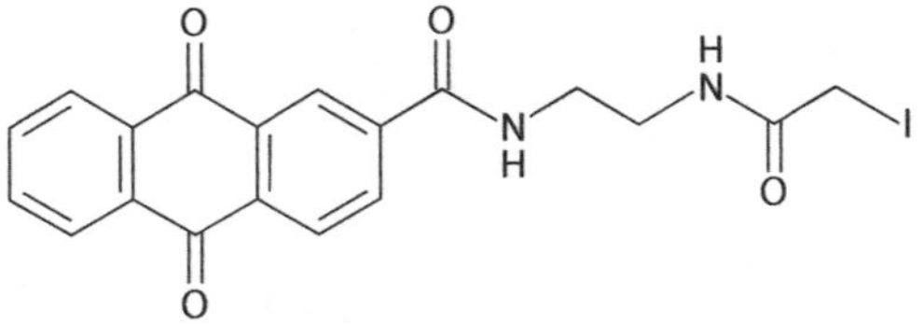

81

CXX

N-(3-Trifluoroethanesulfonyloxy-propyl)anthraquinone-2-carboxamide (NTPAC)

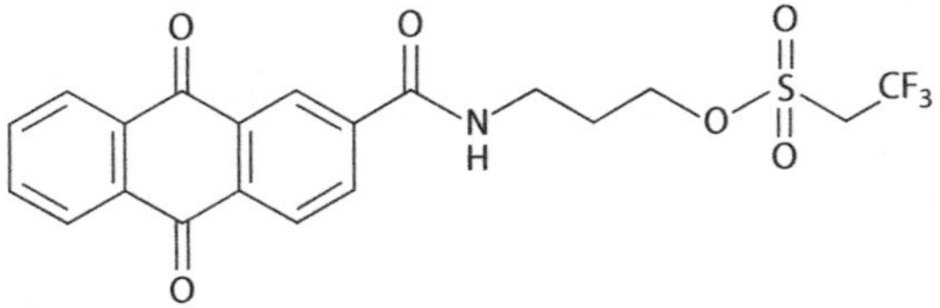

82

CXXI

N-(4-Azidobenzoyl)-N,N′-dimethyl-N-[((8-methoxy)psoralen-5-yl) methyl]cystamine

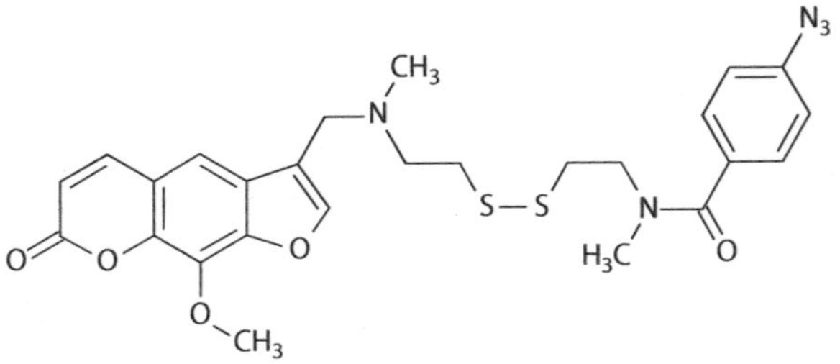

83

CXXII

N-(4-Azidobenzoyl)-*N*,*N'*-dimethyl-*N'*-[3-(8-psoralenyloxy)propyl]
cystamine

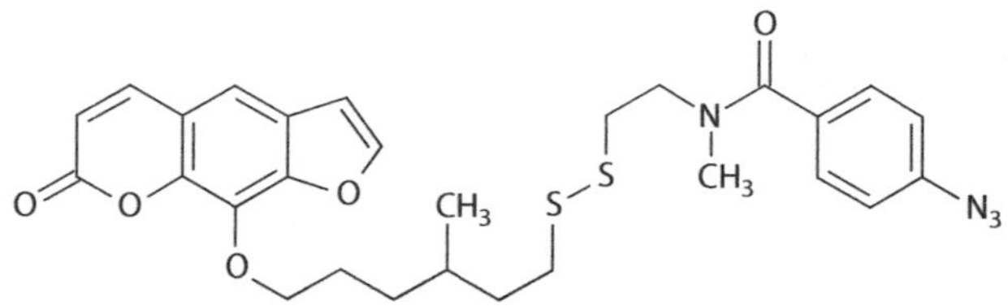

83

CXXIII

N-(4-Azidobenzoyl)-*N*,*N'*-dimethyl-[((4,8,5'-trimethyl)psoralen-4'-
yl)-methyl]cystamine

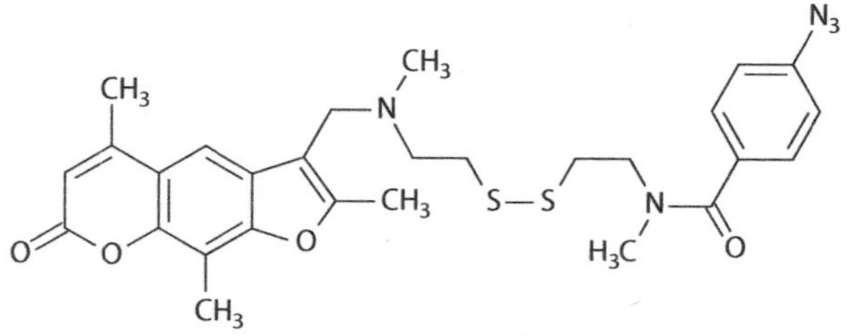

83

C. Guanidinyl Group–Anchored Photoactivatable Reagent

CXXIV

4-Azidophenylglyoxal (APG)

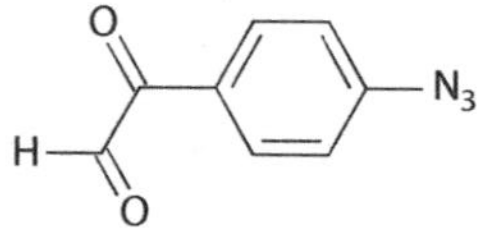

1,84

D. Carboxyl-, Carboxamide-, and Carbonyl-Group-Anchored Photoactivatable Reagents

CXXV

N-(4-Azido-2-nitrophenyl)ethylene-diamine;
N-(β-aminoethyl)-4-azido-2-nitroaniline

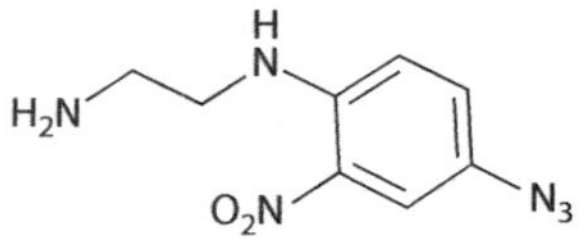

85

(continued)

(Continued)

Name (Abbreviation)	Structure	References
CXXVI		
N-(5-Azido-2-nitrophenyl)ethylene-diamine	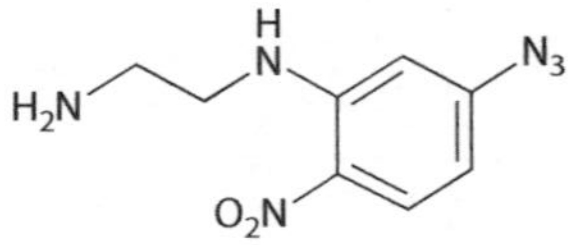	86
CXXVII		
N-(4-Azido-2-nitrophenyl) putrescine (ANP)	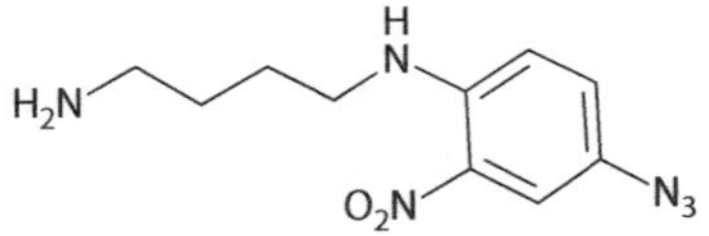	87
CXXVIII		
N-(β-(β-Aminoethyldithioethyl)]-4-azido-2-nitroaniline	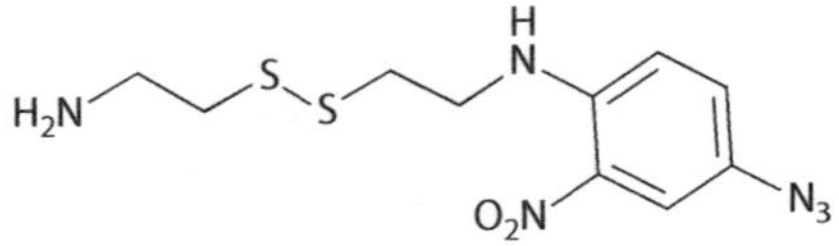	85
CXXIX		
N-(4-Azido-2-nitrophenyl-β-amino-ethyl)-N'-(β-aminoethyl)tartramide	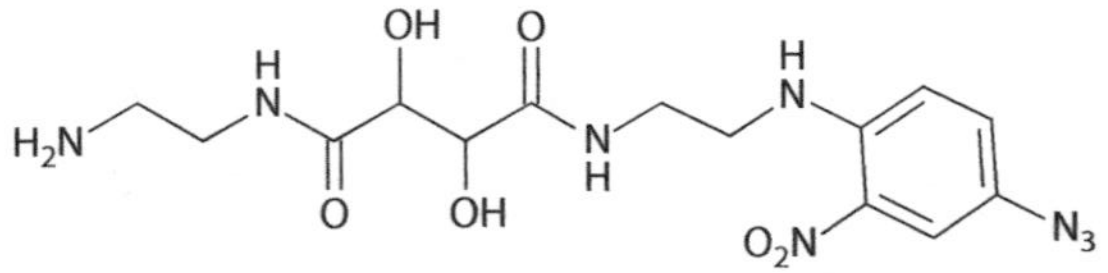	85
CXXX		
N-(4-Azidobenzoyl)-ethylenediamine	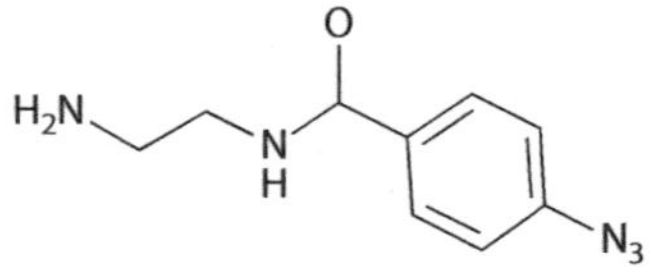	88

CXXXI

N-(4-Azidobenzoyl)-diaminopropane

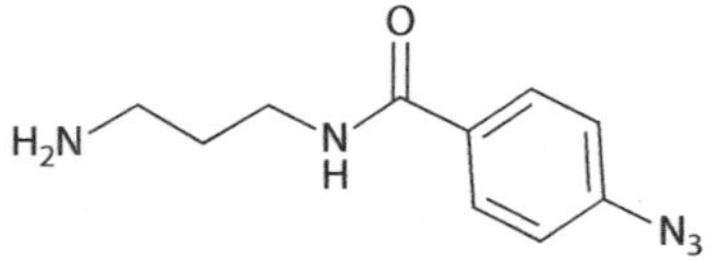

88

CXXXII

N-(4-Azidobenzoyl) putrescine (ABP)

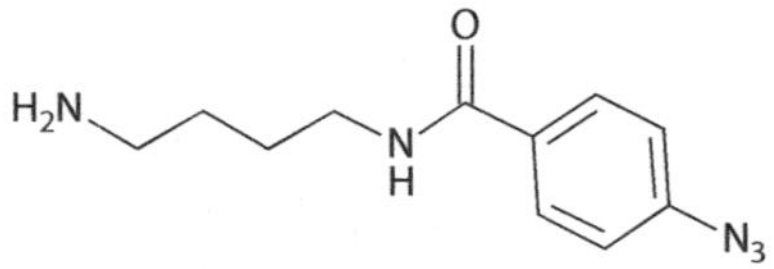

88

CXXXIII

4-(p-Azidosalicylamido)butylamine (ASBA)

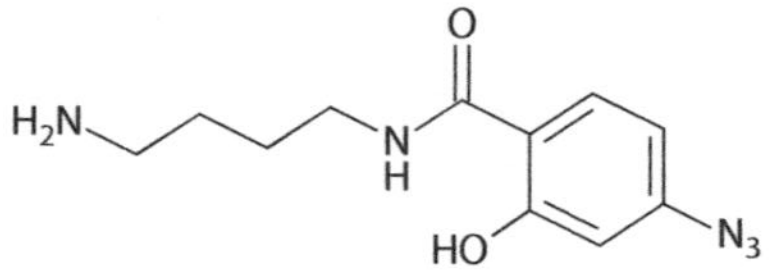

89

CXXXIV

p-Azidobenzoylhydrazide (ABH)

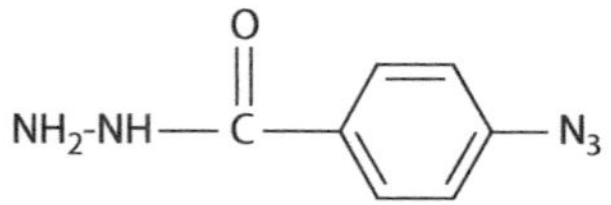

90

E. Photoaffinity Labels

CXXXV

2-[(4-Azido-2-nitrophenyl)amino]ethyl triphosphate (NANTP)

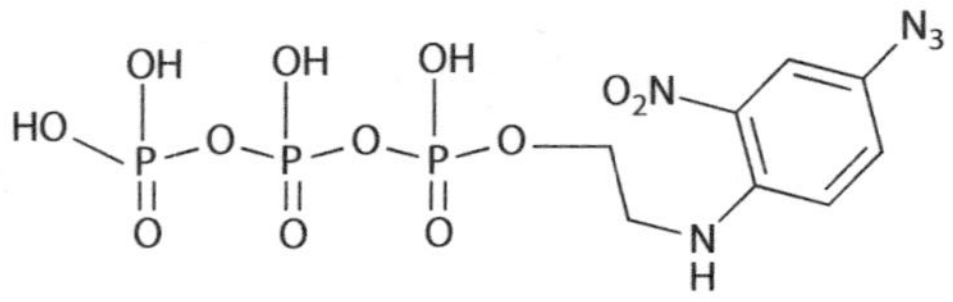

91

(continued)

(Continued)

Name (Abbreviation)	Structure	References
CXXXVI		
5'-(p-Fluorosulfonylbenzoyl)-8-azidoadenosine (FSBAzA)	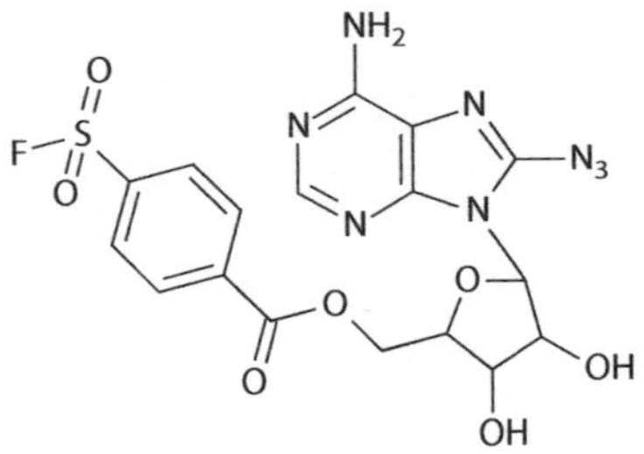	92
CXXXVII		
2-Azidoadenosine monophosphate (2-N₃AMP)	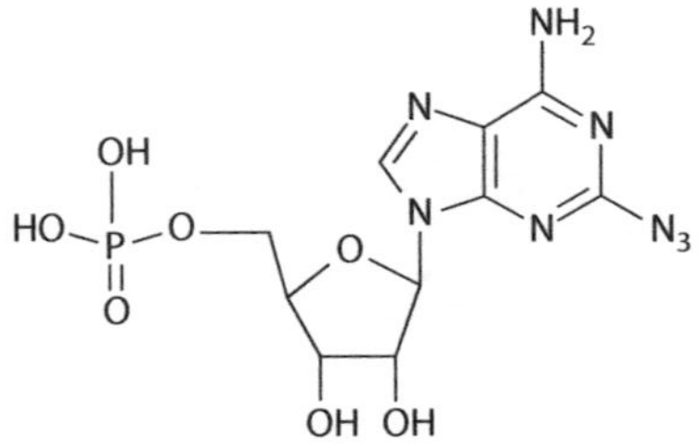	93
CXXXVIII		
2-Azidoadenosine diphosphate (2-N₃ADP)	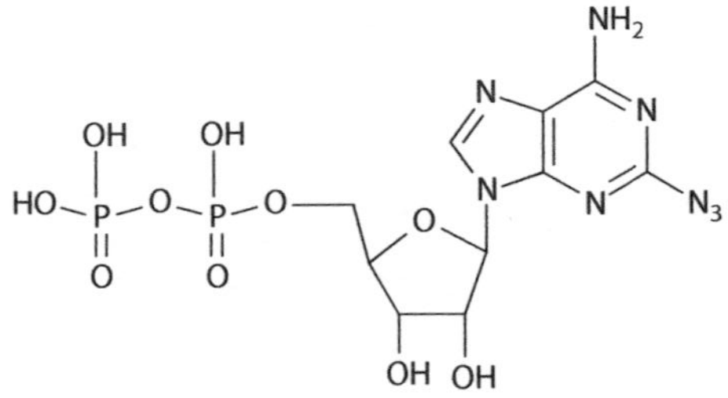	94

95

96

96

(continued)

CXXXIX

3′-Arylazido-β-alanine-δ-azido-ATP; 3′-*O*-[3-[*N*-azido-(2-nitrophenyl)-amino]propionyl]-8-azioadenosine-5′-triphosphate (DiN₃ATP)

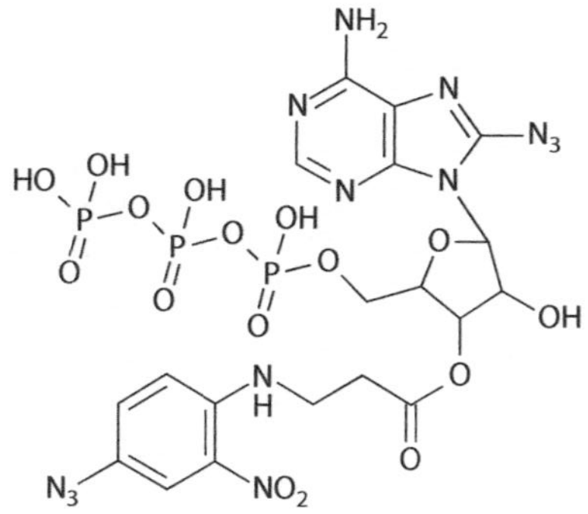

CXL

Adenosine 5′-diphosphate *N*-(4-(benzoyl)phenyl]methyl) phosphoramide (ATP-BP, 5-BzATP)

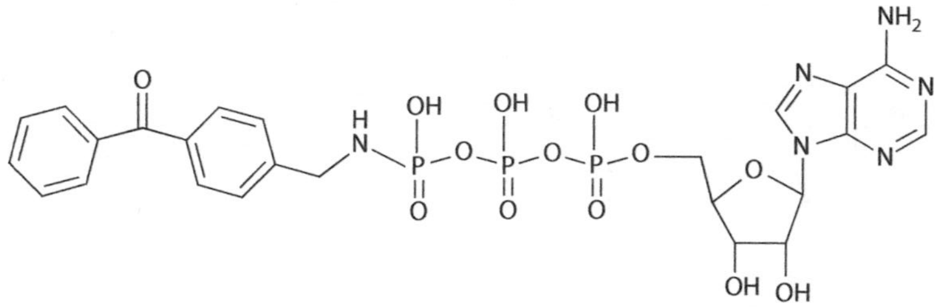

CXLI

Guanosine 5′-diphosphate *N*-(4-(benzoyl)phenyl]methyl) phosphoramide (GTP-BP)

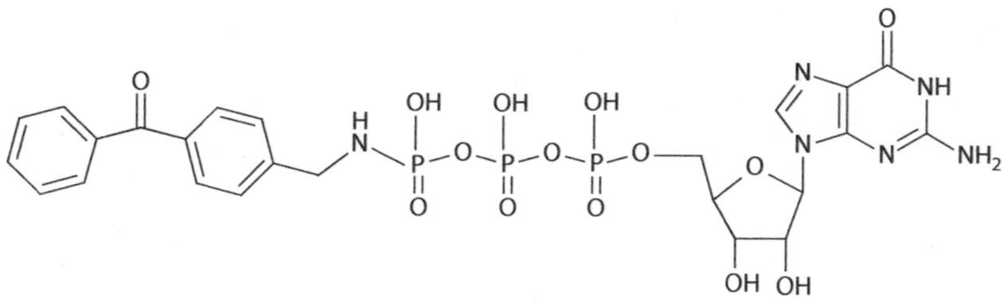

(Continued)

Name (Abbreviation)	Structure	References
CXLII 8-Azido-adenosine 5′-diphosphate N-(4-(benzoyl)phenylmethyl) phosphoramide (8-N$_3$-ATP-BP)		96
CXLIII 8-Azido-guanosine 5′-diphosphate N-(4-(benzoyl)phenylmethyl) phosphoramide (8-N$_3$-GTP-BP)		96
CXLIV 3′-O-(4-Benzoyl)benzoyl-adenosine 5′-triphosphate (3-BzATP); 3′(2′)-O-(4-benzoyl)benzoyladenosine 5′-triphosphate (Bz$_2$ATP)		97,98

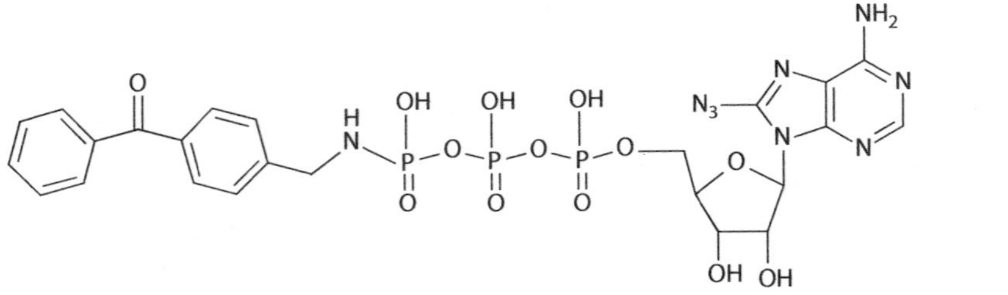

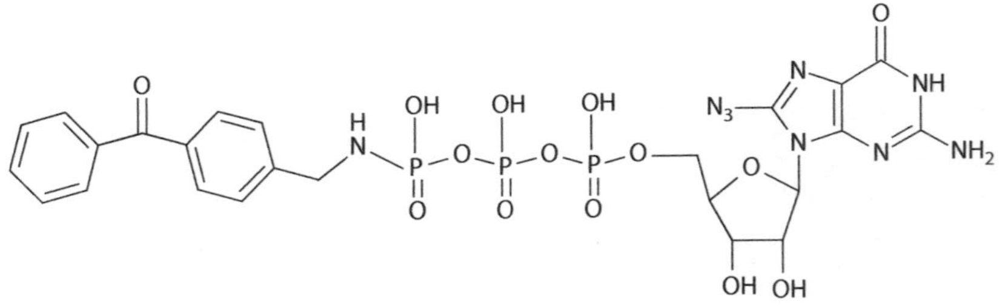

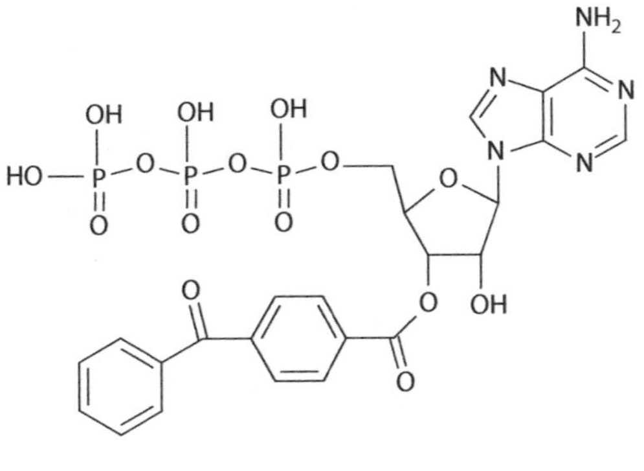

CXLV

9-Arylazide-*N*-acetylneuraminic acid (9-AAz-NeuAc)

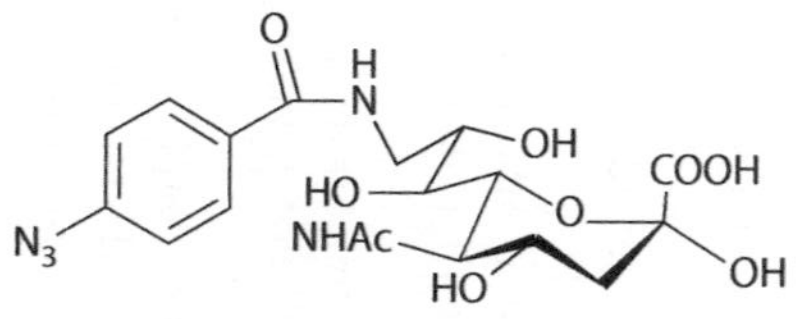

99

CXLVI

O-Penta-acetyl-5-*N*-acyldiazirine-amino-1-methylsialate (Ac$_5$-5-SiaDAz)

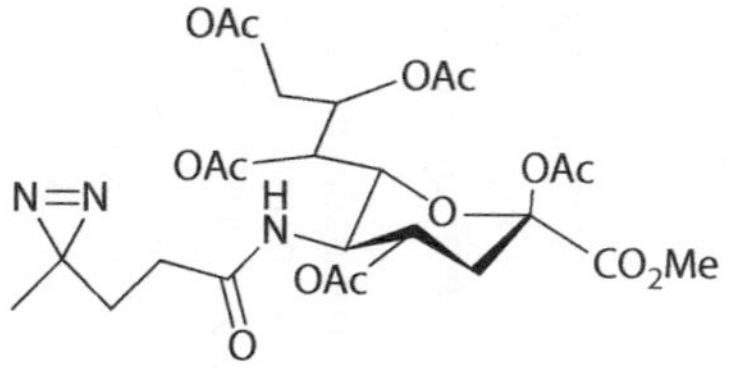

100

CXLVII

O-Tetraacetyl-*N*-acyldiazirine 2-aminomannose (Ac$_4$-ManNDAz)

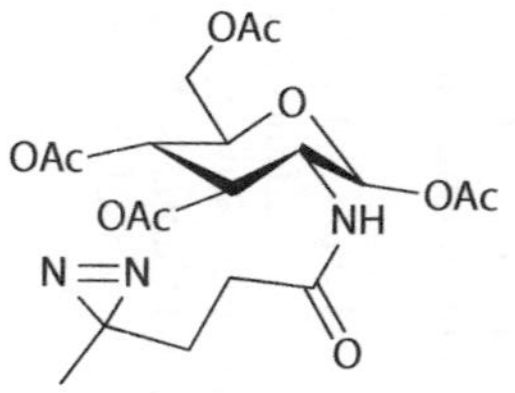

100

CXLVIII

[15-^{3}H]-3-Diazo-4-oxo-10,13-ethano-11-*cis*-retinal

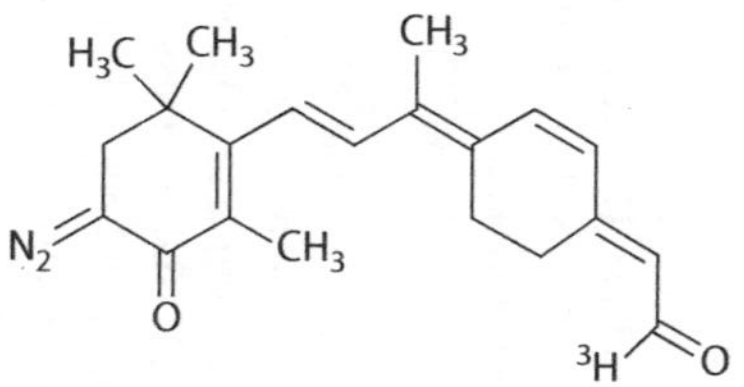

101

CXLIX

o-Dimethyl-*p*-trifluoromethyldiazirine phenyl *cis*-retinal

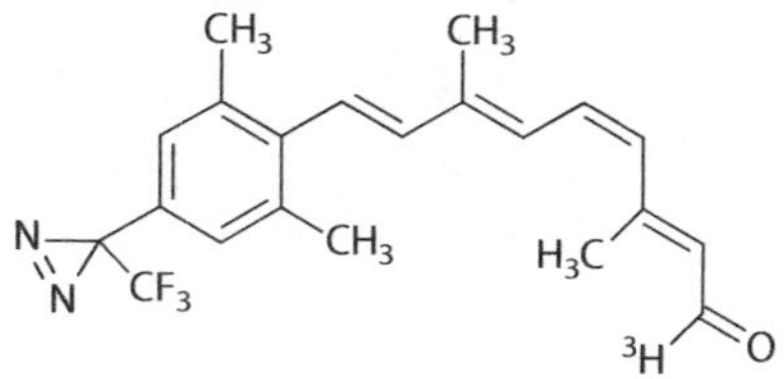

102

REFERENCES

1. Han, K.-K., Richard, C., and Delacourte, A., Chemical cross-links of proteins by using bifunctional reagents, *Int. J. Biochem.*, 16, 129–252, 1984.
2. Wood, C. L. and O'Dorisio, M. S., Covalent cross-linking of vasoactive intestinal polypeptide to its receptors on intact human lymphoblasts, *J. Biol. Chem.*, 260, 1243–1247, 1985.
3. Lee, J. H. and Hoover, T. R., Protein crosslinking studies suggest that *Rhizobium meliloti* C4-dicarboxylic acid transport protein D, a sigma 54-dependent transcriptional activator, interacts with sigma 54 and the beta subunit of RNA polymerase, *Proc. Natl. Acad. Sci. U. S. A.*, 92, 9702–9706, 1995.
4. Ji, T. H. and Ji, I., Macromolecular photoaffinity labeling with radioactive photoactivable heterobifunctional reagents, *Anal. Biochem.*, 121, 286–289, 1982.
5. Schwartz, M. A., A 125I-radiolabel transfer crosslinking reagent with a novel cleavable group, *Anal. Biochem.*, 149, 142–152, 1985.
6. Lewis, R. V., Roberts, M. F., Dennis, E. A., and Allison, W. S., Photoactivated heterobifunctional cross-linking reagents which demonstrate the aggregation state of phospholipase A2, *Biochemistry*, 16, 5650–5654, 1977.
7. Ji, I. and Ji, T. H., Both alpha and beta subunits of human choriogonadotropin photoaffinity label the hormone receptor, *Proc. Natl. Acad. Sci. U. S. A.*, 78, 5465–5469, 1981.
8. Schwartz, I. and Ofengand, J., *E coli* tRNAPhe modified at the 3-(3-amino-3-carboxypropyl) uridine with a photoaffinity label is fully functional for aminoacylation and for ribosomal interaction, *Biochim. Biophys. Acta*, 697, 330–335, 1982.
9. Yaqub, M. and Guire, P., Covalent immobilization of L-asparaginase with a photochemical reagent, *J. Biomed. Mater. Res.*, 8, 291–297, 1974.
10. Schmidt, R. R. and Betz, H., Cross-linking of beta-bungarotoxin to chick brain membranes. Identification of subunits of a putative voltage-gated K + channel, *Biochemistry*, 28, 8346–8350, 1989.
11. Witzemann, V., Muchmore, D., and Raftery, M. A., Affinity-directed cross-linking of membrane-bound acetylcholine receptor polypeptides with photolabile alpha-bungarotoxin derivatives, *Biochemistry*, 18, 5511–5518, 1979.
12. Yan, M., Cai, S. X., Wybourne, M. N., and Keana, J. F., N-Hydroxysuccinimide ester functionalized perfluorophenyl azides as novel photoactive heterobifunctional cross-linking reagents. The covalent immobilization of biomolecules to polymer surfaces, *Bioconjug. Chem.*, 5, 151–157, 1994.
13. Pandurangi, R. S., Lusiak, P., Desai, S., and Kuntz, R. R., Chemistry of bifunctional photoprobes: Part 4: Synthesis of the chromogenic, cleavable, water soluble and heterobifunctional (*N*-methyl amino perfluoroaryl azide benzamido)-ethyl-1,3-dithiopropionyl sulfosuccinimide: An efficient protein crosslinking agent, *Bioorg. Chem.*, 26, 201–212, 1998.
14. Vanin, E. F. and Ji, T. H., Synthesis and application of cleavable photoactivatable heterobifunctional reagents, *Biochemistry*, 20, 6754–6760, 1981.
15. Cheng, M. and Guillory, R. J., Sulfo-SADP (sulfosuccinimidyl[4-azidophenyldithio]propionate) an active site directed reagent inhibiting the NADPH dependent O_2-generation of leukocyte cytochrome b(558), *J. Biochem. Mol. Biol. Biophys.*, 6, 177–1784, 2002.
16. Schwartz, M. A., Das, O. P., and Hynes, R. O., A new radioactive cross-linking reagent for studying the interactions of proteins, *J. Biol. Chem.*, 257, 2343–2349, 1982.
17. Wine, R. N., Dial, J. M., Tomer, K. B., and Borchers, C. H., Identification of components of protein complexes using a fluorescent photo-cross-linker and mass spectrometry, *Anal. Chem.*, 74, 1939–1945, 2002.
18. Jaffe, C. L., Lis, H., and Sharon, N., New clevable photoreactive heterobifunctional cross-linking reagents for studying membrane organization, *Biochemistry*, 19, 4423–4429, 1980.
19. Denny, J. B. and Blobel, G., 125I-labeled crosslinking reagent that is hydrophilic, photoactivatable, and cleavable through an azo linkage, *Proc. Natl. Acad. Sci. U. S. A.*, 81, 5286, 1984.
20. Imai, N., Kometani, T., Crocker, P. J., Bowdan, J. B., Demir, A., Dwyer, L. D., Mann, D. M., Vanaman, T. C., and Watt, D. S., Photoaffinity heterobifunctional cross-linking reagents based on N-(azidobenzoyl) tyrosines, *Bioconjug. Chem.*, 1, 138–143, 1990.
21. Imai, N., Dwyer, L. D., Kometani, T., Ji, T., Vanaman, T. C., and Watt, D. S., Photoaffinity heterobifunctional cross-linking reagents based on azide-substituted salicylates, *Bioconjug. Chem.*, 1, 144–148, 1990.
22. Bomgarden, R. D., Etienne, C., Beach, Z., Deshpande, A. M., and Kaboord, B. J., New diazirine-based photoreactive crosslinkers, heterobifunctional crosslinkers for studying extracellular and intracellular protein–protein interactions, *Previews Pierce Protein Res. Prod.*, 12, 2–3, 2008.

23. Yakovlev, A. A., Crosslinkers and their utilization for studies of intermolecular interactions, *Neurochem. J.*, 3, 139–144, 2009.

24. Lozito, T. P. and Tuan, R. S., Mesenchymal stem cells inhibit both endogenous and exogenous MMPs via secreted TIMPs, *J. Cell. Physiol.*, 226, 385, 2011.

25. Gomes, A. F. and Gozzo, F. C., Chemical cross-linking with a diazirine photoactivatable cross-linker investigated by MALDI- and ESI-MS/MS, *J. Mass. Spectrom.*, 45, 892–899, 2010.

26. Osswald, M., Döring, T., and Brimacombe, R., The ribosomal neighbourhood of the central fold of tRNA: Cross-links from position 47 of tRNA located at the A, P or E site, *Nucleic Acids Res.*, 23, 4635–4641, 1995.

27. Bochkariov, D. E. and Kogon, A. A., Application of 3-[3-(3-(trifluoromethyl)diazirin-3-yl)phenyl]-2,3-dihydroxypropionic acid, carbene-generating, cleavable cross-linking reagent for photoaffinity labeling, *Anal. Biochem.*, 204, 90–95, 1992.

28. Dolder, M., Michel, H., and Sigrist, H., 3-(Trifluoromethyl)-3-(m-isothiocyanophenyl)diazirine: Synthesis and chemical characterization of a heterobifunctional carbene-generating crosslinking reagent, *J. Protein Chem.*, 9, 407–415, 1990.

29. Mchedlidze, M. T., Sumbatian, N. V., Bondar', D. A., Taranenko, M. V., and Korshunova, G. A., New photoreactive cleavable reagents with (trifluoromethyl)diazirine group, *Bioorg. Khim.*, 29, 200–207, 2003.

30. Ji, T. H., A novel approach to the identification of surface receptors. The use of photosensitive heterobifunctional cross-linking reagent, *J. Biol. Chem.*, 252, 1566–1570, 1977.

31. Middaugh, C. R., Vanin, E. F., and Ji, T. H., Chemical crosslinking of cell membranes, *Mol. Cell Biochem.*, 50, 115–141, 1983.

32. Fink, G., Fasold, H., Rommel, W., and Brimacombe, R., Reagents suitable for the crosslinking of nucleic acids to proteins, *Anal. Biochem.*, 108, 394–401, 1980.

33. Maassen, J. A., Cross-linking of ribosomal proteins by 4-(6-formyl-3–3-azidophenoxy)butyrimidate, a heterobifunctional, cleavable cross-linker, *Biochemistry*, 18, 1288–1292, 1979.

34. Das, M., Miyakawa, T., Fox, C. F., Pruss, R. M., Aharonov, A., and Herschman, H. R., Specific radiolabeling of a cell surface receptor for epidermal growth factor, *Proc. Natl. Acad. Sci. U. S. A.*, 74, 2790–2794, 1977.

35. Young, E. F., McKee, M. J., Ferguson, D. G., and Kranias, E. G., Structural characterization of phospholamban in cardiac sarcoplasmic reticulum membranes by cross-linking, *Membr. Biochem.*, 8, 95–106, 1989.

36. Harris, R. and Findlay, J. B., Investigation of the organisation of the major proteins in bovine myelin membranes. Use of chemical probes and bifunctional crosslinking reagents, *Biochim. Biophys. Acta*, 732, 75–82, 1983.

37. Fleet, G. W., Knowles, J. R., and Porter, R. R., The antibody binding site. Labelling of a specific antibody against the photo-precursor of an aryl nitrene, *Biochem. J.*, 128, 499–508, 1972.

38. Erecińska, M., Vanderkooi, J. M., and Wilson, D. F., Cytochrome c interactions with membranes. A photoaffinity-labeled cytochrome c, *Arch. Biochem. Biophys.*, 171, 108–116, 1975.

39. Sigrist, H., Allegrini, P. R., Kempf, C., Schnippering, C., and Zahler, P., 5-Isothiocyanato-1-naphthalene azide and rho-azidophenylisothiocyanate. Synthesis and application in hydrophobic heterobifunctional photoactive cross-linking of membrane proteins, *Eur. J. Biochem.*, 125, 197–201, 1982.

40. Dormán, G. and Prestwich, G. D., Benzophenone photophores in biochemistry, *Biochemistry*, 33, 5661–5673, 1994.

41. Mariano, P. S., Glover, G. I., and Wilkinson, T. J., Photochemistry of modified proteins. Benzophenone-containing bovine serum albumin, *Photochem. Photobiol.*, 23, 147–154, 1976.

42. Mourey, R. J., Estevez, V. A., Marecek, J. F., Barrow, R. K., Prestwich, G. D., and Snyder, S. H., Inositol 1,4,5-trisphosphate receptors: Labeling the inositol 1,4,5-trisphosphate binding site with photoaffinity ligands, *Biochemistry*, 32, 1719–1726, 1993.

43. Bogdanov, A., Jr., Tung, C. H., Bredow, S., and Weissleder, R., DNA binding chelates for nonviral gene delivery imaging, *Gene Ther.*, 8, 515–522, 2001.

44. Chowdhry, V., Vaughan, R., and Westheimer, F. H., 2-Diazo-3,3,3-trifluoropropionyl chloride: Reagent for photoaffinity labeling, *Proc. Natl. Acad. Sci. U. S. A.*, 73, 1406–1408, 1976.

45. Pascual, A., Casanova, J., and Samuels, H. H., Photoaffinity labeling of thyroid hormone nuclear receptors in intact cells, *J. Biol. Chem.*, 257, 9640–9647, 1992.

46. Shafer, J., Baronowsky, P., Laursen, R., Finn, F., and Westheimer, F. H., Products from the photolysis of diazoacetyl chymotrypsin, *J. Biol. Chem.*, 241, 421–427, 1966.

47. Harrison, J. K., Lawton, R. G., and Gnegy, M. E., Development of a novel photoreactive calmodulin derivative: Cross-linking of purified adenylate cyclase from bovine brain, *Biochemistry*, 28, 6023–6027, 1989.

48. Trommer, W. E., Kolkenbrock, H., and Pfleiderer, G., Synthesis and properties of a new selective bifunctional cross-linking reagent, *Hoppe Seylers Z. Physiol. Chem.*, 356, 1455–1458, 1975.

49. Aggeler, R., Chicas-Cruz, K., Cai, S. X., Keana, J. F., and Capaldi, R. A., Introduction of reactive cysteine residues in the epsilon subunit of Escherichia coli F1 ATPase, modification of these sites with tetrafluorophenyl azide-maleimides, and examination of changes in the binding of the epsilon subunit when different nucleotides are in catalytic sites, *Biochemistry*, 31, 2956–2961, 1992.

50. Hixson, S. H. and Hixson, S. S., P-Azidophenacyl bromide, a versatile photolabile bifunctional reagent. Reaction with glyceraldehyde-3-phosphate dehydrogenase, *Biochemistry*, 14, 4251–4254, 1975.

51. Mattson, G., Conklin, E., Desai, S., Nielander, G., Savage, M. D., and Morgensen, S., A practical approach to crosslinking, *Mol. Biol. Rep.*, 17, 167–183, 1993.

52. Muramoto, K. and Ramachandran, J., Photoreactive derivatives of corticotropin. 2. Preparation and characterization of 2-nitro-4(5)-azidophenylsulfenyl derivatives of corticotrophin, *Biochemistry*, 19, 3280–3286, 1980.

53. Moreland, R. B., Smith, P. K., Fujimoto, E. K., and Dockter, M. E., Synthesis and characterization of N-(4-azidophenylthio)phthalimide: A cleavable, photoactivable crosslinking reagent that reacts with sulfhydryl groups, *Anal. Biochem.*, 121, 321–326, 1982.

54. Huang, C. H. and Richards, F. M., Reaction of a lipid-soluble, unsymmetrical, cleavable, cross-linking reagent with muscle aldolase and erythrocyte membrane proteins, *J. Biol. Chem.*, 252, 5514–5521, 1977.

55. Mikkelsen, R. B. and Wallach, D. F., Photoactivated cross-linking of proteins within the erythrocyte membrane core, *J. Biol. Chem.*, 251, 7413–7416, 1976.

56. Peletskaya, E. N., Boyer, P. L., Kogon, A. A., Clark, P., Kroth, H., Sayer, J. M., Jerina, D. M., and Hughes, S. H., Cross-linking of the fingers subdomain of human immunodeficiency virus type 1 reverse transcriptase to template-primer, *J. Virol.*, 75, 9435–9445, 2001.

57. Cai. K., Itoh, Y., and Khorana, H. G., Mapping of contact sites in complex formation between transducin and light-activated rhodopsin by covalent crosslinking: Use of a photoactivatable reagent, *Proc. Natl. Acad. Sci. U. S. A.*, 98, 4877–4882, 200.

58. Ebright, Y. W., Chen, Y., Kim, Y., and Ebright, R. H., S-[2-(4-Azidosalicylamido)ethylthio]-2-thiopyridine: Radioiodinatable, cleavable, photoactivatible cross-linking agent, *Bioconjug. Chem.*, 7, 380–384, 1996.

59. Gautam, A., Mulugu, S., Alexander, K., and Bastia, D., A single domain of the replication termination protein of *Bacillus subtilis* is involved in arresting both DnaB helicase and RNA polymerase, *J. Biol. Chem.*, 276, 23471–23479, 2001.

60. Chong, P. C. and Hodges, R. S., A new heterobifunctional cross-linking reagent for the study of biological interactions between proteins. II. Application to the troponin C–troponin I interaction, *J. Biol. Chem.*, 256, 5071–5076, 1981.

61. Dhanasekaran, N., Wessling-Resnick, M., Kelleher, D. J., Johnson, G. L., and Ruoho, A. E., Mapping of the carboxyl terminus within the tertiary structure of transducin's alpha subunit using the heterobifunctional cross-linking reagent, 125I-N-(3-iodo-4-azidophenylpropionamido-S-(2-thiopyridyl) cysteine, *J. Biol. Chem.*, 263, 17942–179450, 1988.

62. Zecherle, G. N., Oleinikov, A., and Traut, R. R., The C-terminal domain of Escherichia coli ribosomal protein L7/L12 can occupy a location near the factor-binding domain of the 50S subunit as shown by cross-linking with N-[4-(p-azidosalicylamido)butyl]-3-(2′-pyridyldithio)propionamide, *Biochemistry*, 31, 9526–9532, 1992.

63. Traut, R. R., Dey, D., Bochkariov, D. E., Oleinikov, A. V., Jokhadze, G. G., Hamman, B., and Jameson, D., Location and domain structure of *Escherichia coli* ribosomal protein L7/L12: Site specific cysteine crosslinking and attachment of fluorescent probes, *Biochem. Cell Biol.*, 73, 949–958, 1995.

64. Friebel, K., Huth, H., Jany, K. D., and Trommer, W. E., Semi-reversible cross-linking. Synthesis and application of a novel heterobifunctional reagent, *Hoppe Seylers Z. Physiol. Chem.*, 362, 421–428, 1981.

65. Brunner, J. and Semenza, G., Selective labeling of the hydrophobic core of membranes with 3-(trifluoromethyl)-3-(m-[125I]iodophenyl)diazirine, a carbene-generating reagent, *Biochemistry*, 20, 7174–7182, 1981.

66. Nassal, M., 4′-(1-Azi-2,2,2-trifluoroethyl)phenylalanine, a photolabile carbene-generating analog of phenylalanine, *J. Am. Chem. Soc.*, 106, 7540–7545, 1984.

67. Halbfinger, E., Gorochesky, K., Lévesque, S. A., Beaudoin, A. R., Sheihet, L., Margel, S., and Fischer, B., Photoaffinity labeling on magnetic microspheres (PALMm) methodology for topographic mapping: Preparation of PALMm reagents and demonstration of biochemical relevance, *Org. Biomol. Chem.*, 1, 2821–2832, 2003.

68. Döring, T., Mitchell, P., Osswald, M., Bochkariov, D., and Brimacombe, R., The decoding region of 16S RNA; a cross-linking study of the ribosomal A, P and E sites using tRNA derivatized at position 32 in the anticodon loop, *EMBO J.*, 13, 2677–2685, 1994.

69. Peletskaya, E. N., Kogon, A. A., Tuske, S., Arnold, E., and Hughes, S. H., Nonnucleoside inhibitor binding affects the interactions of the fingers subdomain of human immunodeficiency virus type 1 reverse transcriptase with DNA, *J. Virol.*, 78, 3387–3397, 2004.

70. Chevolot, Y., Bucher, O., Léonard, D., Mathieu, H. J., and Sigrist, H., Synthesis and characterization of a photoactivatable glycoaryldiazirine for surface glycoengineering, *Bioconjug. Chem.*, 10, 169–175, 1999.

71. Resek, J. F., Bhattacharya, S., and Khorana, H. G., A new photo-crosslinking reagent for the study of protein–protein interactions, *J. Org. Chem.*, 58, 7598, 1993.

72. von Ballmoos, C., Appoldt, Y., Brunner, J., Granier, T., Vasella, A., and Dimroth, P., Membrane topography of the coupling ion binding site in Na+-translocating F1F0 ATP synthase, *J. Biol. Chem.*, 277, 3504–3510, 2002.

73. Nagase, T., Nakata, E., Shinkai, S., and Hamachi, I., Construction of artificial signal transducers on a lectin surface by post-photoaffinity-labeling modification for fluorescent saccharide biosensors, *Chemistry*, 9, 3660–3669, 2003.

74. Henkin, J., Photolabeling reagent for thiol enzymes. Studies on rabbit muscle creatine kinase, *J. Biol. Chem.*, 252, 4293–4297, 1977.

75. Paquatte, O., Fried, A., and Tu, S. C., Delineation of bacterial luciferase aldehyde site by bifunctional labeling reagents, *Arch. Biochem. Biophys.*, 264, 392–399, 1988.

76. Husain, S.S., Ferguson, J. B., and Fruton, J. S., Bifunctional inhibitors of pepsin, *Proc. Natl. Acad. Sci. U. S. A.*, 68, 2765–2768, 1971.

77. Qvit, N., Monderer-Rothkoff, G., Ido, A., Shalev, D. E., Amster-Choder, O., and Gilon, C., Development of bifunctional photoactivatable benzophenone probes and their application to glycoside substrates, *Biopolymers*, 90, 526–536, 2008.

78. Jelenc, P. C., Cantor, C. R., and Simon, S. R., High yield photoreagents for protein crosslinking and affinity labeling, *Proc. Natl. Acad. Sci. U. S. A.*, 75, 3564–3568, 1978.

79. Nielsen, P. E., Hansen, J. B., and Buchardt, O., Photochemical cross-linking of protein and DNA in chromatin. Synthesis and application of a photosensitive cleavable derivative of 9-aminoacridine with two photoprobes connected through a disulphide-containing linker, *Biochem. J.*, 223, 519–526, 1984.

80. Choithani, J., Sethi, D., Kumar, P., and Gupta, K. C., Preparation of oligonucleotide microarrays on modified glass using a photoreactive heterobifunctional reagent, 1-*N*-(maleimidohexanoyl)-6-*N*-(anthraquinon-2-oyl) hexanediamine (MHAHD), *Surf. Sci.*, 602, 2389–2394, 2008.

81. Patnaik, S., Swami, A., Sethi, D., Pathak, A., Garg, B. S., Gupta, K. C., and Kumar, P., N-(Iodoacetyl)-N′-(anthraquinon-2-oyl)-ethylenediamine (IAED): A new heterobifunctional reagent for the preparation of biochips, *Bioconjug. Chem.*, 18, 8–12, 2007.

82. Kumar, P., Agarwal, S. K., and Gupta, K. C., N-(3-Trifluoroethanesulfonyloxypropyl)anthraquinone-2-carboxamide: A new heterobifunctional reagent for immobilization of biomolecules on a variety of polymer surfaces, *Bioconjug. Chem.*, 15, 7–11, 2004.

83. Elsner, H., Buchardt, O., Møller, J., and Nielsen, P. E., Photochemical crosslinking of protein and DNA in chromatin. II. Synthesis and application of psoralen-cystamine-arylazido photocrosslinking reagents, *Anal. Biochem.*, 149, 575–581, 1985.

84. Ngo, T. T., Yam, C. F., Lenhoff, H. M., and Ivy, J., p-Azidophenylglyoxal. A heterobifunctional photoactivable cross-linking reagent selective for arginyl residues, *J. Biol. Chem.*, 256, 11313–11318, 1981.

85. Gorman, J. J. and Folk, J. E., Transglutaminase amine substrates for photochemical labeling and cleavable cross-linking of proteins, *J. Biol. Chem.*, 255, 1175–1180, 1980.

86. Darfler, F. J. and Marinetti, G. V., Synthesis of a photoaffinity probe for the beta-adrenergic receptor, *Biochem. Biophys. Res. Commun.*, 79, 1–7, 1977.

87. Hegyi, G., Mák, M., Kim, E., Elzinga, M., Muhlrad, A., and Reisler, E., Intrastrand cross-linked actin between Gln-41 and Cys-374. I. Mapping of sites cross-linked in F-actin by N-(4-azido-2-nitrophenyl) putrescine, *Biochemistry*, 37, 17784–17792, 1998.

88. Hegyi, G., Michel, H., Shabanowitz, J., Hunt, D. F., Chatterjie, N., Healy-Louie, G., and Elzinga, M., Gln-41 is intermolecularly cross-linked to Lys-113 in F-actin by N-(4-azidobenzoyl)-putrescine, *Protein Sci.*, 1, 132–144, 1992.

89. Korutla, L. N., Stewart, G. J., Lasz, E. C., Maione, T. E., and Niewiarowski, S., Evaluation of recombinant platelet factor 4 and protamine sulfate for heparin neutralization: Clotting and clearance studies in rat, *Thromb. Haemost.*, 71, 609–114, 1994.

90. Praetorius, J., Backlund, P., Yergey, A. L., and Spring, K. R., Specific lectin binding to beta1 integrin and fibronectin on the apical membrane of madin-darby canine kidney cells, *J. Membr. Biol.*, 184, 273–281, 2001.

91. Nakamaye, K. L., Wells, J. A., Bridenbaugh, R. L., Okamoto, Y., and Yount, R. G., 2-[(4-Azido-2-nitrophenyl)amino]ethyl triphosphate, a novel chromophoric and photoaffinity analogue of ATP. Synthesis, characterization, and interaction with myosin subfragment 1, *Biochemistry*, 24, 5226–5235, 1985.

92. Dombrowski, K. E. and Colman, R. F., 5′-(p-Fluorosulfonyl)benzoyl-8-azidoadenosine: A new bifunctional affinity label for nucleotide binding sites in proteins, *Arch. Biochem. Biophys.*, 275, 302–308, 1989.

93. Riquelme, P. T. and Czarnecki, J. J., Conformational and allosteric changes in fructose 1,6-bisphosphatase upon photoaffinity labeling with 2-azidoadenosine monophosphate, *J. Biol. Chem.*, 258, 8240–8245, 1983.

94. Grammer, J. C., Kuwayama, H., and Yount, R. G., Photoaffinity labeling of skeletal myosin with 2-azidoadenosine triphosphate, *Biochemistry*, 32, 5725–5732, 1993.

95. Schäfer, H. J. and Dose, K., Photoaffinity cross-linking of the coupling factor 1 from *Micrococcus luteus* by 3′-arylazido-8-azido-ATP, *J. Biol. Chem.*, 259, 15301–15306, 1984.

96. Rajagopalan, K., Chavan, A. J., Haley, B. E., and Watt, D. S., Synthesis and application of bidentate photoaffinity cross-linking reagents. Nucleotide photoaffinity probes with two photoactive groups, *J. Biol. Chem.*, 268, 14230–14238, 1993.

97. Williams, N. and Coleman, P. S., Exploring the adenine nucleotide binding sites on mitochondrial F1-ATPase with a new photoaffinity probe, 3′-O-(4-benzoyl)benzoyl adenosine 5′-triphosphate, *J. Biol. Chem.*, 257, 2834–2841, 1982.

98. Mahmood, R., Cremo, C., Nakamaye, K. L., and Yount, R. G., The interaction and photolabeling of myosin subfragment 1 with 3′(2′)-O-(4-benzoyl)benzoyladenosine 5′-triphosphate, *J. Biol. Chem.*, 262, 14479–14486, 1987.

99. Han, S., Collins, B. E., Bengtson, P., and Paulson, J. C., Homomultimeric complexes of CD22 in B cells revealed by protein-glycan cross-linking, *Nat. Chem. Biol.*, 1, 93–97, 2005.

100. Tanaka, Y. and Kohler, J. J., Photoactivatable crosslinking sugars for capturing glycoprotein interactions, *J. Am. Chem. Soc.*, 130, 3278, 2008.

101. Nakanishi, K., Zhang, H., Lerro, K. A., Takekuma, S., Yamamoto, T., Lien, T. H., Sastry, L. et al., Photoaffinity labeling of rhodopsin and bacteriorhodopsin, *Biophys. Chem.*, 56, 13, 1995.

102. Nakayama, T. A. and Khorana, H. G., Orientation of retinal in bovine rhodopsin determined by cross-linking using a photoactivatable analog of 11-cis-retinal, *J. Biol. Chem.*, 265, 15762–15769, 1990.

Index